IONIC LIQUIDS: APPLICATIONS AND PERSPECTIVES

Edited by **Alexander Kokorin**

Ionic Liquids: Applications and Perspectives
Edited by Alexander Kokorin

Published by InTech
Janeza Trdine 9, 51000 Rijeka, Croatia

Publishing Process Manager Ana Nikolic

Technical Editor Teodora Smiljanic

Cover Designer InTech Design Team

Additional hard copies can be obtained from orders@intechopen.com

Ionic Liquids: Applications and Perspectives, Edited by Alexander Kokorin
p. cm.
ISBN 978-953-307-248-7

For EU product safety concerns:
IN TECH d.o.o., Prolaz Marije Krucifikse Kozulić 3, 51000 Rijeka, Croatia,
info@intechopen.com or visit our website at intechopen.com.

Meet the editor

Alexander I. Kokorin: born: 1947, Moscow; DSc., PhD; Principal Research Fellow (Research Professor) of Department of Kinetics and Catalysis, N. Semenov Institute of Chemical Physics, Russian Academy of Sciences, Moscow. Area of research interests: physical chemistry of complex-organized molecular and nanosized systems, including polymer-metal complexes; the surface of doped oxide semiconductors. He is an expert in structural, absorptive, catalytic and photocatalytic properties, in structural organization and dynamic features of ionic liquids, in magnetic interactions between paramagnetic centers. The author or co-author of 3 books, over 200 articles and reviews in scientific journals and books. He is an actual member of the International EPR/ESR Society, European Society on Quantum Solar Energy Conversion, Moscow House of Scientists, of the Board of Moscow Physical Society.

Contents

Preface

This book is the second in the series of publications on this field by this publisher, and contains a number of latest research developments on ionic liquids (ILs), first of all on room temperature ILs. It is a promising new area that has received a lot of attention during the last 20 years. Readers will here find 30 chapters on recent applications of ILs in polymer sciences, material chemistry, catalysis, nanotechnology, biotechnology and electrochemical applications, which are collected in 6 sections. Also modern trends and perspectives are discussed. The authors of each chapter are scientists and technologists with strong expertise in their respective fields. This book offers an international forum for exchanging states of the arts and knowledge in ILs applications. The readers will be able to perceive a trend analysis and examine recent developments in different areas of ILs chemistry and technologies. I hope that the book will help in systematization of knowledges in ILs science, creation of new approaches in this field and further promotion of ILs technologies and engineering for the future.

Prof. Dr. Alexander Kokorin
N.Semenov Institute of Chemical Physics RAS,
Moscow
Russian Federation

Part 1

Polymers

1

Advanced Applications of Ionic Liquids in Polymer Science

Elaheh Kowsari
Amirkabir University of Technology
Islamic Republic of Iran

1. Introduction

During past few years, ionic liquids have kept attracting much attention as "green and designer" media for chemical reactions. Room-temperature ionic liquids have emerged as a potential replacement for organic solvents in catalytic processes on both laboratory and industrial scales (Holbrey & Seddon, 1999b). Literature reports on a wide range of reactions including advances in alkylation reactions (Earle et al., 1998), Diels-Alder cyclizations (Earle et al., 1999; Jaeger & Tucker, 1989), and the development of commercially competitive processes for dimerization, oligomerization, and polymerization of olefins (Abdul-Sada et al., 1995a; 1995b; Ambler et al., 1996; Chauvin et al., 1988; 1989). Effectively, Ionic liquids, among a unique set of chemical and physical properties (Chauvin, 1996; Chauvin & Mussmann 1995; Seddon, 1997), have no measurable vapor pressure, which lends them as ideal replacements for volatile, conventional organic solvents. The wide and readily accessible range of room-temperature ionic liquids with corresponding variations in physical properties, prepared by simple structural modifications to the cations (Gordon et al., 1998; Holbrey & Seddon, 1999a) or changes in anions (Bonhoˆte et al., 1996; Wilkes & Zaworotko, 1992), offers the opportunity to design an ionic liquid-solvent system optimized for particular processes. In other words, these ionic liquids can be considered as "designer solvents" (Freemantle, 1998).

Applications of ionic liquids as solvents for polymerization processes have widely been reviewed in literature (Kubisa, 2004; Shen & Ding, 2004; Lua et al., 2009). Ionic liquids have been used in polymer science, mainly as polymerization media in several types of polymerization processes, including conventional free radical polymerization (Sarbu, & Matyjaszewski, 2001), living/controlling radical polymerizations (such as atomtransfer radical polymerizations (ATRP) (Ding et al., 2005; Shen & Ding., 2004; Biedron & Kubisa., 2001; Biedron & Kubisa., 2002; Biedron & Kubisa., 2003), reversible addition-ragmentation transfer (RAFT) (Perrier & Davis, 2002), as well as in ionic and coordination polymerizations (Chiefari et al., 1998; Vijayaraghavan & MacFarlane et al., 2004). When radical polymerizations are conducted in an ionic liquid, a significant increase of kp/kt ratio is normally observed in comparison to those carried out in other polar/coordinating solvents. As solvents for ATRP and RAFT, ionic liquids facilitate separation of the polymer from residual catalyst and reduce the extent of side-reactions.

The use of ionic liquids in polymer science is not limited to their application as solvents. Ionic liquids are also used as additives, including plasticizers, components of polymer electrolytes, and porogenic agents) to polymers. More recently, properties of polymers containing chemically bound ionic liquid moiety (polymeric ionic liquids) are studied and the possibilities of their applications are being explored. Ionic liquids are also investigated as components of the polymeric matrixes (such as polymer gels), templates for porous polymers, and novel electrolytes for electrochemical polymerizations (Przemysław & Kubisa, 2009). This chapter focuses on the recent developments and achievements gained from applications of ionic liquids in the preparation of functional polymers as well as properties modification of polymers caused by ionic liquids.

2. Radical polymerization

2.1 Radical polymerization (ATRP) of acrylates in ionic liquids

It is known, that properties of ILs commonly depend on the structure of the cation (the symmetry and the length of alkyl substituents, the presence of hydrophobic groups, etc.) as well as on the degree of anion charge delocalization. To elucidate the influence of the nature of ILs on the yield and molecular weight of vinyl polymers, the radical polymerization of appropriate monomers (MMA) in different ionic liquids has been studied (Yakov et. al, 2007).For the MMA polymerization, the dependence between polymer η_{inh} and the length of alkyl substitute in asymmetrical 1-methyl-3-alkylimidazolium ILs was examined (Table 1). The results presented in Table 1 (entries 1–9) show, that increase of carbon chain length of imidazolium ILs leads to the reduction of polymer molecular weight, especially in the case of tetrafluoroborate ILs (Table 1, entries 5–9). Table 1 (entries 10, 11) shows, that a polymer having relatively high η_{inh} value was obtained in both tetrafluoroborate IL: [1,3-Bu_2Im]BF_4 and [1-Bu-3-(*iso*-Bu)Im]BF_4, but in the case of 1,3-di-n-butylimidazolium tetrafluoroborate (entry 10) the η_{inh} value of the polymer is higher (3.13 and 2.30 dl g^{-1}, respectively).

The influence of the anion nature on the radical polymerization of MMA can be revealed by comparison of the data, obtained in ILs with ordinary used [1-Me-3-BuIm] cation (Table 1, High values of polymer molar mass and an increase in the rate of free radical polymerization of MMA in ILs can be assigned to the strong effect of ionic media on the chain propagation (activation energy decrease) and chain termination, for the reason of high viscosity of the reaction system, the so-called gel effect. (Yakov et. al, 2007).

2.2 Polymerized ionic liquids: synthesis and applications

Recently, much attention has been paid to polymerized ionic liquids or polymeric ionic liquids, which are macromolecules obtained from polymerizing ionic liquid monomers (Lu et. al, 2009). Their potential applications involve polymeric electrolytes (Galiński et. al, 2006; Ricks-Laskoski& Snow, 2006; Sato et. al, 2007; Susan et. al, 2005), catalytic membranes (Carlin & Fuller 1997), ionic conductive materials (Hirao et. al, 2000; Washiro et. al, 2004; Matsumi et. al, 2006), CO_2 absorbing materials (Tang et. al, 2005a; Tang et. al, 2005b; Tang et. al, 2005c; Tang et. al, 2005d), microwave absorbing materials (Tang et. al, 2008; Amajjahe, & Ritter 2009), and porous materials (Yan & Texter, 2006; Yan et. al, 2007).

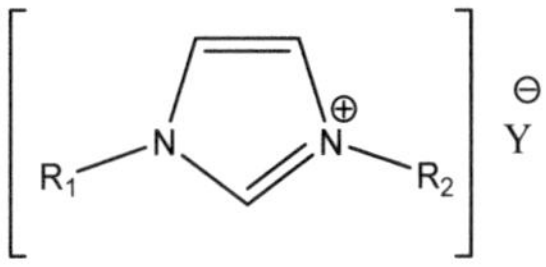

				PMMA	
Entry[a]	R_1	R_2	Y	Yield (%)	η_{inh}[b] (dlg^{-1})
1	CH_3	C_2H_5	$(CF_3SO_2)_2N$	98	3.06
2	CH_3	C_4H_9	$(CF_3SO_2)_2N$	98	2.81
3	CH_3	C_2H_5	CF_3SO_3	96	4.09 [c]
4	CH_3	C_4H_9	CF_3SO_3	94	2.97
5	CH_3	C_3H_7	BF_4	93	3.00
6	CH_3	C_4H_9	BF_4	92	2.70
7	CH_3	C_5H_{11}	BF_4	91	2.86
8	CH_3	C_6H_{13}	BF_4	90	1.58
9	CH_3	C_7H_{15}	BF_4	95	1.55
10	C_4H_9	C_4H_9	BF_4	92	3.13
11	C_4H_9	*i*-C_4H_9	BF_4	96	2.30
12	CH_3	C_4H_9	PF_6	97	3.89
13	CH_3	C_4H_9	SbF_6	98	2.92
14	CH_3	C_6H_{13}	$(CF_3CF_2)_3PF_6$	98	3.29
15	$[P^+(C_6H_{13})C_{14}H_{29}]BF_4^-$			96	3.01
16	$[P^+(C_6H_{13})C_{14}H_{29}]PF_6^-$			98	3.31
17	$[P^+(C_6H_{13})C_{14}H_{29}]C^-$			98	3.48 [d]
18	$[P^+(C_8H_{17})_4]B^-$			95	3.32
19	Benzene [e]			41	0.36

[a] Polymerization parameters: [AIBN]= 0.5 wt%, [MMA] =50 wt%, reaction time = 4 hr, reaction temperature, T = 60 °C.
[b]For the solutions of 0.05 g of PMMA in 10.0 ml of $CHCl_3$ at 25.0 °C.
[c]M_w = 5,770,000 g/mol (determined by static light scattering in acetone).
[d]Mw = 4,100,000 g/mol (determined by static light scattering in acetone).
[e]For comparison. (Reproduced from Vygodskii1, et. al (2007) *Polym. Adv. Technol.* 3, 18, 50–63, Copyright (2007), with permeation from John Wiely & Sons)

Table 1. IL's nature effect upon the radical polymerization of MMA

A variety of polymers having imidazolium moieties in the side chains have been reported, including poly (meth) acrylate (Washiro et. al, 2004; Ding et. al, 2004; Nakashima et. al, 2007; Juger et. al, 2009), polystyrene (Tang et. al, 2005c; Tang et. al, 2005e), and poly(*N*-vinylimidazolium) derivatives (Amajjahe, & Ritter 2009; Leddet et. al, 2001; Marcilla et. al, 2004; Marcilla et. al, 2005), and most of these poly(ionic liquid)s were prepared by conventional radical polymerizations. Free radical polymerizations of various *N*-vinylimidazolium derivatives were reported to proceed in the presence of conventional radical initiator, and various copolymers involving the imidazolium group were also synthesized by this method (Mu, et. al, 2005; Sugimura et. al, 2007).

2.2.1 Microwave-absorbing ionic liquid polymer

Microwave-absorbing materials are applicable to reducing electromagnetic interference from personal computers, stealth aircraft technology, microwave cookware, and microwave darkroom protection (Petrov & Gagulin, 2001; Bregar, 2004; Yoshihiro et. al, 2002; Saib et. al, 2006; Zou et. al, 2006). Microwave absorption is usually achieved by combining dielectric and magnetic loss. Most microwave absorbing materials are polymer composites with conductive fillers, such as graphite, carbon black, and metals, or magnetic fillers, such as ferrites and carbonyl iron powders. These fillers make the microwave-absorbing materials black and hard to fabricate, for example, into precise parts or thin films (Peng et. al, 2005; Bosman et. al, 2003). Conductive polymers such as polyaniline, polypyrrole, polyalkylthiophenes, and poly(4,4'-diphenylene diphenylvinylene) were also reported as microwave-absorbing materials (Truong et. al, 2003; Chandrasekhar & Naishadham, 1999; Olmedo et. al, 1995; Wan, et. al, 2001; Phang, et. al, 2005). The structures of poly(ionic liquid)s studied by Tang and coworker are shown in Fig. 1.

Since the poly(ionic liquid)s exhibit no magnetic loss and are insulators (the reported ionic conductivity of a poly(ionic liquid) with a similar structure to the polymers reported here is about 10^{-8}s m^{-1}[34]), their microwave absorptions are solely due to the dielectric loss.

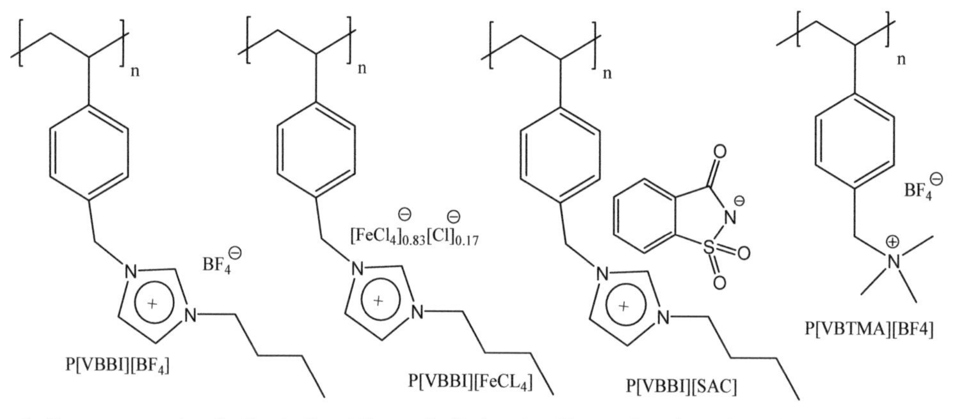

Fig. 1. Structures of poly(ionic liquid)s: poly[1-(*p*-vinylbenzyl)-3-butylimidazolium tetrafluoroborate] (P[VBBI][BF_4]), poly[1-(*p*-vinylbenzyl)- 3-butylimidazolium tetrachloroferrate] (P[VBBI][$FeCl_4$]), poly[1-(*p*-vinylbenzyl)-3-butylimidazolium o-benzoicsulphimide] (P[VBBI][Sac]), and poly[*p*-vinylbenzyltrimethylammonium tetrafluoroborate] (P[VBTMA][BF_4]).
(Reproduced from Tang et. al, (2008) *Macromolecules*, 41, 2, 493-496, Copyright (2008) with permeation from American Chemical Society)

The dielectric constant έ is shown in Fig. 2 as a function of frequency. At frequencies lower than 1 GHz, the έ decreases with increasing frequency. At higher frequencies, the έ remains essentially constant. The poly(ionic liquid)s with the imidazolium cations have similar dielectric constants (έ ≈ 4). In contrast, P[VBTMA][BF_4] with the ammonium cations has a higher dielectric constant (έ ≈ 5.2). This is not surprising because this material is more polar than poly(ionic liquid)s with imidazolium cations, which is in line with results of other ionic liquids (Tokuda et. al, 2006).

To probe the effect of anion on the dielectric loss, the dielectric losses of three poly(ionic liquid)s, P[VBBI][Sac], P[VBBI][BF_4], and P[VBBI][$FeCl_4$], with the same backbone and cations but different anions are compared. Sac^- is a mostly used organic anion while BF_4^- is a widely used inorganic anion in ionic liquids. A poly(ionic liquid) with $FeCl_4^-$ anions containing transition metal ions Fe^{3+} is also synthesized to test whether such kinds of anions can further increase the loss factor (Tang et. al, 2008).

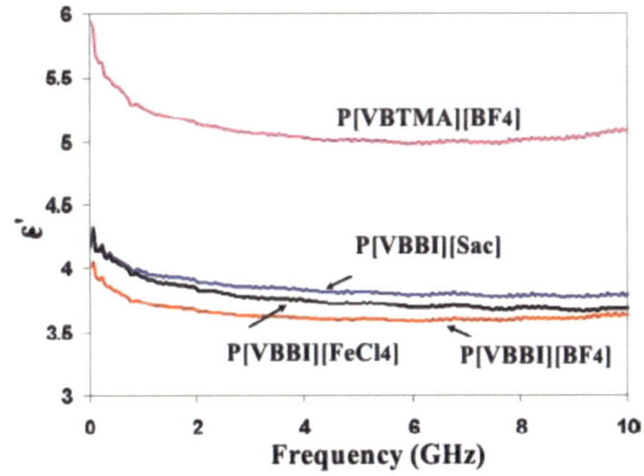

Fig. 2. Dielectric constant ἐ (the real part of complex permittivity) as a function of frequency (Reproduced from Tang et. al, (2008) *Macromolecules*, 41, 2, 493-496, Copyright (2008), with permeation from American Chemical Society)

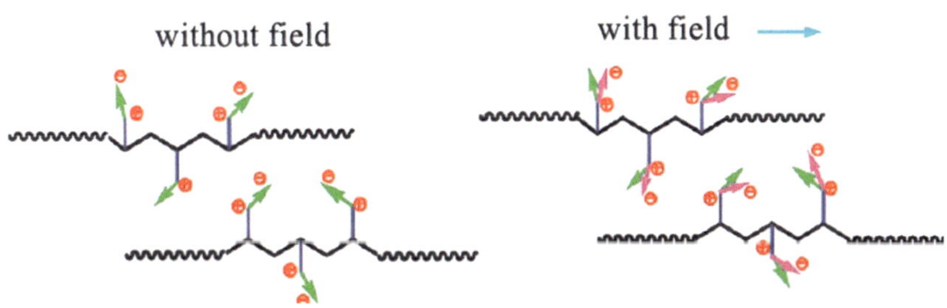

Fig. 3. Orientation of ion-pair dipoles in poly(ionic liquid) without and with an electric field (green vector vs purple vector). (Tang et. al, (2008) *Macromolecules*, 41, 2, 493-496, reprinted with permeation from American Chemical Society)

As sketched in Fig. 3, poly(ionic liquid)s have strong dipole moments created by the permanent ion pairs. In the absence of an external field, these dipoles are randomly oriented (randomly point in different directions) and continually jump from one orientation to another as a result of thermal agitation. In an external field, these dipoles orient themselves in the direction of the applied field (Fig. 3). (Tang et. al, 2008)

2.2.2 Electrowetting of a new ionic liquid monomer and polymer system

Synthesis and electrowetting of a new ionic liquid monomer and polymer system was reported by Holly and coworkers (Holly et. al, 2008). The formation of the monomeric ionic liquid salt and its polymerization is depicted in Fig. 4.

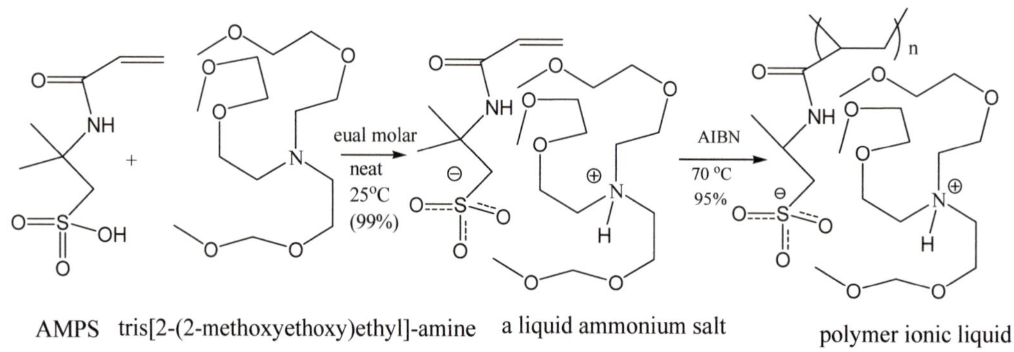

Fig. 4. Reaction Scheme for AMPS oxyethylene ammonium salt monomer and polymer (Reproduced from Ricks-Laskoski & Snow (2006) *J. Am. Chem. Soc.* 128, 38, 12402-12403, Copyright (2006), with permeation from American Chemical Society)

The uniqueness of the oxyethylene amine in the formation of the ammonium cationic species contributes to both the ionic and liquid nature of the monomer and polymer. Even more remarkable is the ability of this polymer to maintain its liquid nature as a macromolecule and to wet a substrate, showing preference for one polarity based upon the makeup of the ionic backbone of the polymer formed (Ricks-Laskoski & Snow, 2006).

Polymerizable ionic liquids and their actuation in an electric field are a combination of material and properties with unique potential to display structural and fluid dynamics above that found for small molecule ionic liquids.

Electrowetting is an electrostatically driven surface effect where a liquid droplet's spreading on a hydrophobic surface is modulated by application of a voltage to the droplet and an underlying conducting substrate (Quilliet &. Berge 2001). A schematic of this effect is illustrated in Figure 5.

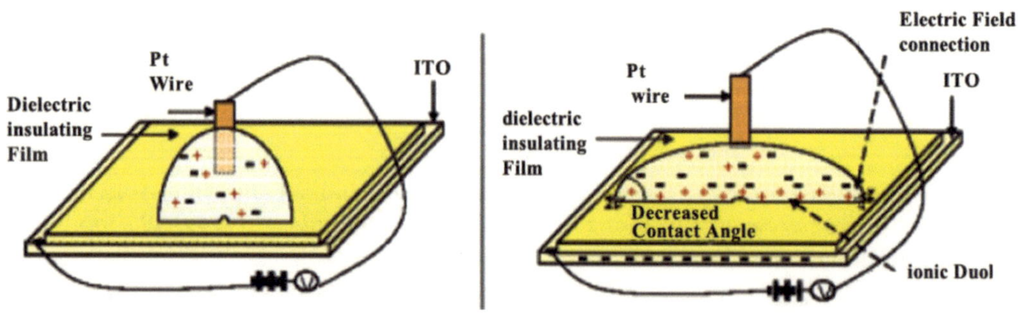

Fig. 5. Depiction of an electrowetting actuation electrode setup with (right) and without (left) an induced electric field (Reproduced from Ricks-Laskoski & Snow (2006) *J. Am. Chem. Soc.* 128, 38, 12402-12403, Copyright (2006), with permeation from American Chemical Society)

The droplet rests on a very thin low-dielectric insulating film (Teflon AF) which is supported on a conducting substrate and is contacted at the top by a very fine wire contact. Application of a voltage builds up a layer of charge on both sides of the interface with the dielectric film and decreases the interfacial energy.

2.2.3 Polymerized ionic liquids: solution properties and electrospinning

The solution properties and electrospinning of a polymerized ionic liquid was explored by Chen and Elabd (Chen & Elabd, 2009) Polymerized ionic liquids are synthesized from polymerizing ionic liquid monomers, where ionic liquids are of great interest due to their unique physiochemical properties. Compared to other polyelectrolyte solutions, this polymerized ionic liquid solution exhibits similar viscosity scaling relationships in the semidilute unentangled and semidilute entangled regimes. However, the electrospraying-electrospinning transition occurs at similar polymer solution concentrations compared to neutral polymers, where electrospinning produced beaded fibers and defect free fibers at ~1.25 and ~2 times the entanglement concentration, respectively. Due to high solution conductivities, electrospinning produces fibers approximately an order of magnitude smaller than neutral polymers at equivalent normalized solution concentrations. In addition, a high ionic conductivity of the solid-state fiber mat was observed under dry conditions and even higher conductivities were observed for polyelectrolyte fiber mats produced from electrospinning polyelectrolyte-ionic liquid solutions, where both anion and cation are mobile species. Structure of polymerized iIonic liquid poly(MEBIm-BF_4) is shown in Fig. 6.
Fig. 7 shows the morphology of the fibers at various ionic liquid contents. Instead of reduced fiber sizes, the existence of ionic liquid results in larger fibers with a ribbon structure. This can be attributed to the nonvolatility of ionic liquid that hinders the solidification of fibers to smaller sizes. Moreover, the liquid component in the fiber collapses the fiber into a ribbon structure.
With the increase of ionic liquid content, more ribbons were observed in the fiber mat (Fig 7).

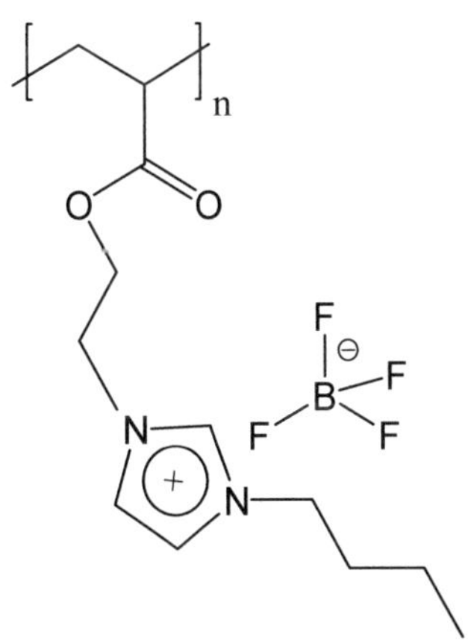

Fig. 6. Structure of polymerized ionic liquid poly(MEBIm-BF_4) (Reproduced from Chen et al., (2009) *Macromolecules* 42, 3368-3373, Copyright (2009), with permeation from American Chemical Society)

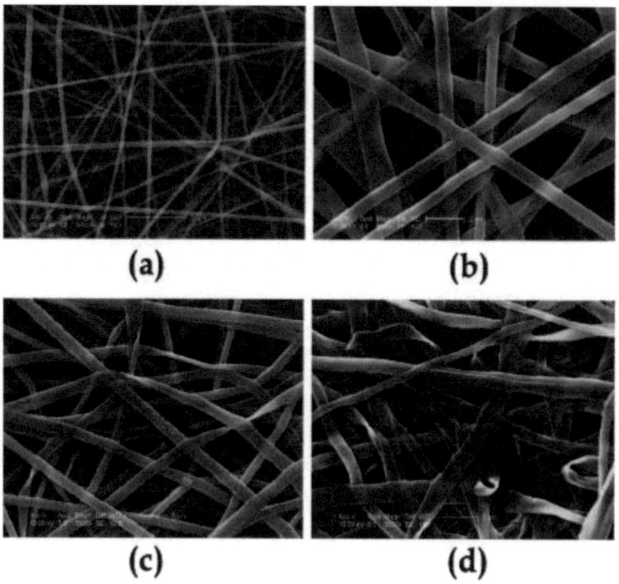

Fig. 7. Field emission scanning electron microscope images of electrospun Nafion-PAA-BMIm-BF4 blend at ionic liquid weight fraction of (a) 0%, (b) 10%, (c) 20%, (d) 30%. The weight ratio of Nafion:PAA is 3:2 at a 10 wt % total polymer concentration (Reproduced from Chen et al., (2009) *Macromolecules* 42, 3368-3373, Copyright (2009), with permeation from American Chemical Society)

2.3 Ultrasound and ionic-liquid-assisted synthesis method

Currently, the study of physical and chemical effects of ultrasound irradiation is a rapidly growing research area. When liquids are irradiated with high-intensity ultrasound irradiation, acoustic cavitations (the formation, growth, and implosive collapse of bubbles) provide the primary mechanism for sonochemical effects. During cavitation, bubble collapse produces intense local heating, high pressures, and extremely rapid cooling rates (Suslick et.al 1991, Suslick 1988). These transient, localized hot spots can drive many chemical reactions, such as oxidation, reduction, dissolution, decomposition, and promotion of polymerization (Suslick 1988). One of the most important recent aspects of sonochemistry has been its application in the synthesis of nanodimensional materials (Suslick & Price, 1999). Ultrasound irradiation offers a very attractive method for the preparation of novel materials with unusual properties and has shown very rapid growth in its application in materials science due to its unique reaction effects and ability to induce the formation of particles of much smaller sizes (Suslick 1988, Suslick & Price, 1999). The advantages of the sonochemical method include a rapid reaction rate, controllable reaction conditions, and the ability to form nanoparticles with uniform shapes, narrow size distributions, and high purity.

2.3.1 Ultrasound and ionic-liquid-assisted synthesis and characterization of polyaniline/Y_2O_3 nanocomposite with controlled conductivity

A sonochemical method has been employed to prepare polyaniline-Y_2O_3 (PANI/Y_2O_3) nanocomposite with controlled conductivity with the assistance of an ionic liquid by

Kowsari and Faraghi (Kowsari & Faraghi, 2010) Ultrasound energy and the ionic liquid replace conventional oxidants and metal complexes in promoting the polymerization of aniline monomer. Ionic liquids (with unique properties) can act as morphology templates for the synthesis of PANI/ Y_2O_3 nanocomposite with novel or improved properties. Here task specific acidic ionic liquids with different counter ions induce different template and different PANI morphology. The structures of ionic liquids studied by Kowsari Faraghi are shown in Fig. 8.

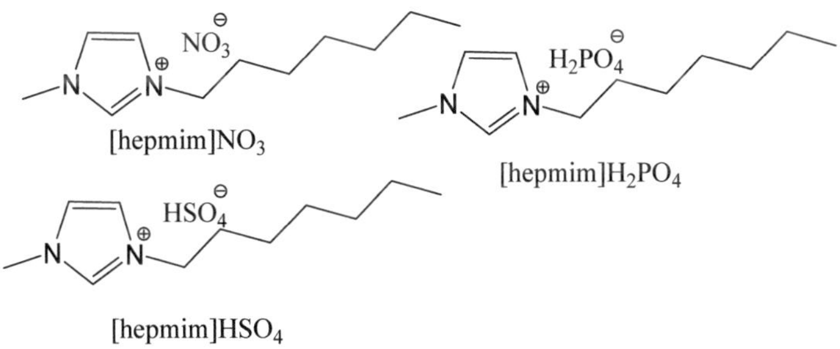

Fig. 8. The structures of ionic liquids in this study

Fig. 9a–d exhibits the morphology of PANI/Y_2O_3 nanocomposite in the presence of different type of ionic liquid. Interesting morphologies were obtained for PANI/Y_2O_3 nanocomposites. To compare the effects of ionic liquid additives on the properties of the resulting PANI/ Y_2O_3 nanocomposite, the monomer concentration ratio was kept constant. The differences in the structure of PANI/ Y_2O_3nanocomposite prepared, in the presence of ionic liquids are clearly visible. At the same magnification PANI/ Y_2O_3 nanocomposite (with [hepmim]. H_2PO_4 ionic liquid) reveals an interesting ribbon structure (Fig. 9a). The nanosheet structure of PANI/ Y_2O_3 in the presence of [hepmim] HSO_4, ionic liquid is shown in Fig. 9b. In Fig. 9b the smooth surface of PANI/ Y_2O_3 is visible. Smooth surfaces of it can be a reason for the better conductivities of these samples compared to Y_2O_3-free PANI. The SEM study shows that the presence of ionic liquid additives in polymerization strongly affects the morphology of PANI/ Y_2O_3. ionic liquids play a key role in tailoring the resultant conducting PANI/ Y_2O_3 structures. It was found that in the presence of [hepmim] NO_3, the products were regular solid microspheres covered with some nanoparticles.(Fig 9c) .TEM image of Y_2O_3 is shown in Fig 9d.

The influence of three different ionic liquid counter anions, namely HSO_4^-, $H_2PO_4^-$ and NO_3^- on conductivity was investigated by preparing composites in the presence of these anions. It was observed that the conductivity was strongly influenced by the type of anion, although the yields of the respective composites were subject to variation. Since the PANI in the composite is in its emeraldine salt form irrespective of the acid used, the variation in conductivity most probably stems from the differences in the size and nature of the dopant anions.

As can be seen in Fig. 10, the conductivities may be classified in the decreasing order NO_3^- < HSO_4^-< $H_2PO_4^-$

The effect of the concentration of ionic liquid = [hepmim]HSO_4) on PANI-Y_2O_3 morphology is shown is Fig 11. As is vivid, at 0.2 M concentration morphology is a mixture of fibers and plates. When the concentration increases to 0.4, fibers gradually disappear and change into plates. Also, when the concentration reaches 0.6, the plates become thin in terms of

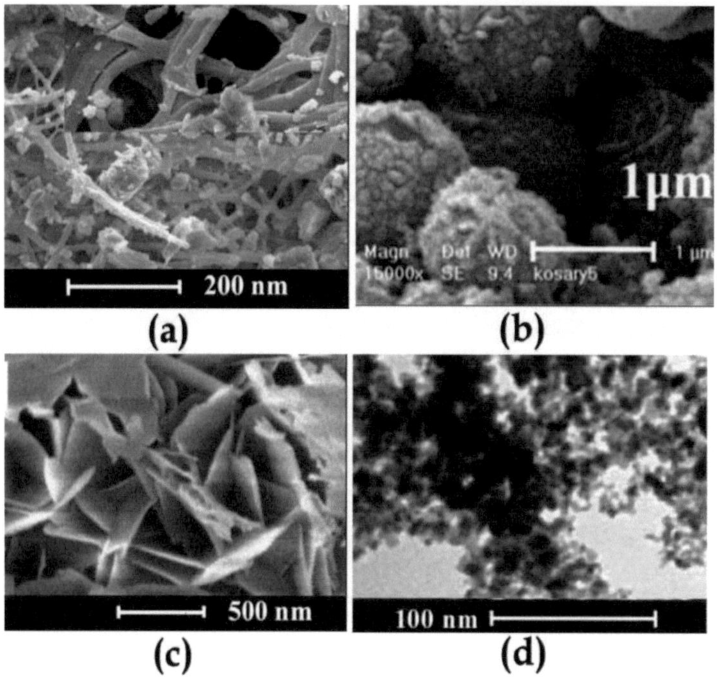

Fig. 9. SEM images of PANI/Y_2O_3 composite at different type of IL: (a) PANI/ Y_2O_3 composite (Y2O3 = 30%, aniline = 0.2 M, and IL = [hepmim]•H_2PO_4 = 0.6 M), (b) PANI/ Y_2O_3 composite (Y_2O_3 = 30%, aniline = 0.2 M, and IL = [hepmim]•HSO_4 = 0.6 M), (c) PANI/ Y_2O_3 composite (Y_2O_3= 30%, aniline = 0.2 M, and IL = [hepmim]•NO_3 = 0.6 M). (d) TEM image of Y_2O_3 (Reproduced from Kowsari & Faraghi (2010) *Ultrason. Sonochem.* 17, 4, 718-725, Copyright (2010), with permeation from Elsevier)

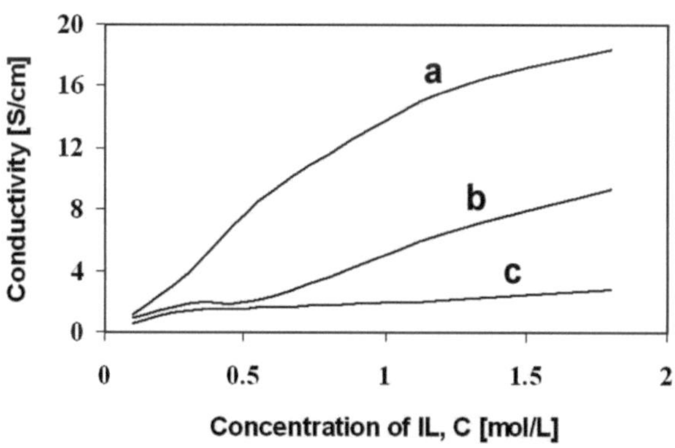

Fig. 10. The conductivities of PANI/Y_2O_3 composites at different concentrations of ILs: (a) PANI/Y_2O_3 composite (Y_2O_3 = 30%, aniline = 0.2 M, and IL = [hepmim]·H_2PO_4 = 0.6 M), (b) PANI/Y_2O_3 composite (Y_2O_3 = 30%, aniline = 0.2 M, and IL = [hepmim]·HSO_4 = 0.6 M), (c) PANI/Y_2O_3 composite (Y_2O_3 = 30%, aniline = 0.2 M, and IL = [hepmim]·NO_3 = 0.6 M). (Reproduced from Kowsari & Faraghi (2010) *Ultrason. Sonochem.* 17, 4, 718-725, Copyright 2010), with permeation from Elsevier)

thickness. As for both mentioned ionic liquids, increase in concentration brings about higher conductivity and yield, as well.

In the present study, when there is an increase in a frequency from 20 to 40 kHz, the morphology evolves as sphere shapes. In additions, the yield would be 55 wt %, 76 wt %, and 80 wt % at 20, 30, and 40 kHz for frequency.

Ultrasonic irradiation at 40 kHz, yields constantly higher degradation efficiencies compared with that at 20 kHz for all ionic liquids. Since, in the present study, the ionic liquid replaces conventional oxidant and metal complexes for polymerization, the increase of ionic liquids degradation leads to the increase of the concentration of alkyl radicals and the increase of aniline polymerization and, therefore, the increase of yield of product.

2.4 Ziegler-Natta polymerisation of ethylene

Ziegler-Natta polymerisation is used extensively for the polymerisation of simple olefins (e.g. ethylene, propene and 1-butene) and is the focus of much academic attention, as even small improvements to a commercial process operated on this scale can be important. Ziegler-Natta catalyst systems, which in general are early transition metal compounds used in conjunction with alkylaluminium compounds, lend themselves to study in the chloroaluminate(III) ionic liquids, especially the ones with an acidic composition.

During studies into the behaviour of titanium(IV) chloride in chloroaluminate(III) ionic liquids Carlin *et al* carried out a brief study to investigate if Ziegler-Natta polymerisation was possible in an ionic liquid (Carlin et. al 1990)

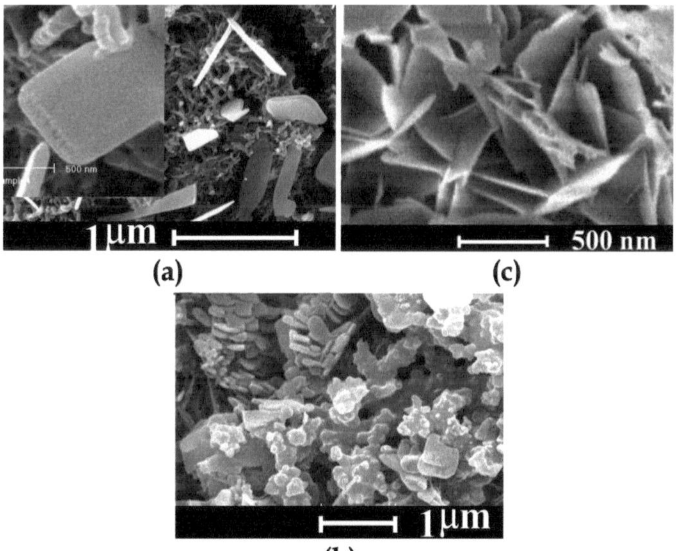

Fig. 11. SEM images of PANI/Y_2O_3 composite (Y_2O_3 = 30%, aniline = 0.2 M, and IL = [hepmim]·HSO_4) at different concentrations of ILs (a) [IL] = 0.2 M, (b) [IL] = 0.4 M, (c) [IL] = 0.6 M (Reproduced from Kowsari & Faraghi (2010) *Ultrason. Sonochem.* 17, 4, 718-725, Copyright (2010), with permeation from Elsevier)

2.5 Microemulsion polymerization by ionic liquids

Microemulsions are thermodynamically stable dispersions containing two immiscible liquids stabilized by surfactants at the liquid-liquid interface. Compared with "classic" water-inoil microemulsions, ionic liquid-in-oil microemulsions, in which an ionic liquid dispersed in an oil-continuous phase by suitable surfactants, are of great interest due to the unique features of both ionic liquid and microemulsion system (Eastoe et al., 2005; Binks et al., 2003; He et al., 2006; Atkin et al., 2007; Gao et al., 2004; Li et al., 2007). Microemulsions composed of a room temperature ionic liquid, 1-butyl-3-methylimidazolium tetrafluoroborate ([Bmim][BF_4]), cyclohexane, and a surfactant, Triton X-100, have been recently studied by Han (Gao et al., 2004) and Eastoe (Eastoe et al., 2005). Small angle neutron scattering data clearly demonstrated the formation of ionic liquid nanodroplets dispersed in cyclohexane (Eastoe et al., 2005). More recently, microemulsions composed of an ionic liquid dispersed in a variety of organic oils, such as toluene, (Li et al., 2007; Gao et al., 2006) xylene (Gao et al., 2006b) and benzene, (Gao et al., 2007) have been investigated by several groups. These microemulsions behaviors are consistent with water-in-oil microemulsions and provide nanosized ionic liquid domains as reaction or extraction media which should avoid contacting with water. However, potential uses of these ionic liquid - based microemulsions have not been intensively studied so far.

2.5.1 Microemulsions with an ionic liquid surfactant

The formation of amphiphilic association structures in and with ionic liquids, such as micelles, vesicles, microemulsions and liquid crystalline phases has been reviewed recently (Hao & Zemb, 2007, Qiu, & Texter, 2008). The formation of micelles formed by common surfactants in EAN was first documented in 1982 by Evans and coworkers (Evans et al, 1982, 1983), whereas the formation of liquid crystals of lipids in ethyl ammonium nitrate (EAN) (Evans, & Kaler, 1983) as well as lyotropic liquid crystals of non-ionic surfactants were observed in the same room temperature ionic liquid (Araos & Warr, 2005). The formation of liquid crystalline phases of binary mixtures of [C_{16}mim][Cl] and EAN has also been described lately (Zhao et al, 2009). The self-aggregation of common ionic and non-ionic surfactants in imidazolium based RTIL was also reported. (Anderson et al, 2003; Patrascu et al, 2006; Hao et al, 2005)

As imidazolium based ionic liquids (ILs) with long-chain hydrocarbon residues exhibit surfactant properties in water (Thomaier& Kunz, 2007; (Bowers et al, 2004; Miskolczy et al, 2004; Sirieix-Plenet et al, 2004; Kaper & Smarsly, 2006; Seoud et al, 2007), Thomaier and coworkers could recently demonstrate that the surfactant like ionic liquid (SLIL) 1-hexadecyl-3-methylimidazolium chloride ([C_{16}mim][Cl]) forms colloidal structures in EAN as well. They found a critical aggregation concentration (cac) approximately ten times higher than the critical micelle concentration (cmc) in the corresponding aqueous system. (Thomaier & Kunz, 2007) This gap is in accordance with the results of Evans *et al.*, who found that the cacs of classical surfactants in EAN are between 5 and 10 times higher than in water. (Evans et al, 1982)

2.5.2 Polymerization of ionic liquid-based microemulsions: The synthesis of polymer electrolytes

S. Yu and coworkers (Yu et al, 2008) reported the first example of polymerization of microemulsions comprising surfactant stabilized ionic liquid nanodomains. Polymerization of these ionic liquid-based microemulsions yielded free-standing, flexible, and transparent polymer electrolytes even though the resulting vinyl polymers are incompatible with ionic liquid cores. The obtained ionic liquid/ polymer composites show high conductivity at both room temperature and elevated temperature. This facile yet and effective method provides a versatile platform for the preparation of semisolid polymer electrolytes containing ionic liquids. Although They only demonstrated this method with vinyl monomers of styrene, methyl methacrylate, vinyl acetate, and *N,N*-dimethylacrylamide and only one surfactant as examples, they believe that this method should be extendable to other liquid monomers with similar polarity and ionic liquids with desirable properties. The elucidation of interaction between the polymeric matrix and ionic liquids need to be further explored in future work.

2.6 Aqueous/ionic liquid interfacial polymerization

Interfacial polymerization (IP) involves step polymerization of two reactive monomers or agents, which are dissolved respectively in two immiscible phases and the reaction takes place at the interface of the two liquids. IP has been used to prepare various polymers, such as poly(urea), poly(amide), and poly(ester) capsules.

It allows the synthesis of polymers at low temperature with limited side reactions, and can avoid the use of catalysts or phase transfer agents. The relative ease of IP has made it the preferred technique in many fields, ranging from microencapsulation of pharmaceutical products to preparing conducting polymers. IP is commonly done with volatile organic solvents as the organic phase, such as benzene (Chu et al, 2002), chloroform (Ashgarian et al, 1999), and toluene (Lu et al, 2002). The nonvolatile nature of ionic liquids gives them significant advantage in minimizing solvent consumption. They are good solvents for organic, inorganic, and polymeric compounds, and have been used as media for chemical reactions (Dupont et al, 2002), including polymerization (Harrisson et al, 2003) and ionic liquid/CO_2 biphasic reactions. Poly aniline (PANI) has been synthesized with water/ionic liquid IP by Kowsari and Yavari (Kowsari & Yavari, 2009). This method has some obvious advantages, such as PANI nanofiber can be fabricated and both solvents used are environmentally benign.

3. Polycondensation processes in ionic liquids

Polycondensation is typically conducted at relatively high temperature thus non-volatile and thermally stable ionic liquids seem to be suitable solvents for polycondensation processes. Research in this area has mainly been directed towards synthesis of polyamides, polyimides and polyesters. On the other hand, in some studies of polycondesation catalytic effect of ionic liquids was observed. More recently, two step procedures (involving post polycondensation in ionic liquid) was applied to polycondensation of sebacic, adipic, and succinic acid with aliphatic diols. Aliphatic polyesters with Mw up to 6×104 were obtained and once more it was noted that solubility of polyesters in ionic liquid was a limiting factor. Solubility depends on structure of IL (nature of cation and anion) and correlation was found between the miscibility of aliphatic polyester/ionic liquid system and the extent to which their solubility parameters matched (Kubisa, 2009).

3.1 Ionic liquids as novel solvents and catalysts for the direct polycondensation

The direct polycondensation of *N,N*'-(4,4'-oxydiphthaloyl)- bis-L-phenylalanine diacid with various aromatic diamines was performed in ionic liquid media by Mallakpour and Kowsari

(Mallakpour & Kowsari, 2005). The influence of various reaction parameters, including the nature of the ionic liquid cations and anions, the monomer structures, the reaction temperature, and the reaction time, on the yields and inherent viscosities of the resulting optically active poly(amide imide)s (PAIs) were investigated. Direct polycondensation preceded in ionic liquids and triphenyl phosphite (a condensing agent) without any additional extra components, such as LiCl and pyridine, which are used in similar reactions in ordinary molecular solvents. Therefore, ionic liquids can act as both solvents and catalysts. Various high-molecular weights, optically active PAIs were obtained in high yields with inherent viscosities ranging from 0.54 to 0.88 dL/g. This method was also compared with three other classical methods for the polycondensation of the aforementioned monomers.

3.2 Polycondensation processes in ionic liquid under microwave irradiation

Application of ionic liquids as media to microwave-assisted reactions offers several advantages. Typical organic solvents are frequently flammable and volatile, which is a safety hazard for high-temperature and closed-vessel applications using microwaves. In contrast, ionic liquids have high boiling-points, low vapor pressures, and high thermal stabilities. In addition, typical ionic liquids have moderately high dielectric constants (in the range of 10–15), and relatively low heat capacities (in the range of 1–2 J/g K). This combination allows ionic liquids to absorb microwaves efficiently. Owing to these advantages, ionic liquids have been investigated as solvents in a number of microwave-mediated reactions. Microwave heating in ionic liquids was used also for polycondensation reactions leading to polyamides (Mallakpour& Kowsari, 2006). Certain advantages have been indicated, but until now only slight improvement of reactions conditions (more efficient heating, higher rates) has been achieved.

4. Ionic polymerization

4.1 Cationic polymerization

Ability of ionic liquids to dissolve wide range of inorganic compounds was exploited in the study in which organoborate acids (HBOB) (bisoxalatoboric acid, bissuccinatoboric and bisglutaratoboric acids) were used as initiators of the cationic polymerization of styrene in pyrollidonium, imidazolium and phosphonium bis (trifluromethanesulfonyl) amide ionic liquids. In another study, cationic polymerization of styrene initiated with $AlCl_3$ in ionic liquid ([bmim][PF6]), supercritical CO_2 and organic solvent (CH_2Cl_2) was investigated. The only conclusion was that in ionic liquids rates and molecular weights are higher than in organic solvent (Kubisa, 2009).

Studies on the dimerisation and hydrogenation of olefins with transition metal catalysts in acidic chloroaluminate(III) ionic liquids report the formation of higher molecular weight fractions consistent with cationic initiation (Chauvin et. al 1990; Ellis et. al 1999). These studies ascribe the occurrence of the undesired side reaction to both the Lewis acid and the proton catalysed routes. Their attempts to avoid these side reactions led to the preparation of alkylchloroaluminate(III) ionic liquids and buffered chloroaluminate(III) ionic liquids (Chauvin et. al 1990; Ellis et. al 1999).

Attempts to bring the benefits of ionic liquid technology by drawing on the inherent ability of the chloroaluminate(III) ionic liquids to catalyse cationic polymerisation reactions, as opposed to minimising them, were patented by Ambler et al of BP Chemicals Ltd in 1993 (Ambler, 1993). They used acidic [EMIM][Cl-$AlCl_3$] ($X(AlCl_3)$ = 0.67) for the polymerisation of butene to give products that find application as lubricants. The polymerisation can be carried out by bubbling butene through the ionic liquid. The product formed a separate layer that floats upon the ionic liquid and was isolated by a simple process. Alternatively, the polymerisation was carried out by injecting the ionic liquid into a vessel charged with butene. After a suitable settling period the poly(butene) was isolated in a similar fashion. The products from these reactions must be best described as oligomers as opposed to polymers as the product is still in the liquid form. Chain transfer to impurities, ionic liquid, monomer and polymer will terminate the propagation reaction resulting in the low-mass products.

4.1.1 Cationic polymerization of styrene in $scCO_2$ and [bmim][PF_6]

Bueno and coworkers (Bueno et. al 2009) presented a study on the cationic polymerization of styrene in different reaction media, supercritical CO_2 ($scCO_2$), [bmim][PF_6], dichloromethane, and $scCO_2$ plus [bmim][PF_6] using $AlCl_3$ as initiator at temperatures from 273 to 333 K. The reactions were analyzed in relation to the monomer conversion rate, polymer structures, average molecular weights and molecular weight distribution. In all cases, the styrene polymerization produced oligomers. Reactions using ionic liquid as a solvent led to higher molecular weight and monomer conversion rate. Monomer conversion rates of about 100% and weight average molecular weight (*Mw*) of 2400 at 298 K were obtained. Reactions with $scCO_2$ as a solvent yielded low monomer conversion rates (around 50%) and a *Mw* = 2000 at 298 K. The oligolystyrenes presented rr syndiotatic-rich sequences in the microstructure in all reaction conditions. The use of ionic liquid and $scCO_2$ results in better yields than the use of the other organic solvents.

4.2 Anionic polymerization

4.2.1 Anionic polymerization of methyl methacrylate in an ionic liquid

Anionic polymerization reactions of methyl methacrylate (MMA) in ionic liquids were carried out by utilizing alkyl lithium initiators such as n-butyl lithium (n-BuLi) and diphenylhexyl lithium (DPHLi) by H. Kokubo and Watanabe (Kokubo & Watanabe, 2008). The polymerization in ionic liquids having bis(trifluoromethyl sulfonyl)amide ([NTf_2]) anion did not yield poly(MMA) (PMMA), because of the deactivation of the initiator due to an attack on the trifluoromethyl group. By using 1-butyl-3-methylimidazolium hexafluorophosphate ([C_4mim][PF_6]), the polymerization reactions proceeded. The PMMA prepared in [C_4mim][PF_6] gave low yields (5–9%), as compared with those (15–62%) obtained forPMMAprepared in tetrahydrofuran (THF), and the prepared PMMA had large polydispersity indices (≈ 2.0). The results obtained can be attributed to the high reaction temperature (0 °C), in comparison to the common anionic polymerization temperature of MMA, -78 °C, and the reaction between the initiator and the imidazolium cation. The initiator was considered to be deactivated because the hydrogen atom at the 2-position of the imidazolium ring was withdrawn by the alkyl lithium initiator. The tacticity of the obtained PMMA, prepared in [C_4mim][PF_6] by utilizing DPHLi, was rich in *mm* triads, similar to that polymerized in toluene

5. Miscellaneous application of ionic liquids in polymer chemistry

5.1 Polymer electrolytes: Compatible system between polymers and ionic liquids (Ion gels)

Polymer electrolytes containing room temperature ionic liquids were first reported on such a system by Noda & Watanabe (Noda & Watanabe, 2000). Successively, the study was expanded to polymer electrolytes containing chemically stable ionic liquids (Noda & Watanabe, 2000; Watanabe et. al 1993; Watanabe et. al 1995; Watanabe & Mizumura 1996; Ogata et. al 1995; Susan et. al 2005). Watanabe and coworkers found that common vinyl monomers were widely soluble in common ionic liquids and that they could be polymerized by free radical polymerization (Noda & Watanabe 2000; Susan et. al 2005). In certain cases, surprisingly good compatibility of the resulting polymers with the ionic liquids could be achieved irrespective of the polymer concentration and temperature (Fig. 12).

Typically, the polymerization of methyl methacrylate (MMA) in a common ionic liquid, 1-ethyl-3-methyl imidazolium bis(trifluoromethane sulfone)imide ([C_2mim][NTf_2]) in the presence of a small amount of a cross-linker gives self standing, flexible, and transparent polymer gels. The polymer and ionic liquid composite gels, which they term "ion gels", show a single T_g for a given range of [C_2mim][NTf_2] composition, and the T_g decreases with an increase in the mole fraction of [C_2mim][NTf_2] (Fig 13) (Susan et. al 2005; Ueki &.Watanabe 2008).

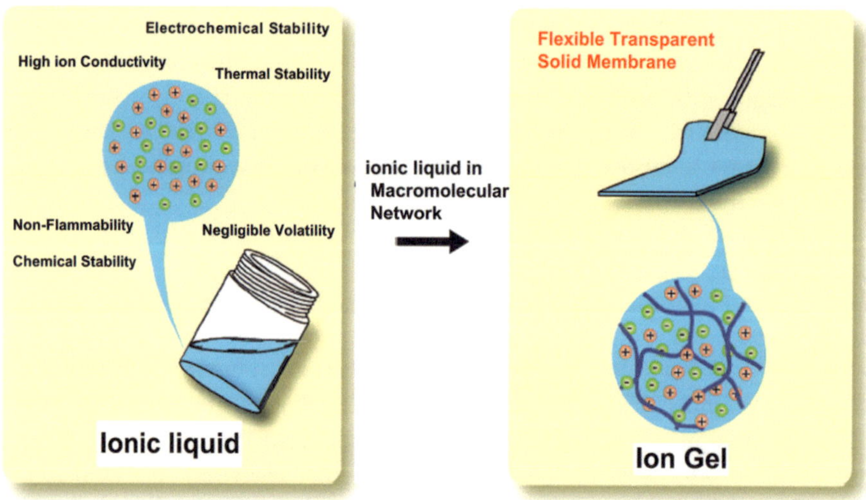

Fig. 12. Preparation of ion gels (compatible binary systems between polymer networks and ionic liquids) by a variety of methods such as in situ radical polymerization of vinyl monomers in ionic liquids and sol–gel transition of macromolecules in ILs. The flexibility of molecular design of ionic liquids can facilitate various interesting applications of ion gels, particularly as ion-conducting polymer electrolytes (Reproduced from Ueki & Watanabe (2008) *Macromolecules*, 41, 11, 3739-3748, Copyright (2008) ,with permeation from American Chemical Society)

5.2 Ion gel gated polymer thin-film transistors

Lee and coworkers (Lee et. al 2007) demonstrate that a gel electrolyte (a so-called "ion gel") based on a mixture of an ionic liquid and a block copolymer can provide both large specific capacitance (>10 μF/cm2) and greatly improved polarization response times (1 ms) when used as the gate dielectric in a polymer TFT (thin-film transistor). The improved properties allow transistor operation at frequencies greater than 100 Hz, significantly faster than what has been demonstrated previously and opening the door to a broader range of applications. Furthermore, the ion gel material is solution processible, making it potentially compatible with high throughput patterning methods (e.g., ink jet printing). Unlike traditional solid polymer electrolytes, which are obtained by the dissolution of salts by ion-coordinating polymers, the ion gel electrolyte is formed by gelation of poly(styrene-block-ethylene oxide-block-styrene) (SOS) triblock copolymer in 1-butyl-3-methylimidazolium hexafluorophosphate ([BMIM][PF_6]), a room temperature ionic liquid; the structures of these materials are shown in Fig. 14.

6. Conclusions

Further progress in the area of ionic liquids and specifically polymerization mechanisms is inevitable. Due to distinctive features and "green" applications of ionic liquids in industry, researchers have found this research topic an interesting and promising one. Very attractive initial applications for the incorporation of ionic liquids monomers into polymers have been reported, but the range of possible ionic liquid structures is much larger than has so far been explored. What seems to be necessary and needed is careful selection of the systems studied where real synthetic advantages of applying ionic liquids are offered or new insights into polymerization mechanisms are provided. These research requirements are opposed to mere mentioning, although new, examples of processes that can be carried out with regard to ionic liquids. This way, achievements in the field can be foreseen.

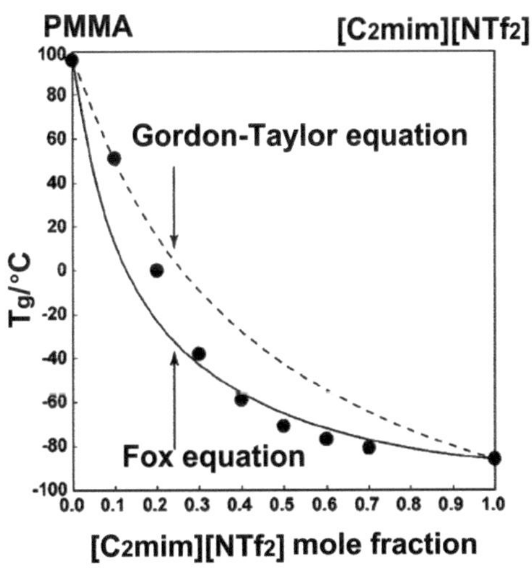

Fig. 13. Glass transition temperatures (*T*g) of PMMA network polymers with dissolved [C_2mim][NTf_2] as a function of [C_2mim][NTf_2] mole fraction. The points denote experimental *T*g results, the broken lines denote the fitted profiles by the Gordon-Taylor equation, and the solid lines denote those by the Fox equation. (Reproduced from Susan et. al (2005) *J Am. Chem. Soc.* 2005, 127, 4976, Copyright (2005), with permeation from American Chemical Society)

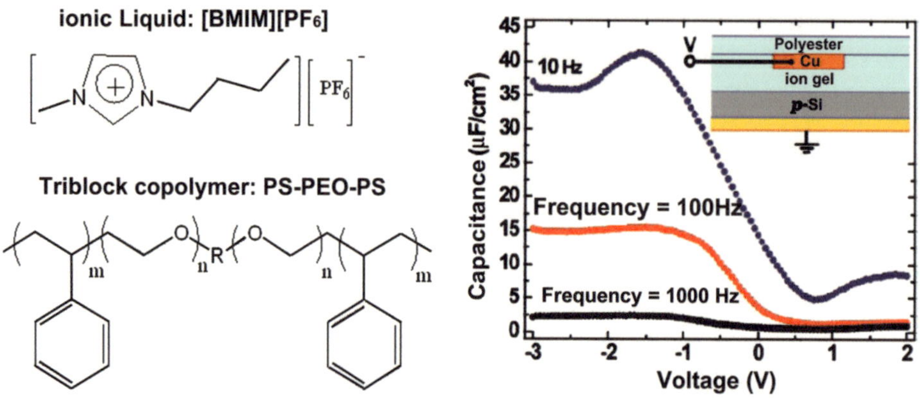

Fig. 14. Structure of the ionic liquid and triblock copolymer ion gel components (left) and C-V characteristics of a p-Si/ion gel/Cu test structure (see inset) at three frequencies (right). The C-V curves indicate large frequency-dependent hole accumulation in the Si at negative bias on the top contact (Reproduced from Lee et. al (2007) *J. Am. Chem. Soc.* 129, 4532-4533, Copyright (2007), with permeation from American Chemical Society)

7. Acknowlegment

The author wishes to express her gratitude to National Elite Foundation for the financial support.

8. Reference

Abdul-Sada, A. K.; Ambler, P.W.; Hodgson, P. K. G., Seddon, K. R. & Stewart N. J. (1995a). Ionic liquids. *World patent*, WO95/21871.

Abdul-Sada, A. K.; Atkins, M. P.; Ellis, B.; Hodgson, P. K. G.; Morgan, M. L. M.& Seddon K. R. (1995b). Alkylation process. *World patent*, WO95/21806.

Amajjahe, S. & Ritter, H. (2009). Microwave-Sensitive foamable poly(ionic liquids) bearing tert-butyl ester groups: Influence of counterions on the ester pyrolysis. *Macromol. Rapid Commun*, 30, 2, 94–98, 1521-3927

Ambler, P. W.; Hodgson, P.K. G & Stewart, N.J. (1993) *European Patent*, EP, 0558187.

Ambler, P. W.; Hodgson, P. K. G. & Stewart, N. J. (1996). Butene polymers. *European patent application, EP*/0558187A0558181.

Anderson, J. L.; Pino, V.; Hagberg, E. C.; Sheares, V. V. & Armstrong, D. W. (2003). Surfactant solvation effects and micelle formation in ionic liquids. *Chem. Commun*, 19 2444-2445, 1359-7345

Araos, M. U. & Warr, G. G. (2005). Self-assembly of nonionic surfactants into lyotropic liquid crystals in ethylammonium nitrate, a room-temperature ionic liquid. *J. Phys. Chem. B*, 109, 30, 14275-14277, 1520-6106

Ashgarian, B.; Cadenhead, D. A. (1999). nterfacial Compression and Polymerization of 4- and 9-((Butoxycarbonyl)methyl)urethane−Diacetylene Monomolecular Films: A Fluorescence Microscopy Study. *Macromolecules*, 32, 17, 5575-5580, 0024-9297

Atkin, R.; Warr, G. G. (2007). Phase behavior and microstructure of microemulsions with a room-temperature ionic liquid as the polar phase. *J. Phys. Chem. B,* 111, 31, 9309–9316, 1520-6106

Biedron, T. & Kubisa, P. (2001). Atom-transfer radical polymerization of acrylates in an ionic liquid. *Macromol Rap Commun,* 22, 15, 1237–1242, 1521-3927

Biedron, T. & Kubisa P. (2002). P. Atom transfer radical polymerization of acrylates in an ionic liquid: synthesis of block copolymers. *J Polym Sci Part A Polym Chem,* 40, 16, 2799–809, 1099 0518

Biedron, T. & Kubisa, P. (2003) Ionic liquids as reaction media for polymerization processes: atom transfer radical polymerization (ATRP) of acrylates in ionic liquids. *Polym Int,* 52, 10, 1584–8, 1097-0126

Binks, B. P.; Dyab, A. K. F.; Fletcher, P. D. I. (2003) Novel emulsions of ionic liquids stabilised solely by silica nanoparticles. *Chem. Commun.,* 20, 2540–2541, 1359-7345

Bosman, Y.; Lau, Y.; Gilgenbach, R. M. (2003) Microwave absorption on a thin film. *Appl. Phys. Lett., 82,* 9, 1353-1355, 0003-6951

Bowers, J.; Butts, C. P.; Martin, P. J.; Vergara-Gutierrez, M. C. & Heenan, R. K. (2004). Aggregation behavior of aqueous solutions of ionic liquids. *Langmuir,* 20, 6, 2191-2198, 0743-7463

Bregar, V. B. (2004). Advantages of ferromagnetic nanoparticle composites in microwave absorbers. *IEEE Trans. Magn.,* 40, 3, 1679-1684, 0018-9464

Bueno, C.; Cabral,V. F.; Cardozo-Filho, L.; Dias, M.L. & Antunes.O. A. C. (2009). Cationic polymerization of styrene in $scCO_2$ and [bmim][PF_6], *J. of Supercritical Fluids,* 48, 2, 183–187, 0896-8446

Carlin, R. T.; Osteryoung, R. A.; Wilkes, J. S. & Rovang, J. (1990). Studies of titanium(IV) chloride in a strongly Lewis acidic molten salt: electrochemistry and titanium NMR and electronic spectroscopy. *Inorg. Chem,* 29, 16, 3003-3009, 0020-1669

Carlin, R. T. & Fuller, J. (1997) Ionic liquid–polymer gel catalytic membrane. *Chem. Commun.,* 15, 1345–1346, 1359-7345

Chandrasekhar, P.; Naishadham, K. (1999). Broadband microwave absorption and shielding properties of a poly(aniline). *Synth. Met,* 105, 2, 115- 120, 0379-6779

Chauvin, Y. (1996). Two-phase catalysis in nonaqueous ionic liquids. *Actualite chimique* v, 44–46, 0151-9093

Chauvin, Y.; Commereuc, D.; Hirschauer, A.; Hugues, F. & Saussine, L. (1988). Process and catalyst for the dimeriation or codimerization of olefins. *French Patent, FR* 2,611,700.

Chauvin, Y.; Commereuc, D.; Hirschauer, A.; Hughes, F. & Saussine, L. (1989). Process and catalyst for the alkylation of isoparaffins with alkenes. *French Patent, FR* 2,611,572.

Chauvin, Y.; Gilbert, B. & Guibard, I. (1990) Catalytic dimerization of alkenes by nickel complexes in organochloroaluminate molten salts. *J. Chem. Soc., Chem. Commun,* 23, 1715-1716, 0022-4936

Chauvin, Y.; Mussmann, L.& Olivier, H. (1995). A novel class of versatile solvents for two-phase catalysts: Hydrogenation, isomerization, and hydroformylation of alkenes catalyzed by rhodium complexes in liquid 1,3-dialkylimidazolium salts. *Angew Chem Int Ed Engl,* 34, 23-24, 2698–2700, 1521-3773

Chen, H.; Yossef A. Elabd,Y. A.(2009). Polymerized ionic liquids: solution properties and electrospinning. *Macromolecules, 42*,9, 3368-3373, 0024-9297

Chiefari, J.; Chong, Y. K.; Ercole, F., Krstina, J.; Jeffery, L.; Mayadunne, R. T. A. (1998). Living free-radical polymerization by reversible additionfragmentation chain transfer: the RAFT process. *Macromolecules,* 31, 16, 5559–5562, 0024-9297

Chu, L. Y.; Park, S. H.; Yamaguchi, T.; Nakao, S. I. (2002). Preparation of micron-sized monodispersed thermoresponsive core−shell microcapsules. *Langmuir,* 18, 5, 1856-1864, 0743-7463

Ding, S.; Tang, H.; Radosz, M. & Shen, Y. (2004). Atom transfer radical polymerization of ionic liquid 2-(1- butylimidazolium-3-yl) ethyl methacrylate tetrafluoroborate. *J. Polym. Sci., Part A: Polym. Chem,*. 42, 22, 5794–5801, 1099-0518

Ding, S.; Radosz, M. & Shen, Y. (2005). Ionic liquid catalyst for biphasic atom transfer radical polymerization of methyl methacrylate. *Macromolecules,* 38, 14, 5921–5928, 0024-9297

Dupont, J.; de Souza, R. F.; Suarez P. A. Z. (2002). Review ionic liquid (Molten Salt) phase organometallic catalysis. *Chem Rev,* 102, 10, 3667-3692, 0009-2665

Earle, M. J.; McCormac, P. B. & Seddon, K. R. (1999). Diels-Alder reactions in ionic liquids. *Green Chem,* 1, 23–25, 1463-9262

Eastoe, S.; Gold, S. E.; Rogers, A.; Paul, T.; Welton, R. K.; Heenan, I. G.(2005). Ionic liquid-in-oil microemulsions. *J. Am. Chem. Soc,* 127, 20, 7302–7303, 0002-7863

Ellis,B.; Keim,W. & Wasserscheid, P.(1999). Linear dimerisation of but-1-ene in biphasic mode using buffered chloroaluminate ionic liquid solvents. *J. Chem. Soc., Chem. Commun,* 4, 337-338, 0022-4936

Elseoud, O. A.; Pires, P. A. R.; Abdel-Moghny, T.; Bastos E. L. (2007). Synthesis and micellar properties of surface-active ionic liquids: 1-alkyl-3- methylimidazolium chlorides. *J. Colloid Interface Sci.*, 313, 1, 296-304, 0021-9797

Evans, D. F.; Yamauchi, A.; Roman R.& Casassa E. Z. (1982). Micelle formation in ethylammonium nitrate, a low-melting fused salt. *J. Colloid Interface Sci.*, 88, 1, 89-96, 0021-9797

Evans,D. F.; Kaler, E. W. & Benton; W. J. (1983). Liquid crystals in a fused salt: distearoylphosphatidylcholine in *N*-ethylammonium nitrate *J. Phys Chem,* 87, 18, 533-535

Evans, D. F.; A. Yamauchi, A.; Wei, G. J.; Bloomfield, V. A. (1983). Micelle size in ethylammonium nitrate as determined by classical and quasi-elastic light scattering. A, *J. Phys. Chem. 87*, 18, 3537-3541

Flannigan, J. D; Hopkins, D. S. D.; Suslick, K. S. (2005). Sonochemistry and sonoluminescence in ionic liquids, molten salts, and concentrated electrolyte solutions. *J. Organomet. Chem,* 690,15, 3513–3517, 0022-328X

Freemantle, M. (1998). Designer solvents—Ionic liquids may boost clean technology development. *Chem Eng News*, 76, 32–37, 0009-2347

Galiński, M.; Lewandowski, A. & Ste-pniak, I. (2006) Ionic liquids as electrolytes. *Electrochim. Acta*, 51, 26, 5567–5580, 0013-4686

Gao, H. X.; Li, J. C.; Han, B. X.; Chen, W. N.; Zhang, J. L.; Zhang, R.; Yan, D. D. (2004) Microemulsions with ionic liquid polar domains. *Phys. Chem. Chem. Phys.* 6, 2914–2916, 1463-9076

Gao, Y. A.; Wang, S. Q.; Zheng, L. Q.; Han, S. B.; Zhang, X.; Lu, D. M.; Yu, L.; Ji, Y. Q.; Zhang, G. Y. (2006) Microregion detection of ionic liquid microemulsions. *J. Colloid Interface Sci.*, 301, 2, 612–616, 0021-9797

Gao, Y. A.; Zhang, J.; Xu, H. Y.; Zhao, X. Y.; Zheng, L. Q.; Li, X. W.; Yu, L. (2006). Structural Studies of 1-Butyl-3-methylimidazolium Tetrafluoroborate/TX-100/ *p*-Xylene Ionic Liquid Microemulsions. *Chem Phys Chem* , 7, 7, 1554–1561, 1439-7641

Gao, Y.; Li, N.; Zheng, L.; Bai, X.; Yu, L.; Zhao, X.; Zhang, J.; Zhao, M.; Li, Z. (2007). Role of solubilized water in the reverse ionic liquid microemulsion of 1-Butyl-3-methylimidazolium Tetrafluoroborate/TX-100/Benzene. *J. Phys. Chem. B,* 111, 10, 2506–2513, 1520-6106

Gordon, C. M., Holbrey, J. D.; Kennedy, A. & Seddon, K. R. (1998). Ionic liquid crystals: Hexafluorophosphate salts. J Mater Chem, 8, 2627–2636, 0959-9428

Hansen, F. K. & Rødsrud, G. (1991). Surface tension by pendant drop: I. A fast standard instrument using computer image analysis. *J. Colloid Interface Sci.*, 141,1, 1-9, 0021-9797

Hao, J.; Song, A.; Wang, J.; Chen, X.; Zhuang, W.; Shi, F.; Zhou, F. & Liu, W. (2005). Self-assembled structures in room-temperature ionic liquids, *Chem. Eur. J.*, 11,13, 3936-3940, 1521-3765

Hao, J.; Zemb,T. (2007). Self-assembled structures and chemical reactions in room-temperature ionic liquids, *Curr. Opin. Colloid Interface Sci.*, *12*, 3, 129-137, 1359-0294

Harrisson, S., Mackenzie, S. R.; Haddleton, D. M. (2003). Pulsed laser polymerization in an ionic liquid: Strong solvent effects on propagation and termination of methyl methacrylate. *Macromolecules*, 36, 14, 5072-5075, 0024-9297

He, Y.; Li, Z.; Simone, P.; Lodge, T. P. (2006). Self-assembly of block copolymer micelles in an ionic liquid. *J. Am. Chem. Soc.*, 128, 8, 2745–2750, 0002-7863

Hirao, M.; Ito, K. &; Ohno, H. (2000). Preparation and polymerization of new organic molten salts; *N*-alkylimidazolium salt derivatives. *Electrochim. Acta,* 45, 8-9, 1291- 1294. 0013-4686

Jaeger, D. A. & Tucker, C. E. (1989). Diels-Alder reactions in ethylammonium nitrate, a low-melting fused salt. *Tetrahedron Lett,* 30, 1785–1788, 0040-4039

Holbrey, J. D. & Seddon, K. R. (1999a). The phase behaviour of 1-alkyl-3-methylimidazolium tetrafluoroborates: ionic liquid and ionic liquid crystals. *J Chem Soc Dalton Trans*, 13, 2133–2139, 1472-7773

Holbrey, J. D. & Seddon, K. R. (1999b). Ionic liquids. *Clean Prod Proc*, 1, 223–237, 1435-2974

Bonho^te, P., Dias, A. P.; Papageorgiou, N., Kalyanasundaram, K. & Gra¨tzel, M. (1996). Hydrophobic, highly conductive ambient-temperature moltensalts. *Inorg Chem,* 35, 5, 1168–1178, 0020-1669

Holly L. Ricks-Laskoski, and Arthur W. Snow (2006). Synthesis and electric field actuation of an ionic liquid polymer. *J. Am. Chem. Soc.*, 128, 38, 12402-12403, 0002-7863

Juger, J.; Meyer, F.; Vidal, F.; Chevrot, C. & Teyssie, D. (2009). Synthesis, polymerization and conducting properties of an ionic liquid-type anionic monomer. *Tetrahedron Lett.*, 50, 1, 128–131, 0040-4039

Kaper, H. & SmarslyB. (2006). Templating and phase behavior of the long chain ionic liquid C_{16}mimCl,. *Phys. Chem, 220,* 10, 1455-1471, 0942-9352

Kato, Y.; Satoshi, S.; Sugimoto, S.; Shinohara, K., Tezuka, N., Kagotani, T. & Inomata, K. (2002). Magnetic Properties and Microwave Absorption Properties of Polymer-Protected Cobalt Nanoparticles. *Mater. Trans,* 43, 3, 406-409, 1345-9678

Kokubo, H. & Watanabe M. (2008). Anionic polymerization of methyl methacrylate in an ionic liquid. *Polym. Adv. Technol,* 19,10, 1441– 1444, 1099-1581

Ku, C. C.; Liepins, R. (1987). *Electrical Properties of Polymers: Chemical Principles,* Hanser Publishers, 978-3446142800, New York

Kubisa, P. (2004). Application of ionic liquids as solvents for polymerization processes. *Prog Polym Sci,* 29, 1, 3–12, 0079-6700

Kubisa, P. (2009) Ionic liquids as solvents for polymerization processes—progress and challenges. *Prog Polym Sci* , 34, 12, 1333–1347, 0079-6700

Kowsari, E.; Yavari, I. (2009) Proceedings of the Polymer Processing Society Europe/Africa Regional Meeting ~ PPS-2009 ~ October 18-21, 2009 Larnaca (Cyprus)

Kowsari, E.; Faraghi, G. (2010). Ultrasound and ionic-liquid-assisted synthesis and characterization of polyaniline/Y_2O_3 nanocomposite with controlled conductivity. *Ultrason. Sonochem,* 17, 4, 718-725, 1350-4177

Leddet, C. & Fischer, A.; Brembilla, A.; Lochon, P. (2001). Influence of the alkyl-chain size on the amphiphilic behaviour of poly(3-alkyl-1-vinylimidazolium) bromide in aqueous medium . *Polym. Bull.,* 46, 1 , 75–82, 0170-0839

Lee, J; Panzer, M. J.; He, Y.; Lodge, T. P & Daniel Frisbie, C. (2007) Ion gel gated polymer thin-film transistors, *J. Am. Chem. Soc.* , 129, 4532-4533, 0002-7863

Li, N.; Gao, Y.; Zheng, L.; Zhang, J.; Yu, L.; Li, X. (2007) Studies on the micropolarities of bmimBF_4/TX-100/Toluene ionic liquid microemulsions and their behaviors characterized by UV−Visible spectroscopy. *Langmuir,* 23, 3, 1091–1097, 0743-7463

Lu, X. F.; Bian, X. K., Shi, L. Q. (2002) Preparation and characterization of NF composite membrane. *J. Membr Sci* 210, 1, 3-11, 0376-7388

Lu, J.; Yan, F. & Texter, J. (2009) Advanced applications of ionic liquids in polymer science *Prog. Polym. Sci.,* 34, 5, 431–448, 0079-6700

Mallakpour, S., Kowsari, E. (2005). Ionic liquids as novel solvents and catalysts for the direct polycondensation of *N,N'*-(4,4'-oxydiphthaloyl)-bis-L-phenylalanine diacid with various aromatic diamines. *J. Polym. Sci. Part A: Polym. Chem,* 43, 24, 6545-6553, 1099-0518

Mallakpour, S; Kowsari, E. (2006). Microwave heating in conjunction with ionic liquid as a novel method for the fast synthesis of optically active poly(amide-imide)s derived from *N,N'*-(*4,4'*-hexafluoroisopropylidenediphthaloyl-(bis-*L*methionine and various aromatic diamines. *Iranian Polym., J.* 15, 3, 239-247, 1026-1265

Marcilla, R.; Blazquez, J. A.; Rodriguez, J.; Pomposo, J. A. & Mecerreyes, D. *J.* (2004). Tuning the solubility of polymerized ionic liquids by simple anion-exchange reactions. *Polym. Sci., Part A: Polym. Chem.,* 42, 1, 208–212, 1099-0518

Marcilla, R.; Blazquez, J. A.; Fernandez, R.; Grande, H.; Rodriguez, J.; Pomposo, J. A. & Mecerreyes, D. (2005) Synthesis of novel polycations using the chemistry of ionic liquids. *Macromol. Chem. Phys.*, 206, 2, 299–304, 1521-3935

Matsumi, N.; Sugai, K.; Miyake, M. & Ohno, H. (2006). Polymerized ionic liquids via hydroboration polymerization as single ion conductive polymer electrolytes. *Macromolecules*, 39, 20, 6924–6927, 0024-9297

Miskolczy, Z.; Sebok-Nagy, K.; Biczok, L. & Goektuerk, S. (2004). Aggregation and micelle formation of ionic liquids in aqueous solution. *Chem. Phys. Lett*,. 400, 1-6, 296-300, 0009-2614

Mu, X. d.; Meng, J. q.; Li, Z. C.; Kou, Y. (2005). Rhodium nanoparticles stabilized by ionic copolymers in ionic liquids: long lifetime nanocluster catalysts for benzene hydrogenation. *J. Am. Chem. Soc*,. 127, 27, 9694–9695, 0002-7863

Nakashima, T.; Sakashita, M.; Nonoguchi, Y. & Kawai, T. (2007). Sensitized photopolymerization of an ionic liquid-based monomer by using CdTe nanocrystals. *Macromolecules*, 40, 18, 6540–6544, 0024-9297

Noda, A. & Watanabe, M. (2000). Highly conductive polymer electrolytes prepared by in situ polymerization of vinyl monomers in room temperature molten salts. *Electrochim. Acta*,, 45, 8-9, 1265-1270, 0013-4686

Ogata, N.; Sanui, K.; Rikukawa, M.; Yamada, S.;Watanabe, M.; Mizumura, T. (1996). *Solid State Ionics* 353, 86.

Olmedo, L.; Hourquebie, P.; Jousse, F. (1995). Microwave properties of conductive polymers. *Synth. Met.*, 69, 1-3, 205- 208, 0379-6779

Patrascu, C.; Gauffre, F.; Nallet, F.; Bordes, R.; Oberdisse, J.; de Lauth-Viguerie, N. & Mingotaud, C. (2006). Micelles in ionic liquids: aggregation behavior of alkyl poly(ethyleneglycol)- ethers in 1-butyl-3-methyl-imidazolium type ionic liquids, *Chem. Phys. Chem*,. 7, 99- 101, 1439-7641, 1439-7641

Peng, C. H.; Hwang, C. C.; Wan, J.; Tsai, J. S.; Chen, S. Y. (2005). Microwave-absorbing characteristics for the composites of thermal-plastic polyurethane (TPU)-bonded NiZn-ferrites prepared by combustion synthesis method. *Mater. Sci. Eng., B* , 117,1, 27-36, 0921-5107

Perrier, S.; Davis, T. P.; Carmichael, A. J.; Haddleton, D. M. (2002). First report of reversible addition-fragmentation chain transfer (RAFT) polymerization in room temperature ionic liquids. *Chem Commun*, 19, 2226–2227, 1359-7345

Petrov, V. M. & Gagulin, V. V. (2001). Microwave absorbing materials. *Inorg. Mater.*, 37,2, 93-98, 0020-1685

Phang, S.W.; Daik, R; Abdullah, M. H. (2005). Poly(4,4'-diphenylene diphenylvinylene) as a non-magnetic microwave absorbing conjugated polymer. *Thin Solid Films*, 477, 1-2, 125- 130, 0040-6090

Qiu, Z. & Texter, J. (2008). Ionic liquids in microemulsions, *Curr. Opin. Colloid Interface Sci*, 13, 4, 252-262, 1359-0294

Quilliet, C. & Berge, B. (2001). Electrowetting: a recent outbreak. *Curr. Opin. Colloid Interface Sci.*, 6, 1, 34-39, 1359-0294

Ricks-Laskoski, H. L. & Snow, A. W. (2006). Synthesis and electric field actuation of an ionic liquid polymer. *J. Am. Chem. Soc.*, 128, 38, 12402–12403, 0002-7863.

Saib, A.; Bednarz, L.; Daussin, R.; Bailly, C.; Lou, X.; Thomassin, J. M.; Pagnoulle, C.; Detrembleur, C.; Jerome, R. & Huynen, I. (2006). Carbon nanotube composites for broadband microwave absorbing materials. *IEEE Trans. Microwave Theor. Tech,* 54, 6, 2745-2754, 0018-9480

Sarbu, T. & Matyjaszewski, K. (2001). ATRP of methyl methacrylate in the presence of ionic liquids with ferrous and cuprous anions. *Macromol Chem Phys;* 202, 17, 3379–91, 1022-1352

Sato, T.; Marukane, S.; Narutomi, T. & Akao, T. (2007). High rate performance of a lithium polymer battery using a novel ionic liquid polymer composite. *J. Power Sources,* 164,1, 390–396. 0378-7753

Seddon, K. R. (1997). Ionic liquids for clean technology. *J Chemical Tech Biotech* 68, 351–356, 0268-2575

Shen, Y. & Ding, S. (2004). Catalyst separation in atom transfer radical polymerization. *Prog Polym Sci,* 29, 10, 1053–1078, 0079-6700

Sirieix-Plenet, J.; Gaillon, L. & Letellier, P. (2004). Behavior of a binary solvent mixture constituted by an amphiphilic ionic liquid, 1-decyl-3-methylimidazolium bromide and water. Potentiometric and conductimetric studies, *Talanta, 63,* 4, 979-986, 0039-9140

Sugimura, R.; Qiao, K.; Tomida, D.; Kume, Y. & Yokoyama, C. (2007) immobilization of palladium acetate on ionic liquid copolymerized polystyrene: A way to eliminate inhibiting effect of imidazolium chloride and enhance catalytic performance. *Chem. Lett,.* 36, 7, 874–875, 0366-7022

Susan, M. A. B. H.; Kaneko, T.; Noda, A.; Watanabe, M. (2005) Ion gels prepared by in situ radical polymerization of vinyl monomers in an ionic liquid and their characterization as polymer electrolytes. *J. Am. Chem. Soc.* 127, 13, 4976-4983, 0002-7863

Suslick, K. S. (1988). *Ultrasound,* VCH, Weinhein, 0-89573-328-5, Germany

Suslick, K. S.; Choe, S. B.; Cichowlas, A. A.; Grinstaff, M. W. (1991). Sonochemical synthesis of amorphous iron. *Nature,* 353, 6343, 414-416, 0028-0836

Suslick, K. S.; Price G. J. (1999) Application of ultrasound to materials chemistry. *Ann. Rev. Mater. Sci,.* 29, 295-326, 0084-6600

Tang, J.; Tang, H.; Sun, W.; Plancher, H.; Radosz, M.; Shen, Y. (2005). Poly(ionic liquid)s: a new material with enhanced and fast CO_2 absorption. *Chem. Commun,* 26, 3325–3327. 1359-7345

Tang, H.; Tang, J.; Ding, S.; Radosz, M. & Shen, Y. (2005). Atom transfer radical polymerization of styrenic ionic liquid monomers and carbon dioxide absorption of the polymerized iIonic liquids. *J. Polym. Sci., Part A: Polym. Chem,.* 43, 7, 1432–1443, 1099-0518

Tang, J.; Sun, W.; Tang, H.; Radosz, M.; Shen, Y. (2005). Enhanced CO_2 absorption of poly(ionic liquid)s. *Macromolecules,* 38,6, 2037–2039, 0024-9297

Tang, J.; Tang, H.; Sun, W.; Radosz, M. & Shen, Y. (2005) Low-pressure CO_2 sorption in ammonium-based poly(ionic liquid)s .*Polymer,* 46, 26, 12460–12467, 0032-3861, 0024-9297

Tang, J.; Tang, H.; Sun, W.; Radosz, M. & Shen, Y. (2005). Poly(ionic liquid)s as new materials for CO_2 Absorption. *J. Polym. Sci., Part A: Polym. Chem,.* 43, 22, 5477–5489, 1099-0518

Tang, J.; Radosz, M.; Shen, Y. (2008) Poly(ionic liquid)s as optically transparent microwave-absorbing materials. *Macromolecules,* 41, 2, 493– 496, 0024-9297

Thomaier, S. & Kunz, W. (2007) Aggregates in mixtures of ionic liquids, *J. Mol. Liq,* 130, 1-3, 104-107, 0167-7322

Tokuda, H.; Tsuzuki, S., Hayamizu, M. K.; Watanabe, M. (2006). How ionic are room-temperature ionic liquids? An indicator of the physicochemical properties. J. Phys. Chem. B, 110, 39, 19593-19600, 1520-6106

Truong, V. T.; Riddell, S. Z.; Muscat, R. F. (1998). Polypyrrole based microwave absorbers. *J. Mater. Sci,* 33, 20, 4971-4976, 0022-2461

Ueki, T. & Watanabe, M. (2008). Macromolecules in ionic liquids: progress, challenges, and opportunities. *Macromolecules* , 41, 11, 3739-3748, 0024-9297

Vijayaraghavan, R. & MacFarlane, D. R. (2004) Living cationic polymerisation of styrene in an ionic liquid. *Chem. Commun,*6,700-701

Vijayaraghavan, R.; Pringle, J. M. & MacFarlane D. R. (2008). Anionic polymerization of styrene in ionic liquids. *European Polymer Journal,* 44 , 6,1758–1762, 0014-3057

Vygodskii1,Y. S.; Mel'nik1,O. A.; Lozinskaya1, E. I.; Shaplov1, A. S.; I. A. Malyshkina, I. A.; Gavrilova,n. D.; Lyssenko1,k. A.; Antipin1,M. Y. Golovanov, D. G.; G.; Korlyukov1, A. A.; Ignat'ev, N .& Welz-Biermann, U. (2007). The influence of ionic liquid's nature on free radical polymerization of vinyl monomers and ionic conductivity of the obtained polymeric materials. *Polym. Adv. Technol,* 3, 18, 50–63, 1099-1581

Wan, M. X.; Li, J. C.; Li, S. Z. (2001). Microtubules of polyaniline as new microwave absorbent materials. *Polym. Adv. Technol.,* 12, 11-12, 651- 657, 1099-1581

Wang, Y. Y.; Li,W. & Dai, L. Y. (2007). Cationic ring-opening polymerization of 3,3-bis(chloromethyl)oxacyclobutane in ionic liquids, *Chinese Chemical Letters,* 18, 10, 1187–1190, 1001-8417

Washiro, S.; Yoshizawa, M.; Nakajima, H. & Ohno, H. (2004). Highly ion conductive flexible films composed of network polymers based on polymerizable ionic liquids. *Polymer,* 45, 5 1577–1582, 0032-3861

Watanabe, M.; Yamada, S.; Sanui, K.; Ogata, N. J. (1993). High ionic conductivity of new polymer electrolytes consisting of polypyridinium, pyridinium and aluminium chloride. *J Chem. Soc Chem. Commun.* , 11, 929-931, 0022-4936

Watanabe, M.; Yamada, S. & Ogata, N. (1995) Ionic conductivity of polymer electrolytes containing room temperature molten salts based on pyridinium halide and aluminium chloride. *Electrochim. Acta* , 40,13-14, 2285-2288, *0013-4686*

Wilkes, J. S. & Zaworotko, M. J. (1992). Air and water stable 1-ethyl-3-methylimidazolium based ionic liquids. *J Chem Soc Chem Commun,* 13, 965–967, 0022-4936

Yan, F. & Texter, J. (2006). Surfactant ionic liquid based microemulsions for polymerizatiom. *Chem. Commun,* 2696–2698. 1359-7345

Yan, F. & Texte J. (2007) Solvent-reversible poration in ionic liquid copolymers. *Angew. Chem., Int. Ed,* 46, 14, 2440–2443, 1521-3773

Yu, S.; Yan, F.; Zhang, X.; You, J.; Wu,P.; Lu, J.; Xu, Q.; Xia, X. & Ma, G. (2008). Polymerization of ionic liquid-based microemulsions: A versatile method for the synthesis of polymer electrolytes. *Macromolecules,* 41, 10, 3389-3392, 0024-9297

Zhao, Y.; Chen, X.; WengX. (2009). Liquid crystalline phases self-organized from a surfactant-like ionic liquid C_{16}mimCl in ethylammonium nitrate, *J. Phys. Chem B,* 113,7, 2024-2030, 1520-6106

Zou, Y. H.; Liu, H. B.; Yang, L.; Chen, Z. Z. (2006). The influence of temperature on magnetic and microwave absorption properties of Fe/graphite oxide nanocomposites. *J. Magn. Magn. Mater, 302*, 2, 343-347, 0304-8853

2

Ionic Liquids in Biphasic Ethylene Polymerisation

Wioletta Ochędzan-Siodłak
Faculty of Chemistry, University of Opole, Poland

1. Introduction

Over the past several years there has been a considerable increase in the interest of ionic liquids as versatile reaction media for a wide variety of synthetic processes. The popularity of ionic liquids has been related primarily to their unique physical and chemical properties, which can be tailored by the judicious selection of cations and anions. These properties make it possible to use ionic liquids as solvents for many organic, inorganic, and organometallic compounds, including catalysts and co-catalysts, for various types of reactions. In particular, the polar but weakly coordinating character of ionic liquids enables them to be applied as mediums in catalysis, to immobilize various kinds of catalyst precursors which contain transition metals. Recently, ionic liquids have also been viewed as a "green" replacement to conventional organic solvents (Olivier-Bourbigou, 2002; Wasserscheid et al., 2004; Wilkes, 2004; Welton, 2004; Kubisa, 2004; Jain, 2005 ; Parvulescu et al., 2007; Chowdhury et al., 2007; Ochędzan-Siodłak, 2009).

Amongst many reactions carried out in an ionic liquids environment, dimerisation and oligomerisation reactions catalyzed by nickel complexes should be noted. In the majority of cases, an improvement in the activity and selectivity of the processes as well as the limitation of undesirable side reactions were achieved. Due to the good solubility of transition metal compounds in the ionic liquid phase, and simultaneously, a poor miscibility of the produced olefins, the reaction is commonly performed in a biphasic mode. The product creates a separate phase over the ionic liquid phase containing the dissolved catalyst. Thus, it becomes possible to easily separate the product from the reaction mixture, to minimize the waste of the expensive transition metal compound, and to use the catalyst repeatedly (recycling) (Chauvin et al.,1990; 1995; 1997; Einloft et al. 1996; Simon et al., 1998; Ellis et al., 1999; Pinheiro et al., 2001; Wasserscheid et al., 2001; Bernando-Gusmão et al., 2003; Wasserscheid et al., 2004).

The successful application of ionic liquids in the oligomerisation of 1-olefins inspired us to investigate the biphasic technique in polymerisation reactions using metallocene catalysts, which constitute the next generation of organometallic catalysts for olefin polymerization. Metallocenes in the past gained a considerable interest due to their high activities - up to several tons of the polymer product per gram of the transition metal. Unfortunately, it was found that such high activities can only be obtained for homogenous catalyst systems dissolved in carcinogenic solvents and activated by a great excess of expensive methylaluminoxane (MAO). Furthermore, the homogenous systems reveal low stabilities

during the polymerisation reactions, reaction fouling, and a lack of control of the polymer morphology (Kaminsky, Laban, 2001; Ewen, 2000).
To overcome these undesirable features, the metallocene systems were subjected to an anchorage on a solid carrier, for example SiO_2, Al_2O_3, $MgCl_2$, as well as polymer compounds. The obtained heterogenisation of the metallocene systems makes it possible to control the product morphology, as well as resulting in an absence of reactor fouling, and a decrease in the amount of the MAO activator required. Unfortunately, the anchorage of the catalyst to the solid carrier results in a decrease of the activity (10-60%) as compared to the homogeneous analogue. The heterogenisation process is technically difficult to proceed, time consuming, and the polymer product reveals a considerable degree of inhomogeneity (Chen, 2000; Hlatky, 2000; Severn et al., 2005; Ochędzan-Siodłak, Nowakowska, 2005).
Therefore, there is a need to search for new, non-conventional performance methods of olefin polymerisation using metallocene catalysts. It seems that the immobilization of the metallocene catalyst in ionic liquids could be a potential alternative to its heterogenisation on a solid carrier.

2. Ethylene polymerisation

The ethylene polymerisation performed using the [C_2-mim][$AlCl_4$] ionic liquid as a medium of the $TiCl_4$ catalyst activated by the $AlEtCl_2$ alkylaluminium compound was described for the first time by Carlin and Osteryoung. The polymerisation was carried out in the single phase mode (in the ionic liquid). Although only small amounts of the polyethylene product were obtained (0.23 kgPE/(molTi · 10 min)), the results indicated that ionic liquids can be applied to olefin polymerisation (Carlin et al., 1990). It was further found that amongst the Cp_2MCl_2 metallocene catalysts (M = Ti, Hf, Zr) activated by the $AlCl_{3-x}R_x$ alylaluminium compounds (R = Me, Et), only the titanocene catalyst turned out to be active (Carlin, Wilkes, 1990).

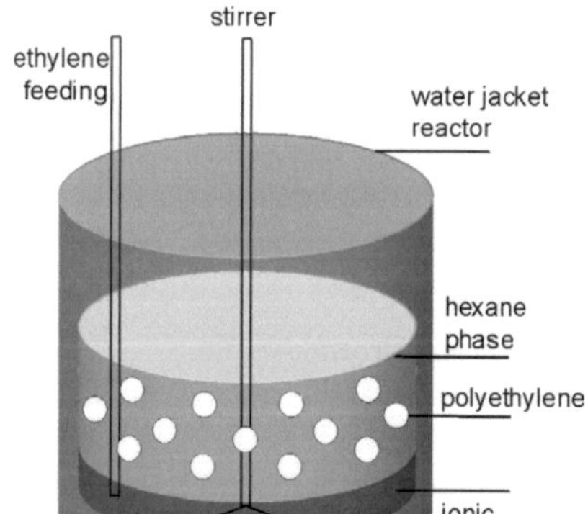

Fig. 1. The schematic representation of the studied biphasic ionic liquid/hexane polymerisation.

In our work, the polymerisation reaction was performed in the biphasic mode (Figure 1). The polymerisation of the simplest olefin, ethylene, was investigated as a model. The lower phase consisted of the ionic liquid, in which the Cp_2TiCl_2 catalyst and the alkylaluminium activators were dissolved. Hexane constituted the upper phase. For the ionic liquids, 1-*n*-

alkyl-3-methylimidazolium and 1-*n*-alkyl-4-methylpyridinium chloroaluminates substituted by alkyl chains of various length (n = butyl, hexyl, octyl) were applied (Ochędzan-Siodłak et al., 2007; 2008; 2009).

2.1 Biphasic process

For all biphasic ionic liquid/hexane polymerisation experiments performed, the following facts have been observed. Two phases, hexane and ionic liquid, are clearly visible from the beginning to the end of the reaction. At the beginning of the reaction the ionic liquid phase becomes white and swells considerably as the polyethylene appears. At the same time, the hexane phase remains colourless and transparent. After 5-10 minutes, the hexane phase becomes a white suspension as the polyethylene is progressively transferred from the ionic liquid phase. The polyethylene is obtained in form of a powder. It is white when obtained from the hexane phase and off-white when obtained from the ionic liquid phase. The polyethylene from the hexane phase can be easily isolated and subjected to analysis.

It is very important for the biphasic catalytic process to maintain the catalyst in the selected phase. This has been shown by experiments where the ionic liquid phase contained the Cp_2TiCl_2 catalyst and the alkylaluminium compound was washed with hexane. Catalyst leakage was not detected, which indicates that the titanocene is firmly immobilised in the ionic liquid phase and remains during the entire duration of the polymerisation (Ochędzan-Siodłak, Pawelska, 2008).

During the course of the polymerisation reaction, the polyethylene is progressively transferred to the hexane phase. At the same time, the amount of polyethylene in the ionic liquid phase decreases, however, the total polyethylene yield increases with the polymerisation time (Figure 2). This confirms that the polymerisation reaction takes place in the ionic liquid phase. It also indicates that the catalyst is stable.

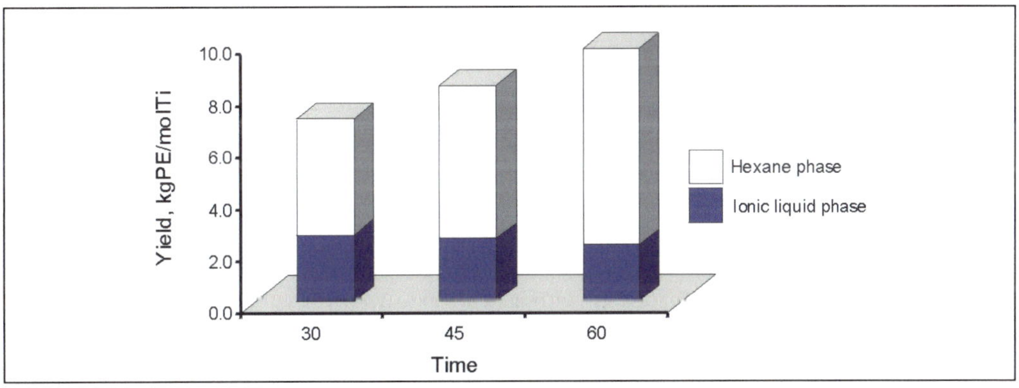

Fig. 2. The influence of the reaction time on the PE yield in the biphasic ethylene polymerisation. Polymerisation conditions: ionic liquid - $[C_4\text{-mim}][AlCl_4]$, catalyst - Cp_2TiCl_2, activator-$AlEt_2Cl$, Al/Ti = 33, standard pressure, room temperature.

The stability of the catalyst was also proved by the re-use of the ionic liquid phase containing the catalyst and the activators in the consecutive polymerisation reactions (Figure 3). The experiments consisted of three cycles, which were successfully performed, although the polymerisation yield gradually decreased. The catalyst was maintained in the ionic liquid phase whereas the polyethylene product was removed after each cycle with the hexane phase. Thus, catalyst recycling is possible. Also, it seems to be a very important way to increase catalyst performance in the biphasic processes (Ochędzan-Siodłak et al., 2008).

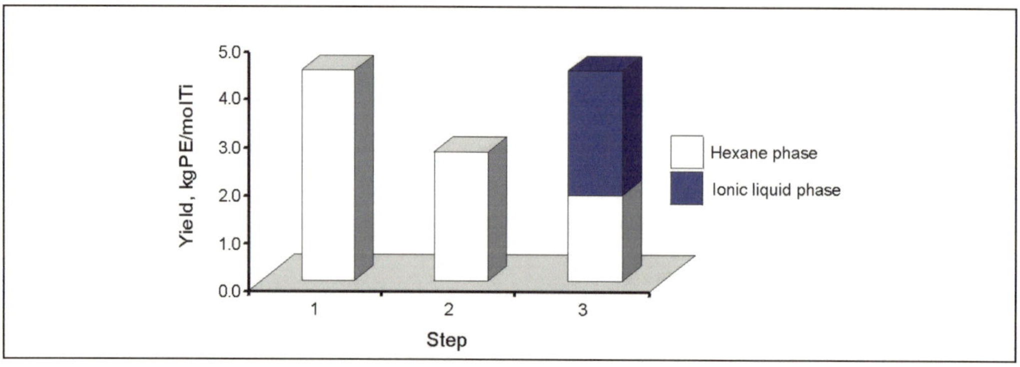

Fig. 3. The 3-step catalyst recycling in the biphasic ethylene polymerisation. Polymerisation conditions: ionic liquid - $[C_4\text{-mim}][AlCl_4]$, catalyst - Cp_2TiCl_2, activator - $AlEt_2Cl$, Al/Ti = 33, standard pressure, room temperature, reaction time - 30 min.

The results presented show that the application of the ionic liquid as a medium of the titanocene catalyst as well as the performance of the polymerisation reaction in the biphasic mode offer the following advantages; the polymer product can be easily separated from the reaction mixture using a simple decantation technique; the product is characterised by high purity; the product is transferred to the hexane phase, whereas the catalyst and activator remain in the ionic liquid phase. In addition, the catalyst presented in the ionic liquid phase is stable and catalyst leakage is not observed. Thus, recycling of both the catalyst and the ionic liquid is possible. It should be also noted that reaction fouling (common for the metallocene/MAO systems) is not present (Ochędzan-Siodłak et al, 2007; 2008).

2.2 Activators

The great advantage of ionic liquids over classic solvent is the fact that their properties can be tailored by the choice of suitable cations and anions. Applying a titanocene catalyst to the ethylene polymerisation limits the choice to the chloroaluminate ionic liquids because so far it is only this anion which makes the use of the alkylaluminium compounds as necessary catalyst activators possible. It should be noted that introduction of the alkylaluminium compound inevitably influences the anionic part of the ionic liquid. The kind and the amount of the alkylaluminium compound used are important because a range of mixtures of the chloroaluminate anions can be created. Also, the Lewis acidity of the ionic liquid is changed. This has considerable influence on the course and the performance of the ethylene polymerisation carried out in the biphasic mode.

The following alkylaluminium compounds: $AlEtCl_2$, $AlEt_2Cl$, $AlEt_3$, and MAO were investigated as activators of the Cp_2TiCl_2 catalyst (Figure 4). The best results were obtained for the compounds containing the chlorine atom, $AlEtCl_2$ and $AlEt_2Cl$. The more chlorine atom the activator contains, the higher the total yield, with the greater amount of polyethylene product transferred to the hexane phase. Interestingly, the worst results were obtained using methylaluminoxane (MAO), which is the best activator for the homogeneous metallocene catalysts. Therefore, the application of the ionic liquid to the ethylene polymerisation eliminates the need for the expensive MAO as the activator, which can be superseded by traditional, cheaper alkylaluminium compounds (Ochędzan-Siodłak et al., 2007; 2008).

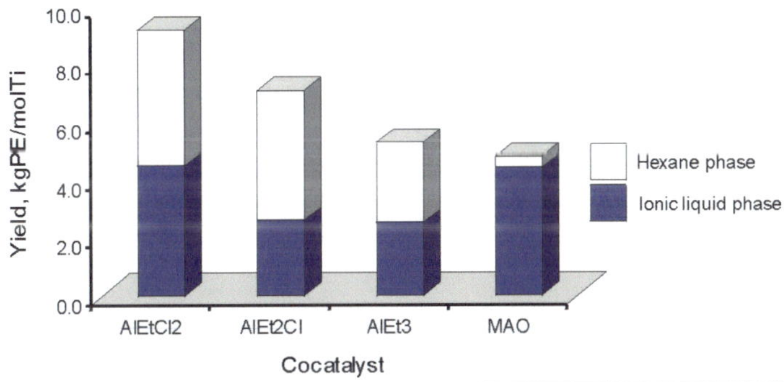

Fig. 4. The influence of various kinds of alkylaluminium activators on the performance of the biphasic ethylene polymerisation. Polymerisation conditions: ionic liquid - [C_4-mim][$AlCl_4$], catalyst - Cp_2TiCl_2, Al/Ti = 33, standard pressure, room temperature, reaction time - 30 min.

2.3 Imidazolium ionic liquids

The further investigations focused mainly on the modification of the cationic part of the ionic liquid. Substituting the imidazolium and pirydinium cations by alkyl chains of various lengths makes it possible to tailor some of the physical properties of the ionic liquids. The two most effective activators, $AlEtCl_2$ and $AlEt_2Cl$, were applied (Ochędzan-Siodłak et al., 2007; 2008; 2009). The study performed using 1-*n*-alkyl-3-methylimidazolium ionic liquids, [C_n-mim][$AlCl_4$], show that the length of the alkyl chain at the ionic liquid cation and the concentration of the alkylaluminium compound influence the performance of the biphasic polymerisation reaction, i.e. the total polymerisation yield, the amount of the polyethylene product gathered in the hexane phase, and the phase separation (Figure 5). For each studied ionic liquid, there is an optimal concentration of the alkylaluminium compound, in which the yield of the reaction is the highest and the majority of the product transfers to the hexane phase. For [C_4-mim][$AlCl_4$] and [C_6-mim][$AlCl_4$], ionic liquids with shorter alkyl chains, the activator/catalyst molar ratio (Al/Ti) reaches optimum at the value 100. A higher ratio results in a decrease of the yield and the majority of the product remains in the ionic liquid phase. For [C_8-mim][$AlCl_4$], the ionic liquid with a longer alkyl chain, the optimal activator/catalyst molar ratio is higher (Al/Ti = 133) resulting in a total PE yield 120 kgPE/molTi·h. An even higher yield can be obtained at the activator/catalyst molar ratios 167 and 200. However, the majority of the product remains in the ionic liquid phase, which is disadvantageous to the biphasic process. Therefore, if two criteria for the performance of the biphasic process are the amount of the product gathered in the hexane phase and the ratio between the amount of product in both phases, the ionic liquid with the longest alkyl chain is the better medium for the titanocene catalyst as it results in the greatest amount of the pure polymer product from the hexane phase. It was also found that prolongation of the reaction time (up to 120 min.) improves the reaction yield. This confirms the stability of the catalyst immobilised in the ionic liquid and makes it possible to obtain the greatest amount of the product from the hexane phase.

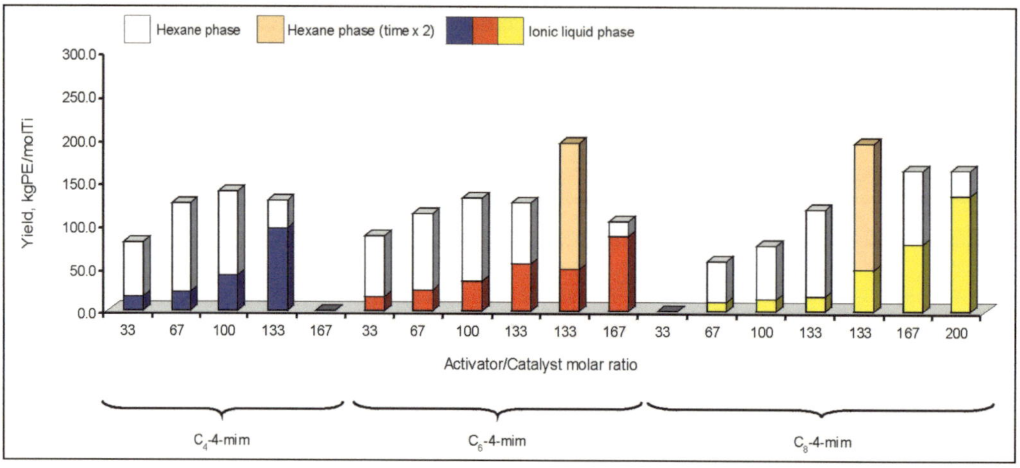

Fig. 5. The influence of the alkyl chain at the cation of the ionic liquids and the concentration of the $AlEtCl_2$ alkylaluminium activator on the performance of the biphasic ethylene polymerisation. Polymerisation conditions: ionic liquid - $[C_n\text{-mim}][AlCl_4]$ (n = 4,6,8), catalyst - Cp_2TiCl_2, activator -$AlEtCl_2$, pressure - 0.5MPa, room temperature, reaction time - 60 min.

When the $AlEt_2Cl$ compound is used instead of $AlEtCl_2$, similar results were obtained (Figure 6). The yield increases with the increase of the concentration of the alkylaluminium compound. For the ionic liquids with shorter alkyl chains, the best results are obtained at the lower activator/catalyst molar ratio, whereas for the ionic liquid with the longer alkyl chain, a higher activator/catalyst molar ratio is required. However, the obtained yield is considerably lower, which indicates that the $AlEt_2Cl$ compound is not as efficient as $AlEtCl_2$.

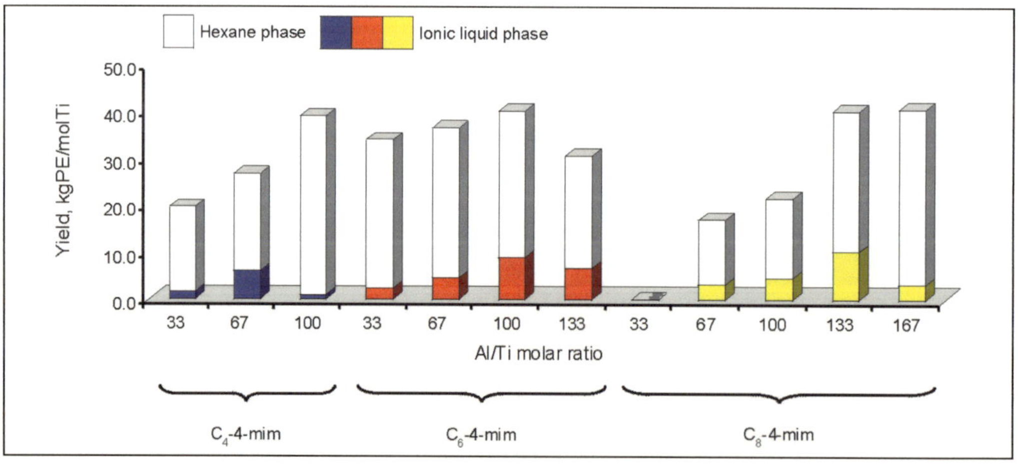

Fig. 6. The influence of the alkyl chain at the cation of the ionic liquids and the concentration of the $AlEt_2Cl$ alkylaluminium activator on the performance of the biphasic ethylene polymerisation. Polymerisation conditions: ionic liquid - $[C_n\text{-mim}][AlCl_4]$ (n = 4,6,8), catalyst - Cp_2TiCl_2, activator - $AlEt_2Cl$, pressure - 0.5MPa, room temperature, reaction time - 60 min.

2.4 Pyridinium ionic liquids

Similar studies were carried out for the pirydinium ionic liquids. The pyridinium cation not only differs in its structure in comparison to the imidazolium cation, but also the position of the alkyl chain can be changed. 1-*n*-Alkyl-4-methylpyridinium chloroaluminates [C_n-4-mpy][$AlCl_4$] were investigated at similar polymerisation conditions as for the imidazolium ionic liquids (Figure 7). It was found that the yield increases with the increase of the alkylaluminium concentration. Nevertheless at higher activator/catalyst molar ratios, the majority of the product remains in the ionic liquids, which is a disadvantageous phenomenon. Only at a lower activator/catalyst molar ratio (Al/Ti = 67) is the majority of the polymer product gathered in the hexane phase. In contrast to the imidazolium ionic liquids, the best results were obtained for the pyridinium cations with shorter alkyl chains ([C_4-4-mpy][$AlCl_4$]). This can particularly be seen at lower Al/Ti molar ratios.

The change of the position of the alkyl chain at the pyridinium cation was investigated on the [C_8-4-mpy][$AlCl_4$] and [C_8-3-mpy][$AlCl_4$] ionic liquids. In both cases the polymerisation yield and the transfer of the polymer to the hexane phase is slightly better for the ionic liquid having the alkyl chain at the *para* position at the pyridinium cation. This indicates that the position of the alkyl chain at the ionic liquid cation influences the performance of the biphasic process, although the effect is moderate. Generally, the application of pyridinium ionic liquids to the biphasic polymerisation results in higher yields in comparison to imidazolium analogues at similar reaction conditions.

To improve the performance of the biphasic polymerisation, the reaction time was extended twice (up to 120 min.). Advantageous increases in the amount of the product in the hexane phase as well as in the total polymerisation yield were observed. Again, the best result was obtained for the [C_4-4-mpy][$AlCl_4$] ionic liquid with the shortest alkyl chain.

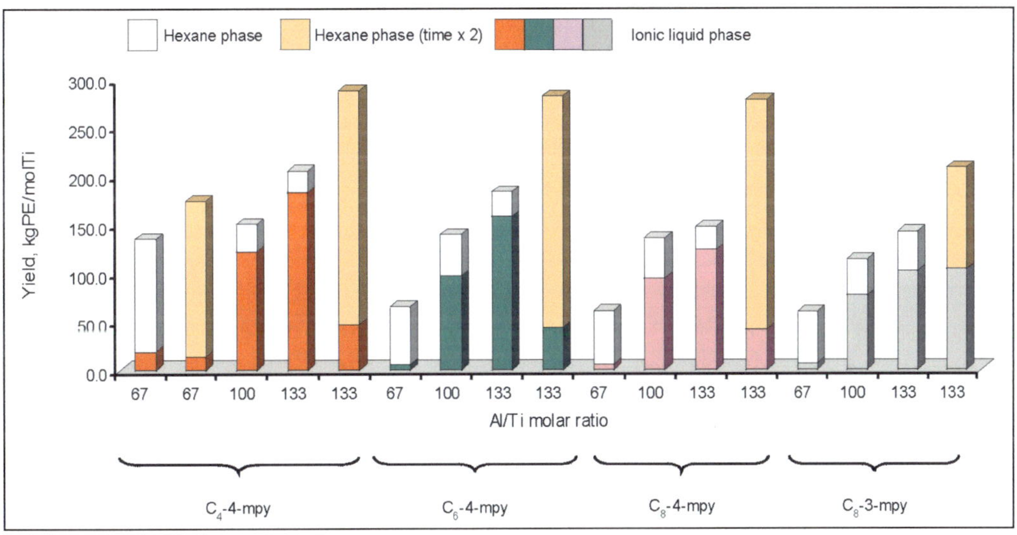

Fig. 7. The influence of the alkyl chain at the pyridinium cation and the concentration of the $AlEtCl_2$ alkylaluminium activator on the performance of the biphasic ethylene polymerisation. Polymerisation conditions: ionic liquids: [C_n-4-mpy][$AlCl_4$] (n = 4,6,8); [C_8-3-mpy][$AlCl_4$], catalyst - Cp_2TiCl_2, activator -$AlEtCl_2$, pressure - 0.5MPa, room temperature, reaction time - 60 min.

By contrast, the length of the alkyl chain at the pyridinium cations in comparison to imidazolium analogues has a negative effect, which can clearly be seen using the $AlEt_2Cl$ activator instead of $AlEtCl_2$ (Figure 8). Similarly to the imidazolium ionic liquids, the application of the $AlEt_2Cl$ activator results in a considerable decrease of the reaction yield. However, it should be noted that almost the entire amount of the polymer product is gathered in the hexane phase. This phenomenon is of great importance because it results in a product of high purity. Furthermore, the negligible amount of the product in the ionic liquid phase makes it possible to apply this phase in consecutive multi-step reactions. Thus, the recycling of both the catalyst and the ionic liquid can easily be done. It seems that the $AlEt_2Cl$ activator could be successfully applied in a continuous flow process, whereas the $AlEtCl_2$ activator is better in a single reaction step.

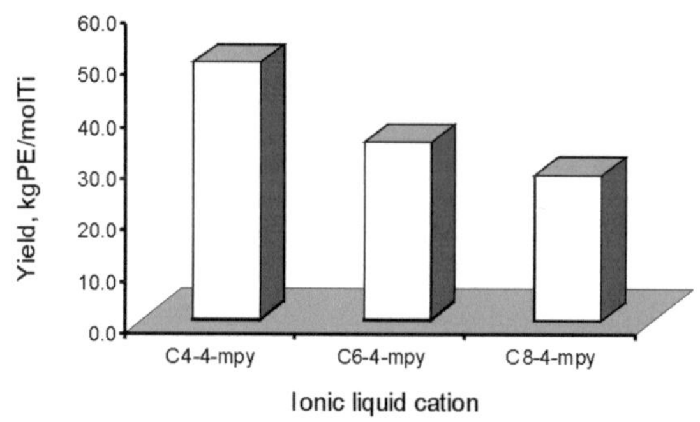

Fig. 8. The influence of the alkyl chain at the pyridinium cation and the concentration of the $AlEt_2Cl$ alkylaluminium activator on the performance of the biphasic ethylene polymerisation. Polymerisation conditions: ionic liquid - $[C_n\text{-4-mpy}][AlCl_4]$ (n = 4,6,8), catalyst - Cp_2TiCl_2, activator - $AlEt_2Cl$, Al/Ti = 133, standard pressure - 0.5MPa, room temperature, reaction time - 60 min.

2.5 Ionic liquids properties

The presented results show that the structure of the ionic liquid cation has a considerable influence on the performance of the biphasic polymerisation. The structure of the cation influences the physical properties of the ionic liquids, such as density and viscosity (Olivier-Bourbigou, 2010).

Generally, an increase in the alkyl chain length leads to a decrease in the density and an increase in the viscosity of the ionic liquid. In biphasic ethylene polymerization the viscosity and the density of the ionic liquid play a key role in the dispersion of both phases, which influences the mass transfer of the product from the ionic liquid phase to the hexane phase. This can be seen the most clearly in the case of the polymerisations carried out in the imidazolium ionic liquids (Figure 5). In the case of the pyridinium analogues, the mass transfer seems to be more hindered, and this effect is much smaller.

For polymerisation performed at lower alkylaluminium (activator) concentration, the majority of the product is gathered in the hexane phase. This can be explained by the

presence of a relatively small number of active sites. In consequence, a smaller amount of the polymer is produced, but this does not exceed the amount of product which the system is able to transfer from the ionic liquid phase where the product is created to the hexane phase where the product is gathered. For the polymerisation performed at a higher alkylaluminium concentration, a greater number of active sites appears, which results in a greater amount of the polymer produced than the system is able to transfer to the hexane phase. Thus, a considerabe amount of the product remains in the ionic liquid phase. This phenomenon is observed for all the ionic liquids studied. The polymerisation performed using the $AlEt_2Cl$ compound support this conclusion. $AlEt_2Cl$ is an activator of lower efficiency than $AlEtCl_2$. It creates a smaller number of active sites, and thus, a smaller mass of the polymer product, but this is easily transferred to the hexane phase.
The optimum activator/catalyst molar ratio observed in the case of the polymerisation carried out in the imidazolium ionic liquids is another interesting phenomenon. Too low or too high a concentration of the alkylaluminium compound is disadvantageous and leads to a decrease of the reaction yield. This phenomenon is connected with the acidity of the ionic liquid phase. Carlin and Wilkes proved that only ionic liquids of the Lewis acid character are useful for the ethylene polymerisation reaction. The acid character of the chloroaluminate ionic liquid can be obtained by adding the correct amount of $AlCl_3$. Usually, during the synthesis of the ionic liquids from $AlCl_3$ and the given chloride of the organic cation, an excess of $AlCl_3$ is added $AlCl_3/[C_n\text{-mim}]Cl>1$). However, at a higher concentration of the $AlCl_4^-$ and $Al_2Cl_7^-$ anions, complexation of the titanocene catalyst to either the $Cp_2Ti(AlCl_4)_2$ or $Cp_2TiCl(Al_2Cl_7)$ forms can occur (Carlin, Wilkes, 1990).

$$Cp_2TiCl_2 + Al_2Cl_7^- \leftrightarrow Cp_2TiCl(AlCl_4) + AlCl_4^- \quad (1)$$

$$Cp_2Ti(AlCl_4) + Al_2Cl_7^- \leftrightarrow Cp_2Ti(AlCl_4)_2 + AlCl_4^- \quad (2)$$

$$Cp_2Ti(AlCl_4) + Al_2Cl_7^- \leftrightarrow Cp_2Ti(Al_2Cl_7) + AlCl_4^- \quad (3)$$

The proper acidity of the ionic liquid can be obtained using alkylaluminium compounds containing chlorine atoms. These compounds, such as the $AlEtCl_2$ applied in this work, lead to the creation of various anions in the reaction medium: $AlCl_4^-$, $Al_2Cl_7^-$, $Al_2EtCl_6^-$, $AlEtCl_3^-$, $Al_2Et_2Cl_5^-$. The optimum parameters of the chloroaluminate ionic liquid applied as a medium of the catalyst for the polymerisation depends on the amount and kinds of the anions. Higher concentrations of the $AlCl_4^-$ and $AlEtCl_3^-$ anions, of neutral character, considerably decrease the acidity of the chloroaluminate ionic liquid. In contrast, higher concentrations of the $Al_2Cl_7^-$ and $Al_2EtCl_6^-$ anions result in the ionic liquid being too acidic, which also leads to a decrease of the polymerisation yield. Therefore, the judicious selection of the kind and amount of the alkylaluminium Lewis acid is very important when tailoring the acidity of the ionic liquid. The presented results indicate that the application of the $AlEtCl_2$ compound, which reveals a higher Lewis acidity in comparison to $AlEt_2Cl$, results in a better performance of the ionic liquid in the polymerisation reaction (Chauvin, 1990; Thiele, de Souza, 2007).
The optimum activator/catalyst molar ratio not only enables the reduction of the titanium compound, but also creates the appropriate environment for the polymerisation reaction, which results from the correct proportion of the chloroaluminate anions being present in the ionic liquid medium.

It is known that in the homogeneous metallocene systems, the active site is represented by the cationic form of the metallocene; Cp_2MR^+. Thus, the necessary condition for the polymerisation reaction is:

i. detachment of the $AlCl_4^-$ group and creation of the unsaturated Cp_2TiR^+, according to the equation:

$$Cp_2TiR(AlCl_4) \leftrightarrow Cp_2TiR^+ + AlCl_4^- \quad (4)$$

ii. exchange of the $AlCl_4^-$ group by ethylene molecule and creation of the $Cp_2TiR(CH_2CH_2)^+$complex:

$$Cp_2TiR(AlCl_4)+CH_2CH_2 \leftrightarrow Cp_2TiR(CH_2CH_2)^+ +AlCl_4^- \quad (5)$$

In both cases, the centre of the polymerisation reaction is the cationic complex, created by the break of the M-Cl bond, which is stronger for Zr and Hf and weaker for Ti. Therefore, the inactivity of the Zr and Hf compounds results from the stronger M-Cl bonds, which make it impossible to create the active cationic form [Cp_2MR^+] (Carlin, Wilkes, 1990). It should be noted, however, that the chloroaluminate ionic liquid participates in the creation of the active centre, which is supported by literature (Chauvin, 1997; Simon, 1998; Bernardo-Gusmao, 2003).

3. Polyethylene properties

The presented biphasic polymerisation experiments show that the polyethylene product is present in both the ionic liquid and hexane phases. According to the concept of biphasic polymerisation, the product gathered in the hexane phase is the most important. Therefore only the properties of this polyethylene have been analysed. The polyethylene from the ionic liquid phase is contaminated by the products of decomposition of the catalyst, activator, and ionic liquids, which means that it is impossible to determine the PE properties correctly.

The data presented in Table 1 show the influence of the imidazolium ionic liquids on the properties of the PE produced. The obtained polyethylene was linear in character. On average, 3.0-9.4 methyl groups on the 1000 methylene groups occur. The melting temperature (~133°C) is relatively narrow and typical for commercial HDPE (Vasilie , Pascu, 2005). The high crystallity degree (~86%) should be noted. This value varies depending on the length of the alkyl chain at the ionic liquid cation; the lowest is for [C_4-mim][$AlCl_4$] (~82%), and the highest for [C_8-mim][$AlCl_4$] (~87%). Another advantageous feature of the polyethylene product is a high value of the bulk density, which again varies depending on the ionic liquid applied. Again, the lowest value was obtained for [C_4-mim][$AlCl_4$] (~230 g/dm^3) and the highest for [C_8-mim][$AlCl_4$] (~470 g/dm^3). Such high values of the PE bulk density can be obtained only for the heterogeneous metallocene catalyst (Kang, 1999; Razavi, 2000; Jang, 2003; Wei, 2004; Ochędzan-Siodłak, 2005; Hammawa, 2006). In the case of the homogeneous systems, the bulk density usually does not exceed 100 g/dm^3 (Soares, 2000; Lee, 2000). The high values of the crystallity degree and bulk density of the polyethylene obtained show a high level of the molecular order and indicate good mechanical parameters of the polymer product. Nevertheless, the polyethylene obtained has a relatively small molecular weight (M_w) in the range of 70 000–170 000 g/mol. The molecular weight decreases with the increase of the activator/catalyst molar ratio. Therefore, an increase of the activator concentration results in shorter PE chains. It indicates that the alkylaluminium compound participates, most probably, in the chain termination reaction. The molecular weight distribution is monomodal (MWD from 4.3 to 7.2). This is characteristic for the polymer obtained using supported metallocene catalysts. The polyethylene obtained using the $AlEt_2Cl$ activator has a similar melting temperature, crystallity degree, molecular weight, molecular weight distribution, and linearity to the analogue polymer obtained using the $AlEtCl_2$ activator, regardless of the ionic liquid used. However, the bulk density is lower when $AlEt_2Cl$ is applied.

Item	Activator/Catalyst molar ratio	T_m (°C)	Crystallinity (DSC) (%)	Mw × 10^{-3} (g/mol)	MWD	Bulk density (g/dm^3)	CH_3/ 1000CH_2
$[C_4mim]^+[AlCl_4]^-$							
1	67	134	86.8	159.6	4.4	243	3.7
2	100	133	83.5	86.7	5.8	220	6.3
3	133	137	81.6	71.7	4.3	240	3.0
$[C_6mim]^+[AlCl_4]^-$							
4	67	133	85.9	121.5	6.2	-	7.6
5	100	132	86.4	110.0	6.8	370	7.3
6	133	132	86.6	45.9	4.8	440	7.0
7[a]	100	133	85.5	116.3	4.3	170	9.4
$[C_8mim]^+[AlCl_4]^-$							
8	67	135	89.2	165.6	7.2	-	-
9	100	134	88.6	129.2	6.8	340	6.2
10	133	133	85.8	82.5	5.3	470	8.3
11[a]	100	134	80.5	77.8	5.4	160	7.0
[a] $AlEt_2Cl$ activator							

Table 1. Selected properties of the polyethylene produced using imidazolium ionic liquids in the biphasic polymerisation

The application of the pyridinium ionic liquid in the biphasic polymerisation results in a polyethylene product with slightly different properties than that obtained using the imidazolium ionic liquids (Table 2). It is a linear polyethylene and the number of branches (from 4.3 to 12.9 of the CH_3 groups per 1000 CH_2 groups) is comparable for all the samples analysed. The number of branches increases with an increase in the activator concentration, regardless of the ionic liquid used. The melting temperatures (127-130°C) are lower but the molecular weights (M_w = 100 000–270 000 g/mol) are higher than for the PE samples obtained in the biphasic polymerisation using imidazolium ionic liquids. It should be noted that the lowest M_w have PE samples obtained using $[C_8\text{-3-mpy}][AlCl_4]$. The molecular weight distributions (MWDs) of the studied samples are relatively broad, from 3.1 to 6.9. Again, the distinguishing feature of the PE produced is the high bulk density, which reaches up to a value of 520 g/dm^3, which corresponds to that of the polyethylenes obtained over supported metallocene catalysts. A decrease of the bulk density is observed with the increase of the activator/catalyst molar ratio. The position of the alkyl chain at the pyridinium cation has an influence on the bulk density. The polyethylene produced using the $[C_8\text{-4-mpy}][AlCl_4]$ ionic liquid has a higher bulk density than that obtained using $[C_8\text{-3-mpy}][AlCl_4]$ (Al/Ti=67 and 100).

All the studied PE samples reveal an exceptionally high crystallinity degree, with an average value of 93%, as determined by the DSC method. Such a high crystallinity is atypical for PE obtained using a metallocene catalyst (Prasad, 1999; Wei, 2004) and it corresponds to that of the commercial HDPE (crystallinity = 77-90%) obtained using Ziegler-Natta catalysts (Doak, 1986).
The polyethylene samples obtained using the $AlEt_2Cl$ activator revealed considerable purity due to the efficient mass transfer to the hexane phase of the polymer produced. They reveal much lower molecular weights, bulk densities, and crystallinity - similar to those of the polyethylene obtained in the biphasic process using imidazolium ionic liquids (Ochędzan-Siodłak, 2008). Generally, the polyethylene obtained using $AlEtCl_2$ as the activator is characterised by better properties than that obtained using $AlEt_2Cl$.

Item	Activator/Catalyst molar ratio	T_m (°C)	Crystallinity (DSC) (%)	Mw × 10^{-3} (g/mol)	MWD	Bulk density (g/dm^3)	CH_3/ 1000CH_2
	[C_4-4-mpy][$AlCl_4$]						
1	67	128	94.0	261.6	5.9	370	5.9
2	100	126	95.0	233.0	6.7	215	8.9
3	133	127	95.1	151.5	5.4	–	9.9
4[a]	133	129	94.2	142.9	6.7	165	9.1
	[C_6-4-mpy][$AlCl_4$]						
5	67	129	94.9	260.0	5.5	444	5.2
6	100	128	93.1	229.3	6.7	283	7.1
7	133	126	95.1	150.9	6.1	136	12.9
8[a]	133	127	90.6	130.1	6.9	145	11.1
	[C_8-4-mpy][$AlCl_4$]						
9	67	129	96.0	266.8	6.9	520	4.4
10	100	128	95.0	222.9	5.4	395	5.3
11	133	133	93.0	158.6	5.9	210	6.2
12[a]	133	130	87.4	122.5	5.6	137	9.8
	[C_8-3-mpy][$AlCl_4$]						
13	67	128	96.0	141.3	3.4	481	5.0
14	100	128	98.0	138.2	3.9	357	6.0
15	133	127	98.0	111.5	3.1	-	8.1

[a] $AlEt_2Cl$ activator

Table 2. Selected properties of the polyethylene produced using pyridinium ionic liquids in the biphasic polymerisation

The presented results show that the polyethylene product obtained in the biphasic process joins the properties of the polyethylenes obtained using the homogenous and heterogeneous metallocene catalysts (a low molecular weight as well as broad molecular weight distribution and high bulk density, respectively). Moreover, the product is characterised by a high crystallity degree resulting from the high level of the macromolecular order, which is particularly observed for the samples obtained from the pyridinium ionic liquids. Properties such as molecular weight, bulk density, and crystallity degree can be modified by both the kind of the alkyl chain at the ionic liquid cation and the concentration of the alkylaluminium compound.

4. Conclusion

The presented studies describe the biphasic ionic liquid/hexane polymerisation of ethylene using a titanocene catalyst. For the ionic liquids, imidazolium and pyridinium chloroaluminates were applied. It was found that the catalyst is stable in the polymerisation reaction and firmly immobilized in the ionic liquid. Therefore, application of ionic liquids as media of the catalyst seems to be an interesting alternative to the catalyst's heterogenisation on the solid carrier. The studied biphasic mode makes it possible to re-use (recycle) both the catalyst and the ionic liquid phase. The traditional alkylaluminium compounds are more suitable activators for the titanocene catalyst than methylaluminoxane, which is considered as the most effective activator for metallocene catalysts. Polyethylene (PE) formed in the ionic liquid phase is progressively transferred to the hexane phase. Thus, the polymer is characterized by a high purity and can easily be separated from the reaction mixture by a simple decantation technique.

It is possible to influence the performance of the biphasic polymerisation and the properties of the polymer product by changing the character of the ionic liquid. This can be done by the modification of the cationic and anionic parts of the ionic liquid. In the presented work the 1-*n*-alkyl-3-methylimidazolium and 1-n-alkyl-4-methylpyridinium cations with alkyl chains of various length were investigated. The chloroaluminate anions can be modified by the choice of the kind and the concentration of the alkylaluminium compound, which is also applied as the catalyst activator. The following alkylaluminium compounds were investigated: $AlEt_3$, $AlEt_2Cl$, $AlEtCl_2$, and MAO. The results obtained using compounds containing the chlorine atom, $AlEtCl_2$ and $AlEt_2Cl$ were better than when methylaluminoxane was applied. The $AlEtCl_2$ compound turned out to be the most suitable for the biphasic ethylene polymerisation as regardless of the ionic liquid cation used, the highest yields were obtained. It should be noted that the reaction yield depends on the concentration of the alkylaluminium compound. This can be explained by the balance of the various chloroaluminate anions created, which influence the acidity of the ionic liquids, which is necessary for the polymerisation reaction. The concentration of the alkylaluminium compound, described in the from of activator/catalyst molar ratio (Al/Ti), determines the amount of the polymer product gathered in the hexane phase. Generally, the lower the Al/Ti molar ratio, the greater amount of product present in the hexane phase and a correspondingly smaller amount remaining in the ionic liquid phase. Relatively low activator/catalyst molar ratios result in a low concentration of the active sites, and the amount of the polymer produced does not exceed the ability of the biphasic system to transfer the product from the ionic liquid to the hexane phase. Therefore, for each ionic liquid an optimum concentration of the alkylaluminium compound can be determined. For the imidazolium ionic liquids, the longer the alkyl chain at the cation the better the mass transfer of the product to the hexane phase. The length of the alkyl chain at the imidazolium cation changes the physical properties of the ionic liquids, particularly density and viscosity, which influence the dispersion and separation of both phases. For the pyridinium ionic liquids such a tendency is not observed. The transfer of the product to the hexane phase is best with a shorter alkyl chain at the pyridinium cation and only at a lower concentration of the alkylaluminium compound. Although the application of pyridinium ionic liquids results in better total yield (from both the ionic liquid and hexane phases), the mass transfer of the product to the hexane phase is hindered at higher $AlEtCl_2$ concentrations. This can be overcome, however, by extending the reaction time.

The analysis of the physical and molecular properties of the polyethylene product obtained in the biphasic process shows that regardless of the ionic liquid used, the product displays unique properties, particularly a non-typical, very high crystallinity degree - up to 95%. It is also possible to control the selected properties of the polymer product, e.g. the melting point, crystallinity degree and bulk density (Ochędzan-Siodłak et al., 2007; 2008; 2009; 2010). The presented results indicate that the biphasic polymerisation using ionic liquid as a medium of the catalyst is an interesting option, and even desirable from the point of view of modern green chemistry. It also makes it possible to obtain a polymer product with interesting properties, which can be tailored by the change of the ionic liquid's parameters. Further investigation should be focused on the elimination of the disadvantageous features of the biphasic polymerisation. It is particularly important to improve the catalytic performance of the system, the mass transfer of the product, and the control of the polymer morphology. A promising way to fulfil these requirements could be application of the supported ionic liquid phase (SILP) methodology, where the ionic liquid containing an organometallic catalyst is anchored on a solid carrier (Valkenberg et al., 2001; 2002; Mehnert, 2005; Riisager et al., 2006). This method makes it possible to control, simultaneously, the course of the polymerisation reaction and the physical and chemical properties of the polymer product. Additionally, the amount of the ionic liquid phase can be considerably reduced and the performance of the catalyst can be improved, while the catalyst re-use and the easy separation procedure can be maintained (Gu, Li, 2009).

5. Acknowledgement

This work is supported by The Ministry of Science and Higher Education for Grant No N N209 335337.

6. References

Bernando-Gusmão, K., Queiroz, L.F.T., de Souza, R.F., Leca, F., Loup, C., Réau, R. (2003). Biphasic oligomerisation of ethylene with nickel-1,2-diiminophosphorane complexes immobilized in 1-n-butyl-3-methylimidazolium organochloroaluminate. *J. Catal.*, Vol., 219, pp. 59-62.

Carlin, R.T., Osteryoung, R.A., Wilkes, J.S., Rovang, J. (1990). Studies of titanium(IV) chloride in a strongly Lewis acidic molten salt: Electrochemistry and titanium NMR and electronic spectroscopy. *Inorg. Chem.*, Vol., 29, pp. 3003-3009.

Carlin, R.T., Wilkes, J.S. (1990). Complexation of Cp_2MCl_2 in a chloroaluminate molten salt: relevance to homogeneous Ziegler–Natta catalysis. *J. Mol. Catal.*, Vol., 63, pp. 125-129.

Chauvin, Y., Einloft, S., Olivier, H. (1995). Catalytic dimerization of propene by nickel-phosphine complexes in 1-butyl-3-methylimidazolium chloride/$AlEt_xCl_{3-x}$ (x = 0, 1) ionic liquids. *Ind. Eng. Chem. Res.*, Vol., 34, pp. 1149-1155.

Chauvin, Y., Gilbert, B., Guibard, I. (1990). Catalytic dimerization of alkenes by nickel complexes in organochloroaluminate molton salts. *J. Chem. Soc. Chem. Commun.*, Vol., 23, pp. 1715-1716.

Chauvin, Y., Olivier, H., Wyrvalski, C.N., Kimon, L.C., Souza, R.F. (1997). Oligomerization of n-butenes catalyzed by nickel complexes dissolved in organochloroaluminate ionic liquids. *J. Catal.*, Vol., 165, pp.275-278.

Chauvin, Y., Olivier, H., Wyrvalski, C.N., Kimon, L.C., Souza, R.F. (1997). Oligomerization of n-butenes catalyzed by nickel complexes dissolved in organochloroaluminate ionic liquids. *J. Catal.*, Vol., 165, pp. 275-278.

Chien, J.C.W. (1999). Supported metallocene polymerization catalysis. *Topics in Catal.*, Vol., 7, pp. 23-36.

Chowdhury, S., Mohan, R.S., Scott, J.L. (2007). Reactivity of ionic liquids. *Tetrahedron*, Vol., 63, pp. 2363-2389.

Doak, K.W. (1986). *Encyclopedia of polymer science and engineering*, vol. 6. J.I. Kroschwitz (Ed.), p. 384, Willey, New York.

Einloft, S., Dietrich, F.K., de Souza, R.F., Dupont, J. (1996). Selective two-phase catalytic ethylenedimerization by Ni complexes/$AlEtCl_2$ dissolved in organoaluminate ionic liquids. *Polyhedron*, Vol., 15, pp. 3257- 3259.

Ellis, B., Keim, W., Wasserscheid, P. (1999). Linear dimerisation of but-1-ene in biphasic mode using buffered chloroaluminate ionic liquid solvents. *Chem. Commun.*, pp. 337-338.

Ewen, J.A. (2000). Metallocene polymerization catalysts: Past, Present and Future, In: *Metallocene-based polyolefins*, J. Scheirs, W. Kaminsky (Ed.), pp. 3-31, John Wiley, Sons, Ltd., Chichester.

Gu, Y., Li, G. (2009). Ionic Liquids-Based Catalysis with Solids: State of the Art. *Adv. Synth. Catal.*, Vol., 351, pp. 817–847.

Hammawa, H., Wanke, S.E. (2006). Gas-phase olefin polymerization over supported metallocene/MAO catalysts: influence of support on activity and polydispersity. *Polym. Int.*, Vol., 55, pp. 426-434.

Hlatky, G.G. (2000). Heterogeneous single-site catalysts for olefin polymerization. *Chem. Rev.*, Vol., 100, pp. 1347-1376.

Jain, N. Kumar, A. Chauchan, S. Chauchan, S.M.S. (2005) Chemical and biochemical transformations in ionic liquids. *Tetrahedron*, Vol., 61, pp. 1015-1060.

Jang, Y-J., Nenov, N., Klapper, M., Muller, K. (2003). Organic nanoparticles with polypropyleneoxide chains as support for metallocene catalysts: ethylene homopolymerization and ethylene α-olefin copolymerization. *Polymer Bull.*, Vol., 50, pp. 343-350.

Kaminsky, W., Laban, A. (2001). Metallocene catalysis. *Appl. Catal. A: Gen.*, Vol., 222, pp. 47-61.

Kang, K.K., Oh, J.K., Jeong, Y.T., Shiono, T., Ikeda, T. (1999). Highly active $MgCl_2$-supported $CpMCl_3$ (M= Ti, Zr) catalysts for ethylene polymerization. *Macromol. Rapid Commun.*, Vol., 20, pp. 308-311.

Kubisa, P. (2004). Application of ionic liquids as solvents for polymerization processes. *Prog. Polym. Sci.*, Vol., 29, pp. 3-12.

Lee, B.Y., Oh, J.S. (2000). Preparation of anchored metallocene complexes on dehydroxylated silica and their use in the polymerization of ethylene. *Macromolecules*, 33, 3194-3195.

Mehnert, C.P. (2005). Supported ionic liquid catalysis. *Chem. Eur. J.*, Vol., 11, pp. 50-56.

Ochędzan-Siodłak, W. (2009). Ionic liquids in olefins oligomerization and polymerization (in Polish). *Wiadomości Chemiczne* (Wrocław), Vol., 63, pp. 776-801.

Ochędzan-Siodłak, W., Dziubek, K., Czaja, K. (2009). Comparison of imidazolium and pyridinium ionic liquids as mediums of the titanocene catalyst for biphasic ethylene polymerization. *Polimery (Warsow)*, Vol., 54, pp. 501-506.

Ochędzan-Siodłak, W., Dziubek, K., Siodłak, D. (2008). Biphasic ethylene polymerisation using 1-n-alkyl-3-methylimidazolium tetrachloroaluminate ionic liquid as a medium of the Cp_2TiCl_2 titanocene catalyst. *Eur. Polym. J.*, Vol., 44, pp. 3608-3614.

Ochędzan-Siodłak, W., Nowakowska, M. (2005). Heterogeneous zirconocene catalyst on magnesium support $MgCl_2(THF)_2$ modified by $AlEt_2Cl$ for ethylene polymerisation *Eur. Polym. J.*, Vol., 41, pp. 941-947.

Ochędzan-Siodłak, W., Pawelska, P. (2008). Chloroaluminate ionic liquids as a medium of titanocene catalyst activated by alkyl aluminium compounds for ethylene polymerisation. *Polimery (Warsow)*, Vol., 53, pp. 36-41.

Ochędzan-Siodłak, W., Sacher-Majewska, B. (2007). Biphasic ethylene polymerisation using ionic liquid over a titanocene catalyst activated by an alkyl aluminium compound. *Eur. Polym. J.*, Vol., 43, pp. 3688–3694.

Olivier-Bourbigou, H., Magna, L. (2002). Ionic liquids: perspectives for organic and catalytic reactions. *J. Mol. Catal. A: Chem.*, Vol., 182-183, pp. 419-437.

Olivier-Bourbigou, H., Magna, L., Morvan, D. (2010). Ionic liquids and catalysis: Recent progress from knowledge to applications. *Appl. Catal. A: Gen.*, Vol., 373, pp. 1-56.

Parvulescu, V.I., Hardacre, Ch. (2007). Catalysis in ionic liquids. Chem. Rev., 107, 2615–2665.

Pinheiro, M.F., Mauler, R.S., de Souza, R.F. (2001). Biphasic ethylene polymerisation with a diiminenickel catalyst. *Macromol. Rapid. Commun.*, Vol., 22, No., 6, pp. 425-428.

Prasad, A. (1999). *Polymer data handbook.* J.E. Mark (Ed.), p. 531, Oxford University Press: Inc., New York.

Razavi, A.C.R. (2000). Metallocene catalysts technology and environment. *Chemistry*, Vol., 3, pp. 615-625.

Riisager, A., Fehrmann, R., Haumann, M., Wasserscheid, P. (2006). *Topics Catal.*, 40, 91-102.

Severn, J.R., Chadwick, J.C., Duchateau, R., Friedrich, S. (2005). "Bound but not gagged" - Immobilizing single-site α-olefin polymerization catalysts. *Chem. Rev.*, Vol., 105, pp. 4073-4147.

Simon, L.C., Dupont, J., de Souza, R.F. (1998). Two-phase n-butenes dimerization by nickel complexes in molten salt media. *Appl. Catal. A: Gen.*, Vol., 175, pp. 215-220.

Soares, J.B.P., Kim, J.D.J. (2000). Copolymerization of ethylene and α-olefins with combined metallocene catalysts. I. A formal cwiterion for molecular weight bimodality. *J. Polym. Sci. Chem.*, Vol., 38, pp. 1408-16.

Thiele, D., de Souza, R.F. (2007). The role of aluminum species in biphasic butene dimerization catalyzed by nickel complexes. *J. Mol. Catal. A: Chem.*, Vol., 264, pp. 293-298.

Valkenberg, M.H., de Castro, C., Hölderich, W.F. (2001). Immobilisation of chloroaluminate ionic liquids on silica materials. *Topics in Catal.*, Vol., 14, pp. 139-144.

Valkenberg, M.H., de Castro, C., Hölderich, W.F. (2002). Immobilisation of ionic liquids on solid supports. *Green Chem.*, Vol., 4, pp. 88–93.

Vasilie C., Pascu M. (2005). *Practical Guide of Polyethylene*, p. 32, Rapra Technology Limited, ISBN: 1-85957-493-9, Shawbury, UK Wasserscheid, P., Eichmann, M. (2001). Selective dimerisation of 1-butene in biphasic mode using buffered chloroaluminate ionic liquid solvents—design and application of a continuous loop reactor. *Catal. Today*, Vol., 66, pp.309-316.

Wasserscheid, P., Hilgers, C., Keim, W. (2004). Ionic liquids–weakly-coordinating solvents for the biphasic ethylene oligomerisation to α-olefins using cationic Ni-complexes. *J. Mol. Catal. A: Chem.*, Vol., 214, pp.83-90.

Wei, L., Tang, T., Huang, B. (2004). Synthesis and characterization of polyethylene/clay–silica nanocomposites: a montmorillonite/silica-hybrid-supported catalyst and in situ polymerization. *J. Polym. Sci. Part A: Polym. Chem.*, Vol., 42, pp. 941-949.

Welton, J. (2004). Ionic liquids in catalysis. *Coord. Chem. Rev.*, Vol., 248, pp. 2459-2477.

Wilkes, J.S. (2004). Properties of ionic liquid solvents for catalysis. *J. Mol. Catal. A: Chem.*, Vol., pp. 214, 11-17.

Part 2

Natural Polymers

3

Synthesis of Ionic Liquids, Solubility for Wood and Its Application for Graft Copolymer with Acrylamide

Guo Liying
School of Petrochemical Engineering, Shenyang University of Technology. Liaoyang, 111003, China

1. Introduction

With the increasing awareness in sustainable development of environmental protection, there is a keen aspiration in exploring natural recoverable and environment-friendly materials [1]. Wood is a typical abundant natural material which is environment-friendly and easily-obtained. Cellulose, hemi-cellulose and lignin are the main components of wood powder which build the spongy three-dimension structures [2]. The poor solubility of wood powder in normal solvents strongly limits its application, hence it is important to find out environment-friendly solvents for wood powder.

The conditions of using conventional method to dissolve wood in solvent such as phenol or polyhydric alcohols were very strict and the original structures of wood were destroyed which results in the loss of mechanical properties [3, 4]. In contrast, ionic liquids which were known as green solvents for cellulose [5-9], can partly dissolve wood powder, and improve the reactivity due to the structural laxity [10-12]. The applications of ionic liquids ranged from chemical synthesis, biological catalysis, preparation of functional materials, etc. [13-19]. In addition, as a newly-developed technique which was widely used in various chemical areas, microwave can be applied in dissolving wood powder as a new and high efficient method.

Regarding the importance in preparing environment-friendly polymer materials from natural resources, studies of lignin were reported from several countries such as U.S., Japan, etc for decades. Polyacrylamide, enjoying the reputation of "hundred industry assistants" is used in various fields. But production cost is very high for its application. It has been of inevitable trend and broad prospects to develop multi-functional lignin flocculant using inexpensive lumber [20, 21]. Meanwhile, copolymerizations of lignin or cellulose with vinyl monomers in aqueous medium were reported [22-25], which have the shortages of low conversion of the polymerization, instable final product and difficulty in controlling the reactions. To overcome the disadvantages stated above, here, we report the synthesis of a novel flocculant with wood and acrylamide (AM) by graft copolymerization (AM-g-wood) in ionic liquids in which wood powder was used as alternative to lignin, which can introduce the new concepts of utilizing natural recoverable resources. Besides, graft copolymer of AM-g-wood as a novel modified flocculant, whose production cost was decreased greatly in this research, also could be widely applied in areas such as water treatment, petrochemical industry, mineral processing, papermaking and so on. The related work is being researched.

2. Experimental

2.1 Materials

Materials: N-methylimidazole, N-ethyllimidazole, and 1-vinlylimidazole (Purity≥99%) were obtained from Linhai Kaile Chemical Plant; Allyl Chloride and Chloroethanol were from Sinopharm Chemical Reagent Co. Ltd. and Tianjin Guangfu Fine Chemical Research Institute separately; Potassium Persulfate ($K_2S_2O_8$ or KPS) and acrylamide were received from Sinopharm Chemical Reagent Co. Ltd.; Benzene, Methanol, Acetone, Sodium hydroxide(NaOH), and Diethyl Ether were provide by Shanghai Suyi Chemical Reagent Co. Ltd.; Fir powder was donated by Huangshan Wood Factory. All chemicals and materials were used as received without further purification.

2.2 Synthesis of ionic liquids

Ionic liquids were synthesized from imidazole and allyl chloride (or chloroethanol), and their chemical structures were confirmed by FTIR and ^{1}HNMR. Typical procedure is: into a round bottom flask, imidazole materials and allyl chloride (or chloroethanol) were added. The reaction was conducted under refluxing for 8-48 h. After reaction, the mixture was extracted with ether for several times to remove the residual imidazole. The ionic liquids can be obtained after being dried at 80°C (vacuum: 0.08MPa) for 24h (yield: >90%). The chemical structures of the ionic liquids were shown in Scheme 1.

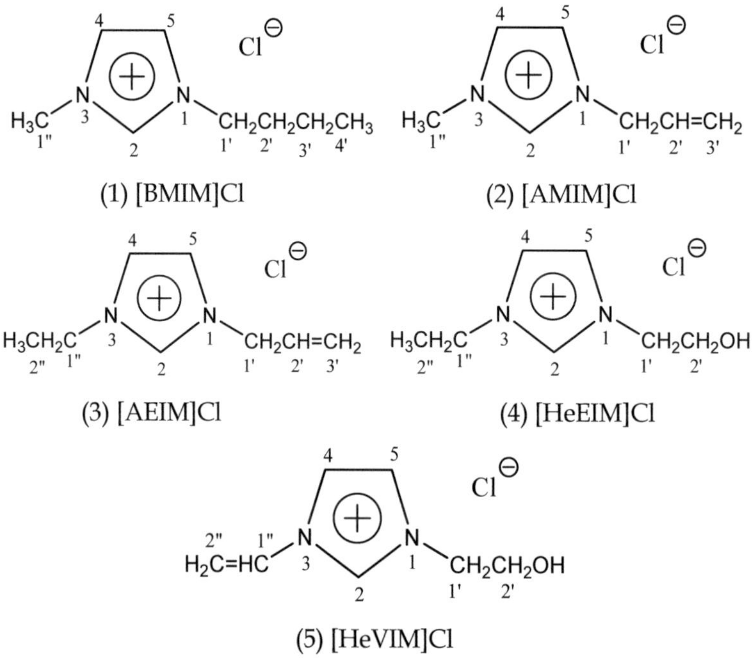

Scheme 1. Chemical structures and abbreviation of the examined ionic liquids

1. 1-butyl-3-methyl imidazolium chloride ([BMIM]Cl) was synthesized according to the general procedure provided by Lu et al.[26] with a slight of modification. FTIR (film): ν=3422 (ν O-H), 2918 and 2843 (ν C-H), 1632 (ν C=C), 1575 (ν C=N), 1472 (β C-H), 1169

(δ C-H), 718 (δ imidazole ring); ^{1}H NMR (DMSO, 25°C): δ=0.84 (3H, t, H-4'), 1.26 (2H, m, H-3'), 1.76 (2H, m, H-2'), 4.00 (3H, s, H-1"), 4.23 (2H, t, H-1'), 7.42 (1H, d, H-4), 7.58 (1H, d, H-5), 10.26 (1H, s, H-2).

2. 1-allyl-3-methyl imidazolium chloride ([AMIM]Cl) was synthesized according to the general procedure provided by Zhang et al.[9] with a slight of modification. FTIR (film): v=3409 (v O-H), 3083 and 2857 (v C-H), 1645 (v C=C), 1573 (v C=N), 1424 (β C-H), 1167 (δ C-H), 998 (ρ C-H), 950 (ρ C-C), 763 (δ imidazole ring); ^{1}H NMR (DMSO, 25°C): δ=3.88 (3H, s, H-1"), 4.80 (2H, dd, H-1'), 5.22 (2H, m, H-3'), 5.81 (1H, m, H-2'), 7.26 (1H, d, H-4), 7.59 (1H, d, H-5), 10.08 (1H, s, H-2).
3. 1-allyl-3-ethyl-imidazolium chloride ([AEIM]Cl) was synthesized from allyl chloride 25.3g (0.33mol) and N-ethylimdiazole 28.8g (0.3mol). The reaction was conducted under refluxing for 8h at 70°C in water bath with stirring under the protection of N_2. FTIR (film): v=3432 (v O-H), 3082 and 2980 (v C-H), 1645 (v C=C), 1563 (v C=N), 1449 (β C-H), 1163 (δ C-H), 997 (ρ C-H), 943 (ρ C-C), 761 (δ imidazole ring); ^{1}H NMR (DMSO, 25°C): δ=1.45 (3H,t, H-2"), 4.20 (2H, q, H-1"), 4.77 (2H, dd, H-1'), 5.34 (2H, m, H-3'), 6.00 (1H, m, H-2'), 7.43 (1H,d, H-4), 7.47 (1H, d, H-5), 8.74 (1H, s, H-2).
4. 1-(2-hydroxylethyl)-3-ethyl imidazolium chloride ([HeEIM]Cl) was synthesized from chloroethanol 16.7mL (0.25mol) and N-ethylimdiazole 28.8g (0.3mol). The reaction was conducted under refluxing for 48h at 80°C in water bath with stirring. FTIR (film): v=3392 (v O-H), 3145 and 2994 (v C-H), 1632 (v C=C), 1566 (v C=N), 1449 (β C-H), 1165 (δ C-H), 1072 (v C-O), 750 (δ imidazole ring); ^{1}H NMR (DMSO, 25°C): δ=1.43 (3H, t, H-2"), 3.73 (2H, m, H-2'), 4.22 (2H, q, H-1"), 4.26 (2H, t, H-1'), 5.48 (1H, br, H-O-H), 7.81 (1H, d, H-4), 7.86 (1H, d, H-5), 9.39 (1H, s, H-2).
5. 1-(2-hydroxylethyl)-3-ethylene imidazolium chloride ([HeVIM]Cl) was synthesized from chloroethanol 16.7mL (0.25mol) and N-ethylene-imdiazole 26.3g (0.28mol). The reaction was conducted under refluxing for 10h at 60°C in water bath with stirring under the protection of N_2. FTIR (film): v=3406 (v O-H), 2994 and 2878 (v C-H), 1651 (v C=C), 1572 (v C=N), 1495 (v_{as} C-H), 1372 (β C-H), 1078 (v C-O), 962 (ρ C-H), 916 (ω O-H), 750 (δ imidazole ring); ^{1}H NMR (DMSO, 25°C): δ=3.60 (2H, m, H-2"), 3.77 (2H, m, H-2'), 4.30 (2H, t, H-1'), 5.47 (1H, br, H-O-H), 7.40 (1H, m, H-1"), 7.97 (1H, d, H-5), 8.32 (1H, d, H-4), 9.80 (1H, s, H-2).

2.3 The pretreatment of wood powder

2.3.1 Extraction of wood powder with benzene and ethanol

Wood powder with particle size at 0.2mm was extracted with a mixture of benzene/ethanol (volume ratio: 1/1) using soxhlet extractor [27]. Then, the powder was dried at 70°C (vacuum: 0.04MPa) for 12h in Vacuum drying oven.

2.3.2 Treatment of wood powder with 25% NaOH

Wood powder with particle size at 0.2mm was treated with 25% NaOH solution in air for 2h. Alternative conditions are: microwave treatment for 5min at 60°C (360w), or treated MCC in 25% NaOH solution (40mL) at 140°C for 2h in an airtight high pressure pot (60mL), which was placed in Electronic constant temperature oven. The mixture was filtered, neutralized and washed with water several times. Then the fir powder was dried at 70°C (0.04 MPa) for 12h in Vacuum drying oven.

2.4 Dissolution of fir powder by microwave heating

Into a flask, ionic liquid and pretreated fir powder (W_0) above (weight ratio: 25:1) were introduced. The flask was placed in microwave oven (NJL 07-3) and heated at 90°C (400w) for 1h with stirring. The parallel experiment was copied. One of them was shifted to polymerization tube for future use. The other mixture was allowed to cool to room temperature followed by the centrifuge separation (2500r/min, 5min). The finial fir powder was dried at 70°C (0.04Mpa) for 12h in order to calculate dissolution rate [11].

2.5 Synthesis and purification of AM-g-Wood

2.5.1 Synthesis of AM-g-Wood

Into a polymerization tube which was placed in water bath at 45°C, the mixture of fir powder/ionic liquid treated by microwave was added. Then air was purged out by N_2 stream before the introduction of initiator $K_2S_2O_8$ (I/M: 1/1000). AM was added dropwise into the mixture, and the dropping rate was controlled to avoid the homo-polymerization. The polymerization was stopped after 3h, and aged for 24h before purification.

2.5.2 Purification of AM-g-Wood

Typical purification procedure of AM-g-Wood is: the copolymer was washed over ethanol several times to remove the ionic liquid and residual monomer, followed by drying at 60°C under vacuum (0.08 MPa) for 24h. Then, the raw product (W_1) was extracted with acetone for 48h using soxhlet extractor to give the refined product, whose mass was W_2 after being dried under the same conditions above. The effect of graft reaction was evaluated by grafting degree (D) and grafting efficiency (E) which can be calculated by the following formulas:

$$\text{Grafting degree (\%)} = (W_2-W_0) / W_2 \times 100\%$$

$$\text{Grafting efficiency (\%)} = (W_2-W_0) / (W_1-W_0) \times 100\%$$

W_0-----Total mass of wood powder
W_1-----Mass of raw product of AM-g-Wood
W_2-----Mass of refined product of AM-g-Wood

2.6 Characterizations

MAGNA-IR750 FT-IR Spectrometer and AVANCE AV-400 Super-conducting Fourier Digital NMR Spectrometer was used to analyze the structures of ionic liquids; MAGNA-IR750 FT-IR Spectrometer, D/Max-rB X-ray diffractometer was used to analyze the structures and the crystallinity of pretreated fir powder and the graft copolymer of AM-g-Wood; ShimazuTGA-50H Thermo-gravimetric Analyzer and JSM6700F Field Emission Scanning Electron Microscope were applied to analyze the thermal stability and apparent morphology of AM-g-Wood respectively.

3. Results and discussion

3.1 Wood pretreatment and dissolution in ionic liquids

3.1.1 Chemical structures of fir powder pretreated with different methods

Fir power was extracted with a mixture of benzene/ethanol or activated with 25% (mass fraction) NaOH under normal temperature and pressure, microwave and high pressure in order to make it dissolve more easily in ionic liquids. Chemical structures of the treated fir power, including the change of wave number, wave form and the interaction of hydrogen bonding, were examined by FTIR and the results were shown in Fig.1.

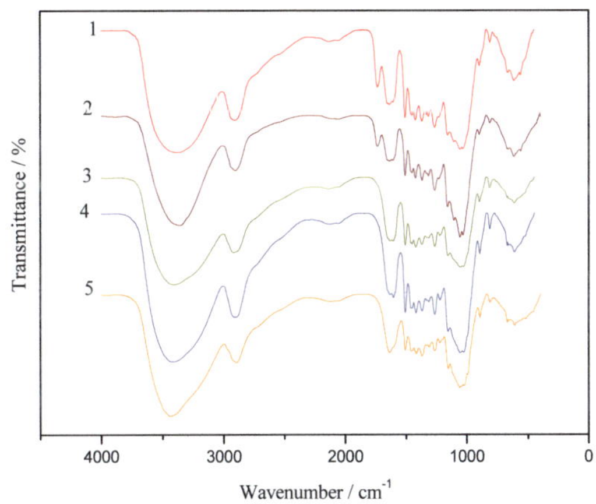

Fig. 1. FTIR spectra of fir powder pretreated with different methods

(1-Original wood; 2-Extracted by benzene/ethanol; 3-Treated with 25% NaOH (mass fraction) at normal temperature and pressure; 4-Treated with 25% NaOH by microwave heating; 5-Treated with 25% NaOH in high pressure pot)

The peak around 3400cm^{-1} is stretching vibration of O-H in hydrogen bonding. The peak originally ascribe to the absorption of lignocelluloses weakened which suggested the breaking of hydrogen bonding in the wood system after extraction and NaOH treatments, and this peak shifted to high wave number by the greatest degree and the waveform was the most narrow after high pressure treatment (Fig.1), which indicated the hydrogen bonding of sample 5 was broken most seriously. Similarly, the shrinkage of peak around 1730cm^{-1} indicated that certain amount of fatty acids and acetyl groups or uronic acids of the hemicelluloses have been removed partly by the extraction, and this peak disappeared after NaOH treatment which suggested that the hemicelluloses of wood was degraded and the fatty acids, uronic acids and acetyl groups, etc, were completely dissolved by NaOH solution. However, absorption peaks of wood powder mainly around 1500cm^{-1}, 1268cm^{-1}, 1060cm^{-1} remained unchanged after extraction and NaOH treatments under different conditions [28, 29]. Followed by the change of wave forms and analysis above, the preliminary conclusion could be reached that it is easier to dissolve fir powder treated with 25% NaOH under high pressure.

3.1.2 Crystallinity and dissolution rate of the pretreated fir powder

Crystallinity of fir powder treated under different technologic conditions were measured by Turley method, which scanned and counted in rule times at diffraction angles of 5°, 18.6°,22.6° and 40°, then wrote down the diffraction peak intensity respectively and calculating the crystallinity. Dissolution rate of pretreated fir powder dissolved in [HeEIM]Cl by microwave heating was studied, and the results were shown in Table 1.

Sample number	Angle and Intensity				Crystallinity (%)	Dissolution rate (%)
	5°	18.6°	22.6°	40°		
1	17177	37523	74608	10498	61.0	6.2
2	16614	21780	38468	4949	60.2	8.4
3	14156	45671	54162	11196	19.6	13.6
4	8171	45386	49822	8773	10.6	16.8
5	8914	42504	43507	10450	2.4	18.3

(1-Original wood; 2-Extracted by benzene/ethanol; 3-Treated with 25% NaOH (mass fraction) at normal temperature and pressure; 4-Treated with 25% NaOH by microwave heating; 5-Treated with 25% NaOH in high pressure pot)

Table 1. Crystallinity and dissolution rate in [HeEIM]Cl of pretreated fir powder

Crystallinity of fir powder, which extracted by mixture of benzene/ethanol (the extract content was 6.1%), had little change while the dissolution rate increased (in Table 1). This is because some impurity and small molecule fattiness of wood powder were extracted by the mixture. When the fir powder was pretreated with 25% NaOH, the crystallinity of sample 5 was decreased and the dissolution rate was increased, especially under high pressure condition, the crystallinity decreased from 61.0% to 2.4%, while the dissolution rate increased from 6.2% to 18.3%. There were two reasons for the result: on the one hand, alkali had a certain effect on decreasing crystallinity and degrading hemicellulose of wood; on the other hand, higher temperature and pressure were helpful to destroy the crystal form of fir powder, causing the crystallinity to decrease and dissolution rate to increase. High pressure method, therefore, was used as the best pretreatment method for wood powder.

3.1.3 Solubility of different ionic liquids

Fir powder pretreated with 25% NaOH under high pressure (Sample 5 in Table 1) was dissolved in different ionic liquids by microwave heating, and the results were shown in Fig.2.

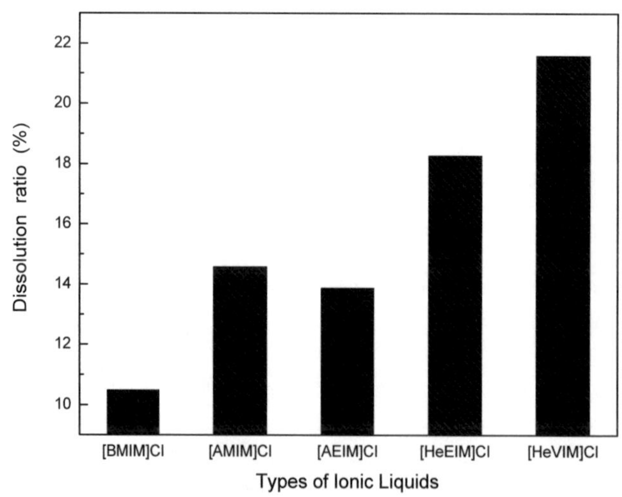

Fig. 2. Influence of different ionic liquids on dissolution rate

Due to the presence of hydroxyl and alkene groups in the side chains which contribute to the destroy of hydrogen bond, the solubility of fir powder using [HeVIM]Cl was the highest (Fig.2). The solubility of ionic liquid with hydroxyl side chain ([HeEIM]Cl) was better than alkene side chain ([AEIM]Cl) and alkyl side chain ([BMIM]Cl) (Fig.2). However, solubility of ionic liquids with shorter side chains ([AMIM]Cl) was better than that of [AEIM]Cl , which is ascribe to the reduced efficient chloride concentration [6]. Ionic liquid of [BMIM]Cl, whose side chain was of the longest alkyl groups, therefore, had the worst solubility. [HeEIM]Cl was used as radium of the following graft copolymerizations instead of [HeVIM]Cl, which had the best solubility, but it was more difficult to be purified.

3.2 Effect of reaction medium on graft copolymerization

AM was polymerized with fir powder (Sample 5 in Table 5) at 45 °C using aqueous and [HeEIM]Cl medium respectively. The results were shown in Fig.3.

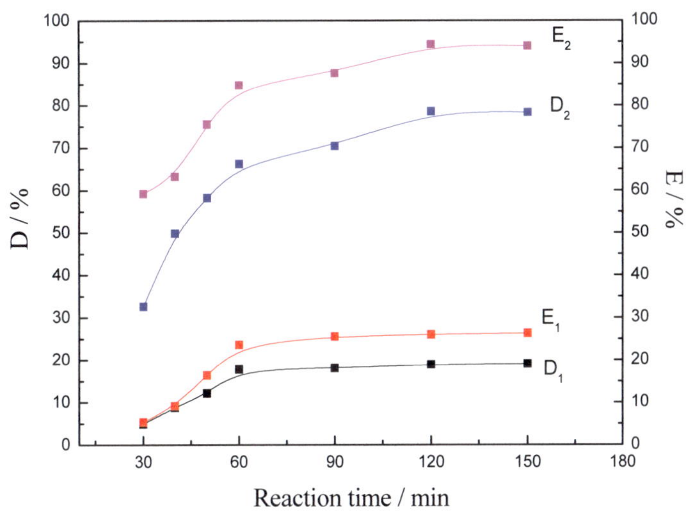

(1-reaction medium of H_2O; 2- reaction medium of [HeEIM]Cl)

Fig. 3. Influence of reaction medium on grafting degree (D) and grafting efficiency (E)

Graft copolymer of AM-g-Wood could be prepared much more easily in [HeEIM]Cl than in aqueous system (Fig.3). Similarly, grafting degree (D) and grafting efficiency (E) of AM-g-Wood prepared in [HeEIM]Cl were much higher than that in aqueous medium. This is because fir powder can be dissolved in ionic liquid before copolymerization which resulted in the breaking of hydrogen bond and enhancing the reactivity of the powder. In Fig.3, grafting degree and grafting efficiency increased very fast during the beginning of the reaction (≤1h), and became stable in aqueous medium, while increased slowly in [HeEIM]Cl thereafter. This phenomenon can be explained by the typical conventional free radical polymerization mechanism: In the first stage, a great number of free radicals were formed and the reaction was very quick. With the reaction went on, propagation slowed down due to the loss of radicals during the process. The best condition for graft copolymerization, therefore, was used [HeEIM]Cl as the reaction medium with the reaction time of 1h at least.

3.3 Orthogonal test

To optimize the technical conditions, orthogonal test was designed based on three factors and levels (Table 2) separately. The effects of different conditions, such as mass fraction of AM, fluid ratio of [HeEIM]Cl to Wood, and mass proportion of wood to AM on graft copolymerization were carefully studied. Details were shown in Table 2 and 3.

Levels	Factors		
	A (Mass fraction of AM)	B (Fluid ratio of [HeEIM]Cl to wood)	C (Mass proportion of wood to AM)
1	20%	25:1	1:3
2	25%	30:1	1:4
3	30%	35:1	1:5

Table 2. Factors and levels of the orthogonal test

No.	A	B	C	Grafting degree (%)	Grafting efficiency (%)
1	1	1	1	61.2	82.2
2	1	2	2	77.5	89.3
3	1	3	3	70.4	69.5
4	2	1	2	84.1	91.4
5	2	2	3	82.5	75.8
6	2	3	1	78.9	89.5
7	3	1	3	74.8	82.6
8	3	2	1	79.1	94.1
9	3	3	2	74.3	90.5
K1	209.1 / 241.0	220.1 / 256.2	219.2 / 265.8		
K2	245.5 / 256.7	239.1 / 259.2	235.9 / 271.2	—	—
K3	228.2 / 267.2	223.6 / 249.5	227.7 / 227.9		
R	36.4 / 26.2	19.0 / 9.7	16.7 / 43.3		

Table 3. Result of orthogonal test

The data in Table 3 showed that factors A, B and C had different effect on grafting degree and grafting efficiency of the synthesized products. The affecting order on grafting degree was A>B>C, while on grafting efficiency was C>A>B. The best comprehensive effects came from $A_2B_2C_2$ sample, which means that the mass ratio of AM was 25%, fluid ratio of [HeEIM]Cl to fir powder was 30:1, and mass proportion of wood to AM was 1:4.

3.4 Chemical structures of original wood and AM-g-Wood

Chemical structures of original wood powder and its graft copolymer of AM-g-Wood, which were prepared in [HeEIM]Cl using the best conditions according to the results of orthogonal test, were confirmed using FTIR. The results were shown in Fig.4.

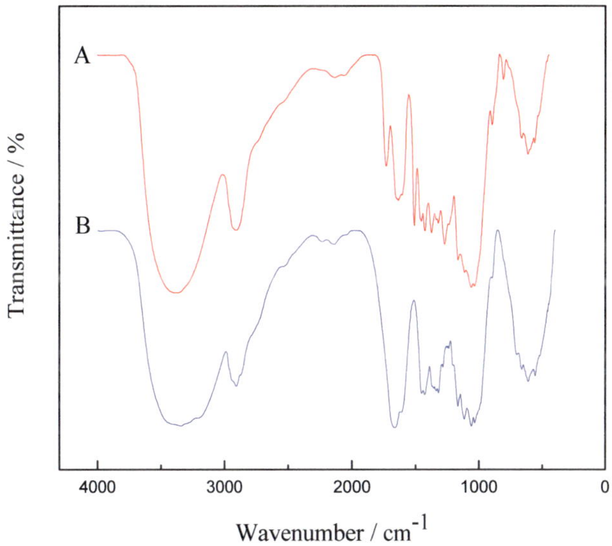

Fig. 4. FTIR spectra of original Wood (A) and graft copolymer AM-g-Wood (B)

For curve A of Fig.4, peak around 1735cm^{-1} was the stretching vibration of C=O in fatty molecules and acetyl groups or uronic acids of the hemicelluloses of original wood powder, peak around 1638cm^{-1} was characteristic vibration of conjugated carbonyl in lignin of wood powder, and peaks around 1509cm^{-1}, 1160cm^{-1}, and 1058cm^{-1} were typical absorption of aromatic skeletal vibration, phenol-ether group and aliphatic ether vibration in lignin of wood, peak around 1268cm^{-1} was the stretching vibration of C-O-C in celluloses and hemicelluloses [28-30]. For curve B, the peak around 1735cm^{-1} disappeared after grafted, which suggested that fatty molecules and hemicelluloses have been dissolved by ionic liquids. The new stronger peak in 1671cm^{-1} came from the typical stretching vibration of -$CONH_2$. There were still existence of peaks around 1160cm^{-1} and 1058cm^{-1} after grafted polymerization which came from the wood. The wave form of peak around 1268cm^{-1} became unnoticeable after grafted, which suggested that the content of cellulose in the wood was decreased. These analysis strongly suggested that AM has been grafted onto the fir powder.

3.5 Crystal form of original wood and AM-g-Wood

Crystal forms of original wood powder and its graft copolymer of AM-g-Wood were measured by X-ray diffractometer and the results were shown in Fig.5.

In Fig 5, diffraction peaks appeared at angles of 16.6°, 22.1°, and 34.8° which demonstrate that original wood powder have the same diffraction results as wood fiber. Crystallinity of the original wood powder was 61% based on the Segal formula calculation [31]. The diffraction peaks from wood powder disappeared after grafted polymerization and the crystal area were completely converted into amorphous region, which suggested that the ordered crystal structures have been destroyed. The change of the crystal type also suggested the graft copolymerization occurred between wood powder and acrylamide (Fig 5).

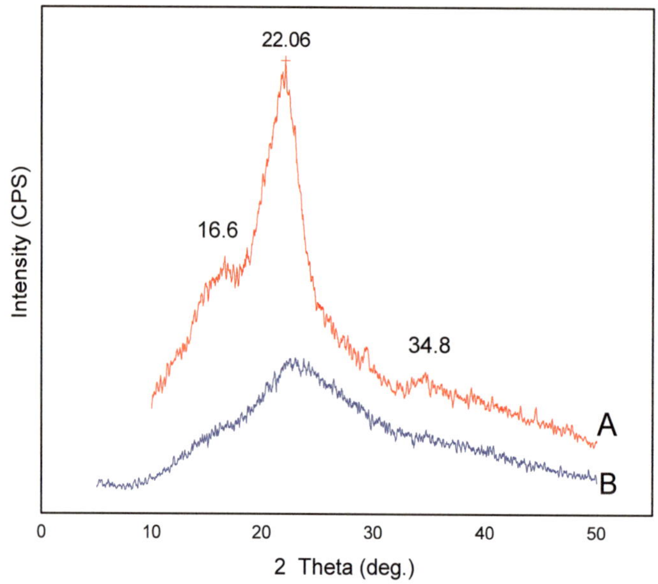

Fig. 5. X-ray spectra of original Wood (A) and graft copolymer AM-g-Wood (B)

3.6 Thermal stability of original wood and AM-g-Wood

Thermal stabilities of original wood powder and its graft copolymer of AM-g-Wood were examined by Thermo-gravimetric analyzer and the results were shown in Fig.6.

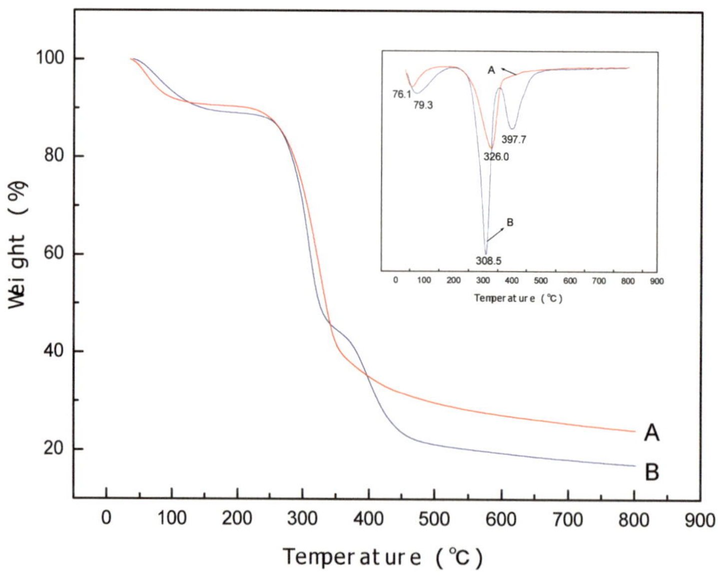

Fig. 6. TGA and DTG curves of original Wood (A) and graft copolymer AM-g-Wood (B)

As shown in Fig. 6, there were two steps for thermal decomposition of original wood: The first one was from 34°C to 175°C, which indicated that wood lost water in this range and the weight loss rate was around 9.5%; The second step was from 175°C to 794°C, responding to the thermal decomposition of wood and the weight loss rate was 66.2%. Actually, the original fir powder began to decompose and the chains of macromolecules were broken at 175°C. The temperature at 326°C refers to the largest weight loss rate. The residual weight rate was 24.4% after the temperature reached 800°C.

However, the thermal decomposition of the graft copolymer AM-g-Wood had three steps: The first step was from 38°C to 189°C, indicating the lost of water from copolymer with weight loss rate around 10.3%. The second step was from 190°C to 343°C, suggesting the weight loss (43.5%) resulted from imino reactions between amide groups of the PAM in the copolymer. The third step was from 343°C to 800°C with weight loss of 28.4%, and this is because the graft chains of PAM were broken during this temperature range. The final residual weight rate was 17.8% after the temperature reached 800°C. As such, the initial thermal decomposition temperature of the grafted wood was 343°C which was much higher than that of the original wood (175°C). Meanwhile, the temperature of the maximum weight loss rate for the grafted wood was 398°C, which was also higher than original wood (326°C). These results demonstrated that AM has been successfully grafted onto fir powder and the thermal stability of the wood after copolymerization was greatly improved.

3.7 Morphology of wood and AM-g-Wood

SEM results of the wood powder before dissolved (A), wood residue after dissolved in [HeEIM]Cl (B) and graft copolymer of AM-g-Wood (C) which was synthesized in [HeEIM]Cl were shown in Fig.7.

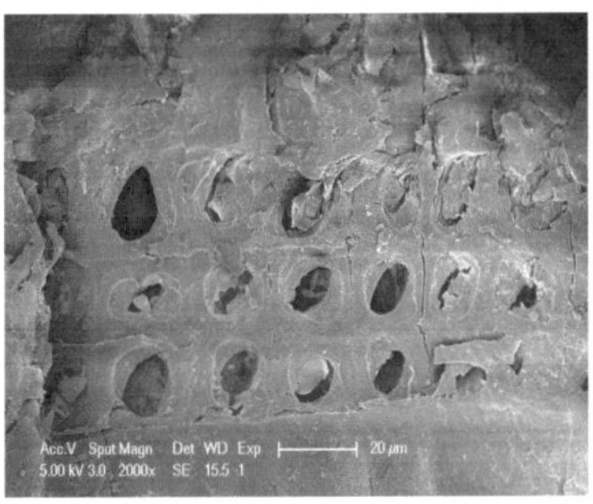

(A) wood before dissolved

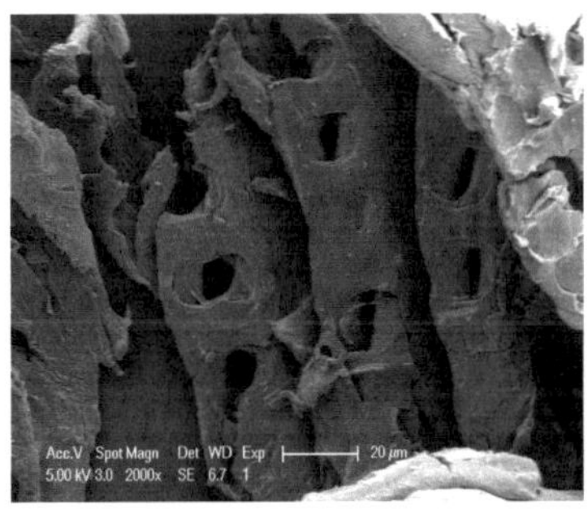

(B) wood residue after dissolved

(C) graft copolymer of AM-g-Wood

Fig. 7. SEM micrographs of wood and its graft copolymer

Microstructure of the wood powder before dissolved in [HeEIM]Cl was shown in A (Fig.7) which was loose and porous, while looser structure and bigger porous were found in B for wood after dissolved in [HeEIM]Cl. However, the loose and porous structures completely disappeared after polymerization (C) and it was difficult to detect fibril, which is very different from A and B. This difference should be ascribe to the existence of AM-g-wood structure which was synthesized from the graft copolymerization of AM and wood. Followed by the changes of wood appearance, therefore, a graft copolymerization occurred between fir powder and AM, and grafted copolymer of AM-g-Wood should be prepared more easily with wood residue as its looser structure and bigger porous.

4. Conclusions

Various ionic liquids with different functional groups of imidazole side chains have been synthesized, and their solubility for fir powder has been discussed. AM-g-Wood was synthesized in [HeEIM]Cl with $K_2S_2O_8$ as initiator and the conditions of graft copolymerization were optimized by orthogonal test. The specific conclusions were listed as follows:

1. The structures of synthesized ionic liquids have been confirmed by FT-IR and ^{1}HNMR. It was suggested that the nature of side groups greatly affected the solubility of the ionic liquids. The ionic liquid of [HeVIM]Cl which contains hydroxyl and alkene groups in imidazole side chains was the best solvent for fir powder, but it was difficult to be purified.
2. The fir powder which has been pretreated with 25% NaOH solution under high pressure has the lowest crystallinity and the highest dissolution rate.
3. AM-g-Wood could be prepared readily in [HeEIM]Cl. Compared to aqueous medium, AM-g-Wood prepared in [HeEIM]Cl had advantages such as much better grafting degree and grafting efficiency, higher yields and so on.
4. Orthogonal test suggested that the best conditions for graft copolymerization were: mass fraction of AM was 25%, fluid ratio of [HeEIM]Cl to wood was 30:1, and mass proportion of wood to AM was 1:4.
5. Results of FT-IR and SEM demonstrated that AM has been successfully grafted onto the wood powder. The crystal structure of the wood has been completely destroyed after graft copolymerization, and the thermal stability has been greatly improved.

5. Reference

[1] Wang Y, He J, Lu T, Wu YY, Fan YM (2004) Research on graft copolymerization of cellulose. Fine and Specialty Chemicals 12:18-21

[2] Vogli E, Sieber H, Greil P (2002) Biomorphic SiC-ceramic prepared by Si-vapor phase infiltration of wood. Journal of the Europe Ceramic Society 22: 2663-2668

[3] Lin LZ, Yoshioka M, Yao YG, Shiraishi N (1994) Liquefaction of wood in the presence of phenol using phosphoric acid as a catalyst and the flow properties of the liquefied wood. Journal of Applied Polymer Science 52: 1629-1636

[4] Yamada T, Ono H (2001) Characterization of the products resulting from ethylene glycol liquefaction of cellulose. Journal of Wood Scicence 47: 458-464

[5] Cuissinat C, Navard P, Heinze T (2008) Swelling and dissolution of cellulose. Part IV: Free floating cotton and wood fibres in ionic liquids. Carbohydrate Polymers 72: 590-596
[6] Swatloski RP, Spear SK, Holbrey JD, Rogers RD (2002) Dissolution of cellulose with ionic liquids. J Am Chem Soc 124:4974-4975
[7] Liu LY, Chen HZ (2006) Enzymatic hydrolysis of cellulose materials treated with ionic liquid [BMIM]Cl. Chinese Science Bulletin 51: 2432-2436
[8] Guo LY, Shi TJ, Li Z, Duan YP, Wang YG (2008) Synthesis of novel and functionalized ionic liquid [HeEIM]Cl and its solubility for cotton fibre. Chemical Journal of Chinese Universities 29:1901-1907
[9] Zhang H, Wu J, Zhang J, He JS (2005) 1-allyl-3-methylimidazolium chloride room temperature ionic liquid: A new and powerful nonderivatizing solvent for cellulose. Macromolecules 38: 8272-8277
[10] Fort DA, Remsing RC, Swatloski RP, Moyna P, Moyna G, Rogers RD (2007) Can ionic liquids dissolve wood? Processing and analysis of lignocellulosic materials with 1-n-butyl-3-methylimidazolium chloride. Green Chemistry 9: 63-69
[11] Guo LY, Shi TJ, Li Z, Duan YP (2008) Solubility of two kinds of imidazolium ionic liquids for fir powder. Journal of Chemical Industry and Engineering 59:1299-1304
[12] Kilpeläinen I, Xie HB, King A, Granstrom M, Heikkinen S, Argyropoulos DS (2007) Dissolution of wood in ionic liquids. J Agric Food Chem 55: 9142-9148
[13] Jain N, Kumar A, Chauhan S, Chauhan SMS (2005) Chemical and biochemical transformations in ionic liquids. Tetrahedron 61:1015-1060
[14] Kubisa P (2004) Application of ionic liquids as solvent for polymerization Processes. Progress in Polymer Science 29: 3-12
[15] Scurtoa M, Leitner W (2006) Expanding the useful range of ionic liquids: melting point depression of organic salts with carbon dioxide for biphasic catalytic reactions. Chemistry Communication 35: 3681-3683
[16] Lu XB, Hu JQ, Yao X, Wang ZP, Li JH (2006) Composite system based on chitosan and room-temperature ionic liquid: Derict electrochemistry and electrochemistry and electrocatalysis of hemoglobin. Biomacromolecules 7: 975-980
[17] Weuster-Botz D (2007) Process intersification of whole-cell biocatalysis with ionic liquids. Chem Rec 7: 334-340
[18] Zein El-Abedin S, Endres F (2006) Electrodeposition of metals and semiconductors in air- and water-stable ionic liquids. Chemphyschem 7: 58-61
[19] Lu XB, Zhang Q, Zhang L, Li JH (2006) Direct electron transfer of horseradish peroxidase and its biosensor based on chitosan and room temperature ionic liquid. Electrochemistry Communications 8: 874-878
[20] Mckague AB (1974) Flocculating agent derived from kraft lignin. Appl Chem Biotechnol 24: 607-615
[21] Dizhbite T, Telyshera G (1998) Multifunctional activity of lignin derivatives in pulp/ filler compositions. Adv Lignocellul Chem Ecol Friendly Pulping Bleaching Technol, Eur Workshop lignocellul Pulp 97-100
[22] Liu S, Sun G (2008) Radical graft functional modification of cellulose with allyl monomers: Chemistry and structure characterization. Carbohydrate Polymers 71:614-625

[23] Canché-Escamilla G., Pacheco-Catalán DE, Andrade-Canto SB (2006) Modification of properties of rayon fibre by graft copolymerization with acrylic monomers. J Mater Sci 41:7296-7301

[24] Seven P, Coskun M, Demirelli K (2008) Synthesis and characterization of two-armed graft copolymers prepared with acrylate and methacrylate using atom transfer radical polymerization. Reactive & Functional Polymers 68: 922-930

[25] Yang F, Li G, Ren FX, Yu XQ, Wang GX, Zhu SB (2007) Study on Preparation and Properties of Poly (AM -co- CMC). Polymer Materials Science and Engineering 23:78-81

[26] Lu ZX, Yuan X, Wu J, Wang LJ, LuoHA (2005) Synthesis and spectroscopic characterization of N, N'-dialkylimidazolium-based ionic liquids. Chemical World 46:148-150

[27] Peng YH, Zeng XQ, Lu HM (2006) Purification and application of lignin in urea-formaldehyde resin. Contemporary Chemical Industry 35: 377-379

[28] Faix O, Bremer J, Schmidit O, Stevanovic T (1991) Monitoring of chemical changes in white-rot degraded beech wood by pyrolysis-gas chromatography and Fourier transform infrared spectroscopy. Journal of Analytical and Applied Pyrolysis 21:147–162

[29] Pandey KK, Pitman AJ (2003) FTIR studies of the changes in wood chemistry following decay by brown-rot and white-rot fungi. International Biodeterioration & Biodegradation 52:151-160

[30] Faix O (1988) Practical uses of FTIR spectroscopy in wood science and technology. Mikrochim Acta [Wien] 1:21-25

[31] Xue ZH, Zhao GJ (2007) Influence of different treatments on wood crystal properties. Journal of Northwest Forestry University 22:169-171

4

Selective Breakdown of (Ligno)cellulose in Ionic Liquids

Haibo Xie and Zongbao K. Zhao
Dalian National Laboratory for Clean Energy and Dalian Institute of Chemical Physics, CAS, Dalian 116023, PR China

1. Introduction

(Ligno)cellulose is mainly composed of cellulose, lignin, hemicellulose, and extractives, which represents an abundant carbon-neutral renewable resource. Increasing concerns about global warming and diminishing fossil fuel reserves have stimulated much effort to use (ligno)cellulose as an alternative to petroleum for the production of fuels and chemicals. From a sustainability point of view, shifting society's dependence away from fossil-based energy resources to renewable alternatives can be regarded as an important contribution towards the establishment of sustainable and "Low Carbon" economy for all nations (Ragauskas et al. 2006; Corma et al. 2007). It is estimated that by 2025, up to 30% of raw materials for the chemical industry will be produced from renewable sources. To achieve this goal, innovative processing technologies, separation and depolymerization processes, as well as catalytic conversion systems are in high demand. In general, it is difficult to dissolve (ligno)cellulose in its native form because of the three-dimensional cross-linked lignin network and strong hydrogen bonds among the polymeric matrix. The inaccessibility of a reliable medium for dissolution of (ligno)cellulose has severely hampered the efficient utilization of lignocellulosic biomass (Kamm 2007).

The emerging ionic liquid-based technologies have paved an environment-friendly and homogenous manner to use (ligno)cellulose. Recent reviews have presented the progresses in bio-materials derived from biopolymers with the ionic liquids platform (Zhu et al. 2006; Pinkert et al. 2009), we thus will not cover this area here. The primary focus of this chapter is to provide an up-to-date overview on the selective breakdown of (ligno)cellulose into platform molecules with catalytic strategies in ionic liquids and to address ongoing in-depth understanding of key issues that bridge the carbohydrates chemistry and petroleum chemistry through ionic liquids. In this regard, those studies on the chemistry of monosaccharides in ionic liquids will be mentioned only if it is essential to clarify some issues related to transformation of (ligno)cellulosic materials.

2. Dissolution of (ligno)cellulose in ionic liquids

For a long time, the dissolution and processing of (ligno)cellulose and cellulose are one of the biggest obstacles for their utilization. Rogers et al (Swatloski et al. 2002) reported that imidazolium-based ionic liquids could dissolve cellulose (up to 25 wt%), and an amorphous cellulose can be produced with addition of anti-solvents (e.g. water, ethanol). Since then, a large library of ionic liquids have been synthesized to develop a more efficient and greener solvent system for dissolution of cellulose (Fukaya et al. 2006; Kilpelainen et al. 2007; Zavrel et al. 2009; Zakrzewska et al. 2010) and lignocellulosic materials (Kilpelainen et al. 2007). Recently, dissolution of cellulose in ionic liquids has been reviewed, and the structures of ionic liquids and their solubility to different types of carbohydrates have been collected (Zakrzewska et al. 2010). Some representative structures of ionic liquids are shown in Figure 1.

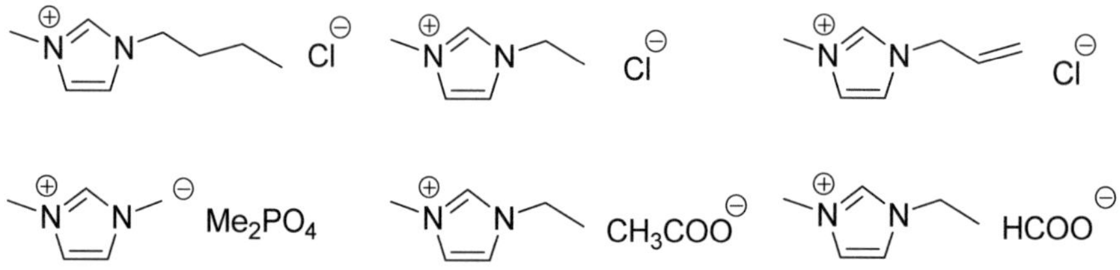

Fig. 1. Representative ionic liquids for biomass dissolution

Despite the increasing number of known ionic liquids capable of dissolving biomass, the mechanism of this dissolution process remains not well understood. There have been a few theoretical and experimental studies, including molecular dynamic studies and NMR analyses. For example, ^{13}C and $^{35/37}Cl$ NMR relaxation measurements have been done for cellulose solution in ionic liquids at varied temperatures and concentrations to investigate the interaction of the chloride ion of the ionic liquids with cellulose, and it was found that the solvation of cellulose by the ionic liquid 1-*n*-butyl-3-methylimidazolium chloride ([C_4mim]Cl) involved hydrogen-bonding between the carbohydrate hydroxyl proton and the chloride ion in a 1 : 1 stoichiometry (Remsing et al. 2006). Further multinuclear NMR spectroscopy experiments demonstrated that the anions in ionic liquids were involved in specific interactions with the solutes, and thus controlled the solvation process (Remsing et al. 2007). On the other hand, the ^{13}C relaxation rate of the imidazolium carbons showed no strong correlation with sugar concentration, regardless of the structure of either the cation or the cellulose, indicating that their role was not essential to the solvation process. Yet, molecular dynamics simulations showed that weak hydrogen bonding interactions with ionic liquids cation occurred through the acidic hydrogen at the C(2) position on the imidazolium ring (Youngs et al. 2007).

To provide an in-depth rationale for selecting ionic liquids capable of dissolving cellulose, further molecular dynamics simulations with an all-atom force field were performed on a system consisting of 1-ethyl-3-methyl imidazolium acetate ([Emim][OAc]) (Remsing et al. 2008; Liu et al. 2010). Results showed that the interaction energy between the polysaccharide chain and the ionic liquids was stronger (3 times) than that for either water or methanol. In addition to the anion acetate forming strong hydrogen bonds with hydroxyl groups of cellulose, some of the cations were found to be in close contact with the polysaccharides through hydrophobic interactions. It was also observed that the preferred β-(1,4)-glycosidic linkage conformation of the cellulose was altered when the cellulose was dissolved in [Emim][OAc] as compared to that found in crystalline cellulose dispersed in water. Furthermore, this kind of changes was not reversible during the regeneration process. They hypothesized that this conformational change in the cellulose was one of the key factors in the observed increase in enzymatic hydrolysis rates after ionic liquids pretreatment (Liu et

al. 2010). Although the fundamental molecular knowledge of these systems elucidated the role of cation during the dissolution process, recently publications of non-imidazolium based ionic liquids demonstrated good solubility to cellulose (Yang et al. 2010), which argued against the key role of anions to a certain extent. Such molecular level knowledge of ionic liquids system should enable the design and rapid computational screening of a wide range of ionic liquids for biomass pretreatment and facilitate the development of new efficient and economic ionic liquids-based biorefinery technologies (Pinkert et al. 2009).
In the traditional petrochemical industry, crude oil is fractionated and refined to produce various grades of liquid transportation fuels, and hydrocarbon feedstocks are functionalized to produce intermediates and specialty chemicals with advanced catalytic strategies (Gallezot 2008; Cheng et al. 2009; Ong et al. 2010). The concept of biorefinery is similar in terms of the utilization of a renewable source of carbon to produce heat, biomaterials, fuels and value-added chemicals. The overall strategy in the production of hydrocarbon fuels and chemicals from biomass is (1) to de-polymerize biopolymers, (2) to reduce the oxygen content of the parent feedstocks or de-polymerized intermediates and, (3) to create C-C bonds between biomass-derived intermediates to increase the molecular weights of the end products (Alonso et al. 2010). Ionic liquids have been regarded as 'greener' solvents, partially because their properties can be designed through tuning the structures of their constitutive ions (Rogers et al. 2003). The dissolution of (ligno)cellulose in ionic liquids resulted in a full release of all the functional groups and bonds from the well-organized strong hydrogen bonding matrix (Xie et al. 2007; Xie et al. 2009). Thus, biomass dissolved in ionic liquids was more susceptible to chemical attack by external reagents and catalysts. Bearing these in mind, a lot of efforts have been devoted to producing valuable chemicals from biomass *via* hydrolysis, dehydration, hydrogenolysis, etc., taking the advantages of this new homogenous platform.

3. Selective breakdown of (ligno)cellulose in ionic liquids

3.1 Hydrolysis of (ligno)cellulose

Hydrolysis of cellulose to fermentable sugars is virtually an essential step in any practical cellulosic biofuel production *via* a biological route. Furthermore, more chemical processes are seen to convert monosaccharides into fuels and value-added chemicals (Huber et al. 2006; Chheda et al. 2007). Two methods including acid hydrolysis and enzymatic hydrolysis are currently known for biomass hydrolysis. To make cellulosic materials more susceptible to hydrolysis, a pretreatment process is needed to reduce the crystallinity and increase the porosity. Traditional pretreatment technologies using oxidants, organic solvents, lime, or mineral acids, remain imperfect in terms of process greenness and eco- and environmental friendliness (Stephanopoulos 2007). Therefore, the development of advanced pretreatment technologies is one of the main issues in biorefinery. The inherent of dissolution of (ligno)cellulose in ionic liquids provided a new pretreatment technology and an inert homogeneous platform for the hydrolysis of (ligno)cellulose.

3.1.1 Enzymatic hydrolysis of (ligno)cellulose

Presently, two types of research have been carried out to hydrolyze the lignocellulose into fermentable sugars in ionic liquids. One is enzymatic hydrolysis of cellulosic materials regenerated from ionic liquids with the addition of an anti-solvent, and the other is *in situ* enzymatic hydrolysis in ionic liquids.

Shortly after regeneration of cellulose was demonstrated, efforts have been dedicated to characterizing the regenerated material, finding new applications, and further optimizing the process. It was confirmed that the crystalline structure of cellulose was destroyed (*vide ante*), which should allow a greater accessibility for the hydrolytic enzymes to rapidly penetrate and hydrolyze the (ligno)cellulose. Liu et al firstly investigated the enzymatic hydrolysis of cellulosic samples that had been pretreated with ionic liquids (Liu et al. 2006). It was found that the hydrolysis rates of wheat straw and steam-exploded wheat straw treated with [C_4mim]Cl were improved significantly. For example, wheat straw samples pretreated with [C_4mim]Cl gave a 70% conversion, while those pretreated with water had only a 42% conversion. Further WXRD and morphology study by SEM of the regenerated (ligno)cellulose material showed that diffraction signals from the crystalline regions of spruce sawdust disappeared after the dissolution-regeneration process (Kilpelainen 2007). Those were in an amorphous and porous morphology (Figure 2). For the spruce thermomechanical pulp (TMP) samples regenerated from 1-methyl-3-alkylimidazolium ([Amim]Cl) solution by precipitation upon water addition, glucose yield was 60% by the enzymatic hydrolysis, while it was only 12% for the untreated sample.

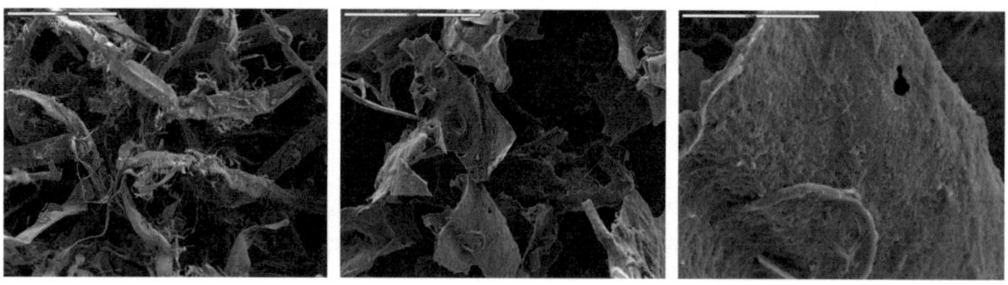

Fig. 2. Morphology study of spruce TMP treated by the ionic liquid [Amim]Cl

As noted earlier, such kinds of crystalline structure and morphology transformation, as well as β to α conformational change are anticipated to allow a greater accessibility for the hydrolytic enzymes. An extended research demonstrated an initial clue and potential to produce bioethanol from wheat straw with the ionic liquid pretreatment technology (Li et al. 2009). The fermentability of the enzymatic hydrolysates was evaluated using the yeast *Saccharomyces cerevisiae*, and the ethanol production was 0.43 g/g glucose after 26 h of fermentation. Results indicated that the pretreatment with the ionic liquid 1-methyl-3-ethylimidazolium diethylphosphate ([Emim]DEP) had negligible affects on the downstream ethanol production. The reusability of ionic liquid was also investigated in this study, and the yields of reducing sugars were higher than 52% even though the ionic liquids were recycled for 5 times. Although biomass pretreatment using ionic liquid is technically attractive, issues about process-costs and the ecological and environmental impacts require further scrutinize.

Recently ionic liquids have also been considered as solvents for biocatalysis (Zhao 2010), and examples of bioreactions in ionic liquids were seen with higher selectivity, faster reaction rate and even greater enzyme stability (Moniruzzaman et al. 2010). It is reasonable to perceive direct enzymatic hydrolysis of cellulose using ionic liquids as the reaction media. However, those conventional cellulose-dissolving ILs are composed of anions such as Cl^-, dca^-, HCO_2^-, or $CH_3CO_2^-$. While these anions facilitate the dissolving of cellulose through the formation of strong hydrogen bonds, they also tend to damage molecular interaction networks in a protein leading to denaturation of enzyme. In addition, high viscosity of ionic liquids may also limit the effectiveness of enzymatic transformations. All of these limitations hindered the enzymatic conversion of cellulose in ionic liquids. Therefore, the design and synthesis of ionic liquids capable of dissolving cellulose, and enzyme-compatible, especially compatible with cellulase as well, have been paid much attention to pursue an *in situ* enzymatic saccharification of cellulose (Zhao 2010). In 2008, Kamiya et al firstly reported their efforts on this topic (Kamiya et al. 2008). They dissolved cellulose sample in the ionic liquid [Emim]DEP, and then mixed the ionic liquid solution with different volumes of citrate buffer (10 mM, pH 5.0), followed by the addition of cellulase directly at 40 °C. Little cellulose activity was observed when the volumetric ratio of ionic liquid to water was higher than 3:2. However, decreasing the ratio enhanced enzymatic activity remarkably, and a reducing sugar yield over 70% (composed of 50% glucose + 20% cellobiose) was achieved after 24 h when the ionic liquid to water ratio was 1:4. Importantly, glucose formation in the aqueous-ionic liquid mixture was approximately 2-fold more than that of the aqueous system under identical conditions, indicating that the ionic liquid was truly enzyme-compatible. In comparison, when the ionic liquid [Emim][OAc] was applied under these conditions, cellulase activity was only approximately half of that with [Emim]DEP, suggesting that the anion structure played a key role in terms of the cellulase activity in ionic liquids.

The activity and stability of a commercial cellulase samples in eight ionic liquids have been investigated by optical and calorimetric techniques to gain further insights into the *in situ* enzymatic saccharification process (Bose et al. 2010). Some important issues concerning the cellulase in ionic liquids have been elucidated. Among those ionic liquids, hydrolysis was observed only in 1-hydrogen-3-methylimidazolium chloride and tris-(2-hydroxyethyl) methyl ammonium methylsulfate (HEMA). Interestingly, hydrolysis at 65 °C had a higher initial rate but reached a plateau after 2 h in the aqueous buffer, whereas the reaction in these two ionic liquids proceeded with a lower initial rate but went monotonically over time. Further investigation of relationship of enzymatic activity and thermal stability with viscosity and reaction temperature demonstrated that this difference in the rate of hydrolysis was largely attributed to two factors, the higher viscosity of the ionic liquids and enzyme stability. The inherent high viscosity of the ionic liquids was a retarding factor for the enzymatic hydrolysis, since it slowed the diffusion of the enzyme to its substrate resulting in a lower observed activity. Reversibility of the unfolding process was studied by gradual cooling of the denatured enzyme from temperature higher than the $T_{1/2}$. It was found that cellulase unfolding was irreversible in buffer and accompanied by precipitation of the enzyme. In HEMA, folding and unfolding processes were reversible even at up to 120 °C, which indicated that excellent cellulase thermo stability can be achieved in particular ionic liquids. However, in 1-hydrogen-3-methylimidazolium chloride, the unfolding process was completely irreversible. The potential of enzymatic saccharification process based on ionic liquids technology can be envisioned, however, it remains challenging to design an ionic liquid that compromises these opposing properties of dissolving cellulose, retaining the activity of the enzymes, yet having a low viscosity. From an eco- and environment-friendly point of view, efforts should be put on reducing operation costs and synthesis of novel ionic liquids.

3.1.2 Acid hydrolysis of (ligno)cellulose

Many attempts to hydrolyze cellulose into fermentable sugars catalyzed by mineral acids have been reported. These processes were usually done under heterogeneous and harsh conditions in which the (ligno)cellulose raw material was typically not solubilized in the reaction medium. Thus, the supramolecular and crystalline structure of cellulose imposed important restrictions on the kinetics of the hydrolysis. Consequently, traditional acid hydrolysis of (ligno)cellulose was inefficient and cost-intensive. Noting that the full dissolution of cellulose in ionic liquids could break internal and external supramolecular structures among the cellulosic fibers, and we envisioned that it should provide an opportunity to revisit the acid-catalyzed hydrolysis technology. In 2007, we firstly reported the hydrolysis behavior of cellulose in ionic liquids in the presence of mineral acids (Li and Zhao 2007). It was interesting to find that catalytic amounts of acid were sufficient to drive the hydrolysis reaction. For example, when the acid/cellulose mass ratio was set to 0.46, yields of total reducing sugar (TRS) and glucose were 64% and 36%, respectively, after 42 min at 100 °C. In fact, excess acid loading in the ionic liquid system was detrimental in terms of sugar yields because side reaction tended to occur that consumed the hydrolysis products. Preliminary kinetic study indicated that the cellulose hydrolysis catalyzed by H_2SO_4 followed a consecutive first-order reaction sequence, where k_1 for TRS formation and k_2 for TRS degradation were 0.073 min^{-1} and 0.007 min^{-1}, respectively. Our further study on the hydrolysis behavior of (ligno)cellulose in ionic liquids demonstrated that hydrochloric acid was also an effective catalyst (Li et al. 2008). TRS yields were up to 66%, 74%, 81% and 68% for hydrolysis of corn stalk, rice straw, pine wood and bagasse, respectively, in the presence of only 7 wt% catalyst at 100 °C under an atmospheric pressure within 60 min. Under those conditions, the constants for k_1 and k_2 were 0.068 min^{-1} and 0.007 min^{-1}, respectively, for the hydrolysis of corn stalk.

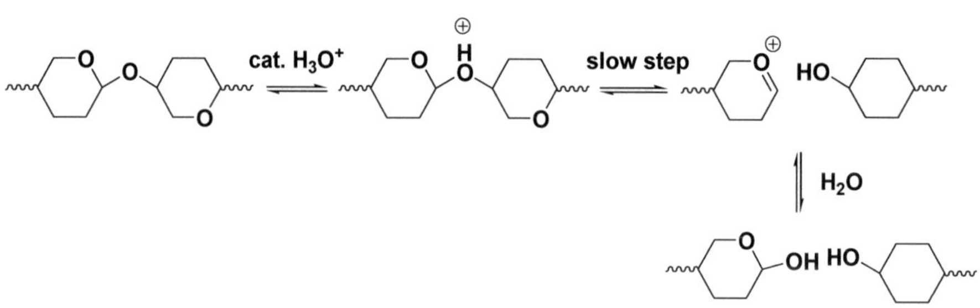

Fig. 3. Proposed mechanism for hydrolysis of cellulose. Hydrogen, hydroxyl, and hydroxymethyl groups are omitted for better clarity (Adopted from the reference (Rinaldi et al. 2010)).

Further study showed that the rates of these two competing reactions varied with acid strength and that for acids with an aqueous pKa below zero (Ka < 1), and the hydrolysis reaction was significantly faster than the degradation of glucose, thus allowing hydrolysis to be performed with a high selectivity in glucose (Vanoye et al. 2009). Although the acidity in ionic liquids is a concept still in its infancy, the relative acidity order of several organic and inorganic acids in ionic liquids was found identical to that in water (MacFarlane et al. 2006). The proposed hydrolysis of cellulose usually consists of three consecutive processes, (1) protonation of the glycosidic oxygen; (2) formation of a cyclic carbocation; (3) nucleophilic attack on the cyclic carbocation species by a water molecule (Figure 3). Although the second one is the rate-limiting step, the protonation of the glycosidic oxygen is difficult in some electron deficient acetals on the basicity of the acetal O-site and the hydroxyl O-sites and the acidic strength. Therefore, although some ionic liquids containing weak basic anions (e.g.

acetate, organophosphates) are capable of dissolving cellulose, they are not good solvents for hydrolysis of cellulose, because a relative weak acid formed by addition of a strong acidic catalyst (Rinaldi et al., 2010).

A recent work on pretreatment of wood species in ionic liquids in the presence of mineral acids suggested that a significant amount of lignin was also degraded, together with hydrolysis of cellulose and hemicellulose (Li et al. 2010). This was based on the results of GC-MS and ^{31}P NMR analysis of the ethyl acetate extracted samples produced during the pretreatment process. In fact, authors demonstrated the presence of compounds such as 5-hydroxymethylfurfural, furan-2-carboxylic acid, catechol, methylcatechol, methylguaiacol, acetoguaiacone, and acetol, among which those phenolic compounds and acetol should be originated from lignin.

The facts that the higher the degree of polymerization (DP) value of cellulose, the longer was the reaction time required for a satisfactory glucose yield, and that the longer reaction time favored glucose formation, while the shorter reaction time produced more TRS, implied that cellulose hydrolysis in ionic liquids catalyzed by mineral acids most likely followed a random hydrolysis mechanism, as observed with the concentrated acid system (Li et al. 2007). During the hydrolysis process, both endoglycosidic and exoglycosidic scission occurred, but the endoglycosidic product, oligoglucoses, was the major one in the initial stage, which usually was observed in traditional heterogeneous hydrolytic systems. Such phenomena were later confirmed in the study of solid acid (Amberlyst 15DRY) catalyzed hydrolysis of cellulose and (ligno)cellulose in ionic liquids (Rinaldi et al. 2008; Rinaldi et al. 2010). In those studies, depolymerized cellulose was precipitated and recovered by addition of water to the hydrolytic system, and the DP value was estimated by gel-permeation chromatography. It was found that the size of recovered cellulose fibers became successively smaller over time, resulting in a colloidal dispersion for the material recovered after 5 h. The depolymerization of cellulose proceeded progressively, resulting in the formation of soluble oligosaccharides if the reaction was carried out over a long time. For example, cellooligomers consisted of approximately ten anhydroglucose units (AGU) were seen after 5 h.

There is an intriguing question in the case of hydrolysis of cellulose catalyzed by solid acids in ionic liquids, where does the hydrolysis reaction take place, on the catalyst surface or in the bulk solution? The observation of induction period for the production of glucose and titration results of the ionic liquids separated from a suspension of Amberlyst 15DRY in [Bmim]Cl suggested that proton was progressively released into the bulk liquid within an hour upon through an ion-exchange process involving $[Bmim]^+$ of the ionic liquid and H^+ species of the solid acid. The true catalytic species in the case of Amberlyst 15DRY-catalyzed hydrolysis was the released H^+ species, and an activation process by washing it with a H_2SO_4 solution is required for the reuse of this solid catalyst (Rinaldi et al. 2010).

A full depolymerization of cellulose into fermentable sugar or other derivatives in ionic liquids challenges separation technology, because those ionic liquids *per se* are excellent solvents for monosaccharides, polar, water soluble and unstable under high temperature. Up to now, an efficient and economic method has not been achieved (Binder and Raines 2010; Brennan et al. 2010), which maybe the severe barrier of the ionic liquids based technologies for a practical use point of view. In this regard, the controlled depolymerization of cellulose into water insoluble oligosaccharides followed by an enzymatic hydrolysis seemed the most interesting transformation. For example, cellooligomers with a DP value around 30 were precipitated in 90% yield by the addition of water after microcrystalline cellulose was hydrolyzed in ionic liquids for 1.5 h. Another general strategy is to use ionic liquids as pretreatment agents to facilitate the hydrolysis process (*vide ante*). For example, pretreatment of cellulose with [Bmim]Cl led to loosen crystalline cellulose through partial transformation of cellulose I to cellulose II, which could facilitate the attack of proton during the subsequent hydrolysis over Nafion NR50 (Kim et al. 2010).

Other acids, such as trifluoroacetic acid, Brönsted acid ionic liquids, have also been used and satisfactory results were obtained in hydrolysis of cellulose in ionic liquids (Amarasekara and Owereh 2009; Amarasekara and Owereh 2010). In 2010, a more environment-friendly process was reported for biomass conversion in ionic liquid-water mixture (Zhang et al. 2010). Under relatively mild conditions ($\leq$ 140 °C, 1 atm) and in the absence of acid catalysts typically employed in biomass conversion, the dissolved cellulose in [Emim]Cl can be converted into reducing sugars in up to 97% yield. The combined study of experimental methods and *ab initio* calculations demonstrated the *Kw* value of water in the mixture was up to 3 orders of magnitude higher than the pure water under ambient conditions. Such high *Kw* values are typically achievable under high temperature or subcritical conditions, which is responsible for the remarkable performance without adding acid catalysts. They hypothesized that the increased [H^+] was due to the enhanced water auto ionization by ionic liquids. This process will be affected by the electrostatic environment of the solution, the broad dielectric medium of the solvent and the temperature. Comparative *ab initio* calculations based on the thermodynamic cycle showed that IL-water mixture exhibited higher concentrations of both [H^+] and [OH^-] than pure water, thus enabling the acid- and base-catalyzed reactions.

3.2 Catalytic conversion of (ligno)cellulose into furfurals

Given the chemical differences between biomass feedstocks and conventional fuels and chemicals, biomass should obviously be subjected to deoxygenation/dehydration in order to reduce the oxygen content. A desired intermediate derived from biomass, usually referred as platform molecules, should afford a great degree of flexibility in downstream processes and applications (Alonso et al. 2010). In 2004, the National Renewable Energy Laboratory proposed twelve building block chemicals that can be readily produced from sugars *via* biological or chemical transformations (Werpy and Petersen 2004). These twelve building blocks can be subsequently converted into a number of high-value chemicals or bulky materials. Among them, 5-hydroxymethylfurfural (HMF), is recognized as a versatile intermediate in biofuel chemistry and the petroleum industry (Figure 4). Production of HMF from fructose by dehydration is relatively easy. Although various conditions were published for this transformation, we will not discuss it. Instead, we will focus on production of HMF from cellulose and lignocellulosic materials. We also cover a few examples where HMF was produced using glucose because it had a tight connection to our main focus.

Although there were early studies on HMF production using glucose as the feedstock, yields were not appreciable. In 2007, a system of metal chloride in ionic liquids was introduced to convert glucose into HMF (Zhao et al. 2007). Specifically, a record high yield of 68% was achieved in the presence of $CrCl_2$ in [Emim]Cl. It was proposed that glucose was isomerized through a $CrCl_2$-associated enediol intermediate to fructose followed by dehydration reaction. It was interesting to note that an addition of strong ligands, 2,2′-bipyridine or glyceraldehydes, essentially shut down the reaction: HMF was receive in less than 2% yield in the presence of 2,2′-bipyridine. The authors concluded that the metal catalyst interacted with the hemiacetal portion of glucopyranose, and that there was little interaction with polyalcohol portion of the sugar. The authors also proposed that the $CrCl_3^-$ anion played an important role in proton transfer, facilitating mutarotation of glucose to fructose.

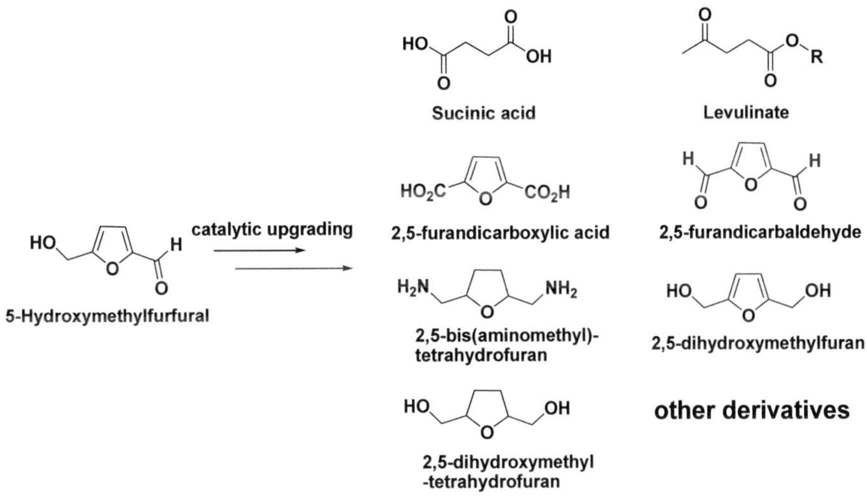

Fig. 4. Versatile derivatization of 5-hydroxymethyl furfural

Given these advancements on production of HMF from monosaccharides, the direct conversion the most abundant of cellulose and (ligno)cellulose into HMF remains challenging. We were the pioneer in the direct transformation of cellulose and (ligno)cellulose into furans (Li 2008). We demonstrated that $CrCl_3$ in ionic liquid was an outstanding system for HMF production from cellulose under microwave irradiation conditions. Thus, treatment of Avicel cellulose (100 mg) and $CrCl_3 \cdot 6H_2O$ (10 mg) in $[C_4mim]Cl$ (2.0 g) under microwave irradiation (400 W) for 2 min gave HMF in 62% yield (Li et al. 2009). Interestingly, when these components were heated at 100 °C over an oil-bath for 240 min, HMF yield was only 17%, but TRS yield was 45%. These results suggested that microwave irradiation was the key to high yields of HMF and $CrCl_3 \cdot 6H_2O$ was capable of catalyzing hydrolysis of cellulose. It should also mention that this system afforded HMF in over 90% yield when glucose was employed. Following up this work, we also demonstrated that furans production under similar conditions using biomass as the feedstocks (Zhang et al. 2010). Corn stalk, rice straw and pine wood treated under typical reaction conditions produced HMF and furfural in yields of 45–52% and 23–31%, respectively, within 3 min. In a separate study, a pair of metal chlorides ($CuCl_2$ and $CrCl_2$) was used as the catalysts for direct conversion of cellulose into HMF in [Emim]Cl under mild conditions (Su et al. 2009). For example, HMF yields were around 55% when the reactions were held at 80–120 °C for 8 h in the presence of the paired catalyst $CuCl_2/CrCl_2$ (χ_{CuCl2} = 0.17). Authors also demonstrated that the catalytic performance of recovered [Emim]Cl and the catalysts was maintained in repeated uses once the reaction was done in an extractive separation manor using methylisobutylketone as the extracting agent. It was interesting to note that single metal chlorides at the same total loading showed considerably less activity under similar conditions. This results were in sharp contrast to our work where high HMF yield was achieved using $CrCl_3 \cdot 6H_2O$ along as the catalyst (Li et al. 2009).

In parallel, chemical transformation of lignocellulosic biomass into furans in the DMA/ionic liquids mixture catalyzed by acid catalysts was also developed (Binder and Raines 2009). In this system, the ionic liquids were used as additives. A broad range of substrates including cellulose and (ligno)cellulose were tested, and good results were achievable. For example, in the case of cellulose, HMF yield was 22% in the DMA/LiCl-HCl system in the presence of $CrCl_2$, while the yield reached 54% if ionic liquids were added to 60 wt% concentration. Interestingly, a mineral acid was required to hydrolyze the cellulose into glucose for a satisfactory HMF yield in this mixed solvent system, while in the pure ionic liquid system, $CrCl_2$ acted as a catalyst both for the hydrolysis process and for the muturonation and dehydration processes. In 2010, conversion of cellulose into HMF in up to 89% yield was reported using $CrCl_2$ as a co-catalyst in [Emim]Cl-water mixture under relatively mild conditions (≤ 140 °C, 1 atm) (Zhang et al. 2010).

Given the significant achievements in the conversion hexoses into HMF in ionic liquid in the presence of metal chlorides, in-depth information is still needed for a creditable mechanism. Back to 1970s, the conversion of glucose into HMF in acidic water was studied (Harris and Feather 1974; Harris and Feather 1975; Farber et al. 1989). Although the yields were low, mechanistic insights were collected from isotopic labeling experiments. It was believed that an 1,2-hydride shift was a major contributor in the process of formation of fructose and HMF from glucose. Inspired by the enzyme-catalyzed process, in which the initial proton transfer is catalyzed by a basic residue (His53) located in the cavity of the enzyme active site, Pidko and co-workers reported the molecular-level details of the unique reactivity of chromium (II) chloride towards selective glucose dehydration in ionic liquids through combining kinetic experiments, in situ X-ray absorption spectroscopy and density functional theory calculations (Pidko et al. 2010). They proposed that the chloride anions of the ionic liquids play the role of basic mediators upon formation of a hydrogen-bonding network with the hydroxyl groups of the carbohydrate, thus facilitating the proton transfer; the rate-limiting H-shift reaction of the open form of the carbohydrate is facilitated by the transient self-organization of the Lewis acidic Cr^{2+} centers into a binuclear complex. This research provided a primary cooperative nature of the Cr complexes and the presence of moderately basic sites in the ionic liquids.

Accordingly, two variations for the mechanism were proposed recently (Binder and Raines 2009; Binder and Raines 2010). The enolization pathway that goes through an enediol(ate) intermediate that is protonated at C-1 to yield a ketose assisted by Cr^{2+}, which is dehydrated into HMF *via* a furanosyl oxocarbenium ion. The hydride shift pathway that goes through a chromium assisted 1,2-hydride shift to form the ketose from the aldose in a single step. Their further study using isotopic labeled glucose demonstrated the ketose formation occurring *via* 1,2-hydride shift, similar to that found in the catalytic mechanism of xylose isomerase. We considered that both in these two pathways, the cooperative action of Cr^{2+} and halide anion contributed the formation of the enediol intermediate and the 1,2-hydride shift to form ketose from aldose. Collectively, a putative mechanism for the chromium-catalyzed conversion of glucose into HMF was presented in Figure 5. It is worthy to mention that although $CrCl_2$ has been very successful in terms of HMF production from glucose and cellulose, $CrCl_3$ or $CrBr_3$ gave similar, even better results for the same chemistry (Binder and Raines 2009; Li et al. 2009; Binder and Raines 2010). We believe that the good performance of $CrCl_2$ might be due to the oxidation of $CrCl_2$ into $CrCl_3$ by residual oxygen in the reaction system. Apart from the metal catalysts for the efficient conversion of glucose into HMF in ionic liquids, heteropoly acids were also very efficient for the dehydration of glucose to

HMF, particularly in the presence of acetonitrile as a cosolvent (Chidambaram and Bell 2010). For example, a glucose conversion of 98% with a HMF selectivity of 99% can be achieved using 12-molybdophosphoric acid (12-MPA) as a catalyst in [Emim]Cl and acetonitrile after 3 h at 393 K. The authors proposed that the high selectivity was ascribed to the stabilization of the 1,2-enediol and other intermediates involved in the dehydration of glucose, the avoidance of the 2,3-endiol intermediate leading to furylhydroxymethyl ketone and the suppression of the formation of humans from glucose by the added acetonitrile.

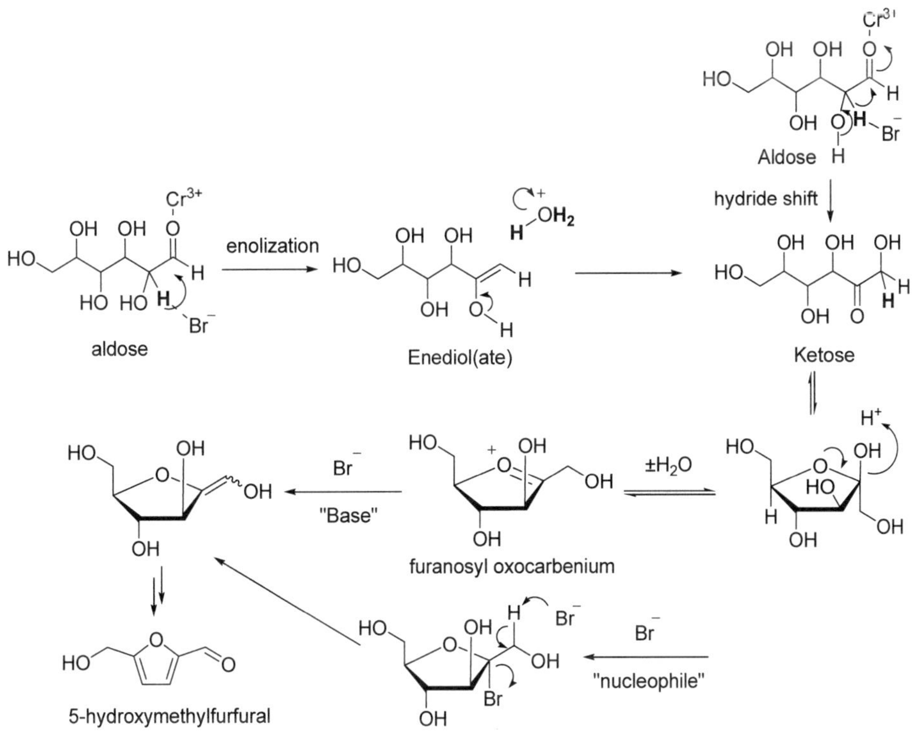

Fig. 5. Possible mechanisms for the dehydration of glucose into HMF in the case of bromide based ionic liquids in the presence of Cr^{2+} (Formulated and improved from that in references (Binder and Raines 2009; Binder and Raines 2010)).

3.3 Reactions leading to miscellaneous products

As expected, a successful lignocellulosic biorefinery can be realized through a combination of different catalytic technologies and biomass processing strategies in ionic liquids for the flexible production of fuels, chemicals and materials (Ragauskas et al. 2006; Corma et al. 2007; Gallezot 2008; Alonso et al. 2010). Sugar alcohols, a family of important bioproducts, are usually produced from the hydrogenation of monosaccharides. Given that cellulose can be dissolved in ionic liquids, it is logical to develop processes for direct hydrogenolysis of cellulose into sugar alcohols. Reductive depolymerization of cellulose was recently reported in the presence of hydrogen gas catalyzed by the combination of a heterogeneous metal catalyst and a homogeneous ruthenium catalyst (Ignatyev et al. 2010). Authors began their study with the hydrogenation of 1,1-diethoxycyclohexane using a combination of the heterogeneous catalysts Rh/C and Pt/C in the presence of Lewis acid $BF_3 \cdot Et_2O$ at 20 °C

under 2 MPa hydrogen gas, and ethoxycyclohexane was obtained in 100% yield. Further study found that the catalytic system was not effective for the hydrogenation of cellobiose. However, a much better result, 43% yield of sorbitol was obtained when the heterogeneous catalysts were replaced by a homogeneous catalyst precursor ($HRuCl(CO)(PPh_3)_3$). Under optimized conditions, sorbitol yield was 51% using cellulose as the feedstock. Additional control experiments confirmed that each component of the catalytic mixture, such as hydrogen, $HRuCl(CO)(PPh_3)_3$, Pt/C, and water, was necessary for the reaction and that the role of the ruthenium complex was to act as a hydrogen transport agent in the ionic liquids *via* formation of hydride compounds, thus supplying Pt/C catalyst with hydrogen.
Transition metal nanoparticles as catalysts for organic transformation has been received extensively attention due to their superior catalytic activities and selectivity relative to their corresponding bulk catalytic materials (Johnson 2003). It is generally received that their small particle size and high surface structure dominated their better performance. Zhu and co-workers (Zhu et al. 2010) designed an ionic liquid capable reversible binding cellulose with the consideration of the ability to stabilize transition metal nanoparticles and the reversible reaction of a boronic acid with sugars (Figure 6). They anticipated that boronic acids could break up the crystal packing of cellulose by reversible binding with the multiple hydroxyl groups, thereby improving solubility and catalytic activity. It was found that in the absence of the designed ionic liquids, a 15% conversion was achieved for cellulose hydrogenation to hexitols using a Ru nanocluster catalyst in [Bmim]Cl. However, the cellulose can be successfully converted into hexitols in high yields ranging from 76% to 93%. It is also interesting to find that cellulose was smoothly hydrolyzed to glucose using the designed ionic liquids with a 95% yield after 5 h at 80 °C. Furthermore, the nanoparticle catalyst was easily recovered from the reaction mixture, and the activity was maintained after 5 runs. Further study by ^{11}B NMR supported the complex formations between the ionic liquid with cellulose, which facilitated the solubilization of cellulose.

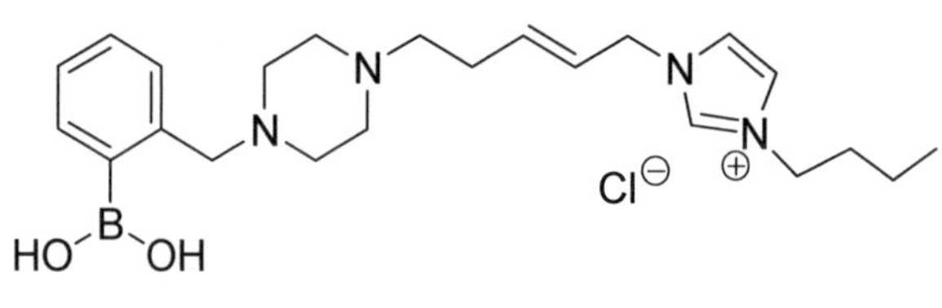

Fig. 6. Boronic acid functionalized ionic liquids

Long-chain alkyl glycosides are no-ionic compounds with excellent surfactant properties, low toxicity and good biodegradability. They have a couple of applications as cosmetics and detergents, food emulsifiers and pharmaceutical dispersing agents. One-pot catalytic conversion of cellulose into biodegradable surfactants was reported in ionic liquids under mild conditions (Villandier and Corma 2010). It was found that a sulfuric resin can catalyze a direct conversion of cellulose into alkyl glucosides with mass yields up to 82% (Figure 7). A lot of alcohols including butanol, hexanol, and octanol were used in the study, and total yields of surfactants were ranged from 7.2% to 91%. The amount of water in the reaction medium played an important role. During the hydrolysis process, water was a reactant inhibiting the formation of HMF as well as hemiacetals between the aldehyde group of the monosaccharide and the alcohols. They found that a removal of water from the system by reducing the reaction pressure could facilitate the glycosidation significantly.

A very similar work was also reported where cellulose was converted into the related alkyl glycosides in [Bmim]Cl in the presence of an acidic resin Amberlyst 15DRY (Ignatyev et al. 2010).

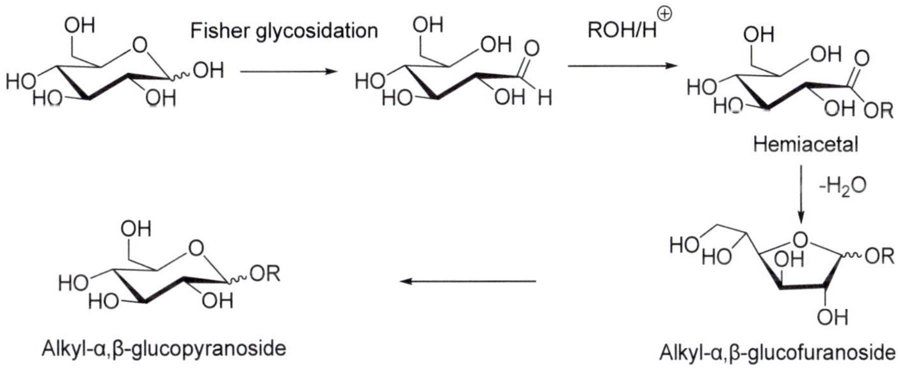

Fig. 7. Conversion of glucose into alkyl-α,β-glycoside surfactants

3.4 Greener pathways for the conversion of (ligno)cellulose in ionic liquids

The development of a sustainable chemical industry requires sustainable raw materials and technologies for the conversion of raw materials into end products. The design of a "green" compound or 'green' technology, whether the role is as a solvent, reagent or catalyst should ideally address issues such as low toxicity and ready biodegradability without the generation of toxic, persistent metabolites (Kamm 2007) . The conversion of biomass into value-added chemicals with the ionic liquids platform already have met two of the 12 green chemistry principles, using green solvents and biorenewable resources as raw materials. With more studies in this field, researchers are also realizing the potential problems, such as potential biotoxicity of ionic liquids, low biodegradability, cost-intensive separation process and energy consumption (Coleman 2010; Thi et al. 2010). With the development of green technology and green chemistry, researchers are also introducing new technologies and concepts into this field (Anastas and Zimmerman 2003) (Figure 8).

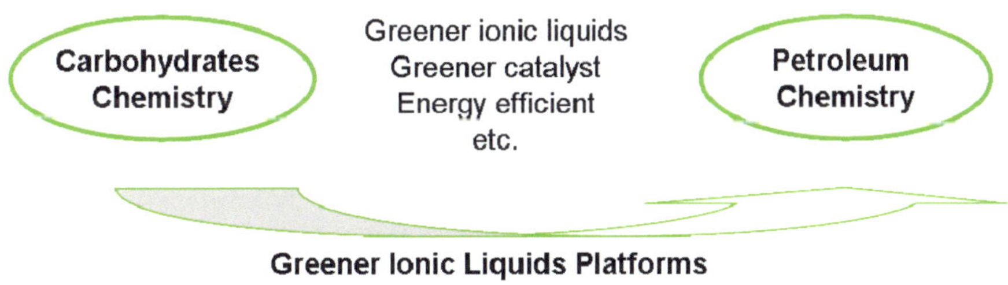

Fig. 8. Greener pathways for the conversion of (ligno)cellulose in ionic liquids

Microwave irradiation is regarded as a more energy efficient heating method and it has been widely used in organic synthesis. In 2009, we demonstrated that microwave irradiation could drastically improve the $CrCl_3$-catalyzed HMF production form cellulose in ionic liquids (Li et al. 2009). In fact, ionic liquids have excellent dielectric properties which can facilitate the energy transformation to heat the reaction rapidly, volumetrically and simultaneously. Other microwave effects, such as lowering activation energy or increasing the preexponential factor in the Arrhenius law due to orientation effect of polar species in an electromagnetic field might also contributed.

With increasing concerns about the environmental effects, the design, synthesis and application of low toxic and biodegradable ionic liquids have been obtained much attention. In this regard, some renewable materials-based ionic liquids have been prepared to investigate their effectiveness in the conversion of fructose into HMF (Hu et al. 2008; Hu et al. 2009). Some ionic liquids prepared entirely from cheap and renewable materials, such as ChoCl/malonic acid, ChoCl/oxalic acid and ChoCl/citric acid were found very effective for the reaction. For example, HMF yield was 83% where the ionic liquid ChoCl/citric acid was used as the solvent. It was also found that HMF was rather stable in these ionic liquids. In order to improve the HMF yield and selectivity in ChoCl/citric acid, and to reuse the ionic liquid, ethyl acetate was added as an extracting solvent to extract HMF out of the system, which gave a HMF yield of 86%, about 9% higher than that in the absence of the extract solvent. The ionic liquids-catalyst mixture was successfully recycled after pouring out the ethyl acetate phase. The most important aspects of this work are, 1) ionic liquids synthesized from biorenewable feedstock, and 2) the reaction performed under continuous reaction-extraction fashion. These are two important factors determining the practical potential of a technology in a large scale. Further study from the same group also showed that the system was also appropriate for the direct conversion of inulin to HMF in 64% yield under mild conditions (Hu et al. 2009).

With the consideration of the toxicity of chromium ion, researchers are devoting to discovering a cheap, non-toxic, easily handled catalysts for the selective conversion of (ligno)cellulose into furans. $SnCl_4$ is a common, cheap, easy handling Lewis acid, and has a much lower toxicity. Hu and co-workers investigated its catalytic activity in several ionic liquids for the conversion of glucose into HMF. The activity mostly varied with the anions in ionic liquids, and a best result of 61% yield was obtained in $[Emim]BF_4$. It is interesting to mention that although the Cl^- is essential for the Cr^{2+} catalytic system, it does not work well in this system (Hu et al. 2009). Inspired by a catalytic effect on glucose dehydration by lanthanide chlorides in supercritical water, the catalytic effect of lanthanide catalysts on the conversion of glucose into HMF in ionic liquids was investigated (Stahlberg et al. 2010). However, HMF yields were moderate or relatively low (less than 30%). Germanium (IV) chloride is an environment-friendly Lewis acid. We recently investigated the direct conversion of carbohydrates into HMF in $[C_4mim]Cl$ (Zhang et al. 2010). It was found that monosaccharides D-fructose and D-glucose, disaccharides sucrose and maltose, and even polysaccharides cellulose were successfully converted into HMF with good yields ranging from 30% to 92% under mild conditions. Thus, a HMF yield of 92% was achieved in 5 min in the case of fructose. Interestingly, we noticed that a higher content of water content in the system resulted in a lower HMF yield. However, addition of 5 Å molecular sieves during the dehydration of glucose afforded an increase in HMF yield from 38% to 48%.

4. Conclusions and prospects

Clearly, the conversion of biomass into value added chemicals in ionic liquids is just kicking start, and the complex challenge will certainly require the integration of advanced catalytic and separation technologies. This chapter has sought to highlight the up-to-date progress in this area, especially molecular level understanding of the dissolution of biomass in ionic liquids, strategies for the conversion of (ligno)cellulose into fermentable sugars and furans, and related chemicals.

Undoubtedly, many milestone-results have been obtained so far. For example, a remarkable TRS yield of 97% was obtained for hydrolysis of cellulose in the water system under mild conditions, which should provide cheaper fermentable feedstocks. If the separation of monosaccharides from ionic liquids were realized economically, the cost of the second generation of bioenergy is expected to be reduced significantly. Indeed, there are at least two references about sugar recovery after the hydrolysis of cellulose in ionic liquids. One is to use ion-exclusion chromatography to simultaneously recover the ionic liquids and the sugars from the reaction mixture, and the obtained sugar solution was proved suitable for the production of ethanol by ethanologenic bacterium or yeast (Binder and Raines 2010). Another method is to recover the sugars from ionic liquids by extraction based on the chemical affinity of sugars to boronates such as phenyl boronic acid and naphthalene-2-boronic acid (Brennan et al. 2010). It was shown that boronate complexes could extract up to 90% of mono- and disaccharides from aqueous ionic liquids solutions, pure ionic liquids systems, or hydrolysates of corn stover containing ionic liquids. Both of these methods delivered a solution of fermentable sugars, reduced toxic byproducts, and presented technical approaches to the challenging separation problem in the ionic liquid-based biorefinery. The integration of these processes into the traditional fermentable processes should provide better chances for the production of the second generation of biofuels.

The furans, especially HMF, are regarded as important intermediates bridging the carbohydrate chemistry and petroleum chemistry. For a long time, the direct and efficient conversion of cellulose and glucose into HMF is not successful. With the ionic liquids technology, HMF can now be produced in over 70% yields directly from cellulose under mild conditions. The continual development of downstream products from furans remains challenging. In the future, efforts should also be paid to the synthesis, structural characterization and properties study of new bioproducts or biofuels from these platform molecules. The increasing end product market will promote the related field developing.

Bearing all of these significant progresses in mind, technological and scientific challenges, however, remain to be addressed, such as, economic synthesis at a large scale, definition of recycling pathways, efficient products/ionic liquids separation, development of new efficient catalysts, production of new value-added chemicals, biodegradation/bioaccumulation studies as well as toxicity and handling considerations of ionic liquids. The future is prosperous, while opportunities always are sided with challenges on the road to industrial applications. Further research in the area of selective breakdown of (ligno)cellulose, or more general speaking, utilization of biomass incorporating ionic liquids technology is expected as below:

1. Development of new catalytic systems for more efficient transformations, such as depolymerization, dehydration, hydrogenolysis, or alcoholysis, of (ligno)cellulose and related materials;
2. Exploitation of novel chemistry of (ligno)cellulose in ionic liquids or ionic liquid-containing media;
3. Integration of the ionic liquids processes with other chemical or biological processes;
4. Design and preparation of cheaper, greener and process-compatible ionic liquids;

5. Developing new strategies for ionic liquids recycling and product recovery;
6. Catalytic conversion of lignin into value-added chemicals.

5. References

Amarasekara, A. S. and Owereh, O. S. (2009). Hydrolysis and Decomposition of Cellulose in Bronsted Acidic Ionic Liquids under Mild Conditions. Industrial & Engineering Chemistry Research 48(22): 10152-10155.

Amarasekara, A. S. and Owereh, O. S. (2010). Synthesis of a Sulfonic Acid Functionalized Acidic Ionic Liquid Modified Silica Catalyst and Applications in the Hydrolysis of Cellulose. Catalysis Communications 11(13): 1072-1075.

Alonso, D. M., Bond, J. Q. and Dumesic, J. A. (2010). Catalytic Conversion of Biomass to Biofuels. Green Chemistry, 12(9):1493-1513

Anastas, P. T. and Zimmerman, J. B. (2003). Design through the 12 Principles of Green Engineering. Environmental Science & Technology 37(5): 94A-101A.

Binder, J. B., Cefali, A. V., Blank, J. J. and Raines, R. T. (2010). Mechanistic Insights on the Conversion of Sugars into 5-Hydroxymethylfurfural. Energy & Environmental Science 3(6): 765-771.

Binder, J. B. and Raines, R. T. (2009). Simple Chemical Transformation of Lignocellulosic Biomass into Furans for Fuels and Chemicals. Journal of the American Chemical Society 131(5): 1979-1985.

Binder, J. B. and Raines, R. T. (2010). Fermentable Sugars by Chemical Hydrolysis of Biomass. Proceedings of the National Academy of Sciences of the United States of America 107(10): 4516-4521.

Bose, S., Armstrong, D. W. and Petrich, J. W. (2010). Enzyme-Catalyzed Hydrolysis of Cellulose in Ionic Liquids: A Green Approach toward the Production of Biofuels. Journal of Physical Chemistry B 114(24): 8221-8227.

Brennan, T. C. R., Datta, S., Blanch, H. W., Simmons, B. A. and Holmes, B. M. (2010). Recovery of Sugars from Ionic Liquid Biomass Liquor by Solvent Extraction. Bioenergy Research 3(2): 123-133.

Cheng, S. M. and Zhu, S. D. (2009). Lignocellulosic Feedstock Biorefinery-the Future of the Chemical and Energy Industry. Bioresources 4(2): 456-457.

Chheda, J. N., Huber, G. W. and Dumesic, J. A. (2007). Liquid-Phase Catalytic Processing of Biomass-Derived Oxygenated Hydrocarbons to Fuels and Chemicals. Angewandte Chemie-International Edition 46(38): 7164-7183.

Chidambaram, M. and Bell, A. T. (2010). A Two-Step Approach for the Catalytic Conversion of Glucose to 2,5-Dimethylfuran in Ionic Liquids. Green Chemistry 12(7): 1253-1262.

Coleman, D. and Gathergood, N. (2010). Biodegradation Studies of Ionic Liquids. Chemical Society Reviews 39(2): 600-637.

Corma, A., Iborra, S. and Velty, A. (2007). Chemical Routes for the Transformation of Biomass into Chemicals. Chemical Reviews 107(6): 2411-2502.

Farber, G. K., Glasfeld, A., Tiraby, G., Ringe, D. and Petsko, G. A. (1989). Crystallographic Studies of the Mechanism of Xylose Isomerase. Biochemistry 28(18): 7289-7297.

Fukaya, Y., Sugimoto, A. and Ohno, H. (2006). Superior Solubility of Polysaccharides in Low Viscosity, Polar, and Halogen-Free 1,3-Dialkylimidazolium Formates. Biomacromolecules 7(12): 3295-3297.

Gallezot, P. (2008). Catalytic Conversion of Biomass: Challenges and Issues. ChemSusChem 1(8-9): 734-737.

Harris, D. W. and Feather, M. S. (1974). Intramolecular C-2- C-1 Hydrogen Transfer-Reactions During Conversion of Aldoses to 2-Furaldehydes. Journal of Organic Chemistry 39(5): 724-725.

Harris, D. W. and Feather, M. S. (1975). Studies on Mechanism of Interconversion of D-Glucose, D-Mannose, and D-Fructose in Acid Solution. Journal of the American Chemical Society 97(1): 178-182.

Hu, S. Q., Zhang, Z. F., Song, J. L., Zhou, Y. X. and Han, B. X. (2009). Efficient Conversion of Glucose into 5-Hydroxymethylfurfural Catalyzed by a Common Lewis Acid $Sncl_4$ in an Ionic Liquid. Green Chemistry 11(11): 1746-1749.

Hu, S. Q., Zhang, Z. F., Zhou, Y. X., Song, J. L., Fan, H. L. and Han, B. X. (2009). Direct Conversion of Inulin to 5-Hydroxymethylfurfural in Biorenewable Ionic Liquids. Green Chemistry 11(6): 873-877.

Hu, S. Q., Zhang, Z. F., Zhou, Y. X., Han, B. X., Fan, H. L., Li, W. J., Song, J. L. and Xie, Y. (2008). Conversion of Fructose to 5-Hydroxymethylfurfural Using Ionic Liquids Prepared from Renewable Materials. Green Chemistry 10(12): 1280-1283.

Huber, G. W., Iborra, S. and Corma, A. (2006). Synthesis of Transportation Fuels from Biomass: Chemistry, Catalysts, and Engineering. Chemical Reviews 106(9): 4044-4098.

Ignatyev, I. A., Van Doorslaer, C., Mertens, P. G. N., Binnemans, K. and de Vos, D. E. (2010). Reductive Splitting of Cellulose in the Ionic Liquid 1-Butyl-3-Methylimidazolium Chloride. ChemSusChem 3(1): 91-96.

Ignatyev, I. A., Mertens, P. G. N., Van Doorslaer, C., Binnemans, K. and de Vos, D. E. (2010). Cellulose Conversion into Alkylglycosides in the Ionic Liquid 1-Butyl-3-Methylimidazolium Chloride. Green Chemistry 12(10): 1790-1795.

Johnson, B. F. G. (2003). Nanoparticles in Catalysis. Topics in Catalysis 24(1-4): 147-159.

Kamiya, N., Matsushita, Y., Hanaki, M., Nakashima, K., Narita, M., Goto, M. and Takahashi, H. (2008). Enzymatic *in Situ* Saccharification of Cellulose in Aqueous-Ionic Liquid Media. Biotechnology Letters 30(6): 1037-1040.

Kamm, B. (2007). Production of Platform Chemicals and Synthesis Gas from Biomass. Angewandte Chemie-International Edition 46(27): 5056-5058.

Kilpelainen, I., Xie, H., King, A., Granstrom, M., Heikkinen, S. and Argyropoulos, D. S. (2007). Dissolution of Wood in Ionic Liquids. Journal of Agricultural and Food Chemistry 55: 9142-9148.

Kim, S. J., Dwiatmoko, A. A., Choi, J. W., Suh, Y. W., Suh, D. J. and Oh, M. (2010). Cellulose Pretreatment with 1-n-Butyl-3-Methylimidazolium Chloride for Solid Acid-Catalyzed Hydrolysis. Bioresource Technology 101(21): 8273-8279.

Li, B., Filpponen, I. and Argyropoulos, D. S. (2010). Acidolysis of Wood in Ionic Liquids. Industrial & Engineering Chemistry Research 49(7): 3126-3136.

Li, C. Z. (2008). Application of Ionic Liquids in Biomass Conversion and Friedel-Crafts Acylation Reactions. Ph.D. Thesis, Dalian Institute of Chemical Physics, CAS.

Li, C. Z., Wang, Q. and Zhao, Z. K. (2008). Acid in Ionic Liquid: An Efficient System for Hydrolysis of Lignocellulose. Green Chemistry 10(2): 177-182.

Li, C. Z., Zhang, Z. H. and Zhao, Z. K. (2009). Direct Conversion of Glucose and Cellulose to 5-Hydroxymethylfurfural in Ionic Liquid under Microwave Irradiation. Tetrahedron Letters 50(38): 5403-5405.

Li, C. Z. and Zhao, Z. K. (2007). Efficient Acid-Catalyzed Hydrolysis of Cellulose in Ionic Liquid. Advanced Synthesis & Catalysis 349: 1847-1850.

Li, Q., He, Y. C., Xian, M., Jun, G., Xu, X., Yang, J. M. and Li, L. Z. (2009). Improving Enzymatic Hydrolysis of Wheat Straw Using Ionic Liquid 1-Ethyl-3-Methyl Imidazolium Diethyl Phosphate Pretreatment. Bioresource Technology 100(14): 3570-3575.

Liu, H. B., Sale, K. L., Holmes, B. M., Simmons, B. A. and Singh, S. (2010). Understanding the Interactions of Cellulose with Ionic Liquids: A Molecular Dynamics Study. Journal of Physical Chemistry B 114(12): 4293-4301.

Liu, L. Y. and Chen, H. Z. (2006). Enzymatic Hydrolysis of Cellulose Materials Treated with Ionic Liquid Bmim Cl. Chinese Science Bulletin 51(20): 2432-2436.

MacFarlane, D. R., Pringle, J. M., Johansson, K. M., Forsyth, S. A. and Forsyth, M. (2006). Lewis Base Ionic Liquids. Chemical Communications (18): 1905-1917.

Moniruzzaman, M., Nakashima, K., Kamiya, N. and Goto, M. (2010). Recent Advances of Enzymatic Reactions in Ionic Liquids. Biochemical Engineering Journal 48(3): 295-314.

Ong, Y. K. and Bhatia, S. (2010). The Current Status and Perspectives of Biofuel Production *Via* Catalytic Cracking of Edible and Non-Edible Oils. Energy 35(1): 111-119.

Pidko, E. A., Degirmenci, V., van Santen, R. A. and Hensen, E. J. M. (2010). Glucose Activation by Transient Cr^{2+} Dimers. Angewandte Chemie-International Edition 49(14): 2530-2534.

Pinkert, A., Marsh, K. N., Pang, S. S. and Staiger, M. P. (2009). Ionic Liquids and Their Interaction with Cellulose. Chemical Reviews 109(12): 6712-6728.

Ragauskas, A. J., Williams, C. K., Davison, B. H., Britovsek, G., Cairney, J., Eckert, C. A., Frederick, W. J., Hallett, J. P., Leak, D. J., Liotta, C. L., Mielenz, J. R., Murphy, R., Templer, R. and Tschaplinski, T. (2006). The Path Forward for Biofuels and Biomaterials. Science 311(5760): 484-489.

Remsing, R. C., Liu, Z. W., Sergeyev, I. and Moyna, G. (2008). Solvation and Aggregation of N,N'-Dialkylimidazolium Ionic Liquids: A Multinuclear NMR Spectroscopy and Molecular Dynamics Simulation Study. Journal of Physical Chemistry B 112(25): 7363-7369.

Remsing, R. C., Swatloski, R. P., Rogers, R. D. and Moyna, G. (2006). Mechanism of Cellulose Dissolution in the Ionic Liquid 1-N-Butyl-3-Methylimidazolium Chloride: A C-13 and Cl-35/37 NMR Relaxation Study on Model Systems. Chemical Communications (12): 1271-1273.

Remsing, R. C., Wildin, J. L., Rapp, A. L. and Moyna, G. (2007). Hydrogen Bonds in Ionic Liquids Revisited: Cl-35/37 NMR Studies of Deuterium Isotope Effects in 1-N-Butyl-3-Methylimidazolium Chloride. Journal of Physical Chemistry B 111(40): 11619-11621.

Rinaldi, R., Palkovits, R. and Schuth, F. (2008). Depolymerization of Cellulose Using Solid Catalysts in Ionic Liquids. Angewandte Chemie-International Edition 47(42): 8047-8050.

Rinaldi, R., Meine, N., vom Stein, J., Palkovits, R. and Schuth, F. (2010). Which Controls the Depolymerization of Cellulose in Ionic Liquids: The Solid Acid Catalyst or Cellulose? ChemSusChem 3(2): 266-276.

Rogers, R. D. and Seddon, K. R. (2003). Ionic Liquids - Solvents of the Future? Science 302(5646): 792-793.

Stahlberg, T., Sorensen, M. G. and Riisager, A. (2010). Direct Conversion of Glucose to 5-(Hydroxymethyl)furfural in Ionic Liquids with Lanthanide Catalysts. Green Chemistry 12(2): 321-325.

Stephanopoulos, G. (2007). Challenges in Engineering Microbes for Biofuels Production. Science 315(5813): 801-804.

Su, Y., Brown, H. M., Huang, X. W., Zhou, X. D., Amonette, J. E. and Zhang, Z. C. (2009). Single-Step Conversion of Cellulose to 5-Hydroxymethylfurfural (HMF), a Versatile Platform Chemical. Applied Catalysis A-General 361(1-2): 117-122.

Swatloski, R. P., Spear, S. K., Holbrey, J. D. and Rogers, R. D. (2002). Dissolution of Cellose with Ionic Liquids. Journal of the American Chemical Society 124(18): 4974-4975.

Thi, P. T. P., Cho, C. W. and Yun, Y. S. (2010). Environmental Fate and Toxicity of Ionic Liquids: A Review. Water Research 44(2): 352-372.

Vanoye, L., Fanselow, M., Holbrey, J. D., Atkins, M. P. and Seddon, K. R. (2009). Kinetic Model for the Hydrolysis of Lignocellulosic Biomass in the Ionic Liquid, 1-Ethyl-3-Methyl-Imidazolium Chloride. Green Chemistry 11(3): 390-396.

Villandier, N. and Corma, A. (2010). One Pot Catalytic Conversion of Cellulose into Biodegradable Surfactants. Chemical Communications 46(24): 4408-4410.

Werpy, T., Petersen, G. Eds. (2004). Top Value Added Chemicals from Biomass, Report: Doe/Go-102004-1992;. U.S. Department of Energy.

Xie, H., Kilpeläinen, I., King, A., W., T., Järvi, P. and Argyropoulos, D. S., (2009). Opportunities with Dissolved Wood. Cellulose Solvents: For Analysis, Shaping and Chemical Modification. Liebert, T., et al., ACS Symposium Series, 2009, Chapter 19, 343-363.

Xie, H., King, A., Kilpelainen, I., Granstrom, M. and Argyropoulos, D. S. (2007). Thorough Chemical Modification of Wood-Based Lignocellulosic Materials in Ionic Liquids. Biomacromolecules 8(12): 3740-3748.

Yang, F., Li, L. Z., Li, Q., Tan, W. G., Liu, W. and Xian, M. (2010). Enhancement of Enzymatic *in Situ* Saccharification of Cellulose in Aqueous-Ionic Liquid Media by Ultrasonic Intensification. Carbohydrate Polymers 81(2): 311-316.

Youngs, T. G. A., Hardacre, C. and Holbrey, J. D. (2007). Glucose Solvation by the Ionic Liquid 1,3-Dimethylimidazolium Chloride: A Simulation Study. Journal of Physical Chemistry B 111(49): 13765-13774.

Zakrzewska, M. E., Bogel-Lukasik, E. and Bogel-Lukasik, R. (2010). Solubility of Carbohydrates in Ionic Liquids. Energy & Fuels 24: 737-745.

Zavrel, M., Bross, D., Funke, M., Buchs, J. and Spiess, A. C. (2009). High-Throughput Screening for Ionic Liquids Dissolving (Ligno-)Cellulose. Bioresource Technology 100(9): 2580-2587.

Zhang, Y. T., Du, H. B., Qian, X. H. and Chen, E. Y. X. (2010). Ionic Liquid-Water Mixtures: Enhanced Kw for Efficient Cellulosic Biomass Conversion. Energy & Fuels 24: 2410-2417.

Zhang, Z. H., Wang, Q., Xie, H. B., Dong, K., Liu, W. J. and Zhao, Z. K. (2010). Catalytic Conversion of Carbohydrates into 5-Hydroxymethylfurfural by Germanium (IV) Chloride in Ionic Liquids. ChemSusChem, DOI: 10.1002/cssc.201000279.

Zhang, Z. H. and Zhao, Z. K. (2010). Microwave-Assisted Conversion of Lignocellulosic Biomass into Furans in Ionic Liquid. Bioresource Technology 101(3): 1111-1114.

Zhao, H. (2010). Methods for Stabilizing and Activating Enzymes in Ionic Liquids - a Review. Journal of Chemical Technology and Biotechnology 85(7): 891-907.

Zhao, H. B., Holladay, J. E., Brown, H. and Zhang, Z. C. (2007). Metal Chlorides in Ionic Liquid Solvents Convert Sugars to 5-Hydroxymethylfurfural. Science 316(5831): 1597-1600.

Zhu, S. D., Wu, Y. X., Chen, Q. M., Yu, Z. N., Wang, C. W., Jin, S. W., Ding, Y. G. and Wu, G. (2006). Dissolution of Cellulose with Ionic Liquids and Its Application: A Mini-Review. Green Chemistry 8(4): 325-327.

Zhu, Y. H., Kong, Z. N., Stubbs, L. P., Lin, H., Shen, S. C., Anslyn, E. V. and Maguire, J. A. (2010). Conversion of Cellulose to Hexitols Catalyzed by Ionic Liquid-Stabilized Ruthenium Nanoparticles and a Reversible Binding Agent. ChemSusChem 3(1): 67-70.

5

Chemical Modification of Cellulose with Succinic Anhydride in Ionic Liquid with or without Catalysts

CF Liu[1], AP Zhang[2,3], WY Li[1] and RC Sun[1,4]
[1]*State Key Laboratory of Pulp and Paper Engineering, South China University of Technology,*
[2]*Institute of New Energy and New Material, South China Agricultural University,*
[3]*Biomass Energy Key Laboratory of Guangdong Regular Higher Education Institutions, South China Agricultural University,*
[4]*Institute of Biomass Chemistry and Technology, Beijing Forestry University, China*

1. Introduction

Currently, environmental problems such as global warming and dwindling energy resources are becoming increasingly critical issues. Utilization of biomass, especially inedible lignocelluloses to create fuels, energy, chemicals, composites and a host of other products is highly desirable for the construction of sustainable society (Zhang et al., 2007; Ogaki et al., 2009; Zhang et al., 2010). U.S. Department of Energy has targeted to achieve 10% of basic chemical building blocks arising from lignocellulosic biomass by 2020, and achieve a further increase to 50% by 2050 (Mohanty et al., 2002). At the same time, the European Union has developed a scenario for supplying 25% of the E.U.'s gasoline demand with biofuels by 2030 (Himmel et al., 2007).

The integrated utilization and multi-product biorefinery of lignocellulosic biomass is becoming the significant issue and development tendency. In this promising pattern, the lignocellulosic biomass is firstly fractionated to three main components, that is, lignin, hemicelluloses, and cellulose, then the isolated components are independently utilized to produce different products according to their own properties. Cellulose, the most abundant natural polymer, is a homopolymer composed of D-glucopyranose units linked by β (1→4) glycosidic bonds (Pandey et al., 2000). It is non-toxic, renewable, biodegradable and modifiable, and has great potential for an excellent industrial material (Richardson & Gorton, 2003). The promising applications of cellulose include biofibers, biopolymers, biofuels, and biocomposites (Reddy & Yang, 2005; Clark et al., 2009). Due to three hydroxyl groups available within one anhydroglucose units (AGU), a great variety of chemical modifications of cellulose are possible (Potthast et al., 2006). Chemical modifications of cellulose can introduce functional groups into the macromolecules in heterogeneous phase or homogeneous phase to improve the overall utilization of cellulosic polymers. Homogeneous functionalization has been one focus of cellulose research for a long time because more uniform and stable products can be obtained in homogeneous phase than heterogeneous phase (Regiani et al., 1999; El Seoud et al.,

2000; Satge et al., 2004). The intrinsic lack of solubility of native cellulose in water and most organic solvent systems constitutes a major obstacle for cellulose utilization. The efficient dissolution of cellulose is a long-standing goal in cellulose research and development and still of great importance (Heinze et al., 2005).

Recently, due to their unique physico-chemical properties, such as chemical and thermal stability, non-flammability and immeasurably low vapor pressure, ionic liquids (ILs) have been considered as the most potential green solvents in the future, which have attracted a great deal of scientific attention in many fields (Forsyth et al., 2002; Turner et al., 2003). Many kinds of ILs with a variety of structures have been reported as novel solvents for cellulose dissolution, including a series of alkylimidazolium salts containing chloride (Swatloski et al., 2002; Luo et al., 2005; Heinze et al., 2005; Zhang et al., 2005), formate (Fukaya et al., 2006), acetate (Hermanutz et al., 2007; Kosan et al., 2008; de Maria & Martinsson, 2009), and alkylphosphate (Fukaya et al., 2006; Fukaya et al., 2008; Kamiya et al., 2008; Mazza et al., 2009).

Based on the dissolution of cellulose in ILs, the investigation into the use of ILs as reaction media for cellulose functionalization has also been reported. Modification of cellulose represents one of the most versatile transformations as it provides access to a variety of biobased materials with valuable properties. Acylation of cellulose with linear chain acylation reagents such as anhydride or chloride is the most common method to produce cellulosic bioproducts. Because of the wide application of cellulose acetate, cellulose acetylation with acetic anhydride or acetyl chloride in ILs has been extensively studied, and cellulose acetates with high degree of substitution (DS) were easily prepared (Wu et al., 2004; Abbott et al., 2005; Heinze et al., 2005; Barthel & Heinze, 2006; Granstrom et al., 2008). Furthermore, ILs were also be reported as reaction media for cellulose modification with other liquid reagents, such as carbanilation with phenyl isocyanate (Barthel & Heinze, 2006; Schlufter et al., 2006), acylation with lauroyl chloride (Barthel & Heinze, 2006), and perpropionylation with propionic anhydride (Schlufter et al., 2006). However, an undesired byproduct, namely, the corresponding carboxylic acid, is undoubtedly produced. This acid must be removed from the reaction after modification.

On the other hand, cellulose derivatives obtained with solid reagents such as cyclic anhydrides have also been widely used in various applications such as water absorbents for soil in agriculture, natural absorbents for the removal of heavy metal ions in wastewater treatment, medicine for drug delivery systems, and thermoplastic materials (Hadano et al., 2003; Yoshimura et al., 2006). More importantly, modification with cyclic anhydrides such SA does not yield such a byproduct. Moreover, the reaction results in a pendant carboxylic moiety attached to the cellulose via a covalent ester bond, providing a site upon which further reactive chemistry is possible. In the present study, we investigated the possibility of cellulose succinoylation in ILs without any catalysts. Three catalysts were also explored to increase cellulose modification efficiency. The cellulosic derivatives were then characterized by Fourier transform infrared (FT-IR), and solid-state CP/MAS ^{13}C nuclear magnetic resonance (NMR) spectroscopies as well as thermal analysis.

2. Experimental

2.1 Materials

Rice straw (RS) was obtained from Guangzhou, China. It was dried in sunlight and then cut into small pieces. The cut RS was ground and screened to prepare 20-40 mesh size particles

(450-900 μm). Ionic liquid [C_4mim]Cl was obtained from the Chemer Chemical Co., Ltd., Hangzhou, China, and used as received. All of other chemicals used were of analytical grade and purchased from Sigma-Aldrich, Guangzhou, China.

2.2 Isolation of cellulose from RS

The dried RS powder was first dewaxed with toluene-ethanol (2:1, v/v) in a Soxhlet apparatus for 6 h. The dewaxed RS was oven-dried and then soaked in distilled water (800 mL) at 80 °C for 2 h. The insoluble residue was then dewaxed and water-soluble-free sample was delignified with sodium chlorite at pH 3.8-4.0, adjusted by 10% acetic acid, at 75 °C for 2 h. The residue was collected by filtration, washed with distilled water and ethanol, and then dried in a cabinet oven for 16 h at 50 °C. The holocellulose obtained was extracted with 10% KOH (600 mL) at 20 °C for 10 h. After filtration, the residue was washed thoroughly with distilled water until the filtrate was neutral, and then suspended in 95% ethanol (liquor to solid ratio 25:1) for 10 h. After filtration, the solid residue was dried in an oven at 50 °C for 16 h.

2.3 Succinoylation of cellulose in ionic liquid

The succinylated cellulosic derivatives containing free carboxylic groups were prepared by homogeneous reaction of RS cellulose without any catalysts or with three different catalysts including iodine (I_2), *N*-bromosuccimide (NBS), and 4-dimethylaminopyridine (DMAP) in the solvent system containing [C_4mim]Cl and DMSO. Dried cellulose was added to [C_4mim]Cl in three-necked flask continuously purged with gaseous N_2, and the mixture of cellulose/[C_4mim]Cl was stirred at 80 °C up to 3 h to guarantee the complete dissolution of cellulose. To reduce viscosity and achieve suitable mixing, 5 mL DMSO was added to this mixture. The succinoylation reaction was carried out according to the following procedures.

2.3.1 Succinoylation without any catalysts

To the cellulose solution in [C_4mim]Cl/DMSO system, SA, previously dissolved in 5 mL DMSO, was added at the corresponding reaction temperature with the 4:1 molar ratio of SA/AGU in cellulose in Table 1. The mixture was heated in oil bath under N_2 atmosphere with magnetic stirring for the desired time. After the required time, the resulting mixture was slowly poured into 150 mL of isopropanol with stirring, washed thoroughly with isopropanol to eliminate ILs, unreacted anhydride, and byproducts, and then dried in a vacuum at 50 °C for 16 h.

2.3.2 Succinoylation catalyzed with iodine

To the cellulose solution in [C_4mim]Cl/DMSO system, SA and I_2, previously dissolved in 5 mL DMSO, were added at the corresponding reaction temperature with the 4:1 molar ratio of SA/AGU in cellulose and the corresponding weight ratio of I_2/SA shown in Table 2. The mixture was heated in oil bath under N_2 atmosphere with magnetic stirring for the desired time. Then the reaction mixture was treated with a saturated solution of sodium thiosulfate (2 mL) with stirring, and then vigorously shaken for 2 min to guarantee complete transformation of iodine to iodide. The resulting mixture was slowly poured into 150 mL of isopropanol with continuous stirring. The solid was filtered, washed thoroughly with isopropanol to eliminate ILs, un-reacted anhydride, and byproducts, and then dried in vacuum at 50 °C for 16 h.

2.3.3 Succinoylation catalyzed with *N*-bromosuccimide

To the cellulose solution in [C_4mim]Cl/DMSO system, SA and NBS, previously dissolved in 5 mL DMSO, were added at the corresponding reaction temperature with the 4:1 molar ratio of SA/AGU in cellulose and the corresponding weight ratio of NBS/SA shown in Table 2. The mixture was heated in oil bath under N_2 atmosphere with magnetic stirring for the desired time. After the required time, the resulting mixture was slowly poured into 150 mL of isopropanol with stirring, washed thoroughly with isopropanol to eliminate ILs, unreacted anhydride, NBS, and byproducts, and then dried in a vacuum at 50 °C for 16 h.

2.3.4 Succinoylation catalyzed with 4-dimethylaminopyridine

To the cellulose solution in [C_4mim]Cl/DMSO system, SA and DMAP, previously dissolved in 5 mL DMSO, were added at the corresponding reaction temperature with the 4:1 molar ratio of SA/AGU in cellulose and the corresponding weight ratio of DMAP/SA shown in Table 2. The mixture was heated in oil bath under N_2 atmosphere with magnetic stirring for the desired time. After the required time, the resulting mixture was slowly poured into 150 mL of isopropanol with stirring, washed thoroughly with isopropanol to eliminate ILs, unreacted anhydride, DMAP, and byproducts, and then dried in a vacuum at 50 °C for 16 h.

2.4 Determination of degree of substitution

The DS of cellulosic preparations was determined by direct titration method (Jeon et al., 1999; Stojanovic et al., 2005). A known weight of the sample was dissolved in 10 mL of DMSO by stirring at 75 °C for 30 min. After cooling, 5-6 drops of phenolphthalein indicator were added. This solution was titrated against 0.01 mol/L standard NaOH solution until a permanent pale pink color was seen. The DS was calculated by using the following equation:

$$DS = \frac{162 \times (V_{NaOH} \times c_{NaOH})}{m - 100 \times (V_{NaOH} \times c_{NaOH})}$$

where: 162 g/mol is the molar mass of an AGU, 100 g/mol is the net increase in the mass of an AGU for each succinoyl substituted, m is the weight of sample analyzed, V_{NaOH} is the volume of standard NaOH solution consumed in the titration, and c_{NaOH} is the molarity of standard NaOH solution.

2.5 Characterization of the native and succinylated cellulose

Viscosity of the cellulose was measured by British Standard Methods for determination of limiting viscosity number of cellulose in dilute solutions, Part 1, cupriethylenediamine (CED) method (BS 6306, Part 1, 1982). The viscosity average DP (degree of polymerization) of cellulose was estimated from their intrinsic viscosity [η] in CED hydroxide solution, $P^{0.90}=1.65[\eta]$, where P is an indeterminate average DP (Evans & Wallis, 1989). Molecular weight (Mw) of cellulose was then calculated from P by multiplied by 162, the Mw of an AGU. The FT-IR spectra of the cellulose and succinylated cellulosic derivatives were recorded on an FT-IR spectrophotometer (Nicolet 510) from finely ground samples (1%) in KBr pellets in the range 4000-400 cm^{-1}. Thirty-two scans were taken for each sample with a resolution of 2 cm^{-1} in the transmission mode. The solid-state CP/MAS ^{13}C NMR spectra were obtained on a Bruker DRX-400 spectrometer with 5 mm MAS BBO probe employing both Cross Polarization and Magic Angle Spinning and each experiment was recorded at ambient temperature. The spectrometer operated at 100 MHz. Acquisition time was 0.034 s, the delay time 2 s, and the proton 90° pulse time 4.85 μs. Each spectrum was obtained with an accumulation of 5000 scans.

Thermal analysis of the cellulose and succinylated cellulosic samples was performed using thermogravimetric analysis (TGA) and differential thermal analysis (DTA) on a simultaneous thermal analyzer (SDT Q600, TA Instrument). The apparatus was continually flushed with nitrogen. The sample weighed between 8 and 12 mg and the scans were run from room temperature to 550°C at a rate of 10°C per minute.

3. Results and discussion

3.1 Modification without any catalysts

Delignification of dewaxed and water-soluble-free RS followed by alkaline extraction yielded 44.6% cellulose (based on the dry weight of RS). The intrinsic viscosity, degree of polymerization, and molecular weight of the native cellulosic preparation were determined to be 348 mL · g^{-1}, 1165, and 188730 g · mol^{-1}, respectively.

Due to three hydroxyl groups available within one anhydroglucose units (AGU), a great variety of chemical modifications of cellulose are possible. In the present study, succinoylation of the cellulose isolated from RS in [C_4mim]Cl/DMSO system was investigated with or without catalysts. Usually, SA reacts with cellulose O-H groups to form the monoester as shown in Fig. 1. After cellulose was dissolved in [C_4mim]Cl, the viscous solution obtained was diluted by DMSO to achieve a suitable and clear mixiture. Solid SA and catalyst previously dissolved in DMSO were added to the solution to achieve homogeneous reaction. Table 1 shows the effects of reaction conditions including the molar ratio of SA/anhydroglucose units (AGU) in cellulose, reaction time, and reaction temperature on DS of the cellulose succinylated without any catalysts.

As shown in Table 1, an increase of molar ratio of SA/AGU in cellulose from 1:1 to 2:1, 3:1, 4:1, 6:1, 8:1, and 10:1 resulted in the improvement of the DS of succinylated cellulose from 0.02 to 0.04, 0.07, 0.11, 0.15, 0.22, and 0.31, respectively, which was due to the greater availability of succinic anhydride molecules in the proximity of the cellulosic molecules at higher concentration of succinic anhydride. The DS increased from 0.08 to 0.12, 0.15, 0.16, and 0.20 with the enhancement of reaction temperature from 90 to 100, 110, 120, and 130°C, respectively. The reason for this enhancement of succinoylation was presumed to be due to the favourable effect of temperature on succinoylation reaction. A higher reaction temperature would enhance the mobility of the molecules, the diffusion of succinic anhydride into the cellulosic molecules, and the swell ability of cellulose, therefore, increasing succinoylation rates. The rate and time of collisions of succinic anhydride with

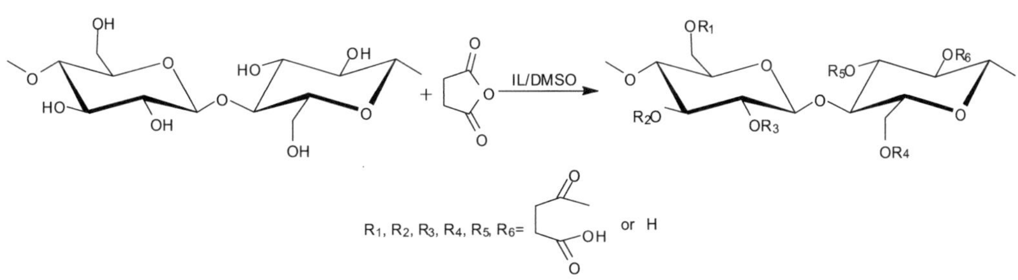

Fig. 1. Reaction of cellulose with succinic anhydride in ionic liquid.

Succinoylation conditions				Succinylated cellulose	
Cellulose/[C_4mim]Cl(%)	Molar ratio (SA/AGU)	T(°C)	t(min)	Sample	DS
2	1:1	110	60	1	0.02
2	2:1	110	60	2	0.04
2	3:1	110	60	3	0.07
2	4:1	110	60	4	0.11
2	6:1	110	60	5	0.15
2	8:1	110	60	6	0.22
2	10:1	110	60	7	0.31
2	6:1	90	60	8	0.08
2	6:1	100	60	9	0.12
2	6:1	120	60	10	0.16
2	6:1	130	60	11	0.20
2	6:1	110	15	12	0.04
2	6:1	110	30	13	0.11
2	6:1	110	90	14	0.17
2	6:1	110	120	15	0.18

Table 1. The DS of succinylated cellulose without any catalysts in [C_4mim]Cl/DMSO system.

cellulosic molecules increased with the improvement of reaction time. The DS of succinylated cellulose was 0.11 within 30 min, 0.15 within 60 min, 0.17 in 90 min and 0.18 in 120 min. The low succinoylation was propably due to the increased chain of succinic anhydride compared with acetylation. The similar reduced carboxymethylation efficiency of cellulose was also reported (Heinze et al., 2005).

3.2 Iodine-catalysed succinoylation

Recently, it has been demonstrated that iodine (I_2) is an effective catalyst for acetylation of alcohols and polysaccharides without solvents (Biswas et al., 2005; Ren et al., 2007). In the present study, we investigated the possibility of cellulose succinoylation catalyzed with I_2 in [C_4mim]Cl/DMSO system to increase cellulose modification efficiency. As shown in Table 2, the increase of the dosage of I_2 from 2% to 4%, 6%, 8%, 10%, and 15% resulted the improved DS from 0.74 to 0.92, 1.13, 1.25, 1.29, and 1.30, respectively. In comparison to the sample 5 obtained without any catalyst under the same conditions (DS=0.15), the succinoylation efficiency clearly increased. The possible mechanism of succinoylation and the actual role of iodine are not clear. However, a plausible explanation is that iodine might be ionized into I^+ and I^- in ILs. I^+ in turn activates the carbonyl groups of succinic anhydride to form as the acylation reagent for further reaction, as shown in Fig. 2.

3.3 NBS-catalysed succinoylation

It has been demonstrated that NBS is a novel, highly effective catalyst for acetylation of alcohols under mild reaction conditions. It is an inexpensive and commercially available reagent that is traditionally used as an oxidizing agent or brominating agent (Karimi & Seradj, 2001). In the present study, the possibility of cellulose succinoylation catalyzed with NBS was explored in [C_4mim]Cl/DMSO system to increase cellulose modification efficiency. As shown in Table 2, the DS was enhanced from 1.14 to 1.76, 1.95, 2.11, 2.17, and 2.18 with the improvement of NBS dosage from 1% to 3%, 5%, 7%, 10% and 15%, respectively, indicating the highly efficient succionylation catalysed with NBS compared with that without any catalyst. The possible mechanism of NBS-catalysed succinoylation is shown in Fig. 3. NBS might act as a source for Br^+, which in turn activates the carbonyl groups of SA to produce the highly reactive acylating agent. This acylating agent reacts with hydroxyl groups of cellulose, which upon elimination of NBS, produce succinoylated cellulose.

Succinoylation conditions					Succinylated cellulose	
SA/AGU	Catalyst	Catalyst/SA(%)	T/°C	t/min	Sample	DS
6:1	I_2	2	110	60	16	0.74
6:1	I_2	4	110	60	17	0.92
6:1	I_2	6	110	60	18	1.13
6:1	I_2	8	110	60	19	1.25
6:1	I_2	10	110	60	20	1.29
6:1	I_2	15	110	60	21	1.30
6:1	NBS	1	110	60	22	1.14
6:1	NBS	3	110	60	23	1.76
6:1	NBS	5	110	60	24	1.95
6:1	NBS	7	110	60	25	2.11
6:1	NBS	10	110	60	26	2.17
6:1	NBS	15	110	60	27	2.18
6:1	DMAP	1	110	60	28	1.26
6:1	DMAP	2	110	60	29	1.54
6:1	DMAP	3	110	60	30	1.85
6:1	DMAP	5	110	60	31	1.92
6:1	DMAP	8	110	60	32	1.99
6:1	DMAP	15	110	60	33	2.27

Table 2. The DS of succinylated cellulose with different catalysts in $[C_4mim]Cl$/DMSO system.

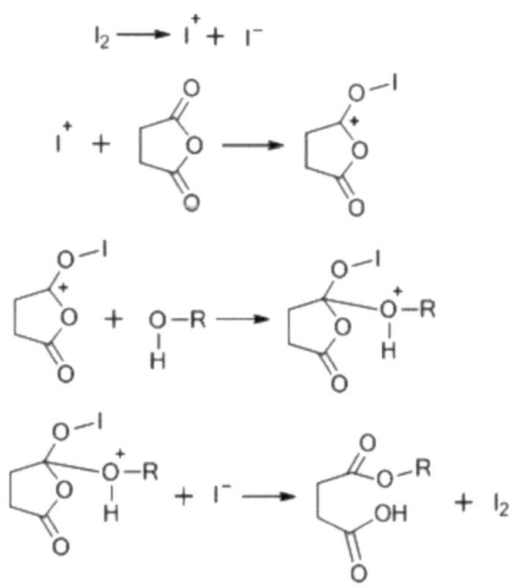

Fig. 2. Possible mechanism of succinoylation of cellulose using I_2 as a catalyst.

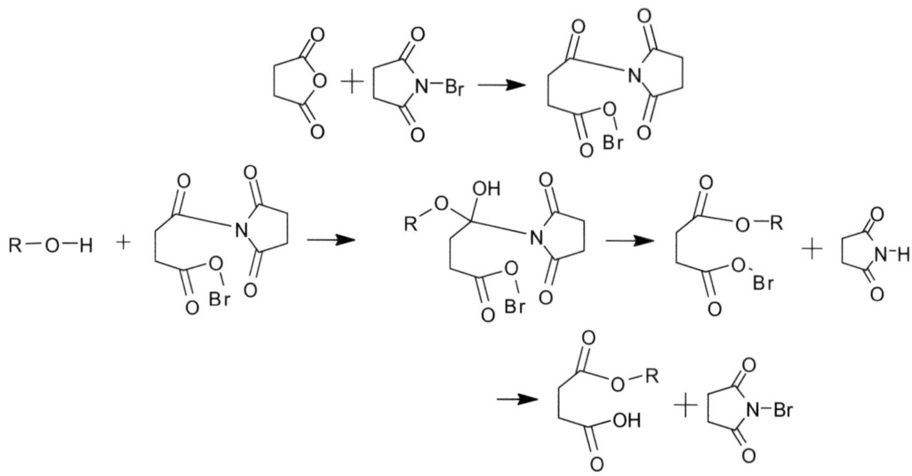

Fig. 3. Possible mechanism of succinoylation of cellulose using NBS as a catalyst.

3.4 DMAP-catalysed succinoylation

Pyridine has been found to be effective in the modification of wood with various long chain anhydrides, because it serves not only to swell the wood structure, thereby permiting effective ingress of reagent, but also catalyses the reaction via nucleopilic mediated catalysis (Sun et al., 2001). In comparison to pyridine, DMAP was found to be approximately 10^4 times more active when used as acylation catalyst (Hill et al., 2000). In the present study, we studied the succinoylation of cellulose in [C_4mim]Cl/DMSO system using DMAP as a catalyst. As shown in Table 2, the improvement of DMAP dosage from 1% to 2%, 3%, 5%, 8%, and 15% led to the increment in DS of succinylated cellulose from 1.26 to 1.54, 1.85, 1.92, 1.99, and 2.27, respectively. The possible mechanism of DMAP-catalysed succinoylation is shown in Fig. 3. The nucleofilic attack of DMAP on a carbonyl group of succinic anhydride leads to intermediate, which could react with cellulose hydroxyl groups and produce cellulose ester.

3.5 FT-IR

Fig. 4 illustrates the FT-IR spectra of native cellulose (spectrum 1) and succinylated samples 9 (spectrum 2, without any catalysts) and 24 (spectrum 3, 5% NBS). In spectrum 1, the absorbances at 3431, 2923, 1629, 1381, 1165, and 1048 cm^{-1} are associated with native cellulose. The strong adsorption at 3431 cm^{-1} is due to the stretching of hydroxyl groups and that at 2923 cm^{-1} to the C-H stretching. The band at 1629 cm^{-1} relates to the bending mode of the absorbed water. The peak at 1381 cm^{-1} is attributed to the O-H bending. The absorption band at 1165 cm^{-1} corresponds to C-O antisymmetric bridge stretching. A strong peak at 1048 cm^{-1} arises from C-O-C pyranose ring skeletal vibration.

Compared with spectrum 1, the spectra 2 and 3 of succinylated cellulose provide evidence of succinoylation by the occurrence of the absorbance at 1732 and 1573 cm^{-1}. In general, the absorption by carbonyl bonds in esters gives peaks at 1750 cm^{-1}, and one in carboxylic acids exhibits a band at 1712 cm^{-1} (Jayakumar et al., 2000). The overlapping band at 1732 cm^{-1} is indicative of absorption by carbonyl group in carboxyl and esters. The absorption band at

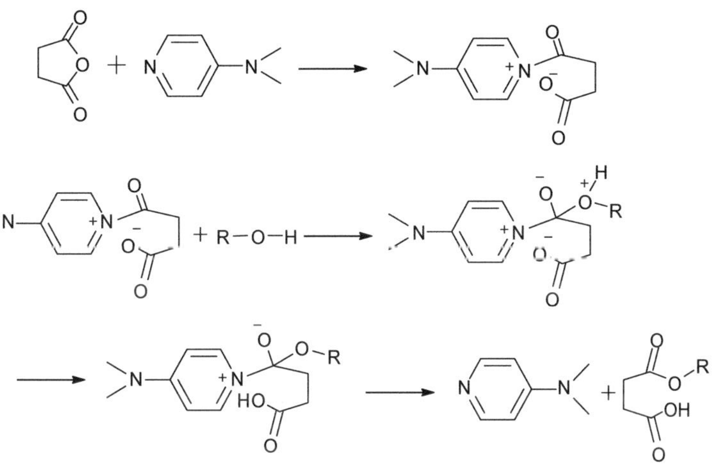

Fig. 4. Possible mechanism of succinoylation of cellulose using DMAP as a catalyst.

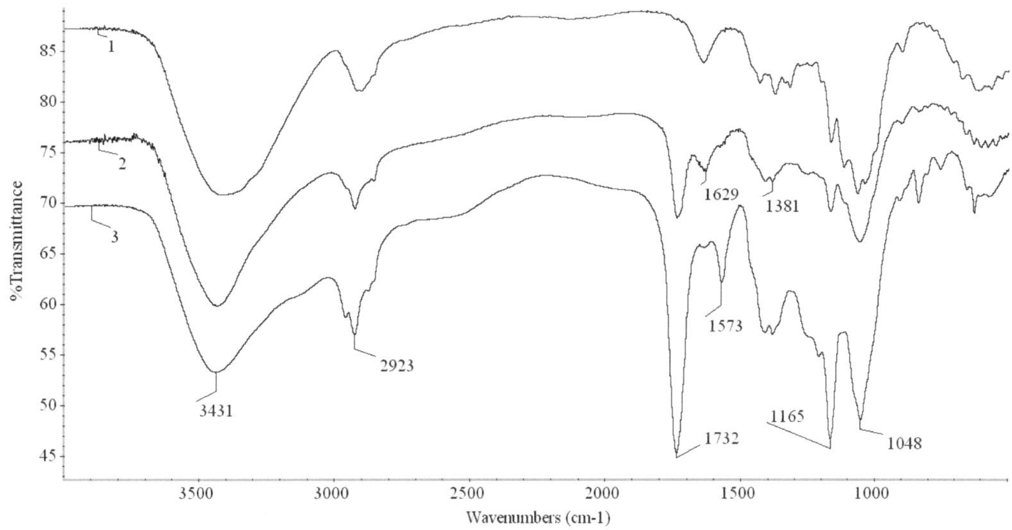

Fig. 5. FT-IR spectra of native cellulose (spectrum 1) and succinylated cellulose samples 9 (spectrum 2) and 24 (spectrum 3)

1573 cm^{-1} is associated with the antisymmetric stretching of carboxylic anions (Yoshimura et al., 2006), indicating the formation of monoester. In addition, the intensity of the absorption band at 1165 cm^{-1} for C-O antisymmetric stretching increased after succinoylation, suggesting that the esterification reaction does occur. These significant changes of the absorption bands indicated that the monoester of succinylated cellulose was formed in [C_4mim]Cl/DMSO system under the conditions given. It should be noted that the intensities of the bands at 1732, 1573, and 1165 cm^{-1} are significantly improved from spectrum2 to spectrum 3, indicating the enhanced succinoylation of cellulose. It suggested that NBS is a high efficient catalyst of cellulose succinoylation in [C_4mim]Cl/DMSO. As expected, the absence of peaks at 1850 and 1780 cm^{-1} in spectra 2 and 3 for succinylated cellulose confirmed that the products are free of the unreacted SA.

3.6 Solid-state CP/MAS ^{13}C-NMR

CP/MAS ^{13}C-NMR spectroscopy is the most effective of the few methods for the characterization of the lignocellulosic materials directly and is capable of providing detailed information directly on solid samples (Maunu, 2002). In the present study, the succinoylation reaction of cellulose with or without catalysts was also studied by solid-state CP/MAS ^{13}C-NMR spectroscopy, and the correspongding spectra are illustrated in Fig. 6.

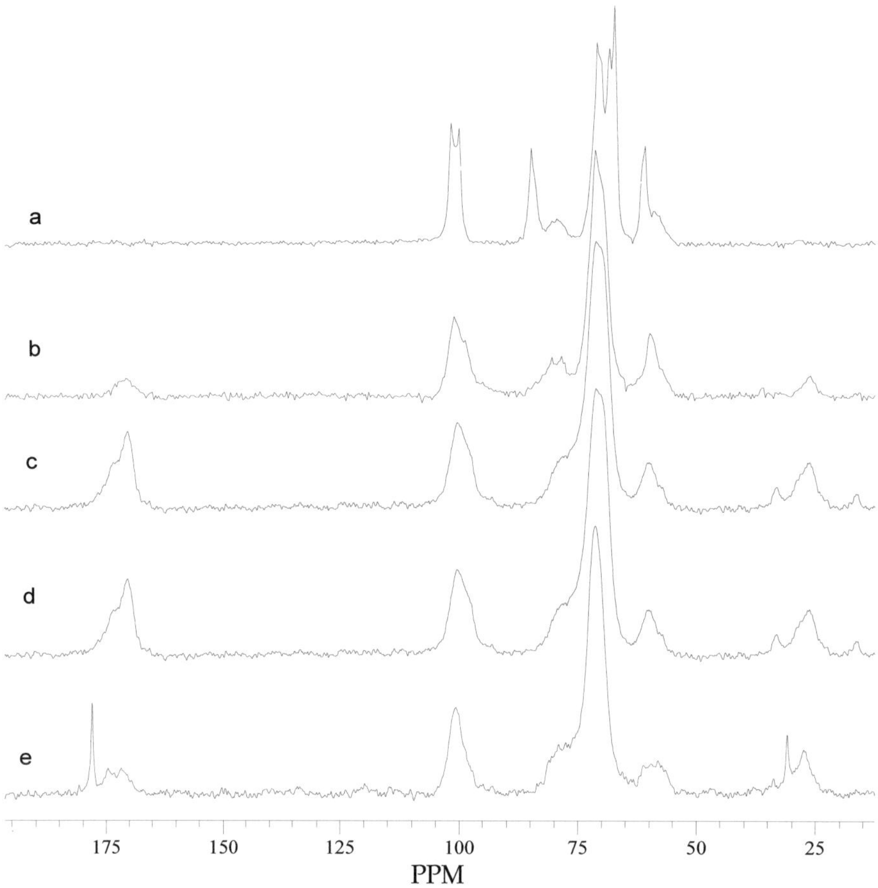

Fig. 6. Solid-state CP/MAS ^{13}C-NMR spectra of native cellulose (spectrum a) and succinylated cellulose without catalysts (sample 5, spectrum b) and with catalyst NBS (sample 24, spectrum c), DMAP (sample 31, spectrum d), and iodine (sample 17, spectrum e).

In spectrum a, all noticeable signals are distributed in the region between 50 and 110 ppm for the carbon atoms of the carbohydrate moiety. The typical signals at 101.4 (C-1), 85.2 (C-4 of crystalline cellulose), 79.7 (C-4 of amorphous cellulose), 71.2 (C-5), 67.8 (C-2 and C-3), 61.1 (C-6 of crystalline cellulose), and 59.8 ppm (C-6 of amorphous cellulose) have all been reported before (Chang & Chang, 2001; Maunu, 2002). Evidently, the presence of the signals of the carboxylic group at 171.0 ppm and the methylene group at 26.4 ppm provided evidence of succinoylation, which indicated that the reaction shown in Fig. 1 does occur. The signals at 85.2 ppm for C-4 of crystalline cellulose and 61.1 ppm for C-6 of crystalline cellulose disappeared in spectra b to e after succinoylation, suggesting the complete disruption of cellulose crystalline structure during the dissolution and functionalization, which indicated that succinoylation reaction occurs in homogeneous phase. The intensity of the signal for C-6 significantly decreased after succinoylation, and that at 67.8 ppm for C-2 and C-3 also decreased, which indicated the succinoylation reaction at C-6, C-2, and C-3 positions does occur. Obviously, the succinoylation reaction is preferred at C-6. The significant decrease of the signal for C-6 indicated that the reactivity of C-6 is higher than that of C-2 and C-3.

3.7 Thermal analysis

Fig. 7 illustrates TGA curves of native cellulose (curve a) and succinylated cellulose without catalyst (sample 6, curve b) and with NBS catalyst (sample 22, curve 3). The native cellulose starts to decompose at 226°C, whereas the regenerated cellulose begins to decompose at 205°C and 202°C for samples 6 and 10, respectively. At 50% weight loss, the decomposition temperature occurs at 338°C for native cellulose, 330°C for sample 6, and 285°C for sample 22. These decreasing trends of decomposition temperature suggested that thermal stability of cellulose decreased after succinoylation, especially true for high DS samples.

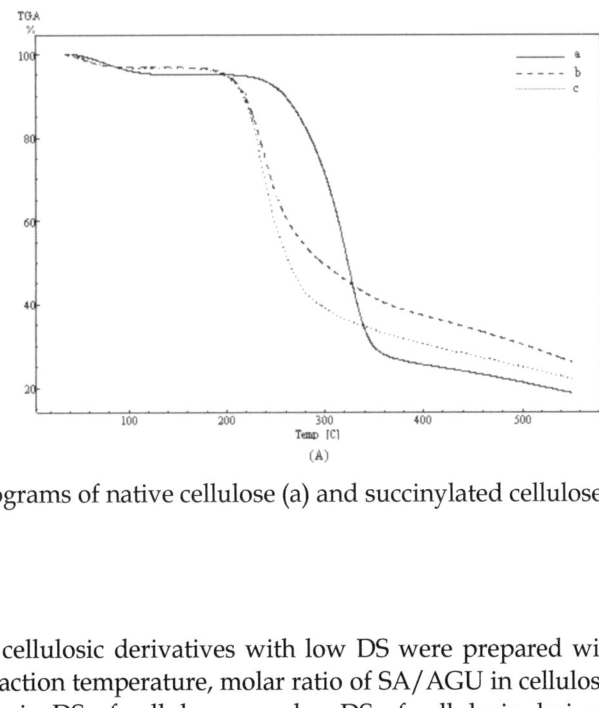

Fig. 7. The thermograms of native cellulose (a) and succinylated cellulose samples 6 (b) and 22 (c)

4. Conclusion

The succinylated cellulosic derivatives with low DS were prepared without any catalysts. The increase of reaction temperature, molar ratio of SA/AGU in cellulose, and reaction time led to an increase in DS of cellulose samples. DS of cellulosic derivatives obtained with iodine, NBS, and DMAP catalysts was significantly improved compared with the products without any catalysts, indicating that these three catalysts were effective catalysts for cellulose succinoylation in ILs. The possible mechanism of homogeneous succinoylation catalyzed with these catalysts and the actual role of these catalysts were also proposed. Fourier transform infrared and solid-state cross-polarization/magic angle spinning ^{13}C-NMR spectroscopies also provided evidence of catalyzed homogeneous succinoylation reaction. The results indicated that the reaction of hydroxyl groups at C-6, C-2, and C-3 positions in cellulose all occurred. The thermal stability of the succinylated cellulose was found to decrease upon chemical modification.

5. Acknowledgements

The authors are grateful for the financial support of this research from the National Natural Science Foundation of China (Nos. 30871994, 30972325, and 30710103906), the Guangdong Natural Science Foundation (No. 8451064101000409), Specialized Research Fund for the Doctoral Program of Higher Education (No. 20070561040), Chinese Universities Scientific Fund (No. 2009ZZ0024), and National Basic Research Program of China (No. 2010CB732201)

6. References

Abbott, A.P., Bell, T.J., Handa, S. & Stoddart, B. (2005). O-acetylation of cellulose and monosaccharides using a zinc based ionic liquid, *Green Chemistry* 7(10): 705-707.

Barthel, S. & Heinze, T. (2006). Acylation and carbanilation of cellulose in ionic liquids, *Green Chemistry* 8(3): 301-306.

Biswas, A., Shogren, R.L. & Willett, J.L. (2005). Solvent-free process to esterify polysaccharides, *Biomacromolecules* 6(4): 1843-1845.

Chang, S.T. & Chang, H.T. (2001). Comparisons of the photostability of esterified wood, *Polymer Degradation and Stability* 71(2): 261-266.

Clark, J.H., Deswarte, F.E.I. & Farmer, T.J. (2009). The integration of green chemistry into future biorefineries, *Biofuels Bioproducts & Biorefining-Biofpr* 3(1): 72-90.

de Maria, D. & Martinsson, A. (2009). Ionic-liquid-based method to determine the degree of esterification in cellulose fibers, *Analyst* 134(3): 493-496.

Evans, R. & Wallis, A.F.A. (1989). Cellulose molecular-weights determined by viscometry, *Journal of Applied Polymer Science* 37: 2331-2340.

Forsyth, S.A., MacFarlane, D.R., Thomson, R.J. & von Itzstein, M. (2002). Rapid, clean, and mild o-acetylation of alcohols and carbohydrates in an ionic liquid, *Chemical Communications* 7(7): 714-715.

Fukaya, Y., Sugimoto, A. & Ohno, H. (2006). Superior solubility of polysaccharides in low viscosity, polar, and halogen-free 1,3-dialkylimidazolium formates, *Biomacromolecules* 7(12): 3295-3297.

Fukaya, Y., Hayashi, K., Wada, M. & Ohno, H. (2008). Cellulose dissolution with polar ionic liquids under mild conditions: Required factors for anions, *Green Chemistry* 10(1): 44-46.

Granstrom, M., Kavakka, J., King, A., Majoinen, J., Makela, V., Helaja, J., Hietala, S., Virtanen, T., Maunu, S.L., Argyropoulos, D.S. & Kilpelainen, I. (2008). Tosylation and acylation of cellulose in 1-allyl-3-methylimidazolium chloride, *Cellulose* 15(3): 481-488.

Hadano, S., Onimura, K., Tsutsumi, H., Yamasaki, H. & Oishi, T. (2003). Syntheses of chemical-modified cellulose obtained from waste pulp, *Journal of Applied Polymer Science* 90(8): 2059-2065.

Heinze, T., Schwikal, K. & Barthel, S. (2005). Ionic liquids as reaction medium in cellulose functionalization, *Macromolecular Bioscience* 5(6): 520-525.

Hermanutz, F., Gaehr, F., Uerdingen, E., Meister, F. & Kosan, B. (2007). New developments in dissolving and processing of cellulose in ionic liquids, *Macromolecular Symposia* 262(1): 23-27.

Hill, C.A.S., Cetin, N.S. & Ozmen, N. (2000). Potential catalysts for the acetylation of wood, *Holzforschung* 54(3): 269-272.

Himmel, M.E., Ding, S.Y., Johnson, D.K., Adney, W.S., Nimlos, M.R., Brady, J.W. & Foust, T.D. (2007). Biomass recalcitrance: Engineering plants and enzymes for biofuels production, *Science* 315(5813): 804-807.

Jayakumar, R., Balaji, R. & Nanjundan, S. (2000). Studies on copolymers of 2-(n-phthalimido)ethyl methacrylate with methyl methacrylate, *European Polymer Journal* 36(8): 1659-1666.

Jeon, Y.S., Viswanathan, A. & Gross, R.A. (1999). Studies of starch esterification: Reactions with alkenyl-succinates in aqueous slurry systems, *Starch-starke* 51(2-3): 90-93.

Kamiya, N., Matsushita, Y., Hanaki, M., Nakashima, K., Narita, M., Goto, M. & Takahashi, H. (2008). Enzymatic in situ saccharification of cellulose in aqueous-ionic liquid media, *Biotechnology Letters* 30(6): 1037-1040.

Karimi, B. & Seradj, H. (2001). N-bromosuccinimide (nbs), a novel and highly effective catalyst for acetylation of alcohols under mild reaction conditions, *Synlett* (4): 519-520.

Kosan, B., Michels, C. & Meister, F. (2008). Dissolution and forming of cellulose with ionic liquids, *Cellulose* 15(1): 59-66.

Luo, H.M., Li, Y.Q. & Zhou, C.R. (2005). Study on the dissolubility of the cellulose in the functionalized ionic liquid, *Polymeric Materials Science & Engineering* 21(2): 233-235.

Maunu, S.L. (2002). NMR studies of wood and wood products, *Progress in Nuclear Magnetic Resonance Spectroscopy* 40(2): 151-174.

Mazza, M., Catana, D.A., Vaca-Garcia, C. & Cecutti, C. (2009). Influence of water on the dissolution of cellulose in selected ionic liquids, *Cellulose* 16(2): 207-215.

Mohanty, A.K., Misra, M. & Drzal, L.T. (2002). Sustainable bio-composites from renewable resources: Opportunities and challenges in the green materials world, *Journal of Polymers and the Environment* 10(1-2): 19-26.

Ogaki, Y., Shinozuka, Y., Hatakeyama, M., Hara, T., Ichikuni, N. & Shimazu, S. (2009). Selective production of xylose and xylo-oligosaccharides from bamboo biomass by sulfonated allophane solid acid catalyst, *Chemistry Letters* 38(12): 1176-1177.

Pandey, A., Soccol, C.R., Nigam, P. & Soccol, V.T. (2000). Biotechnological potential of agro-industrial residues. I: Sugarcane bagasse, *Bioresource Technology* 74(1): 69-80.

Reddy, N. & Yang, Y. (2005). Biofibers from agricultural byproducts for industrial applications, *Trends in Biotechnology* 23(1): 22-27.

Ren, J.L., Sun, R.C., Liu, C.F., Cao, Z.N. & Luo, W. (2007). Acetylation of wheat straw hemicelluloses in ionic liquid using iodine as a catalyst, *Carbohydrate Polymers* 70(4): 406-414.

Richardson, S. & Gorton, L. (2003). Characterisation of the substituent distribution in starch and cellulose derivatives, *Analytica Chimica Acta* 497(1-2): 27-65.

Schlufter, K., Schmauder, H.P., Dorn, S. & Heinze, T. (2006). Efficient homogeneous chemical modification of bacterial cellulose in the ionic liquid 1-n-butyl-3-methylimidazolium chloride, *Macromolecular Rapid Communications* 27(19): 1670-1676.

Stojanovic, Z., Jeremic, K., Jovanovic, S. & Lechner, M.D. (2005). A comparison of some methods for the determination of the degree of substitution of carboxymethyl starch, *Starch-starke* 57(2): 79-83.

Sun, R.C., Sun, X.F. & Zhang, F.Y. (2001). Succinoylation of wheat straw hemicelluloses in n,n-dimethylformamide/lithium chloride systems, *Polymer International* 50(7): 803-811.

Swatloski, R.P., Spear, S.K., Holbrey, J.D. & Rogers, R.D. (2002). Dissolution of cellose with ionic liquids, *Journal of the American Chemical Society* 124(18): 4974-4975.

Turner, M.B., Spear, S.K., Huddleston, J.G., Holbrey, J.D. & Rogers, R.D. (2003). Ionic liquid salt-induced inactivation and unfolding of cellulase from trichoderma reesei, *Green Chemistry* 5(4): 443-447.

Wu, J., Zhang, J., Zhang, H., He, J.S., Ren, Q. & Guo, M. (2004). Homogeneous acetylation of cellulose in a new ionic liquid, *Biomacromolecules* 5(2): 266-268.

Yoshimura, T., Matsuo, K. & Fujioka, R. (2006). Novel biodegradable superabsorbent hydrogels derived from cotton cellulose and succinic anhydride: Synthesis and characterization, *Journal of Applied Polymer Science* 99(6): 3251-3256.

Zhang, H., Wu, J., Zhang, J. & He, J.S. (2005). 1-allyl-3-methylimidazolium chloride room temperature ionic liquid: A new and powerful nonderivatizing solvent for cellulose, *Macromolecules* 38(20): 8272-8277.

Zhang, M.J., Qi, W., Liu, R., Su, R.X., Wu, S.M. & He, Z.M. (2010). Fractionating lignocellulose by formic acid: Characterization of major components, *Biomass & Bioenergy* 34(4): 525-532.

Zhang, Y.H.P., Ding, S.Y., Mielenz, J.R., Cui, J.B., Elander, R.T., Laser, M., Himmel, M.E., McMillan, J.R. & Lynd, L.R. (2007). Fractionating recalcitrant lignocellulose at modest reaction conditions, *Biotechnology and Bioengineering* 97(2): 214-223.

6

Preparation of Polysaccharide-based Materials Compatibilized with Ionic Liquids

Jun-ichi Kadokawa
Kagoshima University
Japan

1. Introduction

Ionic liquids (ILs) are low-melting-point molten salts, defined as which form liquids at room temperature or even at temperatures lower than a boiling point of water. The property is owing to that the liquid state is thermodynamically favorable due to the large size and conformational flexibility of the ions, in which these behaviors lead to small lattice enthalpies and large entropy changes that favor the liquid state (Welton, 1999). In the past more than a decade, ILs have attracted much attention due to their specific characteristics such as a negligible vapor pressure, excellent thermal stabilities, and controllable physical and chemical properties (Plechkova & Seddon, 2008). Beyond these traditional properties of ILs, recently, interests and applications on ILs have been extended to the researches related to biomolecules such as naturally occurring polysaccharides, because of specific good affinities of ILs for them (Seoud et al., 2007).

Polysaccharides are widely distributed in nature and have been regarded as structural materials and as suppliers of water and energy (Schuerch, 1986). Natural polysaccharides such as cellulose, starch, chitin, and other many kinds of them have recently received much attention for use as resources of new green and sustainable materials because of their eco-friendly properties. However, the natural polysaccharides have often exhibited solubility, processability, and feasibility problems due to numerous hydrogen bonds in their polymeric chains, causing difficulty in employing them in a wide variety of materials.

In 1934, Graenacher had already discovered that a molten *N*-ethylpyridinium chloride, in the presence of nitrogen-containing bases, dissolved cellulose (Graenacher, 1934). This was probably the first example of the cellulose dissolution using IL-type solvents. However, this was thought to be of little practical value at the time because the concept of ILs had not been put forward. In 2002, Rogers et al. comprehensively reported that an IL, 1-butyl-3-methylimidazolium chloride (BMIMCl) dissolved cellulose in relatively high concentrations and this research opened up a new way for the development of a class of cellulose solvent systems (Swatloski et al., 2002). Since this publication, ILs have began to be used in the processing of cellulose and other polysaccharides, which mainly concern with the dissolution, homogeneous derivatization and modification, and regeneration (Zhu et al., 2006, Liebert & Heinze, 2008, Feng & Cheng, 2008, Pinkert et al., 2009).

In recent years, ILs have been extensively used as functional components in the polysaccharides-based materials besides the above traditional uses as solvents and media. The concept in these researches is based on the properties of ILs that exhibit good affinities for the polysaccharides, which are suitably lead to showing well-compatibilization. In this chapter, on the basis of these backgrounds and viewpoints, the preparation of polysaccharides-based materials compatibilized with ILs is described, which is divided by the following two sections. A first topic is the preparation of polysaccharide-polymeric IL composite materials by means of in-situ polymerization of polymerizable ILs. As a second topic, the preparation of IL-gel materials from various natural polysaccharides is disclosed.

2. Preparation of polysaccharide-polymeric ionic liquid composites by in-situ polymerization of polymerizable Ionic Liquids

2.1 Polymeric ionic liquids

Polymeric ILs are defined as the polymers obtained by polymerization of ILs having polymerizable groups (polymerizable ILs) (Green et al., 2009). Thus, 'polymeric ILs' are termed just the polymeric forms of ILs, but they are not necessary to show liquid form at room temperature or even at some ambient temperatures. The polymeric ILs, therefore, are often called 'polymerized ILs' too. The major advantages for providing the polymeric ILs are to be enhanced stability, and improved processability and feasibility in application as practical materials. Polymerizable ILs as a source of the polymeric ILs can be available by incorporating the polymerizable groups at both anionic and cationic sites in the IL structures (Fig. 1). In the former case, polymerizable anions are ionically exchanged to some anions of general ILs (Fig. 1(a)), giving the polymerizable ILs. In the latter case, vinyl, meth(acryloyl), and vinyl benzyl groups have typically been appeared as the polymerizable group (Fig. 1(b)). 1-Vinylimidazole is a commercially available reagent. Therefore, the vinylimidazolium-type polymerizable ILs are prepared by quaternization of 1-vinylimidazole with a variety of alkyl halides. The reaction of haloalkyl (meth)acrylates or vinylbenzyl halides with 1-alkylimidazoles gives the corresponding imidazolium-type polymerizable ILs having the (meth)acryloyl or vinyl benzyl polymerizable groups (Fig. 2). Furthermore, when haloalkyl (meth)acrylates or vinylbenzyl halides are reacted with 1-vinylimidazole, the polymerizable ILs having two polymerizable groups are produced. Because these ILs can be converted into insoluble and stable polymeric ILs with the cross-linked structure by the radical polymerization (Fig. 3), they have a highly potential as the source of the components in the practical materials. On the basis of the above viewpoints, the polymerizable ILs having the acrylate or vinylbenzyl group have been employed for the preparation of polysaccharide-polymeric IL composite materials.

2.2 Preparation of cellulose-polymeric IL composites

Cellulose is the most abundant organic substance on the earth, which consists of β-(1→4)-linked glucose repeating units. Cellulose has limited applications because of the strong inter- and intramolecular hydrogen bonds between the hydroxy groups of the glucose residues. Therefore, considerable efforts have been still devoted to the compatibilization of cellulose with synthetic polymers for improvement of the processability of cellulose. In-situ polymerization method, for example, those giving interpenetrating polymer network (IPN) system has been useful for formation of composites from cellulose (Fig. 4). The IPNs are a special class of polymer blends or composites based on two or more polymers, with each polymer chemically cross-linked or at least one network being prepared in the presence of the others. If all polymers are cross-linked, the blend is called a full-IPN. Semi-IPN differs

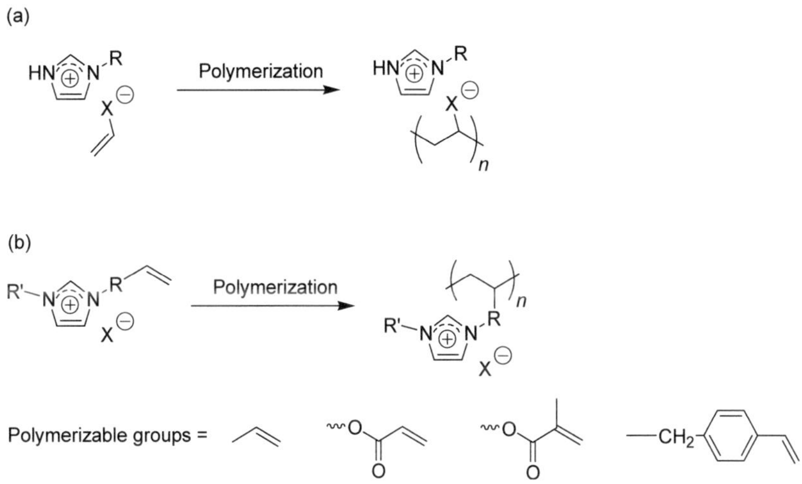

Fig. 1. Polymerization of polymerizable ILs having a polymerizable group at anionic site (a) and cationic site (b).

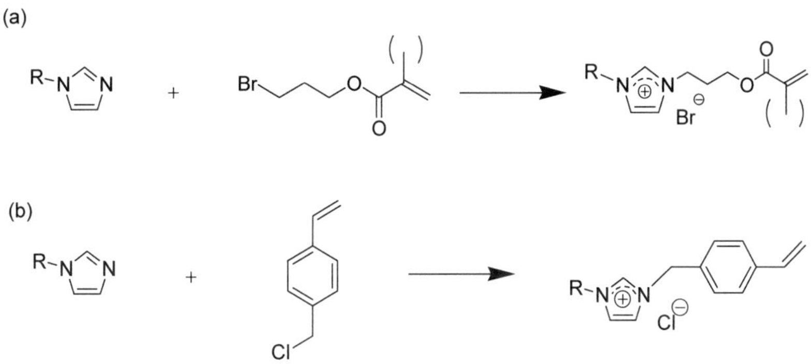

Fig. 2. Typical synthetic schemes for polymerizable ILs having (meth)acrylate (a) and vinylbenzyl (b) groups.

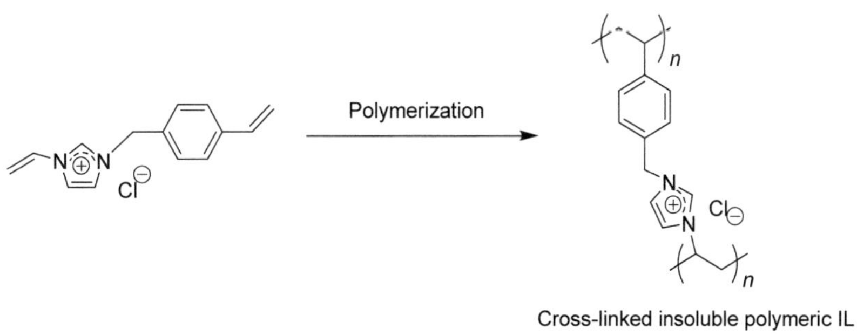

Fig. 3. Polymerization of a polymerizable IL having two polymerizable groups to produce a cross-linked insoluble polymeric IL.

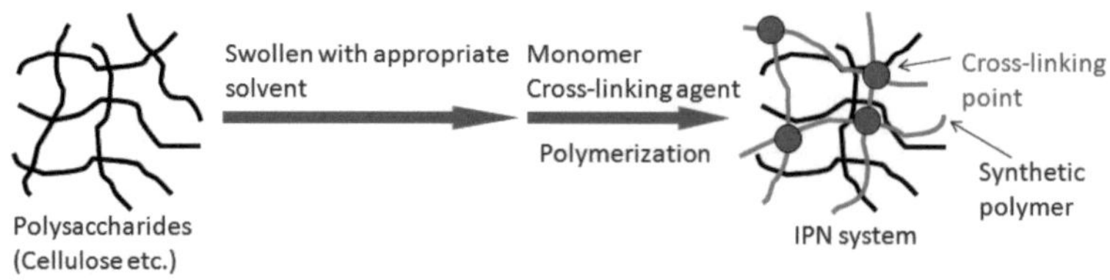

Fig. 4. Schematic image of interpenetrating polymer network (IPN) system.

from the full-IPN, in which it is composed of a non-cross-linked polymer entrapped in another polymer networks. Since the imidazolium-type ILs such as BMIMCl are good solvents for cellulose, it has been interesting to employing the imidazolium-type polymeric ILs as one component for good compatibilization with cellulose. To obtain the IPN-composite of cellulose with the polymeric IL by the in-situ polymerization method, in the first study, a polymerizable IL, 1-(4-acryloyloxybutyl)-3-methylimidazolium bromide (AcMIMBr) was used (Fig. 5) (Murakami et al., 2007).

First, cellulose was dissolved in BMIMCl (9.1 wt%) by heating at 100 °C for 24 h. Then, AcMIMBr (an equimolar amount of the glucose units in cellulose) and a radical initiator, AIBN, were added to the solution and the radical polymerization of the mixture was carried out at 80 °C for 5 h. After the polymerization mixture was washed with acetone and treated further with refluxed acetone, the residue was subjected to Soxhlet extraction with methanol to give the composite. The unit ratio of cellulose to AcMIMBr was calculated by the elemental analysis to be 1.0 : 0.63 and the ^{1}H NMR spectrum of the Soxhlet extract indicated that a part of poly(AcMIMBr) was washed out by the extraction. The IR spectrum of the resulting material suggested the structure composed of cellulose and poly(AcMIMBr). Furthermore, the thermal gravimetric analysis (TGA) and the X-ray diffraction (XRD) data supported the efficient compatibilization of cellulose with poly(AcMIMBr) in the composite. When cellulose and poly(AcMIMBr) were simply mixed in BMIMCl and the obtained mixture was subjected to the same isolation procedure as that for the aforementioned composite, almost of poly(AcMIMBr) was removed out. This result indicated that the simple mixing method of cellulose with poly(AcMIMBr) in BMIMCl was not useful for the efficient compatibilization.

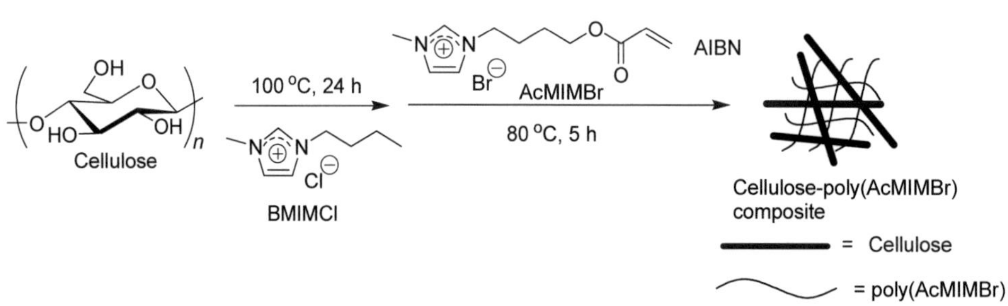

Fig. 5. Preparation of cellulose-polymeric IL composite using two ILs, BMIMCl and AcMIMBr.

Although the aforementioned approach was the efficient for the preparation of the cellulose-polymeric IL composites, the procedure should be improved due to the following disadvantageous points; (1) the necessity of two kinds of ILs and (2) the instability of the obtained composites toward solvents. To solve the second problem, the polymerizable IL having two polymerizable groups was employed, which was polymerized by a radical initiator to give the insoluble and stable cross-linked polymeric IL. Thus, in the improved procedure for the preparation of cellulose-polymeric IL composites in the next paper, a combined use of a polymerizable IL having acryl and vinyl groups (1-(3-acryloyloxypropyl)-3-vinylimidazolium bromide, AcVIMBr) with AcMIMBr (an alkyl group = propyl) was conceived (Fig. 6) (Takegawa et al., 2009). The former acted as a cross-linker in the produced polymeric ILs by the radical copolymerization of these two ILs. Therefore, the property of the polymeric IL can be changed by the feed ratio of the two ILs in the polymerization. Furthermore, it was found that a mixture of these polymerizable ILs mostly disrupted the crystalline structure of cellulose by pre-treatment, indicating the preparation of the composites to exclusion of use of the solvent IL such as BMIMCl unlike the case in the first study.

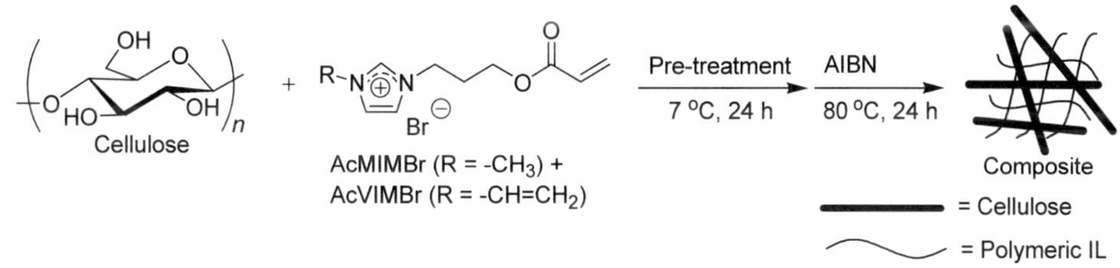

Fig. 6. Preparation of cellulose-polymeric IL composite using acrylate-type polymerizable ILs.

Therefore, for the preparation of the composites, first, cellulose (30 wt% for AcMIMBr + AcVIMBr) was pre-treated in mixtures of AcMIMBr and AcVIMBr with various weight ratios (100 : 0 – 50 : 50) at 7 °C for 24 h. The ability of the ILs that disrupted the crystalline structure of cellulose was evaluated by the XRD and TGA measurements of the pre-treated mixtures. Consequently, the analytical data indicated that the crystalline structure of cellulose was largely disrupted by the pre-treatment. Then, the formation of the composites was performed by the in-situ polymerization of the polymerizable ILs in the pre-treated mixtures. After AIBN was added to the pre-treated mixtures, the mixtures were heated at 80 °C for 24 h, resulting in the composites without any further isolation and purification procedures. The XRD and TGA results of the resulting composites indicated good compatibility of cellulose and the polymeric ILs in the composites in all cases. The compatibility between cellulose and the polymeric ILs in the composites was also evaluated by the SEM observation. The SEM images of the surface and inside area in the composite showed the completely different morphology from those of the standard cellulose, supporting the good compatibility of two components in the composite (Fig. 7). The stability of the composites was examined by washing with methanol by the Soxhlet extraction manner. After washing the composites obtained by using the mixtures of AcMIMBr and AcVIMBr, the unit ratios of cellulose to the polymeric ILs were hardly changed from those before washing. On the other hand, the content of the polymeric IL in the composite resulted by using AcMIMBr alone decreased by washing. The results indicated that the composites from the mixtures of AcMIMBr and AcVIMBr were very stable because of consisting of the cross-linked polymeric IL components formed by the copolymerization of the two polymerizable ILs. In contrast, the component of the polymeric IL in the composite from AcMIMBr alone was washed out, due to its linear structure, which was soluble in methanol. The mechanical property of the composites obtained from the mixtures of AcMIMBr and AcVIMBr with the various weight ratios was evaluated by the stress-strain curves under tensile mode. The composite from AcMIMBr alone showed the more elastic property compared with other composites containing cross-linked poly(AcMIMBr/AcVIMBr). The higher contents of the units from AcVIMBr in the composites affected to sustain the higher stresses until 30 wt% of AcVIMBr in the mixture of the two polymerizable ILs was used.

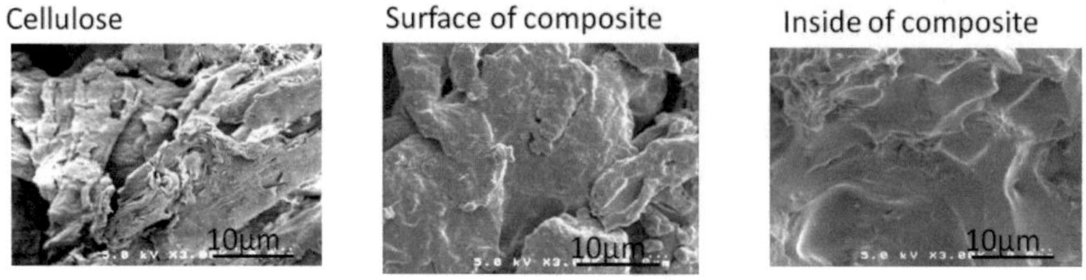

Fig. 7. SEM images of cellulose and composite.

The above in-situ polymerization method was extended to use another polymerizable IL, which was an IL having vinylbenzyl and vinyl polymerizable groups (1-vinyl-3-(4-vinylbenzyl)imidazolium chloride, VVBnIMCl) (Fig. 8) (Kadokawa et al., 2008a). This IL gives the polymeric IL composed of the polystyrene main-chain by radical polymerization. The preparation of the composites composed of cellulose and poly(VVBnIMCl) was carried out by the same procedure as that using AcMIMBr/AcVIMBr. Thus, cellulose (9.1 - 50.0 wt%) was first pre-treated with VVBnIMCl at 7 °C for 24 h. The XRD profile of the pre-treated mixture (33.3 wt%) showed only the slight diminution of the crystalline peaks. This result in comparison with that using AcMIMBr/AcVIMBr indicated that the ability of AcMIMBr/AcVIMBr disrupting the crystalline structure of cellulose was much higher than that of VVBnIMCl. However, the TGA result of the pre-treated mixture (33.3 wt%) using VVBnIMCl exhibited the similar curve as that using AcMIMBr/AcVIMBr, where an onset weight loss at around 250 °C due to the thermal degradation of cellulose was observed, which was ca. 50 °C lower than that of a standard cellulose. The TGA data indicated that the cellulose chains were partially swollen in the pre-treated mixture using VVBnIMCl.

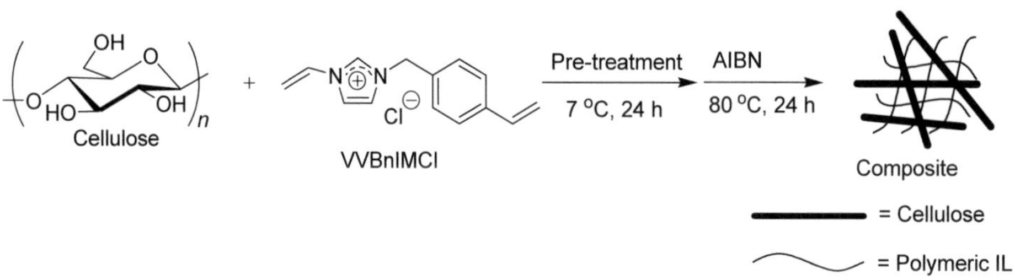

Fig. 8. Preparation of cellulose-polymeric IL composite using vinylbenzyl-type polymerizable IL.

After AIBN was added to the pre-treated mixtures, the mixtures were heated at 80 °C for 24 h to occur the radical polymerization. The composites were thus obtained without any isolation and purification procedures. The XRD and TGA results of all the composites indicated the partial disruption of the crystalline structure of cellulose. The SEM images of the composite supported good compatibility between cellulose and poly(VVBnIMCl) in the composite. The mechanical properties of the composites with 9.1, 33.3, and 50.0 wt% cellulose contents were evaluated by the tensile testing. The composites sustained the stresses in the range 2.5 - 6.5 MPa and the strains at break were 1.3 - 2.0 %. The higher content of cellulose in the composite (50 wt%) caused a brittle property. However, the appropriate content of cellulose affected to sustain higher stress. Consequently, the composite containing cellulose of 33.3 wt% indicated good mechanical property. Because of the presence of poly(VVBnIMCl) with the cross-linked structure in the composite, it showed the good stability toward washing with methanol.

This in-situ polymerization method was applied to the preparation of cellulose-based ionic porous material compatibilized with the polymeric IL (Fig. 9) (Prasad et al., 2010). This was achieved by combining the in-situ polymerization method with the templating technique using the oil/IL emulsion system. The in-situ polymerization of a mixture of AcMIMBr and AcVIMBr was first performed in a solution of cellulose in a solvent of BMIMCl. Then, the sonication of the mixture coexisting with corn oil and sorbitan monooleate, followed by the treatment with the appropriate solvents gave the porous material. The pore sizes in the materials ascertained from the SEM images were found to be around 0.15 - 1.3 μm accompanied with the smaller sizes of 30 – 70 nm (Fig. 10).

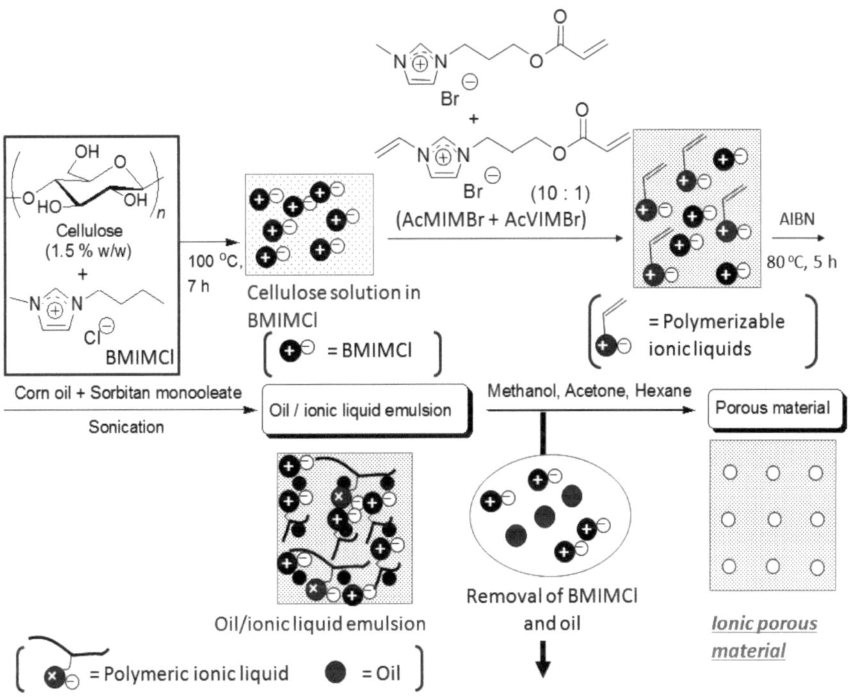

Fig. 9. Schematic image for formation of cellulose-based ionic porous material compatibilized with polymeric IL.

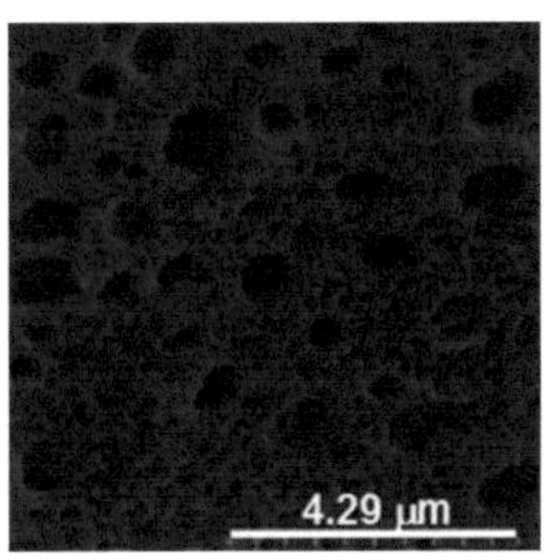

Fig. 10. SEM image of porous material.

2.3 Preparation of carrageenan-polymeric IL composites

Seaweed-originated polysaccharides like agarose and carrageenan have been used to prepare some functional materials such as electrical conductive materials by incorporating salts into their gel matrices as well as by doping with conducting polymers such as polyaniline and polypyrrole. Following the preparation of the aforementioned cellulose composites with the polymeric ILs, the new composite materials using *ι*-carrageenan of a seaweed polysaccharide was also prepared by compatibilizing with the polymeric ILs by means of in-situ polymerization technique (Fig. 11) (Prasad & Kadokawa, 2010). *ι*-Carrageenan is a water-soluble phycocolloid extracted from red algae consisting of alternating 1,3-linked α-D-galactopyranose unit and 1,4-linked β-3,6-anhydro-D-galactopyranose unit with sulfates on C-4 of the galactopyranose unit and C-2 of the anhydro-D-galactopyranose unit (Section 3.2). Due to the presence of the charges, it is expected to form well-miscible composites with the polymeric ILs through cross-linking via ionic exchange, which possibly contributes to improving the mechanical properties of the materials. It has also been considered that the material may exhibit the electrical conductivity due to the ionic nature. Therefore, the preparation of the composite materials composed of *ι*-carrageenan and the polymeric ILs by the in-situ polymerization of the polymerizable ILs was performed. The polymerizable ILs used were AcVIMBr and VVBnIMCl. For the preparation of the composite using AcVIMBr, BMIMCl was used for the dissolution of *ι*-carrageenan because its direct soak in AcVIMBr was not possible due to the high viscosity of AcVIMBr. Thus, *ι*-carrageenan was first dissolved in BMIMCl by heating at 100 °C for 7 h. Then, the radical polymerization of AcVIMBr with AIBN in the solution was performed, followed by the work-up procedure, to give the composite. Because unlike AcVIMBr, VVBnIMCl was no viscous, on the other hand, the composite using VVBnIMCl was prepared without use of any solvent for dissolution of *ι*-carrageenan. The CP-MS ^{13}C NMR and IR spectra of the resulting composites fully supported which were composed of *ι*-carrageenan and the corresponding polymeric ILs.

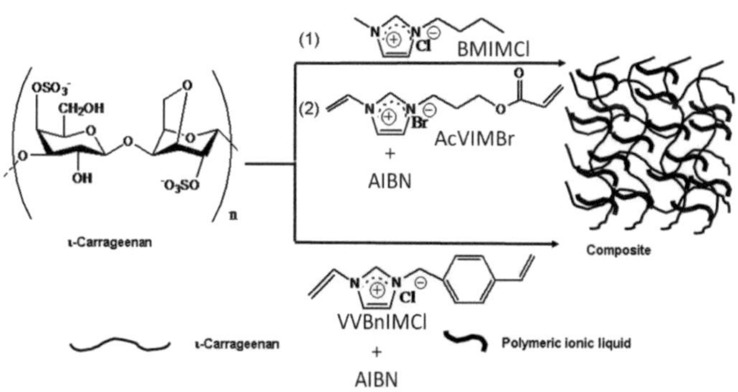

Fig. 11. Preparation of *ι*-carrageenan-polyIL composites.

Immiscible components were not seen in the charge coupled device (CCD) camera views (mm scale) of the surfaces of the composites. However, the SEM images with the nm scale of the two composites exhibited the different phase structures. The SEM image of the composite with poly(AcVIMBr) showed loose structures with nano scale distributions, whereas closer packed nano structures were observed in the SEM image of the composite with poly(VVBnIMCl). The morphology for the former material was probably caused by the removal of the solvent, BMIMCl, during the work-up procedure.

The mechanical properties of the composites of ι-carrageenan with poly(AcVIMBr) and poly(VVBnIMCl), which were evaluated under compressive mode, showed the fracture stresses of 108.07 and 102.0 MPa with the fracture strains of 25.0 and 10.7 %, respectively. These values were larger than those of the respective cellulose composite with each polymeric IL. The good mechanical properties of the present composites were reasonably explained by the formation of the good miscibility as observed in the SEM images, which were owing to the cross-linking through the ionic exchange between ι-carrageenan and the polymeric ILs. The electrical conductivities of the composites with poly(AcVIMBr) and poly(VVBnIMCl) were measured to be 6.2 x 10^{-4} and 1.6 x 10^{-4} S cm^{-1}, respectively, which were comparable to the conductance of semiconductor. On the other hand, cellulose composites with poly(AcVIMBr) and poly(VVBnIMCl) showed the electrical conductivity values of ca. 1.3 x 10^{-5} and 2.8 x 10^{-5} S cm^{-1}, respectively. The higher electrical conductivity of the present composites indicated the role of the effective ionic exchange during the formation of the composites, due to the presence of charges in ι-carrageenan.

3. Preparation of ionic liquid-gels of polysaccharides

3.1 Preparation of ionic liquid-gels of cellulose

When a solution of cellulose in BMIMCl (13.0 wt%) was left standing at room temperature for 7 days, it was found that the formation of a gel gradually took place with exclusion of excess BMIMCl from the gel matrix (Fig. 12) (Kadokawa et al., 2008b). Thus, the gel was purified by washing with ethanol to remove the excluded BMIMCl and dried under reduced pressure. The resulting gel had flexible and manipulatable natures and was characterized by the elemental analysis, XRD, and TGA. The XRD result of the gel indicated that the crystalline structure of cellulose was mostly disrupted in the gel. The elemental analysis and TGA data suggested that gel was composed not only of cellulose and BMIMCl, but also of water. On the basis of the above analytical results, it was considered that the gel was produced by the formation of non-crystalline aggregates of cellulose in the solution during gradually absorbing water owing to the strong hygroscopic nature of BMIMCl (Fig. 13). The aggregates probably acted as cross-linking points for the gelation.

When the gel was heated at 120 °C, it was gradually softened, and converted into a fluid at 150 °C. The soft material, which was obtained by heating the gel at 120 °C for 2 h, could be transferred to the gel form again by standing it at room temperature for 2 days. The regenerated gel was still flexible and was more transparent than the primary formed gel. Because the aggregates of cellulose as the cross-linking points in the regenerated gel were probably dispersed well by the heating-cooling process compared with those in the primary gel, the gel after regeneration was more transparent than before.

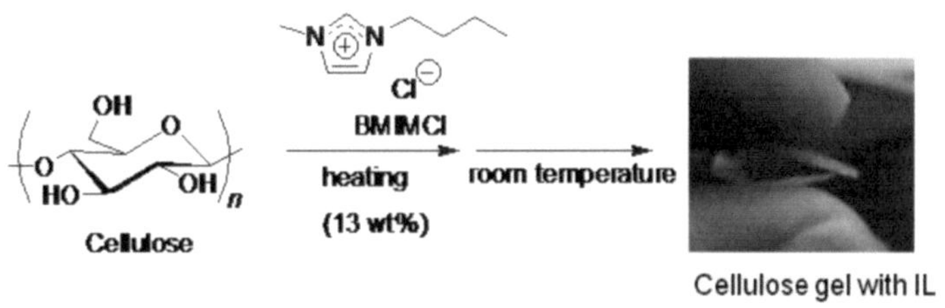

Fig. 12. Procedure for preparation of cellulose gel with IL.

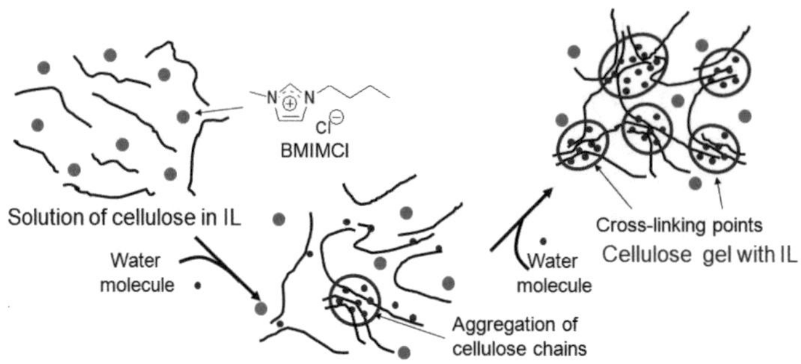

Fig. 13. Schematic image for procedure of formation of cellulose gel with IL.

This gelation technique was extended to the preparation of cellulose/starch composite gel with BMIMCl (Kadokawa et al., 2009). When the mixture of cellulose (9.1 wt%) and starch (4.8 wt%) in BMIMCl was heated at 100 °C for 24 h, a homogeneous liquid was formed, which was left standing at room temperature for 4 - 6 days. Consequently, the excess BMIMCl was excluded from the liquid to take place the gelation. Thus, the resulting gel was purified by washing with methanol to remove the excluded excess BMIMCl and dried under reduced pressure. In the XRD profile of the gel, the diffraction peaks due to the crystalline structures of the standard cellulose and starch were not obviously exhibited, indicating that the crystalline structures of the polysaccharides were largely disrupted in the gel. The results of the elemental analysis and TGA indicated that the gel was composed of the polysaccharides, BMIMCl, and water, as same as the aforementioned gel from cellulose and BMIMCl. From the data, the similar gelation process was conceived as that for the gel of cellulose with BMIMCl.

The homogeneous mixture of cellulose and starch (9.1 wt% each) in BMIMCl, obtained by heating as same as above, had a very viscous nature, and thus, could be raised from the surface. The resulting fine linear material was subjected to the formation of the composite fibrous material by soaking into acetone. The SEM picture of the obtained material showed the fiber form with ca. 100 - 200 μm diameters, but the crystalline and particle images, which were observed in the SEM pictures of the standard cellulose and starch, were not seen. The XRD result of the material indicated that the crystalline structures of cellulose and starch chains were not regenerated well, suggesting the good compatibilization of the two polysaccharides in the fibrous material.

It was found that a clear liquid of chitin (4.8 wt%) with ionic liquid, 1-allyl-3-methylimidazolium bromide (AMIMBr) was obtained by heating a mixture of chitin with AMIMBr at 100 °C for 48 h (Prasad, Murakami et al., 2009). The disappearance of the chitin powder in the liquid was confirmed by the CCD camera view and the SEM observation. The IR spectrum, TGA curve, and XRD profile of the regenerated chitin from the liquid as well as the analytical data of its hexanoyl derivative supported that the degradation and decreasing the molecular weight of chitin did not frequently occur during the dissolution process. On the other hand, when a mixture of 6.5 wt% chitin was soaked in AMIMBr at room temperature for 24 h, heated at 100 °C, and then cooled to room temperature, it turned into the gel-like form. The dynamic rheological measurements showed that both 4.8 wt% and 6.5 wt% chitins with AMIMBr behaved as the weak gels. However, it was seen that the obtained 6.5 wt% chitin with AMIMBr did not flow upon leaning a test tube, while the 4.8 wt% chitin with AMIMBr started to flow upon leaning (Fig. 14). The much lower yield stress of 4.8 wt% chitin with AMIMBr than that of 6.5 wt% chitin with AMIMBr supported that 4.8 wt% chitin with AMIMBr can flow under gravitation.

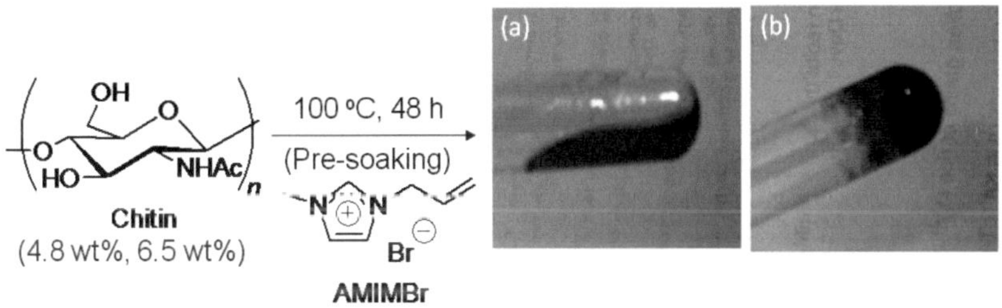

Fig. 14. Dissolution and gelation of chitin with IL; 4.8 wt% (a) and 6.5 wt% (b).

Thus, the preparation of the chitin/cellulose composite gel and film using the two ionic liquids, AMIMBr and BMIMCl, was performed (Fig. 15) (Takegawa et al., 2010). First, chitin and cellulose were dissolved in each appropriate ionic liquid. Then, the two liquids were mixed in the desired ratios at 100 °C to give the homogeneous mixtures. The gels were obtained by standing the mixtures at room temperature for 4 days and the films were obtained by casting the mixtures on a glass plate, followed by soaking in water and drying. The resulting gels and films were characterized by the XRD and TGA measurements, which showed relatively good miscibility among the polysaccharides and the ionic liquids in the materials. The mechanical properties of the gels and films were changed depending on the ratios of chitin to cellulose in the materials.

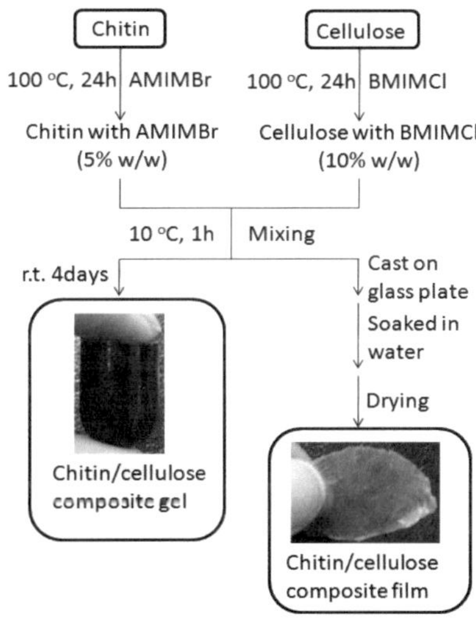

Fig. 15. Preparation procedure for chitin / cellulose composite gel and film with ILs.

The chitin/cellulose composite gel with ionic liquids was employed as the novel electrolyte for an electric double layer capacitor (EDLC) (Yamazaki et al., 2009, Yamazaki et al., 2010a, 2010b). First, the gel was treated with an aqueous 2.0 mol/L H_2SO_4 solution for 3 h. Electrochemical characteristics of the obtained acidic chitin/cellulose composite gel electrolyte were investigated by galvanostatic charge-discharge measurements. The test cell with the composite gel electrolyte showed a specific capacitance of 162 F/g at room temperature, which was higher than that for a cell with an H_2SO_4 electrolyte (155 F/g). The composite gel electrolyte exhibited the excellent high-rate discharge capability in a wide range of current densities as well as an aqueous H_2SO_4 solution. Moreover, the discharge capacitance of the test cell retained over 80 % of its initial value in 10^5 cycles even at a high current density of 5000 mA/g. The temperature dependence of the ionic conductivity for the composite gel electrolyte was the Arrhenius-type and similar to an aqueous H_2SO_4 solution. Furthermore, the self-discharge of the EDLC cell with the composite gel electrolyte was advantageously slow in comparison with the cell containing only the aqueous H_2SO_4 solution. The above results indicated that the acidic chitin/cellulose composite gel electrolyte had a practical applicability to an advanced EDLC with the excellent stability and working performance.

3.2 Preparation of ionic liquid-gels of food hydrocolloids

Besides the abundant polysaccharides like cellulose, starch, and chitin, many kinds of natural polysaccharides from various sources such as plants, animals, seaweeds, and bacteria have been known. For example, some polysaccharides are used as hydrocolloids for a stabilizer, a viscous agent, and a structure provider in food industries (Fig. 16). However, there have not been many reports on the use of these food hydrocolloids as a source of practical materials.

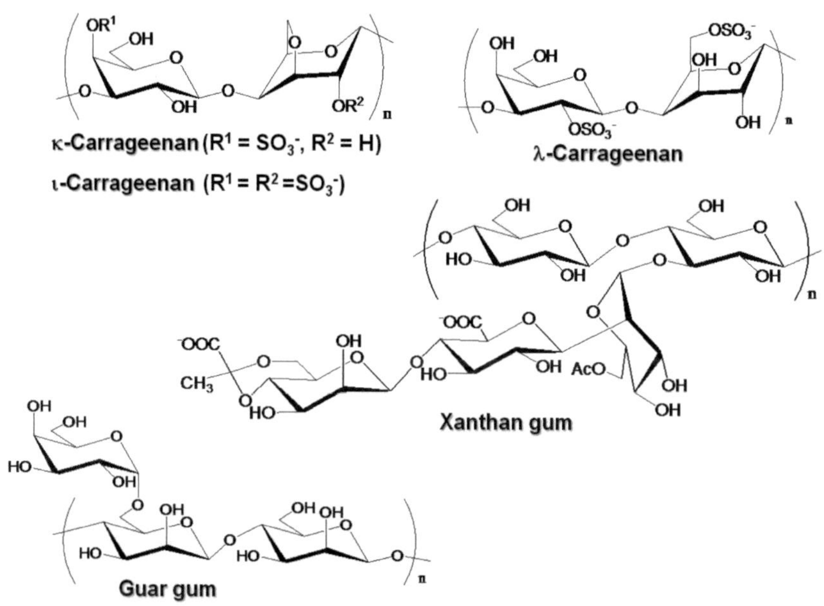

Fig. 16. Structures of typical food hydrocolloids, carrageenans, xanthan gum, and guar gum.

Carrageenan is a water soluble phycocolloid extracted from red algae and is used mostly as a stabilizer and a structure provider in the food and ice cream industries. Three major varieties of carrageenans are known, which are κ-, ι-, and λ-carrageenans. They differ in the number and position of sulfate groups on the repeating garactopyranose units. To efficiently employ the food hydrocolloids as a source of the materials, the gelling system of κ-, ι-, and λ-carrageenans with BMIMCl was reported (Prasad, Kaneko et al., 2009). First, 13.0 wt% carrageenan was dissolved in BMIMCl by heating at 100 °C for 10 h with stirring. After the solution was cooled to room temperature, a gel was formed, which could suitably be handled after 12 h of standing upon addition of acetone and excess ionic liquid leaching out from the gel was washed out with ethanol. κ-Carrageenan gave the formation of hard gel while other two carrageenans gave the formation of softer gels with BMIMCl (Fig. 17). A schematic representation for the formation of the gels is shown in Fig. 18. At elevated temperatures, carrageenans are soluble in BMIMCl, and thus exist as random coils in the solutions. Upon cooling, three dimensional networks develop, leading to the formation of junction zones in the polysaccharide chains (Phase 1). The subsequent treatment of the mixture with acetone causes further aggregation and stabilization of the junction zones, particularly for the parts close to the surface, resulting in a stronger gelling system with the exclusion of BMIMCl (Phase 2). The excluded excess BMIMCl was washed out with ethanol to give the formation of the compact gels. The composite gels of carrageenans and cellulose with BMIMCl could be obtained by the same procedure as that described above. The mechanical properties of the carrageenan gels, as well as the composite gels with cellulose and BMIMCl, under the compressive mode were better than those of the hydrogels of κ- and ι-carrageenans and the cellulose gel with BMIMCl.

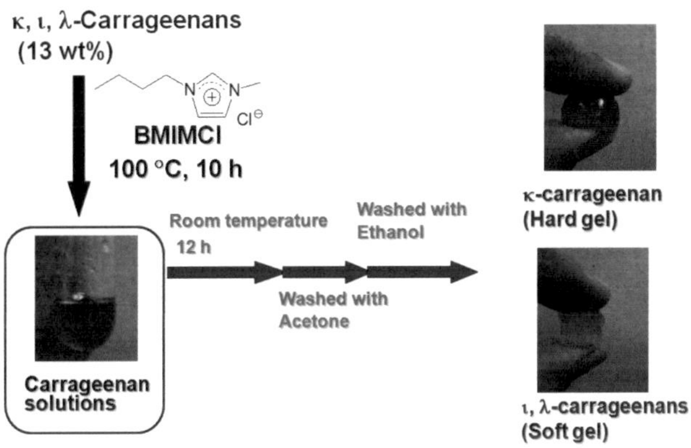

Fig. 17. Preparation of carrageenan gels with IL.

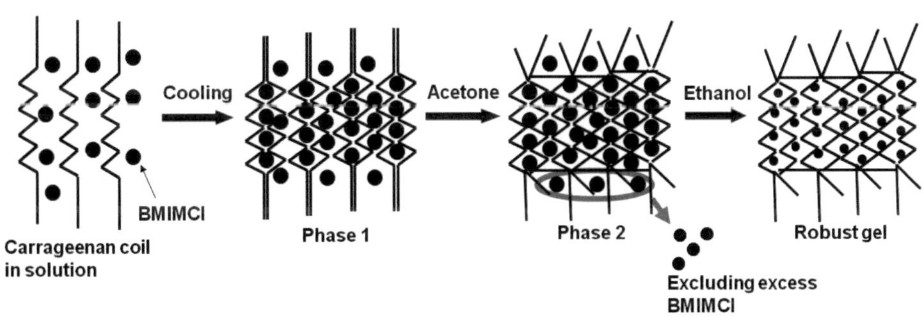

Fig. 18. Schematic representation for formation of carrageenan gels with IL.

Xanthan gum, which is a water soluble polysaccharide produced by *Xanthomonas campestris*, is also a useful food hydrocolloid. It has a cellulose main-chain with anionic trisaccharide side chains attached to alternating glucose units in the main-chain. It was found that xanthan gum/BMIMCl gels in various contents were obtained from the 9.1 - 50 wt% solutions of xanthan gum in BMIMCl (Izawa et al, 2009, Izawa & Kadokawa, 2010). When the solutions were left standing at room temperature for 30 min, they turned totally into the gel form (Fig. 19). The process of gelation was completely different from that for the aforementioned gels of cellulose and carrageenan with BMIMCl because long duration or treatment with appropriate organic solutions was required to generate the stable gels in their cases. It was considered that the ion-exchange of carboxylate metal cations in the original xanthan gum with BMIMCl took place during the gelation to form imidazolium-carboxylate pair in xanthan gum (Fig. 20). All the gels exhibited good mechanical properties under compressive mode. The 9.1 wt% gel showed elastic nature and the properties gradually changed to harder with increasing the contents of xanthan gum in the gels. It should be additionally noted that these gels did not have apparent fracture points. Interestingly, the gels had the sufficient strength to apply to the tensile test. The 9.1 and 16.7 wt% gels showed good elasticity, in which they were elongated to strain values near by 400%. In contrast, the 50 wt% gel had a hard nature and a fracture stress was 2.1 MPa. Furthermore, the xanthan gum/BMIMCl gels exhibited thermally induced shape-memory effect. First, a desired permanent shape of the gel was prepared by heating-cooling process using an appropriate mold. Subsequently, the gel was softened and deformed by heating at around 50 °C, and then the intended temporary shape was fixed by cooling the gel at room

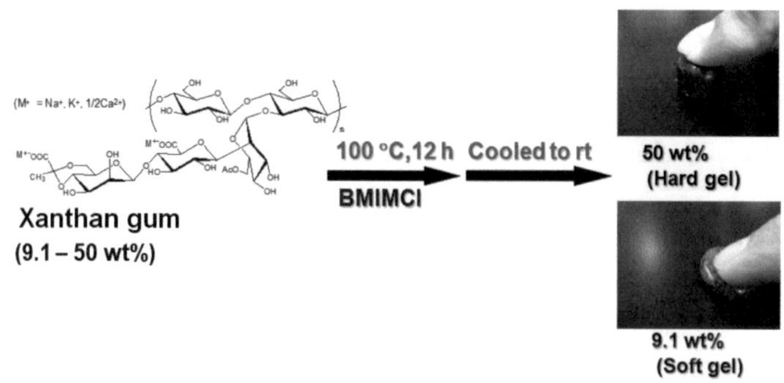

Fig. 19. Preparation of xanthan gum gels with IL.

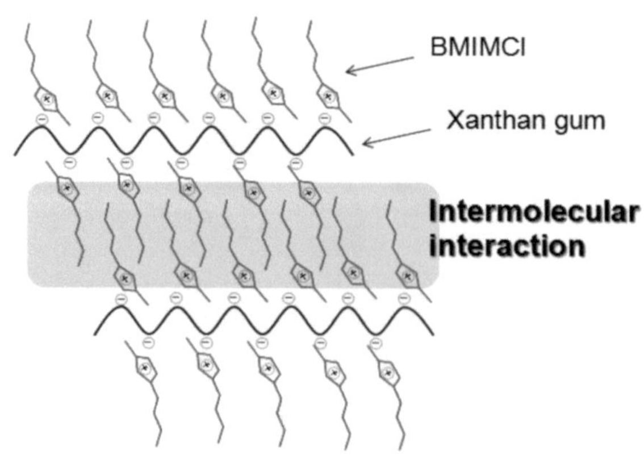

Fig. 20. Ionic exchange between xanthan gum and BMIMCl.

temperature. The temporary shape of the gel gradually returned back to the permanent shape by heating at around 50 °C. Consequently, the gel was almost recovered from temporary shape to permanent shape. It was supposed that the shape memory behavior of the gel was probably generated by the effect of ion-exchange during the gel formation and the rigid nature of the xanthan gum main-chain (Fig. 20).

The 9.1 wt% xanthan gum/BMIMCl gel was converted into a hydrogel of xanthan gum by soaking it (1.15 g) in water (100 mL) (Fig. 21). A water content of the obtained hydrogel (primary form) was estimated to be 94.1 wt%, indicating the formation of the xanthan gum hydrogel with high water content. This material was completely different from the aqueous xanthan gum dispersion because that was thixotropic gel-like paste. The elemental analysis data showed that the gel still contained 2.4 equivalents of the imidazolium groups to a repeating unit of xanthan gum. This is probably due to the presence of ion pair between the imidazolium and carboxylate formed by ion-exchange of carboxylate metal salts in the original xanthan gum with BMIMCl during the formation of the xanthan gum/BMIMCl gel. Further ion-exchange of the imidazolium groups in the primary formed hydrogel with Ca^{2+} was attempted by soaking it in 0.2 mol/L $CaCl_2$ aqueous solution to obtain an ionically cross-linked xanthan gum hydrogel with divalent cations (Ca^{2+} form) (Fig. 21). The water content of the resulting hydrogel was estimated to be 85.2 wt%, indicating shrinkage of this material compared with the primary hydrogel. In addition, the ionically cross-linked

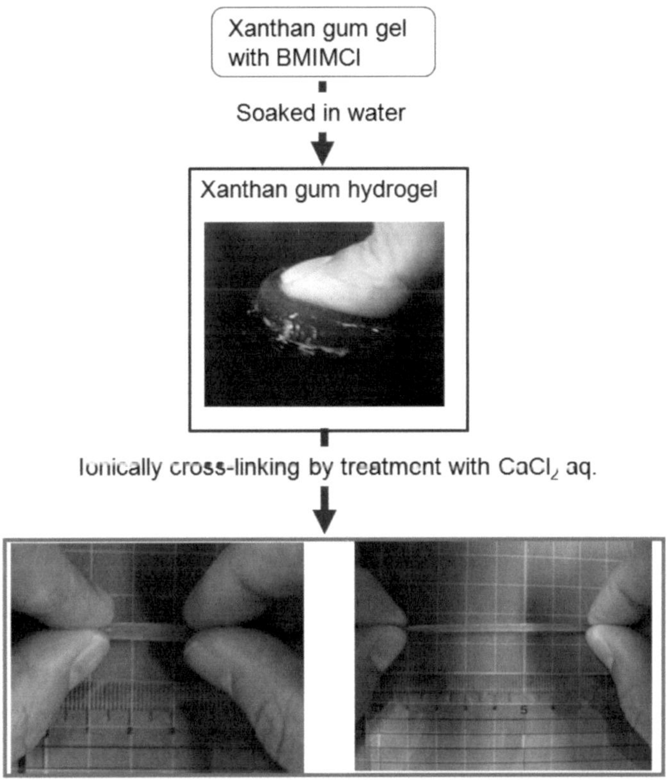

Fig. 21. Preparation of xanthan gum hydrogels from xanthan gum gel with IL.

hydrogel showed a reversible swelling-shrinking behavior. When the ionically cross-linked hydrogel was soaked in water for 1 day, it was gradually swollen to be 134.2% weight, which contained water of 24.4 times weights of xanthan gum. Soaking the swollen hydrogel in 0.2 mol/L $CaCl_2$ aqueous solution for 1 day caused rapid shrinkage to be 7.7% weight. The swelling-shrinking cycle of the hydrogel was repeated further twice. These results indicated that the present hydrogel exhibited salt concentration-induced responsive property. This behavior was probably caused by the conformation change of xanthan gum, which was induced by ionic strength of the aqueous solution. Under compressive mode, the hydrogel in the Ca^{2+} form showed much better mechanical property than the primary formed hydrogel. The primary hydrogel was too weak to be subjected to a tensile test, whereas the hydrogel in the Ca^{2+} form had sufficient strength for tensile testing, which exhibited elongation up to 550%. These results suggested that cross-linking with Ca^{2+} in the hydrogel strongly contributed to enhancement of mechanical properties. The ion-exchanged hydrogels with Na^+ and Fe^{3+} were also prepared by soaking the primary hydrogel in NaCl and $FeCl_3$ aqueous solutions, respectively.

The gel formation of food hydrocolloids with BMIMCl was not limited to the anionic polysaccharides, but extended to a neutral polysaccharide, which was guar gum (GG) (Prasad, Izawa et al., 2009). Guar gum (GG) is a galactomannan extracted from the seed of the leguminous shrub *Cyamopsis tetragonoloba* and consists of a (1→4)-linked β-D-mannopyranose main-chain with a branched α-D-garactopyranose unit at 6 position. As a typical procedure for the preparation of the gel of GG with BMIMCl (Fig. 22), first, GG (13.0 wt%) was dissolved in BMIMCl by heating at 100 °C for 5 h. After the solution was cooled to room temperature, it formed a gel, which was suitably handled after 12 h of standing upon addition of acetone and the excess BMIMCl leaching out from the gel was washed out with ethanol. However, the obtained gel started to leach out BMIMCl again after 12 h of preparation unlike other polysaccharide gels with BMIMCl described above. When the gel was further soaked in ethanol for 12 h, a stable hard material of GG with BMIMCl was obtained. The elemental analysis data of the material indicated that during the soaking process, most of BMIMCl was leached out from the gel matrices and subsequent aggregation of the GG chains existing on the surface area of the gel probably took place, causing stabilization of the material, because BMIMCl was soluble, but GG was insoluble in ethanol. The resulting hard material was further applied to the film formation. The hard material with an appropriate size was put on a compression machine and compressed applying gradually variable forces up to a maximum of ca. 10 MPa. Since expulsion of some BMIMCl from the material due to the compression was observed, the film thus obtained was further soaked in ethanol for 2 h to remove the excluded BMIMCl and dried at room temperature (Fig. 20). The film was stable and did not leach out BMIMCl even after long storage under ambient conditions. The tensile testing of the film showed a fracture stress of 17 MPa with a fracture strain of 18.2%. It was observed that the film became hard upon heating and hence it could be arranged to give a shape at higher temperature (Fig. 23). For example, the film was twisted to a desired shape and was heated to a temperature of ca. 70 – 80 °C. The film became hard immediately and retained the shape for a few minutes after the heating was stopped. When the film was then left standing at room temperature, it gradually became soft and returned to the original shape after ca. 15 minutes.

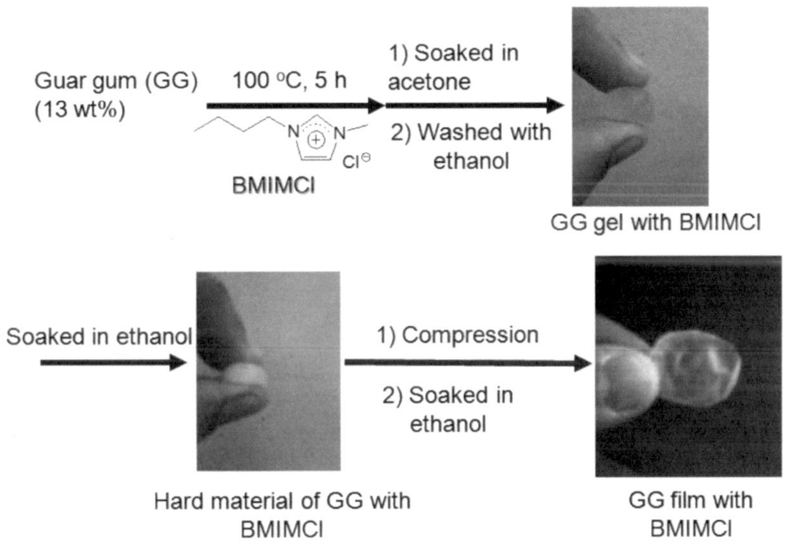

Fig. 22. Preparation of guar gum gel and film with IL.

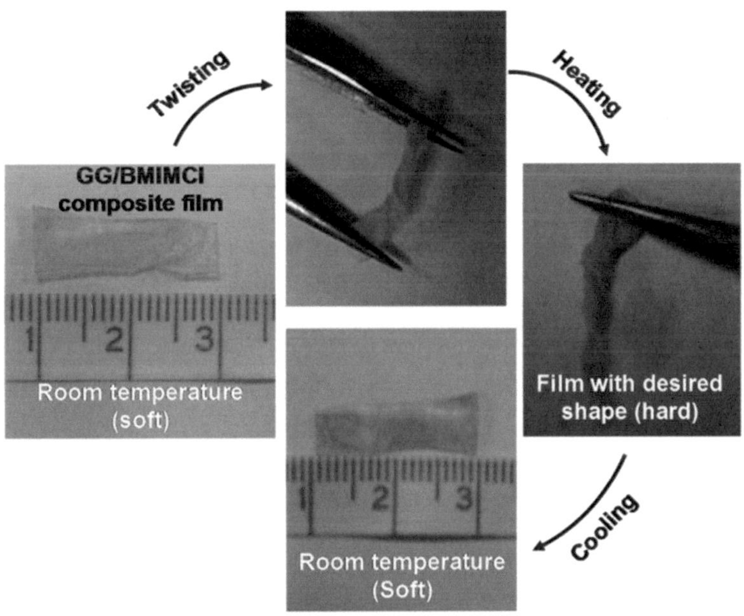

Fig. 23. Temperature-induced shaping ability of guar gum film with BMIMCl.

4. Conclusion

This chapter overviewed the preparation of polysaccharide-based materials compatibilized with ILs. Because of the solubility, processability, and feasibility problems of the polysaccharides due to their numerous hydrogen bonds, they have often showed difficulty in employing in a wide variety of materials. On the basis of the viewpoint that the imidazolium-type ILs have good affinity with the polysaccharides, the well-compatible materials of the polysaccharides with the ILs have been produced from not only abundant polysaccharides but also food hydrocolloids. This type of the research described herein provides new polysaccharide-based functional materials, which leads to efficient usage of the polysaccharides as an organic resource in the practical materials. Therefore, the studies in this research field will increasingly develop with attracting the much attention, which contributes to construction of green and sustainable society in the future.

5. References

Feng, L. & Chen. Z. -I. (2008). Research progress on dissolution and functional modification of cellulose in ionic liquids. *Journal of Molecular Liquids*, 142, 1-5, 0167-7322

Graenacher, C. (1934). Cellulose solution. *US patent*, 1943176

Green, O.; Grubjesic, S.; Lee, S. & Firestone, M. A. (2009). The design of polymeric ionic liquids for the preparation of functional materials. *Journal of Macromolecular Science, Part C: Polymer Reviews*, 49, 339-360, 1558-3724

Izawa, H.; Kaneko, Y. & Kadokawa, J. (2009). Unique gel of xanthan gum with ionic liquid and its conversion into high performance hydrogel. *Journal of Materials Chemistry*, 19, 6969-6972, 0959-9428

Izawa, H. & Kadokawa, J. (2010). Preparation and characterizations of functional ionic liquid-gel and hydrogel of xanthan gum. *Journal of Materials Chemistry*, 20, 5235-5241, 0959-9428

Kadokawa, J.; Murakami, M. & Kaneko, Y. (2008a). A facile method for preparation of composites composed of cellulose and a polystyrene-type polymeric ionic liquid using a polymerizable ionic liquid. *Composite Science and Technology*, 68, 493-498, 0266-3538

Kadokawa, J.; Murakami, M. & Kaneko, Y. (2008b). A facile preparation of gel materials from a solution of cellulose in ionic liquid. *Carbohydrate Research*, 343, 769-772, 0008-6215

Kadokawa, J.; Murakami, M.; Takegawa, A. & Kaneko, Y. (2009). Preparation of cellulose-starch composite gel and fibrous material from a mixture of the polysaccharides in ionic liquid. *Carbohydrate Polymers*, 75, 180-183, 0144-8617

Liebert, T. & Heinze, T. (2008). Interaction of ionic liquids with polysaccharides 5. Solvents and reaction media for the modification of cellulose. *BioResources*, 3, 576-601, 1930-2126

Murakami, M.; Kaneko, Y. & Kadokawa, J. (2007). Preparation of cellulose-polymerized ionic liquid composite by in-situ polymerization of polymerizable ionic liquid in cellulose-dissolving solution. *Carbohydrate Polymers*, 69, 378-381, 0144-8617

Pinkert, A.; Marsh, K. N.; Pang, S. & Staiger, M. P. (2009). Ionic liquids and their interaction with cellulose. *Chemical Reviews*, 109, 6712-6728, 0009-2665

Plechkova, N. V. & Seddon, K. R. (2008). Applications of ionic liquids in the chemical industry. *Chemical Society Reviews*, 37, 123-150, 0306-0012

Prasad, K.; Izawa, H.; Kaneko, Y. & Kadokawa, J. (2009). Preparation of temperature-induced shapeable film material from guar gum-based gel with an ionic liquid. *Journal of Materials Chemistry*, 19, 4088-4090, 0959-9428

Prasad, K.; Kaneko, Y. & Kadokawa, J. (2009). Novel gelling systems of κ-, ι- and λ carrageenans and their composite gels with cellulose using ionic liquid. *Macromolecular Bioscience*, 9, 376-382, 1616-5187

Prasad, K.; Murakami, M.; Kaneko, Y.; Takada, A.; Nakamura, Y. & Kadokawa, J. (2009). Weak gel of chitin with ionic liquid, 1-allyl-3-methylimidazolium bromide. *International Journal of Biological Macromolecules*, 45, 221-225, 0141-8130

Prasad, K. & Kadokawa, J. (2010). Preparation of composite materials composed of ι-carrageenan and polymeric ionic liquids. *Polymer Composites*, 31, 799-806, 0272-8397

Prasad, K.; Mine, S.; Kaneko, Y. & Kadokawa, J. (2010). Preparation of cellulose-based ionic porous material compatibilized with polymeric ionic liquid. *Polymer Bulletin*, 64, 341-349, 0170-0839

Schuerch, C. (1986). Polysaccharides. *Encyclopedia of polymer science and engineering, 2nd edition*, Vol. 13, pp. 87-162, John Wiley & Sons, New York

Seoud, O. A. E.; Koschella, A.; Fidale, L. C.; Dorn, S. & Heinze, T. (2007). Applications of ionic liquids in carbohydrate chemistry: A windows of opportunities. *Biomacromolecules*, 8, 2629-2647, 1525-7797

Swatloski, R. P.; Spear, S. K.; Holbrey, J. D. & Rogers, R. D. (2002). Dissolution of cellulose with ionic liquids. *Journal of the American Chemical Society*, 124, 4974-4975, 0002-7863

Takegawa, A.; Murakami, M.; Kaneko, Y. & Kadokawa, J. (2009). A facile preparation of composites composed of cellulose and polymeric ionic liquids by in situ polymerization of ionic liquids having acrylate groups. *Polymer Composites*, 30, 1837-1841, 0272-8397

Takegawa, A.; Murakami, M.; Kaneko, Y. & Kadokawa, J. (2010). Preparation of chitin/cellulose composite gels and films with ionic liquids. *Carbohydrate Polymers*, 79, 85-90, 0144-8617

Welton, T. (1999). Room-temperature ionic liquids. Solvents for synthesis and catalysis. *Chemical Reviews*, 99, 2071-2083, 0009-2665

Yamazaki, S.; Takegawa, A.; Kaneko, Y.; Kadokawa, J.; Yamagata, M. & Ishikawa, M. (2009). An acidic cellulose-chitin hybrid gel as novel electrolyte for an electric double layer capacitor. *Electrochemistry Communications*, 11, 68-70, 1388-2481

Yamazaki, S.; Takegawa, A.; Kaneko, Y.; Kadokawa, J.; Yamagata, M. & Ishikawa, M. (2010a). Performance of electric double-layer capacitor with acidic cellulose-chitin hybrid gel electrolyte. *Journal of Electrochemical Society*, 157, A203-A208, 0013-4651

Yamazaki, S.; Takegawa, A.; Kaneko, Y.; Kadokawa, J.; Yamagata, M. & Ishikawa, M. (2010b). High/low temperature operation of electric double layer capacitor utilizing acidic cellulose-chitin hybrid gel electrolyte. *Journal of Power Sources*, 195, 6245-6249, 0378-7753

Zhu, S.; Wu, Y.; Chen, Q.; Yu, Z.; Wang, C.; Jin, S.; Ding, Y. & Wu, G. (2006). Dissolution of cellulose with ionic liquids and its application: a mini-review. *Green Chemistry*, 8, 325-327, 1463-9262

Polymerization of Cyclodextrin-Ionic Liquid Complexes for the Removal of Organic and Inorganic Contaminants from Water

Mphilisi M Mahlambi, Tshepo J Malefetse,
Bhekie B Mamba and Rui WM Krause
University of Johannesburg, Department of Chemical Technology, Doornfontein, South Africa

1. Introduction

The removal of toxic contaminants from water still remains a huge challenge for water supplying companies and municipalities. Both organic and inorganic (mostly heavy metal) pollutants are often present in water distribution networks as a result of agricultural, domestic and industrial operations. To remove these pollutants from water distribution networks, effective techniques need to be developed. In our laboratories, research that involves the use of polymers of cone-shaped cyclodextrins (CDs) interconnected by linkers has been successfully demonstrated in the removal of organic pollutants from water. We have now undertaken a programme whereby polymers possessing moieties capable of removing both organic and inorganic contaminants from water, the so-called multifunctional complexes will be synthesised. To access these multifunctional complexes, CDs were attached to salts of organic cations (i.e. imidazolium and pyridinium) in a manner similar to the formation of ionic liquids (ILs). These cyclodextrin-ionic liquid (CD-IL) derivatives were then polymerised with bifunctional linkers with the expectation that the resulting polymers will be able to perform the dual role of removing a wide range of both organic and inorganic pollutants from water.

Pollutants from many industries often end up in water systems; other pollutants emanate from domestic and agricultural run-offs. Organic pollutants such as organic solvents, pesticides, fungicides and inorganic pollutants such as nitrates, cyanides and dissolved gases are present in many water systems (Schwarzenbach et al, 2003). When these pollutants are introduced into the environment, they are subjected to physical, chemical and biological processes, leading in the generation of derivatives possessing different environmental behaviour and effects (Schwarzenbach et al, 2003). These contaminants may leach into the soil and contaminate ground water. They also tend to bioaccumulate in the bodies of organisms. The small size and chemical inertness (or unreactiveness) of these pollutants make their removal from water using the current water treatment methods difficult (Li and Ma, 1999). Also, they have health effects such as kidney and liver damage, the destruction of nerve and circulatory systems and other psychological effects detrimental to human health. Since communities need ultra pure water, there is a need to remove pollutants even at very low concentrations to ensure the health and safety of water users.

Recent studies in our laboratories have involved the synthesis and application of standard (i.e unfunctionalised) and functionalised CD polymers in the removal of organic contaminants from water at concentration levels as low as ppb (μgL^{-1}). CDs can form inclusion complexes with organic compounds and thereby facilitating their removal. However, to apply CDs in the removal of organic pollutants from water, they need to be converted to water-insoluble polymers. Ionic liquids (ILs) on the other hand have demonstrated the ability to extract heavy metal ions from water. Although technologies for removing inorganic species from water are well known, to the best of our knowledge there is no general technique yet developed that is capable of removing organic and inorganic species from water simultaneously. Therefore, the use of CDs linked to ionic liquid appendages is envisaged to provide a dual system capable of removing both organic and inorganic pollutants from water at the same time. While the CD moiety in the polymer will retain its traditional role of absorbing organic contaminants, the IL component will perform the role of extracting the heavy metals present in water.

This work seeks to extend the application of cyclodextrin polymers to incorporate the removal of inorganic pollutants from water. The objectives of this study were the following:

i. To synthesise and characterise 6-deoxy-6-(*p*-tosyl)-*β*-cyclodextrin (CDOTs) and 6-deoxy-6-iodo-*β*-cyclodextrin (CDI) precursors. The aim here is to establish and optimize a reliable reaction path for the synthesis of CD-IL precursors. Characterisation of the CDOTs and the CDI precursors were carried out mainly by IR and NMR spectroscopy.
ii. To synthesise and characterise CD-IL complexes by reacting CDOTs and CDI precursors synthesised in (i) with 3-alkylimidazole and pyridine. The CD-IL complexes were characterised with IR and NMR spectroscopy.
iii. To polymerise the CD-IL complexes generated in (ii) with suitable bifunctional linkers such as hexamethylene diisocyanate and toluene-2,4-diisocyanate. IR spectroscopy was used for the characterisation of the polymers.
iv. To study the thermal stability of the CD-IL polymers using thermogravimetric analysis (TGA) and differential scanning calorimetry (DSC).
v. To study the morphology of the polymers using scanning electron microscopy (SEM) and Brunauer-Emmet-Teller (BET).
vi. To test the ability of the CD-IL polymers to remove organic and inorganic pollutants from water by performing model contamination experiments. These experiments involve deliberately placing the polymers into an aqueous solution containing known concentrations of the contaminants. GC-MS and UV-Visible spectroscopy were used to quantify the degree of absorption of the organic contaminants. Extraction of the inorganic contaminants was quantified by using atomic absorption (AA) spectroscopy.

A schematic representation for the synthesis of CD-IL complexes and CD-IL polymers are shown in Scheme 1.1.

2. Literature review

2.1 Organic pollutants

The problems associated with the use of the conventional water treatment techniques for the removal of these pollutants from the water systems and the approaches undertaken in our laboratories to address these problems are also described. The chapter then concludes with an overview of the applications of cyclodextrins and ionic liquids in the attempt to remove organic and inorganic contaminants from drinking water systems. Organic pollutants in

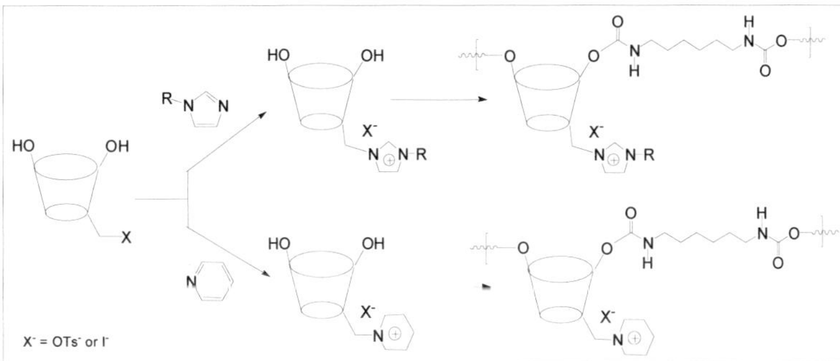

Scheme 1.1. Synthetic pathway for the preparation of CD-IL polymers

water pose a major threat to human health even when present at low concentrations. Some organic pollutants are referred to as persistent organic pollutants (POPs) because they remain in the environment for long periods of time. POPs can be absorbed by plants or animals thereby finding their way to poisoning food chains of living organisms (Oleszczuk and Baran, 2004). They have also been linked to adverse human effects such as cancer, nervous system damage, reproductive disorders, as well as the disruption of the immune system (EPA, 2006). POPs released in one part of the world can be transported globally *via* oceans and atmospherically, and their effects can be felt in regions which are distant from where they originate. POPs are mostly from human activity but there are other organics that are also of major concern.

Also, drinking water disinfection by-products pose a health risk to humans. Chlorine is still widely used for water disinfection. A residual amount of free chlorine (0.5-1.0 ppm) can keep levels of micro organisms such as bacteria to their lowest (Halmowitz, 1991). However, during chlorination, chlorine combines with water to form hypochlorous acid. Hypochlorous acid in turn reacts with NOM to form DBPs. The low volatility of DBPs, combined with their extreme stability, has led to their high persistence both in the environment and in the bodies of organisms after exposure (Gardio et al., 2006). They are mainly found in fatty tissues and bioaccumulate in food chains. DBPs are known to induce or aggravate adverse health problems in humans such as cancer, liver damage, failure of immune systems and the disruption of hormonal functions (Gardio et al., 2006). Due to the problems associated with DBPs, it is imperative to find better techniques to remove both NOM and its by-products from drinking water systems.

Phenolic compounds are common constituents of many industrial and commercial wastes (Rogers et al., 1999; Wikström and Marklund 2000). They may be introduced into the environment by accidental spills, illegal release of industrial and municipal wastewater, and excessive use of pesticides (Park et al., 1999). Phenolic compounds can be easily incorporated into the soil organic matter or dissolved humic materials because they are highly soluble in water. The presence of minute amounts of some phenolic compounds in water causes bad tastes and odours, which are greatly aggravated by the usual addition of chlorine for germicidal purposes (Niegowski, 1952; Kawaguchi 2004). Chlorinated and nitrated phenols pose a health hazard to both humans and animals. For instance, *para*-nitrophenol (PNP), though slightly soluble in water, is regarded as a possible mutagen to humans. 2,4,6-Trichlorophenol (TCP) is also known to affect the respiratory system and is a recognised carcinogen (NPI, 2007). Examples of important nitrated and chlorinated phenols are shown in Figure 2.1.

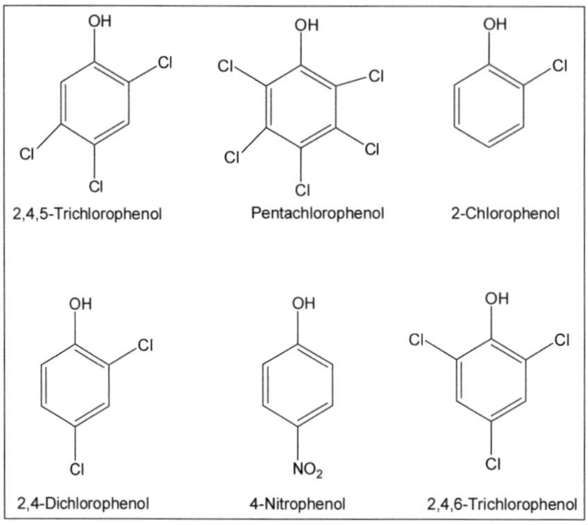

Fig. 2.1. Structures of chlorinated and nitrated phenols

The removal of phenolic compounds from industrial wastewater is currently an environmental and economic challenge. This effluent is resistant to conventional water treatment techniques due to the poor biodegradability and toxicity of phenolic compounds (Santos et al., 2007; Sabio et al., 2006). Moreover, it is also expensive to use these water treatment technologies to remove these organic pollutants at ppb levels ($\mu g.L^{-1}$).

2.2 Current water treatment methods for organic pollutants

Water treatment technologies used for the removal of organic species around the world include filtration using activated carbon, reverse osmosis, zeolites, and the use of dendrimers (Li and Ma, 1999). However, these techniques often do not remove organic contaminants to accepted levels. Due to the failure of most of the aforementioned water treatment techniques to remove organic pollutants from water, the world is facing challenges in meeting the rising demand for clean water. This is because the sources of available fresh water supplies are decreasing due to extended droughts, population growth and more stringent health-based regulations (Savage and Diallo, 2005). Also, water scientists are questioning the viability of current water treatment practices in meeting the water demands of all users according to regulations by the World Health Organisation (WHO). Since the current water treatment techniques fail to remove pollutants to the required levels due to their diversity and varying chemical properties, better technologies need to be developed. In our laboratories the use of cyclodextrin polyurethanes has been studied. These CD polymers have been found to have a high efficiency for the removal of organic pollutants from water, even at ppb levels (Mhlanga et al., 2007). Cyclodextrin polyurethanes are discussed in the following sections.

2.3 Cyclodextrins

Cyclodextrins (CDs) were first discovered by Villiers in 1891 (Szeijtli, 1998). CDs are cyclic oligomers formed by the enzymatic hydrolysis of starch by *Bacillus macerans*. The glucose units in a CD molecule are linked by α-(1,4) linkages (Bender and Komiyana, 1978; Croft and Bartsch 1983). Figure 2.2 illustrates the *a*-(1, 4) linkages and the numbering system in a typical CD moiety.

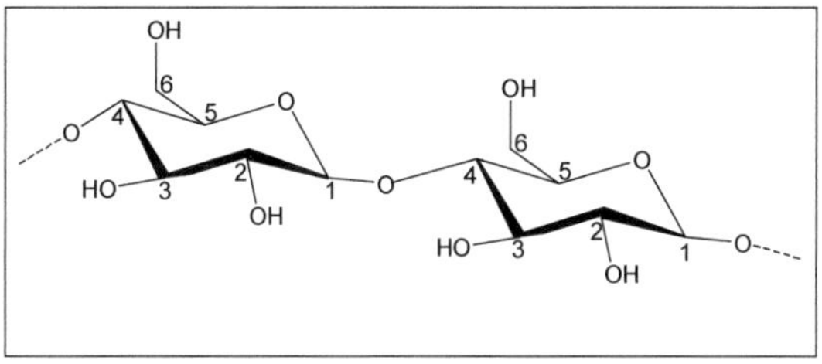

Fig. 2.2. α-(1, 4) linkages and the numbering system in a CD molecule

2.3.1 Physical properties of cyclodextrins

The most common CDs are made of six, seven or eight glucose units and are called α-, β- and γ-CDs, (Szeijtli, 1998) respectively. The β-CD has been extensively studied by our research group and other researchers and was as a result used in this study. Higher molecular weight CDs are known, however their use is not common even in industry (Martin, 2004). Figure 2.3 shows the three common types of CDs.

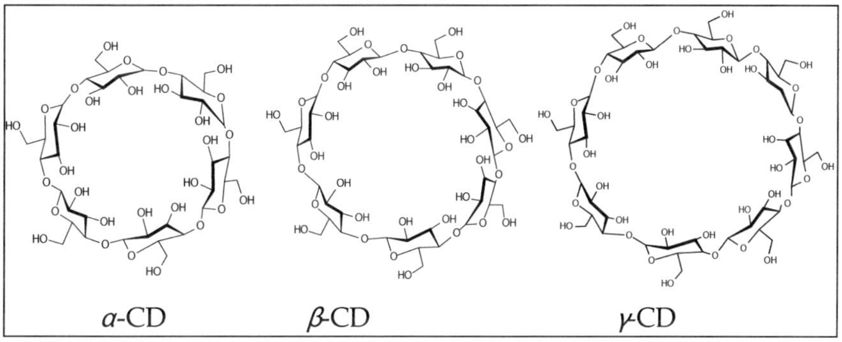

Fig. 2.3. Structures of α-, β- and γ-CDs, respectively showing the glucose units

CDs have a characteristic toroidal shape that form well defined cavities (Bender and Komiyana, 1978). These cavities are about 8Å deep. Depending on the number of the glucose units, the diameter of these cavities range between 5 and 10 nm. They have a hydrophobic interior because of the presence of carbon and hydrogen atoms and this feature allows them to host several compounds in their cavities. Their exterior cavities are hydrophilic because of the presence of hydroxyl groups and this makes them soluble in water. Figure 2.4 shows a representation of CD moiety with a hydrophilic exterior and a hydrophobic cavity.

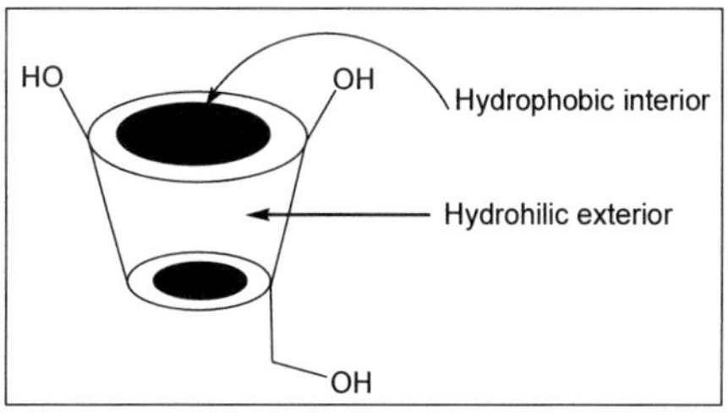

Fig. 2.4. Hydrophobic interior and hydrophilic exterior of a CD

2.3.2 Formation of inclusion complexes

Cyclodextrins can form inclusion complexes with various compounds (guests) of low molecular weight (Figure 2.5) (Li and Tan, 2001; Phan et al., 2002). Examples of the guest molecules include acids, apolar aliphatic and aromatic hydrocarbons and amines. The CD moiety harbours these small, suitably shaped organic compounds in its tubular cavities by shielding the bound species from the surrounding aqueous environment (Weinz et al., 1997). This phenomenon is a result of the hydrophobic-hydrophobic interaction between the host CD and the organic species (Liu et al., 2003; Weinz et al., 1997). The formation of inclusion complexes does not involve the formation of bonds but is an attraction between the host and guest as a result of their polarities (Szeijtli, 1998).

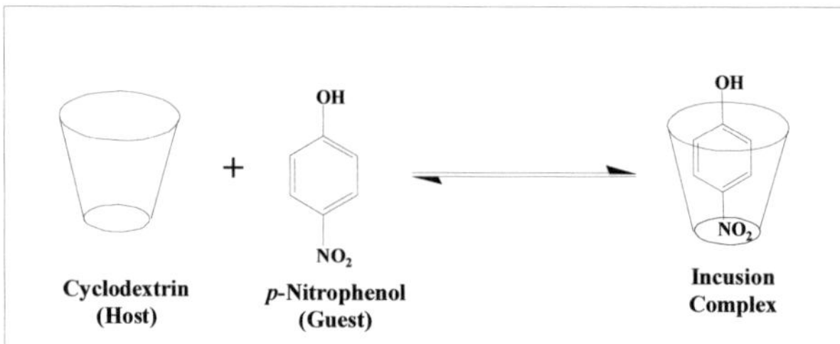

Fig. 2.5. Schematic representation of an inclusion complex

Because the lengths of the diameter of the CD vary, the organic species must be able to fit (at least partly) into the CD cavity (compatible geometry). The size of the organic compound and the type of the CD used is important for the formation of inclusion complexes. Direct exploitation of this phenomenon for water treatment is however limited by the solubility of CDs in water. Hence, making insoluble CD derivatives and polymers is a prerequisite if CDs are to be used in water treatment.

2.3.3 Cyclodextrin derivatives

The presence of hydroxyl groups in the outer cavity can be exploited in order to alter the physical and chemical properties of CDs (Li and Ma, 1999; Szeijtli, 1998). For example, functionalising CDs can decrease their solubility, increase their stability in the presence of light, heat and oxidizing agents and also make them more volatile (Harada, 1997; Liu et al., 2003). The primary hydroxyl groups of a CD molecule (C-6 OHs) are more reactive compared to their secondary hydroxyl groups (C-2, C-3 OHs). These hydroxyl groups are shown in Figure 2.6. The C-2 hydroxyl groups are acidic. The C-3 hydroxyl groups are less

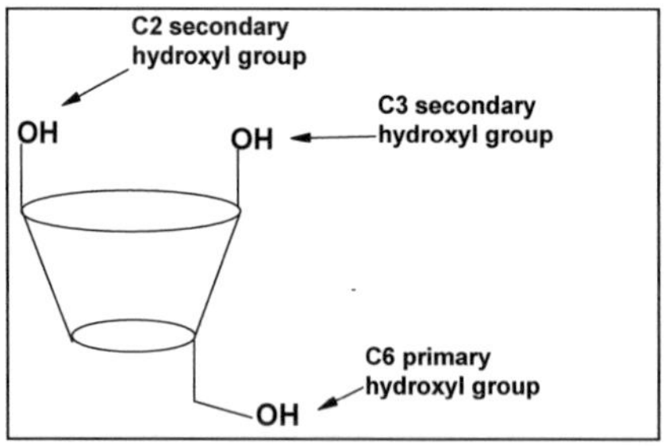

Fig. 2.6. The structure of a CD showing the three different hydroxyl groups.

reactive because they are not easily accessible due to sterric hindrance. The primary hydroxyl groups at C-6 are basic hence an acid is used to remove the H in the OH, i.e.

$$R{-}OH + OH^{-} \rightarrow R{-}O^{-} + H_2O \quad (1)$$

and are the most easily accessible thus often more reactive during nucleophilic substitution reactions.

2.3.3.1 Monofunctionalisation of CDs

The purpose of functionalising CDs is to modify their physico-chemical properties and also to introduce groups with specific activity. The two most common types of CD substitution reactions are mono- and per-functionalisation reactions. Monofunctionalisation refers to the reaction of only one hydroxyl group while per-functionalisation involves the whole set of hydroxyl groups. In monofunctionalisation, regioselective reactions should be the target (Sutyagin et al., 2002). Regioselective reactions usually involve the protection of other OH groups to prevent them from participating in the reaction while exposing the target group to reactions. Monofunctionalisation at C-2 and C-3 is well documented (Rong and D'Souza, 1990; Fugita et al., 1984). In this study monofunctionalisation at C-6 was mainly exploited. Monofunctionalisation of C-6 can be achieved because in the presence of a base (e.g. NaOH), the proton at C-6 is removed to make O^{-} which is a nucleophile that can be easily substituted. Thus when a CD is reacted with a tosyl group in a basic medium, a monofunctionalised CD derivative (CDOTs) is afforded (Byun et al., 1998; Muderawan et al., 2005; Zhong et al., 1998). The CDOTs is an important precursor for all C-6 CD derivatives because the tosylate is a good leaving group and can be easily replaced by other nucleophiles (e. g. halogens, azides and acetates). In this study, the CDOTs was also used as a precursor for the CD derivatives that were synthesised.

2.3.4 Cyclodextrin polymers

The solubility of CDs and their derivatives limits their application in water treatment. Converting them to water insoluble entities is therefore crucial if CDs are to be utilised in the removal of contaminants from water. To attain this property, CD derivatives have been polymerised with suitable bifunctional linkers in our laboratories to produce water insoluble polymers (Mhlanga et al., 2007; Mamba et al., 2007; Salipira et al., 2007). The presence of the OH groups in the CDs or CD-derivatives allows a reaction in which the monomers react with linkers to produce insoluble polymers. A reaction scheme for the polymerisation of CDs with diisocyanate linkers (hexamethylene diisocyanate (HDI)) is shown in Figure 2.7 (Li and Ma, 1999).

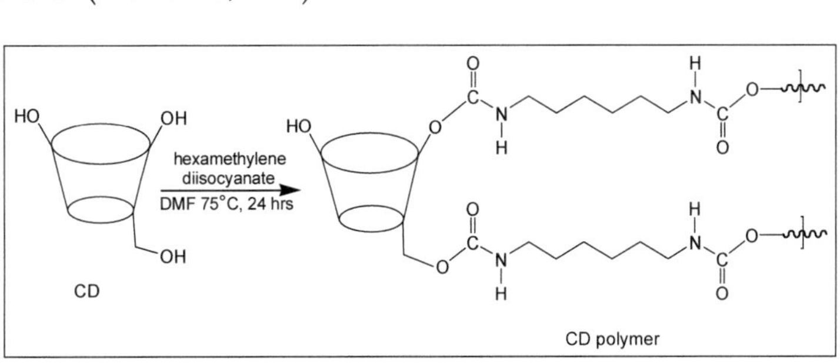

Fig. 2.7. Polymerisation of CDs with an isocyanate linker (HDI)

The CD polyurethanes have demonstrated a high efficiency for the absorption of organic pollutants in water (Mhlanga et al., 2007; Mamba et al., 2007; Salipira et al., 2007). However, in real water systems organic and inorganic contaminants co-exist; the challenge is to craft a water treatment system that is capable of simultaneously removing these contaminants from water. In an attempt to overcome this challenge, we functionalised CDs with ionic liquids (ILs). Since ILs are known to chelate heavy metals from aqueous media and CDs are able to remove organic pollutants from water, a system which combines CDs and ionic liquids is expected to absorb both organic and inorganic pollutants from water.

2.4 Inorganic pollutants in water

Inorganic contaminants are compounds that typically do not contain the element carbon in their structure (waterborne, 2006). Inorganic pollutants are introduced into the environment naturally, for example, from weathering of parent rocks, industrial wastes and agricultural run-offs (waterborne, 2006; Petanen and Romantschuk, 2002). Inorganic pollutants include dissolved gases (e. g. oxygen, carbon dioxide, nitrogen and radon), heavy metals (e. g. Hg^{2+}, Cd^{2+}, Cr^{6+}) and anions such as fluorides (F^-), nitrates (NO_3^-) and cyanides (CN^-). These contaminants also pose health risks to living organisms because they bioaccumulate in the body resulting in chronic health effects (Morales et al., 1999). Inorganic contaminants are not biodegradable; hence they can typically poison the food chains of living organisms (waterborne, 2006).

2.5 Ionic liquids

Ionic liquids (ILs) are salts of organic cations which have aroused an increasing interest for their promising role as alternative media for volatile solvents (Wei et al., 2003). Ionic Liquids (ILs) are low melting point molten salts composed of organic cations (e.g. pyridinium and imidazolium ions) and mostly inorganic anions like Cl^-, Br^-, $[PF_6]^-$ and $[BF_4]^-$. Compared to conventional salts, ILs are liquid at room temperature (Wei et al., 2003). The most common types of organic cations are the imidazolium and pyridinium ions. Examples of the structures of ionic liquids are depicted below in Figure 2.8.

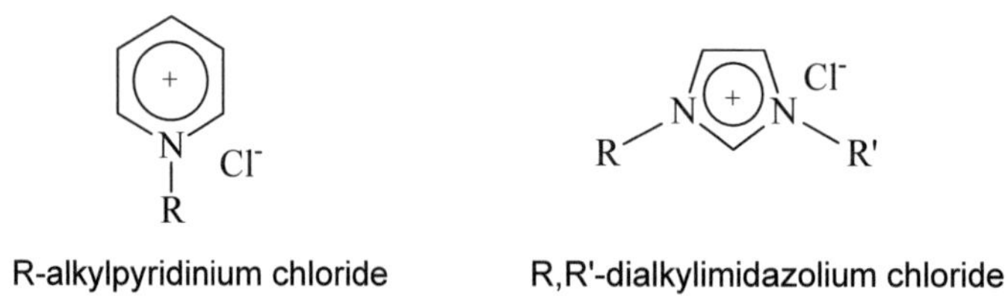

Fig. 2.8. Examples of ionic liquids

The organic cationic component and the alkyl substituents are known to be responsible for the low melting points associated with these compounds. The ability to dissolve organic solvents has also been linked to the cationic component and the alkyl substituents since both of them are organic in character (Welton, 1999). The alkyl group is usually an alkane chain whereby increasing the length of the alkane chain one can change the resulting properties (e.g., viscosity, hydrophobicity, and melting point) (Ferreira et al., 2000). Besides pyridinium and imidazolium cations, other less common organic cations have been used for the preparation of ionic liquids. These include ammonium, pyrrolidium, sulfonium and phosphonium ions (Figure 2.9) (Pereiro et al., 2007).

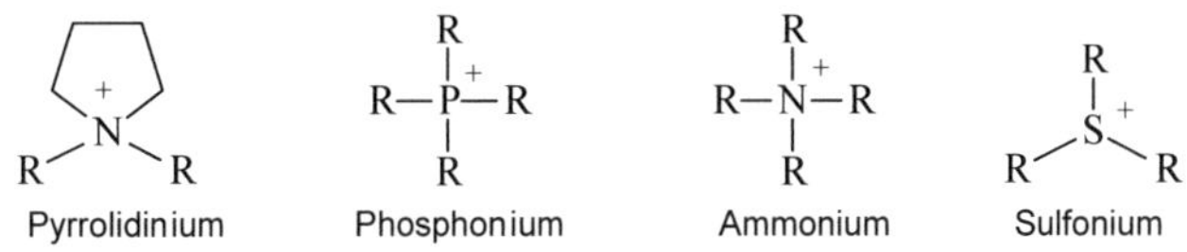

Fig. 2.9. Less common organic cations used for the synthesis of ILs.

2.5.1 Synthesis of ILs

Ionic liquids are made up of both a cation and an anion, whose forces of attraction are not sufficiently strong to hold them together as solids at ambient temperatures. It is therefore possible, by proper choice of starting material, to synthesize ionic liquids that are liquid at or below room temperature (Ferreira et al., 2000). Volatile organic chemicals (VOC's) are used to prepare non-volatile ionic liquids which have negligible vapour pressure; therefore new and better methods need to be researched since many processes used for cleaning ionic liquids involve washing with water and VOC's. This will also help in terms of eliminating or decreasing the hazards in the system and eliminate the waste before it is produced. The two most common methods of synthesising ionic liquids are metathesis reaction involving a halide salt and acid-base neutralisation reactions (Welton, 1999). For example, the metathesis reaction of 1-ethyl-3-methylimidazolium iodide with ammonium tetrafluoroborate is shown in Scheme 2.1.

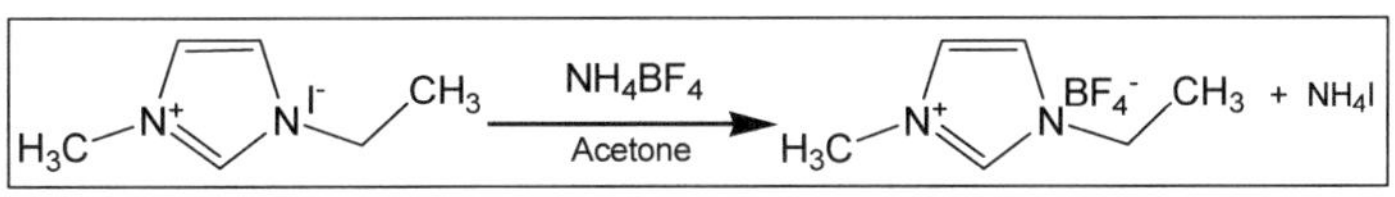

Scheme 2.1. Synthetic pathway for 1-ethyl-3-methylimidazolium tetrafluoroborate

Also, a reaction an alkyl halide with pyridinium and imidazolium cations yields ionic liquids, as shown in Scheme 2.2.

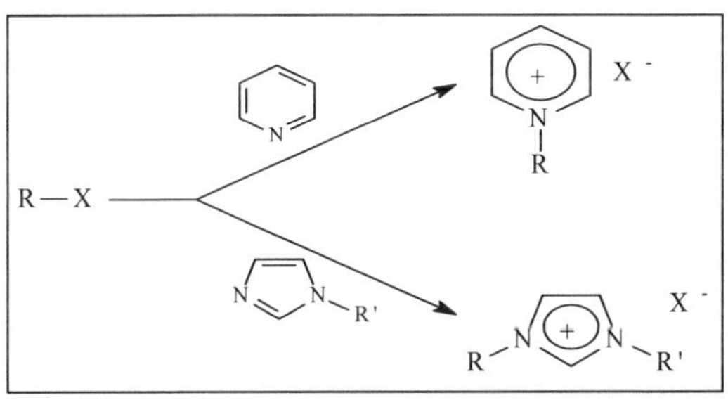

Scheme 2.2. Pathway for the synthesis of pyridinium and imidazolium based ILs

The anions, to a great extent influence both the chemical and physical properties of the ionic liquids. For example, although both anions are air- and moisture-stable, $[BF_4]^-$ and $[PF_6]^-$ anions are used for the preparation of water-miscible and water-immiscible ionic liquids, respectively (Ferreira et al., 2000). Besides anions such as hexafluorophosphate, tetrafluoroborate and halide ions, other common inorganic anions that are used for the preparation of ILs include $[SbF_6]$ and $[(CF_3SO_2)_2N]^-$. The anion need not necessarily have to be inorganic; ILs possessing organic anions such as alkylsulfate, tosylate and methanesulfonate are known (Figure 2.10) (Sigma-aldrich, 2007).

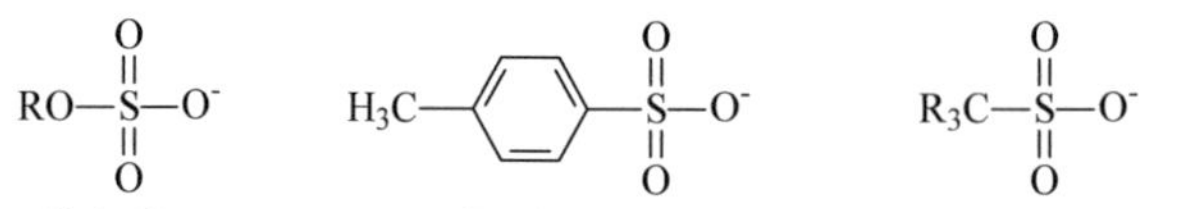

Fig. 2.10. Examples of organic anions used for the synthesis of ILs.

An example of such an IL (ethylpyridinium tosylate) is shown in Figure 2.11.

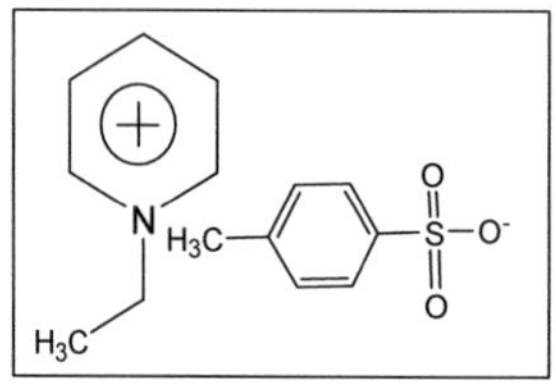

Fig. 2.11. Structure of ethylpyridinium tosylate

2.7.2 Physical properties of ILs

The physical properties of ionic liquids depend on the nature and the size of their cationic and anionic species (Visser and Rogers, 2004). These properties relate to the choice of the two ionic components (Ferreira et al., 2003; Visser and Rogers, 2004). Hence, ILs are sometimes referred to as "designer solvents". ILs are gaining a wide recognition in analytical chemistry because they possess several unique properties (Liu et al., 2004). These physical properties include negligible vapour pressure (Liu et al., 2003), an ability to dissolve a wide range of organic and inorganic materials (Welton, 1999), high thermal stability (they have low melting points and high boiling points, for example, some ILs are liquid at 400°C, while others are liquid at -96°C) (Liu et al 2005) as well as variable viscosity and miscibility with water and other organic solvents (Liu et al., 2003). These physical properties have triggered an interest in applying ionic liquids in separation chemistry and synthesis. In addition, these properties form the basis for the utilisation of ILs in replacing traditional toxic, flammable and volatile organic compounds (VOCs), which are not environmentally friendly. To this end, ILs have the potential to be used as alternative reaction media for "Green Chemistry" (Wei et al., 2003).

2.7.2.1 Hydrophobic and hydrophilic ILs

The hydrophilic and hydrophobic nature of ILs depends on the type of anion used in the synthesis of the target ionic liquid (Welton, 1999; Cocalia et al., 2006). This characteristic feature determines the extent of miscibility of ILs with water and other organic solvents. It also provides an alternative for the use of ILs in nonaqueous, polar phase separation systems, i.e. liquid-liquid extraction (Welton, 1999). For instance, hydrophobic ILs can be used as immiscible polar phases with water. Examples of common hydrophobic and hydrophilic anions are shown in Figure 2.11. In this study, one hydrophobic (tosylate) and one hydrophilic (iodide) anion were utilised in the synthesis of ionic liquids.

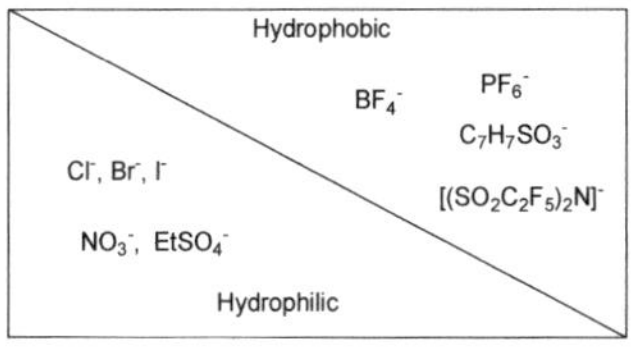

Key: $[PF_6]^-$ = hexafluorophosphate, $[BF_4]^-$ = tetrafluoroborate, $[C_7H_7SO_3]^-$ = tosylate $[(SO_2C_2F_5)_2N]^-$ = bis (trifyl) imide and $[EtSO_4]^-$ = ethylsulfonate.

Fig. 2.11. Examples of hydrophobic and hydrophilic IL anions

2.7.3 Chemical properties of ILs

Ionic liquids have been widely considered as greener solvents which are suitable for application in a wide range of organic chemical reactions (Chowdhury et al., 2007). Applications of ILs include control of product distribution, enhanced reactivity rates, recyclability, catalysis and ease of recovering products from a reaction mixture (Chowhury et al., 2007). Further, they have shown to possess much needed properties such as inflammability and less volatility. Thus they are emerging as novel replacements for the widely used VOC's which are detrimental to the environment (Liu et al., 2003; Cruz and Moens, 2000). ILs are generally referred to as polar solvents because they are able to dissolve and stabilise dipolar and charged solutes (Welton, 1999). The localisation of the anions and the delocalisation of the cationic charge also influence the polarity of ILs. Ionic liquids can be applied in battery industries since they are good electrolytes and good solvents for Friedel-Crafts alkylations, hydrogenations and Diels-Alder reactions, to mention just a few (Welton, 1999; Cruz and Moens, 2000). However, the reactivity of ILs is also affected by the acidity and basicity of the IL under investigation.

2.7.3.1 Acidic and basic ILs

The acidic and basic nature of ionic liquids is largely determined by the ability of the salt to act as a hydrogen bond donor (acid) or acceptor (base) (Welton, 1999; Chowdhury et al., 2007). Incorporation of an aliphatic carboxylic group to the cationic component of an IL can result in the formation of an acidic ionic liquid. Also, reacting the cationic component with an acidic anion, e. g. $[HSO_4^-]$, results in an acidic IL (Chowdhury et al., 2007). Increasing the acidity of the anion has been found to improve the catalytic activity of the IL. On the other hand, basic ILs are a result of combining the cationic component of the IL with a basic anion, e. g. $[OH^-]$. Knowing the acidic and basic nature of an IL is important because it allows for the proper choice of an IL for a specific application. This has led to a tailored synthesis of task specific ionic liquids (TSILs).

2.7.4 Applications and uses of ILs

The application of any ionic liquid depends on the purpose for which it is designed. ILs that have been specifically designed for use in catalysis, chromatography, spectrometry, electrochemical applications, as chemo sensors and extraction solvents have been reported in a number of publications (Visser and Rogers, 2004; Liu et al., 2004; Cruz and Moens, 2000). Task specific ionic liquids that have been synthesised for the purposes of metal extraction from aqueous media have also been reported (Pereiro et al., 2007; Liu et al., 2005; Visser et al., 2002). Metal ions such as Hg^{2+}, Cd^{2+}, Sr^{2+} and Cs^{+} have been removed from aqueous media using ILs (Visser and Rogers, 2004). The ability of ILs to form complexes with most metal-halides, especially transition metal-halides, is the basis for their application in metal extraction from aqueous media (Welton, 1999). Metal ions form complexes with both the anion and cation of the IL with a specific coordination. The coordination mechanism depends on the ionisation state of the metal ion. For instance, metals with ionisation state of two, [M (II)], form complexes with bidentate coordination (Welton, 1999). To enhance the metal ion extraction capability of ionic liquids, metal ligating groups such as urea and sulphur, are used (Visser et al., 2002). Affiliating such ligands on the IL cations improve the extraction efficiency of the targeted metal ions.

2.8 Problems associated with current water treatment technologies

Current water treatment techniques can either absorb organic or inorganic pollutants from an aqueous medium, but not both. In addition, these technologies fail to remove pollutants to the required levels and this is attributed to the diversity and varying chemical properties of organic pollutants. Furthermore, in real water systems organic and inorganic pollutants co-exist hence techniques that can remove both contaminants simultaneously would be ideal. Cyclodextrin polyurethanes have been applied in the removal of organic pollutants from water (Mhlanga, 2006). The new challenge is to functionalise these polymers such that they are capable of removing both organic and inorganic pollutants present in water. One possibility is to functionalise the CD moieties with salts of organic cations known as ionic liquids. On their own, ionic liquids have demonstrated the ability to remove inorganic contaminants (heavy metal ions such as Hg^{2+}, As^{3+} and Cr^{3+}) from an aqueous solution (Visser et al., 2002; Tran and De Paoli, 2002). Since cyclodextrin polyurethanes and ionic liquids have independently shown the capability to remove organic and inorganic contaminants, respectively, it is expected that cyclodextrin polymers functionalised with ionic liquids can simultaneously remove organic and inorganic contaminants from water.

2.9 Summary

From the literature overview in this section, the health risk associated with the presence of organic and inorganic pollutants in water has been highlighted. The failure of conventional water treatment technologies to remove organic and inorganic pollutants to ppb levels remains an ongoing research concern. Water insoluble CD polymers have demonstrated the ability to remove organic pollutants at ppb levels. On the other hand ILs are able to remove metal ions from water. To the best of our knowledge, there is no known technique that simultaneously removes organic and inorganic contaminants from water. Water insoluble polymers of CDs functionalised with ILs should provide a dual system that is capable of removing both organic and inorganic pollutants from water simultaneously

3. Experimental methodology

3.1 Reagents and solvents

Unless otherwise specified, all chemicals and reagents were obtained from suppliers and used without further purification. All reactions were carried out under inert conditions of argon or nitrogen. All solvents used in this study were dried over calcium hydride for 2 days and distilled before being used. The distilled solvents were stored under molecular sieves.

3.2 Synthetic procedures

The synthesis of monosubstituted cyclodextrins mainly takes place at the C-6 hydroxyl group. A strong nucleophilic reagent (i.e. sulfonic anhydride (Ts_2O)), was used for the displacement of the hydroxyl group at the C-6 position. Monotosylated cyclodextrins are generally good precursors for derivatives of CDs due to the ease of formation of the mono-tosylate and ease with which the tosyl leaving group can be replaced. Therefore, in order to access a CD monotosylate, Ts_2O had to be first synthesised.

3.2.1 Synthesis and characterisation of *p*-toluene sulfonic anhydride (Ts_2O)

p-Toluene sulfonyl chloride (2.00g, 10.4mmol) was dissolved in dichloromethane (12.5mL) and *p*-toluene sulfonic acid (TsOH) (0.52g, 2.63mmol) was added gradually with vigorous stirring under nitrogen atmosphere. The resulting mixture was stirred overnight at room temperature. The milky reaction mixture was then filtered through silica gel to remove the unreacted *p*-toluene sulfonic acid. Hexane (50mL) was added to the clear filtrate and a precipitate was afforded. The crystals of Ts_2O (1) were collected as white needle-shaped granular/solids after drying overnight under vacuum.

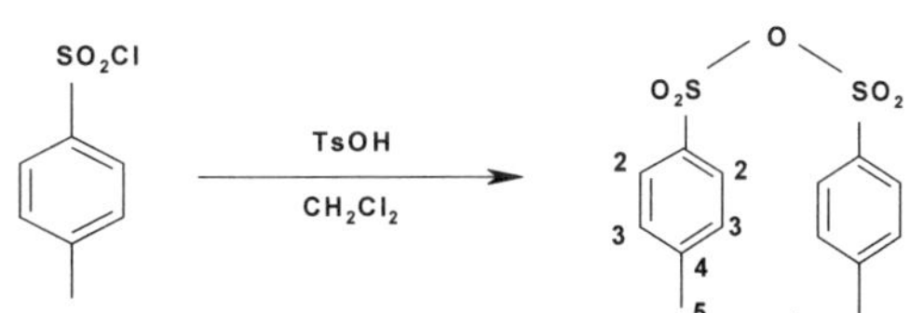

1

Scheme 3.1. Synthetic pathway for the preparation of *p*-toluene sulfonic anhydride.

IR/KBr, cm^{-1}: 3095 (CH, Ar), 3058 (CH, Ar), 2925 (C-H), 1928, 1809, 1652, 1592, 1488 (SO_2), 1373, 1305, 1925, 1175 (SO_2), 1121, 1079, 1015. ^1H NMR, DMSO-$_{d6}$: 7.40 (d, 4H, H-2), 7.0 (d, 4H, H-3), 2.0 (s, 6H, CH_3). ^{13}C NMR, DMSO-$_{d6}$: 145 (C-1), 139 (C-2), 129.0 (C-3), 126.0 (C-4), 21.5 (C-5). Yield: 5.89g, 18.1mmol, 73%. M.pt = 124-126°C. TLC: Eluent: 2:1 hexane/ethylacetate, R_f: 0.80.

3.7.2 Synthesis of monofunctionalised CD precursors

3.7.2.1 Synthesis and characterisation of mono-6-deoxy-6-(*p*-tosyl)-*β*-cyclodextrin (2) (*β*-CDOTs)

A solution of *β*-CD (2.86g, 2.53mmol), *p*-toluene sulfonic anhydride (1.24g, 3.80mmol) and distilled water (60 mL) was stirred at room temperature under an inert atmosphere. A solution of NaOH (1.25g, 31.3mmol) in water (13 mL) was then added after 2 hours. After 10 minutes, unreacted Ts_2O was removed by filtration through silica gel. The pH of the filtrate was adjusted to ~ 8 by the addition of ammonium chloride (3.36g). Overnight refrigeration of the resultant precipitate followed by drying under reduced pressure led to the isolation of 2 as a white powdery solid (2.25g, 1.75mmol, 68%).

3.7.2.2 Synthesis and characterisation of mono-6-iodo-*β*-cyclodextrin (*β*-CDI)

Dry *β*-CDOTs (1.00g, 0.88mmol) was dissolved in anhydrous DMF (40 mL). To this solution, sodium iodide (1.20g, 7.76mmol, 10 equiv), was added under nitrogen with vigorous stirring which was continued for 5 hours at 90°C in an inert atmosphere. Addition of acetone (25 mL) led to the formation of a precipitate. The mixture was filtered and thereafter

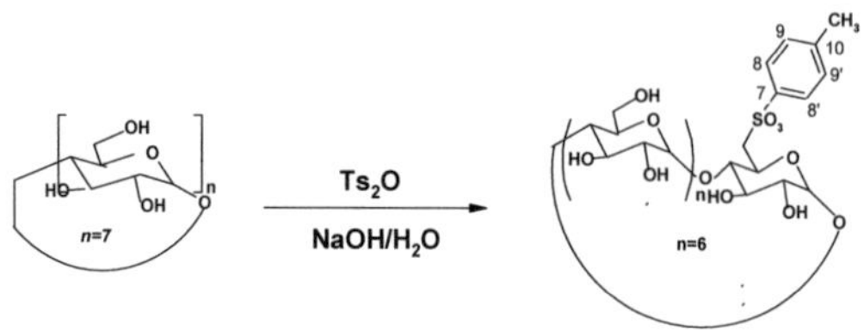

2

Scheme 3.2. A synthetic pathway for the synthesis of β-CDOTs

IR/KBr, cm^{-1}: 3377 (O-H), 2934 (C-H), 1635 (C=C), 1599 (C-C), 1357 (SO_2, Assy.), 1156 (SO_2, Sym), 1037 (C-O). ^1H NMR/ppm, DMSO-$_{d6}$: 7.53(d, H_{Ar}, 2H), 7.21 (d, H_{Ar}, 2H), 4.55 (s, OH-6), 5.40-5.80 (m, H-6, 2H), 4.0 (m, H-6), 3.20-3.55 (m, H-3, H-5, H-6), 5.40-5.80 (br, OH-2, OH-3) , 2.90-3.20 (m, H-2, H-4) 4.63 (d, H-1, 7H), 2.21, (s, $-CH_3$, 3H). ^{13}C NMR/ppm, DMSO-$_{d6}$: 145.0 (C_{Ar}), 132.8 (C_{Ar}), 129.9 (C_{Ar}), 127.6 (C_{Ar}), 101.9 (C-1), 81.5 (C-4), 72.4 (C-2), 73.1 (C-3), 72.2 (C-5), 59.9 (C-6), 21.2 ($-CH_3$). Yield = 2.25g, 1.75mmol, 69%.M.pt = 171-174°C. TLC: R_f = 0.49.

2 3

Scheme 3.3. Synthetic pathway for the preparation of β-CDI.

FTIR/KBr, cm^{-1}: 3346 (O-H), 2924 (C-H), 1027 (C-O), 586 (C-I). ^1H NMR/ppm, DMSO-$_{d6}$: 5.40-5.63 (m, OH-2, OH-3), 4.62 (d, H-1), 4.30 (br, OH-6), 4.30(m, H-6), 3.20-3.55 (m, H-3, H-5, H-6), 3.00-3.20 (m, H-2, H-4), 1.90 (CH_2-I). ^{13}C NMR/ppm, DMSO-$_{d6}$: 102.0 (C-1), 81.6 (C-4), 73.1 (C-3), 72.4 (C-2), 72.1 (C-5), 60.0 (C-6), 30.9 (C-6′) M.pt = 206-209°C. TLC: R_f = 0.43.

washed with acetone (100 mL). The acetone was then removed under reduced pressure (rotary evaporator). To ensure complete removal of unreacted NaI, this procedure (washing and evaporation of acetone) was repeated three times. The resulting solid was dissolved in deionised water (50 mL) and precipitated by the addition of copious amounts of acetone. The precipitate was filtered off and dried in a desiccator to give 3 as white powdery solid (0.79g, 0.63 mmol, 82%).

3.7.3 General procedure for the synthesis of CD-IL complexes

Typically an alkyl imidazole or pyridine (10 equiv.) was added dropwise to a stirred solution of dry β-CDI or β-CDOTs (1.00g, 0.88mmol) dissolved in anhydrous DMF (40 mL). Stirring was continued at elevated temperatures (80°C and 90°C for β-CDI an β-CDOTs, respectively) under nitrogen for a further 24 hours. After cooling to room temperature, acetone (25 mL) was added to precipitate the product. The reaction mixture was then stirred for 30 minutes. Evaporation of organic solvents using a rotary evaporator afforded a white solid. The solid was finally dissolved in deionized water (50 mL) and precipitated by the addition of acetone (200 mL). The precipitate was filtered off and dried under vacuum to yield a white powdery solid.

3.7.3.1 Synthesis and characterisation of mono-6-deoxy-6-(3-methylimidazolium)-*β*-cyclodextrin tosylate (4) (*β*-CD-MIMOTs)

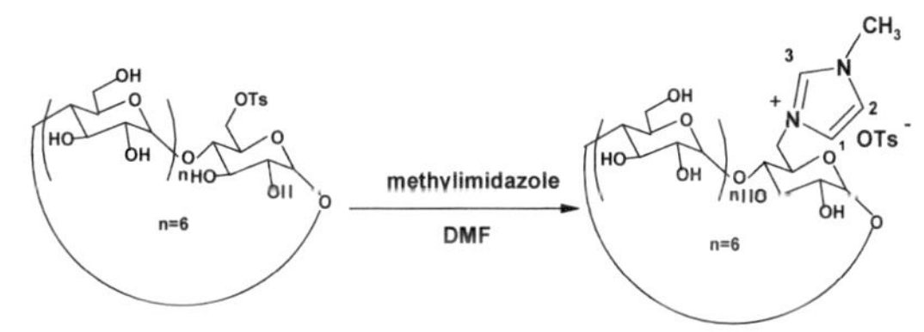

2 4

Scheme 3.4. Synthetic pathway for the preparation of *β*-CD-MIMOTs

IR/KBr, cm^{-1}: 3388 (O-H), 2929 (C-H), 1027 (C-O), 1641 (C=C), 1156 (C-N). ^1H NMR/ppm, $DMSO_{d6}$: 8.99 (s, $=CH_{im}$), 7.92 (s, $=CH_{imid}$), 7.65 (s, $=CH_{imid}$), 7.45 (d, $H_{Ar(ortho)}$), 7.09 (d, $H_{Ar(meta)}$), 5.20-6.21 (OH-2, OH-4), 4.76 (s, H-1), 4.27 (t, OH-6) 3.87 (s, H-6), 3.40-3.66 (m, H-3, H-5, H-6), 3.20-3.41 (m, H-2, H-4), 2.25 (s, $-CH_{3imid}$), 2.04 (s, $-CH_{3OTs}$). ^{13}CNMR/ppm, $DMSO_{d6}$: 145.4 ($C_{Ar(meta)}$), 137.9 ($C_{Ar(ortho)}$), 128.3 ($C_{Ar(ipso)}$), 125.6 ($C_{Ar(para)}$), 137.2 (C_{im}) 123.17 (C_{im}), 123.0 (C_{im}), 102.0 (C-1), 81.6 (C-4), 73.1 (C-3) 72.5 (C-2), 72.1 (C-5), 60.1 (C-6), 48.76 (C-6′). Yield = 0.91g, 0.66 mmol, 97%. M.pt = 226-229°C, TLC: = 0.58.

3.7.3.2 Synthesis and characterisation of mono-6-deoxy-6-(3-butylimidazolium)-*β*-cyclodextrin tosylate(5) (*β*-CD-BUTIMOTs)

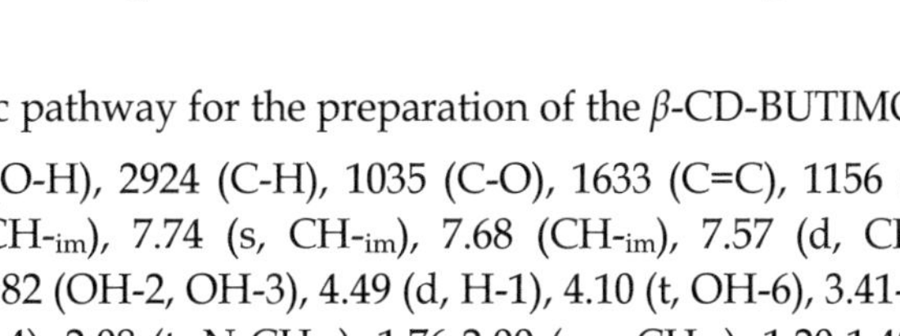

2 5

Scheme 3.5. Synthetic pathway for the preparation of the *β*-CD-BUTIMOTs

IR/KBr, cm^{-1}: 3346 (O-H), 2924 (C-H), 1035 (C-O), 1633 (C=C), 1156 (C-N). ^1HNMR/ppm $DMSO\text{-}_{d6}$: 9.06 (s, $CH\text{-}_{im}$), 7.74 (s, $CH\text{-}_{im}$), 7.68 ($CH\text{-}_{im}$), 7.57 (d, $CH_{Ar(ortho,OTs)}$), 7.08 (d, $CH_{Ar(meta,OTs)}$), 5.52-5.82 (OH-2, OH-3), 4.49 (d, H-1), 4.10 (t, OH-6), 3.41-3.72 (m, H-3, H-5, H-6), 3.10-3.41 (H-2, H-4), 2.08 (t, $N\text{-}CH_2\text{-}$), 1.76-2.00 (m, $-CH_2-$), 1.20-1.40 (m, $-CH_2\text{-}CH_3$) 0.84 ($CH_{3\text{-}OTs}$), 0.005 (s, $CH_{3\text{-}im}$). ^{13}CNMR/ppm $DMSO\text{-}_{d6}$: 145.4 ($C\text{-}2'_{OTs}$), 137.1 ($C\text{-}2_{OTs}$), 128.2 ($C\text{-}3_{OTs}$), 125.6 ($C\text{-}3'_{OTs}$), 138.0 ($C\text{-}_{8but}$), 123.5 ($C\text{-}_{10but}$), 123.1 ($C\text{-}_{9but}$), 102.0 (C-1), 81.6 (C-4), 73.1 (C-3), 72.49 (C-2), 72.09 (C-5), 60.00 (C-6), 48.71 (C-6′) ,31.55 ($N\text{-}CH_2$), 31.2 ($-CH_2-$), 21.29 - $CH_2\text{-}CH_3$, 19.3 ($CH_{3\text{-}OTs}$), 13.75 ($CH_{3\text{-}im}$). Yield: 0.96g, 0.67 mmol, 98%, M.pt: 240-243°C. TLC: R_f= 0.61.

3.7.3.3 Synthesis of mono-6-deoxy-6-pyridinium-*β*-cyclodextrin tosylate (6) (*β*-CD-PYROTs)

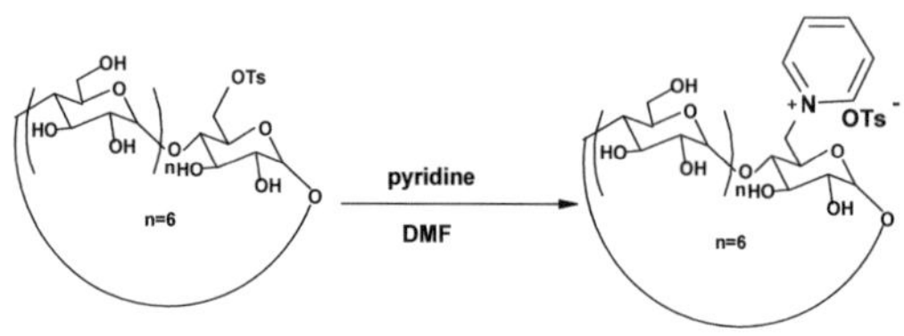

2 6

Scheme 3.6. A synthetic pathway for the preparation of *β*-CD-PYROTs

IR/KBr, cm^{-1}: 3395 (O-H), 2933 (C-H), 1033 (C-O), 1636 (C=C), 1159 (C-N). ^{1}H NMR/ppm, DMSO-$_{d6}$: 9.08 (d, $=CH_{pyr}$), 8.20 (t, $=CH_{pyr}$), 7.80 (dd, $=CH_{pyr}$), 7.57 (br d, $H_{Ar(ortho)}$), 7.21 (d, $H_{Ar(meta)}$), 5.01 (s, H-1), 4.40-4.82 (m,OH-2, OH-3), 3.98 (t, OH-6), 3.52-3.88 (m, H-3, H-5, H-6), 3.21-3.52 (m, H-2, H-4), 2.15 (d, $-CH_{3OTs}$). ^{13}CNMR/ppm, DMSO-$_{d6}$: 147.69 ($C_{py(C-8)}$), 145.16($C_{py(C-9)}$), 125.66 ($C_{py(C-10)}$), 143.00 ($C_{OTs(C-1)}$), 138.25 ($C_{OTs(C-2)}$), 137.92 ($C_{OTs(C-3)}$), 128.38 ($C_{OTs(C-4)}$), 102.08 (C-1), 81.68 (C-4), 73.22 (C-3), 72.53 (C-2), 72.17 (C-5), 60.09 (C-6), 45.42 (C-6′). Yield = 2.49g, 0.70 mmol, 90.7%. M.pt = 242-245°C. TLC: R_f = 0.55.

3.7.3.4 Synthesis and characterisation of mono-6-deoxy-6-(3-methylimidazolium)-*β*-cyclodextrin iodide (7) (*β*-CD-MIMIOD)

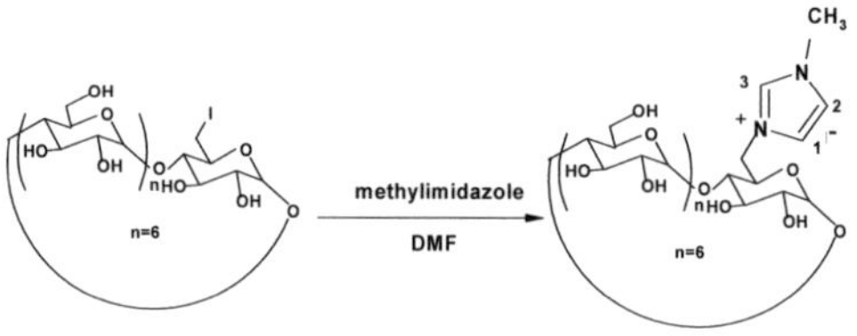

3 7

Scheme 3.7. A synthetic scheme for the preparation of *β*-CD-MIMIOD

FTIR/KBr, cm^{-1}: 3393 (O-H), 2919 (C-H), 1027 (C-O), 1645 (C=C), 1156 (C-N). ^{1}H NMR/ppm, DMSO-$_{d6}$: 5.60-5.80 (br m, OH-2, OH-3), 4.81 (s, H-1), 4.54 (br, OH-6), 3.41-3.80 (m, H-3, H-5, H-6), 3.22-3.41 (m, H-2, H-4), 7.48 (d, H_{Ar}-1), 7.10 (d, H_{Ar}-2), 7.92 (s, H_{Ar}-3), 2.06 (s, $-CH_3$). ^{13}C NMR/ppm, DMSO-$_{d6}$: 128.4($C_{Ar-imid}$), 125.7($C_{Ar-imid}$), 102.2 (C-1), 81.8 (C-4), 73.2 (C-3), 72.6 (C-2), 72.2 (C-5), 60.1 (C-6), 30.9 ($-CH_{3-imid}$). Yield = 0.86g, 0.63 mmol, 92%, M.pt: 197-202°C. TLC: R_f = 0.54.

3.7.3.5 Synthesis and characterisation of mono-6-deoxy-6-(3-butylimidazolium)-*β*-cyclodextrin iodide (8) (*β*-CD-BUTIMIOD)

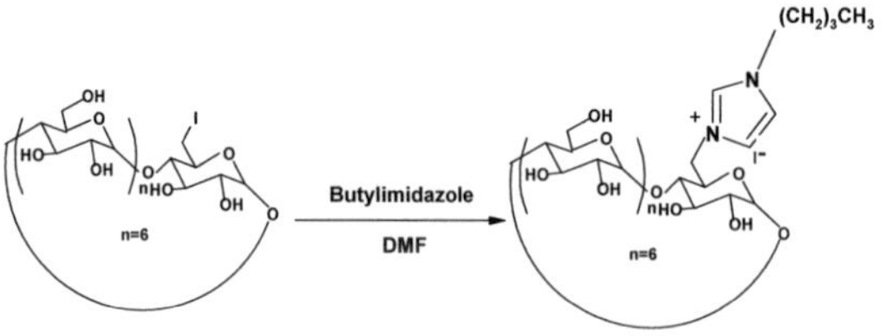

3 8

Scheme 3.8. Synthetic pathway for the preparation of *β*-CD-BUTIMIOD

FTIR/KBr, cm^{-1}: 3378 (O-H), 2924 (C-H), 1032 (C-O), 1655 (C=C), 1151 (C-N). ^1H NMR/ppm, DMSO-$_{d6}$: 7.10 (d, H_{Ar}), 7.46 (d, H_{Ar}), 7.92 (s, H_{Ar}), 5.60-5.81 (m, OH-2, OH-3), 4.81 (s, H-1), 4.43-4.62 (br, OH-6), 3.29-3.65 1.73 (s, $CH_{2\text{-im}}$), 1.21 (s, $CH_{2\text{-im}}$), 0.85 (s, $CH_{2\text{-im}}$), (m, H-3, H-5, H-6), 3.15-3.39 (m, H-2, H-4), 2.27 (m), 2.06 (s), 1.93 (s). ^{13}C NMR/ppm, DMSO-$_{d6}$: 128.3($C_{Ar\text{-imid}}$), 125.6($C_{Ar\text{-imid}}$), 102.1 (C-1), 81.7 (C-4), 73.2 (C-3), 72.5 (C-2), 72.2 (C-5), 60.1 (C-6), 30.9(-$CH_{3\text{-imid}}$), 36.0(-$CH_{2\text{-imid}}$), 20.9. Yield = 0.82g, 0.58 mmol, 85%. M.pt: 219-221°C. TLC: R_f = 0.57.

3.7.3.6 Synthesis and characterisation of mono-6-deoxy-6-pyridinium *β*-cyclodextrin iodide (9) (*β*-CD-PYRIOD)

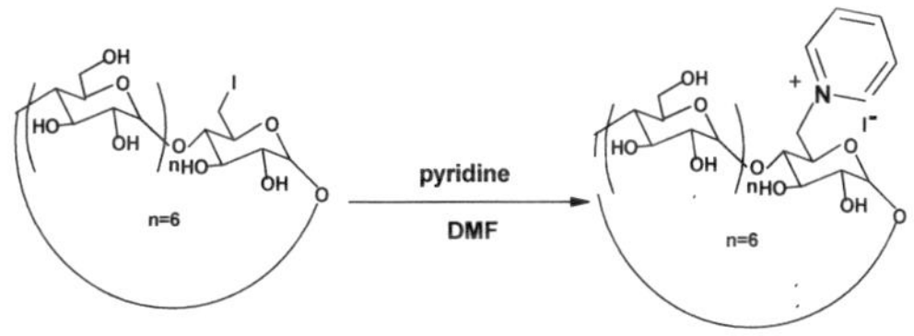

3 9

Scheme 3.9. A schematic pathway for the preparation of *β*-CD-PYRIOD

FTIR/KBr, cm^{-1}: 3388 (O-H), 2929 (C-H), 1027 (C-O), 1656 (C=C), 1151 (C-O). ^1H NMR/ppm, DMSO-$_{d6}$: 9.01 (d, H_{Ar}), 7.48 (d, H_{Ar}), 7.12 (t, H_{Ar}), 5.50-5.84 (m, OH-2, OH-3), 4.81 (s, H-1), 4.51 (t, OH-6), 3.42-3.78 (m, H-3, H-5, H-6), 2.98-3.42 (m, H-2, H-4). ^{13}C NMR/ppm, DMSO-$_{d6}$: 128.2 ($C_{Ar\,(ortho)}$), 125.6 ($C_{Ar\,(meta)}$), 120.3 ($C_{Ar\,(para)}$), 102.4 (C-1), 81.9 (C-4), 73.6 (C-3), 72.8 (C-2), 72.4 (C-5), 60.4 (C-6) 48.3 (C-6′). Yield = 0.855g, 0.63 mmol, 88%, M.pt: 216-218°C. TLC: R_f = 0.51

3.8 Synthesis of CD-IL polymers

The CD-IL complexes (both tosylate and iodide derivatives) were reacted with bifunctional diisocyanate linkers to produce CD-IL polymers. Typically, the *β*-CD-IL complex (1.00g, 0.88 mmol) was dissolved in DMF (20 mL), and reacted with the bifunctional linker (10 equiv.) in DMF at 75°C for 18-24 hours. Two bifunctional linkers were used for the polymerisation of the CD-IL complexes. These are hexamethylene diisocyanate (HDI) and toluene-2,4-diisocyanate (TDI). The polymerisation reaction was monitored by IR spectroscopy. The completion of the polymerisation was confirmed by the total disappearance of the isocyanate peak at 2270 cm^{-1} after 18-24 hours (Li and Ma, 1999). The reaction mixture was then precipitated by the addition of acetone (100 mL). The solid formed was then left to settle in acetone for 10 minutes to allow for the removal of residual DMF from the polymers. To remove traces of DMF which may have been present, the polymers were filtered and washed with copious amounts of acetone (100 mL). The polymers were then dried overnight under reduced pressure. The examples of the synthesised polymers are shown in Figures 3.10 and 3.11.

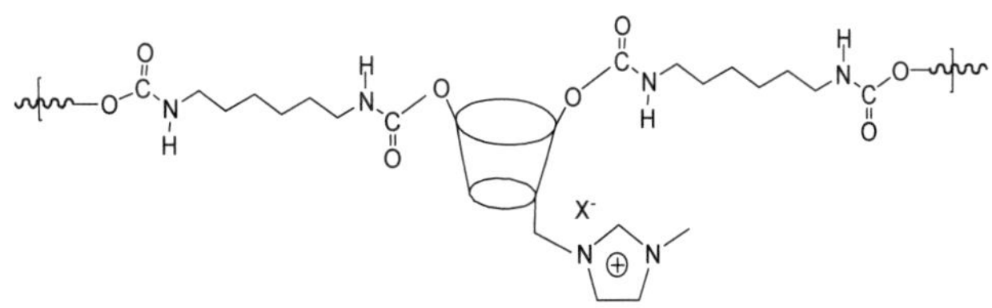

Fig. 3.1. Structure of CD-IL hexamethylene diisocyanate polymer

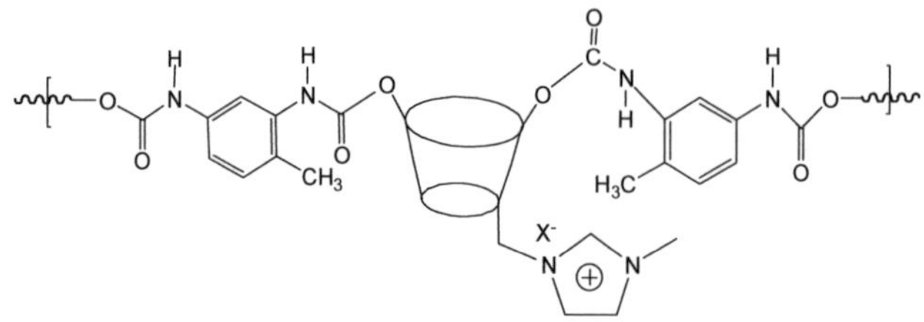

Fig. 3.2. Structure of CD-IL toluene-2,4-diisocyanate polymer

3.9 Absorption studies of the CD-IL polymers

The synthesized CD-IL polymers were found to be insoluble in water and other organic solvents, an important feature for their application in the removal of organic and inorganic pollutants from water. Using model contamination experiments, the ability of the polymers to absorb organic and inorganic pollutants from an aqueous medium was studied. In these model experiments, a known concentration of the pollutant in water was passed through the polymers. Quantification of organic pollutants was achieved by using UV-visible spectroscopy and GC-MS analysis. Inorganic pollutants (heavy metals) were quantified using AAS.

3.9.1 UV-Visible spectroscopy absorption experiments

To measure the UV absorbance of *para*-nitrophenol (PNP), a Cary 50 UV-visible spectrophotometer was used. PNP was chosen because phenolic compounds are prevalent as pollutants in most aquatic systems. Compounds that can be analysed by this technique must have chromophores (i.e. covalently bonded but unsaturated groups such as NO_2, C=C and C=O) that absorb electromagnetic radiation in the ultraviolet and visible regions of the spectrum (Field et al., 1995). An added advantage of using PNP is that it can be easily monitored by UV-visible spectroscopy since it strongly absorbs at 318 nm (visible region of the spectrum). A 10 mm quartz cuvet cell was used as a sample holder. The instrument was calibrated using PNP standards (2, 5, 10, 15 and 20 ppm) and a calibration curve was plotted and used to determine the amount of the PNP absorbed by the polymers. For the actual analysis of the pollutants, the pollutant under investigation, (30 mL, 10 $mg.L^{-1}$) was typically passed through a CD-IL polymer (300 mg) (i.e. HDI and TDI polymers) packed in a C-18 SPE chromatography column. PNP was allowed to pass through the polymers at a flow rate of 5 mL per minute. It was observed that after the extraction process that the polymers assumed the colour of the pollutant (i.e. turned yellow) while the filtrate became clearer. The filtrate was then analysed for the amount of PNP present after it had been passed through the polymer. The residual concentration of PNP absorbed by the polymers was determined from a calibration curve drawn from standards.

3.9.2 GC-MS analysis

The GC-MS uses gas chromatography and mass spectroscopy as a combined analytical technique for analysing mixtures of compounds (Frederic, 2004). GC analysis separates the components in a sample and provides a representative spectral output while MS identifies the substances by electrically charging the specimen molecules; this entails accelerating them through a magnetic field thus breaking them into charged fragments and detecting the mass ions. By combining these techniques, a solution containing a number of compounds can be analysed qualitatively and quantitatively. The sample is injected into the injector port and passed through the column by a mobile phase. Helium was used as a carrier gas in all experiments in this study. The gas sweeps the analyte into a column packed with a 0.25 μm coating of a polymeric stationary phase in order to promote maximum separation. As the compounds emerge from the column, a signal is created whenever a compound is detected. The time duration that a compound is retained in the GC column is referred to as the retention time (T_R). Though the retention time can assist in differentiating between compounds, it is sometimes not a reliable determiner since similar compounds can have similar retention times. Therefore, more information other than retention times is required to identify a particular compound. As the compounds elute, the MS identifies the substances by electrically charging the specimen molecules, accelerating them through a magnetic field, breaking them into charged fragments and thus detecting the different charges. These fragments are charged ions with a specific mass. The mass of the fragments divided by the charge is called mass to charge ratio (*m/z*). The mass spectrum can thus be used to confirm the identity of a given compound using *m/z* values. Also, an internal library system can be used to relate an unknown compound by comparing the mass spectrum of a compound with mass spectra of compounds having similar fragmentation patterns and *m/z* values.

3.9.2.1 Preconcentration of water samples

The concentration of water samples when performing analyses at low concentrations is crucial because it ensures that samples are within the detection limit of the instrument (Valerie, 2003). The method used for this procedure was EPA Method 525.2, which is an EPA method for the analysis of semi-volatile organic compounds in drinking water samples (Zymark Coporation, 2001). It utilises solid phase extraction (SPE) for isolation and concentration of the analyte and GC-MS for the determination of the preconcentrated analyte. For the purposes of this study, this method was used with some modifications. For instance, a 30 mL water sample, instead of a 1000 mL, was loaded onto a conditioned C-18 SPE cartridge. The cartridge was then dried and eluted with ethyl acetate (EtOAc) and dichloromethane (CH_2Cl_2). After this the sample was concentrated under nitrogen to about 1 mL before analysis. The first step involved in the SPE method is the conditioning of the column with methanol. The second step involves loading of the sample in the solid sorbent. Thirdly, the sorbent is washed with a solvent that has low elution strength and lastly, the desired analyte is eluted with a suitable solvent

3.9.2.1.1 2,4,6-Trichlorophenol absorption tests

2,4,6-Trichlorophenol (TCP) was selected as an additional organic pollutants to be studied because it is known to affect the respiratory system and is a recognised carcinogen. For the preparation of 50 ppm TCP standard solution (1000 mL), TCP (0.500 g) was dissolved in acetone (2 mL) and then made up to the mark by adding CH_2Cl_2. Serial dilutions were then carried out to prepare 1, 2, 5, 10, 15 and 20 ppm standard solutions. Residual TCP was determined using a calibration curve obtained from these standards. An empty C-18 SPE cartridge was filled with the polymer under investigation (0.300 g). The cartridge was then connected to a filtering unit comprising of separating funnel and a receiving Erlenmeyer flask. The Erlenmeyer flask was in turn connected to a suction pump (see Figure 3.3 for the manual set up of the SPE) to enhance a smooth and controlled flow of the water samples through the polymer. The polymer was conditioned by first passing through CH_2Cl_2 (5 mL) after which a spiked water sample (30 mL, 10 ppm) was passed through the polymer at a flow rate that was maintained between 3 and 5 mL per minute. Residual TCP was extracted from the eluent with CH_2Cl_2.

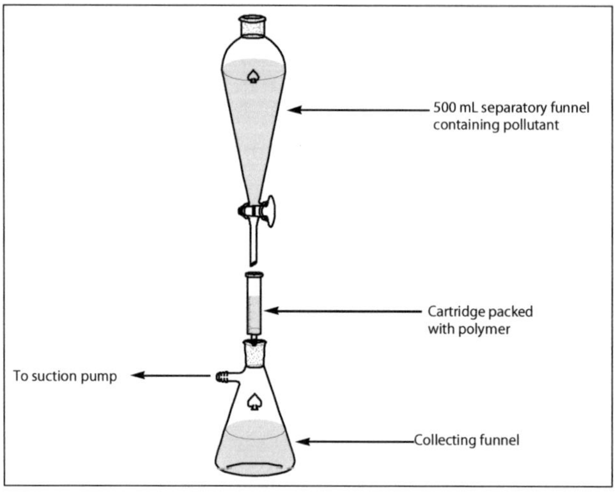

Fig. 3.3. Diagram showing the manual SPE used in this project

3.10 Extraction of heavy metals

Chromium and cadmium were used as model pollutants for the extraction study of heavy metals from water by the polymers. Besides being available in our laboratory, these pollutants were selected because they are toxic and lethal to the human body, particularly the central nervous system. Apparently these metals bind to sulphur bonds of enzymes resulting in the malfunctioning of the enzymes. A chromium standard (1000 $mg.L^{-1}$, 1000 mL) was prepared by dissolving potassium dichromate ($K_2Cr_2O_7$) (2.84 g), which was first pre-dried in an oven at 110°C for 2 hours, in distilled water (200 mL). HNO_3 (1.5 mL) was then added. To prepare cadmium standard solution (1000 $mg.L^{-1}$, 1000 mL), cadmium metal (1.00 g) was dissolved in a 1:1 solution of distilled water/HCl. $La(NO_3)_2$ (1.00g) was added to stabilise the cadmium standards.

3.10.1 AAS analysis

For AAS analysis, serial dilutions were made from the 1000 $mg.L^{-1}$ standard chromium and cadmium solutions. This led to the preparation of 2, 4, 5, 6 and 8 ppm for chromium standards and 0.4, 0.6. 0.8 1, 1.5 and 2 ppm for cadmium standards. These concentrations were chosen based on the detection limits for chromium and cadmium. These are 0.2-10 ppm and 0.05-2 ppm, respectively. Absorption of chromium was determined at 5 ppm while cadmium was measured at 1 ppm. Distilled water was used to zero the instrument as a reference sample. All measurements were performed at room temperature.

4. Results and discussions

4.1 CD-IL complexes/derivatives

The CD-IL derivatives were successfully synthesized and their physical parameters are summarised in Table 4.1.

CD and CD-IL derivatives	Melting point (°C)	Colour and form	Yield (%)
β-CDOTs (2)	171-174	White fine powder	68
β-CDI (3)	206-209(dec)	White fine powder	82
β-CDMIMOTs (4)	229-229(dec)	White solid	98
β-CDBUTIMOTs (5)	240-242	White powder	97
β-CDPYROTs (6)	242-245(dec)	White solid	91
β-CDMIMIOD (7)	197-202(dec)	White solid	92
β-CDBUTIMIOD (8)	219-221(dec)	White powder	85
β-CDPYRIOD (9)	216-218(dec)	White powder	88

β-CDOTs: mono-6-tosyl, .. β-CDMIMOTs: mono-6-methyl imidazolium tosylate
β-CDBUTIMOTs: mono-6-butyl imidazolium tosylate,
β-CDPYROTs: mono-6-pyridinium tosylate, β-CDI: mono-6-CD iodide,
β-CDMIMIOD: mono-6-methyl imidazolium iodide
β-CDBUTIMIOD: mono-6-butyl imidazolium iodide,
β-CDPYRIOD: mono-6-pyridinium iodide. Dec = decomposes at its melting point.

Table 4.1. Yields and physical properties of the β-CD-IL complexes

The melting points of the synthesised CD-IL derivatives were found to range between 190 and 245°C. With the exception of the β-CDBUTIMOTs derivative, all the CD-IL complexes melted with decomposition. These compounds were found to be soluble in water and several organic solvents (e.g. DMF, ethanol, DMSO and ethyl acetate). For application in water treatment, it was therefore necessary to polymerise the CDI-IL complexes into water-insoluble polymers. The synthetic reactions and characterisations of these CD-IL polymers are discussed in the section that follows.

4.2 Synthesis and characterisation of the CD-IL polymers

Polymerisation of the CD-IL complexes was performed as per the procedure described in section 3. Each CD-IL complex was polymerised with bifunctionalised linkers; hexamethylene diisocyanate (HDI) and toluene 2,4-diisocyanate (TDI). The polymerisation reactions were monitored using infrared spectroscopy. Figures 4.12a and 4.12b show the gradual disappearance of the isocyanate peak (N-C=O) at 2270 cm^{-1} as the polymerisation reaction proceeded. The complete disappearance of the isocyanate peak at 2270cm^{-1} (after 24 hrs) marked the end of the polymerisation reaction.

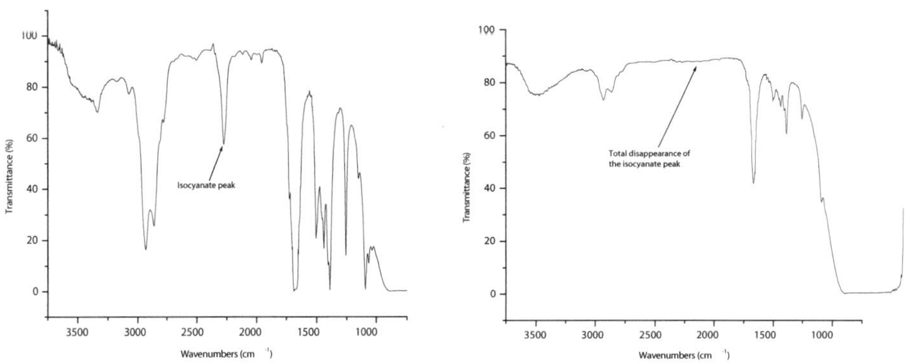

Fig. 4.12. IR spectrum showing isocyanate peak at 00 mins and after 24 hrs

4.3 Physical properties of the β-CD-IL polymers

The physical properties (colour, texture and the yields) of the synthesised CD-IL polymers are shown in Table 4.2.

CD-IL polymer	Texture	Colour	Yield (%)
B-CD/HDI	Powder	White	100
B-CD/TDI	Powder	White	98
β-CDMIMOTs/HDI	Powder	White	97
β-CDMIMOTs/TDI	Powder	Light brown	99
β-CDBUTIMOTs/HDI	Powder with granules	Crème white	95
β-CDBUTIMOTs/TDI	Powder	Light brown	96
β-CDPYROTs/HDI	Granulated powder	White	93
β-CDPYROTs/TDI	Powder	Yellowish brown	96
β-CDMIMIOD/HDI	Granulated powder	Crème white	100
β-CDMIMIOD/TDI	Granulated powder	Light brown	97
β-CDBUTIMIOD/HDI	Powder	White	100
β-CDBUTIMIOD/TDI	Powder with granules	Light brown	95
β-CDPYRIOD/HDI	Crystalline with powder	Beige	100
β-CDPYRIOD/TDI	Crystalline	Yellow	94

β-CDMIMOTs: mono-6-methyl imidazolium tosylate
β-CDBUTIMOTs: mono-6-butyl imidazolium tosylate
β-CDPYROTs: mono-6-pyridinium tosylate β-CDI: mono-6-CD iodide
β-CDMIMIOD: mono-6-methyl imidazolium iodide
β-CDBUTIMIOD: mono-6-butyl imidazolium iodide
β-CDPYRIOD: mono-6-pyridinium iodide.

Table 4.2. Physical properties of the β-CD-IL polymers

While these polymers were produced in high yields, it should be noted that they are insoluble in water, an important feature for their application in water treatment. Unfortunately, NMR studies could not be performed due to the unavailability of solid-state NMR equipment. However, to study the morphology of these polymers SEM and BET surface area analyses were carried out. SEM is important because it gives us an idea of the physical morphology of the polymers while BET analysis provides an insight into the effect of the pore volume and surface area on the absorption efficiency of the polymers.

4.4 Scanning electron microscopy (SEM) analysis of the polymers

SEM was utilised in order to determine and compare the surface features of both the native β-CD and the CD-IL polymers. The SEM micrographs of both the β-CD/HDI and the CD-IL/HDI polymers are shown from Figure 4.13a to Figure 4.13e.

SEM images of the TDI linked polymers were also investigated and the SEM micrographs are presented in Figures 4.13 h-n.

From these SEM micrographs it can be observed that the SEM micrographs of the native β-CD/HDI or TDI polymers and the CD-IL polymers are different. Some of the polymers exhibit a spongy surface (e.g. β-CDPYROTs/HDI and β-CDMIMOTs/TDI) while others are crystalline (e.g. β-CDBUTIOD/HDI and β-CDMIMIOD/TDI). From these results it can be concluded that the type of cation (pyridine or alkyl imidazole) or anion (iodide or tosylate) have an effect on the physical surface of the polymers. However, it is not clear at this stage whether these differences are as a result of some discrepancies during the processing of the polymers, thus BET analysis was carried out to further elucidate their physical properties.

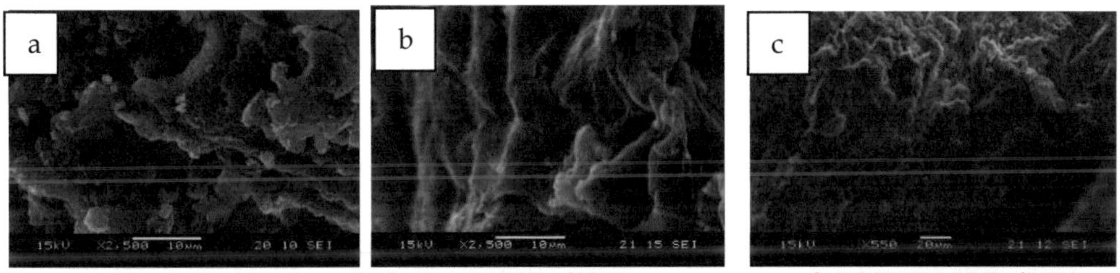

Figure 4.13a: β-CD/HDI, b: β-CDMIMOTs/HDI, c: β-CDBUTIMOTs/HDI

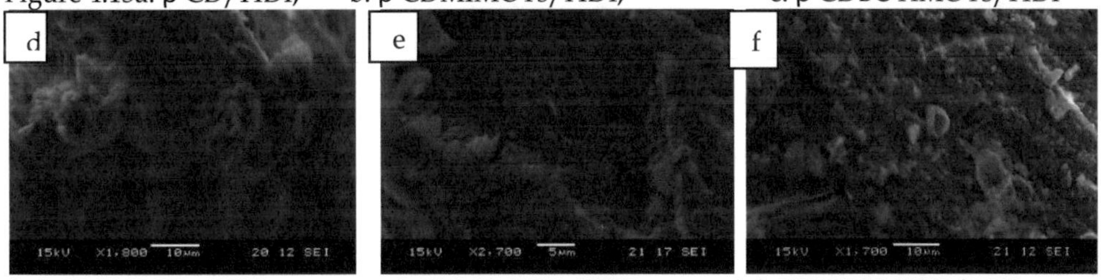

d: β-CDPYROTs/HDI, e: β-CDMIMIOD/HDI, f: β-CDBUTIMIOD/HDI

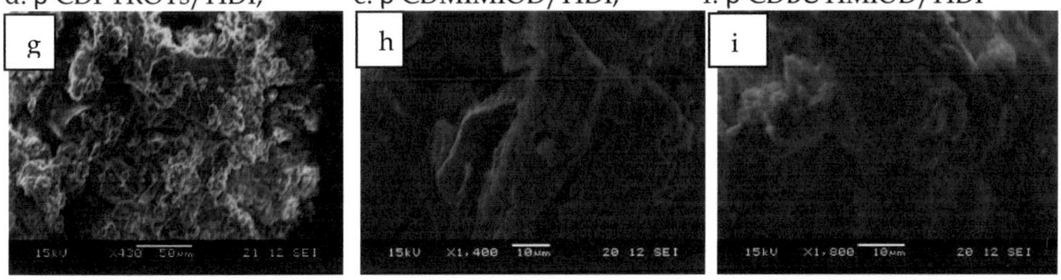

g: β-CDPYRIOD/HDI, h: β-CD/TDI, i: β-CDMIMOTs/TDI

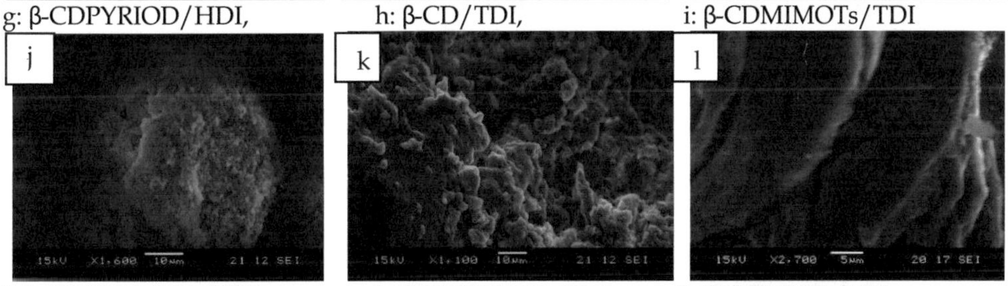

j: β-CDBUTIMOTs/TDI, k: β-CDPYROTs/TDI, l: β-CDMIMIOD/TDI

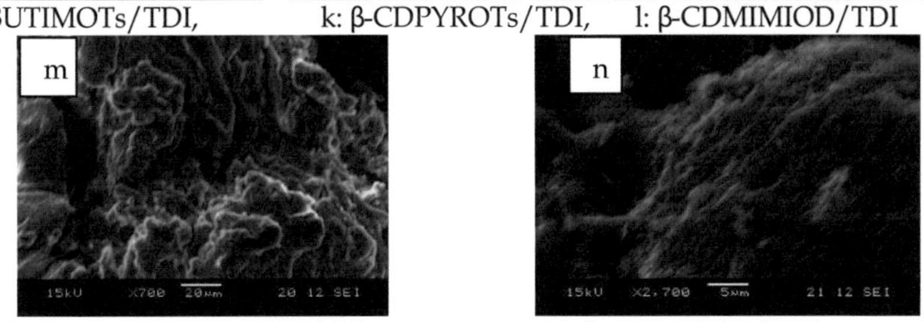

m: β-CDBUTIMIOD/TDI, n: β-CDPYRIOD/TDI,

Fig. 4.13.

4.5 BET analysis of the polymers

Polymers	HDI		TDI	
	Surface area (m^2/g)	Pore volume (cm^3/g)	Surface area (m^2/g)	Pore volume (cm^3/g)
Β-CD	2.53	0.009	1.706	0.009
β-CDMIMOTs	23.26	0.084	22.71	0.11
β-CDBUTIMOTs	2.99	0.014	10.51	0.023
β-CDPYROTs	2.89	0.063	1.64	0.026
β-CDMIMIOD	2.51	0.012	1.81	0.080
β-CDBUTIMIOD	10.88	0.058	1.78	0.0066
β-CDPYRIOD	0.4427	0.0036	7.00	0.030

Table 4.3. BET results of the polymers

These BET results show that there is no pattern observed between the surface area and the incorporation of the imidazolium or pyridinium ring onto the CD backbone. For example, the highest surface area recorded was for the *β*-CDMIMOTs/HDI polymer (23.26 m^2/g) while the lowest was 0.4427 m^2/g (*β*-CDPYRIOD/HDI polymer). However, the surface area significantly increased for the *β*-CDMIMOTs (both HDI and TDI polymers.

4.6 Thermal analysis of the polymers

To determine the stability, melting and decomposition temperatures of the polymers, thermal analysis of the polymers was carried out using TGA and DSC. The findings of this study are presented in the following sections.

4.6.1 TGA analysis of the polymers

The polymers generally had similar heating profiles. Representative TGA graphs of the native *β*-CD/HDI, *β*-CDMIMIOD/HDI and *β*-CDPYROTS/TDI polymers are shown in Figure 4.14. Other TGA thermographs are shown in Appendix B.

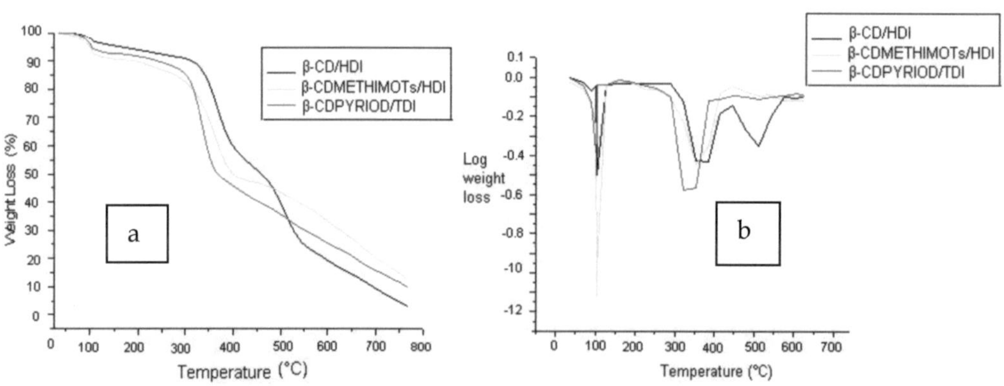

Fig. 4.14. (a) TGA thermographs and (b) derivative curves of the native *β*-CD and CD-IL polymers

The TGA thermographs (Figure 4.14(a) showed some differences between the native *β*-CD polymers and the CD-IL polymers. The polymers showed a three step degradation process;

the first step at about 100°C (loss of water), the second step at about 350°C (melting point of the polymer) and lastly the decomposition of the polymers at about 520°C. However, to fully study the thermal behaviour of these polymers derivative curves of both the β-CD polymers and the CD-IL polymers were done. These derivative curves are presented in Figure 4.14(b). Analysis of the derivative curves confirmed that the native β-CD polymers showed three different regions which were characterised by the loss of water at 100°C, the melting point at about 350°Cand the decomposition of the polymers at about 520°C. The imidazolium CD-IL polymers on the other hand showed only two regions; loss of moisture at 100°C and the melting point was accompanied by degradation at about 350°C.However, the pyridinium based CD-IL polymers had lower melting points at 320°C.

4.6.2 DSC analysis of the polymers

DSC analysis was also performed in order to compare the thermal properties of the polymers with or without the attachment of the imidazolium or pyridinium moieties. Noteworthy is that the DSC thermograms are generally similar thus Figure 4.16 shows representative DSC thermographs of the polymers.

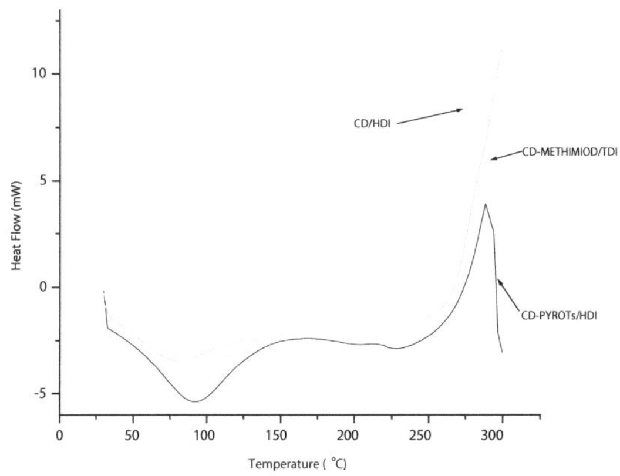

Fig. 4.16. DSC chromatograms of the native β-CD and CD-IL polymers

The DSC thermograms all showed an endothermic peak associated with the evaporation of water and other organic solvents at about 95°C. The thermograms also showed an extra endothermic peak at 250°C which corresponds to the melting points of these polymers. For all three polymers degradation occurred at around 300°C.

4.7 Absorption studies of the polymers

Four pollutants (two organic and two inorganic) were used as model pollutants for the study of the absorption capabilities of the polymers. These pollutants are *para*-nitrophenol (PNP), 2,4,6-trichlorophenol (TCP), chromium VI and cadmium II and were readily available in our laboratory.

4.7.1 Absorption of *para*-nitrophenol (PNP) and 2,4,6-trichlorophenol (TCP)

p-Nitrophenol (PNP) was chosen as a model pollutant because phenolic compounds are prevalent as pollutants in most aquatic systems. Also, PNP can be easily monitored by UV-Visible spectroscopy because of its yellow colour. Polymers formed by cross-linking the synthesised CD-IL complexes with HDI and TDI linkers were investigated for their ability to absorb PNP. *p*-Nitrophenol (30 mL, 10mg.L^{-1}) was passed through the polymers (300 mg) as described in section 3. The absorbance values were recorded in triplicates at room temperature. TCP, which is known to affect the respiratory system and is carcinogenic, was also used as a model pollutant and was studied using GC-MS. TCP standard solution (30 mL, 10 mg.L^{-1}) was used to determine the absorption efficiencies of the polymers. The residual amount of the TCP was extracted by liquid-liquid extraction with CH_2Cl_2 as outlined in section 3 after it was passed through the polymers. Results obtained from these analyses are summarised in Tables 4.3. The native *β*-CD polymer was included in the study for comparison purposes.

Polymers	Amount of PNP and TCP removed (%)			
	HDI linker		TDI linker	
	PNP	TCP	PNP	TCP
β-CD	64	54	68	58
β-CDMIMOTs	80	69	77	69
β-CDBUTIMOTs	68	69	64	77
β-CDPYROTs	15	63	40	60
β-CDMIMIOD	65	75	55	74
β-CDBUTIMIOD	63	67	71	71
β-CDPYRIOD	24	69	48	61

Table 4.3. Percentage absorption efficiency of PNP by HDI and TDI polymers

As mentioned in section 2, the amount of the pollutant absorbed can be affected by the type of cation or anion, or the length of the alkyl chain attached to the imidazolium or pyridinium ring. It was, therefore, necessary to study the effect of these properties on the absorption efficiency of PNP and TCP by the polymers. Also, the effect of the linker on the overall performance of the polymer was investigated. These variations are discussed in the section that follows.

4.7.1.1 Effect of the type of the cation on the extraction of the organic pollutants

It is very clear from Table 4.3 that the incorporation of the imidazolium ring onto the CD backbone generally enhances the absorption efficiencies of the polymers (i.e. with respect to PNP). In fact, the highest absorption (80%) was recorded for *β*-CDMIMOTs/HDI. However, the pyridinium-based showed less affinity for PNP. For example, the absorption efficiencies of *β*-CDPYROTs/HDI and *β*-CDPYRIOD/HDI were a mere 15% and 24%, respectively. The corresponding TDI polymers followed a similar pattern. In contrast, both the imidazolium- and pyridinium-based polymers are better at absorbing TCP when compared to the native cyclodextrin polymers. It is not clear at this stage why the extraction of PNP is influenced by the type of cation.

4.7.1.2 Effect of the imidazolium alkyl chain on the extraction of the organic pollutants

The effect of the length of the alkyl chain attached to the imidazolium ring on the absorption efficiency of the polymers was investigated. According to previous studies (Visser et al, 2001; Pandey, 2006), the longer the alkyl chain the less effective is the ionic liquid for metal ion extraction. In our study, two substituents (methyl and butyl) have been incorporated onto the ionic liquid component (imidazolium) of the polymer and this allowed us to test this concept. Results presented in Table 4.3 indeed confirmed that the alkyl chain had an effect on the amount of pollutant absorbed by the polymers. As expected, the methylimidazolium polymers were generally found to have a higher absorption efficiencies compared to their butylimidazolium counterparts.

4.7.1.3 Effect of the type of anion on the extraction of the organic pollutants

To further study the absorption efficiencies of the polymers, the effect of the counter-anion (tosylate and iodide) on the absorption capacity of the polymers was investigated. It was observed that the amount of pollutant absorbed and the type of anion present in the polymer were not well correlated. As summarised in Figure 4.17a and Figure 4.17b, the type of the anion does not seem to significantly affect the absorption of both PNP and TCP.

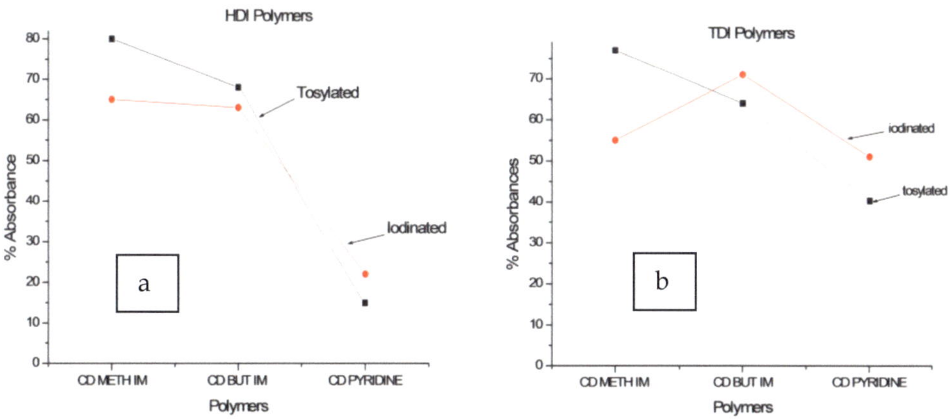

Fig. 4.17. Comparison of PNP absorbed by tosylated versus iodinated HDI-linked (a) and (b) TDI-linked polymers.

4.7.1.4 Effect of the linker on the extraction of the organic pollutants

Figures 4.17a and 4.17b were also used to determine how the linkers affected the absorption efficiency of the polymers. According to these results, the tosylated polymers showed better absorption efficiencies than the TDI linked polymers (especially the imidazolium polymers) for the HDI-linked polymers. However, this was not the case with the pyridinium polymers where higher absorption efficiency was observed for the iodinated polymers. For the TDI-linked polymers, it was observed that the iodinated polymers had better absorption efficiencies compared to the tosylated polymers with the exception of β-CD-methylimidazolium tosylate polymer.

4.7.2 Absorption of chromium (Cr^{6+}) and cadmium (Cd^{2+}) by the polymers

Chromium is a toxic inorganic pollutant (heavy metal), especially when present in water as Cr^{6+}. This metal ion can penetrate the skin causing irritation, liver and kidney damage as well as a decrease in male sperm counts (Li and Ma, 1999; Morales et al., 1999). Chromium standards were prepared as outlined in Chapter 3 and a Cr^{6+} concentration of 5 mg.L^{-1} (30 mL) was passed through the polymers (300 mg). The detection limit of Cr^{6+} by AAS is 0.002 mg.L^{-1}. A working range of 0.2-10 mg.L^{-1} was recommended thus 5 mg.L^{-1} was used for its analysis. Cadmium ion (Cd^{2+}) was also used as a model inorganic pollutant to study the absorption efficiencies of the CD-IL polymers. Cadmium was studied because it is toxic and even lethal to the human body, particularly the central nervous system (Morales et al., 1999). This metal ion has a great affinity for sulphur hence it attacks the sulphur bonds of enzymes making them to function improperly (Lewis and Evans, 2001). Cadmium standards were prepared as outlined in section 3 and a concentration of Cd^{2+} (1 mg.L^{-1}, 30 mL) was passed through the polymers (300 mg). The detection limit of Cd^{2+} by AAS is 0.002 mg.L^{-1}. Since a working range of 0.005-2 mg.L^{-1} is recommended, a 1 mg.L^{-1} was used for Cd^{2+} analysis. After passing the Cr^{6+} and the Cd^{2+} spiked water sample through the CD-IL polymers, AAS analysis of the eluent was performed. Results obtained are shown in Table 4.4. Native β-CD polymers were also used to see if they can absorb inorganic pollutants from an aqueous medium.

Polymers	Amount of Cr^{6+} and Cd^{2+} removed (%)			
	HDI linker		TDI linker	
	Cr^{6+}	Cd^{2+}	Cr^{6+}	Cd^{2+}
β-CD	34	18	60	24
β-CDMIMOTs	100	33	100	28
β-CDBUTIMOTs	98	32	97	23
β-CDPYROTs	100	24	95	20
β-CDMIMIOD	69	15	43	18
β-CDBUTIMIOD	51	21	95	17
β-CDPYRIOD	34	10	60	13

Table 4.4. Comparison of absorption efficiency of Cr^{6+} by HDI and TDI polymers

It is clear from Table 4.4 that the incorporation of the ionic liquid component onto the CD backbone resulted in high absorption of Cr^{6+}. Absorption of Cd^{2+} was however significantly reduced. For example, β-CDMIMOTs/HDI polymer absorbed 100% of the Cr^{6+} but only 33% of the Cd^{2+} was absorbed. This can be the attributed to the fact that ionic liquids are selective in the absorption of heavy metal ions (Welton, 1999; Chowbury et al., 2007). For example, the incorporation of sulphur or urea in the alkyl chain of the cation has been reported to greatly enhance the absorption of Cd^{2+} and Hg^{2+} from an aqueous medium (Visser et al., 2002). This observation suggests that, to improve the extraction of Cd^{2+}, further studies that incorporate sulphur or urea on the cations need to be carried out. A discussion on how variations of cation, anion, alkyl chain as well as the linker impact on the extraction efficiency of the polymers follows in the next section.

4.7.2.1 Effect of the type of the cation on the extraction of the inorganic pollutants

The incorporation of the imidazolium or pyridinium rings onto the CD backbone was found to have no bearing on the amount of the pollutant absorbed by the polymers. Some imidazolium-based polymers (e.g. β-CDMIMOTs) and some pyridinium-based polymers (e.g. β-CDPYROTs) had 100% absorption efficiencies for Cr^{6+} while some imidazolium-based polymers (e.g. β-CDBUTIMIOD/HDI) and some pyridinium-based polymers (e.g. β-CDPYRIOD/HDI) had low absorption efficiencies of 51 and 34%, respectively. The same polymers, however, showed less efficiency in the absorption of Cd^{2+}.

4.7.2.2 Effect of the imidazolium alkyl chain on the extraction of the inorganic pollutants

The length of the alkyl chain of the imidazolium ring was also investigated to determine its effect on the absorption efficiency of the polymers. From Table 4.4 it was observed that increasing the alkyl chain of the imidazolium ring did not have much bearing on the amount of either Cr^{6+} or Cd^{2+} absorbed by the polymers. No pattern could be established between the amount of pollutant (either Cr^{6+} or Cd^{2+}) absorbed and the length of the alky chain attached to the imidazolium ring.

4.7.2.3 Effect of the type of anion on the extraction of the inorganic pollutants

The type of anion of the ionic liquid seems to have an effect on the amount of Cr^{6+} and Cd^{2+} absorbed by the polymers (Table 4.4). The tosylated polymers (both HDI and TDI) showed high absorption efficiencies of Cr^{6+} and Cd^{2+} than the iodinated polymers (Figure 4.18). The tosylate anion enhances the chelation (hexavalent dentition) of the chromium with the cationic component of the CD-IL polymers because it is a hydrophobic anion (Welton, 1999), while the iodide anion is a hydrophilic anion; hence the tosylate anion enhances the chelation of the heavy metal by repelling the water molecules from the cationic component (Cocalia et al., 2006).

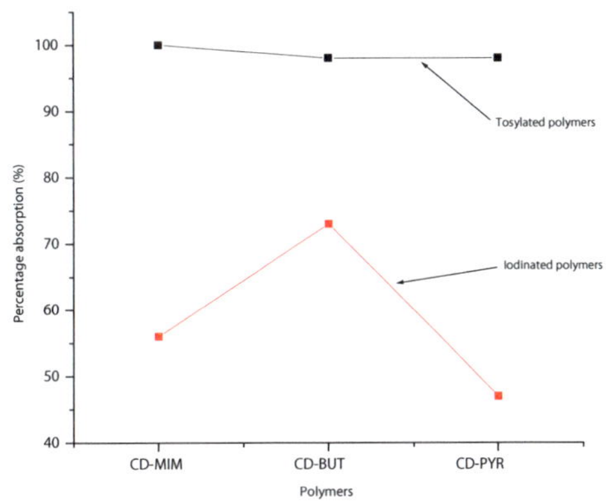

Fig. 4.18. Comparison of Cr^{6+} absorbed by average tosylated versus average iodinated polymers (both (TDI and HDI)

4.7.2.4 Effect of the linker on the extraction of the inorganic pollutants

For the absorption of Cd^{2+} and the Cr^{6+}, there appears to be no correlation between the amount of pollutant absorbed and the type of linker used. For instance, the β-CDMIMOTs polymer absorbed 100% of Cr^{6+} irrespective of the linker used. However, it is noteworthy that the native β-CD polymers were able to absorb some chromium and cadmium from water. The HDI and TDI polymers showed 34% and 60% of Cr^{6+}, respectively, while 18 and 24% absorption efficiencies were respectively observed for the Cd^{2+} polymers. This observation is attributed to the ability of cyclodextrins to form complexes with chromium (Cardas et al., 2005). Similar findings (Brusseau et al., 1997), confirm that not only can modified cyclodextrins form complexes with organic compounds, but they can also complex with cationic heavy metals.

4.8 Comparison of absorption and surface area

After attaining the absorption efficiencies of the polymers, it was necessary to see if there was any correlation between the amount of pollutant absorbed and the surface area of the polymers. The results obtained are summarised in Table 4.5.

Sample	HDI					TDI				
	Surface area (m^2/g)	% PNP absorbed	% TCP absorbed	% Cr^{6+} absorbed	%Cd^{2+} absorbed	Surface area (m^2/g)	% PNP absorbed	% TCP absorbed	% Cr^{6+} absorbed	% Cd^{2+} absorbed
β-CD	2.53	64	54	34	18	1.706	68	58	60	24
β-CDMIMOTs	23.26	80	69	100	33	22.71	77	69	100	28
β-CDBUTIMOTs	2.99	68	69	98	32	10.51	64	77	97	23
β-CDPYROTs	2.89	15	63	100	24	1.64	40	60	95	20
β-CDMIMIOD	2.51	65	75	69	15	1.81	55	74	43	18
β-CDBUTIMIOD	10.88	63	67	51	21	1.78	71	71	95	17
β-CDPYRIOD	0.4427	24	69	34	10	7.00	48	61	60	13

Table 4.5. Comparison of surface area versus percentage absorption

From these results, it was observed that there is no direct relationship between the amount of pollutant absorbed and the surface areas of the polymers. For example, both *β*-CDMIMOTs/HDI and *β*-CDPYROTs/HDI polymers showed 100% absorption efficiency for PNP yet their surface areas are significantly different, 23.26 m^2/g and 2.89 m^2/g, respectively.

4.9 Absorption efficiencies of the polymers on a mixture of organic and inorganic pollutants

In real water samples, organic compounds and heavy metals co-exist (Brusseau et al., 1997). It was, therefore, necessary to study the absorption efficiency of the CD-IL polymers in a synthetic mixture containing both organic and inorganic pollutants. These experiments were carried out to determine if the organic and inorganic pollutants would compete for the absorption sites of the CD-IL.

4.9.1 PNP and chromium

From the absorption studies of PNP and Cr^{6+}, it was evident that the CD-IL polymers exhibited enhanced absorption efficiencies on exclusive absorption tests; hence studies to investigate whether they would still exhibit such enhanced efficiencies were carried out. PNP and Cr^{6+} were thus chosen to study the effect of the presence of both organic and inorganic pollutants on the absorption efficiency of the polymers (Table 4.6).

Polymers	HDI linked		TDI linked	
	PNP absorbed	Cr^{6+} absorbed	PNP absorbed	Cr^{6+} absorbed
β-CD	63 (64)	14 (34)	60 (68)	40 (60)
β-CDMIMOTs	62 (80)	72 (100)	34 (77)	36 (100)
β-CDBUTIMOTs	13 (68)	21 (98)	34 (64)	50 (97)
β-CDPYROTs	27 (15)	23 (100)	40 (40)	29 (95)
β-CDMIMIOD	16 (65)	36 (69)	20 (55)	14 (43)
β-CDBUTIMIOD	69 (63)	25 (51)	47 (71)	27 (95)
β-CDPYRIOD	14 (24)	10 (34)	43 (48)	50 (60)
Values in parenthesis refer to individual absorption of the pollutant by the polymer				

Table 4.6. Comparison of HDI and TDI linked CD-IL polymers on the absorption efficiencies of a mixture of PNP (10 ppm) and Cr^{6+} (5 ppm)

From the results in Table 4.6, it was observed that the presence of both organic and inorganic pollutants generally resulted in reduction in the absorption efficiency of the pollutants. This is an indication that there is competition between the organic and inorganic pollutants for complexing sites on the polymers. However, for β-CDBUTIMIOD/HDI and β-CDPYROTs/HDI polymers, the presence of Cr^{6+} enhanced the absorption efficiency of PNP.

4.8.2 PNP and cadmium

Since the CD-IL polymers showed low absorption efficiency for the mixture of PNP and chromium spiked water samples (though separate analysis showed high absorption efficiencies), a mixture of PNP and cadmium spiked water samples was passed through the polymers in order to determine if the same observations would be seen. The results for this analysis are shown in Table 4.7.

Polymers	HDI linked		Amount removed Cd^{2+} (%)	
	% PNP absorbed	% Cd^{2+} absorbed	% PNP absorbed	% Cd^{2+} absorbed
β-CD	24 (64)	12 (18)	42 (68)	17 (24)
β-CDMIMOTs	48 (80)	33 (33)	23 (77)	28 (28)
β-CDBUTIMOTs	15 (68)	32 (32)	25 (64)	23 (23)
β-CDPYROTs	18 (15)	24 (24)	51 (40)	20 (20)
β-CDMIMIOD	16 (65)	15 (15)	23 (55)	18 (18)
β-CDBUTIMIOD	66 (63)	21 (21)	43 (71)	17 (17)
β-CDPYRIOD	13 (24)	10 (10)	37 (48)	13 (13)
Values in parenthesis refer to individual absorption of the pollutant by the polymer				

Table 4.7. Comparison of HDI and TDI linked CD-IL polymers on the absorption efficiencies of PNP (10 ppm) and Cd^{2+} (1 ppm)

From Table 4.7 it was observed that the absorbances of the CD-IL polymers remain the same, (low regardless of the type of heavy metal ion used (i.e. whether Cd^{2+} or Cr^{6+})). It was observed that the absorption efficiency of the polymers generally decreased in the presence of both organic and inorganic pollutants an indication that Cd^{2+} retards the absorption of both organic and inorganic pollutants from water. However, a thorough investigation needs to be carried out to ascertain these observations.

5. Conclusions

Based on the aims and objectives of this investigation as set out earlier in the chapter, it can be concluded that this study was successful and the following conclusions can therefore be drawn:

Alkyl imidazolium and pyridinium based CD-IL complexes were successfully synthesised from monotosylated and monoiodinated CDs in relatively high yields and they were successfully characterised using IR, ^1H NMR and ^{13}C NMR spectroscopic techniques. The CD-IL complexes were successfully polymerised using hexamethylene diisocyanate (HDI) and toluene-2,4-diisocyanate to form CD-IL polymers. Unlike the water soluble CD-IL complexes, the synthesised CD-IL polymers were insoluble in water and common organic

solvents such as DMF, CH_2Cl_2 and CH_3OH. This property is very important because it is usually exploited thus allowing their application in water treatment. Further analysis of the polymers using SEM revealed that some of the polymers (e.g. β-CDPYROTs/HDI and β-CDMIMOTs/TDI) had a spongy surface while others were crystalline (e.g. β-CDMIMIOD/TDI and β-CDBUTIOD/HDI). However, it could not be determined whether this phenomenon was due to the incorporation of the pyridinium or the imidazolium rings. Also, BET results were not conclusive because no relationship could be established between the pore volume, surface area and the amount of pollutant absorbed by the polymers.
Thermal analyses of the polymers revealed that not only were CD-IL polymers more stable than the native CD polymers, the thermal stability of the CD-IL polymers were affected by the type of cation. It was established that the polymers containing pyridinium-based polymers are less stable than the imidazolium-based polymers. The CD-IL polymers were effective at the removal of organic and inorganic pollutants from water. The β-CDMIMOTs/HDI polymer showed high absorption efficiency for both organic (80% with respect to PNP) and inorganic (100% with respect to Cr^{6+}) pollutants. The CD-IL polymers showed very little affinity for cadmium. This implies that the CD-IL polymers we synthesised are not suitable for the absorption of Cd^{2+} from an aqueous media. From the results obtained from this study, it was found that the absorption efficiency of the CD-IL polymers was generally retarded when a synthetic mixture of organic and inorganic pollutants was analysed. This therefore revealed that the organics and inorganics compete for complexing sites of the CD-IL polymers.

6. Future work

Also, as part of future work on this subject, the following is under study for future work
This study only explored the use of imidazolium and pyridinium cations in the synthesis of CD-IL complexes. Synthesis of CD-IL complexes using other cations like ammonium and phosphonium cations need to be explored. Other anions such as hexafluorophosphate (PF_6), tetrafluoroborate (BF_4) and trioctylmethylammonium thiosalicylate (TOMATS) have also not been tried. Conversion of the tosylate or iodide anions (of the CD-IL complexes we synthesised) into hydrophobic anions (e.g. the less toxic TOMATS) would generate new and insoluble CD-IL complexes. This would thus eliminate the need for the preparation of the CD-IL polymers. Attachment of thio, thioether, thiourea and urea on the cationic components of the ionic liquid has been reported to have high complexation capabilities for cadmium and mercury (Visser et al., 2002). Such attachments, especially on the imidazolium ring, need to be explored.
Above that, absorption efficiencies of the polymers were done only at high concentrations (ppm levels) and using only four pollutants. It would be important to test these polymers at lower concentrations (ppb levels) and on a wide range of organic and inorganic pollutants. The preliminary studies in our laboratory have shown that there is competition for complexation sites between the anion and the cation, conditions that would decrease this competition need to be found to exploit the full extent of the absorption capabilities of the polymers. Since the effectiveness of the polymers was studied for only one cycle, more multiple cycles need to be studied in order to test the strength of the polymers and their ability to still absorb pollutant species after a number of cycles.

7. References

Bender M. L. and Komiyana M. (1978) *Cyclodextrin Chemistry* p2.

Brusseau M. L., Wang X. & Wang W. (1997) Enhanced Solubilization of a Metal−Organic Contaminant Mixture (Pb, Sr, Zn, and Perchloroethylene) by Cyclodextrin. *Environ. Sci. Technol.* 31 1087-1092.

Byun H. S., Zhong N. and Bittman R. (2000) *Organic Syntheses* 77 225-228.

Cardas S., Foray M-F., Gadelle A., Gerbaud G. & Bardet M. (2005) High-resolution solid state ^{13}C NMR study of per(3,6-anhydro)-*a*-cyclodextrin based polymers and of their chromium complexes *Carbohydrate Polymers* 61 88-94.

Chowdhury S., Mohan R. S. & Scott J. L. (2007) Reactivity of ionic liquids *Tetrahedron* 63 2363-2389.

Cocalia V. A., Holbrey J. D., Gutowski K. E., Bridges N. J. & Rogers R. D. (2006) Use of ionic liquids in the study of fruit ripening by high-resolution ^{13}C NMR spectroscopy: 'green' solvents meet green bananas *Tshinghua Science and Technology* 11183-193.

Croft A. P. & Bartsch R. (1983) Synthesis of chemically modified cyclodextrins *Tetrahedron* 39 1417-1474.

Cruz D. M. P. & Moens L. (2004) Thermal Analysis and Oxidation Reactions with Room-Temperature Ionic Liquids. DOE ERULF Program, University of Puerto Rico, National Renewable Energy Laboratory, Golden, Colorado

Ferreira P., Goncalves I. S., Pillinger M., Rocha J., Santos P. & Teixeira-Dias J. J. C. (2000) Modifications of *β*-cyclodextrin with ferrocenyl groups by ring opening of an encapsulated [1] ferrocenophane *Organometallics* 19 1455-1457.

Field L. D., Sternhell S. & Kalman J. R. 1995 *Organic Structures from Spectra,* John Wiley and Sons, New York, p6

Frederic D. GC/MS Analysis. *http://www.scientific.org/tutorials/articles/gcms.html* Accessed 19.05.2004.

Fujita K., Nagamura S. & Imoto T. (1984) Convenient preparation and effective separation of the C-2 and C-3 tosylates of α-cyclodextrin *Tetrahedron Letters* 25 5673-5676.

Garrido Frenich A., Martinez Vidal J. L., Cruz Sicilia A. D. Gonzalez-Rodriguez M. J., & Plaza Bolanos P. (2006) Multiresidue analysis of organochlorine and organophosphorus pesticides in muscle of chicken, pork and lamb by gas chromatography−triple quadrupole mass spectrometry *Analytica Chimica* 558 45-52.

Halmowitz S. (1991) *Chemical Engineering* 75-80.

Harada A. (1997) Construction of supramolecular structures from cyclodextrins and polymers. *Carbohydrate Polymers* 34 183-192.

http://www.epa.gov/pesticides Accessed 01-05-06.

http://www.npi.gov.au/database/substance-info/profiles/90.htm Accessed 28-05-07.

http://www.sigma-aldrich.com/chemicalsynthesis. ChemFiles 5 (6) 1-24.

http://www.waterbornne.org.doc;factsheets.doc;inorganic.htm Accessed 26-04-06.

Kawaguchi M., Ishii I., Okanouchi N., Sakni N., Ito R. & Inowe K. (2004) *Organochlorine Compounds* 66 181-185.

Lewis R. & Evans W. *Chemistry* 2nd Ed. Palgrave Macmillan, Hampshire United Kingdom 2001, pp 415.

Li D. Q. & Ma M. (1999) New organic nanoporous polymers and their inclusion complexes *Chem. Mater.* 11 872-874.

Li D. Q. & Ma M. (1999) Nanosponges: From inclusion chemistry to water purifying technology *Chemical Science and Technology* 26-28.

Li D. Q. & Ma M. (1999) From inclusion chemistry to water purification technology *Chemtech.* 35 31-37.

Li J. & Tan D. (2001) Inclusion complexes formation between cyclodextrins and poly(1,3-dioxolane) *Macromolecules* 34 1542-1544.

Liu J., Chi Y., Jiang J., Tai C., Peng J. & Hu J. (2004) Ionic liquid-based liquid-phase microextraction, a new sample enrichment procedure for liquid chromatography *Journal of Chromatography A.* 1026 143-147.

Liu J., Jiang G., Chi Y., Cai Y., Zhou Q. & Hu J. (2003) Use of ionic liquids for liquid-phase microextraction of polycyclic aromatic hudrocarbons *Journal of Anal. Chem.* 75 5870-5876.

Liu J., Johnsson J. A. & Jiang G. (2005) Application of ionic liquids in analytical chemistry. *Trends in Analytical Chemistry* 24 20-27.

Mamba B. B., Krause R. W., Malefetse T. J. & E. N. Nxumalo (2007) Monofuctionalized cyclodextrin polymers for the removal of organic pollutants from water. *Environ. Chem. Lett.* 5 79-84.

Martin Del Ville E. M. (2004) Cyclodextrins and their uses: a review *Process Biochemistry* 39 1033-1046.

Mhlanga S. D., Mamba B. B., Krause R. W. & Malefetse T. J. (2007) Removal of organic contaminants from water using nanosponge cyclodextrin polymers *J. Chem. Technol. Biotechnol.* 82 382-388.

Mhlanga S. D. MTech Dissertation (2006) University of Johannesburg.

Morales M. M., Marti P., Llopis A., Campos L. & Sagrado S. (1999) An environmental study by factor analysis of surface seawaters in the Gulf of Valencia (Western Mediterranean) *Analytica Chimica Acta* 394 109-117.

Muderawan W., Ong T., Tang W., Yong D., Ching C. & Ng S. (2005) Synthesis of ammonium substituted β-cyclodextrins for enantioseparation of anionic analytes. *Tetrahedron Letters* 46 1747-1749.

Niegowski S. J. (1952) Destruction of phenols by oxidation with ozone. *Industrial and Engineering Chemistry* 45 632-634.

Oleszczuk P. & Baran S. (2004) Application of solid-phase extraction to determination of polycyclic aromatic hydrocarbons in sewage sludge extracts. *Journal of Hazardous Materials* 113 237-245.

Pandey S. (2006) Analytical applications of room-temperature ionic liquids: A review of recent efforts *Analytica Chimica Acta* 556 38-45.

Park J., Dec J., Kim J. & Bollag J. (1999) Effect of Humic Constituents on the Transformation of Chlorinated Phenols and Anilines in the Presence of Oxidoreductive Enzymes or Birnessite *Environ. Sci. Technol.* 33 2028-2039.

Pereiro A. B., Legido J. L. & Rodriguez A. (2007) Physical Properties of Ionic Liquids Based on 1-Alkyl-3-methylimidazolium Cation and Hexafluorophosphate as Anion and Temperature Dependence *J. Chem. Thermodynamics* 391219-1230.

Petanen T & Romantschuk M. (2002) Use of bioluminescent bacterial sensors as an alternative method for measuring heavy metals in soil extracts *Analytica Chimica Acta* 456 55-56.

Phan T. N. T., Bacquet M. & Morcellet M. The removal of organic pollutants from water using new silica-supported β-cyclodextrin derivatives *Reactive and Functional Polymers* 52 (2002) 117-125.

Rogers J. D., Jedral W. & Bunce N. J. (1999) Electrochemical Oxidation of Chlorinated Phenols *Environ. Sci. Technol.* 33 1453-1457.

Rong D. & D'Souza V. T. A convenient method for functionalization of the 2-position of cyclodextrins *Tetrahedron Letters* 31 (1990) 4275-4287.

Sabio E., Zamora F., Gonzalez J. F., Gonzalez-Garcia C. M., Roman S. & Al-Kassir (2006) Characterisation under static and dynamic conditions of commercial activated carbons for their use in wastewater plants *Applied Surface Science* 6058-6063.

Salipira K. L., Mamba B. B., Krause R. W., Malefetse T. J. & Durbach S. H. (2007) Carbon nanotubes and cyclodextrin polymers for the removal of organic pollutants from drinking water. *Environ. Chem. Lett.* 55 13-17.

Santos A., Yustos P., Rodriguez S., Simon E. & Garcia-Ochoa F. (2007) Abatement of phenolic mixtures by catalytic wet oxidation enhanced by Fenton's pretreatment: Effect of H_2O_2 dosage and temperature *Journal of Hazardous Materials* 146 595-601.

Savage N. & Diallo M. S. (2005) Nanomaterials and water purification: Opportunities and challenges *Journal of Nanoparticle Research* 7 476-486.

Schwarzenbach R. P., Gschweld P. M. & Imboden D. M. *Environmental Organic Chemistry* 2nd Ed 2003, 4-7 .

Sutyagin A. A., Glazyrin A. E., Kurochkina G. I., Gracher K. K. & Nifant'ev E. E. (2002) Regioselective Acetylation of β-Cyclodextrin *Russian Journal of Chemistry* 72 147-150.

Szeijtli J. (1998) Introduction and General Overview of Cyclodextrin Chemistry *Chem. Rev.* 98 1743-1753.

Tran C. D. & De Paoli Lacendra S. H. (2002) Determination of Binding Constants of Cyclodextrins in Room-Temperature Ionic Liquids by Near-Infrared Spectrometry. *Anal. Chem.* 74 5337-5341.

US Environmental Protection Agency (EPA), (2000) *Chemical Precipitation,* US EPA, Washington, DC, (EPA832-f-00-018).

Valerie C. (2003) Solid phase extraction of trace elements. *Spectrochimica Acta Part B:* 58 1177-1233.

Visser A. E. & Rogers R. D. (2004) Influence of structural variation in room-temperature ionic liquids on the selectivity and efficiency of competitive alkali metal salt extraction by a crown ether. *Ind. Eng. Chem. Res.* 39 3596-3604.

Visser A. E., Swatloski R. P., Griffin S.T., Hartmen D. H. & Rogers R. D. (2001) Liquid/Liquid Extraction of Metal Ions in Room Temperature Ionic Liquids *Sep. Sci. Technol.* 36 785-804.

Visser A. E., Swatloski R. P., Reichert W. M., Mayton M., Sheff M., Wierzbicki A., Davis Jr. J. H. & Rogers R. D. (2002) Task-Specific Ionic Liquids Incorporating Novel Cations for the Coordination and Extraction of Hg^{2+} and Cd2+: Synthesis, characterization and Extraction Studies. *Environ. Sci. Technol.* 36 2523-2529.

Wei G.T., Yang Z. & Chen C. J. (2003) Room temperature ionic liquid as a novel medium for liquid/liquid extraction of metal ions. *Analytica Chimica Acta* 488 183-192.

Weinz G., Bons Steinbrum M. & Landfester K. (1997) Solid state polycondensation within cyclodextrin channels leading to watersoluble polyamide rotaxanes. *Tetrahedron* 5315575-15592.

Welton T. (1999) Room-Temperature Ionic Liquids: Solvents for Synthesis and Catalysis. *Chem. Rev.* 99 2071-2083.

Wikstrom E. & Marklund S. (2000) Secondary Formation of Chlorinated Dibenzo-*p*-dioxins, Dibenzofurans, Biphenyls, Benzenes, and Phenols during MSW Combustion. *Environ. Sci. Technol.* 34 604-609.

Zhong N., Byun H. S. & Bittman R. (1998) An improved synthe- sis of 6-O-monotosyl-6-deoxy-β-cyclodextrin. *Tetrahedron Letters* 39 2919-2920.

Zymark Corporation. Application Brief (2001) Method 252.2, *Semi-Volatile Organics in Waterby Solid Phase Extraction and GC/MS Detection, Zymark Corporation.*

Part 3

Nanotechnology

8

Ionic Liquid as Novel Solvent for Extraction and Separation in Analytical Chemistry

Li Zaijun[1], Sun Xiulan[2] and Liu Junkang[1]
[1]*College of Chemical and Materials Engineering, Jiangnan University, Wuxi 214122,*
[2]*State Key Laboratory of Food Science and Technology, Wuxi 214122,*
China

1. Introduction

Direct analysis of samples, regardless their origin, is desirable although for the majority of cases unfeasible on account of the complexity of the sample matrix, inadequate concentration of the target analytes, or even incompatibility with the detector. In these cases, a sample pretreatment step is required for interference removal and analytes separation/preconcentration. Despite valuable advances developed in separation science, the traditional solvent extraction is widely used for samples preparation. Main drawback of the solvent extraction is the requirement of large amounts of high-purity solvents that are expensive and toxic and result in the production of hazardous waste. Therefore, the search for new solvent is a key trend in solvent extraction evolution. In this sense, ionic liquid, which is ionic media resulting from the combination of organic cations and various anions, has attracted much attention taking into account their special features like: low-vapor pressure, high viscosity, dual natural polarity, good thermal stability and a wide range of miscibility with water and other organic solvents, hence many environmental and safety problems associated with organic solvents are avoided. As a result of their unique chemical and physical properties, the ionic liquid has aroused increasing interest for their promising role as alternative medium for classical solvent extraction [1-7] and organic synthesis [8-10]. Recently, ionic liquids were rapidly developed as environment-friendly acceptor phase for various microextraction to sample preparation in analytical chemistry such as liquid phase microextraction, single drop liquid phase microextraction, solid phase microextraction, dispersive liquid phase microextraction and cold induced aggregation microextraction. The microextraction with ionic liquid often exhibited better extraction efficiency and enrichment factor than that with conventional solvent extractions. Among the microextractions, dispersive liquid phase microextraction is very rapid and effective. Since the ionic liquid is dispersed completely into aqueous phase and the analytes will more easily migrate into the ionic liquid phase because of the much large contact area, the procedure for dispersive liquid phase microextraction may be complete within several minutes. In recent years, the ionic liquid extraction coupled with different analytical technology has widely been applied to determination of ultra trace analytes in water or other samples such as high performance liquid chromatography (HPLC) and gas chromatography mass spectrometry (GC-MS) for organic compounds, and flame atomic absorption spectrometry (AAS) and graphite furnace atomic absorption spectrometry (ETAAS) for metal ions. In general, the extractions with ionic liquid have increasingly become the most important samples preparation technologies in analytical chemistry.

2. Application of ionic liquids as medium for classical solvent extraction

2.1 Metal ion extraction and its application

For metal ion extraction, one needs to find the right combination of ionic liquid and ligand to obtain similar behavior as in organic solvents or one needs to use the differences from organic solvents for achieving more advantageous separations. Although the metal ion extraction in the ionic liquid/aqueous two phase system often indicated high efficiency and selectivity that excelled most of organic solvents, the metal ion partitioning always rely on the species of the ionic liquid, metal ion and ligand seriously. For examples, Bartsch group reported on the influence of structural variations in room temperature ionic liquid on the selectivity and efficiency of competitive alkali metal salt extraction by crown ethers [11]. They showed that extraction efficiency generally diminished as the length of the 1alkyl group on the ionic liquid was increased. It was demonstrated that metal ion affinity for the hydrophobic phase necessitates the presence of an extractant [12]. Professor Rogers has investigated Cyaex-923 as an ligand for actinide ions in liquid/liquid separation using the hydrophobic ionic liquid $[C_{10}mim][Tf_2N]$ as the extracting phase, a mixture of tri-alkyl-phosphine oxides of varying mixture of tri-alkyl-phosphine oxides of varying chain lengths (hexyl- and octyl-) and has been used for actinide extraction into molecular solvents. Distribution ratio measurements for Pu^{4+} and Am^{3+} showed little acid dependency for Pu^{4+}, but strong acid dependency studies seem to support 1:1 metaltoligand binding for Pu^{4+}, but significantly the notion that ionic liquids can provide different coordination environments [13]. In such application, imidazolium cations were also derived to include the task specific functionalities, such as metal ligating groups that when used as part of the solvent or doped in the less expensive ionic liquids could dramatically enhance the extraction of targeted metal ions [14]. Moreover, the effect of pH on the extraction efficiencies showed a big difference due to its differences in solvent behavior of ionic liquid and conventional organic solvents. After metal ion was extracted to ionic liquid, a strong solution such as nitric acid was applied to stripping metal ion from the ionic liquid. We have investigated the conditions of stripping metal ions from ionic liquid too, the experiments indicated metal ion is easy to be stripped from ionic liquid phase and enter aqueous phase if the neutral complex was formed in a base medium. Otherwise, metal ion stripping required a relatively strong base solution. However, the stripping process is slow very much at room temperature. In order to overcome the problem, a new stripping procedure was developed for stripping metal ion from ionic liquid, in which the equal volume of acetone and the ionic liquid containing metal ion complex mixed before a strong solution was added, whole process can complete within several minutes. At the same, above stripping method may be used to renew of the ionic liquids.

Recently, we attempt that metal ion extraction in the ionic liquid/aqueous two phase system combined with spectrometry technology including AAS and spectrophotometry to determine ultra trace amounts of metal elements in environmental samples. It was found that the extraction procedure can decrease the detect limit and improve the selectivity of analytical method remarkably. For atomic absorption spectrometry, the metal ion extraction must be back-extracted into aqueous solution, then was introduced into atomic absorption spectrometer for determination. For example, a new room temperature ionic liquid

$[C_4tmsim][PF_6]$ was developed as a novel medium for liquid/liquid extraction of inorganic mercury with CDAA, the extraction and back-extraction efficiencies were 99.9 and 100.1 % for 5.0 μg.L^{-1} standard mercury in 1000 mL of water solution, respectively, the detection limit is 0.01 μg.L^{-1} mercury in real water sample, proposed method has been used to the determination of trace inorganic mercury in natural water with satisfactory results [2]. If the ionic liquid phase containing metal complex is transparent, it can be directly used to spectrophotometric determination. So that, metal ion extraction in the ionic liquid/aqueous two phase system in which some excellent chromogenic reagent was selected as ligand to form metal complex is very selective, this may overcome the drawback in sensitivity of conventional spectrophotometric methods. For example, we reported a highly sensitive and selective extraction spectrophotometric method for determination of trace germanium in natural water with a new chromogenic reagent methybenzeneazosalicylfluorone abbreviated as MBASF, in which ionic liquid $[C_4mim][PF_6]$ was used as novel medium for liquid/liquid extraction of germanium (IV). In the presence of TritonX-100, MBASF reacted with germanium (IV) to form a red complex rapidly, the complex was then extracted into the $[C_4mim][PF_6]$ phase (shown in Fig.1), absorbance of the complex in ionic liquid at 496 nm was recorded and used to determine trace Ge^{4+}. The apparent molar absorptivity of the complex and the detection limit for the real sample were found to be 3.12×10^6 L.moL^{-1}.cm^{-1} and 0.2 ng.mL^{-1} respectively. The absorbance of the complex at 496 nm increases linearly with the concentration up to 4mg of Ge^{4+} in 250 mL of aqueous solution. The interference study show the determination of germanium is free from the interference of almost all positive and negative ions found in the natural water samples. The determination of Ge^{4+} in natural water was carried out by the present method and ETAAS. The results were

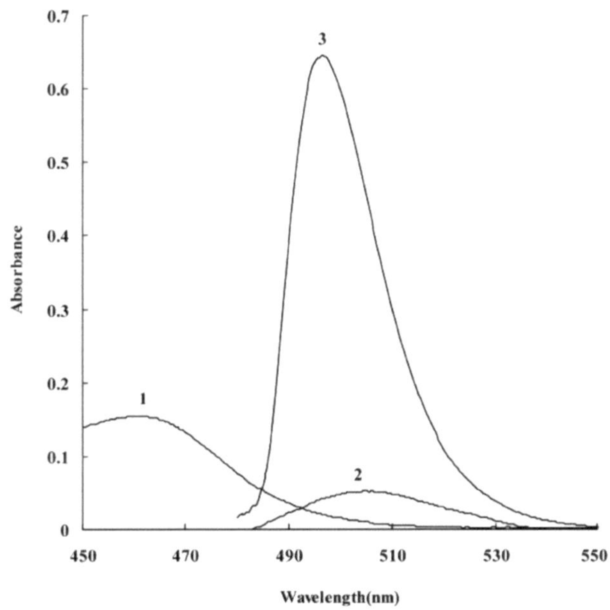

Fig. 1. Absorption spectrum of the reagent against water (1), the complex in the aqueous phase (2), and in the $[C_4mim][PF_6]$ (3), germanium (IV): 4.0mg in 250 mL of aqueous solution, $[C_4mim][PF_6]$

satisfactorily comparable so that the applicability of the proposed method was confirmed using the real samples [40]. Metal ion extraction in ionic liquid/aqueous two phase system and their application were listed in Table 1.

Metal ion	Ionic liquid	Ligand	Detection technology /detection limit	Ref.
Ni^{2+}	$[C_4mim][PF_6]$,	PAN	FI-FAAS/12.5 μg.L^{-1}	15
Fe^{3+}, Cu^{2+}, Ni^{2+}, Mn^{2+}	trioctylmethylammonium salicylate			16
Cd^{2+}	ionic liquid functionalized 2-mercaptobenzothiazole	ammonium pyrrolidine dithiocarbamate and diethydithiocarbamate	AAS	17
Pb^{2+}, Cu^{2+}, Hg^{2+}, Cd^{2+}	$[C_4mim][PF_6]$	2-aminothiophenol		18
Co^{2+}	1-hexylpyridinium hexafluorophosphate	1-phenyl-3-methyl-4-benzoyl-5-pyrazolone	AAS/0.70 mug.L^{-1}	19
Cd^{2+}	$[C_4mim][PF_6]$	2-mercaptobenzothiazole	AAS/4.6 ng.L^{-1}.)	20
La^{3+}	$[C_4mim][Tf_2N]$	18-Crown-6 and 2-thenoyltrifluoroacetone		21
Sr^{2+}	$[C_4mim][PF_6]$	DCH-18-C-6		22
Am^{3+}	$[C_4mim][Tf_2N]$	tri-n-butylphosphate and n-octyl(phenyl)-N,N-diisobutylcarbamoylmethylphosphine oxide		23
UO_2^{+}	$[C_4mim][PF_6]$	tri-n-butyl phosphate		24
Mn^{2+}, Co^{2+}, Ni^{2+}, Zn^{2+}, Cd^{2+}	$[C_nmim][PF_6]$, $[C_nmim][Tf_2N]$			25
Pb^{2+}, Cd^{2+}	$[C_4mim][PF_6]$	tetra-n-propylammonium iodide	differential-pulse anodic stripping voltammetry/0.001 mug.mL^{-1}	26
Th^{4+}	$[C_8mim][PF_6]$	primary amine N-1923		27
UO_2^{2+}	$[C_4mim][Tf_2N]$	ri-n-butylphosphate		28
Y^{3+}	$[C_8mim][PF_6]$	Cyanex-923 and EDTA		29
Pu^{4+}	$[C_8mim][PF_6]$	carbamoyl methyl phosphine oxide		30
La^{3+}	$[C_4mim][Tf_2N]$	2-thenoyltrifluoroacetone and 18-crown-6		31
Ni^{2+}	$[C_4mim][PF_6]$	2-pyridinealdoxime		32
Sr^{2+}, Cs^{+}	tri-butylmethylammonium bis((trifluoromethyl)sulfonyl)imide	ionophores dicyclohexano-18-crown-6 and calix[4]arenebis-(tert-octylbenzo-crown-6)		33
Pb^{2+}	$[C_4mim][PF_6]$	dithizone	ETAAS/1.0 ng.L^{-1}	5
Al^{3+}	$[C_4mim][PF_6]$	3,5-ditertbutylsalicylfluorone	ETAAS/0.06 mug.L^{-1}	6
Zn^{2+}	Butyl-methyl-pyrrolidinium	2,4-dinitrophenolate		34

	trifluoromethylsulfonate			
Pb^{2+}	$[C_8mim][PF_6]$	dithizone	ETAAS/1.0 ng.L^{-1}	7
Sc^{3+}	$[C_8mim][PF_6]$	Cyanex -925		35
M^{2+}	$[C_4mim][PF_6]$	benzoyltrifluoroacetone, 2-naphtoyltrifluoroacetone and trifluoroacetylacetone		36
Pd^{2+}	tri-n-octylmethylammonium chloride and tri-n-octylmethylammonium nitrate	Aliquat-336		37
Cu^{2+}	$[C_4mim][PF_6]$	1-methylimidazole		38
La^{3+}	$[C_4mim][PF_6]$	N,N,N ',N '-tetrakis(2-pyridylmethyl)ethylenediamine		39
Ge^{4+}	$[C_4mim][PF_6]$	methybenzeneazosalicylfluorone	Spectrophotometry/0.2 ng.mL^{-1}	40
UO_2^{2+}, Am^{3+}, Nd^{3+}, Eu^{3+}	$[C_9mim][Tf_2N]$	dialkylphosphoric or dialkylphosphinic acids		41
UO_2^{2+}	$[C_4mim][PF_6]$	tri-n-butylphosphate		42
Cu^{2+}	$[C_4mim][Tf_2N]$	beta-diketonate		43
Cu^{2+}	$[C_4mim][PF_6]$	dithizone, 8-hydroxyquinoline, and 1-(2-pyridylazo)-2-naphthol		44

Table 1. Reviews in extraction of metal ions with ionic liquid as medium

2. Organic compound/gas extraction and separation

Ionic liquids have been widely applied to organic compound/gas extraction and separation too. Basing on the ionic liquid various unique solvent characteristics and extraction performance, room temperature ionic liquids may be used as liquid membrane materials to separation organic compound and gas. For example, $[C_4mim][PF_6]$ as membrane solution can be utilized for hydrocarbon separation by using a supported liquid membrane. Although the permeation rates through the membrane based on the ionic liquids were less than those of water, the selectivity of aromatic hydrocarbons was greatly improved [45]. Ionic liquid as special solution combined with classical separation technologies such as liquid/solid phase microextraction was used to beforehand separation and enrichment of the sample for high performance liquid chromatography, this may improve the separation effect of analytical technology remarkably. Ionic liquid as electrolyte can enhance electric capability to capillary zone electrophoresis. For example, professor Qi reported the ionic liquid used as running electrolyte in capillary zone electrophoresis with cyclodextrin (β-CD) as modifier for the separation of anthraquinones extract of Chinese herb Paedicalyx attopevensis Pierre ex Pitard [46]. Ionic liquid as ion pair reagent was applied to reversedphase high performance liquid chromatography. Moreover, utilization of supercritical carbon dioxide ($scCO_2$) in concert with various popular ionic liquids is shown to have modified and enhanced the overall extraction capabilities and applications of these neoteric solvents [47-50]. Brennecke group have investigated the recovery of organic products from ionic liquids using $scCO_2$. The results indicated the recovery rates of various aromatic and aliphatic solutes from [C_4mim][PF_6] in the presence of environmentally benign $scCO_2$ are presented. In a subsequent study, interesting outcomes showing utility of $scCO_2$ as a separation switch for ionic liquid/organic mixtures were published by the same group [47, 48]. In general, we conjecture ionic liquid will rapidly be applied to many new fields of the extraction and separation with understanding on the extraction performance of room temperature ionic liquid.

3. Application of ionic liquids in microextractions

3.1 Liquid phase microextraction

Ionic liquid has been widely applied to various liquid phase microextractions for sample preparation. Commonly, volume of the ionic liquid for liquid phase microextraction is often lower than 25 μL in order to obtain high enrichment factor. To fine mix the ionic liquid with aqueous phase, a relatively long extraction time and strong stir process are strongly required. However, so small volume of the ionic liquid is difficult to be precisely collected from aqueous phase due to its dissolution in water and adhere to wall of the vial (Table 2). To overcome the problems, some new types of liquid phase microextraction have been reported in literatures. For examples, single drop microextraction (SDME) (Table 3) and dispersive liquid microextraction (DLME) (Table 4).

Analyte	Ionic liquid	Sample	Remark	Ref.
Steroidal antiinflammatory drugs	[C_4mim][PF_6]	Human urine	Enrichment factors varied between 10.69 for naproxen and 13.93for fenbufen. HPLC system was employed for analysis. A dynamic liquid-phase microextraction using an ionic liquid as acceptor phase is reported.	51
Aromatic and aliphatic hydrocarbons	[C_4mim][PF_6]	Water	GC-MS was coupled with determination of anlyte; for the first time hollow fiber-protected ionic liquid supported three-phase (liquid–liquid–liquid) microextraction; enrichment factor and extraction time are 210 and 40min.	52
4-Nonylphenol, 4-tert-octylphenol	[C_4mim][PF_6]	Water	HPLC was employed for analysis; enrichment factors are between 19.5 and 33.8, and extraction time was about at 30 min.	53
Chlorpromazine, promethazine, levomepromazine, prochlorperazine,	[C_4mim][PF_6]	Human urine	HPLC was employed for analysis.	54

trifluoperazine, fluphenazine and thioridazine				
Phorate, parathion and poxim	$[C_4mim][PF_6]$	Water	HPLC was employed for analysis; enrichment factors are in the range from 553 to 665.	55
Benzene, toluene, ethylbenzene, and xylene), polycyclic aromatic hydrocarbons, phthalates, phenols, aromatic amines, herbicides, organotin, and organomecury	$[C_4mim][PF_6]$, $[C_8mim][PF_6]$	Water	HPLC, AAS and or cold-vapor atomic fluorescence spectrometry determination was employed for analysis, respectively; the enrichment factors determined were in the range of 5-168 for 15 min extraction by $[C_4mim][PF_6]$ and 4-178 for 30 min extraction by $[C_8mim][PF_6]$.	56
Chlorpromazine, promethazine, levomepromazine, prochlorperazine, trifluoperazine, fluphenazine and thioridazine	$[C_4mim][PF_6]$	Human urine	The proposed method is a valuable alternative for the analysis of these drugs in urine within the concentration range 0.07-10μg.mL^{-1}. Limits of detection were in the range from 21 ng.mL^{-1} for thioridazine to 60ng.mL^{-1} for levomepromazine.	102
Cobalt	$[C_6mim][PF_6]$	Environmental and biological samples	ETAAS was employed for detection of cobalt.	111
Triazines	$[C_6mim][PF_6]$	water	HPLC was employed for detection of triazines, the detection limits are for the triazine range from 0.36 to 1.41mug.L^{-1}.	116
Palladium	$[C_6mim][PF_6]$	food additive	The limit of detection was 0.2ng.mL^{-1}.	117
Cadmium	$[C_4mim][PF_6]$	saline	o,o-Diethyldithiophosphate was used as a ligand. FAAS was employed for detection of cadmium, the limit of detection (3s) and the enhancement factor were 0.07 mug.L^{-1} and 78, respectively.	127

Table 2. Reviews for common liquid phase microextraction using ionic liquid as acceptor phase

SDME includes direct-immersion liquid phase microextraction and headspace liquid phase microextraction. Because during headspace liquid microextraction the ionic liquid has no loss because it lack effective vapor pressure even at high temperature, headspace liquid phase microextraction often provide a higher enrichment factor for most of analytes than direct-immersion liquid phase microextraction. A typical experimental set-up of headspace liquid phase microextraction is illustrated in Fig.2 [59]. Before each extraction, the microsyringe was rinsed with methanol to avoid analyte carry-over and air bubble

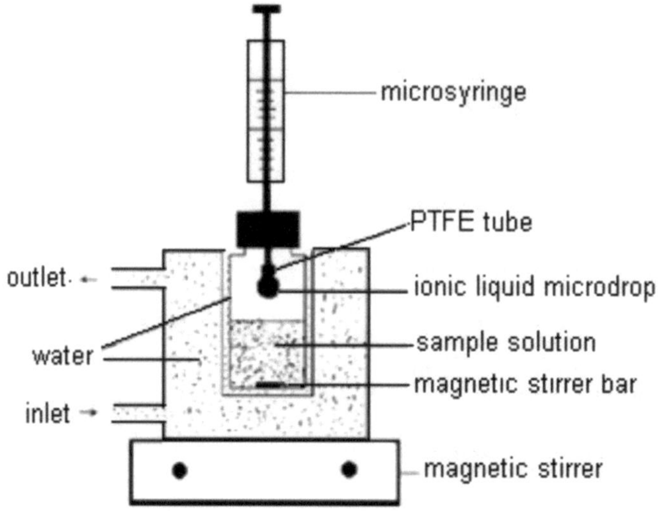

Fig. 2. Schematic diagram of SDME

formation. For the extraction an aliquot of the sample solution was placed in a vial immersed in the recirculating cell. After the uptake of the 10-25 μL of the ionic liquid, the microsyringe needle was inserted through the septum and then tightly fitted with a long polytetrafluoroethylene tube, and the syring to begin with the extraction process, the syringe plunger was depressed completely to expose an ionic liquid drop and the magnetic stirrer was turned. After extracting, the magnetic stirrer was switched off, and then the microdrop was retracted back into the microsyringe and injected directly into detector for analysis. SDME is based on the equilibrium of the three phases: the condensed aqueous phase, the headspace phase and the microdrop phase. Mass transfer in the headspace is assumed to be a faster process since diffusion coefficients in the gas phase are typically 10^4 times greater than corresponding diffusion coefficients in the condensed phase. Thus, both slow mass transfers in the aqueous phase and diffusion of analyte into the microdrop are limiting steps in the overall extraction process was clamped so that the tip of the needle was located in a constant position in the headspace of the vial. The penetration theory indicates the aqueous phase mass transfer coefficient of solute increases with increasing stirring rate, and heating improves the mass transfers in the aqueous phase and induces convection in the headspace. Therefore, consequently the equilibrium between the aqueous and vapor phase can be attained more rapidly. Moreover, an addition of salt (such as NaCl and Na_2SO_4 etc) to the sample can decrease the solubility of analytes in the aqueous phase and enhance their partitioning into the organic phase, the extraction is often enhanced with increasing salt concentration and polarity of compound. Thus, a salt concentration of about 30 % was used in most SDME.

Analyte	Ionic liquid	Sample	Remark	Ref.
Chlorobenzenes	$[C_4mim][PF_6]$	Water	HPLC was coupled with analysis; extraction time is 37min; an obvious salting-out effect was observed.	57
Formaldehyde	$[C_8mim][PF_6]$	Shiitake mushroom	2,4-Dinitrophenylhydrazine was used as derivative agent; HPLC was coupled	58

			with analysis; extraction time is 30min	
Dichlorodiphenyltrichloroethane and its metabolites	$[C_4mim][PF_6]$	Water	HPLC was coupled with analysis; an obvious salting-out effect was observed; extraction time is 30min	59
Phenols	$[C_8mim][PF_6]$	Water	GC was used for detection of phenols; effect of salting-out is obvious.	60
Phenols	$[C_4mim][PF_6]$	Water	HPLC was employed for analysis; enrichment factors were in the range from 17.2–160.7.	61
Chloroform, bromodichloromethane, dibromochloromethane and bromoform	$[C_8mim][PF_6]$	Water	GC-MS was employed for analysis; extraction time was about 30min; an obvious salting-out effect was observed well.	62
Chlorinated anilines	$[C_4mim][PF_6]$	Water	HPLC was employed for analysis; extraction temperature and salting-out have seriously influenced on the extraction efficiencies.	63
Benzophenone-3	$[C_6mim][PF_6]$	Urine	HPLC was employed for analysis.	64
Dichloromethane, *p*-xylene, and *n*-undecane	$[C_4mim][PF_6]$	Water	GC-MS was employed for the analysis; a new removable interface that enables the introduction of the extracted analytes into the GC system, while preventing the ionic liquid from entering the column, has been developed	65
Benzene, toluene, ethylbenzene and xylene isomers	$[C_4mim][PF_6]$	Waters	GC-MS was employed for the analysis; extraction temperature and salting-out effect on the extraction performance.	66
Cobalt, mercury and lead	$[C_4mim][PF_6]$	Biological and environmental samples	Electrothermal vaporization inductively coupled plasma mass spectrometry was employed for determination; after 10 min of extraction, the enrichment factors were 350 (cobalt), 50 (mercury) and 60 (lead); 1-(2-pyridylazo)- 2-naphthol as both extractant and chemical modifier.	67
Chlorobenzenes	$[C_6mim][PF_6]$	Water	A 20 min of extraction time was required; a microwave was used for assist extraction.	68
Mmanganese	$[C_4mim][PF_6]$	Water	ETAAS was employed for analysis; 1-(2-thiazolylazo)-2-naphthol was used for chelator of manganese; the drop of the ionic liquid was directly. injected into the graphite furnace.	69
1-Naphthylamine, N,N-dimethylaniline and diphenylamine	$[C_4mim][PF_6]$	Water	Enrichment factors are in the range from 13.7 to 116.3.	93
4-Dichlorophenol, 2-naphthol, 2-nitrophenol and 4-chlorophenol	$[C_4mim][PF_6]$	Water	The effect of complex matrices natural water samples could be resolved with addition of sodium ethylene diamine tetracetate into the samples.	71

4-Nonylphenol, 4-tert-octylphenol	$[C_6mim][PF_6]$	Water	HPLC was employed for analysis; the proposed procedure has a detection limit and enrichment factor of 0.3mug.L^{-1} and 163 for 4-nonylphenol, and 0.7mug.L^{-1} and 130 for 4-tert-octylphenol, respectively.	72
chlorobenzenes	$[C_6mim][PF_6]$		Limits of detection ranging between 1 and 4ng.L^{-1}.	103
mercury species	$[C_6mim][PF_6]$	water	Limits of detection were between 1.0 and 22.8 mug.L^{-1} for the four species of mercury.	104
lead	$[C_4mim][PF_6]$	certified reference materials	Lead was complexed with ammonium pyrroldinedithiocarbamate; The limit of detection and the enhancement factor were 0.015 and 76µg.L^{-1}.	105
Trihalomethanes	$[C_4mim][PF_6]$	water	The limits of the detection range between 0.1 ng/mL (bromoform) and 0.9 ng.mL^{-1} (chloroform).	106
Tributyltin and Triphcnyltin	$[C_4mim][PF_6]$	water	Detection limits are between 1 and 100 mug.mL^{-1}.	108
	$[C_4mim][PF_6]$	Chinese Medicines	The limit of detection was 0.39mug.L^{-1}.	115
phenols	$[C_4mim][PF_6]$		capillary electrophoresis was employed for analysis, detection limits to three phenols were less than 0.05µg.L^{-1}.	120
carbonyl compounds	$[C_8mim][PF_6]$	water	Enrichment factor is up to 150. HPLC was employed for detection of carbonyl compounds, the limit of detection was in the range of 0.04-2.03ng.mL^{-1}.	124
cobalt	$[C_6mim][PF_6]$	water	ETAAS was employed for detection of cobalt, the detection limit was 0.04mug.mL^{-1}, and the enrichment factor was 120.	131
2-hydroxy-4-methoxybenzophenone, isoamyl-4-methoxycinnamate, 3-(4'-methylbenzylidene)camphor, 2-ethylhexyl 2-cyano-3,3-diphenylacrylate, 2-ethylhexyl 4-dimethylaminobenzoate and 2-ethylhexyl 4-methoxycinnamate	$[C_6mim][PF_6]$	water	HPLC was employed for detection, limits of detection were found in the low mug.L^{-1} range.	128
lead	Tetradecyl(trihexyl)phosphonium chloride	water	2-(5-bromo-2-pyridylazo)-5-diethylaminophenol was used as ligand. ETAAS was employed for detection of lead, its detection limit is 4.5 mug.L^{-1}.	129

Table 3. Reviews for single drop microextraction using ionic liquid as acceptor phase

Recently, a novel liquid phase microextraction technology termed dispersive liquid phase micro-extraction (DLME) was developed and validated (Table 4). From the name, it is obviously that it was based on the temperature change or using organic disperser making ionic liquid completely disperse in the aqueous phase and increase the chance of mass transfer into ionic liquid phase, in which the analytes will more easily migrate into the ionic liquid phase because of the much larger contact area than that of conventional single drop liquid micro-extraction, and the ionic liquid was concentrated into one drop by cooling and centrifugation. Experimental indicated that this system could be tuned to a great extent because ionic liquids can be designable and the method does not suffer from the limitations of that in conventional solvent micro-extraction. DLME has been applied to detect trace organophosphorus pesticides [73], fipronil, chlorfenapyr, buprofezin, hexythiazo [74] and mercury [75] with high extraction efficiencies. In general, DLME is more simple and rapid than SDME.

Analyte	Ionic liquid	Sample	Remark	Ref.
Organophosphorus pesticides	$[C_6mim][PF_6]$	water	HPLC was coupled with analysis; a 70℃ of heating and 30min of the centrifugation process were required strongly in order to form a fine dispersive phase and extraction efficiency; serious salting-out effect was also observed.	73
Fipronil, chlorfenapyr, buprofezin, and hexythiazo	$[C_4mim][PF_6]$	Water	HPLC was employed for analysis; enrichment factors are in the range from 209 to 276.	74
Hg^{2+}	$[C_6mim][Ff_2N]$	Water	Spectrophotometry was employed for analysis; 4,4'-bis(dimethylamino)-thiobenzophenone was used as chelator of mercury; enrichment factor is about to 18.8.	75
Sm^{3+}, Eu^{3+}, Gd^{3+}, Dy^{3+}	$[C_4mim][PF_6]$	uranium dioxide powder	1-hydroxy-2,5-pyrrolidinedione was used as agent.	90
polycyclic aromatic hydrocarbons	$[C_8mim][PF_6]$	tap, bottled, fountain, well, river, rainwater, treated and raw wastewater	Analysis of extracts was performed by HPLC coupled with fluorescence detection; Quantification limits obtained for all of these considered compounds are between 0.1 and 7 $ng.L^{-1}$.	91
Zn^{2+}	1-hexylpyridinium hexafluorophosphate	water and milk	Zinc was complexed with 8-hydroxyquinoline; The ionic liquid was proposed for the	92

			preconcentration of trace levels of zinc as a prior step to determination by flame atomic absorption spectrometry; the limit of detection (3s) and the enhancement factor were 0.22 mug.L^{-1} and 71.	
non-steroidal anti-inflammatory drugs	$[C_4mim][PF_6]$	urine	The suitability of the proposal is evaluated by means of the determination of non-steroidal anti-inflammatory drugs in urine by liquid chromatography/ultraviolet detection; Limits of detection were in the range from 8.3 ng.mL^{-1} for indomethacin to 32 ng.mL^{-1} for ketoprofen.	93
Cd^{2+}	$[C_6mim][PF_6]$	water	Dispersive liquid-liquid microextraction followed by electrothermal atomic absorption spectrometry; The enrichment factor of the method is 67. The detection limit was 7.4 ng.L^{-1}.	94
Dichlorodiphenyltrichloroethane and its main metabolites	$[C_6mim][PF_6]$		The experimental results indicated that the detection limits obtained for p,p'-DDD, p,p'-DDT, o,p'-DDT and p,p'-DDE were 0.24, 0.24, 0.45, 0.24 ng.mL^{-1}, respectively	95
Thiophanate-methyl, carbofuran, carbaryl, tebuconazole, iprodione, oxyfluorfen, hexythiazox and fenazaquin	$[C_6mim][PF_6]$	water	This paper describes a dispersive liquid-liquid microextraction procedure using room temperature ionic liquids coupled to high-performance liquid chromatography with diode array detection capable of quantifying trace amounts of eight pesticides; Limits of detection are between 0.320-4.66μg.kg^{-1}).	96
phenols	$[C_4mim][PF_6]$	water	The settled phase was injected into HPLC for separation and detection of phenols; The main advantages of the proposed method are high speed, high	97

			recovery, good repeatability and environmental friendliness.	
aromatic compounds	$[C_4mim][PF_6]$	water	The limits of detection for seven polycyclic aromatic hydrocarbons varied from 0.02 to 0.3 mug.L^{-1}.	98
dicofol and DDT	1-isooctyl-3-methylimidazolium hexafluorophosphate	water	The enrichment factors were 532 for dicofol and 540 for DDT; Proposed extraction combined with gas chromatography mass spectrometry was a valuable alternative for analysis of dicofol and DDT in environmental water samples within the concentration range 0.005-3.0 mug.L^{-1}. The detection limits were 1.3 ng.L^{-1} for DDT and 3.2 ng.L^{-1} for dicofol.	99
parathion, phoxim, phorate and chlorpyifos	$[C_8mim][PF_6]$	water	Enrichment factor of analytes is 200 and the limits of detection calculated at a signal-to-noise ratio of 3 were in the range of 0.1-5.0 mug.L^{-1}.	100
thiophanate-methyl, carbofuran, carbaryl, tebuconazole, iprodione, oxyfluorfen, hexythiazox, and fenazaquin	$[C_6mim][PF_6]$	fruit	Limits of detection were in the range 0.651-5.44 mug/kg for table grapes and 0.902-6.33 A mug.kg^{-1} for plums.	101
three triazine and two phenylurea herbicides	$[C_6mim][PF_6]$	water	HPLC was used for detection of five herbicides, their limit of detections were between 0.46 and 0.89µg.L^{-1}.	109
triclosan and triclocarban	$[C_6mim][PF_6]$	water	HPLC-MS was used for detection of triclosan and triclocarban, limits of detection in the range from 0.040 to 0.58 µg.L^{-1}	110
Pb^{2+}, Cd^{2+}	$[C_6mim][PF_6]$	saline	FAAS was employed for detection of lead and cadmium, detection limits are 0.6µg/L for lead and 0.03µg.L^{-1} for cadmium	112
triclosan and triclocarban	$[C_6mim][PF_6]$	water	HPLC-ESI-MS-MS was employed fortriclosan and triclocarb, detection limits are in the range from 0.040 to 0.58 mµg.L^{-1}.	113

organophosphorus pesticides	1,3-dibutylimidazolium hexafluorophosphate	water and fruit	Enrichment factors are more than 300 times. HPLC was employed for detection of pesticides, Limits of detection were 0.01-0.05 μg.L^{-1} for four analytes.	121
V^{5+}	[C_4mim][PF_6]	environmental and biological samples	2-(5-bromo-2-pyridylazo)-5-diethylaminophenol was used as ligand. ETAAS was employed for detection of vanadium. The detection limit achieved after preconcentration of 5 mL of sample solution, was 4.8 ng.L^{-1}	122
triclosan and triclocarban	[C_6mim][PF_6]	water	HPLC-ESI-MS/MS was employed for detection of triclosan and triclocarban, detection limit is 0.04 and 0.3 mug.L^{-1}	125
anthraquinones	[C_6mim][PF_6]	Radix	HPLC was employed for detection of anthraquinones, the limits of detection for all target analytes were ranged from 0.50 to 2.02 μg.L^{-1}.	126
Cr^{6+}	[C_6mim][PF_6]	lake and tap water	ETAAS was employed for detection of chromium, the limit of detection of the proposed method was 0.07ng.mL^{-1}.	130

Table 4. Reviews for dispersive liquid phase microextraction using ionic liquid as acceptor phase

3.2 Solid phase microextraction

Solid-phase microextraction (SPME) has developed rapidly and it has been widely applied in sample analysis. This is because this simple and solvent-free sample preparation technique, which combines extraction, concentration and sample introduction in one step, is portable, sensitive and convenient to couple with various analytical instruments, especially GC. Almost a dozen different fiber coatings have been commercialized. Among them, the polydimethylsiloxane shows excellent selectivity for non-polar compounds while the polyacrylate coating facilitates the extraction of relatively polar analytes. To date, most of these fiber coatings suffered from limited life span, relatively low operating temperatures in GC applications, and incompatibility with the organic solvents in HPLC mobile phases. In recent years, ionic liquid was widely applied to coating for SPME (Table 4). A typical experimental set-up of headspace SPME mode was illustrated in Fig.3 [79]. The SPME fiber was coated with IL prior to every extraction, the analytes were then extracted and desorbed on the injection port of gas chromatography, and finally the ionic liquid coating on the fiber was washed out with solvents. The coating and washing out of the ionic liquid from the fiber can be finished in a few minutes. Experimental demonstrated that the relatively large

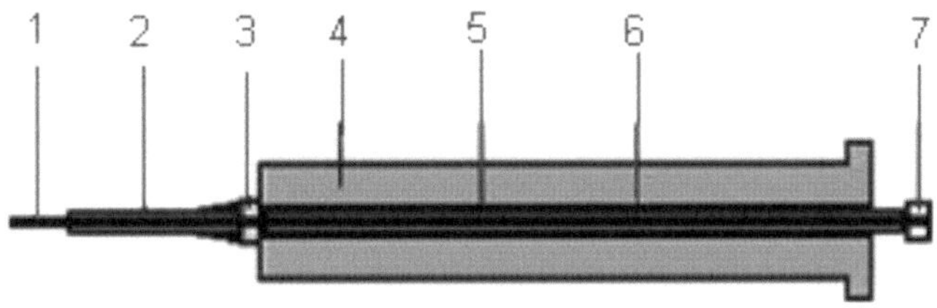

Fig. 3. Schematic diagram of the home-made SPME device: (1) ionic liquid-coated fiber tip; (2) stainless steel needle; (3) silicone rubber O-ring; (4) glass syringe body; (5) moveable stainless steel tubing; (6) stainless steel fiber or polyimide coated fused-silica fiber; (7) localizer.

viscosity of ionic liquid aided its coating on the SPME fiber and permitted easy washing out after sample injection. On the other hand, the involatility and good thermal stability of the ionic liquid will keep it staying on the fiber when the analytes were thermally desorbed in the injection port of GC, though loss of some of the ionic liquid film occurred due to the simple coating procedure. This proposed disposable ionic liquid-coated fibers show reproducibility comparable with the widely used commercially available SPME fibers, but its sensitivity is lower due to the relatively thin coating. The major advantage of the disposable ionic liquid-coated fiber is the very low cost because only a very low amount of ionic liquid is consumed for each determination. Another advantage of this disposable coating is the carryover was avoided as new fiber coating is used for every determination. However, the detection limits were higher than those obtained using commercially available SPME fibers due to the relative thin film of ionic liquid that can be coated on the fiber. Therefore, this proposed method is only applicable to samples with relatively high levels of analytes. Considering that ionic liquid possess good extractability for various organic compounds and metals ions, and task-specific ionic liquid can be designed and synthesized for selective extraction of target analytes, disposable SPME/LPME coating ionic liquid might has great potential in sample preparation.

Baghdadi group reported a new microextraction with ionic liquid termed cold-induced aggregation microextraction procedure (CIAME) [85]. In this method, very small amounts of 1-hexyl-3-methylimidazolium hexafluorophosphate and 1-hexyl-3-methylimidazolium bis (trifluoromethylsulfonyl)imide (as extractant solvents) were dissolved in a sample solution containing Triton X-114 (as an anti-sticking agent). Afterwards, the solution was cooled in the ice bath and a cloudy solution was formed. After centrifuging, the fine droplets of extractant phase were settled to the bottom of the conical-bottom glass centrifuge tube. CIAME is a simple and rapid method for extraction and preconcentration of metal ions from water samples and can be applied for the sample solutions containing high concentration of salt and water miscible organic solvents. Furthermore, this technique is much safer in comparison with the organic solvent extraction. Performance of the technique was evaluated by determination of the trace amounts of mercury as a test analyte in several real water samples. Michler thioketone was chosen as a complexing agent. Analysis was carried out using spectrophotometric detection method. Type and amount of IL and the surfactant, temperature and the other parameters were optimized. Under the optimum conditions, the limit of detection of the method was 0.3 ng mL^{-1} and the relative standard deviation was 1.32% for 30 ng.mL^{-1} mercury.

Analyte	Ionic liquid	Sample	Remark	Ref.
	trihexyltetradecylphosphonium, N-butyl-4-methylphyridinium tetrafluoroborate		Ionic liquid-mediated sol-gel hybrid organic-inorganic materials present enormous potential for effective use in analytical microextraction; Detection limits ranging from 4.9 to 487ng.L^{-1}.	76
Polycyclic aromatic hydrocarbons	[C_6mim]Br, 1-hexadecyl-3-butylimidazolium bromide and 1,3-didodecylimidazolium bromide	sediment BCR-535	GC-MS was employed for the analysis; an aqueous ionic liquid aggregate was used as acceptor; the extraction tube was introduced into the microwave cavity.	77
Esters	polymeric ionic liquids	Water	For the fist time polymeric ionic liquids was used as fiber coatings for extraction; GC was employed for analysis; long lifetime and high thermal stability of the polymeric ionic liquid solid phase microextraction coatings may provide them particular advantages in GC–MS applications involving highly selective ester and FAME extractions from complex matrices.	78
benzene, toluene, ethylbenzene, and xylenes	[C_8mim][PF_6]	Paint	GC was employed for analysis; the SPME fiber was coated with ionic liquid prior to every extraction; an extraction time of 30min was selected; an obvious salting-out effect was observed well.	79
ethanol	[C_4mim]Cl and [C_2mim]acetate		The application of ionic liquids as nonderivatizing solvents for the pretreatment and regeneration of cellulose is a growing area of research; GC was employed for analysis.	80
nonane, methyl caproate, *o*-cresol, *sec*-phenethyl-alcohol	six dicationic ionic liquids	Water	GC was employed for analysis; the partitioning behavior of six imidazolium-based ionic liquid aggregates in water has been studied for the first time. Monocationic and dicationic ionic liquids-aggregates behave as traditional surfactant systems; the obtained partition coefficients show that hydrophobic analytes are preferably extracted with ionic liquids-aggregates compared to more polar analytes.	81
4-Chlorophenol, 3-	[C_8mim][PF_6]	Water	HPLC was employed for analysis; a 60 min of extraction time was required; an obvious salting-out effect was observed.	82

chorophenol, 2,4-dichlorophenol and 2,4,6-trichlorophenol				
Aromatic compounds	$[C_{16}mim]Br$, 1-hexadecyl-3-butylimidazolium bromide, 1,3-didodecylimidazolium bromide	water	HPLC was employed for analysis.	83
Polycyclic aromatic hydrocarbons	$[C_8mim][PF_6]$, 1-benzyl-3-methylimidazolium trifluoromethanesulfonate:and 1-methyl-3-phenylpropylimidazolium trifluoromethanesulfonate	Water	nafion membrane was used to support the ionic liquid; GC-MS was employed for analysis.	84
short-chain alcohols, polar and basic amines	$[C_4mim][PF_6]$	water	two ionic liquids contained styrene units that allowed for polymerization and higher carbon content of the bonded silica particles; The sensitivities of the two ionic liquid fibers differing by the anion were similar. Their efficacy compares favorably to that of commercial fibers for polar analytes. The mechanical strength and durability of the polymeric ionic liquid fibers were excellent.	87
Methamphetamine and amphetamine	1-Ethoxyethyl-3-methylimidazloium bis(trifluoromethane)sulfonylimide	human urine	combined with gas chromatography/mass spectrometry working in selected ion monitoring mode, the new method showed good linearity in the range of 20-1500μg/ L and low detection limits (0.1μg.L^{-1} for MAP and 0.5 μg.L^{-1} for AP.	88
polycyclic aromatic hydrocarbons	$[C_4mim][PF_6]$	mosquito coil incense	A simple and cost effective solid-phase microextraction device has been developed	89
Sulfadiazine, sulfamerazine sulfamethazine, sulfadimethox	$[C_4mim][PF_6]$	water	By using ionic liquid as membrane liquid and tri-n-octylphosphine oxide as additive, hollow fiber supported liquid phase microextraction was developed for the determination of five sulfon-amides in environmental water samples by high performance liquid chromatography with ultraviolet	107

ine and sulfamethoxazole		detection; Detection limits are between 0.1 and 0.4 mug.L-1.	
Acetonitrile	Bonded ionic liquid polymeric material	GC was employed for detection.	114
CO_2	poly(1-vinyl-3-hexylimidazolium)-bis[(trifluoromethyl)sulfonyl]imide and poly(1-vinyl-3-hexylimidazolium) taurate	GC was employed for detection of CO_2.	118
polycyclic aromatic hydrocarbons and substituted phenols	poly(1-vinyl-3-hexadecylimidazolium) bis((trifluoromethyl)sulfonyl)imide	GC-MS was employed for analysis, detection limits oscillate from 0.005 ng.mL-1 for fluoranthene to 4.4 ng/mL for 4-chloro- 3-methylphenol.	119
lead and nickel	$[C_8mim][PF_6]$	Ammonium pyrroldinedithiocarbamate was used as ligand. ETAAS was employed for detection lead and nickel, the detection limits of the method were 0.03 and 0.02 mug.mL-1, for Ni and Pb, respectively	123

Table 5. Reviews for solid phase microextraction using ionic liquid as acceptor phase

4. The combination of analytical technologies with the extraction

Many modern analytical technologies have been applied to couple with microextraction using ionic liquid as extracting media for different analytical purpose such as AAS for metal ions, HPLC, GC and GC-MS for organic compounds. When ionic liquids are employed as medium in various LPME and SPME, HPLC is preferred to GC as separation technique since this ionic liquid is incompatible with GC due to the non-volatility of the ionic liquid. Consequently, the utilizations of ionic liquid in GC are limited to their use as stationary phases, as coaters of the surface of fibers in solid-phase microextraction or as solvents in various liquid phase microextraction including SDME and DLME etc. In order to overcome the problem, some effective attempts have been reported recently in the literatures. For example, Miguel Valcarcel group has successfully developed a new interface for the direct coupling between ionic liquid-SDME and GC-MS [65] (shown in Fig.4). Using this interface, no ionic liquid can reach the GC system dirtying the chromatographic column and a proper volatilization and subsequent transference of the substances to the GC is achieved. Therefore, the advantages of the use of ionic liquids with SDME, avoiding the irreproducibility associated with the use of organic solvents to form the drop, are added to the analytical possibilities of GC-MS. Furthermore, using the interface analytes of an extensive range of polarity/volatility can be determined on account of the fact that no solvent delay is necessary in the detection step. Moreover, Zeng Baizhao group also

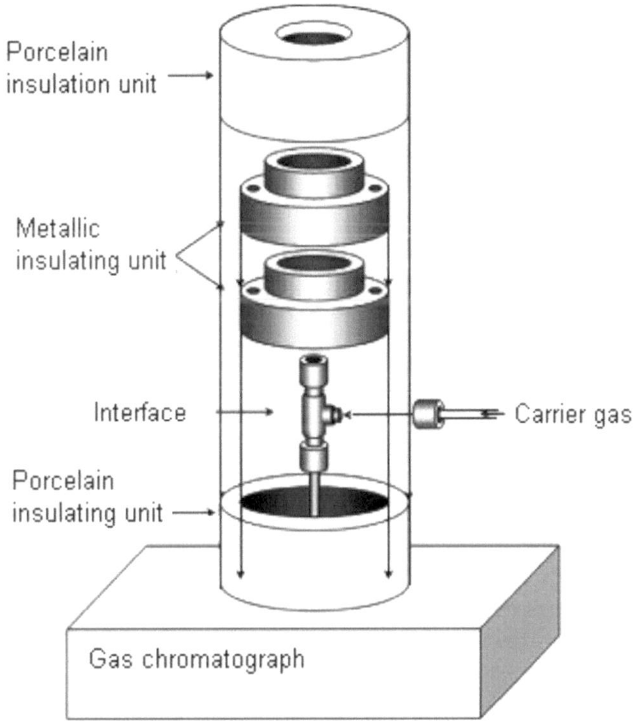

Fig. 4. Scheme of the household unit of the interface for the direct introduction of analytes extracted in a drop of ionic liquid into a gas chromatograph

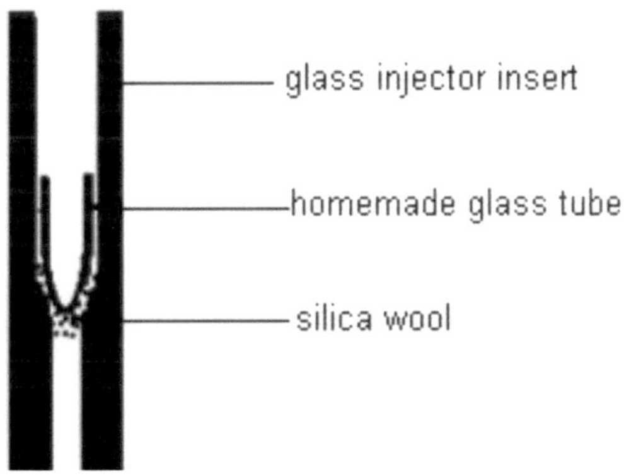

Fig. 5. The modified sample injection part of GC instrument. The dimension of the homemade tube: outer diameter, 2.0 mm; length, about 15 mm

provided anther scheme to avoid ionic liquid leaking into the chromatographic column (shown in Fig.5). In the scheme through placing a small glassy tube in the sample injection part, ionic liquid no long enters into the chromatographic column, thus ionic liquid based microextraction technique can be coupled with GC. To evaluate the proposed method, five chlorobenzene derivatives are used as model. Results show that the chlorobenzenes can be extracted effectively by the ionic liquid and the extractant exhibits good chromatographic behavior. The linear ranges are up to 1.5mg.L^{-1} and the detection limits range from 0.1 to 0.5 µg.L^{-1}. As ionic liquids have many unique properties, ionic liquid based microextraction technique coupled with GC has potential application in the detection of pollutants etc [86]. In our laboratory, above two schemes have been investigated too, the results exhibited above two schemes can not avoid the ionic liquid entering into chromatographic column.

5. Effect of solvent properties on the extraction performances

Although several 1-alkyl-3-methylimidazolium hexafluorophosphate room temperature ionic liquids, abbreviated as [C$_n$mim][PF$_6$], have been widely developed as friendly acceptor phase for various extractions in literatures, their application also suffers from many drawbacks. Firstly, these ionic liquids are difficult to be dispersed into aqueous phase and tend to tightly adhere on vessel wall due to high viscosity. The special property results that a long heating process (or mechanical agitation) and extract operation were required strongly in order to obtain good extraction efficiency. Secondly, most of the ionic liquids have relatively high water-solubility, the aqueous phase in the extraction system contains large amounts of imidazolium and hexafluorophosphate. The salting-out will effect on the extraction performance and following analysis in present of high ionic strength for complex samples. Moreover, the imidazolium dissolved in water is of good surface activity, the property may result in producing serious emulsification phenomenon during extraction and prolong phase separation time. In order to resolve above problems, some new ionic liquids have been designed, synthesized and applied to microextraction in the laboratory. For examples, 1,3-diisooctylimidazolium hexafluorophosphate (shown in Fig.6) was developed as a novel solvent for DLPME. In this study, dicofol and its three degradation products were selected as model compounds to optimize the factors relevant to the extraction and evaluate the applicability of the method. Experimental showed the proposed method offers the advantage of simplicity of operation, rapidity, high extraction efficiency and enrichment factor. The temperature-controlled ionic liquid 1-isooctyl-3-methylimidazolium hexafluorophosphate has been developed as novel medium for DLPME. This new mode of LPME with temperature-controlled ionic liquid increases the rapidity due to good hydrophobic property and relatively high melting point of the ionic liquid, the stability of extraction because of the larger volume of extraction solvent that may be used, enhances the sensitivity because of the much larger contact area than that of conventional LPME and SPME. Comparing with room temperature ionic liquid, proposed method has all performance during liquid/liquid and co-precipitation extraction, it can extract various polar or non-polar analyte from aqueous phase into the ionic liquid with high extraction efficiency. Moreover, new extraction can easily avoid the degradation of the pesticide during enrichment step due to no use a long heating time. Thirdly, the incompatibility between the extraction with ionic liquid and GC or GC-MS has become a problem in analytical chemistry due to ionic liquid's decomposition and leakage. Recently, Zhao group carried out a beneficial attempt to resolve above difficulty [78]. In this study, a new class of stationary phase coatings for SPME based on polymeric ionic liquids is presented. Proposed SPME coatings exhibit exceptional film stability, high thermal stability, reproducible extraction efficiencies, and long lifetimes. Moreover, the fiber coatings were used to extract esters and fatty acid methyl esters from aqueous solution followed by gas chromatographic separation and flame ionization detection. To examine the effect of the matrix on the coatings, extractions were carried out in a synthetic wine solution followed by recovery experiments in two real wine samples.

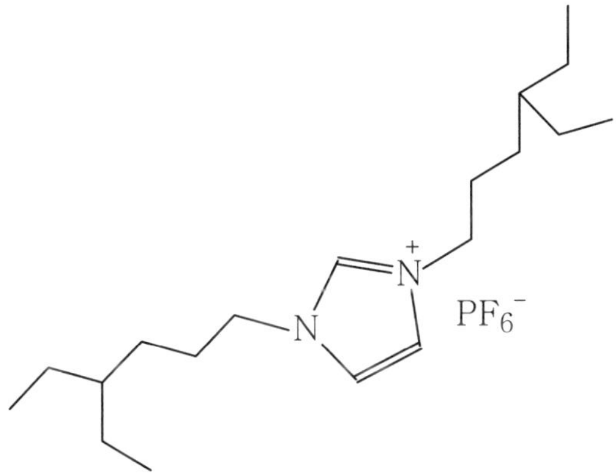

Fig. 6. Molecular structure of 1,3-diisooctylimidazolium hexafluorophosphate

When carrying out the extractions in aqueous solution, the detection limits for most analytes ranged from 2.5 to 50mug.L^{-1} whereas lower detection limits were obtained for larger fatty acid methyl esters. Recovery experiments carried out in red and white wines ranged from 70.2 to 115.1% using the poly(ionic liquid) fibers compared to 61.9 to 102.9% using a commercial polydimethylsiloxane fiber of similar film thickness. The structural tuning capability of these new coating materials makes them widely amendable to performing task-specific microextractions.

In selection of solvents for extraction and extractive distillation processes knowledge of the activity coefficients at infinite dilution (γ^{∞}) is very important and useful. A large number of works concerns measurements of activity coefficients at infinite dilution of organic solvents in ionic liquids. In separation processes both properties of extractant are important, namely selectivity S and capacity k which can be directly calculated from activity coefficients at infinite dilution for different separation problems. Activity coefficients at infinite dilution are also very helpful for characterizing the behaviour of liquid mixtures, estimation of mutual solubilities, fitting the excess molar energy (GE) model parameters, predicting the existence of an azeotrope, analytical chromatography, calculation of Henry constant and partition coefficients, development of thermodynamic models based on the group contribution methods such as mod. Typical solvents used in separation of aromatic hydrocarbons from aliphatic hydrocarbons are polar components such as sulfolane and N-methylpyrrolidone (NMP). Replacement of these volatile solvents in the separation processes by the nonvolatile ionic liquids can offer advantages such as less complex processes, more simple regeneration of the extractant using techniques as simple as flash distillation or stripping and the environmental protection. Extraction of aromatics from aromatic/aliphatic mixture with ionic liquids is expected to require less process steps and less energy consumption than extraction with conventional solvents due to negligible vapor pressure of ionic liquids. Andrzej Marciniak has made the analysis of cation and anion structure of the ionic liquid and effect of the temperature on the selectivity and the capacity for aliphatics/aromatics and n-hexane/hex-1-ene separation problems [132]. It was found the highest values of selectivity show ionic liquids with less aliphatic character of the anion and the cation, e.g. based on following cations 1,3-dimethylimidazolium, 1-ethyl-3-methylimidazolium, 1-ethylpyridinium, triethylsulfonium and with -C≡N group in the structure, like [C_3mim]CN. Unfortunately always when the ionic liquid reveals high values of the selectivity, the capacity takes low values. Therefore a selection of proper anion and cation is required. The highest values of capacity have Tf_2N^- and $FeCl_4^-$ anions. As was shown most of ionic liquids with high values of both selectivity and capacity is based on Tf_2N^- anion. 1-Butyl-4-methyl-pyridinium bis(trifluoromethylsulfonyl)amide have higher values of selectivity than sulfolane and NMP and much higher value of capacity than these two mentioned entrainers for the n-hexane/benzene separation. The highest selectivity for the n-hexane/benzene separation has [C_4mim][SCN] above 4 times higher than for sulfolane with a slightly better capacity. There are a large number of ionic liquids with selectivity and capacity better than sulfolane. A large number of ionic liquids have comparable or better both of selectivity and capacity values than NMP, e.g.: [1-ethyl-3-methyl-imidazolium bis(trifluoromethylsulfonyl)amide, triethyl-sulfonium bis(trifluoromethylsulfonyl)-amide, 1-ethyl-2,3-dimethylimidazolium bis(trifluoromethylsulfonyl)amide, hexyl-trimethyl-ammonium bis(trifluoromethylsulfonyl)amide, tributylmethylphosphonium methylsulfate, 1-hexyl-oxymethyl-3-methylimidazolium bis(trifluoromethylsulfonyl)amide and [C_6mim][Tf_2N]. For n-hexane/hex-1-ene the 3-(3-(Acryloyloxy)propyl)-1-methyl-imidazolium bromide has the highest selectivity of 15.2. For this separation process majority of ionic liquids have selectivity on the level of 2. To improve selectivity and capacity lower temperature is favourable.

From the values of selectivities and capacities, obtained for the large number of ionic liquids, it can be concluded that majority of ionic liquids may replace conventional entrainers applied for the separation processes of aliphatic/aromatic hydrocarbons and alkanes/alkenes.

6. Extraction mechanism

There are many investigations concerning extraction mechanism, these mainly are solvent ion pair extraction, ion exchange, and simultaneous combination of these. Among these, ion exchange mode was widely applied to understand the separation performance. In general, the ion exchange mode think metal ions will replace imidazolium cation or the anion combined with the ionic liquid in ionic liquid phase. For example, Dietz's group has suggested that high distributions in ionic liquid/aqueous system were due to a change in extraction equilibria in comparison to the octanol/aqueous system. In octanol, Sr^{2+} is extracted as a strontium nitratocrown ether complex, whereas in ionic liquids, the two axial bound nitrates are substituted with two water molecules bound to the metal ion, and the charged $Sr(CE)(H_2O)_2^{2+}$ species, where CE is dicyclohexano-18-crown-6-ether, is transferred to the ionic liquid phase, the neutrality of the system is maintained through the migration of two imidazolium cations to the aqueous phase for each extracted metal ion. Therefore, the loss of the ionic liquid components in the aqueous phase would be a major limitation for practical applications of ionic liquids in separations and enrichments [143]. However, the nature of the ionic liquid is not the only factor which determines the metal extraction pathway, ligand type also has a very important contribution. Professor Jensen has suggested La^{3+} was shown to follow an anion exchange mechanism when extracted in an ionic liquid/aqueous system, in contrast to the cation exchange observed in molecular solvents. Here, changing the lipophilicity of the ionic liquid did not change the partitioning mechanism [144]. Recently, the ion exchange theory was also applied to explain the extraction performance of amino acids using a room temperature ionic liquid 1butyl3methylimidazolium hexafluorophosphate with dicyclohexano-18-crown-6 system

[145]. The indication of an ion exchange mechanism is the significant leakage of the dialkylimidazolim cation into the aqueous phase on extraction of amino acids-confirmed by monitoring the corresponding UV band. After extraction of amino acids the clear absorption peak of the dialkylimidazolium cation appeared in the spectrum of the aqueous phase at ca. 300 nm. Increasing the initial amino acid concentration in the aqueous phase from 1×10^{-3} to 1×10^{-2} mol L^{-1} results in an increase of the dialkylimidazolium concentration in water after extraction. Clearly this is caused by ion exchange of the dialkylimidazolium cation with the amino acid cation. Thus, the absence of any appreciable anion effect and distribution of dialkylimidazolium cation into water, which accompanies extraction, strongly suggest that extraction of amino acids with $DC_{18}C_6$ into the ionic liquid is an ion exchange process. However, above all extraction mechanism mode did not consider the extraction functions of the surface charge on the ionic liquid phase and the ionic liquid dissolved in water. In our previous work, the surface tension and Zeta potential of the ionic liquid solution were investigated in detail in order to understand the extraction performance in the room temperature ionic liquid [C_4tmsim][PF_6] with CDAA system. It was found the surface tension of the solution decreased with increase of the concentration of [C_4tmsim][PF_6] solution, its critical micelle concentration and surface tension of saturated [C4tmsim][PF_6] solution are 1.59×10^{-3} mol L^{-1} and 65.95 Nm/m, respectively. The data showed ionic liquid has surface activity, this property allow the ionic liquid to fully interact with the complex or CDAA in aqueous phase and easy to extract the complex into the ionic liquid phase due to micelle extraction function, which is unique property of surfactant. On the other hand, the Zeta potential value of saturated [C_4tmsim][PF_6] solution was found to be +2.1mV. The result showed the double electron layer at the [C_4tmsim][PF_6] water interface existed the positive charge. The positive charge will produce strong interaction between Hg-CDAA complex or CDAA species and ionic liquid surface. Above two factors allow ionic liquid has a higher the extraction efficiency than ordinary organic solvent. It should be stated that Zeta potential may be positive or negative value for different ionic liquid, that decide various ionic liquid has different extraction performance. In order to test suggested extraction mechanism, the extraction efficiencies of CDAA in different acidity was researched in detail. The result is very interesting that the order of the extraction efficiency of CDAA is in base > neutral > acid medium. Because CDAA is a weak organic acid, its dissociation product three reagent species in water, these are RH_2, RH^- and R^{2-} respectively, in which CDAA was replaced by RH_2 in order to simple. In strong base solution, reagent specie is R_2 mainly, it is easy to extract into ionic liquid due to its strong negative electric charge, the order accord with suggested extraction mechanism [2]. Recently, professor Yan Yongsheng suggested a new extraction mechanism by use of ionic liquid salt aqueous two phase system (shown in Fig.7). Aqueous two-phase system consisting of a hydrophilic ionic liquid (1-butyl-3-methylimidazolium tetrafluoraborate, [C_4mim][BF_4]) and Na_2CO_3, which is a novel, simple, non-toxic and effective sample pretreatment technique coupled with molecular fluorescence spectrophotometry, was developed for the simultaneous separation, enrichment and rapid analysis of roxithromycin. The extraction yield of roxithromycin (ROX) in [C_4mim][BF_4]-Na_2CO_3 is influenced by the types of salts, concentrations of Na_2CO_3 and [C_4mim][BF_4], as well as the extracting temperature. The mechanism of ionic liquid-salt aqueous two-phase system formation was discussed by hydration theory, and the extraction mechanism of the [C_4mim][BF_4]-salt was investigated by FT-IR spectroscopy and UV-vis spectroscopy. The results demonstrate that no chemical (bonding) interactions are observed between ionic liquid and roxithromycin (ROX), while the nature properties of the ROX are not altered [146].

It should be pointed the extraction performance of the ionic liquids are very complex due to its complex solvent characteristics, the theory modes concerning extraction mechanism need improve and confirmed by reliable experiments technologies in the future.

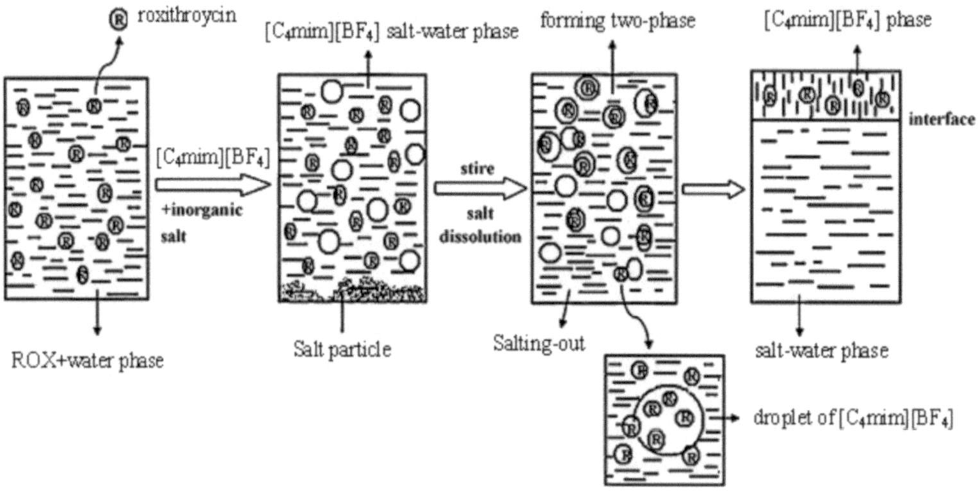

Fig. 7. The schematic diagram of separating ROX in ionic liquid-salt aqueous two phase system

7. References

[1] H.X. Shan, Z.J. Li, Spectroscopy and Spectral Analysis, 2008, 28, 214-217.
[2] Z.J. Li, Q. Wei, R. Yuan, X. Zhou, H.Z. Liu, H.X. Shan, Q.J. Song, Talanta, 2007, 71, 68-72.
[3] Z.J. Li, J. Chang, H.X. Shan, J.M. Pan, Reviews in Analytical Chemistry, 2007, 26, 109-153.
[5] Z.J. Li, Q.P. Peng, H.X. Shan, Journal of AOAC International, 2007, 90, 1191-1196.
[6] Z.J. Li, N.P. Lu, X. Zhou, Q.J. Song, Journal of Pharmaceutical and Biomedical Analysis, 2007, 43, 1609-1614.
[7] H.X. Shan, Z.J. Li, M. Li, Microchimica Acta, 2007, 159, 95-100.
[8] K. Sekiguchi, M. Atobe, T. Fuchigami, Electrochemistry communications, 2002, 4, 881-885.
[9] I.A. Ansari, R. Gree, Organic letters, 2002, 4, 1507-1509.
[10] H.X. Shan, Z.J. Li, M. Li, G.X. Ren, Y.J. Fang, Journal of Chemical Technology and Biotechnology, 2008, 83, 886-891.
[11] S.K.Chun, S.V.Dzyuba and R.A.Bartsch, Anal. Chem., 2001, 73(15), 3737.
[12] A.E.Visser, W.M.Swatloski, S.T.Griffin,D.H.Hartman and R.D.Rogers, Sep. Sci. Technol., 2001, 36, 785.
[13] V.A.Cocalia, J.D.Holbrey, K.E.Gutowski, N.J.Bridges and R.D.Rogers, Tsinghua Science and Technology,2006, 11(2), 188.
[14] A.Visser, R. Swatloski, W. Matthewreichert, Rebeccamayton, Seansheff, Andrzejwierzbicki, J.R. Jamesh Davis and R.D.Rogers, Environmental Science and Technology, 2002, 36 (11), 2523.
[15] D. Shayesteh, S. Ali Mohammad Haji, B. Mahboubeh Shirani, J. Abbas Ali, Journal of Hazardous Materials, 2010, 173 (1-3), 534-538.

[16] V.M. Egorov, D.I. Djigailo, D.S. Momotenko, D.V. Chernyshov, I.I. Torocheshnikova, S.V. Smirnova, I.V. Pletnev, Talanta, 2010, 80(3), 1177-1182.
[17] N. Li, G.Z. Fang, B. Liu, J. Zhang, L.J. Zhao, S. Wang, Analytical Sciences, 2010, 26(4), 455-459.
[18] R. Lertlapwasin, N. Bhawawet, A. Imyim, S. Fuangswasdi, Separation and Purification Technology, 2010, 72(1), 70-76.
[19] H. Abdolmohammad-Zadeh, E. Ebrahimzadeh, Central Europen Journal of Chemistry, 2010, 8(3), 617-625.
[20] N. Pourreza, K. Ghanemi, Journal of Hazardous Materials, 2010, 178 (1-3), 566-571.
[21] H. Okamura, N. Hirayama, K. Morita, K. Shimojo, H. Naganawa, H. Imura, Analytical Sciences, 2010, 26 (5), 607-611.
[22] C. Xu, X.H. Shen, Q.D. Chen, H.C. Gao, Science in China Series B-Chemistry, 2009, 52 (11), 1858-1864.
[23] A. Rout, K.A. Venkatesan, T.G. Srinivasan, P.R.V. Rao, Radiochimica Acta, 2009, 97(12), 719-725.
[24] J.S. Wang, C.N. Sheaff, B. Yoon, R.S. Addleman, C.M. Wai, Chemistry-A Europan Journal, 2009, 15(17), 4458-4463.
[25] K. Kidani, N. Hirayama, H. Imura, Analytical Sciences, 2008, 24 (10), 1251-1254.
[26] A. Kamio, Y. Nagaosa, Analytical Sciences, 2008, 24 (10), 1363-1367.
[27] Y. Zuo, J. Chen, D.Q. Li, Separation and purification Technology, 2008, 63 (3), 684-690.
[28] P. Giridhar, K.A. Venkatesan, S. Subramaniam, T.G. Srinivasan, P.R.V. Rao, Journal of Alloys and Compounds, 2008, 448 (1-2), 104-108.
[29] X.Q. Sun, B. Peng, J. Chen, D.Q. Li, F. Luo, Talanta, 2008, 74 (4), 1071-1074.
[30] K.V. Lohithakshan, S.K. Aggarwal, Radiochimica Acta, 2008, 96 (2), 93-97.
[31] N. Hirayama, H. Okamura, K. Kidani, H. Imura, Analytical Sciences, 2008, 24 (6), 697-699.
[32] K. Kozaka, K. Fulinaga, J. Noro, Y. Watanabe, Y. Komatsu, Bunseki Kagaku, 2008, 57(8), 593-597.
[33] P.Y. Chen, Electrochimica Acta, 2007, 52 (17), 5484-5492.
[34] N.V. Prondzinski, A. Babai, A.V. Mudring, K. Merz, Zettschrift Fur Anorganische Und Allgemeine Chemie, 2007, 633 (9), 1490-1492.
[35] X.Q. Sun, D.B. Wu, J. Chen, D.Q. Li, Journal of Chemical Technology and Biotechnology, 2007, 82 (3), 267-272.
[36] N. Hirayama, M. Deguchi, T. Honjo, Solvent Extraction Research and Devlopment-Japan, 2006, 13, 83-88.
[37] P. Giridhar, K.A. Venkatesan, T.G. Srinivasan, P.R.V. Rao, Hydrometallurgy, 2006, 81(1), 30-39.
[38] H.L. Huang, H.P. Wang, G.T. Wei, T.W. Sun, J.F. Huang, Y.W. Yang, Environmental Science and Technology, 2006, 40 (15), 4761-4764.
[39] K. Shimojo, H. Naganawa, F. Kubota, M. Goto, Chemicstry Letters, 2006, 35(5), 484-485.
[40] X. Zhou, Z.J. Li, R. Yuan, H.Z. Liu, Analytical Letters, 2006, 39(5), 863-877.
[41] V.A. Cocalia, M.P. Jensen, J.D. Holbrey, S.K. Spear, D.C. Stepinski, R.D. Rogers, Dalton Transactions, 2005, (11), 1966-1971.
[42] P. Giridhar, K.A. Venkatesan, T.G. Srinivasan, P.R.V. Rao, Journal of Radioanalytical and Nuclear Chemistry, 2005, 265 (1), 31-38.
[43] S. Mekki, C.M. Wai, I. Billard, G. Moutiers, C.H. Yen, J.S. Wang, A. Ouadi, C. Gaillard, P. Hesemann, Green Chemistry, 2005, 7 (6), 421-423.
[44] G.T. Wei, J.C. Chen, Z.S. Yang, Journal of the Chinese Chemical Society, 2003, 50(6),1123-1130.

[45] M. Matsumoto, Y. Inomoto and K. Kondo, J. Membrane Sci., 2005, 246, 77.
[46] S.D.Qi, S.Y.Cui, X.G.Chen and Z.D.Hu, J. Chromatogr. A, 2004, 1059, 191.
[47] L.A. Blanchard and J.F. Brennecke, Ind. Eng. Chem. Res., 2001, 40, 287.
[48] A.M. Scurto, S.N.V.K. Aki and J.F. Brennecke, J. Am. Chem. Soc.,2002, 124, 10276.
[49]J.F. Brennecke, L.A. Blanchard, J.L. Anthony, Z.Y. Gu, I. Zarraga and D.T. Leighton, Clean solvents, ACS Symp. Ser.,2002, 819, 82.
[50] M. Solinas, A. Pfaltz, P.G. Cozzi and W. Leitner, J. Am. Chem. Soc.,2004, 126, 16142.
[51] M. Cruz-Vera, R. Lucena, S. Cardenas, M. Valcárcel, Journal of Chromatography A, 2008, 1202, 1-7.
[52] C. Basheer, A.A. Alnedhary, B.S. Madhava Rao, R. Balasubramanian, H.K. Lee, Journal of Chromatography A, 2008, 1210, 19-24.
[53] X.M. Fu, S.G. Dai, Y. Zhang, International Journal of Environmental Analytical Chemistry, 2006, 86, 985-993.
[54] M. Cruz-Vera, R. Lucena, S. Cardenas, M.Valcarcel, Journal of Chromatography B, 2009, 877, 37-42.
[55] H.X. Xie, L.J. He, Y. Wu, K. Lu, X.Z. Si, Chinese Journal of Analytical Chemistry, 2007, 35, 187-190.
[56] J.F. Liu, Y.G. Chi, G.B. Jiang, Journal of Separation Science, 2005, 28, 87-91.
[57] L. Vidal, E. Psillakis, C.E. Domini, N. Grane, F. Marken, Analytica Chimica Acta, 2007, 584, 189-195.
[58] J.F. Liu, J.F. Peng, Y.G. Chi, G.B. Jiang, Talanta, 2005, 65, 705-709.
[59] C.L. Ye, Q.X. Zhou, X.M. Wang, Analytica Chimica Acta, 2006, 572, 165-171.
[60] F.Q. Zhao, J. Li, B.Z. Zeng, Journal of Separation Science, 2008, 31, 3045-3049.
[61] C.L. Ye, Q.X. Zhou, X.M. Wang, J.P. Xiao, Journal of Separation Science, 2007, 30, 42-47.
[62] Eva Aguilera-Herrador, Rafael Lucena, Soledad Cardenas, Miguel Valcarcel. Journal of Chromatography A, 2008, 1209, 76-82.
[63] J.F. Peng, J.F. Liu, G.B. Jiang, C. Tai, M.J. Huang, Journal of Chromatography A, 2005, 1072, 3-6.
[64] L. Vidal, A. Chisvert, A. Canals, A. Salvador, Journal of Chromatography A, 2007, 1174, 95-103.
[65] E. Aguilera-Herrador, R. Lucena, S. Cardenas, M. Valcarcel, Analytical Chemistry, 2008, 80, 793-800.
[66] E. Aguilera-Herrador, R. Lucena, S. Cardenas, M.Valcarcel, Journal of Chromatography A, 2008, 1201, 106-111.
[67] L.B. Xia, X. Li, Y.L. Wu, B. Hu, R. Chen, Spectrochimica Acta Part B, 2008, 63, 1290-1296.
[68] L. Vidal, C.E. Domini, N. Grane, E. Psillakis, A. Canals, Analytica Chimica Acta, 2007, 592, 9-15.
[69] J.L. Manzoori, M. Amjadi, J. Abulhassani, Talanta, 2009, 77, 1539-1544.
[70] Q.X. Zhou, C.L. Ye, Microchimica Acta, 2008, 162, 153-159.
[71] Q.X. Zhou, J.P. Xiao, C.L. Ye, X.M. Wang, Chinese Chemical Letters, 2006, 17, 1073-1076.
[72] J.F. Liu, Y.G. Chi, G.B. Jiang, C. Tai, J.F. Peng, J.T. Hu, Journal of Chromatography A, 2004, 1026, 143-147.
[73] Q.X. Zhou, H.H. Bai, G.H. Xie, J.P. Xia, Journal of Chromatography A, 2008, 1188, 148-153.
[74] Y. Liu, E.C. Zhao, W.T. Zhu, H.X. Gao, Z.Q. Zhou, Journal of Chromatography A, 2009, 1216, 885-891
[75] M. Gharehbaghi, F. Shemirani, M. Baghdadi, International Journal of Environmental Analytical Chemistry, 2009, 89, 21-33.

[76] A.M. Shearrow, G.A. Harris, L. Fang, P.K Sekhar, L.T. Nguyen, E.B. Turner, S. Bhansali, Journal of Chromatography A, 2009, 1216, 5449-5458.
[77] L. Guerra-Abreu, V. Pino, J.L. Anderson, A.M. Afonsoa, Journal of Chromatography A, 2008, 1214, 23–29.
[78] F. Zhao, Y.J. Meng, J.L. Anderson, Journal of Chromatography A, 2008, 1208, 1-9.
[79] J.F. Liu, N. Li, G.B. Jiang, J.M. Liu, J.A. Jonsson, M.J. Wen, Journal of Chromatography A, 2005, 1066, 27-32.
[80] F. Zhao, T.K. Ponnaiyan, C.M. Graham, C.A. Schall, S. Varanasi, J.L. Anderson, Analytical Bioanalytical Chemistry, 2008, 392, 1271-1275.
[81] V. Pino, Q.Q. Baltazar, J.L. Anderson, Journal of Chromatography A, 2007, 1148, 92-99.
[82] J.F. Peng, J.F. Liu, X.L. Hu, G.B. Jiang, Journal of Chromatography A, 2007, 1139, 165-170.
[83] C. Yao, V. Pino, J. L. Anderson, Journal of Chromatography A, 2009, 1216, 948-55.
[84] Y.N. Hsieh, P.C. Huang, I.W. Sun, T.J. Whang, C.Y. Hsu, H.H. Huang, C.H. Kuei, Analytica Chimica Acta, 2006, 557, 321-328.
[85] M. Baghdadi, F. Shemirani, Analytica Chimica Acta, 2008, 613, 56-63.
[86] F.Q. Zhao, S. Lu, W. Du, B.Z. Zeng, Microchim Acta, 2009, 165, 29-33.
[87] E. Wanigasekara, S. Perera, J.A. Crank, L. Sidisky, R. Shirey, A. Berthod, D.W. Armstrong, Analytical and Bioanalytical Chemistry, 2010, 396, 511-524.
[88] Y. He, J. Pohl, R. Engel, L. Rothman, M. Thomas, Journal of Chromatography A, 2009, 1216, 4824-4830.
[89] K.P. Huang, G.R. Wang, B.Y. Huang, C.Y. Liu, Analytica Chimica Acta, 2009, 645, 42-47.
[90] M.H. Mallah, F. Shemirani, M.G. Maragheh, Environmental Science and Technology, 2009, 43, 1947-1951.
[91] M.T. Pena, M.C. Casais, M.C. Mejuto, R. Cela, Journal of Chromatography A, 2009, 1216, 6356-6364.
[92] H. Abdolmohammad-Zadeh, G.H. Sadeghi, Analytica Chimica Acta, 2009, 649, 211-217.
[93] M. Cruz-Vera, R. Lucena, S. Cardenas, M. Valcarcel, Journal of Chromatography A, 2009, 1216, 6459-6465.
[94] S.Q. Li, S. Cai, W. Hu, H. Chen, H.L. Liu, Spectrochimica Acta Part B-Atomic Spectroscopy, 2009, 64, 666-671.
[95] H.H. Bai, Q.X. Zhou, G.H. Xie, J.P. Xiao, Analytica Chimica Acta, 2009, 651, 64-68.
[96] L.M. Ravelo-Perez, J. Hernandez-Borges, M. Asensio-Ramos, M. Angel Rodriguez-Delgado, Journal of Chromatography A, 2009, 1216, 7336-7345.
[97] Y.C. Fan, M.L. Chen, C. Shen-Tu, Y. Zhu, Journal of Analytical Chemistry, 2009, 64, 1017-1022.
[98] C. Yao, J.L. Anderson, Analytical and Bioanalytical Chemistry, 2009, 395, 1491-1502.
[99] Z.J. Li, P.P. Chen, C.P. Yu, Y.J. Fang, Z.Y. Wang, M. Li, H.X. Shan, Current Analytical Chemistry, 2009, 5, 324-329.
[100] L.J. He, X.L. Luo, H.X.Xie, C.J. Wang, X.M. Jiang, K. Lu, Analytical Chimica Acta, 2009, 655, 52-59.
[101] L.M. Ravelo-Perez, J. Hernandez-Borges, A.V. Herrera-Herrera, Analytical and Bioanalytical Chemistry, 2009, 395, 2387-2395.
[102] P. Berton, E. M. Martinis, L. D. Martinezc, R. G. Wuilloud, Analytica Chimica Acta, 2009, 640, 40-46
[103] A. Chisvert, I.P. Roman, L. Vidal, Journal of Chromatography A, 2009, 1216, 1290-1295.
[104] F. Pena-Pereira, I. Lavilla, C. Bendicho, L. Vidal, A. Canals, Talanta, 2009, 78, 537-541.
[105] J.L. Manzoori, M. Amjadi, J. Abulhassani, Analytica Chimica Acta, 2009, 644, 48-52.
[106] E. Aguilera-Herrador, R. Lucena, S. Cardenas, M. Valcarcel, Journal of Chromatography A, 2009, 1216, 5580-5587.

[107] Y. Tao, J.F. Liu, X.L. Hu, H.C. Li, T. Wang, G.B. Jiang, Journal of Chromatography A, 2009, 1216, 6259-6266.
[108] H. Sheikhloie, M. Saber-Tehrani, P. Abrumand-Azar, S. Waqif-Husain, Acta Chromatographica, 2009, 21, 577-589.
[109] S.L. Wang, L.P. Ren, C.Y. Liu, J. Ge, F.M. Liu, Analytical and Bioanalytical Chemistry, 2010, 397 (7), 3089-3095.
[110] R.S. Zhao, X. Wang, J. Sun, S.S. Wang, J.P. Yuan, X.K. Wang, Analytical and Bioanalytical Chemistry, 2010, 397(4), 1627-1633.
[111] P.Berton, RG. Wuilloud, Analytica Chimica Acta, 2010, 662 (2), 155-162.
[112] Yousefi, Seyed Reza; Shemirani, Farzaneh, Analytica Chimica Acta, 2010, 669(1-2), 25-31.
[113] RS. Zhao, X. Wang, J. Sun, S.S.Wang, J.P. Yuan, X.K.Wang, Analytical and Bioanalytical Chemistry, 2010, 397(4), 627-1633.
[114] E. Wanigasekara, S. Perera, J.A.Crank, L. Sidisky, R. Shirey, A. Berthod, D.W. Armstrong, Analytical and Bioanalytical Chemistry, 2010, 396 (1), 511-524.
[115] H. Xiao-Wen, J.Ye, L. Yan, Chinese Journal of Analytical Chemistry, 2010, 38 (5), 727-730.
[116] Y Wang, JY. You, CL Bao, H.Q. Zhang, A.M. Yu, Y.Yu, Chinese Journal of Chemistry, 2010, 28 (5), 785-790.
[117] M. Vaezzadeh, F. Shemirani, B. Majidi, Food and Chemical Toxicology, 2010, 48 (6), 1455-1460.
[118] QC. Zhao, J.L.Anderson, Journal of Chromatography A, 2010, 1217 (27), 4517-4522.
[119] J. Lopez-Darias, V. Pino, J.L. Anderson, C.M. Graham, A.M. Afonso, Journal of Chromatography A, 2010, 1217 (8), 1236-1243.
[120] Q. Wang, H.D. Qiu, J. Li, X. Liu, S.X. Jiang, Journal of Chromatography A, 2010, 1217 (33), 5434-5439.
[121] L.J. He, X.L. Luo, X.M. Jiang, L.B. Qu, Journal of Chromatography A, 2010, 1217 (31), 5013-5020.
[122] P. Berton, E.M.Martinis, R.G. Wuilloud, Journal of Hazardous Materials, 2010, 176 (1-3), 721-728.
[123] J. Abulhassani, J.L. Manzoori, M. Amjadi, Journal of Hazardous Materials, 2010, 176 (1-3), 481-486.
[124] Q.Y. Liu, Y.J. Liu, S.Z. Chen, Q.J. Liu, Journal of Separation Science, 2010, 33 (15), 2376-82.
[125] R.S. Zhao, X. Wang, J. Sun, J.P. Yuan, S.S. Wang, X.K. Wang, Journal of Separation Science, 2010, 33 (12), 1842-1848.
[126] H.F. Zhang, Y.P. Shi, Talanta, 2010, 82 (3), 1010-1016.
[127] S. Mahpishanian, F. Shemirani, Talanta, 2010, 82 (2), 471-476.
[128] L.Vidal, A.Chisvert, A. Canals, A. Salvador, Talanta, 2010, 81 (1-2), 549-555.
[129] E.M. Martinis, P. Berton, J.C.Altamirano, U. Hakala, R.G. Wuilloud, Talanta, 2010, 80 (5), 2034-2040.
[130] H. Chen, P. Du, J. Chen, S.H. Hu, S.Q. Li, H.L. Liu, Talanta, 2010, 81 (1-2), 176-179.
[131] M. Amjadi, J.L. Manzoori, J. Abulhassani, Journal of AOAC International, 2010, 93 (3), 985-991.
[132] A. Marciniak, Fluid Phase Equilibria, 2010, 294, 213-233.
[133] M.L.Dietz and J.A.Dzielawa, Chem. Commun.,2001, 20, 2124.
[134]M.P.Jensen, J.Neuefeind, J.V.Beitz, S.Skanthakumar and L.Soderholm, J. Am. Chem.Soc.,2003, 125, 15466.
[135] V.S.Svetlana, I.T.Irina,A.F.Andrey and V.P.Igor,Anal. Bioanal. Chem.,2004, 378, 1369.
[136] C.X. Li, J. Han, Y. Wang, Y.S. Yan, X.H. Xu, J.M. Pan, Analytica Chimica Acta, 2009, 653, 178-183.

9

Sample Treatments Based on Ionic Liquids

Eva Aguilera-Herrador, Rafael Lucena,
Soledad Cárdenas and Miguel Valcárcel
Department of Analytical Chemistry, University of Córdoba
Spain

1. Introduction

The term ionic liquid (IL) is nowadays used to define a broad class of semi-organic salts or salt mixtures composed entirely by ions which are liquid in the range 180-600 K. They are hydrogen bonded substances with strong interionic interactions which condition their crystallographic structure. The presence of at least one constituent ion with a delocalized charge prevents the formation of a stable crystal lattice (Palacio & Bhushan, 2008).

Although ILs are known since 1914, their expansion arrives with the availability of air and moisture stable ones. This second generation of ILs combines the unique properties of the former with the ease of handling of the latter. Their most relevant attributes or properties are as follows: negligible vapor pressure, good thermal stability, tunable viscosity and miscibility with water, inorganic and organic substances, a wide electrochemical window, high conductivity, high heat capacity and suitability to control reactions.

The electronic and steric features of the cation and anion condition their physical-chemical properties. Their synthetic flexibility permits the modulation or adaptation of certain properties of the IL for a particular objective. Density, viscosity and solubility (crucial in any extraction procedure) can be cited among the more tuneable features. Moreover, the nature of cation and anion structures influences the acid/basic and solvent properties and raises the differences with conventional molecular solvents. Finally, the potential introduction of functional groups capable of having complementary interactions with other solvents or molecules opens up new variability on ILs properties and behavior.

A huge number of potential cation-anion combinations are possible, which gives rise to an unusual large number of different compounds. The most popular high weight organic cations are imidazolium, pyridinium, pyrrolidinium quaternary ammonium and tetraalkylphosphonium. The most used by far are the 1,3-dialkylimidazolium salts as the likely result of their easily tailored properties. Regarding the most preferred organic anions, perfluorated (e.g. trifluoromethanesulphonate) and halogen–free (e.g. n-alkylsulphates) are among the most reported in the literature.

The presence of impurities is a very remarkable aspect in the use of any IL, as its properties can be dramatically altered by the presence of trace amounts of compounds usually arising from unreacted starting material or water. This fact has led to irreproducibility between reactions and procedures. The overcoming of such limitation is crucial for the implementation of ILs in analytical science and their consolidation as an alternative to traditional molecular organic solvents.

2. Ionic liquids in analytical chemistry

Ionic liquids can be successfully used in different steps of the (bio)chemical measurement process. In addition to selectivity and sensitivity improvement of existing methodologies, ILs can offer to the analytical scientists original alternatives to face up new applications.
Probably, their use in the preliminary steps is of advantage as they can markedly reduce the sample pretreatment step while improving its efficiency. This section presents a current picture of the contribution of ILs to improve the analytical process. It must be taken into consideration that the following sections will be exclusively devoted to the role of ionic liquids in sample preparation.

2.1 Extraction

Ionic liquids have been extensively used for the liquid-liquid extraction of metals (aided by complexation reagent), small organic molecules (hydrocarbons, PAHs) and also large biomolecules such as proteins and DNA thanks to hydrophobic interactions helped by electrostatic and also salting out effects. This wide applicability derives from the possibility of ad hoc design of ionic liquid used as extractant. In this way, the selectivity and efficiency can be modulated from the specific application as well as the solubility and miscibility between the extractant and the sample matrix. Several studies can be found in the literature dealing with this topic. For example, it has been established that the distribution coefficients for different metal cations were maximum for shorter 1-alkyl-3-methyl imidazolium ILs (C_4 expected). On the contrary, the extraction of proteins is favored with the use of ILs presenting a long alkyl chain in the cation. Particularly interesting is the applicability of ionic liquids to break a variety of azeotropic systems and chiral separations.
Liquid-liquid extraction has been moved to the miniaturized version of the technique in order to overcome its negative connotations. As expected, ILs have played a crucial role in the development of several modalities of liquid phase microextraction (LPME). 1-octyl-3-methylimidazolium-PF_6 demonstrated excellent performance for the extraction of PAHs from waters in both, direct immersion and headspace LPME. Since then, ILs have been used in dynamic LPME, hollow fiber protected LPME, single drop microextraction and dispersive liquid-liquid microextraction.
The applicability of ILs in solid-phase microextraction (SPME) was established by Liu *et al.* (Liu *et al.*, 2005) for the determination of benzene, toluene, ethylbenzene and xylene in paints. The thermal stability, extraction capacity and the easy immobilization on fused silica capillaries are among the most remarkable advantages of this coating. Their usefulness has been demonstrated in both headspace and direct immersion approach.
All these extraction modalities will be deeply commented in section 4.

2.2 Chromatography

Room temperature ionic liquids have been used as novel stationary phases in a variety of chromatographic separations. The research in gas chromatographic uses of ILs has been leaded by Armstrong´s group, who studied the behavior of $[C_4MIM][PF_6]$ and $[C_4MIM][Cl]$ as coatings in fused silica capillary columns (Armstrong *et al.*, 1999). The most valuable features in this application field are the high viscosity and thermal stability together with the easy immobilization on the inner surface of the fused silica capillary. The versatile combinations of ILs permit the efficient separation of polar and non-polar compounds, although long retention times and tailing peaks can be obtained for alcohols and organic

acids and even irreversible interaction of alkylamines has been reported. Notwithstanding this, the design of specific ionic liquids can overcome this limitation (Anderson & Armstrong, 2003).
In liquid chromatography, ILs can be used either as mobile phase modifiers or as stationary phases. The high viscosity and UV absorption of ILs limits their use as mobile phases. However, their presence at low concentrations (as additives) results in a decreased peak tailing, reduced band broadening and thus improved resolution. This effect is justified by a competitive reaction between imidazolium cations and the polar groups of the analytes for the silanol groups of the silica surface. Additional hydrophobic interactions between the alkyl chain of the cation and the C_{18} groups can also be behind this favorable effect of the ionic additive. More recently, ionic liquid-based stationary phases with immobilized imidazolium based ILs have been synthesized. In most cases, they work under strong anion exchange character and reversed phase interactions.

2.3 Electrophoresis

Ionic liquids have been employed to improve the electrophoretic separation in non aqueous capillary electrophoresis as their use in the conventional mode is hindered by their high viscosity and conductivity. ILs have been used to separate water-insoluble dyes due to the modification of the overall electrophoretic mobility of the system and the dissociation of the analytes in presence of the ILs. Further research has demonstrated the capability of ILs to improve the electrophoretic separation of chiral and achiral compounds aided by polymeric surfactants (Mwongela *et al.*, 2003).
Ionic liquids have also been covalently bonded to the inner surface of the fused silica capillary to reduce the analytes adsorption and reverse the electrosmotic flow (Qin & Li, 2002). Biomolecules (DNA and basic proteins), drugs, inorganic cations and alkylphosphonic acids have been successfully separated using capillaries modified with ILs. The reusability of the capillary (up to 96 runs) and the reproducibility improvement have been reported as additional advantages of this modality.

2.4 Sensors

The applicability of ionic liquids in electrochemistry has been boosted by the high ionic conductivity, low volatility and good redox robustness. However, most of the references deal with the electrodes construction and voltammetric data generation, being the analytical applications very limited yet. The high viscosity of ILs has been exploited for the development of membrane-free electrode modified with a thin layer of IL (Buzzeo *et al.*, 2004). They would be especially useful under extreme temperature conditions owing to their thermal stability and low volatility. The combination of hydrophobic ionic liquids, such as 1-ethyl-3-methylimidazolium trifluoromethylsulfonate, with membranes (e.g. Nafion) has been proposed as alternative to water for the construction of electromechanical actuators and sensors (Bennett & Leo, 2004). Although they are more stable operating in air, their response is rather slow in comparison with the use of water as consequence of the higher viscosity of the medium. Moreover, irreversible redox reactions can also occur.
More relevant is the use of ionic liquids in sensing approaches for the determination of organic vapors and gases. In this regard, variation on the viscosity of ionic liquids has been successfully employed as sensing material for the fabrication of quartz crystal microbalances (Liang *et al.*, 2002).

3. Ionic liquids as solvents

The solvent power of ILs, recently reviewed (Poole & Poole, 2010), has attracted much attention not only for extraction but also for the use of ILs as solvents in synthesis. The use of ILs as solvents is based on their peculiar physical and solvation properties. Among physical characteristics, we can highlight their high density, viscosity and conductivity, their low vapor pressure, and their miscibility with other organic solvents and water. ILs involve a great variety of compounds, in the range of 10^4 (Chiappe & Pieraccini, 2005), which clearly indicates their versatility. Moreover, some of their characteristics are tunable and therefore task-specific ILs (Davis, 2004) can be synthesized for a defined application. The environmental friendly nature of ILs, which is under scrutiny (Couling *et al.*, 2006), is also a positive aspect. In fact their recycling is easier compared to conventional organic solvents due to their special physical properties.

ILs have been extensively used in liquid-liquid extraction (LLE) for the isolation from aqueous samples of a great variety of compounds. In this sense, they have been selected for the extraction of metals following three different procedures (Zhao *et al.*, 2005):

- Extraction of metals by using crown ethers. The resulted crown- ether complexes present a high hydrophobicity but they still remain a residual electric charge. The use of ILs for extraction of these complexes presents efficiencies up to 10^4 better compared to conventional organic solvents (Dai *et al.* 1999).
- Extraction of metals as neutral complexes using anionic ligands (Wei *et al.*, 2003) (Hirayama *et al.* 2005). This approach has been extensively used for the extraction of different metals such as mercury (Li *et al.*, 2007a), aluminium (Li *et al.* 2007b) or nickel (Dadfarnia *et al.*, 2010).
- Extraction of metals using task-specific ILs which present a cationic interchangeable group in their structure (Visser *et al.*, 2001).

The chemical nature of ILs (tunable polarity and miscibility with water) allows also their use for the extraction of organic compounds from water (Simone *et al.*, 2010).

In summary, ILs have become an important tool in sample preparation, especially in microextraction techniques. Their peculiar characteristics have allowed the improvement of the existing techniques and moreover the development of new ones. These aspects will be considered in depth in the following section of the chapter.

4. Ionic liquids in microextraction techniques

Sample pre-treatment is a key step in the analytical process in order to improve the sensitivity (by the preconcentration of the target analytes) and selectivity (by avoiding the presence of potential interferences) as well as to overcome the incompatibility with the detection system. The ideal sample treatment should be: a) simple, reducing the number of steps; b) automatic, reducing operator manipulation; c) suitable for miniaturization, reducing sample and solvent consumption; d) rapid, to allow a high sample throughput in order to resolve problems in real time; e) inexpensive; and f) safe to the operators and environment (Lucena *et al.* 2009).

Microextraction techniques, including solid phase microextraction (SPME) and liquid phase microextraction (LPME), emerged in the 1990´s as a consequence of analytical chemistry evolution towards automation, miniaturization and simplification. In this context, the unusual physical properties of ionic liquids as well as their excellent solvent properties have been exploited in a large extent.

4.1 Ionic liquid in liquid phase microextraction

Liquid-liquid extraction (LLE), a classical sample treatment technique, is based on the partitioning (distribution) of a chemical substance between two immiscible phases. In the usual approach the aqueous sample containing the analytes is mixed with an appropriate organic solvent which presents a chemical affinity towards the target analyte at the operating conditions (temperature, pH, ionic strength ...etc). After a vigorous shaken both phases are conveniently mixed, the transference of the analytes from the donor to the acceptor phase taking place. Despite the different approaches developed, LLE remains as a manual technique involving different steps (subsequent extractions, back-extractions, evaporation of the extracts ...etc) which are a source of potential contamination or analyte losses, directly affecting the reproducibility of the final measurements. Moreover, the technique is tedious, time consuming and it requires the use of large volumes of pure organic solvents which are expensive and hazardous to operators and environment.
In this scenario, LPME emerged as an alternative to classical LLE. Dasgupta (Liu & Dasgupta, 1996) proposed in 1996 for the very first time the use of solvents in the low microliter range as extractants in LLE. Apart from the obvious economical and safety connotations, the reduction of the volume of extractant has allowed an evident improvement in the enrichment factors since the sample/extractant volumes ratio is dramatically increased.
From this origin, LPME has evolved spectacularly and it has become a reality in routine laboratories. In this sense, a large number of sub-techniques have been developed in order to solve different analytical problems. Taking into account that this chapter is focused on the role of ILs in this context, only those techniques which make use of the peculiar characteristic of ILs will be considered in depth.

4.1.1 Ionic liquids in single drop microextraction

Single drop microextraction technique (SDME) was firstly proposed by Jeannot and Cantwell in 1996 (Jeannot & Cantwell, 1996). Despite the utility of this contribution, the widespread use of SDME is due basically to the works published in 1997, almost simultaneously, by Jeannot and Cantwell (Jeannot & Cantwell, 1997) and He and Lee (He & Lee, 1997). In these articles, the authors proposed the use of a conventional microsyringe as extraction manifold for SDME. Theoretically, SDME is based on the transference by passive diffusion of the analytes from the sample to a micro-drop of extractant which is located in the tip of a microsyringe. The technique involves in the same step the isolation and preconcentration of the target analytes allowing also the direct injection of the extracts in the analytical instrument. Depending on the phases involved, two different extractions schemes are possible. In the direct immersion mode (DI-SDME), the microsyringe tip is inserted into the aqueous phase and the drop is directly exposed to the sample. In the headspace mode (HS-SDME) the extractant is exposed to the vapor phase generated from the sample, the extraction involving three different phases (aqueous, headspace and organic extractant).
In both modes, conventional organic solvents have been successfully applied for the resolution of diverse analytical problems. However, ILs present special characteristics like tunable polarity and miscibility with water as well as exceptional physical (viscosity, surface tension...) and chemical (extraction capabilities) properties which make them so attractive as alternatives to conventional solvents. Moreover, ILs have been considered as green solvents due to their negligible vapor pressure compared with volatile organic solvents, making them an interesting tool with respect to the safety to operators and environment.

Liu *et al.* (Liu *et al.*, 2003) proposed for the first time the use of ILs in SDME (in both extraction modes) for the extraction of polycyclic aromatic hydrocarbons from water samples. The IL employed, 1-octyl-3-methyl imidazolium hexafluorophospate, ([C_8MIM][PF_6]), presented superior capabilities compared to 1-octanol in terms of enrichment factors. On the one hand, the adequate viscosity and surface tension of ILs allow the suspension of larger drops in the tip of the microsyringe. On the other hand, their negligible vapor pressure makes the microdrop more stable (avoiding potential evaporation during the extraction) in the headspace mode.

DI-SDME is the preferred technique for the isolation of non-volatile and/or thermolabile compounds from aqueous samples. The solvents employed in these procedures should fulfil some requirements like immiscibility with water (in order to avoid extractant losses), affinity towards the analytes (to achieve good preconcentration factors) and good stability on the microsyringe tip. 1-alkyl-3-methyl imidazolium hexafluorophospate ([C_nMIM][PF_6] with n=4,6,8) ILs have been extensively used in DI-SDME for the extraction of organic compounds as well as inorganic species. Vidal *et al.* have proposed the use of [C_6MIM][PF_6] for the extraction of typical UV-filters in water samples (Vidal *et al.*, 2007a) (Vidal *et al.* 2010). The chemical compatibility of these ILs with the chromatographic columns allows the separation and determination of the extracted analytes by liquid chromatography coupled to UV detection. Preconcentration factors up to 98 (for 3-benzophenone) were obtained. In these articles, the authors present an interesting study of the influence of different chemical variables on the extraction of a given analyte.

In the same way, ILs have been employed for the extraction of inorganic species such as mercury, manganese or lead. According to their ionic nature, these analytes cannot be directly isolated in the ILs phase, a previous complexation with an appropriate ligand being necessary. Pena-Pereira *et al.* used [C_6MIM][PF_6] for the extraction of mercury species, including inorganic and organic ones, making speciation possible by HPLC-UV (Pena-Pereira *et al.*, 2009). In other sense, for the total determination of metals electrothermal atomic absorption spectrometry (ETAAS) is preferred due to its selectivity and sensitivity levels. Manzoori *et al.* proposed for the first time the direct coupling of DI-SDME using ILs with ETAAS (Manzoori *et al.*, 2009a). This coupling, which is possible due to an adequate selection of the pyrolysis temperature and time, allows the determination of manganese in the low part per billion range using [C_4MIM][PF_6] as extractant. The same procedure has been proposed for the determination of lead in water and food samples with similar success (Manzoori *et al.*, 2009b). Although the use of [C_nMIM][PF_6] ILs is predominant, Martinis *et al.* have proposed the use of tetradecyl(trihexyl)phosphonium chloride as extractant for the determination of lead in waters (Martinis *et al.*, 2010).

DI-SDME effectiveness is restricted by the stability of the drop in the microsyringe tip, which limits the stirring of the sample solution and involves the use of drops into the 2-10 μL range. Different approaches have been developed in order to overcome these limitations. Cycle flow configuration (Xia *et al.* 2005) allows the free flowing of the sample through the IL microdop avoiding its potential detaching during the extraction. This approach has allowed the determination of cobalt, mercury an lead in biological and environmental samples by electrothermal vaporization inductively coupled plasma mass spectrometry (ETV-ICP-MS) (Xia *et al.*, 2008). In other sense, dynamic liquid-phase microextraction has become an excellent alternative when higher volumes of ILs (in the range of 50 μL) are required for liquid chromatographic analysis (Cruz- Vera *et al.* 2008). This approach makes use of a special manifold, completely mechanized, which performs all the steps involved in the extraction. The obtained enrichment factors, in the range from 10 to 20, are acceptable for the resolution of different analytical problems such as the determination of non steroidal anti-inflammatory drugs or phenotiazines derivatives in biological samples (Cruz-Vera *et al.* 2009a). In samples of biological concern, ILs seem to present an excellent selectivity towards the target analytes, clean extracts being obtained after sample pretreatment.

The non-volatility of ILs, due to their low vapor pressure, makes them superior solvents than the conventional organic ones in HS-SDME. In fact, the evaporation of the solvent during the extraction is a critical limitation since it affects the extraction effectiveness and reproducibility. Therefore, ILs have been extensively used in this context for the determination in environmental samples of several contaminants such as chlorinated anilines (Peng *et al.*, 2005), organochlorine pesticides (Ye *et al.*, 2006), phenols (Ye *et al.* 2007), chlorobenzenes (Vidal *et al.*, 2007b) and aromatic amines (Zhou & Ye, 2008). Although HS-SDME is usually performed in conventional furnaces or heating baths, the use of microwave energy can be employed improving the expeditiousness of the extractions (Vidal *et al.*, 2007c). Moreover, in the latter approach the design allows the heating of the sample while the ILs remains unheated, which clearly improves the efficiency of the extraction.

All the above described analytical methods use liquid chromatography or atomic techniques for the determination of the extracted analytes, even when volatile compounds are under study. This fact is ascribed to the non-volatility of ILs which is a clear limitation in gas chromatography (GC). In fact, the presence of ILs dirties the chromatograph and even blocks the analytical column making the analysis unfeasible. Some research groups have focused their efforts in the development of interfaces and/or strategies that overcomes this shortcoming making possible the use of IL in GC. The first approach, which was proposed in 2008, made use of a special removable interface directly coupled to the chromatograph which allowed analytes desorption from the ILs avoiding the entering of the solvent in the system (Aguilera-Herrador *et al.*, 2008a). The excellent performance of the interface has been exploited in some applications which use ILs in HS-SDME for the determination of monoaromatic hydrocarbons (Aguilera-Herrador *et al.*, 2008b) and halogenated hydrocarbons (Aguilera-Herrador *et al.*, 2008c) in waters. The versatility of the interface makes possible its application with other instrumental techniques such as ion-mobility spectrometry (Aguilera-Herrador *et al.*, 2009).

Zhao and co-workers proposed another alternative for making compatible the use of ILs with GC. In this case, a minimal modification of the injection liner and a special injection procedure are required (Zhao *et al.*, 2008a). The microsyringe with the IL containing the extracted analytes is inserted in the injector of the chromatograph and the microdrop is exposed to the carrier gas in the liner, which is employed as desorption chamber. Once injected the analytes in the chromatographic column, the IL is retracted into the syringe avoiding its introduction in the chromatograph. A similar approach has been recently proposed for the determination of pesticides is soil samples (Zhang & Lee, 2010). Finally, the application of a commercially available thermal desorption unit especially employed for stir bar sorptive extraction (SBSE) has been also proposed for this coupling (Chisvert *et al.* 2009). The commercial desorption unit is modified including an inner tube in order to increase the sample throughput.

In all the described applications, both in DI-SDME and HS-SDME, pure ILs are used as extractants. However, special combinations of IL with chemical modifiers can be also employed. Micellar ionic liquids, which are the result of combining ILs with surfactants, have been recently evaluated for the extraction of aromatic compounds from aqueous samples (Yao *et al.*, 2010). The obtained results indicate the good extractability of micellar ILs as well as a special selectivity towards the evaluated analytes.

4.1.2 Ionic liquids in hollow fiber protected liquid phase microextraction

The instability of the drop in the syringe tip may cause its detachment during the extraction, which is undesirable. Hollow fiber protected liquid phase microextraction (HF-LPME) faces up this shortcoming using a polymeric cylindrical membrane as container for the extraction solvent (Pedersen-Bjergaard & Rasmussen, 1999). The extraction solvent is located in the lumen of a hollow fiber of minimal dimensions (600 µm I.D., 200 µm of wall thickness and variable porosity) usually made in polypropylene. According to the phases involved in the extraction, two different modalities are possible. In the two phases mode (2P-HF-LPME) the extraction solvent is located in the lumen of the fiber as well as in the pores. This type of extraction, which is similar to the classical LLE, is ideal for hydrophobic analytes. In the three phases mode (3P-HF-LPME) the organic solvent is located in the pores of the membrane acting as physical barrier (supported liquid membrane, SLM) between two aqueous phases, namely: the donor (sample) and the acceptor (extractant). This extraction mode, similar to classical LLE with back-extraction, is the alternative of choice for hydrophobic ionizable compounds. In fact, in 3P-HF-LPME the pH gradient established at both sides of the liquid membrane is the driving force of the extraction. Conventional organic solvents (like toluene, undecane, dihexylether and octanol) have been successfully applied in both extraction modes although their tendency to evaporation is a critical aspect being a key source of irreproducibility. Once again, ILs are a good alternative to this solvents due to their peculiar characteristics.

Fortunato *et al.* suggested the potential of $[C_nMIM]$-based ILs as supported liquid membranes due to their high viscosities and interfacial tensions and their low solubility in water when certain types of anions are employed (Fortunato *et al.*, 2004). This usefulness has been later on demonstrated by different research groups. Peng *et al.* proposed the use of $[C_8MIM][PF_6]$ as supported liquid membrane for the extraction of chlorophenols from water samples (Peng *et al.*, 2007). The effective immobilization of the IL in the pores of the hollow fiber was demonstrated by the authors using scanning electron microscopy (SEM). The SEM pictures obtained before and after the impregnation process are presented in Figure 1. The

(a) (b)

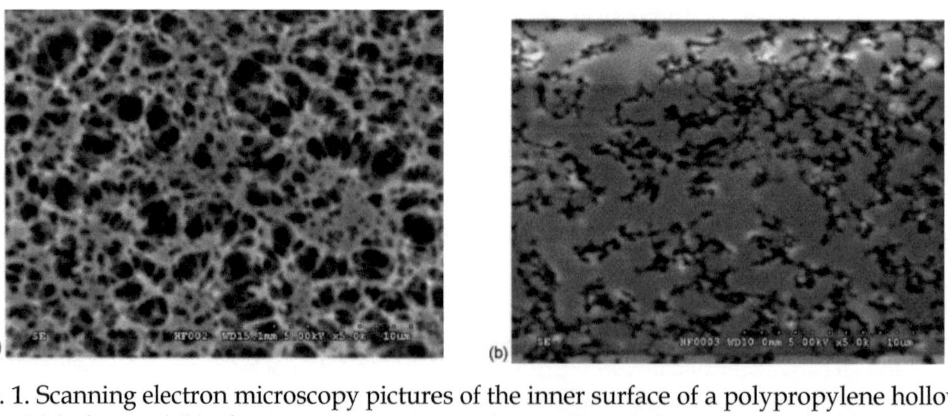

Fig. 1. Scanning electron microscopy pictures of the inner surface of a polypropylene hollow fiber (a) before and (b) after its impregnation with ionic liquid. Picture magnified by 5000 times. Figure from Peng *et al.* (2007) and reproduced with permission of Elsevier.

potential of ILs as SLM can be also ascribed to their chemical characteristics which make possible the development of special applications. On the one hand, ILs present higher polarities compared to the conventional organic solvents used as SLM. This fact has been exploited for the extraction of sulphonamides in environmental samples which are hard to extract with conventional solvents (Tao *et al.*, 2009). On the other hand, ILs present variable solubility in water and organic solvents. In this sense Basheer *et al.* proposed a special 3P-HF-LPME with the IL acting a SLM between an aqueous phase (the sample) and an organic phase (the acceptor). The method allows the determination of aliphatic and aromatic hydrocarbons in storm water by GC since the organic extracts can be directly injected in the chromatograph (Basheer *et al.*, 2008). Finally, the good performance of ILs in the 2P-HF-LPME has been recently demonstrated by the determination of lead and nickel using ETAAS (Abulhassani *et al.* 2010).

4.1.3 Ionic liquids in dispersive liquid-liquid microextraction

Microextraction techniques are surface dependent processes. Due to the small contact area between sample and extractant, the kinetic of the extraction process is too low and the time necessary to achieve the distribution equilibrium is unpractical for analysis purposes. *Dispersive liquid-liquid microextraction* (DLLME) faces up this shortcoming using an innovative approach (Rezaee *et al.*, 2006). In DLLME the extractant solvent is dispersed physically, chemically or assisted by an external energy source, in the sample in order to obtain a cloud of fine extractant drops, increasing dramatically the contact surface with the sample. DLLME allows the rapid extraction, almost instantaneously, of the analytes providing therefore excellent enrichment factors since the distribution equilibrium is achieved. In classical DLLME the extraction solvent should fulfil some general requirements such as: (a) immiscibility with water; (b) good chemical affinity towards the target analytes and (c) low vapor pressures. Moreover, taking into account that the extractant is recovered by centrifugation after the extraction, the solvent must be denser than water. In the light of these requirements, ILs are excellent solvents for DLLME.

Fan *et al.* proposed for the first time the use of ILs in DLLME for the determination of aromatic amines in water samples (Fan *et al.*, 2008). The IL, $[C_4MIM][PF_6]$, is directly added to the sample and the mixture is aspirated in a 1mL-microsryringe and later on discharged in a test tube. The process is repeated twice producing the physical dispersion of the IL into the sample and providing a cloudy solution of fine microdrops. Enrichment factors in the range of 31 to 269 are obtained allowing the determination of the analytes in the low μg/L level. Physical dispersion has been also applied for the determination of phenols in waters (Fan *et al.*, 2009).

According to Liu and co-workers (Liu *et al.*, 2009a), chemical dispersion is more effective than physical dispersion when ILs are used for the extraction of heterocyclic insecticides in water samples. In fact, the extraction recoveries were 5 fold better when an organic solvent is employed as disperser. Chemical dispersion is based on the use of an organic solvent (called disperser), which is miscible with the sample (water) and the extractant (IL). A mixture of the disperser and extractant solvents, in the appropriate proportion, is rapidly injected into the sample forming the cloudy solution immediately. After a determined period of time (named extraction time) the resulted mixture is centrifuged in order to recover the IL containing the isolated analytes. This strategy has been employed for the determination of organophosphorus pesticides (He *et al.*, 2009a), polycyclic aromatic

hydrocarbons (Pena *et al.*, 2009) and bactericides (Zhao *et al.*, 2010a) in water samples. As well as other microextraction techniques, DLLME can be used for the extraction and determination of inorganic species in liquid samples. In order to favor the extraction, an appropriate ligand has to be added to the sample to form the corresponding chelate which is the extractable substance. Thus, the extraction of zinc (Abdolmohammad-Zadeh & Sadeghi, 2009), cobalt (Berton & Wuilloud, 2010a) and aluminium (Abdolmohammad-Zadeh & Sadeghi, 2010) have been successfully accomplished. DLLME can also operate with solid samples but a previous treatment is always required. A specific amount of sample is located in a container and the analytes are extracted by leaching with an appropriate solvent (usually acetonitrile). The extracts, which also contain matrix compounds, are later on evaporated and reconstituted in water. Once the aqueous matrix is obtained, the DLLME procedure is developed. In this way the determination of pesticides in bananas (Ravelo-Pérez et al., 2009a) and table grapes and plums (Ravelo-Pérez et al. 2009b) by liquid chromatography have been possible. Moreover, the versatility of DLLME allows the treatment of complex samples. In fact, Wang *et al.* have determined triazines in honey. The sample (2 g) has to be conveniently diluted with 20 mL of water before the extraction. The authors proposed the use of a surfactant (Triton X 114) as disperser for the extractant ([C_6MIM][PF_6]). Both, disperser and extractant, are directly added to the diluted sample and the mixture is vigorously shaken for 10 min. After the extraction, the IL is recovered by centrifugation and finally analyzed by HPLC (Wang *et al.*, 2010).
The use of IL-based microextraction techniques for the treatment of samples with a high content of electrolytes is usually problematic since the solubility of a given IL in water dramatically depends on the ionic strength. At high levels of ionic strength, the IL is completely soluble in water and its recovery after the extraction is unaffordable. This phenomenon is caused essentially by an ion exchange process, the so-called metathesis reaction, between the IL and the electrolytes present in the sample. In this way, when a [C_nMIM][PF_6]-based IL is introduced in a sample with a high concentration of chloride, the chloride anion may exchange with the hexafluophosphate. As a consequence of this metathesis reaction a new IL ([C_6MIM][Cl]), completely soluble in water, is produced. Yousefi *et al.* have proposed a strategy in order to minimize this reaction by the external addition of $NaPF_6$ to the sample (Yousefi & Shemirani, 2010). Due to the common ion effect, the solubility of the IL decreases even in the presence of high concentration of electrolytes. This approach has been applied in a DLLME configuration for the determination of lead and cadmium in saline aqueous samples by FAAS. The common ion effect has also supported the development of a novel microextraction technique called *in situ solvent formation microextraction* (ISFME) (Baghdadi & Shemirani, 2009). In ISFME, a water-miscible IL (such as [C_6MIM][BF_4]) is added to the sample, being completely dissolved and therefore maximizing the contact surface with the target analytes. Later on, a common ion reagent (like $NaPF_6$) is added to the extraction vial inducing the metathesis reaction. As a consequence, a water-immiscible IL (in this case [C_6MIM][PF_6]) is formed which is finally recovered by centrifugation. ISFME has been characterized in depth by Yao & Anderson (Yao & Anderson, 2009). The authors have demonstrated the superior features of ISFME compared to traditionally DLLME or DI-SDME, using ([C_4MIM][Cl] and $LiNTf_2$ as IL and common ion reagent, respectively. Enrichment factors in the range of 189 to 753 have been obtained for the determination of 13 aromatic compounds in water samples. ISFME has been also used for the determination of cadmium in saline samples by FAAS providing limits of detection of 0.07 μg/L (Mahpishanian & Shemirani, 2010).

The dispersion of the extraction solvent into the sample is the crucial step in any DLLME and different alternatives, apart from the physical and chemical approaches, have been developed in recent years. In these novel approaches, an external energy source is employed in order to assist the dispersion process. The technique, *temperature controlled ionic liquid dispersive liquid phase microextraction* (TILDLME) was proposed in 2008 by Zhou and co-workers (Zhou *et al.*, 2008a) and it uses an external heat source. TILDLME is based on different well established steps. First of all, a specific amount of IL is added to the sample which is subsequent incubated at a high temperature (in the range of 45-90°C) for an appropriate time. During the heating of the sample, the IL is completely dispersed, even solubilized, into the sample matrix maximizing the contact area. Later on, the vial is introduced in an ice-bath decreasing the temperature of the mixture. In such conditions, the IL becomes insoluble and a cloudy solution appears in the extraction vial. After a proper extraction time, the mixture is centrifuged and the IL, with the extracted analytes, is recovered for its subsequent analysis. In TILDLME three especial properties of ILs are exploited, namely: (a) their good extractability towards a great variety of compounds; (b) their negligible vapor pressure which avoids the evaporation of the extractant during the heating process, and (c) their variable water-solubility which directly depends on the working temperature. Compared to classical DLLME, TILDLME provides better enrichment factors, especially for polar substances, since no chemical disperser is employed. Moreover, the contact area between sample and extractant is dramatically increased. TILDLME has been successfully applied for the determination of organophosphorous pesticides (Zhou *et al.*, 2008b), herbicides (Zhou et al., 2009a) or bactericides (Zhao *et al.*, 2010b). As other extraction techniques previously described, TILDLME can be also employed for the extraction of inorganic species such as vanadium (Berton *et al.*, 2009) or lead (Bai *et al.*, 2010). In 2008, the *cold induced aggregation microextraction* (CIAM) technique is introduced for the first time (Baghdadi & Sheminari, 2008). CIAM is similar to TILDLME but lower temperatures are used in the incubation step. CIAM has been successfully used for the determination of cobalt in water samples (Gharehbaghi *et al.*, 2009).

Ultrasounds can also be used to assist DLLME originating a new technique which is called *ultrasound-assisted ionic liquid dispersive liquid-liquid phase microextraction* (USA-DLLME) (Zhou *et al.*, 2009b). In USA-DLLME the IL is dispersed by using an ultrasound source and once the IL is dispersed, the extraction vessel is introduced in an ice-bath inducing the aggregation of the IL. The resultant cloudy solution is finally centrifuged and the IL, with the extracted analytes, is recovered for its subsequent analysis. Conventional or high energy ultrasounds (Mao *et al.*, 2009) can be employed. As the previously described techniques, the versatility of USA-DLLME allows its application for the resolution of diverse analytical problems. In this sense, it has been used for the extraction of cadmium from water samples (Li *et al.*, 2009).

In the light of the results, DLLME is a powerful technique for the extraction of different analytes in a great variety of sample matrixes. DLLME allows an efficient miniaturization and simplification of the sample pre-treatment although its automation is clearly difficult. This difficulty arises from the unavoidable centrifugation step which is a typical off-line process. Some efforts have been made to avoid this step in order to open up a new horizon in DLLME automation. Berton *et al.* (Berton *et al.*, 2010b) have recently proposed an automatic TILDLME method, based on a flow injection manifold, for the determination of vanadium species in water samples. With no centrifugation the IL is recovered after the extraction using a florisil column, where the IL remains due to its high viscosity. The centrifugation step is also avoided in the technique *one-step in-syringe ionic liquid-based dispersive liquid-liquid microextraction* (Cruz-Vera *et al.;* 2009b). In this technique the complete DLLME procedure is developed in a conventional plastic syringe. Once the cloudy solution is formed, the IL is recovered by the slight movement of the syringe plunger. This separation is possible due to the tendency of the IL to interact with the plastic walls of the syringe.

4.2 Ionic liquids in solid phase microextraction

Solid phase microextraction (SPME) was introduced by Pawliszyn and co-workers (Berladi & Pawliszyn, 1989) (Arthur & Pawliszyn, 1990) as a simple, time efficient and solvent-free technique. SPME integrates the sampling, isolation and preconcentration of the target analytes in one step as well as it allows the direct coupling with different instrumental techniques. SPME is based on the partition of the analytes of interest between the sample and a small amount of extracting phase dispersed and immobilized in a solid support and the subsequent thermal or solvent-based desorption (Cruz-Vera et al., 2009c). SPME has been in a continuous evolution and different approaches, based on the sorptive phase minimization, have been developed.

Classical SPME used a fiber with an appropriate sorptive coating as extraction unit. The fiber is integrated in a syringe manifold which permits the development of the extraction as well as the subsequent injection of the analytes in the instrument, commonly a gas chromatograph. The fiber coating plays a key role on the extraction of the analytes. The ideal coating must fulfil some general requirements such as: (a) good affinity towards the target analytes in order to allow a good preconcentration; (b) thermal stability, especially when analytes desorption is assisted by an external heat source; (c) chemical stability, in particular when the desorption is performed by means of a chemical elution; (d) long life span; (e) low memory effect to avoid carry-over effects, (f) appropriate batch to batch reproducibility and (g) low price. As it is evident, the ideal coating does not exist and it should be selected according to the analytical problem under study. Moreover, the development of new coatings has turned into the goal of an extensive research in recent years. In this context, ILs can play an important role due to their characteristics that have been widely described in this chapter.

Liu and co-workers proposed for the first time the use of IL in a hybrid LPME-SPME approach (Liu *et al.*, 2005). In fact, $[C_8MIM][PF_6]$ is employed to cover a fused silica-fiber producing a disposable IL-coating. The IL-coating is of physical nature, without a covalent bonding, and it is performed by dipping a pretreated fused silica fiber into an IL solution in dichloromethane. After dipping, the fiber is dried removing the excess of dichloromethane, leaving a coating of $[C_8MIM][PF_6]$. The dipping and evaporating cycle is repeated three times in order to achieve the desired coating thickness. The use of IL in this type of coating presents some advantages such as: (a) due to its liquid nature, the coating can be easily removed using appropriate solvents; (b) due to its liquid nature, the coating presents higher diffusion coefficients which involve faster sorption kinetics and (c) due to its high viscosity, the IL can cover easily the fiber. Using this approach, the determination of benzene, toluene, ethylbenzene and xylene in paints has been achieved. Taking into account that the IL is only physically coated into the fiber, the volume of IL is reduced and therefore the sorption capacity limited. Hsieh *et al.* have proposed the use of nafion membranes as inner coating in order to maximize the volume of IL (Hsieh *et al.*, 2006). This polymeric membrane presents some anionic groups that may interact with imidazolium cations increasing the volume of immobilized IL and providing a more stable coating. The proposed configuration used [C_8MIM][Tfo] as extractant for the determination of ultra traces of PAHs in water samples. The physical coating can be also employed in coated fused silica capillary previously etched with an appropriate solution. Etching process produces a rough capillary surface which is able to immobilize a higher volume of IL compared to the original bare fused silica capillary (Huang *et al.*, 2009). The use of elastomers has been proposed to improve coating consistency. For this purpose a mixture of IL and elastomers in a volatile solvent is placed in a beaker and heated until a suitable viscosity is achieved. In this moment, a fused silica fiber is dipped in this solution. Later on, the fiber is drawn and dried until the solvent is completely evaporated. After a cycle of dipping and evaporation, a coated fiber can be obtained. The IL is trapped in the polymeric coated but it remains in its liquid form. A 50 μm in thickness polymeric coating containing [C_4MIM][NTf_2] has been successfully applied to the determination of methamphetamine and amphetamine in forensic samples (He *et al.*, 2009b).

All the mentioned approaches correspond to a hybrid LPME-SPME strategy since typically SPME devices are employed while a liquid phase is used as extractant. The liquid nature of the extractant provides evident advantages (e.g faster sorption kinetics) but it limits the stability and reusability of the coatings. Polymeric ionic liquids (PILs) allow the development of classical SPME fibers providing better chemical and thermal stability compared to the above mentioned approaches (Zhao *et al.* 2008b). Moreover, PILs coatings present a longer life span (they can be used up to 150 times) and good reproducibility between fibers. This type of coatings are prepared with a dipping-evaporation process, similar to those described previously, using a solution of PILs in a volatile organic solvent. When the solvent is evaporated, a solid PIL-coating is obtained. SEM pictures of PILs coating are presented in Figure 2. PIL-based SPME fibers have been successfully employed for the extraction of a great variety of organic substances (López-Darias *et al.* 2010) or carbon dioxide (Zhao & Anderson, 2010).

ILs have been covalently bonded to a silica support in order to obtain a fiber coating (Wanigasekara *et al.* 2010). The IL is previously derivatized to obtain an appropriate reagent which is able to react with a silica subtract. As a result, the liquid state of the IL is lost but their inherent characteristics (such as low vapor pressure, chemical stability or affinity towards organic compounds) remain unaltered. This IL-silica coated fiber has been used for the extraction of polar compounds.

In the SPME context, ILs can be used as solvents for the synthesis of stationary phases for capillary microextraction (CME), a SPME mode based on the location of the sorptive phase in the inner volume of a fused- silica capillary. The IL may play a double role since it can act as a reaction solvent and/or porogen solvent. The use of IL as reaction solvent has some advantages such as: (a) the solvent can be recycled; (b) the solvent is thermo-stable in the temperature range in which the reaction takes place and (c) the reagents present a higher stability in ILs compared to other organic solvents. On the other hand, the use of ILs as porogens provides the final polymer with a higher porosity and therefore with a higher extraction efficiency. Moreover, the IL employed can be removed by a thermal treatment releasing a pure polymeric coating (Shearrow *et al.*, 2009a). The synthesis process is reproducible, both in run-to-run and capillary-to-capillary conditions. This approach has been successfully used for the fabrication of coatings focused on the extraction of polar substances, which is a challenge in the analytical chemistry context (Shearrow *et al.*, 2009b).

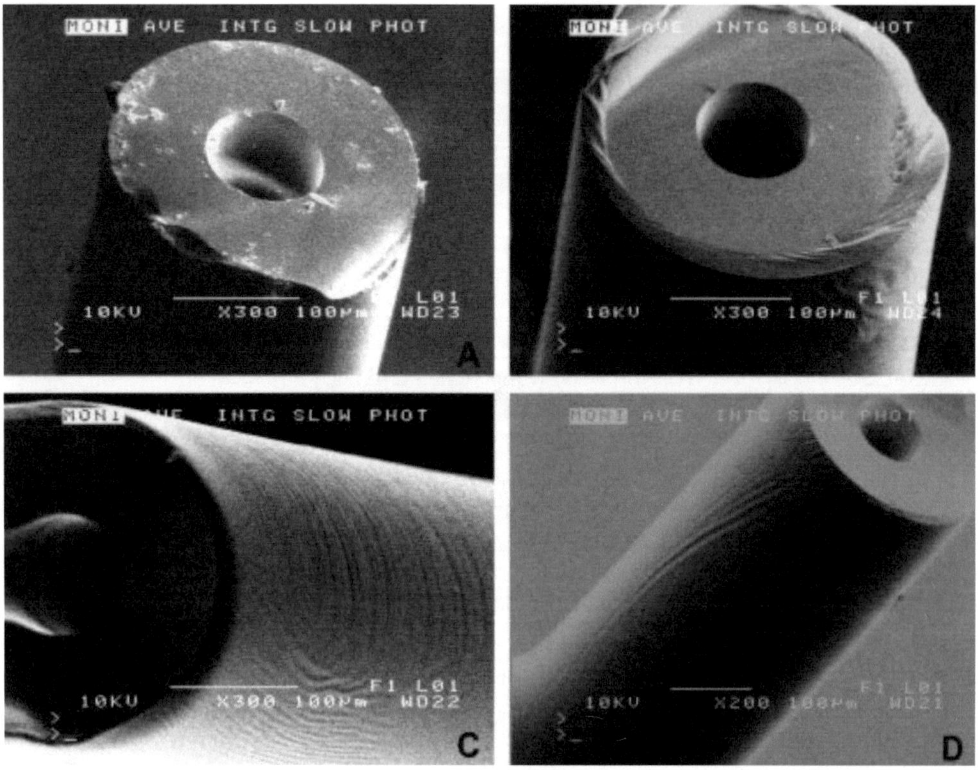

Fig. 2. Scanning electron micrographs of a 100-μm inner diameter bare fused silica support (A) and various angles of the fused silica support coated with PIL (B–D). Figure from Zhao *et al.* (2008b) and reproduced with permission of Elsevier.

5. Ionic liquid-based sorbents for solid phase extraction

The use of a variety of sorbents in solid phase (micro)extraction techniques has several advantages compared to liquid phase (micro)extraction for sample clean-up and/or analyte(s) preconcentrations. In a previous section (see section 4.2) the main contributions of ILs to solid phase microextraction have been highlighted. However, ILs have also contributed to improve the sorbent capability of conventional sorbent materials in several ways, the most relevant of which are commented on.

The use of a soft functional material arising from the combination of ILs and carbon nanotubes, allowed by the specific interaction between imidazolium ion components of the IL and the p-electronic nanotube surface, has been described. The resultant gelatinous material, consisting of highly electroconductive nanowires and fluid electrolytes, has a high capacity to absorb/desorb analytes from a variety of liquid samples. This capacity is the result of the synergistic effect between ILs solvation power and the adsorption capacity of carbon nanotubes. This new sorbent has been successfully applied to the determination of traces of PAHs in river waters (Polo-Luque *et al.*, in press. a).

A conventional filter can be modified when a dissolution of carbon nanotubes in ILs is passed through it, leading to the stable retention of the nanotubes. This modified nanofilter has been applied to organic interference removal for the determination by non-aqueous capillary electrophoresis (NACE) of imidazolium and pyridinium-based ionic liquids in river water (Polo-Luque *et al.*, in press. b). In a similar way, typical C_{18} sorbents can be modified using a dissolution of carbon nanotubes in ILs; this sorbent has been used for the preconcentration and determination of sulphonamides in river water by capillary electrophoresis (Polo-Luque *et al.*, in press. c).

6. Ionic liquids in sample dissolution

Dissolution is the classical beginning in the analysis of solid and solid-liquid samples (Kellner *et al.*, 2005). The general objectives of this substep of the preliminary operations of analytical processes are: (a) to make the next detection step possible; (b) to improve sample homogeneity and, thus, to enhance reproducibility (and quality) of the results; (c) to enhance sensitivity; and (d) to contribute to the simplification. In many real cases, it is not an easy task to distinguish between dissolution and extraction (see part 4 of this chapter). In this context, solvent changeover can also be considered.

The capacity of dissolution of ILs is undeniable (Weingärtner, 2008) (see part 3 of this article). They have been involved as solvents in both, physico-chemical studies and analytical processes. The most relevant applications in the field of chemical analysis are commented on below.

6.1 Ionic liquids as matrixes for mass spectrometric analysis

The application of ILs as replacements for the conventional liquid or solid matrixes in Matrix Assisted Laser Desorption/Ionization Mass Spectrometry (MALDI-MS) has recently been recognized (Baker *et al.*, 2005). MALDI-MS is a powerful technique for determining high-molecular-weight compounds (e.g. biomacromolecules, synthetic polymers) as well as low-molecular weight compounds. The requirements of an effective matrix for MALDI-MS are as follows (Liu *et al.*, 2009b):

- To dissolve (liquid matrix) or co-crystallize (solid matrix) with the sample;
- To contain a chromophore to absorb laser light and promote ionization of the analyte;
- To remain in the condensed phase under high-vacuum conditions;
- To minimize or avoid thermal and chemical sample degradation.

As can be seen in Figure 3, the applications of ILs as matrixes for MALDI-MS satisfy almost all of these requirements, but their ability to promote ionization of the analyte is not accomplished in all cases since it depends on the type of analyte and IL. In addition, the great variety of ILs makes not possible to extend their general features to their role as matrixes in MALDI-MS analysis. A recent review (Tholey & Heinzle, 2006) compiles the most relevant aspects and fields of application of the topic.

One of the first attempts to use ILs as matrixes in MALDI-MS was reported by Armstrong's group using peptides, proteins and poly(ethylene glycol) (PEG-2000) as model analytes (Armstrong *et al.*, 2001). All of the IL-based matrixes showed excellent solubilizing properties and vacuum stability compared to other frequently employed solid and liquid matrixes. There are other recent papers aimed to compare ILs with conventional solid and liquid matrixes for a variety of sample-analyte-pairs (Mank *et al.*, 2004) (Zabet-Moghaddam *et al.*, 2004). Besides the properties pointed out by Armstrong in 2001 (Armstrong *et al.*, 2001) (see Figure 3), it has been demonstrated that liquid consistency of IL-matrix sample preparations considerably enhances MALDI-MS analysis in terms of shot-to-shot reproducibility, which is a bottle-neck in this technique.

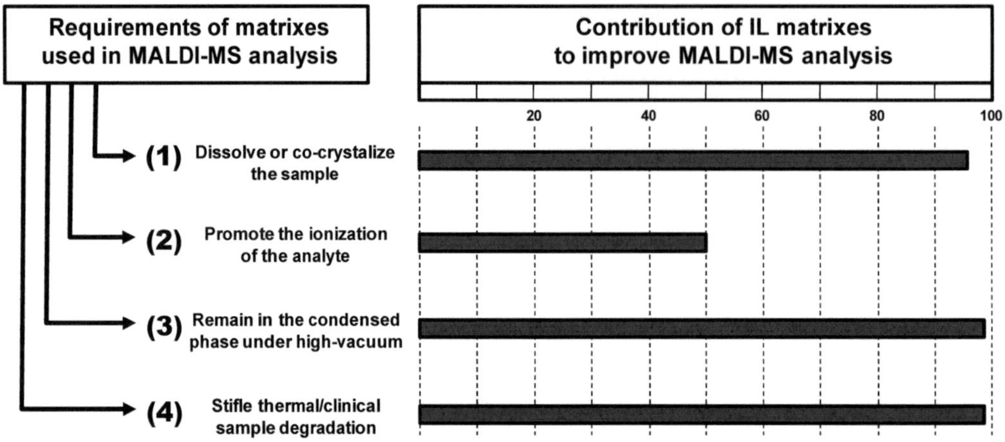

Fig. 3. Fulfillment by ILs of the basic requirements of matrixes in MALDI-MS analysis. For details, see text.

ILs as matrixes show also a synergy with the technique Matrix-Assisted Laser Desorption/Ionization-Time-of-Flight Mass Spectrometry (MALDI-TOF-MS) especially for screening purposes. Quantitative MALDI-TOF-MS using isotopic labeled internal standards and IL-based matrixes have been applied for the screening of 10 pyranose oxidase variants using aliquots of enzyme reaction mixtures without previous purification steps, which can be considered a relevant approach (Bungert *et al.*, 2004). In some cases, this technique is combined with a previous separation step. Such is the case of the fast screening of a mixture of low molecular weight compounds (three alkaloids, two anesthetics and one antibiotic) based on Thin-Layer Chromatography (TLC) followed by direct on-shot MALDI-TOF-MS identification with nearly matrix-free mass spectra using a UV-absorbing IL matrix. This technique is fast and sensitive and it requires little sample preparation and manipulation (Santos *et al.*, 2004).

6.2 Ionic liquids as solvents in molecular spectroscopy

Since 1986, ionic liquids have been demonstrated to be able to solvate a wide range of chemical species including organic, inorganic, metal complexes and organometallic compounds (Appleby *et al.*, 1986). The main advantages of the use of ILs as solvents in spectroscopy are (Koel, 2005): (a) they offer stability to a variety of species, particularly metal complexes; (b) they circumvent problems arising from solvation and solvolysis; and (c) they permit reliable UV-visible and IR solution spectra. Enhanced resolution and significant spectra shifts compared to those of the standards published have been observed. The spectra of the substance of interest obtained in ILs resemble to those recorded in solvents of low dielectric constants such as dichloromethane. The majority of publications on this topic are related to physico-chemical studies, but analytical processes can have benefit from having the analyte dissolved in an appropriate solvent, which can be achieved using extraction (see part 4 of this chapter) or solvent changeover by dissolution of the solid residue after the dryness of the first solvent employing heating or under a N_2 stream.

6.3 Ionic liquids as solvents in headspace gas chromatography

Owing to the inherent characteristics of ILs, such as non-volatility and heat stability, they can be used advantageously as solvents in headspace GC. The study performed by André

and co-workers (André *et al.*, 2005) is based on the use of three acidic, neutral and basic model analytes with boiling points above 200 °C which were dissolved in ILs; the analytes could be determined at the low-ppm level.
This approach has been successively applied to the determination of traces of low vapor pressure residual solvents in pharmaceutical drug products, a crucial aspect of their quality. Using the ionic liquid $[C_4MIM][BF_4]$ as the matrix medium, residual solvents such as N-methyl-2-pyrrolidone and dimethylformamide in commercially-available pharmaceutical preparations have been determined using static headspace GC (Liu & Jiang, 2007).

6.4 Ionic liquids as solvents for nanoparticles

ILs are suitable solvents for the dissolution (extraction) of a great variety of nanomaterials both organic (e.g. carbon nanoparticles) and inorganic (e.g. gold nanoparticles, quantum dots, etc.). Solubility of nanoparticles in ILs can have different foundations, the most relevant being salting out effects and cation exchange processes, among others.
ILs can be exploited as useful tools for the analysis of environmental and biological samples for the determination of the pollution caused by manufactured nanomaterials and for clarification of their toxicity in humans, animals and plants. Some specific developments in this context have been described in the last years (Wei *et al.*, 2004) (Huang *et al.*, 2006) (Nakashima & Kawai, 2005).
In addition, ILs can be used for the characterization of nanomaterials. The fact that they can be easily solubilized in ILs facilitates the analysis. This is the case of the characterization of a commercial mixture of single-walled carbon nanotubes that can be characterized by Raman spectrometry as a single component. Other alternative consists of the initial sample dissolution in an IL ($[C_4MIM][BF_4]$) using ultrasonic energy, and the subsequent encapsulation of the nanotubes in sodium dodecylsulphate micelles. An aliquot of the dissolution is introduced in a capillary electrophoretic equipment and an electropherogram of 8 peaks (in less than 4 minutes) is obtained, which demonstrates the non-homogeneity of the original sample of carbon nanotubes (López-Pastor *et al.*, 2008).

7. Final remarks

Due to the fact that the majority of the preliminary operations of the analytical process rely on the use of solvents (e.g. dissolution, extraction, solvent changeover, etc), any contribution to minimize or to avoid the negative connotations of the conventional ones, both organic (e.g. hexane) or inorganic (e.g. HNO_3) should be welcome. Such is the case of ILs which, since two decades ago, have open up new perspectives and promising approaches in this field, enhancing the analytical properties (top, basic and productive) through the improvement of the sample preparation steps (Valcárcel & Rios, 1993). The main advantages of ILs as solvents in comparison with the traditional ones can be justified by their unique properties that have been described in part 1 of this chapter, being the most relevant for such purpose: solvent capability and tunability, thermal stability and very low vapor pressure, as well as their viscosity, density, surface tension, immiscibility with other solvents, etc.
There are presently several drawbacks that limit the use of ILs in Analytical Sciences, the most relevant being the following: (a) hygroscopy, which can led to the hydrolysis of the anion components; (b) impurity, traces of other compounds (e.g.water) that change the expected properties of ILs to be used for a fixed fit-for-purpose; (c) instability, arising from the decomposition of ILs under heating, ultrasonication or microwave energy, frequently

used in sample preparations; (d) alteration of the structure of the dissolved analytes (e.g. biomolecules); and (e) toxicity, despite the reputation of ILs as green solvents, recent studies evidence the relative toxicity of some of the ILs used in practice. The environmental fate and toxicity of ILs have been recently reviewed by Pham *et al* (Pham *et al.* 2010).

8. Acknowledgements

Financial support from the Junta de Andalucía (FQM-02300) and the Spanish DGICyT (Grant CTQ2007-60426) is gratefully acknowledged

9. References

Abdolmohammad-Zadeh, H. & Sadeghi, G. H. (2009). A novel microextraction technique based on 1-hexylpyridinium hexafluorophosphate ionic liquid for the preconcentration of zinc in water and milk samples. *Anal. Chim. Acta,* 649 (September, 2009) 211-217, ISSN: 0003-2670.

Abdolmohammad-Zadeh, H. & Sadeghi, G. H. (2010). Combination of ionic liquid-based dispersive liquid-liquid micro-extraction with stopped-flow spectrofluorometry for the pre-concentration and determination of aluminum in natural waters, fruit juice and food samples. *Talanta,* 81 (May, 2010) 778-785, ISSN: 0039-9140.

Abulhassani, J.; Manzoori, J. L. & Amjadi, M. (2010). Hollow fiber based-liquid phase microextraction using ionic liquid solvent for preconcentration of lead and nickel from environmental and biological samples prior to determination by electrothermal atomic absorption spectrometry. J. Hazard. Mater. 176 (April, 2010) 481-486, ISSN: 0304-3894.

Aguilera-Herrador, E.; Lucena, R.; Cárdenas, S. & Valcárcel, M. (2008a). Direct coupling of ionic liquid based single-drop microextraction and GC/MS. *Anal. Chem.* 80 (February, 2008) 793-800, ISSN: 0003-2700.

Aguilera-Herrador, E.; Lucena, R.; Cárdenas, S. & Valcárcel, M. (2008b). Ionic liquid-based single-drop microextraction/gas chromatographic/mass spectrometric determination of benzene, toluene, ethylbenzene and xylene isomers in waters. *J. Chromatogr. A,* 1201 (August, 2008), 106-111, ISSN: 0021-9673.

Aguilera-Herrador, E.; Lucena, R.; Cárdenas, S. & Valcárcel, M. (2008c). Determination of trihalomethanes in waters by ionic liquid-based single drop microextraction/gas chromatographic/mass spectrometry. *J. Chromatogr. A,* 1209 (October, 2008), 76-82, ISSN: 0021-9673.

Aguilera-Herrador, E.; Lucena, R.; Cárdenas, S. & Valcárcel, M. (2009). Ionic liquid-based single drop microextraction and room-temperature gas chromatography for on-site ion mobility spectrometric analysis. *J. Chromatogr. A,* 1216 (July, 2009), 5580-5587, ISSN: 0021-9673.

Anderson, J. L. & Armstrong. D. W. (2003). High-stability ionic liquids. A new class of stationary phases for gas chromatography. *Anal. Chem.,* 75 (September, 2003) 4851-4858, ISSN: 0003-2700.

Andre, M.; Loidl, J.; Laus, G.; Schottenberger, H.; Bentivoglio, G.; Wurst, K. & Ongania, K. H. (2005). Ionic liquids as advantageous solvents for headspace gas chromatography of compounds with low vapor pressure. *Anal. Chem.,* 77 (January, 2005) 702-705, ISSN: 0003-2700.

Appleby, D.; Hyssey, C. L.; Seddon, K. R. & Turp, J. E. (1986). Room-temperature ionic liquids as solvents for electronic absorption spectroscopy of halide complexes. *Nature*, 323 (October, 1986) 614-615, ISSN: 0028-0836.

Armstrong, D. W.; He, L. & Liu Y. S. (1999). Examination of ionic liquids and their interaction with molecules, when used as stationary phases in gas chromatography. *Anal. Chem.* 71 (September, 1999) 3873-3876, ISSN: 0003-2700.

Armstrong, D. W.; Zhang, L. K; He, L. & Gross, M. L. (2001) Ionic liquids as matrixes for matrix-assisted laser desorption/ionization mass spectrometry. *Anal. Chem.* 73 (August, 2001) 3679-3686, ISSN: 0003-2700.

Bai, H.; Zhou, Q.; Xie, G. & Xiao, J. (2010). Temperature-controlled ionic liquid-liquid-phase microextraction for the pre-concentration of lead from environmental samples prior to flame atomic absorption spectrometry. *Talanta*, 80 (March, 2010) 1638-1642, ISSN: 0039-9140.

Baghdadi; M. & Sheminari, F. (2008). Cold-induced aggregation microextraction: a novel sample preparation technique based on ionic liquids. *Anal. Chim. Acta*, 613 (April, 2008) 56-63, ISSN: 0003-2670.

Baghdadi; M. & Sheminari, F. (2009). In situ solvent formation microextraction based on ionic liquids: A novel sample preparation technique for determination of inorganic species in saline solutions. *Anal. Chim. Acta*, 634 (February, 2009) 186-191, ISSN: 0003-2670.

Baker, G. A.; Baker, S. N.; Pandey, S. & Bright, F. V. (2005). An Analytical View of Ionic Liquids. *Analyst*, 130 (June, 2005) 800-808, ISSN: 0003-2654.

Basheer, C.; Alnedhary, A. A.; Madhava Rao, B. S.; Balasubramanian, R. & Lee, H. K. (2008). Ionic liquid supported three-phase liquid-liquid-liquid microextraction as a sample preparation technique for aliphatic and aromatic hydrocarbons prior to gas chromatography-mass spectrometry. *J. Chromatogr. A*, 1210 (November, 2008), 19-24, ISSN: 0021-9673.

Bennett, M. D. & Leo, D. J. (2004). Ionic liquids as stable solvents for ionic polymer transducers. *Sensor Actuator A. Phys.*, 115 (September, 2004) 79-90, ISSN 0924-4247.

Berladi, R. P. & Pawliszyn, J. B. (1989). The application of chemically modified fused silica fibers in the extraction of organics from water matrix samples and their rapid transfer to capillary columns. *Water Pollut. Res. J. Can.* 24 (1989) 179-191. ISSN: 0197-9140.

Berton, P.; Martinis, E. M.; Martinez, L. D. & Wuilloud, R. G. (2009). Room temperature ionic liquid-based microextraction for vanadium species separation and determination in water samples by electrothermal atomic absorption spectrometry. *Anal. Chim. Acta*, 640 (April, 2009) 40-46, ISSN: 0003-2670.

Berton, P. & Wuilloud, R. G. (2010a). Highly selective ionic liquid-based microextraction method for sensitive trace cobalt determination in environmental and biological samples. *Anal. Chim. Acta*, 662 (March, 2010) 115-162, ISSN: 0003-2670.

Berton, P.; Martinis, E. M. & Wuilloud R. G. (2010b). Development of an on-line temperature-assisted ionic liquid dispersive microextraction system for sensitive determination of vanadium in environmental and biological samples. *J. Hazard Mater.* 176 (April, 2010) 721-728, ISSN: 0304-3894.

Bungert, D.; Bastian, S.; Heckmann-Pohl, D. M.; Giffhorn, F.; Heinzle, E. & Tholey, A. (2004). Screening of sugar converting enzymes using quantitative MALDI-TOF mass spectrometry. *Biotech. Letters*, 26 (July, 2004) 1025-1030, ISSN: 0141-5492.

Buzzeo, M. C.; Hardacre, C. & Compton, R. C. (2004). Use of room temperature ionic liquids in gas sensor design. *Anal. Chem.*, 76 (July, 2004) 4583–4588, ISSN: 0003-2700.

Catherine L. Arthur, Janusz Pawliszyn. (1990). Solid phase microextraction with thermal desorption using fused silica optical fibers. *Anal. Chem.*, 62 (October, 1990) 2145–2148, ISSN: 0003-2700.

Chiappe, C. & Pieraccini, D. (2005). Ionic liquids: solvent properties and organic reactivity. *J. Phys. Org. Chem.*, 18 (April, 2005) 275-297, ISSN: 0894-3230.

Chisvert, A.; Román, I. P.; Vidal, L. & Canals, A. (2009). Simple and commercial readily-available approach for the direct use of ionic liquid-based single-drop microextraction prior to gas chromatography determination of chlorobenzenes in real water samples as model analytical application. *J. Chromatogr. A*, 1216 (February, 2009), 1290-1295, ISSN: 0021-9673.

Couling, D. J.; Bernot, R. J.; Docherty, K. M.; Dixon, J. K. &. Maginn, E. J. (2006). Assessing the factors responsible for ionic liquid toxicity to aquatic organisms via quantitative structure–property relationship modeling. *Green Chem.* 8 (August, 2006) 82-90, ISSN: 1463-9262.

Cruz-Vera, M.; Lucena, R.; Cárdenas, S. & Valcárcel, M. (2008). Ionic liquid-based dynamic liquid-phase microextraction: application to the determination of anti-inflammatory drugs in urine samples. *J. Chromatogr. A*, 1202 (August, 2008) 1-7, ISSN: 0021-9673.

Cruz-Vera, M.; Lucena, R.; Cárdenas, S. & Valcárcel, M. (2009a). Determination of phenothiazine derivatives in human urine by using ionic liquid-based dynamic liquid-phase microextraction coupled with liquid chromatography. *J. Chromatogr. B*, 877 (January, 2009) 37-42, ISSN: 1570-0232.

Cruz-Vera, M.; Lucena, R.; Cárdenas, S. & Valcárcel, M. (2009b). One-step in-syringe ionic liquid-based dispersive liquid-liquid microextraction. *J. Chromatogr. A*, 1216 (September, 2010) 6459-6465, ISSN: 0021-9673.

Cruz-Vera, M.; Lucena, R.; Cárdenas, S. & Valcárcel, M. (2009c). Sorptive microextraction for liquidchromatographic determination of drugs in urine. *Trends Anal. Chem.*, 28 (month, 2009) 1164- 1173, ISSN: 0165-9936.

Dadfarnia, S.; Shabani, A. M.; Bidabadi, M. S. & Jafari, A. A. (2010). A novel ionic liquid/micro-volume back extraction procedure combined with flame atomic absorption spectrometry for determination of trace nickel in samples of nutritional interest. *J Hazard Mater.*, 173 (January, 2010) 534-538, ISSN: 0304-3894.

Dai, S.; Ju, Y. H. & Barnes, C. E. (1999) Solvent extraction of strontium nitrate by a crown ether using room-temperature ionic liquids. *J. Chem. Soc. Dalton Trans.*, (month, 1999) 1201-1202, ISSN: 1477-9226.

Davis, J. H. (2004). Task specific ionic liquids. *Chem. Letters*, 33 (September, 2004) 1072-1077, ISSN: 0366-7022.

Fan, Y. C.; Hu, Z. L.; Chen, M. L.; Tu, C. S. & Zhu, Y. (2008). Ionic liquid based dispersive liquid–liquid microextraction of aromatic amines in water samples. *Chinese Chem. Lett.*, 19 (August, 2008), 985-987, ISSN 1001-8417.

Fan, Y.C.; Chen, M. L.; Tu, C. S. & Zhu, Y. (2009). A ionic liquid for dispersive liquid-liquid microextraction of phenols. *J. Anal. Chem.*; 64 (October, 2009) 1017-1022, ISSN: 1061-9348.

Fortunato, R.; Afonso, C. A. M.; Reis, M. A. M. & Crespo, J. G. (2004). Supported liquid membranes using ionic liquids: study of stability and transport mechanisms. *J. Membr. Sci.*, 242 (October, 2004) 197-2009, ISSN: 0376-7388.

Gharehbaghi, M.; Shemirani, F. & Farahani, M. D. (2009). Cold-induced aggregation microextraction based on ionic liquids and fiber optic-linear array detection spectrophotometry of cobalt in water samples. *J. Hazard. Mater.* 165 (June, 2009) 1049-1055, ISSN: 0304-3894.

He, Y. & Lee, H. K. (1997). Liquid-phase microextraction in a single drop of organic solvent by using a conventional microsyringe. *Anal. Chem.*, 69 (November, 1997) 4634-4640, ISSN: 0003-2700.

He, L.; Luo, X.; Xie, H.; Wang, C.; Jiang, X. & Lu, K. (2009a). Ionic liquid-based dispersive liquid-liquid microextraction followed high-performance liquid chromatography for the determination of organophosphorus pesticides in water sample. *Anal. Chim. Acta*, 655 (November, 2009) 52-59, ISSN: 0003-2670.

He, Y.; Pohl, J.; Engel, R.; Rothman, L. & Thomas, M. (2009b). Preparation of ionic liquid based solid-phase microextraction fiber and its application to forensic determination of methamphetamine and amphetamine in human urine. *J. Chromatogr. A.*, 1216 (June, 2009) 4824-4830, ISSN: 0021-9673.

Hirayama, N.; Deguchi, M.; Kawasumi, H. & Honjo, T. (2005). Use of 1-alkyl-3-methylimidazolium hexafluorophosphate room temperature ionic liquids as chelate extraction solvent with 4,4,4-trifluoro-1-(2-thienyl)-1,3-butanedione. *Talanta*, 65 (January, 2005) 255-260, ISSN: 0039-9140.

Hsieh, H.; Huang, P.; Sun, I.; Whang, T.; Hsu, C.; Huang, H. & Kuei, C. (2006). Nafion membrane-supported ionic liquid–solid phase microextraction for analyzing ultra trace PAHs in water samples. *Anal. Chim. Acta*, 557 (January, 2006) 321-328, ISSN: 0003-2670.

Huang, H. L.; Wang, H. P.; Wei, G. T.; Sun, I. W.; Huang, J. F. & Yang, Y. W. (2006). Extraction of nanosize copper pollutants with an ionic liquid. *Environ. Sci. Technol.*, 40 (August, 2006) 4761-4764, ISSN: 0013-936X.

Huang, K. P.; Wang, G. R.; Huang, B. Y. & Liu, C. Y. (2009). Preparation and application of ionic liquid-coated fused-silica capillary fibers for solid-phase microextraction. *Anal. Chim. Acta*, 645 (July, 2009) 42-47, ISSN: 0003-2670.

Jeannot, M. A. & Cantwell, F. (1996). Solvent microextraction into a single drop. *Anal. Chem.*, 68 (July, 1996) 2236-2240, ISSN: 0003-2700.

Jeannot, M. A. & Cantwell, F. (1997). Mass transfer characteristics of solvent extraction into a single drop at the tip of a syringe needle. *Anal. Chem.*, 69 (January, 1997) 235-239, ISSN: 0003-2700.

Kellner, R.; Mermet, J. M.; Otto, M.; Valcárcel, M. & Widmer, H. M. (2005). *Analytical Chemistry (2nd edition)*, Wiley-VCH, ISBN: 3-527-3059-4, Wheinheim.

Koel, M. (2005). Ionic Liquids in Chemical Analysis. *Crit. Rew. Anal. Chem.*, 35 (January, 2005) 177-192, ISSN: 1040-8347.

Li, Z.; Wei, Q.; Yuan, R.; Zhou, X.; Liu, H.; Shan, H. & Song, Q. (2007a). A new room temperature ionic liquid 1-butyl-3-trimethylsilylimidazolium hexafluorophosphate as a solvent for extraction and preconcentration of mercury with determination by cold vapor atomic absorption spectrometry. *Talanta*, 71 (January, 2007) 68-72, ISSN: 0039-9140.

Li, Z.; Lu, N.; Zhou, X. & Song, Q. (2007b). Extraction spectrophotometric determination of aluminum in dialysis concentrates with 3,5-ditertbutylsalicylfluorone and ionic liquid 1-butyl-3-trimethylsilylimidazolium hexafluorophosphate. *J. Pharm Biomed Anal.*, 43 (April, 2007) 1609-1614, ISSN: 0731-7085.

Li, S.; Cai, S.; Hu, W.; Chen, H. & Liu, H. (2009). Ionic liquid-based ultrasound-assisted dispersive liquid–liquid microextraction combined with electrothermal atomic absorption spectrometry for a sensitive determination of cadmium in water samples. *Spectrochim. Acta Part B,* 64, (July, 2009) 666-671, ISSN: 0584-8547.

Liang, C.; Yuan, C. Y.; Warmack, R. J.; Barnes, C. E. & Dai, S. (2002). Ionic liquids: a new class of sensing materials for detection of organic vapors based on the use of a quartz crystal microbalance. *Anal. Chem.*, 74, (May, 2002) 2172-2176, ISSN: 0003-2700.

Liu, H. & Dasgupta, P. K. (1996). Analytical Chemistry in a Drop. Solvent Extraction in a Microdrop. *Anal. Chem.*, 68, (June, 1996) 1817-1821, ISSN: 0003-2700.

Liu, J.; Jiang, G.; Chi, Y.; Cai, Y.; Zhou, Q. & Hu, J. (2003). Use of ionic liquids for liquid-phase microextraction of polycyclic aromatic hydrocarbons. *Anal. Chem.,* 75, (November, 2003) 5870-5876, ISSN: 0003-2700.

Liu, J. F.; Li, N.; Jiang, G. B.; Liu, J. M.; Jönsson, J. A. & Wen, M. J. (2005). Disposable ionic liquid coating for headspace solid-phase microextraction of benzene, toluene, ethylbenzene, and xylenes in paints followed by gas chromatography-flame ionization detection. *J. Chromatogr. A,* 1066(February, 2005) 27-32, ISSN: 0021-9673.

Liu, F. H. & Jiang, Y. (2007). Room temperature ionic liquid as matrix medium for the determination of residual solvents in pharmaceuticals by static headspace gas chromatography. *J. Chromatogr. A,* 1167 (October, 2007) 116-119, ISSN: 0021-9673.

Liu, Y.; Zhao, E.; Zhu, W.; Gao, H. & Zhou, Z. (2009a). Determination of four heterocyclic insecticides by ionic liquid dispersive liquid-liquid microextraction in water samples. *J. Chromatogr. A,* 1216, (February, 2009) 885-891, ISSN: 0021-9673.

Liu, R.; Liu, J. F.; Yin, Y. G.; Hu, X. L. & Jiang, G. B. (2009b). Ionic liquids in sample preparation. *Anal. Bioanal. Chem.,* 393 (February, 2009) 871-883, ISSN: 1618-2642.

López-Darias, J.; Pino, V.; Anderson, J. L.; Graham, C. M. & Afonso, A. M. (2010). Determination of water pollutants by direct-immersion solid-phase microextraction using polymeric ionic liquid coatings. *J. Chromatogr. A,* 1217, (February, 2010) 1236-1243, ISSN: 0021-9673.

López-Pastor, M.; Domínguez-Vidal, A.; Ayora-Cañada, M. J.; Simonet, B. M.; Lendl, B. & Valcarcel, M. (2008). Separation of Single-Walled Carbon Nanotubes by Use of Ionic-Liquid-Aided Capillary Electrophoresis. *Anal. Chem.,* 80 (April, 2008) 2672-2679, ISSN: 0003-2700.

Lucena, R.; Cruz-Vera, M.; Cárdenas, S. & Valcárcel, M. (2009). Liquid phase microextraction in bioanalytical sample preparation. *Bioanalysis,* 1, (April, 2009) 135-149, ISSN 1757-6180.

Mahpishanian, S. & Shemirani, F. (2010). Preconcentration procedure using in situ solvent formation microextraction in the presence of ionic liquid for cadmium determination in saline samples by flame atomic absorption spectrometry. *Talanta,* 82 (July, 2010) 471-476, ISSN: 0039-9140.

Mank, M.; Stahl, B. & Boehm, G. (2004). 2,5-Dihydroxybenzoic acid butylamine and other ionic liquid matrixes for enhanced MALDI-MS analysis of biomolecules. *Anal. Chem.* 76 (May, 2004) 2938-2950, ISSN: 0003-2700.

Manzoori, J. L.; Amjadi, M. & Abulhassani, J. (2009a). Ionic liquid-based single drop microextraction combined with electrothermal atomic absorption spectrometry for the determination of manganese in water samples. *Talanta,* 77, (February, 2009) 1539-1544, ISSN: 0039-9140.

Manzoori, J. L.; Amjadi, M. & Abulhassani, J. (2009b). Ultra-trace determination of lead in water and food samples by using ionic liquid-based single drop microextraction-electrothermal atomic absorption spectrometry. *Anal. Chim. Acta,* 644, (June, 2009) 48-52, ISSN: 0003-2670.

Mao, T.; Hao, B.; He, J.; Li, W.; Li, S. & Yu, Z. (2009). Ultrasound assisted ionic liquid dispersive liquid phase extraction of lovastatin and simvastatin: a new pretreatment procedure. *J. Sep. Sci.* 32 (September, 2009) 3029-3033, ISSN: 1615-9306.

Martinis, E. M.; Bertón, P.; Altamirano, J. C.; Hakala, U. & Wuilloud, R. G. Tetradecyl(trihexyl)phosphonium chloride ionic liquid single-drop microextraction for electrothermal atomic absorption spectrometric determination of lead in water samples. *Talanta*, 80 (March, 2010) 2034-2040, ISSN: 0039-9140.

Mwongela, S. M.; Numan, A.; Gill, N. L.; Agbaria, R. A. & Warner, I. M. (2003). Separation of Achiral and Chiral Analytes Using Polymeric Surfactants with Ionic Liquids as Modifiers in Micellar Electrokinetic Chromatography. *Anal. Chem.* 75, (November, 2003) 6089-6096, ISSN: 0003-2700.

Nakashima, T. & Kawai, T. (2005). Quantum dots-ionic liquid hybrids: efficient extraction of cationic CdTe nanocrystals into an ionic liquid. *Chem. Commun.*, 28 (March, 2005) 1643-1645, ISSN: 1359-7345.

Palacio, M. & Bhushan, B. (2008). Ultrathin Wear-Resistant Ionic Liquid Films for Novel MEMS/NEMS Applications. *Adv. Mater.*, 20 (March, 2008) 1194-1198, ISSN: 0935-9648.

Pham, T. P.; Cho, C. W. & Yun, Y. S. (2010). Environmental fate and toxicity of ionic liquids: a review. *Water Res.* 44 (January, 2010) 352-372, ISSN: 0043-1354.

Pedersen-Bjergaard, S. & Rasmussen, K. E. (1999). Liquid-liquid-liquid microextraction for sample preparation of biological fluids prior to capillary electrophoresis. *Anal. Chem.* 71, (July, 1999) 2650-2656, ISSN: 0003-2700.

Pena, M. T.; Casais, M. C.; Mejuto, m. C. & Cela, R. (2009). Development of an ionic liquid based dispersive liquid-liquid microextraction method for the analysis of polycyclic aromatic hydrocarbons in water samples. *J. Chromatogr. A.*, 1216 (September, 2009) 6356-6364, ISSN: 0021-9673.

Pena-Pereira, F.; Lavilla, I.; Bendicho, C.; Vidal, L. & Canals, A. (2009). Speciation of mercury by ionic liquid-based single-drop microextraction combined with high performance liquid chromatography-photodiode array detection. *Talanta*, 78, (April, 2009) 537-541, ISSN: 0039-9140.

Peng, J.; Liu, J.; Jiang, G.; Tai, C. & Huang M. (2005). Ionic liquid for high temperature headspace liquid-phase microextraction of chlorinated anilines in environmental water samples. *J. Chromatogr. A*, 1072, (April, 2005) 3-6, ISSN: 0021-9673.

Peng, J.; Liu, J.; Hu, X. & Jiang, G. (2007). Direct determination of chlorophenols in environmental water samples by hollow fiber supported ionic liquid membrane extraction coupled with high-performance liquid chromatography. *J. Chromatogr. A*, 1139, (January, 2007) 165-170, ISSN: 0021-9673.

Polo-Luque, M. D.; Simonet, B. M. & Valcárcel, M. (in press. a). New soft materials of ionic liquids with carbon nanotubes as sorbents for SPE. *Anal. Chem.*, in press.

Polo-Luque, M. D.; Simonet, B. M. & Valcárcel, M. (in press. b). Combination of carbon nanotubes-modified filters for the NACE determination of trace levels of ionic-liquids in water. *J. Chromatogr. A.*, in press.

Polo-Luque, M. D.; Simonet, B. M. & Valcárcel, M. (in press. c). Solid-phase extraction-capillary electrophoretic determination of sulfonamide residues in surface water by using C18-carbon nanotubes as novel sorbent material. *Electrophoresis*, in press.

Poole, C. F. & Poole, S. K. (2010) Extraction of organic compounds with room temperature ionic liquids. *J. Chromatogr. A., 1217* (April, 2010) 2268-2286, ISSN: 0021-9673.

Qin, W. & Li, S. F. Y. (2002). An ionic liquid coating for determination of sildenafil and UK-103,320 in human serum by capillary zone electrophoresis-ion trap mass spectrometry. *Electrophoresis*, 23 (December, 2002) 4110-4116, ISSN: 1522-2683.

Ravelo-Pérez, L. M.; Hernández-Borges, J.; Asensio-Ramos, M. & Rodríguez-Delgado, M. A. (2009a). Ionic liquid based dispersive liquid-liquid microextraction for the extraction of pesticides from bananas. *J. Chromatogr. A*, 1216 (October, 2009) 7336-7345, ISSN: 0021-9673.

Ravelo-Pérez, L. M.; Hernández-Borges, J.; Herrera-Herrera, A. V. & Rodríguez-Delgado, M. A. (2009b). Pesticide extraction from table grapes and plums using ionic liquid based dispersive liquid-liquid microextraction. *Anal. Bioanal. Chem.*, 395 (December, 2009) 2387-2395, ISSN: 1618-2642.

Rezaee, M.; Assadi, Y.; Milani Hosseini, M. R.; Aghaee, E.; Ahmadi, F. & Berijani, S. (2006). Determination of organic compounds in water using dispersive liquid-liquid microextraction. *J. Chromatogr. A*, 1116, (May, 2006) 1-9, ISSN: 0021-9673.

Santos, L. S.; Haddad, R.; Höehr, N. F.; Pilli, R. A. & Eberlin, M. N. (2004). Fast screening of low molecular weight compounds by thin-layer chromatography and "on-spot" MALDI-TOF mass spectrometry. *Anal. Chem.*, 76 (April, 2004) 2144-2147, ISSN: 0003-2700.

Shearrow, A. M.; Harris, G. A.; Fang, L.; Sekhar, P. K.; Nguyen, L. T.; Turner, E. B.; Bhansali, S. & Malik, A. (2009a). Ionic liquid-mediated sol-gel coatings for capillary microextraction. . *J. Chromatogr. A*, 1216, (July, 2009) 5449-5458, ISSN: 0021-9673.

Shearrow, A. M.; Bhansali, S. & Malik, A. (2009b). Ionic liquid-mediated bis[(3-methyldimethoxysilyl)propyl] polypropylene oxide-based polar sol-gel coatings for capillary microextraction. *J. Chromatogr. A*, 1216, (September, 2009) 6349-6355, ISSN: 0021-9673.

Simoni, L. D.; Chapeaux, A.; Brennecke, J. F. & Stadtherr, M. A. (2010). Extraction of biofuels and biofeedstocks from aqueous solutions using ionic liquids. *Computers and chemical engineering*, 34 (September, 2010) 1406-1412, ISSN 0098-1354.

Tao, Y.; Liu, J.; Hu, X.; Li, H.; Wang, T. & Jiang, G. (2009). Hollow fiber supported ionic liquid membrane microextraction for determination of sulfonamides in environmental water samples by high-performance liquid chromatography. *J. Chromatogr. A*, 1216, (August, 2009) 6259-6266, ISSN: 0021-9673.

Tholey, A. & Heinzle, E. (2006) Ionic (liquid) matrices for matrix-assisted laser desorption/ionization mass spectrometry-applications and perspectives. *Anal. Bioanal. Chem.*, 386 (September, 2006) 24-37, ISSN: 1618-2642.

Valcárcel, M. & Ríos, A. (1993). The Hierarchy and Relationships of Analytical Properties. *Anal. Chem.*, 65 (September, 1993) 781A-787A, ISSN: 0003-2700.

Vidal, L.; Chisvert, A.; Canals, A. & Salvador, A. (2007a). Sensitive determination of free benzophenone-3 in human urine samples based on an ionic liquid as extractant phase in single-drop microextraction prior to liquid chromatography analysis. *J. Chromatogr. A*, 1174, (December, 2007) 95-103, ISSN: 0021-9673.

Vidal, L.; Psillakis, E.; Domini, C. E.; Grané, N.; Marken, F. & Canals, A. (2007b). An ionic liquid as a solvent for headspace single drop microextraction of chlorobenzenes from water samples. *Anal. Chim. Acta*, 584, (February, 2007) 189-195, ISSN: 0003-2670.

Vidal, L.; Domini, C. E.; Grané, N.; Psillakis, E. & Canals, A. (2007c). Microwave-assisted headspace single-drop microextration of chlorobenzenes from water samples. *Anal. Chim. Acta*, 592, (May, 2007) 9-15, ISSN: 0003-2670.

Vidal, L.; Chisvert, A.; Canals, A. & Salvador, A. (2010). Ionic liquid-based single-drop microextraction followed by liquid chromatography-ultraviolet spectrophotometry detection to determine typical UV filters in surface water samples. *Talanta,* 81, (April, 2010) 549-555, ISSN: 0039-9140.

Visser, A. E.; Swatloski, R. P.; Reichert, W. M.; Davis, J. H.; Rogers, R. D.; Mayton, R.; Sheff, S. & Wierzbicki, A. (2001). Task-specific ionic liquids for the extraction of metal ions from aqueous solutions. *Chem. Commun.*, (month, 2001), 135 - 136, ISSN: 1359-7345.

Wang, Y.; You, J.; Ren, R.; Xiao, Y.; Gao, S.; Zhang, H. & Yu, A. (2010). Determination of triazines in honey by dispersive liquid-liquid microextraction high-performance liquid chromatography. *J. Chromatogr. A.*, 1217 (June, 2010) 4241-4246, ISSN: 0021-9673.

Wanigasekara, E., Perera, S.; Crank, J. A.; Sidisky, L.; Shirey, R.; Berthod, A. & Armstrong, D. W. (2010). Bonded ionic liquid polymeric material for solid-phase microextraction GC analysis. *Anal. Bioanal. Chem.*, 396 (January, 2010) 511-524, ISSN: 1618-2642.

Wei, G. T.; Yang, Z. & Chen, C. J. (2003) Room temperature ionic liquid as a novel medium for liquid/liquid extraction of metal ions. *Anal. Chim. Acta,* 488 (July, 2003) 183-192, ISSN: 0003-2670.

Wei, G. T.; Yang, Z.; Lee, C. Y.; Yang, H. Y. & Wang, C. R. (2004). Aqueous-organic phase transfer of gold nanoparticles and gold nanorods using an ionic liquid. *J. Am. Chem. Soc.,* 126 (April, 2004) 5036-5037, ISSN: 0002-7863.

Weingärtner, H. (2008). Understanding ionic liquids at the molecular level: facts, problems, and controversies. *Angew. Chem. Int. Ed.,* 47 (January, 2008) 654-670, ISSN: 1433-7851.

Xia, L.; Hu, B.; Jiang, Z.; Wu, Y.; Li, L. & Chen, R. (2005). 8-Hydroxyquinoline–chloroform single drop microextraction and electrothermal vaporization ICP-MS for the fractionation of aluminium in natural waters and drinks. *J. Anal. Atom. Spectrom., 20,* (May, 2005) 441-446, ISSN: 0267-9477.

Xia, L.; Li, X.; Wu, Y.; Hu, B. & Chen, R. (2008). Ionic liquids based single drop microextraction combined with electrothermal vaporization inductively coupled plasma mass spectrometry for determination of Co, Hg and Pb in biological and environmental samples. *Spectrochim. Acta Part B,* 63, (November, 2008) 1290-1296, ISSN: 0584-8547.

Yao, C. & Anderson, J. L. (2009). Dispersive liquid-liquid microextraction using an in situ metathesis reaction to form an ionic liquid extraction phase for the preconcentration of aromatic compounds from water. *Anal. Bioanal. Chem.* (November, 2009) 1491-1502, ISSN: 1618-2642.

Yao, C.; Twu, P. & Anderson, J. L. (2010). Headspace Single Drop Microextraction Using Micellar Ionic Liquid Extraction Solvents. *Chromatographia,* 72, (September, 2010) 393-402, ISSN 0009-5893.

Ye, C.; Zhou, Q. & Wang, X. (2006). Headspace liquid-phase microextraction using ionic liquid as extractant for the preconcentration of dichlorodiphenyltrichloroethane and its metabolites at trace levels in water samples. *Anal. Chim Acta.* 572 (July, 2006) 165-171, ISSN: 0003-2670.

Ye, C.; Zhou, Q.; Wang, X. & Xiao, J. (2007). Determination of phenols in environmental water samples by ionic liquid-based headspace liquid-phase microextraction coupled with high-performance liquid chromatography. *J. Sep. Sci.,* 30 (January, 2007) 42-47, ISSN: 1615-9306.

Yousefi, S. R. & Shemirani, F. (2010) Development of a robust ionic liquid-based dispersive liquid-liquid microextraction against high concentration of salt for preconcentration of trace metals in saline aqueous samples: application to the determination of Pb and Cd. *Anal. Chim. Acta*, 669 (June, 2010) 25-31, ISSN: 0003-2670.

Zabet-Moghaddam, M.; Heinzle, E. & Tholey, A. (2004) Qualitative and quantitative analysis of low molecular weight compounds by ultraviolet matrix-assisted laser desorption/ionization mass spectrometry using ionic liquid matrices. *Rapid Commun. Mass Spectrom.*, 18 (January, 2004) 141-148, ISSN: 0951-4198.

Zhang, J. & Lee, H. K. (2010). Headspace ionic liquid-based microdrop liquid-phase microextraction followed by microdrop thermal desorption-gas chromatographic analysis. *Talanta,* 81 (April, 2010) 537-542, ISSN: 0039-9140.

Zhao, H.; Xia, S. & Ma, P. (2005). Use of ionic liquids as green solvents for extractions. *J. Chem. Technol. Biotechnol.* 80 (October, 2005) 1089-1096, ISSN: 0268-2575.

Zhao, F.; Li, J. & Zeng, B. (2008a). Coupling of ionic liquid-based headspace single-drop microextraction with GC for sensitive detection of phenols. *J. Sep. Sci.,* 31 (September, 2008) 3045-3049, ISSN: 1615-9306.

Zhao, F.; Meng, Y. & Anderson, J. L. (2008b). Polymeric ionic liquids as selective coatings for the extraction of esters using solid-phase microextraction. *J. Chromatogr. A,* 1208 (October, 2008) 1-9, ISSN: 0021-9673.

Zhao, R.; Wang, X.; Sun J.; Wang, S.; Yuan, J. & Wang, X. (2010a). Trace determination of triclosan and triclocarban in environmental water samples with ionic liquid dispersive liquid-phase microextraction prior to HPLC–ESI-MS–MS. *Anal. Bioanal. Chem.*, 397 (June, 2010) 1627-1633, ISSN: 1618-2642.

Zhao, R. S.; Wang, X.; Sun, J.; Yuan, J. P.; Wang, S. S. & Wang, X. K. (2010b). Temperature-controlled ionic liquid dispersive liquid-phase microextraction for the sensitive determination of triclosan and triclocarban in environmental water samples prior to HPLC-ESI-MS/MS. *J. Sep. Sci.* 33 (June, 2010) 1842-1848, ISSN: 1615-9306.

Zhao, Q. & Anderson, J. L. (2010c). Selective extraction of CO2 from simulated flue gas using polymeric ionic liquid sorbent coatings in solid-phase microextraction gas chromatography. *J. Chromatogr. A*, 1217 (July, 2010) 4517- 4522, ISSN: 0021-9673.

Zhou, Q. & Ye, C. (2008). Ionic liquid for improved single-drop microextraction of aromatic amines in water samples. *Microchim. Acta,* 162 (July, 2008) 153-159, ISSN: 0026-3672.

Zhou, Q.; Bai, H.; Xie, G. & Xiao, J. (2008a). Temperature-controlled ionic liquid dispersive liquid phase micro-extraction. *J. Chromatogr. A,* 1177 (January, 2008) 43-49, ISSN: 0021-9673.

Zhou, Q.; Bai, H.; Xie, G. & Xiao, J. (2008b). Trace determination of organophosphorus pesticides in environmental samples by temperature-controlled ionic liquid dispersive liquid-phase microextraction. *J. Chromatogr. A,* 1188 (April, 2008) 148-153, ISSN: 0021-9673.

Zhou, Q.; Bai, H.; Xie, G. & Xiao, J. (2009a). Temperature-controlled ionic liquid-dispersive liquid-phase microextraction for preconcentration of chlorotoluron, diethofencarb and chlorbenzuron in water samples. *J. Sep. Sci.* 32 (November, 2009) 3945-3950, ISSN: 1615-9306.

Zhou, Q.; Zhang, X. & Xiao, J. (2009b). Ultrasound-assisted ionic liquid dispersive liquid-phase micro-extraction: a novel approach for the sensitive determination of aromatic amines in water samples. *J Chromatogr A.* 1216 (May,2009) 4361-4365, ISSN: 0021-9673.

10

Cold-Induced Aggregation Microextraction: A Novel Sample Preparation Technique Based on Ionic Liquids for Preconcentration of Cobalt Prior to Its Determination by Fiber Optic-Linear Array Detection Spectrophotometry in Real Water Samples

Maysam Gharehbaghi, Farzaneh Shemirani,
Malihe Davudabadi Farahani and Majid Baghdadi
University of Tehran
Iran

1. Introduction

Despite good development in modern analytical instruments, direct determination of trace analytes at low concentrations is often a problem for analytical chemists and, as a result, a sample-preparation step is required. The sample preparation in analytical process typically consists of an extraction–separation step, which offers not only the ability to isolate target analytes from the matrix solution, thus reducing or even eliminating originally present interferences but also the opportunity for these analytes to be preconcentrated and determined at very low levels. Liquid–liquid extraction (LLE), based on the partitioning of an analyte from one liquid phase to another immiscible liquid phase, is a still widely used traditional sample preparation technique. However, this method is time-consuming and tedious. Also large amounts of high purity solvents, which are expensive and toxic, are inevitable to be used. The continuous quest for novel sample preparation procedures has led to a development of new methods, which have main advantages like rapidity and reduction of solvents consumption, such as cloud point extraction (CPE), solid-phase microextraction (SPME), single drop microextraction (SDME), hollow fiber-liquid phase microextraction (HF-LPME), dispersive liquid–liquid microextraction (DLLME) and others.

Cloud point extraction is considered to be a green extraction method because the extraction is performed while using surfactants, which may exhibit lower toxicity, volatility and flammability compared with organic solvents used in other extraction techniques. This technique is widely used in extraction of metals (Bezerra et al., 2005; Dallali et al., 2009), organic compounds (Carabias-Martinez et al., 2000) and proteins (Saitoh & Hinze, 1995). Despite many benefits when using CPE, in rather high content of salt, background is increased since the enrichment phase is composed of a little aqueous sample. Also in the presence of more than 3 % of water-miscible organic solvents such as acetone the phase separation does not occur and the extraction system is destroyed (Manzoori & Karim-

Nezhad, 2003). In addition, the application of this technique is limited by the influence of temperature, pressure, pH, and selection of specified surfactants (Paleologos et al., 2005).

Solid-phase microextraction (SPME), a widely used sample preparation technique, is a solventless, simple and convenient method, which combines extraction, preconcentration and sample introduction in one step (Risticevic et al., 2009). However, it is expensive; its fiber is fragile and has a limited lifetime. Unfortunately, large amounts of eluents are used in this method. In addition, the carryover between extractions is also problematic when determining some analytes at very low concentrations.

Single-drop microextraction is a type of solvent microextraction (SME) technique that is performed by exposing a single drop of solvent to the headspace or directly into the matrix of the sample (Genfa & Dasgupta, 2000; He & Lee, 1997; Jeannot & Cantwell, 1996; Liu & Dasgupta, 1995). It was developed as a solvent-minimized sample preparation procedure, which is inexpensive and advantageous. Since very little amount of the solvent is used, there is a minimal exposure to toxic organic solvents (Ahmadi et al., 2006; Jeannot & Cantwell, 1997). In comparison with solid phase microextraction, it has advantages like wide choice of extraction solvents, cheaper devices and simpler operation and no analyte carryover occurs. However, relatively small volumes of extractant drop lead to lower sensitivity. In addition, it is time-consuming and at fast stirring rates, the organic drop is broken up and air bubbles are formed.

Hollow-fiber liquid-phase microextraction is a sample preparation technique, in which the analytes are extracted into an organic layer filled in pores of a suspended hollow fiber or further extracted back into an aqueous phase inside the fiber, depending on whether a two-phase or three-phase system is applied (Pedersen-Bjergaard & Rasmussen, 1999; Rasmussen & Pedersen-Biergaard, 2004). Because of the larger contact area between the analyte aqueous solution and extraction phase, the extraction efficiency obtained with this method is higher than with SDME. However, the extraction procedure is tedious and in most cases, the extraction equilibrium is not attained during a short time. Moreover, carryover and contamination of the hollow fiber can be an outcome.

Dispersive liquid-liquid microextraction (DLLME), another type of SME which was developed in 2007, is simple, fast and inexpensive. However, the amount of used disperser solvent is relatively high, therefore it may lead to decrease of the extraction recovery of less hydrophobic species.

In recent years, room-temperature ionic liquids (RTILs) have attracted increasing interest and are applied more and more as the extraction solvent replacing the volatile solvent in sample preparation (Huddleston et al., 1998), due to their unique chemical and physical properties, such as negligible vapour pressure, non-flammability, good extractability for various organic compounds and metal ions as a neutral or charged complexes, as well as tunable viscosity and miscibility with water and organic solvents. However, most of them were conducted in the liquid–liquid extraction and as a result a large volume of ionic liquid was required (Smirnova et al., 2004; Xia et al., 2006), which is tedious and costly. Then the microextraction based on ILs was developed by Liu et al. (Liu et al., 2003). Subsequently, due to a combination of the viscosity and the expense of ionic liquids, we used an ionic liquid (IL) at the microscale for the analytical extraction and the determination of mercury ions in water samples by using spectrophotometric detection (Gharehbaghi et al., 2009).

In 2008 Baghdadi and Shemirani developed a new method called cold-induced aggregation microextraction (CIAME), which was based on the use of ILs in homogeneous liquid–liquid microextraction (HLLME) (Baghdadi & Shemirani, 2008). In this method, a very small amount of hydrophobic IL as an extractant solvent is dissolved in the sample solution containing Triton X-114. Triton X-114 prevents IL from sticking onto the surface of a centrifuge tube wall, therefore we have named it "anti-sticking agent". One should be aware that there is no interface between the water and the extractant phases; as a result mass transfer from aqueous phase into separated phase has no important effect on the extraction step. After dissolving, the solution is cooled in the ice bath, a cloudy solution is formed due to the solubility reduction of IL and fine droplets of it are formed. During the formation of fine droplets of the extractant phase, the hydrophobic species are collected by the extractant molecules and the extraction process is completed after the formation of the droplets. After centrifuging, the fine droplets of extractant phase settle to the bottom of the conical-bottom glass centrifuge tube.

This method is simple, rapid, safe and robust against high content of salt and water-miscible organic solvents. In comparison with CPE, background is too low in case of saline solutions, because of very low solubility of water in ILs. Also, owing to high viscosity of ILs, removal of bulk aqueous phase is easier. This method is more suitable for the extraction of heat-susceptible species in comparison with CPE. CIAME provides a high recovery and has a low toxicity since only very small amounts of IL as a "green extraction solvent" is used.

2. Importance

2.1 Cobalt; the occurrence, benefits and applications, exposure and toxicity and methods of determination

Cobalt is a natural element found throughout the environment, used in pigment manufacture and used to make superalloys (alloys that maintain their strength at high temperatures approaching their melting points). It can benefit or harm human health. Cobalt is an essential element in people's lives, as a constituent of vitamin B_{12}. It has also been used as a treatment for anemia, because it stimulates red blood cell production. The significance of cobalt as a transition metal in its wide spectrum of applications is covering many frontier areas of study, particularly in medicine. We can be exposed to low levels of cobalt by breathing air, eating food or drinking water. Food and drinking water are the largest sources of exposure to cobalt for general population.

Toxicological effects of large amounts of cobalt include vasodilatation, flushing and cardiomyopathy in men and animals (Yuzefovsky et al., 1994). Even though cobalt is not considered to be as toxic as most heavy metals, it is an equally harmful element. Hence owing to the importance of cobalt, its determination from associated elements by extractive spectrophotometry has been of a considerable importance.

A wide variety of chelating agents has been reported for the spectrophotometric determination of cobalt (Appadoo & Bhagwat, 1994; Carvalho et al., 1996; Chaudhari & Sawant, 1993; Dasilva & Martins, 1992; Kalika et al., 1993; Kamburova et al., 1994; Khambekar & Sawant, 1997; Lokhande et al., 1996; Maheshwari & Balasubramanian, 1995; More & Sawant, 1992; Pillai & Shinde, 1995; R.A. Chaudhari & A.D. Sawant, 1993; Reddy & Sarma, 1994; Sharma & Dave, 1997; Shen et al., 1995; Taher & Puri, 1995; Toral et al., 1993; Umebayashi & Ishiguro, 1996). However, these methods suffer from limitations such as critical pH (Dasilva et al., 1992; Maheshwari et al., 1995; Pillai et al., 1995), long color development time (Umebayashi et al., 1996) and interference with some ions (Chaudhari et al., 1993; Kalika et al., 1993; Lokhande et al., 1996; Maheshwari et al., 1995; More et al., 1992; Pillai et al., 1995; Reddy et al., 1994; Toral et al., 1993; Umebayashi et al., 1996).

In addition, the preconcentration and determination of cobalt (II) has been studied in various extraction methods such as: liquid-liquid extraction (LLE), flow injection extraction (FIE), cloud point extraction (CPE), solid phase extraction (SPE) and dispersive liquid-liquid microextraction (DLLME).

New LLE methods based on liquid membranes (Kakoi et al., 1998; Ribeiro et al., 2004) and hollow fiber contactors (Soldenhoff et al., 2005) were used for extraction and preconcentration of cobalt. However, it is time-consuming, tedious and uses large amounts of high purity solvents, which are expensive and toxic. The continuous quest for novel sample preparation procedures has led to a development of new methods, among which main advantages belong their speed and negligible volume of used solvents. Initial efforts to address the problem of large solvent consumption have led to a development of the FIE method. Some of these methods have been applied for cobalt extraction (Andac et al., 2001; Cao et al., 1999; Cassella et al., 2001; Dadfarnia & Jafarzadeh, 1999; Dzherayan et al., 2002; Fujimoto et al., 1999; Li et al., 2006a; Li et al., 2006b; Nogueira et al., 1998; Shabani et al., 2003; Song et al., 2006; Tsakovski et al., 2002). FIE has some advantages compared with LLE like low cost, high extraction speed and reduced solvent and sample consumption. However, the amount of used solvent is still in the order of several hundred micro liters per analysis.

Also preconcentration of cobalt in some CPE methods has been reported (Bezerra et al., 2007; Chen & Teo, 2001; Donati et al., 2006; Ghaedi et al., 2008; Gil et al., 2008; Nascentes & Arruda, 2003; Safavi et al., 2004; Shemirani & Shokoufi, 2006; Shokoufi et al., 2007b). Despite many benefits from the use of CPE, in high content of salt, background is increased since the enrichment phase is composed of a little aqueous sample and in very high content of salt, the density of the sample becomes equal or even higher than that of micelles, therefore they can not be settled. Also in the presence of more than 3 % of water-miscible organic solvents such as acetone phase separation does not occur.

Several SPE preconcentration procedures for cobalt have been reported using various sorbents (Blitz-Raith et al., 2007; Ghaedi et al., 2007; Praveen et al., 2005; Yang et al., 2002).

DLLME is a type of solvent microextraction (SME), which has been developed in 2007. A combined method including fiber optic-linear array detection spectrophotometry (FO-LADS) and DLLME was developed for preconcentration and determination of cobalt (Shokoufi et al., 2007a). In this method 1,2-dichlorobenzene (a harmful solvent) was used as an extractant. Another DLLME method, which was applied to preconcentrate trace levels of cobalt as a prior step to its determination by spectrophotometric detection, was reported (Gharehbaghi et al., 2008). This method is simple, fast and inexpensive. However, the amount of used disperser solvent is relatively high, therefore it may happen that recoveries of relatively less hydrophobic species decrease.

Homogeneous liquid-liquid microextraction (HLLME) utilizes the phase separation phenomenon from a homogeneous solution and the target solutes are extracted into a separated phase. However, HLLME has some problems; for instance, sometimes it is not compatible with some instrumental analysis and it also requires an addition of a reagent such as acid, base, salt, etc. As a result, some interesting compounds are probably destroyed; moreover, the addition of the reagent causes a release of heat during the extraction.

2.2 Vantage points of cold-induced aggregation microextraction method

In this research the performance of CIAME was investigated with the determination of cobalt in water samples using FO-LADS. The effects of various experimental parameters on the extraction were investigated. This method is simple, rapid and viscosity independent for extraction and preconcentration of metal ions and organic compounds from water samples, and is robust against high content of salt and water-miscible organic solvents. Additionally, in comparison with the organic solvent extraction, it is much safer since only small amounts of surfactant and IL are used which are being considered as "green solvents" for various separation processes and in comparison with IL-based DLLME, the extraction recoveries of CIAME are relatively high (Gharehbaghi et al., 2009).

Another important capability of this method is that it can be operated in a continuous mode. If the solution is cooled during centrifuging, fine droplets of extractant phase are continuously formed and analytes are extracted and transferred to the bottom of the centrifuge tube. In order to have a good recovery in case of microextraction in which the volume of the extractant solvent is very small, a large distribution coefficient is necessary, unless the recovery is low and continuous microextraction is required. This project is currently being performed in our laboratory.

3. Experimental

3.1 Instrumentation

A UV–vis light source, optical fiber and spectrograph model 2048 were prepared from Avantes (Eerbeek, Netherlands) for FOLADS. A 50 μL quartz cylindrical micro-cell (Hellma, Mullheim, Germany) was used as a determination cell. A Universal 320R refrigerated centrifuge equipped with an angle rotor (6-place, 9000 rpm, Cat. No.1620A) and a Jeio Tech BW-05G water bath were obtained from Hettich (Kirchlengern, Germany).

3.2 Reagents

All used reagents were of analytical grade. Triply distilled water was used throughout the experiment. 1-(2-pyridylazo)-2-naphthol (PAN), acetone, ethanol, acetic acid, 1-hexyl-3-methylimidazolium hexafluorophosphate [Hmim][PF_6] and 1- hexyl-3-methylimidazolium bis (trifluoromethylsulfonyl) imide [Hmim][Tf_2N] and all used salts were obtained from Merck (Darmstadt, Germany). Triton X-100 and Triton X-114 were purchased from Fluka (Buchs, Switzerland). A stock solution of cobalt (II) (1000 mg L^{-1}) was prepared by dissolving appropriate amount of $CoCl_2 \cdot 6H_2O$ and working standard solutions were obtained by appropriate stepwise dilution of the stock standard solutions. All stock and working standard solutions were stored in plastic (polypropylene) bottles with leak proof screw cap, which were cleaned before being used by soaking in 10% nitric acid solution for at least 24 h and then rinsed thoroughly with triply distilled water. A 10^{-3} mol L^{-1} PAN solution was prepared by dissolving the appropriate amount in absolute ethanol. The viscosity of ILs is high and their handling is difficult, therefore working solutions ([Hmim][PF_6], 0.8 mg μL^{-1} and [Hmim][Tf_2N], 0.5 mg μL^{-1}) were prepared in acetone.

3.3 Cold-induced aggregation microextraction procedure

The sample or standard solution containing Co(II) in the range of 1.5–65 ng mL^{-1}, PAN (2.5×10^{-5} mol L^{-1}), Triton X-114 (0.05 %, w/v), sodium nitrate (0.25 %, w/v), oxalate (4.0×10^{-3} mol L^{-1}) was poured in a 10 mL conical-bottom glass centrifuge tube. After 1 or 2 min, while greenish complex of cobalt was formed, pH of the solution was adjusted to 1 by concentrated HCl. [Hmim][PF_6] (64mg) and [Hmim][Tf_2N] (5 mg) were added and the volume of the solution was adjusted to 10.0 mL by triply distilled water. After that, tube was kept in a thermostated bath at 35 °C for 4 min. After shaking, it was placed in an ice-water bath for 10 min and a cloudy solution was formed. Then, the mixture was centrifuged for 5 min at 5000 rpm. As a result, fine droplets of IL settled at the bottom of the centrifuge tube (about 8 µL). The bulk aqueous phase was removed simply by inverting the tubes. The extraction steps are illustrated in Fig. 1. Afterwards, IL-phase was dissolved in 50 µL of pure ethanol and transferred to quartz cylindrical micro-cell. The absorbance of the complex was measured at 578 nm.

4. Results and discussion

It is necessary to investigate the effect of all parameters that can probably influence the extraction performance. In this methodology these parameters are the kind and amount of IL and anti-sticking agent, ligand concentration, pH, salt concentration, temperature and centrifuge conditions, which were investigated and optimized in order to achieve a high recovery and enrichment factor. In all optimization steps concentration of cobalt was 30 ng mL^{-1}.

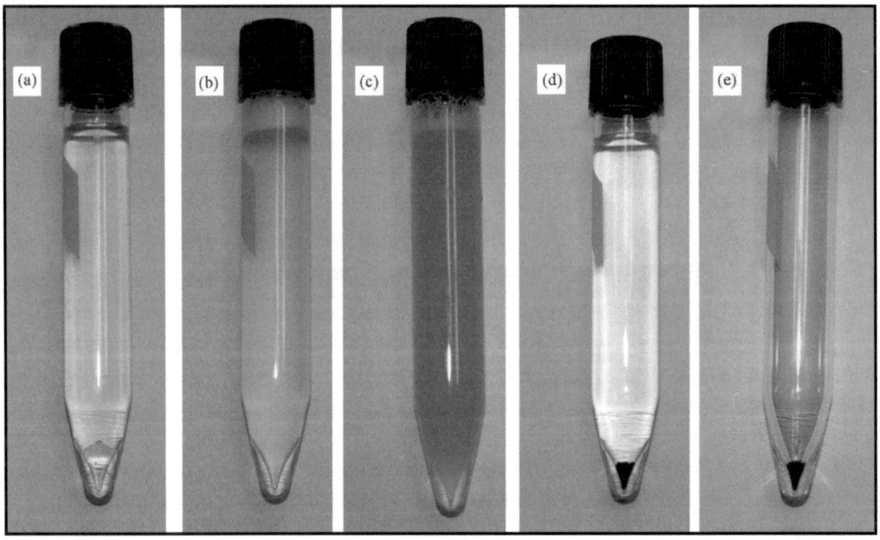

Fig. 1. Photography of different steps in CIAME: (a) after adding IL in the sample solution, (b) after shaking and dissolving the IL, (c) after cooling and phase separation, (d) after centrifuging and (e) after removing the bulk aqueous phase.

4.1 Selection of ionic liquid

For the selection of a suitable IL some considerations were recommended, which were mentioned in the previous work (Baghdadi et al., 2008). In this work, we focused on ILs containing imidazolium cation. Imidazolium-ILs containing PF_6^- as an anion are hydrophobe, relatively inexpensive and liquid in experimental conditions, therefore they are suitable for LLE. From [Bmim][PF_6], [Hmim][PF_6] and [Omim][PF_6] according to some physicochemical properties like density, viscosity and water solubility (Liu et al., 2005) and with regard to the point that the sample volume was 10 mL, [Hmim][PF_6] was chosen as an extractant. Thus, about 75 mg of this IL was solved in the aqueous sample solution.

4.2 Selection of anti-sticking agent

Like in the previous work, after centrifugation, it was observed that some of the IL-phase stick on the wall of the centrifuge tube. In order to overcome this problem, a non-ionic surfactant was added to the sample solution (Baghdadi et al., 2008). In the presence of non-ionic surfactant and during the phase separation, molecules of the surfactant surround the fine droplets of IL. Hence, interactions of IL with the wall of the centrifuge tube decrease and consequently, IL-phase does not stick on the wall of the centrifuge tube. The effects of two non-ionic surfactants (Triton X-114 and Triton X-100) were investigated and compared. In the presence of Triton X-100 absorbance decreased, while in the case of using Triton X-114 absorbance increased. Therefore, the Triton X-114 was chosen as the anti-sticking agent at the optimum concentration of 0.05 % (w/v).

4.3 Effect of pH and PAN concentration

In CIAME method pH plays a unique role in the metal-chelate formation and the subsequent extraction. The effect of pH on the extraction of cobalt from water samples was studied in the range of 0.5–6 after Co–PAN complex formation. Results reveal that the absorbance is slightly reduced by increasing pH. As well as at low pH, cations are less likely to precipitate; pH 1 seems a proper choice for extraction.

The effect of PAN concentration on the absorption is shown in Fig. 2. As it is well expected, in accordance with ML_2 stoichiometry of the complex the absorption is higher when increasing the PAN concentration. We investigated PAN concentration in the range of 1.5×10^{-7} to 4×10^{-5} mol L^{-1}. The maximum absorbance was obtained at a concentration of 2.5×10^{-5} mol L^{-1} of the ligand and after that, the absorbance stays approximately constant.

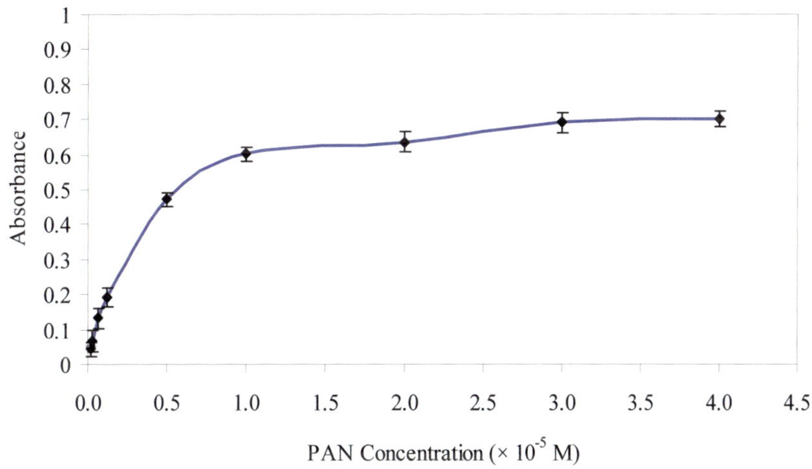

Fig. 2. Effect of PAN concentration on the absorbance of complex. Utilized conditions: Cobalt 30 ng mL^{-1}, pH = 1, $NaNO_3$ 0.2 % (w/v), Triton X-114 0.05 % (w/v), [Hmim][PF_6] 60 mg, [Hmim][Tf_2N] 5 mg, diluting agent 50 µL.

4.4 Effect of IL amount

In the presence of high content of salt, the solubility of [Hmim][PF_6] increases and phase separation does not occur. But according to the common ion effect, solubility decreases in the presence of IL with a common ion (such as [Hmim][Tf_2N]). This is one of the interesting properties of ILs. The effect of [Hmim][PF_6] in the presence and absence of [Hmim][Tf_2N] is shown in Fig. 3. In the presence of [Hmim][Tf_2N], lower amount of [Hmim][PF_6] was required for the phase separation. The optimum amount of [Hmim][PF_6] for the complete extraction was 70 and 60 mg in the absence and presence of [Hmim][Tf_2N], respectively.

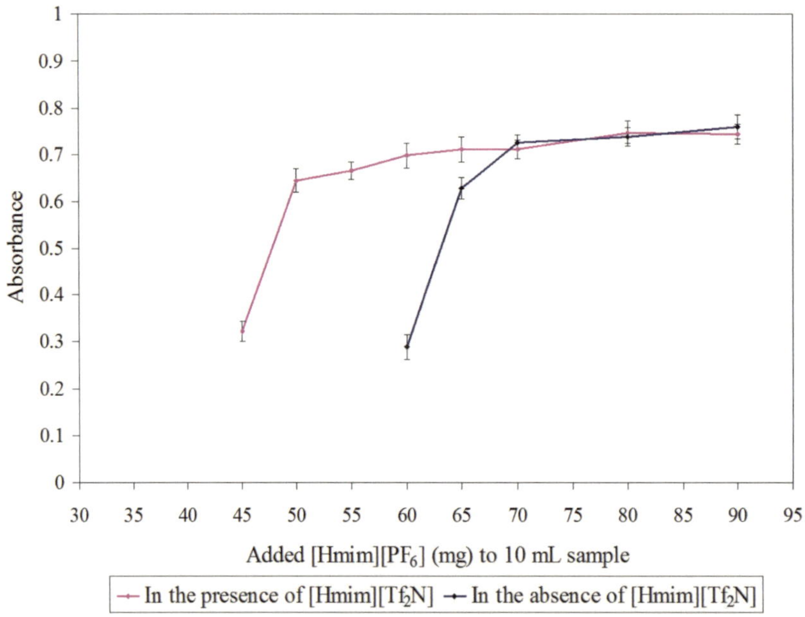

Fig. 3. Effect of the amount of [Hmim][PF_6] on the absorbance of Co-PAN complex. Utilized conditions: Cobalt 30 ng mL^{-1}, PAN 2.5×10^{-5} mol L^{-1}, pH = 1, $NaNO_3$ 0.2 % (w/v), Triton X-114 0.05 % (w/v), [Hmim][Tf_2N] 5 mg, diluting agent 50 μL.

4.5 Effect of salt concentration and water-miscible organic solvents

$NaNO_3$ was chosen in order to study the salt effect in the range of 0–1 % (w/v). The absorbance slightly increased as a result of the salting out effect. A concentration of 0.2% (w/v) $NaNO_3$ was selected for subsequent experiments in order to increase recovery. Also to investigate the robustness of the method against salty solutions, various solutions with the concentration up to 10% were examined. As shown in Fig. 4, in the absence of [Hmim][Tf_2N], absorbance decreased by increasing the salt concentration as a result of increasing in solubility of [Hmim][PF_6], but in the presence of [Hmim][Tf_2N] phase separation occurred up to 10% $NaNO_3$.

Since acetone and ethanol were selected as the solvents for the IL and ligand, their effects were investigated, too. In the presence of acetone, the absorbance was nearly constant up to 10 % and afterwards, it decreased and in the presence of ethanol, the absorbance was almost constant up to 12 %. This is one of the advantages of CIAME compared to CPE. In case of CPE, the extraction system is destroyed in the presence of over 3% water-miscible organic solvents (Manzoori et al., 2003). Even in case of DLLME based on common organic solvents like chloroform, carbon tetrachloride, etc., the extraction efficiency was clearly reduced for solutions containing water-miscible organic solvents.

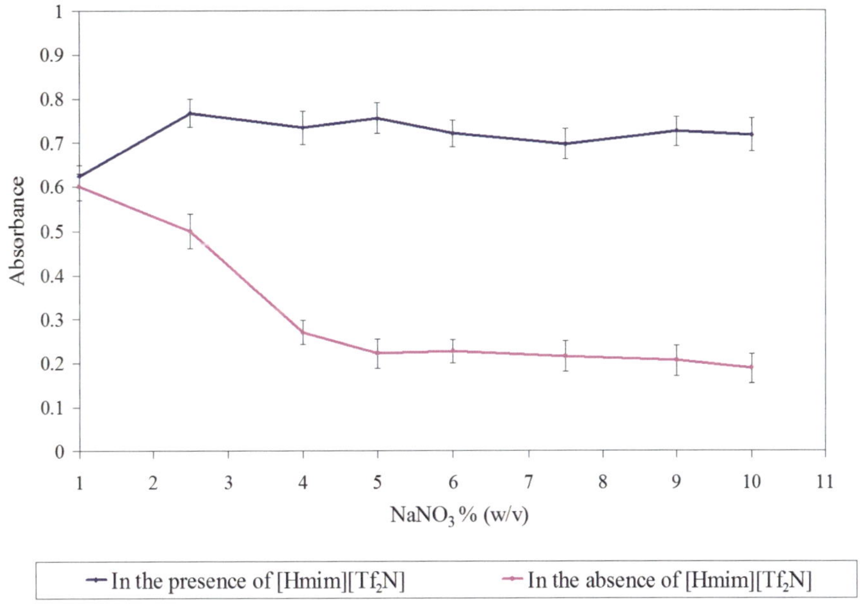

Fig. 4. Effect of $NaNO_3$ on the absorbance of Co-PAN complex in the range of 1-10 % (w/v). Utilized conditions: Cobalt 30 ng mL^{-1}, PAN 2.5×10^{-5} mol L^{-1}, pH = 1, Triton X-114 0.05 % (w/v), [Hmim][PF_6] 64 mg, [Hmim][Tf_2N] 5 mg, diluting agent 50 µL.

4.6 Effect of temperature

Before shaking the solutions containing IL, they were heated in the range of 20-60 °C. In this range the absorbance was constant; warming of the solution had no effect on the extraction process. This is another advantage of CIAME in comparison with CPE, especially if the species are heat-sensitive. Since ILs are dissolved more easily at the temperatures above 30 °C, a temperature of 35 °C was chosen for the heating step.

After dissolving of IL, samples were cooled in the temperature range of 0-25 °C. As the temperature decreased, the absorbance increased, as a result of decreasing IL solubility. Hence, a temperature of 0 °C was applied in all experiments by placing the centrifuge tubes in an ice bath for 10 min.

4.7 Effect of centrifuge conditions

The effect of centrifugation rate on the absorbance was studied in the range of 1000 - 7000 rpm. It was found, that over 4000 rpm IL-phase completely settled, therefore a rate of 5000 rpm was selected as the optimum point.

At the optimum rate, the absorbance was investigated as a function of the centrifugation time. Over 2 min, absorbance was constant, indicating complete transfer of IL-phase to the bottom of centrifuge tube. Therefore, the optimum centrifugation time was chosen as 3 min.

4.8 Effect of coexisting ions

The effects of coexisting ions in real water samples on the recovery of cobalt were also studied. In these experiments, 10mL solutions containing 30 ng mL^{-1} of cobalt and various amounts of interfering ions were treated. A given species was considered to interfere if it resulted in a ± 5% variation of the absorbance. The obtained results are given in Table 1. Most of the examined cations and anions did not interfere with the extraction and determination of Co^{2+}. However, some of the tried species, such as Fe^{3+}, Cu^{2+} and Pd^{2+} interfered with the determination of Co^{2+} ion. These interferences were eliminated in the presence of proper masking agents, such as 1.5×10^{-4} mol L^{-1} acetylacetone for Pd^{2+}, 4.0×10^{-3} mol L^{-1} oxalate for Fe^{3+} (and Ca^{2+}) and 2.5×10^{-3} mol L^{-1} ascorbic acid and 4.0×10^{-3} mol L^{-1} iodide for Cu^{2+}. In the presence of the masking agents, no interference was observed for Ca^{2+} up to 1000, Fe^{3+} and Cu^{2+} up to 100 and Pd^{2+} up to 10 times relative to Co^{2+} concentration. Thus, a quantitative extraction of Co^{2+} was possible.

Interferent	Interferent/Co(II) (weight ratio)	Recovery (%)	Interferent	Interferent/Co(II) (weight ratio)	Recovery (%)
Ba^{2+}	1000	98	Cu^{2+}	100	-[c]
Mg^{2+}	1000	100		100	105[d]
Mn^{2+}	1000	99	Hg^{2+}	100	92
Cr^{3+}	1000	104		50	99
Cd^{2+}	1000	100	Pd^{2+}	10	188
Zn^{2+}	1000	103		10	95[e]
Ca^{2+}	1000	98[a]	I^-	1000	100
Ag^+	500	97	SO_4^{2-}	1000	101
Pb^{2+}	500	101	NO_3^-	1000	98
Al^{3+}	500	102	PO_4^{3-}	1000	98
Fe^{3+}	100	165	CH_3COO^-	1000	100
	100	98[b]	NO_2^-	1000	98

Table 1. Effect of interferents on the recovery of 30 μg L^{-1} Co(II) in water sample using CIAME

[a]Ca^{2+} masked with 4.0×10^{-3} mol L^{-1} oxalate .

[b]Fe^{3+} masked with 4.0×10^{-3} mol L^{-1} oxalate.

[c]In the presence of this interferent the shape of spectrum is changed completely.

[d]Cu^{2+} masked with 2.8×10^{-3} mol L^{-1} ascorbic acid and 3.9×10^{-3} mol L^{-1} iodide.

[e]Pd^{2+} masked with 1.9×10^{-4} mol L^{-1} acetylacetone.

4.9 Figures of merit

Table 2 summarizes analytical characteristics of the optimized method, including limit of detection, reproducibility and enhancement factor. The limit of detection (LOD = 0.14 ng mL^{-1}) was calculated as $3\ S_b/m$ (S_b: standard deviation of the blank signals; m: slope of calibration curve after the preconcentration). A good correlation coefficient (r = 0.9997) was obtained and only small deviations between sequential determinations (RSD = 2.32 %) were found. The calibration curve was investigated up to 65 ng mL^{-1} which was linear. Enhancement factor (EF = 165) was obtained from the slope ratio of calibration curve after and before the preconcentration.

Parameter	Analytical Feature
Linear range (ng mL^{-1})	1.5-65
r^2	0.9995
Limit of detection (ng mL^{-1}) (3σ, n = 5)	0.14
Repeatability (RSD[a], %) (n = 5)	2.32
Enhancement factor[b] (EF)	165
Sample volume (mL)	10
Extraction time (min)	<20

Table 2. Analytical characteristics of CIAME for determination of Co(II)
[a]RSD was obtained for 30 µg L^{-1} concentration of Co(II) .
[b]Enhancement factor is the slope ratio of calibration graph after and before extraction.

The equation of the calibration curve after the preconcentration procedure is given in Eq. (1). Also the equation of the calibration curve before the preconcentration procedure is given at below in Eq. (2). (These standard samples of cobalt were chosen up to 200 ng mL^{-1} with execs PAN to measure their absorbance and obtain the equation of the calibration curve accurately).

$$A= 2.30 \times 10^{-2} C_{(Co)} + 0.0094 \quad (r^2 = 0.9995, r = 0.9997) \tag{1}$$

$$A= 1.39 \times 10^{-4} C_{(Co)} + 0.0323 \quad (r^2 = 0.9983, r = 0.9991) \tag{2}$$

4.10 Accuracy

For evaluating the accuracy of the method, a certified reference material (CRM) was analyzed. The Reference Standard "SCP-ES-L-1" (SCP SCIENCE Canada, ground water (Low level)) is a ground water that has been spiked with metals. The certified value and analytical results are presented in Table 3. The result indicates effectiveness and accuracy of the proposed method.

Certified reference material	Certified[a] (ng mL^{-1})	Found[b] (ng mL^{-1})	Found[c] (ng mL^{-1})	Recovery (%)
SCP-ES-L-1	51 ± 8	48.6 ± 0.2 [d]	49.5 ± 0.6 [d]	101.8

Table 3. Determination of cobalt in a standard reference material using DLLME
[a]Results after dilution 1 : 500.
[b]Determined by GFAAS.
[c]Determined by CIAME/FO-LADS.
[d]Mean value ± standard deviation based on five replicate measurements.

4.11 Analysis of natural water

The proposed methodology was applied for the determination of cobalt in different water samples. Damavand mineral water, Tajan river water and Caspian Sea water samples were collected from the north of Iran and analyzed by CIAME as a prior step to its determination by FO-LADS. No concentration of cobalt in the tap and mineral water samples was detected. Moreover, the robustness of the proposed method was checked by performing recovery tests on a saline serum and a synthetic sample (containing Mn^{2+}, Zn^{2+}, Mg^{2+}, Cd^{2+}, Cr^{3+} of 1000 ng mL^{-1} and Na^+, K^+ of 3000 ng mL^{-1}). Each type of water was spiked with variable amounts of Co(II) to assess matrix effects. The results are shown in Table 4. The relative recoveries of cobalt from mentioned water samples at various spiking levels were between 96.0 and 103.0 %. These results demonstrated that matrices of these water samples, under presented conditions, had little effect on CIAME of cobalt.

Sample	Spiked (ng mL^{-1})	Found (ng mL^{-1})[a]	Recovery (%)
Mineral water	-	N.D[b]	-
	10	10.1 ± 0.15	101.0
	20	20.3 ± 0.32	101.5
Tap water[c]	-	N.D[b]	-
	10	10.2 ± 0.22	102.0
	15	14.7 ± 0.35	98.0
River water	-	4.3 ± 0.39	-
	10	13.9 ± 0.23	96.0
	25	28.5 ± 0.33	96.8
Sea water	-	1.7 ± 0.27	-
	10	12.0 ± 0.32	103.0
	25	27.1 ± 0.55	101.6
Saline Serum[d]	25	25.6 ± 0.2	102.4
	45	43.9 ± 0.4	97.6
Synthetic sample	10	10.1 ± 0.18	101.0
	15.0	14.9 ± 0.23	99.3

Table 4. Determination of cobalt in real and synthetic samples
[a]Mean ± S.D. (n = 5).
[b]Not detected.
[c]From drinking water system of Tehran, Iran.
[d]Produced by Daru Paksh Company for injection.

4.12 Comparison of CIAME with other methods

A comparison of the represented method with other reported preconcentration methods is given in Table 5. In comparison with other reported methods, CIAME has low LOD (0.14 ng mL^{-1}), high enrichment factor (165) and relative short extraction procedure (less than 20 min). All these results indicate that CIAME is a reproducible, simple and low cost technique that can be used for the preconcentration of metal ions like cobalt from water samples.

5. Conclusions and future work

In this study, a new mode of HLLME based on IL, named cold-induced aggregation microextraction (CIAME), was developed for preconcentration of cobalt from water samples as a prior step to its determination by FO-LADS. Cobalt was employed as a test analyte and 1-(2-pyridylazo)-2-naphthol (PAN) as a complexing agent to assess the extraction procedure.

Method	LOD[a] (ng mL^{-1})	RSD.[b] (%)	Enhance-ment factor	Sample consumpt-ion (mL)	Calibration range (ng mL^{-1})	References
CPE/Spectrophotometry	7.5	2.2	10[c]	10	20–200	(Safavi et al., 2004)
CPE/FAAS	1.06	5.41	28.5	12.5	25-200	(Nascentes et al., 2003)
CPE/FAAS	0.24	2.1	57[d]	50	0-120	(Chen et al., 2001)
SPE/Spectrophotometry	10	2.23	100[e]	250	10-400	(Yang et al., 2002)
DLLME/FO-LADS	0.2	<4	165	10	1-70	(Shokoufi et al., 2007a)
DLLME/Spectrophotometry	0.5	2.5	125	50	2-60	(Gharehbaghi et al., 2008)
CPE/FAAS	5	1.71	20[f]	10	0-200	(Giokas et al., 2001)
SPE/ETAAS	0.004	4.5	87	11.5	0.01–0.25	(Sant'Ana et al., 2002)
CIAME/FO-LADS	0.14	2.32	165	10	1.5-65	This work

Table 5. Characteristic performance data obtained by using CIAME and other techniques for determination of cobalt

[a]Limit of detection.

[b]Relative standard deviation.

[c]Ratio of the aqueous phase to final volume of surfactant-rich phase.

[d]The enhancement factor, as the ratio of absorbance of preconcentrated sample to that obtained without preconcentration.

[e]Ratio of the aqueous phase to final volume of eluent phase.

[f]Preconcentration factor.

This method is simple, rapid, safe and robust against high content of salt and water-miscible organic solvents. In comparison with CPE, background is too little in case of saline solutions, because of very low solubility of water in ILs. Also owing to high viscosity of ILs removing the bulk aqueous phase is easier. This method is more suitable for extraction of heat-susceptible species in comparison with CPE. CIAME provides high recovery and has low toxicity since only very small amounts of an IL as a "green extraction solvent" is used.

In addition, the proposed method offers good sensitivity in comparison with other combination methods, which were used, FAAS or conventional UV–vis spectrophotometer as the detection technique.

Our research group is currently working on CIAME in a continuous mode and as a future work would like to develop a microextraction technique in which the volume of the extractant solvent is very small. In order to have a good recovery, in this case a large distribution coefficient is necessary unless a continuous microextraction, in which fine droplets of extractant phase are continuously formed and analytes are extracted and transferred to the bottom of the centrifuge tube, is applied.

Financial support from University of Tehran, as well as proofreading by Barbora Ehrlichová is gratefully acknowledged.

6. References

Ahmadi, F.;Assadi, Y.;Hosseini, S., & Rezaee, M. (2006). Determination of organophosphorus pesticides in water samples by single drop microextraction and gas chromatography-flame photometric detector. *Journal of Chromatography A*, 1101(1-2), 307-312.

Andac, M.;Asan, A.;Isildak, I., & Cesur, H. (2001). Flow-injection spectrophotometric determination of cobalt(II) at low mu g 1(-1) levels with 4-benzylpiperidinedithiocarbamate. *Analytica Chimica Acta*, 434(1), 143-147.

Appadoo, C., & Bhagwat, V. W. (1994). Spectrophotometric determination of cobalt with m-(mercaptoacetamido) phenol. *Asian Journal of Chemistry*, 6, 703–705.

Baghdadi, M., & Shemirani, F. (2008). Cold-induced aggregation microextraction: A novel sample preparation technique based on ionic liquids. *Analytica Chimica Acta*, 613(1), 56-63.

Bezerra, M. A.;Maeda, S. M. D.;Oliveira, E. P.;de Carvalho, M. D. B., & Santelli, R. E. (2007). Internal standardization for the determination of cadmium, cobalt, chromium and manganese in saline produced water from petroleum industry by inductively coupled plasma optical emission spectrometry after cloud point extraction. *Spectrochimica Acta Part B-Atomic Spectroscopy*, 62(9), 985-991.

Bezerra, M. D.;Arruda, M. A. Z., & Ferreira, S. L. C. (2005). Cloud point extraction as a procedure of separation and pre-concentration for metal determination using spectroanalytical techniques: A review. *Applied Spectroscopy Reviews*, 40(4), 269-299.

Blitz-Raith, A. H.;Paimin, R.;Cattrall, R. W., & Kolev, S. D. (2007). Separation of cobalt(II) from nickel(II) by solid-phase extraction into Aliquat 336 chloride immobilized in poly(vinyl chloride). *Talanta*, 71(1), 419-423.

Cao, Q. E.;Zhao, Y. K.;Cheng, X. G.;Hu, Z. D., & Xu, Q. H. (1999). Sequential fluorescent determination of copper(II) and cobalt(II) in food samples by flow injection analysis. *Food Chemistry*, 65(3), 405-409.

Carabias-Martinez, R.;Rodriguez-Gonzalo, E.;Moreno-Cordero, B.;Perez-Pavon, J. L.;Garcia-Pinto, C., & Laespada, E. F. (2000). Surfactant cloud point extraction and preconcentration of organic compounds prior to chromatography and capillary electrophoresis. *Journal of Chromatography A*, 902(1), 251-265.

Carvalho, M. S.;Fraga, I. C. S.;Neto, K. C. M., & Silva, E. Q. (1996). Selective determination of cobalt using polyurethane foam and 2-(2-benzothiazolylazo)-2-p-cresol as a spectrophotometric reagent. *Talanta*, 43(10), 1675-1680.

Cassella, R. J.;Salim, V. A.;Jesuino, L. S.;Santelli, R. E.;Ferreira, S. L. C., & de Carvalho, M. S. (2001). Flow injection determination of cobalt after its sorption onto polyurethane foam loaded with 2-(2-thiazolylazo)-p-cresol (TAC). *Talanta*, 54(1), 61-67.

Chaudhari, R. A., & Sawant, A. D. (1993). Extraction and spectrophotometric determination of cobalt. *Asian Journal of Chemistry*, 5, 1-5.

Chen, J. R., & Teo, K. C. (2001). Determination of cobalt and nickel in water samples by flame atomic absorption spectrometry after cloud point extraction. *Analytica Chimica Acta*, 434(2), 325-330.

Dadfarnia, S., & Jafarzadeh, M. H. (1999). Online trace enrichment and determination of cobalt ion as an anionic complex by flow injection atomic absorption spectrometry. *Microchemical Journal*, 63(2), 226-234.

Dallali, N.;Zahedi, M. M.;Yamimi, Y., & Agrawal, Y. K. (2009). Simultaneous cloud point extraction and determination of zn, co, ni, and pb by flame atomic absorption spectrometry. *Reviews in Analytical Chemistry*, 28(2), 125-136.

Dasilva, J. F., & Martins, W. (1992). Extraction of Fe(III), Cu(II), Co(II), Ni(II) and Pb(II) with thenoyltrifluoroacetone using the ternary solvent system water ethanol methylisobutylketone. *Talanta*, 39(10), 1307-1312.

Donati, G. L.;Nascentes, C. C.;Nogueira, A. R. A.;Arruda, M. A. Z., & Nobrega, J. S. (2006). Acid extraction and cloud point preconcentration as sample preparation strategies for cobalt determination in biological materials by thermospray flame furnace atomic absorption spectrometry. *Microchemical Journal*, 82(2), 189-195.

Dzherayan, T. G.;Shkinev, V. M.;Shpigun, L. K.;Kamilova, P. M., & Geckeler, K. E. (2002). Water-soluble polymers for spectrophotometric and flow injection determination of cobalt with nitroso-R-salt. *Talanta*, 57(1), 7-13.

Fujimoto, T.;Teshima, N.;Kurihara, M.;Nakano, S., & Kawashima, T. (1999). Catalytic determination of cobalt at sub-nanogram levels using the oxidative coupling of 3-methyl-2-benzothiazolinone hydrazone with N-ethyl-N-(2-hydroxy-3-sulfopropyl)-3,5-dimethoxyaniline by manual and flow-injection methods. *Talanta*, 49(5), 1091-1098.

Genfa, Z., & Dasgupta, P. K. (2000). A continuous film-recirculable drop gas-liquid equilibration device. Measurement of trace gaseous ammonia. *Analytical Chemistry*, 72(14), 3165-3170.

Ghaedi, M.;Ahmadi, F., & Soylak, M. (2007). Preconcentration and separation of nickel, copper and cobalt using solid phase extraction and their determination in some real samples. *Journal of Hazardous Materials*, 147(1-2), 226-231.

Ghaedi, M.;Shokrollahi, A.;Ahmadi, F.;Rajabi, H. R., & Soylak, M. (2008). Cloud point extraction for the determination of copper, nickel and cobalt ions in environmental samples by flame atomic absorption spectrometry. *Journal of Hazardous Materials*, 150(3), 533-540.

Gharehbaghi, M.;Shemirani, F., & Baghdadi, M. (2008). Dispersive liquid-liquid microextraction and spectrophotometric determination of cobalt in water samples. *International Journal of Environmental Analytical Chemistry*, 88(7), 513-523.

Gharehbaghi, M.;Shemirani, F., & Baghdadi, M. (2009). Dispersive liquid-liquid microextraction based on ionic liquid and spectrophotometric determination of mercury in water samples. *International Journal of Environmental Analytical Chemistry*, 89(1), 21-33.

Gil, R. A.;Gasquez, J. A.;Olsina, R.;Martinez, L. D., & Cerutti, S. (2008). Cloud point extraction for cobalt preconcentration with on-line phase separation in a knotted reactor followed by ETAAS determination in drinking waters. *Talanta*, 76(3), 669-673.

Giokas, D. L.;Paleologos, E. K.;Tzouwara-Karayanni, S. M., & Karayannis, M. I. (2001). Single-sample cloud point determination of iron, cobalt and nickel by flow injection analysis flame atomic absorption spectrometry - application to real samples and certified reference materials. *Journal of Analytical Atomic Spectrometry*, 16(5), 521-526.

He, Y., & Lee, H. K. (1997). Liquid phase microextraction in a single drop of organic solvent by using a conventional microsyringe. *Analytical Chemistry*, 69(22), 4634-4640.

Huddleston, J. G.;Willauer, H. D.;Swatloski, R. P.;Visser, A. E., & Rogers, R. D. (1998). Room temperature ionic liquids as novel media for 'clean' liquid-liquid extraction. *Chemical Communications*(16), 1765-1766.

Jeannot, M. A., & Cantwell, F. F. (1996). Solvent microextraction into a single drop. *Analytical Chemistry*, 68(13), 2236-2240.

Jeannot, M. A., & Cantwell, F. F. (1997). Mass transfer characteristics of solvent extraction into a single drop at the tip of a syringe needle. *Analytical Chemistry*, 69(2), 235-239.

Kakoi, T.;Ura, T.;Kasaini, H.;Goto, M., & Nakashio, F. (1998). Separation of cobalt and nickel by liquid surfactant membranes containing a synthesized cationic surfactant. *Separation Science and Technology*, 33(8), 1163-1180.

Kalika, K. C.;Kumar, A.;Rajeev, P., & Rajeev, S. (1993). Simultaneous spectrophotometric estimation of cobalt(II) and nickel(II) with phenylazobenzaldoxime. *Asian Journal of Chemistry*, 5, 559–562.

Kamburova, M.;Alexandrov, A., & Trifonov, K. (1994). Extraction-spectrophotometric determination of cobalt in soils and plants using tetrazolium violet. *Chemia Analityczna*, 39(5), 639-643.

Khambekar, A. M., & Sawant, A. D. (1997). Extractive and spectrophotometric determination of Co(II) using p-nitroisonitrosoacetophenone. *Indian Journal of Chemistry Section a-Inorganic Bio-Inorganic Physical Theoretical & Analytical Chemistry*, 36(5), 459-460.

Li, B. X.;Wang, D. M.;Lv, H., & Zhang, Z. J. (2006a). Chemometrics-assisted simultaneous determination of cobalt(II) and chromium(III) with flow-injection chemiluminescence method. *Spectrochimica Acta Part a-Molecular and Biomolecular Spectroscopy*, 65(1), 67-72.

Li, B. X.;Wang, D. M.;Lv, H. G., & Zhang, Z. J. (2006b). Flow-injection chemiluminescence simultaneous determination of cobalt(II) and copper(II) using partial least squares calibration. *Talanta*, 69(1), 160-165.

Liu, J.;Jiang, G. B.;Chi, Y. G.;Cai, Y. Q.;Zhou, Q. X., & Hu, J. T. (2003). Use of ionic liquids for liquid-phase microextraction of polycyclic aromatic hydrocarbons. *Analytical Chemistry*, 75(21), 5870-5876.

Liu, J. F.;Jonsson, J. A., & Jiang, G. B. (2005). Application of ionic liquids in analytical chemistry. *Trac-Trends in Analytical Chemistry*, 24(1), 20-27.

Liu, S. R., & Dasgupta, P. K. (1995). Liquid droplet - a renewable gas sampling interface. *Analytical Chemistry*, 67(13), 2042-2049.

Lokhande, R. S.;Jaywant, A. S., & Barhate, V. D. (1996). Extractive spectrophotometric determination of cobalt(II) with p-methylisonitrosoacetophenonehydrazone. *Asian Journal of Chemistry*, 8(4), 610-612.

Maheshwari, V., & Balasubramanian, N. (1995). Spectrophotometric determination of cobalt with nitrosochromotropic acid. *Fresenius Journal of Analytical Chemistry*, 351, 333-335.

Manzoori, J. L., & Karim-Nezhad, G. (2003). Selective cloud point extraction and preconcentration of trace amounts of silver as a dithizone complex prior to fame atomic absorption spectrometric determination. *Analytica Chimica Acta*, 484(2), 155-161.

More, P. S., & Sawant, A. D. (1992). Extractive and spectrophotometric determination of Co(II) at trace levels using isonitroso-4-methyl-2-pentanone. *Indian Journal of Chemistry Section a-Inorganic Bio-Inorganic Physical Theoretical & Analytical Chemistry*, 31(12), 984-985.

Nascentes, C. C., & Arruda, M. A. Z. (2003). Cloud point formation based on mixed micelles in the presence of electrolytes for cobalt extraction and preconcentration. *Talanta*, 61(6), 759-768.

Nogueira, A. R. A.;Souza, G. B.;Batista, L. A. R.;Arruda, M. A. Z., & Miyazawa, M. (1998). Catalytic determination of cobalt by flow injection spectrophotometry: application in animal faeces analysis using hydrochloric acid as extracting solution. *Animal Feed Science and Technology*, 74(1), 79-84.

Paleologos, E. K.;Giokas, D. L., & Karayannis, M. I. (2005). Micelle-mediated separation and cloud-point extraction. *Trac-Trends in Analytical Chemistry*, 24(5), 426-436.

Pedersen-Bjergaard, S., & Rasmussen, K. E. (1999). Liquid-liquid-liquid microextraction for sample preparation of biological fluids prior to capillary electrophoresis. *Analytical Chemistry*, 71(14), 2650-2656.

Pillai, V. P., & Shinde, V. M. (1995). Synergistic extraction of copper(II), cobalt(II) and nickel(II) with trifluoroacetylacetone and pyridine. *Indian Journal of Chemistry Section a-Inorganic Bio-Inorganic Physical Theoretical & Analytical Chemistry*, 34(5), 407-409.

Praveen, R. S.;Daniel, S., & Rao, T. P. (2005). Solid phase extraction preconcentration of cobalt and nickel with 5,7-dichloroquinone-8-ol embedded styrene-ethylene glycol dimethacrylate polymer particles and determination by fame atomic absorption spectrometry (FAAS). *Talanta*, 66(2), 513-520.

R.A. Chaudhari & A.D. Sawant (1993). Extraction and spectrophotometric determination of cobalt. *Asian Journal of Chemistry*, 5, 1-5.

Rasmussen, K. E., & Pedersen-Biergaard, S. (2004). Developments in hollow fibre-based, liquid-phase microextraction. *Trac-Trends in Analytical Chemistry*, 23(1), 1-10.

Reddy, B. R., & Sarma, P. (1994). Extractive spectrophotometric determination of cobalt using cyanex-272. *Talanta*, 41(8), 1335-1339.

Ribeiro, C. P.;Costa, A. O. S.;Lopes, I. P. B.;Campos, F. F.;Ferreira, A. A., & Salum, A. (2004). Cobalt extraction and cobalt-nickel separation from a simulated industrial leaching liquor by liquid surfactant membranes using Cyanex 302 as carrier. *Journal of Membrane Science*, 241(1), 45-54.

Risticevic, S.;Niri, V. H.;Vuckovic, D., & Pawliszyn, J. (2009). Recent developments in solid-phase microextraction. *Analytical and Bioanalytical Chemistry*, 393(3), 781-795.

Safavi, A.;Abdollahi, H.;Nezhad, M. R. H., & Kamali, R. (2004). Cloud point extraction, preconcentration and simultaneous spectrophotometric determination of nickel and cobalt in water samples. *Spectrochimica Acta Part a-Molecular and Biomolecular Spectroscopy*, 60(12), 2897-2901.

Saitoh, T., & Hinze, W. L. (1995). Use of surfactant-mediated phase-separation (cloud point extraction) with affinity ligands for the extraction of hydrophilic proteins. *Talanta*, 42(1), 119-127.

Sant'Ana, O. D.;Oliveira, L. G.;Jesuino, L. S.;Carvalho, M. S.;Domingues, M. D. F.;Cassella, R. J., & Santelli, R. E. (2002). A novel sampling strategy for cobalt determination by ETAAS using polyurethane foam. *Journal of Analytical Atomic Spectrometry*, 17(3), 258-262.

Shabani, A. M. H.;Dadfarnia, S., & Dehghan, K. (2003). On-line preconcentration and determination of cobalt by chelating microcolumns and flow injection atomic spectrometry. *Talanta*, 59(4), 719-725.

Sharma, R. R., & Dave, D. P. (1997). Solvent extraction and spectrophotometric determination of cobalt(II) with isonitrosopropiophenone. *Indian Journal of Chemistry Section a-Inorganic Bio-Inorganic Physical Theoretical & Analytical Chemistry*, 36(7), 629-630.

Shemirani, F., & Shokoufi, N. (2006). Laser induced thermal lens spectrometry for cobalt determination after cloud point extraction. *Analytica Chimica Acta*, 577(2), 238-243.

Shen, H. X.;Tang, Y. P.;Xiao, X. L.;Zhang, S. F., & Liu, R. X. (1995). Synthesis and studies on the analytical functions of a highly selective spectrophotometric reagent 2-(8-quinolylazo)-5-n,n'-dimethylaminobenzoic acid. *Analyst*, 120(5), 1599-1602.

Shokoufi, N.;Shemirani, F., & Assadi, Y. (2007a). Fiber optic-linear array detection spectrophotometry in combination with dispersive liquid-liquid microextraction for simultaneous preconcentration and determination of palladium and cobalt. *Analytica Chimica Acta*, 597(2), 349-356.

Shokoufi, N.;Shemirani, F., & Memarzadeh, E. (2007b). Fiber optic-linear array detection spectrophotometry in combination with cloud point extraction for simultaneous preconcentration and determination of cobalt and nickel. *Analytica Chimica Acta*, 601(2), 204-211.

Smirnova, S. V.;Torocheshnikova, II;Formanovsky, A. A., & Pletnev, I. V. (2004). Solvent extraction of amino acids into a room temperature ionic liquid with dicyclohexano-18-crown-6. *Analytical and Bioanalytical Chemistry*, 378(5), 1369-1375.

Soldenhoff, K.;Shamieh, M., & Manis, A. (2005). Liquid-liquid extraction of cobalt with hollow fiber contactor. *Journal of Membrane Science*, 252(1-2), 183-194.

Song, Z. H.;Yue, Q. L., & Wang, C. N. (2006). Flow injection chemiluminescence determination of femtogram-level cobalt in egg yolk, fish tissue and human serum. *Food Chemistry*, 94(3), 457-463.

Taher, M. A., & Puri, B. K. (1995). Column preconcentration of cobalt with the ion-pair of 2-nitroso-1-naphthol-4-sulfonic acid tetradecyldimethylbenzylammonium chloride supported on naphthalene using 2nd-derivative spectrophotometry. *Analyst*, 120(5), 1589-1592.

Toral, M. I.;Richter, P.;Silva, L., & Salinas, A. (1993). Determination of mixtures of cobalt and iron by 1st derivative spectrophotometry. *Microchemical Journal*, 48(2), 221-228.

Tsakovski, S.;Benkhedda, K.;Ivanova, E., & Adams, F. C. (2002). Comparative study of 8-hydroxyquinoline derivatives as chelating reagents for flow-injection preconcentration of cobalt in a knotted reactor. *Analytica Chimica Acta*, 453(1), 143-154.

Umebayashi, Y., & Ishiguro, S. I. (1996). Spectrophotometric study of thiocyanato complexation of cobalt(II) and nickel(II) ions in micellar solutions of a nonionic surfactant Triton X-100. *Journal of Solution Chemistry*, 25(8), 731-746.

Xia, Z.;Li, Z.;Rui, Y., & Liu, H. (2006). A novel room temperature ionic liquid extraction spectrophotometric determination of trace germanium in natural water with methybenzeneazosalicylfluorone. *Analytical Letters*, 39(4/6), 863-877.

Yang, G. Y.;Huang, Z. J.;Hu, Q. F., & Yin, J. Y. (2002). Study on the solid phase extraction of Co(II)-QADEAB chelate with C-18 disk and its application to the determination of trace cobalt. *Talanta*, 58(3), 511-515.

Yuzefovsky, A. I.;Lonardo, R. F.;Wang, M. H., & Michel, R. G. (1994). Determination of ultra-trace amounts of cobalt in ocean water by laser-excited atomic fluorescence spectrometry in a graphite electrothermal atomizer with semi online flow-injection preconcentration. *Journal of Analytical Atomic Spectrometry*, 9(11), 1195-1202.

11

Applications of Ionic Liquids in Azeotropic Mixtures Separations

Ana B. Pereiro[1,2] and Ana Rodríguez[1]
[1]*Department of Chemical Engineering, University of Vigo, 36310 Vigo,*
[2]*Present address: Instituto de Tecnología Química e Biológica, UNL,*
Av. República 127 2780-157 Oeiras,
[1]*Spain*
[2]*Portugal*

1. Introduction

The increasing concern about environmental issues, as well as the establishment of new regulations, has recently directed the attention of the scientific community to novel processes based on greener technologies. In many areas of industry, solvent mixtures accumulation occurs due to the hardness of recycling. The separation of these mixtures into the pure components is necessary so that they can be reused. However, most solvent mixtures contain azeotropes and thus, their separation by simple distillation becomes impossible.

Extractive distillation is the separation process most widely used to remove one of the components in the azeotropic system. In this process, the addition of a new solvent (entrainer) is used to interact more favourably with one component of the original mixture altering their relative volatilities. This obvious advantage is constrained to the high energy costs necessary to achieve a fluid phase system. Within this context, liquid–liquid separation, based on the immiscibility between two liquid phases at room temperature, emerges as a beneficial alternative to reduce the energy consumption and the environmental impact.

Ionic liquids (ILs) have become one of the growing "green" media for engineers not only due to their remarkable physicochemical properties but also for their recyclability. The most outstanding reason of interest in these neoteric solvents is their negligible vapor pressure at room temperature (Earle et al., 2006), which decreases the risk of worker exposure and the loss of solvent to the atmosphere. Moreover, ILs can be tailored for a specific application by accurately selecting the cation and the anion (Huddleston et al., 2001). This feature is very attractive for industry since fine-tuning of solvent properties permit the optimization of the chemical engineering needs in terms of efficiency and cost of the processes.

During the last years, ILs have exhibited the ability to separate azeotropes including ethanol + water (Jork et al., 2004; Seiler et al., 2004), tetrahydrofuran (THF) + water (Jork et al., 2004; Seiler et al., 2004; Hu et al., 2006), benzene + heptane (Letcher & Deenadayalu, 2003; Gonzalez et al., 2009), ethyl acetate + ethanol (Zhang et al., 2008), and ethyl tert-butyl ether

(ETBE) + ethanol (Arce et al., 2007). Although the increasing number of publications addressing azeotropic separations with ILs, these studies only analyze the liquid-liquid equilibria (LLE), vapor-liquid equilibria or simulation of the extractive distillation process.

The present work considers the use of ILs formed by 1-alkyl-3-methylimidazolium cation ($[C_nMIM^+]$) and several anions as solvents in a bench-scale extraction process. One of the selected anions was hexafluorophosphate $[PF_6^-]$ because it has been commonly investigated and considered historically one of the most important anion families; despite the fact that this anion can undergo hydrolysis producing hydrofluoric acid when in contact with water and at high temperatures (Swatloski et al., 2003; Najdanovic-Visak et al., 2002). Consequently, its industrial application has been restricted to those processes under water free conditions and moderate temperatures. Other anions chosen are methyl sulfate $[MeSO_4^-]$ and ethyl sulfate $[EtSO_4^-]$, since they display the most promising potential for application in industrial processes (Holbrey et al., 2002) because they can be easily synthesized in an halide-free manner at reasonable cost, they are chemically and thermally stable, and they have low melting points and relatively low viscosities. Taking into account all these features, these ILs were considered good candidates to be tested as extracting solvents or entrainers in the separation of azeotropic mixtures.

In order to evaluate such a possibility, the liquid–liquid equilibria of different azeotropic mixtures with ionic liquids at 298.15 K and atmospheric pressure were accomplished. The experimental data were successfully correlated by applying the NRTL equation (Renon & Prausnitz, 1968), thus facilitating their implementation and use in computerized applications. The capacity of the selected ionic liquids as azeotrope breakers in liquid extraction processes was evaluated by means of the selectivity and the solute distribution ratio. This capacity was compared with other extracting solvents referred in literature.

From the analysis of all extraction capacities, the systems with the best results were selected to carry out the lab-scale extraction processes incorporating a solvent recycling stage. The operation conditions of the lab-scale extraction process were optimized by using HYSYS software. The optimized conditions were assessed in practice in a laboratory-scale packed column and the extraction efficiencies of the extraction processes in the packed column were calculated.

2. Experimental

2.1 Chemicals

1-Ethyl-3-methylimidazolium ethyl sulfate ($[C_2MIM]$ $[EtSO_4]$) was purchased from Solvent Innovation with a purity ≥ 99 wt%. The others ionic liquids were synthesized according to procedure described in previous research (Pereiro et al., 2006a; Pereiro et al., 2007a; Pereiro et al., 2007b). NMR and positive FAB mass spectra were performed and the results are in agreement with literature. The ionic liquids were always used directly following the reduction of its water content to a mass fraction < 0.03%, determined by Karl Fischer titration. Hexane (from Aldrich, ≥ 99.0 wt%), heptane (from Aldrich, ≥ 99.0 wt%), 2-butanone (from Merck, 99.5 wt%), 2-propanol (from Merck, 99.7 wt%), ethyl acetate (from Fluka, 99.8 wt%), cyclohexane (from Fluka, 99.8 wt%) and ethanol (from Merck, ≥ 99.8 wt%) were dried over 4 Å molecular sieves (supplied by Aldrich) for several weeks before use, and the purities were verified by means of gas chromatography.

2.2 Equipment and techniques

2.2.1 Experimental liquid- liquid equilibrium procedure

Ternary liquid-liquid equilibrium data were determined in a glass cell (Fig. 1a-b) containing a magnetic stirrer and thermostatted by a water jacket connected to a bath controlled to ± 0.01 K. The temperature in the cell was measured with an ASL F200 digital thermometer with an uncertainty of ± 0.01 K (Fig. 1c).

Two techniques were used to study the ternary liquid-liquid equilibrium. The first based on the experimental determination of the binodal curve, estimating the immiscible area. The second is grounded on the determination of the tie-lines, which calculates the composition of each layer.

The binodal curve was determined by adding known quantities of the three components corresponding to the immiscible area into the equilibrium cell. Then, we slowly added known quantities of solute maintaining the stirring until the "cloud point" disappears (Fig. 1a).

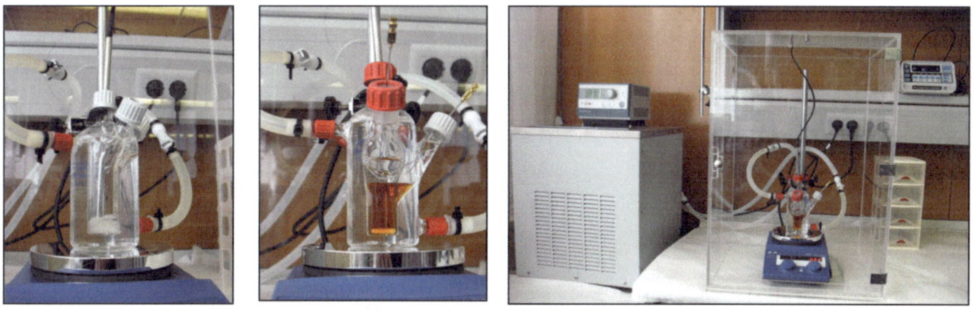

a) b) c)

Fig. 1. (a)-(b) Liquid-liquid equilibria cells. (c) Experimental equipment used to determine of the liquid-liquid equilibria.

For the determination of the tie-lines, 30 ml of ternary mixture of known composition was added to the cell, the temperature was brought to 298.15 ± 0.01 K, and the mixture was stirred vigorously for 1 h and left to settle for 4 h (Fig. 1b). Then, samples of both layers were taken with a syringe and the densities and refractive indices were determined. Lastly, the phase compositions were inferred by means of calibration curves which had been previously constructed at 298.15 K. These curves were obtained by fitting the composition on the binodal curve by means of refractive indices and densities at 298.15 K. The uncertainty of the phase composition is ± 0.004 in mass fraction. All weight measurements were performed in a Mettler AX-205 Delta Range balance with an uncertainty of ± 10^{-4} mass fraction. Also, densities were measured with an Anton Paar DSA-48 digital vibrating tube densimeter with an uncertainty of ± 2×10^{-4} g cm^{-3}. Finally, refractive indices were calculated via a Dr. Kernchen ABBEMAT WR automatic refractometer with an uncertainty of ± 4×10^{-5}.

2.2.2 Packed-column experiments

The practical performance of ionic liquids as azeotrope breakers in extraction processes was researched by using ionic liquids for continuous countercurrent separation of the azeotropic mixture in a 54 × 1585 mm glass extraction column packed with 8 × 8 mm Raschig rings to a height of 1475 mm (Fig. 2). The experiment was carried out at room temperature under steady-state conditions. First, the azeotropic mixture and solvent streams were pumped in, the extract stream was pumped out (with FMI QV laboratory pumps) and the raffinate stream came out of the column under gravity. Then, samples from the top and bottom of the column were taken periodically and the composition was determined by analyzing their density and refractive index. At last, the ionic liquid was recovered from the extract stream and was recycled into the packed column after regeneration.

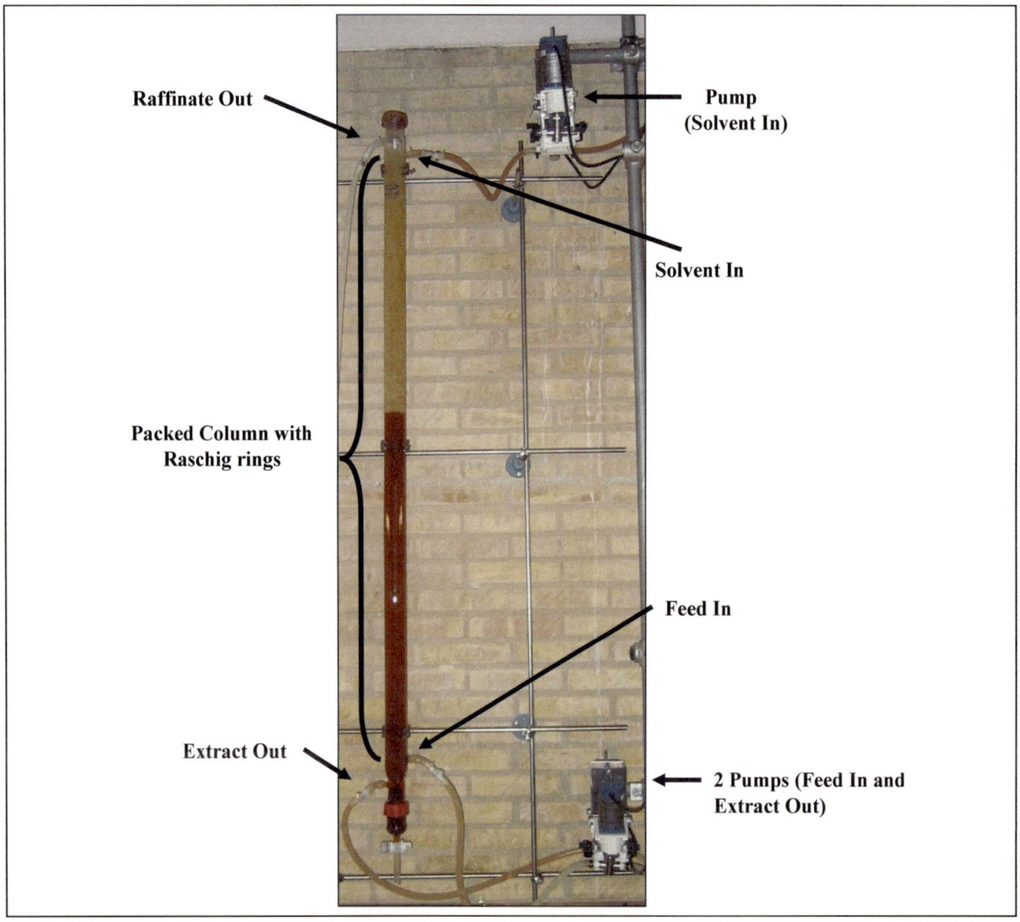

Fig. 2. Snapshot of the lab-scale packed countercurrent extraction column. The feed inlet, the raffinate and extract outlets are depicted, as well as the two pumps employed in the experimental set-up.

2.2.3 Regeneration of the Ionic Liquids

The ionic liquids used during these experiments were recovered and purified from the extract stream by removing the rest of its components in a Büchi R 3000 rotary evaporator with a vacuum controller. This operation is straightforward due to the fact that vapor pressure of ionic liquids is lower than the one of the rest of the components. The purity of the recovered ILs was verified by comparing its density at 298.15 K and NMR with the density and NMR of the freshly synthesized ones.

3. Results and discussion

3.1 Liquid-liquid separation of the azeotropic mixtures

The implementation of a separation process requires the accurate knowledge and careful control of the thermodynamic properties of the mixture, especially the phase boundaries. In this study, ternary liquid-liquid equilibria of different azeotropic mixtures with ionic liquids as extraction agents (Table 1) were measured at 298.15 K and atmospheric pressure. The selected azeotropic mixtures are commonly used in different processes of coating industry, petrochemical and food industry. The separation of each azeotropic mixture is discussed in the following sections.

Azeotropic mixture	**Ionic Liquids**	**Azeotropic mixture**	**Ionic Liquids**
Ethanol + 2-Butanone	$[C_1MIM][MeSO_4]$	Hexane+Ethanol	$[C_1MIM][MeSO_4]$
	$[C_4MIM][PF_6]$		$[C_4MIM][MeSO_4]$
2-Propanol+2-Butanone	$[C_1MIM][MeSO_4]$		$[C_2MIM][EtSO_4]$
	$[C_4MIM][PF_6]$		$[C_6MIM][PF_6]$
2-Propanol+Ethyl acetate	$[C_1MIM][MeSO_4]$		$[C_8MIM][PF_6]$
	$[C_4MIM][PF_6]$	Heptane+Ethanol	$[C_1MIM][MeSO_4]$
	$[C_6MIM][PF_6]$		$[C_4MIM][MeSO_4]$
Hexane+Ethyl acetate	$[C_6MIM][PF_6]$		$[C_2MIM][EtSO_4]$
	$[C_8MIM][PF_6]$		$[C_6MIM][PF_6]$
Cyclohexane+2-Butanone	$[C_6MIM][PF_6]$		$[C_8MIM][PF_6]$
	$[C_8MIM][PF_6]$		

Table 1. Azeotropic mixture + ionic liquid systems analyzed by liquid-liquid equilibria at 298.15 K and atmospheric pressure.

Values of solute distribution ratio, β, and selectivity, S, are widely used parameters in assessing the solvent feasibility the in liquid–liquid extraction. The solute distribution ratio supplies the amount of ionic liquid required for the process, related to the capacity of the IL and the selectivity evaluate the efficiency of the ionic liquid used as solvent. These parameters are defined as follows:

$$\beta = \frac{x_2^{\mathrm{II}}}{x_2^{\mathrm{I}}} \tag{1}$$

$$S = \left(\frac{x_1^{\mathrm{I}}}{x_1^{\mathrm{II}}}\right) \cdot \left(\frac{x_2^{\mathrm{II}}}{x_2^{\mathrm{I}}}\right) \tag{2}$$

where x is the mole fraction, subscripts 1 and 2 refer to inert and solute, respectively, and superscripts I and II indicate the organic (raffinate) and ionic liquid (extract) phases, respectively.

In order to perform simulation studies and process design, the experimental data were fitted with NRTL equation. The parameters were adjusted to minimize the difference between the experimental and calculated mole fraction defined as:

$$\text{O.F.} = \sum_{i=1}^{n}\left[\left(x_{1i}^{I} - x_{1i}^{I}(\text{calc})\right)^2 + \left(x_{2i}^{I} - x_{2i}^{I}(\text{calc})\right)^2 + \left(x_{1i}^{II} - x_{1i}^{II}(\text{calc})\right)^2 + \left(x_{2i}^{II} - x_{2i}^{II}(\text{calc})\right)^2\right] \quad (3)$$

where x_{1i}^{I}, x_{2i}^{I}, x_{1i}^{II}, x_{2i}^{II} are the experimental mole fraction; $x_{1i}^{I}(\text{calc})$, $x_{2i}^{I}(\text{calc})$, $x_{1i}^{II}(\text{calc})$ and $x_{2i}^{II}(\text{calc})$ are the calculated mole fraction; and superscripts I and II indicate the organic (raffinate) and ionic liquid (extract) phases, respectively.
The deviations were calculated by applying the following expression:

$$\sigma = \left(\frac{\sum_{i}\left(x_{ilm}^{exp} - x_{ilm}^{calc}\right)^2}{6k}\right)^{1/2} \quad (4)$$

where x is the mole fraction and the subscripts i, l and m provide the component, the phase and the tie - line, respectively. The k value refers to the number of experimental tie-lines.

3.1.1 Separation of alcohols + 2-Butanone

Traditionally, 2-butanone has been used as a solvent in paints and resin adhesives. A mixture of different alcohols with this ketone that form azeotropes is a very common product. Given the wide diversity of alcohol + ketone mixtures, study of the binary mixtures ethanol or 2-propanol with 2-butanone has been considered (Pereiro & Rodriguez, 2007c; Pereiro & Rodriguez, 2007d).
At present, the separation of these azeotropic mixtures is made by azeotropic distillation (Berg, 1995; Berg, 1999) using as entrainers: amyl acetate, methyl formate, 2,2-dimethyl butane or 2,3-dimethyl butane for the azeotrope ethanol + 2-butanone; and 3-methylpentane, amyl ether or acetonitrile for the azeotrope 2-propanol + 2-butanone.
The evaluation of $[C_4MIM][PF_6]$ and $[C_1MIM][MeSO_4]$ as potential solvents in liquid –liquid extraction for the recovery of alcohols from the azeotropes was carried out through the analysis of liquid-liquid equilibrium data. Binodal curves and the tie lines were obtained for the mixtures of ethanol + 2-butanone + $[C_4MIM][PF_6]$, 2-propanol + 2-butanone + $[C_4MIM][PF_6]$, 2-butanone + ethanol + $[C_1MIM][MeSO_4]$ and 2-butanone + 2-propanol + $[C_1MIM][MeSO_4]$ at 298.15 K and atmospheric pressure.
The values of the selectivity for the studied ternary systems as a function of the solute composition in the organic phase are plotted in Figure 3. A comparison with conventional organic extractive solvents (Katayama et al., 1998; Katayama & Amano, 2005) was made and also depicted in the Figure 3.
All the selectivity values for all areas of the binodal curves are higher than unity, from which it can be inferred that the extraction of the solute from the azeotropic system is indeed possible. A comparison between the selectivity values for the ternary systems shows that the $[C_1MIM][MeSO_4]$ obtains higher values than $[C_4MIM][PF_6]$ for the ethanol + 2-butanone and the opposite behavior is observed for 2-propanol + 2-butanone. In Figure 3a, the selectivity values at low concentration of ethanol for the ethanol + 2-butanone separation are similar in the $[C_1MIM][MeSO_4]$ and glycerol and higher than they are in water and $[C_4MIM][PF_6]$. On the other hand, for 2-propanol + 2-butanone, the selectivity values of the $[C_4MIM][PF_6]$ are higher than for the other solvents and ionic liquid.

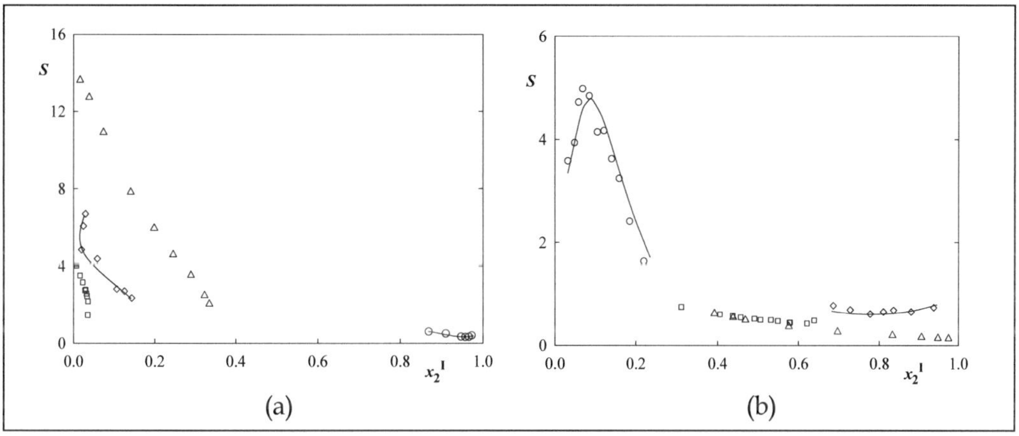

Fig. 3. Selectivity (S) of the systems presenting azeotrope: (a) ethanol (1) + 2-butanone (2); (b) 2-propanol (1) + 2-butanone (2) where: O, [C_4MIM][PF_6] this work; ◇, [C_1MIM][$MeSO_4$] this work; Δ, glycerol (Katayama et al., 1998); □, water (Katayama & Amano, 2005); solid line, NRTL correlation as a function of the solute mole fraction (ethanol or 2-butanone) in the organic phases at 298.15 K.

3.1.2 Separation of 2-Propanol + Ethyl Acetate

The azeotropic mixture ethyl acetate + 2-propanol is present in the solvent extraction of edible oils (Bera et al., 2006). The liquid-liquid equilibrium data were obtained for the ternary mixtures of ethyl acetate + 2-propanol + [C_1MIM][$MeSO_4$], 2-propanol + ethyl acetate + [C_4MIM][PF_6] and 2-propanol + ethyl acetate + [C_6MIM][PF_6] at 298.15 K and atmospheric pressure (Pereiro & Rodriguez, 2007e). The binodal curve for ternary mixture 2-propanol + ethyl acetate + [C_8MIM][PF_6] was also determined (Fig. 4).

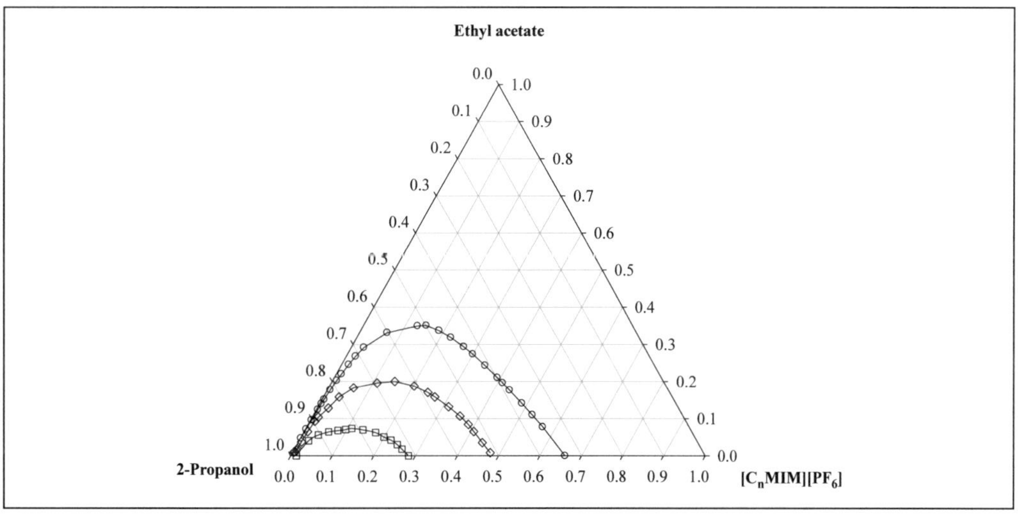

Fig. 4. Experimental binodal curves of the ternary systems 2-propanol (1) + ethyl acetate (2) + [C_nMIM][PF_6] (3) at 298.15 K, where: O, [C_4MIM][PF_6]; ◇, [C_6MIM][PF_6]; and □, [C_8MIM][PF_6].

In Figure 4, it is observed how the immiscibility region decreases if the length of the alkyl chain in the imidazolium ring increases when the [C_nMIM][PF_6] was used as solvent in the liquid-liquid extraction.
The analysis of the LLE data (Pereiro & Rodriguez, 2007e) indicates that the alkyl chain length of the imidazolium ring plays a negative role in the capability of the [C_nMIM][PF_6] to purify the alcohol. This negative role may be caused mainly by the hydrophobic steric effect of the alkyl group which reduces the polar character of the secondary -OH group of 2-propanol. From the LLE data it is also verified that the use of [C_1MIM][$MeSO_4$] as solvent leads to higher values of solute distribution ratio and selectivity than the ionic liquids involving PF_6 as the anion due to the fact that the $MeSO_4$ ionic liquid contains just two -CH_3 groups and the steric alkyl effect is reduced.

3.1.3 Separation of Hexane + Ethyl Acetate

The ILs [C_6MIM][PF_6] and [C_8MIM][PF_6] have been chosen for the separation of the azeotropic mixture hexane + ethyl acetate. This azeotrope is present in the process for purifying grafted polyolefins (Gupta & Carey, 2006), and its separation is made by azeotropic batch distillation (Rodriguez-Donis et al., 2005) with heterogeneous entrainers such as methanol, acetonitrile, water, and nitromethane. The evaluation of these two ILs as extraction solvents for the recovery of ethyl acetate from its mixture with hexane was carried out through the analysis of LLE (Pereiro & Rodriguez, 2008a).

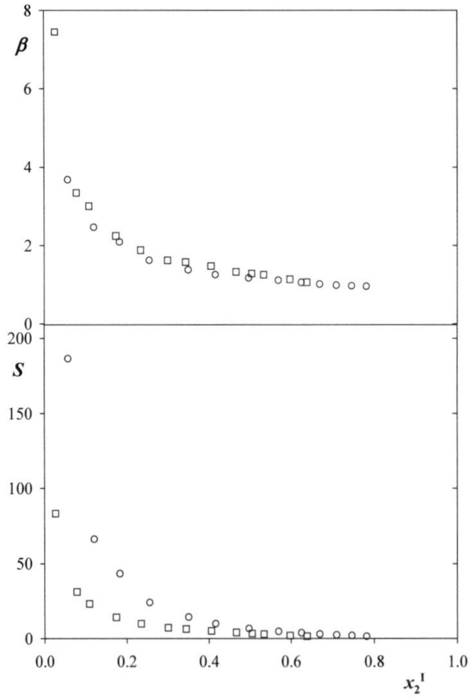

Fig. 5. Solute distribution ratio (β) and selectivity (S) of the systems presenting azeotrope hexane (1) + ethyl acetate (2) with: O, [C_6MIM][PF_6] and □, [C_8MIM][PF_6] versus ethyl acetate mole fraction in the organic phases at 298.15 K.

The solute distribution ratio (Eq. 1) provides the solvent capacity of the ionic liquid, related to the amount of ionic liquid required for the process. The selectivity (Eq. 2) is an important parameter to assess the efficiency of the ionic liquid used as solvent in the selective extraction of the solute from the azeotropic system. The values of the solute distribution ratio and selectivity for the studied ternary systems as a function of the solute composition in the organic phase are plotted in Figure 5.

A comparison between the selectivity values for the ternary systems shows that the $[C_6MIM][PF_6]$ obtains better values than $[C_8MIM][PF_6]$ for the removal of hexane from its azeotropic mixture with ethyl acetate.

3.1.4 Separation of Cyclohexane + 2-Butanone

The azeotropic mixture cyclohexane + 2-butanone occurs in the process for purifying grafted polyolefins (Gupta & Carey, 2006) and its separation is made by membrane in conjunction with a dephlegmation (Wijmans et al., 2005).

Liquid-liquid equilibrium data were obtained for the mixtures of cyclohexane + 2- butanone + $[C_6MIM][PF_6]$ and cyclohexane + 2-butanone + $[C_8MIM][PF_6]$ at 298.15 K and atmospheric pressure (Pereiro & Rodriguez, 2008b).

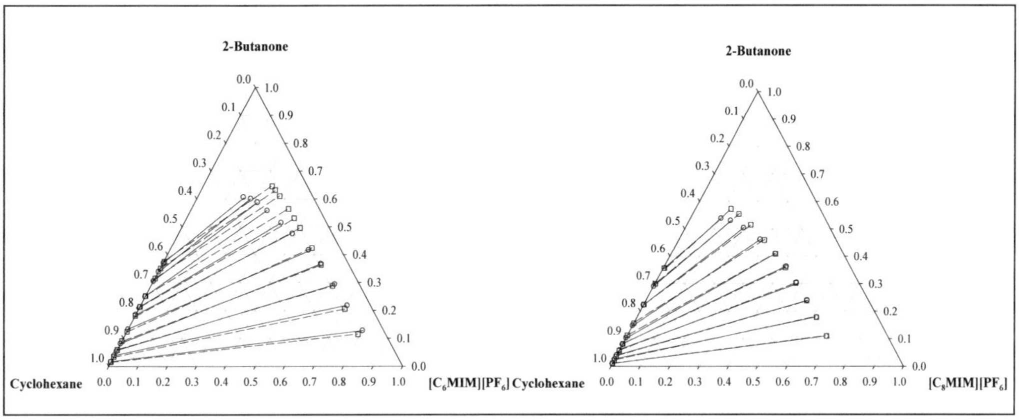

Fig. 6. Experimental tie-lines of the ternary systems at 298.15 K: (a) cyclohexane (1) + 2-butanone (2) + $[C_6MIM][PF_6]$ (3); (b) cyclohexane (1) + 2-butanone (2) + $[C_8MIM][PF_6]$ (3) where: O and solid line, experimental data; □ and dashed line, NRTL correlation.

The corresponding triangular diagrams with the experimental tie-lines for the studied systems are shown in Figure 6. An examination of this figure indicates a clear idea of the shape and the size of the immiscibility region of the systems. The positive slope shows that the solute goes preferentially to the solvent-rich phase. Another significant aspect is the fact that the ionic liquid does not enter in the organic-rich phase, *i.e.* the presence of $[C_6MIM][PF_6]$ and $[C_8MIM][PF_6]$ was not detected in the upper phase.

The use of $[C_6MIM][PF_6]$ as solvent leads to higher values of selectivity (Pereiro & Rodriguez, 2008b) than the $[C_8MIM][PF_6]$, indicating that it would be a better choice as solvent for this azeotropic separation. The analysis of the data indicate that the increase of alkyl chain length of the imidazolium decreases the capability of the $[C_nMIM][PF_6]$ to purify the cyclohexane from the azeotropic mixture.

3.1.5 Separation of Alkanes + Ethanol

As a result of the reduction of lead in gasoline, a growing number of processes in which alkanols and alkanes co-exist to produce oxygenated additives for gasolines are under development or have already reached the industrial production stage (Pucci, 1989). The azeotropic mixtures of either hexane or heptane with ethanol are chosen due to the difficulty that lies in separating them. The liquid–liquid separation leads to an environmentally friendly extraction process of these azeotropic mixtures as an alternative to azeotropic distillation (Laroche et al., 1991; Marwil, 1984), pervaporation (Okada & Matsuura, 1988) and reverse osmosis (Laatikainen & Lindstrom, 1986) which are procedures used for the separation of these azeotropes.

The [C_1MIM][$MeSO_4$], [C_4MIM][$MeSO_4$], [C_2MIM][$EtSO_4$], [C_6MIM][PF_6] and [C_8MIM][PF_6] ILs have been selected to act as azeotrope breakers for the alkanes + ethanol separation. Ternary liquid-liquid equilibria of theses mixtures were determined at 298.15 K and atmospheric pressure (Pereiro et al., 2006b; Pereiro & Rodriguez, 2008c; Pereiro & Rodriguez, 2008d;Pereiro & Rodriguez, 2009a; Pereiro & Rodriguez, 2009b; Pereiro & Rodriguez, 2009c; Pereiro et al., 2010).

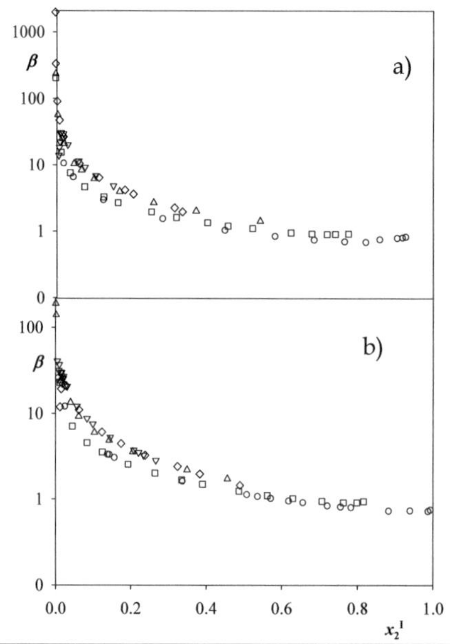

Fig. 7. Solute distribution ratio (β) of the systems presenting azeotrope: (a) hexane (1) + ethanol (2); (b) heptane (1) + ethanol (2) where: ○, [C_6MIM][PF_6]; □, [C_8MIM][PF_6]; △, [C_1MIM][$MeSO_4$]; ▽, [C_4MIM][$MeSO_4$]; and ◇, [C_2MIM][$EtSO_4$] as a function of the solute mole fraction (ethanol) in the organic phase at 298.15 K.

The corresponding values for the solute distribution ratio of the studied ternary systems are plotted in Figure 7 as a function of the ethanol mass fraction in the organic-rich phase. High values of these parameters are desired and all ionic liquids are suitable for extraction processes. Although, by analysing the selectivity values (Pereiro & Rodriguez, 2009c) we can conclude that shorter alkyl chain on the imidazolium cation increases selectivity, being favourable for alkane/ethanol separation.

3.2 Lab-Scale extraction process

From the analysis of all extraction capacities, the systems composed by alkanes (hexane or heptane) + ethanol + ionic liquids with the methyl sulfate anion ([C_1MIM][$MeSO_4$] and [C_4MIM][$MeSO_4$]) were selected to carried out the lab-scale extraction process incorporating a solvent recycling stage (Pereiro & Rodriguez, 2008d; Pereiro & Rodriguez, 2009a; Pereiro & Rodriguez, 2009b; Pereiro & Rodriguez, 2009c). The use of alkylsulfate ILs as azeotrope breakers amplifies the selectivities and solute distribution ratio, encouraging us to test these ILs in our lab-scale extraction process.

3.2.1 Selection of column operation conditions

Operating conditions for simulations and packed column experiments were selected in order to lessen cost while respecting the requirement of an elevated purity of the raffinate. Both cost and purity rise when the solvent/feed flow ratio in the column and/or the purity of the solvent stream increase. Moreover, the cost of solvent recovery grows when the purity is as high as desired. For that reason, this study has been carried out for solvent purities (ILs) of 70-100%. The restrictions applied on solvent stream were raffinate purities >85 wt% and solvent/feed ratio lower than 2.

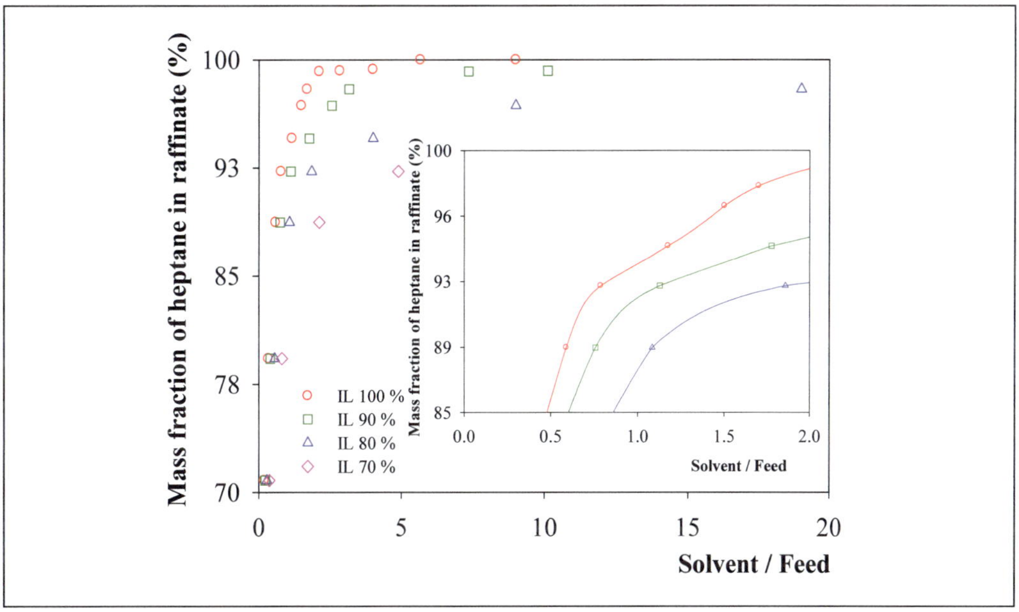

Fig. 8. Effect of solvent purity ([C_1MIM][$MeSO_4$]) on the raffinate purity for the ternary system heptane + ethanol + [C_1MIM][$MeSO_4$] at 298.15 K.

Figure 8 shows plots of raffinate purity alongside solvent/feed ratio calculated from the experimental tie line data for the ternary system heptane + ethanol + [C_1MIM][$MeSO_4$] as illustrated in Figure 9, where is showed an example of a tie-line. Each point in Figure 8 corresponds to the crossing between the tie-line and the line that combine the feed (azeotrope) with solvent stream (IL stream with different purities). More details on this procedure can be found elsewhere (Pereiro & Rodriguez, 2008d).

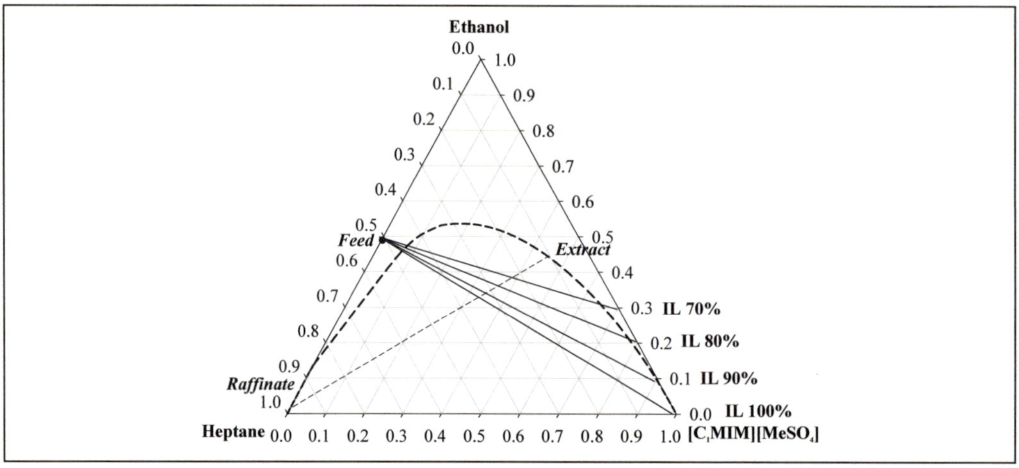

Fig. 9. Ternary diagram depicting the phase compositions for the *Raffinate* and *Extract* streams when an azeotropic mixture of heptane and ethanol (*Feed*) is mixed with $[C_1MIM][MeSO_4]$ containing: 0% (100% IL), 10% (90% IL), 20% (80% IL), or 30% (70% IL) of ethanol in ratios corresponding to the crossing with the tie-line (dashed line).

3.2.2 Simulation results

The design of the extraction process was accomplish via HYSYS v.3.2 (Aspen Technology Inc., Cambridge, MA, USA) with the NRTL equation fitted to experimental tie-line data. The simulation model is illustrated schematically in Figure 10, where a liquid–liquid extractor with one equilibrium stage models the packed column and a short-cut distillation process models solvent recovery. As simulation constraints, the solvent and feed compositions were kept constant (in the region of the chosen the theoretical operation conditions, as mentioned in the above sub-section), and the flow rates were optimized in order to maximize raffinate purity.

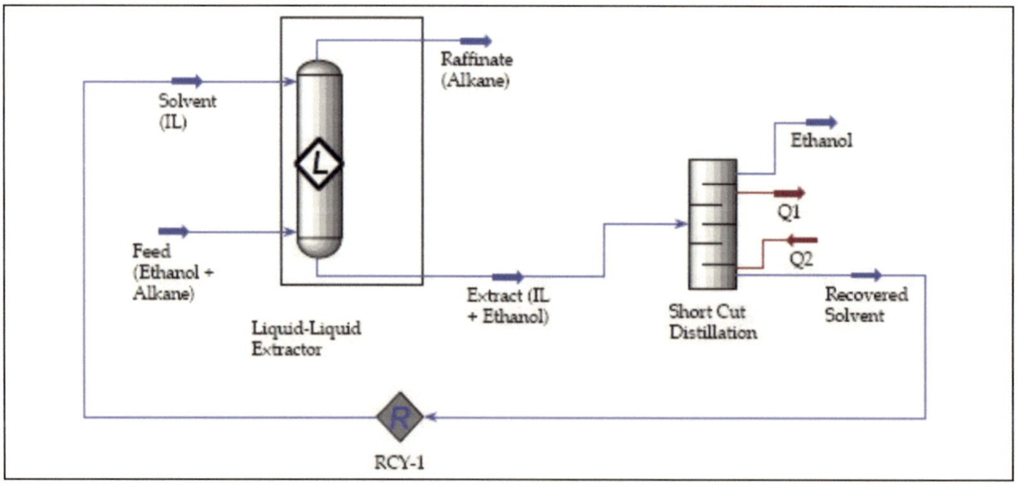

Fig. 10. Flowsheet simulation for the extraction process of ethanol + alkanes azeotropic mixtures using ionic liquid as solvent.

The ability of ionic liquids to act as azeotrope breakers in liquid-liquid extraction processes for the separation of the mixture ethanol + alkanes has been clearly proven by the performed simulations.

3.2.3 Results of packed-column experiments

The evolution of the alkane content observed in raffinate stream of the countercurrent column extraction experiment, carried out under operating conditions attained from the simulation results, is plotted in Figure 11. The amount of time needed to attain steady state, raffinate purity and the extract mass composition under steady state are outlined in Table 2.

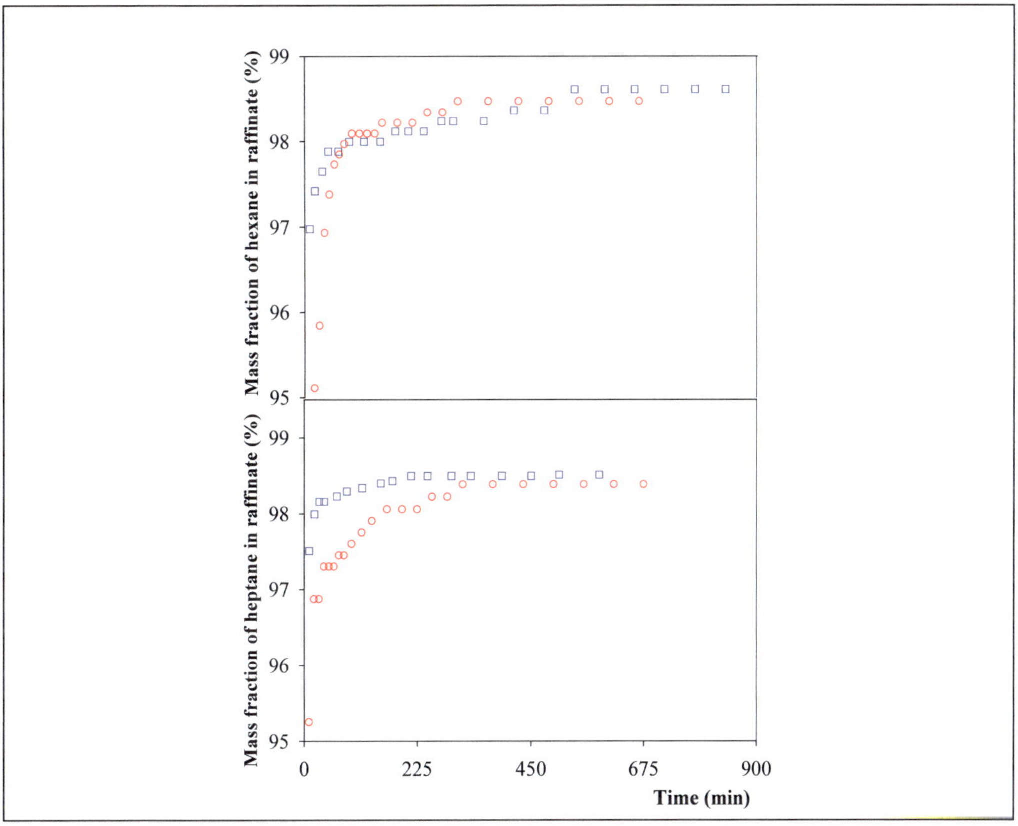

Fig. 11. Time dependence of the mass fractions of heptane in raffinate following start-up of the extraction column of Fig. 2 for: ○, [C_1MIM][$MeSO_4$]; and □, [C_4MIM][$MeSO_4$].

The extraction of alkane with a purity ≥ 98 wt. % was feasible by using the packed extraction column and the above mentioned ILs ([C_1MIM][$MeSO_4$] and [C_4MIM][$MeSO_4$]) as solvent. A comparison between packed column experimental data and theoretical data from experimental tie-lines is also depicted in Table 2. The performance of the experimental column exceeds the theoretical (from LLE) and simulation-based expectations, certainly due to the superior mixing between the feed and the solvent.

Figure 12 compares the compositions of the initial feed and the two outlet streams of the countercurrent packed column for the two ionic liquids (([C_1MIM][$MeSO_4$] and

[C$_4$MIM][MeSO$_4$]). The results indicate clearly that the [C$_1$MIM][MeSO$_4$] has a lower amount of alkane in the extract stream.

	Raffinate				Extract		
	% w_1^I	% w_2^I	% w_3^I		% w_1^{II}	% w_2^{II}	% w_3^{II}
Hexane + Ethanol + [C$_1$MIM][MeSO$_4$]							
t = 305 min	98.5	1.5	0.0	t = 485 min	6.2	36.8	57.0
one equilibrium stage	89.8	9.9	0.3	one equilibrium stage	3.7	32.6	63.7
Hexane + Ethanol + [C$_4$MIM][MeSO$_4$]							
t = 535 min	98.6	1.4	0.0	t = 655 min	9.4	36.4	54.2
one equilibrium stage	96.2	3.5	0.3	one equilibrium stage	10.0	38.1	51.9
Heptane + Ethanol + [C$_1$MIM][MeSO$_4$]							
t = 315 min	98.4	1.6	0.0	t = 495 min	3.6	38.4	58.0
one equilibrium stage	88.7	10.6	0.7	one equilibrium stage	4.7	40.1	55.2
Heptane + Ethanol + [C$_4$MIM][MeSO$_4$]							
t = 4 min	98.1	1.9	0.0	t = 8 min	11.9	46.0	42.1
one equilibrium stage	92.8	7.2	0.0	one equilibrium stage	10.5	44.0	45.5

Table 2. Comparison of experimental data under steady state for the extraction processes with theoretical data (from tie-lines) for the studied ternary systems.

In order to contrast the two ionic liquids for extraction processes in the packed column, the extraction efficiency, *E*, was calculated. This parameter indicates the ability of extraction solvent ([C$_1$MIM][MeSO$_4$] and [C$_4$MIM][MeSO$_4$]) to remove solute (ethanol) from the azeotropic mixtures (ethanol + alkanes) in the extraction column. This parameter is defined as follows:

$$E = \frac{w_1^{\mathrm{F}} - w_1^{\mathrm{R}}}{w_1^{\mathrm{F}} - w_1^{\mathrm{Eq}}} \tag{5}$$

where w is the mass fraction, subscript 1 refer to inert component (alkane) and superscripts F, R and Eq indicate the feed stream, raffinate stream and one equilibrium stage (theoretical data), respectively.

Azeotropic mixture	[C$_1$MIM][MeSO$_4$]	[C$_4$MIM][MeSO$_4$]
Hexane+Ethanol	1.81	1.14
Heptane+Ethanol	1.26	1.13

Table 3. Extraction efficiencies for the separation of the azeotrope alkanes + ethanol in the packed column.

In the separations of the azeotropes alkanes (hexane and heptane) + ethanol, the extraction efficiencies are listed in Table 3. The values obtained for [C$_1$MIM][MeSO$_4$] are better in comparison with those for [C$_4$MIM][MeSO$_4$]. Moreover, [C$_1$MIM][MeSO$_4$] has both lower viscosity and synthesized cost.

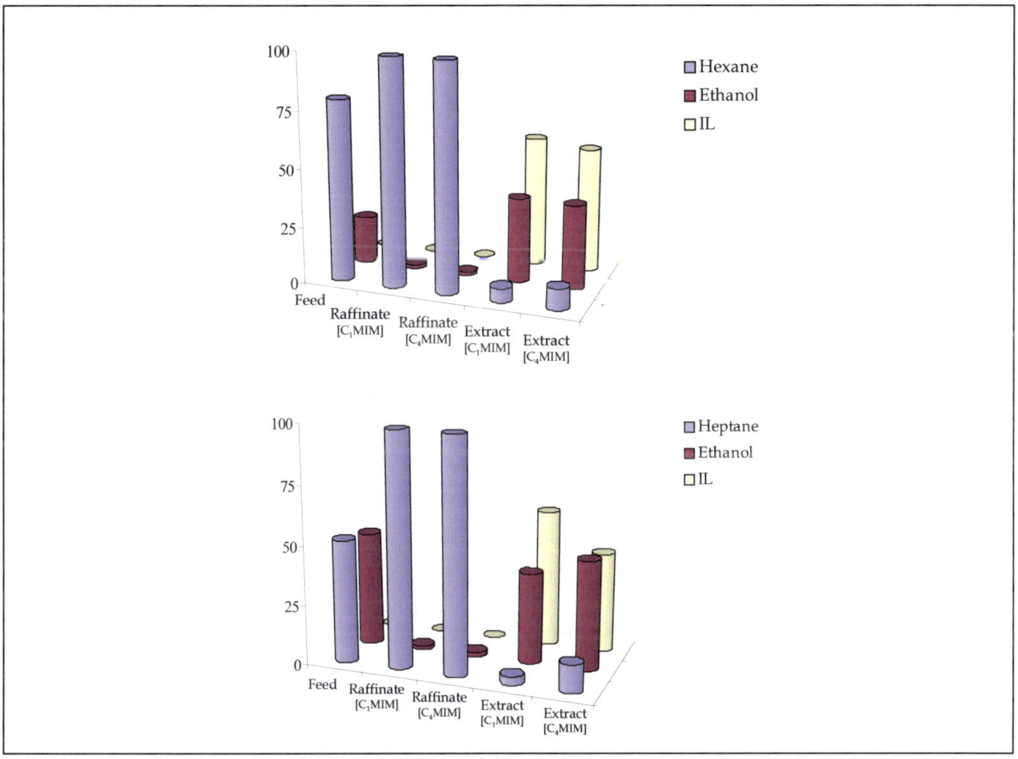

Fig. 12. Mass composition (%) of the feed, raffinate and extract streams of the packed column (Figure 2) under steady state conditions for $[C_1MIM][MeSO_4]$ and $[C_4MIM][MeSO_4]$.

4. Conclusion

The ability of different ILs as solvents for the separation of azeotropes by liquid-liquid extraction was demonstrated. The first approach for designing any extraction process should be addressed on the basis of several points: i) determination of LLE data for the ternary systems azeotropic mixtures + ILs, ii) correlation of the experimental data by means of theoretical equations, iii) evaluation of the extraction capacity with decisive parameters such as the solute distribution ratios and selectivities, iv) computational simulation to optimize the operating conditions, v) experimental lab-scale extraction process.

The alkyl sulfate-based ILs are the most promising alternative as solvents in the separation of the azeotropes studied in this work. The increase of the alkyl side chain length of the imidazolium ring plays a negative role in the extraction capability of the ionic liquids.

The extraction processes for the separation of the azeotropes ethanol plus alkanes were carried out with methyl sulfate-based ILs ($[C_1MIM][MeSO_4]$ and $[C_4MIM][MeSO_4]$). As it was proposed, the LLE data enabled the identification of theoretically appropriate operating conditions for countercurrent continuous extraction process at room-temperature including a solvent recycling stage. The operation conditions were subsequently optimized by simulation techniques. Experiments with a laboratory-scale packed column under steady-state conditions achieved a raffinate purity of over 98 wt %. The lab-scale experiments also confirmed the possibility of ready online recovery of selected ILs.

To contrast the different extraction processes in the packed column, the extraction efficiency was calculated. The values acquired for $[C_1MIM][MeSO_4]$ are better than those for $[C_4MIM][MeSO_4]$. Moreover, $[C_1MIM][MeSO_4]$ is considered the best candidate due to its lower viscosity and cost. If the raffinate purity of alkane obtained in the extraction process is taken into account, scaling up for industrial application seems viable. Some likely issues to be addressed in the near future include the scaling-up of the process in a pilot plant, to check the viability of the industrial application.

The use of ILs as azeotrope breakers should reduce the effects of the conventional solvents (VOCs) in the global climate change. This approach is an excellent opportunity to achieve the enhanced goals on reducing emissions of greenhouse gases established in the last United Nations Climate Conference (COP15).

5. References

Arce, A.; Rodriguez, H.; Soto, A. (2007). Use of a green and cheap ionic liquid to purify gasoline octane boosters. *Green Chem.*, 9, 3, March 2007 247–253, 1463-9262

Bera, D.; Lahiri, D.; De Leonardis, A.; De, K. B.; Nag, A. (2006). A Novel Azeotropic Mixture for Solvent Extraction of Edible Oils. *Agricultural Engineering International: the CIGR Ejournal.*, Manuscript FP 06 005, 8, April 2006 1–6, 1682-1130

Berg, Lloyd. (1995). Separation of 2-butanone from isopropanol by azeotropic distillation. *U.S. Patent* 5423954, June 1995

Berg, Lloyd. (1999). Separation of methyl ethyl ketone from ethanol by azeotropic distillation. *U.S. Patent* 5868907, February 1999

Earle, M. J.; Esperança, J. M. S. S.; Gilea, M. A.; Lopes, J. N. C.; Rebelo, L. P. N.; Magee, J. W.; Seddon, K. R.; Widegren, J. A. (2006). The distillation and volatility of ionic liquids. *Nature,* 439, 7078, February 2006 831–834, 0028-0836

Gonzalez, E. J.; Calvar, N.; Gonzalez, B.; Dominguez, A. (2009). (Liquid plus liquid) equilibria for ternary mixtures of (alkane plus benzene plus [EMpy] $[ESO_4]$) at several temperatures and atmospheric pressure. *J. Chem. Thermodyn.*, 41, 11, November 2009 1215-1221, 0021-9614

Gupta, V. P.; Carey, E. P. (2006). Extraction of Impurities from Grafted Polyolefins. *U.S. Patent* 7060787, June 2006

Holbrey, J. D.; Reichert, W. M.; Swatloski, R. P.; Broker, G. A.; Pitner, W. R.; Seddon, K. R.; Rogers, R. D. (2002). Efficient, halide free synthesis of new, low cost ionic liquids: 1,3-dialkylimidazolium salts containing methyl- and ethyl-sulfate anions. *Green Chem.* 4, 5, October 2002, 407–413, 1463-9262

Hu, X.; Yu, J.; Liu, H. (2006). Separation of THF and water by room temperature ionic liquids. *Water Sci. Tech.*, 53, 11, 2006 245–249, 0273-1223

Huddleston, J. G.; Visser, A. E.; Reichert, W. M.; Willauer, H. D.; Broker, G. A.; Rogers, R. D. (2001). Characterization and comparison of hydrophilic and hydrophobic room temperature ionic liquids incorporating the imidazolium cation. *Green Chem.*, 3, 4, August 2001 156–164, 1463-9262

Jork, C.; Seiler, M.; Beste, Y.-A.; Arlt, W. (2004). Influence of ionic liquids on the phase behavior of aqueous azeotropic systems. *J. Chem. Eng. Data*, 49, 4, July-August 2004 852–857, 0021-9568

Katayama, H.; Hayakawa, T.; Kobayashi, T. (1998). Liquid-liquid equilibria of three ternary systems: 2-propanone-glycerol-methanol, 2-butanone-glycerol-ethanol, and 2-

butanone-glycerol-2-propanol in the range of 283.15 to 303.15 K. *Fluid Phase Equilib.*, 144, 1-2, February 1998 157-167, 0378-3812

Katayama, H.; Amano, K. (2005). Liquid-liquid equilibria of three ternary systems: Water plus 2-butanone plus aliphatic alcohols (Ethanol, 2-propanol and 1-propanol). *J. Chem. Eng. Japan*, 38, 7, July 2005 459-464, 0021-9592

Laatikainen, M.; Lindstrom, M. (1986). Separation of methanol-ethanol and ethanol-normal-heptane mixtures by reverse-osmosis and pervaporation. *Acta Polytech Scandinavica Chem. Tech. Metall. Series*, 175, 1986 1-61, 1239-0518

Laroche, L.; Bekiaris, N.; Andersen, H. W.; Morari, M. (1991). Homogeneous azeotropic distillation - comparing entrainers. *Can. J. Chem. Eng.*, 69, 6, December 1991 1302-1319, 0008-4034

Letcher, T. M.; Deenadayalu, N. (2003). Ternary liquid-liquid equilibria for mixtures of 1-methyl-3-octyl-imidazolium chloride plus benzene plus an alkane at T-298.2 K and 1 atm. *J. Chem. Thermodyn.*, 35, 1, January 2003 67-76, 0021-9614

Marwil, S. J. (1984). Separation of hydrocarbon and alcohol azeotropic mixtures by distillation with anhydrous ammonia. *U.S. Patent* 4437941, March 1984

Najdanovic-Visak, V.; Esperanca, J. M. S. S.; Rebelo, L. P. N.; Nunes da Ponte, M.; Guedes, H. J. R.; Seddon, K. R.; Szydlowski, J. (2002). Phase Behaviour of Room Temperature Ionic Liquid Solutions: an Unusually Large Co-Solvent Effect in (Water + Ethanol). *Phys. Chem. Chem. Phys.*, 4, 10, 2002 1701–1703, 1463-9076

Okada, T.; Matsuura, T. (1988). A study on the pervaporation of ethyl alcohol/heptane mixtures by porous cellulose membranes. *Proc. Int. Conf. Pervaporation Processes Chem. Ind. 3*, pp. 224–230

Pereiro, A. B.; Santamarta, F.; Tojo, E.; Rodriguez, A.; Tojo, J. (2006a). Temperature dependence of physical properties of ionic liquid 1,3-dimethylimidazolium methyl sulfate. *J. Chem. Eng. Data*, 51, 3, May 2006 952-954, 0021-9568

Pereiro, A. B.; Tojo, E.; Rodriguez, A.; Canosa, J.; Tojo, J. (2006b). $HMImPF_6$ ionic liquid that separates the azeotropic mixture etanol + heptane. *Green Chem.*, 8, 3, March 2006 307-310, 1463-9262

Pereiro, A. B.; Legido, J. L.; Rodriguez, A. (2007a). Physical properties of ionic liquids based on 1-alkyl-3-methylimidazolium cation and hexafluorophosphate as anion and temperature dependence. *J. Chem. Thermodyn.*, 39, 8, August 2007 1168-1175, 0021-9614

Pereiro, A. B.; Verdia, P.; Tojo, E.; Rodriguez, A. (2007b). Physical properties of 1-butyl-3-methylimidazolium methyl sulfate as a function of temperature. *J. Chem. Eng. Data*, 52, 2, March-April 2007 377-380, 0021-9568

Pereiro, A. B.; Rodriguez, A. (2007c). Ternary Liquid-Liquid Equilibria Ethanol + 2-Butanone + 1-Butyl-3-Methylimidazolium Hexafluorophosphate, 2-Propanol + 2-Butanone + 1-Butyl-3-Methylimidazolium Hexafluorophosphate, and 2-Butanone + 2-Propanol + 1,3-Dimethylimidazolium Methyl Sulfate at 298.15 K. *J. Chem. Eng. Data*, 52, 6, November 2007 2138-2142, 0021-9568

Pereiro, A. B.; Canosa, J.; Rodriguez, A. (2007d). Liquid-liquid equilibria of 1,3-dimethylimidazolium methyl sulfate with ketones, dialkyl carbonates and acetates. *Fluid Phase Equilib.*, 254, 1-2, June 2007 150-157, 0378-3812

Pereiro, A. B.; Rodriguez, A. (2007e). Ternary (liquid + liquid) equilibria of the azeotrope (ethyl acetate + 2 - propanol) with different ionic liquids at T = 298.15 K. *J. Chem. Thermodyn.*, 39, 12, December 2007 1608-1613, 0021-9614

Pereiro, A. B.; Rodriguez, A. (2008a). Phase Equilibria of the Azeotropic Mixture Hexane + Ethyl Acetate with Ionic Liquids at 298.15 K. *J. Chem. Eng. Data,* 53, 6, June 2008 1360-1366, 0021-9568

Pereiro, A. B.; Rodriguez, A. (2008b). Measurement and correlation of liquid - liquid equilibria of the azeotrope (cyclohexane + 2 - butanone) with different ionic liquids at T = 298.15 K. *J. Chem. Thermodyn.*, 40, 8, August 2008 1282-1289, 0021-9614

Pereiro, A. B.; Rodriguez, A. (2008c). A study on the liquid-liquid equilibria of 1-alkyl-3-methylimidazolium hexafluorophosphate with ethanol and alkanes. *Fluid Phase Equilib.*, 270, 1-2, August 2008 23-29, 0378-3812

Pereiro, A. B.; Rodriguez, A. (2008d). Azeotrope-breaking using [BMIM] [$MeSO_4$] ionic liquid in an extraction column. *Sep. Purif. Technol.*, 62, 3, September 2008 733-738, 1383-5866

Pereiro, A. B.; Rodriguez, A. (2009a). Separation of Ethanol-Heptane Azeotropic Mixtures by Solvent Extraction with an Ionic Liquid. *Ind. Eng. Chem. Res.*, 48, 3, February 2009 1579-1585, 0888-5885

Pereiro, A. B.; Rodriguez, A. (2009b). Purification of hexane with effective extraction using ionic liquid as solvent. *Green Chem.*, 11, 3, March 2009 346-350, 1463-9262

Pereiro, A. B.; Rodriguez, A. (2009c). Effective extraction in packed column of ethanol from the azeotropic mixture ethanol + hexane with an ionic liquid as solvent. *Chem. Eng. J.*, 153, 1-3, November 2009 80-85, 1385-8947

Pereiro, A. B.; Deive, F. J.; Esperança¸ J. M. S. S.; Rodriguez, A. (2010). Alkylsulfate-based ionic liquids to separate azeotropic mixtures. *Fluid Phase Equilib.*, 291, 1, April 2010 13-17, 0378-3812

Renon, H.; Prausnitz, J. M. (1968). Local compositions in thermodynamic excess functions for liquid mixtures. *AIChE J.*, 14, 1, 1968 135-144, 0001-1541

Pucci, A. (1989). Phase-equilibria of alkanol alkane mixtures in new oil and gas process-development. *Pure Appl. Chem.*, 61, 8, August 1989 1363-1372, 0033-4545

Rodriguez-Donis, I.; Acosta-Esquijarosa, J.; Gerbaud, V.; Pardillo-Fondevila, E.; Joulia, X. (2005). Separation of n-hexane-ethyl acetate mixtures by azeotropic batch distillation with heterogeneous entrainers. *Chem. Eng. Process.*, 44, 1, January 2005 131-137, 0255-2701

Seiler, M.; Jork, C.; Kavarnou, A.; Arlt, W.; Hirsch, R. (2004). Separation of azeotropic mixtures using hyperbranched polymers or ionic liquids. *AIChE J.*, 50, 10, October 2004 2439–2454, 0001-1541

Swatloski, R. P.; Rogers, J. D.; Rogers, R. D. (2003). Ionic Liquids are not always Green: Hydrolysis of 1-Butyl-3-methylimidazolium Hexafluorophosphate. *Green Chem.*, 5, 4, August 2003 361–363, 1463-9262

Wijmans, J. G.; Baker, R. W.; Mairal, A. P. (2005). Separation of organic mixtures using gas separation or pervation and dephlegmation. *U.S. Patent* 6899743, May 2005

Zhang, D. L.; Deng, Y. F.; Li, C. B.; Chen, J. (2008). Separation of ethyl acetate-ethanol azeotropic mixture using hydrophilic ionic liquids. *Ind. Eng. Chem. Res.*, 47, 6, March 2008 1995–2001, 0888-5885

12

Application of Ionic Liquids in Liquid Chromatography

Jolanta Flieger
Medical University of Lublin
Poland

1. Introduction

Ionic liquids (ILs) are salts formed by a bulky nitrogen- or phosphorous-containing dissymmetrical organic cations such as alkyl-imidazolium, pyridinium or pyrrolidinium, ammonium or phosphonium and a variety of organic anions: triflate, dicyanamide, acetate, trifluoroacetate, trifluoromethylsulfate or inorganic: bromide, chloride, nitrate, perchlorate, chloroaluminate, tetrafluoroborate, or hexafluorophosphate. The number of possible combinations of anion-cation amounts 10^{18}, which makes this class of substances one of the biggest in chemical chest of substances. Properties of ILs depend on their structure. In order to obtain ionic liquid of desirable properties, with the specific melting point, viscosity, density, miscibility with water and organic solvents, one should select appropriate ions. For this reason ionic liquids are often called "designer" solvents.

There are certain common characteristics which definitely distinguish this class of chemical compounds. They have the ability of dissolving inorganic and organic compounds, including enzymes, simultaneously retaining their activity. They are stable chemically and thermally. The difference between the boiling point and of melting is of 300 °C. They possess low vapor pressure, which is why ionic liquids don't produce volatile environmental pollution. The most specific feature of these compounds is the low melting temperature -below 100 °C. If melting point is below room temperature (25 °C), the IL is called room-temperature ionic liquid (RTIL).

ILs found applications as new non-molecular types of solvents with unique properties in different areas of chemistry like organic synthesis, electrochemistry, extraction, spectroscopy and mass spectrometry. The growing interest in ILs has been observed also in chromatographic methods: gas chromatography (GC), capillary electrophoresis (CE) and liquid chromatography (LC). From a point of view of separation techniques, an important property of ILs is a possibility of generating strong donor-acceptor interactions and ability of undergoing coulombian interactions (chromatography), good electrical conduction (capillary electrophoresis), the great viscosity, the surface tension, the thermal stability (gas chromatography), excellent solvent properties (extraction). Worth noting are critical reviews emphasizing specific advantageous of ILs application in chromatographic and electromigration techniques (Buszewski & Studzińska, 2008; Marszałł & Kaliszan, 2007; Berthod et al., 2008; Pandey, 2006; Koel, 2005; Stalcup & Cabovska, 2004; Poole, 2007; Sun & Armstrong, 2010; Poole, 2004).

A major problem associated with using ILs, as solvents, in liquid chromatography is their high viscosity, many times bigger than one of methanol or acetonitrile. A completely different situation concerns using ionic liquids as small additives (2-60 mM) to water-organic mobile phases. In such minute concentration, ionic liquids lose their unique properties and they simply become salts. Apart from that, small concentration of the additive doesn't change the eluent viscosity drastically and can be applied in conditions both of high pressure chromatography (HPLC) as well as thin-layer chromatography (TLC). The addition of ionic liquids is particularly beneficial in analysis of basic compounds on the reversed-phase columns obtained on the basis of silica, where silanol interactions were usually a disadvantage. Activity of reducing of silanol interactions through ionic liquids exceeds commonly used tertiary amines as well as bivalent cations, frequently applied for this reason.

Modification of the stationary phase with the help of ILs relies on covalent bonding of silanol groups of silica gel with the cation of an ionic liquid. Obtained phases change its properties into ion-exchanging able to separate inorganic as well as organic ionic analytes.

This review focuses on application of ILs in liquid chromatography (HPLC and TLC), and electromigration techniques. Control of ILs by these techniques is also discussed according to increases of their popularity in industry.

2. Physicochemical properties of ionic liquids

Physicochemical properties of ILs depend on their components. The asymmetric organic cations and large polyatomic anions are responsible for lowering of the lattice energy and finally lowering of the melting point of resulting compound. For instance the melting point of archetypical sodium chloride equals 800 °C. Displacement of sodium by imidazolium cation lowers the melting point ten times. Increasing popularity of ILs in chemical industry is connected with their low vapor pressure, which is lower than the one of water, mercury and nonacosane - least volatile among alkanes. In the room temperature, the pressure achieves non-measurable values thus they appear to be friendly for environment. Besides that, they exhibit a wide range of solubility, viscosity and density depending on the kind of anion and cation in their structure. To find proper composition for exact analytical application, physicochemical properties of almost all existing ILs are collected in data banks available on the following websites: Dortmund Data Bank, Germany (Data Bank of Ionic Liquids, DECHEMA, DETHERM on the WEB, Thermophysical Properties of Pure Substances & Mixtures, http://www.ddbst.de/new/Default.htm or another one: NIST Boulder Colorado, USA (Ionic Liquids Database (IL Thermo), Thermodynamics Research Center, National Institute of Standards and Technology, Boulder, CO, USA, http://ilthermo.boulder.nist.gov/ILThermo/mainmenu.uix.

One of the advantages of ILs is their high thermal robustness. They possess thermal stability in the range approximately from -40°C to 200°C, which is much wider than other solvents like water, ammonium, acetone, methanol, benzene or nitrobenzene. ILs could be hydrophobic or hydrophilic depending on the properties of their ions. Hydrophilic ILs similarly to regular salts dissociate into ions which are hydrated in aqueous solutions. Ions exerting strong interactions with water molecules cause increase of the structuring of water and they are known as ''structure-makers'' or 'kosmotropes'. Hydrophobic ions partially or completely immiscible with water which decrease the structuring of water are known as 'chaotropes. meaning disorder-makers'. These terms originated from the Hofmeister series,

which orders the ion abilities of proteins precipitation. Comparing several thermodynamic parameters like the viscosity *B*-coefficients obtained from the Jones-Dole empirical equation, structural entropies (ΔS^0), structural volumes and ion mobility expressed as the limiting Walden product $\Lambda_m\eta$ quantification of the ILs ions kosmotropicity/chaotropicity can be evaluated (Zhao, 2006). According to Zhao investigation, it appears that smaller organic cations including $[MMIM]^+$, $[EMIM]^+$, $[MePy]^+$, $[EtPy]^+$ and $[Me_4N^+]$ are chaotropes, while those of medium size (such as $[BuPy]^+$ and $[Et_4N^+]$ are borderline ions, in turn larger ions such as $[BMIM]^+$, $[Pr_4N^+]$, $[Bu_4N^+]$, and $[Pe_4N^+]$ possess kosmotropic properties. The anions can be kosmotropes or chaotropes depending on the position in Hofmeister series of salts.
The properties of a modern ILs could be characterized by the following features (Johnson, 2007):viscosity: < 100 cP, dielectric constant: < 30, specific conductivity: < 10 mScm^{-1}, molar conductivity: < 10 Scm^2 mol^{-1}, electrochemical window > 2V.
Comparing properties of ILs and other liquids according to their molarities and molar conductance, it is clear that modern ILs consist of not only ions but also un-dissociated molecules in a form of ion-pairs. From a point of view of liquid chromatography such properties as viscosity, refractivity, solubility, E_T^N - Reichardt polarity index are the most important. Table 1 contains properties of ILs common in chromatographic techniques.

Symbol	Name	Melting point (°C)	Density (25 °C)	Refrac. index	Visc. (25 °C) (cP)	Sol. in water	E_T^N (×100)
EMIM BF_4	1-Ethyl-3-methyl imidazolium tetrafluoroborate	6	1.248		66	s	
BMIM BF_4	1-Butyl-3-methyl imidazolium tetrafluoroborate	−82	1.208	1.429	233	s	67
BMMIM BF_4	1-Butyl-2,3-dimethyl imidazolium tetrafluoroborate	37	1.2		780	s	
HMIM BF_4	1-Hexyl-3-methyl imidazolium tetrafluoroborate	−82	1.208		310	p	
OMIM BF_4	1-Octyl-3-methyl imidazolium tetrafluoroborate	−79	1.11		440	p	54
DMIM BF_4	1-Decyl-3-methyl imidazolium tetrafluoroborate	−25	1.072		930		63
BMIM PF_6	1-Butyl-3-methyl imidazolium hexafluorophosphate	10	1.373	1.411	400	18 g L^{-1}	67
HMIM PF_6	1-Hexyl-3-methyl imidazolium hexafluorophosphate	−61	1.304		800	n	

OMIM PF_6	1-Octyl-3-methyl imidazolium hexafluorophosphate	−40	1.2		810	n	
EMIM Cl	1-Ethyl-3-methyl imidazolium chloride	89	1.12*	Solid	Solid	s	
HMIM Cl	1-Hexyl-3-methyl imidazolium chloride	−75	1.05		7500	s	
OMIM Cl	1-Octyl-3-methyl imidazolium chloride	0	1		16,000	s	
$(C_7)_4$ N Cl	Tetraheptyl ammonium chloride	−9	0.882		598	s	
BMIM Cl	1-Butyl-3-methyl imidazolium chloride	65	1.10*	Solid	Solid	s	
BMIM Br	1-Butyl-3-methyl imidazolium bromide					s	
BMIM I	1-Butyl-3-methyl imidazolium iodide	−72			1110		
EMIM Act	1-Ethyl-3-methyl imidazolium acetate	−20	1.03		91	s	
BMIM Act	1-Butyl-3-methyl imidazolium acetate	−20	1.06		525	s	
EA NO_3	Ethylammonium nitrate	12.5	1.122	1.4537	32	s	95
PA NO_3	Propylammonium nitrate	4	1.157	1.4561	67	s	92
TBA NO_3	Tributylammonium nitrate	21.5	0.918	1.4627	640	s	80
EMIM MS	1-Ethyl-3-methyl imidazolium methylsulfate	5	1.24		80	s	
BMIM MS	1-Butyl-3-methyl imidazolium methyl sulfate	−20	1.2		180	s	

Table 1. Physicochemical properties of selected ionic liquids (Berthod et al., 2008).

3. Chromatography

ILs were applied firstly in gas chromatography as new liquid stationary phases owing to their low volatility, high viscosity and non-flammability (Barber et al., 1959). Barber and co-workers proposed GC stationary phases modified with molten stearates of Mn, Co, Ni, Cu, and Zn for separation of alcohols and amines. Later on, GC columns based on ethylammonium nitrate and ethylpyridinium bromide and tetraalkylphosphonium salts were prepared and tested (Pachloec et al., 1982; Pacholec & Poole, 1983). ILs-based stationary phases exhibited selectivity towards polar and hydrogen bond forming analytes. However, they possessed some limitations concerning narrow liquid range, poor thermal stabilities, poor efficiency and lack of selectivity according to n-hydrocarbons. Since imidazoium ILs: 1-Butyl-3-methylimidazolium hexafluorophosphate [BMIM][PF_6] and chloride [BMIM][Cl] salts started to be coated onto silica capillaries, GC systems have achieved satisfactory efficiency useful for separation either nonpolor or polar compounds (Armstrong et al., 1999; Berthod et al., 2001; Anderson & Armstrong, 2003).

In liquid chromatography, ILs have not been widely applied so far. However, the number of publications dealing with this subject has recently visibly increased. Generally, ILs are too viscous to be used as solvents and exhibit poor transmittance of low-wavelength UV light, worsening detection limits. However, when ILs are diluted, properties characteristic for the neat ones are lost and ILs become common organic salts. Among whole group, only pyridinium and imidazolium hexafluorophosphates or tetrafluoroborates are the most commonly applied in solution as mobile phase modifiers. In this case ILs can play different roles: coating residual silanols, modifying the stationary phase and acting as ion-pairing reagents.

3.1 Thin-layer chromatography

The first paper concerning application of ILs in thin-layer chromatography (TLC) was published by Kaliszan's group (Kaliszan et al., 2004). The imidazolium classes of tetrafluoroborate ILs covering [EMIM] [BF_4], [HxMIM] [BF_4], [Hx-HpOMIM] [BF_4] were tested as mobile phase modifiers in analysis of several basic compounds on an octadecyl bound-silica stationary phase. It appeared that proposed modifiers were more effective in chromatographic separation of chosen basic analytes in comparison to other standard additives such as triethylamine (TEA), dimethyloctylamine (DMOA) or ammonia. Whereas TEA, DMOA and ammonia showed negligible effect on investigated compounds retention, ILs appeared to be the most effective suppressor of attractive interactions between acidic free silanols present on the stationary phase surface and analyzed basic drugs causing drastically retention decreases, improvement of selectivity and spots shape. Effects of investigated additives on chromatographic results are compared in Fig.1.

Dependences of ionic liquid concentration in eluent system versus obtained experimental retardation factors (R_F) exhibit typical Langmuir shapes. Independently on the kind of stationary phase (silica or the octadecyl silica-covered plates), mobile phase composition (pure acetonitrile, water-acetonitrile) and IL used, saturation level was achieved at ca. 0.5% v/v of ionic liquid content in eluent system. Comparison of the effectiveness of investigated silanol screening agents on the basis of reciprocal of retardation factor vs. reciprocal of the additive concentration relationship (Fig.2), proved the most effective adsorption of exemplary ionic liquid [EMIM] [BF_4].

Addition of imidazolium tetrafluoroborates to mobile phase improved also linearity of the classical Snyder-Soczewiński relationship of R_M (log1-R_F/R_F) versus organic modifier concentration in the mobile phase. This suggests that chromatographic systems modified with ILs could be applied for much more precise determination of chromatographic lipophilicity parameters (R_M^0, φ_0) by extrapolation and interpolation graphical techniques. Using of IL as the mobile phase modifier is useful also in normal-phase TLC systems. Kaliszan and his group (Bączek et al., 2005) described behavior of a homologous series of peptides analysed on a silica support by the use of the eluent system enriched with [EMIM] [BF_4]. It appeared that experimental dependences of the retention coefficient, R_M on the volume percentage of organic modifier in the mobile phase containing ionic liquid were

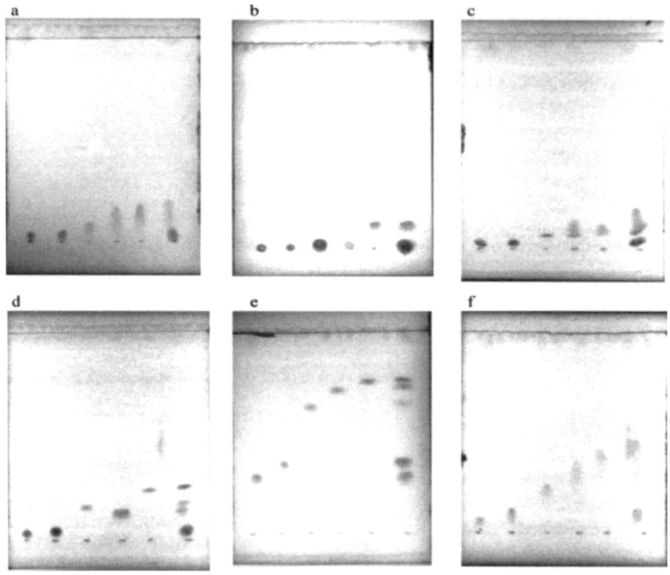

Fig. 1. Chromatograms of thioridazine, trifluoropromazine, phenazoline, naphazoline, tiamenidine and the mixture of the drugs, as spotted from left to right, on RP-18 F254 plates, developed with water-acetonitrile 40:60 v/v eluent either pure or containing 1.5% v/v of various additives: a-no additive; b-NH4OH; c-DMOA; d-TEA; e-[EMIM][BF4]; f-buffer of pH 2.87. (Kaliszan et al., 2004).

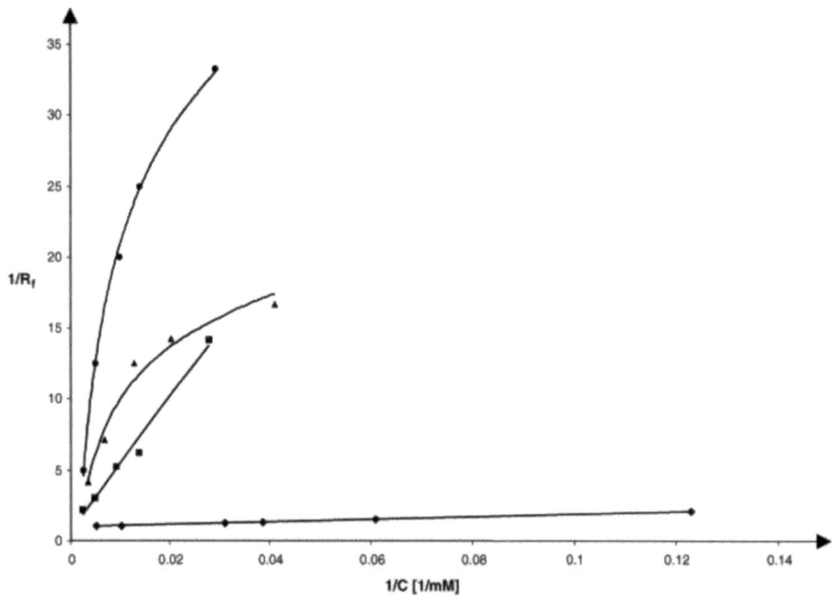

Fig. 2. Plots of reciprocal of retardation factor of tiamenidine on octadecylsilica plates with acetonitrile as eluent vs. the reciprocal of concentration of additive in the mobile phase. Additives are denoted as follows: IL 1 (✦), TEA (▪), DMOA (▴) and NH_4OH (·).(Kaliszan et al., 2004).

nonlinear and could be described with a third-degree polynomial function. Whereas quadratic equation was well fitted to dependences obtained for chromatographic systems without addition of IL.

RTILs were examined also as matrixes for online separation and identification of peptides, proteins, and poly(ethylene glycol) (PEG-2000) with matrix-assisted laser desorption/ionization-mass spectrometry (MALDI-MS) (Armstrong at el., 2001; Mehl & Hercules, 2000). The method described hybrid TLC-MALDI plates. Authors concluded that tested ILs, owing to good solubilizing properties and vacuum stability, in comparison to other commonly used liquids and solid matrixes, should be further investigated as potential MALDI matrixes.

3.2. High-performance liquid chromatography

3.2.1 Ionic liquids as mobile phase solvents

In the 80s of the last century, Poole et al. (Poole et al., 1986) elaborated on the synthesis of six alkylammonium salts of the general formula $R_{n-a}\ NH^{+}_{a}\ X^{-}$, where R = ethyl, propyl, butyl or sec.−butyl, X^{-} = nitrate or thiocyanate, $n < 4$ and a = 0, 1,2 or 3. Authors measured their physical properties including melting point, refractive index, UV spectra, miscibility with common organic solvents, aqueous solution pH, density and viscosity and concluded that obtained liquid organic salts forming immiscible solvent pairs with non-polar organic solvents appear to be suitable for liquid—liquid chromatography and liquid—liquid extraction. Some of them, such as the alkylammonium nitrate or the thiocyanate salts mixed with a second solvent of low viscosity such as water, methanol, acetonitrile, tetrahydrofuran and dichloromethane could be used as mobile phases in liquid chromatography. Studies of their solvent properties in microcolumn RPLC demonstrated that the back pressure on the column was too high, in spite of the fact of a small flow of 0.05 ml/min, a small internal diameter of the column (1mm) and the least viscous of ILs applied (Shetty et al., 1987). Additionally, applied aqueous solutions of thiocyanates corroded metal parts of the apparatus and demonstrated poor baseline stability. In turn tetra-n-alkylammonium sulfonates were too basic and caused silica-based columns degradation process. Therefore, the project was abandoned.

A couple of years after that, Waichigo applied a few monoalkyloammonium nitrates and acetates adding them, as modifiers, to water in concentrations of 20-60 % using classical C18 columns or the copolymer-styrene-divinylobenzene (Waichigo et al., 2005; Waichigo & Danielson, 2006 a; Waichigo & Danielson, 2006 b; Waichigo et al., 2007). It appears that ethylammonium nitrate possesses similar solvent strength to tetrahydrofuran. Elongation of alkyl chain in ammonium salt structure is connected with increasing of elution strength of IL in chromatographic system. In turn, the solvent strength of ethylammonium acetate is similar to methanol. Waichigo observed the lower number of theoretical plates (20%) and the twice higher back pressure on the column compared with the conventional arrangement of reversed phases and water-organic eluent. One should emphasize it that majority of authors applying alkylammonium salts achieved worse effects comparing with classical RP-HPLC as for: the transparency in UV, viscosities triggering back pressure, worsened efficiency. Taking into account the fact that alkylammonium salts possess the lowest viscosity in comparison to other ILs, it is difficult to expect that they would displace currently applied organic modifiers like methanol or acetonitrile. However, the non-molecular nature of ILs could give meaningful improvement of selectivities (Fig.3)

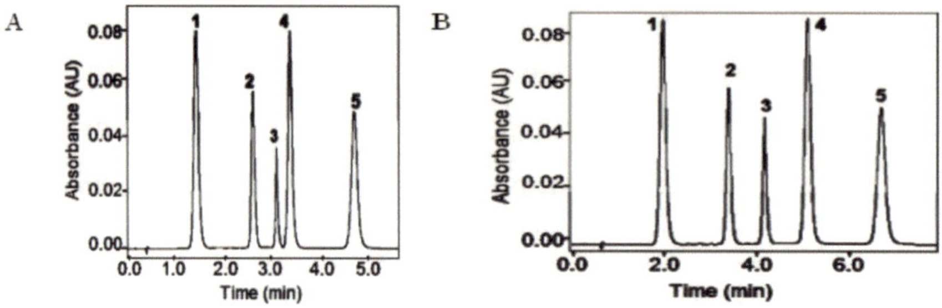

Fig. 3. Chromatogram of 1-ascorbic acid, 2-nicotinic acid, 3-thiamine, 4-pyridoxine, and 5-niacinamide using the Aquasep C8 column with a)30%methanol-70%water. a)30% ionic liquid ethylammonium acetate-70%acetate. Flowrate:1.0mLmin-1. UV detection at 254 nm. (Waichigo et al., 2005).

3.2.2 Ionic liquids as mobile phase modifiers.

ILs were found to be more useful when added to the mobile phase at millimolar concentration, however all their unique properties were lost and they were just salts adjusting the solvent strength. By dilution of IL with water or organic-aqueous solvent or by increasing the column temperature, the viscosity of ionic liquids can be drastically reduced. Several advantages associated with the use of ionic liquids as mobile phase additives should be emphasized. First of all, these compounds don't damage the silica based columns, do not change radically pH of the mobile phase, and could replace the less advantageous alkylamines with aim of silanol screening.

Many papers demonstrate the usefulness of ILs at these conditions in terms of improvement of efficiency and peak symmetry (He et al., 2003; Zhang et al., 2003; Zhang et al., 2004; Marszałł et al., 2005,; Marszałł et al., 2006 a; Ruiz-Angel et al., 2006). Fig.4 presents comparison of prochlorperazine peaks depending on the kind of modifier added to the mobile phase (Flieger, 2009).

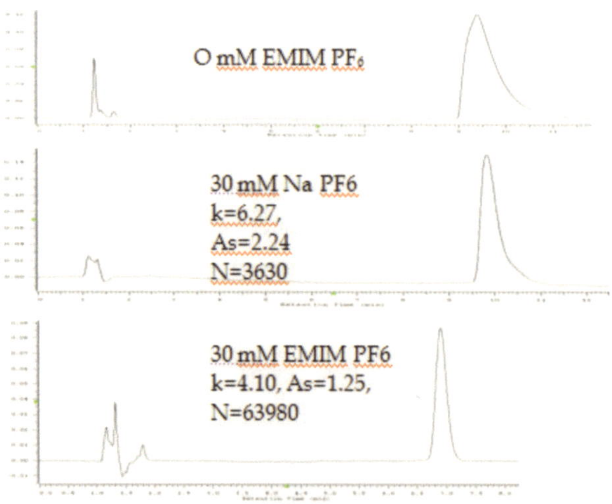

Fig. 4. The effect of different mobile phase additives on chromatographic parameters of prochlorperazine. Eluent: 40%ACN/10mM phosphate buffer (pH2.8), column: Zorbax Extend-C18 (Flieger, 2009).

To understand the influence of the particular ions of added reagents, the eluent systems containing the following components: ACN/phosphate buffer pH=2.8, ACN/phosphate buffer (pH=2.8)/$NaPF_6$ and ACN/phosphate buffer (pH=2.8)/[EMIM] [PF_6] were compared. As it can be seen, unacceptable parameters in system containing only organic-aqueous mobile phase acidified by phosphate buffer being a consequence of the ionic nature of the analyte, cause the creation of tailing peaks with not satisfactory symmetry. Addition of chaotropic additive $NaPF_6$ to the mobile phase improves the peak symmetry and efficiency of the system, nevertheless A_S factor is still out of an acceptable range. The best parameters of peak symmetry and efficiency were obtained for ionic liquid [EMIM] [PF_6] as the mobile phase additive. Change of Na^+ cation, which is not adsorbed into hydrophobic surface into the enlarged imidazole cation possessing the structure enabling to penetrate the space between the hydrocarbon chains, appears to be the most advantageous.

However, special attention should be devoted to appropriate concentration of IL in the mobile phase to avoid dissymmetry and multiplification of peak as a result of existing analyte in different forms: ion-pair and free protonated base. Fig.5 shows how peaks profile of nortryptyline can be improved according to increasing concentration of [BMIM][PF_6] (Flieger & Czajkowsa-Żelazko, 2010).

Improving peak symmetry, efficiency expressed as theoretical plates number and visible in decreasing peaks tailing and reducing band broadening, are connected with changing of the retention mechanism in the presence of ILs in a chromatographic system. Both ions contribute to the retention of analytes. They can affect obtained results either antagonistically or synergistically.

Many authors confirm suppression of silanophilic interaction by ILs which coat the surface and interact with free silanol groups of silica based supports (Marszałł et al.,2006 a; Marszałł et al.,2005; Ruiz-Angel et al.,2006; Flieger, 2009). As new masking agents enabling reduction of deleterious silanol effect, ILs appeared to be more advantageous in comparison to traditional additives like amino quenchers (triethylamine (TEA), dimethyloctylamine (DMOA), cyclohexylamine, quaternary ammonium ions) applied so far for this reason. It was demonstrated that either anions can be adsorbed on hydrophobic stationary phase according to their chaotropicity (Pan et al., 2004; Gritti & Guiochon, 2004) or ionic liquids' cations following their lyotropy related to the lengths of the alkyl chains (Berthod et al.,2005; Zhao, 2006). Ions can also move together with the mobile phase and interact with ionic analytes. Protonated basic analytes could interact with anions creating neutral ion-pairs with longer retention. In this case ion-associated complexes are formed only with that ones exhibiting chaotropic properties ($H_2PO_4^- < HCOO^- < CH_3SO_3^- < Cl^- < NO_3^- < CF_3COO^- < BF_4^- < ClO_4^- < PF_6^-$) (LoBrutto et al., 2001; Flieger, 2006). So far, adsorption of ILs on adsorbent surface and ion-pairing mechanism has been proposed to explain retention mechanism in chromatography applying ILs as modifiers.

Firstly Jiang (Jiang et al., 2004) described addition of 2–50 mM IL to aqueous mobile phases as silanol suppressing agents in the separation of basic and strong hydrophilic analytes ephedrines, catecholamines, and amino acids. Later on Marszałł et al. evaluated the silanol-suppressing potency of ILs (Marszałł et al., 2006 a) applying dual retention model proposed by Nahum and Horvath:

$$\frac{[A]}{k_0 - k} = \frac{1}{k_2 K_A} + \frac{[A]}{k_2} \qquad (1)$$

where k_0 indicates retention factor at the absence of any additives and k is the retention factor obtained at the concentration [A] of the silanol suppressor.

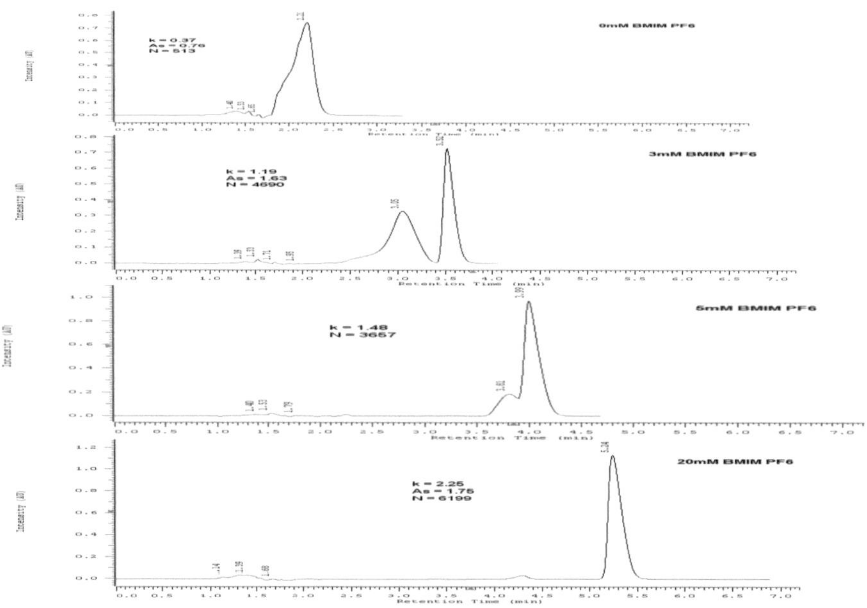

Fig. 5. The effect of different concentration of BMIM PF_6 on retention, peak symmetry and efficiency of nortryptyline. Column: on Zorbax Eclipse- XDB-C18, Eluent system: 40% ACN / 20 mM phosphate buffer pH = 2.9 (Flieger & Czajkowska-Żelazko, 2010).

The effect of different imidazolium based ILs was established on the basis of calculated binding constant of the silanol complex (K_A). It appears that the higher values were obtained for chloride and bromide ionic liquids than for tetrafluoroborates. Additionally, it should be emphasized that observed screening effect was achieved at lower concentration of these additives. Determined stability constant of the silanol complexes together with recommended additive concentration are presented in Table 2.

Advantages connected with suppression of silanophilic interaction appear simultaneously with the changes in analytes' retention factor. The increasing K_A values of alkylimidazolium halides, reflecting the silanol-masking potency, are connected with decreases of the basic analytes retention and selectivity of their separation.

In the light of conducted experiments, it is obvious that the cation of ionic liquid is responsible for the silanol screening effect, whereas the retention factor primarily depends on the properties of its anion. Only more chaotropic character of ILs anion, meaning not only hydrophobic nature but also entirely ion-pairing properties, leads to drastic retention increase. When anion of IL does not create ion-associated complex with anlayte in ionic form, decrease in retention is observed. Summarizing, the retention mechanism in chromatographic system modified with ILs additives is complex covering the creation of ion pairs as well as ion exchange and hydrophobic partitions.

Generally, the retention of the basic compounds increases with the increase of the ionic liquids with chaotropic anion concentration in the mobile phase in reversed-phase systems (Flieger, 2009). However, these dependences differ in the shapes especially at their initial sectors. Fig. 6 illustrates the effect of concentration of 1- ethyl-3-methyl-imidazolium hexafluorophospate [EMIM][PF_6], 1-butyl-3-methyl-imidazolium hexafluorophosphate [BMIM][PF_6] added to acetonitrile-aqueous eluent system on retention of basic compounds

Investigated ionic liquids	Benzyltriethylammonium hydrochloride		Thioridazine hydrochloride		Chlorpromazine hydrochloride	
	K_A x 10^{-3} [M^{-1}]	C_A^* [mM]	K_A x 10^{-3} [M^{-1}]	C_A^* [mM]	K_A x 10^{-3} [M^{-1}]	C_A^* [mM]
[EMIM][BF_4]	2.15	4.65	7.11	1.41	7.40	1.35
[PMIM][BF_4]	2.37	4.22	7.67	1.30	8.68	1.15
[BMIM][BF_4]	2.38	4.20	7.92	1.26	9.45	1.06
[HMIM][BF_4]	2.71	3.69	9.39	1.07	17.95	0.56
[OMIM][BF_4]	3.12	3.20	13.52	0.74	18.43	0.54
TEA	1.13	8.86	0.52	19.31	0.61	16.49
[MMIM][MSO_4]	2.45	4.07	8.53	1.17	14.31	0.69
[EMIM][ESO_4]	2.61	3.82	6.99	1.43	7.09	1.41
[BMIM][OSO_4]	1.90	5.26	1.05	9.52	0.94	10.63
[EMIM][Br]	3.57	3.34	10.31	0.96	8.24	1.21
[EMIM][Cl]	2.99	2.80	9.7	1.03	7.90	1.26
[EMIM][Ts]	0.26	38.46	5.44	1.83	6.36	1.57
[BMIM][Br]	2.59	3.86	7.08	1.41	7.05	1.41
[BMIM][Cl]	2.89	3.46	6.54	1.52	6.94	1.43
[OMIM][Cl]	5.58	1.79	13.81	0.72	16.28	0.61

Table 2. Stability constant of the silanol complex, K_A and the recommended additive concentration, C_A^*, in 50:50 v/v acetonitrile/water pH 3 when using 1-alkyl-3-methylimidazolium ionic liquids. (Marszałł et al., 2006a).

belonging to phenothiazine derivatives on an octadecyl bounded-silica stationary phase. This class of isotherms possesses characteristic Langmuir shape related to horizontal orientation of the adsorbate molecules on the adsorbent surface. At higher concentration of ionic liquids, the shape of the isotherms differs, depending on the type of the additive. Quite different are also the concentrations provided to column saturation. The addition of [BMIM] [PF_6] ensures obtaining of the plateau at 2mM, and the shape of this dependence is a typical one of the monolayer adsorption (the basic type of Langmuir adsorption). In case of [EMIM] [PF_6], the shape of the above mentioned dependence suggests the incidence of a multi-layer adsorption. Quite different is the shape of the isotherm for the additive containing anion of a neutral character [BMIM] [Cl]. In this case, retention time of bases decreases with increasing ionic liquid concentration. This effect could be a consequence of repulsion forces between the adsorbed imidazolium cation on the adsorbent surface and protonated analytes.

Xiaohua obtained similar results (Xiaohua et al., 2004) analyzing the series of amines: benzylamine, benzidine, N-ethylaniline, N, N′-dimethylaniline and *k* o-,m-,p-phthalic acids in the presence of [BMIm][BF4] in the mobile phase. Decreases of basic and acidic compounds retention proved that the part of ionic liquids coat on the reversed-phase surface and simultaneously anion of ionic liquid added does not interact with analytes. Kaliszan and his group also observed the retention decrease of neuroleptic basic drugs on octadecylsilica support and mobile phase comprising acetonitrile: phosphoric buffer (pH 3) 70/30 (v/v) modified with 1-ethyl-3-methylimidazolium tetrafluoroborate and 1,3-dimethylimidazolium methyl sulfate (Marszałł et al., 2005).

Owing to addition of ILs at minute concentration to the mobile phase we can improve also separation selectivity. Xiaohua and co-workers (Xiaohua et al., 2004) showed improvement of amines mixture separation covering benzylamine, benzydine, N,N-dimethyloaniline, N-ethyloanilinecan on C18 column and water with addition of 1-ethyl-3-methylimidazolium tetrafluoroboran as an eluent system.

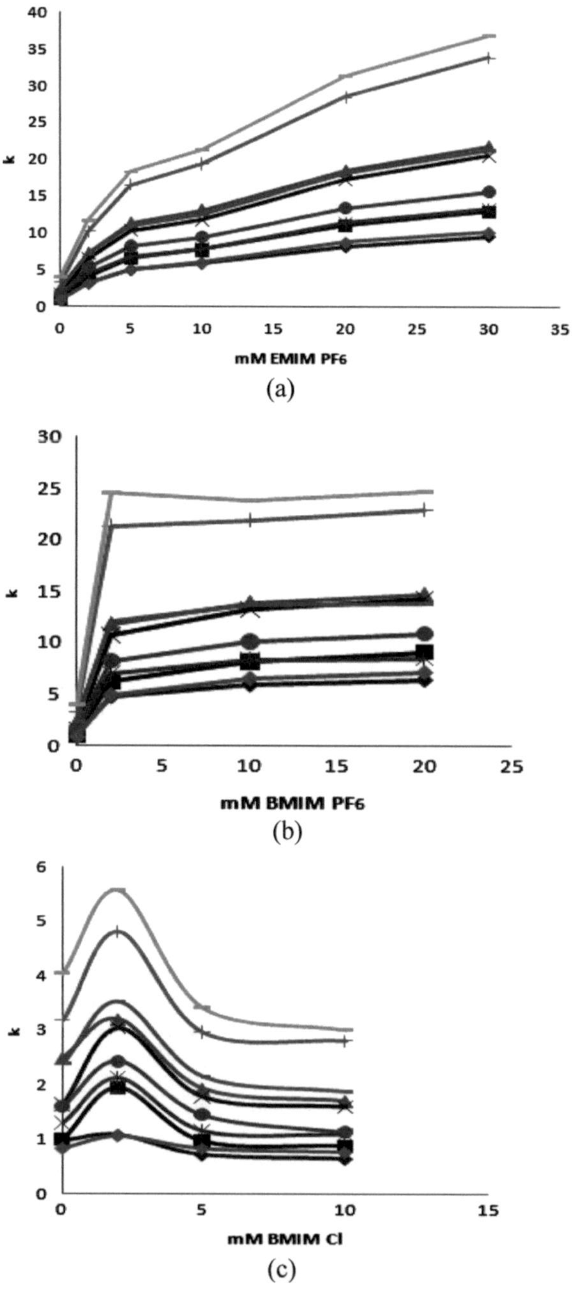

Fig. 6. Experimental retention factor for the investigated neuroleptic drugs vs. concentration of ILs in mobile phase containing 30% ACN/10 mM phosphate buffer (pH=2.8). A-EMIM PF_6, B-BMIM PF_6, C-BMIM Cl. (Flieger, 2009).

In my latest studies, some new aspects devoted to application of ionic liquids as mobile phase modifier were revealed (Flieger & Czajkowska-Żelazko, 2010). Analyzing chosen basic drugs in reversed-phase systems on two columns: Zorbax XDB-C18 and Zorbax SB-Phenyl in acetonitrile aqueous mobile phase acidified and modified with [BMIM] [Cl] or [BMIM][PF_6], visible differences in chromatographic behavior were detected. Fig.7 presents the retention dependences of pindolol on the concentration of investigated ionic liquids in 10%ACN buffered eluent system.

As it could be seen, pindolol shows an increase of retention with increasing concentration of ionic liquid possessing chaotropic anion [PF_6^-], as a result of ion-pair creation between protonated basic molecule [BH^+] and the counter-anion [PF_6^-]. This process achieves saturation level after which ionic liquid concentration does not influence the retention. In turn, the presence of neutral anion [Cl^-] causes deceasing of retention factor. Moreover, at this graph unexpected retention increase on Zorbax SB-Phenyl at low concentration (1 mM) of [BMIM] [Cl] in 10% ACN mobile phase could be observed. This deviation is the consequence of differences in structure of interphases created in compared systems.

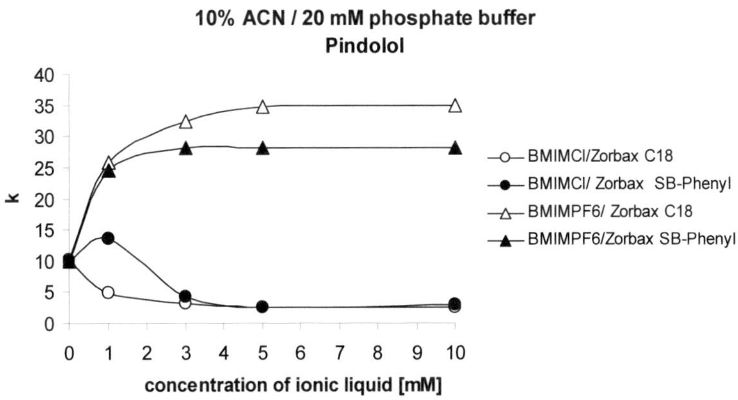

Fig. 7. Experimental retention factors for pindolol versus ionic liquids (1-butyl-3-methyl-imidazolium hexafluorophosphate-[BMIM] [PF_6], 1-butyl-3-methyl-imidazolium chloride-[BMIM] [Cl] concentration in 10% ACN / 20 mM phosphate buffer pH = 2.9 obtained for different hydrophobic columns: Zorbax Eclipse- XDB-C18 and Zorbax SB-Phenyl. Flow rate: 0.9 ml/min., temperature: 20°C, UV detection at 254 nm; sample: 3 µl injection. (Flieger & Czjkowska-Żelazko, in press).

Multilayer of adsorbed acetonitrile molecules forming on the adsorbent surface is a good medium for solubility of ionic liquids as a liophilic species. Assuming that free silanols are in ionized state, the addition of even small amount of ionic liquid (1 mM for Zorbax XDB-C18 or 1-3 mM for Zorbax SB-Phenyl) could be enough to reduce free silanols and to increase of stationary phase hydrophobicity owing to electrostatic interactions occurring accordingly to the following scheme:

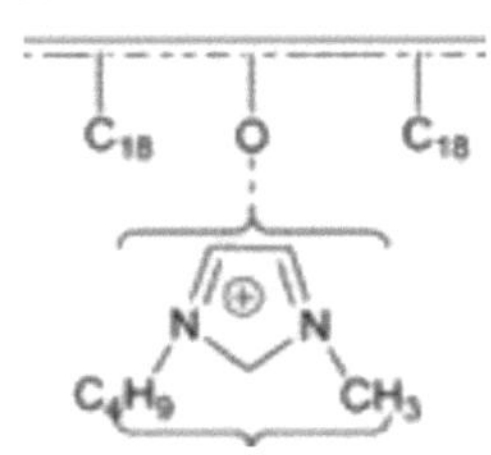

This effect could be useful in practice for improvement of separation selectivity by retention enhancement. Fig.8 presents example of improvement of 1,4 benzodiazepine separation by addition of 3 mM [BMIM][Cl] to acetonitrile-water mobile phase on Zorbax SB-Phenyl column.

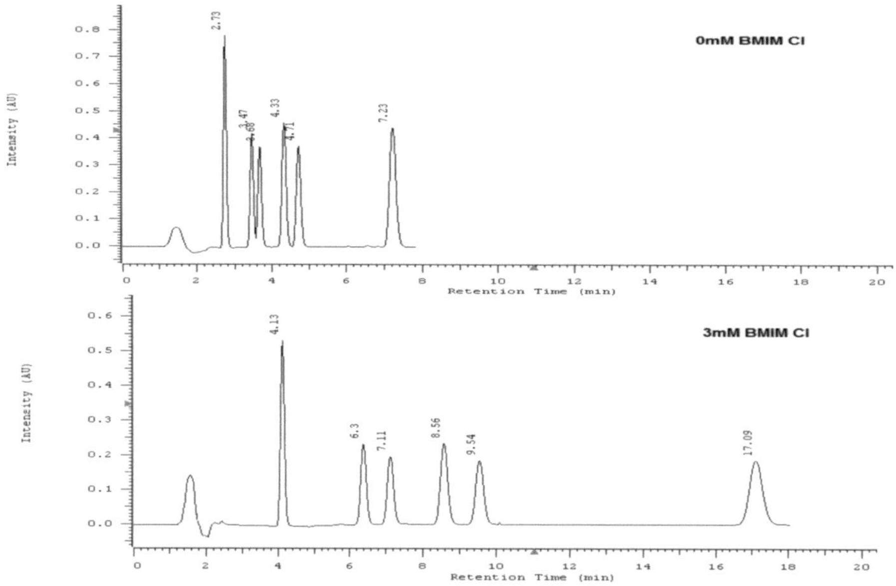

Fig. 8. Separation of 1,4-benzodiazepines mixture on Zorbax SB-Phenyl column and eluent system: 50% ACN / 20 mM phosphate buffer pH = 2.9 with addition of 3 mM [BMIM] [Cl]. For HPLC conditions see Experimental part. The peaks order: dechlordiazepam, oxazepam, lorazepam, nitrazepam, demethyldiazepam, diazepam.

Further increase of [BMIM] [Cl] concentration in the mobile phase causes decrease in retention of protonated amines probably owing to excess adsorption of liophilic ionic liquid cation with polar part oriented toward polar eluent environment in agreement with the following scheme:

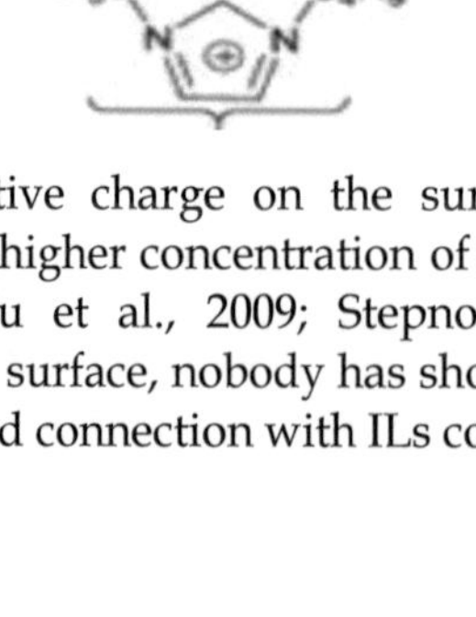

According to this scheme, positive charge on the surface is responsible for retention decrease of protonated amines at higher concentration of ionic liquid chloride in the mobile phase.Though, some papers (Hu et al., 2009; Stepnowski et al., 2006) demonstrated adsorption of ILs on hydrophobic surface, nobody has shown before real reflectance of these speculations in retention factor and connection with ILs concentration in the mobile phase.

Addition of ILs to the mobile phase in RP-HPLC systems appears to be especially advantageous for determination of the lipophilicity parameters of ionogenic basic analytes owing to supression of silanophilic interactions and improvement of linearity of log k vs. volume fraction of organic modifier in eluent system:

$$\log k = -S\varphi + \log k_w \qquad (2)$$

where log k indicates the isocratic capacity factor, φ the volume fraction of organic modifier in the mobile phase, log k_w the retention factor extrapolated to pure water, S the slope of the regression curve.

The correlation between the reference lipiophilicity parameter from the n-octanol-water partition system (log P_{oct}) and chromatographically determined lipophilicity parameters (S, log k_w, φ_0) is statistically significant mainly for congeneric group of neutral solutes. Application of ionic liquids additives improved evidently this correlation for compounds exhibiting acid–base behavior. Kaliszan (Kaliszan et al 2004) proposed [EMIM] [BF_4] for this purpose. Fig.9 illustrates improvement of correlations between the chromatographic lipophilicity parameters, log k_w, and log P_{oct} obtained for a series of basic drugs by gradient HPLC at the presence of [EMIM] [BF_4] in the mobile phase.

For acid compounds, the mechanism for separation and the effects of ionic liquids appear to be different. Hu et al (Hu et al., 20) applied 1-butyl-3-methylimidazolium chloride [C_4MIM][Cl], 1-octyl-3-methylimidazolium chloride [C_8MIM][Cl], and 1-decyl-3-methylimidazolium chloride [C_{10}MIM][Cl] as mobile phase additives in reversed-phase separation phenoxy acid herbicides and phenols at neutral pH on conventional C18 column. It was demonstrated that [BMIM][Cl], was the most efficient. The retention time of the target

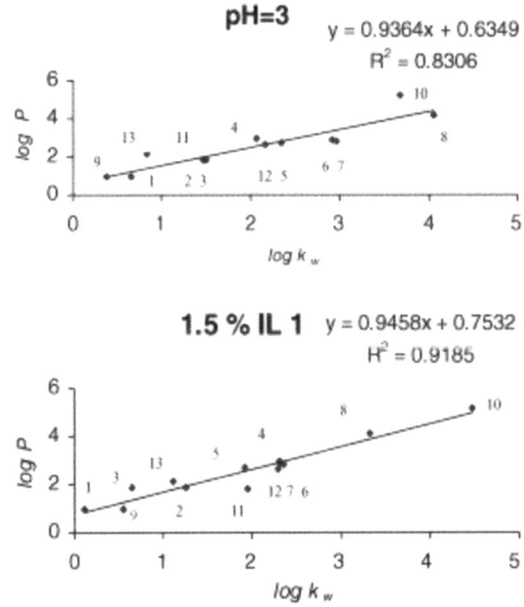

Fig. 9. Relationships between log P and the lipophilicity parameter log k_w, determined by gradient HPLC with a buffered at pH 2.87 water–methanol eluent not comprising (upper plot) and comprising (bottom plot) 1.5% (v/v) of IL (Kaliszan et al., 2004).

acids was shortened with the increase of the alkyl chain length and the concentrations of ionic liquids. The delocalization of the positive charge on the imidazolium cation, the repulsion between chlorine ions of ionic liquids and the acid compounds, as well as the stereo-hindrance effect are probably responsible for the observed effects.

3.2.3 Ionic liquids as liquid chromatography stationary phases

The newest approach in liquid chromatography is stationary phase modification by the use of ILs. Initial experiments were performed by the dynamic coating of the silica support using alkylammonium nitrate ILs in chlorinated hydrocarbons or ethers in hexane containing up to 75% (v/v) of the polar organic solvent as the mobile phase (Shetty et al. 1990). In this way, ILs formed the so-called supported liquid phases generating liquid-liquid separation mechanism. Obtained biphasic systems were effective for identifying systems for liquid–liquid extraction of complex mixtures and determination of partition coefficients. This method, owing to favorable features including speed, accuracy and minimal consumption of reagents, appears to be suitable in spite of some disadvantages concerning UV absorbance limitations caused by ILs. This idea found application in countercurrent chromatography (CCC) requiring two immiscible liquid phases as separation medium. Berthod et al. (Berthod et al. 2004) constructed biphasic liquid system consisting of water and [BMIM] [PF_6] with addition of acetonitrile reducing viscosity for determination of the distribution constant of various aromatic compounds containing basic, acidic and neutral substituents. A preparative high-speed counter-current chromatography method for isolation and purification of flavonoid compounds from *Oroxylum indicum* was successfully established by using ionic liquids as the modifier of the two-phase solvent system. Two flavonoid compounds including baicalein-7-*O*-diglucoside and baicalein-7-*O*-glucoside were purified from the crude extract of *O. indicum* by using ethyl acetate–water–[BMIM][PF_6] (5:5:0.2, v/v) as two-phase solvent system (Liu et al., 2010).

Liu and his group was among the first who examined new LC stationary phases based on covalently bonded ILs. They immobilized a vinyl-hexylimidazolium tetrafluorobarate IL at the surface of porous silica particles activated with a silane-coupling agent (mercaptopropyl trimethoxyxilane). The obtained column was applied for separation of an ephedrine mixture (Jiang et al. 2004). The initial results weren't satisfactory. However, the next study of the same team concerning covalent bonding between silanol groups and the cation of the ILs turned out to be a considerable success (Qiu et al., 2006 a; Qiu et al., 2006 b). Owing to the reaction of silanised with 3-chloropropyltrimethoxysilane gel with imidazolium salt, stationary phase possessing anion-exchanging properties able to separate inorganic as well as organic anions was obtained. A schematic diagram of SilprMim synthesis and separation results is shown in Fig.10 and Fig.11.

Next, Stalcup and co-workers (Sun et al. 2005) received new column thanks to the reaction of silica gel with 1-butyl-3 heptylimidazolium bromide. IL's cation was covalently immobilized through an n-alkyl ether chain. Properties of obtained column were studied under reversed-phase conditions for the separation of a group of neutral aromatic solutes applying the linear solvation energy relationship approach. Retention properties were remarkably similar to conventional phenyl-based stationary phases despite the presence of significant charge on the surface. The slope of the log $k_{butylimidazolium}$ versus log k_{phenyl} is close to unity (0.958) indicating additionally that residual alkyl linker does not play a large role in the retention process (Fig.12).

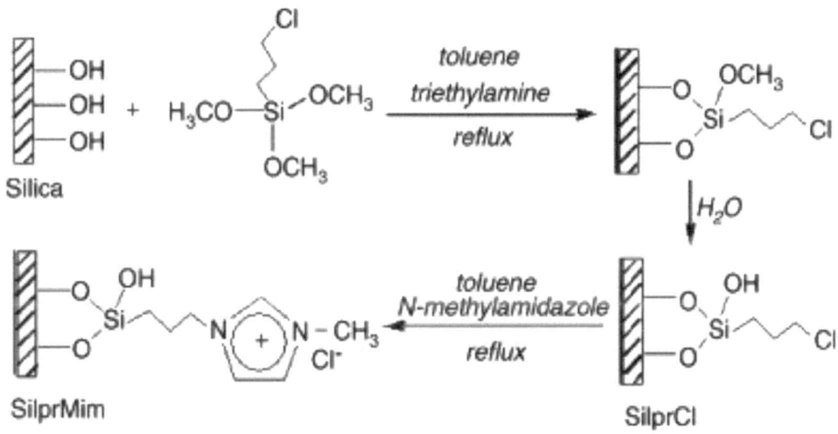

Fig. 10. Preparation of N-methylimidazolium functionalized silica. (Qiu et al. 2006 b)

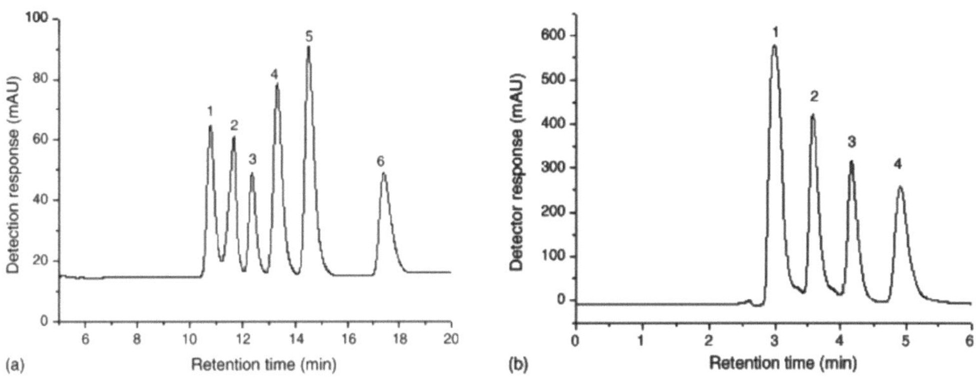

Fig. 11. Chromatogram of separation of test mixture composed of: (a) iodate(1) chloride(2); bromide(3); nitrate(4); iodide(5); thiocyanate(6). Mobile phase 20mmoll^{-1}KH_2PO_4 (pH 4.6); (b) N-ethylaniline(1); N,N-dimethylaniline(2); o-nitroaniline(3);andbenzidine(4); - Mobile phase was 50mmoll^{-1} KH_2PO_4 (pH 4.6) (Qiu et al. 2006 b).

Recently, consecutive study describing other stationary phase has been published (Wang et al. 2006). Trimethoxysilane "ionosilane" derivatives of RTILs based on alkylimidazolium bromides were bonded to 3 μm diameter silica particles. The ionic liquid moiety was predominantly attached to the silica surface through two siloxane bonds of the ionosilane derivative. The obtained structure was confirmed by thermogravimetric analysis (TGA) and ^{13}C and ^{29}Si NMR spectroscopies. Properties of the column were studied by using aromatic carboxylic acids as model compounds. It appears that the separation mechanism involves multiple interactions including ion exchange, hydrophobic interaction, and other electrostatic interactions. Authors of the study emphasize it that a clear advantage of the new columns is a possibility of reducing the content of organic solvent in the mobile phase and a diversified selectivity depending on the spatial orientation of bounded ligands (Fig.13).

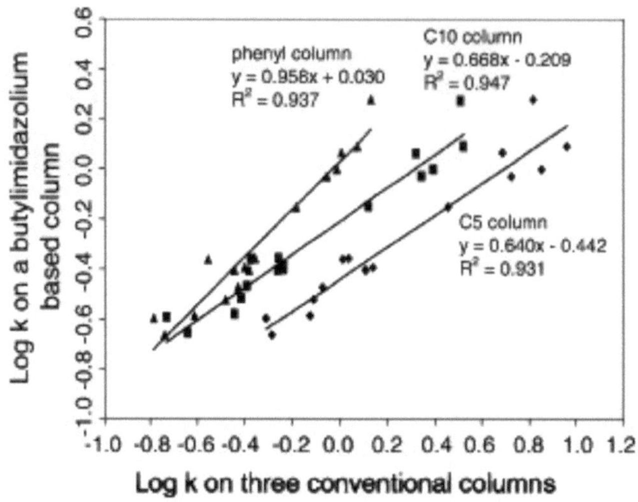

Fig. 12. Plot of log *k* obtained using 60% MeOH on the butylimidazolium column versus log *k* obtained on conventional aliphatic reversed phase columns under the same mobile phase conditions (Sun et al. 2005).

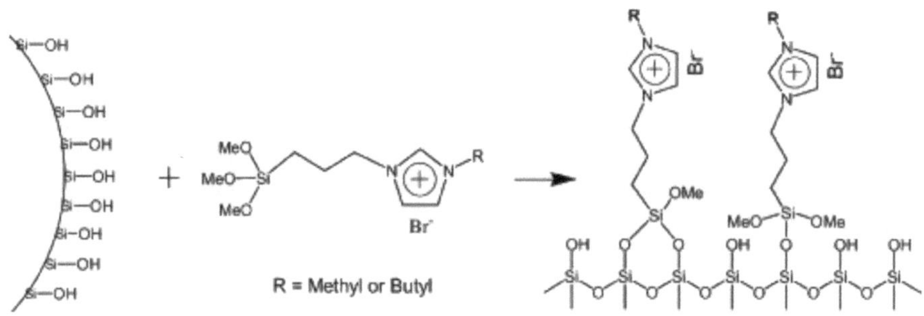

Fig. 13. Schematic procedure of ILs bonded stationary phase synthesis (Wang et al. 2006).

4. Liquid chromatography of ionic liquids

In the recent years, ILs have gained great popularity in chemical industry. Their determination is very important because of high probability of their migration into the environment. Unfortunately, this subject has not been extensively explored so far. There are only several papers dealing with analytical methods suitable for ILs determination. Among them chromatographic techniques, especially reversed-phase liquid chromatography, appear to be the most suitable for analysis of ILs as well as determination of their properties or identifying systems suitable for their liquid-liquid extraction. Different stationary phases such as cholesterol ligands bonded chemically to silica, alkylamide, butyl, octyl, octadecyl, phenyl, aryl, mixed (containing octyl, octadecyl, phenyl, aminopropyl, cyanopropyl) were tested for this purpose (Buszewski et al., 2006). Due to the non-molecular nature of ILs, the retention mechanism is not as simple as the one of molecular species. The retention mechanism combines hydrophobic and ionic interactions (ion-pairing, ion-exchange). In agreement with hydrophobic nature of retention mechanism, retention of ILs increases with

increasing hydrophobicity of the stationary phase. However, for packings containing protonized functional groups, the lowering of the retention factor was observed owing to electrostatic repulsion forces between entities and charged surface. The main interactions, predominant in the case of aryl stationary phase, are dispersive ones and of π-π type.
The best results according to separation selectivity towards ILs were achieved on the butyl and silica monolith with bonded octadecyl groups probably owing to the lowest heterogeneity of these ligands (Buszewski et al., 2006). Routinely, liquid chromatography with the use of an octyl stationary phase could be applied for isolation, identification and quantification of 1-alkyl and 1-aryl-3-methylimidazilium based ILs (Stepnowski et al., 2003). It should be stressed that when a salt is added to the mobile phase, only ILs differing with the cation can be separated (Ruiz-Angel & Berthod, 2006). To improve IL peak shape and enhance retention time, $NaPF_6$ or $NaClO_4$ could be added to the mobile phase. These anionic additives, possessing chaotropic properties, form ion-pairs with the IL cations. In turn, mobile phases containing no salts are able to separate ILs differing with the anion. Their elution order depends on the lyotropy of the anions. Berthod investigated eleven 1-alkyl-3-methyl imidazolium ionic liquid salts in reversed phase mode with a Kromasil C_{18} column. It was shown that it is possible to separate different ILs depending on the mobile phase composition. Additionally, working with salt-free mobile phases, fronting peaks for all ILs were observed, associated with the IL/C_{18} stationary phase interactions responsible for concave isotherms. With mobile phase containing 0.01 M of salt, tailing IL peaks corresponding to convex adsorption isotherms were obtained, as could be seen in Fig.14.
Taking into account the fact that short chain ILs, as the early eluting compounds in reversed-phase systems, are not satisfactorily resolved, Stepnowski and Mrozik (Stepnowski & Mrozik, 2005) applied the SCX stationary phase with strong cation exchange properties. They investigated the chromatographic behavior of 8 Ils: 7 homologues of 1-alkyl-3-methylimidazolium and 4-methyl-*N*-butylpyridinium. Depending on the mobile phase composition, different interactions such as cation exchange, nonspecific hydrophobic interactions, and adsorption affected the retention process. In comparison to the chromatographic behavior of Ils in the classical reversed phase system much higher selectivity of highly hydrophilic compounds was noticed.
In order to take advantage of the potential π–π interactions, an ether-linked phenyl stationary phase with polar endcapping was also proposed in the separation of imidazolium and pyridinium ILs (Stepnowski et al., 2006). With phenyl-bonded phases, very good separations of the cations were achieved however; gradient elution was required to achieve separation at appropriate time (Fig.15). Once again special attention was paid to the short-chain hydrophilic entities, known to be poorly separated on conventional reversed-phase columns. Authors declare the applicability of the proposed method for monitoring degraded cations of 1-butyl-3-methylimidazolium salts obtained in the Fenton process.
The aim of this study was also investigation of the nature of interactions occurring in the chromatographic system. Possible interaction between imidazolium IL and phenyl-bonded stationary phase is illustrated schematically in Fig.16. In case of acetonitrile, which is able to form electrono-donor-acceptor complexes, suppression of π–π interactions between the solute and the aromatic moiety of the stationary phase could be assumed. Thus at higher acetonitrile concentration in the mobile phase only hydrophobic and hydrogen-bonding interactions would be responsible for the separation mode. Recently, Studzińska and Buszewski proved that ILs are bonded much stronger to variety of stationary phase surfaces when pH of mobile phase is less acidic (Studzińska & Buszewski 2010).

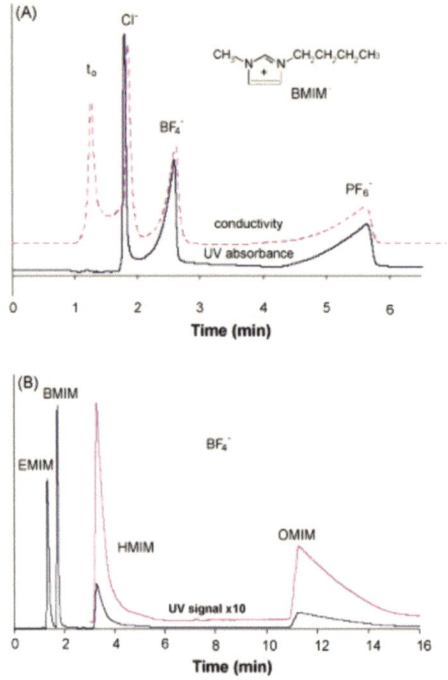

Fig. 14. RPLC separation of ionic liquids. (A) Separation of three BMIM ILs; continuous line: UV 254 nm detection; dotted line: conductimetric detection; salt-free mobile phase: acetonitrile/water 10/90% (v/v), 1 mL/min. (B) Separation of EMIM Cl, BMIM BF_4, HMIM BF_4 and OMIM PF_6; mobile phase: acetonitrile/water 0.01 M NaCl 30/70% (v/v), 1 mL/min; UV detection 254 nm; column Kromasil C_{18} 150 mm × 4.6 mm i.d. (Ruiz-Angel & Berthod, 2006).

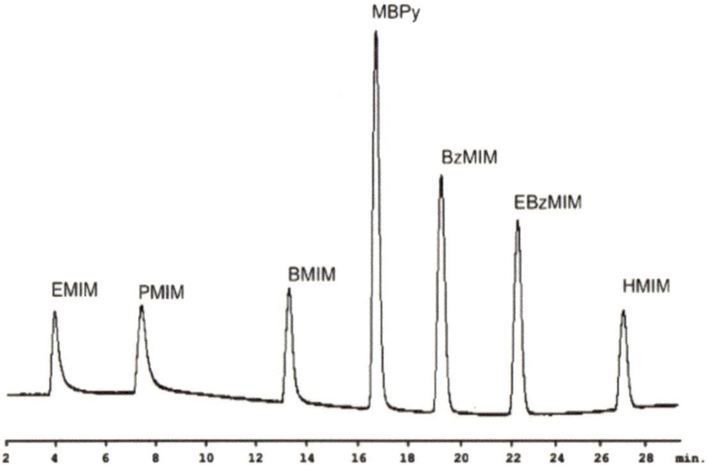

Fig. 15. Gradient separation of selected imidazolium and pyridinium ionic liquids under the conditions investigated. Mobile phase: methanol (2–20%)-water (5mM K_2HPO_4).pH 3.0.Column: Synergi Polar RP (Stepnowski et al., 2006).

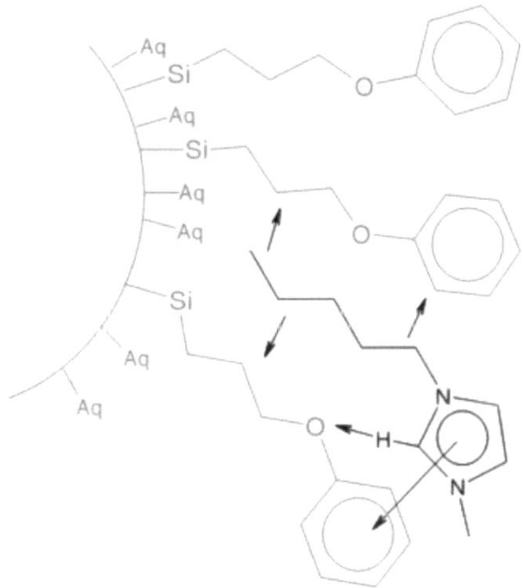

Fig. 16. Possible interactions of imidazolium ionic liquid cation and phenyl-bonded stationary phase. Arrows indicate hydrogen-bonding, hydrophobic and π–π interactions (Stepnowski et al., 2006).

Chromatographic behavior of ionic liquid cations in view of quantitative structure-retention relationship (QSRR) was studied by Molíková and co-workers (Molíková et al., 2010). The most significant QSRR equation between experimentally determined log k_w values and the structural descriptors of test analytes from molecular modeling (the solvent accessible surface: *SAS*, the differences in the energies of the highest occupied and the lowest unoccupied molecular orbitals: *diffHL*) was obtained in case of columns with the highest carbon loads and larger specific surface areas, i.e. Symmetry C18 and Symmetry C8. In turn, the column ACE 5 C18 allowed to obtain the best separations of the ILs according to cations kind. As the authors declare, identification of QSRRs can provide the knowledge regarding hydrophobicity and retention mechanism of ILs. It could be further helpful in prediction of their biological effects and the way of these chemicals into the environment.

5. Electromigration techniques.

Ils, due to good electrical conductivity and appropriate viscosity, could be used as electrolytes, additives in the run buffer or reagents for dynamic coating or covalent attachment to the capillary walls in capillary electrophoresis (CE). The main disadvantage of this chromatographic technique is an electroosmotic flow (EOF) provoked by silanol groups on the inner surface of silica capillaries. To reverse this process ILs can be covalently bonded to a fused-silica capillary surface. Qin and Li (Qin &Li, 2002; Qin et al. 2003) covalently bonded propyl methyl imidazolium chloride [PMIM][Cl] to a silica capillary. Due to the positive charge of the bonded moieties and reduction of interaction between analytes and the capillary wall, separation of sildefanil and its metabolites in human serum and DNA analysis with MS detection were possible. The capillary wall modified with 1-ethyl-3-methylimidazolium was used for separation of positively charged inorganic ions: NH_4^+ and K^+in human urine, the separation of K^+, Na^+, Li^+, Ca^{2+}, Mg^{2+} and Ba^{2+} ions in red wine. The main advantage of those capillaries appears to be increases in the working pH range of the running buffer with good efficiency and the compatibility with mass spectrometry (Qin & Li, 2002; Qin & Li, 2003; Qin et al., 2003).

Fused-silica capillaries could be chemically modified with an analogue of the imidazole-based IL (1-butyl-3-(3-triethoxysilylpropyl)-4,5-dihydroimidazolium tetrafluoroborate) as well as zwitterionic salt (*N* -3-(3-triethoxysilylpropyl)-4,5-dihydroimidazole) (Borissowa et al., 2007).

It appeared that the electroosmotic flow in the capillary coated with an ionic liquid analogue was found to be anodic and dependent on the pH of the separation buffer. Whereas in the case of a zwitterionic capillary, the electroosmotic flow was cathodic and its velocity remained almost constant in the pH range of 4–7. Schematic illustration of ionic liquid bonding to the capillary wall is presented in Fig.17.

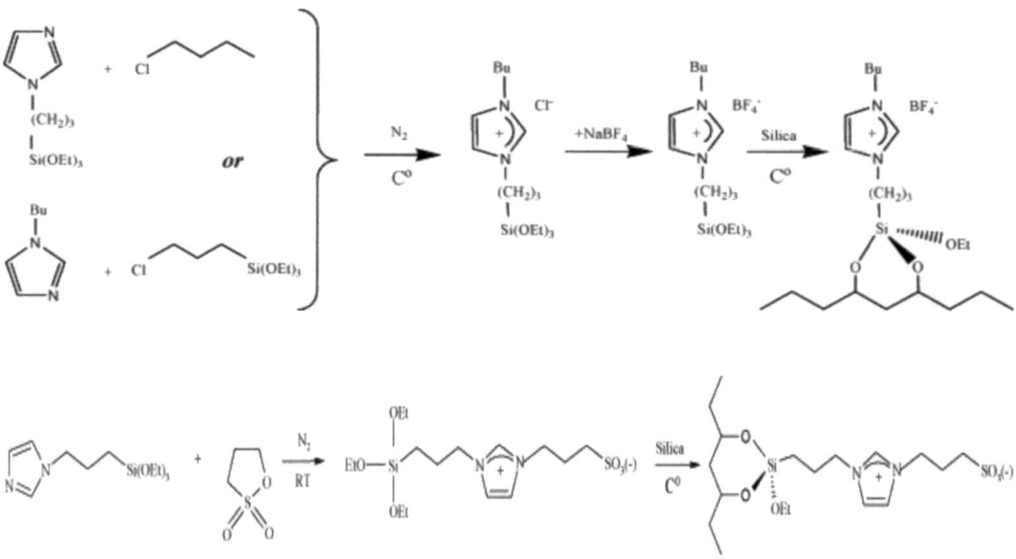

Fig. 17. Scheme representing the IL and zwitterionic salt coating procedure (Borissowa et al., 2007).

ILs could be applied also as background electrolytes in aqueous as well as non-aqueous elecytophoresis (NACE). In NACE, the EOF does not change its direction by addition of IL indicating that the dynamic coating process is meaningless. That is why interactions between the solute and ILs ions and significant changes in the electrophoretic mobility are mainly responsible for enhancement of separation selectivity. Effect of ionic liquids on EOF in NACE was monitored by contactless conductivity detection (Seiman et al., 2008). Dialkylimidazolium ILs were used in NACE for separation of water insoluble dyes (Vaher et al., 2000), phenols and aromatic acids (Vaher et al., 2002a), polyphenols (Vaher et al., 2002b).

In turn, ILs used as the electrolytes in aqueous capillary electrophoresis, enable the dynamic coating of the capillaries wall. IL cations, as well as anions, influence the migration behavior of analytes but the resolving power of the system is much more affected by activity of the IL cation. Analytes can interact either with the ILs cations adsorbed on the capillary wall or with the free IL cations in the electrolyte solution. Not meaningless is the suppression of the EOF and changing eletrophoretic properties of capillary wall by IL's cationic molecule. It makes the separation mechanism more complex. Fig.18 illustrates mechanism of polyphenol separation using 1-alkyl-3-methylimidazolium based IL in the background electrolyte (Yanes et al., 2001).

ILs as dynamic coating materials were used for separation of different groups of compounds. For instance, dimethyldinonylammonium bromide was used for determination of carboxylates as copper complexes (Laamanen et al.,2005), 1-alkyl-3-methylimidazolium-based ionic liquids were used as the running electrolytes for separation of acidic and basic proteins (Jiang et al.,2003; Wu et al., 2008) and flavone derivatives (Qi et al.,2006), nicotinic, isonicotinic and picolinic acids (Marszałł et al., 2006 b), monocationic and dicationic phosphonium-based ILs were tested for separation of inorganic and small organic anions (Krizek et al.,2009). Dicationic ion paring reagent (*N,N'*-dibutyl 1,1'-pentylenedipyrrolidinium) was applied for determination of several anions in the positive mode using CE-ESI-MS method achieving greater sensitivity, in comparison to the direct MS detection in the negative mode (Lin et al., 2009).

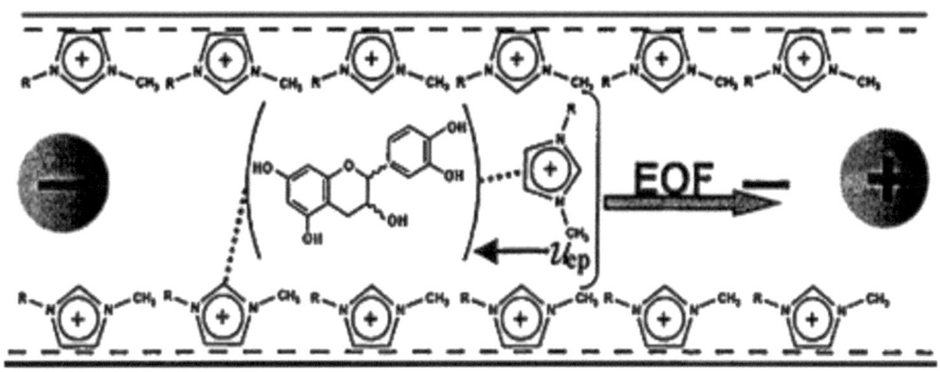

Fig. 18. Mechanism of polyphenols' separation using 1-alkyl-3-methylimidazolium-based ionic liquids. (Yanes et al., 2001).

Separation of both neutral and charged analytes is possible owing to micellar capillary electrophoresis (MCE). In this chromatographic technique, long chain alkylimidazolium ILs could be used as surfactants and short chain ones as modifiers. 1-butyl-3-methylimidazolium tetrafluoroborate [BMIM][BF_4] was added as modifier to the micellar system containing SDS resulting in the complete separation of active components of lignans in medicinal herbs (Tian et al., 2005). As the authors claimed, ILs affected the size and shape of the micelles, altering the separation. Long chain alkylimidazolium ILs added to the aqueous buffer were used as surfactants for separation of neutral methylresorcinol isomers and benzene derivatives (Borissova et al., 2008). 1-Butyl-3-methylimidazolium dodecanesulfonate (BAS) was applied in microchip MCE for the separation of proteins (Xu et al., 2008).

Chiral ILs could be used for enantiomeric separation in electromigation techniques. S-[3-(chloro-2-hydroxypropyl)trimethylammonium] NTf_2 was used as a chiral selector to separate various enantiomeric pharmaceutical compounds in CE (Tran& Mejac, 2008), undecenoxycarbonyl-L -pryrrolidinol bromide, and undecenoxycarbonyl-L -leucinol bromide to separate enantiomeric acidic analytes were used in MCE (Rizvi & Shamasi 2006). Sometimes ILs chiral selector requires another chiral selector to achieve enantioseparation. For instance, two chiral ILs: (ethyl- and phenylcholine of bis(trifluoromethylsulfonyl)imide) acting synergistically with cyclodextrins were able to improve the enantioselectivity for 2-arylpropionic acids (Francois et al.,2007).

6. Concluding remarks

Growth of ionic liquids applications in analytical chemistry has been observed during the last years. It appears to be reasonable, considering a wide range of their physico-chemical properties. As green solvents they could find many industrial applications in future. That is why elaborating on new sensitive methods of their determination in environmental and industrial samples appears to be a dire must. Taking into account chromatographic techniques the most promising progress connected with good solvent properties of ionic liquids, their low volatility and electrical conductivity, is observed in liquid chromatography as well as electromigration techniques. The next generation of ionic liquids also with chiral properties will be synthesized and will find other applications. Special precaution should be taken as for the toxicity, stability and purity of these reagents. Long term stability devoted especially to ionic liquids containing perfluorinated anions should be taken into account in connection with the degradation products which could damage the silica based adsorbents. Understanding ionic liquids properties in chromatographic system is still unclear. Some authors assume that an ionic liquid should not be used as an alternative to a normal salt because the distinction between liquid and solid is lost in solution and ILs properties become the same as those of organic salts. However, other authors satisfactorily demonstrate special behavior of ionic liquids in chromatographic systems, stressing unique stationary phase modification and advantageous interactions with investigated compounds. Interpretation of the separation mechanisms when ionic liquids are used still appears to be an interest of the scientific forum.

7. References

Armstrong, D.W., He, L. & Liu, Y.S.(1999). Examination of ionic liquids and their interactions with molecules, when used as stationary phase in gas chromatography, *Anal.Chem.* Vol. 71 (No. 17): 3873-3876.

Armstrong, D.W., Zhang, Li-K., He, L. & Gross, M.L. (2001). Ionic Liquids as Matrixes for Matrix-Assisted Laser Desorption/Ionization Mass Spectrometry, *Anal. Chem.* Vol.73 (No.15): 3679–3686.

Anderson, J.L. & Armstrong, D.W. (2003). High-stability ionic liquids. A new class of stationary phases for gas chromatography, *Anal.Chem.* Vol. 75 (No. 19): 4851-4858.

Bączek, T., Marszałł, M.P., Kaliszan, R., Walijewski, Ł., Makowiecka, W., Sparzak, B., Grzonka, Z., Wiśniewski, K.& Juszczyk, P. (2005). Behavior of peptides and computer-assisted optimization of peptides separations In a normal-phase thin-layer chromatography system with and without the addition of ionic liquid In the eluent, *Biomed.Chromatogr.* Vol. 19 *(No. 1): 1-8.*

Barber, D.W., Phillips, C.S.G., Tusa G. F. & Verdin, A. (1959). The chromatography of gases and vapours. Part VI. Use of the stearates of bivalent manganese, cobalt, nickel, copper, and zinc as column liquids in gas chromatography, *J. Chem. Soc.*18-24.

Berthod, A., He, L. & Armstrong, D,W. (2001). Ionic liquids as stationary phases solvents for methylated cyclodextrines in gas chromatography, *Chromatographia* Vol. 53 (No.1-2): 63-68.

Berthod, A. & Carda-Broch, S. (2004). Use of the ionic liquid 1-butyl-3-methylimidazolium hexafluorophosphate in countercurrent chromatography, *Anal. Bioanal. Chem.* Vol. 380 (No.1): 168-177.

Berthod, A., Ruiz-Angel, M.J.& Huguet, S.(2005). Nonmolecular Solvents in Separation Methods: Dual Nature of Room Temperature Ionic Liquids, *Anal. Chem.* Vol. 77 (No. 13): 4071-4080.

Berthod, A., Ruiz-Angel, M.J. & Carda-Broch, S. (2008). Ionic liquids in separation techniques, *Journal of Chromatography A* Vol. 1184 (No. 1-2): 6-18.

Borissova, M.; Vaher, M.; Koel, M.& Kaljurand, M. (2007). Capillary zone electrophoresis on chemically bonded imidazolium based salts, *J. Chromatogr. A* Vol. 1160 (No. 1-2):320–325.

Borissova, M., Palk, K.& Koel, M. (2008). Micellar electrophoresis using ionic liquids, *J. Chromatogr. A* Vol. 1183 (No. 1-2):192-195.

Buszewski,B., Kowalska, S.& Stepnowski, P. ((2006). Influence of stationary phase properties on the separation of ionic liquid cations by RP-HPLC, *J.Sep.Sci.* Vol. 29 (No. 8): 1116-1125.

Buszewski, B. & Studzińska, S. (2008). A review of ionic liquids in chromatographic and electromigration techniques, *Chromatographia* Vol. 68 (No. 1-2): 1-10.

Flieger, J. (2006). The effect of chaotropic mobile phase additives on the separation of selected alkaloids in reversed-phase high-performance liquid chromatography, *J. Chromatogr. A* Vol. 1113 (No. 1-2): 37-44.

Flieger, J. (2009). Effect of ionic liquids as mobile phase additives on chromatographic parameters of neuroleptic drugs in reversed-phase high-performance liquid chromatography, *Anal..Lett.* Vol. 42 (No. 11): 1632–1649.

Flieger, J.& Czajkowska-Żelazko, A. (2010). Ionic liquids as mobile phase additives in reversed phase high performance liquid, submitted to the *Journal of Separation science.*

Francois, Y., Varenne, A., Juillerat, E., Villemin, D. & Gareil P. (2007). Evaluation of chiral ionic liquids as additives to cyclodextrins for enantiomeric separations by capillary electrophoresi,s *J. Chromatogr. A* Vol. 1155 (No. 2):134-141.

Gritti, F.& Guiochon, G. (2004). Retention of Ionizable Compounds in Reversed-Phase Liquid Chromatography. Effect of the Ionic Strength of the Mobile Phase and the Nature of the Salts Used on the Overloading Behavior, *Anal.Chem.* Vol. 76 (No. 16): 4779-4789.

He, L., Zhang, W., Zhao, L., Liu, X. & Jiang, S. (2003). Effect of 1-alkyl-3-methylimidazolium-based ionic liquids as the eluent on the separation of ephedrines by liquid chromatography, *J. Chromatogr. A* Vol. 1007 (No. 1-2):39-45.

Hu, X., Peng,J., Huang, Y., Yin, D.& Liu,J. (2009). Ionic liquids as mobile phase additives for high-performance liquid chromatography separation of phenoxy acid herbicides and phenols, *J.Sep.Sci.* Vol.32 (No.23-24):4126-4132.

Jiang, T., Gu, Y., Liang, B., Li, J., Shi, Y.& Ou, Q. (2003). Dynamically coating the capillary with 1-alkyl-3-methylimidazolium-based ionic liquids for separation of basic proteins by capillary electrophoresis, *Anal. Chim. Acta* Vol. 479 (No. 2):249-254.

Johnson, K.E. (2007). What's an ionic liquid?, *The Electrochemical Society Interface Spring 2007:38-41.*
www.electrochem.org/dl/interface/spr/spr07/spr07_p38.pdf

Kaliszan, R., Marszałł, M.P., Markuszewski, M.J., Bączek, T. & Pernak, J. (2004). Suppresion of deleterious effects of free silanols in liquid chromatography by imidazolium tetrafuoroborate ionic liquids. *J. Chromatogr. A* Vol. 1030 (No. 1-2): 263-271.

Koel, M., (2005). Ionic liquids in chemical analysis, *Critical Review in Analytical Chemistry* Vol.35 (No.3): 177-192.

Krizek, T., Breitbach, Z.S., Armstrong, D.W., Tesarova, E.& Coufal, P. (2009). Separation of inorganic and small organic anions by CE using phosphonium-based mono- and dicationic reagent, *Electrophoresis* Vol. 30 (No.22):3955-3963.

Laamanen, P., Busi, S., Lahtinen, M.& Matilainen, R. (2005). A new ionic liquid dimethyldinonylammonium bromide as a flow modifier for the simultaneous determination of eight carboxylates by capillary electrophoresis, *J. Chromatogr. A* Vol. 1095 (No. 1-2):164-171.

Lin, X., Gerardi, A.R., Breitbach, Z.S., Armstrong, D.W.& Colyer, C.L. (2009). CE-ESI-MS analysis of singly charged inorganic and organic anions using a dicationic reagent as a complexing agent, *Electrophoresis* Vol. 30 (No.22):3918-3925.

Liu, S.J., Zhou, F., Xiao, X.H., Zhao, L., Liu, X. & Jiang, S.X. (2004). Surface confined ionic liquid - A new stationary phase for the separation of ephedrines in high-performance liquid chromatography, *Chin. Chem. Lett.* Vol. 15 (No. 9): 1060-1062.

Liu, R., Xu, L., Li,A. & Sun,A. (2010). Preparative isolation of flavonoid compounds from *Oroxylum indicum* by high-speed counter-current chromatography by using ionic liquids as the modifier of two-phase solvent system, *J.Sep.Sci.*Vol.33 (No.8):1058-1063.

LoBrutto, R., Jones, A. & Kazakevich, Y.V. (2001). Effect of counter-anion concentration on retention in high-performance liquid chromatography of protonated basic analytes, *J. Chromatogr. A* Vol. 913 (No. 1-2): 189-196.

Marszałł, M.P., Baczek, T.& Kaliszan, R. (2005). Reduction of silanophilic interactions in liquid chromatography with the use of ionic liquids, *Anal. Chim. Acta* Vol. 547 (No. 2): 172-178.

a.Marszałł, M.P., Baczek, T.& Kaliszan, R. (2006). Evaluation of the silanol-suppressing potency of ionic liquids, *J. Sep. Sci.* Vol. 29 (No. 8):1138-1145.

b.Marszałł, M.P., Markuszewski, M.J.& Kaliszan, R. (2006). Separation of nicotinic acid and its structural isomers using 1-ethyl-3-methylimidazolium ionic liquid as a buffer additive by capillary electrophoresis, *J. Pharm. Biomed. Anal.* Vol. 41 (No. 1):329–332.

Marszałł, M.P.& Kaliszan, R., (2007). Application of ionic liquids In liquid chromatography, *Critical Reviews in Analytical Chemistry* Vol. 37 (No.14): 127-140.Mehl, J.T.& Hercules, D.M., (2000). Direct TLC-MALDI coupling using a hybrid plate. *Anal.Chem.*Vol. 72 (No. 1): 68-73.

Molíková,M., Markuszewsk,i M.J., Kaliszan,R.& Jandera,P.(2010). Chromatographic behaviour of ionic liquid cations in view of quantitative structure-retention relationship, *J.Chromatogr. A* Vol.1217 (No.8):1305-1312.

Pacholec, F., Butler, H.T.& Poole, C.F. (1982). Molten organic salt phase for gas-liquid chromatography, *Anal. Chem.*, Vol. 54 (No. 12): 938-1941.

Pacholec, F.& Pool,e C.F. (1983). Stationary phase properties of the organic molton salt ethylpyridinium bromie In gas chromatohraphy, *Chromatographia* Vol. 17 (No. 7): 370-374.

Pan, L., LoBrutto, R., Kazakevish, Y.V.& Thomson, R. (2004). Influence of inorganic mobile phase additives on the retention, efficiency and peak symmetry of protonated basic compounds in reversed-phase liquid chromatography, *J. Chromatogr. A* Vol. 1049 (No.1-2):63-73.

Pandey S (2006). Analytical applications of room-temperature ionic liquids: A review of recent efforts, *Anal. Chim. Acta* Vol. 556 (No.1):38–45.

Poole, C.F., Kersten, B.R., Ho,S.S.J., Coddens, M.E.,& Furton, K.G. (1986). Organic salts, liquid at room temperature, as mobile phases in liquid chromatography, *J.Chromatogr.A* Vol. 352 (No. 1-2): 407-425.

Poole, C.F. (2004). Chromatographic and spectroscopic methods for the determination of solvent properties of room temperature ionic liquids, *Journal of Chromatography A* Vol. 1037 (No.1-2): 49-82.

Poole, C.F., (2006). Applications of Ionic Liquids in Extraction, Chromatography, and Electrophoresis, in Grushka, E., Grinberg, N., (ed.), *Advances in Chromatography,* Vol. 45, CRC Press, Boca Raton, Fl, pp. 89-125.

Qi, S., Li, Y., Deng, Y., Cheng, Y., Chen, X. & Hu, Z. (2006). Simultaneous determination of bioactive flavone derivatives in Chinese herb extraction by capillary electrophoresis used different electrolyte systems—Borate and ionic liquids, *J. Chromatogr. A* Vol. 1109 (No. 2):300-306.

Qin, W. & Li, S.F.Y. (2002).An ionic liquid coating for determination of sildenafil and UK-103,320 in human serum by capillary zone electrophoresis-ion trap mass spectrometry, *Electrophoresis* Vol. 23 (No. 24):4110-4116.

Qin, W.& Li, .SF.Y. (2003). Electrophoresis of DNA in ionic liquid coated capillary, *Analyst* Vol.128 (No. 1): 37–41.

Qin, W., Wei, H. & Li, S.F.Y. (2003). 1,3-Dialkylimidazolium-based room-temperature ionic liquids as background electrolyte and coating material in aqueous capillary electrophoresis, *J. Chromatogr. A* Vol. 985 (No. 1-2): 447–454.

a.Qiu H., Jiang S., Liu X., (2006). *N*-Methylimidazolium anion-exchange stationary phase for high-performance liquid chromatography *J. Chromatogr. A*, Vol. 1103, (No. 2), p. 265-270.

b.Qiu H., Jiang S., Liu X., Zhao L.,(2006). Novel imidazolium stationary phase for high-performance liquid chromatography *J. Chromatogr. A* Vol. 1116, (No.1-2):46-50.

Rizvi, S.A.A.& Shamsi, S.A. (2006). Synthesis, Characterization, and Application of Chiral Ionic Liquids and Their Polymers in Micellar Electrokinetic Chromatography, *Anal. Chem.* Vol. 78 (No. 19):7061-7069.

Ruiz-Angel, M.J.& Berthod, A.(2006). Reversed phase liquid chromatography of alkyl-imidazolium ionic liquids, *J. Chromatogr. A* Vol. 1113 (No. 1-2):101- 108.

Ruiz-Angel, M.J., Carda-Broch, S.& Berthod, A. (2006). Ionic liquids versus triethylamine as mobile phase additives in the analysis of β-blockers, *J. Chromatogr. A* Vol. 1119 (No. 1-2):202-208.

Seiman, A., Vaher, M.& Kaljurand, M. (2008). Monitoring of the electroosmotic flow of ionic liquid solutions in non-aqueous media using thermal marks, *J. Chromatogr. A* Vol. 1189 (No. 1-2):266-273.

Shetty, P.H., Poole, S.K. & Poole. C.F. (1990). Applications of ethylammonium and propylammonium nitrate solvents in liquid-liquid extraction and chromatography, *Anal. Chim. Acta* Vol. 236 (No.1):51-61.

Shetty, P.H., Youngberg, P.J., Kersten, B.R.& Poole, C.F., (1987). Solvent properties of liquid organic salts used as mobile phases in microcolumn reversed-phase liquid chromatography, *J. Chromatogr.* Vol. 411 (No.1-2): 61-79.

Stalcup, A.M. & Cabovska, B. (2004). Ionic liquids in chromatography and capillary electrophoresis, *Journal of Liquid Chromatography and Related Technologies* Vol. 27 (No. 7-9): 1443-1459.

Stepnowski, P., Müller, A., Behrend, P., Ranke, J., Hoffmann, J.& Jastorff, B. (2003). Reversed-phase liquid chromatographic method for the determination of selected room-temperature ionic liquid cations, *J.Chromatogr.A* Vol. 993 (No.1-2): 173-178.

Stepnowski, P. & Mrozik, W. (2005). Analysis of selected ionic liquid cations by ion exchange chromatography and reversed-phase high performance liquid chromatography, *J.Sep.Sci.*, Vol. 28 (No. 2): 149-154.

Stepnowski, P., Nichthauser, J.,Mrozik, W. & Buszewski, B.(2006). Usefulness of π...π aromatic interactions in the selective separation and analysis of imidazolium and pyridinium ionic liquid cations, *Analytical & Bioanalytical Chemistry* Vol. 385 (No.8): 1483-1491.

Studzińska, S.& Buszewski, B. (2010). Study of retention mechanism of imidazolium-based ionic liquids in HPLC, *J.Sep.Sci.* Vol.33 (No.9): 1264-1273.

Sun,Y., Cabovska B., Evans C.E., Ridgway T.H., Stalcup A.M., (2005). Retention characteristics of a new butylimidazolium-based stationary phase, *Anal. Bioanal. Chem.* Vol.382 (No.3), p. 728-734.

Sun,P., Armstrong, D.W. (2010). Ionic liquids in analytical chemistry, *Anal.Chim.Acta* Vol.661 (No. 1): 1-16.

Tian, K., Qi, S., Cheng, Y., Chen, X.. & Hua, Z. (2005). Separation and determination of lignans from seeds of Schisandra species by micellar electrokinetic capillary chromatography using ionic liquid as modifier, *J. Chromatogr. A* Vol.1078 (No. 1-2):181–187.

Tran, C.D.& Mejac, I.. (2008). Chiral ionic liquids for enantioseparation of pharmaceutical products by capillary electrophoresis, *J. Chromatogr. A* Vol. 1204 (No.2): 204-209.

Vaher, M., Koel ,M.& Kaljurand, M. (2000). Non-aqueous capillary electrophoresis in acetonitrile using lonic-liquid buffer electrolytes, *Chromatographia* Vol.53 (No.5-6):S302–S306.

a.Vaher, M., Koel, M.& Kaljurand, M. (2002). Application of 1-alkyl-3-methylimidazolium-based ionic liquids in non-aqueous capillary electrophoresis, *J. Chromatogr. A* Vol. 979 (No.1-2):27-32.

b.Vaher, M., Koel, M.& Kaljurand, M. (2002). Ionic liquids as electrolytes for nonaqueous capillary electrophoresis, *Electrophoresis* Vol.23 (No.3) :426-430.

Waichigo M.M., Riechel T.L. & Danielson N.D., (2005). Ethylammonium Acetate as a Mobile Phase Modifier for Reversed Phase Liquid Chromatography, *Chromatographia* Vol. 61 (No. 1-2):17-23.

a.Waichigo, M.M.& Danielson, N.D. (2006). Ethylammonium Formate as an Organic Solvent Replacement for Ion-Pair Reversed-Phase Liquid Chromatography, *J. Chromatogr. Sci.*, Vol. 44 (No.10): 607-614.

b.Waichigo, M.M.& Danielson, N.D., (2006). Comparison of ethylammonium formate to methanol as a mobile-phase modifier for reversed-phase liquid chromatography, *J. Sep. Sci.* Vol. 29 (No5):599-606.

Waichigo, M.M., Hunter, B.M., Riechel, T.L.& Danielson, N.D. (2007). Alkylammonium Formate Ionic Liquids as Organic Mobile Phase Replacements for Reversed-Phase Liquid Chromatography, *J. Liq. Chromatogr. Rel. Technol.* Vol. 30 (No. 2):165-184.

Wang Q., Baker G.A., Baker S.N., Colón L.A. (2006). Surface confined ionic liquid as a stationary phase for HPLC, *Analyst*, Vol.131 (No.9)1000-1005.

Wu, X., Wei, W., Su, Q., Xu, L.& Chen, G. (2008).Simultaneous separation of basic and acidic proteins using 1-butyl-3-methylimidazolium-based ion liquid as dynamic coating and background electrolyte in capillary electrophoresis *Electrophoresis* Vol. 29 (No. 11):2356-2362.

Xiaohua, X., Liang, Z., Xia, L., & Shengxiang, J. (2004). Ionic liquids as additives in high performance liquid chromatography: Analysis of amines and the interaction mechanism of ionic liquids, *Anal. Chim. Acta* Vol. 519 (No. 2): 207–211.

Xu, Y., Li, J. & Wang, F. (2008). Eukaryotic expression and antimicrobial spectrum determination of the peptide tachyplesin II, *J. Chromatogr. A* Vol. 1207 (No. 2):175-183.

Yanes E.G., Gratz S.R., Baldwin M.J., Robison S.E.& Stalcup A.M. (2001). Capillary Electrophoretic Application of 1-Alkyl-3-methylimidazolium-Based Ionic Liquids, *Anal. Chem.* Vol.73 (No. 16): 3838-3844.

Zhang, W., He, L., Gu, Y., Liu, X. & Jiang, S.(2003). Archaeal lipid biomarkers and isotopic evidence of anaerobic methane oxidation associated with gas hydrates in the Gulf of Mexico, *Anal. Lett.* Vol. 36 (No. 6): 827-836.

Zhang, W., He, L., Liu, X. & Jiang, S. (2004). Generalized nonlinear models and variance function estimation, *Chin. J. Chem.* Vol. 22 (No. 3): 549-570.

Zhao, H., (2006). Are ionic liquids kosmotropic or chaotropic? An evaluation of available thermodynamic parameters for quantifying the ion kosmotropicity of ionic liquids, *J.Chem.Technol.Biotechnol.* Vol. 81 (No.6): 877-891.

13

Selection of Ionic Liquid Solvents for Chemical Separations Based on the Abraham Model

William E. Acree, Jr.[1], Laura M. Grubbs[1] and Michael H. Abraham[2]
[1]University of North Texas,
[2]University College London
[1]United States
[2]United Kingdom

1. Introduction

Room temperature ionic liquids (RTILs) have generated considerable interest in the past decade because of their unique physical and chemical properties. Each year the number of published applications employing RTILs as solvent media has increased. New generation RTILs are a popular solvent choice for manufacturing applications involving nano-materials and new pharmaceutical drug molecules, as high-temperature lubricants for metal-to-metal contacts, as reservoirs for the controlled release of drug molecules in pharmaceutical formulations, as chromatographic stationary phases for gas chromatographic separations, as gas absorption agents, and as an extraction solvent system for the removal of aromatic nitrogen and sulfur compounds from coal and petroleum feedstocks. RTILs are usually made by combining a poorly coordinating cation and anion to give a highly polar ionic liquid. Ionic liquids are often immiscible with supercritical carbon dioxide, saturated linear hydrocarbons and several acyclic organic solvents. Liquid immiscibility makes RTILs ideally suited for synthetic methods involving biphasic catalysis.

Currently synthetic procedures are available for preparing more than 300 different ionic liquids (ILs), including dication and tricationic *bis*/*tris*-imidazolium-based ILs, polymeric ionic liquids (PILs), and chirial ionic liquids. Methods have also been developed for introducing polar functional groups to the end of an alkyl CH_2-chain. The overall physical and solubilizing properties of ILs result from the composite properties of the cation and anion. The anion generally controls the extent to which the RTIL is miscible with water. The cation of an IL is usually a bulky organic structure (alkylimidazolium, alkylpyridinium, alkylpyrrolidinium, tetraalkylphosphonium and tetraalkylammonium - See Figure 1 for molecular structures) with low molecular symmetry. Cation type and size/symmetry affect the IL's melting point temperature. The melting point temperature is important because it represents the lower limit of liquidity, and when combined with thermal stability, it defines the temperature interval over which the IL can be used as a liquid solvent. Experimental studies have found that ILs having dications have a greater temperature interval of thermal stability than their monocation counterparts.

The solubilizing properties of an IL can be modified by changing the cation-anion combination. Our understanding of the properties of ILs has improved considerably in recent years, to the point where we can now begin to suggest possible IL solvents for achieving a desired chemical separation. One specific application involves the addition of an IL (as an entrainer) to an azeotropic system whose components are not separable by ordinary distillation methods. The added ionic liquid entrainer interacts with the components of the azeotropic system, and alters the component's relative volatilities. Verma and Banerjee (2010) examined the various IL combinations involving 10 cations (imidazolium, pyridinium and quinolium) and 24 anions as entrainers for water + ethanol, water + 2-propanol and water + tetrahydrofuran systems with the 1,3-dimethylimidazolium cation, $[M_2Im]^+$, in combination with the acetate, $[Ac]^-$, chloride $[Cl]^-$, and bromide, $[Br]^-$, anions giving the highest relative volatility. Chemicals in azeotropic mixtures have been separated also by liquid-liquid extraction using an IL solvent as schematically illustrated in Figure 2. Pereiro *et al.* (2010) explored 1-ethyl-3-methylimidazolium ethyl sulfate IL as an extraction solvent for the removal of ethanol from ethanol + heptane and ethanol + hexane azeotropic mixtures. Ionic liquid entrainers have been used successfully in the separation of alkanes from alkenes (propane versus propene (Mokrushina *et al.*, 2010), hexane versus 1-hexene (Lei *et. al.*, 2006), alkanes from aromatic compounds (cyclohexane versus benzene (Zhu *et al.* 2004), and of acetylene from ethylene (Palgunadi *et al.*, 2010). Recent review articles (Soukup-Hein *et al.*, 2009; Lei *et al.* 2003; Pandey, 2006; Poole and Poole, 2010) have discussed the advances that have been made in chemical separations using IL solvents.

Gas-liquid chromatography (glc) and high-performance liquid chromatography (hplc) afford convenient methods for separating organic compounds in mixtures prior to quantification of mixture composition. The IL can serve as the stationary phase, or in the case of hplc can be an organic modifier added to the mobile phase to affect the solute's partitioning characteristics. The solute partitions between the stationary and mobile phases as it passes through the chromatographic column. The elution time is governed by the partition coefficient, which is defined as

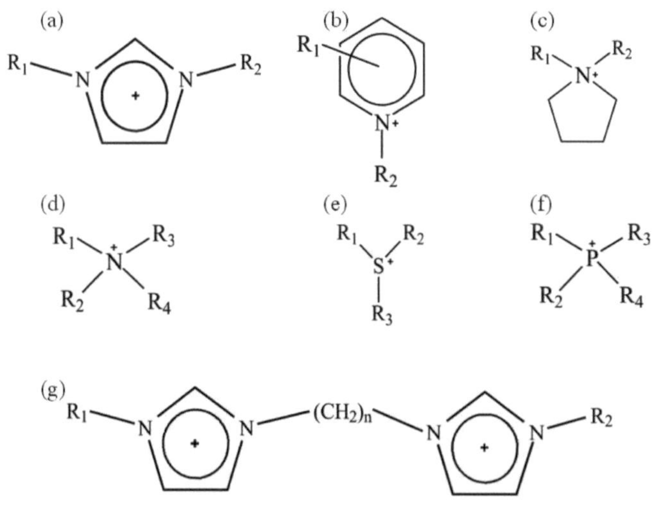

Fig. 1. Molecular Structures for 1,3-disubstituted imidazolium (a), substituted pyridinium (b), substituted pyrrolidinium (c), tetraalkylammonium (d), trialkylsulfonium (e), tetraalkylphosphonium (f), and *bis*(1,3-disubstituted imidazolium) cations (g). The substitutes are alkyl or functionalized alkyl chains and are denoted as R_1, R_2, R_3 and R_4.

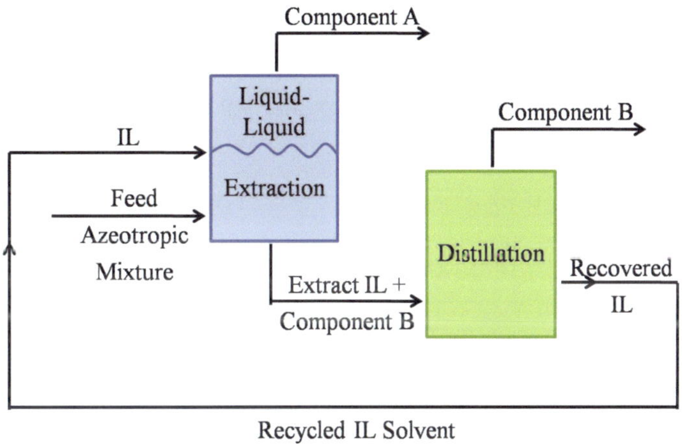

Fig. 2. Process for the separation of Component A and Component B from an azeotropic mixture using an IL solvent

$$P = \frac{C_{solute,mobile\ phase}}{C_{solute,stationary\ phase}} \tag{1}$$

the ratio of the molar concentration of the solute in the respective mobile and stationary phases. The partition coefficient is a relative measure of the affinity that the solute exhibits for each phase. Each solute is expected to interact differently with the two phases.

Numerous publications have reported using IL solvents in conjunction with chromatographic analyses (Soukup-hein *et al.*, 2009; Pandey, 2006; Poole and Poole, 2010). For example, Seely *et al.* (2008) studied chemical separations of complex mixtures on a two-dimensional gas chromatographic column. The system consisted of a trihexyl(tetradecyl)phoshonium *bis*(trifluoromethane)-sulfonamide IL primary column coupled to a (5 % diphenyl + 95 % dimethyl)polysiloxane secondary column. The authors found that the two-dimensional system separated the major components of diesel fuel into three narrow bands: saturated hydrocarbons, monoaromatics and diaromatics. Yao *et al.* (2009) employed various ILs containing the *tris*(pentafluoroethyl)trifluorophosphate anion. $[FAP]^-$, as extraction solvents for direct immersion single drop micro-extraction studies prior to hplc analysis. Ionic liquids tend to form larger and more stable microdroplets than do the more traditional organic solvents, and their elution time in hplc is very short and does not affect the chemical separation. The largest enrichment factors for compounds with high molar masses and fused rings were obtained with trihexyl(tetradecyl)phoshonium b*is*(pentafluoroethyl)trifluorophosphate. 1-Methyl-3-hexyl-imidazolium [FAP] gave the better results for the smaller molecules studied.

Headspace micro-extraction methods (Zhao *et al.* 2008 and 2009; Aguilera-Herrador *et al.* 2008; Liu *et al.*, 2003) have been developed for the partitioning of volatile and semivolatile compounds into an exposed IL drop. In the micro-extraction setup depicted in Figure 3 a syringe needle is inserted through the sample vial septum. The syringe needle is slowly depressed to expose the IL microdrop to the vapor above the aqueous sample. The volatile solutes then partition into the ionic liquid solvent from the vapor phase. After a predetermined equilibration time, the microdrop is retracted back into the syringe needle.

The syringe is removed and inserted into the injection port of a gas chromatograph. The syringe plunger is depressed in order to expose, but not inject, the IL microdrop. The volatile solutes thermally desorb from the IL drop at the higher temperature. Solute concentrations are quantified by gas chromatographic analysis.

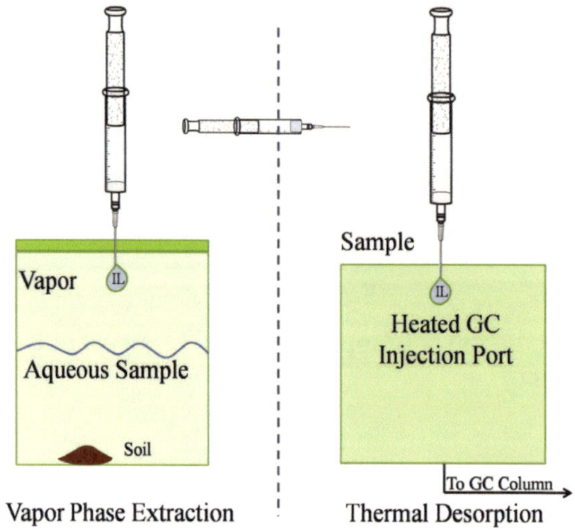

Fig. 3. Simple headspace ionic liquid-based microextraction setup for extracting volatile organic compounds from a contaminated soil sample. Organic contaminates on the soil dissolve into the aqueous phase. The organic vapors above the aqueous solution partition into the ionic liquid droplet.

The fore-mentioned experimental studies document the application of IL solvents to separation problems encountered by chemicals and engineers. Our knowledge of the solubilizing properties of ionic liquids has now matured to the point where researchers can make educated guesses concerning which IL solvent is most likely to give the desired separation. This chapter will discuss predictions based upon the Abraham general solvation model (Abraham, 1993a,b; Abraham *et al.*, 2004).

2. Thermodynamic properties and chemical separations

The thermodynamic basis for chemical separations using IL solvents is governed by solute partitioning between two immiscible or partly miscible phases. In the case of gas-liuqid chromatography the measured adjusted retention time, t_r', is related to the solute's infinite dilution activity coefficient, γ_{solute}^{∞}, (Mutelet *et al.*, 2010)

$$\ln \gamma_{1,2}^{\infty} = \ln\left(\frac{n_{solvent}.R.T}{V_N.P_1^0}\right) - P_1^0.\frac{B_{11}-V_1^0}{RT} + \frac{2.B_{13}-V_1^{\infty}}{RT}.J.P_0 \qquad (2)$$

$$V_N = \frac{3}{2}x\frac{[(\frac{P_{inlet}}{P_{outlet}})^2-1]}{[(\frac{P_{inlet}}{P_{outlet}})^3-1]}U_o t_r' x\frac{T_{col}}{T_{room}}(1-\frac{P_{water}}{P_{outlet}}) \qquad (3)$$

and gas-to-liquid partition coefficient, K,

$$\log K = \log \left(\frac{RT}{\gamma_{solute}^{\infty} P_{solute}^{o} V_{solvent}}\right) \tag{4}$$

through standard thermodynamic relationships. In Eqns. 2 - 4, T_{col} is the column temperature, U_o is the flow rate of the carrier gas (mobile phase) measured at ambient room temperature (T_{room}) with a soap-bubble flowmeter, and P_{water} is the vapor pressure of water at T_{room}. P_{inlet} and P_{outlet} denote inlet and outlet pressures, respectively. Rigorous calculations do account for vapor phase nonideality, B_{11} is the second virial coefficient of the solute (component 1) in the gaseous state at temperature T, B_{13} is the mutual coefficient between the solute and mobile phase carrier gas, and P_1^o is the solute's vapor pressure at temperature T, R denotes the universal gas constant, $V_{solvent}$ refers to the molar volume of the solvent at temperature T, and $n_{solvent}$ is the number of moles of solvent inside the column.

Chemical separation is achieved whenever two eluting solutes have sufficiently different retention times. Assuming an isothermal chromatographic separation, and that the column conditions remain constant during the course of the chemical separation, one can algebraically manipulate equations 2-4 to obtain the following expression

$$\frac{t_{r,solute\,B}{}'}{t_{r,solute\,A}{}'} = \left(\frac{\gamma_A^{\infty}}{\gamma_B^{\infty}}\right) x \left(\frac{P_A^{o}}{P_B^{o}}\right) \tag{5}$$

relating the ratio of adjusted retention times and the infinite dilution activity coefficients. The ratio of vapor pressures represents the separation (separation factor is $\alpha = \gamma_A^{\infty} P_A^{o} / \gamma_B^{\infty} P_B^{o}$) that would be expected from the vapor pressure differences for the two solutes. The infinite dilution activity coefficient measures the enhanced separation that one could get from solute interactions with the liquid phase solvent. Solution models that accurately predict infinite dilution activity coefficients facilitate the design of manufacturing processes and analytical methods employing azeotropic distillations, gas stripping and gas chromatographic separations.

Chemical separations can also be achieved through solute partitioning between two condensed phases. Practical examples include liquid-liquid extraction and hplc. These partitioning processes are described by a partition coefficient as well

$$P = \frac{C_{solute,\, phase\, \alpha}}{C_{solute,\, phase\, \beta}} \tag{6}$$

For notation purposes gas-to-liquid partition coefficients are denoted as K in the chapter, while condensed phase-to-condensed phase partition coefficients are referred to as P. The notational distinction is needed because the Abraham solvation parameter model uses different linear free energy relationships (LFER) to predict log K and log P values.

From thermodynamic considerations it can be shown that the solute's partition coefficient is

$$P = \frac{\gamma_{solute,phase\,\alpha}^{\infty} V_{phase\,\alpha}}{\gamma_{solute,phase\,\beta}^{\infty} V_{phase\,\beta}} \tag{7}$$

a ratio of the product of the solute's infinite dilution activity coefficient in each phase ($\gamma_{solute, phase\ \alpha}^{\infty}$ or $\gamma_{solute, phase\ \beta}^{\infty}$) times the molar volumes of each respective phase corrected for solvent α - solvent β mutual saturation. The solute activity coefficients in Eqn. 7 pertain to the equilibrated phases in physical contact with one another, and thus are not necessarily the values that would be obtained for the solute dissolved in each pure solvent.

Practical partition coefficients should not be confused with indirect, hypothetical partitioning processes. The latter are also thermodynamic transfer processes. The Gibbs energy of transfer is

$$\Delta G_{trans} = - RT \ln \text{Partition coefficient} \quad (8)$$

It is possible to calculate ΔG_{trans} values for solute transfer even when the two condensed phases are not in direct physical contact with each other. A hypothetical transfer process can be set up

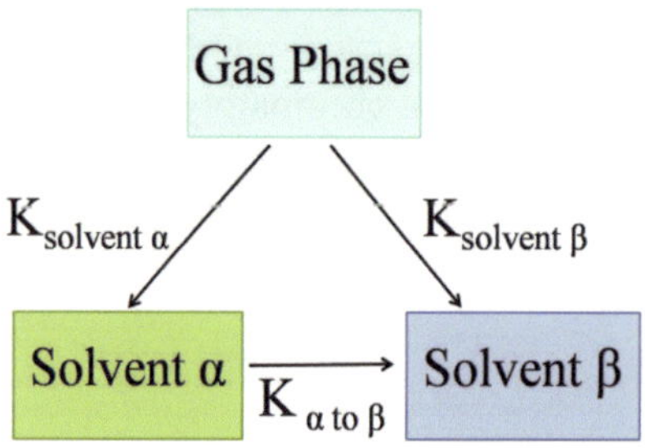

where the "hypothetical" solvent α-to-solvent β partition coefficient can be calculated from

$$\log P_{\alpha\text{-}\beta} = \log K_{solvent\ \beta} - \log K_{solvent\ \alpha} \quad (9)$$

the gas-to-solvent α and gas-to-solvent β partition coefficients. Abraham model correlations have been developed for predicting "hypothetical" water-to-anhydrous IL solvents. Even though hypothetical these predicted log P values for water-to-anhydrous ILs are useful. The predicted values can be converted back to γ_{solute}^{∞} values for solutes dissolved in the anhydrous IL using Eqns. 4 and 10.

$$\log P = \log K_{IL} - \log K_W \quad (10)$$

In Eqn. 10 K_W is the solute's gas-to-water partition coefficient.

Most of the Abraham model correlations that have been developed for predicting partition coefficients of solutes into IL solvents pertain to 298 K (Acree and Abraham, 2006; Abraham and Acree, 2006; Proctor *et al.*, 2008; Sprunger *et al.*, 2007b; Sprunger *et al.*, 2008; Sprunger *et al.*, 2009a,b,c; Sprunger *et al.*, 2010; Abraham *et al.*, 2009). Not all chemical separations take place at 298 K or even under isothermal conditions. There is a need to convert estimated partition coefficients for 298 K to other temperatures. From a thermodynamic standpoint, the gas-to-condensed phase partition coefficient, K, and water-to-organic solvent partition coefficient, P, can be estimated at other temperatures from measured partition coefficient at 298.15 K and the solute's enthalpy of solvation, ΔH_{solv}, or the enthalpy of transfer, ΔH_{trans}, between the two condensed phases from Eqn. 11 or Eqn. 12.

$$\log K(at\,T) - \log K(at\,298.15K) = \frac{-\Delta H_{Solv}}{2.303R}(1/T - 1/298.15) \quad (11)$$

and

$$\log P(at\,T) - \log P(at\,298.15K) = \frac{-\Delta H_{trans}}{2.303R}(1/T - 1/298.15) \quad (12)$$

The enthalpy of transfer needed in Eqn. 12 is defined as

$$\Delta H_{trans} = \Delta H_{Solv,Org} - \Delta H_{Solv,W} \quad (13)$$

the difference in the enthalpy of solvation of the solute in the specified organic solvent minus its enthalpy of solvation in water. The above equations assume zero heat capacity changes. Abraham model correlations have been developed for estimating ΔH_{solv} for organic vapors and gases in IL solvents using ion-specific equation coefficients.

3. The Abraham model: IL-specific correlations

The Abraham general solvation model is one of the more useful approaches for the analysis and prediction of Gibbs energies of solute transfer in chemical and biochemical systems. Published applications include the partitioning of solutes into organic and IL solvents both from the gas phase and from water (Acree and Abraham, 2006; Abraham and Acree, 2006; Proctor *et al.*, 2008; Sprunger *et al.*, 2007b; Sprunger *et al.*, 2008; Sprunger *et al.*, 2009a,b,c; Sprunger *et al.*, 2010; Abraham *et al.*, 2009), partitioning of volatile organic compounds and drug molecules between human/rat blood and select body organs/tissues (Abraham *et al.*, 2006; Abraham *et al.*, 2007; Abraham *et al.*, 2008), partitioning of solutes into humic acid (Mintz *et al.*, 2008a), sorption of gases and organic solutes onto polydimethylsiloxane solid-phase microextraction surfaces, (Sprunger *et al.*, 2007c) and the distribution of solutes between water and sodium dodecyl sulfate (SDS) micelles (Sprunger *et al.*, 2007a).
The method relies on two linear free energy relationships (lfers), one for transfer processes occurring within condensed phases (Abraham, 1993a,b; Abraham *et al.*, 2004):

$$SP = c + e \cdot \mathbf{E} + s \cdot \mathbf{S} + a \cdot \mathbf{A} + b \cdot \mathbf{B} + v \cdot \mathbf{V} \quad (14)$$

and one for processes involving gas-to-condensed phase transfer

$$SP = c + e \cdot \mathbf{E} + s \cdot \mathbf{S} + a \cdot \mathbf{A} + b \cdot \mathbf{B} + l \cdot \mathbf{L} \quad (15)$$

The dependent variable, SP, is some property of a series of solutes in a fixed phase, which in the present study will be the logarithm of solute partition coefficient between two immiscible (or partly miscible) phases or the enthalpy of solvation. The independent variables, or descriptors, are solute properties as follows: **E** and **S** refer to the excess molar refraction and dipolarity/-polarizability descriptors of the solute, respectively, **A** and **B** are measures of the solute hydrogen-bond acidity and basicity, **V** is the McGowan volume of the solute and **L** is the logarithm of the solute gas phase dimensionless Ostwald partition coefficient into hexadecane at 298 K. The first four descriptors can be regarded as measures of the tendency of the given solute to undergo various solute-solvent interactions. The latter two descriptors, **V** and **L**, are both measures of solute size, and so will be measures of the solvent cavity term that will accommodate the dissolved solute. General dispersion interactions are also related to solute size, hence, both **V** and **L** will also describe the general solute-solvent interactions. Solute descriptors are available for more than 4,000 organic, organometallic and inorganic solutes. No single article lists all of the numerical values;

however, a large compilation is available in one published review article (Abraham *et al.*, 1993a), and in the supporting material that has accompanied several of our published papers (Abraham *et al.*, 2006; Abraham *et al.*, 2009; Mintz *et al.*, 2007). Solute descriptors can be obtained by regression analysis using various types of experimental data, including water-to-solvent partitions, gas-to-solvent partitions, solubility data and chromatographic retention data as discussed elsewhere (Abraham *et al.*, 2010; Zissimos *et al.*, 2002a,b). There are also commercial software packages (Pharma Algorithms, 2006) and several published estimation schemes (Mutelet and Rogalski, 2001; Arey *et al.*, 2005; Platts *et al.*, 1999; Abraham and McGowan, 1987) for calculating the numerical values of solute descriptors from molecular structural information if one is unable to find the necessary partition, chromatographic and/or solubility data. For any fully characterized system/process (those with calculated values for the equation coefficients) further values of SP can be estimated for solutes with known values for the solute descriptors.

The usefulness of Eqns. 14 and 15 in the characterization of solvent phases is that he coefficients *e, s, a, b, l* and *v* are not just curve-fitting constants. The coefficients reflect particular solute-solvent interactions that correspond to chemical properties of the solvent phase. The excess molar refraction, **E**, is defined from the solute refractive index, and hence the *e* coefficient gives a measure of general solute-solvent dispersion interactions. The **V** and **L** descriptors were set up as measures of the endoergic effect of disrupting solvent-solvent bonds. However, solute volume is always well correlated with polarizability and so the *v* and *l* coefficients will include not only an endoergic cavity effect but also exoergic solute-solvent effects that arise through solute polarizability. The **S** descriptor is a measure of dipolarity and polarizability and hence the *s* coefficient will reflect the ability of a solvent to undergo dipole- and dipole-induced dipole interactions with the solute. The **A** descriptor is a measure of solute hydrogen bond acidity, and hence the *a* coefficient will reflect the complementary solvent hydrogen bond basicity. Similarly the *b* coefficient will be a measure of solvent hydrogen bond acidity. All this is straightforward for gas-to-solvent partitions because there are no interactions to consider in the gas phase. For partition between solvents, the coefficients in Eqn. 14 then refer to differences between the properties of the two phases.

Listed in Tables 1 and 2 are the Abraham model equation coefficients that have been reported (Grubbs *et al.*, 2010; Revelli *et al.*, 2009; Mutelet *et al.*, 2010; Revelli *et al.*, 2010) or calculated for the chapter for describing solute transfer from the gas phase (log K) and from water (log P) into 1-methyl-3-ethylimidazolium *bis*(trifluoromethyl-sulfonyl)imide, ($[MEIm]^+[(Tf)_2N]^-$), 1-methyl-3-butylimidazolium *bis*(trifluoromethylsulfonyl)-imide, ($[MBIm]^+[(Tf)_2N]^-$), 1-methyl-3-hexyl-imidazolium *bis*(trifluoromethylsulfonyl)imide, ($[MHIm]^+[(Tf)_2N]^-$), trimethylbutylammonium *bis*(trifluoromethyl-sulfonyl)imide, ($[M_3BAm]^+[(Tf)_2N]^-$), hexyltrimethylammonium *bis*(trifluoromethylsulfonyl)imide, ($[HexM_3Am]^+[(Tf)_2N]^-$), 1,3-dimethoxyimidazolium *bis*((trifluoromethyl)sulfonyl)imide ($[(Meo)_2Im]^+[(Tf)_2N]^-$), 1-ethanol-3-methylimidazolium *bis*(trifluoromethyl)sulfonylimide, ($[EtOHMIm]^+[(Tf)_2N]^-$), trihexyltetradecylphosphonium *bis*(trifluoromethyl-sulfonyl)imide, ($[H_3TdP]^+[(Tf)_2N]^-$), 1-methylethylether-3-methyl-imidazolium *bis*((trifluoromethyl)sulfonyl)imide, ($[MeoeMIm]^+[(Tf)_2N]^-$), 1-methyl-3-butylimidazolium tetrafluoroborate, ($[MBIm]^+[BF_4]^-$), 1-methyl-3-octylimidazolium tetrafluoroborate, ($[MOIm]^+[BF_4]^-$), 1-methyl-3-butyl-imidazolium hexafluorophosphate, ($[MBIm]^+[PF_6]^-$), 1-methyl-3-ethylimidazolium ethylsulfate, ($[MEIm]^+[EtSO_4]^-$), 1-methyl-3-butylimidazolium octylsulfate, ($[MBIm]^+[OtSO_4]^-$), 1-methyl-3-butylimidazolium trifluoromethanesulfonate,

Solvent	c	e	s	a	b	l	N[a]	SD	R^2
($[MBIm]^+[BF_4]^-$)	-0.600	0.356	2.534	3.312	0.284	0.604	66	0.099	0.997
	(0.026)	(0.075)	(0.068)	(0.100)	(0.074)	(0.011)			
($[EtOHMIm]^+[(Tf)_2N]^-$)	-0.793	0.139	2.404	2.587	1.353	0.581	81	0.100	0.993
	(0.047)	(0.061)	(0.065)	(0.074)	(0.077)	(0.011)			
($[MBIm]^+[PF_6]^-$)	-0.460	-0.191	2.747	2.228	0.363	0.663	91	0.154	0.994
	(0.033)	(0.081)	(0.086)	(0.097)	(0.101)	(0.015)			
($[MBIm]^+[(Tf)_2N]^-$)	-0.394	0.089	1.969	2.283	0.873	0.696	104	0.111	0.994
	(0.029)	(0.068)	(0.085)	(0.104)	(0.113)	(0.009)			
($[MHIm]^+[(Tf)_2N]^-$)	-0.348	-0.240	2.060	2.184	0.561	0.754	77	0.117	0.993
	(0.032)	(0.102)	(0.102)	(0.136)	(0.113)	(0.010)			
($[MOIm]^+[BF_4]^-$)	-0.409	-0.049	1.562	2.911	0.803	0.778	61	0.140	0.987
	(0.050)	(0.115)	(0.135)	(0.125)	(0.155)	(0.013)			
($[MEIm]^+[(Tf)_2N]^-$)	-0.486	0.068	2.296	2.278	0.988	0.651	65	0.094	0.996
	(0.041)	(0.058)	(0.052)	(0.052)	(0.126)	(0.067)			
($[M_3BAm]^+[(Tf)_2N]^-$)	-0.457	0.000	2.188	2.375	0.663	0.668	58	0.120	0.990
	(0.048)		(0.150)	(0.198)	(0.197)	(0.013)			
($[MBIm]^+[Trif]^-$)	-0.666	0.179	2.264	3.856	0.567	0.698	51	0.105	0.989
	(0.057)	(0.124)	(0.130)	(0.140)	(0.118)	(0.018)			
($[MeomMIm]^+[(Tf)_2N]^-$)	-0.508	0.000	2.635	2.379	0.418	0.602	52	0.108	0.981
	(0.076)		(0.087)	(0.129)	(0.093)	(0.018)			
($[Meo)_2Im]^+[(Tf)_2N]^-$)	-0.762	-0.013	2.557	2.427	1.157	0.584	48	0.084	0.989
	(0.063)	(0.046)	(0.072)	(0.106)	(0.077)	(0.015)			
($[CNPrMIm]^+[C(CN)_2]^-$)	-1.489	-0.418	3.089	4.807	0.626	0.644	45	0.121	0.987
	(0.098)	(0.117)	(0.115)	(0.163)	(0.135)	(0.025)			
($[MEIm]^+[C(CN)_2]^-$)	-0.888	0.361	2.833	4.690	0.364	0.596	74	0.109	0.989
	(0.055)	(0.082)	(0.075)	(0.104)	(0.074)	(0.014)			
($[HexM_3Im]^+[(Tf)_2]^-$)	-0.462	0.000	2.073	2.022	0.637	0.684	50	0.123	0.968
	(0.090)		(0.081)	(0.150)	(0.103)	(0.021)			
($[H_3TdP]^+[(Tf)_2]^-$)	-0.406	-0.576	1.602	2.358	-0.009	0.959	59	0.112	0.982
	(0.058)	(0.123)	(0.125)	(0.156)	(0.147)	(0.018)			
($[MBIm]^+[OtSO_4]^-$)	-0.228	-0.287	1.940	4.862	-0.302	0.880	56	0.116	0.984
	(0.081)	(0.093)	(0.103)	(0.168)	(0.140)	(0.023)			
($[MEIm]^+[EtSO_4]^-$)	-0.677	0.000	2.557	5.327	0.000	0.588	53	0.125	0.986
	(0.062)		(0.072)	(0.072)		(0.015)			

[a] Statistical information: N is the number of data points, SD is the standard deviation, and R^2 is the squared correlation coefficient.

Table 1. Abraham model correlation equation coefficients for the transfer of solutes from gas to the anhydrous (dry) ionic liquid solvents (log K correlation).

Solvent	c	e	s	a	b	v	N[a]	SD	R^2
($[MBIm]^+[BF_4]^-$)	-0.082	0.454	0.541	-0.427	-4.583	2.961	66	0.132	0.992
	(0.049)	(0.092)	(0.095)	(0.134)	(0.099)	(0.057)			
($[EtOHMIm]^+[(Tf)_2N]^-$)	-0.402	0.304	0.470	-1.082	-3.510	2.977	79	0.133	0.990
	(0.078)	(0.078)	(0.091)	(0.099)	(0.103)	(0.063)			
($[MBIm]^+[PF_6]^-$)	-0.056	0.193	0.737	-1.351	-4.526	3.109	86	0.154	0.988
	(0.046)	(0.080)	(0.087)	(0.100)	(0.102)	(0.059)			
($[MBIm]^+[(Tf)_2N]^-$)	-0.018	0.416	0.153	-1.312	-4.187	3.347	101	0.131	0.994
	(0.044)	(0.084)	(0.105)	(0.131)	(0.139)	(0.039)			
($[MHIm]^+[(Tf)_2N]^-$)	-0.065	0.010	0.260	-1.476	-4.313	3.587	75	0.115	0.996
	(0.042)	(0.105)	(0.103)	(0.135)	(0.111)	(0.039)			
($[MOIm]^+[BF_4]^-$)	-0.115	0.210	0.000	-0.511	-4.338	3.617	59	0.159	0.994
	(0.076)	(0.130)		(0.143)	(0.177)	(0.063)			
($[MEIm]^+[(Tf)_2N]^-$)	0.029	0.351	0.202	-1.684	-3.585	3.059	64	0.119	0.993
	(0.048)	(0.103)	(0.121)	(0.163)	(0.137)	(0.055)			
($[M_3BAm]^+[(Tf)_2N]^-$)	0.047	-0.051	0.356	-1.262	-4.400	3.209	57	0.120	0.996
	(0.060)	(0.133)	(0.154)	(0.202)	(0.198)	(0.049)			
($[MBIm]^+[Trif]^-$)	-0.217	0.195	0.497	0.062	-4.310	3.282	51	0.130	0.992
	(0.091)	(0.143)	(0.166)	(0.176)	(0.145)	(0.084)			
($[MeomMIm]^+[(Tf)_2N]^-$)	-0.140	-0.067	0.818	-1.185	-4.523	3.105	48	0.120	0.993
	(0.120)	(0.097)	(0.122)	(0.172)	(0.116)	(0.117)			
($[(Meo)_2Im]^+[(Tf)_2N]^-$)	-0.412	-0.104	0.761	-1.124	-3.776	3.055	46	0.130	0.993
	(0.133)	(0.123)	(0.124)	(0.171)	(0.118)	(0.106)			
($[CNPrMIm]^+[C(CN)_2]^-$)	-0.928	0.373	1.224	1.042	-4.307	3.046	44	0.150	0.988
	(0.156)	(0.146)	(0.154)	(0.207)	(0.166)	(0.125)			
($[MEIm]^+[C(CN)_2]^-$)	-0.404	0.344	0.945	0.987	-4.526	2.957	70	0.126	0.992
	(0.105)	(0.095)	(0.100)	(0.133)	(0.088)	(0.087)			
($[HexM_3Im]^+[(Tf)_2]^-$)	-0.275	0.000	0.407	-1.478	4.320	3.510	48	0.140	0.994
	(0.136)		(0.115)	(0.176)	(0.114)	(0.111)			
($[H_3TdP]^+[(Tf)_2]^-$)	-0.155	-0.163	-0.029	-1.271	-5.042	4.246	59	0.136	0.996
	(0.092)	(0.146)	(0.157)	(0.192)	(0.179)	(0.085)			
($[MBIm]^+[OtSO_4]^-$)	-0.050	0.198	0.179	1.146	-5.154	4.008	55	0.179	0.986
	(0.155)	(0.132)	(0.175)	(0.296)	(0.228)	(0.135)			
($[MEIm]^+[EtSO_4]^-$)	-0.079	-0.021	0.554	1.491	-4.944	2.895	53	0.147	0.992
	(0.090)	(0.158)	(0.183)	(0.228)	(0.211)	(0.068)			

[a] Statistical information: N is the number of data points, SD is the standard deviation, and R^2 is the squared correlation coefficient.

Table 2. Abraham model correlation equation coefficients for the transfer of solutes from water to the anhydrous (dry) ionic liquid solvents (log P correlation).

($[MBIm]^+[Trif]^-$), 1-methyl-3-ethyl-imidazolium dicyanamide, ($[MEIm]^+[C(CN)_2]^-$), and 1-(3-cyanopropyl)-3-methylimidazolium dicyanamide, ($[CNPrMIm]^+[C(CN)_2]^-$). The IL-specific equation coefficients were obtained by regressing infinite dilution activity coefficients and Henry's law constants into the respective IL in accordance with Eqns. 14 and 15. Solubility data for gaseous solutes are often reported in the literature as Henry's law constants, which are related to log K by

$$\log K = \log\left(\frac{RT}{K_{Henry}\ V_{solvent}}\right) \tag{16}$$

The statistics of each derived correlation are quite good as reflected by the low standard deviations (SD) and near unity values for the squared correlation coefficients (R^2). The standard error in each equation coefficient is denoted in parenthesis immediately below the coefficient. The remaining statistical information is as follows: N denotes the number of experimental data points for the given IL data set.

The log P correlations do pertain to solute transfer from water-to-anhydrous IL solvent. As noted above these represent a hypothetical partitioning processes, and calculated values based on the equation coefficients listed in Table 2 may be different than the experimental log P values determined by direct partitioning of the solute between water and the IL solvent. At the present time there has been few practical water-to-IL systems studied. Abraham *et al.* (2003) reported log P correlations for 1-butyl-3-methylimidazoium hexafluorophosphate, ($[BMIm]^+[PF_6]^-$)

$$\text{Log P} = -0.17 + 0.45\,\mathbf{E} + 0.23\,\mathbf{S} - 1.76\,\mathbf{A} - 1.83\,\mathbf{B} + 2.15\,\mathbf{V} \tag{17}$$

and for 1-hexyl-3-methylimidazolium hexafluorophosphate, ($[HMIm]^+[PF_6]^-$)

$$\text{Log P} = -0.13 + 0.05\,\mathbf{E} + 0.40\,\mathbf{S} - 1.48\,\mathbf{A} - 2.11\,\mathbf{B} + 2.30\,\mathbf{V} \tag{18}$$

Room temperature ionic liquids have been used as a solvent media for organic synthesis. Most (if not all) of the classic synthetic methods have been performed in IL solvents. Much higher product yields and shorter reaction times have been reported for ILs than for the more conventional organic solvents. Product removal from the IL reaction media can often be accomplished by carbon dioxide supercritical fluid extraction. Many imidazolium-based ionic liquids are nearly insoluble in supercritical carbon dioxide, even at very high pressures. Roth (2009) recently reviewed the partitioning of organic compounds between IL solvents and supercritical fluids, with particular emphasis on supercritical carbon dioxide.

There have been very few attempts to correlate/predict the partition coefficients of organic solutes in IL solvents - supercritical fluid carbon dioxide systems. Planeta and coworkers (Planeta and Roth 2005; Planeta *et al.*, 2007; Planeta *et al.*, 2009) correlated the relative partition coefficients of solutes in the ($[BMIm]^+[BF_4]^-$)/CO_2 , ($[BMIm]^+[MeSO4]^-$)/CO_2 and ($[ThtdP]^+[Cl]^-$)/CO_2 with the Abraham model

$$\log\frac{P_{solute,A}}{P_{naphthalene}} = eE_A + sS_A + aA_A + bB_A + vV_A \tag{19}$$

The authors used naphthalene as the reference solute, and the descriptors in Eqn. 19 pertain to solute A. The derived correlations did provide a satisfactory description of the experimental partition coefficient data, suggesting that the Abraham model could be used in the preliminary design stage of real-world extraction processes employing IL/CO_2 systems. The solubilizing properties of supercritical carbon dioxide are temperature and pressure-dependent. Correlations derived for one set of experimental conditions might not apply at other operating temperatures and processes.

4. Abraham model: Ion-specific equation coefficients

The physical and chemical properties of ILs can be modified by changing the cation-anion combination. There are currently synthetic procedures for making more than 300 different ILs and the number is likely to grow in number in future years as more applications are discovered. It would be impractical to develop an Abraham model correlation for each known IL. To address this concern, Sprunger and coworkers (Sprunger *et al.*, 2007b; Sprunger *et al.*, 2008; Spurnger *et al.*, 2009c; Grubbs *et al.*, 2010) modified the basic Abraham solvation parameter model for the gas-to-IL partition coefficient

$$\log K = c_{cation} + c_{anion} + (e_{cation} + e_{anion})\,\mathbf{E} + (s_{cation} + s_{anion})\,\mathbf{S} + (a_{cation} + a_{anion})\,\mathbf{A} + (b_{cation} + b_{anion})\,\mathbf{B} + (l_{cation} + l_{anion})\,\mathbf{L} \quad (20)$$

and water-to-IL partition coefficient

$$\log P = c_{cation} + c_{anion} + (e_{cation} + e_{anion})\,\mathbf{E} + (s_{cation} + s_{anion})\,\mathbf{S} + (a_{cation} + a_{anion})\,\mathbf{A} + (b_{cation} + b_{anion})\,\mathbf{B} + (v_{cation} + v_{anion})\,\mathbf{V} \quad (21)$$

by rewriting each of the six solvent equation coefficients as the summation of their respective cation and anion contribution. In separating the equation coefficients it was assumed that the soltue's interaction with a given cation is not influenced by the surrounding IL anion, and that the solute's interaction with a given anion is unaffected by the surrounding IL cation. Once calculated, the ion-specific equation coefficients can be combined to build the Abraham model correlation for the desired cation-anion pair.

During the three years since the modified version was first suggested we have periodically updated the numerical values of the coefficients as new experimental data became available. For this chapter we have reanalyzed our large log K and log P databases for solutes dissolved in ILs to yield the following correlations

$$\log K = \sum_{cation} (c_{cation} + e_{cation}E + s_{cation}S + a_{cation}A + b_{cation}B + l_{cation}L) + \sum_{anion} (c_{anion} + e_{anion}E + s_{anion}S + a_{anion}A + b_{anion}B + l_{anion}L) \quad (22)$$

$(N = 2084, R^2 = 0.998, R^2_{adj} = 0.998, SD = 0.112, F = 5810)$

and

$$\log P = \sum_{cation} (c_{cation} + e_{cation}E + s_{cation}S + a_{cation}A + b_{cation}B + v_{cation}V) + \sum_{anion} (c_{anion} + e_{anion}E + s_{anion}S + a_{anion}A + b_{anion}B + v_{anion}V) \quad (23)$$

$(N = 2054, R^2 = 0.997, R^2_{adj} = 0.996, SD = 0.139, F = 2720)$

The calculated cation-specific and anion-specific equation coefficients are listed in Tables 3 and 4, along with their respective standard errors. For the most part, the larger standard errors are noted in the equation coefficients for those ions for which experimental data is limited. The number of data points for the individual ions range from a minimum of 22 log K values for the $[B(CN)_4]^-$ anion to more than 400 log K values for the $[BMIm]^+$ and $[EMIm]^+$ cations and $[(Tf)_2N]^-$ anion, which is more than sufficient for the regression analysis. The 23 cation-specific equation coefficients and 13 anion-specific equation coefficients that have been reported in Tables 3 and 4 can be combined to give predictive log K and log P correlations for a total of 299 (23 times 13) IL solvents. Predicted log K and log P values can be converted to infinite dilution activity coefficients, γ_{solute}^{∞}, through Eqns.4 and 10. Sprunger *et al.* found that very little (if any) loss in descriptive/predictive ability resulted from splitting the equation coefficients into the individual ion contributions.

One note regarding the computation methodology that we used in calculating of the ion-specific equation coefficients. The cation-specific and anion-specific coefficients are paired in that each cation-specific coefficient goes together with its anion-specific counterpart to make up a summed value that the five solute descriptors are multiplied by. If one were to perform a regression analysis on Eqns. 20 - 23 the statistical software would generate numerical equation coefficients based on some reference point. The reference point would likely depend on the particular database used and the software's built-in convergence routine. Calculation of additional ion values at some later time would be difficult as there would be no guarantee that the next regression analyses would find the same reference point. In accordance with the computation methodology suggested by Sprunger *et al.* (2007b) the anion-specific equation coefficients of $[(Tf)_2N]^-$ were set equal to zero. Setting fixed reference points is not uncommon in thermodynamics. For example in calculating the chemical potentials of individual ions, the chemical potential of the hydrogen ion in water is defined to be zero and the values of all other ions are computed relative to this defined thermodynamic reference state.

Ion	c_{ion}	e_{ion}	s_{ion}	a_{ion}	b_{ion}	l_{ion}
CATIONS[a]						
$[MEIm]^+$	-0.502	0.050	2.421	2.266	0.875	0.650
(N=420)[b]	(0.026)	(0.073)	(0.088)	(0.112)	(0.100)	(0.009)
$[BMIm]^+$	-0.407	0.064	2.071	2.279	0.761	0.699
(N=402)	(0.023)	(0.053)	(0.062)	(0.077)	(0.078)	(0.007)
$[MHIm]^+$	-0.339	-0.135	2.034	2.365	0.505	0.757
(N=238)	(0.027)	(0.078)	(0.093)	(0.108)	(0.100)	(0.008)
$[MOIm]^+$	-0.229	-0.246	1.292	1.808	1.146	0.836
(N=114)	(0.032)	(0.090)	(0.114)	(0.108)	(0.127)	(0.010)
$[M_3BAm]^+$	-0.457	-0.005	2.188	2.375	0.663	0.668
(N=58)	(0.045)	(0.123)	(0.141)	(0.187)	(0.185)	(0.012)
$[M_2EIm]^+$	-0.611	0.188	2.380	2.101	0.899	0.667
(N=39)	(0.072)	(0.134)	(0.142)	(0.175)	(0.138)	(0.022)
$[4\text{-}BMPy]^+$	-0.479	0.141	2.293	2.364	0.672	0.713

(N=76)	(0.055)	(0.115)	(0.132)	(0.138)	(0.121)	(0.017)
[3-BMPy]+	-0.263	0.000	2.368	2.265	0.271	0.688
(N=36)	(0.115)		(0.260)	(0.253)	(0.246)	(0.034)
[NEP]+	-0.668	0.246	2.399	2.403	0.936	0.672
(N=31)	(0.076)	(0.192)	(0.180)	(0.209)	(0.150)	(0.028)
[PM_2Im]+	-0.822	0.780	2.357	3.432	0.926	0.526
(N=34)	(0.120)	(0.169)	(0.196)	(0.227)	(0.207)	(0.035)
[$HexM_3Am$]+	-0.459	-0.039	2.096	2.021	0.624	0.684
(N=50)	(0.083)	(0.102)	(0.095)	(0.138)	(0.100)	(0.020)
[HexdMIm]+	0.019	-0.452	0.821	1.810	0.523	0.997
(N=31)	(0.142)	(0.157)	(0.165)	(0.245)	(0.187)	(0.044)
[HxomMIm]+	-0.463	-0.394	2.478	2.428	0.337	0.786
(N=34)	(0.109)	(0.212)	(0.272)	(0.205)	(0.217)	(0.033)
[$(Hxom)_2Im$]+	-0.314	-0.479	2.076	2.376	0.287	0.835
(N=34)	(0.109)	(0.212)	(0.272)	(0.205)	(0.217)	(0.033)
[$(Meo)_2Im$] +	-0.762	-0.013	2.557	2.427	1.154	0.584
(N=48)	(0.083)	(0.104)	(0.096)	(0.140)	(0.101)	(0.020)
[EtOHMIm] +	-0.843	0.098	2.438	2.684	1.402	0.578
(N=151)	(0.039)	(0.059)	(0.062)	(0.073)	(0.074)	(0.010)
[H_3TdP]+	-0.406	-0.576	1.602	2.338	-0.009	0.959
(N=59)	(0.058)	(0.124)	(0.126)	(0.157)	(0.149)	(0.018)
[MeoeMIm]+	-0.507	-0.015	2.644	2.378	0.413	0.602
(N=52)	(0.078)	(0.020)	(0.094)	(0.135)	(0.100)	(0.019)
[BMPyrr]+	-0.570	-0.075	2.687	2.338	0.570	0.711
(N=31)	(0.076)	(0.300)	(0.391)	(0.273)	(0.282)	(0.025)
[CNPrMIm]+	-1.103	0.108	2.677	2.383	1.136	0.699
(N=45)	(0.110)	(0.155)	(0.159)	(0.216)	(0.177)	(0.029)
[PrOHPy]+	-0.895	-0.065	3.006	3.013	1.483	0.673
(N=38)	(0.117)	(0.202)	(0.250)	(0.240)	(0.247)	(0.034)
[E_3S]+	-0.606	-0.196	2.992	2.444	0.355	0.690
(N=31)	(0.112)	(0.278)	(0.368)	(0.241)	(0.254)	(0.033)
[MO_3Am]+	-0.387	-0.130	1.460	2.327	0.501	0.927
(N=32)	(0.064)	(0.222)	(0.353)	(0.336)	(0.463)	(0.020)
ANIONS[c]						
[$(Tf)_2N$]-	0.000	0.000	0.000	0.000	0.000	0.000
(N=860)						
[BF_4]-	-0.203	0.217	0.370	1.093	-0.408	-0.068
(N=307)	(0.027)	(0.066)	(0.073)	(0.091)	(0.086)	(0.009)
[PF_6]-	-0.033	-0.184	0.647	0.007	-0.362	-0.085

(N=169)	(0.029)	(0.068)	(0.075)	(0.090)	(0.092)	(0.010)
$[EtSO_4]^-$	-0.176	-0.035	0.123	3.048	-0.862	-0.063
(N=53)	(0.062)	(0.143)	(0.162)	(0.205)	(0.189)	(0.016)
$[Trif]^-$	-0.301	0.030	0.329	1.712	-0.321	0.011
(N=163)	(0.044)	(0.114)	(0.133)	(0.147)	(0.128)	(0.015)
$[F_3Ac]^-$	-0.305	-0.191	0.499	3.265	-0.336	0.019
(N=32)	(0.055)	(0.233)	(0.343)	(0.686)	(0.725)	(0.019)
$[OtSO_4]^-$	0.277	-0.232	-0.150	2.402	-1.015	0.120
(N=58)	(0.072)	(0.099)	(0.117)	(0.175)	(0.156)	(0.018)
$[SCN]^-$	-0.727	0.284	0.783	2.607	-0.288	-0.006
(N=91)	(0.069)	(0.153)	(0.202)	(0.153)	(0.167)	(0.020)
$[C(CN)_2]^-$	-0.387	0.310	0.412	2.424	-0.510	-0.055
(N=119)	(0.063)	(0.113)	(0.118)	(0.156)	(0.126)	(0.017)
$[E_2PO_4]^-$	0.089	0.145	-0.184	5.186	-0.966	0.064
(N=38)	(0.105)	(0.175)	(0.182)	(0.237)	(0.184)	(0.035)
$[FAP]^-$	0.179	-0.015	0.063	-1.314	0.238	-0.053
(N=144)	(0.050)	(0.099)	(0.117)	(0.129)	(0.136)	(0.015)
$[B(CN)_4]^-$	0.243	-0.347	0.445	0.034	-0.225	-0.005
(N=23)	(0.157)	(0.469)	(0.607)	(0.839)	(0.775)	(0.047)
$[NO_3]^-$	-0.211	0.624	0.577	2.578	-0.788	-0.191
(N=27)	(0.224)	(0.200)	(0.203)	(0.328)	(0.192)	(0.067)

[a] Cation abbreviations: $[4\text{-BMPy}]^+$ is 4-methyl-N-butylpyridinium cation; $[3\text{-BMPy}]^+$ is 3-methyl-N-butylpyridinium cation; $[MEIm]^+$ is 1-methyl-3-ethylimidazolium cation ; $[M_2EIm]^+$ is 1,2-dimethyl-3-ethylimidazolium cation; $[MBIm]^+$ is 1-methyl-3-butylimidazolium cation; $[MHIm]^+$ is1-methyl-3-hexylimidazolium cation; $[M_3BAm]^+$ is trimethylbutylammonium cation; $[MOIm]^+$ is 1-methyl-3-octylimidazolium cation; $[NEPy]^+$ is N-ethylpyridinium cation; $[PM_2Im]^+$ is 1-propyl-2,3-dimethylimidazolium cation; $[(Meo)_2Im]+$ is 1,3-dimethoxy-imidazolium cation; $[EtOHMIm]^+$ is 1-ethanol-3-methylimidazolium cation; $[H_3TdP]^+$ is trihexyltetradecylphosphonium cation; $[MeoeMIm]^+$ is 1-methylethylether-3-methyl-imidazolium cation; $[HxomMIm]^+$ is 1-hexyloxymethyl-3-methylimidazolium; $[(Hxom)_2Im]^+$ is 1,3-dihexyloxymethylimidazolium cation; $[PrOHPy]^+$ is 1-(3-hydroxypropyl)pyridinium; $[BMPyrr]^+$ is 1-butyl-1-methylpyrrolidinum cation; $[CMPrMIm]^+$ is 1-(3-cyanopropyl)-3-methylimidazolium cation; $[HexM_3Am]^+$ is hexyltrimethylammonium cation; $[HexdMIm]^+$ is 1-hexadecyl-3-methylimidazolium cation; $[E_3S]^+$ is triethylsulfonium cation; and $[MO_3Am]^+$ is methyl(trioctyl)ammonium cation.

[b] Number of experimental data points associated with the specified ion.

[c] Anion abbreviations: $[(Tf)_2N]^-$ is *bis*(trifluoromethylsulfonyl)imide anion; $[BF_4]^-$ is tetrafluoroborate anion; $[PF_6]^-$ is hexafluorophosphate anion; $[EtSO_4]^-$ is ethylsulfate anion; $[Trif]^-$ is the trifluoromethanesulfonate anion; $[F_3Ac]^-$ is the trifluoroacetate anion;$[OtSO_4]^-$ is octylsulfate anion; $[SCN]^-$ is thiocyanate anion; $[C(CN)_2]^-$ is dicyanamide anion; $[E_2PO_4]^-$ is diethylphosphate anion; $[B(CN)_4]^-$ is tetracyanoborate anion; $[NO_3]^-$ is nitrate anion; and $[FAP]^-$ is *tris*(pentafluoroethyl)trifluorophosphate anion.

Table 3. Cation-Specific and Anion-Specific Equation Coefficients for the Abraham Model Gas-to-RTIL Correlation (Eqn.22)

Ion[a]	c_{ion}	e_{ion}	s_{ion}	a_{ion}	b_{ion}	v_{ion}
CATIONS						
[MEIm]+	-0.037	0.152	0.554	-1.422	-4.011	3.150
(N=415)[b]	(-0.043)	(0.091)	(0.113)	(0.142)	(0.125)	(0.045)
[BMIm]+	-0.028	0.448	0.180	-1.399	-4.160	3.332
(N=392)	(0.036)	(0.065)	(0.079)	(0.099)	(0.098)	(0.034)
[MHIm]+	-0.040	0.252	0.165	-1.263	-4.375	3.586
(N=236)	(0.045)	(0.098)	(0.116)	(0.136)	(0.124)	(0.041)
[MOIm]+	-0.011	0.009	-0.150	-1.641	-3.98	3.872
(N=112)	(0.055)	(0.111)	(0.143)	(0.137)	(0.159)	(0.052)
$[M_3BAm]^+$	0.047	-0.051	0.356	-1.262	-4.400	3.209
(N=57)	(0.071)	(0.156)	(0.180)	(0.236)	(0.231)	(0.058)
$[M_2EIm]^+$	-0.095	0.292	0.443	-1.681	-4.024	3.174
(N=39)	(0.115)	(0.156)	(0.187)	(0.224)	(0.172)	(0.103)
[4-BMPy]+	-0.128	0.207	0.591	-1.349	-4.283	3.433
(N=76)	(0.087)	(0.133)	(0.169)	(0.176)	(0.151)	(0.079)
[3-BMPy]+	0.134	0.170	0.678	-1.204	-4.934	3.324
(N=36)	(0.151)	(0.309)	(0.317)	(0.325)	(0.258)	(0.173)
[NEP]+	-0.322	0.323	0.552	-1.234	-3.951	3.370
(N=31)	(0.130)	(0.222)	(0.231)	(0.273)	(0.189)	(0.140)
$[PM_2Im]^+$	-0.499	0.685	0.568	-0.238	-4.006	2.907
(N=34)	(0.174)	(0.192)	(0.256)	(0.284)	(0.258)	(0.151)
$[HexM_3Am]^+$	-0.278	0.013	0.401	-1.476	-4.315	3.512
(N=48)	(0.139)	(0.127)	(0.129)	(0.176)	(0.123)	(0.111)
[HexdMIm]+	0.167	-0.218	-0.448	-1.884	-4.597	4.461
(N=31)	(0.212)	(0.182)	(0.227)	(0.311)	(0.234)	(0.190)
[HxomMIm]+	-0.039	-0.645	1.184	-1.374	-4.779	3.609
(N=34)	(0.163)	(0.255)	(0.335)	(0.256)	(0.275)	(0.142)
$[(Hxom)_2Im]^+$	0.107	-0.628	0.747	-1.441	-4.808	3.750
(N=34)	(0.163)	(0.255)	(0.335)	(0.256)	(0.275)	(0.142)
$[(Meo)_2Im]^+$	-0.412	-0.104	0.761	-1.124	-3.776	3.055
(N=34)	(0.142)	(0.131)	(0.132)	(0.183)	(0.126)	(0.113)
[EtOHMIm]+	-0.404	0.229	0.517	-1.026	-3.493	2.931
(N=148)	(0.061)	(0.070)	(0.080)	(0.092)	(0.091)	(0.052)
$[H_3TdP]^+$	-0.155	-0.164	-0.055	-1.317	-4.985	4.250
(N=59)	(0.093)	(0.148)	(0.154)	(0.189)	(0.168)	(0.086)
[MeoeMIm]+	-0.161	0.014	0.658	-1.282	-4.262	3.125
(N=50)	(0.131)	(0.026)	(0.127)	(0.173)	(0.124)	(0.104)
[BMPyrr]+	-0.182	-0.760	2.188	-1.449	-5.039	3.423

(N=31)	(0.127)	(0.368)	(0.487)	(0.344)	(0.355)	(0.120)
$[CNPrMIm]^+$	-0.596	0.172	0.724	-1.497	-3.509	3.284
(N=44)	(0.179)	(0.194)	(0.210)	(0.277)	(0.220)	(0.148)
$[PrOHPy]^+$	-0.302	-0.299	1.476	-0.503	-3.760	3.125
(N=38)	(0.175)	(0.249)	(0.311)	(0.299)	(0.310)	(0.151)
$[E_3S]^+$	-0.062	-1.347	2.716	-1.550	-5.274	3.242
(N=31)	(0.170)	(0.348)	(0.458)	(0.302)	(0.325)	(0.148)
$[MO_3Am]^+$	-0.092	0.000	-0.146	-1.129	-4.609	4.114
(N=32)	(0.103)		(0.135)	(0.116)	(0.274)	(0.095)
ANIONS						
$[(Tf)_2N]^-$	0.000	0.000	0.000	0.000	0.000	0.000
(N=843)						
$[BF_4]^-$	-0.084	0.129	0.240	1.108	-0.401	-0.296
(N=305)	(0.045)	(0.080)	(0.094)	(0.116)	(0.108)	(0.043)
$[PF_6]^-$	0.000	-0.299	0.544	0.088	-0.321	-0.264
(N=164)		(0.083)	(0.096)	(0.115)	(0.114)	(0.048)
$[EtSO_4]^-$	-0.042	-0.173	0.000	2.913	-0.933	-0.256
(N=53)	(0.095)	(0.174)		(0.258)	(0.236)	(0.078)
$[Trif]^-$	-0.193	-0.339	0.506	1.520	-0.293	-0.047
(N=162)	(0.072)	(0.136)	(0.166)	(0.187)	(0.161)	(0.068)
$[F_3Ac]^-$	-0.288	-0.771	0.891	2.255	0.332	0.049
(N=32)	(0.094)	(0.270)	(0.426)	(0.854)	(0.904)	(0.096)
$[OtSO_4]^-$	0.013	-0.258	0.000	2.537	-1.001	0.678
(N=56)	(0.126)	(0.121)		(0.249)	(0.201)	(0.109)
$[SCN]^-$	-0.566	-0.757	1.878	2.390	-0.843	-0.120
(N=91)	(0.105)	(0.191)	(0.251)	(0.193)	(0.212)	(0.190)
$[C(CN)_2]^-$	-0.287	0.179	0.351	2.350	-0.525	-0.255
(N=116)	(0.105)	(0.139)	(0.156)	(0.200)	(0.158)	(0.094)
$[E_2PO_4]^-$	0.059	0.136	-0.120	5.217	-1.029	0.196
(N=38)	(0.164)	(0.194)	(0.242)	(0.302)	(0.231)	(0.156)
$[FAP]^-$	0.132	-0.171	0.121	-1.314	0.244	-0.107
(N=144)	(0.079)	(0.122)	(0.148)	(0.162)	(0.170)	(0.069)
$[B(CN)_4]^-$	0.346	-1.011	1.195	-0.105	-0.676	-0.070
(N=23)	(0.237)	(0.618)	(0.766)	(1.053)	(0.952)	(0.205)
$[NO_3]^-$	-0.037	0.322	0.376	2.539	-0.801	-0.676
(N=27)	(0.303)	(0.215)	(0.283)	(0.403)	(0.243)	(0.268)

[a] Cation and anion abbreviations are given in Table 3.
[b] Number of experimental data points associated with the specified ion.

Table 4. Cation-Specific and Anion-Specific Equation Coefficients for the Abraham Model Water-to-RTIL Correlation (Eqn.23)

Both LFERs are statistically very good, and describe experimental log K and log P databases that cover a 12.5 log unit and 8.7 log unit range to within standard deviations of 0.112 log units (Eqn. 22) and 0.139 log units (Eqn. 23) as shown in Figures 4 and 5. Based on the limited number of replicate independent activity coefficient measurements that have been

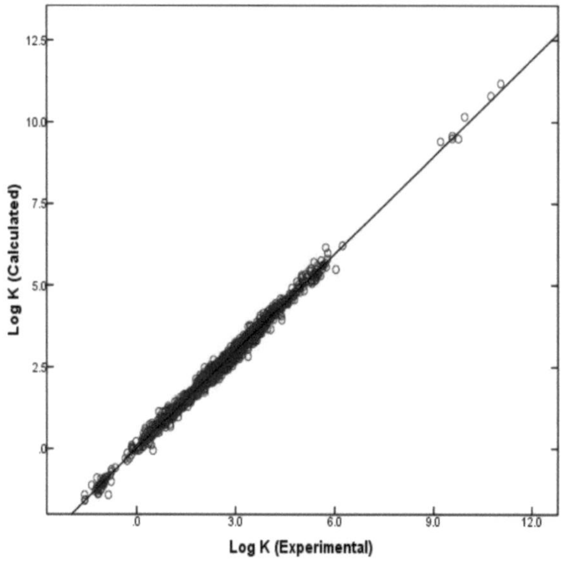

Fig. 4. Comparison between the 2084 experimental log K data points and predicted values based on Eqn. 22.

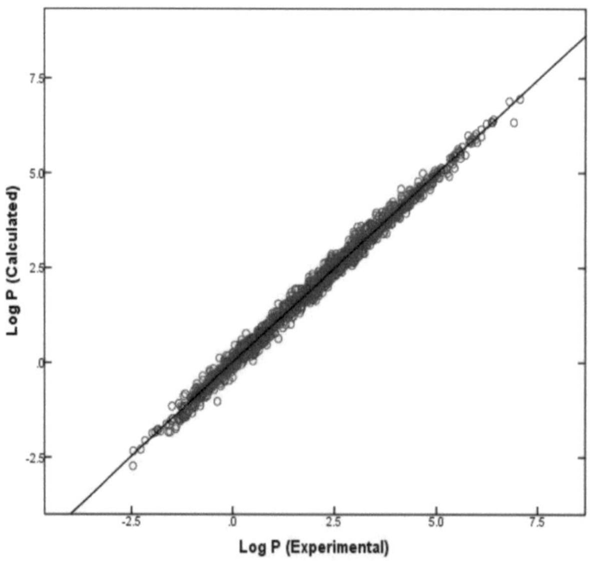

Fig. 5. Comparison between the 2054 experimental log P data points and predicted values based on Eqn. 23.

performed for solutes dissolved in ILs we believe that 0.05 to 0.15 log units would be a reasonable guesstimate of the uncertainty associated with the experimental log K values. Slightly larger standard deviations are expected for the log P correlation which also includes the experimental uncertainties in the log K_w data used to convert the log K values to log P. Careful examination of the individual residuals between the calculated and observed values revealed that Equation 22 described 70.1 % (1461 of 2084 values) of the gas-to-IL partition coefficient data to within 0.1 log units, 92.0 % (1918 of 2084 values) to within 0.2 log units, and 98.1 % (2045 of 2084 values) to within 0.3 log units of observed values. Only 1.9 % of the predicted log K values fell more than 0.3 log units from the experimental value, with the largest residual being -0.55 log units. Similar results were noted for Eqn. 23; 56.8 % of the back-calculated water-to-IL partition coefficients differed from the observed value by less than 0.1 log units, 85.8 % differed by less than 0.2 log units, and 98.7 % differed by less than 0.3 log units. Less than 1.3 % of the predicted log P values were more than 0.3 log units from the observed value. The largest residual for the log P calculations is -0.61 log units. These values should reflect the predictive ability that Eqns. 22 and 23 would exhibit in terms of predicting partition coefficients for new compounds dissolved in ILs containing the 23 cations and 13 anions given in Tables 3 and 4, provided that the solute descriptors of the compounds fall within the area of predictive chemical space defined by the calculated equation coefficients: **E** = 0.000 to **E** = 1.500; **S** = 0.000 to **S** = 1.720; **A** = 0.000 to **A** = 1.030; **B** = 0.000 to **B** = 1.280; **V** = 0.109 to **V** = 1.799; and **L** = -1.200 to **L** = 7.833. A few of the ion-specific data sets spanned a slightly smaller range of solute descriptors. Predicted activity coefficients can be converted to infinite dilution activity coefficients, γ_{solute}^{∞}, through Eqns.4 and 10.

The major advantage of splitting the equation coefficients into individual cation-specific and anion-specific contributions is that one can make predictions for more RTILs. Normally one needs partition coefficient data for 40-50 solutes dissolved in a given RTIL to develop a RTIL-specific Abraham model correlation. By combining all of the experimental data for a RTIL containing lets say either a 1-methyl-3-octylimidazolium cation, $[MOIm]^+$, and trifluoromethane-sulfonate anion, $[Trif]^-$, it may be possible to calculate $[MOIm]^+$-specific and $[Trif]^-$-specific equation coefficients where there was too few data points for the $[MOIm]^+[Trif]^-$ ionic liquid to develop a meaningful correlation. Revelli *et al.* (2010a) further extended the Abraham model's predictive capability for IL solvents by further splitting the cation-specific equation coefficient into functional group values

$$\log K = \sum_{group} c_i n_i + \sum_{group} e_i n_i E + \sum_{group} s_i n_i S + \sum_{group} a_i n_i A + \sum_{group} b_i n_i B + \sum_{group} l_i n_i L + (c_{anion} + e_{anion} E + s_{anion} S + a_{anion} A + b_{anion} B + l_{anion} L) \quad (24)$$

and

$$\log P = \sum_{group} c_i n_i + \sum_{group} e_i n_i E + \sum_{group} s_i n_i S + \sum_{group} a_i n_i A + \sum_{group} b_i n_i B + \sum_{group} v_i n_i V + (c_{anion} + e_{anion} E + s_{anion} S + a_{abion} A + b_{anion} B + v_{anion} V) \quad (25)$$

where n_i is the number of times a particular functional group appears in the cation. The authors were able to mathematically describe the 1450 available gas-to-IL partition coefficients (log K values) and 1410 water-to-IL partition coefficients (log P values) with 21 groups: 12 functional groups characterizing the cations (CH_3, CH_2, N_{cyclic}, C_{cyclic}, *etc.*) and 9 individual anions ($[PF_6]^-$, $[BF_4]^-$, $[SCN]^-$, *etc.*) to within 0.15 and 0.17 log units, respectively. The number of functional group values is likely to increase in future years as more experimental data becomes available for functionalized IL solvents. In selecting a predictive method, we suggest that the first choice by the IL-specific Abraham equation if available, followed by the ion-specific Abraham model, and then the Abraham group contribution model of Revelli *et al.* For many IL solvents the group contribution model will be the only option available at the present time.

5. Abraham model: Enthalpy of solvation correlations

Abraham model correlations constructed from the ion-specific equation coefficients in Tables 3 and 4 pertain to 298.15 K. A method is needed for extrapolating the predicted log K and log P values to other temperatures for applications involving azeotropic distillation and temperature-programmed gas-liquid chromatographic separation. The majority of experimental data used in the log K and log P regressions came from published gas-liquid chromatographic retention measurements. As part of the experimental studies the authors determined the infinite dilution coefficients of volatile solutes in IL solvents at several temperatures. For the majority of measurements the median temperature corresponded to 323 ± 5 K. Solutes studied include inert gases, diatomic gas molecules, linear and cyclic alkanes and alkenes (up to dodecane), alkylbenzenes, linear and branched alcohols, linear and cyclic monoethers (plus 1,4-dioxane), chlorinated methanes and a few of the smaller aldehydes and ketones. The solute descriptor space defined by these compounds would be: **E** = 0.000 to **E** = 0.850; **S** = 0.000 to **S** = 0.900; **A** = 0.000 to **A** = 0.430; **B** = 0.000 to **B** = 0.650; **V** = 0.109 to **V** = 1.800; and **L** = -1.200 to **L** = 5.700. The predictive area of chemical space spanned by this set of solutes is smaller than the range covered by the log K (Eqn. 22) and log P (Eqn. 23) correlations.

The solute's molar enthalpy of solution in the IL is calculated from the variation of γ_{solute}^{∞} with temperature, *i.e.*, $\Delta H^{ex,\infty} = R\, \partial \ln \gamma_{solute}^{\infty} / \partial(1/T)$. Enthalpies of solution determined in this fashion assume that $\Delta H^{ex,\infty}$ is independent of temperature over the range of the experimental measurements. Enthalpies of solvation are calculated by

$$\Delta H_{Solv} = \Delta H^{ex,\infty} - \Delta H_{Vap} \quad (26)$$

$$\Delta H_{Solv} = \Delta H^{ex,\infty} - \Delta H_{Sub} \quad (27)$$

the solute's enthalpy of vaporization (ΔH_{Vap}, liquid solute) or enthalpy of sublimation (ΔH_{Sub}, solid solute) from the molar enthalpy of solution. Large compilations of ΔH_{Vap} and ΔH_{Sub} data are available in the published literature (Chickos and Acree, 2002; Chickos and Acree, 2003) for use in converting $\Delta H^{ex,\infty}$ to ΔH_{Solv}, and *vice versa*.

Mintz *et al.* (Mintz *et al.*, 2007; Mintz *et al.*, 2008b; Mintz *et al.*, 2009) and Sprunger *et al.* (2009a) demonstrated that the Abraham model does provide an accurate mathematical description of enthalpies of solvation of organic vapors and gases in water, $\Delta H_{Solv,W}$,

$$\Delta H_{Solv,W}\ (\text{kJ/mole}) = -13.310(0.457) + 9.910(0.814)\mathbf{E} + 2.836(0.807)\mathbf{S}$$
$$- 32.010(1.102)\mathbf{A} - 41.816(0.781)\mathbf{B} - 6.354(0.200)\mathbf{L}$$
$$(\text{with } N = 368,\ SD = 3.68,\ R^2 = 0.964,\ R^2_{adj} = 0.964,\ F = 1950.5) \quad (28)$$

and

$$\Delta H_{Solv,W}\ (kJ/mole) = -\ 6.952(0.651) + 1.415(0.770)\ \mathbf{E} - 2.859(0.855)\ \mathbf{S}\ -\ 34.086(1.225)\ \mathbf{A}$$

$$-42.686(0.850)\ \mathbf{B} - 22.720(0.800)\ \mathbf{V}$$

$$\text{(with N = 369, SD = 4.04, } R^2 = 0.959,\ R^2_{adj} = 0.958,\ F = 1688.2) \qquad (29)$$

as well as into organic solvents and ionic liquids. From a thermodynamic standpoint Eqn. 28 is the enthalpic temperature derivative of the Abraham model's gas-to-condensed phase transfer equation. Equation 29 might be more useful in some predictive applications in instances where the **L**-descriptor is not known. Equation 29 uses the McGowan volume, **V**-descriptor, which is easily calculable from the individual atomic sizes and numbers of bonds in the molecule (Abraham and McGowan, 1987).

To provide as complete set of ion-specific equation coefficients as possible we have reanalyzed our enthalpy of solvation data for solutes dissolved in IL solvents at 323 K in accordance with the Eqns. 14 and 15 of the Abraham model. The derived correlations

$$\Delta H_{solv} = \sum_{cation} (c_{cation} + e_{cation} E + s_{cation} S + a_{cation} A + b_{cation} B + l_{cation} L) + \sum_{anion} (c_{anion} + e_{anion} E + s_{anion} S + a_{anion} A + b_{anion} B + l_{anion} L) \qquad (30)$$

$$(N = 977,\ R^2 = 0.998,\ R^2_{adj} = 0.998,\ SD = 1.602,\ F = 3058)$$

and

$$\Delta H_{solv} = \sum_{cation} (c_{cation} + e_{cation} E + s_{cation} S + a_{cation} A + b_{cation} B + v_{cation} V) + \sum_{anion} (c_{anion} + e_{anion} E + s_{anion} S + a_{anion} A + b_{anion} B + v_{anion} V) \qquad (31)$$

$$(N = 977,\ R^2 = 0.998,\ R^2_{adj} = 0.997,\ SD = 1.655,\ F = 2860)$$

describe the observed enthalpy of solvation to within standard deviations of SD = 1.602 kJ/mole and SD = 1.655 kJ/mole using the ion-specific equation coefficients tabulated in Tables 5 and 6. Standard errors in the equation coefficients are given in parenthesis immediately below the respective coefficient. For the most part, the larger standard errors were noted in the equation coefficients for those ions for which experimental data were limited. Large standard errors were similarly noted in the ion-specific equation coefficients for our initial log K and log P Abraham model correlations. The standard errors did decrease in magnitude when additional log *K* and log P values were added to the database. We expect the same to happen with the ΔH_{solv} correlations.

Comparsions of experimental ΔH_{Solv} data versus back-calculated values based on the two Abraham model correlations are graphically depicted in Figures 6 and 7. Each calculation uses only the values for the cation and anion in the given RTIL. For example, to predict the enthalpies of solvation of organic vapors and gases in 1-(3-hydroxypropyl)pyridinium *tris*(pentafluoroethyl)trifluorophosphate one would use only the six equation coefficients for the $[PrOHPy]^+$ and the six equation coefficients for the $[FAP]^-$ anion. The 13 cation-specific equation coefficients and 10 anion-specific equation coefficients listed in Tables 5 and 6 can be used to predict enthalpies of solvation of organic vapors and gases in 130 (13 x 10) different IL solvents. The number of ionic liquids that one can make ΔH_{Solv} predictions for is far smaller than the 299 ILs for which log K and log P predictions can be made.

Predicted values of enthalpies of solvation based on Eqns. 30 and 31 can be used to correct predicted log K and log P values at 298.15 K to other temperatures. For the log P corrections one will need an experimental or predicted value for the enthalpy of solvation of the solute in water, $\Delta H_{Solv,W}$, as well (see Eqns. 12 and 13). The Abraham model correlation derived by Mintz *et al.* can be used to estimate $\Delta H_{Solv,W}$ if an experimental value is not available. An error/uncertainty of ± 2 kJ/mole in the enthalpy of solvation results in an error of slightly less than 0.04 log units in extrapolating a log K value measured at 298.15 K to a temperature of 313.15. This level of predictive error will be sufficient for most practical chemical separation applications. For gas-liquid chromatographic separations Abraham model correlations have been reported (Breitbach and Armstrong, 2008; Anderson *et al.*, 2002; Baltazar *et al.*, 2008; Zhao *et al.*, 2009) for predicting solute retention factors on a few select mono-cationic and di-cationic IL stationary phases at higher temperatures.

Ion[a]	c_{ion}	e_{ion}	s_{ion}	a_{ion}	b_{ion}	l_{ion}
CATIONS						
$[MEIm]^+$	-6.007	3.914	-15.247	-20.601	-7.607	-7.153
(N=257)[b]	(0.821)	(1.563)	(1.749)	(2.147)	(1.704)	(0.249)
$[BMIm]^+$	-6.223	-1.307	-8.699	-20.815	-9.869	-7.072
(N=179)	(0.642)	(1.407)	(1.926)	(2.169)	(2.328)	(0.173)
$[MHIm]^+$	-5.543	0.111	-6.842	-19.484	-10.780	-7.795
(N=101)	(0.685)	(1.577)	(2.013)	(2.362)	(2.357)	(0.187)
$[MOIm]^+$	-7.758	1.088	-4.778	-21.886	-6.993	-7.448
(N=86)	(0.907)	(2.153)	(2.631)	(2.686)	(2.588)	(0.274)
$[M_3BAm]^+$	-7.255	1.352	-9.405	-22.715	-7.191	-6.761
(N=51)	(0.969)	(2.041)	(2.724)	(3.067)	(3.151)	(0.237)
$[M_2EIm]^+$	-0.592	4.122	-15.849	-23.339	-7.193	-8.667
(N=37)	(1.570)	(2.097)	(2.221)	(2.629)	(2.015)	(0.442)
$[4\text{-}BMPy]^+$	-2.796	5.802	-19.743	-27.272	-2.956	-8.378
(N=69)	(1.127)	(2.094)	(2.554)	(2.343)	(2.210)	(0.334)
$[3\text{-}BMPy]^+$	-6.188	5.800	-18.460	-20.302	-2.050	-7.565
(N=36)	(1.992)	(3.406)	(4.291)	(4.378)	(4.375)	(0.589)
$[E_3S]^+$	-2.913	15.064	-32.880	-15.755	-5.551	-7.845
(N=28)	(1.642)	(4.601)	(7.428)	(10.973)	(12.526)	(0.495)
$[BMPyr]^+$	-5.800	10.249	-25.046	-16.107	-3.629	-7.191
(N=30)	(1.710)	(4.440)	(5.683)	(4.725)	(4.739)	(0.539)
$[HxomMIm]^+$	-3.748	13.318	-24.942	-27.250	1.260	-8.645
(N=34)	(1.589)	(3.132)	(4.019)	(3.056)	(3.335)	(0.479)
$[(Hxom)_2Im]^+$	-4.547	13.240	-21.422	-28.533	0.734	-8.833

(N=34)	(1.589)	(3.132)	(4.019)	(3.056)	(3.235)	(0.479)
$[PrOHPy]^+$	-25.941	-15.090	28.322	-15.026	-33.898	-2.429
(N=35)	(2.861)	(7.406)	(9.520)	(12.678)	(11.759)	(0.844)
ANIONS						
$[(Tf)_2N]^-$	0.000	0.000	0.000	0.000	0.000	0.000
(N=394)						
$[BF_4]^-$	-0.520	3.920	3.391	1.969	-1.162	0.494
(N=136)	(0.751)	(1.805)	(2.224)	(2.303)	(2.144)	(0.227)
$[PF_6]^-$	-5.420	12.391	-24.092	13.322	14.997	0.806
(N=54)	(0.813)	(2.647)	(3.369)	(3.463)	(3.616)	(0.254)
$[EtSO_4]^-$	-3.021	-2.344	6.454	-16.911	1.412	1.106
(N=42)	(1.551)	(2.634)	(3.036)	(3.747)	(3.276)	(0.379)
$[Trif]^-$	2.171	2.471	-5.152	-19.614	6.577	-0.696
(N=132)	(1.253)	(2.224)	(2.796)	(3.195)	(3.132)	(0.378)
$[F_3Ac]^-$	5.481	9.027	-6.060	-12.187	-16.266	-1.763
(N=27)	(1.849)	(4.864)	(7.649)	(11.189)	(12.648)	(0.561)
$[NO_3]^-$	-3.591	0.499	-0.226	-14.195	5.620	0.520
(N=28)	(3.254)	(2.953)	(2.980)	(4.528)	(2.842)	(0.970)
$[SCN]^-$	7.978	14.895	-23.862	-26.264	12.733	-2.220
(N=84)	(1.103)	(2.444)	(3.255)	(2.689)	(2.961)	(0.316)
$[B(CN)_4]^-$	20.462	5.485	-24.845	-39.242	21.911	-4.524
(N=22)	(2.434)	(7.100)	(9.553)	(13.453)	(12.973)	(0.729)
$[FAP]^-$	25.824	17.977	-48.802	-4.930	10.119	-4.938
(N=58)	-2.405	(6.922)	(8.926)	(12.306)	(11.333)	(0.713)

[a] Cation and anion abbreviations are given in the footnote to Table 3.
[b] Number of experimental data points associated with the specified ion.

Table 5. Cation-Specific and Anion-Specific Equation Coefficients for the Abraham Model Enthalpy of Solvation Correlation (Eqn.30)

Ion[a]	c_{ion}	e_{ion}	s_{ion}	a_{ion}	b_{ion}	v_{ion}
CATIONS						
$[MEIm]^+$	-0.522	-8.710	-15.755	-24.770	-7.602	-25.280
(N=257)[b]	(1.027)	(1.469)	(1.838)	(2.270)	(1.765)	(0.914)
$[BMIm]^+$	0.671	-10.742	-13.978	-23.150	-10.347	-26.260
(N=179)	(0.794)	(1.409)	(2.004)	(2.238)	(2.395)	(0.646)
$[MHIm]^+$	0.103	-12.846	-7.474	-16.018	-15.782	-27.380
(N=101)	(0.851)	(1.584)	(2.083)	(2.421)	(2.412)	(0.697)
$[MOIm]^+$	-3.094	-14.937	-1.923	-21.742	-10.249	-25.306
(N=86)	(1.147)	(2.122)	(2.745)	(2.799)	(2.672)	(1.021)
$[M_3BAm]^+$	-1.158	-7.091	-15.020	-25.483	-6.221	-24.671

(N=51)	(1.197)	(2.059)	(2.867)	(3.202)	(3.259)	(0.891)
$[M_2EIm]^+$	4.006	-5.855	-21.604	-28.854	-3.145	-29.304
(N=37)	(1.881)	(1.962)	(2.463)	(2.829)	(2.100)	(1.576)
$[4\text{-}BMPy]^+$	1.375	-11.008	-16.169	-28.921	-5.933	-27.874
(N=69)	(1.372)	(2.024)	(2.654)	(2.454)	(2.280)	(1.195)
$[3\text{-}BMPy]^+$	1.324	-6.735	-20.274	-23.121	-4.582	-28.299
(N=36)	(2.435)	(3.415)	(4.442)	(4.505)	(4.529)	(2.107)
$[E_3S]^+$	4.009	-10.891	-17.029	-23.710	-12.059	-28.593
(N=28)	(2.124)	(4.789)	(7.631)	(11.458)	(12.920)	(1.860)
$[BMPyr]^+$	2.509	-6.665	-20.110	-20.750	-9.554	-27.793
(N=30)	(2.141)	(4.169)	(5.749)	(4.885)	(4.842)	(1.961)
$[HxomMIm]^+$	2.168	-5.927	-19.612	-27.547	-4.531	-30.045
(N=34)	(1.977)	(3.112)	(4.111)	(3.166)	(3.397)	(1.732)
$[(Hxom)_2Im]^+$	1.063	-6.379	-15.883	-28.570	-5.017	-30.303
(N=34)	(1.977)	(3.112)	(4.111)	(3.166)	(3.397)	(1.732)
$[PrOHPy]^+$	-24.316	-6.193	7.321	-15.580	-23.480	-8.586
(N=35)	(3.552)	(8.005)	(9.947)	(13.201)	(11.998)	(3.057)
ANIONS						
$[(Tf)_2N]^-$	0.000	0.000	0.000	0.000	0.000	0.000
(N=394)						
$[BF_4]^-$	0.525	3.258	-5.021	-0.272	3.289	0.404
(N=136)	(0.968)	(1.774)	(2.346)	(2.427)	(2.225)	(0.862)
$[PF_6]^-$	-3.055	14.791	-27.034	9.240	18.214	0.108
(N=54)	(1.074)	(2.662)	(3.515)	(3.608)	(3.733)	(1.006)
$[EtSO_4]^-$	-1.724	5.034	-1.145	-17.689	4.040	2.461
(N=42)	(1.964)	(2.584)	(3.309)	(4.005)	(3.406)	(1.449)
$[Trif]^-$	0.513	-0.513	-2.324	-16.775	5.462	-0.738
(N=132)	(1.525)	(2.158)	(2.908)	(3.281)	(3.220)	(1.340)
$[F_3Ac]^-$	7.964	-7.797	13.304	-17.312	-24.326	-7.359
(N=27)	(2.404)	(5.003)	(7.884)	(11.705)	(13.049)	(2.124)
$[NO_3]^-$	-6.235	4.613	-3.847	-10.664	8.528	3.247
(N=28)	(3.616)	(2.622)	(3.381)	(4.656)	(2.911)	(3.178)
$[SCN]^-$	9.842	1.469	-11.336	-31.443	8.565	-8.113
(N=84)	(1.422)	(2.512)	(3.355)	(2.831)	(3.130)	(1.205)
$[B(CN)_4]^-$	23.574	-15.339	-4.703	-39.309	10.023	-15.787
(N=22)	(3.047)	(7.629)	(9.887)	(14.040)	(13.192)	(2.647)
$[FAP]^-$	29.378	-4.429	-27.315	-4.750	-3.066	-17.347
(N=58)	(3.001)	(7.540)	(9.341)	(12.818)	(11.546)	(2.585)

[a] Cation and anion abbreviations are given in the footnote to Table 3.
[b] Number of experimental data points associated with the specified ion.

Table 6. Cation-Specific and Anion-Specific Equation Coefficients for the Abraham Model Enthalpy of Solvation Correlation (Eqn.31)

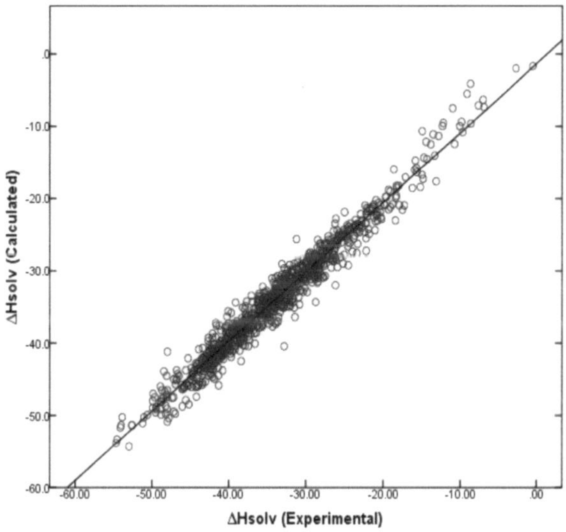

Fig. 6. Comparison between the 977 experimental ΔH_{Solv} data points and predicted values based on Eqn. 30.

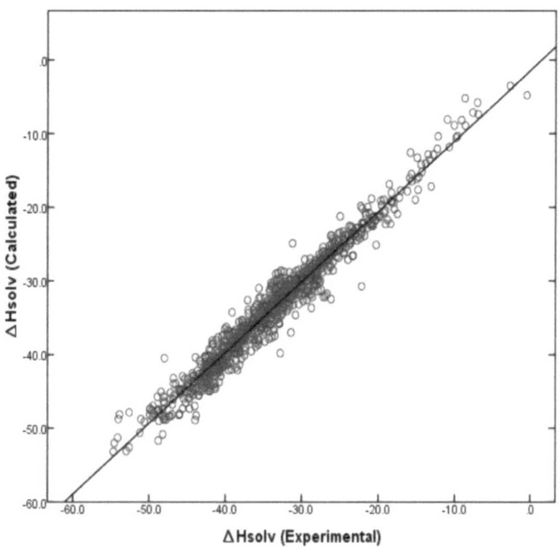

Fig. 7. Comparison between the 977 experimental ΔH_{Solv} data points and predicted values based on Eqn. 31

6. Conclusion

The Abraham general solvation provides a reasonably accurate mathematical description of the thermodynamic properties governing the solute transfer into anhydrous ionic liquid solvents from both water and from the gas phase. Derived expressions based on the Abraham model allow one to estimate the estimate the log K and log P values at 298 K for many solute-IL systems based on published ion-specific equation coefficients and/or published group contribution values. Comparison of experimental versus predicted values for more than 30 IL solvents suggest that the predicted log P and log K values should fall within 0.15 log units of the observed value in most cases. The model contains provisions for correcting the predicted log K and log P values to other temperatures not too far removed from 298 K.

7. References

Abraham, M. H. & McGowan, J. C. (1987) The use of characteristic volumes to measure cavity terms in reversed phase liquid chromatography. *Chromatographia* 23 (4) 243–246.

Abraham, M. H. (1993a) Scales of solute hydrogen-bonding: their construction and application to physicochemical and biochemical processes. *Chemical Society Reviews* 22 (2),73-83.

Abraham, M. H. (1993b) Application of solvation equations to chemical and biochemical processes.*Pure and Applied Chemistry* 65 (12), 2503-12.

Abraham, M. H.; Acree, W. E., Jr. (2006) Comparative analysis of solvation and selectivity in room temperature ionic liquids using the Abraham linear free energy relationship. *Green Chemistry* 8 (10), 906-915.

Abraham, M. H.; Zissimos, A. M.; Huddleston, J. G.; Willauer, H. D.; Rogers, R. D. & Acree, W. E., Jr.(2003) Some Novel Liquid Partitioning Systems: Water-Ionic Liquids and Aqueous Biphasic Systems. *Industrial & Engineering Chemistry Research* 42, 413-418.

Abraham, M. H.; Ibrahim, A. & Zissimos, A. M. (2004) Determination of sets of solute descriptors from chromatographic measurements. *Journal of Chromatography, A* 1037 (1-2), 29-47.

Abraham, M. H.; Ibrahim, A.; Zhao, Y.; Acree, W. E., Jr. (2006) A data base for partition of volatile organic compounds and drugs from blood/plasma/serum to brain, and an LFER analysis of the data. *Journal of Pharmaceutical Sciences* 95 (10), 2091-2100.

Abraham, M. H.; Ibrahim, A. & Acree, W. E. Jr. (2007) Air to liver partition coefficients for volatile organic compounds and blood to liver partition coefficients for volatile organic compounds and drugs . *European Journal of Medicinal Chemistry* 42 (6), 743-751.

Abraham, M. H.; Ibrahim, A. & Acree, W. E. Jr. (2008) Air to lung partition coefficients for volatile organic compounds and blood to lung partition coefficients for volatile organic compounds and drugs. *European Journal of Medicinal Chemistry* 43 (3), 478-485.

Abraham, M. H.; Acree, W. E., Jr. & Cometto-Muniz, J. E. (2009) Partition of compounds from water and from air into amides. *New Journal of Chemistry* 33 (10), 2034-2043.

Abraham, M. H.; Smith, R. E.; Luchtefeld, R.; Boorem, A. J.; Luo, R.; Acree, W. E., Jr. (2010) Prediction of solubility of drugs and other compounds in organic solvents. Journal of Pharmaceutical Sciences 99 (3), 1500-1515.

Acree, W. E., Jr. & Abraham, M. H. (2006) The analysis of solvation in ionic liquids and organic solvents using the Abraham linear free energy relationship. *Journal of Chemical Technology and Biotechnology* 81 (8), 1441-1446.

Aguilera-Herrador, E.; Lucena, R.; Cardenas, S.; Valcarcel, M. (2008) Ionic liquid-based single-drop microextraction/gas chromatographic/mass spectrometric determination of benzene, toluene, ethylbenzene and xylene isomers in waters. *Journal of Chromatography, A* 1201 (1), 106-111.

Anderson, J. L.; Ding, J.; Welton, T. & Armstrong, D. W. (2002) Characterizing ionic liquids on the basis of multiple solvation interactions. *Journal of the American Chemical Society* 124 (47), 14247-14254.

Arey, J. S.; Green, W. H., Jr. & Gschwend, P. M. (2005) The electrostatic origin of Abraham's solute polarity parameter. *J. Phys. Chem. B 109*, 7564-7573.

Baltazar, Q. Q.; Leininger, Suzette K.; Anderson, J. L. (2008) Binary ionic liquid mixtures as gas chromatography stationary phases for improving the separation selectivity of alcohols and aromatic compounds. *Journal of Chromatography, A*, 1182 (1), 119-127.

Breitbach, Z. S.; Armstrong, D. W. (2008) Characterization of phosphonium ionic liquids through a linear solvation energy relationship and their use as GLC stationary phases. *Analytical and Bioanalytical Chemistry* 390 (6), 1605-1617.

Chickos, James S.; Acree, William E., Jr. (2003) Enthalpies of vaporization of organic and organometallic compounds, 1880-2002. *Journal of Physical and Chemical Reference Data* 32 (2), 519-878.

Chickos, J. S. & Acree, W. E., Jr. (2002) Enthalpies of sublimation of organic and organometallic compounds. 1910-2001. *Journal of Physical and Chemical Reference Data* 31 (2), 537-698.

Grubbs, L. M.; Saifullah,M.; De La Rosa, N. E.; Acree, W. E., Jr.; Abraham, M. H.; Zhao, Q. & Anderson, J. L. Cation-specific and anion-specific Abraham model correlations for solute transfer into ionic liquids. *Global Journal of Physical Chemistry,* 1 (1), 1-19.

Lei, Z.; Li, C. & Chen, B. (2003) Extractive distillation: a review. *Separation and Purification Reviews* 32 (2), 121-213.

Lei, Z.; Arlt, W.; Wasserscheid, P. (2006) Separation of 1-hexene and n-hexane with ionic liquids. *Fluid Phase Equilibria,* 241 (1-2), 290-299.

Liu, J.-F.; Jiang, G.-B.; Chi, Y.-G.; Cai, Y.-Q.; Zhou, Q.-X.; Hu, J.-T. (2003) Use of ionic liquids for liquid-phase micro-extraction of polycyclic aromatic hydrocarbons. *Analytical Chemistry* 75 (21), 5870-5876.

Mintz, C. & Acree, W. E Jr. (2007) Partition coefficient correlations for transfer of solutes from gas phase and water to room temperature ionic liquids. *Physics and Chemistry of Liquids* 45 (3), 241-249.

Mintz, C.; Clark, M.; Acree, W. E., Jr. & Abraham, M. H. (2007) Enthalpy of solvation correlations for gaseous solutes dissolved in water and in 1-octanol based on the Abraham model. *Journal of Chemical Information and Modeling* 47 (1), 115-121.

Mintz, C.; Ladlie, T.; Burton, K.; Clark, M.; Acree, W. E., Jr. & Abraham, M. H. (2008a) Characterization of the partitioning of gaseous solutes into humic acid with the Abraham model and temperature-independent equation coefficients. *QSAR & Combinatorial Science* 27 (4), 483-491.

Mintz, C.; Ladlie, T.; Burton, K.; Clark, M.; Acree, W. E., Jr. & Abraham, M. H. (2008b) Enthalpy of solvation correlations for gaseous solutes dissolved in alcohol solvents based on the Abraham model. *QSAR & Combinatorial Science*, 27 (5), 627-635.

Mintz, C.; Burton, K.; Ladlie, T.; Clark, M.; Acree, W.E.; Abraham, M. H. (2009) Enthalpy of solvation correlations for organic solutes and gases dissolved in N,N-dimethylformamide and *tert*-butanol. *Journal of Molecular Liquids* 144 (1-2), 23-31.

Mokrushin, V.; Assenbaum, D.; Paape, N.; Gerhard, D.; Mokrushina, L.; Wasserscheid, P.; Arlt, W.; Kistenmacher, H.; Neuendorf, S. & Goeke, V. (2010) Ionic liquids for propene-propane separation. *Chemical Engineering & Technology*, 33 (1), 63-73.

Mutelet, F. & Rogalski, M. (2001) Experimental determination and prediction of the gas-liquid n-hexadecane partition coefficients. *J. Chromatogr., A 923*, 153-163.

Mutelet, F.; Revelli, A.-L.; Jaubert, J.-N.; Sprunger, L. M.; Acree, W. E., Jr. & Baker, G. A. (2010) Partition Coefficients of Organic Compounds in New Imidazolium and Tetralkyl-ammonium Based Ionic Liquids Using Inverse Gas Chromatography. *Journal of Chemical & Engineering Data* 55 (1), 234-242.

Palgunadi, J.; Kim, H. S.; Lee, J. M. & Jung, S. (2010) Ionic liquids for acetylene and ethylene separation: Material selection and solubility investigation. *Chemical Engineering and Processing* 49 (2), 192-198.

Pandey, S. (2006) Analytical applications of room-temperature ionic liquids: A review of recent efforts. *Analytica Chimica Acta* 556 (1), 38-45.

Pereiro, A. B.; Deive, F. J.; Esperança, J. M. M. S. & Rodríguez, A. (2010) Alkylsulfate-based ionic liquids to separate azeotropic mixtures. *Fluid Phase Equilibr.*, 291 (1), 13-17.

Pharma Algorithms, (2006) ADME Boxes, Version 3.0; Pharma Algorithms Inc.: Toronto, Ontario, Canada, 2006.

Planeta, J. & Roth, M. (2005) Solute Partitioning between the Ionic Liquid 1-n-Butyl-3-methylimidazolium Tetrafluoroborate and Supercritical CO2 from Capillary-Column Chromatography. *Journal of Physical Chemistry B* 109 (31), 15165-15171.

Planeta, J.; Karasek, P. & Roth, M (2007) Limiting partition coefficients of solutes in biphasic trihexyltetradecyl-phosphonium chloride ionic liquid-supercritical CO2 system: measurement and LSER-based correlation. *Journal of Physical Chemistry B* 111 (26), 7620-7625.

Planeta, J.; Karasek, P. & Roth, M. (2009) Distribution of organic solutes in biphasic 1-n-butyl-3-methylimidazolium methyl sulfate-supercritical CO2 system. *Journal of Physical Chemistry B*, 113 (28), 9520-9526.

Platts, J. A.; Butina, D.; Abraham, M. H.; Hersey, A. (1999) Estimation of molecular linear free energy relation descriptors using a group contribution approach. *J. Chem. Inf. Comput. Sci.* 1999, *39*, 835-845.

Poole, C. F. & Poole, Salwa K. (2010) Extraction of organic compounds with room temperature ionic liquids. *Journal of Chromatography, A* 1217(16), 2268-2286.

Proctor, A. Sprunger, L. M.; Acree, W. E., Jr.; Abraham, M. H. (2008) LFER correlations for the solubilising characterisation of room temperature ionic liquids containing trifluoromethanesulfonate and trifluoroacetate anions. *Physics and Chemistry of Liquids* 46 (6), 631-642.

Revelli, A.-L.; Sprunger, L. M.; Gibbs, J.; Acree, W. E.; Baker, G. A. & Mutelet, F. (2009) Activity coefficients at infinite dilution of organic compounds in trihexyl(tetradecyl)phosphonium bis(trifluoromethylsulfonyl)imide using inverse gas chromatography. *Journal of Chemical and Engineering Data* 54 (3), 977-985.

Revelli, A.-L.; Mutelet, F. & Jaubert, J.-N. (2010a) Prediction of partition coefficients of organic compounds in ionic liquids: use of a linear solvation energy relationship with parameters calculated through a group contribution method. *Industrial & Engineering Chemistry Research* 49 (8), 3883-3892.

Revelli, A.-L.; Mutelet, F.; Jaubert, J.-N.; Garcia-Martinez, M.; Sprunger, L. M.; Acree, W. E., Jr. & Baker, Gary A. (2010b) Study of ether-, alcohol-, or cyano-functionalized ionic liquids using inverse gas chromatography. *Journal of Chemical & Engineering Data,* 55 (7), 2434-2443.

Roth, M. (2009) Partitioning behaviour of organic compounds between ionic liquids and supercritical fluids. *Journal of Chromatography,* A 1216 (10), 1861-1880.

Seeley, J. V.; Seeley, S. K.; Libby, E. K.; Breitbach, Z. S. & Armstrong, D. W. (2008) Comprehensive two-dimensional gas chromatography using a high-temperature phosphonium ionic liquid column. *Analytical and Chemistry* 390(1), 323-332.

Soukup-Hein, R. J.; Warnke, M. M. & Armstrong, D. W. (2009) Ionic liquids in analytical chemistry. *Annual Review of Analytical Chemistry* 2, 145-168.

Sprunger L.; Acree W. E., Jr. & Abraham M. H. (2007a) Linear free energy relationship correlation of the distribution of solutes between water and sodium dodecyl sulfate (SDS) micelles and between gas and SDS micelles. *Journal of Chemical Information and Modeling* 47 (5), 1808-17.

Sprunger, L.; Clark, M.; Acree, W. E., Jr. & Abraham, M. H. (2007b) Characterization of room-temperature ionic liquids by the Abraham model with cation-specific and anion-specific equation coefficients. *Journal of Chemical Information and Modeling* 47 (3), 1123-1129.

Sprunger, L.; Proctor, A.; Acree, W. E. Jr. & Abraham, M. H. (2007c) Characterization of the sorption of gaseous and organic solutes onto polydimethyl siloxane solid-phase microextraction surfaces using the Abraham model. *Journal of Chromatography, A* 1175 (2), 162-173.

Sprunger, L. M.; Proctor, A.; Acree, W. E., Jr. & Abraham, M. H. (2008) LFER correlations for room temperature ionic liquids: Separation of equation coefficients into individual cation-specific and anion-specific contributions. *Fluid Phase Equilibria* 265 (1-2), 104-111.

Sprunger, L. M.; Achi, S. S.; Acree, W. E. & Abraham, M. H. (2009a) Linear Free Energy Relationship Correlations for Enthalpies of Solvation of Organic Solutes into Room-Temperature Ionic Liquids Based on the Abraham Model with Ion-Specific Equation Coefficients.. *Industrial and Engineering Chemistry Research* 48 (18), 8704-8709.

Sprunger, L. M.; Gibbs, J.; Baltazar, Q. Q.; Acree, W. E., Jr.; Abraham, M. H.; Anderson, J. L. (2009b) Characterisation of room temperature ionic liquid chromatographic stationary phases by combining experimental retention factor and partition coefficient data into a single model. *Physics and Chemistry of Liquids* 47 (1), 71 83.

Sprunger, L. M.; Gibbs, J.; Proctor, A.; Acree, W. E., Jr.; Abraham, M. H.; Meng, Y.; Yao, C. & Anderson, J. L. (2009c) Linear free energy relationship correlations for room temperature ionic liquids: revised cation-specific and anion-specific equation coefficients for predictive applications covering a much larger area of chemical space. *Industrial and Engineering Chemistry Research* 48 (8), 4145-4154.

Verma, V. K. & Banerjee, T. (2010) Ionic liquids as entrainers for water+ethanol, water+2-propanol, and water+THF systems: A quantum chemical approach. *Journal of Chemical Thermodynamics* 42 (7), 909-919.

Yao, C.; Pitner, W. R. & Anderson, J. L. (2009) Ionic liquids containing the tris(pentafluoroethyl)trifluorophosphate anion: a new class of highly selective and ultra hydrophobic solvents for the extraction of polycyclic aromatic hydrocarbons using single drop microextraction. *Analytical Chemistry* 81 (12), 5054-5063.

Zhao, Q.; Eichhorn, J.; Pitner, W. R. & Anderson, J. L. (2009) Using the solvation parameter model to characterize functionalized ionic liquids containing the tris(pentafluoroethyl)trifluorophosphate (FAP) anion. *Analytical and Bioanalytical Chemistry* 395 (1), 225-234.

Zhao, F.-Q.; Li, J.; Zeng, B.-Z. (2008) Coupling of ionic liquid-based headspace single-drop microextraction with GC for sensitive detection of phenols. *Journal of Separation Science* 31 (16-17), 3045-3049.

Zhao, F.; Lu, S.; Du, W. & Zeng, B. (2009) Ionic liquid-based headspace single-drop microextraction coupled to gas chromatography for the determination of chlorobenzene derivatives. *Microchimica Acta* 165(1-2), 29-33.

Zhu, J.; Chen, J. & Fei, W. (2004) Separation of aromatic hydrocarbons or olefins from paraffins using new ionic liquids. *Huagong Xuebao* (Chinese Edition), 55, 2091-2094.

Zissimos, A. M.; Abraham, M. H.; Barker, M. C.; Box, K. J.; Tam, K. Y. (2002a) Calculation of Abraham descriptors from solvent-water partition coefficients in four different systems; evaluation of different methods of calculation. *Journal of the Chemical Society, Perkin Transactions 2* (3), 470-477.

Zissimos, A. M.; Abraham, M. H.; Du, C. M.; Valko, K.; Bevan, C.; Reynolds, D.; Wood, J.; Tam, K. Y. (2002b) Calculation of Abraham descriptors from experimental data from seven HPLC systems; evaluation of five different methods of calculation. *Journal of the Chemical Society, Perkin Transactions 2* (12), 2001-2010.

Part 4

Materials Chemistry

14

DBU Derived Ionic Liquids and Their Application in Organic Synthetic Reactions

Xinzhi Chen[1] and Anguo Ying[2]
[1]*Department of Chemical and Biochemical Engineering, Zhejiang University, Hangzhou,*
[2]*School of Pharmaceutical and Chemical Engineering, Taizhou University, Taizhou,*
P.R. China

1. Introduction

With the continuing depletion of natural resources and the growing environmental awareness, current and future chemists are being trained to develop synthetic routes in economically beneficial manner. One strategy to realize these green processes in organic synthesis is the replacement of toxic volatile organic solvents with environmentally more benign reaction mediums. Among the novel green solvents that have been reported, ionic liquids (ILs) have been one of the most active areas of green chemistry over the past decade, due to their excellent chemical and thermo properties such as good thermal stability, negligible vapor pressure, ease of handling, potential for recycling, good coordinating and dissolving capability. Thus, ILs were widely used in various organic transformations(Miao & Chan, 2006; Ying et al., 2008; Ying et al., 2008). However, most of the ILs studied by chemists are structurally based on imidazole which is inert for organic reactions. So the need for the developments of novel task-specific ionic liquids still exists.

On the other hand, 1,8-diazabicyclo[5.4.0]undec-7-ene(2,3,4,6,7,8,9,10- octakydropyrimido [1,2-α]azep -ine, DBU) was found to be superior to other tertiary amines as catalyst, base or promoter. For example, using DBU as catalyst for Balylis-Hillman reaction and aza-Michael addition was reported by Aggarwal and Kim respectively(Aggarwal & Mereu, 1999; Yeom et al., 2007). DBU-mediated CO2-fixation reaction(Yoshida et al., 2008), DBU-promoted, S-N-type Smiles and Ireland-Claisen rearrangement reactions(Ma et al., 2007; Li et al., 2007), DBU-catalyzed addition reactions of sulfonylimidates(Matsubara & Kobayashi, 2008), DBU-promoted chemoselective cleavage of acetylenic TMS group(Yeom et al., 2008), and DBU-assisted unusual dehydrogenation(Kim et al., 2008) were all found to be very efficient. Considering the special role that DBU played in organic synthesis, ionic liquids with a DBU moisture (Scheme 1) were developed(Tolstikoua & Shainyan, 2006). Because of its large size, low symmetry of cationic moiety and charge delocalization over the N-C=N triad render these type of ionic liquids low melting points.

However, the salts in Scheme 1 are inert and can only been used as reaction solvents. Then, we attempted to develop a new class of task-specific DBU-derived ionic liquids which could not only serve as reaction medium but also as catalyst or promotor. The preparation of the five novel ILs can be simply realized by neutralization reaction of DBU and the corresponding carboxyl acids, including acetic acid, lactic acid, propionic acid and butyric acid (Scheme2)(Ying et al., 2009; Ying et al., 2010; Ying, 2010). Among these ILs, [DBU][Tfa]

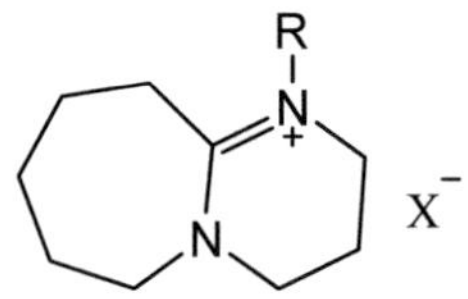

R=Me, Et, Bu, $PhCH_2$, $H(CF)_nCH_2$;

X=Cl, Br, OTs, OTf.

Scheme 1

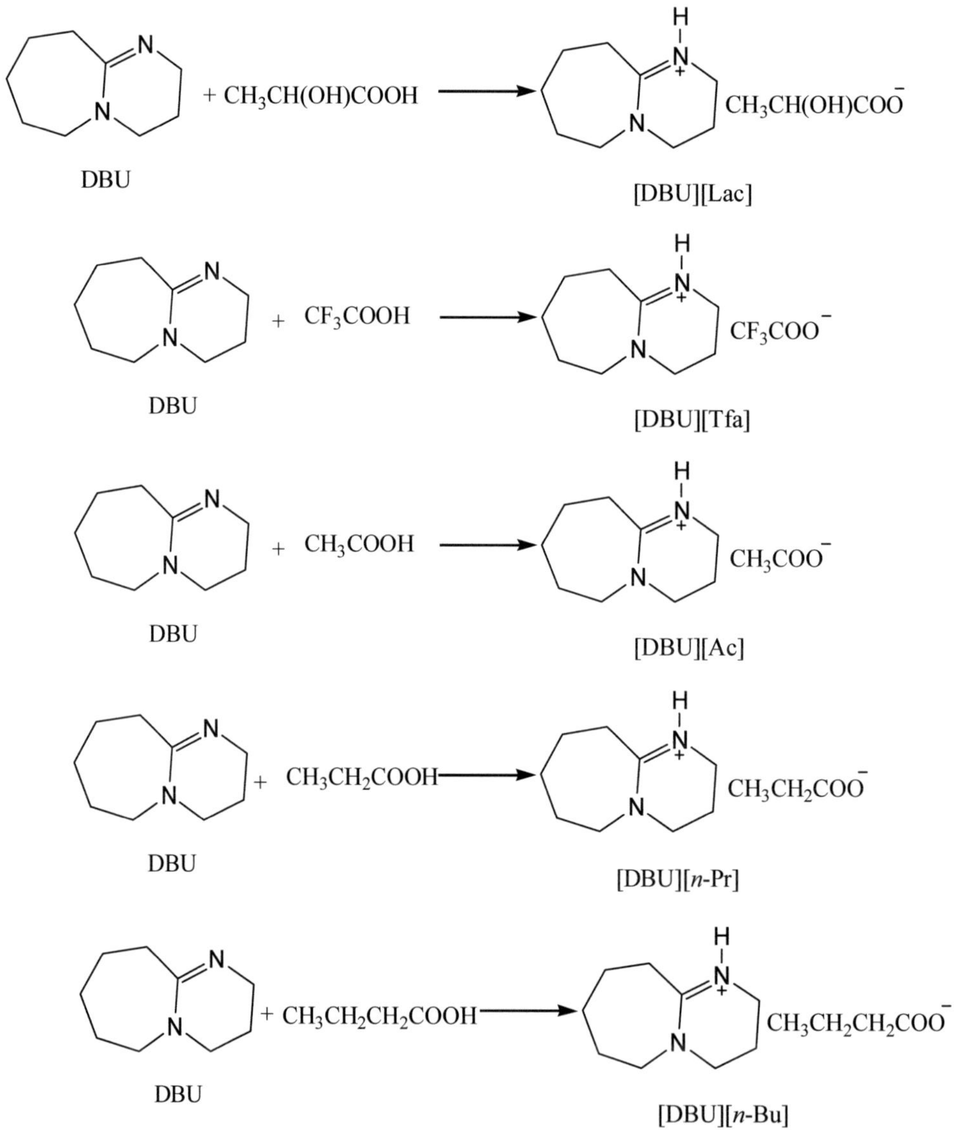

Scheme 2

is liquid with weak acidicity at room temperature, while the other four ionic liquids, [DBU][Ac], [DBU][Lac], [DBU][n-Pr] and [DBU][n-Bu] show property of weak basicity[11]. Next, we will investigate the applications of the novel DBU-based ionic liquids in Michael addition and Knoevenagel condensation. To improve the recycling capability, ionic liquids immobilized on Magnetic nanoparticles (MNP) will also been summarized in this chapter.

2. Application in Michael addition

2.1 Michael addition (aliphatic amines as Michael donors)

Carbon-nitrogen bond forming reaction is one of the most important methodologies in synthetic organic chemistry for preparation of β-aminocarbonyl compounds, which not only constitute a component of biologically active natural products but also serve as a key intermediate for the syntheisis of β-aminoalcohols, β-lactams and β-amino acids(Kleinmann, 1991; Georg, 1993; Corey, 1989). Manich reaction between enolates and imines provide a classic route for the construction of β-amino carbonyl compounds(Arend et al., 1998; Kobayashi & Ishitani, 1999). However, this type of reaction often requires harsh reaction conditions and long reaction time. Compared with the Manich reaction, the aza-Michael addition of amines to electron deficient alkenes has attracted considerable attention as an alternative protocol for C-N bond formation due to its atomic economy and operational simplicity. Most aza-Michael reactions are usually carried out with a strong acid or a base, which would lead to by-products or undesired harmful residues. Thus, milder Lewis acidic catalysts such as $LiClO_4$(Azizi & Said, 2004), $Yb(OTf_3)$(Jenner, 1995; Matsubara et al., 1994), $Bi(NO_3)$(Srivastava & Banik, 2003), $FeCl_3.6H_2O$(Xu et al., 2004), $CeCl_3.7H_2O$(Bartoli et al., 2001), $InCl_3$(Loh & Wei, 1998), SmI_2(Reboule et al., 2005), $Cu(OTf)_2$(Xu et al., 2005), and so on are employed in the Michael protocol. Recently, some novel reagents used as catalyst or promoter in conjugate addition have been reported, including β-cyclodextrin(Surendra et al., 2006), bromodimethylsulfonium bromide(Khan et al., 2007), boric acid in water(Chaudhuri et al., 2005), $ZrOCl_2.8H_2O$ on montmorillonite K10(Hashemi et al., 2006), imidazoliHum-based polymer supported CuI(Alleti et al., 2008), KF/Al_2O_3(Kantam et al., 2008), $[HP(HNCH_2CH_2)_3N]NO_3$(Yang et al., 2005), etc. However, many of the above methods suffered from some drawbacks, such as the requirement for a large excess of reagents, substrate-selective for some catalysts, and often involvement of some toxic solvents such as 1,2-dichloroethane or acetonitrile.

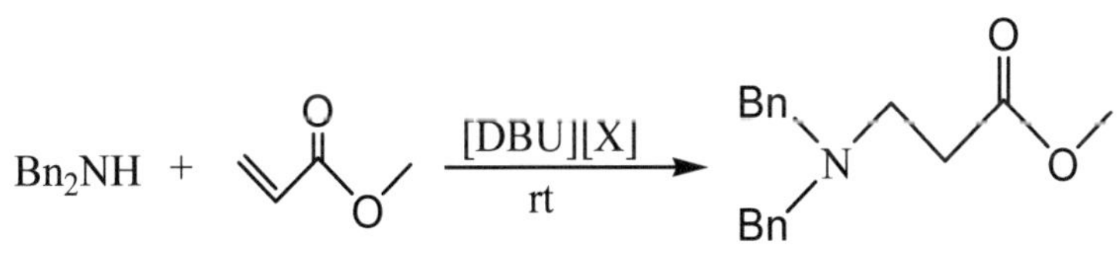

Among the five ionic liquids, [DBU][Ac] showed slightly higher catalytic activity than the other ionic liquids(Table 1, entry 7 and entries 10-13). For the comparison with DBU as promoter for aza-Michael addition reported by Kim et al., 50 mol % of the ionic liquid [DBU][Ac] was employed for the model reaction between dibenzylamine and methyl acrylate in acetonitrile at room temperature(Yeom et al., 2007). The reaction catalyzed by [DBU][Ac] was slightly faster than that promoted by its parent base DBU (Table 1, entries 1-2), indicating the rationality of [DBU][Ac] as catalyst for the reaction. Other solvents such as methanol, toluene, and dichloromethane were also tested and they were all effective

Entry	Catalyst	Solvent	Catalyst amount (mmol)	Time (h)	Yield (%)
1b	DBU	Acetonitrile	0.5	6	95
2	[DBU][Ac]	Acetonitrile	0.5	5	94
3	[DBU][Ac]	Methanol	0.5	5	86
4	[DBU][Ac]	Toluene	0.5	5	89
5	[DBU][Ac]	Dichloromethane	0.5	5	90
6	[DBU][Ac]	Solvent-free	0.5	4.5	96
7	[DBU][Ac]	Solvent-free	0.3	4.5	95
8	[DBU][Ac]	Solvent-free	0.1	4.5	72
9	[DBU][Ac]	Solvent-free	0.01	4.5	43
10	[DBU][Tfa]	Solvent-free	0.3	4.5	83
11	[DBU][Lac]	Solvent-free	0.3	4.5	92
12	[DBU][*n*-Pr]	Solvent-free	0.3	4.5	89
13	[DBU][*n*-Bu]	Solvent-free	0.3	4.5	83

Table 1. Aza-Michael reaction of dibenzylamine (1 mmol) with methyl acrylate (1.3 mmol) under **various reaction conditions**

[a] Isolated yields. [b] reference (Yeom et al., 2007): using DBU as catalyst.

The main aim of the work is to examine if the non-solvent addition of aliphatic amines to electron deficient olefins can proceed smoothly with DBU based task-specific ionic liquids as catalysts.

Reaction medium for the model reaction (entries 3-5). However, because of the toxicity of organic solvents and our pursuit for the establishment of the environmentally benign process for organic transformations, we attempted to conduct the reaction of dibenzylamine and methyl acrylate under solvent-free condition and to our delight, good yield (96 %) was obtained within 4.5 h (Table 1, entry 6). So we chose solvent-free conditions for further study. Next, the amount of catalyst [DBU][Ac] was reduced to 0.3 equiv and almost no decrease of yield was observed (entry 7). However, the product yield decreased obviously when the amount was further reduced to 0.01 equiv (Table 1, entries 8-9). As a result, we adopted 0.3 equiv of [DBU][Ac] under solvent-free condition at room temperature for the following investigations.

With the optimal catalytic system in hand, we investigated the suitability of a wide range of nitrogen nucleophiles for the [DBU][Ac] catalyzed aza-Michael reactions using methyl acrylate as substrate, aliphatic secondary amines such as morpholine, piperidine, 1-methylpiperazine, 1-ethylpiperazine and pyrazole underwent conjugate reaction with methyl acrylate favourably and excellent yields of Michael adducts were obtained at ambient temperature under solvent-free conditions within short reaction time (Table 2, entries 1-4). Morpholine, piperidine and pyarzole reacted faster and afforded higher products yields in the presence of ionic liquid [DBU][Ac] than those promoted by DBU (entries1, 4 and 5). It is worthy to note that when imidazole were treated as Michael donor with methyl acrylate, it disppeared in 2 hours and no Michael product was detected with the formation of acylation product. Judging from the disappearance of OCH3 group of products from ^{1}H NMR spectrum, it may provides a possible method for acylation reaction of Michael acceptors. However, 2-isopropylimidazole could react with methyl acrylate and 93 % Michael adduct yield was obtained in 10 hours. From the above, steric bulkiness of reagent can assist the formation of Michael adduct and is detrimental to acylation process. Primary amine, benzylamine, were also treated with methyl acrylate and 20 % yield of di-substituted product was formed which was the same with that catalyzed by DBU(Yeom et al., 2007) (Table 2, entry 8).

Having obtained favourable results with methyl acrylate, we studied the solventless reaction of morpholine with other electron-deficient olefins under the same conditions. Other various vinyl esters, acrylonitrile and acrylamide were effective substrates to give desired products in good to excellent yields (Table 2, entries 9-11 and 13-14). Among the acrylate esters, the increasing of carbon number acrylates would lead to decreased reactivity (entries 9 and 10), as reported by Lin et al.(2007) α-Methyl and β-bezene substituted ester, ethyl cinnamate were also tested as acceptors respectively and were found that the former gave the rational yield in short time and the latter exhibited relatively inert reaction activity due to steric hindrance of benzene at β position (entries 11 and 12).

Entry	Nitrogen nucleophile	Michael acceptor	Reaction time/Ref. 19 (h)	Product	Yieldb/Ref. 19 (%)
1	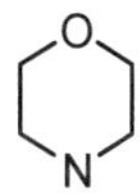	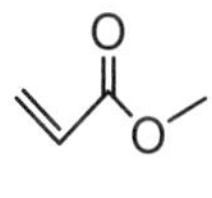	1.5/4	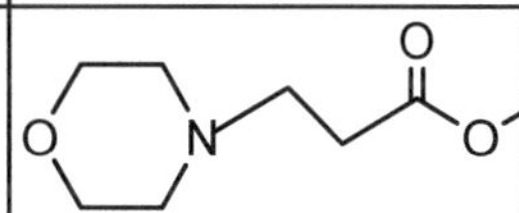	93/90
2	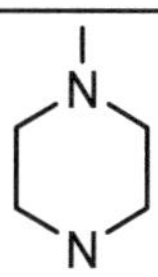	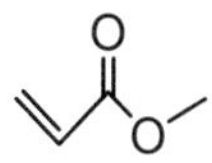	1.5	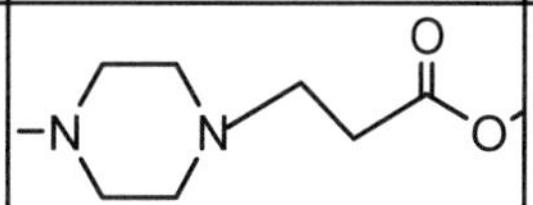	90
3	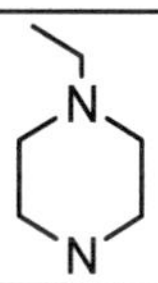	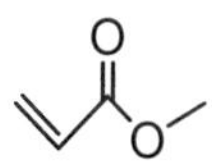	1.5	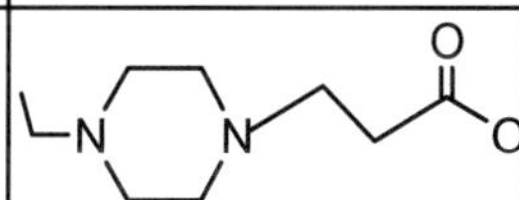	87
4	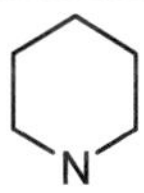	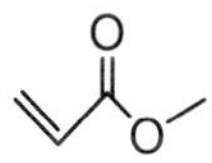	1.5/4	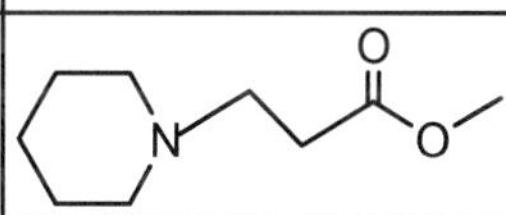	92/88
5	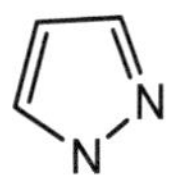	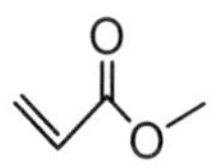	5/14	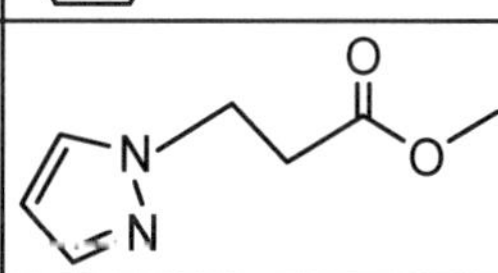	93/95
6	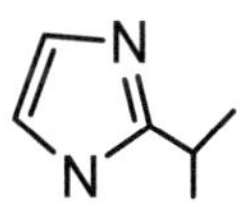	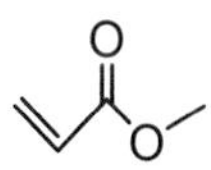	10	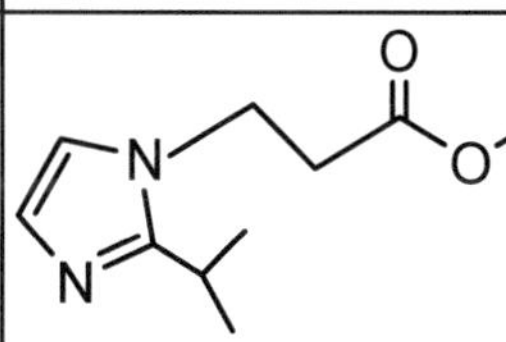	93
7	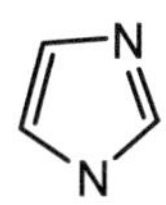	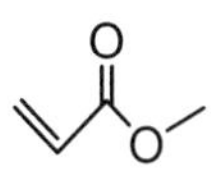	2/14	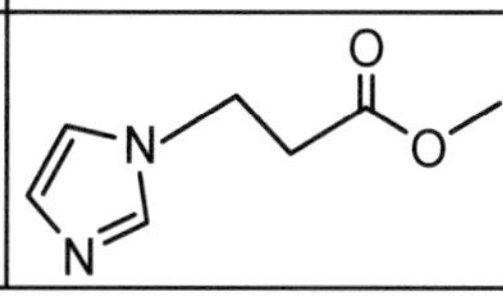	0/95

8	BnNH2	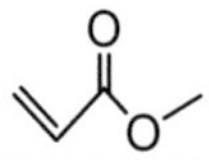	3/3	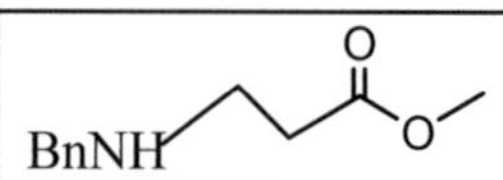	80/75
9	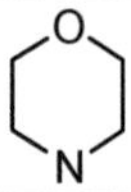	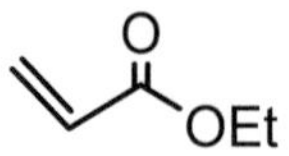	3	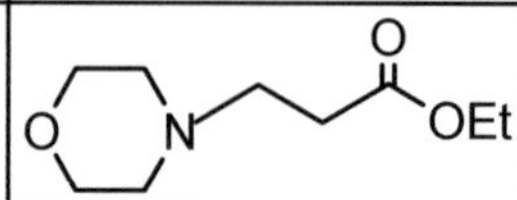	90
10	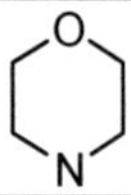	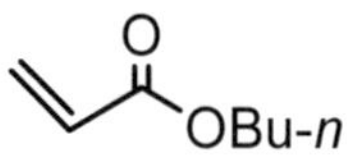	5	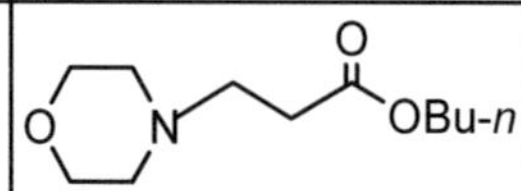	93
11	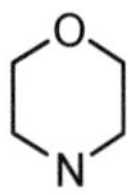	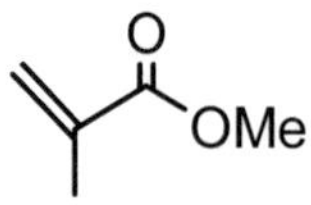	4	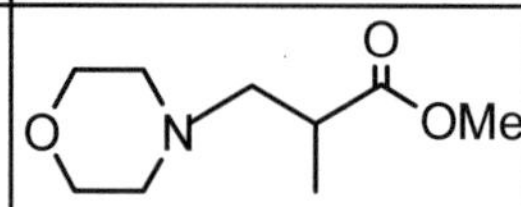	90
12	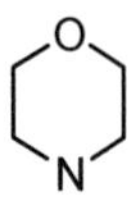	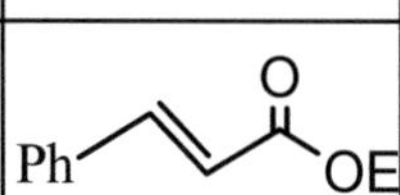	15	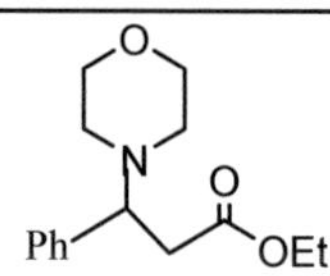	65
13	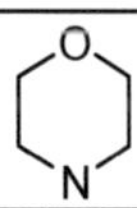	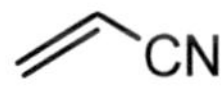	3	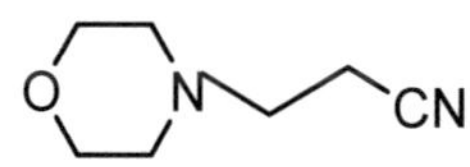	88
14	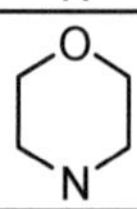	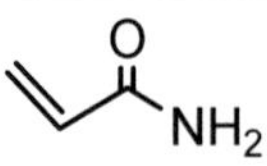	3	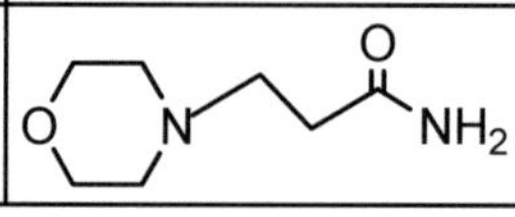	93

Table 2. Results of aza-Michael addition of various aliphatic amines to electron-deficit alkenes using [DBU][Ac] as catalyst under solvent-free conditions[a]

[a] Reactions were carried out on 1.0 mmol scale of substrate with 1.3 equiv of α,β-unsaturated compounds in the presence of 0.3 equiv ionic liquid at room temperature;

[b] Yields of isolated products.

In order to demonstrate the industrial applicability of this methodology, the solvent-free aza-Michael condensation of piperidine and methyl acrylate was carried out on a larger scale (100 mmol). The reaction was completed in 2 hours, excellent yield of 94% for the conjugate product was obtained. On the same scale, the recyclability of catalytic system was investigated using the same reaction as model reaction. Upon the completion of the reaction, the product was isolated through vacuum distillation while the residue ionic liquid was dried to remove water at 60 °C under vacuum. The recovered ionic liquid was reused in subsequent reactions. As shown in Figure 1, the ionic liquid [DBU][Ac] can be recycled for six times without considerable decrease of activity and the used ionic liquid remained intact (^{1}H NMR).

Although [DBU][Ac] has weaker basicity than DBU, its catalytic property for aza-Michael reaction is better than that of DBU. The reason, we speculated, is that amines exhibited higher nucleophilicity in the presence of ionic liquids than in organic solvents(Xu et al., 2007; Kim et al., 2002; Crowhurst et al., 2006; Meciarova et al., 2006).

In conclusion, we have developed a mild, simple and efficient methodology using a new basic ionic liquid [DBU][Ac] as catalyst for the conjugate reaction of various aliphatic and aromatic amines with a variety of structurally diverse olefins. The reactions are conducted at

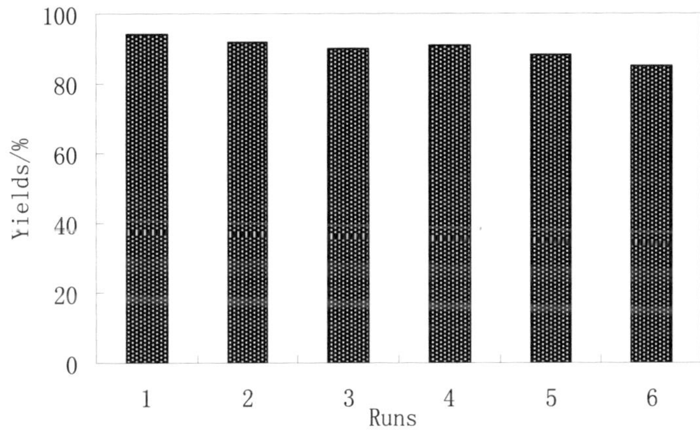

Fig. 1. Reuse of ionic liquid for aza-Michael reaction between piperidine (100 mmol) and methyl acrylate under solvent-free conditions

room temperature under solvent-free conditions and in most cases, from high to excellent yields of the desired 1,4-adducts were obtained. Also, the protocol could be scaled up to 100 mmol and proceeded smoothly, showing the potential for industrial applicability. Upon completion of the reaction, the catalyst [DBU][Ac] could be recovered by drying under vacuum at 60 °C and reused for six times without significant loss of activity. The applicability of the task-specified ionic liquid [DBU][Ac] in other fields of organic transformation are underway in our laboratory.

2.2 Michael addition (aromatic amines as Michael donors)

Because of the inertness of arylamines compared with aliphatic amines, most of the available methods suitable for Michael addition of aliphatic amines are not successful with aromatic amines. Duan and his co-workers used CAN (ceric ammonium nitrate) as effective promoter for aza-Michael addition of aromatic and aliphatic amines to α,β-unsaturated electrophiles in absence of solvent under ultrasound irradiation(Duan et al., 2006). Very recentzly, Bhanage et al. reported $Y(NO_3)_3{\cdot}6H_2O$ catalyzed aza-conjugate reaction between weakly nucleophilic aromatic amines and various Michael acceptors such as esters, nitriles and amides under solvent-free conditions(Bhanushali et al., 2008). However, both of the two procedures utilized transition metal catalysts, which limited their application from economic and environmental viewpoints, and more importantly, they have not been involved in the aza-Michael reaction with α,β-unsaturated ketones as Michael acceptors. Basic ionic liquid [bmim]OH (1-butyl-3-methylimidazolium hydroxide) was first introduced as catalyst for the conjugate reaction of aromatic amines and to our disappointment, however, relatively long reaction times were required(Yang et al., 2006). Thus, the development of the efficient and green protocol for aza-Michael addition of aromatic amines to electron-deficient ketones still remains a challenging task and is highly desirable. Herein we wish to used our recently developed task-specific ionic liquids rather than the reported inert ionic liquids(Tolstikova & Shainyan, 2006) derived from DBU for aza-Michael addition of aromatic amines to α,β-unsaturated ketones at room temperature under solvent-free conditions.

The reaction of aniline and ethyl vinyl ketone (EVK) was selected as model to optimize the reaction conditions. Firstly, using [bmim]OH which is highly efficient in Michael reaction of carbon nucleophiles and aliphatic amines as catalyst(Yang et al., 2007), 8 h was required for consumption of the starting material (Table 3, entry 6)(Yang et al., 2007). Inert ionic liquid [bmim]BF_4 (1-butyl-3-methylimidazolium tetrafluoroborate) was also tested as promoter of the reaction and only 55 % yield of the desired product was obtained in 40 min (Table 3, entry 7). For comparison, some Lewis bases such as DMAP, PPh_3, and DBU were subjected to the model reaction and bis-adduct was formed thereby decreasing the product yield (Table 3, entries 8-10). Especially, with DBU, the bis-addition product was produced as much as 32 % yield while using its corresponding ionic liquids [DBU][Ac] , [DBU][Lac], [DBU][n-Pr], [DBU][n-Bu] and [DBU][Tfa] no bis-addition reaction was observed and 89-99 % yields were afforded within the same period, demonstrating good chemo- and regio-selectivity of these ionic liquids in the reaction of aniline and ethyl vinyl ketone (Table 3, entries 1-5). Among the three ionic liquids, [DBU][Lac] performed better than the other ionic liquids, [DBU][Ac], [DBU][n-Pr], [DBU][n-Bu] and [DBU][Tfa] (entries 1-5). So we chose [DBU][Lac] as catalyst for further investigations. A blank experiment was also carried out to demonstrate the catalytic activity of the DBU-based task-specific ionic liquids (Table 3, entry 11). To optimize the details of the reaction conditions, solvent effect was studied in the same model reaction between aniline and EVK (Table 3, entry 1). Toluene, CH_2Cl_2, CH_3OH, and CH3CN were all found to be effective media for the reaction and relatively lower reactant concentration in organic solvent than that under solvent-free conditions decreased the reaction rate. We then selected the solvent-free conditions rather than using organic solvents on consideration of the environmental effect and reaction rate.

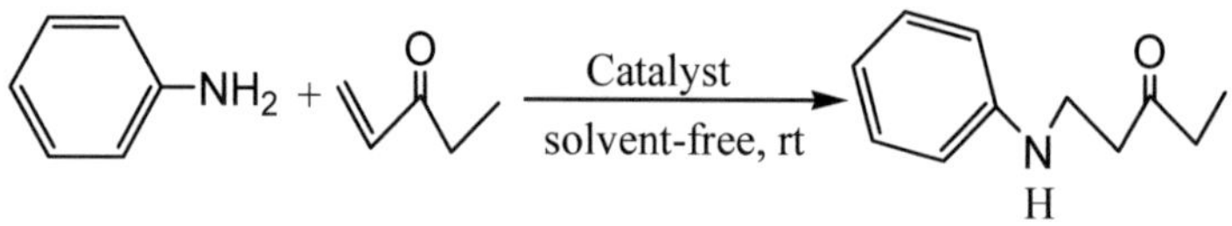

Entry	Catalyst	Time (min)	Yieldb (%)
1	[DBU][Lac]	30	99 (82c, 75d, 76e, 70f)
2	[DBU][Ac]	40	96
3	[DBU][n-Pr]	40	90
4	[DBU][n-Bu]		89
5	[DBU][Tfa]	40	96
6g	[bmim]OH	480	9823
7	[bmim]BF4	40	55
8	DMAP	40	64
9	PPh3	40	59
10	DBU	40	43
11	-	40	37

Table 3. Aza-Michael Reaction of Aniline with Ethyl Vinyl Ketone under Various Reaction Conditions[a]

[a] Reactions conditions: aniline (1 mmol), ethyl vinyl ketone (1.5 mmol), catalyst (30 mol %), solvent-free conditions, rt. [b] GC yield. [c] Toluene. [d] CH_2Cl_2. [e] CH_3OH. [f] CH_3CN.
[g] Reference: Yang et al., 2007.

With the efficient catalytic system in hand, EVK was treated with other aromatic amines. The results are summarized in Table 4 (entries 1-5). Aromatic amines with electron-donating group at bezene ring were effective nucleophiles to react smoothly with EVK and excellent yields were afforded in very short reaction time (Table 4, entries 1-2), while 2-Methyl substituted amine, due to its steric hindrance, reacted relatively slowly (entry 3). p-NO_2 substituted amine could not react all (entry 5). Arylamine with weakly electron-drawing substitution Cl was also tested with EVK under the above reaction conditions and 95 % GC yield was obtained in 1 h (entry 4).

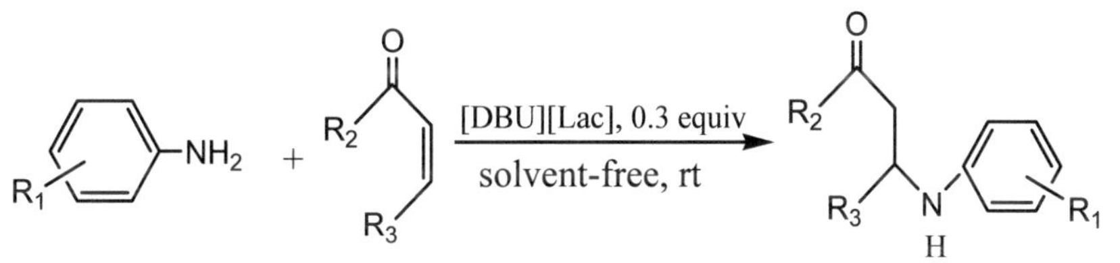

Entry	R1	R2	R3	Time (h)	GC yield (%)	Isolated yield (%)
1	4-MeO	Et	H	0.5	98	95
2	4-Me	Et	H	0.1/8c	97/98c	90
3	2-Me	Et	H	0.8	92	85
4	4-Cl	Et	H	1/8c	95/97c	90
5	2-NO2	Et	H	2	NR	NR
6	4-H	$-(CH_2)_3-$		5	89	82
7	4-Cl	$-(CH_2)_3-$		6	90	79
8b	3-Cl	$-(CH_2)_3-$		5	75	68
9	2-Me	$-(CH_2)_3-$		8	85	77
10	4-Me	$-(CH_2)_3-$		5/24c	95/71c	92
11[b]	2-Cl	$-(CH_2)_2-$		8	70	65
12[b]	2-Br	$-(CH_2)_2-$		8	75	69
13[b]	2-Me	$-(CH_2)_2-$		8	86	78
14[d]	4-H	Ph	Ph	5	92	85
15[d]	4-Cl	Ph	Ph	12	75	72
16[d]	4-MeO	Ph	Ph	8	88	83
17[d]	4-Me	Ph	Ph	8	91	87

Table 4. [DBU][Lac] catalyzed aza-Michael reactions of various aromatic amines with α,β-unsaturated ketones at room temperature[a]

[a] Reaction conditions: aromatic amines (1 mmol), α,β-unsaturated ketones (1.5 mmol), [DBU][Lac] (30 mol %), solvent-free conditions, rt. [b] Reaction conducted at 60 °C.

[c] Results obtained from reference: Yang et al., 2007, reaction conditions: [bmim]OH (0.3 equiv.), aromatic amines (1.0 mmol), acceptors (1.5 mmol), rt, solvent-free. [d] 1.0 Equiv of [DBU][Lac] required for dissolving chalcone and aromatic amines.

In order to evaluate the generality of the ionic liquid [DBU][Lac] as catalyst for aza-Michael reaction of aromatic amines, 2-cyclohexen-1-one, 2-cyclopenten-1-one and chalcone were tested under the same conditions (Table 4, entries 6-17). To our delight, 2-cyclohexen-1-one and chalcone were effective Michael acceptors to react with various arylamines, giving good to excellent isolated yields (entries, 6-10, 14-17). However, 2-cyclopenten-1-one showed relatively lower reactivity. Accordingly, higher temperature (60 °C) was required (entries, 11-13). For further comparison with [bmim]OH, [DBU][Lac] was used as promoter for the additions of p-methyl aniline to EVK and p-methyl aniline to 2-cyclohexen-1-one respectively (entries, 2 and 10). With comparable or higher products yields obtained, the two reactions catalyzed by [DBU][Lac] proceeded much faster than those promoted by [bmim]OH. Moreover, chalcone reacted smoothly with various aromatic amines using [DBU][Lac] as catalyst (entries, 14-17) while no reaction occurred in the presence of the ionic liquid [bmim]OH(Yang et al., 2007). All cases summarized in Table 4 obviously demonstrate that the novel ionic liquid [DBU][Lac] has excellent catalytic activity for aza-Michael reaction of aromatic amines and α,β-unsaturated ketones.

The recyclability of the ionic liquid [DBU][Lac] was then studied using the reaction of aniline and 2-cyclohexen-1-one as model. The results are shown in Table 5. Upon the completion of the reaction, the reaction solution was extracted with ethyl acetate and purified by flash chromatography. The addition product was identified by ^{1}H NMR, ^{13}C NMR and MS. The residual ionic liquid was washed with ethyl ether, dried under vacuum at 60 °C for 2 h and reused for subsequent reactions. As shown in Table 5, the recovered ionic liquid could be used for 8 times without obvious loss of catalytic activity. It is worthwhile to note that the ionic liquid [DBU][Lac] used for 8 runs remained intact, judging from its ^{1}H NMR spectrum.

Cycle	t (h)	GC yieldb (%)
1	5	89
2	5	90
3	5	88
4	5	89
5	5	83
6	5	83
7	6	90
8	6	90

Table 5. Recycling and Reusability of the catalyst [DBU][Lac] in the Reaction between aniline and 2-cyclohexen-1-one[a]

[a] Reaction conditions: 1 mmol of aniline, 1.5 mmol of 2-cyclohexen-1-one, 0.3 mmol of [DBU][Lac] without solvents at rt. [b] GC yields.

As for the role of DBU-based ionic liquids for the activation in the aza-Michael addition of aromatic amines to α,β-unsaturated ketones, [DBU][Lac], [DBU][Ac], [DBU][*n*-Pr] and [DBU][*n*-Bu] were regarded as Brønsted bases to promote the addition and [DBU][Tfa] played the role as Brønsted acid for the this type of aza-Michael reaction. The reason for relatively lower catalytic activity of [DBU][*n*-Pr] and [DBU][*n*-Bu] might be their much higher hindrance of anions. The reason for slightly higher catalytic activity of [DBU][Lac] than those of the other four ionic liquids may be the activation of carbonyl group of Michael acceptors by the hydroxyl group at the α-position of the carboxyl in lactate anion.

In conclusion, we have developed five task-specific ionic liquids, [DBU][Lac], [DBU][Ac], [DBU][*n*-Pr], [DBU][*n*-Bu] and [DBU][Tfa]. The ionic liquids were then used as catalysts for aza-Michael addition of aniline to EVK. Among the five ionic liquids, [DBU][Lac] exhibited the best catalytic activity. Using [DBU][Lac] as catalyst, various aromatic amines were subjected to 2-cyclohexen-1-one, 2-cyclopenten-1-one and chalcone, from good to excellent yields were obtained. This protocol also has some advantages, such as readily work-up of the reactions, excellent reaction selectivity and good recyclability of the ionic liquid (reused for 8 times without significant loss of activity).

3. Application in the Knoevenagel condensation

3.1 Condensation under solvent-free conditions

Knoevenagel condensation is one of the most important methodologies for carbon-carbon double bond formation in synthetic chemistry(Freeman, 1980; Tietze, 1996). The condensation products are the key intermediates for synthesis of natural and therapeutic drugs, polymer, cosmetics and perfumes(Tietze, 2004; Yu et al., 2000). Generally, Knoevenagel reactions are carried out by condensation of active methylene compounds with aldehydes using some organic bases with their salts as catalysts, including dimethylamino pyridine, piperidine, guanidine, ethylenediamine(Narsaiah et al., 2004; Han et al., 2008), and so on. Also, alternative protocols for Knoevenagel condensations catalyzed by Lewis acids such as $ZnCl_2$(Shanthan Rao & Venkataratnam, 1991), $CuCl_2$(Attanasi et al., 1983), $TiCl_4$(Green et al., 1985), $LaCl_3$(Narsaiah & Nagaiah, 2003), $Mg(ClO_4)_2$(Bartoli et al., 2008) and various heterogeneous solid bases, including Zeolites(Saravanamurugan et al., 2006; Martns et al., 2008), sulfate-ion promoted Zirconia(Reddy et al., 2006), clay(Bigi et al., 1999), and layered double hydroxides (LDHs)(Kantam et al., 2006) have been reported in literatures. However, many of those procedures required the use of large amount of organic solvents as reaction medium, long reaction time, harsh reaction conditions and had difficulties in the reuse of catalysts, which prompt chemical researchers for further development of more environmentally benign, efficient and operationally simple Knoevenagel protocols.

Ionic liquids, due to their unique properties such as good solvating ability, negligible vapor pressure, variable polarity, nonflammability and recyclability have been widely used as catalyst as well as reaction medium in Knoevenagel condensations with more or less success(Paun et al., 2007; Hu et al., 2005; Gao et al., 2007; Yue et al., 2008; Santamarta et al., 2008; Cai et al., 2006; Yeom et al., 2007). However, the costs of those methods increased because large excess amount of ionic liquids were required, which would greatly limit their industrial applications. From both environmental and economical points of view, Knoevenagel reaction promoted by catalytic amount of ionic liquid without solvent may be an attractive catalytic reaction system with many advantages such as lower cost, reduced pollution and simple operation.

Very recently, Kakade et al. investigated the use of DBU as catalyst for Knoevenagel reaction of 2-chloroquinoline-3-carbaldehyde and ethyl cyanoacetate under ultrasonic irradiation(Kakade et al., 2008). Encouraged by this, we tested the reaction of benzaldehyde with ethyl cyanocetate catalyzed by 20 mol % amount of DBU without solvent (Table 6, entry 10) and were pleased to found that 91 % yield of condensation product was obtained within 20 min. However, using DBU as catalyst has some problems that it could not be reused and has unplesant flavour during the operational process which also exists in other reaction with organic bases as promoters for synthetic transformation. In order to overcome these problems and promped by our wish for invention of new task-specified ionic liquid used for organic transformations, we would like to report a novel, basic, task specific ionic liquids, [DBU][X](X=Tfa, Lac, Ac, *n*-Pr, *n*-Bu) and their use as catalysts for Knoevenagel reactions between aromatic aldehydes and active methylene ingredients.

Then we examined novel DBU derived ionic liquids used as catalyst for the reaction of benzaldehyde with ethyl cyanoacetate under solvent-free conditions at room temperature. The results shows that [DBU][Lac] demonstrates the most excellent catalytic activity(Table 6, entry 3 and entries 6-9). A remarkable yield of desired product was obtained (Table 6, entry 3). Some literatures reported that water could promote the Knoevenagel coupling(Shaabani et al., 2007; El-Rahman et al., 2007; Deb & Bhuyan, 2005). Thus, we carried out the reaction in water for the purpose of comparison with that under solvent-free conditions and found that solvent-free condensation is much faster than aqueous reaction (Table 6, entries 3, 4).

A blank experiment to demonstrate the catalytic ability of [DBU][Lac] was carried out. Only 37 % of the desired product was obtained in absence of [DBU][Lac] even with long reaction time (Table 6, entry 11). This result clearly indicated the effective catalytic role of the new ionic liquid in Knoevenagel condensation. To find a optimal amount of catalyst used for the model reaction of benzaldehyde and ethyl cyanoacetate, the amount of [DBU][Lac] was reduced from 0.5 equiv to 0.01 equiv. The results are collected in Table 5. The reactivity for lower loading of [DBU][Lac] (0.01 equiv) decreased obviously. Increasing the amount of [DBU][Lac] from 20 mol % to 50 mol %, expected improvement in yield of electrophilic alkene has not been observed (Table 6, entries 1-4). Thus, 0.2 equiv was the optimal amount of catalyst for further experiments.

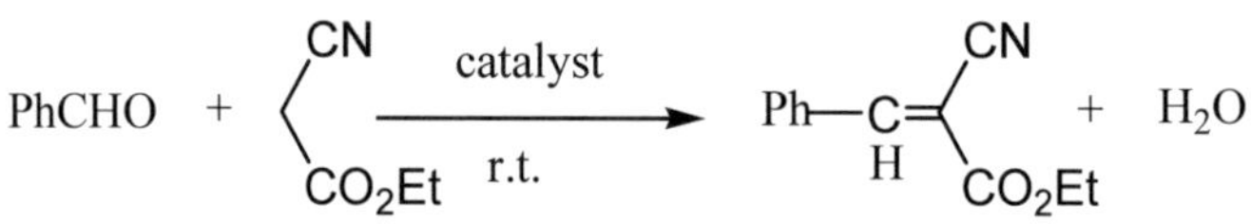

Entry	Catalysts (mol %)	Time (min)	Yieldsa (%)
1	[DBU][Lac] (0.1)	30	59
2	[DBU][Lac] (1)	30	73
3	[DBU][Lac] (20)	20	93
4b	[DBU][Lac] (20)	120	95
5	[DBU][Lac] (50)	20	94
6	[DBU][Ac] (20)	20	90
7	[DBU][n-Pr] (20)	20	85
8	[DBU][n-Bu] (20)	20	83
9	[DBU][Tfa] (20)	20	78
10	DBU (20)	20	91
11	-	10	37

Table 6. Results of varying the amounts of [DBU][Lac] in the solvent-free Knoevenagel condensation of benzaldehyde and ethyl cyanoacetate at room temperature
[a] Isolated yields of products [b] Reaction in water.

With the best catalytic system in hand, we investigated the Knoevenagel reaction of various aromatic aldehydes with ethyl cyanoacetate and malononitrile. The results are shown in Table 7. A variety of structurally diverse aromatic aldehydes reacted favorably with active nucleophilic reagents to give the desired products in good to excellent yields. The products could be isolated by simple filtration. Some crude products need recrystallization in ethanol. All products were verified by their melting point, ^{1}H NMR and ^{13}C NMR spectroscopy which were in good agreement with literatures.

Aromatic aldehydes bearing withdrawing groups reacted with ethyl cyanoacetate, affording lower yields of desired products, while the reaction time is shorter than those with electron-donating substituents (Table 7, entries 1-6). Such results are mainly attributed to the strong activities of electro-deficient aromatic aldehydes, which led to some side reactions and thus resulted in lower yield. When malononitrile was treated with aromatic aldehydes carrying electron donating or withdrawing groups, the reactions proceeded within short time to achieve very high isolated yields (Table 7, entries 1-5). It is notable that electron-donating or electron-withdrawing substituents on aromatic rings have less effect on Knoevenagel reaction presumably due to the strong activity of acidic malononitrile. It can also been observed from Table 7 that more sluggish methylene compounds, acetylacetone can not converse completely and moderate yield afforded (Table 7, entry 17). The order of reaction rate as well as efficiency of Knoevenagel reaction is as follows: malononitrile>ethyl cyanoacetate> acetylacetone, which is accordance with the acidic activity of three nucleophiles. In addition, Knoevenagel condensation of hetero aromatic aldehydes such as 2-furaldehyde and 3- pyridinecarboxaldehyde with active methylene compounds also underwent smoothly at room temperature (Table 7, entries 7, 8, 14, 15). It is worthy to note that all products obtained are E-geometry exclusively and no subsequent Michael adduct is detected. Unfortunately, almost no desired condensation product was found when aliphatic aldehydes and ketones were used as the substrates under the above reaction conditions due to the inertness of these two types of compounds.

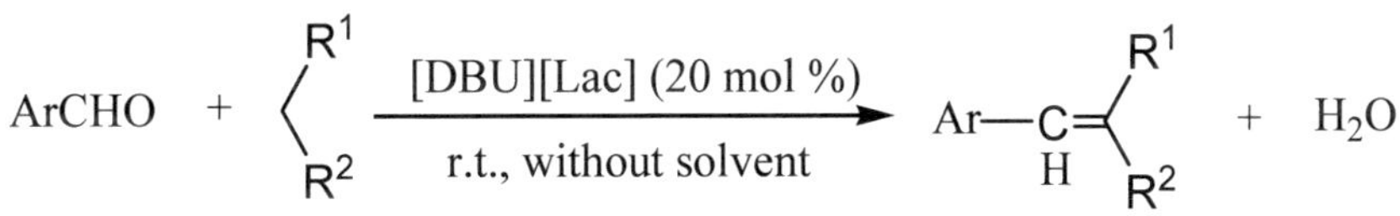

Entry	Ar	R1	R2	Time (min)	Products	Yields a (%)
1	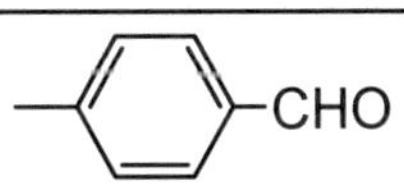	CN	CO2Et	30	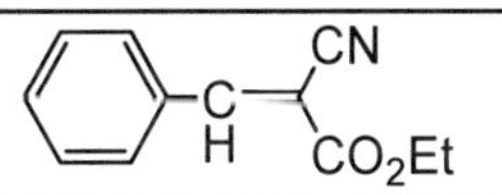	95
2	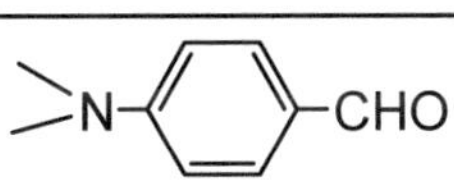	CN	CO2Et	60	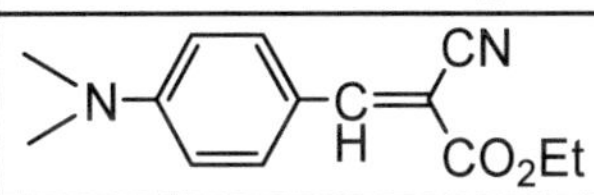	95,
3	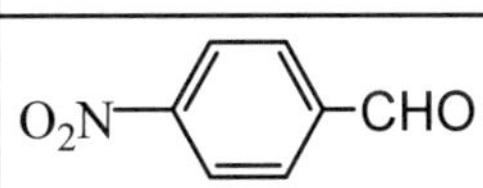	CN	CO2Et	3	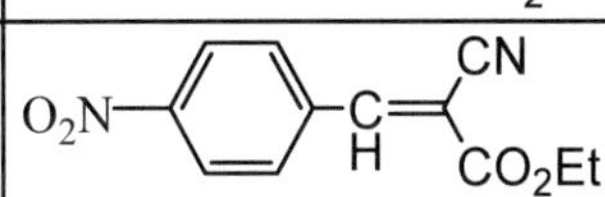	80
4	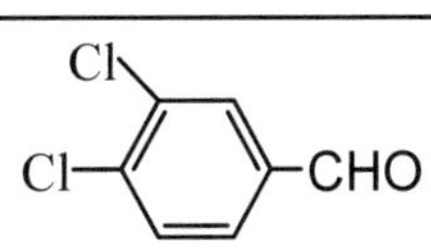	CN	CO2Et	3	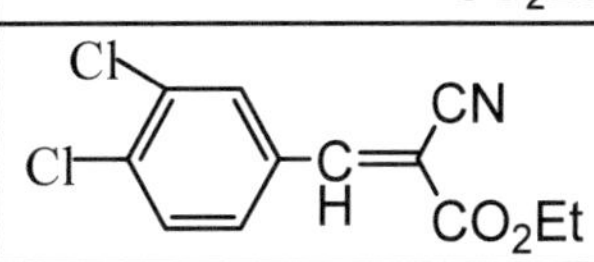	83

5	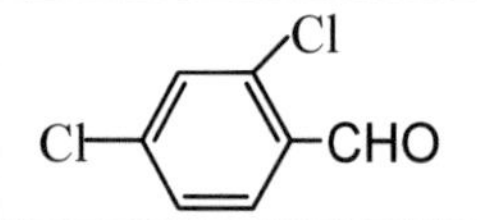	CN	CO2Et	10	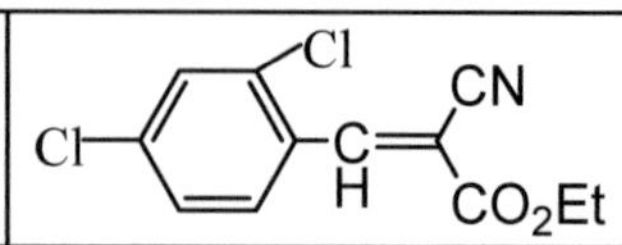	86
6	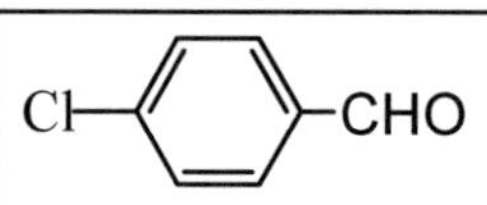	CN	CO2Et	5	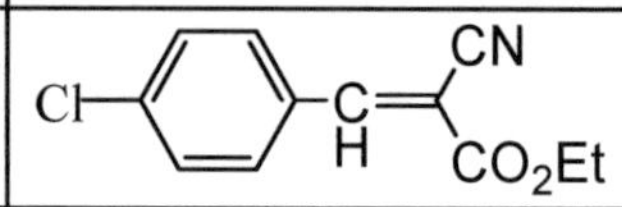	93
7	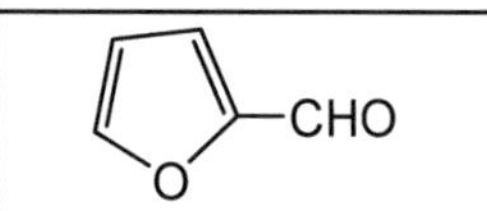	CN	CO2Et	1	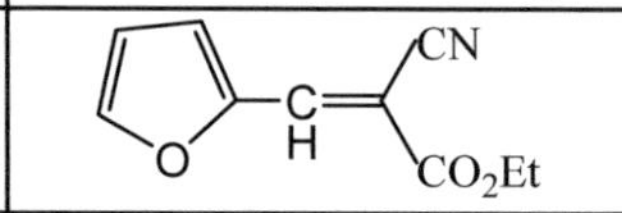	90
8	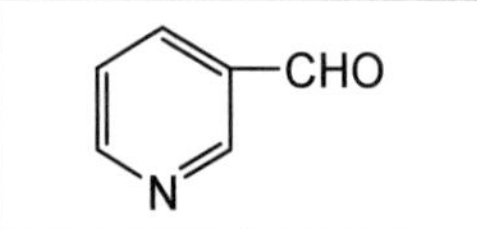	CN	CO2Et	20	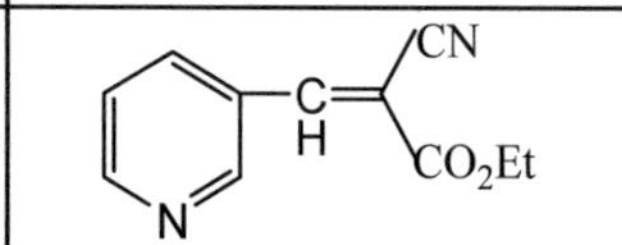	75
9	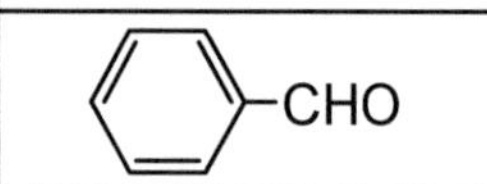	CN	CN	2	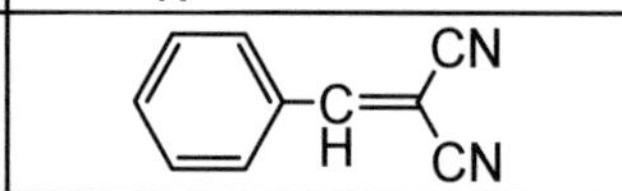	93
10	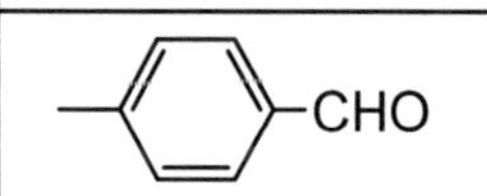	CN	CN	1	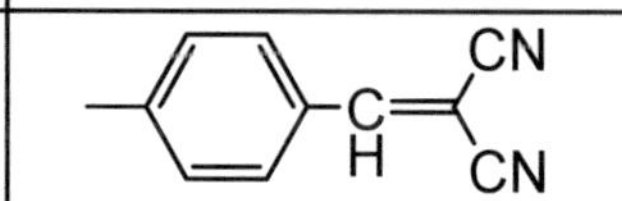	92
11	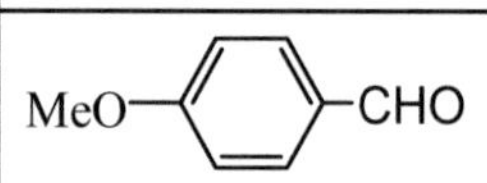	CN	CN	2	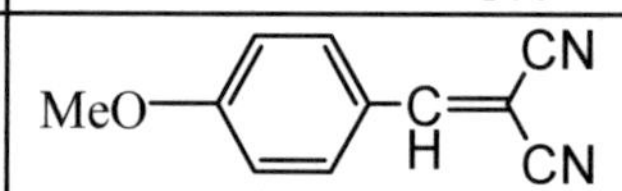	90
12	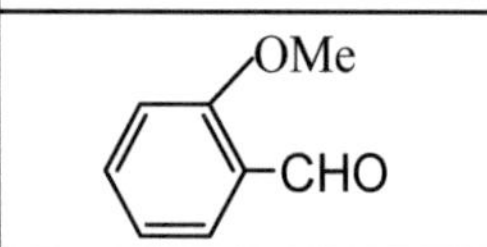	CN	CN	5	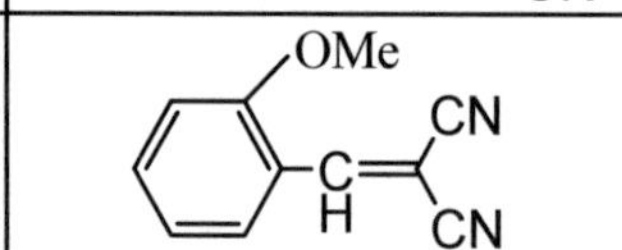	95
13	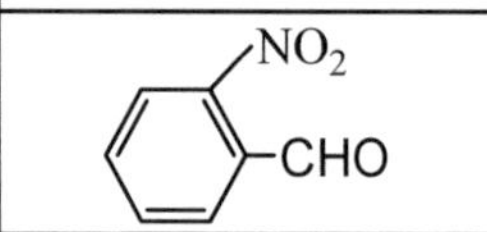	CN	CN	2	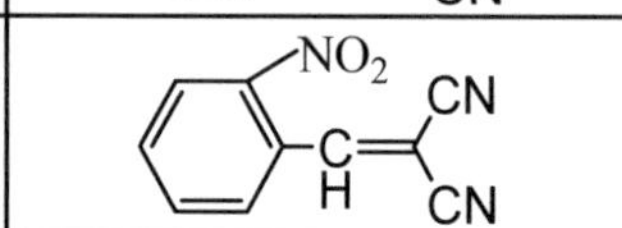	86
14	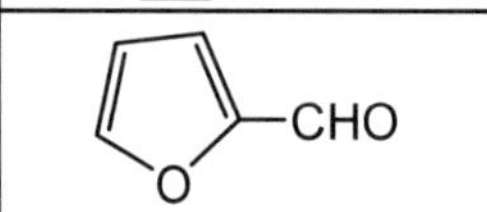	CN	CN	5	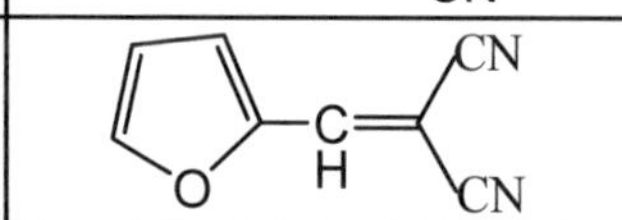	90
15	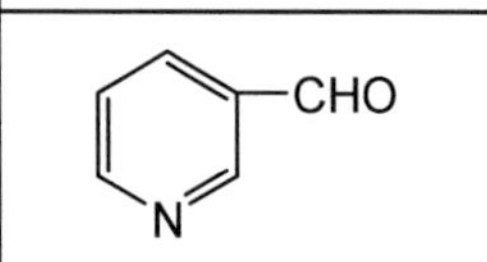	CN	CN	10	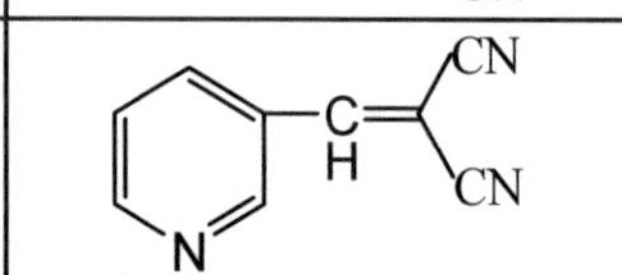	85
16	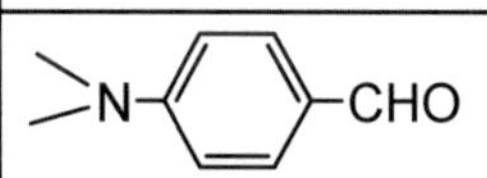	CN	CN	3	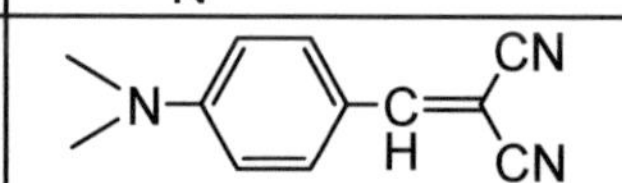	93
17c	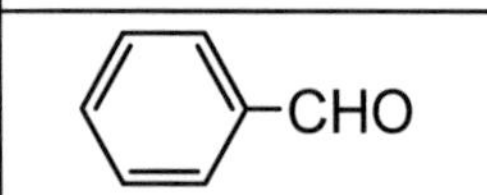	COMe	COMe	300	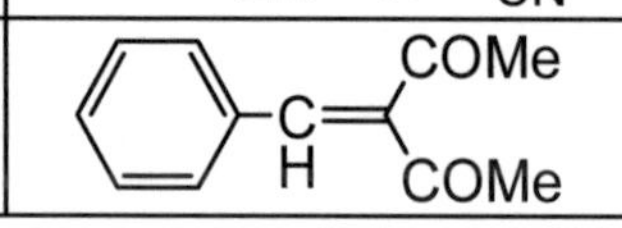	65

Table 7. Results of Knoevenagel condensation between various aromatic aldehydes and methylene active compounds at room temperature under solvent-free conditions[b]
[a] Isolated yields of desired products; [b] Reaction conditions: aromatic aldehydes (1 mmol),

active methylene compounds (1mmol) stired with 20 mol % [DBU][Lac] as catalyst without solvent at room temperature; [c] Reactions at 60°C;

For the purpose of comparison with other methodologies on the catalytic efficiency, we carried out the reaction of slugguish substrate 4-(dimethylamino) benzaldehyde with ethyl cyanoacetate. As shown in Table 8, only reaction in guanidinium lactate provided the comparable yield (97.2 %) while it required 2.5 equiv amount of ionic liquid, which would limit its scale-up (Table 8, entry 4-5). There was a problem about recyclability of catalytic system in the condensation reaction promoted by phase-transfer catalyst TEBA (Table 8, entry 3). Other methods suffered from the longer reaction times as well as lower yields of products (Table 8, entries 1-2). All the results show that the present catalytic system (Table 8, entry 5) is very efficient and economic for Knoevenagel condensation.

Entry	Reaction conditions	Yields (%)	Ref.
1	Perfluoroalkylated pyridine catalyzed in n-octane, at 80℃ for 10 h.	78	Yi & Cai, 2008
2	PEG as solvent, catalyzed by L-proline, reaction time: 210 min	81.2	Liu et al., 2008
3	Catalyst: TEBA, reaction time: 15 min	86	Rong et al., 2006
4	2.5 equiv amount of cyclic guanidinium lactate as solvent, reaction time: 3 min	97.2	Liang et al., 2008
5	60 min for reaction, 0.2 equiv [DBU][Lac] as catalyst	95	-

Table 8. Cmparison of the present catalytic system with other reported protocols in the model reaction between 4-(dimethylamino) benzaldehyde with ethyl cyanoacetate.

Finally, in order to demonstrate the industrial applicability of this methodology, the solvent-free Knoevenagel condensation of benzaldehyde and ethyl cyanoacetate was carried out on a larger scale (100 mmol). The reaction completed in 20 min., excellent yield of 94 % for the condensation product was achieved. On the same scale, the recyclibility of catalytic system was investigated using the same reaction as model reaction. Upon the completion of the reaction, the product was isolated via readily filtration, washed with water and dried under vacuum without further separation, while the filtrate containing [DBU][Lac] was dried to remove water at 60 °C under vacuum. The recycled ionic liquid was reused in subsequent reactions. As shown in Figure 2, the ionic liquid [DBU][Lac] can be recycled for six times without considerable decrease of activity and the used ionic liuid remained intact (^1H NMR). As for the role of [DBU][Lac] in Knoevenagel condensation, we speculate that the basicity of the ionic liquid can attract the proton of the nucleophile to form carbon anion firstly, and secondly, catalytic amount of ionic liquid can improve the nucleophilicity of carbon anion to facilitate the condensation reaction, as same as the behavior observed in larger amount of ionic liquid(Kim et al., 2002; Kim et al., 2003; Liang et al., 2008). However, more work need to be done to clarify the exact mechanism of the [DBU][Lac] mediated Knoevenagel reaction.

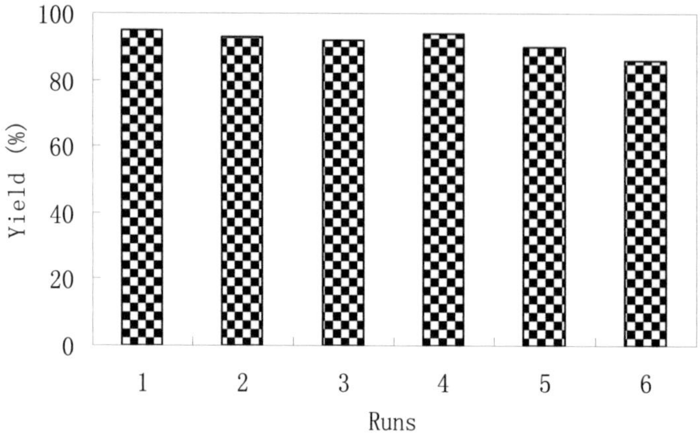

Fig. 2. Reuse of catalyst for Knoevenagel condensation between benzaldehyde and ethyl cyanoacetate (100 mmol) under solvent-free conditions

In conclusion, we have established a simple, mild and efficient methodology for solvent-free Knoevenagel condensation between aromatic aldehydes and methylene compounds at ambient temperature with ionic liquid [DBU][Lac] as catalyst. This protocol has notable advantages, such as high product yields, reuse of ionic liquid on large scale.

3.2 Condensation reaction in water

In the last decades, there has been incredible growth in research involving the use of water as a green, environmentally benign replacement for a wide range of processes that currently rely on conventional organic solvents from both environmental and economical points of view(Breslow, 1991; Li, 1993; Ribe & Wipf, 2001). More recently, Song and co-workers employed water as reaction solvent for Knoevenagle condensation between aromatic aldehydes and malononitirle(Cai et al., 2006). Compared with organic solvents media, the advantage of water is very obvious from points of reaction efficiency or environmental "greeness".

Then we examined DBU based ionic liquids as catalysts for the reaction of benzaldehyde with ethyl cyanoacetate in water at room temperature. Among the five ionic liquids, [DBU][Ac] has the best efficient catalytic activity(Table 9, entries 3 and 5-8). An excellent yield of desired product was obtained (Table 9, entry 3). The result is comparable with that catalyzed by parent catalyst DBU. Bhuyan et al. publicated a method for Knoevenagel reaction in aqueous medium without any catalyst(Deb & Bhuyan, 2005). However, the method was substrate-selective and the aqueous reaction of benzaldehyde with ethyl cyanoacetate in absence of [DBU][Ac] is very slow with low yield obtained (Table 9, entry 9). The result indicated that the ionic liquid [DBU][Ac] played an important role as catalyst during the reaction process.

To find a optimal loading amount of catalyst for the model reaction of benzaldehyde and ethyl cyanoacetate, the amount of [DBU][Ac] was reduced from 0.5 equiv to 0.01 equiv. The results are collected in Table 9. Considering both catalytic activity and the cost of catalyst, we chose 0.2 equiv as optimal amount of catalyst for further examinations.

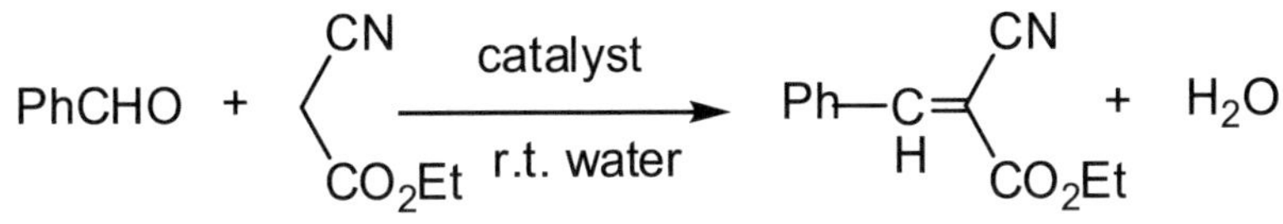

Entry	Catalysts (mol %)	Time (h)	Yieldsa (%)
1	[DBU][Ac] (1)	3	59
2	[DBU][Ac] (5)	3	73
3	[DBU][Ac] (20)	2	95
4	[DBU][Ac] (50)	2	94
5	[DBU][Lac] (20)	2	92
6	[DBU][n-Pr] (20)	2	90
7	[DBU][n-Bu] (20)	2	85
8	[DBU][Tfa] (20)	2	85
9	-	10	36
10	DBU (20)	2	93

Table 9. Results of DBU and varying the amounts of [DBU][Ac] in the aqueous Knoevenagel condensation of benzaldehyde and ethyl cyanoacetate at room temperature
[a] Isolated yields of products

With the rational catalytic system in hand, we studied the Knoevenagel reaction of various aromatic aldehydes with malononitrile and ethyl cyanoacetate. The results are shown in Table 10. A variety of structurally diverse aromatic aldehydes reacted favorably with active nucleophilic reagents, giving desired products in good to excellent yields. The products could be isolated by simple filtration without further tedious purification process. All products were verified by their melting point, ^{1}H NMR and ^{13}C NMR spectroscopy which were in good agreement with the data in other published literatures.

In order to demonstrate the industrial applicability of this methodology, the Knoevenagel condensation of benzaldehyde and ethyl cyanoacetate was carried out on a larger scale (100 mmol) in water (100 mmL). The reaction completed in 2 h, excellent yield of 97% for the condensational product was achieved. On the same scale, the recyclability of catalytic system was investigated using the same reaction as model. Upon the completion of the reaction, the high pure product was afforded by readily filtration without further separation, while the remaining aqueous medium containing [DBU][Ac] was reused directly without additional recovery. As shown in Figure 3, the catalytic system of [DBU][Ac] in water can be reused for ten times without noticeable loss of activity.

In summary, we have established a simple, mild and efficient methodology for Knoevenagel condensation between aromatic aldehydes and methylene compounds at ambient temperature in water with ionic liquid [DBU][Ac] as catalyst. This protocol has notable advantages, such as high product yields, aqueous reaction medium, ease of work-up, reuse of ionic liquid on large scale, which make it more efficient and cleaner with industrial potential.

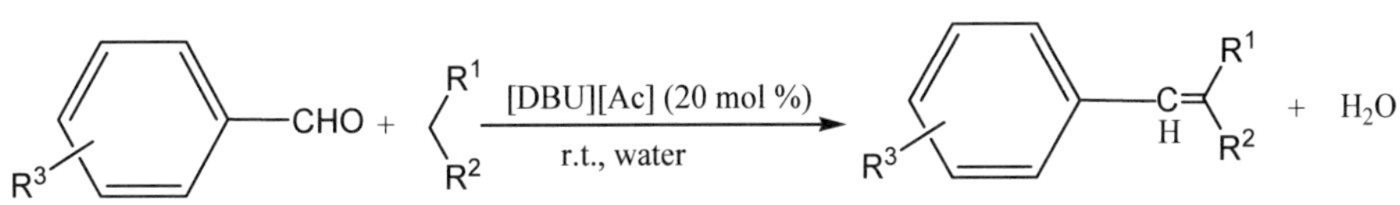

Entry	R3	R1	R2	Time (h)	Yieldsa (%)
1	H	CN	CN	0.3	93
2	4-Me	CN	CN	0.3	94
3	4-MeO	CN	CN	0.3	94
4	2-MeO	CN	CN	0.3	90
5	2-NO2	CN	CN	0.2	88
6d	4-Dimethyl	CN	CO2Et	6	96
7	4-Me	CN	CO2Et	4	95
8	4-NO2	CN	CO2Et	1	90
9	3,4-Dichloro	CN	CO2Et	3	82
10	2,4-Dichloro	CN	CO2Et	2	75
11	4-Cl	CN	CO2Et	1.5	95
12d	H	COMe	COMe	10	63b
13	2-Furaldehyde	CN	CN	1	86
14	3-Pyridinecarboxaldehyde	CN	CN	1	96
15	2-Furaldehyde	CN	CO2Et	3	92
16	3-Pyridinecarboxaldehyde	CN	CO2Et	3	94

Table 10. Results of Knoevenagel condensation between various aromatic aldehydes and methylene active compounds at room temperature in aqueous media[c]
[a] Isolated yields of desired products [b] Starting materials detected
[c] Reaction conditions: aromatic aldehydes (1 mmol), active methylene compounds (1mmol) stirred in water (1mL) with 20 mol % [DBU][Ac] as catalyst at room temperature
[d] Reactions at 60°C

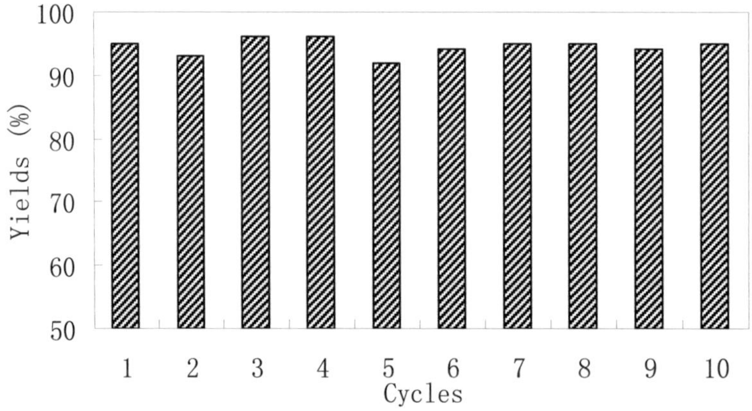

Fig. 3. Reuse of catalyst for Knoevenagel condensation between benzaldehyde and ethyl cyanoacetate (100 mmol) in 100 mL water

4. Ionic liquids immobilized on MNP and their applications in organic synthesis

In order to improve the recycling and reuse property of ionic liquids, immobilization of ionic liquids on polymers is frequently employed(Welton, 2004). However, the reduced activities of these type of ionic liquids catalysts are often observed because of poor dispersion of supported ionic liquids in the reaction system. Thus, novel supports that ensure both good recyclability and catalytic activity are still desirable.

MNP have recently appeared as a new type of catalyst support because of their easy preparation and functionalization, large surface area ratio, facile separation via magnetic force as well as low toxicity and price(Yoon et al., 2003; Stevens et al., 2005; Lee et al., 2006; Abu-Reziq et al., 2006; Luo et al., 2008). These features have made MNP a promising alternate of porous/mesoporous catalyst supports.

Combined advantages of ionic liquids with those of MNP, MNP-supported ionic liquid catalysts (MNP-ILs) were developed by some famous research groups. Luo and his co-workers(Zheng et al., 2009) developed a series of novel MNP-ILs and successfully utilized these catalyst in CO2 cycloaddition reactions (Scheme 3). The activity of the supported catalyst is comparable with that of the free ILs catalysts for this reaction. The catalysts could be easily recycled using a magnetic force and reused for up to 11 times with essentially no loss of activity.

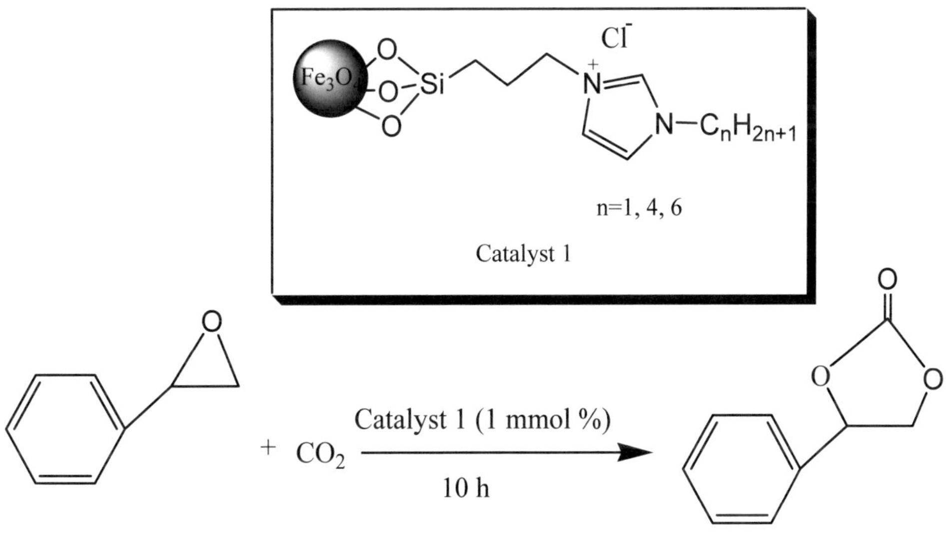

Scheme 3

Pd-NHC-ionic liquid matrix was immobilized into ionic liquid layers coated on the surface of Fe_3O_4(Taher et al., 2009). The catalyst 2 has shown excellent catalytic activity and high stability for the Suzuki coupling reaction in water (Scheme 4). This heterogeneous catalyst can be recycled for five times without significant loss of the catalytic activity. Furthermore, recovery of the catalyst by an external permanent magnet is facile and efficient.

Shan and his co-workers(Wang et al., 2010) prepared a novel magnetic nano-solid acid catalyst, which includes grafting ionic liquid onto Fe_3O_4 nanoparticles, followed by the sulfonation of phenyls groups in the ionic liquid. The catalyst 3 shows an excellent performance in the acetalization of the carbonyl under mild reaction conditions (Scheme 5), and it can be recycled without obvious loss of catalytic activity.

In summery, MNP-supported ionic liquid catalysts (MNP-ILs) have been developed, which overcome the disadvantages of both magnetic nano-particles and ionic liquids. The three examples have showed the excellent catalytic activities and facile work-up of the MNP-ILs.

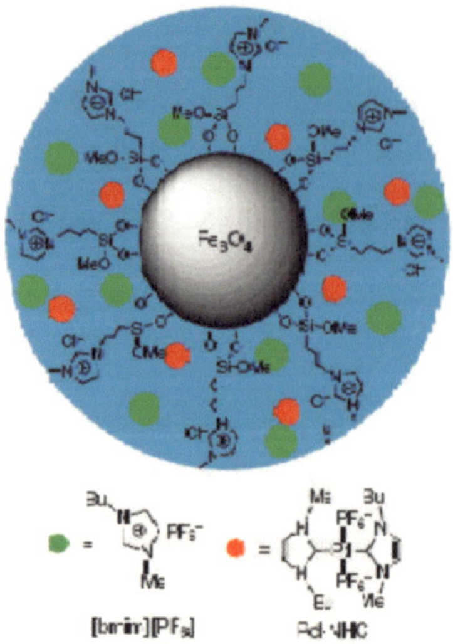

Catalyst 2

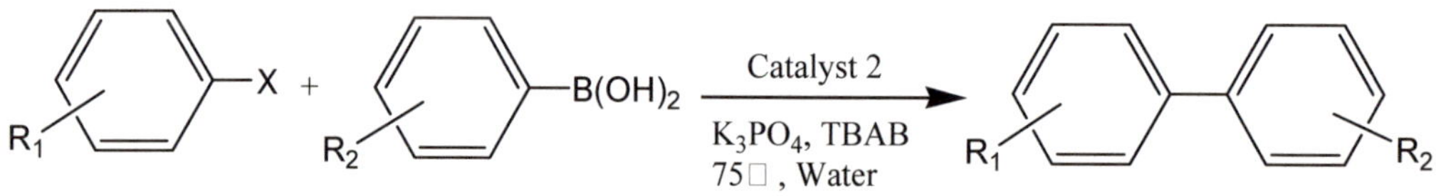

Scheme 4

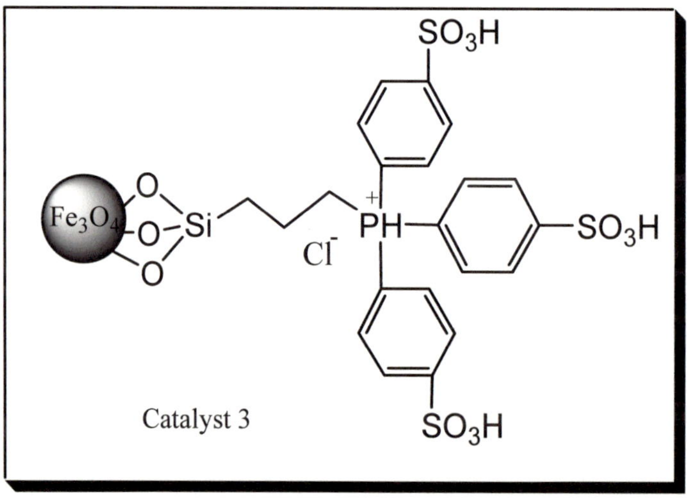

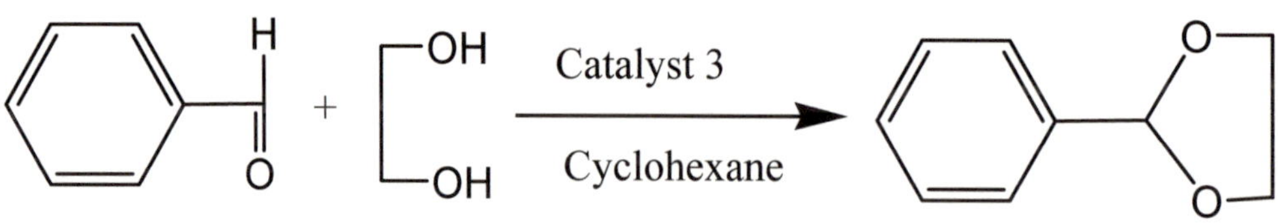

Scheme 5

5. Conclusion

This chapter has mainly described our recently prepared task-specific ionic liquids derived from DBU, [DBU][Tfa], [DBU][Lac], [DBU][Ac], [DBU][*n*-Pr], and [DBU][*n*-Bu], and their applications in aza-Michael addition and Knoevenagel condensation. Compared with some conventional catalysts, the novel DBU based ionic liquid catalysts demonstrate relatively higher catalytic activities, much better chemo- and stero- selectivity, readily recovery property and excellent recyclability. However, the application of DBU based ionic liquids in other organic transformations including asymmetric organic synthesis need to be further developed.

Magnetic nanoparticles (MNP) are of great interest in researchers because of their good stability, easy preparation and functionalization, large surface ratio and facile separation. Considering the features of ionic liquids, MNP-supported ionic liquid catalysts (MNP-ILs) have been developed by chemists. Very recently, MNP-ILs used as catalysts have been successfully utilized in cycloaddition reactions, Suzuki coupling reaction and acetalization reaction. The results show that MNP-ILs can smoothly catalyze reaction, be readily recovery from reaction solution *via* magnetic force, as well as good recyclability and stability. It is hoped that this review will help stimulate the relevant research of MNP-ILs to move forward.

6. Acknowledgement

We are grateful to the generous financial support from the National Natural Science Foundation of China (21076183) and that of the Zhejiang Province(Y4090045, R4090358)

7. References

Abu-Reziq, R.; Alper, H.; Wang, D. & Post, M. L. (2006). Metal supported on dendronized magnetic nanoparticles: highly selective hydroformylation catalysts *J. Am. Chem. Soc.* Vol. (128) : 5279-5282.

Aggarwal, V. K. & Mereu, A. (1999). Superior amine catalysts for the Baylis-Hillman reaction : the use of DBU and its implication *Chem. Commun.* 2311-2312.

Alleti, R.; Woon, S. O.; Perambuduru, M.; Ramena, C. V. & Reddy, V. P. (2008). Imidazolium-based polymer supported gadolinium triflate as a heterogeneous recyclable Lewis acid catalyst for Michael additions *Tetrahedron Lett.* Vol.(49):3466-3470.

Arend, M.; Westermann, B. & Risch, N. (1998). Modern variants of the Mannich reactions Angew. *Chem. Int. Ed.* Vol.(37):1044-1070.

Attanasi, O; Fillippone, P & Mei A. (1983). Effect of metal ions in organic synthesis. Part XVI. Knoevenagel condensations of aldehydes and tosylhydrazones with 2,4-pentanedione by copper(II) chloride-catalyzed reaction *Synth Commun.* Vol. (13):1203-1208.

Azizi, N. & Said, M. R. (2004). LiClO4 Accelerated Michael addition of amines to α,β-unsaturated olefins under solvent-free conditions *Tetrahedron* Vol.(60):383-387.

Bartoli, G.; Bosco, M. & Carlone A. (2008). Magnesium perchlorate as efficient Lewis acid for the Knoevenagel condensation between β-diketones and aldehydes *Tertrahedron Lett.* Vol. (49):2555-2557.

Bartoli, G.; Bosco, M.; Marcantoni, E.; Petrini, M.; Sanbri, L. & Torregiani, E. (2001). Conjugate Addition of Amines to α,β-Enones Promoted by CeCl3·7H2O-NaI System Supported in Silica Gel *J. Org. Chem.* Vol.(66):9052-9055.

Bhanushali, M. J.; Nandurkar, N. S.; Jagtap, S. R. & Bhanage, B. M. (2008). Y(NO3)3·6H2O catalyzed aza-Michael addition of aromatic/hetero-aromatic amines under solvent-free conditions *Catal. Commun.* Vol. (9):1189-1195.

Bigi, F.; Chesini, L. & Maggi R. (1999). Montmorillonite KSF as an Inorganic, Water Stable, and Reusable Catalyst for the Knoevenagel Synthesis of Coumarin-3-carboxylic Acids *J. Org. Chem.* Vol. (64):1033-1035.

Breslow, R.(1991). Hydrophobic effects on simple organic reactions in water *Acc. Chem. Res.* Vol. (24):159-164

Cai, Y. Q.; Peng, Y. Q. & Song, G. H. (2006). Amino-functionalized ionic liquid as an efficient and recyclable catalyst for Knoevenagel reactions in water *Catal. Lett.* Vol. (109):61-64.

Chaudhuri, M. K.; Hussain, S.; Kantam, M. L. & Neelima, B. (2005). Boric acid: a novel and safe catalyst for aza-Michael reactions in water *Tetrahedron Lett.* Vol.(46):8329-8331.

Corey, E. J. & Reichard, G. A. (1989). Enantioselective and practical syntheses of R- and S-fluoxetines *Tertrahedron Lett.* Vol.(30):5207-5210.

Crowhurst, L.; Lancaster, N. L.; Arlandis, J. M. P. & Welton, T. (2006). Manipulating Solute Nucleophilicity with Room Temperature Ionic Liquids *J. Am. Chem. Soc.* Vol. (126):11549-11555.

Deb, M. L. & Bhuyan, P. J. (2005). Uncatalyzed Knoevenagel condensation in aqueous medium at room temperature *Tetrahedron Lett.* Vol. (46):6453-6456.

Duan, Z.; Xuan, X.-J.; Li, T.; Yang, C.-F. & Wu, Y.-J. (2006). Cerium(IV) ammonium nitrate (CAN) catalyzed aza-Michael addition of amines to α,β-unsaturated electrophiles *Tetrahedron Lett.* Vol. (47):5433-5436.

EI-Rahman, N. M. A. ; EI-Kateb, A. A. & Mady, M. F. (2007). Simplified approach to the uncatalyzed Knoevenagel condensation and Michael addition reactions in water using microwave irradiation *Synth. Commun.* Vol. (37):3961-3970.

Freeman, F. (1980). Properties and reactions of ylidenemalononitriles *Chem. Rev.* Vol. (80):329-350.

Gao, G. H.; Lu, L.; Zou, T., et al. (2007). Basic ionic liquid: A reusable catalyst for Knoevenagel condensation in aqueous media *Chem. Res Chinese Univ.* Vol. (23):169-172.

Georg, G. I. (1993). *The Organic Chemistry of β-Lactams VCH*, New York.

Green, B.; Grane, R. I. & Khaidem I. S. (1985). Synthesis of steroidal 16,17-fused unsaturated δ-lactones *J. Org. Chem.* Vol. (50):640-644.

Han, J. J.; Xu Y. F. & Su Y. P. (2008). Effect of hydrophobic and hydrophilic properties of solid acid catalysts on the esterification of maleic anhydride with ethanol *Catal. Commun.* Vol. (9):2077-2079.

Hashemi, M. M.; Eftekhari-Sis, B.; Abdollahifar, A. & Khalili, B. (2006). ZrOCl2·8H2O on montmorillonite K10 accelerated conjugate addition of amines to α,β-unsaturated alkenes under solvent-free conditions *Tetrahedron* Vol.(62):672-677.

Hu, Y.; Chen, J.; Le Z. G., et al. (2005). Organic reactions in ionic liquids. Ionic liquids ethylammonium nitrate-promoted Knoevenagel condensation of aromatic aldehydes with active methylene compounds *Synth. Commun.* Vol. (35):739-744.

Jenner, G. (1995). Catalytic high pressure synthesis of hindered β-amino ester *Tetrahedron Lett.* Vol.(36):233-236.

Kakade, G. K.; Madje, B. R.; Pokalwar, R. U., et al. (2008). DBU-catalyzed Knoevenagel condensation under ultrasonic irradiation *Indian J. Het. Chem.* Vol. (17):379-380.

Kantam, M. L.; Ravindra, A. & Reddy C. V. (2006). Layered double hydroxides-supported diisopropylamide: synthesis, characterization and application in organic reactions *Adv. Synth. Catal.* Vol. (348):569-578.

Kantam, M. L.; Roy, M.; Roy, S.; Sreedhar, B. & De, R. L. (2008). Polyaniline supported CuI: An efficient catalyst for C-N bond formation by N-arylation of N(H)-heterocycles and benzylamines with aryl halides and arylboronic acids, and aza-Michael reactions of amines with activated alkenes *Catal. Commun.* Vol.(9):2226-2230.

Khan, A. T.; Parvin, T.; Gazi, S. & Choudhury, L. H. (2007). Bromodimethylsulfonium bromide-mediated Michael addition of amines to electron deficient alkenes *Tetrahedron Lett.* Vol.(48):3805-3808.

Kim, D. W.; Song, C. E. & Chi, D. Y. (2003). Significantly Enhanced Reactivities of the Nucleophilic Substitution Reactions in Ionic Liquid *J. Org. Chem.* Vol. (68):4281-4285.

Kim, D. W.; Song, C. E. & Chi, D. Y. (2002). New method of fluorination using potassium fluoride in ionic liquid: significantly enhanced reactivity of fluoride and improved selectivity *J. Am. Chem. Soc.* Vol. (124):10278-10279.

Kim, S. C.; Lee, K. Y.; Lee, H. S. & Kim, J. N. (2008). Synthesis of poly-substituted benzenes starting from Baylis-Hillman adducts: DBU-assisted unusual dehydrogenation *Tetrahedron* Vol.(64):103-109.

Kleinmann, E. F. (1991). *Comprehensive Organic Synthesis*; Pergamon: New York, NY.

Kobayashi, S. & Ishitani, H. (1999). Catalytic Enantioselective Addition to Imines *Chem. Rev.* Vol.(99):1069-1094.

Lee, D.; Lee, J.; Lee, H.; Jin, S.; Hyeon, T. & Kim, B. M. (2006). Filtration-free recyclable catalytic asymmetric dihydroxylation using a ligand immobilized on magnetic mesocellular mesoporous silica *Adv. Synth. Catal.* Vol. (348):41-46.

Li, C. J. (1993). Organic reactions in aqueous media - with a focus on carbon-carbon bond formation *Chem. Rev.* Vol.(93):2023-2035.

Li ,C. J. & Chen, T. H. (1997). *Organic Reacions in Aqueous Media*, Wiley, New York.

Li, P.; Wang, L.; Zhang, Y. & Wang, G. (2008). Silica gel supported pyrrolidine-based chiral ionic liquid as recyclable organocatalyst for asymmetric Michael addition to nitrostyrenes *Tetrahedron* Vol. (64):7633-7638.

Li, Y. X.; Goeke, A.; Wang, R. Y.; Wang, Q. R. & Fráter, G. (2007). DBU-mediated Ireland-Claisen rearrangement of allyl alk-3-enoates :an efficient synthesis of 2-ethylidene-γ,δ-unsatuated carboxylic acids *Tetrahedron* Vol.(63):9605-9613.

Liang, D. P.; Xin, X.; Duan, H. F., et al. (2008). Tetramethylguanidium lactate ionic liquid: catalyst-free medium for Michael addition reaction *Chem. Res. Chinese Univ.* Vol. (24):36-41.

Liu, Y.; Liang, J.; Liu, X. H., et al. (2008). Polyethylene glycol (PEG) as a benign solvent for Knoevenagel condensation *Chinese Chem. Lett.* Vol. (19):1043-1046.

Loh, T. P. & Wei, L. L. (1998). Indium trichloride-catalyzed conjugate addition of amines to α,β-ethylenic compounds in water *Synlett* 975-976.

Luo, S. Z.; Zheng, X. X. & Cheng, J. P. (2008). Asymmetric bifunctional primary aminocatalysis on magnetic nanoparticles *Chem. Commun.* 5719-5721.

Ma, C.; Zhang, Q.; Ding, K.; Xin, L. & Zhang, D. J. (2007). A novel DBU-promoted S-N-type Smiles rearrangement reaction under controlled microwave heating *Tetrahedron Lett.* Vol.(48):7476-7479.

Martns, L.; Vieira K. M. & Rios L. M. (2008). Basic catalyzed Knoevenagel condensation by FAU zeolites exchanged with alkylammonium cations *Catal. Today* Vol. (706):133-135.

Matsubara, R. & Kobayashi, S. (2008). DBU-catalyzed addition reactions of sulfonylimidates *Synthesis* 3009-3011.

Matsubara, S.; Yoshiyoka, M. & Utimoto, K. (1994). Lanthanoid triflate catalyzed conjugate addition of amines to α ,β-unsaturated esters. A facile route to optically active β-lactam *Chem. Lett.* 827-830.

Meciarova, M.; Toma, S. & Kotrusz, P. (2006). Michael addition of thiols to □ -enones in ionic liquids with and without organocatalysts *Org. Biomol. Chem.* Vol. (4):1420-1424.

Miao, W. W. & Chan T. H. (2006). Ionic-Liquid-Supported Synthesis: A Novel Liquid-Phase Strategy for Organic Synthesis *Acc. Chem. Res.* Vol.(39):897-908.

Narsaiah A.V.; Basak A.K. & Visali B. (2004). An eco-friendly synthesis of electrophilic alkenes catalyzed by dimethylaminopyridine under solvent-free conditions *Synth. Commun.* Vol. (16):2893-2901.

Narsaiah, A. V. & Nagaiah, K. (2003). An efficient Knoevenagel condensation catalyzed by LaCl3.7H2O in heterogeneous medium *Synth. Commun.* Vol. (33):3825-3832.

Paun, C.; Barklie, J. ; Goodrich P. et al. (2007). Supported and liquid phase task specific ionic liquids for base catalyzed Knoevenagel reactions *J. Mol. Catal. A: Chem.* Vol. (269):64-71.

Reboule, I.; Gil, R. & Collin, J. (2005). Aza-Michael reactions catalyzed by samarium diiodide *Tetrahedron Lett.* Vol.(46):7761-7764.

Reddy, B. M.; Patil, M. K. & Rao K. N. (2006). An easy-to-use heterogeneous promoted zirconia catalyst for Knoevenagel condensation in liquid phase under solvent-free conditions *J. Mol. Catal. A: Chem.* Vol. (258):302-307.

Ribe, S. & Wipf, P. (2001). Water-accelerated organic transformations *Chem. Commun.* 299-307.

Rong, Y. C.; Li, X. Y.; Wang, H. Y., et al. (2006). Efficient green procedure for the Knoevenagel condensation under solvent-free conditions *Synth. Commun.* Vol. (36):2407-2412.

Santamarta, F.; Verdía, P. & Tojo E. (2008). A simple, efficient and green procedure for Knoevenagel reaction in [MMIm][MSO4] ionic liquid *Catal. Commun.* Vol. (9):1779-1781.

Saravanamurugan, S.; Palanichamy, M. & Hartmann M. (2006). Knoevenagel condensation over □ and Y zeolites in liquid phase under solvent free conditions *Appl. Catal. A: General*, Vol. (298):8-15.

Shaabani, A.; Samadi, S. & Rahmati A. (2007). One-pot, three-component condensation reaction in water. An efficient and improved procedure for the synthesis of pyran annulated heterocyclic systems *Synth. Commun.* Vol. (37):491-499.

Shanthan Rao P. & Venkataratnam R.V. (1991). Zinc chloride as a new catalyst for Knoevenagel condensation *Tetrahedron Lett.* Vol. (32):5821-5822.

Srivastava, N. & Banik, B. K. (2003). Bismuth nitrate-catalyzed versatile Michael reactions *J. Org. Chem.* Vol.(68):2109-2114.

Stevens, P. D.; Fan, J.; Gardimalla, H. M. R.; Yen, M. & Gao, Y. (2005). Superparamagnetic Nanoparticle-Supported Catalysis of Suzuki Cross-Coupling Reactions *Org. Lett.* Vol. (7):2085-2088.

Surendra, K.; Krishnaveni, N. S.; Sridhar, R. & Rao, K.R. (2006). β-Cyclodextrin promoted aza-Michael addition of amines to conjugated alkenes in water *Tetrahedron Lett.* Vol.(47):2125-2127.

Taher, A.; Kim, J. B.; Jung, J. Y.; Ahn, W. S. & Jin, M. J. (2009). Highly Active and Magnetically Recoverable Pd-NHC Catalyst Immobilized on Fe3O4 Nanoparticle-Ionic Liquid Matrix for Suzuki Reaction in Water *Synlett* 2477-2482.

Tietze, L.F. (1996). Domino Reactions in Organic Synthesis *Chem. Rev.* Vol. (96):115-136.

Tietze, L. F. (2004). Domino reactions in the synthesis of heterocyclic natural products and analogs *Pure. Appl. Chem.* Vol. (76):1967-1983.

Tolstikoua, L. L. & Shainyan, B. A. (2006). Ionic liquids on the basis of 2,3,4,6,7,8,9,10 - octahydropyrimido-[1,2-a]azepine (1,8-diazabicyclo[5.4.0]undec-7-ene) *Russian J. Org. Chem.* Vol.(42):1068-1074.

Wang, P.; Kong, A. G.; Wang, W. J.; Zhu, H. Y. & Shan, Y. K. (2010). Facile Preparation of Ionic Liquid Functionalized Magnetic Nano-Solid Acid Catalysts for Acetalization Reaction *Catal. Lett.* Vol. (135):159-164.

Welton, T. Coord. (2004). Ionic liquids in catalysis *Chem. Rev.* Vol. (248):2459-2477.

Xu, D. Q.; Luo, S. P.; Wang, Y. F.; Xia, A. B.; Yue, H. D.; Wang, L. P. & Xu, Z. Y. (2007). Organocatalysts wrapped around by poly(ethylene glycol)s (PEGs): A unique host-guest system for asymmetric Michael addition reactions *Chem. Commun.* 4393-4395.

Xu, J. M.; Qian, C.; Liu, B. K.; Wu, Q. & Lin. X. F. (2007). A fast and highly efficient protocol for Michael addition of N-heterocycles to α,β-unsaturated compound using basic ionic liquid [bmIm]OH as catalyst and green solvent *Tetrahedron* Vol.(63):986-990.

Xu, L. W.; Li, L. & Xia, C. G. (2004). Transition-metal-based Lewis acid catalysis of aza-type Michael additions of amines to α,β-unsaturated electrophiles in water *Helv. Chim. Acta.* Vol.(87):1522-1526.

Xu, L. W.; Li, J. W.; Xia, C. G.; Zhou, S. L. & Hu, X. X. (2005). Efficient copper-catalyzed chemoselective conjugate addition of aliphatic amines to α,β-unsaturated compounds in water *Synlett* 2425-2427.

Yang, L.; Xu, L. W. & Xia, C. G. (2005). Highly efficient KF/Al2O3-catalyzed versatile hetero-Michael addition of nitrogen, oxygen, and sulfur nucleophiles to α,β - ethylenic compounds *Tetrahedron Lett.* Vol.(46):3279-3282.

Yang, L.; Xu, L.-W.; Zhou, W.; Li, L. & Xia, C.-G. (2006). Highly efficient aza-Michael reactions of aromatic amines and N-heterocycles catalyzed by a basic ionic liquid under solvent-free conditions *Tetrahedron Lett.* Vol. (47):7723-7726.

Yeom, C. E.; Kim, M. J.; Choi, W. & Kim, B. M. (2008). DBU-promoted facile, chemoselective cleavage of acetyletynic TMS group *Synlett.* 565-568.

Yeom, C. E.; Kim, M. J. & Kim, M. B. (2007). 1,8-Diazabicyclo[5.4.0]undec-7-ene(DBU) promoted efficient and versatile aza-Michael addition *Tetrahedron* Vol.(63):904-909.

Yi ,W. B. & Cai C. (2008). Perfluoroalkylated pyridine catalyzed Knoevenagel condensation: an important complement of fluorous catalysis without fluorous solvent *Catal. Commun.* Vol. (9):1291-1296.

Ying, A. G. (2010). Preparation of DBU derived task-specific ionic liquids and the study of their applications in the Michael additions and Knoevenagel condensations; Dissertation for Doctor degree, Zhejiang university, pp. 42-50.

Ying, A. G.; Chen, X. Z.; Ye, W. D.; Zhang, D. F.; Liu, L. & Chen, J. H. (2008). Application of ionic liquids in organic synthesis promoted by microwave irradiation *Prog. Chem.* Vol.(20):1642-1650.

Ying, A. G.; Liu, L.; Wu, G. F.; Chen, G.; Chen, X. Z. & Ye, W. D. (2009). Aza-Michael addition aliphatic or aromatic amines to α,β-unsaturated compounds catalyzed by a DBU-derived ionic liquid under solvent-free conditions *Tetrahedron Lett.* Vol.(50):1653-1657.

Ying, A. G.; Liu, L.; Wu, G. F.; Chen, X. Z.; Ye, W. D.; Chen, J. H. & Zhang, K. Y. (2009). Knoevenagel condensation catalyzed by DBU brönsted ionic liquid without solvent *Chem. Res. Chinese Universities* Vol.(25):876-881.

Ying, A. G.; Wang, L. M.; Deng, H. X.; Chen, J. H.; Chen, X. Z. & Ye, W. D. (2009). Green and efficient aza-Michael additions of aromatic amines to α,β-unsaturated ketones catalyzed by DBU based task-specific ionic liquids without solvent *Arkivoc*, Vol.(XI):288-298.

Ying, A. G.; Wang, L. M.; Wang, L. L.; Chen, X. Z. & Ye, W. D. (2010). Green and efficient Knoevenagel condensation catalysed by a DBU based ionic liquid in water *J. Chem. Res.* 30-33.

Ying, A. G.; Ye, W. D.; Liu, L.; Wu, G. F.; Chen, X. Z.; Qian, S. & Zhang, Q. P. (2008). Progress in the applicatioin of ionic liquids to organic synthesis *Chinese J. Org. Chem.* Vol.(28):2081-2094.

Yoon, T. J.; Lee, W.; Oh, Y. S. & Lee, J. K. (2003). Magnetic nanoparticles as a catalyst vehicle for simple and easy recycling *New J. Chem.* Vol. (27):227-229.

Yoshida, M.; Komatsuzaki, Y. & Ihara, M. (2008). Synthesis of 5-vinylideneoxazolidin-2-ones by DBU-mediated CO_2 fixation reaction of 4-(benzylamino)-2-butynyl carbonates benzoates *Org. Lett.* Vol.(10):2083-2086.

Yu. N.; Aramini J. M. & Germann M. W. (2000). Reactions of salicylaldehydes with alkyl cyanoacetates on the surface of solid catalysts: syntheses of 4H-chromene derivatives *Tertrahedron Lett.* Vol. (41):6993-6996.

Yue, C. B.; Mao, A. Q.; Wei Y. Y., et al. (2008). Knoevenagel condensation reaction catalyzed by task-specific ionic liquid under solvent-free conditions *Catal. Commun.* Vol. (9):1571-1574.

Zheng, X. X.; Luo, S. Z.; Zhang, L. & Cheng, J. P. (2009). Magnetic nanoparticle supported ionic liquid catalysts for CO2 cycloaddition reactions *Green Chem.* Vol. (11):455-458.

15

Hydrogenation in Ionic Liquids

Mukund Ghavre, Saibh Morrissey and Nicholas Gathergood
Dublin City University
Ireland

1. Introduction

One of the principal present-day challenges facing the field of transition metal catalysis is the efficient recycling and reuse of catalysts and ligands. The use of ILs is rapidly advancing in this discipline. Due to their tunable physico-chemical properties, which differ markedly from those of conventional organic and aqueous media, ILs can provide a means of catalyst immobilization. The non-nucleophilic and weakly co-ordinating nature of many classes of ionic liquid provides an inert reaction medium that can extend the lifetime of a catalyst. Recyclability of the catalyst system is a key attribute of IL media and it is this enhancement of catalyst performance that is driving research in this field. Low-polarity compounds, for example diethyl ether and *n*-hexane, are poorly soluble in common ILs, providing a suitable accompanying solvent for biphasic catalysis. The positive aspects of homogeneous and heterogeneous catalysis are combined using a biphasic system, in which the catalyst resides in the IL, but the substrates/products are retained in the alternate phase. Thus, the biphasic system provides a cost-effective way to successfully separate the desired product by simple decantation, leaving the catalyst immobilised in the IL and ready for reuse. Product isolation is often simplified even in monophasic catalysis if the substrate is soluble in the IL medium, but the product can be separated by simple extraction or distillation, due to the low vapour pressure of the IL. The reduced polarity of the hydrogenated products in comparison with the substrate can also be exploited for separation from the IL/catalyst phase. Increasing the difference in polarity between the IL and the hydrogenated product can also render the product insoluble in the IL, thus allowing facile decantation of the product from the IL, leaving the IL/catalyst phase ready for the next reaction.

Research into catalytic hydrogenations in ILs began in 1995 with the almost simultaneous work of Chauvin[1] and Dupont.[2] Since then this field has been extended from conventional hydrogenation using transition metal catalysts to transfer hydrogenation and the effect of nanoparticles on hydrogenation reactions in ILs. Supported ionic liquid catalysis (SILC) is also a relatively new field. Using this method, the added benefit of selectivity provided by the homogeneous catalyst can be combined with the attributes of heterogeneous biphasic catalysis. The homogeneous catalyst is, in effect, immobilised on a heterogeneous support. One of the first investigations in this area was carried out by Mehnert *et al.*[3] in 2002, with a flurry of papers ensuing from 2007-2009.

It is intended that this chapter should cover recent progress in hydrogenation reactions carried out in ILs. Wasserscheid and Schulz[4] contributed a chapter in 'The Handbook of Homogeneous Hydrogenation' covering homogeneous hydrogenation in ILs which covers advances up to 2005. More recent summaries of hydrogenation in ILs can be found in 'Ionic Liquids in Synthesis' as part of the chapter on Transition Metal Catalysis in Ionic Liquids[5] and within several reviews covering the wider area of IL catalysis.[6, 7, 8] Moreover, outside the scope of this chapter is the concept of tailored ILs used in hydrogenation reactions, which was covered in 2008 by Sebesta *et al.*[9] For each hydrogenation reaction reviewed in this chapter, the catalyst and IL are noted, together with the genre of substrate. Particular consideration is paid to conversion and turnover numbers obtained using ILs as alternatives to commonly used volatile organic solvents. A section describing kinetic and thermodynamic studies of reactions in ILs is also presented. ILs are not always the preferred choice of solvent, and cases where the IL was found to be less effective than a conventional molecular solvent are included. Throughout this chapter abbreviations used are as follows:

[emim]: 1-ethyl-3-methylimidazolium, [bmim]: 1-*n*-butyl-3-methylimidazolium, [omim]: 1-*n*-octyl-3-methylimidazolium, [dmim]: 1-*n*-decyl-3-methylimidazolium, [dodecylmim]: 1-*n*-dodecyl-3-methylimidazolium, [bdmim]: 1-*n*-butyl-2,3-dimethylimidazolium, [TEA]: tetraethylammonium, [TBA]: tetrabutylammonium, [mbpy]: 4-methyl-*N*-butyl-pyridinium, [DAMI]: 1,3-di(*N,N*-dimethylaminoethyl)-2-methylimidazolium, [C_8Py]: *N*-octylpyridinium, [BMPL]: *N*-butyl-*N*-methylpyrrolidinium, [B3MPYR]: *n*-butyl-3-methylpyridinium, [bmmim]: 1-*n*-butyl-2,3-dimethylimidazoliium, [bmimOH]: hydroxyl-functionalized butyl-3-methylimidazolium, [BF_4]: tetrafluoroborate, [PF_6]: hexafluorophosphate, [NTf_2]: trifluoromethanesulfonimide, [OTf]: triflate, [$N(CN)_2$]: dicyanamide, [NO_3]: nitrate, [HSO_4]: hydrogen sulphate, [$EtOSO_3$]: ethyl sulphate, [$BuOSO_3$]: butyl sulphate, [$HexOSO_3$]: hexyl sulphate, [$OctOSO_3$]: octyl sulphate, TOF: turn over frequency, TON: turn over number, BINAP: 2,2'-*bis*(diphenylphosphino)-1,1'-binaphthyl, COD: 1,5-cyclo-octadiene, ee: enantiomeric excess, CIL: chiral ionic liquid, $scCO_2$: supercritical carbon dioxide, IPA: isopropanol, SILC: supported ionic liquid catalysts, SILP: supported ionic liquid phase, SSILP: structured supported ionic liquid-phase, SCILL: solid catalyst with ionic liquid layer, CTH: catalytic transfer hydrogenation, OSN: organic solvent nanofiltration, TSIL: task-specific ionic liquid, COE: cyclooctene, COA: cyclooctane, PVP: poly(*N*-vinyl-2-pyrrolidone), [N_{6222}]: triethylhexylammonium, [N_{22212}]: triethyldodecylammonium, [N_{22214}]: triethyltetradecylammonium, MAA: methyl acetoacetate, Ts-DPEN: *N*-(*p*-toluenesulfonyl)-1,2-diphenylethylenediamine, [$P_{14,6,6,6}$]: trihexyltetradecylphosphonium, [$P_{4,4,4,4}$]: tetrabutylphosphonium, [P1,*i*4,*i*4,*i*4]: tri-*iso*-butylmethylphosphonium, ECOENG 500™: Peg-5 cocomonium methosulfate.

2. Kinetic and Thermodynamic properties:

A major advantage of ionic liquids as the solvent in hydrogenation reactions is the ability to fine tune the properties of the solvent by altering the structure. Hence it becomes very important for a chemist to investigate the kinetic and thermodynamic aspects of the reactions. Although only few detailed studies of the kinetics of hydrogenation reactions in ionic liquids have been reported, from available results it can be shown that the reaction kinetics, product selctivity, reaction yields, TOFs of catalysts are greatly influenced by the cations and anions of ionic liquids and their concentration in the reaction system.

It is also important to compare the trends of H_2 gas solubilities in ionic liquids and molecular solvents, while studying reaction kinetics. Table 1 summarizes the H_2 solubilities in various solvents, along with densities and viscosities. [10]

Solvent	Henry's Constant, k_H/MPa[a]	$10^3[H_2]$ (M)	Density (g/Ml)	Viscosity (cP)	Ref.
Water[b]	6.8 x 10^3	0.81	0.9982[11]	0.89[c,][12]	[13]
Methanol[b]	6.6 x 10^2	3.75	0.7914[11]	0.55[c,][12]	[14]
Ethanol[b]	5.9 x 10^2	2.98	0.7893[11]	1.06[c,][12]	[14]
Toluene[b]	2.69 x 10^2	3.50	1.4961[11]	0.45[b,][12]	[14]
Benzene[b]	4.47 x 10^2 (4.39 x 10^2)	2.54 (2.57)	0.878[11]	0.60[c,][12]	[10,13]
Cyclohexane[b]	2.57 x 10^2 (2.55 x 10^2)	3.63 (3.66)	0.777[11]	1.62[b,][12]	[10,13]
[bmim][BF_4][b]	5.8 x 10^2 (1.63 x 10^2)	0.86[d] (3.0)	1.12[15]	219[c,][15]	[10,16]
[bmim][PF_6][b,c]	6.6 x 10^2 (5.38 x 10^2)	0.73[d] (0.88)	1.363[15]	450[c,][15]	[10,16]
[bmim]Tf_2N[c]	4.5 x 10^2	0.77[d]	1.433	69[c,][15]	[10]
[bm_2im][Tf_2N][c,e]	3.8 x 10^2	0.86[d]	1.421	97.1[c,][17]	[10]
[bupy][Tf_2N][c,f]	3.9 x 10^2	0.89[d]	1.449	57[c,][18]	[10]
[bmpy][Tf_2N][c,g]	3.7 x 10^2	0.90[d]	1.387	85[c,][19]	[10]
[bmim][SbF_6][c]	4.9 x 10^2	0.93[d]	1.699	95	[10]
[bmim][CF_3CO_2][c]	4.9 x 10^2	0.98[d]	1.198	73[b,][20]	[10]
[hmim][BF_4][c,h]	5.7 x 10^2	0.79[d]	1.14[21]	314.0[b,][22]	[10]
[omim][BF_4][c]	6.4 x 10^2	0.62[d]	1.106	135.0[b,][22]	[10]
[bmim][CF_3SO_3][c]	4.6 x 10^2	0.97[d]	1.290[20]	90[b,][20]	[10]
[$P(C_6H_{13})_3(C_{14}H_{29})$] [$PF_3(C_2F_5)_3$][c]	0.7 x 10^2	1.84[d]	1.196	498[b,][21]	[10]

[a] $k_H = P_{H_2}/X_{H_2}$, where the partial pressure of hydrogen is expressed in MPa. [b] 293 K.
[c] 298 K. [d] Calculated from the solubility under 10.1 MPa, supposing that it changes linearly with the partial pressure. [e] $[bm_2im]^+$ = 1,2-Dimethyl-3-butylimidazolium.
[f] $[bupy]^+$ = *N*-Butylpyridinium. [g] $[bmpy]^+$ = *N*-Butyl-*N*-methylpyrrolidinium.
[h] $[hmim]^+$ = 1-Hexyl-3-methylimidazolium.

Table 1. Solubility of H_2 in water, organic solvents and ionic liquids, at 0.101 MPa (1 atm)

Table 1 shows that H_2 solubility in ionic liquids is typically much lower than in molecular solvents, which can lead to low reaction rates. Mass transfer effects associated with low gas solubility play a key role in hydrogenation, and may be critical when processes such as catalytic asymmetric hydrogenation are carried out in ionic liquids. Blackmond and co-workers[23] described in detail the key kinetic parameters affecting enantioselectivity in asymmetric hydrogenations, namely the concentration of molecular H_2 in the liquid phase, itself related to the pressure of the system, the rate of mass transfer and the intrinsic kinetics of competing reactions. Blackmond also determined that in cases where the enantioselectivity decreases with increasing H_2 pressure, the system can benefit from H_2-starved conditions. In other words a diffusion-limited regime could be beneficial because the rate of consumption of H_2 by the reaction would be higher than the rate of diffusion of H_2 in the liquid phase. For other reactions requiring high H_2 concentrations, low solubility problems can be solved by carrying out reactions at elevated pressures, which raises the H_2 solubility.[16] Interestingly in many cases product selectivities are achieved due to solubility differences between the intermediate and fully hydrogenated products.[24] For example, in the partial hydrogenation of 1,3-butadiene using Pd(0) nanoparticles in [bmim][BF_4], it was observed that 1,3-butadiene is at least three times as soluble in the ionic liquid as the intermediate butenes, inhibiting further hydrogenation and leading to product selectivity.[25]

The high viscosity of ionic liquids can also be a limitation for in hydrogenation reactions as diffusion of reactants through the medium is restricted. Temperature also plays an important role in hydrogenation, following the usual trend that at higher temperatures a high reaction rate is observed. However, the viscosity of the ionic liquid also decreases as temperature increases, facilitating mass transport of the reactants.

The choice of the anion is another crucial aspect of selecting an ionic liquid for hydrogenation studies. For example, in heterogeneous catalytic hydrogenations, the reaction occurs at the surface of the catalyst and the solvent cannot directly affect the energy of the activated complex. Nevertheless, solvent polarity (which varies from one anion to another) still plays an important role because polar solvents facilitate the adsorption of nonpolar substrates on the catalyst, while non-polar solvents have the opposite effect.[26] Gas solubility *(vide infra)* also varies with the anion, and it can be seen from Table 1 that ionic liquids with ditriflimide [NTf_2^-] as the counter anion have greater H_2 solubility than ILs with tetrafluoroborate [BF_4^-]. Anions can also have more specific interactions with the catalyst, which can control the conversion and enantiomeric excess of the product.[27]

Kinetic studies into the heterogeneous catalytic hydrogenation of cyclohexene in ionic liquid-alcohol mixtures have also been carried out by Khodadadi-Moghaddam *et al.*[28] with a Pt/Al_2O_3 catalyst in the IL, 2-hydroxy ammonium formate and the alcohols, methanol, ethanol or propan-2-ol at 25 °C. Mass transfer limitation effects of H_2 solubility and solvent polarity on reaction rates were studied. The rotation speed of the reaction mass was used to determine the mass transfer barriers as the hydrogen transfers from gas phase to liquid phase. It was found that there is a linear relationship between observed rate constant and rotation speed up to 700 rpm, after which the rate constant becomes independent of the rotation speed up to 1250 rpm, which suggests that the reaction is under kinetic control and takes place without external mass transfer limitations. The H_2 solubility was measured for the IL-alcohol mixtures which shows that increase in mole fraction of the IL increases the solubility and ultimately in pure IL the solubility was maximum (0.01 mL of H_2 gas in 20 g IL at 25 °C and 1 atm). A rate expression was derived considering the RDS as dissociative adsorption of hydrogen on catalyst as follows,

$$K_{obs} = K_{app}\left(\frac{H_{H_2}}{H_T}\right) C_{H_2}$$

Where k_{obs} = observed rate constant

k_{app} = apparent rate constant (incudes other concentrations and parameters)

H_{H_2} = Henry's law constant for H_2 in the solvent.

H_T = Henry's law constant for the transition state.

C_{H_2} = Concentration of H_2 in liquid phase.

Based on this equation the rate constant was found to be proportional to H_2 concentration in liquid phase and the reaction was first order with respect to H_2. (Figure 1)

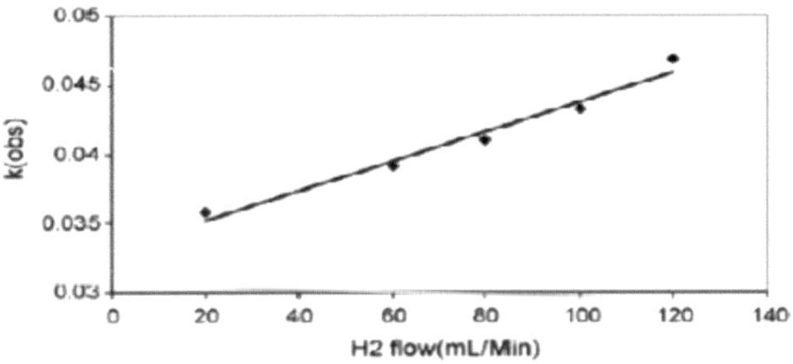

Fig. 1. Dependence of k_{obs} on hydrogen flow in ionic liquid at 25 °C (solvent 20 g, cyclohexene 0.1 g, rotation speed 1250 rpm and catalyst 0.02 g).

Furthermore the rate constants were calculated for three mixtures of IL-alcohol with increasing mole fractions of the IL (Table 2). The results indicated that with increasing mole fraction of IL, π* (dipolarity/polarizability) increases, leading to an increase in rate constant. In heterogeneous catalysis the reaction occurs on the catalyst surface, and polar solvents facilitate non-polar substrate adsorption on the catalyst.[26] Hence, Kishida linked the increase in rate constants to increased solvent polarity.

X_{IL}	IL-methanol mixture					IL-ethanol mixture					IL-propan-2-ol mixture				
	k x 10^{-2} s^{-1} g^{-1}	E^N	Π*	β	α	k x 10^{-2} s^{-1} g^{-1}	E^N	Π*	β	α	k x 10^{-2} s^{-1} g^{-1}	E^N	Π*	β	α
0.0	0.80	0.76	0.57	0.81	1.16	1.03	0.65	0.51	0.91	0.97	0.26	0.54	0.49	1.07	0.74
0.1	1.20	0.88	0.71	0.82	1.31	1.17	0.84	0.63	0.91	1.27	0.35	0.80	0.56	1.04	1.23
0.2	2.43	0.89	0.80	0.80	1.25	1.29	0.85	0.68	0.93	1.27	0.50	0.81	0.60	1.02	1.24
0.3	2.97	0.89	0.87	0.78	1.20	2.41	0.85	0.72	0.91	1.22	0.62	0.82	0.66	0.98	1.21
0.4	3.31	0.89	0.88	0.77	1.20	3.17	0.85	0.79	0.87	1.18	0.97	0.82	0.71	0.95	1.18
0.5	3.60	0.89	0.95	0.74	1.15	3.24	0.86	0.81	0.89	1.18	1.32	0.83	0.76	0.91	1.16
0.6	4.03	0.89	1.01	0.67	1.10	3.24	0.85	0.89	0.81	1.12	1.59	0.84	0.80	0.89	1.15
0.7	4.23	0.89	1.04	0.66	1.09	3.12	0.87	0.94	0.79	1.11	3.00	0.84	0.87	0.83	1.11
0.8	4.62	0.90	1.05	0.67	1.11	3.30	0.87	1.01	0.72	1.07	5.73	0.86	0.95	0.76	1.09
0.9	5.19	0.90	1.03	0.71	1.11	3.31	0.89	1.04	0.71	1.08	6.59	0.88	0.98	0.74	1.10
1.0	7.25	0.89	1.15	0.59	1.01	7.25	0.89	1.15	0.59	1.01	7.25	0.89	1.15	0.59	1.01

K = Rate constant, E^N = Normalized polarity parameter, Π* = Dipolarity/polarizability,
β = Hydrogen-bond acceptor basicity, α = Hydrogen-bond donor acidity.

Table 2. First-order rate constant of the reaction in the RTIL mixed with methanol, ethanol or propan-2-ol, together with solvatochromic parameters for the media

Similarly Fonseca *et al.*[29] carried out hydrogenation of 1-decene on Ir(0) nanoparticles in [bmim][PF_6] at 75 °C with varying pressures of H_2. They observed an increase in initial reaction rates with increase in H_2 pressures upto 4 atm, which was expected. However, above 4 atm the reaction rate is independent of the H_2 pressure, which can be ascribed to the surface saturation of the Ir(0) nanoparticles with H_2 gas. Thus at elevated pressures, (> 4 atm), a monomolecular mechanism is proposed (Figure 2) for hydrogenation occurring at the surface of catalytic nanoparticles within the IL.

$$S + C \underset{K_{a-1}}{\overset{K_{a1}}{\rightleftharpoons}} SC \xrightarrow{K_C} P$$

Fig. 2. Monomolecular mechanism

Where S = substrate, C = activated catalyst, P = final product, K_a = adsorption rate constant, K_{a-1} = desorption rate constant and K_c = catalytic rate constant.
When K_c was calculated at various pressures, it was found that for pressures ≥ 4 atm, K_c is almost identical (0.45 ± 0.06 min^{-1}) and at low pressures (2 atm) reaction rates decreased significantly, indicating that below 4 atm the hydrogenation is controlled by a mass transfer process.
In the case of fibre-supported Rh catalysts the rate of homogeneous hydrogenation was found to be dependent on the ligand to metal ratio and acid to IL ratio. Ruta *et al.*[30] when carried out gas phase hydrogenation of 1,3-butadiene using $SMF_{Inconel}$ supported $[Rh(nbd)Cl]_2$ catalyst, PPh_3 as ligand and an acid (H_3PO_4 or HBF_4) in ionic liquids ([bmim][BF_4] and [bmim][PF_6]) , found that with [bmim][BF_4]:HBF_4 of 0.5 and PPh_3:Rh of 8, the reaction was fast *i.e.* 285 h^{-1}. Further investigation showed that addition of excess acid favoured the formation of a cationic dihydride species (Figure 3).

$$[Rh(H)_2L_n]^+ \rightleftharpoons RhHL_n + H^+$$

Fig. 3. Acid equilibrium for cationic dihydride species

Excellent work was carried out by Kernchen *et al.*[24] in which the hydrogenation of 1,4-cyclooctadiene was performed using a [bmim][n-$C_8H_{17}OSO_3$] coated Ni catalyst in *n*-dodecane. This IL/ *n*-dodecane biphasic system allows effective partitioning of the intermediate (in this case cyclooctene) into the hydrocarbon layer, preventing over-reduction to cyclooctane. Table 3 depicts the values of partition coefficients of COD, COE and COA.

T in °C	K_N in (kg i/kg IL)/(kg i/kg *n*-dodecane) With i = COD, COE or COA		
	COD	COE	COA
20	0.30	0.25	0.20
50	0.37	0.31	0.25

Table 3. Nernst participation coefficients K_N for COD, COE and COA in biphasic system *n*-dodecane and ionic liquid [bmim][n-$C_8H_{17}OSO_3$]

For the hydrogenation of COD with uncoated Ni catalyst at 50 °C Kernchen obtained 40 % yield for COE (c.f. IL coated catalyst under the same conditions, 70 % yield). The K_N values from Table 3 offer a reasonable explanation for the results obtained. The low solubility of COE in the ionic liquid facilitates transfer to the n-dodecane layer and hence over-reduction is avoided. Further studies showed that the reaction follows first order kinetics with respect to COD and is zero order with respect to H_2 gas. An example of arene reductions by Rh nanoparticles supported on an ionic liquid-like co-polymer was reported by Zhao *et al.*[31] In this case most of the intermediate dienes were found to be soluble in [bmim][BF_4] which was used as a solvent, hence hydrogenation did not stop at the diene and further hydrogenated products predominate. Dyson *et al.*[32] have also performed reductions of benzene, exploiting solubility differences using K_2PtCl_4/[*N*-octyl-3-picolinium][$AlCl_4$]. The group also put forward a mechanism to account for the superior activity of K_2PtCl_4 (99 % yield of cyclohexane, at 100 °C, 50 bar H_2 in 240 mins) over other catalysts screened (Figure 4) in which a trimetal face is coordinated by the aromatic π-system of benzene.

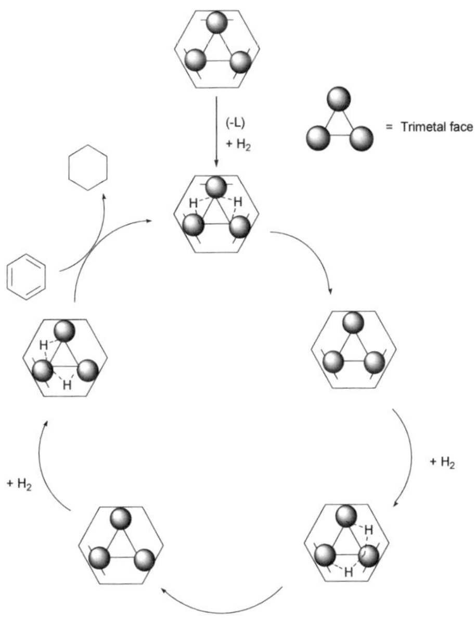

Fig. 4. The arene-exchange mechanism that could operate at a metal face

Interestingly when Silveira *et al.*[33] studied the partial hydrogenation of benzene to cyclohexene using Ru(0) nanoparticles, under solvent-free conditions and in ionic liquids ([bmim][PF_6] and [bmim][BF_4]) at 75 °C and 4 atm H_2, it was found that in the absence of a solvent, the reactions were faster (Table 4). Similar results were observed for the reduction of cyclohexanone using Ir(0) nanoparticles[34] and for other arenes using Ir(0) and Rh(0) nanoparticles.[35] The authors indicate that the reactions in ionic liquids may have been slowed by mass transfer effects. Correspondingly, when the reduction of 1,3-butadiene was attempted using Pd(0) particles at 40 °C and 4 atm H_2[25] under solvent-free conditions, 1,3-butadiene was consumed in under 2 h, compared to 6 h when [bmim][BF_4] was used as the solvent.

Mass transfer limitations have been a major challenge in many ionic liquid-based hydrogenations. For example, even in the biphasic reduction of caffeate in [bmim][PF_6] and tetradecane although the initial reaction rate was high (880 mmol kg^{-1} h^{-1}), after 1 hour this decreased to 12.4 mmol kg^{-1} h^{-1}[36]. Wolfson *et al.*[37] have overcome this problem by addition of water in their biphasic asymmetric hydrogenation reactions which were carried out with 2-acetamidoacrylate using Rh-EtDuPHOS as a catalyst in [bmim][PF_6]. When the reaction was performed with the ionic liquid in isolation, no conversion was observed. However, when water was included as a cosolvent, the product was isolated in 68 % yield with 96 % ee, with a 50 % v/v ratio of ionic liquid to water giving optimal reaction rates (TOF ≈ 1000 h^{-1} at 20 °C and 5 bar H_2).

Entry	Medium	Substrate	t [h]	Conv. [%]	TON[a]	TOF [h^{-1}][b]
1	-	1-hexene	0.7	> 99	500	714
2	[bmim][BF_4]	1-hexene	0.6	> 99	500	833
3	[bmim][PF_6]	1-hexene	0.5	> 99	500	1000
4	-	cyclohexene	0.5	> 99	500	1000
5	[bmim][BF_4]	cyclohexene	5.0	> 99	500	100
6	[bmim][PF_6]	cyclohexene	8.0	> 99	500	62
7	-	2,3-dimethyl-2-butene	1.2	76	380	316
8	-	benzene	5.5	90	450	82
9	[bmim][BF_4]	benzene	17.3	30	150	9
10	[bmim][PF_6]	benzene	18.5	73	365	20
11	[bmim][CF_3SO_3]	benzene	17.5	50	240	14
12	-	benzene[c]	2.0	> 99	250	125
13	-	toluene[c]	5.6	> 99	250	45
14	-	isopropylbenzene[c]	6.4	> 99	250	39
15	-	*tert*-butylbenzene[c]	14.1	> 99	250	18
16	-	anisole	18	< 1	-	-

[a] Turnover number TON = mol of hydrogenated product/mol of Ru.
[b] Turnover frequency TOF = TON/h. [c] Arene/Ru = 250.

Table 4. Hydrogenation of alkenes and arenes by Ru^0 nanoparticles under multiphase and solventless conditions (75 °C and 4 atm, constant pressure, substrate/Ru = 500)

While studying the asymmetric hydrogenation of acetophenone Fow *et al.*[38] investigated supported Ru and Rh-based catalysts containing either BINAP or chiraphos ligands immobilised in phosphonium ionic liquids, with the bases, K_2CO_3 and K_3PO_4 as additives. Fow proposed a kinetic model to rationalise the results of these experiments (Figure 5). For full details for catalysts structure and preparation see Fow *et al.*[38]

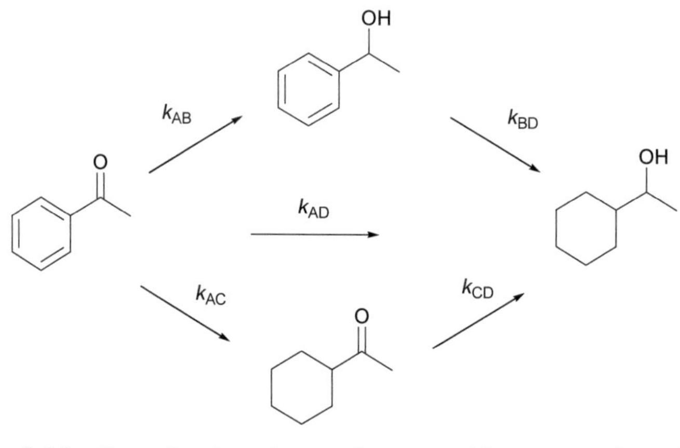

Fig. 5. Kinetic model for the reduction of acetophenone with supported catalysts.

Using a series of supported Ru and Rh catalysts immobilised in ionic liquids, reductions were carried out at between 30 and 80 °C (results Table 5) and rate constants were calculated (Table 6).

Catalyst	Temp. (°C)	Time (h)	Conv. (%)	Selectivity[a] (%)	ee[b] (%)	Configuration[b]
Ru/dec/PO_4	50	15	22	35	49	*S*
Ru/tos/CO_3	50	17	36	19	55	*S*
Rh/dec/PO_4	30	30	95	41	13	*R*
Rh/dec/PO_4	50	15	90	70	30[c]	*R*
Rh/dec/PO_4	80	4	100	45	5	*R*
Rh/dbp/PO_4	30	29	20	58	19	*R*
Rh/tos/CO_3	50	22	100	6	74	*S*
Rh(chi)/tos/CO_3	50	24	63	38	2	*S*

[a] Chemoselectivity to 1-phenylethanol, [b] Enantiomeric excess (ee) and configuration of 1-phenylethanol were determined by gas chromatography, [c] 49 % ee at 10 % conversion

Table 5. Conversion and selectivity in the hydrogenation of acetophenone

Catalyst	Temp (°C)	k_{AB} (h^{-1})	k_{AC} (h^{-1})	k_{BD} (h^{-1})	k_{CD} (h^{-1})	k_{AD} (h^{-1})	Ratio (k_{AB}/k_{AC})	Ratio $(k_{AB} + k_{CD})/(k_{AC} + k_{BD})$
Rh/dec/PO_4	30	0.064	0.038	0.018	0.043	0.006	1.7	1.9
Rh/dec/PO_4	50	0.169	0.046	0.008	0.065	0.005	3.7	4.3
Rh/dec/PO_4	80	0.678	0.177	0.062	0.038	0.299	3.8	3.0
Rh/dbp/PO_4	30	0.185	0.099	0	0.391	0.036	1.9	5.8
Rh/tos/CO_3	50	0.136	0.117	0.137	0.017	0	1.2	0.6
Ru/dec/PO_4[a]	50	0.108	0.109	0.100	0.012	0.045	1.0	0.6
Ru/tos/CO_3[a]	50	0.044	0.145	0	0.356	0.036	0.3	2.8

[a] Catalyst deactivation was taken into account.

Table 6. Rate constants for the hydrogenation of acetophenone with supported catalysts

As expected the reaction rates are temperature dependent. A higher ratio for k_{AB}/k_{AC} supports the obtained selectivities of 2-phenylethanol. It was found that Rh/dec/PO_4 proved a more effective catalyst for the reduction at 50 °C in terms of reaction rate, conversion and selectivity. The activation energies calculated for reaction AB (42 $kJmol^{-1}$) were measured higher than reactions AC and BD (28 and 24 $kJmol^{-1}$ respectively) which again supports the higher selectivities at high temperatures.

When chemoselective hydrogenation of cinnamaldehyde was carried out by Kume *et al.*[39] using Pd nanoparticles at 80 °C and 5 MPa of H_2, they found that the reaction which uses Pd nanoparticles immobilized on ionic liquid modified silica gel in *m*-xylene has high reaction rates than that of reactions with $Pd(OAc)_2$ immobilized in ionic liquids in terms of TOFs (Table 7). In case of Pd/SiO_2-IL (PF_6) the reaction took only 20 minutes for completion with TOF > 47,000 h^{-1} whereas the analogous reaction in [bmim][PF_6] using $Pd(OAc)_2$ as catalyst took 6 h to complete with TOF 18 h^{-1}. The increase in reaction rates are proposed to be due to the high surface area of the silica gel which promotes adsorption of substrate and H_2 on to the Pd nanoparticles. The low yields in cases when ionic liquids were used as the solvent were ascribed to low solubility of H_2 in the medium.

Run	Catalyst	Time	Yield (%)	TOF (h^{-1})
1[b]	Pd/[bmim][Cl]	6 h	56.2	18
2[b]	Pd/[bmim][PF_6]	6 h	100	33
3[b]	Pd/[bmim][BF_4]	6 h	75.3	23
4[b]	Pd/[bmim][NO_3]	6 h	90.1	29
5[c]	Pd/SiO_2-IL[Cl]	20 min	20.7	270
6[d]	Pd/SiO_2-IL[PF_6]	20 min	100	> 47,000
7[e]	Pd/SiO_2-IL[BF_4]	20 min	70.8	24,260
8[e]	Pd/SiO_2-IL[NO_3]	20 min	100	> 33,000
9[e]	Pd/SiO_2	20 min	64.8	22,200

[a] Temperature 80 °C; H_2 5 MPa; cinnamaldehyde 2.7 mmol; *m*-xylene 2 g, [b] $Pd(OAc)_2$ 0.0134 mmol; ionic liquid 5 mmol., [c] Pd 0.108 mg, [d] Pd 0.016 mg, [e] Pd 0.022 mg.

Table 7. Hydrogenation of cinnamaldehyde catalyzed by Pd catalyst

While discussing reaction rates, one has to consider the rates of formation of metal nanoparticles as well. When Scheeren *et al.*[40] studied the hydrogenation of cyclohexene by catalyst precursor PtO_2 dispersed in [bmim][PF_6] at 75 °C, they proposed a mechanism of formation of Pt(0) nanoparticles and calculated the rate of their formation. This mechanism suggests nanoparticles are formed in four stages, nucleation, autocatalytic surface growth, agglomeration and autocatalytic agglomeration to form large agglomerates (Figure 6).

(a) $A \xrightarrow{k_1} B$

(b) $A + B \xrightarrow{k_2} 2B$

(c) $B + B \xrightarrow{k_3} C$

(d) $B + C \xrightarrow{k_4} 1.5C$

* The four equations correspond to: (a) slow nucleation of catalyst precursor A to a nanocluster B, (b) autocatalytic surface growth, (c) agglomeration step leading to the formation of bulk metal C, and (d) autocatalytic agglomeration of smaller nanoparticles with larger bulk metal particles.

Fig. 6. Four-step mechanism for transition-metal nanocluster nucleation, growth and agglomeration

Rate constants were calculated for the four step formation of nanoparticles. The kinetic data was found to be consistent with formation of large agglomerates of bulk metal catalyst (Table 8) and shows that the autocatalytic surface growth is faster than other steps.

Floris *et al.*[41] have discussed the effect of ion pairs of the catalyst and ionic liquid in the asymmetric hydrogenation of methyl acetoacetate at 60 °C and a hydrogen pressure of 50 bars. The [PF_6^-] based ionic liquids showed around 50 % lower activity in terms of TOF than [NTf_2^-] salts. (Table 9)

PtO_2/ C_6H_{12} (molar ratio)	Equations in kinetic model[a]	k_1 (h^{-1})[b]	k_2 ($M^{-1}h^{-1}$)[b]	k_3 ($M^{-1}h^{-1}$)[b]	k_4 ($M^{-1}h^{-1}$)[b]
1/4000	a, b	0.143	204.00	-	-
1/4000	a, b, c	0.184	329.88	15.96	-
1/1000	a, b	0.094	58.10	-	-
1/1000	a, b, c	0.086	78.72	7.26	-
1/1000	a, b, c, d	0.677	901.39	9.40	146.17

[a] From Figure 6. [b] Rate constants corrected for reaction stoichiometry

Table 8. Kinetic constants for hydrogenation of cyclohexene by PtO_2 in [bmim][PF_6][a]

IL	TOF_{90} (h^{-1})	ee_{90} (%)	S_{90} (%)	$ee^{re\text{-}use}$ (%)
MeOH	1100	98	79	-
[N_{6222}][NTf_2]	390	93	87	54
[bmim][NTf_2]	340	97	91	54
[N_{6222}][PF_6]	210	55	86	-
[bmim][PF_6]	160	78	88	32

Reaction conditions: 2 g MAA, 17 mL IL-MeOH 1/1 wt., S/C = 1580, 333 K, 50 bar H_2.

Table 9. The ion pair effect

There are two rate limiting factors which can be considered. First is the hydrogen gas solubility, and second is the structural modification of active catalytic centre by anion pairing. Figure 7 depicts plausible catalyst-IL interactions in the case of the [NTf_2^-] based IL.

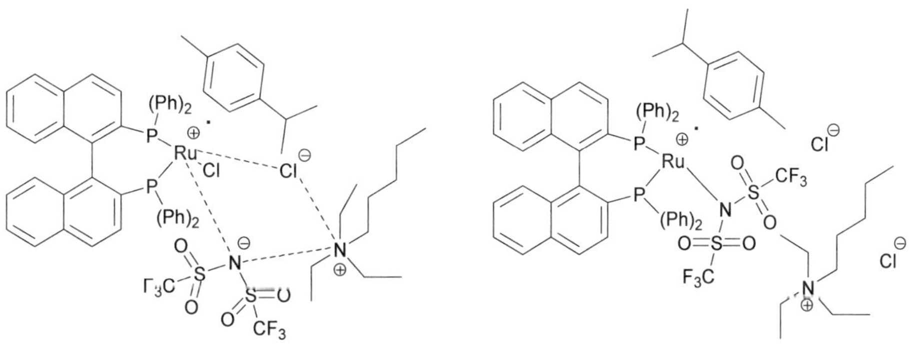

Fig. 7. Plausible catalyst-IL interactions in an [NTf_2^-] IL

3. Transition metal catalysis in ILs

Commonly used heterogeneous catalysts such as palladium or platinum on solid supports are among the catalysts employed for the hydrogenation of substrates in ILs. Although increased temperature and pressure may be a requirement when using the IL, classic palladium, platinum and ruthenium catalysts have been shown to give superior results when used in an IL compared with a common organic solvent.

Xu *et al.*[42] used a range of imidazolium ILs containing [BF_4] and [PF_6] anions for the catalytic heterogeneous hydrogenation of halonitrobenzenes to the corresponding haloanilines (Figure 8).

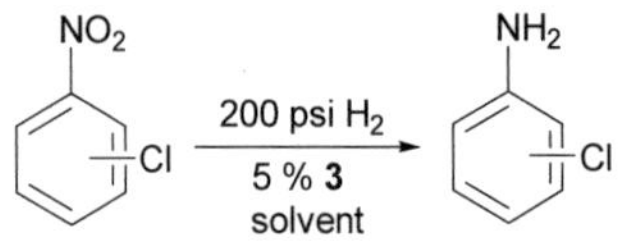

Dehalogenation (%)	
[bmim][BF_4]	Methanol
ortho: 0	*ortho*: 22.7
meta: 0.8	*meta*: 29.0
para: 2.5	*para*: 44.2

Fig. 8. Hydrogenation of halonitrobenzenes to haloanilines

Raney nickel (**1**), platinum on carbon (**2**) and palladium on carbon (**3**) were employed as metal catalysts, and methanol was used as a reference organic solvent due to its wide application in heterogeneous catalytic hydrogenations. Although increased temperatures and pressures were required for the IL systems (100 °C, 31.0 bar (**1**), 13.8 bar (**2** and **3**)) in comparison with the methanol systems, i.e. 80 °C, 13.8 bar for (**1**) and 30 °C, 2.8 bar for (**2**) and (**3**), the ILs performed better as solvents for these reactions, with the undesirable dehalogenation being greatest for all substrates tested in methanol rather than the ILs (Figure 8). Taking for example *ortho-*, *meta-* and *para-*chloronitrobenzene, and 5 % (**3**), for which the greatest differences in results between IL and organic solvent were evident, in [bmim][BF_4] dehalogenation ranged from as little as 0 % with *o*-chloronitrobenzene to 0.8 % with the *meta* derivative and at most 2.5 % with the *para* substituted derivative. However, when methanol was used as the solvent, dehalogenation ranged from 22.7 % for *ortho* to 44.2 % for the *para*-chloro isomer (Figure 8). The same trend was evident using 5 % (**2**) and (**1**) as catalyst, albeit to a lesser extent. The reaction rates were found to be lower in [bmim][BF_4] than in methanol, which was attributed to mass transfer processes.

Anderson *et al.*[43] selected the α,β-unsaturated aldehydes, citral and cinnamaldehyde (Figure 9 and Figure 10), to demonstrate the superior selectivity obtained using pyridinium, imidazolium and ammonium ILs over common organic solvents in hydrogenation reactions. A palladium on carbon catalyst (**3**) was used for the reactions.

In the case of cinnamaldehyde, although the temperature was increased for the reaction carried out in the IL (60 °C), superior selectivity towards hydrocinnamaldehyde was obtained (78-100 %) compared with several conventional organic solvents (78-89 %). Worth noting for these hydrogenations is the variation of selectivities across a series of [bmim] ILs. [Bmim][PF_6] showed a selectivity of 100 %, [bmim][OTf] 91 %, and [bmim][OAc] 78 % for formation of hydrocinnamaldehyde. Recycling of the [bmim][BF_4] system showed catalyst activity to decrease by 50 % upon the first recycle but remained constant thereafter for five successive reactions. The selectivity however remained almost constant for all recycles carried out. The authors note that if the IL system without a substrate is treated with hydrogen gas for one hour prior to the reaction, the recycling ability of the system can be improved. In the case of citral hydrogenation, similar trends were observed. The selectivity towards citronellal obtained using ILs ([bmim][PF_6], [bmim][BF_4], [C_8Py][BF_4], [C_6mim][NTf_2], and [emim][NTf_2]) ranged from 81-100 %, with the organic solvents giving only 62-77 % selectivity (cyclohexane: 62 %, toluene: 77 %, and dioxane: 77 %). The authors concluded that due to the high viscosity of the ILs the rate of diffusion of aldehyde was reduced compared with conventional solvents and reactions rates were correspondingly lower. Using the dicyanamide IL [bmim][$N(CN)_2$] Arras *et al.*[44] also achieved the selective

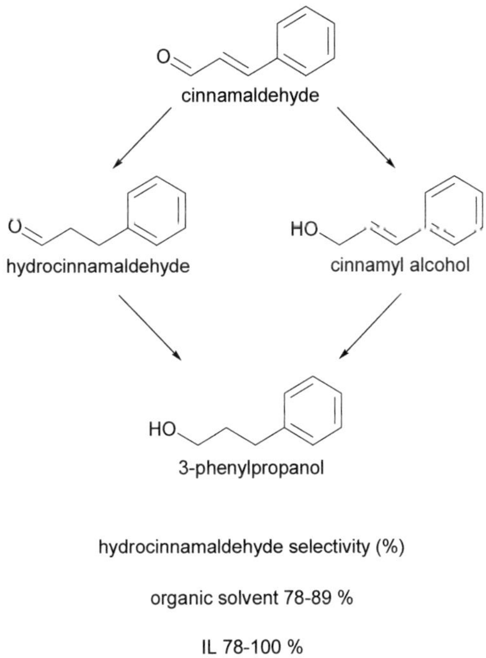

Fig. 9. Reaction pathway of cinnamaldehyde hydrogenation

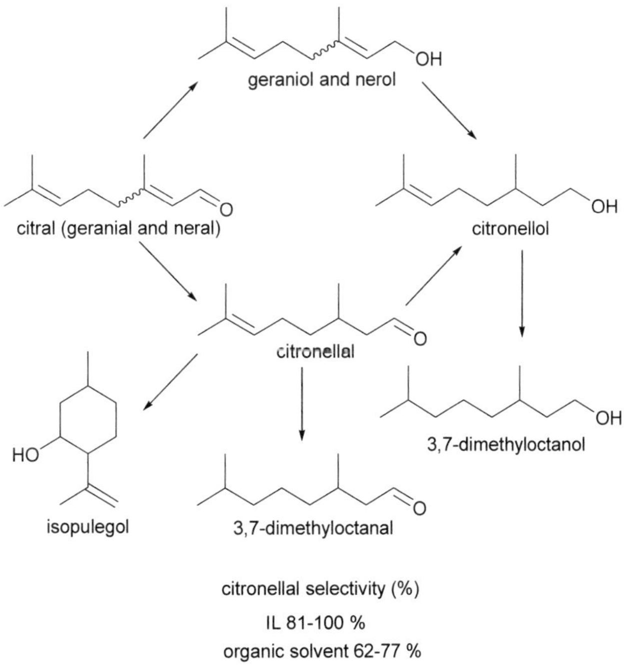

Fig. 10. Reaction pathway of citral hydrogenation

hydrogenation of citral to citronellal using 10 % (**3**) at 50 °C and 1.0 MPa H_2. In the IL-free system 100 % conversion and 41 % selectivity towards the desired product was obtained. They found that with the IL as bulk solvent, 100 % conversion was obtained with 97 % selectivity. However, using the catalyst coated with IL the selectivity increased to > 99 % with 100 % conversion. When the IL was only present as an additive a reduction in the conversion (42 %) was observed while the selectivity remained high (> 99 %).

The same group continued this work by investigating the use of ILs as additives and coatings on palladium supported catalysts for the hydrogenation of citral.[45] They compared the results obtained by using ILs containing perfluorinated anions ([bmim][NTf_2], [bmim][PF_6] and [BMPL][NTf_2]) and the dicyanamide anion ([bmim][$N(CN)_2$], [BMPL][$N(CN)_2$] and [B3MPYR][$N(CN)_2$]). The performance of the dicyanamide ILs as either additives or coatings improved in comparison the ditriflimide or hexafluorophosphate ILs regardless of the heterocyclic cation. Using Pd/SiO_2 (**4**) as catalyst, at 50 °C and 2.0 MPa H_2, the highest conversions and selectivities towards citronellal were achieved using the dicyanamide ILs. Conversions ranging from 63 - 75 % and selectivities from 59 - 62 % were obtained using the ditriflimide or hexafluorophosphate ILs as a catalyst coating. The results for conversion obtained with [bmim][NTf_2] and [BMPL][NTf_2] were comparable with those obtained in the absence of an IL under the same reaction conditions, although the selectivity obtained was higher (79 % conversion, 45 % selectivity). Using [bmim][$N(CN)_2$], [BMPL][$N(CN)_2$] and [B3MPYR][$N(CN)_2$] conversion was > 99 % and selectivity ranged from 81 - 99 %. Almost quantitative yield of citronellal was obtained using [BMPL][$N(CN)_2$] (conversion = 100 %, selectivity = 99 %). The authors attribute the improved results of the hydrogenation using dicyanamide ILs to the sensitivity of the hydrogenation reaction to halide impurities from the fluorinated anions and also an electronic interaction between the dicyanamide anion and the palladium catalyst. Curiously the low hydrogen solubility in ILs did not affect the conversion, indicating no mass transport limitations. When the research was extended to Ru/Al_2O_3 catalyzed hydrogenation of citral in ditriflimide ILs[46], it was determined that addition of the ionic liquid reduces the initial TOF of the catalyst. For [bmim][NTf_2] the initial TOF decreased from 14.0×10^{-2} s^{-1} (initial TOF for reaction without IL) to 7.2×10^{-2} s^{-1}.

Geldbach *et al.*[32] investigated the generation of catalysts from metal chlorides in the Lewis acidic IL, [*N*-octyl-3-picolinium]$AlCl_4$ (Figure 11), by dissolving a series of metal chlorides in the IL and adding benzene as the substrate. The hydrogenation of benzene to cyclohexane is an important petrochemical process and research in this area is in continuous evolution.[47,48,49]

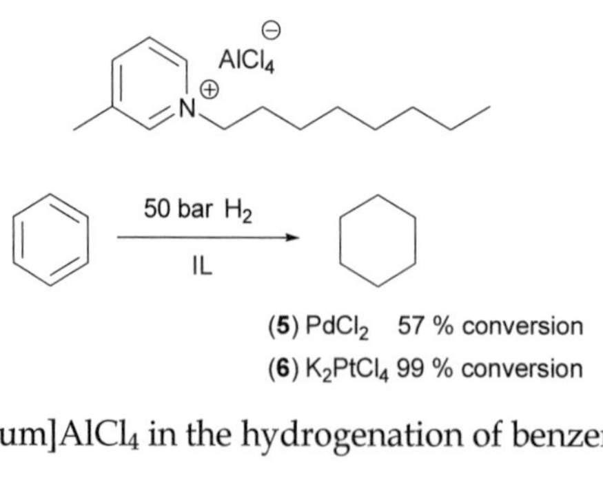

Fig. 11. [*N*-octyl-3-picolinium]$AlCl_4$ in the hydrogenation of benzene

Hydrogen was added to the biphasic mixture and the conversions to cyclohexane recorded. Of all the metal chlorides used, only the palladium catalyst, $PdCl_2$, (**5**) and platinum catalyst, K_2PtCl_4, (**6**) showed any significant activity, with 57 and 99 % conversion respectively. K_2PtCl_4 was investigated further in relation to catalyst concentration and temperature due to the impressive result. Generally, higher K_2PtCl_4 catalyst loadings gave increased conversion, and raising the temperature also led to an increase in conversion (e.g. 0.0071 mol% (**6**): 18 % conversion at 20 °C and 80 % conversion at 75 °C; and 0.14 mol% (**6**): 67 % conversion at 20 °C and > 99 % conversion at 75 °C). This groups research also extends to the examination of a ruthenium cluster catalyst in [bmim][BF_4] for the same hydrogenation reactions however no activity was observed using the IL.[23] The importance of the IL-promoted reduction can be appreciated when one considers that much higher temperatures have been recommended to achieve conversion of benzene to cyclohexane under conventional conditions. Jasik *et al.*[47] using a temperature of 175 °C to achieve complete hydrogenation and Bakar *et al.*[48] recommended 200 °C for maximum conversion.

Deshmukh *et al.*[50] also used a Lewis acidic IL ([bmim][$AlCl_4$]) for the hydrogenation of a selection of arenes, in the presence of a Pd/C catalyst. Although the hydrogenation of benzene can require harsh conditions, this group combined the concept of the activation of aromatics by Lewis acids (using the IL) and the activation of molecular hydrogen by Pd/C to achieve > 99 % conversion of benzene to cyclohexane under ambient conditions (1 bar H_2, RT). Under the same conditions they were able to achieve > 99 % conversion of biphenyl, naphthalene and napthacene and 97 % conversion of anthracene. Deshmukh[50] also tackled the problem of the hydrogenation of fullerene with their novel system. Harsh conditions are usually necessary for hydrogenation of this substrate (120 bar H_2, 400 °C),[51] however hydrogenation was achieved with only 5 bar H_2 at RT using this system.

Although ILs has many 'green' attributes, in hydrogenation reactions there are some cases, in which conventional organic solvents are preferred. Using the bimetallic catalyst system, Ag-In/SiO_2 (**7**) for the hydrogenation of citral to selectively form the acyclic/allylic terpene alcohols, geraniol and nerol, Steffan *et al.*[52] showed the non-polar solvent hexane to be superior to [bmim][NTf_2]. The chemoselective hydrogenation of citral to geraniol and nerol was lower in the IL compared to the organic solvent. Steffan explained the lower conversion of citral in ILs compared with hexane by suggesting that the lower solubility of hydrogen in the IL (0.77×10^{-3} $molL^{-1}$ at 298 K and 0.1 MPa of H_2, estimated from the H_2 solubility at 10.1 MPa assuming a linear relationship with partial pressure)[10] was the limiting factor. Information pertaining to hydrogen solubility in ILs can be found in recent sources.[53,54,55]

While investigating mass transfer effects in the hydrogenation of phenylacetylene to styrene and ethyl benzene (Figure 12) using a rotating disc reactor, Hardacre *et al.*[56] found [bmim][NTf_2] to give lower reaction rates than a non-polar hydrocarbon solvent.

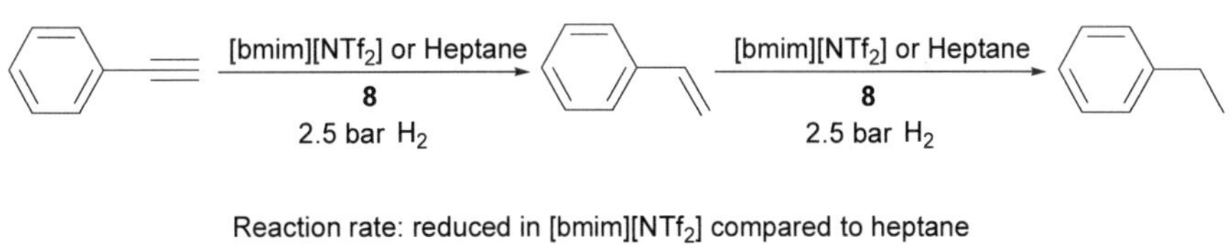

Fig. 12. Hydrogenation of phenylacetylene

Using palladium on calcium carbonate (**8**) as a catalyst, they investigated several parameters in [bmim][NTf_2] and heptane, including the effect of phenyl acetylene concentration in the solvent and the rotation speed. The rate of reaction was reduced in the IL (0.942 mmol min^{-1} at 6.0 bar) compared with the organic solvent (12.976 mmol min^{-1} at 5.5 bar), postulated to be due to the varying rate of diffusion of gaseous hydrogen through the liquid medium to the catalyst surface. The liquid to solid mass transfer coefficient ($k_{LS}a$) was calculated to vary from 0.144 to 0.150 s^{-1}, over a range of phenylacetylene concentrations and hydrogen pressures. These values indicate that the reaction is limited by liquid to solid mass transfer process, in particular the transport of dissolved hydrogen. This fact was supported by the calculated activation energies for conversion of phenylacetylene to ethyl benzene in heptane and [bmim][NTf_2] obtained between 9 and 33 kJ mol^{-1}.

Recently Khodadadi-Moghaddam *et al.*[28] investigated the kinetic parameters of the hydrogenation of cyclohexene, in mixtures of 2-hydroxyethylammonium formate and various alcohols (methanol, ethanol and IPA). Using a Pt-Al_2O_3 (**9**) catalyst, the rate constant for the reaction carried out in the IL/IPA mixture was twenty eight times higher than when IPA was used as the reaction medium. From studying solvent effects on the reaction the authors explain the discrepancy in rate to be due to the varying polarities of solvent and substrate - specifically that because of the polarity of the IL, the non-polar cyclohexene is more abundant on the catalyst surface, promoting the reaction. Furthermore, Khodadadi-Moghaddam and co-workers state that low gas solubility in many ILs is compensated by fast gas diffusion in reactions involving hydrogen gas.[57] This group extended their work to investigate this effect using acetone as the hydrogenation reaction substrate.[58] They postulate that the presence of the polar carbonyl group in acetone lowers the extent of adsorption on the catalyst surface compared with cyclohexene. The first-order rate constant of the hydrogenation reaction of cyclohexene to cyclohexane was found to be 8.7 times higher using the IL as solvent than compared to IPA. The rate constant of the hydrogenation reaction of acetone to propan-2-ol was also higher in the IL than IPA, albeit only 3.3 times.

Biphasic reaction conditions are one important method for hydrogenations using homogeneous catalysts when efficient recycling of catalyst is of importance. Hydrogenation reactions have been carried out using rhodium and ruthenium catalysts in biphasic systems using imidazolium based ILs ([bmim], [hmim] and [omim] [BF_4]). With the use of a rhodium catalyst ([Rh(η^4-C_7H_8)(PPh_3)$_2$][BF_4]) (**10**), Dyson *et al.*[59] demonstrated a biphasic hydrogenation of an alkyne using [omim][BF_4] containing the catalyst and an aqueous phase containing the substrate, 2-butyne-1,4-diol (Figure 13).

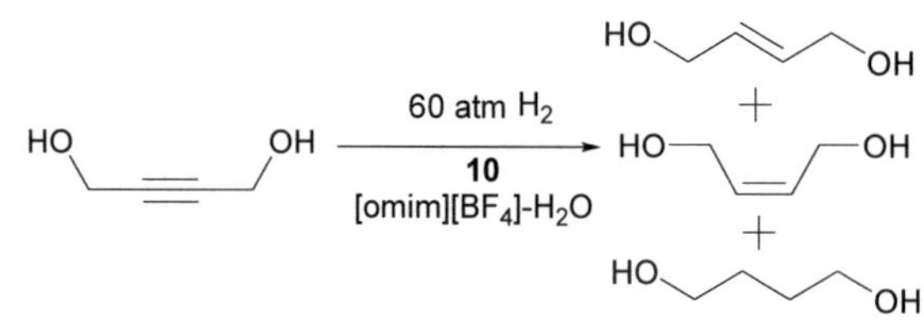

Fig. 13. Hydrogenation of butyne-1,4-diol

At room temperature, the phases were immiscible; however at the reaction temperature of 80 °C homogeneity was attained. Hydrogenation reactions were carried out under 60 atm H_2 with facile separation of the reduced products from the catalyst/IL phase being achieved simply by cooling the reaction. The products dissolved in the aqueous layer were isolated and reuse of the IL/catalyst system demonstrated. The limitations of this system were shown with maleic acid, when the reduced product, succinic acid, was found to be soluble

in both the IL and aqueous phase. This type of cationic rhodium catalyst has been also used by Esteruelas *et al.*[60] employing an organic solvent to selectively hydrogenate phenylacetylene to styrene ([Rh(2,5-norbornadiene)(PPh_3)$_2$] (**11**) in DCM at 25 °C and 1 atm. H_2 pressure). At 50 °C and 39.48 atm. H_2 with $scCO_2$ as reaction solvent (157.91 atm. CO_2) Zhao *et al.*[61] selectively hydrogenated the same substrate as Dyson *et al.*,[59] 2-butyne-1,4-diol, to butane-1,4-diol (84 % at 100 % conversion) using a stainless steel reactor wall (SUS 316) to promote the reaction with no catalyst.
Wolfson *et al.*[37] used [bmim][PF_6] as reaction medium in the hydrogenation of 2-cyclohexen-1-one with Rh(PPh_3)$_3$Cl (**12**) (Wilkinson's catalyst) (Figure 14) and methyl 2-acetamidoacrylate with Rh-EtDuPHOS (**13**).

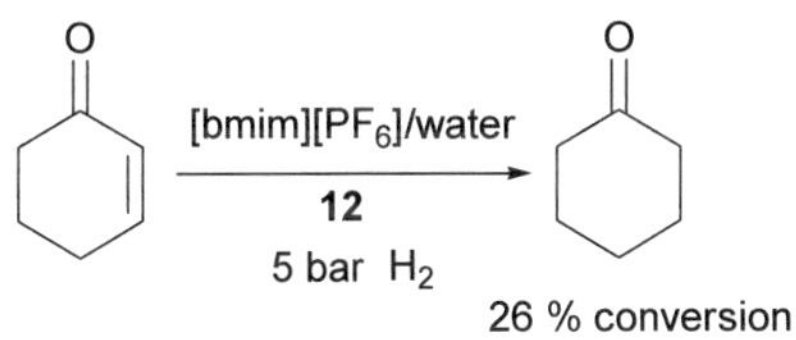

Fig. 14. Hydrogenation of 2-cyclohexen-1-one

As water was shown by this group to enhance the activity of Wilkinson's catalyst (**12**), they studied this parameter in the biphasic hydrogenation of 2-cyclohexen-1-one. Diethyl ether and hexane were screened but demonstrated low hydrogenation activity. The conversion to cyclohexanone increased from 4 % (100 % selective) in diethyl ether and 7 % (100 % selective) in hexane as co-solvent to 26 % in water (90 % selective). In the analogous homogeneous reaction with Wilkinson's catalyst (**12**) and only ethanol as solvent, 100 % conversion was achieved, albeit with low selectivity (27 %). Increased selectivity was achieved using DCM as solvent (100 %), although a compromise in conversion was observed (17 %).
Water was also used as solvent in conjunction with the [bmim][PF_6] for the biphasic hydrogenation of methyl 2-acetamidoacrylate at 5 bar H_2 and 20 °C. 68 % conversion was obtained (66 % upon re-use) with 96 % ee (97 % upon re-use). The group proposes that the use of water as the second solvent in biphasic IL reactions has a beneficial effect on activities due to the creation of a well mixed 'emulsion-like' system. Using methanol as the sole phase in a homogeneous reaction, 54 % conversion with 97 % selectivity was obtained. However, using the IL as sole reaction phase, no reaction was observed.
Scurto *et al.*[62] used biphasic hydrogenation conditions with $scCO_2$ and a rhodium catalyst (**14**) for the hydrogenation of 2-vinyl-naphthalene (Figure 15).

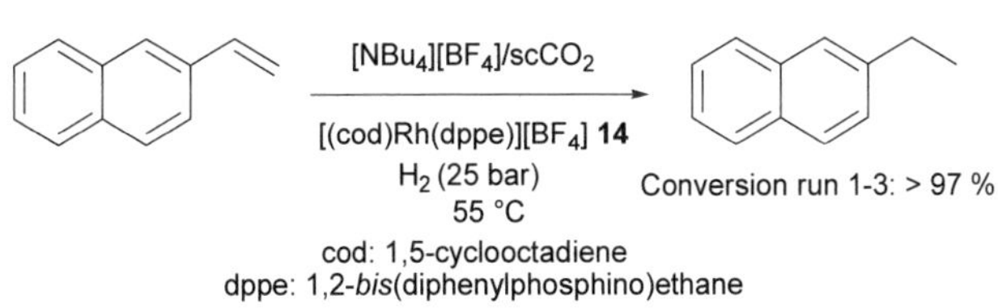

Fig. 15. Hydrogenation of 2-vinyl-naphthalene

[TBA][BF_4] was pressurised with CO_2 to give a high melting point depression of the salt for subsequent use as a reaction solvent in the liquid phase. Conversions for the first three runs using the IL were high (> 93 %). The authors explained that the drop in conversion to 62 % by the fifth run may have been due to accidental oxygen introduction or loss of catalyst during the recycling procedure.

Suarez *et al.*[63] used a ruthenium catalyst ($RuCl_2(TPPMS)_3(DMSO)$ (**15**); TPPMS: triphenylphosphine monosulfonate) immobilised in [bmim][PF_6] for the biphasic hydrogenation of 1-hexene (Figure 16).

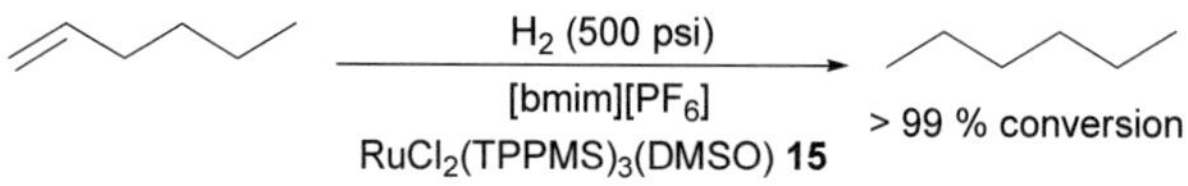

Fig. 16. Hydrogenation of 1-hexene

They investigated the effect of different parameters on the hydrogenation rate and conversion. It was observed that with increasing temperature, the viscosity of the IL decreased, therefore the conversion rate increased. However at the upper limit of temperature, 120 °C, decomposition of the catalyst was observed. Increasing the pressure also increased the conversion, until it levelled off at pressures higher than 500 psi. Overall, greater than 99 % conversion was observed for the hydrogenation of 1-hexene in the IL. However, upon recycling, the total conversion decreased (70 % after six reuses). The catalytic activity of the system using this catalyst (**15**) with the IL is, however, lower in comparison to using (**15**) in a toluene/water biphasic system.[64] Two other substrates were also investigated (cyclohexene and crotonaldehyde). 34 % conversion was achieved with cyclohexene, while only 25 % was achieved using crotonaldehyde, although good selectivity was attained with 1-butanol formed as the only product.

The tailoring of ILs to carry out a specific role together with acting as reaction medium is emerging as an efficient way to limit the number of reagents required in a chemical reaction.[9] One example is the synthesis of imidazolium IL (1-(*N,N*-dimethylaminoethyl)-2,3-dimethylimidazolium trifluoromethanesulfonate, [mammim][OTf]) for the hydrogenation of carbon dioxide to form formic acid[65] (Figure 17). This reaction is thermodynamically unfavourable because the standard Gibbs free energy ΔG_{298}° of reaction is + 32.9 kJ mol^{-1}.

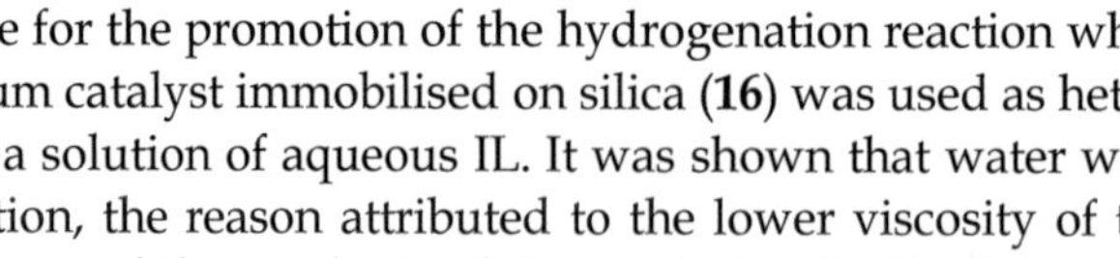

Fig. 17. Hydrogenation of carbon dioxide

The IL acts as a base for the promotion of the hydrogenation reaction which was carried out at 60 °C. A ruthenium catalyst immobilised on silica (**16**) was used as heterogeneous catalyst phase dispersed in a solution of aqueous IL. It was shown that water was necessary for the success of the reaction, the reason attributed to the lower viscosity of the IL. The basic IL [mammim][OTf] favoured the synthesis of the product, with the formation of a formate salt driving the reaction. TOFs as high as 103 h^{-1} were reached. Higher TOFs were, however, observed using other ruthenium catalysts in organic solvents (e.g. acetone, NEt_3, [Ru_2(μ-

CO)(CO)$_4$(μ-dppm)$_2$], 38.11 atm. H_2 and CO_2, TOF = 130 h^{-1})[66] and scCO_2 ([$RuH_2(PMe_3)_4$], 80 – 85 atm H_2, TOF = 630 h^{-1})[67] for the hydrogenation of carbon dioxide to formic acid. An increase in H_2 and CO_2 pressure together with increasing the amount of IL was found to favour higher TOFs. The process was shown to be suitable for recycling procedures, with no significant reduction in TOF being observed after four recycles. Based on the assumption that the efficiency of CO_2 hydrogenation would increase if the IL contained more than one basic group, Zhang[68] synthesised a novel IL containing two tertiary amino groups, 1,3-di(*N,N*-dimethylaminoethyl)-2-methylimidazolium trifluoromethanesulfonate [DAMI][OTf] (Figure 18). It was indeed shown that increasing the number of basic groups in the IL increased the formation of formic acid. A formic acid to IL ratio of 0.145:1 HCO_2H/[mammim][OTf] (wt/wt) was achieved in one reaction cycle. Using [DAMI][OTf] containing two basic groups this increased to 0.246:1.

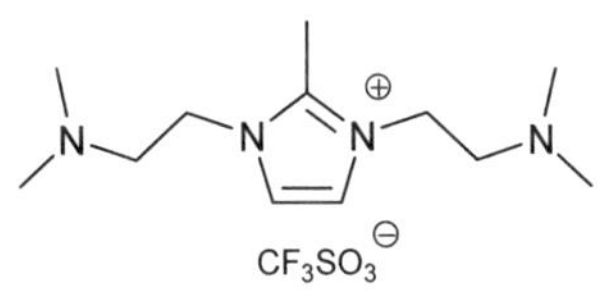

[DAMI][OTf]

Fig. 18. [DAMI][OTf]

Obert *et al.*[69] used a ruthenium on carbon catalyst for the selective hydrogenation of propionitrile to propylamine under biphasic reaction conditions. In 1998 Dow Chemical Co. patented a process for this hydrogenation using a biphasic system of water and organic solvent where selectivities towards propylamine of > 80 % were achieved.[70] The work carried out by this group was based on two approaches encompassing the use of Brønsted-acidic ILs (dimethylcyclohexylammonium hydrogensulfate and 1-butylimidazolium hydrogensulfate) or neutral ILs ([emim][$EtOSO_3$]). The Brønsted-acidic IL was used for protonation of the primary amine in order to prevent its subsequent over-reduction to di- and tri-propylamine. The neutral IL [emim][$EtOSO_3$] was used as a medium from which the primary amine could be extracted by an organic solvent. Control experiments were carried out using one phase consisting of an organic solvent. At full conversion up to almost 50 % selectivity for propylamine was observed using a system of 1,2,4-trichlorobenzene (48.3 %), methanol (26.8 %), and cyclohexane (34.8 %). The best results were achieved using a biphasic system of 1,2,4-trichlorobenzene and the Brønsted-acidic IL, 85 % selectivity was obtained at full conversion. Using the biphasic neutral IL, the selectivity decreased to 70 %. Recycling of the neutral IL biphasic system was again demonstrated. However, the use of the Brønsted-acidic IL system required a laborious basic aqueous work-up.

Bouquillon *et al.*[71] used novel ILs, including a readily biodegradable IL for the hydrogenation of phenoxyocta-2,7-diene (Figure 19). The work for the preparation of these biodegradable solvents began with Gathergood and Scammells in 2002[72] where the same principles that are used in the synthesis of biodegradable surfactants were applied to the design of environmentally friendly ILs. Subsequent studies showed the presence of an ester linkage in the side chain of the IL cation promoted biodegradation.[73] The counterion was also a significant factor, with the octylsulfate anion present in examples which were readily biodegradable. Impressive conversions were obtained by Bouquillon *et al.*[71] for the hydrogenation of phenoxyocta-2,7-diene using a palladium catalyst (**17**). The biodegradable octylsulfate imidazolium IL displayed superior conversion (85 %) to the ditriflimide derivative (75 %). The potential for reuse of the IL/catalyst system was exemplified by the recycling of the octylsulfate system, albeit with a significant decrease in conversion to 55 % observed.

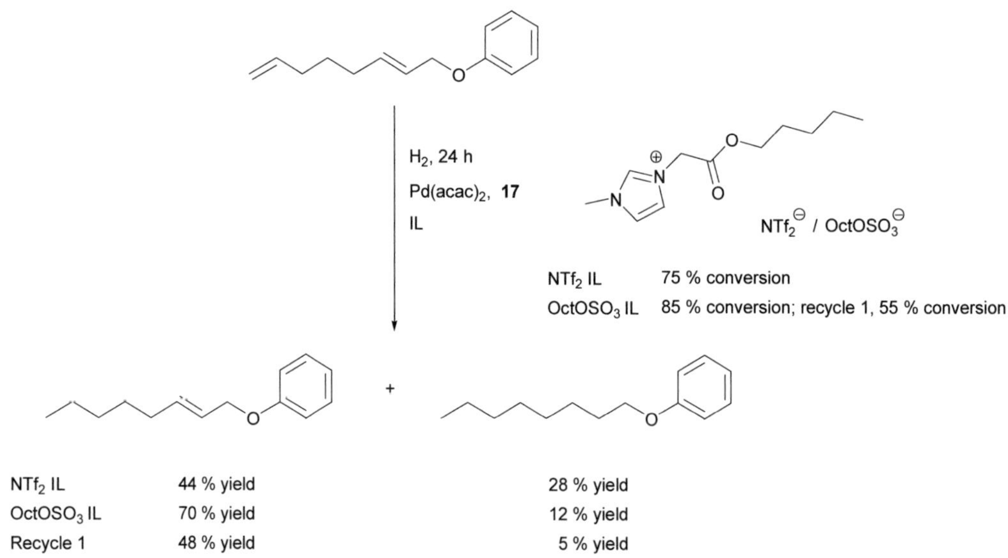

Fig. 19. Hydrogenation using ester-functionalised ILs

The potential of ILs with ester functionalities as solvents for hydrogenation reactions has been recently shown to cover a wider range of substrates. In their communication, Morrissey *et al.*[74] used the previously mentioned imidazolium ILs together with other novel ILs containing ester groups for the hydrogenation of *trans*-cinnamaldehyde, and a range of cinnamate esters. The use of these novel ILs was shown to give superior selectivity towards hydrocinnamaldehyde in the reduction of *trans*-cinnamaldehyde when compared with commercially available [bmim][NTf_2] and [bmim][$OctOSO_3$] ILs and the common organic solvent, toluene. Using a Pd/C catalyst [10 % (**3**)] at 1 atm H_2 pressure, impressive conversions (97 - 100 %) and selectivities (88 - 100 %) were obtained using the ester functionalised imidazolium ILs (Figure 20). A comprehensive review of the hydrogenation of α,β-unsaturated aldehydes comprising *trans*-cinnamaldehyde has been published by Gallezot *et al.*[75]

The hydrogenation of benzyl cinnamate was investigated by the same group, who found that selective hydrogenation of the olefin in the presence of the benzyl group was not possible using either the commercial ILs ([bmim][NTf_2] and [bmim][$OctOSO_3$]) or common organic solvents (e.g. THF, ethyl acetate, and methanol). Instead, complete hydrogenolysis of the benzyl group was observed under these conditions. However, by using the ester-functionalised ILs, 3-methyl-1-(propoxyethoxycarbonylmethyl)imidazolium [NTf_2^-] and [$OctOSO_3^-$] hydrogenation of the olefin occurred with complete selectivity, leaving the benzyl ester intact (Figure 21).

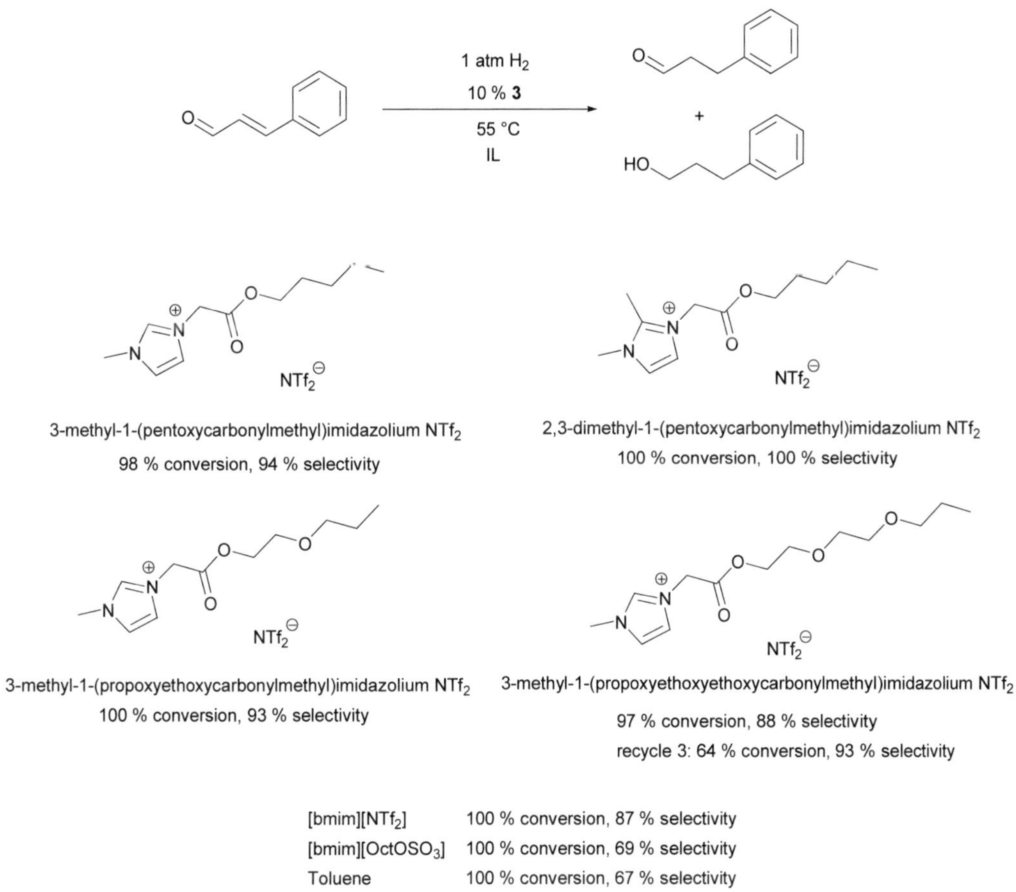

Fig. 20. Hydrogenation of *trans*-cinnamaldehyde using ester-functionalised ILs

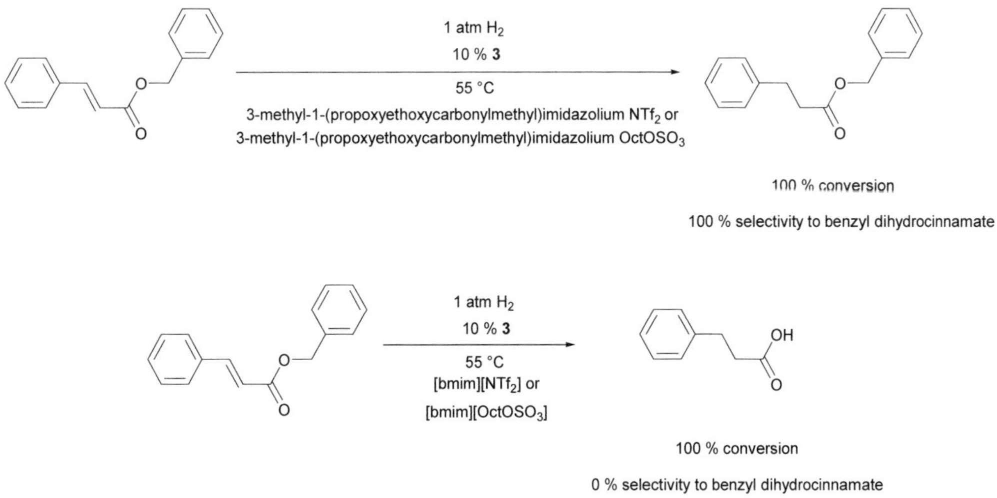

Fig. 21. Hydrogenation of benzyl cinnamate using ester-functionalised ILs

4. Asymmetric Hydrogenation in ILs

Asymmetric hydrogenation is one of the most reliable methods for the synthesis of enantiomerically pure products.[76] For the greater part, the source of chiral induction originates from chiral ligands coordinated to a metal catalyst.[77,78] Extensive research into the hydrogenation of prochiral substrates in ILs has been carried out in recent years. The majority of this work has employed ruthenium or rhodium based catalysts.

4.1 Hydrogenation using rhodium catalysts

Rhodium catalysts, DiPFc-Rh (**18**), and EtDuPHOS-Rh (**13**) were compared by Boyle *et al.*[79] for the hydrogenation of α-benzamido cinnamate in [bmim][BF_4] and [emim][OTf] (Figure 22). The conversion was negligible using the IL [bmim][BF_4] (0 - < 2 %), and [emim][OTf] became the focus of the following reactions, giving a conversion of 95 % and 89 % ee using 1 mol% (**13**) as catalyst (60 psi, 50 °C). Enamide esters were hydrogenated by Burk *et al.*[80] using the same catalyst (**13**) with methanol as the reaction solvent giving ≥ 99 % ee at 100 % conversion (30 psi H_2, 20 - 25 °C, 0.1 - 0.005 mol% (**13**) catalyst). Ru-BINAP (**19**) was also included in this study and gave the highest enantioselectivity (95 %) albeit with only 16 % conversion (60 psi H_2, RT, 1 mol% catalyst). This may be compared with a conventional organic solvent based system (THF/MeOH, 1:1 mixture) with Ru-BINAP immobilised on silica, which gave 100 % conversion and 85 % ee in the hydrogenation of α-benzamido cinnamate[81] at 29 psi H_2 pressure and 35 °C.

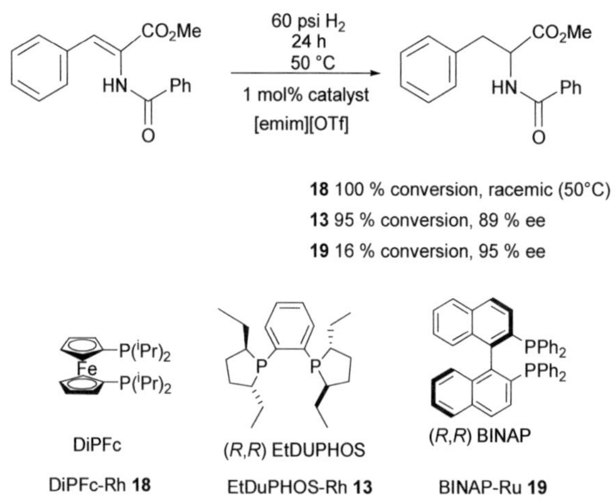

Fig. 22. Hydrogenation of α-benzamido cinnamate and Rh and Ru catalysts

Schmitkamp *et al.*[82] investigated the hydrogenation of dimethyl itaconate and methyl 2-acetamidoacrylate (Figure 23) using ditriflimide CILs derived from L-proline (*L*-prolinium methyl ester NTf_2) and L-valine (L-valinium methyl ester NTf_2) and a rhodium catalyst with tropoisomeric ligands (**20** and **21**).

They investigated the effects of using different CILs and the influence of sulfonated *tropos* ligands on the conversion and enantioselectivity of the reactions. For the hydrogenation of methyl 2-acetamidoacrylate, using (**21**), 49 (*S*) % ee was obtained using L-prolinium methyl ester [NTf_2], where a racemic mixture was obtained under the same conditions using L-valinium methyl ester [NTf_2]. This group thus used (**21**) for all subsequent hydrogenations.

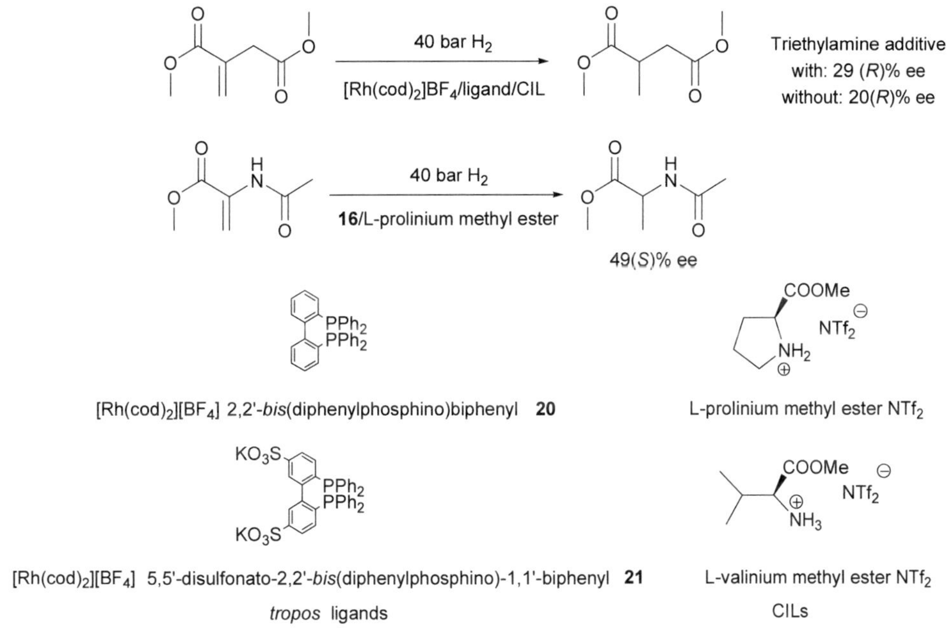

Fig. 23. Hydrogenation of prochiral substrates using CILs and a rhodium catalyst

Concerning the acrylate substrate, good enantiomeric excess (69 %) was obtained for the (*S*) enantiomer when triethylamine was used as an additive in the reaction. Using dimethyl itaconate as the substrate, the amine additive was again found to increase the enantioselectivity (from 20 (*R*) % ee without additive to 29 (*R*) % ee with additive). The sulfonate groups present in the *tropos* ligand were shown to be essential for increased enantioselectivity for both substrates. Enantioselectivity decreased dramatically when the unsulfonated 2,2′-*bis*(diphenylphosphino)biphenyl ligand (**20**) was employed. Changing from (**20**) to (**21**), a drop in enantioselectivity was observed for methyl 2-acetamidoacrylate, from 49 (*S*) to 28 (*S*) % ee, and in combination with triethylamine as an additive, from 69 (*S*) to 52 (*S*) % ee. Recycling of the system was possible by product extraction using $scCO_2$. The recycling procedure showed a reduction in conversion from > 99 % for the first run to 57 % in run three. Enantioselectivity was also moderately compromised during the recycling procedure, decreasing from 69 % to 52 % ee over three cycles.

Sulfonated ligands were also investigated by She *et al* [83] for the hydrogenation of dimethyl itaconate. A chiral rhodium complex containing water soluble BINAPS ligand (**22**) (Figure 24) was used for the reaction in ILs [bmim][BF_4] and [bmim][PF_6].

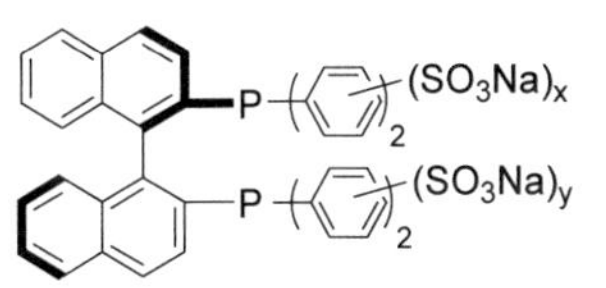

(*R*)-BINAP-*n*SO_3Na, *n* = x + y = 3-4

22

Fig. 24. Water soluble BINAP ligand

An IL/IPA biphasic system was used and conversions up to 100 % were obtained, with moderate enantioselectivities (49 - 70 %). Catalytic activity began decreasing, however, after four runs of recycling the system, but the authors found that the addition of fresh ligand to the catalyst re-established its performance.

[Bmim][PF_6] was used by Wolfson *et al.*,[37] as a reaction medium in the asymmetric hydrogenation of methyl 2-acetamidoacrylate with a rhodium catalyst [Rh-EtDuPHOS, (**13**)] (Figure 25).

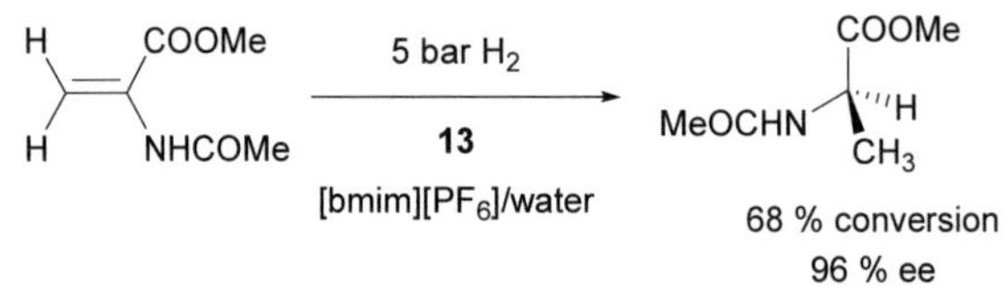

Fig. 25. Hydrogenation of methyl 2-acetamidoacrylate

The reaction did not proceed when performed in the IL alone. To enable the recycling of the catalyst immobilised in the IL, solvents immiscible in the IL were screened. Water gave the highest conversion (68 %) compared to IPA (31 %), diethyl ether (12 %) and hexane (0 %). However, the enantioselectivity remained the same (95 - 96 % ee) for the three solvents that gave conversion. The authors postulate that water is the best co-solvent due to greater mixing with the IL phase, with the water droplets dispersed more effectively than organic solvents in the IL medium.

The hydrogenation of (Z)-α-acetamidocinnamic acid and methyl-(Z)-α-acetamidocinnamate, was carried out in the ILs, [bmim][BF_4], [bmim][PF_6] and [mbpy][BF_4] using a rhodium catalyst ([Rh(COD)(DIPAMP)][BF_4]) (**23**) (Figure 26).[84]

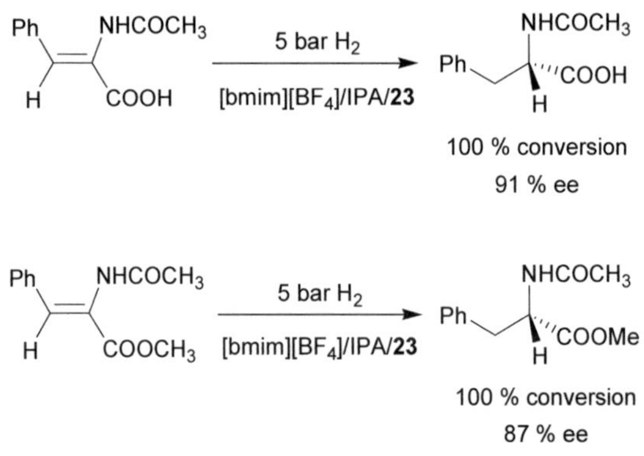

Fig. 26. Hydrogenation of (Z)-α-acetamidocinnamic acid and methyl-(Z)-α-acetamidocinnamate

In this case, IPA was used as the co-solvent in a biphasic system to facilitate recycling of the catalyst phase. A study of the effect of temperature on the enantioselectivity showed this value to peak at 55 °C. At 5 bar H_2 pressure, conversion percentage was above 97 % for the both substrates in [bmim][BF_4] and [bmim][PF_6]. Enantioselectivity was also good, with enantioselectivities between 71 - 92 %. In many asymmetric reactions the enantioselectivity decreases with increasing temperature. However, in the above case the reverse trend was observed, which led Halpern to investigate the mechanism more closely.[85] Frater *et al.*[84]

demonstrated that the catalyst system retained activity up to the fourth recycle, after which the conversion decreased slightly, with enantioselectivity remaining constant for each subsequent recycle.

Shariati *et al.*[86] used a rhodium catalyst (**24**) (Figure 27) for the asymmetric hydrogenation of methyl α-acetamido cinnamate.

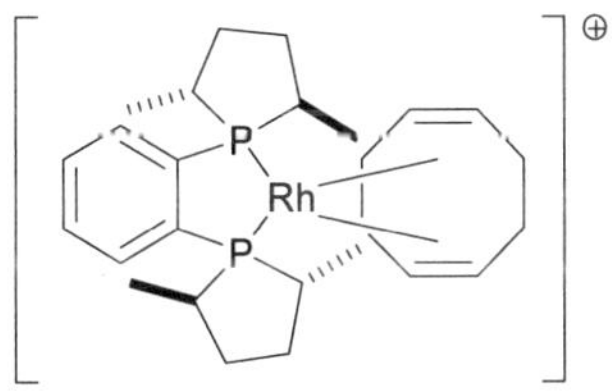

Fig. 27. Rh-MeDuPHOS (**24**)

[Bmim][BF_4] was used as the solvent and the effect of variations in the pressure of H_2 and CO_2 on the conversion and enantioselectivity of the reactions were studied. It was found that when the pressure was increased, conversion increased and selectivity decreased. Increased CO_2 pressure resulted in a decrease in conversion but an increase in selectivity. Using the IL as solvent at 20 bar H_2, 94.2 % conversion was obtained with 91.9 % ee. An increase in the pressure to 50 bar led to an increase in conversion (100 %) but a decrease in enantioselectivity (56.2 %).

4.2 Hydrogenation using ruthenium catalysts

Using methanol as co-solvent, a series of tetradecyl(trihexyl)phosphonium [P_{66614}] ILs were tested for the hydrogenation of dimethyl itaconate to dimethyl methylsuccinate (Figure 28), using the catalyst (*R*)-Ru-BINAP (**25**), at near ambient temperature (35 °C), and 20 bar H_2.[87]

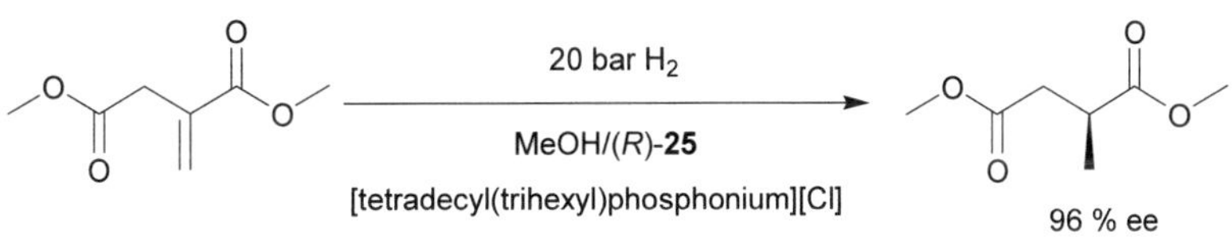

Fig. 28. Hydrogenation of dimethyl itaconate

In order to recycle the system, organic solvent nanofiltration (OSN) was used to separate the catalyst and IL from the product. Recycling of the methanol/[P_{66614}][Cl] system was achieved eight times with no loss in enantioselectivity or catalyst activity. Compared with pure methanol as solvent, the methanol/[P_{66614}][Cl] and methanol/[TBA][Cl] systems showed increased enantioselectivities, ranging from 75 % in pure methanol, to 96 % respectively in the co-solvent systems. The authors demonstrated the dependence of catalytic activity on the anion of the IL. Using [tetradecyl(trihexyl)phosphonium][Cl], good enantioselectivities and yields were obtained. However, in the case of [P_{66614}][decanoate], [P_{66614}][PF_6] and [P_{66614}][BF_4] (Figure 29), no improvement in enantioselectivity or yield was observed.

The hydrogenation of ethyl 4-chloro-3-oxobutyrate to give ethyl 4-chloro-3-hydroxy butyrate (Figure 30) was investigated by Starodubtseva *et al.*[88] using a Ru-BINAP (**25**) catalyst in various IL systems.

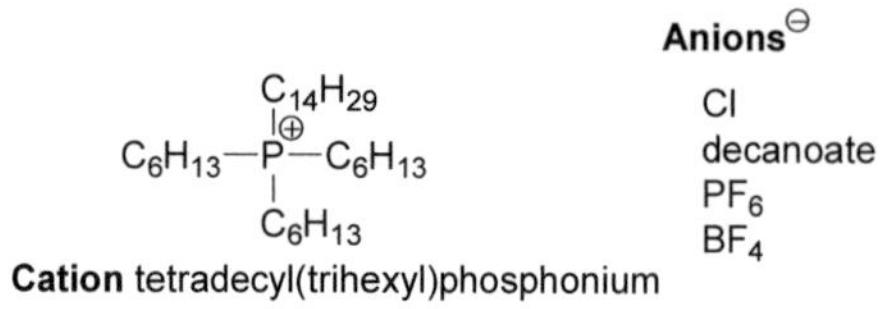

Fig. 29. Tetradecyl(trihexyl)phosphonium ILs

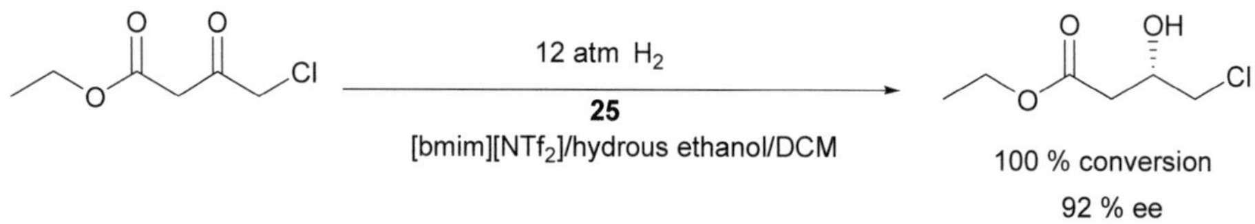

Fig. 30. Hydrogenation of ethyl 4-chloro-3-oxobutyrate

Hydrogenation results were poor using the ILs neat or in combination with an aprotic co-solvent. Starodubtseva then examined the use of [bmim][PF_6], [bmim][NTf_2] and [TEA][Br] with protic solvents, in particular ethanol. It was found that water content was also important, with superior results obtained for wet ethanol compared to anhydrous ethanol. The anhydrous ethanol formed a biphasic system with the IL and catalyst. When wet ethanol was used a homogeneous mixture with the IL/catalyst system formed. 100 % conversion, 100 % selectivity and 92 % enantioselectivity were observed in the case of the [bmim][NTf_2]/hydrous ethanol/DCM system. The conversion however decreased by more than half its original value (to 46 %) upon the third run. Selectivity remained excellent (100 %) for all three runs, and the enantioselectivity only slightly decreased to 85 % for run three. Using the [TEA][Br] IL as an example, the authors showed increasing temperature to be important for enhanced catalyst activity. By increasing the temperature from 30 °C (conversion = 42 %, selectivity = 93 %, ee = 85 %), to 70 °C (conversion = 100 %, selectivity = 100 %, ee = 96 %) improvements in catalytic performance were evident.

With their novel ruthenium catalyst [($RuCl_2$(TPPTS)$_2$]$_2$-(1*S*,2*S*)-DPENDS-KOH; TPPTS: P(*m*-$C_6H_4SO_3Na$)$_3$ and DPENDS: (1*S*,2*S*)-1,2-diphenyl-1,2-ethylene diamine sulfonate disodium) (**26**), Xiong *et al.*[27] carried out the hydrogenation of aromatic ketones (acetophenone, propiophenone, 2-fluoroacetophenone, 2-chloroacetophenone, 2-bromoacetophenone, 2-(trifluoromethyl)acetophenone, 4-(trifluoromethyl)acetophenone, 2-methoxyacetophenone, and 4-methoxyacetophenone) using a selection of ILs ([emim], [bmim], [omim] and [dodecylmim][OTs] and [bmim][BF_4] and [PF_6]) (Figure 31).

The best results were found using the tosylate ILs, with a large decrease in conversion and enantioselectivity obtained with [bmim][PF_6]. The authors attributed this to the hydrophobicity of [bmim][PF_6] hindering the activity of the hydrophilic catalyst. A decrease in enantioselectivity was observed with an increase in alkyl chain length of the cation of the IL. Various parameters were investigated as a function of catalyst activity in the IL that showed the most promising results, namely [bmim][OTs]. Increasing temperature brought about a decrease in enantioselectivity. An increase in the amount of base added (KOH) significantly increased the conversion and the enantioselectivity, as did the addition of (1*S*,2*S*)-DPENDS. The most effective catalyst precursor was shown to be [$RuCl_2$(TPPTs)$_2$]$_2$ with a conversion of 100 % and 79.2 % ee (5 MPa H_2, 50 °C). Nine aromatic ketones were tested, giving good conversions (68.0 - 100 %) and moderate enantioselectivities (40.0 – 80.6

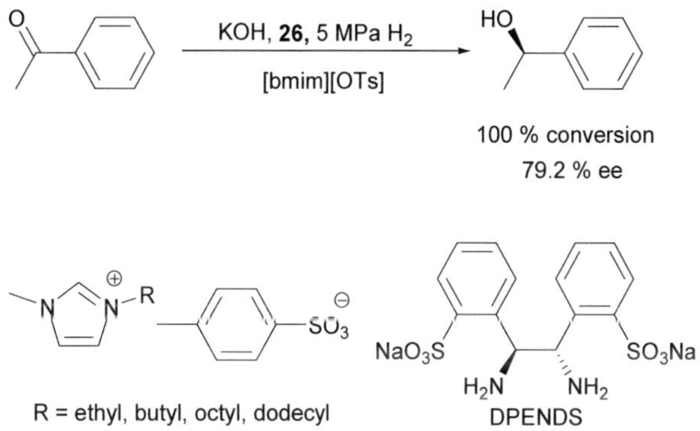

Fig. 31. Hydrogenation of acetophenone using tosylate imidazolium ILs and DPENDS

%). Recycling of the IL/catalyst system over nine runs showed conversions ranging between 100 % and 68.7 %, where even the lowest value of 68.7 % conversion was redeemed by the addition of more KOH. Wang *et al.*[89] used their novel catalyst for the hydrogenation of α,β-unsaturated ketones using ILs. Using benzalacetone as a reference substrate (Figure 32), they found that the lipophilic chains on the cations of the ILs influenced the enantioselectivity.

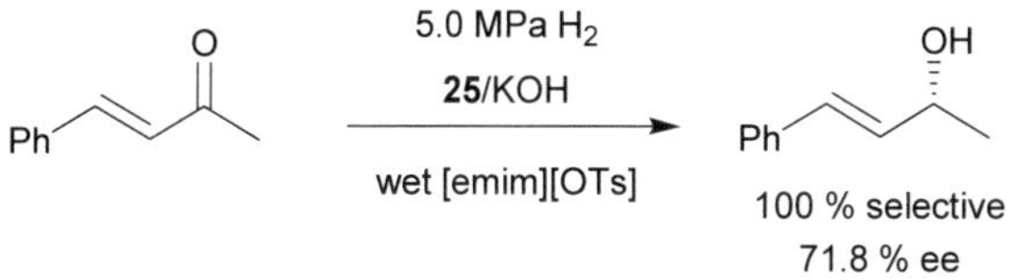

Fig. 32. Selective hydrogenation of benzalacetone

Although selectivity was high for all the unsaturated alcohols screened, enantioselectivity decreased as the ILs alkyl chain length increased (from 71.8 % ee for ethyl to 59.9 % for the dodecyl chain). This group also investigated the effect of base and water content on the reaction. 100 % conversion was reached using strong bases such as NaOH and KOH, while only 4.6 % conversion was obtained in the presence of K_2CO_3. Higher enantioselectivities (70.5 - 71.8 %) were also observed with the hydroxide bases compared to 58.7 % ee for K_2CO_3. Water was found to be a valuable co-solvent, and optimised conditions led to 100 % conversion, 100 % chemoselectivity and 75.9 % ee. Using only water as solvent 79.4 % conversion was obtained with 91.1 % selectivity towards the unsaturated alcohol and 66 % ee (5 MPa H_2, 40 °C). Consistent with Xiong *et al.*[27] favourable results were obtained using $[RuCl_2(TPPTS)_2]_2$ as precatalysts (100 % conversion, 100 % selectivity to unsaturated alcohol, 75.9 % ee in [emim][OTs]). The scope of the reaction was extended to other α,β-unsaturated ketones with good results. Hydrogenation of 2-cyclohexen-1-one gave good conversion (100 %) and chemoselectivity (94.1 % for the unsaturated alcohol), albeit with a moderate ee (48.1 %). 4-Methyl-3-penten-2-one showed good enantioselectivity (84.7 %) and chemoselectivity (84.9 %), however with only poor conversion (29.1 %). The IL/catalyst system was efficiently recycled eight times with conversion dropping from 100 % to 87.9 %. A slight decrease in the chemoselectivity (from 100 % to 99.1 %) was observed and the enantioselectivity remained almost constant for each successive recycle.

Lam *et al.*[90] used a ruthenium catalyst with a dipyridylphosphine ligand (P-Phos) (**27**) for the asymmetric hydrogenation of α- and β-keto esters (Figure 33).

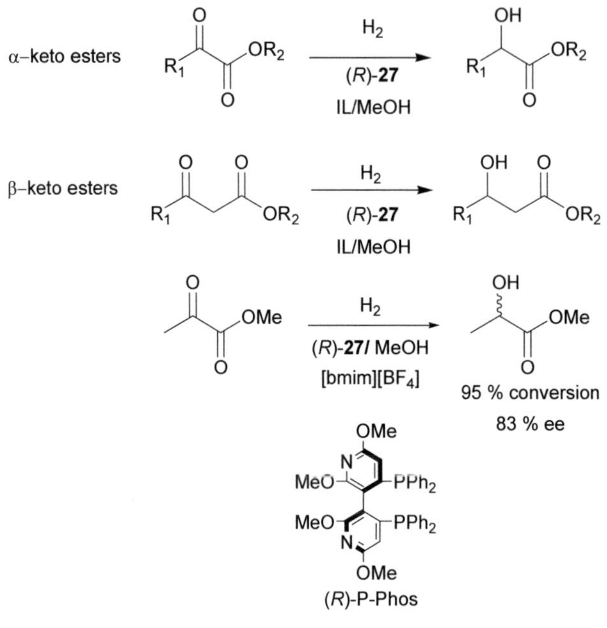

Fig. 33. Asymmetric hydrogenation of α- and β-keto esters using a Ru catalyst

Methyl pyruvate was taken as an example from the α-keto esters and the hydrogenation was carried out using Ru((*R*)-P-Phos)Cl_2 ((*R*)-**27**) and a reference ruthenium BINAP catalyst (**25**). It was found that a co-solvent was crucial for these hydrogenations, as using only the IL as solvent gave negligible conversion. Methanol was used in equal volumes as co-solvent and using (**27**), good conversions were obtained (73 and 95 % conversion for [bmim][PF_6] and [bmim][BF_4] respectively) with good enantioselectivity (86 and 83 % ee). Enantioselectivies decreased when moving to the BINAP ligand, although conversions achieved were higher. The IL used was observed to have an effect on conversion using [bmim][BF_4] and [bmim][PF_6] and methyl 2-oxo-2-phenylacetate as substrate, conversions of 18 and 65 % respectively were obtained with enantioselectivity being relatively unaffected (90 and 93 % respectively). The best conversion (65 %) and enantioselectivity (93 %) were obtained with methyl 2-oxo-2-phenylacetate in [bmim][PF_6]. Methyl acetoacetate was used as a reference substrate for the hydrogenation of β-keto esters. Conversions and enantioselectivities were greater than 98 % with either Ru((*R*)-P-Phos)Cl_2 ((*R*)-**27**) or Ru-BINAP (**25**) as the catalyst. In general, the range of β-keto esters subjected to hydrogenation using (**27**) displayed improved results, with most conversions reaching at least 70 %, and enantioselectivities exceeding 99 %. The hydrogenation of methyl acetoacetate was investigated in [bmim][PF_6] and [bmim][BF_4] for recycling ability. Both IL systems were recycled nine times with similar results although the conversion had dramatically decreased by run nine for both ILs (39 % for [bmim][BF_4] and 49 % for [bmim][PF_6]), enantioselectivities did not fall below 94 % for both ILs over the nine runs.

The hydrogenation of methyl acetoacetate to methyl 3-hydroxybutyrate was investigated by Floris *et al.*[91] using a ruthenium catalyst in a mixed IL/methanol phase. Good stereoselectivity has been obtained with substrates consisting of keto groups β- or γ- to the carboxy function group using catalysts prepared from $[RuCl_2(p\text{-cymene})]_2$,[92] although the authors goal in this work was to demonstrate the effective re-use of the catalyst ((*R*)-[RuCl(BINAP)(*p*-cymene)]Cl) (**28**) immobilised in an IL. The ILs used were a series of *n*-alkyl-triethylammonium ditriflimide $[N_{222n}][NTf_2]$ ILs with varying alkyl chain lengths (n = 6,7,8,10,12, and 14), to investigate their effect on the reaction. The amount of IL required for satisfactory results to be obtained was investigated with the hexyl substituted ionic liquid $[N_{2226}][NTf_2]$. Using only methanol as the reaction solvent, 98 % ee and 79.4 % selectivity and TOF 1091 h^{-1} were obtained. When using a relatively low loading of IL (0.2 wt%) the TOF 850 h^{-1}, was slightly reduced, along with the enantioselectivity, 96.9 %. However the selectivity remained constant at 79.4 %. Increasing the amount of $[N_{2226}][NTf_2]$ added led to a reduction in TOF values and enantioselectivity, although the selectivity increased. A study into the effect of varying alkyl chain length from $[N_{2226}][NTf_2]$ to $[N_{22214}][NTf_2]$ on the reaction showed no apparent trends. A 1:1 ratio of methanol to IL was used for the reactions above. Although the best result in terms of TOF and enantioselectivity was obtained using only methanol as solvent (1091 h^{-1} TOF, 98.0 % ee at 59.85 °C and 5 MPa H_2), this system was deemed unrecyclable by the authors. The highest result for TOF and enantioselectivity for the mixed systems was obtained with the dodecyl example $[N_{22212}][NTf_2]$ (TOF = 509 h^{-1}, 97.5 % ee at 59.85 °C and 5 MPa H_2) and the lowest values with the tetradecyl $[N_{22214}][NTf_2]$ (TOF = 328 h^{-1}, 96.4 % ee at 59.85 °C and 5 MPa H_2). Using the $[N_{22212}][NTf_2]$ in combination with methanol as reaction solvent, the system was successfully recycled with only a slight drop in enantioselectivity upon the second run, from 97.5 % to 93.9 %.

Methyl acetoacetate was again used as a substrate to investigate hydrogenation using mixtures of ILs and methanol with a ruthenium catalyst (**29**).[93] The catalyst was prepared *in situ* in the presence of HBr from one equivalent of [*bis*(2-methylallyl)(1,5-cyclooctadiene) ruthenium(II)] and two equivalents of phenyl-4,5-dihydro-3H-dinaphtho[2,1-c;1′2;-e]phosphepine (phenyl-phosphepine). Various ILs based on ditriflimide (eg. $[bmim][NTf_2]$ and $[N_{1111}][NTf_2]$) were investigated because the mildly coordinating ditriflimide anion had a beneficial effect on the reaction. Although poorer results were obtained using mixed IL systems in comparison with methanol as a sole reaction solvent, the recyclability of the IL systems was demonstrated. Using methanol as solvent for the reduction of this substrate to methyl hydroxybutyrate gave 100 % yield, 100 % conversion and 92.4 % ee. These values however drastically decreased to 35.8 % yield, 85.6 % conversion and 85.3 % ee upon the first recycle of the system. Impressive results for up to 3 recycles were obtained using the IL system of [*bis*(hydroxyethyl)dimethylammonium]$[NTf_2]$. The yield decreased from 99.5 % upon the first run to 93.2 % upon run 3. The conversion decreased from 100 % to 94.2 % and the ee merely decreased from 93.0 to 91.3 % upon the third run.

Zhou *et al.*[94] demonstrated a phosphine-free catalytic system for the hydrogenation of quinolines in the IL $[bmim][PF_6]$ (Figure 34). Using 50 atm. of H_2 at 25 °C, 100 % conversion and 99 % enantioselectivity was obtained for the preparation of (*S*)-1,2,3,4-tetrahydro-2-methylquinoline. The use of the IL facilitated the recycling of the catalyst/IL system even with prolonged exposure to air. Upon recycling of the system by the 8th recycle, the conversion had decreased from 100 to 82 % with only a slight decrease in ee to 97 %. Zhou *et al.* have proposed a mechanism for this reaction. (Figure 35)

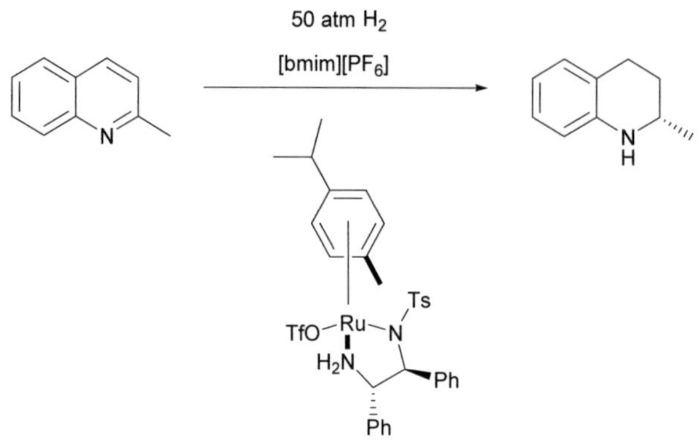

Fig. 34. Hydrogenation of 2-methylquinoline with Ru/Ts-DPEN catalysts [Ts-DPEN: *N*-(*p*-toluenesulfonyl)-1,2-diphenylethylenediamine]

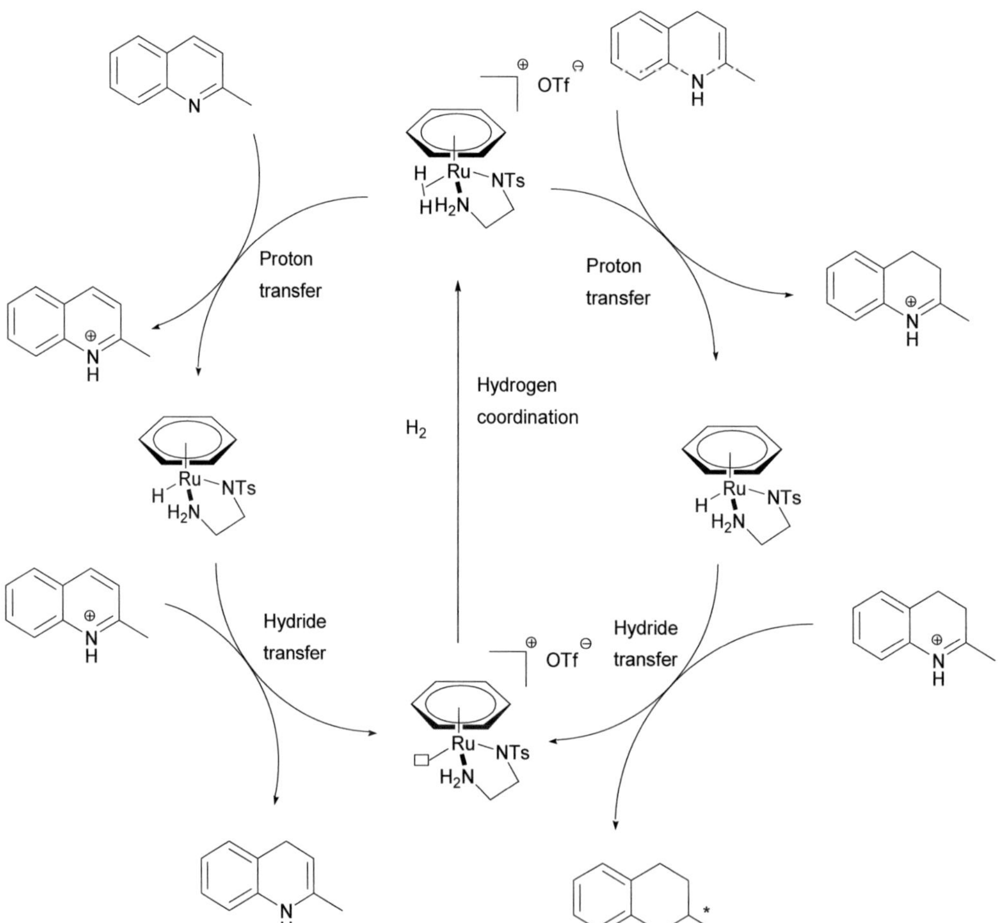

Fig. 35. Proposed mechanism for the reaction in Figure 34

When Dupont *et al.*[16] carried out asymmetric hydrogenation of (Z)-α-acetamido cinnamic acid using Ru-(tolyl-BINAP) in [bmim][BF_4] and IPA mixture at RT, they observed that the reduction was accompanied by kinetic resolution of the substrate

5. Metal nanoparticle catalysed hydrogenation in ILs

Nanoparticles have properties intermediate between bulk and single particles. The advantages of using nanoparticles as catalysts are that they present high catalytic activity. However, these nanoparticles need to be stabilized against aggregation into larger particles and eventually bulk aggregates, in order to retain their catalytic activity. Aggregation occurs due to their extensively large surface area and the main means of their stabilization in solution utilizes electrostatic or steric protection. Surfactants and quaternary ammonium salts have been used for stabilization of these particles, and now this field of nanoparticles used in ILs is emerging. Hydrogenation reactions using nanoparticles in ILs may lead to increased reaction times in comparison to solventless conditions or the use of common organic solvents. IL solvents can, however, have a positive effect on product selectivity and recycling of the nanoparticles can also be facilitated. The formation and stabilization of the nanoparticles used for hydrogenation reactions in ILs is only beginning to be understood,[40] despite the fact that nanoparticles based on various transition metals have now been utilised as catalytic media for hydrogenation reactions in ILs.

5.1 Palladium nanoparticles

Umpierre *et al.*[25] prepared palladium nanoparticles in [bmim][BF_4] and [bmim][[PF_6] for use in the selective hydrogenation of 1,3-butadiene to 1-butene (Figure 36). The selective hydrogenation of 1,3-butadiene is a widely used industrial process and further improvement is required to increase the selectivity towards 1-butene. Palladium on supported catalysts in conjunction with promoters or modifiers[95,96] have also been used, however the selectivity towards 1-butene is impeded by isomerisation of 1-butene and complete hydrogenation to 1-butane.

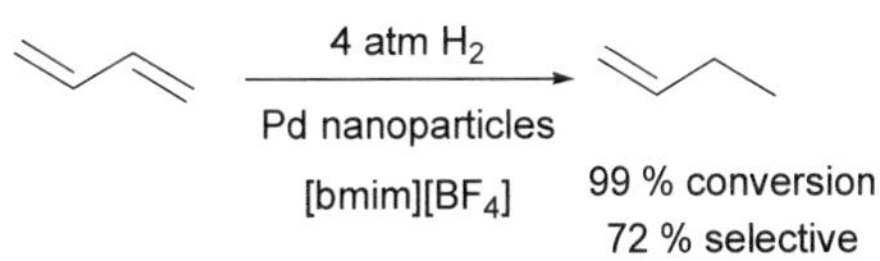

Fig. 36. Selective hydrogenation of 1,3-butadiene

The prepared nanoparticles synthesized by Umpierre *et al.*[25] could be re-dispersed in the ILs or used under solventless conditions. It was found that the reactions carried out in ILs required extended reaction times (6 h) in comparison to reactions carried out under solventless conditions (< 2h). The significant difference is the selectivity achieved in the IL case. Less than 2 % butane was obtained using the IL system, with the 1-butene product formed in high yield. However, even at low conversions and under solventless conditions, a significant amount of butane was formed. The authors suggest the higher selectivity in the IL is due to the lower solubility of butane compared to 1,3-butadiene in the IL. Butenes also did not undergo isomerisation in the ILs, whereas under solventless conditions, isomerisation to 2-butenes occurred after complete conversion of the diene substrate.

Huang *et al.*[97] used phenanthroline ligand-stabilised palladium nanoparticles for the reduction of olefins (cyclohexene, 1-hexene and 1,3-cyclohexadiene) in the IL [bmim][PF_6]. Within the temperature range 20-60 °C and a hydrogen pressure of one bar, conversions of up to 100 % were obtained. For the hydrogenation of 1,3-cyclohexadiene to cyclohexene, a selectivity of 95 % was reached. Aggregation of the palladium nanoparticles was not observed, shown by the recycling of the catalyst media nine times with no loss in catalyst activity. The effect of the ligand on the hydrogenation was investigated, including a study investigating the absence of the ligand on the recyclability of the system. Using the phenanthroline ligand, a conversion of 100 % was reached for the first cycle; the ligand-free system displayed a similar value of 95 %. However, a marked loss in activity was observed upon the first recycle using the ligand-free system (only 35 % conversion). A system without the IL was investigated, using acetic acid as reaction solvent in its place. Aggregation of the nanoparticles was observed during the course of the reaction, and a low conversion of only 5 % was reached.

Hu *et al.*[98] incorporated a bidentate nitrogen ligand into an IL, forming a novel IL capable of stabilizing Pd nanoparticles. The functionalized IL (Figure 37) stabilized nanoparticles were prepared *in situ* in [bmmim][PF_6].

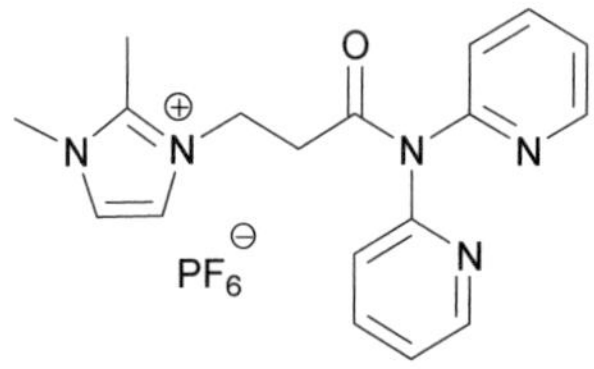

Fig. 37. 2,3-Dimethyl-1-[3-*N,N*-*bis*(2-pyridyl)-propylamido]imidazolium hexafluorophosphate

The stabilized nanoparticles were then used as catalysts for the hydrogenation of cyclohexene, styrene and ethyl acrylate. 100 % yield of reduced product was obtained for each olefin hydrogenation.

This group continued their study of this functionalised IL and investigated the recyclability of the nanocatalysts.[99] Using the IL as ligand for the palladium catalyst in neat [bmmim][PF_6] the selective hydrogenation of functionalised alkenes was investigated. Using this system, the selectivity towards reduction of the C=C bond of α,β-unsaturated aldehydes, esters and ketones was demonstrated. In the case of cinnamaldehyde hydrogenation using a Pd/C catalyst (1.5 MPa H_2 and 35 °C) 55 % conversion was obtained with 35.2 % yield of hydrocinnamaldehyde and 19.4 % yield of 3-phenylpropanol. Using palladium with the functionalised IL as ligand, 75 % conversion was obtained, with 64.5 % yield of hydrocinnamaldehyde and only 10.1 % yield of the fully reduced product, 3-phenylpropanol under the same conditions. The IL-stabilized nanocatalysts were demonstrated to be effective over 7 recycles of the hydrogenation of 2-cyclohexen-1-one. The yield of cyclohexanone remained > 95 % over the recycling procedures.

Chun *et al.*[100] formed palladium nanoparticles deposited on imidazolium-functionalised multi-walled carbon nanotubes (Figure 38) by the reduction of Na_2PdCl_4 with H_2. Direct anion exchange was possible with the bromide functionalised IL on the multi-walled carbon nanotubes, without changing size and distribution of the palladium nanoparticles, to yield hydrophobic [NTf_2^-] and [SbF_6^-] derivatives.

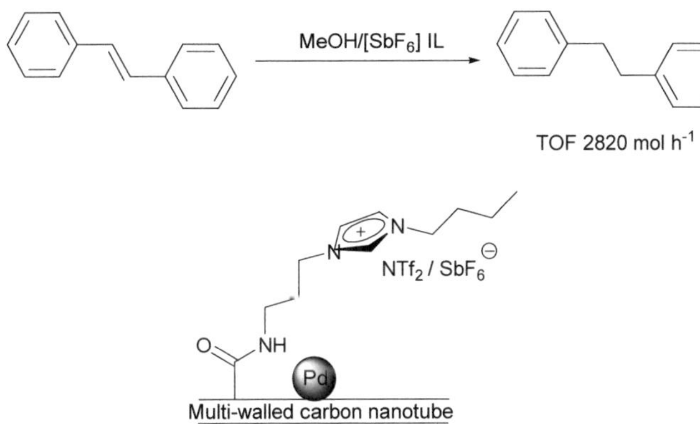

Fig. 38. Imidazolium-functionalised multi-walled carbon nanotubes in the hydrogenation of stilbene

Trans-stilbene was used as a test substrate for the use of the nanoparticles in hydrogenation reactions. Using methanol as a solvent, TOFs up to 2820 mol h^{-1} were attained using the [SbF_6^-] catalyst. When a [bmim][NTf_2]/methanol mixture was used as the solvent, conversion rose from 65 to 100 % for the [NTf_2^-] catalyst and the TOF also rose from 390 to 600 mol/h. The [SbF_6^-] IL was recycled ten times with no loss of catalytic activity. A range of substrates were hydrogenated using this recycled system with a TON of 5000 reached. IL leaching was identified as a minor problem, and additional IL was added upon each cycle of twenty runs. Aggregation of the nanoparticles was postulated to be responsible for the decreased conversion (65 %) for run 50.
Huang *et al.*[101] used palladium nanoparticles immobilized on molecular sieves by the IL, 1,1,3,3-tetramethylguanidinium lactate (Figure 39), for the hydrogenation of olefins (cyclohexene, cyclohexadiene, and 1-hexene).

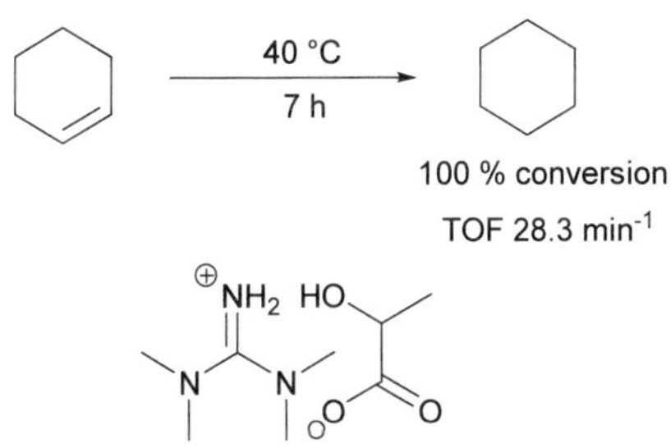

Fig. 39. 1,1,3,3-tetramethylguanidinium lactate and cyclohexene hydrogenation

The synergistic effects of the nanoparticles, molecular sieves and the IL yielded impressive results. Using cyclohexene as a substrate, the system was reused four times with no loss in activity (100 % conversion and 20.0 min^{-1} TOF maintained under a reaction temperature of 20 °C). Cyclohexadiene was selectively reduced to cyclohexene at 20 °C in 3 hours reaching 98 % conversion and 65.3 min^{-1} TOF,
Tao *et al.*[102] also used three guanidinium ILs ([1,1,3,3-tetramethylguanidine][TFA]/[lactic acid]/[acetic acid]) for the preparation of palladium on sepiolite nanoparticles and investigated their use in the hydrogenation of a range of substrates (cyclohexene, styrene, 1-hexene and 1,3-cyclohexadiene). The TFA based IL nanocatalyst was used for most of the

investigative hydrogenations. 1-Hexene and styrene were both hydrogenated to hexane (> 99.0 % conversion, 10,000 h^{-1} TOF) and ethyl benzene (> 99.0 % conversion, 10,000 h^{-1} TOF) under the same conditions (2 MPa H_2, 60 °C, 0.5 h). The selectivity of the catalyst was investigated using the substrate 1,3-cyclohexadiene, where the predominant product obtained was cyclohexene at > 99 % conversion after 3.5 hours. The catalyst system was also applied to the cyclohexene to cyclohexane hydrogenation to probe recyclability. Impressive stability of the catalyst was observed when it was successfully used over five recycling experiments with no loss of catalytic activity (> 99 % conversion, 5,000 h^{-1} TOF). This group showed the superiority of their nanocatalyst system when, using the commercially available 5 % Pd/C for the hydrogenation of cyclohexene, a conversion of only 80.2 % was observed in comparison with 95.0 % conversion obtained under the same conditions using the nanocatalyst. The authors suggested that the cation of the IL was primarily responsible for the successful immobilisation of the palladium on sepiolite as results obtained using lactic and acetic acid derived ILs are almost identical to the TFA-based IL.

Palladium nanoparticles in [bmim][PF_6] were used in combination with gold nanoparticles by Dash *et al.*[103] for the hydrogenation of a range of substrates including allyl alcohol, 1,5-cyclooctadiene, *trans*-cinnamaldehyde and 3-hexyn-1-ol. Even in the presence of poly(vinylpyrrolidene), which was used as a stabilizer, the direct synthesis of the IL led to problems in terms of nanoparticle polydispersity. The preferred alternative mode of synthesis for these nanoparticles was initial preparation in methanol, then switching the solvent to an IL. Gold was found not to induce catalytic activity when used as sole metallic source. TOF results showed a high ratio of palladium to gold with 3:1 furnishing the highest TOF values for all substrates. The lowest catalytic activity was observed for *trans*-cinnamaldehyde and 3-phenylpropanal and 3-phenyl-1-propanol were formed in a 1:1 ratio. The selectivity for the alcohol increased slightly when using the bimetallic catalyst in the 1:1 ratio of Au:Pd. The hydrogenation of 3-hexyn-1-ol furnished 3-hexen-1-ol in significantly higher selectivity compared to 1-hexanol with all catalyst ratios and also using palladium as sole metallic source. Cyclooctene was formed in superior selectivity to cyclooctane for the hydrogenation reactions of 1,3-cyclooctadiene. 100 % selectivity for cycloctene was reached using the bimetallic catalyst in a ratio of 1:3 Au:Pd (Figure 40).

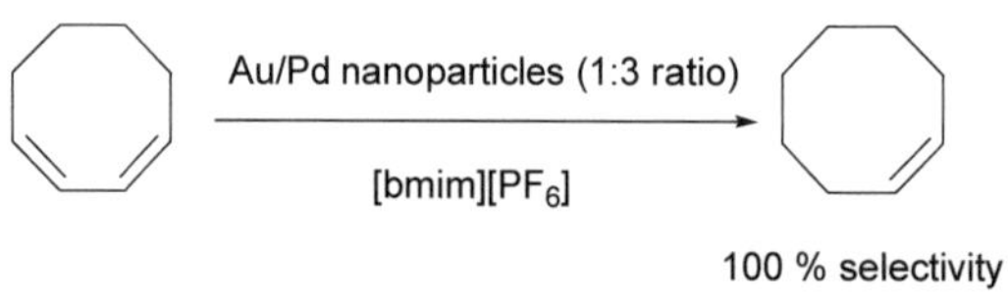

Fig. 40. Hydrogenation of 1,3-cyclooctadiene

The authors report water displays similar TOFs to [bmim][PF_6] (water TOF: 284 h^{-1}, [bmim][PF_6] TOF: 266 h^{-1}), although the IL gave superior recyclability of the system. The catalyst activity was seen to only decrease by less than 4 % for the hydrogenation of allyl alcohol for the first recycle. Several studies have shown that H_2 has low solubility in [bmim] based ILs [10,104], thus hydrogenation reactions in ILs are often mass transfer limited. Under these mass-transfer limited conditions, with respect to H_2 gas, catalyst concentrations were selected in order to optimize both catalyst stability and overall conversion of substrates.

Kume *et al.*[39] prepared palladium nanoparticles immobilized on IL modified-silica gel (Figure 41) for hydrogenation reactions at 80 °C and 5 MPa H_2.

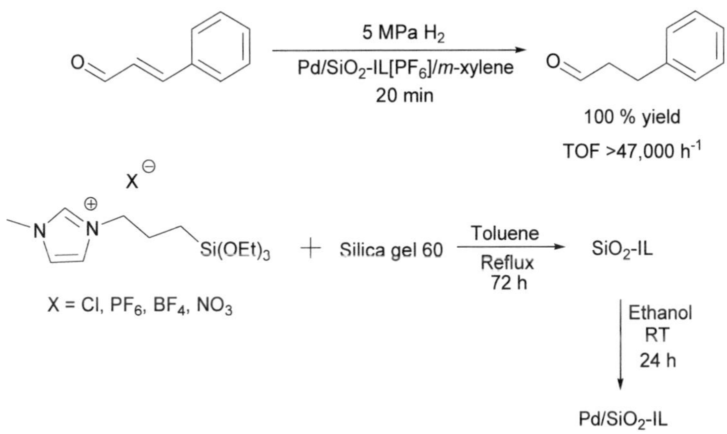

Fig. 41. Preparation of palladium nanoparticles immobilized on IL modified SiO_2 and hydrogenation of cinnamaldehyde

The prepared nanoparticles were used for the hydrogenation of cinnamaldehyde using *m*-xylene and ILs ([bmim][Cl], [bmim][NO_3], [bmim][BF_4], [bmim][PF_6]). Reactions carried out in the neat ILs gave poor catalytic activity, with TOFs of less than 33 h^{-1} after six hours, for all ILs tested. However, TOFs of 47,000 h^{-1} were achieved using the biphasic method in just 20 mins. Using xylene as the only solvent, after 20 minutes a TOF of 22,200 h^{-1} was attained. Hydrocinnamaldehyde was the only product observed for the reaction, and a correlation between the IL anion and TOF was evident. Although the TOF values obtained using neat IL were much less than those obtained using the biphasic system with the IL modified-silica gel, the same trend was evident: TOF values decreased in the order of the following anions used [Cl] < [BF_4] < [NO_3] < [PF_6]. The remarkable improvement in reaction rate when using IL modified-silica gel can be attributed to the high surface area of silica gel, which leads to enhanced contact of substrate and H_2 over the Pd nanoparticles. The silica gel system with the [bmim][PF_6] was successfully recycled nine times with 100 % yield, and a TON of nearly 500,000 achieved.

5.2 Iridium nanoparticles

Fonseca *et al.*[29] formed iridium nanoparticles in [bmim][PF_6] and investigated their application in the catalytic hydrogenation of 1-methylcyclohexene, cyclohexene, 1-decene, and 2,3-dimethyl-2-butene (Figure 42).

They found that increased steric hindrance around the double bond decreased the reaction rate, and increased pressure increased the rate of the reaction up to a certain saturation point. From their studies on the substrate/catalyst concentration ratio they concluded a limiting factor to be also the miscibility of the substrate in the IL. The same group[34] also prepared iridium nanoparticles in [bmim][PF_6] for hydrogenation of ketones. The prepared solid reduced iridium catalyst was removed from [bmim][PF_6] to be used in solventless reactions. One reaction was performed where they re-dissolved the prepared catalyst in [bmim][PF_6]. Comparing the reaction performed under solventless conditions and in [bmim][PF_6], it was seen that the reaction time increased greatly when using the IL (from as low as two hours for solventless conditions, to 17.5 hours for [bmim][PF_6]). The catalyst immobilised in [bmim][PF_6] could, however, be reused fifteen times with almost no loss in

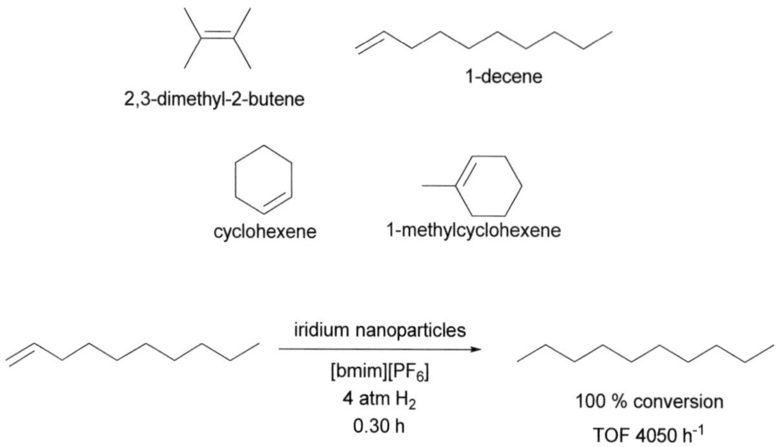

Fig. 42. Substrates used for hydrogenation by Fonseca *et al.*[29] and hydrogenation of 1-decane

activity compared to the catalyst used alone, which showed decreased activity after the third recycle. Increased reaction times were also required when the same group[35] prepared iridium nanoparticles in [bmim][PF_6]. The nanoparticles were subsequently used for the hydrogenation of various arenes by re-dispersion in the IL (reduction of benzene to cyclohexane at 75 °C and 4 atm H_2 after 14 hours, 85 h^{-1} TOF), in an organic solvent (after 2.5 hours in acetone, 200 h^{-1} TOF) or under solventless conditions (after 2 hours, 125 h^{-1} TOF). This was again demonstrated in [bmim][PF_6] hydrogenation reactions. Prolonged reaction times were required, compared with the use of an organic solvent or solvent-free conditions. Also was shown that the iridium nanoparticle catalysts could be reused after the solvent-free hydrogenation reactions (up to seven times) with little loss in catalyst activity. Iridium nanoparticles employed in the [bmim][PF_6] reaction showed a significant loss in activity due to the decomposition of the IL and the nanoparticles.
Dupont *et al.*[105] also prepared iridium nanoparticles in [bmim][PF_6]. These nanoparticles were subsequently used in the hydrogenation of olefins (1-decene, styrene, cyclohexene, methyl methacrylate, 4-vinylcyclohexene) in [bmim][PF_6] (Figure 43).

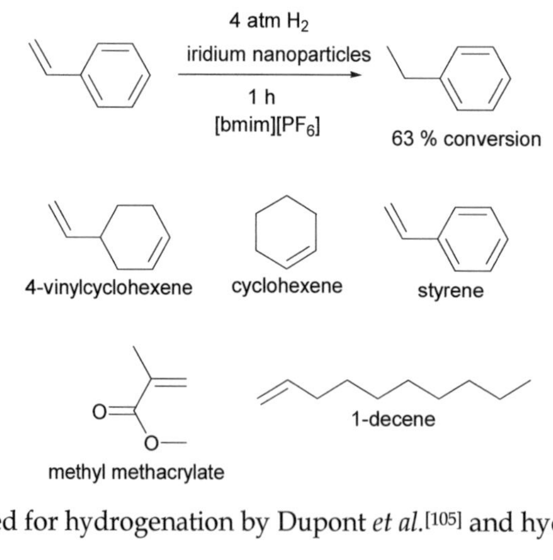

Fig. 43. Substrates used for hydrogenation by Dupont *et al.*[105] and hydrogenation of styrene

Good conversions were obtained (56 - 100 %) within 4 hours. The catalyst system was also shown to maintain efficiency over seven recycles.

5.3 Rhodium nanoparticles

Leger *et al.*[106] employed 2,2′-bipyridine ligands to stabilise zerovalent rhodium nanoparticles for the hydrogenation of aromatic compounds in ILs using biphasic conditions. They investigated the effects of varying anions and cations of the IL on catalytic activity and found the composition of the IL to have a significant influence on selectivity. Using the [bmim] cation, [BF_4] was shown to be the most effective anion favouring hydrogenation of the aromatic double bonds of styrene together with the exocyclic olefinic bond (ethylbenzene/ethylcyclohexane, 8/92 selectivity). With the dicyanamide anion, 100 % selectivity was observed for the product ethylbenzene. Concerning the cationic species of ditriflimide ILs, imidazolium and pyrrolidinium cations displayed identical results (ethylbenzene/ethylcyclohexane, 70/30 selectivity). The pyridinium cation-containing IL displayed superior catalytic activity, with ethylbenzene being formed with 85 % selectivity using [pyridinium][NTf_2]. Leger demonstrated that their catalyst system was viable for a range of aromatic compounds, and they observed that increased steric hindrance due to bulky substituents on the aromatic rings of substrates led to decreased catalytic activity. The catalytic medium was also recycled once with no decrease in catalyst activity.

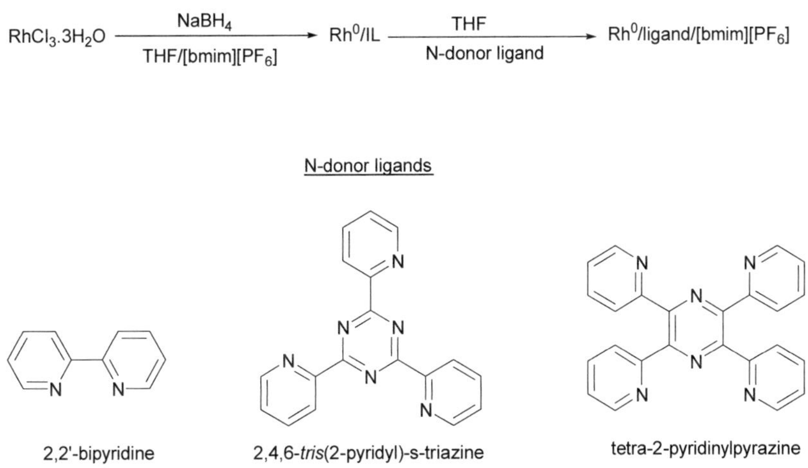

Fig. 44. Synthesis of rhodium ligand-stabilized Rh nanoparticles

Leger *et al.*[107] continued this work by using polynitrogen ligands (Figure 44) in [bmim][PF_6] for the stabilization of zero-valent Rh^0 colloids. When using the tetra-2-pyridinylpyrazine ligand, styrene was selectively hydrogenated to ethylcyclohexane (98 % yield) and ethylbenzene (2 % yield) under 40 bar H_2 at 80 °C. Benzene was reduced to cyclohexane and toluene was reduced to methylcyclohexane with 100 % conversion using either ligands, 2,2′-bipyridine or 2,4,6-*tris*(2-pyridyl)-*s*-triazine. Greater conversion was achieved using 2,4,6-*tris*(2-pyridyl)-*s*-triazine (100 %) for the reduction of ethylbenzene to ethylcyclohexane, than using 2,2′-bipyridine (60 %). Rhodium nanoparticles were stabilised by an IL-like co-polymer (Figure 45) in [bmim][BF_4] and demonstrated high activity for the hydrogenation of various arenes.[31] A record total turnover was obtained (20,000) for the hydrogenation of benzene using this system. Superior results were obtained when using the stabiliser in the IL with the nanoparticles. The solubilities of arenes in reaction media and steric, electronic properties of the substituents on aromatic ring influence the reaction rates.

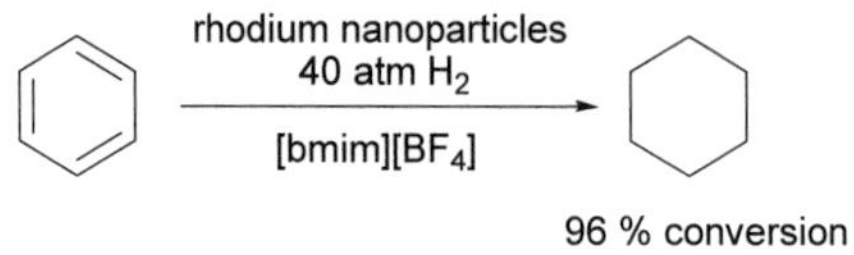

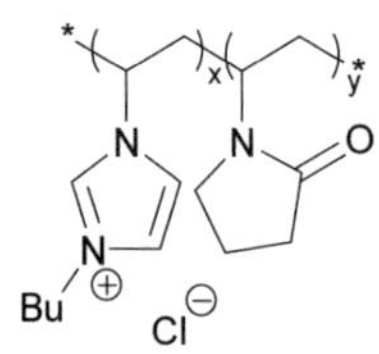

Fig. 45. Poly[(*N*-vinyl-2-pyrrolidone)-*co*-(1-vinyl-3-butylimidazolium chloride)] in the hydrogenation of benzene

Cimpeanu *et al.*[108] used compressed CO_2 to decrease the melting point of ammonium salts for hydrogenation reactions using a Rh nanocatalyst. [Rh(acac)(CO_2)] (**30**) was used as a catalyst precursor and various ammonium bromide salts and organic impurities were subsequently removed by extraction with supercritical CO_2. Together with the desirable decrease in melting points of the ammonium salts to facilitate the dissolution of the rhodium precursor, the supercritical CO_2 is known to enhance the hydrogen availability in the IL.[109] Using cyclohexene and benzene to test their system, results were obtained that were in line with data reported previously using Rh nanoparticles in standard ILs. Interesting selectivity was however obtained for the hydrogenation of (*E*)-2-(benzoylamino)-2-propenoic acid derivatives using the novel system in comparison to commercially available rhodium catalysts. Whereas Wilkinson's catalyst (**12**) shows poor conversion (< 5 % after 3 days) and Rh/Al_2O_3 shows no selectivity between the olefinic and aromatic double bonds, the rhodium nanoparticles stabilised by the ammonium salts demonstrated selectivity between the differing double bonds. Using the rhodium catalyst stabilised in the tetrabutylammonium bromide salt ([Bu_4N][Br]), impressive selectivity was obtained where the phenyl aromatic ring was almost exclusively hydrogenated (Figure 46).

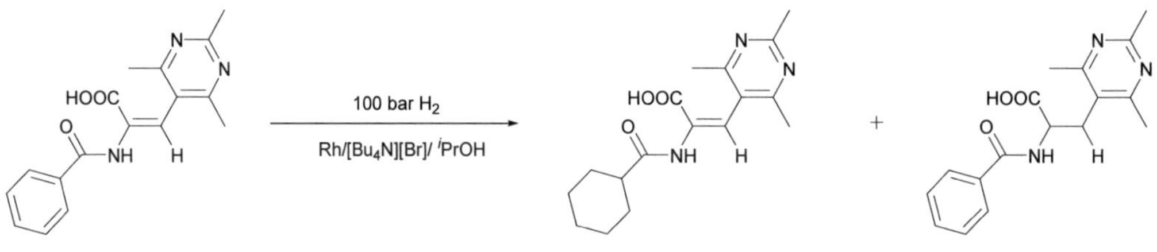

Fig. 46. Selective reduction of aromatic ring

5.4 Platinum nanoparticles

The hydrogenation of *o*-chloronitrobenzene was used to test the performance of Pt(I) and Pt(II) nanoclusters stabilised by an IL-like copolymer in [bmim][BF_4][110] (Figure 47).

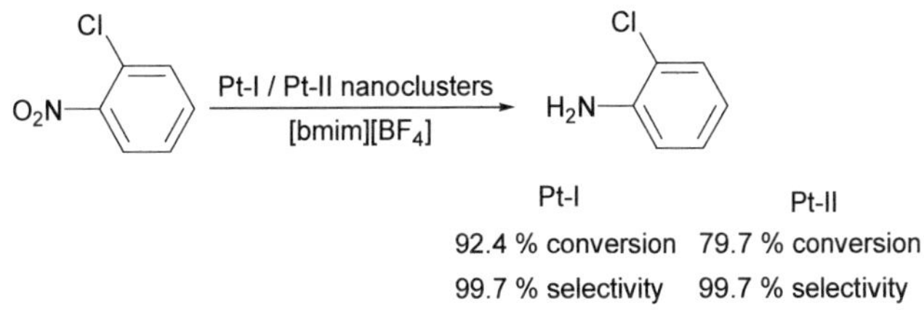

Fig. 47. Selective hydrogenation of *o*-chloronitrobenzene to *o*-chloroaniline

The IL-like copolymer (poly[(*N*-vinyl-2-pyrrolidone)-*co*-(1-butyl-3-vinylimidazolium chloride)]) displayed good solubility in [bmim][BF_4] and was synthesised by Kou and coworkers[110] to surmount the problem of poor solubility of PVP (poly(*N*-vinyl-2-pyrrolidone) polymer. The conversion and selectivity towards *o*-chloroaniline using these catalysts in the IL were superior to results obtained using the conventional PVP-Pt catalyst. The selectivity obtained for *o*-chloroaniline was high using both Pt(I) and Pt(II) nanoclusters (99.7 and 99.1 % respectively), although a drop in conversion was observed moving from Pt(I) (92.4 %) to Pt(II) (79.7 %). From IR studies carried out, the increased selectivity towards *o*-chloroaniline was attributed to interactions between the IL and the substrate.

Scheeren *et al.*[111] prepared platinum nanoparticles for use in the hydrogenation of alkenes and arenes in [bmim][PF_6]. Following the preparation of the nanoparticles, they were re-dispersed in either IL, acetone or used under solventless conditions for the hydrogenation reactions. The prepared nanoparticles proved to be more active than the more commonly used Adam's catalyst. From investigation of the time taken for the hydrogenation of substrates, it was found that the [bmim][PF_6] system displayed the longest reaction time in comparison to the organic solvent system, or solventless conditions. The time taken to reach 100 % conversion, for example, of 1-hexene to hexane was 0.25 hours for acetone and solventless conditions, but 0.4 hours for the IL system. A more dramatic difference in results was shown in the hydrogenation of 2,3-dimethyl-1-butene to 2,3-dimethyl-1-butane, where it took 0.6 hours to achieve 100 % conversion under solventless conditions but 3 hours to obtain only 82 % conversion using [bmim][PF_6]. Only a slight reduction in catalyst activity was noticed upon reuse from solventless conditions. The difference between reaction times was attributed to the typical biphasic conditions of the reactions performed in the ionic liquid, which can be a mass-transfer controlled process.[105]

Abu-Reziq *et al.*[112] used platinum nanoparticles supported on magnetite (Fe_3O_4) modified with functionalised ionic liquids (Figure 48). The ILs were readily synthesised by substitution reaction of the desired alkylimidazole with the alkyl chloride to furnish the alkylimidazolium chloride ILs.

The length of the alkyl chain facilitated regulation of the solubility of the resulting nanoparticles. These supported nanoparticles chemoselectively reduced alkynes to *cis*-alkenes and α,β-unsaturated aldehydes to allyl alcohols, with the added advantage of clean catalyst recovery from the reaction mixture by the application of an external magnetic field. Using their magnetite nanoparticles, impressive chemoselectivity was obtained with a range of substrates; cinnamaldehyde, for example, was selectively hydrogenated to 3-phenylprop-2-en-ol in 99 % yield and diphenylacetylene was hydrogenated to stilbene, the isomeric ratios being *cis*:*trans* 95:5. The latter system was recycled three times, the conversion decreasing a mere 3 %, from 100 % for run one, to 97 % upon the third recycle. The ratio of the *cis* and *trans* product remained almost constant upon recycling. Authors state that the magnetite nano-support can polarize the surface of the platinum nanoparticles and make them partially positive which can lead to selective adsorption and activation of the polar functional groups of the substrate.

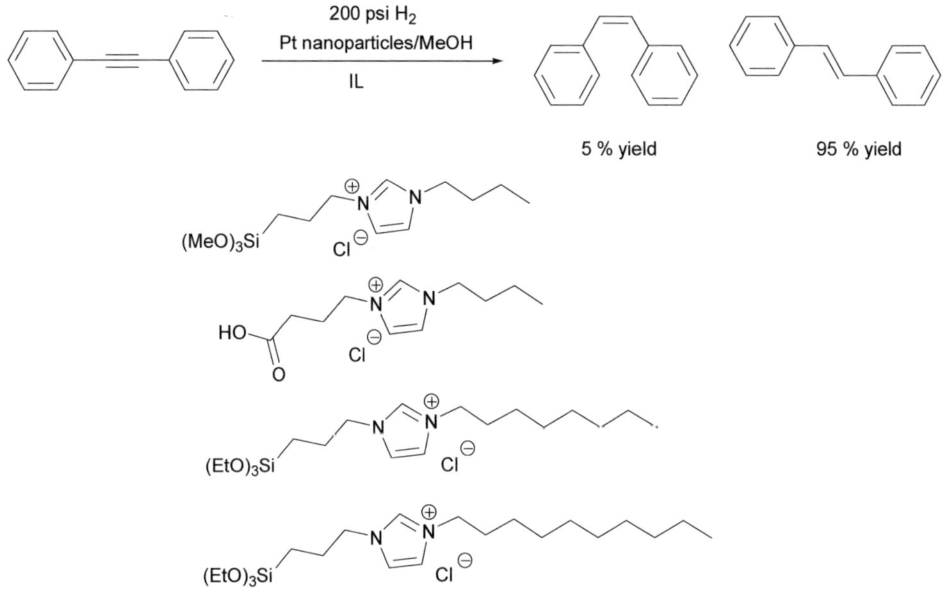

Fig. 48. Functionalised ILs and hydrogenation of diphenylacetylene

5.5 Ruthenium nanoparticles

Silveira *et al.*[33] prepared ruthenium nanoparticles in [bmim][BF_4] and [bmim][NTf_2]. They investigated the hydrogenation of benzene and olefins (1-hexene, cyclohexene, 2,3-dimethyl-2-butene, toluene, *iso*-propylbenzene, *tert*-butylbenzene, and anisole) with these nanoparticle catalysts re-dispersed in the IL and under solventless conditions. Excellent results were obtained for the conversion of hexene, cyclohexene and 2,3-dimethyl-2-butene in [bmim][PF_6] and under solventless conditions (> 99 %). Although the reactions occurred faster under solventless conditions, the catalyst could be reused up to eight times in the IL system with no significant loss in activity. Also investigated was the partial hydrogenation of benzene to cyclohexene, where the IL displayed better selectivity than solventless conditions. For example, at 75 °C and 4 atm H_2 after 2.0 hours at 10 % conversion, 15 % selectivity was obtained for the nanoparticles in the IL. However, under the same conditions after 1.0 hour and at 9 % conversion, solventless conditions furnished cyclohexene in only 4 % selectivity.

Ruthenium nanoparticles were prepared from a ruthenium dioxide precursor in imidazolium ILs.[113] The nanoparticles were subsequently prepared under solventless conditions or under IL biphasic conditions. Although harsh conditions are normally required for benzene hydrogenation, this group successfully hydrogenated benzene to cyclohexane under relatively mild conditions (75 °C, 4 atm H_2) using their ruthenium catalyst. Using [bmim][PF_6] as solvent the highest conversion and TOF (96 % and 49 h^{-1}) of benzene to cyclohexane was obtained in comparison to [bmim][BF_4] (46 % and 15 h^{-1}) or [bmim][OTf] (40 % and 22 h^{-1}). The same catalyst used under solventless conditions displayed 100 % conversion with a TOF of 953 h^{-1} after only 0.7 h. However, selectivity to cyclohexene was obtained using the IL [bmim][BF_4]. This group found that an increase in the temperature of the reaction up to a certain point had a positive effect on the selectivity of the reaction. The kinetic studies completed proved that the hydrogen mass transfer governs the overall reaction rate and product selectivity is not influenced by hydrogen pressure.

Prechtl *et al.*[114] used the ILs [bmim][NTf_2], [bmim][BF_4], [dmim][NTf_2] and [dmim][BF_4] to immobilise Ru(0) nanoparticles for hydrogenation reactions. Biphasic conditions were used for the hydrogenation of arenes at temperatures ranging from 50 – 90 °C at low hydrogen pressure (4 bar). [Bmim][BF_4] containing the nanoparticles showed the lowest conversion (40 %) after 18 hours for the hydrogenation of toluene to methylcyclohexane in comparison with the three other IL systems. Using the [bmim][NTf_2] system, after 24 hours, 50 % conversion was reached. However, the [dmim] ILs better conversion was evident after 18 hours (76 % for [BF_4^-] and 90 % for [NTf_2^-]. After several runs of the nanoparticles in the ILs the particle size and agglomeration state remained nearly unchanged. The systems were also reused successfully several times with little loss in catalytic activity.

Ruthenium nanoparticles were prepared by Prechtl *et al.*[115] by the treatment of [Ru(COD)(2-methylallyl)$_2$] (**31**) with H_2 in a nitrile functionalised IL to form Ru nanoparticles for the hydrogenation of nitrile functionalities. Using benzonitrile as a substrate this group demonstrated the preferential hydrogenation of the nitrile functionality over the aromatic group by the Ru nanoparticles in ILs (Figure 49).

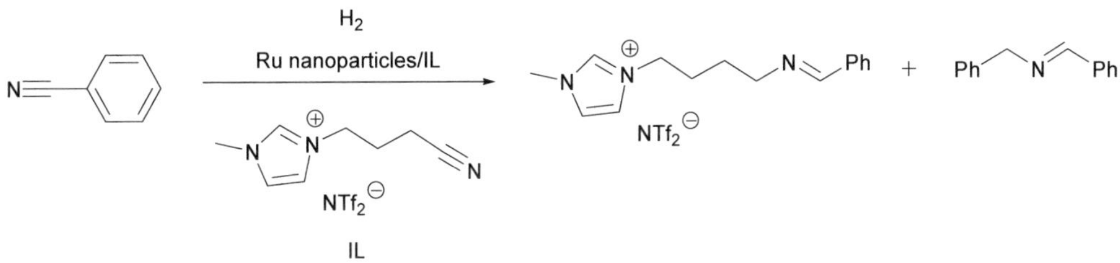

Fig. 49. Selective reduction of nitrile group. (Nucleophilic subsititution with benzyl amine, followed by evolution of ammonia was also observed)

They attribute this to the use of a nitrile functionalised IL giving rise to preferential hydrogenation of the nitrile group whereas arenes are normally hydrogenated by Ru nanoparticles in non-functionalised ILs. Additionally the activation energy for toluene hydrogenation with the ruthenium catalyst in IL is approximately 20 % lower than the activation energy for hydrogenation of nitrile groups in the ILs.[69,114] They found that the selectivity is dependent on the IL support and not on the Ru nanomaterial. For the hydrogenation of benzonitrile to (*E*)-*N*-benzylidene-1-phenylmethanamine, *via* displacement of ammonia, their hypothesis is that the strong coordination of the nitrile group to the ruthenium surface prevents displacement by the aromatic ring of benzonitrile.

5.6 Nickel nanoparticles

Migowski *et al.*[116] demonstrated that the organisation range order of an IL influences the diameter and size distribution of nickel nanoparticles prepared within. 1-Alkyl-3-methylimidazolium [NTf_2] ILs of side chain lengths varying from *n*-butyl to *n*-hexadecyl were used to test this dependency. It was found that an increase in alkyl chain length of the IL up to C_{14} led to a decrease in diameter and size distribution of the nanoparticle, although with side-chain length C_{16} the values for these parameters increased. The IL-Ni colloidal dispersion was used for the biphasic hydrogenation of cyclohexene. TOF values (91 h^{-1}) obtained for the hydrogenation of cyclohexene to cyclohexane in [bmim][NTf_2] at 100 °C and 4 bar H_2 after 14 hours were two orders of magnitude greater than those obtained using conventional nickel supported catalysts at 80 °C and 5 bar H_2.[116] Assuming that the reaction follows a classical monomolecular surface reaction mechanism[29], the catalytic activity expressed as the kinetic constant was calculated to be 9.2×10^{-4} s^{-1}.

5.7 Miscellaneous nanoparticles

Redel *et al.*[117] prepared metal nanoparticles (Co, Rh and Ir) in ILs by decomposition of their corresponding metal carbonyls ($Co_2(CO)_8$, $Rh_6(CO)_{16}$ and $Ir_4(CO)_{12}$) and subsequently used them as catalysts for the hydrogenation of cyclohexene to cyclohexane in ILs ([N_{1114}][NTf_2], [bmim][BF_4] and [bmim][OTf). The catalyst activity observed for the nanoparticles prepared by this route was superior to that obtained by standard metal nanocatalysts in [bmim] ILs. Using [bmim][BF_4] as reaction solvent, the iridium nanoparticles demonstrated the best catalytic activity (1940 mol product (mol metal)$^{-1}$ h^{-1} after 1 hour at 97 % conversion) in comparison to the rhodium nanoparticles (380 mol product (mol metal)$^{-1}$ h^{-1} after 2.5 hours at 95 % conversion), with the cobalt nanoparticles being particularly poor (0.16 mol product (mol metal)$^{-1}$ h^{-1} after 3 hours at 0.8 % conversion). The superior activity of the iridium in comparison with rhodium was postulated to be due to the smaller particle size of the Rh nanoparticles and therefore their accompanying larger surface-to-volume ratio. Even with the reasonably mild conditions used (4 bar H_2, 75 °C) for the hydrogenation reactions, the results obtained were superior to previous work carried out using metal nanoparticles in ILs (eg. RuO_2/Ru/[bmim][PF_6]: 943 mol product (mol metal)$^{-1}$ h^{-1}). The authors suggest the preparation route of the nanoparticles to be the determining factor for the enhanced activity of their nanoparticles. Nanoparticles prepared by the route of reduction of the precatalysts can lead to impurities residing in the IL (eg HCl) which may lower the stabilization effect of the IL and in turn result in catalyst deactivation. However, preparation of the nanoparticles by the decomposition route is safer for the catalyst, because it avoids disruption of the IL phase.

6. Hydrogenation using supported ionic liquid catalysis

An economical means of using ILs in hydrogenation catalysis has emerged. ILs containing a catalyst can be loaded onto a solid support and used for catalytic reactions, thus combining the advantages of using a solid support with the merits of using an IL media. This method of catalysis reduces the amount of IL needed in comparison with conventional IL catalytic systems.

Using [tricaprylmethylammonium][PF_6], [bmim][PF_6] and [bmim][BF_4], Mikkola *et al.*[118] investigated the hydrogenation of citral using a palladium catalyst and SILC (Supported Ionic Liquid Catalysis) technology. Citronellal formation was favoured using the reaction system containing [bmim][PF_6], whereas dihydrocitronellal was favoured using [bmim][BF_4]. [Tricaprylmethylammonium][PF_6] also showed increased formation of dihydrocitronellal. The authors suggest the limiting factor to be the solubility of H_2 in the IL medium, with H_2 solubility being less in [bmim][PF_6] than [bmim][BF_4].[16,119] The cation was also shown to contribute to solubility factors as the dihydrocitronellal formation in [tricaprylmethylammonium][PF_6] was greater than in [bmim][BF_4]. When cyclohexane was used as solvent by Hao *et al.*[120] for the hydrogenation of citral using a Pd/C catalyst (5 MPa H_2, 50 °C) 60 % conversion was achieved with 78 % selectivity towards citronellal. The effect of a Lewis acid and a Brønsted acid on SILC for the one pot synthesis of menthol from citral was investigated by Virtanen *et al.*[121]. In Virtanen's system, supported Pd nanoparticles were immobilized in [*N*-butyl-4-methylpyridinium][BF_4] together with a Lewis or Brønsted acid as an ionic modifier on active carbon cloth (ACC). The initial reaction rate for the conversion of citral decreased on addition of $ZnCl_2$, and the selectivity towards dihydrocitronellal also decreased.

Virtanen *et al.*[122] used SILC technology to hydrogenate citral and cinnamaldehyde, using a palladium catalyst immobilised in [*N*-butyl-4-methylpyridinium][BF_4], [bmim][BF_4], [bmim][PF_6], [N_{1888}][PF_6] and [N_{1888}][HSO_4], tethered to an active carbon support. The stirring rate during each experiment was high (1500 rpm) so that external mass transfer limitations were eliminated. In the case of citral reduction, with all ILs tested, the percentage conversion was high with the main reduced product dihydrocitronellal observed. The notable exceptions were the [PF_6] ILs, where the principle product was citronellal and with a significant decrease in conversion (39 %). The authors contribute this compromise in performance due to lower solubility of H_2 in the [PF_6] ILs. In the case of cinnamaldehyde reduction, [*N*-butyl-4-methylpyridinium][BF_4] and [bmim][PF_6] were studied and for both ILs the main product was hydrocinnamaldehyde. Better conversion results were obtained using the [*N*-butyl-4-methylpyridinium][BF_4] (87 - 100 %) in comparison with the [bmim][PF_6] (23 - 93 %). Slightly superior selectivity towards hydrocinnamaldehyde was obtained using the [bmim][PF_6] (82 - 94 %) in comparison with the [*N*-butyl-4-methylpyridinium] [BF_4] (80 - 88 %). A clear trend of catalyst deactivation was seen from the results of different hydrogenation experiments. The low activity of the palladium catalyst with ionic liquid [bmim][PF_6] was ascribed to poor solubility of hydrogen which was also supported by kinetic studies. When the same reactions were studied to compare the intial reaction rates, it was found that for the catalyst containing ionic liquid [bmim][PF_6] has reaction order close to zero and for the ionic liquid [*N*-butyl-4-methylpyridinium][BF_4] it was 0.7 with respect to hydrogen pressure showing that hydrogen pressure does not make difference in former case.[123] Five different SILC (supported ionic liquid catalyst) compounds have also been used by Virtanen *et al.*[124] to investigate the kinetics of citral hydrogenation (Figure 50) and kinetic modelling was based on the Langmuir-Hinshelwood-Hougen-Watson concept for citral hydrogenation. Palladium nanoparticles in the IL layer were immobilised on an ACC and the results of the hydrogenation compared with conventional palladium on ACC (**32**). The bulk solvent used was *n*-hexane as no leaching of the IL was observed using this solvent, except in the case of [bmim][BF_4]. Of the imidazolium, pyridinium and ammonium ILs studied, the pyridinium [BF_4] IL-containing catalyst displayed the highest TOF observed for all ILs (> 140 mol/h). Selectivity was in most cases highest with dihydrocitronellal, for which pyridinium [BF_4] exhibited the best result (89 % selectivity). All SILCs achieved greater TOF values than the palladium on ACC (**32**) reference standard.

Lou *et al.*[125] used supported ionic liquid catalysis with a ruthenium complex ($RuCl_2(PPh_3)$(*S,S*-DPEN), DPEN = 1,2-diphenylethylenediamine) (**33**) for the hydrogenation of acetophenone using mesoporous materials modified with an imidazolium IL (1-methyl-3-(3-triethoxysilylpropyl)imidazolium [BF_4]) (Figure 51). The ruthenium complex immobilised in [bmim][BF_4] was confined to the surface of the mesoporous material for the hydrogenation reactions (3 MPa H_2, RT, 10 h) in IPA.

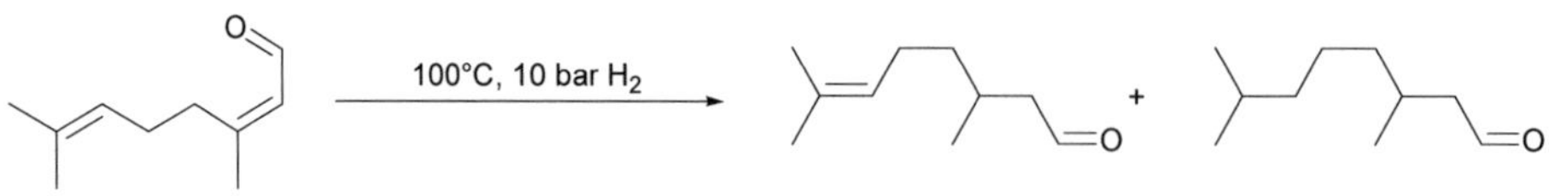

Catalyst and IL	Conv. (%)	Citronellal (%)	Dihydrocitronellal (%)
Pd(acac) in [bmim][PF_6] on ACC[118]	95 %	major product	-
Pd(acac) in [bmim][BF_4] on ACC[118]	97 %	-	major product
Pd SILCA in [bmim][PF_6][122]	95 %	major product	-
Pd SILCA in [bmim][BF_4][122]	97 %	-	major product

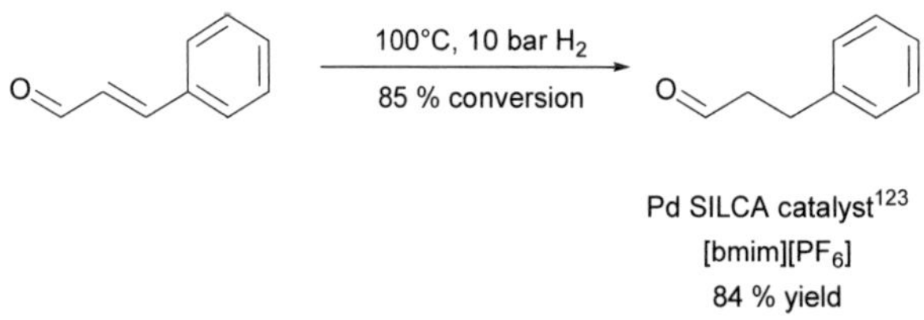

Fig. 50. Hydrogenation of citral and cinnamaldehyde[118,122,123]

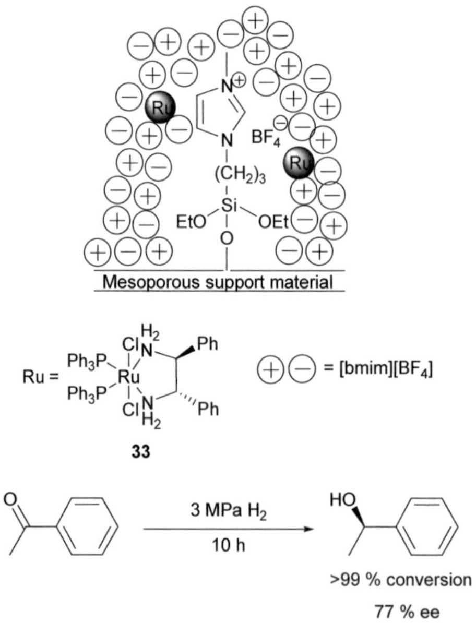

Fig. 51. Ruthenium complex used in hydrogenation of acetophenone promoted by IL-modified mesoporous materials

Lou compared the activity of their supported systems with that of the corresponding homogeneous system, with almost identical results. The homogeneous system gave a conversion of > 99 % with 78 % ee. SiO_2 showed the best activity in terms of recycling ability, compared with other mesoporous materials. The system was recycled five times with no decrease in conversion, (99 %), and little variation in enantioselectivity (79 - 75 % ee). The authors postulate that the larger pore size and complex structure of SiO_2 are the major reasons for the impressive recycling results, due to the prevention of channel blockage during recycling.

Gelesky *et al.*[126] used zerovalent rhodium nanoparticles in [bmim][BF_4] immobilized in a silica network, prepared by the acid or base catalysed sol-gel method, for the hydrogenation of alkenes (1-decene and cyclohexene). This supported ionic liquid phase (SILP) technology was compared with a commercial 5 % Rh/C catalyst and also isolated Rh(0) nanoparticles. The SILP example was found to increase catalytic activity, compared with isolated Rh(0) nanoparticles. The supported catalysts prepared by the acid catalyzed sol-gel method displayed higher IL content in the silica network, and also contained gels of larger pore diameter. Increased catalytic activity resulting from catalysts prepared *via* this route was ascribed to their relatively large pore diameter.

Fow *et al.*[38] used chiral ruthenium and rhodium catalysts immobilised in phosphonium ILs (Figure 52) supported on silica for the hydrogenation of acetophenone.

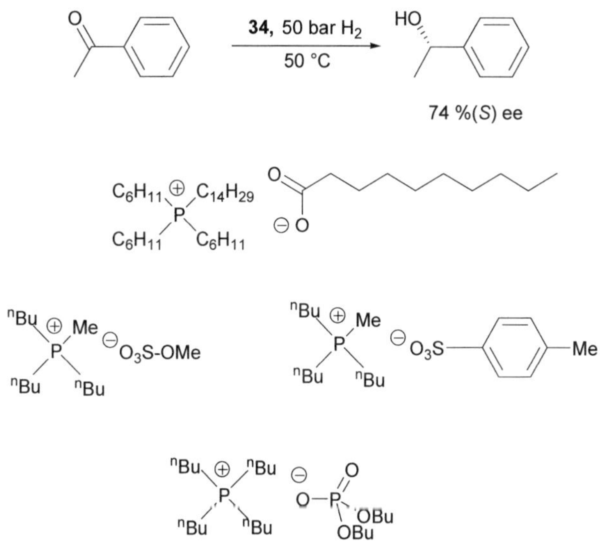

Fig. 52. Phosphonium ILs and hydrogenation of acetophenone

Basic reaction conditions were found to be essential for high conversions. Good enantioselectivity was achieved using their rhodium catalyst, [Rh((*S*)-BINAP)(COD)]$ClO_4 \cdot$ THF (**34**) (74 % ee). No enantioselectivity was observed using the corresponding homogeneous system. The authors describe formation of solvent cages in IL systems as the reason for enantioselectivity promotion with the SILCs.

Chen *et al.*[127] discovered efficient metal scavenging abilities of their task-specific ILs (TSILs) supported on a polystyrene backbone. The materials were formed by the ionic pair coupling of the imidazolium cation tethered to a polystyrene support with L-proline. The application of their novel material soaked with a palladium catalyst was used for the hydrogenation of styrene (Figure 53), where, under mild and solvent-free conditions, a good TON (5,000) and TOF (250 h^{-1}) were obtained.

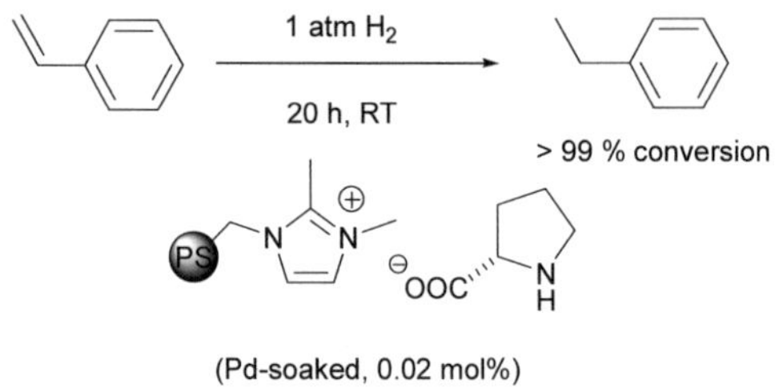

Fig. 53. Hydrogenation of styrene

Ruta *et al.*[30] demonstrated the catalytic activity of their rhodium-based structured supported ionic liquid-phase (SSILP) catalyst with the hydrogenation of 1,3-cyclohexadiene. The IL containing the metal catalyst was confined to a structured support surface consisting of sintered metal fibres. In order to obtain a homogeneous coverage of the support by the IL, these metal fibres were coated with a layer of carbon nanofibres. The hydrogenation of the selected substrate was carried out in the gas phase reaching a TOF of 150 - 250 h^{-1} and selectivity towards cyclohexene of greater than 96 %. The presence of acid and an excess of a triphenylphosphine ligand were essential for the catalytic reaction in the gas phase. Further kinetic studies showed that up to 12 % of catalyst loading the reaction is independent of gas diffusion through the IL film. Lercher *et al.*[128] reported the use of SiO_2 supported Pt nanoparticles immobilized in ionic liquid for the hydrogenation of ethylene gas. This catalyst showed similar activity to a SiO_2 supported Pt nanoparticle catalyst.

7. Transfer hydrogenation

The use of a hydrogen source other than gaseous hydrogen for hydrogen reactions in ILs has recently been explored; to the best of our knowledge, only a few publications have dealt with this subject so far.[129-134]

Catalytic transfer hydrogenations under microwave irradiation were carried out in [bmim][PF_6], using ammonium formate or triethylammonium formate as hydrogen source and 10 % palladium on carbon (**3**) as catalyst by Berthold *et al.*[129] (Figure 54) Substrates containing a wide range of functional groups were investigated, with impressive results obtained in IL systems. Hydrogenation of 4-nitrobenzoic acid methyl ester showed an increase in the yield of 4-aminobenzoic acid methyl ester from 70 % to 92 % when changing from propane-1,3-diol to [bmim][PF_6].

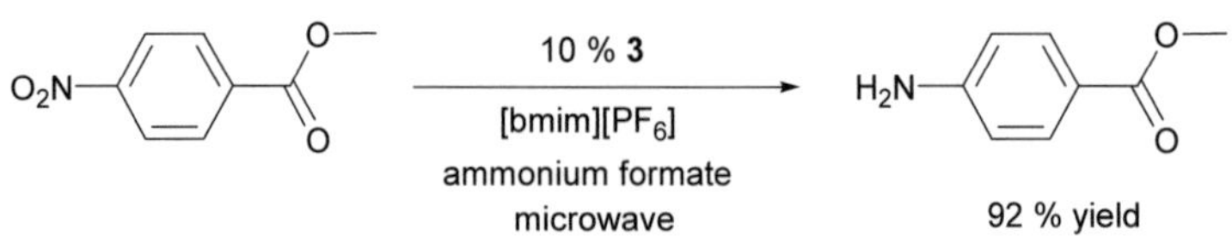

Fig. 54. Hydrogenation of 4-nitrobenzoic acid methyl ester

Baan *et al.*[130] screened six ILs ([bmim][BF_4], [bmim][PF_6], [bmim][Cl], [emim][PF_6], ECOENG™ 212, ECOENG™ 500) in the homogeneous transfer hydrogenation of cinnamic acid using a palladium catalyst (**35**) and ammonium formate as hydrogen donor. (Figure 55) Although only 2 % yield was obtained for the reduction of cinnamic acid using the $[PF_6^-]$ ILs, > 99 % yield was obtained when $[bmim][BF_4]$ and ECOENG™ 212 and 500 were investigated. Under similar conditions in ethanol, toluene, or chloroform negligible yields were observed. The successful reaction with the $[bmim][BF_4]$ was then extended to other α- and aryl-substituted cinnamic acids displaying excellent results (> 99 % yield for five out of nine substrates studied).

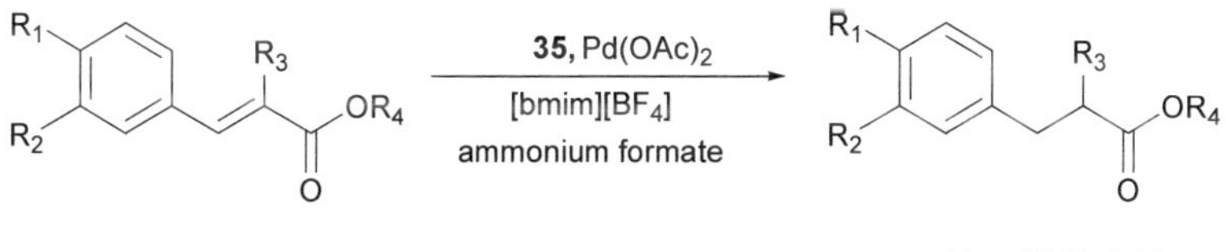

R_1: H, OH, OMe
R_2: H, OH, OMe
R_3: H, Me, Ph, -NHCOCH$_3$-, -NCH(CH$_3$)-
R_4: H, Me

Fig. 55. Hydrogenation of substituted cinnamic acids

Cinnamic acid derivatives were hydrogenated by Baan *et al.*[131] (Figure 56) using catalytic transfer hydrogenation (CTH). Using a basic heterogeneous palladium carrier, magnesium-lanthanum mixed oxide (**36**), in $[bmim][BF_4]$, a variety of substrates were hydrogenated in up to 99 % conversion using HCO_2NH_4. A selection of imidazolium and phosphonium ILs were investigated for the hydrogenation of cinnamic acid. Use of the imidazolium ILs led to increased activity compared with the phosphonium ILs. In $[Emim][EtOSO_3]$ it was found that the source of palladium effected the reaction, with Pd^0 displaying increased conversion (85 %) in comparison with Pd^{II} (53 %). Recyclability was also shown possible by the reuse three times of the catalyst used for the hydrogenation of cinnamic acid, with no loss in activity being observed. Also in $[emim][EtOSO_3]$ and an azeotropic mixture of triethylamine and formic acid and a palladium on magnesium-lanthanum hydrotalcite catalyst, the hydrogenolysis of a variety of *para*-substituted bromo and chlorobenzenes was performed. Overall, the *p*-bromobenzenes were found to be more active than their chlorinated counterparts.

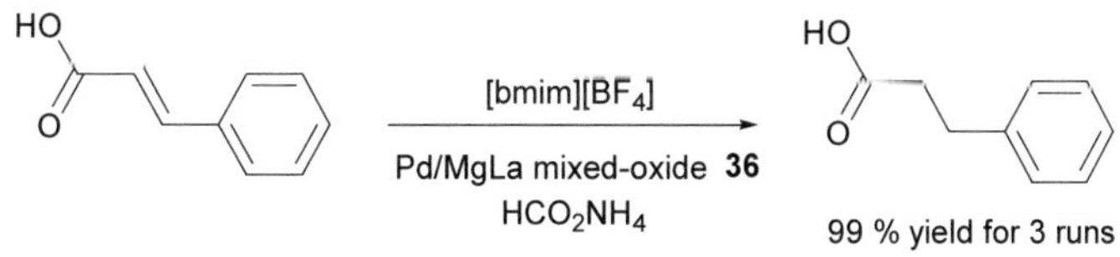

Fig. 56. Hydrogenation of cinnamic acid

Joerger *et al.*[132] used [Ru(arene)(diamine)] catalysts (**37** and **38**) for the asymmetric transfer hydrogenation of acetophenone using a range of ILs, with formic acid as the hydrogen donor. (Figure 57)

In the case of (**37**), catalyst activity was shown to be inhibited by hydrophilic ILs, for example $[bmim][BF_4]$ (conversion < 1 %, 40 hours), $[bmim][MeSO_4]$ (conversion 19 %, 48 hours) and [emim][OTf] (conversion 0 %, 24 hours). The best results using this catalyst were

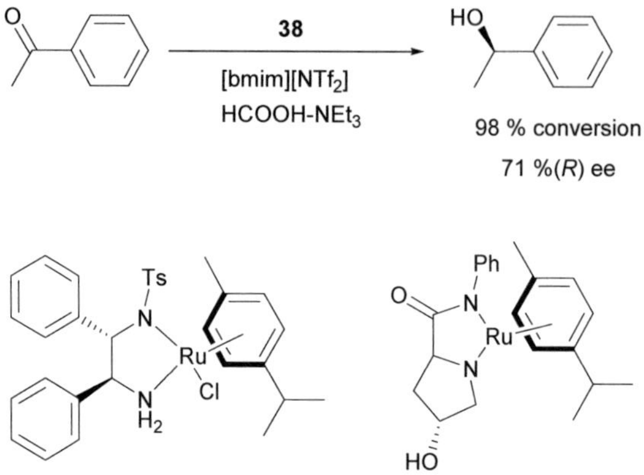

37 38

Fig. 57. [Ru(arene)(diamine)] catalysts and hydrogenation of acetophenone

thus obtained using hydrophobic ILs. Conversions of up to 99 % were reached and 97 % enantioselectivity for the hydrogenation of acetophenone to 1-phenylethanol using ILs [trimethylbutylammonium][NTf_2] (N_{1114}) and [methyltributylphosphonium][NTf_2] (P_{1444}). Without no IL 99 % conversion and 97 % ee was obtained. For the case of [N_{1114}][NTf_2], good recyclability was demonstrated, as long as the reaction time was increased (9 hours for run 1 to 30 hours for run 3), decreasing only from 99 % conversion in run one to 82 % conversion on the third run. A slight decrease was reported when recycling this system in terms of selectivity, from 97 % ee in run 1 to 96 % ee in run 3. Using catalyst (**38**) the highest conversion was observed using hydrophobic ILs [bmim][PF_6] and [bmim][NTf_2], with selectivity of 71 % obtained using both ILs. However, when the cation was changed from [bmim] to *N*-butylpyridinium, conversion decreased to 50 % with the lowest enantioselectivity of 65 % being observed. Upon the fifth run of [bmim][PF_6], the catalyst activity decreased slightly again from 72 - 68 % ee, although excellent conversion was maintained (95 %) throughout the recycling process. Using DCM as reaction solvent, 99 % conversion was achieved and 71 % ee.

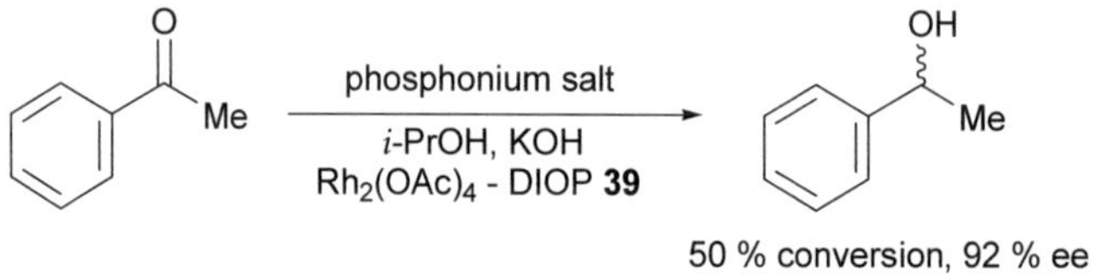

Fig. 58. Hydrogenation of acetophenone to *sec*-phenylethyl alcohol

The rhodium catalyst precursor, dirhodium tetraacetate [$Rh_2(OAc)_4$] was used by Comyns *et al.*[133] for the transfer hydrogenation of acetophenone to *sec*-phenylethyl alcohol using IPA as the hydrogen donor (Figure 58). As reaction solvents this group used a series of tetraalkyl/aryl phosphonium tosylate salts with varying melting points (Figure 59).

The authors describe the advantages of using these higher melting salts over lower melting point ionic liquids. With the ILs that were solid at room temperature, the products could be separated by simple decantation, rather than extraction during work-up. Also the inherent stability of higher melting ILs to harsher conditions provides an additional advantage.

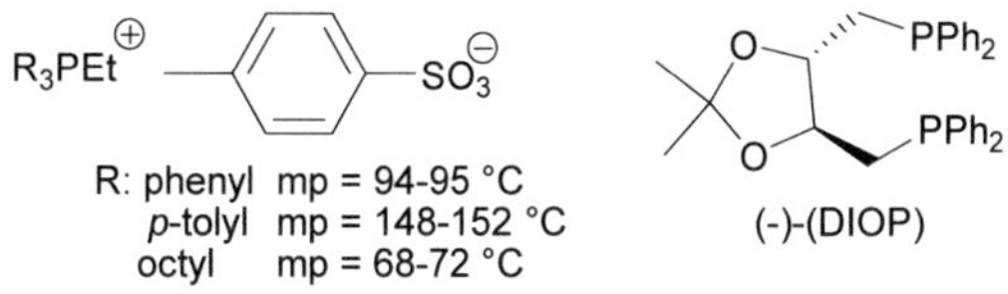

Fig. 59. Phosphonium tosylates salts and (−)-(DIOP) ligand

Reaction temperatures for the transfer hydrogenation ranged from 120-150 °C giving moderate conversions (28-50 %). Using the (−)-(DIOP) (2,3-O-isopropylidene-2,3-dihydroxy-1,4-*bis*(diphenylphosphino)butane) (Figure 59) ligand with the ruthenium catalyst (**38**) impressive enantiomeric excesses of 92 % and 50 % conversion were obtained.

Kantam *et al.*[134] used transfer hydrogenation conditions to selectively reduce carbonyl compounds to the alcohol derivatives using Ru nanoparticles stabilized on magnesium oxide by the incorporation of a basic IL, choline hydroxide (Figure 60). Catalyst preparation involved the treatment of the magnesium oxide crystals with choline hydroxide to yield CHNAP-MgO (CH: Choline Hydroxide, NAP: nanocrystalline aerogel-prepared). CHNAP-MgO was then stirred with $RuCl_3$ solution to obtain Ru(III)-CHNAP-MgO, which in turn was reduced to yield the final catalyst, Ru(0)-CHNAP-MgO.

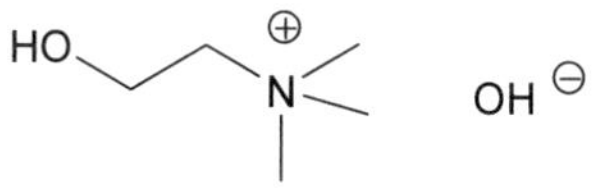

Fig. 60. Choline hydroxide

Under reflux conditions using 2-propanol and KOH in 2-propanol a number of carbonyl functionalised compounds were selectively reduced in high yields (Figure 61).

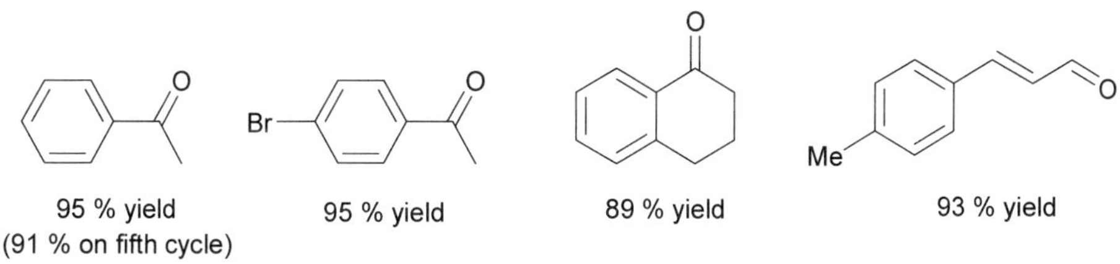

Fig. 61. Carbonyl compounds selectively hydrogenated to alcohols

Worth noting also are the high yields that were obtained for the selective reduction of the pharmaceutically important 4-*iso*-butylacetophenone (94 %) and 6-methoxy-2-acetophenone (94 %), which are intermediate compounds in the synthesis of ibuprofen and naproxen (Figure 62).

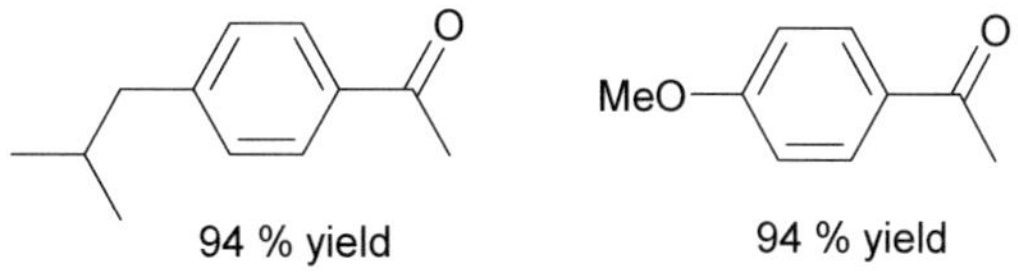

Fig. 62. Pharmaceutically important intermediates

When Hermecz *et al.*[135] carried out hydrogenation of chalcone using Wilkinson's catalyst, they found that the reactions using only ionic liquids as solvents gave better selectivity towards 1,3-diphenylpropan-1-one than reactions performed in a mixture of molecular solvents and ionic liquids (Figure 63).

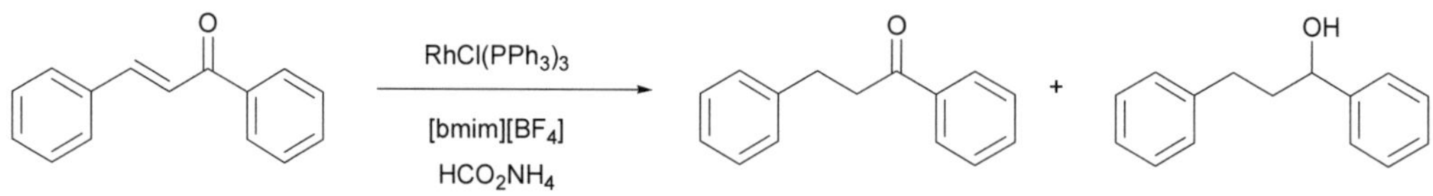

Fig. 63. Hydrogenation of chalcone using Wilkinson's catalyst

The selectivity with molecular solvents ranged from 70 % to 92 % but in ionic liquids was generally > 99 %. Furthermore, the results show that reaction rates are higher in ionic liquids, in particular, [emim][BuOSO$_3$], [emim][HexOSO$_3$] and ECOENG-500™ (Table 10).

Entry	Ionic Liquids	Reaction Time (min)[a]	Conversion %	
			Ketone[b]	Alcohol[c]
1	[emim][EtOSO$_3$]	90	84	0
2	[emim][BuOSO$_3$]	15	> 99	0
3	[emim][HexOSO$_3$]	15	> 99	0
4	[bmim][BF$_4$]	30	> 99	0
5	[bmim]Cl	240	74	26
6	[bmim][AlCl$_4$]	240	0	0
7	[emim][PF$_6$]	240	88	3
8	[bmim][PF$_6$]	240	93	6
9	[hmim][PF$_6$]	240	60	0
10	ECOENG-500™	15	> 99	0
11	[$P_{14,6,6,6}$][Cl]	30	> 99	0
12	[$P_{14,6,6,6}$][PF$_6$]	240	55	0
13	[$P_{14,6,6,6}$][BF$_4$]	240	> 99	0
14	[$P_{4,4,4,4}$][BF$_4$]	90	> 99	0
15	[$P_{1,i4,i4,i4}$][OTs]	240	0	0

[a] Reaction conditions: (2*E*)-1,3-diphenylprop-2-en-1-one (0.2 mmol), RhCl(PPh$_3$)$_3$ (0.02 mmol), NH$_4$CO$_2$H (0.8 mmol), ionic liquid (1 mL), 90 °C, [b] 1,3-diphenylpropan-1-one, [c] 1,3-diphenylpropan-1-ol.

Table 10. Transfer hydrogenation of chalcone in ionic liquids

The group explained the selectivity towards 1,3-diphenylpropan-1-one, by proposing a specific interaction between the ionic liquid and chalcone that prevents the reduction of carbonyl group.

8. Future prospects

This section outlines examples where ILs have been used in hydrogenations, in processes with characteristics which do not fall under the previous headings. These cases demonstrate the potential versatility of ILs and illustrate how their use can lead to innovative approaches to problems in a variety of areas.

ILs have been investigated as buffers for hydrogenation reactions in non-aqueous media.[136] Xu *et al.* demonstrated that the selectivity of the hydrogenation of *trans*-cinnamaldehyde could be modulated by using ILs with different buffering characteristics. Using the ruthenium catalyst, $[RuCl_2(PPh_3)_3]$ (**40**) in DMF at 60 °C under 2 MPa H_2 pressure, Xu demonstrated a reversal of selectivity between the major reduction products, the unsaturated alcohol and saturated aldehyde. A range of IL buffers were synthesised by reacting basic [Rmim][OH] salts with a series of binary or polybasic acids to form IL buffers. These were then used for the hydrogenation of olefins (1-hexene, styrene and cyclohexene) to confirm the buffering ability of the IL-buffers and with *trans*-cinnamaldehyde as a substrate to investigate the dependence of selectivity on the buffer used in a non-aqueous medium. The results of the hydrogenation confirmed that the activity of the catalyst was highly buffer dependent. Notably, IL-buffers from [bmim][OH] and H_3PO_4 with $\log_{10}$([base]/[acid]) of –0.073 ([bmim][H_2PO_4]) and 0.232 ($[bmim]_2[HPO_4]$) were found to give opposite selectivities, with [bmim][H_2PO_4] favouring reduction of the olefin over the carbonyl in the hydrogenation of *trans*-cinnamaldehyde, while $[bmim]_2[HPO_4]$ favoured reduction of the carbonyl. When kinetic studies were carried out, the reaction rate was found to be slightly higher in the absence of the IL-buffers and notable differences in selectivity were observed between the two buffer systems. The authors proposed a catalytic mechanism to explain the reversal in selectivity using different buffering systems in non-aqueous media. (Figure 64)

It was proposed that the IL-buffer $[bmim]_2[HPO_4]$ ($\log_{10}$([Base]/[Acid]) = 0.232) is responsible for the ultimate formation of the $[RuH_4(PPh_3)_3]$ which would lead to preferential reduction of the carbonyl moiety. However, the likely species formed using the IL-buffer [bmim][H_2PO_4] ($\log_{10}$([Base]/[Acid]) = –0.073), $[RuHCl(PPh_3)_3]$ promotes the reduction of the olefin. The ease of preparation of IL-buffers may well promote an increase in research in this area in order to enhance selectivity for transformations of compounds with multi-functionalised reducible groups.

The new concept of SCILL (Solid catalyst with Ionic Liquid Layer) entails the coating of an IL onto a porous solid which is also a heterogeneous catalyst and thus the drawbacks of biphasic homogeneous catalysis with IL/organic liquid systems are circumvented as the amount of ionic liquid required is reduced and mass transfer limitations are avoided due to the small IL film thickness in the pores. Kernchen *et al.*[24] used this technology for the hydrogenation of cyclooctadiene. (Figure 65)

Using a commercial nickel catalyst coated with the IL [bmim][$OctOSO_3$] they obtained poor conversion of cyclooctadiene to cyclooctane. However, no leaching of the catalyst into the organic phase occurred and the selectivity to cyclooctene increased from 40 % using the system without the IL coating to 70 % using the IL coating.

Ruta *et al.*[137] combined the catalytic activity of palladium nanoparticles with a supported IL phase, for the selective hydrogenation of acetylene to ethylene which is an exothermic reaction (ΔH_{298K} = -172 kJ/mol). Monodispersed Pd nanoparticles were obtained via reduction of $Pd(acac)_2$ dissolved in ILs ([bmim][PF_6] and [bmimOH][NTf_2]). Carbon nanofibres were used as the IL support, which were in turn anchored to sintered metal fibres. Existing problems arising from using supported palladium catalysts for this industrial hydrogenation process are the formation of active-site ensembles resulting from catalyst deactivation due to the oligomerization of ethylene.[138] High selectivity (70 - 80 %)

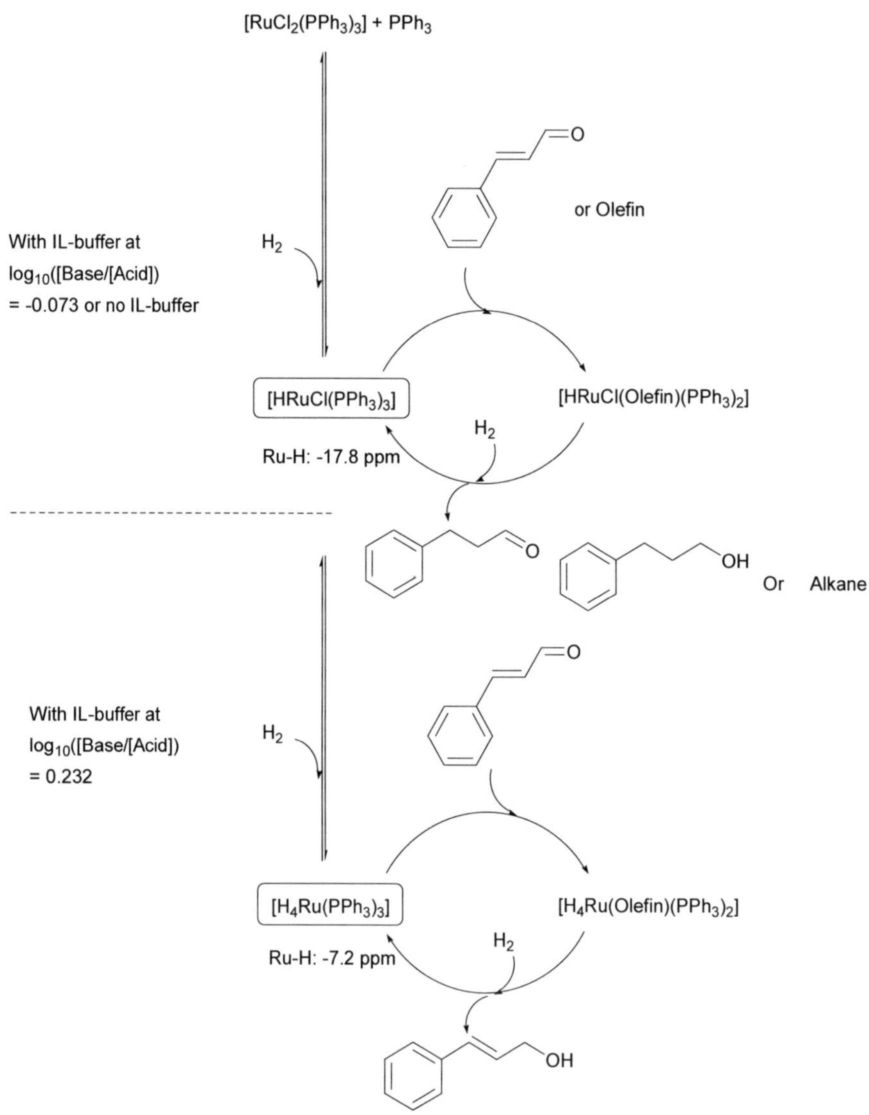

Fig. 64. Hydrogenation using IL-buffers

was obtained using their supported nanoparticle system with a maximum selectivity of 85 % (at 150 °C) being obtained using the IL [bmim][PF_6]. The impressive selectivity is due to the lower solubility of ethylene in the IL compared to acetylene which results in the lowering the subsequent hydrogenation of ethylene to ethane. The systems potential for application in industry was demonstrated by, not only its selectivity, but its long-term stability arising from the ILs preventing the formation of active-site ensembles and therefore reducing catalyst deactivation by this way.

Hamza *et al.*[139] carried out hydrogenation as part of a one-pot multistep process. The process consisted of the hydroformylation of styrene derivatives to yield branched aldehydes. The aldehydes were consequently condensed with reactive methylene compounds (malonitrile, ethyl cyanoacetate) and then hydrogenated. (Figure 66)

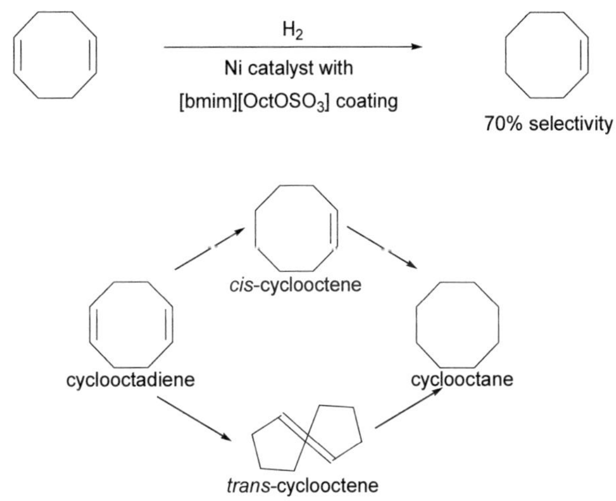

Fig. 65. Hydrogenation of cyclooctadiene

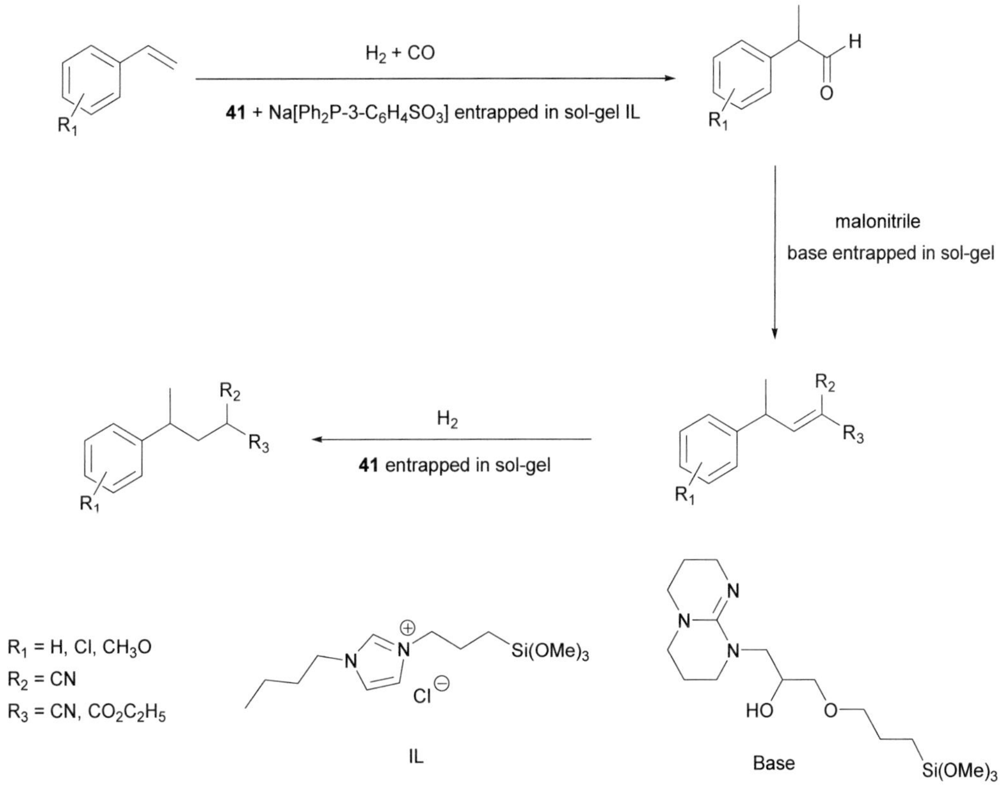

Fig. 66. One-pot multistep process

A silica sol-gel matrix containing co-entrapped $[Rh(cod)Cl]_2$ (**41**) and $Na[Ph_2P\text{-}3\text{-}(C_6H_4SO_3)]$ was confined within an IL (1-butyl-3-[3-(trimethoxysilyl)propyl]imidazolium chloride), The one-pot reaction was carried out at 80 °C with 20.7 bar H_2 and 20.7 bar CO in a mixture of 1:1 1,2-dichoroethane:THF. It was shown that the catalyst could be re-used up to four times, although renewal of the base (1,5,7-triazabicyclo[4.4.0]dec-5-ene modified with (3-glycidoxypropyl)trimethoxysilane), which was separately encaged, was required for each reaction. The authors demonstrated the success of the reaction to be dependent on the Rh catalyst being entrapped within the support and the IL being chemically bound to the sol-gel backbone. Under homogeneous conditions, where the Rh catalyst, base and IL were not entrapped in sol-gel, no reaction proceeded. The role of the IL in the reaction is proposed to be as a carbene ligand for the rhodium complex and also acting as a base that promotes the condensation of the aldehydes with malonitrile. Interesting hydrogenation selectivity was observed for this reaction. The internal double bond of the reaction intermediate was reduced, although no reduction of the external double bond of the styrene derivatives was observed.

Pt(0) and Rh(0) nanoparticles were prepared in $[bmim][BF_4]$ and redispersed in $[bmim][NTf_2]$ with cellulose acetate to create functionalised membrane films.[140] These transition metal containing films were used as catalysts for the hydrogenation of cyclohexene to cyclohexane at 75 °C and 4 atm H_2. The use of cellulose in the catalytic materials was demonstrated to be favourable to catalytic activity. Using the cellulose containing membrane with the Pt(0) nanoparticles and IL, a TOF of 7353 h^{-1} was obtained. However, using the same system without the cellulose led to a dramatic decrease in TOF to 329 h^{-1}. The recycling of their system was shown by the successful reuse of the Pt(0) based IL/cellulose membrane up to 2 times without a decrease in catalytic activity.

Craythorne *et al.*[141] co-entrapped rhodium pre-catalysts ($[RhCl(PPh_3)_3$ (**42**) and $[Rh_2(COD)_2(dppm)(\mu_2\text{-}Cl)]BF_4$ (**43**)) with $[bmim][NTf_2]$ by sol-gel methods and investigated their activity in the hydrogenation of styrene to ethyl benzene. The activity of the catalysts were compared to the same precatalysts prepared by doping of silica glass, the parent homogeneous catalysts, and commercially available Rh catalysts (Rh/C and Rh/alumina). Reactions were carried out using DCM as solvent at 100 °C at 27 bar H_2. In the case of the catalytic system containing the $[RhCl(PPh_3)_3]$ precatalyst, the ionogel outperformed its silica glass counterpart in terms of conversion (98 % in comparison with 75 %) and TOF (32 x 10^2 min^{-1} in comparison to 20 x 10^2 min^{-1}). Irrespective of using the silica doped system or the IL doped system, the TOF values were higher than for the parent homogeneous catalyst (5.5 x 102 min^{-1}). For the systems prepared using $[Rh_2(COD)_2(dppm)(\mu_2\text{-}Cl)]BF_4$ (**43**) the catalyst activity for the catalyst-doped silica glasses (3.8 x 10^2 min^{-1}) and the doped ionogels (3.2 x 10^2 min^{-1}) approached that of the homogeneous catalyst (4.5 x 10^2 min^{-1}). Using heterogeneous rhodium catalysts (Rh/C and Rh/Alumina) the TOF values obtained were significantly lower in comparison to the homogeneous catalytic systems (1.2 x 10^2 min^{-1} (Rh/C) and 0.40 x 10^2 min^{-1} (Rh/Alumina)). This group investigated the capacity of their catalyst system for recycling and found no significant loss in catalytic activity, whereas the heterogeneous rhodium catalysts showed considerable loss in activity over five runs. Levoglucosenone was used to test the catalytic ionogel system using $[RhCl(PPh_3)_3]$ (**42**) in order to determine if the catalytic activity could be extended to more complex substrates. Indeed, at 110 °C and 25 bar H_2, 90 % conversion and 100 % selectivity to the olefinic bond reduced product was observed. (Figure 67)

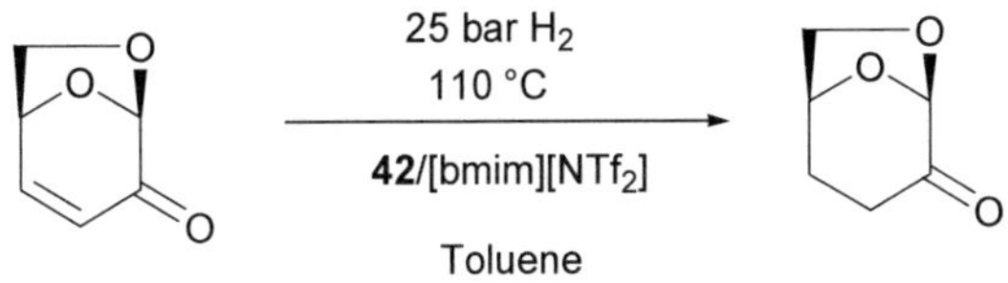

Fig. 67. Hydrogenation of levoglucosenone

Second and third generations of the catalyts were prepared by extracting the used IL with DCM and using it to prepare subsequent catalytic systems. First, second and third generations all demonstrated similar activity.

Cationic dendritic pyrphos-rhodium(norbornadiene) ([pyrphos-Rh(NBD)] (**44**)) complexes were synthesised using poly(propyleneimine) (PPI) and poly(amido amine) (PAMAM) dendrimers as supports, as catalysts for hydrogenation of (Z)-methyl α-acetamidocinnamate in methanol and a biphasic system ([bmim][BF_4]/IPA).[142] [Pyrphos-Rh(NBD)] complexes were also tethered to hyperbranced poly(ethylene imines) (PEI) for use as catalytic systems for the same reaction. Using methanol as reaction solvent at 25 °C and 30 bar H_2 the activity and selectivity of the PPI, PAMAM and hyperbranched PEI-bound pyrphos-Rh(NBD) complexes decreased with increasing molecular size. Studying the catalytic effects in the biphasic system, the reaction temperature was raised to 55 °C in order to facilitate a homogeneous reaction system. PPI, PAMAM and PEI-fixed systems displayed a strong negative 'dendritic effect' (Positive dendritic effect: the activity and/or selectivity of the dendritic catalyst increases in comparison to the mononuclear catalyst) with increasing molecular size in relation to catalyst activity, stereoinduction and reusability. However, the reusability of the IL/IPA biphasic system was demonstrated when the first and second generation PPI-bound pyrphos-Rh(I) complexes showed no loss of activity or selectivity after recycling twice.

Janiak *et al.*[143] reported the hydrogenation of cyclohexene by Ru and Rh nanoparticles in [bmim][BF_4] at 90 °C and 10 bar H_2 pressure, where the nanoparticles were decomposed using corresponding carbonyl complexes and immobilized in [bmim][BF_4]. The results shows that catalyst activity increases with each recycle. In case of the Ru catalysed reaction, for the first run activity was 293 [(mol of product) (mol of metal)$^{-1}$ h^{-1}] which raised to 522 [(mol of product) (mol of metal)$^{-1}$ h^{-1}] at seventh run. Authors have ascribed the raise in activity to surface restructuring of catalyst.

9. Conclusion

With increased environmental awareness throughout the chemical industry, the use of hydrogen gas for hydrogenation reactions is especially popular as it is a clean reducing agent. Coupled with the use of ILs as safer alternatives to VOCs as reaction solvents, hydrogenation reactions can be very attractive as clean, 'green' synthetic methods. It has been demonstrated that ILs can be successfully used as reaction media for a wide variety of substrates for hydrogenation reactions, including transfer hydrogenation reactions. Recyclability of the catalyst system is a major factor in the potential use of the IL. ILs provide a stabilising medium for catalysts and facilitate their immobilisation, thereby facilitating recycling procedures. IL/catalyst systems have been shown to be easily recycled in numerous cases while retaining their activity. Biphasic hydrogenations have also demonstrated recyclability, the substrates and products residing in a separate phase to the

IL and catalyst. A particularly efficient method that has been outlined is where the substrate is dissolved in the IL phase, and the reduction products form a second phase, thus facilitating clean, simple decantation of the desired product from the IL phase. However, if the products are also soluble in the IL phase and a second organic solvent is needed for extraction from the mixture, the requirement for the harmful VOC solvent detracts from the benefit of using the IL unless the recycling ability of the IL is considerable. Although ILs are duly attracting intense attention, their benefits must be balanced with their limitations, and these elements investigated if the solvents are to replace volatile organic solvents on an industrial scale. It has been depicted throughout this review that the viscosity of the ILs poses a problem in terms of increased reaction times. Although, this is an inconvenience in a research laboratory, the increase in cost associated with the heating of the solvent to decrease viscosity or the associated increased reaction times could pose a significant problem at an industrial scale.

Although the principal ILs studied for hydrogenation reactions have been the popular [bmim][BF_4] and [bmim][PF_6], novel ILs have been synthesised and studied as reaction media for these reactions. The cost of the synthesis of novel ILs should be foremost in our minds if these solvents are to be used on a large scale. A predominant factor contributing to cost reduction, and also a possible way forward, may be the use of supported systems, which require a smaller quantity of the IL.

Recycling of the IL is also important, and has significant implications for the original outlay for the material. An IL which can only maintain its required performance for 2-5 cycles has limited use. A significantly higher number of effective recycles is one of the major goals for this research area.

The 'greenness' of ILs has been disputed due to their possible persistence and toxicity in the environment. In our opinion, it is not only the cost and performance that should be a component in designing ILs for use in hydrogenation technologies, but toxicity, bioaccumulation and biodegradability should be given equal merit to the process selection before IL development and use in large scale chemical synthesis

The replacement of a VOC with an ionic liquid does not automatically define the synthetic method as 'green'. Important factors include atom high economy, catalysis, selectivity, recycling, combined with low toxicity and biodegradable chemicals. Catalytic hydrogenations in ionic liquids have the potential to meet all these criteria.

10. References

[1] Y. Chauvin, L. Mussmann and H. Olivier, *Angew. Chem. Int. Ed.*, 1995, *15*, 2698-2700.

[2] A. Z. P. Suarez, E. L. J. Dullius, S. Einloft, F. R. De Souza and J. Dupont, *Polyhedron,* 1996, *15*, 1217-1219.

[3] C. Mehnert, E. Mozeleski and R. Cook, *Chem. Commun.*, 2002, 3010-3011.

[4] P. Wasserscheid and P. Schulz, in *The Handbook of Homogeneous Hydrogenation,* ed. J. de Vries and C. Elsevier, 2006, p.1389. Wiley-VCH

[5] P. Wasserscheid and P. Schulz, in *Ionic Liquids in Synthesis*, ed. P. Wasserscheid and T. Welton, Wiley-VCH, 2008, *vol. 2*, ch. 5, p. 390.

[6] V. I. Parvulescu and C. Hardacre, *Chem. Rev.*, 2007, *107*, 2615-2665.

[7] J. Durand, E. Teuma and M. Gomez, *C. R. Chimie*, 2007, *10*, 152-177.

[8] J. Dupont, R. F. de Souza and P. A. Z. Suarez, *Chem. Rev.*, 2002, *102*, 3667-3692.

[9] R. Sebesta, I. Kmentova and S. Toma, *Green Chem.*, 2008, *10*, 484-496.
[10] P. Dyson, G. Laurenczy, A. Ohlin, J. Vallance, T. Welton, *Chem. Comm.*, 2003, 2418-2419.
[11] *CRC Handbook of Chemistry and Physics*, ed. R. C. Weast, 53rd edn, CRC, OH, 1972-1973.
[12] *Physical Properties of Chemical Compounds*, ACS, 1955.
[13] *IUPAC Solubility Data Series*, ed. C. L. Young, Pergamon Press, Oxford, UK, 1981, *Vol. 5-6.*
[14] W. F. Linke and A. Seidell, *Solubilities of Inorganic and Metal-Organic Compounds*, ACS, Washington DC, 1958, *Vol. I*, p. 1075.
[15] J. G. Huddleston, A. E. Visser, W. M. Reichert, H. D. Willauer, G. A. Broker and R. D. Rogers, *Green Chem.*, 2001, *3*, 156-164.
[16] A. Berger, R. F. de Souza, M. R. Delgado and J. Dupont, *Tetrahedron Asymmetry*, 2001, *12*, 1825-1828.
[17] A. J. McLean, M. J. Muldoon, C. M. Gordon and I.R. Dunkin, *Chem. Comm.*, 2002, 1880-1881.
[18] A. Noda, K. Hayamizu and M. Watanabe, *J. Phys. Chem. B*, 2001, *105*, 4603-4610.
[19] D. R. MacFarlane, P. Meakin, J. Sun, N. Amini and M. Forsyth, *J. Phys. Chem. B*, 1999, *103*, 4164-4170.
[20] P. Bonhote, A. P. Dias, N. Papageorgiou, K. Kalyanasumdaram and M. Gratzel, *Inorg. Chem.*, 1996, *35*, 1168-1178.
[21] Merck KGaA.
[22] L. C. Branco, J. N. Rosa, J. J. Moura Ramos and C. A. M. Afonso, *Chem. Eur. J.*, 2002, *8*, 3671-3677.
[23] Y. Sun, R. N. Landau, J. Wang, C. LeBlond and D.G. Blackmond, *J. Am. Chem. Soc.*, 1996, *118*, 1348-1353.
[24] U. Kernchen, B. Etzold, W. Korth and A. Jess, *Chem. Eng. Technol.*, 2007, *30*, 985-994.
[25] A. Umpierre, G. Machado, G. Fecher, J. Morais and J. Dupont, *Adv. Synth. Catal.*, 2005, *347*, 1404-1412.
[26] S. Kishida, S. Teranishi, *J. Catal.*, *12*, 1968, 90-96.
[27] W. Xiong, Q. Lin, H. Ma, H. Zheng, H. Chen and X. Li, *Tetrahedron: Asymmetry*, 2005, *16*, 1959-1962.
[28] M. Khodadadi-Moghaddam, A. Habibi-Yangjeh and M. Reza Gholami, *Appl. Catal., A*, 2008, *341*, 58-64.
[29] G. Fonseca, J. Domingos, F. Nome and J. Dupont, *J. Mol. Catal. A*, 2006, *248*, 10-16.
[30] M. Ruta, I. Yuranov, P. Dyson, G. Laurenczy and L. Kiwi-Minsker, *J. Catal.*, 2007, *247*, 269-276.
[31] C. Zhao, H. Wang, N. Yan, C. Xiao, X. Mu, P. Dyson and Y. Kou, *J. Catal.*, 2007, *250*, 33-40.
[32] T. Geldbach and P. Dyson, *J. Organomet. Chem.*, 2005, *690*, 3552-3557.
[33] E. Silveira, A. Umpierre, L. Rossi, G. Machado, J. Morais, G. Soares, I. Baumvol, S. Teixeira, P. Fichtner and J. Dupont, *Chem. Eur. J.*, 2004, *10*, 3734-3740.
[34] G. Fonseca, J. Scholten and J. Dupont, *Synlett*, 2004, *9*, 1525-1528.
[35] G. Fonseca, A. Umpierre, P. Fichtner, S. Teixeira and J. Dupont, *Chem. Eur. J.*, 2003, *9*, 3263-3269.
[36] A. Lenourry, J. M. Gardiner and G. Stephens, *Biotechnology Letters*, 2005, *27*, 161-165.

[37] A. Wolfson, I. Vankelecom and P. Jacobs, *Tetrahedron Lett.,* 2005, *46,* 2513-2516.

[38] L. K. Fow, S. Jaenicke, T. Muller and C. Sievers, *J. Mol. Catal. A: Chem,* 2008, *279,* 239-247.

[39] Y. Kume, K. Qiao, D. Tomida and C. Yokoyama, *Catal. Commun.,* 2008, *9,* 369-375.

[40] C. W. Scheeren, J. B. Domingos, G. Machado and J. Dupont, *J. Phys. Chem. C,* 2008, *112,* 16463-16469.

[41] T. Floris, P. Kluson, M. J. Muldoon and H. Pelantova, *Catal. Lett.,* 2010, *134,* 279-287.

[42] D. Xu, Z. Hu, W. Li, S. Luo and Z. Xu, *J.Mol. Catal. A,* 2005, *235,* 137-142.

[43] K. Anderson, P. Goodrich, C. Hardacre and D. Rooney, *Green Chem.,* 2003, *5,* 448-453.

[44] J. Arras, M. Steffan, Y. Shayeghi and P. Claus, *Chem. Commun.,* 2008, 4058-4060.

[45] J. Arras, M. Steffan, Y. Shayeghi, D. Ruppert and P. Claus, *Green Chem.,* 2009, *11,* 716-723.

[46] J. Arras, D. Ruppert and P. Claus, *Appl. Catal. A., 371,* 2009, 73-77.

[47] A. Jasik, R. Wojcieszak, S. Monteverdi, M. Ziolek and M. M. Bettahar, *J. Mol. Catal. A: Chem.,* 2005, *242,* 81-90.

[48] N. H. H. Abu Bakar, M. M. Bettahar, M. Abu Bakar, S. Monteverdi, J. Ismail and M. Alnot, *J. Catal.,* 2009, *265,* 63-71.

[49] S. M. Choi, J. S. Yoon, H. J. Kim, S. H. Nam, M. H. Seo and W. B. Kim, *Appl. Catul., A,* 2009, *359,* 136-143.

[50] R. Deshmukh, J. Lee, U. Shin, J. Lee and C. Song, *Angew. Chem. Int. Ed.,* 2008, *47,* 8615-8617.

[51] A. V. Talyzin, Y. O. Tsybin, A. A. Peera, T. M. Schaub, A. G. Marshall, B. Sundqvist, P. Mauron, A. Zu1ttel and W. E. Billups, *J. Phys. Chem. B Lett.,* 2005, *209,* 5403-5405.

[52] M. Steffan, M. Lucas, A. Brandner, M. Wollny, N. Oldenburg and P. Claus, *Chem. Eng. Technol.,* 2007, *30,* 481-486.

[53] A. Sharma, C. Julcour, A. A. Kelkar, R. M. Deshpande and H. Delmas, *Ind. Eng. Chem. Res.,* 2009, *48,* 4075-4082.

[54] N. Ab Manan, C. Hardacre, J. Jacquemin, D. W. Rooney and T. G. A. Youngs, *J. Chem. Eng. Data,* 2009, *54,* 2005-2022.

[55] J. F. Brennecke, A. Lopez-Castillo and B. R. Mellein, ed. M. Koel, *Ionic Liquids in Chemical Analysis,* 2009, 229, CRC Press.

[56] C. Hardacre, E. Mullan, D. Rooney, J. Thompson and G. Yablonsky, *Chem. Eng. Sci.,* 2006, *61,* 6995-7006.

[57] N. Sieffert, G. Wipff, J. Phys. Chem. B., 111, 2007, 4951-4962.

[58] M. Khodadadi-Moghaddam, Z. Habibi-Yangjeh and M. R. Gholami, *J. Mol. Catal. A: Chem.,* 2009, *306,* 11-16.

[59] P. Dyson, D. Ellis and T. Welton, *Can. J. Chem.,* 2001, *79,* 705-708.

[60] M. A. Esteruelas, I. Gonzalez, J. Herrero and L. A. Oro, *J. Organomet. Chem.,* 1998, *551,* 49-53.

[61] F. Zhao, Y. Ikushima and M. Arai, *Green Chem.,* 2003, *5,* 656-658.

[62] A. Scurto and W. Leitner, *Chem. Commun.,* 2006, 3681-3683.

[63] T. Suarez, B. Fontal, M. Reyes, F. Bellandi, R. Contreras, J. Ortega, G. Leon, P. Cancines and B. Castillo, *React. Kinet. Catal. Lett.,* 2004, *82,* 325-331.

[64] T. Suarez, B. Fontal, M. Reyes, F. Bellandi, R. R. Contreras, A. Bahsas, G. Leon, P. Cancines and B. Castillo, *React. Kinet. Catal. Lett.,* 2004, *82 (2),* 317-324.

[65] Z. Zhang, Y. Xie, W. Li, S. Hu, J. Song, T. Jiang and B. Han, *Angew. Chem. Int. Ed.*, 2008, *47*, 1127-1129.
[66] Y. Gao, J. K. Kuncheria, H. A. Jenkins, R. J. Puddephatt and G. P. A. Yap, *J. Chem. Soc., Dalton Trans.*, 2000, 3212-3217.
[67] P. G. Jessop, Y. Hsiao, T. Ikariya and R. Noyori, *J. Am. Chem. Soc.*, 1996, *118*, 344-355.
[68] Z. Zhang, S. Hu, J. Song, W. Li, G. Yang and B. Han, *ChemSusChem*, 2009, *2*, 234-238.
[69] K. Obert, D. Roth, M. Ehrig, A. Schonweiz, D. Assenbaum, H. Lange and P. Wasserscheid, *Appl. Catal., A*, 2009, *356*, 43-51.
[70] M. F. Zuckerman, *Patent application US 4739120*, 1988, Dow Chemical Co.
[71] S. Bouquillon, T. Courant, D. Dean, N. Gathergood, S. Morrissey, B. Pegot, P. J. Scammells and R. Singer, *Aust. J. Chem.*, 2007, *60*, 843-847.
[72] N. Gathergood and P. J. Scammells, *Aust. J. Chem.*, 2002, *55*, 557-560.
[73] (a) N. Gathergood, M. T. Garcia and P. J. Scammells, *Green Chem.*, 2004, *6*, 166-175; (b) M. T. Garcia, N. Gathergood and P. J. Scammells, *Green Chem.*, 2004, *7*, 9-14; (c) N. Gathergood, P. J. Scammells and M. T. Garcia, *Green Chem.*, 2006, *8*, 156-160.
[74] S. Morrissey, I. Beadham and N. Gathergood, *Green Chem.*, 2009, *11*, 466-474.
[75] P. Gallezot and D. Richard, *Catal. Rev.*, 1998, *40*, 81-126.
[76] H .U. Blaser, C. Malan, B. Pugin, F. Spindler, H. Steiner and M. Studer, *Adv. Synth. Catal.*, 2003, *345*, 103-151.
[77] R. Noyori, M. Kitamura and T. Ohkuma, *PNAS*, 2004, *101*, 5356-5362.
[78] (a) F. Bautista, V. Caballero, J. Campelo, D. Luna, J. Marinas, A. Romero, I. Serrano and A. Llobet, *Top. Catal.*, 2006, *40*, 193-205. (b) A. Chan, S. Laneman and C. Day, *Inorg. Chim. Acta*, 1995, *228*, 159-163. (c) K. Wan and M. Davis, *Tetrahedron: Asymmetry*, 1993, *4*, 2461-2468.
[79] K. Boyle, B. Lipsky and C. Kalbery, *Tetrahedron Lett.*, 2006, *47*, 1311-1313.
[80] M. J. Burk, J. E. Feaster, W. A. Nugent and R. L. Harlow, *J. Am. Chem. Soc.*, 1993, *115*, 10125-10131.
[81] A. R. McDonald, C. Muller, D. Vogt, G. P. M. van Klink and G. van Koten, *Green Chem.*, 2008, *10*, 424-432.
[82] M. Schmitkamp, D. Chen, W. Leitner, J. Klankermayer and G. Francio, *Chem. Commun.*, 2007, 4012-4014.
[83] J. She, L. Ye, J. Zhu and Y. Yuan, *Catal. Lett.*, 2007, *116*, 70-75.
[84] T. Frater, L. Gubicza, A. Szollosy and J. Bakos, *Inorg. Chim. Acta*, 2006, *359*, 2756-2759.
[85] C. R. Landis, J. Halpern, *J. Am. Chem. Soc.*, 1987, *109*, 1746-1754.
[86] A. Shariati, R. Sheldon, G. J. Witkamp and C. Peters, *Green Chem.*, 2008, *10*, 342-346.
[87] H. Wong, Y. H. See-Toh, F. C. Ferreira, R. Crook and A. G. Livingston, *Chem. Commun.*, 2006, 2063-2065.
[88] E. V. Starodubtseva, M. G. Vinogradov, V. A. Pavlov, L. S. Gorshkova and V. A. Ferapontov, *Russ. Chem. Bull., Int. Ed.*, 2004, *53*, 2172-2174.
[89] J. Wang, R. Qin, H. Fu, J. Chen, J. Feng, H. Chen and X. Li, *Tetrahderon: Asymmetry*, 2007, 18, 847-850.
[90] K. H. Lam, L. Xu, L. Feng, J. Ruan, Q. Fan and A. Chan, *Can. J. Chem.*, 2005, *83*, 903-908.
[91] T. Floris, P. Kluson, L. Bartek and H. Pelantova, *Appl. Catal., A*, 2009, *366*, 160-165.

[92] (a) H. U. Blaser, H. P. Jallet and F. Spindler, *J. Mol. Catal. A*, 1996, *107*, 85-94. (b) L. Bartek, P. Kluson and L. Cerveny, *Chem. Listy*, 2005, *99*, 157-167. (c) M. Jahjah, M. Alame, P. Pellet-Rostaing and M. Lemaire, *Tetrahedron: Asymmetry*, 2007, *19*, 2305-2312. (d) A. Wolfson, I. F. J. Vankelecom, S. Geresh and P. A. Jacobs, *J. Mol. Catal. A*, 2004, *217*, 21-26.
[93] E. Oechsner, K. Schneiders, K. Junge, M. Beller and P. Wasserscheid, *Appl. Catal., A*, 2009, *364*, 8-14.
[94] H. Zhou, Z. Li, Z. Wang, T. Wang, L. Xu, Y. He, Q.-H. Fan, J. Pan, L. Gu and A. Chan, *Angew. Chem. Int. Ed.*, 2008, *47*, 8464-8467.
[95] D. C. Lee, J. H. Kim, W. J. Kim, J. H. Kang and S. H. Moon, *Appl. Catal., A*, 2003, *244*, 83-91.
[96] M. Schmal, D. A. G. Aranda, R. R. Soares, F. B. Noronha and A. Frydman, *Catal. Today*, 2000, *57*, 169.
[97] J. Huang, T. Jiang, B. Han, H. Gao, Y. Chang, G. Zhao and W. Wu, *Chem. Commun.*, 2003, 1654-1655.
[98] Y. Hu, Y, Yu, Z. Hou, H. Li, X. Zhao and B. Feng, *Adv. Synth. Catal.*, 2008, *350*, 2077-2085.
[99] Y. Hu, H. Yang, Y. Zhang, Z. Hou, X. Wang, Y. Qiao, H. Li, B. Feng and Q. Huang, *Catal. Commun.*, 2009, *10*, 1903-1907.
[100] Y. S. Chun, J. Y. Shin, C. E. Song and S. Lee, *Chem. Commun.*, 2008, 942-944.
[101] J. Huang, T. Jiang, H. Gao, B. Han, Z. Liu, W. Wu, Y. Chang and G. Zhao, *Angew. Chem. Int. Ed.*, 2004, *43*, 1397-1399.
[102] R. Tao, S. Miao, Z. Liu, Y. Xie, B. Han, G. An and K. Ding, *Green Chem.*, 2009, *11*, 96-101.
[103] P. Dash, N. Dehm and R. Scott, *J. Mol. Catal. A: Chem*, 2008, *286*, 114-119.
[104] J. L. Anthony, E. J. Maginn, J. F. Brennecke, *J. Phys. Chem. B.*, 2002, *106*, 7315-7320.
[105] J. Dupont, G. Fonseca, A. Umpierre, P. Fichtner and S. Teixeira, *J. Am. Chem. Soc.*, 2002, *124*, 4228-4229.
[106] B. Leger, A. Denicourt-Nowicki, A. Roucoux and H. Olivier-Bourbigou, *Adv. Synth. Catal.*, 2008, *350*, 153-159.
[107] B. Leger, A. Denicourt-Nowicki, H. Olivier-Bourbigou and A. Roucoux, *ChemSusChem.*, 2008, *1*, 984-987.
[108] V. Cimpeanu, M. Kocevar, V. I. Parvulescu and W. Leitner, *Angew. Chem. Int. Ed.*, 2009, *48*, 1085-1088.
[109] M. Solinas, A. Pfaltz, P. G. Cozzi and W. Leitner, *J. Am. Chem. Soc.*, 2004, *126*, 16142-16147.
[110] C. Xiao, H. Wang, X. Mu and Y. Kou, *J. Catal.*, 2007, *250*, 25-32.
[111] C. Scheeren, G. Machado, J. Dupont, P. Fichtner and S. Texeira, *Inorg. Chem.*, 2003, *42*, 4738-4742.
[112] R. Abu-Reziq, D. Wang, M. Post and H. Alper, *Adv. Synth. Catal.*, 2007, *349*, 2145-2150.
[113] L. M. Rossi and G. Machado, *J. Mol. Catal. A: Chem.*, 2009, *298*, 69-73.
[114] M. Prechtl, M. Scariot, J. Scholten, G. Machado, S. Teixeira and J. Dupont, *Inorg. Chem.*, 2008, *47*, 8995-9001.
[115] M. H. G. Prechtl, J. D. Scholten and J. Dupont, *J. Mol. Catal. A: Chem.*, 2009, *313*, 74-78.

[116] P. Migowski, G. Machado, S. Texeira, M. Alves, J. Morais, A. Traverse and J. Dupont, *Phys. Chem.*, 2007, *9*, 4814-4821.
[117] E. Redel, J. Kramer, R. Thomann and C. Janiak, *J. Organomet. Chem.*, 2009, *694*, 1069-1075.
[118] J. Mikkola, P. Virtanen, H. Karhu, T. Salmi and D. Murzin, *Green Chem.*, 2006, *8*, 197-205.
[119] (a) J. Jacquemin, M.F. Costa Gomes, P. Husson and V. Majer, *J. Chem. Thermodyn.*, 2006, *38*, 490-502. (b) J. Jacquemin, P. Husson, V. Majer and M. F. Costa Gomes, *Fluid Phase Equilib.*, 2006, *240*, 87-95. (c) J. Jacquemin, P. Husson, V. Majer and M. F. Costa Gomes, *J. Solution Chem.*, 2007, *36*, 967-979. (d) J. Jacquemin, P. Husson, V. Majer, A. A. H. Padua and M. F. Costa Gomes, *Green Chem.*, 2008, *10*, 944-950.
[120] J. Hao, C. Xi, H. Cheng, R. Liu, S. Cai, M. Arai and F. Zhao, *Ind. Eng. Chem. Res.*, 2008, *47*, 6796-6800.
[121] P. Virtanen, H. Karhu, G. Toth, K. Kordas and J. Mikkola, *Jornal of catalysis*, 2009, *263*, 209-219.
[122] P. Virtanen, H. Karhu, K. Kordas and J. Mikkola, *Chem. Eng. Sci.*, 2007, *62*, 3660-3671.
[123] P. Virtanen, T. Salmi and J. Mikkola, *Ind. Eng. Chem. Res.*, 2009, *48*, 10335-10342.
[124] P. Virtanen, J. Mikkola and T. Salmi, *Ind. Eng. Chem. Res.*, 2007, *46*, 9022-9031.
[125] L. Lou, X. Peng, K. Yu and S. Liu, *Catal. Commun.*, 2008, *9*, 1891-1893.
[126] M. Gelesky, S. Chiaro, F. Pavan, J. dos Santos and J. Dupont, *Dalton Trans.*, 2007, 5549-5553.
[127] W. Chen, Y. Zhang, L. Zhu, J. Lan, R. Xie and J. You, *J. Am. Chem. Soc.*, 2007, *129*, 13879-13886.
[128] R. Knapp, A. Jentys and J. Lercher. *Green Chem.* 2009, *11*, 656-661.
[129] H. Berthold, T. Schotten and H. Honig, *Synthesis*, 2002, *11*, 1607-1610.
[130] Z. Baan, A. Potor, A. Cwik, Z. Hell, G. Keglevich, Z. Finta and I. Hermecz, *Synth. Commun.*, 2008, *38*, 1601-1609.
[131] Z. Baan, Z. Finta, G. Keglevich and I. Hermecz, *Tetrahedron Lett.*, 2005, *46*, 6203-6204.
[132] J. Joerger, J. Paris and M. Vaultier, *ARKIVOC*, 2006, 152-160.
[133] C. Comyns, N. Karodia, S. Zeler and J. Anderson, *Catal. Lett.*, 2000, *67*, 113-115.
[134] M. Lakshmi Kantam, R. Sudarshan Reddy, U. Pal, B. Sreedhar and S. Bhargava, *Adv. Synth. Catal.*, 2008, *350*, 2231-2235.
[135] Z. Baan, Z. Finta, G. Keglevich and I. Hermecz, *Green Chem.*, 2009, *11*, 1937-1940.
[136] L. Xu, G. Ou and Y. Yuan, *J. Organomet. Chem.*, 2008, *693*, 3000-3006.
[137] M. Ruta, G. Laurenczy, P. J. Dyson and L. Kiwi-Minsker, *J. Phys. Chem.*, 2008, *112*, 17814-17819.
[138] (a) A. Borodzinski and A. Cybulski, *Appl. Catal. Aq.*, 2000, *198*, 51-66. (b) A. Sarkany, *React. Kinet. Catal. Lett.*, 2001, *74*, 299-307. (c) A. Sarkany, L. Guczi and A. Weiss, *Appl. Catal.*, 1984, *10*, 369-388.
[139] K. Hamza, H. Schumann and J. Blum, *Eur. J. Org. Chem.*, 2009, 1502.
[140] M. A. Gelesky, C. W. Scheeren, L. Foppa, F. A. Pavan, S. L. P. Dias and J. Dupont, *Biomacromolecules*, 2009, *10*, 1888-1893.
[141] S. J. Craythorne, K. Anderson, F. Lorenzini, C. McCausland, E. F. Smith, P. Licence, A. C. Marr and P. C. Marr, *Chem. Eur. J.*, 2009, *15*, 7094-7100.

[142] J. K. Kassube and L. H. Gade, *Adv. Synth. Catal.*, 2009, *351*, 739-749.
[143] C. Vollmer, E. Redel, K. Abu-Shandi, R. Thomann, H. Manyar, C. Hardacre, and C. Janiak, *Chem. Eur. J.*, 2010, *16*, 3849-3858.

16

Palladium Nanoscale Catalysts in Ionic Liquids: Coupling and Hydrogenation Reactions

Martin H. G. Prechtl,[1,2]* Jackson D. Scholten[3] and Jairton Dupont[3]
[1]*Universität zu Köln, Köln,*
[2]*Humboldt Universität zu Berlin, Berlin,*
[3]*Universidade Federal do Rio Grande do Sul, Porto Alegre, RS,*
[1,2]*Germany*
[3]*Brazil*

1. Introduction

Palladium nanoparticles (Pd-NPs) stabilised by ionic liquids (ILs) revealed a promising potential to act as recyclable catalysts for plethora reaction types. These include typical homogeneous as well as heterogeneous catalytic reactions such as hydrogenation of multiple bonds, (Dupont et al., 2002; Huang et al., 2003; Scheeren et al., 2003; Kim et al., 2004; Silveira et al., 2004; Umpierre et al., 2005; Migowski & Dupont, 2007; Rossi et al., 2007; Scholten et al., 2007; Hu et al., 2008; Prechtl et al., 2008; Gelesky et al., 2009; Hu et al., 2009; Prechtl et al., 2009; Redel et al., 2009; Rossi & Machado, 2009; Dupont & Scholten, 2010) and carbon-carbon cross-coupling reactions.(Calo et al., 2003; Calo et al., 2003; Calo et al., 2004; Calo et al., 2004; Calo et al., 2005; Calo et al., 2005; Cassol et al., 2005; Calo et al., 2006; Chiappe et al., 2006; Dubbaka et al., 2006; Phan et al., 2006; Calo et al., 2007; Fei et al., 2007; Dieguez et al., 2008; Calo et al., 2009; Calo et al., 2009; Cui et al., 2010) It is well accepted that Pd-NPs serve as reservoir for active molecular Pd species in carbon-carbon cross coupling reactions (Scheme 1), for example. Moreover, Pd-NPs have been employed as efficient catalyst for typical heterogeneous reactions (Scheme 2).(Phan et al., 2006; Migowski & Dupont, 2007; Gu & Li, 2009) In order to avoid the formation of bulk metal, ILs can be used as suitable stabilisers for the synthesis of mono-dispersed NPs.(Prechtl et al., 2008; Prechtl et al., 2009) ILs form also a protective liquid-film on the sensitive and highly active metal surface, preventing the metal surface from oxidation.(Migowski & Dupont, 2007) One additional property of Pd-NPs/ILs systems is the feature for multiphase catalysis that the NPs are immobilised in the dense IL-phase ("stationary phase"), and are, therefore easily recyclable by simple phase separation of the organic "mobile phase", containing the substrates and product.(Dupont et al., 2002) In many cases, the well-immobilised NPs in ILs can be used as catalyst-phase several times, showing the potential to recyclability of this system. General limitations are reached in the following cases: (I) palladium leaching into the organic layer, (Cassol et al., 2005; Consorti et al., 2005) (II) retention of organic molecules in the IL-phase,(Cassol et al., 2007) and (III) a delayed mass transfer between the organic and the IL-layer due to the relatively high viscosity of the IL.(Dupont & Suarez, 2006)

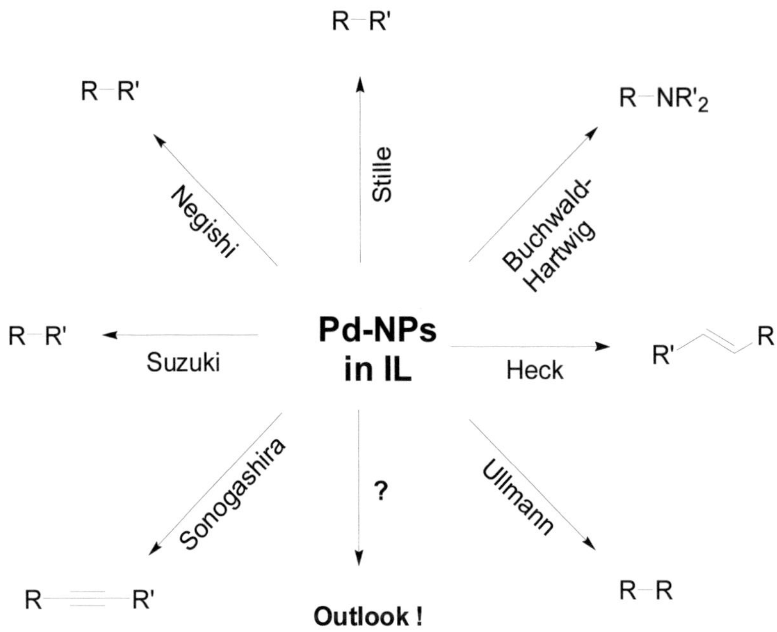

Scheme 1. Exemplary coupling reactions catalysed by metal Pd-NPs in ILs.

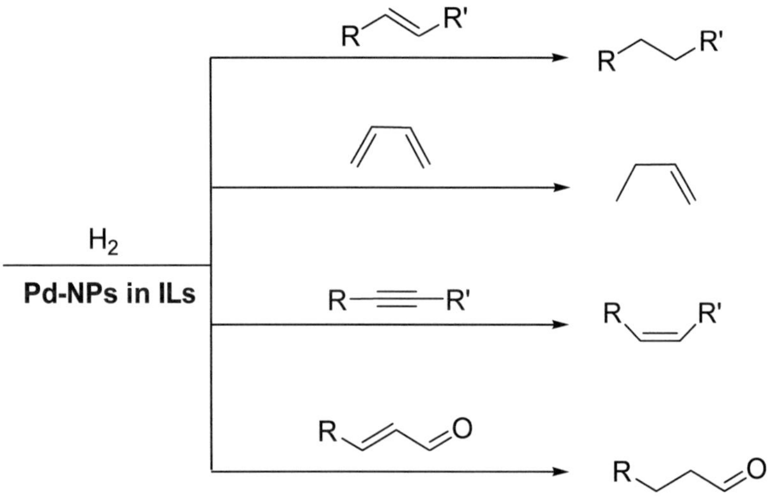

Scheme 2. Hydrogenation reactions catalysed by metal Pd-NPs in ILs.

Pd-based carbon-carbon cross-coupling reactions were discovered more than three decades ago, and remain under the most investigated transition-metal catalysed reactions.(Beletskaya & Cheprakov, 2000; Moreno-Manas & Pleixats, 2003; Phan et al., 2006; Yin & Liebscher, 2007; Liu et al., 2009) These reactions use Pd complexes with strong ligands (e. g. PR_3, NHC), palladacycles, PdX_2 (e. g. X= halide, acetate) or even ligand-free approaches, in which Pd(0) species are catalytically active.(Reetz et al., 1998; de Vries et al., 2002; Bedford, 2003; de Vries et al., 2003; de Vries et al., 2003; Reetz & de Vries, 2004; Dupont et al., 2005) The Heck reaction for instance, with aryl iodides or bromides is catalysed by a plethora of Pd(II) or Pd(0) sources.(Cassol et al., 2005; Migowski & Dupont, 2007) This indicates, at least for ligand-free Pd sources, that soluble Pd-NPs are involved as a reservoir of active species.(Reetz & Westermann, 2000; Rocaboy & Gladysz, 2003; Tromp et al., 2003) The first article of a zero-valent Pd complex suitable for the formation of Pd-NPs, by Takahashi and co-workers was published in 1970, used $Pd(dba)_2$ (dba = dibenzylideneacetone) under thermal decomposition conditions forming metallic palladium and dba in solution.(Takahashi et al., 1970) More studies of Pd-NP synthesis and its applications followed in the 1980s and 1990s by, for example Bönnemann, Reetz and their respective co-workers.(Bonnemann et al., 1990; Bonnemann et al., 1991; Reetz & Helbig, 1994; Reetz & Quaiser, 1995; Reetz et al., 1996; Reetz & Lohmer, 1996) In the last decade, metal NP synthesis in ILs had their breakthrough and numerous detailed studies about Pd-NPs in ILs are available in the literature.

Synthetic approaches using controlled thermal decomposition for the generation of Pd-NPs in ILs use for example: $Pd(dba)_2$,(Takahashi et al., 1970) $Pd(OAc)_2$ or palladium carbene complexes (Scheme 3).(Reetz & Westermann, 2000; Xu et al., 2000; Deshmukh et al., 2001; Calo et al., 2003) Furthermore, palladium salts such as $PdCl_2$, Na_2PdCl_4 can be reduced with metal hydrides or simply dihydrogen. Pd-NPs are also formed starting from palladacycles by reaction with dienes (Scheme 3).(Bonnemann et al., 1990; Cassol et al., 2005; Umpierre et al., 2005)

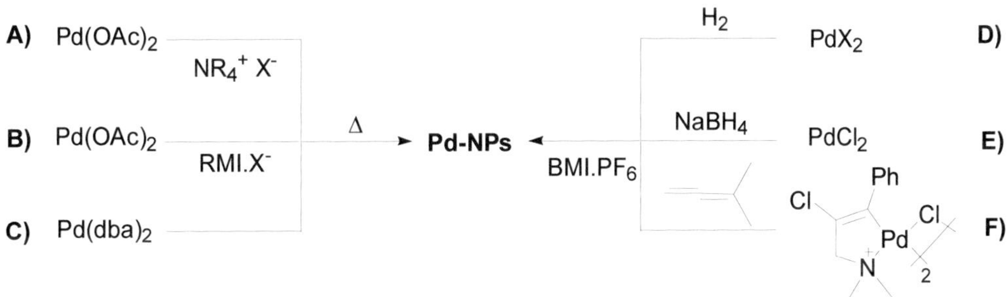

Scheme 3. Metal Pd-NPs formation in ILs by thermal/ultrasonic treatment of Pd salts and Pd carbene complexes where the imidazolium salt (RMI.X) acts as NHC-carbene source (A-C; X = halide or BF_4). Reduction with dihydrogen or hydrides (D-E; X = OAc, halide) or reductive elimination of a palladacycle by reaction with a diene (F). Adapted from references (Takahashi et al., 1970; Reetz & Westermann, 2000; Xu et al., 2000; Deshmukh et al., 2001; Calo et al., 2003; Cassol et al., 2005; Umpierre et al., 2005).

The particle size of Pd-NPs is often strongly related to the type of precursor, and small-sized particles are easily obtained starting from $Pd(OAc)_2$. Besides the Pd-precursor, the state of agglomeration/dispersion also depends on the coordination properties of the IL media and the concentration/solubility of the Pd-precursor in the IL.(Dupont & Scholten, 2010) Especially low concentration and high solubility of the Pd-precursor is quite helpful in obtaining a high dispersion and low agglomeration. It is important to note, that the lifetime of Pd-NPs as catalysts depend, on their stability, which is often related to the preparative protocol used. Also Pd-NPs tend to form large aggregates with smaller surface which often show lower activity. To obtain prolonged catalyst lifetime of the Pd-NPs, these particles may be stabilised by the addition of polymers, which give access for tuning the particle size and their topology.(Yang et al., 2008) A more detailed discussion on the synthesis of metal NPs in ILs can be found in a recently published critical review.(Dupont & Scholten, 2010)

2. Nanoscale Pd-Catalysts in ILs for C-C coupling reactions

2.1 Mizoroki-Heck reaction

The Mizoroki-Heck coupling reaction is one of the best methods in modern organic chemistry.(Beletskaya & Cheprakov, 2000; Moreno-Manas & Pleixats, 2003; Phan et al., 2006; Yin & Liebscher, 2007; Liu et al., 2009) The coupling involves the reaction of an unsaturated halide with olefins catalysed by Pd precursors in organic solvents in presence of a base. Pointing out, that often the active species is not the palladium complex but molecular Pd species derived from Pd-NPs stabilised by the IL serving as reaction medium.(Reetz & Westermann, 2000; Astruc, 2007) These Pd-NPs are resulting from the reduction of the Pd(II) species to Pd(0) in the presence of bases employed in the reaction. In the last two decades, classical solvents have been substituted by ILs in chemical reactions.(Dupont et al., 2002) In fact, ILs are more environmentally benign, which is due to their ability to act as stationary phase for catalysts in recyclable multiphase catalysis. In such multiphase systems, ILs are suitable together with apolar solvents as well as with polar solvents, depending on the polarity of the IL. Moreover, certain ILs such as those with imidazolium cations, show selectivity for a specific product as the ILs are feasible to stabilise ionic transition states due to their inherent physico-chemical properties.(Hardacre et al., 2003; Dupont, 2004; Gozzo et al., 2004; Tsuzuki et al., 2005) One of the first results reported on the role of ILs in Heck reactions was published by Deshmukh and co-workers in 2001. (Deshmukh et al., 2001) For instance, ILs based on the cation [1,3-di-*n*-butylimidazolium] and bearing bromide (BBI.Br) and tetrafluoroborate (BBI.BF_4) as anions promote a significant improve in the rate of Heck reactions. It was observed that under ultrasonic irradiation and in the presence of base, the C-C couplings of several substituted iodobenzenes and alkenes/alkynes at 30 °C were performed successfully affording solely the *trans*-product in high yields (Scheme 4). The authors also proved that a Pd *bis*-carbene took part in the transformation and was reduced to metal Pd-NPs in the process. Moreover, the molecular Pd as the active species is most likely coming from Pd-NPs. Once no reaction was observed for classical organic solvents under the same conditions, the advantage of ILs toward the stabilisation of intermediates can be evidenced in this case.

The formation of similar palladium carbene complexes derived from imidazolium ILs were also reported elsewhere, as well as the deprotonation of imidazolium cations during the catalysis.(Xu et al., 2000; Dupont & Spencer, 2004; Lebel et al., 2004; Bernardi et al., 2009)

In the same context, metal Pd-NPs (1.5–6 nm) dispersed in a tetraalkylammonium salt (tetrabutylammonium bromide, TBAB) could be used as catalysts for the Heck reaction of bromoarenes with 1,1-disubstituted olefins in the presence of tetrabutylammonium acetate (TBAA) as base at 120 °C (entry 1, Table 1).(Calo et al., 2003) In most cases the main products were terminal olefins, which suggest that Pd-hydride species is immediately neutralised by the base avoiding the isomerisation of the olefin. Notably, bromoarenes may couple with less reactive 1,2-disubstituted alkenes, like cinnamates, in presence of Pd-NPs in TBAB at 130 °C.(Calo et al., 2003) Here, $Pd(OAc)_2$ and a Pd *bis*-benzothiazole carbene compound were used as a source of Pd-NPs. The same research group reported recently that aryl chlorides undergo coupling reaction with deactivated alkenes using Pd-NPs in TBAB and TBAA (entry 2, Table 1).(Calo et al., 2009) It has to be pointed out, that it is generally accepted that the true catalyst of the reaction is most likely molecular Pd species detached from the NPs surface.

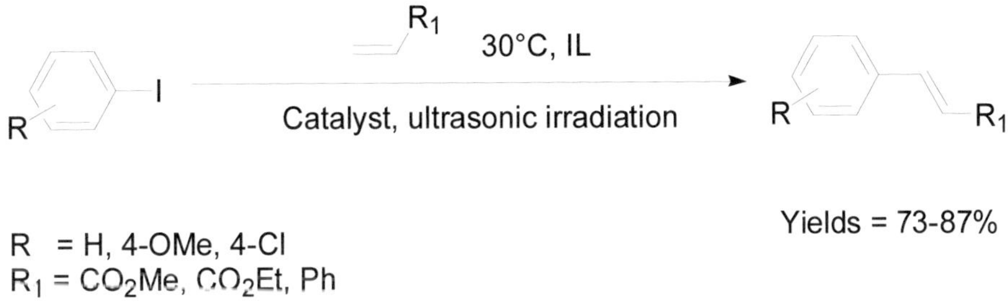

Scheme 4. Selected examples of Heck reaction of substituted iodobenzenes and alkenes in imidazolium-based ILs at 30 °C under ultrasonic conditions. (Deshmukh et al., 2001)

Mechanistic details concerning the role of Pd-NPs in C-C cross-coupling reactions, were investigated by Dupont and co-workers.(Cassol et al., 2005; Consorti et al., 2005) The palladacycle (see Scheme 3) was rapidly transformed into Pd-NPs by reaction with dimethylallene even at room-temperature. These Pd-NPs were suspended in BMI.PF_6 (BMI: 1-*n*-butyl-3-methylimidazolium; PF_6: hexafluorophosphate). Following the analysis of the Pd-powder in IL was conducted with TEM and EDS techniques. The micrographs depict the presence of Pd-NPs and the XRD results of the isolated Pd-NPs confirmed the presence of metallic Pd. The evaluation of the Pd-NPs/IL-system for catalysis was performed in particular with the Heck reaction as benchmark model using with aryl halides and *n*-butyl acrylate (substrate:Pd ratio = 1000:1) at various temperatures and bases (entry 5, Table 1). The addition of $NEt(iPr)_2$ to the dark suspension resulted then in a yellow solution. Contrary with other bases the colour of the suspension remained unchanged. Highest conversions (92-100%) are accessible with aryl iodides and bromides using $NEt(iPr)_2$ as base and lower conversion (<30%) to other bases between 80–130 °C (14 h). The analysis of the organic fraction, after catalysis, by ICP-MS showed leaching of considerable quantities of palladium from the IL into the organic phase. The TEM and ICP-MS data indicate that Pd-NPs in IL simply serve as reservoir for molecular active Pd species. Pointing out, that Pd isolated from the organic layer showed no activity in the Heck reaction also after prolonged period of time in a recycling experiment (20 h). Also attempts to locate Pd-NPs in the organic phase by means of TEM failed. The authors suggested that the reaction pathway starts with oxidative addition of the aryl halide onto the metal surface. The oxidised molecular Pd species is immediately cleaved from the metal surface, which then enters the typical catalytic cycle

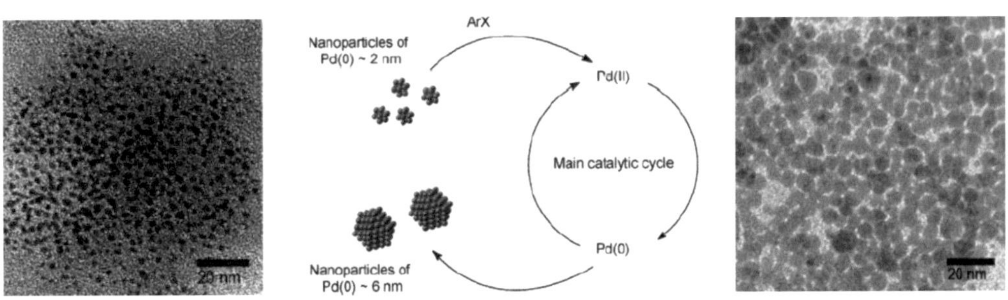

Fig. 1. TEM micrographs of the Pd-NPs in BMI.PF_6: before catalysis (1.7 ± 0.3 nm), left; after catalysis (6.1 ± 0.7 nm), right. The proposed catalytic cycle for the Heck reaction promoted by Pd-NPs is indicated in the middle. Reprinted with permission from reference (Cassol et al., 2005) (Copyright American Chemical Society).

(Figure 1). The molecular Pd species may remain in the catalytic cycle or Pd(0) may agglomerate and precipitate again as Pd-NPs, which stays in agreement with previous reports by de Vries and Reetz for ligand-free Heck reactions.(Reetz & Westermann, 2000; Reetz & de Vries, 2004)

2.2 Suzuki-Miyaura reaction

The Suzuki reaction represents another important example for C-C bond formation. Today, rather few works have been published using Pd-NPs in ILs as catalyst in this reaction.(Pathak et al., 2000; Ramarao et al., 2002; Gopidas et al., 2003; Narayanan & El-Sayed, 2003; Pittelkow et al., 2003; Kim et al., 2004; Liu et al., 2004; Calo et al., 2005; Corma et al., 2005; Fernandez et al., 2007; Durand et al., 2008) For example, Calo and Nacci reported on Pd-NPs in tetraalkylammonium ILs, starting from $Pd(OAc)_2$ in the presence of TBAA at 90 °C, which were used as precursor for the coupling of aryl halides (Scheme 5; entry 6, Table 1).(Calo et al., 2005)

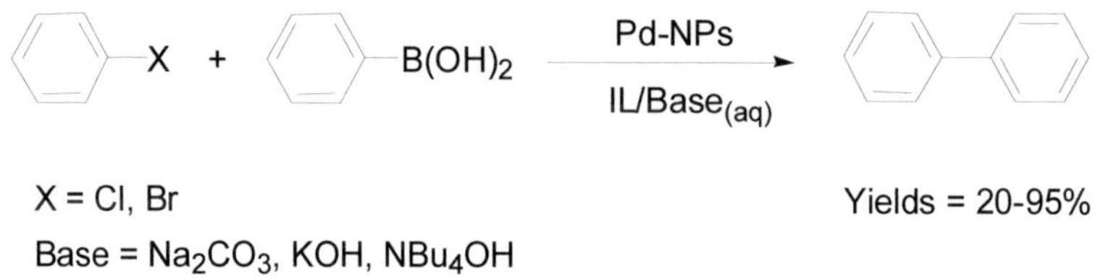

Scheme 5. Suzuki cross-coupling reactions catalysed by Pd-NPs in TBAB or THeptAB (THeptA = tetraheptylammonium) ILs at different temperatures (60-140 °C). (Calo et al., 2005)

Interestingly, with tetrabutylammonium hydroxide as base, the catalytic efficiency increased significantly, thus, the reaction was performed even under milder conditions. This can be explained by the elevated concentration of tetraalkylammonium in water, contributing through partitioning equilibrium, keeping the concentration of the cations in the IL constant. Consequently, the Pd-NPs were effectively stabilised against aggregation. Furthermore, with more hydrophobic IL such as tetraheptylammonium bromide (THeptAB) containing longer side chains than TBAB, the results were improved in the Suzuki reaction, which might be due to the better stabilisation of the Pd-NPs provided by this IL. Moreover, the Pd/THeptAB-system is recyclable for at least three runs. Again, the metal NPs were identified simply as reservoir for the molecular active catalyst in Suzuki reactions, when $Pd(OAc)_2$ as precursor in $BMI.PF_6$ in the presence of functionalised ligands derived from norborn-5-ene-2,3-dicarboxylic anhydride were employed.(Fernandez et al., 2007) Noteworthy is that in organic solvents the molecular catalyst is adequately stabilised by ligands, but in ILs the catalyst is active due to the in situ formation of NPs. Consequently, the formation of Pd-NPs is crucial to obtain an active catalyst in ILs as reaction medium.

2.3 Stille reaction

As another important tool for C-C coupling, the Stille reaction has also been catalysed by Pd-NPs in ILs. It was demonstrated that nitrile-functionalised ILs are superior for Pd-NP stabilisation to non-functionalised ILs (Figure 2).(Zhao et al., 2004; Calo et al., 2005; Chiappe et al., 2006; Fei et al., 2007; Cui et al., 2010) Moreover, tetraalkylammonium bromides are also suitable as reaction medium for the Stille reaction with Pd-NPs catalysts.(Calo et al.,

2005) A library of imidazolium, pyridinium and pyrrolidinium ILs with nitrile groups was designed to improve prolonged catalyst lifetime in the Stille reaction, among other cross-coupling reactions.(Zhao et al., 2004)

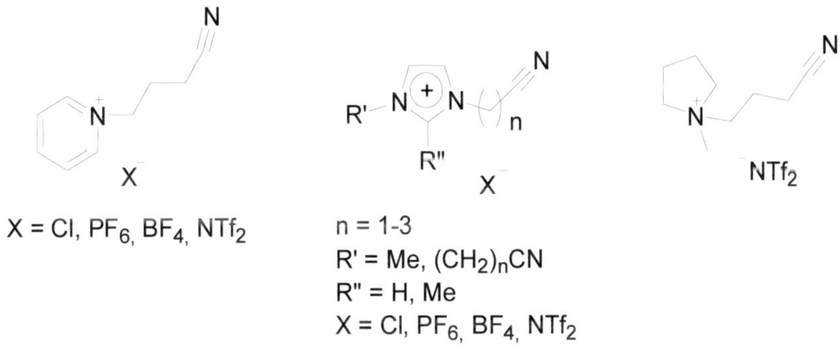

Fig. 2. Selection of ILs with nitrile-groups: pyridinium ((BCN)Py^+; left), imidazolium ((RCN)I^+ or $(RCN)_2I^+$; middle), pyrrolidinium salts ((BCN)P^+; right) with various anions. (Calo et al., 2005; Zhao et al., 2004; Chiappe et al., 2006; Fei et al., 2007; Cui et al., 2010)

In these works an anion-dependence was described using pyridinium ILs with palladium chloride as precursor. It is a direct approach to various molecular palladium species with cations and anions in the coordinating sphere of the palladium. These palladium complexes were tested in C-C coupling reactions in ILs with aliphatic and nitrile side-chains. An encouraging catalytic activity was observed in the coupling of iodobenzene with phenyltributylstannane. And, also recycling with no significant loss of activity was observed exclusively for the nitrile-functionalised IL (entry 8, Table 1). The following facts are characteristic for the nitrile-ILs: (I) the nitrile-group avoids leaching remarkably, (II) Pd-NPs act as reservoir for the active Pd species in the Stille reaction and (III) Pd-NPs were analysed by means of TEM techniques depicted well-dispersed small sized Pd-NPs (5 nm).(Zhao et al., 2004) Similar results were observed with imidazolium ILs including the nitrile-groups. In the entitled reaction, the influences of the cations (BMI, $(RCN)_{1\text{-}2}I$) and anions (BF_4, NTf_2, $N(CN)_2$), and the catalyst source ($Pd(OAc)_2$; $Pd_2(dba)_3$) have been evaluated (entry 10, Table 1).(Chiappe et al., 2006) The nitrile group in the cations and also the cyanamide anion affects the cross-coupling and the catalyst stability. In fact, it was shown that the relative coordination strengths of the ions played a role and that under certain conditions NPs have been observed. In all cases, the nitrile-functionalised ILs are superior to alkylimidazolium ILs for the immobilisation of Pd-catalysts and vinylation of aryl halides with tributylvinylstannane (Scheme 6).

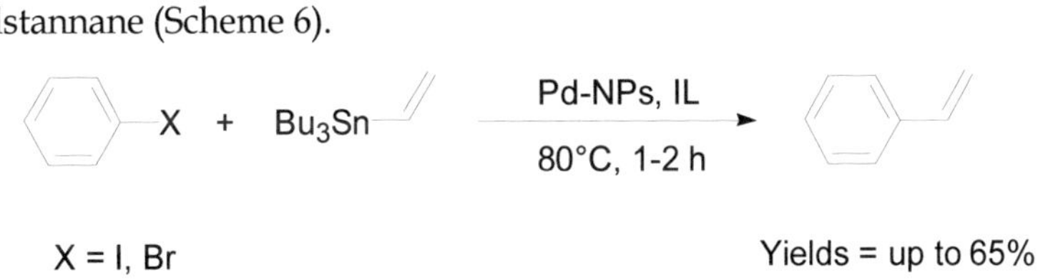

Scheme 6. Example for the Stille reaction of aryl halides with tributylvinylstannane catalysed by Pd-NPs in ILs.(Zhao et al., 2004)

In additional studies, molecular intermediate complexes were found, where (I) the nitrile group is ligated to the metal core, and (II) carbenes (derived from the imidazolium IL reaction media) are coordinated to the central palladium atom.(Fei et al., 2007) Such complexes were tested for C-C coupling reactions, and Pd-NPs were observed in, for instance, the (BCN)MI.BF_4.

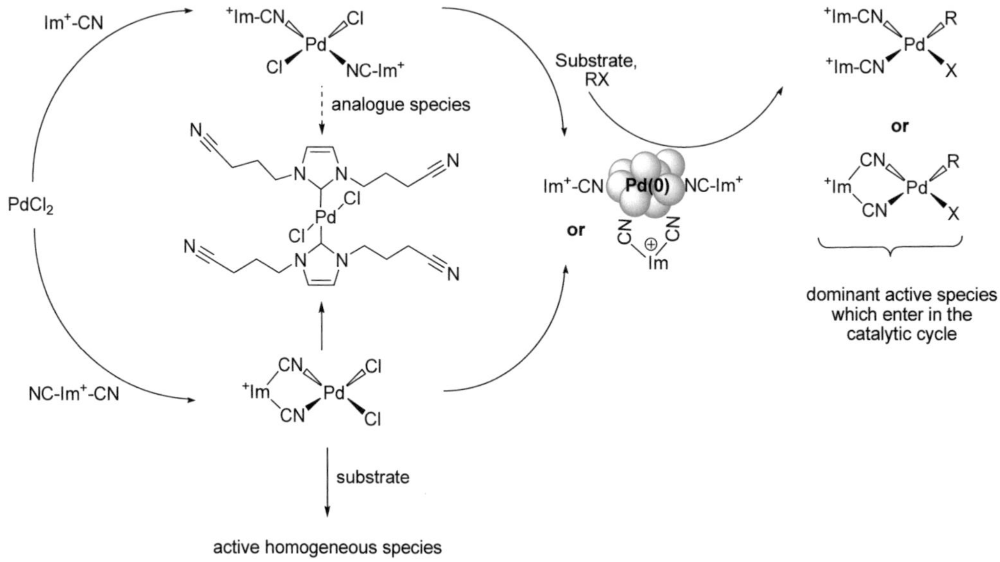

Scheme 7. Proposed mechanism for the formation of nitrile-IL stabilised Pd-NPs. Adapted with permission from reference (Fei et al., 2007) (Copyright American Chemical Society).

However, Pd-NPs are reservoirs for molecular Pd(II) species, the presumably the true active catalyst. The nitrile-functionality and carbenes (derived from the IL) stabilises these molecular intermediates via transient coordination, and they protect Pd-NPs (Scheme 7).(Fei et al., 2007) Most interestingly, palladium leaching into the organic layer was ten times lower in nitrile-ILs than in alkylimidazolium ILs, and the coordinating nitrile-group supports also the solubility of $PdCl_2$. These properties make this catalyst system, achieving high conversions in the Stille reaction, attractive for recycling catalyst.

2.4 Sonogashira reaction

Two different approaches were reported for the Sonogashira reaction with palladium nanocatalysts in IL. (Corma et al., 2005; Gao et al., 2005) One method used Pd-nanowires in IL, and in another protocol a palladacycle was thermally decomposed in IL resulting in Pd-NPs. The palladium nanowires were prepared in a thiol-functionalised IL (TFIL) by use of the seed growth method.(Gao et al., 2005) H_2PdCl_4 was reduced with $NaBH_4$ in a solution of gold colloids (2.2 nm) as seeds in the TFIL. Interestingly, the Pd-nanowires (2-4 nm in diameter) were exclusively obtained with specific ratios of Au and Pd precursors and IL. With low (high) Au concentrations core/shell nanostructures were obtained. The catalytic activity of the Pd nanowires were evaluated for the Sonogashira coupling. They revealed a remarkable activity and catalyst retention with iodoarene and phenyl acetylene as substrates, in presence of CuI and phosphine. Complete conversion was obtained in a few hours at 75 °C using the nanowires (Scheme 8a), contrary to the bimetallic (Pd_{Shell}@Au_{Core})-NPs where a conversion of 82% was obtained.

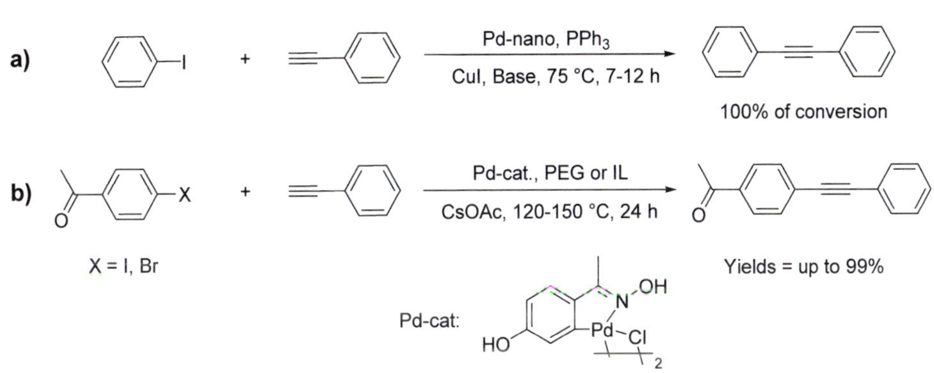

Scheme 8. a) Sonogashira reaction catalysed by Pd-nanowires and b) Pd-catalysed C-C coupling in PEG or ILs (BMI.PF_6 and BM_2I.PF_6).(Corma et al., 2005; Gao et al., 2005)

A palladacycle complex of 4-hydroxyacetophenone oxime was used for a recyclable multiphase system (Scheme 8b; entry 11, Table 1). This palladacycle is known to be a highly active catalyst for C-C bond forming reactions in water.(Corma et al., 2005) The stability of the complex was evaluated at elevated temperature in ILs and in PEG. The palladacycle decomposed in H_2O, BMI.PF_6 and BMI.Cl forming Pd-NPs in water and BMI.PF_6 (2–5 nm) and $PdCl_4^{2-}$ in the latter case. In contrast, the Pd complex remained stable in hot 1-*n*-butyl-2,3-dimethylimidazolium hexafluorophosphate (BM_2I.PF_6) and in PEG. The activity of the complex in PEG was superior to the one in ILs, likely to be related to the stability of the complex. However, the palladacycle also decomposed in PEG during the cross coupling yielding Pd-NPs (2–5 nm) which were stabilised by PEG. This copper- and ligand-free Pd/PEG-system can be applied for Sonogashira coupling on air with moderate to good yields (Scheme 8b). The authors explained that the lower catalytic activity in the ILs is related to the low solubility of CsOAc and unconsidered ILs as suitable media for this Pd-catalyst. Contrary, PEG was identified as more convenient medium for these reactions, due the higher stability of the palladacycle and of the Pd-NPs and the better solubility of cesium acetate.

2.5 Ullmann reaction

One established approach for the synthesis of biaryls is the dimerisation of aryl halides. For this purpose the Ullmann reaction is a traditional method, although the original protocol uses an excess of copper and harsh thermal conditions with temperatures above 200 °C.(Calo et al., 2009) Without doubt, a convenient alternative uses Pd-catalysts for the coupling of aryl halides to symmetrical biaryls. However, reductive conditions are crucial, and reductive agents such as amines, molecular hydrogen, hydroquinone, alcohols, or formic acid salts are commonly used in this approach.(Calo et al., 2009) Protocols introducing improved methods of the Ullmann reaction for recyclable systems, use ILs as reaction media.(Pachon et al., 2006; Calo et al., 2009) Rothenberg, for instance, presented a Pd-NPs catalysed Ullmann reaction based on electroreductive coupling of haloarenes in IL at room temperature (Scheme 9). (Pachon et al., 2006)

The Pd-NPs (2.5 ± 0.5 nm) were released in an electrochemical cell (Pd-anode and Pt-cathode). Here electron-transfer was crucial for closing the catalytic cycle. This system already gives remarkable yields with aryl bromides and iodides as substrates, simply by applying an electric current in water. To improve the electric conductivity and the

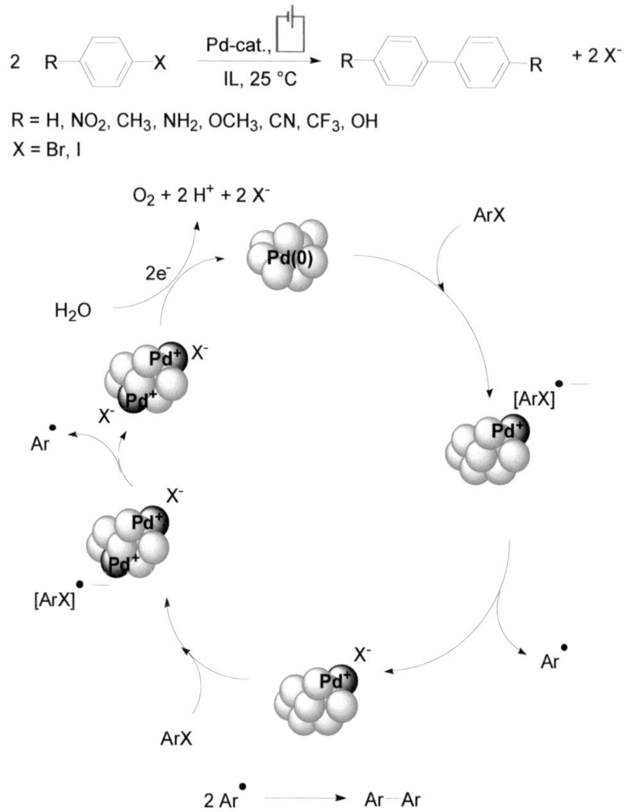

Scheme 9. Ullmann-typed aryl halide coupling with Pd nanocatalysts in IL under electroreductive conditions at 25 °C (top). Proposed catalytic cycle for the electroreductive Pd-NPs catalysed coupling of aryl halides (Pd^+ ions are depicted in dark grey). Noteworthy, the present model includes two single electron transfers from the same cluster, but in general interaction between different clusters is most likely to occur (bottom). Adapted with permission from reference (Pachon et al., 2006) (Copyright Wiley-VCH Verlag GmbH & Co. KGaA).

stabilisation of the Pd-NPs, 1-methyl-3-*n*-octylimidazolium tetrafluoroborate ($OMI.BF_4$) IL was introduced as recyclable solvent. Reaction monitoring at various electrode potentials revealed that a two-electron oxidation of H_2O closes the catalytic cycle by reformation of the Pd(0). Limitations of the system are given for functionalised aryl bromides and iodides with: R = H, NO_2, CH_3, NH_2, OCH_3, CN, CF_3, OH. The conversions vary extremely from 20 to 99% with reaction times from 8 to 24 h at 25 °C, using currents of 10 mA with 1.0–1.6 V. Noteworthy is that aryl chlorides do not undergo homocoupling under the described conditions. This set up is an unique example of electroreductive Pd-NPs catalysis in ILs. These insights into the kinetics suggest the formation of a phenyl radical anion during the reaction (Scheme 9). The advantage of this protocol is that simple electrons and water are the crucial elements for closing the catalytic cycle.(Pachon et al., 2006)
Another very mild approach for reductive homocoupling of aryl, vinyl and heteroaryl towards symmetrical biaryls uses aldehyde as reductant with Pd-NPs in TBAB and TBAA (entry 12, Table 1).(Calo et al., 2009) The IL here is crucial as it acts as base, reaction medium

and it is a stabilising ligand for the Pd-NPs which serve as reservoir for active species. Bromo- and iodo arenes couple to biaryls in absence of other additives at temperatures between 40–90 °C with good conversions (70–90%). In this simple method the substrates and $Pd(OAc)_2$ are added to the IL, providing the catalytically active species directly in situ. One advantage of this approach is that the consumed reductant propanal is converted in presence of TBAA into acrolein (via the enolate ion), which can be simply separated by evaporation. Interestingly, the selectivity of the palladium species is tuneable in dependence of the IL-anion. With TBAB (X = Br) the pathway follows the Heck reaction (not shown) and with TBAA (X = AcO-) the Ullmann reaction (Scheme 10).

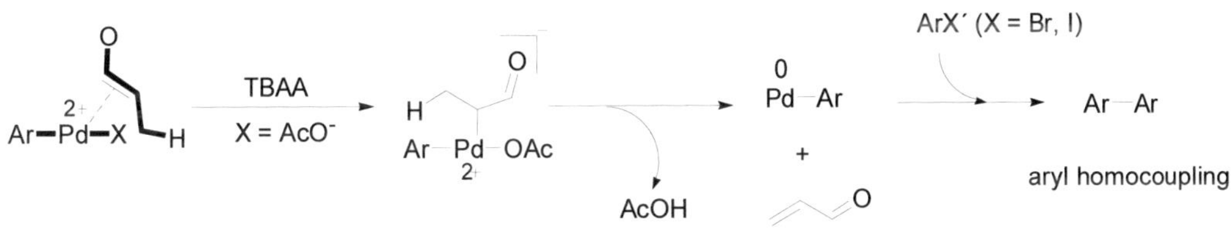

Scheme 10. Pd-NPs catalysed Ullmann reaction in the presence of TBAA. Adapted with permission from reference.(Calo et al., 2009) (Copyright Wiley-VCH Verlag GmbH & Co. KGaA).

2.6 Recyclability of Pd-catalysts in C-C coupling reactions

In summary, convenient recyclable catalyst systems use simple tetraalkylammonium salts as well as imidazolium based ILs. Interestingly, such Pd-NPs in tetraalkylammonium salts or imidazolium ILs are capable to catalyse a whole range of cross-coupling reactions for several recycles in batch reactions of the previously discussed C-C coupling reactions, in particular: Heck, Suzuki, Stille, Sonogashira, and Ullmann reactions. (For references see Table 1 and the citations in the previous segments).

No*	Reaction	IL	Ar-X (X =)	Educt/ Pd	Conv. [%]	Runs
1	Heck	TBAB/TBAA	I, Br, Cl	67 (285)	10-99 (97-99)	n. d. (10)
2		TBAB, TBAA	Cl	67	25-98	n. d.
3		TBAB, BMP.NTf_2	I	200	6-98	5
4		(BCN)MI.NTf_2	I	100	16-99	4
5		BMI.PF_6	I	1000	100	n. d.
6	Suzuki	TBAB, THeptAB	Br, Cl	40	15-99	4
7		TOAB	Br	50	100	n. d.
8	Stille	BPy.NTf_2, (BCN)Py.NTf_2	I	20	44-65 (yield)	9
9		THeptAB	Br, Cl	40	27-98	5
10		(BCN)MI.BF_4	I	20	48-85	4
11	Sonogashira	BMI.PF_6	I	10	57 (yield)	n. d.
12	Ullmann	TBAA	Br	33	81-92	n. d.

BMP: 1-*n*-butyl-1-methylpyrrolidinium, BPy: 1-*n*-butylpyridinium, THeptA: tetraheptylammonium, TOA = tetraoctylammonium. *References: 1. (Calo et al., 2003; Calo et al., 2004) 2. (Calo et al., 2009). 3. (Forsyth et al., 2005). 4. (Fei et al., 2007). 5. (Cassol et al., 2005). 6. (Calo et al., 2005). 7. (Reetz et al., 1996) 8. (Zhao et al., 2004). 9. (Calo et al., 2005). 10. (Chiappe et al., 2006). 11. (Corma et al., 2005). 12. (Calo et al., 2009).

Table 1. Examples for Pd-NPs Catalysed Carbon-Carbon Cross-Coupling Reactions in ILs.

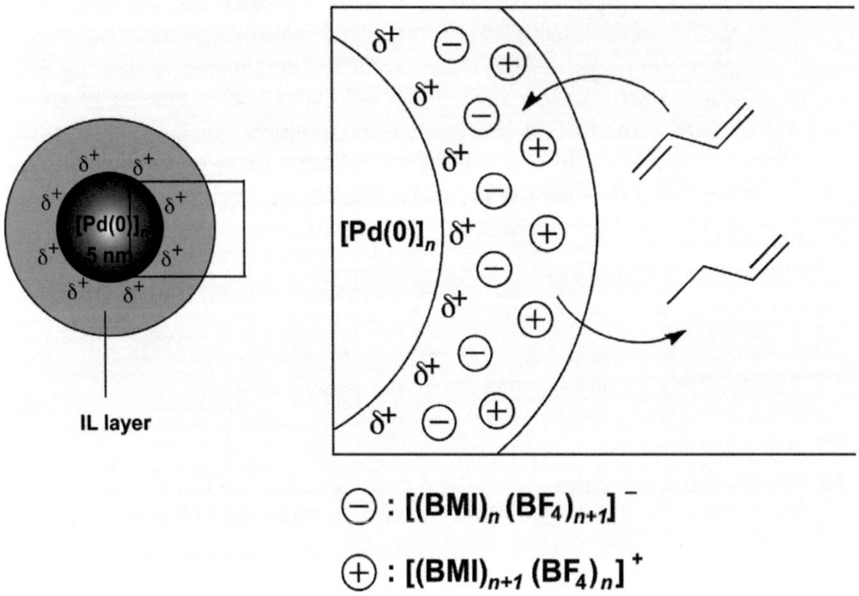

Fig. 3. Hydrogenation of 1,3-butadiene catalysed by metal Pd-NPs in IL. Reprinted with permission from reference (Dupont & Scholten, 2010) (Copyright The Royal Society of Chemistry).

3. Hydrogenation reactions catalysed by Pd-NPs in ILs

There is no doubt that hydrogenation reactions of unsaturated compounds are under the most extensively studied processes in catalysis.(Young et al., 1947; Brown & Brown, 1962; Harmon et al., 1969; Ohkuma et al., 1995; Lu et al., 2008) In this context, the preparation and stabilisation of metal NPs becomes a suitable alternative for the classical homogeneous and heterogeneous systems.(Crooks et al., 2001; Thomas et al., 2003; Astruc et al., 2005; Yan et al., 2010) Particularly, metal NPs prepared in ILs proved to be an outstanding recyclable catalytic-phase for the hydrogenation of different substrates exhibiting high catalytic activities.(Migowski & Dupont, 2007; Dupont & Scholten, 2010) In this section it selected works on the use of Pd(0)-NPs immobilised in ILs as catalyst-phase employed in hydrogenation reactions are briefly discussed (see Table 2).

Stable metal Pd-NPs could be synthesised directly in ILs as unique stabiliser agents or in the presence of additional ligands that, in general, improves the stability and catalytic activity of these nanomaterials.(Prechtl et al., 2010) By use of imidazolium-based ILs as sole stabilising agent, Pd(0)-NPs (4.9 ± 0.8 nm) were generated in $BMI.PF_6$ (or $BMI.BF_4$) from the reduction of $Pd(acac)_2$ (acac = acetylacetonate) by molecular hydrogen (4 atm, constant pressure) at 75 °C.(Umpierre et al., 2005) These NPs dispersed in IL were tested as catalyst on the selective hydrogenation of 1,3-butadiene to butenes. Once that the substrate 1,3-butadiene is at least four times more soluble in $BMI.BF_4$ than butenes, selectivities up to 97% in butenes was achieved at 40 °C and 4 atm of hydrogen (Figure 3). Moreover, 1-butene was the major product obtaining up to 72% of selectivity at 99% of 1,3-butadiene conversion (entry 1, Table 2). These selectivities in butenes could be explained due to the significant difference in solubility of the 1,3-butadiene and butenes, where the products were extracted from the IL layer by the substrate avoiding the subsequent hydrogenation step. Notably, there is no isomerisation of butenes during reaction, suggesting that the Pd(0)-NPs possess pronounced surface-like properties. After the catalytic reactions aggregation of the NPs could be observed, in certain cases.

A "green" method based on the combination of IL and supercritical carbon dioxide was employed as an efficient metal NPs preparation route and product separation.(Jutz et al., 2009) This procedure consists of the removal of the residual ligands after NPs synthesis by the supercritical CO_2 as well as of the extraction of the products at the end of the hydrogenation reaction (Scheme 11). Indeed, Pd-NPs could be prepared in imidazolium ($BMI.PF_6$ and BMI.OTf) or in quaternary ammonium (THAB; THA = tetrahexylammonium) salts by the reduction of $Pd(acac)_2$ with molecular hydrogen. The Pd-NPs synthesised in THAB presented a mean diameter of 3.5 ± 0.6 nm, while the particles in $BMI.PF_6$ or BMI.OTf exhibited considerable agglomeration reaching sizes of 10-30 nm. These nanocatalysts immobilised in ILs were able to catalyse the hydrogenation of acetophenone with good selectivities in 1-phenylethanol (entries 2 and 3, Table 2). The NPs prepared in THAB showed no significant activity for acetophenone hydrogenation. In particular, Pd-NPs in $BMI.PF_6$ proved to be the best catalytic system without loss in its activity after recharges.

When compared to the NPs generated in IL as unique stabiliser agent, a functionalised imidazolium IL, namely 2,3-dimethyl-1-[3-*N,N-bis*(2-pyridyl)-propylamido]imidazolium hexafluorophosphate ($BM_2DPA.PF_6$), was shown to be used as protective ligand to support the synthesis of more stable Pd-NPs in $BM_2I.PF_6$ (BM_2I = 1-*n*-butyl-2,3-dimethylimidazolium).(Hu et al., 2009) Indeed, the functionalised IL displayed an important role in enhancing the stabilisation of NPs. The precursor $Pd(OAc)_2$ could be reduced by molecular hydrogen in $BM_2I.PF_6$ with the presence of $BM_2DPA.PF_6$ producing Pd-NPs (5-6 nm). Notably, the isolation of the NPs and re-dispersion in the IL showed a remarkably effect on the catalytic activity in hydrogenation experiments. In fact, when isolated and re-dispersed in IL, the Pd-NPs provide higher substrate conversions in the catalytic insights by comparison with those non-isolated NPs (entries 4 and 5, Table 2). This can be related to the fact that, when in excess (non-isolated NPs), the ligand could block the active sites of the NP dropping considerably the catalyst's activity. Notably, using 2-cyclohexen-1-one as standard substrate, the system could be re-used efficiently for seven runs (entry 6, Table 2).

Similarly, the use of phenantroline (Phen) as additional ligand on the synthesis of Pd NPs in $BMI.PF_6$ was also demonstrated.(Huang et al., 2003) Protected and well-dispersed NPs (2-5 nm) was obtained from the reduction of $Pd(OAc)_2$ by hydrogen in the IL with the presence of phenantroline. Insights into the catalytic process showed that these ligand-protected Pd-NPs in ILs were very active catalysts for the hydrogenation of olefins, achieving total conversions in the most cases (entries 7 and 8, Table 2). The importance of the additional ligand was proved when cyclohexene hydrogenation was carried out with Pd-NPs prepared only in IL. After the first reaction, a Pd black precipitation occured in the IL and, in the second cycle a drastic decrease in activity was observed. However, using the Phen-protected Pd-NPs in IL, no significant loss in activity was detected for at least 10 recycles during cyclohexene hydrogenation.

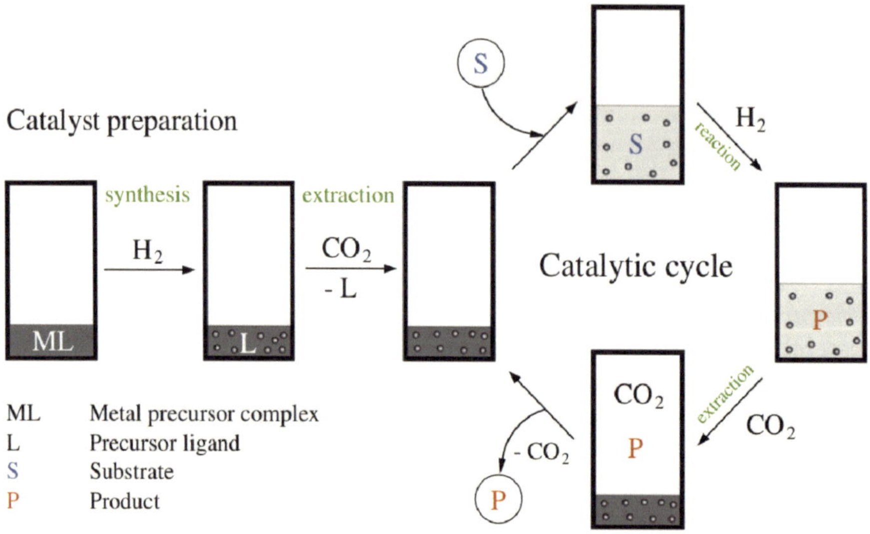

Scheme 11. "Green" method developed to generate metal NPs in ILs and their use as catalyst during hydrogenation reactions. Reprinted with permission from reference (Jutz et al., 2009) (Copyright Elsevier).

In the same context, palladium metal NPs (3 nm) were synthesised from the ethanolic reduction of $PdCl_2$ in the presence of poly(*N*-vinyl-2-pyrrolidone) (PVP) as stabilising ligand.(Mu et al., 2004) Then, these PVP-stabilised NPs were immobilised in BMI.PF_6 and employed as efficient catalyst-phase in olefin hydrogenation under mild conditions (entry 9, Table 2).

As an appropriate capping agent, TBAB was employed for the generation of Pd-NPs.(Le Bras et al., 2004) From a mixture containing a [Pd(II)] precursor, TBAB and tributylamine at 120 °C, metal NPs with mean diameters of 4.1 and 7.5 nm can be prepared using $Pd(OAc)_2$ and $PdCl_2$, respectively. Since the catalytic reactions were performed at room temperature, it was necessary to dissolve the NPs mixture in an organic solvent or in IL. In order to verify the potential recyclability of these systems, BMI.PF_6 was chosen as suitable solvent. The stabilised Pd-NPs (4.1 nm) dispersed in IL present good efficiency and selectivity in hydrogenation of unsaturated compounds (entries 10 and 11, Table 2), even on the recycling tests during five successive reactions with different substrates.

A simple and efficient deposition of Pd-NPs onto an IL functionalised multi-walled carbon nanotubes (IL-MWCNT) support was reported as a proper method to stabilise metal NPs.(Chun et al., 2008) The reduction of an aqueous solution of Na_2PdCl_4 and the IL.Br-MWCNT (bearing a bromide anion) by hydrogen under mild conditions afforded deposited metal Pd-NPs (10 nm) on IL.Br-MWCNTs (Scheme 12). Moreover, the direct exchange of the bromide anion from the system Pd-NPs/IL.Br-MWCNTs to NTf_2 or SbF_6 in water can tune the solubility properties of these materials in other solvents. Interestingly, in a mixture containing an IL and another organic solvent (MeOH or *i*-PrOH), these systems dissolve preferentially in the IL phase. Notably, the Pd-NPs/IL.X-MWCNTs (X = Br, NTf_2, SbF_6) were applied firstly as catalyst on the *trans*-stilbene hydrogenation in MeOH, reaching high TOFs values. In addition, the catalyst containing SbF_6 as anion showed the best results compared to those obtained bearing Br or NTf_2 anion. Thus, Pd-NPs/IL.SbF_6-MWCNTs was

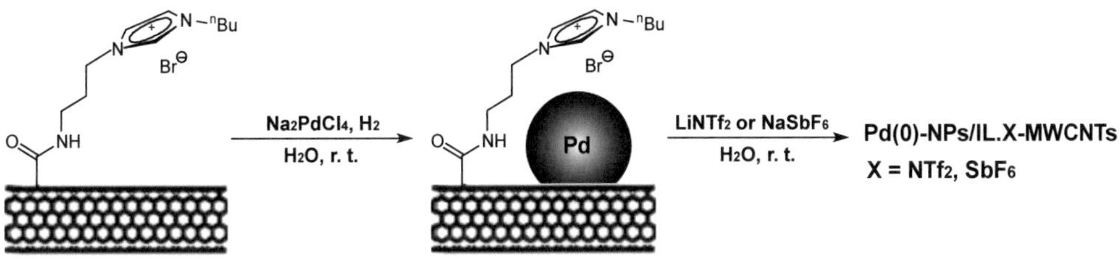

Scheme 12. Synthesis of supported metal Pd-NPs in functionalised IL-MWCNTs. Adapted with permission from reference (Chun et al., 2008) (Copyright The Royal Society of Chemistry).

No*	Solvent	Substrate	Product	T (°C)	t (h)	Conv. (%)	Runs
1	$BMI.BF_4$	1,3-butadiene[a]	1-butene	40	6	99[b]	n.d.
2	$BMI.PF_6$	acetophenone[c]	1-phenylethanol	25	4	98[d]	6
3	BMI.OTf	acetophenone[e]	1-phenylethanol	25	4	33[f]	n.d
4	$BM_2I.PF_6$/ IL-ligand	cinnamaldehyde[g]	aldehyde + alcohol	35	2	75[h]	n.d.
5	$BM_2I.PF_6$/ IL-ligand	cinnamaldehyde[i]	aldehyde	35	2	36[j]	n.d.
6	$BM_2I.PF_6$/ IL-ligand	2-cyclohexen-1-one	cyclohexanone	35	3	100	7
7	$BMI.PF_6$/ Phen-ligand	cyclohexene[k]	cyclohexane	40	5	100	10
8	$BMI.PF_6$/ Phen-ligand	1,3-cyclohexadiene[k]	cyclohexene (cyclohexane)	40	2 (7)	95 (100)[l]	n.d.
9	$BMI.PF_6$/ PVP-ligand	cyclohexene[m]	cyclohexane	40	1	100	n.d.
10	$BMI.PF_6$[n]	benzylpropenoate[o]	benzyl propanoate	25	20	97[p]	n.d.
11	$BMI.PF_6$[n]	1-benzyloxy-2-allylbenzene[o]	1-benzyloxy-2-propylbenzene	25	22	98[p]	n.d.
12	$BMI.SbF_6$ + *i*-PrOH[q]	olefins[r]	alkanes	20	0.08-0.17	100	50

a) Substrate/catalyst (S/C molar ratio) = 1063 and 4 atm of H_2 (constant pressure); b) Selectivity of 72% in 1-butene; c) S/C = 78 and 50 atm of H_2; d) Selectivity of 90% in 1-phenylethanol; e) S/C = 177 and 50 atm of H_2; f) Quantitative selectivity in 1-phenylethanol; g) S/C = 250 and 15 atm of H_2 (isolated and re-dispersed NPs); h) 65% in 3-phenyl propionaldehyde and 10% in 3-phenyl-propan-1-ol; i) S/C = 250 and 15 atm of H_2 (non-isolated NPs); j) Only 3-phenyl propionaldehyde was observed; k) S/C = 500 and 1 atm of H_2 (constant pressure); l) Only cyclohexene was observed at 2 h, but cyclohexane was the product after 7 h of reaction; m) S/C = 250 and 1 atm of H_2; n) The NPs were prepared in TBAB and then isolated and re-dispersed in $BMI.PF_6$ for catalytic insights; o) S/C = 100 and a balloon filled with H_2 was used; p) Yield of product; q) NPs supported in functionalised IL-MWCNTs were dispersed in $BMI.SbF_6$/*i*-PrOH (1/4, v/v) for hydrogenation insights; r) S/C = 100 and 1 atm of H_2 (balloon).

*References: 1. (Umpierre et al., 2005). 2. (Jutz et al., 2009). 3. (Jutz et al., 2009). 4. (Hu et al., 2009). 5. (Hu et al., 2009). 6. (Hu et al., 2009). 7. (Huang et al., 2003). 8. (Huang et al., 2003). 9. (Mu et al., 2004). 10. (Le Bras et al., 2004). 11. (Le Bras et al., 2004). 12. (Chun et al., 2008).

Table 2. Selected hydrogenation reactions of unsaturated compounds catalysed by metal Pd-NPs in ILs. (For references, see in the table caption)

chosen as catalyst to test the potential for recyclability during hydrogenation of several olefins under biphasic conditions (*i*-PrOH/BMI.SbF_6). In fact, this catalytic system could be recycled up to 50 times without loss in activity, indicating to be a robust recyclable catalyst immobilised in the IL phase (entry 12, Table 2).

4. Summary

The summarised contributions about carbon-carbon cross-coupling and hydrogenation reactions with Pd-NPs in ILs show interesting activity and are attractive as recyclable catalyst systems. In both reaction types, the ILs prevented the formation of bulk metal via agglomeration of the NPs. In case of the C-C coupling reactions, the Pd-NPs acted as reservoir for catalytically active molecular Pd species. For hydrogenation reactions, the surface of nanoparticles is catalytically active for heterogeneous reactions. The choice of cations, anions, functional groups and additive bases can influence the Pd-NPs´ reactivity, stability and reaction pathways. Moreover, the choice of palladium precursors for the generation of Pd-NPs most likely plays a minor role, as plethora palladium complexes and palladium salts are known as suitable "pre-catalysts" for the discussed reactions. It is expected that further Pd-catalysed coupling reactions will be more deeply investigated with Pd-NPs in ILs. Additional mechanistic studies might reveal that homogeneous catalytic systems involve NPs as reservoir for molecular species in several reaction types.

5. Acknowledgements

This work is financially supported by the Ministry of Innovation of NRW (MIWFT), Alexander-von-Humboldt Foundation and CNPq.

6. References

Astruc, D. (2007). "Palladium nanoparticles as efficient green homogeneous and heterogeneous carbon-carbon coupling precatalysts: A unifying view." *Inorganic Chemistry* 46(6) 2007: 1884-1894 [0020-1669].

Astruc, D.; Lu, F. & Aranzaes, J. R. (2005). "Nanoparticles as recyclable catalysts: The frontier between homogeneous and heterogeneous catalysis." *Angewandte Chemie-International Edition* 44(48) 2005: 7852-7872.

Bedford, R. B. (2003). "Palladacyclic catalysts in c-c and c-heteroatom bond-forming reactions." *Chemical Communications*(15) 2003: 1787-1796 [1359-7345].

Beletskaya, I. P. & Cheprakov, A. V. (2000). "The heck reaction as a sharpening stone of palladium catalysis." *Chemical Reviews* 100(8) 2000: 3009-3066 [0009-2665].

Bernardi, F.; Scholten, J. D.; Fecher, G. H.; Dupont, J. & Morais, J. (2009). "Probing the chemical interaction between iridium nanoparticles and ionic liquid by xps analysis." *Chemical Physics Letters* 479(1-3) 2009: 113-116 [0009-2614].

Bonnemann, H.; Brijoux, W.; Brinkmann, R.; Dinjus, E.; Joussen, T. & Korall, B. (1991). "Formation of colloidal transition-metals in organic phases and their application in catalysis." *Angewandte Chemie-International Edition in English* 30(10) 1991: 1312-1314 [0570-0833].

Bonnemann, H.; Brijoux, W. & Joussen, T. (1990). "The preparation of finely divided metal and alloy powders." *Angewandte Chemie-International Edition in English* 29(3) 1990: 273-275 [0570-0833].

Brown, H. C. & Brown, C. A. (1962). "New convenient technique for hydrogenation of unsaturated compounds." *Journal Of The American Chemical Society* 84(8) 1962: 1495.

Calo, V.; Nacci, A. & Monopoli, A. (2004). "Regio- and stereo-selective carbon-carbon bond formation in ionic liquids." *Journal Of Molecular Catalysis A-Chemical* 214(1) 2004: 45-56.

Calo, V.; Nacci, A. & Monopoli, A. (2006). "Effects of ionic liquids on pd-catalysed carbon-carbon bond formation." *European Journal Of Organic Chemistry*(17) 2006: 3791-3802.

Calo, V.; Nacci, A.; Monopoli, A. & Cotugno, P. (2009). "Heck reactions with palladium nanoparticles in ionic liquids: Coupling of aryl chlorides with deactivated olefins." *Angewandte Chemie-International Edition* 48(33) 2009: 6101-6103 [1433-7851].

Calo, V.; Nacci, A.; Monopoli, A. & Cotugno, P. (2009). "Palladium-nanoparticle-catalysed ullmann reactions in ionic liquids with aldehydes as the reductants: Scope and mechanism." *Chemistry-a European Journal* 15(5) 2009: 1272-1279 [0947-6539].

Calo, V.; Nacci, A.; Monopoli, A.; Damascelli, A.; Ieva, E. & Cioffi, N. (2007). "Palladium-nanoparticles catalyzed hydrodehalogenation of aryl chlorides in ionic liquids." *Journal Of Organometallic Chemistry* 692 2007: 4397-4401.

Calo, V.; Nacci, A.; Monopoli, A.; Detomaso, A. & Iliade, P. (2003). "Pd nanoparticle catalyzed heck arylation of 1,1-disubstituted alkenes in ionic liquids. Study on factors affecting the regioselectivity of the coupling process." *Organometallics* 22(21) 2003: 4193-4197 [0276-7333].

Calo, V.; Nacci, A.; Monopoli, A.; Fornaro, A.; Sabbatini, L.; Cioffi, N. & Ditaranto, N. (2004). "Heck reaction catalyzed by nanosized palladium on chitosan in ionic liquids." *Organometallics* 23(22) 2004: 5154-5158 [0276-7333].

Calo, V.; Nacci, A.; Monopoli, A.; Ieva, E. & Cioffi, N. (2005). "Copper bronze catalyzed heck reaction in ionic liquids." *Organic Letters* 7(4) 2005: 617-620.

Calo, V.; Nacci, A.; Monopoli, A.; Laera, S. & Cioffi, N. (2003). "Pd nanoparticles catalyzed stereospecific synthesis of beta-aryl cinnamic esters in ionic liquids." *Journal of Organic Chemistry* 68(7) 2003: 2929-2933 [0022-3263].

Calo, V.; Nacci, A.; Monopoli, A. & Montingelli, F. (2005). "Pd nanoparticles as efficient catalysts for suzuki and stille coupling reactions of aryl halides in ionic liquids." *Journal of Organic Chemistry* 70(15) 2005: 6040-6044 [0022-3263].

Cassol, C. C.; Umpierre, A. P.; Ebeling, G.; Ferrera, B.; Chiaro, S. S. X. & Dupont, J. (2007). "On the extraction of aromatic compounds from hydrocarbons by imidazolium ionic liquids." *International Journal Of Molecular Sciences* 8(7) 2007: 593-605.

Cassol, C. C.; Umpierre, A. P.; Machado, G.; Wolke, S. I. & Dupont, J. (2005). "The role of pd nanoparticles in ionic liquid in the heck reaction." *Journal Of The American Chemical Society* 127(10) 2005: 3298-3299 [0002-7863].

Chiappe, C.; Pieraccini, D.; Zhao, D. B.; Fei, Z. F. & Dyson, P. J. (2006). "Remarkable anion and cation effects on stille reactions in functionalised ionic liquids." *Advanced Synthesis & Catalysis* 348(1-2) 2006: 68-74 [1615-4150].

Chun, Y. S.; Shin, J. Y.; Song, C. E. & Lee, S. G. (2008). "Palladium nanoparticles supported onto ionic carbon nanotubes as robust recyclable catalysts in an ionic liquid." *Chemical Communications*(8) 2008: 942-944.

Consorti, C. S.; Flores, F. R. & Dupont, J. (2005). "Kinetics and mechanistic aspects of the heck reaction promoted by a cn-palladacycle." *Journal Of The American Chemical Society* 127(34) 2005: 12054-12065 [0002-7863].

Corma, A.; Garcia, H. & Leyva, A. (2005). "Catalytic activity of palladium supported on single wall carbon nanotubes compared to palladium supported on activated carbon study of the heck and suzuki couplings, aerobic alcohol oxidation and selective hydrogenation." *Journal of Molecular Catalysis a-Chemical* 230 2005: 97-105 [1381-1169].

Corma, A.; Garcia, H. & Leyva, A. (2005). "Comparison between polyethylenglycol and imidazolium ionic liquids as solvents for developing a homogeneous and reusable palladium catalytic system for the suzuki and sonogashira coupling." *Tetrahedron* 61(41) 2005: 9848-9854.

Crooks, R. M.; Zhao, M. Q.; Sun, L.; Chechik, V. & Yeung, L. K. (2001). "Dendrimer-encapsulated metal nanoparticles: Synthesis, characterization, and applications to catalysis." *Accounts of Chemical Research* 34(3) 2001: 181-190.

Cui, Y. G.; Biondi, I.; Chaubey, M.; Yang, X.; Fei, Z. F.; Scopelliti, R.; Hartinger, C. G.; Li, Y. D.; Chiappe, C. & Dyson, P. J. (2010). "Nitrile-functionalized pyrrolidinium ionic liquids as solvents for cross-coupling reactions involving in situ generated nanoparticle catalyst reservoirs." *Physical Chemistry Chemical Physics* 12(8) 2010: 1834-1841 [1463-9076].

de Vries, A. H. M.; Mulders, J.; Mommers, J. H. M.; Henderickx, H. J. W. & de Vries, J. G. (2003). "Homeopathic ligand-free palladium as a catalyst in the heck reaction. A comparison with a palladacycle." *Organic Letters* 5(18) 2003: 3285-3288 [1523-7060].

de Vries, A. H. M.; Mulders, J.; Willans, C. E.; Schmieder-van de Vondervoort, L.; Parlevliet, F. J. & de Vries, J. G. (2003). "Heck reactions with homeopathic palladium." *Abstracts of Papers of the American Chemical Society* 225 2003: 60-ORGN [0065-7727].

de Vries, A. H. M.; Parlevliet, F. J.; Schmieder-van de Vondervoort, L.; Mommers, J. H. M.; Henderickx, H. J. W.; Walet, M. A. M. & de Vries, J. G. (2002). "A practical recycle of a ligand-free palladium catalyst for heck reactions." *Advanced Synthesis & Catalysis* 344(9) 2002: 996-1002 [1615-4150].

Deshmukh, R. R.; Rajagopal, R. & Srinivasan, K. V. (2001). "Ultrasound promoted c-c bond formation: Heck reaction at ambient conditions in room temperature ionic liquids." *Chemical Communications*(17) 2001: 1544-1545 [1359-7345].

Dieguez, M.; Pamies, O.; Mata, Y.; Teuma, E.; Gomez, M.; Ribaudo, F. & van Leeuwen, P. (2008). "Palladium nanoparticles in allylic alkylations and heck reactions: The molecular nature of the catalyst studied in a membrane reactor." *Advanced Synthesis & Catalysis* 350(16) 2008: 2583-2598.

Dubbaka, S. R.; Zhao, D. B.; Fei, Z. F.; Volla, C. M. R.; Dyson, P. J. & Vogel, P. (2006). "Palladium-catalyzed desulfitative mizoroki-heek coupling reactions of sulfonyl chlorides with olefins in a nitrile-functionalized ionic liquid." *Synlett*(18) 2006: 3155-3157.

Dupont, J. (2004). "On the solid, liquid and solution structural organization of imidazolium ionic liquids." *Journal of the Brazilian Chemical Society* 15(3) 2004: 341-350 [0103-5053].

Dupont, J.; Consorti, C. S. & Spencer, J. (2005). "The potential of palladacycles: More than just precatalysts." *Chemical Reviews* 105(6) 2005: 2527-2571.

Dupont, J.; de Souza, R. F. & Suarez, P. A. Z. (2002). "Ionic liquid (molten salt) phase organometallic catalysis." *Chemical Reviews* 102(10) 2002: 3667-3691 [0009-2665].

Dupont, J.; Fonseca, G. S.; Umpierre, A. P.; Fichtner, P. F. P. & Teixeira, S. R. (2002). "Transition-metal nanoparticles in imidazolium ionic liquids: Recycable catalysts for biphasic hydrogenation reactions." *Journal Of The American Chemical Society* 124(16) 2002: 4228-4229.

Dupont, J. & Scholten, J. D. (2010). "On the structural and surface properties of transition-metal nanoparticles in ionic liquids." *Chemical Society Reviews* 39 2010: 1780-1804.

Dupont, J. & Spencer, J. (2004). "On the noninnocent nature of 1,3-dialkylimidazolium ionic liquids." *Angewandte Chemie-International Edition* 43(40) 2004: 5296-5297 [1433-7851].

Dupont, J. & Suarez, P. A. Z. (2006). "Physico-chemical processes in imidazolium ionic liquids." *Physical Chemistry Chemical Physics* 8(21) 2006: 2441-2452.

Durand, J.; Teuma, E.; Malbosc, F.; Kihn, Y. & Gomez, M. (2008). "Palladium nanoparticles immobilized in ionic liquid: An outstanding catalyst for the suzuki c-c coupling." *Catalysis Communications* 9(2) 2008: 273-275 [1566-7367].

Fei, Z. F.; Zhao, D. B.; Pieraccini, D.; Ang, W. H.; Geldbach, T. J.; Scopelliti, R.; Chiappe, C. & Dyson, P. J. (2007). "Development of nitrile-functionalized ionic liquids for c-c coupling reactions: Implication of carbene and nanoparticle catalysts." *Organometallics* 26(7) 2007: 1588-1598 [0276-7333].

Fernandez, F.; Cordero, B.; Durand, J.; Muller, G.; Malbosc, F.; Kihn, Y.; Teuma, E. & Gomez, M. (2007). "Palladium catalyzed suzuki c-c couplings in an ionic liquid: Nanoparticles responsible for the catalytic activity." *Dalton Transactions* 2007: 5572-5581 [1477-9226].

Forsyth, S. A.; Gunaratne, H. Q. N.; Hardacre, C.; McKeown, A.; Rooney, D. W. & Seddon, K. R. (2005). "Utilisation of ionic liquid solvents for the synthesis of lily-of-the-valley fragrance {beta-lilial (r); 3-(4-t-butylphenyl)-2-methylpropanal}." *Journal Of Molecular Catalysis A-Chemical* 231(1-2) 2005: 61-66.

Gao, S. Y.; Zhang, H. J.; Wang, X. M.; Mai, W. P.; Peng, C. Y. & Ge, L. H. (2005). "Palladium nanowires stabilized by thiol-functionalized ionic liquid: Seed-mediated synthesis and heterogeneous catalyst for sonogashira coupling reaction." *Nanotechnology* 16(8) 2005: 1234-1237 [0957-4484].

Gelesky, M. A.; Scheeren, C. W.; Foppa, L.; Pavan, F. A.; Dias, S. L. P. & Dupont, J. (2009). "Metal nanoparticle/ionic liquid/cellulose: New catalytically active membrane materials for hydrogenation reactions." *Biomacromolecules* 10(7) 2009: 1888-1893.

Gopidas, K. R.; Whitesell, J. K. & Fox, M. A. (2003). "Synthesis, characterization, and catalytic applications of a palladium-nanoparticle-cored dendrimer." *Nano Letters* 3(12) 2003: 1757-1760 [1530-6984].

Gozzo, F. C.; Santos, L. S.; Augusti, R.; Consorti, C. S.; Dupont, J. & Eberlin, M. N. (2004). "Gaseous supramolecules of imidazolium ionic liquids: "magic" numbers and intrinsic strengths of hydrogen bonds." *Chemistry-A European Journal* 10(23) 2004: 6187-6193 [0947-6539].

Gu, Y. L. & Li, G. X. (2009). "Ionic liquids-based catalysis with solids: State of the art." *Advanced Synthesis & Catalysis* 351(6) 2009: 817-847.

Hardacre, C.; Holbrey, J. D.; McMath, S. E. J.; Bowron, D. T. & Soper, A. K. (2003). "Structure of molten 1,3-dimethylimidazolium chloride using neutron diffraction." *Journal of Chemical Physics* 118(1) 2003: 273-278 [0021-9606].

Harmon, R. E.; Parsons, J. L.; Cooke, D. W.; Gupta, S. K. & Schoolen.J (1969). "Homogeneous catalytic hydrogenation of unsaturated organic compounds." *Journal Of Organic Chemistry* 34(11) 1969: 3684-3685.

Hu, Y.; Yang, H. M.; Zhang, Y. C.; Hou, Z. S.; Wang, X. R.; Qiao, Y. X.; Li, H.; Feng, B. & Huang, Q. F. (2009). "The functionalized ionic liquid-stabilized palladium nanoparticles catalyzed selective hydrogenation in ionic liquid." *Catalysis Communications* 10(14) 2009: 1903-1907.

Hu, Y.; Yu, Y. Y.; Hou, Z. S.; Li, H.; Zhao, X. G. & Feng, B. (2008). "Biphasic hydrogenation of olefins by functionalized ionic liquid-stabilized palladium nanoparticles." *Advanced Synthesis & Catalysis* 350(13) 2008: 2077-2085.

Huang, J.; Jiang, T.; Han, B. X.; Gao, H. X.; Chang, Y. H.; Zhao, G. Y. & Wu, W. Z. (2003). "Hydrogenation of olefins using ligand-stabilized palladium nanoparticles in an ionic liquid." *Chemical Communications*(14) 2003: 1654-1655.

Jutz, F.; Andanson, J. M. & Baiker, A. (2009). "A green pathway for hydrogenations on ionic liquid-stabilized nanoparticles." *Journal Of Catalysis* 268(2) 2009: 356-366.

Kim, N.; Kwon, M. S.; Park, C. M. & Park, J. (2004). "One-pot synthesis of recyclable palladium catalysts for hydrogenations and carbon-carbon coupling reactions." *Tetrahedron Letters* 45(38) 2004: 7057-7059 [0040-4039].

Le Bras, J.; Mukherjee, D. K.; Gonzalez, S.; Tristany, M.; Ganchegui, B.; Moreno-Manas, M.; Pleixats, R.; Henin, F. & Muzart, J. (2004). "Palladium nanoparticles obtained from palladium salts and tributylamine in molten tetrabutylammonium bromide: Their use for hydrogenolysis-free hydrogenation of olefins." *New Journal of Chemistry* 28(12) 2004: 1550-1553.

Lebel, H.; Janes, M. K.; Charette, A. B. & Nolan, S. P. (2004). "Structure and reactivity of "unusual" n-heterocyclic carbene (nhc) palladium complexes synthesized from imidazolium salts." *Journal of the American Chemical Society* 126(16) 2004: 5046-5047 [0002-7863].

Liu, Y.; Wang, S. S.; Liu, W.; Wan, Q. X.; Wu, H. H. & Gao, G. H. (2009). "Transition-metal catalyzed carbon-carbon couplings mediated with functionalized ionic liquids, supported-ionic liquid phase, or ionic liquid media." *Current Organic Chemistry* 13(13) 2009: 1322-1346 [1385-2728].

Liu, Y. B.; Khemtong, C. & Hu, J. (2004). "Synthesis and catalytic activity of a poly(n,n-dialkylcarbodiimide)/palladium nanoparticle composite: A case in the suzuki coupling reaction using microwave and conventional heating." *Chemical Communications*(4) 2004: 398-399 [1359-7345].

Lu, W. J.; Chen, Y. W. & Hou, X. L. (2008). "Iridium-catalyzed highly enantioselective hydrogenation of the c=c bond of alpha, beta-unsaturated ketones." *Angewandte Chemie-International Edition* 47(52) 2008: 10133-10136.

Migowski, P. & Dupont, J. (2007). "Catalytic applications of metal nanoparticles in imidazolium ionic liquids." *Chemistry-a European Journal* 13(1) 2007: 32-39.

Moreno-Manas, M. & Pleixats, R. (2003). "Formation of carbon-carbon bonds under catalysis by transition-metal nanoparticles." *Accounts of Chemical Research* 36(8) 2003: 638-643 [0001-4842].

Mu, X. D.; Evans, D. G. & Kou, Y. A. (2004). "A general method for preparation of pvp-stabilized noble metal nanoparticles in room temperature ionic liquids." *Catalysis Letters* 97(3-4) 2004: 151-154.

Narayanan, R. & El-Sayed, M. A. (2003). "Effect of catalysis on the stability of metallic nanoparticles: Suzuki reaction catalyzed by pvp-palladium nanoparticles." *Journal of the American Chemical Society* 125(27) 2003: 8340-8347 [0002-7863].

Ohkuma, T.; Ooka, H.; Ikariya, T. & Noyori, R. (1995). "Preferential hydrogenation of aldehydes and ketones." *Journal Of The American Chemical Society* 117(41) 1995: 10417-10418.

Pachon, L. D.; Elsevier, C. J. & Rothenberg, G. (2006). "Electroreductive palladium-catalysed ullmann reactions in ionic liquids: Scope and mechanism." *Advanced Synthesis & Catalysis* 348(12-13) 2006: 1705-1710 [1615-4150].

Pathak, S.; Greci, M. T.; Kwong, R. C.; Mercado, K.; Prakash, G. K. S.; Olah, G. A. & Thompson, M. E. (2000). "Synthesis and applications of palladium-coated poly(vinylpyridine) nanospheres." *Chemistry of Materials* 12(7) 2000: 1985-1989 [0897-4756].

Phan, N. T. S.; Van Der Sluys, M. & Jones, C. W. (2006). "On the nature of the active species in palladium catalyzed mizoroki-heck and suzuki-miyaura couplings - homogeneous or heterogeneous catalysis, a critical review." *Advanced Synthesis & Catalysis* 348(6) 2006: 609-679 [1615-4150].

Pittelkow, M.; Moth-Poulsen, K.; Boas, U. & Christensen, J. B. (2003). "Poly(amidoamine)-dendrimer-stabilized pd(0) nanoparticles as a catalyst for the suzuki reaction." *Langmuir* 19(18) 2003: 7682-7684 [0743-7463].

Prechtl, M. H. G.; Scariot, M.; Scholten, J. D.; Machado, G.; Teixeira, S. R. & Dupont, J. (2008). "Nanoscale ru(0) particles: Arene hydrogenation catalysts in imidazolium ionic liquids." *Inorganic Chemistry* 47(19) 2008: 8995-9001.

Prechtl, M. H. G.; Scholten, J. D. & Dupont, J. (2009). "Tuning the selectivity of ruthenium nanoscale catalysts with functionalised ionic liquids: Hydrogenation of nitriles." *Journal Of Molecular Catalysis A Chemical* 313(1-2) 2009: 74-78.

Prechtl, M. H. G.; Scholten, J. D. & Dupont, J. (2010). "Carbon-carbon cross coupling reactions in ionic liquids catalysed by palladium metal nanoparticles." *Molecules* 15(5) 2010: 3441-3461.

Ramarao, C.; Ley, S. V.; Smith, S. C.; Shirley, I. M. & DeAlmeida, N. (2002). "Encapsulation of palladium in polyurea microcapsules." *Chemical Communications*(10) 2002: 1132-1133 [1359-7345].

Redel, E.; Kramer, J.; Thomann, R. & Janiak, C. (2009). "Synthesis of co, rh and ir nanoparticles from metal carbonyls in ionic liquids and their use as biphasic liquid-liquid hydrogenation nanocatalysts for cyclohexene." *Journal Of Organometallic Chemistry* 694(7-8) 2009: 1069-1075.

Reetz, M. T.; Breinbauer, R. & Wanninger, K. (1996). "Suzuki and heck reactions catalyzed by preformed palladium clusters and palladium/nickel bimetallic clusters." *Tetrahedron Letters* 37(26) 1996: 4499-4502 [0040-4039].

Reetz, M. T. & de Vries, J. G. (2004). "Ligand-free heck reactions using low pd-loading." *Chemical Communications*(14) 2004: 1559-1563 [1359-7345].

Reetz, M. T. & Helbig, W. (1994). "Size-selective synthesis of nanostructured transition-metal clusters." *Journal of the American Chemical Society* 116(16) 1994: 7401-7402 [0002-7863].

Reetz, M. T. & Lohmer, G. (1996). "Propylene carbonate stabilized nanostructured palladium clusters as catalysts in heck reactions." *Chemical Communications*(16) 1996: 1921-1922 [1359-7345].

Reetz, M. T. & Quaiser, S. A. (1995). "A new method for the preparation of nanostructured metal-clusters." *Angewandte Chemie-International Edition in English* 34(20) 1995: 2240-2241 [0570-0833].

Reetz, M. T. & Westermann, E. (2000). "Phosphane-free palladium-catalyzed coupling reactions: The decisive role of pd nanoparticles." *Angewandte Chemie-International Edition* 39(1) 2000: 165-+ [1433-7851].

Reetz, M. T.; Westermann, E.; Lohmer, R. & Lohmer, G. (1998). "A highly active phosphine-free catalyst system for heck reactions of aryl bromides." *Tetrahedron Letters* 39(46) 1998: 8449-8452 [0040-4039].

Rocaboy, C. & Gladysz, J. A. (2003). "Thermomorphic fluorous imine and thioether palladacycles as precursors for highly active heck and suzuki catalysts; evidence for palladium nanoparticle pathways." *New Journal of Chemistry* 27(1) 2003: 39-49 [1144-0546].

Rossi, L. M. & Machado, G. (2009). "Ruthenium nanoparticles prepared from ruthenium dioxide precursor: Highly active catalyst for hydrogenation of arenes under mild conditions." *Journal Of Molecular Catalysis A-Chemical* 298(1-2) 2009: 69-73.

Rossi, L. M.; Silva, F. P.; Vono, L. L. R.; Kiyohara, P. K.; Duarte, E. L.; Itri, R.; Landers, R. & Machado, G. (2007). "Superparamagnetic nanoparticle-supported palladium: A highly stable magnetically recoverable and reusable catalyst for hydrogenation reactions." *Green Chemistry* 9(4) 2007: 379-385.

Scheeren, C. W.; Machado, G.; Dupont, J.; Fichtner, P. F. P. & Texeira, S. R. (2003). "Nanoscale pt(0) particles prepared in imidazolium room temperature ionic liquids: Synthesis from an organometallic precursor, characterization, and catalytic properties in hydrogenation reactions." *Inorganic Chemistry* 42(15) 2003: 4738-4742.

Scholten, J. D.; Ebeling, G. & Dupont, J. (2007). "On the involvement of nhc carbenes in catalytic reactions by iridium complexes, nanoparticle and bulk metal dispersed in imidazolium ionic liquids." *Dalton Transactions* 2007: 5554-5560 [1477-9226].

Silveira, E. T.; Umpierre, A. P.; Rossi, L. M.; Machado, G.; Morais, J.; Soares, G. V.; Baumvol, I. L. R.; Teixeira, S. R.; Fichtner, P. F. P. & Dupont, J. (2004). "The partial hydrogenation of benzene to cyclohexene by nanoscale ruthenium catalysts in imidazolium ionic liquids." *Chemistry-A European Journal* 10(15) 2004: 3734-3740.

Takahashi, Y.; Ito, T.; Sakai, S. & Ishii, Y. (1970). "A novel palladium(o) complex - bis(dibenzylideneacetone)palladium(o)." *Journal of the Chemical Society D-Chemical Communications*(17) 1970: 1065-&.

Thomas, J. M.; Johnson, B. F. G.; Raja, R.; Sankar, G. & Midgley, P. A. (2003). "High-performance nanocatalysts for single-step hydrogenations." *Accounts of Chemical Research* 36(1) 2003: 20-30.

Tromp, M.; Sietsma, J. R. A.; van Bokhoven, J. A.; van Strijdonck, G. P. F.; van Haaren, R. J.; van der Eerden, A. M. J.; van Leeuwen, P. & Koningsberger, D. C. (2003). "Deactivation processes of homogeneous pd catalysts using in situ time resolved spectroscopic techniques." *Chemical Communications*(1) 2003: 128-129 [1359-7345].

Tsuzuki, S.; Tokuda, H.; Hayamizu, K. & Watanabe, M. (2005). "Magnitude and directionality of interaction in ion pairs of ionic liquids: Relationship with ionic conductivity." *Journal of Physical Chemistry B* 109(34) 2005: 16474-16481 [1520-6106].

Umpierre, A. P.; Machado, G.; Fecher, G. H.; Morais, J. & Dupont, J. (2005). "Selective hydrogenation of 1,3-butadiene to 1-butene by pd(0) nanoparticles embedded in imidazolium ionic liquids." *Advanced Synthesis & Catalysis* 347(10) 2005: 1404-1412 [1615-4150].

Xu, L. J.; Chen, W. P. & Xiao, J. L. (2000). "Heck reaction in ionic liquids and the in situ identification of n-heterocyclic carbene complexes of palladium." *Organometallics* 19(6) 2000: 1123-1127 [0276-7333].

Yan, N.; Xiao, C. X. & Kou, Y. (2010). "Transition metal nanoparticle catalysis in green solvents." *Coordination Chemistry Reviews* 254(9-10) 2010: 1179-1218.

Yang, X.; Fei, Z. F.; Zhao, D. B.; Ang, W. H.; Li, Y. D. & Dyson, P. J. (2008). "Palladium nanoparticles stabilized by an ionic polymer and ionic liquid: A versatile system for c-c cross-coupling reactions." *Inorganic Chemistry* 47(8) 2008: 3292-3297.

Yin, L. X. & Liebscher, J. (2007). "Carbon-carbon coupling reactions catalyzed by heterogeneous palladium catalysts." *Chemical Reviews* 107(1) 2007: 133-173 [0009-2665].

Young, W. G.; Meier, R. L.; Vinograd, J.; Bollinger, H.; Kaplan, L. & Linden, S. L. (1947). "Investigations on the stereoisomerism of unsaturated compounds.8. The catalytic hydrogenation of butadiene." *Journal Of The American Chemical Society* 69(8) 1947: 2046-2050.

Zhao, D. B.; Fei, Z. F.; Geldbach, T. J.; Scopelliti, R. & Dyson, P. J. (2004). "Nitrile-functionalized pyridinium ionic liquids: Synthesis, characterization, and their application in carbon - carbon coupling reactions." *Journal of the American Chemical Society* 126(48) 2004: 15876-15882 [0002-7863].

17

Ionic Liquids: Applications in Heterocyclic Synthesis

Clarissa P. Frizzo, Dayse N. Moreira and Marcos A. P. Martins
Federal University of Santa Maria
Brazil

1. Introduction

Ionic liquids (ILs) have become omnipresent in the recent chemical literature; for they can be used as highly customizable solvents for almost any synthetic purpose [Wasserscheid & Welton, 2008]. Especially in the industry, their application goes beyond their use as solvents. The highly diverse properties of these materials make possible a surprising number of applications. In organic reactions, although ionic liquids were initially introduced as alternative green reaction media because of their unique chemical and physical properties of nonvolatility, nonflammability, thermal stability, and controlled miscibility, today they have marched far beyond this boundary, showing their significant role in controlling reactions as solvent or catalysts [Wasserscheid & Welton, 2008]. It is well-known that the microenvironment generated by a solvent can change the outcome of a reaction, in terms of both equilibria and rates [Pârvulescu & Hardacre, 2007]. Since ionic liquids have the potential to provide reaction media that are quite unlike any other available at room temperature, it is possible that they will dramatically affect reactions carried out in them. Undeniably, there have been many claims of great improvements in reaction yields and rates when using ionic liquids [Chiappe & Pieraccini, 2005]. Over the past decade, some authors have manifested interest in providing facts to clarify the question: "how do ionic liquids act in organic reactions?" They have found answers for particular reactions, in that ionic liquids play specific roles depending on the reaction [Martins et al., 2008]. This chapter presents some questions and the best results to afford answers about the role of ILs in the most important reactions involved in heterocyclic synthesis: cyclocondensation and 1,3-dipolar cycloaddition reactions.

Heterocycles form by far the largest of the classical divisions of organic chemistry. Moreover, they are of immense importance not only both biologically and industrially but to the functioning of any developed human society as well. Their participation in a wide range of areas cannot be underestimated. The majority of pharmaceutical products that mimic natural products with biological activity are heterocycles. Most of the significant advances against disease have been made by designing and testing new structures, which are often heteroaromatic derivatives. In fact, in the Comprehensive Medicinal Chemistry (CMC) database, more than 67% of the compounds listed contain heterocyclic rings [Xu & Stevenson, 2000]. Other important practical applications of heterocycles can also be cited, for instance, additives and modifiers in a wide variety of industries including cosmetics, reprography, information storage, plastics, solvents, antioxidants, and vulcanization accelerators. Finally, as an applied science, heterocyclic chemistry is an inexhaustible resource of novel compounds. A huge number of combinations of carbon, hydrogen, and heteroatoms can be designed, providing compounds with the most diverse physical, chemical, and biological properties [Katritzky & Rees, 1984]. It is therefore easy to understand why both the development of new methods and the strategic deployment of known methods for the synthesis of complex heterocyclic compounds continue to drive the field of synthetic organic chemistry.

2. Cyclocondensation reactions

Cyclocondensation (a kind of annulation reaction involving the formation of a ring from one or several acyclic precursors) is a set of condensation reactions in which one-, two-, three-, or multicomponent reactants yield a single main cyclic product with the accompanying formation of some other small molecule(s) [Müller, 1994].

2.1 Characterization of cyclocondensation reactions

Reports of the synthesis of series of three-, five-, six-, and seven-membered heterocyclic rings obtained from cyclocondensation reactions in ILs were found in the literature. These reactions were carried out with different numbers of components, as summarized in **Figure 1**. The functional groups contained in each component can react as electrophiles (E_1, E_2 and E_3) or nucleophiles (Nu_1, Nu_2 and Nu_3). In general, the electrophiles are carbon atoms present in functional groups, such as carbonyl, imine, nitrile, β-carbon of α,β-unsaturated systems, mono- and dihalo-substituted carbons, and acetal and orthoester carbons; and the nucleophiles are either carbon atoms present in the α-position of aldehydes, ketones, enols, enamines, or heteroatoms, such as nitrogen, oxygen, and sulfur.

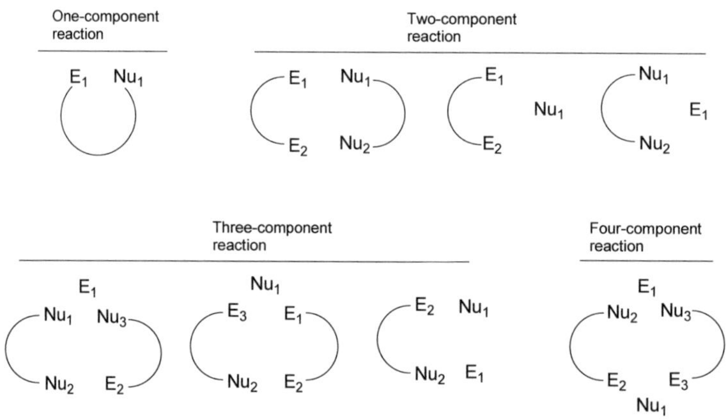

Fig. 1. Number of components in the cyclocondensation reactions in ILs found in the literature.

Table 1 shows the reaction type and building blocks that are found in cyclocondensation reactions in ILs. The first column illustrates the reaction types in accordance with the number of components. The second column explains the number of components, and the third column demonstrates the building blocks of the reactions. Thus, for example, the representation [3 + 2] [CCC + NN] indicates that the heterocycle was formed by two building blocks, one of these building blocks possessing three atoms ([CCC]) and the other possessing two atoms ([NN]). The last column lists the heterocycles obtained. In the one-component cyclocondensation reactions, the formation of one carbon-heteroatom bond was observed. In the two-component cyclocondensation reactions either (i) the formation of two carbon-heteroatom bonds or (ii) the formation of substituted carbons or acetal and orthoester carbons was observed. The formation of carbon-carbon bonds, in general, involves a nucleophilic addition (in most cases, with a second step elimination reaction) of a carbon atom nucleophile (carbonyl α-carbon) to a carbonyl (imine or nitrile) carbon atom or to the β-carbon α,β-unsaturated systems of one carbon-heteroatom and one carbon-carbon bond. In the three-component cyclocondensation reactions there were three possibilities: (i) the formation of three carbon-heteroatom bonds, (ii) the formation of two carbon-heteroatom bonds and one carbon-carbon bond or (iii) the formation of one carbon-heteroatom bond and two carbon-carbon bonds. In the four-component cyclocondensation reactions, either (i) the formation of four carbon-heteroatom bonds or (ii) the formation of two carbon-heteroatom and two carbon-carbon bonds was observed. The formation of carbon-heteroatom bonds, in general, involves either a nucleophilic addition (in most cases, with a second step elimination reaction) of a heteroatom nucleophile (O, N, or S) to a carbonyl (imine or nitrile) carbon atom or to the β-carbon α,β-unsaturated systems or a heteroatom nucleophilic substitution into mono- and dihalo-substituted carbons or acetal and orthoester carbons.

2.2 Ionic liquids in cyclocondensation reactions

The main concern about the use of ILs in cyclocondensation reactions is the origin of catalytic effects. However, the majority of studies in the literature show that ionic liquids in cyclocondensation reactions are at a molar ratio of ≥ 1.0 in relation to substrate. With this in mind arises the question of whether they are catalysts or solvents.

Welton [Welton, 1999] has studied catalytic reactions in ionic liquids and has postulated that the potentially most powerful way in which an ionic liquid can be used in catalysis is as a combination of solvent and catalyst. From this postulate, whenever changing solvent leads to a faster reaction, the new solvent can be considered a catalyst. After all, the reaction has been accelerated, and the solvent has remained unchanged by the process. In this sense, Lee *et al.* [Lee et al., 2010] proposed some pathway to this role of ionic liquids. They suggested that ILs participate in the formation of more reactive catalytic especies, for example, in reactions catalyzed by metal triflates such as $Sc(OTf)_3$, or they stabilize intermediate reactives such as cationic vinyl, arenium intermediates and anionic oxygen radical intermediates. The authors also show the ability of ionic liquids to stabilize the transition state, for example, in the reaction of nucleophilic fluorination. Aiming to respond to the same question, Oliver-Borbigou et al. [Oliver-Borbigou et al., 2010] proposed that ILs can act as solvents and as multifunctional compounds like solvents and ligants, solvents and catalysts and stabilising agents for catalyst intermediates. From this, one might think that the function of ILs differs in different reactions or reaction condition.

Although the solvent properties of ionic liquids are widely described, it appears that their effect is to catalyze reactions. At this point, it is worth remarking that cyclocondensation are generally not catalyzed reactions, or are acid/base catalyzed reactions. Recently, we have published a review about ionic liquids in cyclocondensation reactions to survey the most important contributions and to discuss the role of ionic liquids in these reactions. Here, we have compiled the most important results of that work and we will briefly describe the reactions where ILs had a remarkable role.

No. Components	Reaction Type	Building Blocks	Product
One-component Reactions	[1 + 0]	[CCCCO]	Furans
	[1 + 0]	[NCNOC]	Oxadiazoles
	[1 + 0]	[CCCCCO]	Flavones
Two-component Reactions	[2 + 1]	[CN + C]	Aziridines
	[4 + 1]	[CCCC + N]	Pyrroles
	[3 + 2]	[CCO + CC]	Butenolides
	[4 + 1]	[CCCC + S]	Thiophenes
	[3 + 2]	[CCC + NN]	4,5-Dihydropyrazoles
	[4 + 1]	[NCCN + C]	Imidazoles
	[3 + 2]	[NCN + CC]	Imidazoles
	[3 + 2]	[CCC + NO]	4,5-Dihydroisoxazoles
	[4 + 1]	[NCCO + C]	Oxazoles
	[3 + 2]	[NCS + CC]	2-Thiazoles
	[5 + 1]	[CCNCS + N]	2-Thiazoles
	[4 + 1]	[NCCS + C]	2-Thiazoles
	[4 + 2]	[CCCN + CC]	Quinolines
	[3 + 3]	[CCO + CCC]	Pyrans
	[3 + 3]	[CCC + NCN]	Pyrimidinones
	[5 + 1]	[CCCCN + C]	β-Carbolines
	[5 + 1]	[NCCCO + C]	Oxazines
	[4 + 2]	[NCCS + CC]	Benzothiazines
	[4 + 2]	[NCNC + CN]	Triazines
Three-component Reactions	[2 + 3 + 1]	[CC+ NCN + C]	Pyrimidines
	[2 + 2 +1]	[CC + CO + C]	Furans
	[2 + 2 +1]	[CC + CC + S]	Thiophenes
	[2 + 2 +2]	[CC + CO + CO]	Dioxanes
	[3 + 1 + 1]	[NCN + C + C]	Imidazoles
	[3 + 1 + 1]	[CCO + C + N]	Oxazolidinone
	[3 + 1 + 1]	[CCS + C + N]	4-Thiazolidinones
	[3 + 2 + 1]	[CCN + CC + C]	Pyridines
	[3 + 2 + 1]	[CCC + CC + N]	Pyridines
	[3 + 2 + 1]	[CCN + CC + C]	Quinolines
	[3 + 2 + 1]	[CCO + CC + C]	Pyrans
	[4 + 1 + 1]	[CCCN + C + N]	Quinazolinones
	[4 + 2 + 1]	[NCCN + CC + C]	Benzodiazepines
Four-Component Reactions	[2 + 2 + 1 + 1]	[CC + CC + C + N]	Pyridines
	[2 + 1 + 1 + 1]	[CC + N + C + N]	Imidazoles
	[2 + 2 + 1 + 1]	[CC + CC + C + N]	Acridines

Table 1. Reaction types and building blocks of cyclocondensation reactions in ILs.

The first example shown here is the synthesis of aziridines **3** using ionic liquids from the reaction of imines **1** and EDA (ethyl diazoacetate) **2** (Table 2) [Xia et al., 2003]. The reaction conditions involved equimolar amounts of **1** and **2** in [BMIM][PF_6]. Under these reaction conditions, only the cis-isomer was obtained in a 93% yield. However, when a catalytic amount of [BMIM][PF_6] was used, there was no formation of aziridine **3**. These observation reported by the authors explain the results of entries **4** and **5** in Table 2, where a catalytic amount (0.1 mmol) of ionic liquid was dissolved in co-organic solvents. As summarized in Table 2, arylimines **1**, with either electron-donating or electron-withdrawing groups, reacted readily with **2** in [BMIM][PF_6], affording the corresponding aziridines **3** with high cis selectivities. The remaining ionic liquid was recovered and reused five times with only a gradual decrease in activity observed (93-91% yield). The formation of **3** in ionic liquids proceeded in a shorter reaction time, but it has been suggested to occur in a manner similar to that previously proposed for typical Lewis acids (BF_3•OEt_2) in molecular solvent such as hexane, in which the yield obtained was 93%, after 15 h at 25°C [Xie et al., 1999].

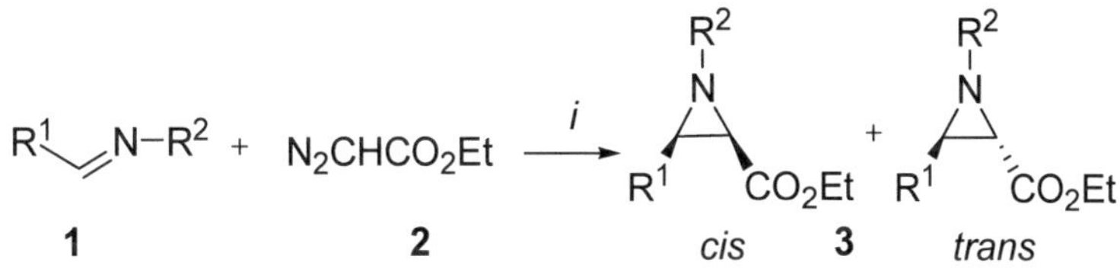

i: IL, r.t., 5 h.

Entry[a]	IL	R[1]	R[2]	Product (Yield %)[b]
1	[BMIM][BF_4][c]	Ph	Ph	82, *cis/trans*; 30:1
2	[BMIM][PF_6][c]	Ph	Ph	95, *cis* only
3	[BMIM][PF_6]	Ph	Ph	93, *cis* only
4	[BMIM][PF_6][d]	Ph	Ph	0
5	[BMIM][PF_6][e]	Ph	Ph	0

[a] All reactions were carried out using 0.5 mmol of imine and 0.5 mmol of EDA in 1.5 mL of ionic liquid for 5 h. [b] The ratio of *cis* and *trans* isomers was determined by GC/MS and ^1H NMR. [c] 1.0 mmol of imine and 0.5 mmol of EDA. [d] 0.5 mmol of imine, 0.5 mmol of EDA and 0.1 mmol of [BMIM][PF_6] in 3 mL of CH_2Cl_2 at room temperature for 7 h. [e] 0.5 mmol of imine, 0.5 mmol of EDA and 0.1 mmol of [BMIM][PF_6] in 3 mL of hexane at room temperature for 7 h.

Table 2. Synthesis of aziridines.

The second example illustrates the role of ILs as liquid support in a cyclocondensation reaction, to synthesize 2-aminothiophenes **9** by the Gewald reaction (Scheme 1) [Hu et al., 2006]. As can be seen, the reaction of an ionic liquid with a minor excess of cyanoacetic acid (1.2 equiv) **5** in the presence of DCC (dicyclohexyl carbodiimide) and a catalytic amount of DMAP (4-dimethylaminopyridine) in dry MeCN produced the functionalized ionic liquid phase bond through ester linkage in **6**. The reactants, ketones or aldehydes **7**, S_8 and EDDA (ethylenediammonium diacetate) were then added. Finally, treatment of the corresponding products **8** with NaOEt in ethanol resulted in a very efficient cleavage of ionic liquid support to provide the 2-aminothiophenes **9** with high purity and without the need for chromatographic purification. Compared to the conventional liquid phase synthesis methods, the ionic liquid phase bond intermediates were easily isolated and purified by simple filtration and washing with Et_2O to remove the few unreacted materials and neutral by-products. As liquid support, the ionic liquid was used at a molar ratio of 1:1 (reactant:IL). The ionic liquid phase was recovered and reused twice with no appreciable decrease in yields. The attainment of thiophenes **9** using molecular solvents such as THF entailed a painstaking and tedious procedure with the addition of $TiCl_4$ at 0°C followed by pyridine and stirring overnight at room temperature [Lütjens et al., 2003]. The yields obtained in THF/$TiCl_4$ were similar to those found in ionic liquid.

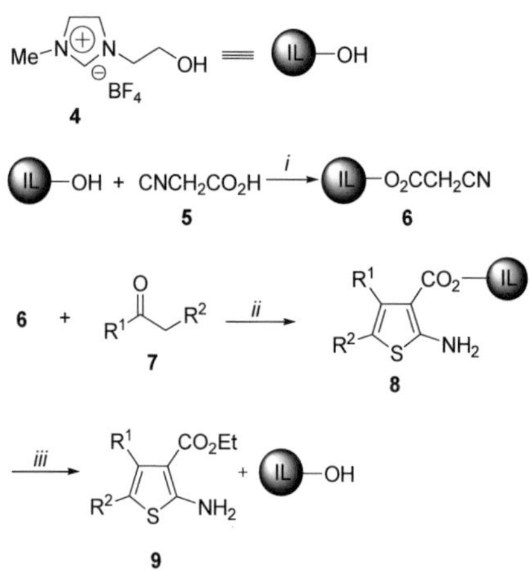

R^1 = H, Me, Et R^2 = Me, Et , CO_2Et, CO_2Me R^1, R^2 =-$(CH_2)_4$-, -$(CH_2)_3$-
i: DCC (1 equiv), DMAP (5%), MeCN, r.t., 12 h;
ii: S_8 (1 equiv), EDDA (1 equiv), 50°C, 3-6h (67-91%);
iii: EtONa (0.5 equiv), EtOH, r.t., 6 h.

Scheme 1

The reaction shown in Scheme 2 is a good example of the use of ILs as solvent in cyclocondensation reactions. The use of ionic liquids as solvent with a molar ratio of 1:10 (reactant:IL) was investigated in the synthesis of imidazo[1,2-*a*]pyrimidines **16** and imidazo[1,2-*a*]pyridines **15**, respectively, from the cyclocondensation of 2-aminopyrimidines **11** or 2-aminopyridines **10** with a suitable α-bromoacetophenone **12-14** [Enguehard et al., 2003]. The authors found that α-tosyloxylation (bromination) of ketones can be performed by treating the ketones with HTIB ([hydroxyl(tosyloxy)iodo]benzene) and 2-aminopyrimidine successively in [BPy][BF_4]. Consequently, the authors reasoned that imidazo[1,2-*a*]pyrimidine **15,16** could be directly prepared by a one-pot procedure. In all of these cases, the ionic liquid was reused four times with a gradual loss of activity (90, 86, 85, 80% yields). The reaction performed in ILs showed rate acceleration and increased yield, when compared with the reaction performed with molecular solvents, such as acetonitrile, where the preparation of imidazo[1,2-*a*]pyridines **15** required refluxing for 6-24 hours and the yield was only 37% [Bienaymé & Bouzid, 1998]. For the preparation of 2-phenylimidazo[1,2-*a*]pyrimidines **16**, refluxing for 6 h in a molecular solvent such as ethanol was necessary [Enguehard et al., 2003].

We consider important to show that task-specific ionic liquids also have applications in heterocyclic synthesis. [HMIM][Tfa] was designed as a protic ionic liquid [Greaves & Drummond, 2008] and Karthikeyan and Perumal [Karthikeyan & Perumal, 2005] proposed

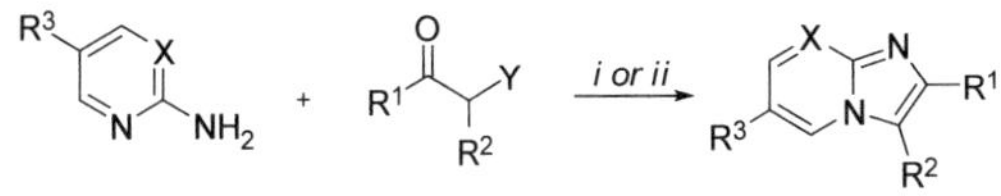

R¹ = Ph, 4-F-Ph, 4-Cl-Ph, 4-Br-Ph, 4-Me-Ph, 4-MeO-Ph, Fur-2-yl, Benzo[*b*]fur-2-yl; R² = H, Me; R³ = H
10,15 (X = CH) **11,16** (X = N) **12** (Y = OTs) **13** (Y =Br) **14** (Y =H)
i: Na_2CO_3, [BPy][BF_4], r.t., 1 h (**15** 56-90%).
ii: HTIB, Na_2CO_3, [BPy][BF_4], r.t., 1 h (**16** 72-85%).

Scheme 2

a methodology using this ionic liquid for the synthesis of pyridines **21,22**, by generating the enaminone from the corresponding β-ketoesters **17,18** for an *in situ* heteroannulation in the Bohlmann-Rahtz reaction. The one-pot, three component reaction of 1,3-dicarbonyl compounds **17,18**, ammonium acetate **20**, and alkynones **19** in [HMIM][Tfa] as solvent gave good results (Scheme 3). Although the reaction time using the ionic liquid was longer than other methods described in the literature, this synthetic route was considered simpler and more convenient. In molecular solvents such as EtOH (temperature 140-160°C) and toluene (it was necessary to add AcOH (5:1) or a Lewis acid such as $ZnBr_2$), the reaction time was 5.5 h (in both cases) [Bagley et al., 2006].

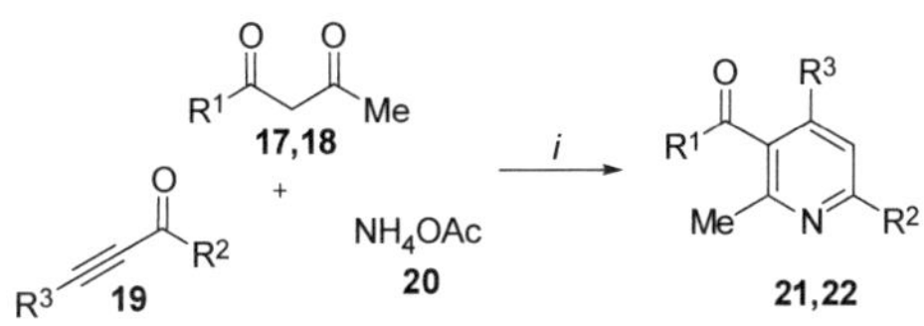

17,21 (R¹ = OMe, OEt) **18,22** (R¹ = Me)
R² = Me, Ph; R³ = H, Ph, $SiMe_3$
i: [HMIM][Tfa], r.t., 24 h (80-94%).

Scheme 3

On the other hand, although some authors do not classify [BMIM][OH] as a basic ionic liquid [MacFarlane et al., 2006], it was used as a base IL in a protocol to synthesize polyfunctionalized pyridines by a cyclocondensation reaction [Ranu et al., 2007]. The conventional method for this reaction involves the condensation of aldehydes **23**, malononitrile **24**, and thiols **25** to afford highly substituted pyridines **26** (Scheme 4). One of the serious limitations of the conventional procedure is the formation of considerable amounts of a side product, enaminonitrile, reducing the yields of the pyridines to 20-48% when using bases such as DABCO (1,4-Diazabicyclo[2.2.2]octane) and Et_3N in ethanol under reflux (2-3 h) [Evdokimov et al., 2006]. Ranu et al. [Ranu et al., 2007] demonstrated that the ionic liquid [BMIM][OH] completely suppressed the side reaction that formed enaminonitrile and raised the (isolated) yields of pyridines to a level of 62-95% (Scheme 4). A wide range of substituted aromatic and heteroaromatic aldehydes **23** as well as several substituted thiophenols **25** underwent this three-component condensation with malononitrile. The ionic liquid was used at a molar ratio of 1:0.5 (reactant:IL) and the authors claimed that the presence of the ionic liquid, [BMIM][OH], was essential, as the reactions did not proceed at all in its absence. The use of other ionic liquids such as

[BMIM][Br] or [BMIM][BF_4] failed to push the reaction to the pyridine stage, and the reaction was stopped at an intermediate step with the formation of arylidenemalononitrile.

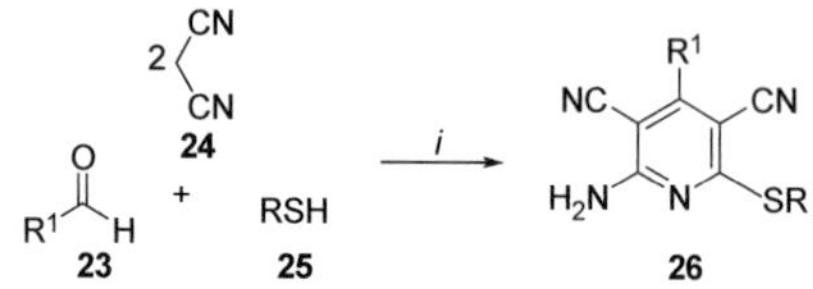

R = Ph, 4-Cl-Ph, Tol-4-yl, Tol-2-yl;
R^1 = Ph, 4-Me-Ph, 4-MeO-Ph, 3-MeO-4-HO-Ph, 2-Br-Ph, 3-Br-Ph, 4-Cl-Ph, 2,6-Cl_2-Ph, 4-O_2N-Ph, 4-MeS-Ph, 4-HO-Ph, Benzodioxol-1,3-yl, Thien-2-yl
i: [BMIM][OH]/EtOH, r.t., 1-2 h (65-95%).

Scheme 4

Protic ionic liquid [HBIM][BF_4] has been reported in the synthesis of quinolines [Palimkar et al., 2003]. The Friedländer heteroannulation protocol was used in ionic liquids at a molar ratio of 1:1 (reactant:IL), which made another catalyst unnecessary for the preparation of **30** (Scheme 5). Two sets of ionic liquids based on BBIM and HBIM salts were used. The capacity of the ionic liquids to promote these heterocyclization reactions was correlated to the basicity of their anions. The authors assumed that the nature of the anion governed the electrophilicity of the imidazolium cation, which in turn had a bearing on the acidity of the ionic liquid. It was observed that the higher the basicity of the anion (increasing pKa of the corresponding acid) the greater the increase in yield. [HBIM][BF_4] afforded the best result and, consequently, all further studies were conducted using this ionic liquid as the reaction medium. The ionic liquid was recovered and reused twice with no appreciable decrease in yield.

Karthikeyan [Karthikeyan & Perumal, 2004] investigated the quinolines synthesis using a [BMIM][Cl]:$ZnCl_2$ melt (1:2 molar ratio), which can act as both a solvent and catalyst on account of its high polarity and Lewis acidity. 2-Aminoketones **27** and ketones/ketoesters **14,17,28,29** were mixed in the [BMIM][Cl]:$ZnCl_2$ melt and stirred at room temperature for 24 h to give quinolines **30** in good to excellent yields (Scheme 5). The ionic liquid was recovered and reused twice with no appreciable decrease in the yield of **30** (89%, 86%). Theoretically, the Friedländer reaction with unsymmetrical ketones such as ethyl methyl ketone can have two possible modes of cyclization giving rise to two regioisomers, 2,3-dimethylquinoline and 2-ethylquinoline, respectively. The reaction path suggested for the Friedländer synthesis involved a sequential formation of the *N*-(2-acylphenyl)-β-enaminone/cyclodehydration reaction. The ionic liquid, promoting the Friedländer reaction with unsymmetrical ketones, regiospecifically afforded the 2,3-dialkylquinolines **30** in excellent yields. The author mentioned that polarity and the large electrochemical window of the ionic liquid may have also contributed to the observed regiospecificity. In the case of 2-aminoacetophenones **27**, the corresponding quinolines **30** were synthesized in excellent yields, that were in fact superior to those reported from conventional procedures using molecular solvent as ethanol under reflux for 12 h [Das et al., 2007].

From these examples, it can be found that for cyclocondensation reactions, ILs have designated present functions of solvent-catalyst, liquid support and co-promoters of the reaction by their task-specific acid or base functions (**Figure 2**). However, in answering the above-mentioned question of whether ILs are catalysts or solvents, based on this important finding and numerous other results collected in our review, we believe that the best approach to it is considering the ionic liquid as a solvent.

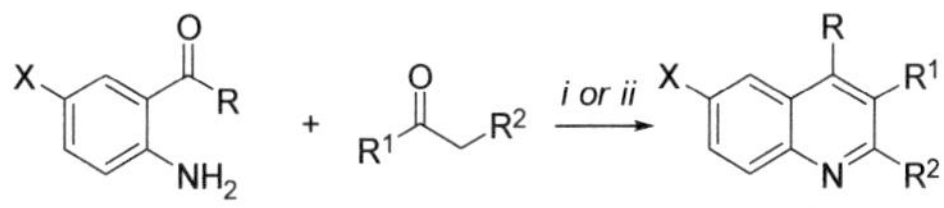

27 **14,17,28,29** **30**

14,30 (R^2 =H) **17,30** (R^2 = CO_2Et) **27,30** (R = Me, Ph; X = H, Cl)
28,30 (R^1 = Me, Ph, 4-Cl-Ph, 4-Br-Ph; R^2 =H, Me, $COCF_3$, $PhCH_2$)
29,30 R^1, R^2 = -$(CH_2)_3$-, -$(CH_2)_4$-, -$(CH_2)_5$-, -$(CH_2)_6$-,
i: [HBIM][BF_4], 100°C, 3-6 h (90-97%).
ii: [BMIM][Cl]:$ZnCl_2$, r.t., 24 h (55-92%).

Scheme 5

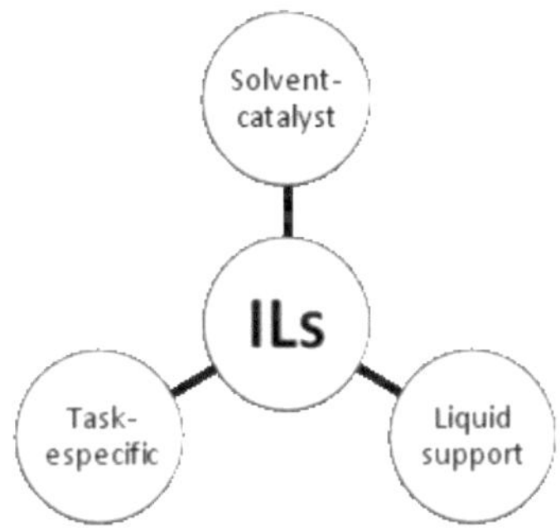

Fig. 2. IL effects in cyclocondensation reactions.

Considering the ionic liquid as a solvent, a single parameter of "polarity", "solvent strength", or "interaction" is not sufficient to explain the variation in experimental results in the many solvent-mediated processes. However, it is reasonable to postulate that the enhanced rate of the reactions is a result of the decrease of activation energy of the slow reaction step, which in turn is most likely due to the general ionic liquid effect. This can be expected for reactions involving highly polar or charged intermediates, such as carbocations or carbanions, and activated complexes, which could become more stable and long-lived in this media [Olivier-Bourbigou & Magna, 2002]. The influence of solvents on rate constants can be understood in terms of transition-state theory. According to this theory, solvents can modify the Gibbs energy of activation (as well as the corresponding activation enthalpies, activation entropies, and activation volumes) by differential solvation of the reactants and the activated complex. The effect of the solvent on reactions was investigated by Hughes and Ingold. They used a simple qualitative solvation model considering only pure electrostatic interactions between ions or dipolar molecules and solvent molecules in initial and transition states [Hughes & Ingold, 1935] and postulated that a change to a more polar solvent will increase or decrease the reaction rate depending on whether the activated reaction complex is more or less dipolar than the initial reactants (**Figure 3**). In this respect, the term "solvent polarity" was used synonymously with the power to solvate solute charges. It was assumed to increase with the dipole moment of the solvent molecules and to decrease with increased thickness of shielding of the dipole charges.
In summary, Welton and Oliver-Borbigou gave a superficial explanation of the role of ILs in organic reactions, asserting that they have a dual action of solvent-catalyst. Lee *et al.* also attempted to explain this, though they limited their explanation to stating that the effects of ILs on organic reactions were due to the stabilization of the reaction transition state. We

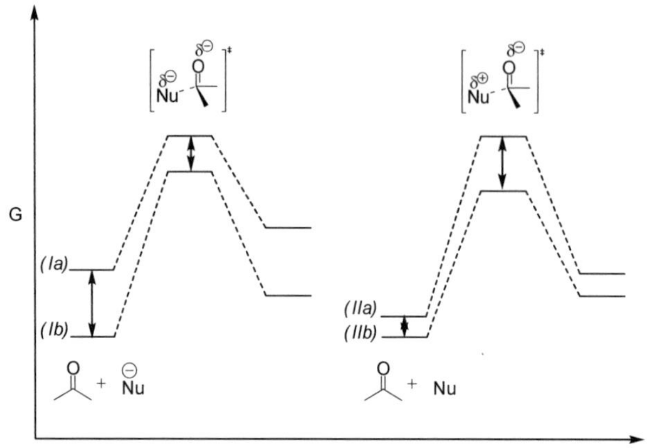

(a) Non polar solvents, and (b) polar solvents.

Fig. 3. Schematic Gibbs energy diagram for a general nucleophilic addition to carbonyl carbon.

have offered a more complete explanation by maintaining that the positive effects of ILs on cyclocondensation reactions are due to the fact that they cause a decrease in the activation energy of the slow reaction and a stabilization of transition states and highly polar or charged intermediates, such as carbocations and carbanions.

In our work, we have also performed an investigation in the Web of Science to show a proliferation of papers in the area of ionic liquids. We found more than 12,000 papers published in the period from 1990 to August 2010, in which more than 95% of the papers were published after the year 2000. These data show the increase in new researchers entering the area. On the other hand, less than 1% of all the papers published in the mentioned period dealt with the application of ionic liquids in heterocyclic synthesis from cyclocondensation reactions! This fact demonstrates that there is a lack in the literature of reports dealing with this theme, in particular on the synthesis of pyrazoles in ionic liquid media. Taking into account the importance of heterocycles, in special pyrazoles, and the environmental and economic need of their obtainment in a highly regioselective manner, and in accordance with works that we have developed in our research for more than twenty years, we decided to contribute to the research on ionic liquid effects in pyrazole synthesis by cyclocondensation reactions.

Our first work to this aim was published together with our review [Martins et al., 2008] and reported the synthesis of 4,5-dihydropyrazoles **33** from the reaction of enones **31** with hydrazine **32** in the presence of the ionic liquid [BMIM][BF_4]. The reaction was performed at 80°C during 1 h. The yields were higher and the reaction time was shorter in comparison to those found for the conventional method (MeOH, reflux, 16 h, 65-73%). (Scheme 6, *i*) [Sanin et al., 1998]. Later, we [Moreira et al., 2008] also employed ILs in the synthesis of 4,5-dihydropyrazoles **35** from the cyclocondensation reaction of enones **31** and cyanoacetohydrazide **34**. The reaction was carried out in [BMIM][BF_4], containing a catalytic amount of HCl conc., at 50°C, during 10-180 min, and the products were obtained in reasonable to good yields. The ionic liquid was used at a molar ratio of 1:1 (reactant:IL). It was possible to affirm that the IL [BMIM][BF_4] allowed the reaction to proceed in a shorter time than when the reaction was carried out in molecular solvents, even when a Brønsted catalyst (HCl) was present.

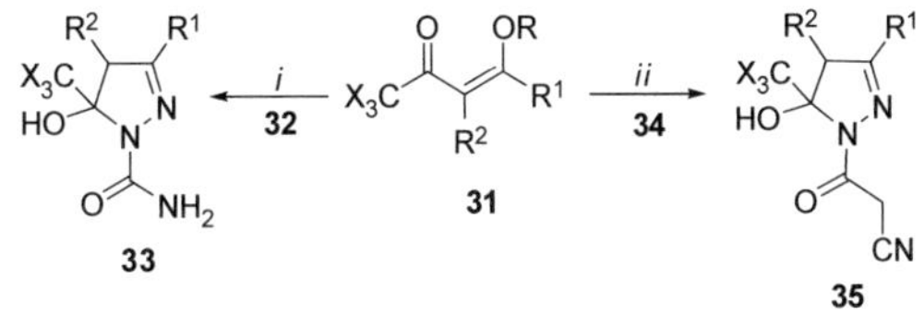

R = Me, Et R^1 = H, Me, Ph R^2 = H, Me
i: $NH_2NHCONH_2$, [BMIM][BF_4], 80°C, 1 h (73-86%)
ii: $NH_2NHCOCH_2CN$, [BMIM][BF_4], HCl, 50°C, 10-180 min (62-95%)

Scheme 6

In our continuing interest to demonstrate the effects of ILs on pyrazole synthesis, we used ionic liquid to promote cyclocondensation reactions to form bis-pyrazoles **38,39** from the reaction of enaminoketones **36** and a series of aryl and alkyl hydrazines **37** [Moreira et al., 2010]. In each case, three ionic liquids ([BMIM][BF_4], [HMIM][HSO_4] and [BMIM][OH]) were evaluated, and the first two proved to be the most suitable for these cyclocondensation reactions. The ionic liquid was used at the same molar ratio of the reactants **36** and **37** and the reactions were performed at 70-90°C (Scheme 7). We found that the catalytic power of [BMIM][BF_4] was improved when the acid HCl was employed. On the other hand, the use of [HMIM][HSO_4], an ionic liquid that combines polar properties with the Brønsted acidic function, promoted these reactions without the need of a co-catalyst. These reactions were performed in molecular solvents such as EtOH, DMF and H_2O, however, the results were unsatisfactory, leading to a lower regioselectivity and a longer reaction time.

In this same context, we published a study where a series of ten imidazolium-based ionic liquids ([BMIM][BF_4], [BMIM][Br], [OMIM][BF_4], [BMIM][PF_6], [DBMIM][Br], [DBMIM][BF_4], [BMIM][OH], [BMIM][SCN], [HMIM][HSO_4] [HMIM][CF_3CO_2]) was evaluted in the promotion of the cyclocondensation reaction of enaminones **40** and *t*-butyl hydrazine **41** (Scheme 8) to form pyrazole **42** [Frizzo et al. 2009]. The best result was achieved when the ionic liquid [BMIM][BF_4] was employed as the reaction media. Either the aryl-, heteroaryl-, alkyl-, or heteroalkyl β-dimethylaminovinyl ketone reacted with *t*-butylhydrazine hydrochloride **41** smoothly at 80°C for 1 h.

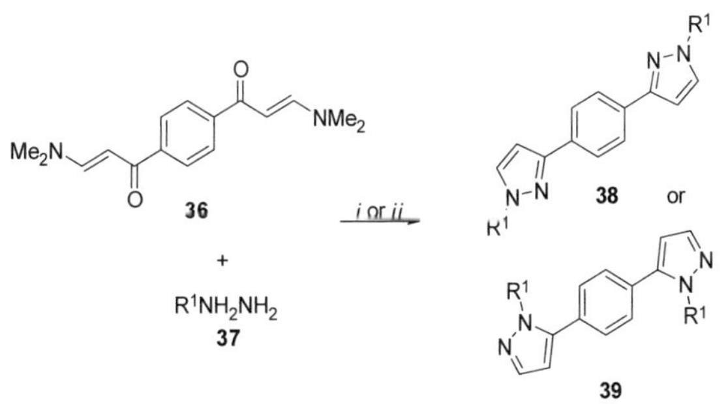

R^1 = H, Ph, 4-O_2N-Ph, 2,4-(O_2N)-Ph, C_6F_5, CO_2Me, $CONH_2$, CH_2CH_2OH, *t*-Bu
i: HCl or BF_3•OEt_2, [BMIM][BF_4], 70-90°C, 0.5-2 h (70-84%)
ii: [HMIM][HSO_4], 90°C, 0.5-3 h (60-88%)

Scheme 7

The reaction of 1-aryl- and 1-heteroaryl-substituted-β-dimethylaminovinyl ketones furnished the products **42** in good to excellent yields. On the other hand, the reaction of 1-hexyl-β-dimethylaminovinyl ketone with *t*-butylhydrazine proceeded, but the product was obtained as a mixture of 1,3- and 1,5-isomers at a ratio of 7:1, respectively. Therefore, it seems that the presence of an alkyl substituent makes the carbonyl carbon of β-dimethyl aminovinyl ketone as reactive as the carbon with carbonyl aryl- and heteroaryl substituents, leading to a mixture of isomers. The ionic liquid was used at the same molar ratio of the reactants. The recyclability and the reuse of the ionic liquid [BMIM][BF_4] was also investigated and it was found that the ionic liquid could be used for several runs without loss of activity. Reaction of **40** and **41** was also performed in ethanol, using the same reaction conditions, however, it led to the products in lower yields than those found with [BMIM][BF_4].

We went on to study the effect of ILs in cyclocondensation reactions to synthesize pyrazole by the evolution of [BMIM][BF_4] in the reaction of enones **31** and *t*-butyl hydrazine **41** to synthesize a series of 3(5)-trifluoromethylpyrazoles **43, 44** (Scheme 9).

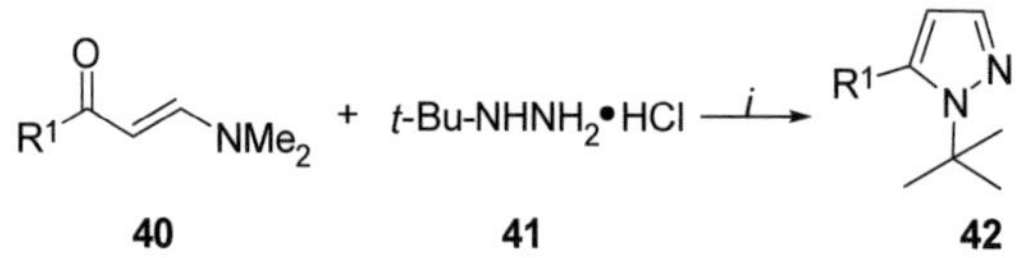

R^1 = Ph, 4-Br-Ph, 4-Cl-Ph, 4-F-Ph, 4-O_2N-Ph, 4-Me-Ph, Fur-2-yl, Thien-2-yl, Pyrrol-2-yl, Pyrid-2-yl, Hexyl.
i: [BMIM][BF_4], 80°C, 1 h (72-96%)

Scheme 8

The reaction was carried out at 78°C during 15 h, using the ionic liquid at a molar ratio of 1:2 in relation to the reactants. The products were obtained as a mixture in good to high yields (70-93%). The *1,5*-isomer was preferentially formed in ILs for most groups in R^1, as shown in Scheme 9. The presence of pyridine was necessary considering the loss of the t-butyl group when this base was not employed.

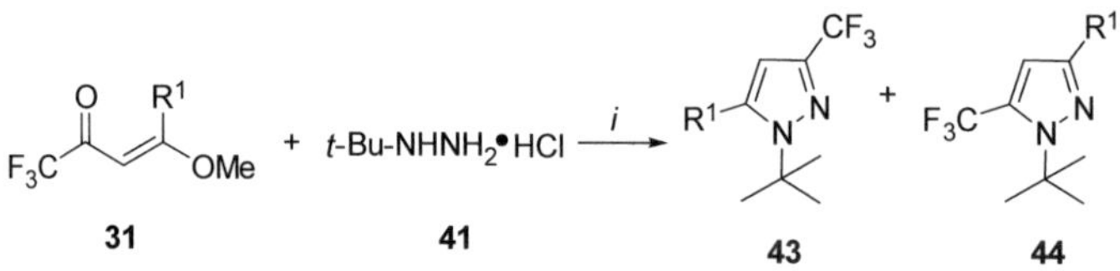

i: Pyridine, [BMIM][BF_4], 78°C, 15 h.

R^1	Molar Ratio **43:44**	Yield (%)[a]
Ph	15:85	85
4-Me-Ph	43:57	72
4-MeO-Ph	44:56	72
4-F-Ph	30:70	81
4-Cl-Ph	36:64	93
4-Br-Ph	39:61	93
Fur-2-yl	25:75	70
Thien-2-yl	57:43	75
Naphth-2-yl	47:57	84

[a] Yields of isolated product

Scheme 9

3. Cycloaddition reactions

A cycloaddition reaction is a reaction in which two or more unsaturated molecules (or parts of the same molecule) combine, with the formation of a cyclic adduct in which there is a net reduction of the bond multiplicity. Cycloadditions provide unsaturated or partially saturated (hetero) cycles with well-defined substitution patterns and often with high stereocontrol. For example, the Diels-Alder reaction has many examples that form heterocyclic as well as carbocyclic ring systems. This half of the literature is about [3+2] cycloaddition reactions which form 5 membered ring heterocyclic systems, in an analogous way to the [4+2] Diels-Alder process which forms 6-membered rings. The reactive partners in this reactions are 1,3-dipoles and dipolarophiles in 1,3-dipolar cycloadditions and diene and dienophile in the Diels-Alder reaction. There are also [2+1] cycloadditions that furnish aziridines and [2+2] cycloadditions that provide β-lactams. The Diels–Alder reaction (DA) [4+2] cycloaddition is one of the most intensively studied organic reactions, not only from a synthetic standpoint [Fringuelli & Taticchi, 2002] [Ibrahim-Ouali, 2009], but also from a theoretical point of view [Apeloig & Matzner, 1995][Imade et al, 1999]. From studies of cycloaddition reactions, in particular Diels-Alder reactions, have established that reactions between dienes (dipoles) and dienophiles (dipolarophiles) fit into the following general profile: (a) it is currently accepted that cycloadditions are concerted processes; they have no distinct intermediates, but the bond formation may be asynchronous; (b) the reaction rates are not influenced much by solvent polarity indicating little change in polarity between reactants and transition state; (c) rates of reaction between dienes (dipoles) and dienophiles (dipolarophiles) vary considerably. This can be explained by the Frontier Molecular Orbital Theory, which considers the interaction between molecular orbitals of the dienes (dipoles) and dienophiles (dipolarophiles).

3.1 Characterization of cycloaddition reactions

Among the cycloaddition reactions, 1,3-dipolar reactions have had an extensively successful history of use in heterocyclic synthesis. The 1,3-dipole is typically represented by closed-shell all-octet valence structures (I). They could be atmospheric components such as ozone (O_3) and nitrous oxide (N_2O), or highly popular azides (N_3R). The [4+2] thermal cycloadditions of 1,3-dipoles with alkene and alkyne dipolarophiles generate six- and five-membered heterocycles and are called 1,3-dipolar cycloadditions because of the dipolar nature of the principal resonance structures and the 1,3-additions that they undergo (**Figure 4**) [Huisgen, 1999].

$$X=\overset{\oplus}{Y}-\overset{\ominus}{Z} \longleftrightarrow \overset{\ominus}{X}-\overset{\oplus}{Y}=Z$$

I

The series depicted in **Figure 4** was constructed from the literature found for 1,3-dipolar cycloaddition reactions in ILs. Reactions were performed with different 1,3-dipoles and dipolarophiles. In general, dipolarophiles, as the C=C block, were enol ethers, alkenes, alkynes, α,β-unsaturated carbonyl or nitrile compounds, and nitriles and imines were the C≡N and C=N blocks, respectively. The blocks that most frequently represented 1,3-dipoles were enamines (C=C-N), ketones (C-C=O), aminoketones (N-C), hydroxylamines (N-O), aldoximes (O-N=C) and azides (N=N-N). In some cases, the 1,3-dipolar cycloaddition had multiple components, with blocks of one atom such as chloroamine-T (N), amine (N), aldehyde (C), enol ether (C) and orthoformate (C). Dipoles vary greatly in stability. Some can be isolated and stored, others are relatively stable, but are usually made on the same day of their use. Others are so unstable they are generated and reacted *in situ*.

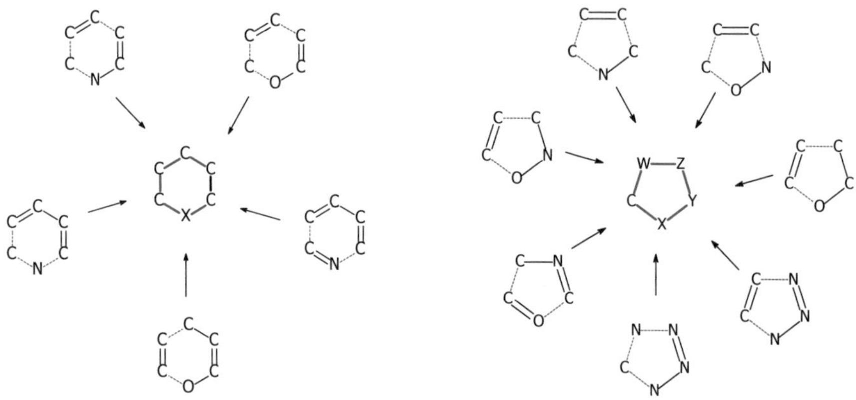

(a) X = O, N **(b)** X = O, N Y, Z, W = C, N

Fig. 4. General 1,3-dipolar cycloaddition and possible combination to form (a) six-membered and (b) five-membered heterocycles.

Table 3 shows the reaction type and building blocks that are found in 1,3-dipolar cycloadditon reactions for heterocyclic synthesis in ILs. The first three columns illustrate the chemical functions of the building blocks of dipolarophiles and 1,3-dipoles and the respective atoms involved in heterocyclic formation. The last column demonstrates the products. We designed Table 3 considering the expanded generalization and classification of 1,3-dipolar cycloadditions. From this classification, for so-called 1,3-dipoles "with a double bond"(I), atoms X and Z can be C, N, or O, while the center atom Y is nitrogen. These are typically referred to as propargylic species and have two sets of degenerate π-orbitals in a linear structure. Dipoles "without a double bond" may have a nitrogen function or oxygen atom at the central position and are isoelectronic with the allyl anion.

3.2 Ionic liquids in cycloaddition reactions

Since Breslow and Rideout [Rideout & Breslow, 1980] evidenced the dramatic accelerating effect of water on cycloaddition reactions in 1980, the solvent effect in this reaction received more attention. Diels–Alder reactions, for example, proceed at an appreciable rate only when either the diene or the dienophile are activated by an electron-donating or electron-withdrawing group, normally characterized by the presence of a heteroatom that can therefore efficiently interact with the solvent. Desimoni et al. [Desimoni et al., 1990] studied the solvent-substrate interaction in Diels–Alder reactions and classified these reactions into three types. Type A is characterized by an increase of the rate constant upon increasing the acceptor number (AN) power of the solvent. This behavior has been attributed to $LUMO_{solvent}$–$HOMO_{solute}$ interactions and considered similar to Lewis acid catalysis. Type B is dominated by the electron donation ability of the solvent, which decreases the reaction rate by soft–soft interactions: $HOMO_{solvent}$–$LUMO_{solute}$ interactions have been considered responsible for this effect. Type C includes all reactions that show a small solvent effect (for example, cyclopentadiene dimerization). In this case, solvent–solvent interactions are dominant.

Blocks	Blocks	Blocks	Product
Enol ether (C=C)	Enamine (C=C-N)	Aldehyde (C)	Tetrahydroquinolines
Enol ether (C=C)	Enamine (C=C-N)	Enol ether (C)	Tetrahydroquinolines
Enol ether (C=C)	Enamine (C=C-N)	Aldehyde (C)	Tetrahydroquinolines
Alkene (C=C)	Enamine (C=C-N)	Aldehyde (C)	Octahydroacridines
Alkene (C=C)	Enamine (C=C-N)	Aldehyde (C)	Tetrahydroquinolines
Alkene (C=C)	Ketone (C-C=O)	Aldehyde (C)	Coumarins
Fullerene (C=C)	Aminoketone (N-C)	Aldehyde (C)	Pyrrolidine
α,β-Unsaturated carbonyl compounds (C=C)	Hydroxylamine (N-O)	Aldehyde (C)	Isoxazolidine
α,β-Unsaturated nitrile compounds (C=C)	Hydroxylamine N-O	Aldehyde (C)	Isoxazolidine
Alkoxydiene (C=C-C=C)	Amine (N)	Aldehyde (C)	Pyridone
Azide (N=N-N)	Amine (N)	Orthoformate (C)	Tetrazole
α,β-Unsaturated carbonyl compounds (C=C)	Aldoxime (O-N=C)	-	2-Isoxazoline
α,β-Unsaturated nitrile compounds (C=C)	Aldoxime (O-N=C)	-	2-Isoxazoline
Heterosubstituted Alkene (C=C)	Aldoxime (O-N=C)	-	2-Isoxazoline
Enol ether (C=C)	Nitrone (O-N=C)		Isoxazolidine
Alkene (C=C)	Chloroamine-T (N)	-	Aziridine
α,β-Unsaturated carbonyl compounds (C=C)	Chloroamine-T (N)	-	Aziridine
Alkene (C=C)	1,4-benzoquinone (C-CO)	-	2,3-Dihydrobenzofuran
Alkene (C=C)	2-Hydroxyaldehyde (C=C-O)	-	Benzopyran
Nitroenamine (C=C)	Arylazide (N=N-N)	-	1,2,3-Triazole
Alkyne (C≡C)	Arylazide (N=N-N)	-	1,2,3-Triazole
Alkyne (C≡C)	Alkylazide (N=N-N)	-	1,2,3-Triazole
Alkyne (C≡C)	Azide (N=N-N)	-	1,2,3-Triazole
Nitrile (C≡N)	Azide (N=N-N)	-	Tetrazole
Nitrile (C≡N)	Azide (N=N-N)	-	Tetrazole
Imine (C=N)	Acid chloride (C-C)	-	Lactam
Imine (C=N)	Diene (C=C-C=C)	-	Tetrahydroquinolines
Imine (C=N)	Alkoxydiene (C=C-C=C)	-	Pyridone
Aldehyde (C=O)	Imidate (C=N-C)	-	Oxazoline

Table 3. Building blocks and products of 1,3-dipolar cycloaddition reactions in ILs.

Actually, the cohesive energy density (ced) (not to be confused with the internal pressure) together with the solvent H-bond acidity (α), has been shown [Gajewski, 1992] to also affect Diels–Alder Type A reactions, whereas in the special case of intramolecular Diels–Alder reactions in highly viscous media, Firestone *et al.* have demonstrated [Firestone & Saffar, 1983] the importance of solvent density. The considerations about the influence of ced on the reaction rate of cycloadditions raise the issue of solvophobic interactions, which are essentially being quantified by the ced. These parameters are also considered to determine the water effect. In terms of TS theory, hydrophobic hydration raises the initial state more than the TS and hydrogen bonding interactions stabilize the TS more than the initial state.

Highly polarizable activated complexes play a key role in these effects [Otto & Engberts, 2000].

Recently, Chiappe *et al.* [Chiappe et al., 2010] published a review about solvent effect on the Diels-Alder reactions in ILs. In this work, the authors discussed the role of ILs in these reactions considering multiparameter linear solvation energy relationships and theoretical analysis. The authors proposed that the endo:exo ratio and associated acceleration observed in the Diels–Alder addition of cyclopentadiene with methyl acrylate was attributed to the ability of the IL to hydrogen bond to the dienophile (methyl acrylate), a process considered to be determined by two competing equilibria. The IL cation ([BMIM]$^+$) can form a hydrogen bond to the anion of the IL (Eqn 1) or to the methyl acrylate (MA) (Eqn 2).

$$[BMIM]^{\oplus} + A^{\ominus} \rightleftharpoons [BMIM]^{\oplus}\cdots A^{\ominus} \quad (1)$$

$$[BMIM]^{\oplus} + MA^{\ominus} \rightleftharpoons [BMIM]^{\oplus}\cdots MA \quad (2)$$

Therefore, the authors proposed that the concentration of the hydrogen-bonded cation–methyl acrylate adduct is inversely proportional to the equilibrium constant for the formation of the cation–anion hydrogen-bonded adduct (K_1). In light of the more recent data on IL structure [Chiappe, 2007] they suppose that the interaction of the cation with reactants and/or the TS implies a reduction of the interactions of this cation with the surrounding anions. The dissolution of a substrate in a solvent, also including ILs, can be represented as follows: a "cavity" is created in the solvent to insert the substrate, and subsequently the reorganization and reorientation of the solvent around the solute occurs [Bruzzone & Chiappe, 2008]. In agreement with the system represented by Eqn 1 and Eqn 2, a strong interaction between IL cations and anions hinders the formation of the cavity, and reduces the rate of reorganization and reorientation of cations and anions around the reactants, thus decreasing the possibility for the cation (or anion) to solvate the reactant and/or the TS. However, the situation is much more complicated than that represented by Eqn 1 and Eqn 2, involving a system more complex than an ion pair, and in which kinetic effects also play a role during the solvent reorganization and reorientation. Also, in this review, Chiappe *et al.* [Chiappe et al., 2010] reported the more important findings of theoretical studies on Diels-Alder reactions in ILs. The main results of this investigation revealed that the Diels–Alder reaction in the presence of the imidazolium cation proceeds via a concerted mechanism similar to the "uncatalyzed" cycloaddition, although the asynchronicity of the process is increased by the presence of the imidazolium cations. The energetic differences between the reactants and the TSs for the endo and exo approaches, calculated for the three dienophiles in the presence of the above-mentioned cations, were qualitatively in agreement with the experimental data and confirmed the high selectivity in favor of the endo path for the reaction of cyclopentadiene with acrolein or methyl acrylate in a [HBIM]-based IL [Chiappe et al., 2010]. On the basis of these data, it was hypothesized that the interaction between the IL cation and the dienophile may be affected by the whole ionic system, and the expression "clamp-effect" was used to define this interaction. More specifically, the IL cation interacts with the dienophile acting as a "clamp", since in an IL, the freedom of motion of the cation is strongly limited by Coulombic interactions with the solvent bulk, which can be considered the clamp support. Therefore, the consequences of cation–dienophile interaction and of the clamp effect on Diels–Alder reactions is, from a simplistic view, that the interaction with the cation determines the polarization of the double bond of the dienophile, increasing its reactivity, whereas the clamp effect blocks one of the reactants, increasing the probability of efficient stacking in the TS. Another finding verified by Chiappe *et al.* [Chiappe et al., 2010] was that the presence of the IL changes the geometry of the TS for all four pathways considered, deforming the diene–dienophile stacking geometry and enhancing the asynchronicity of the reaction when performed in these solvents. Before examining in detail the solvation aspects, it is necessary to recall that the insertion of a solute in a solvent is characterized by a free energy of solvation that can be approximately divided in two parts: the change in electronic energy of the solute given by electrostatic and dispersion interactions with the solvent, and the change of solvent energy due to the necessary reorganization of the solvent molecules in order to embed the solute. The most important solvent effect on the reaction rate emerging from these calculations is given by the solvation free energy, which promotes the aggregation of non-ionic compounds. This "solvophobic" effect, which can be considered similar to that of water, arises from the fact that the (generally positive) solvation free energy of a neutral solute in an IL is dominated by the unfavorable process of creating a cavity of suitable size to accommodate the solute. This process in ILs requires a considerable amount of work due to the lowering of the Coulombic interactions, which cannot be recovered by dipole–ion (or even less efficient) interactions. This discussion is detailed in the work of Chiappe *et al.* [Chiappe et al., 2010].

As we have performed a survey about ionic liquids in heterocyclic synthesis by cyclocondensation reaction, likewise we have addressed cycloaddition reactions. The most important contributions were reported the role of ionic liquids in these reactions, in accordance with Chiappe *et al.* [Chiappe et al., 2010] was discussed.

Here, we have compiled the most important results, and we will briefly describe the reactions where ILs have had remarkable effects, such as rate increase, higher yields and endo/exo selectivity. An important effect of IL in the synthesis of isoxazoline dicarboxylates from the cycloaddition of carboethoxyformonitrile oxide (CEFNO) with different dipolarophiles (e.g., diethyl malonate and acrylonitrile) (Scheme 10), reported by Conti *et al.* [Conti et al., 2003] was the decrease of the by-product furoxane. The reaction conditions involved equimolar amounts of **45**, **46** and $KHCO_3$ in ionic liquid (molar ratio of 1:1.4/reactant:IL). The products were obtained in good yields with electron-rich alkenes and even conjugated dipolarophiles, however the authors also observed the formation of the sub-product furoxane in small amounts, which was formed by dimerization of unstable CEFNO that did not conclude the cycloaddition with acrylate. Probably, the presence of the ionic liquid in basic medium induced the thermodynamically unfavorable migration of the double bond to the terminal position, generating a more reactive alkene that immediately gave rise to cycloaddition. The authors also reported attempts to obtain compounds **47** with diethyl ether as solvent, however this route produced the required products in low yields (20–30%) together with larger amounts of furoxane, even when large amounts of the dipolarophile (45%) were used [Conti et al., 2003].

The better selectivity of cycloaddition reactions was reported by Rosella and Harper [Rosella & Harper, 2009] who performed cycloaddition reactions between benzonitrile oxide, generated *in situ* from the corresponding chloroaldoxime **48** and substituted alkenes (Scheme 11). Initially, the authors performed the cycloaddition reaction with alkene **46** (where, R^1 = Ph; R^2 = CO_2Et), that bear electronic and steric effects, and benzonitrile oxide **48** in three ionic liquids. The authors observed that the reaction in [BMIM][PF_6] furnished a mixture of 4,5-dihydroisoxazoles in a molar ratio of 1:>12 (9:10) and the best conversion

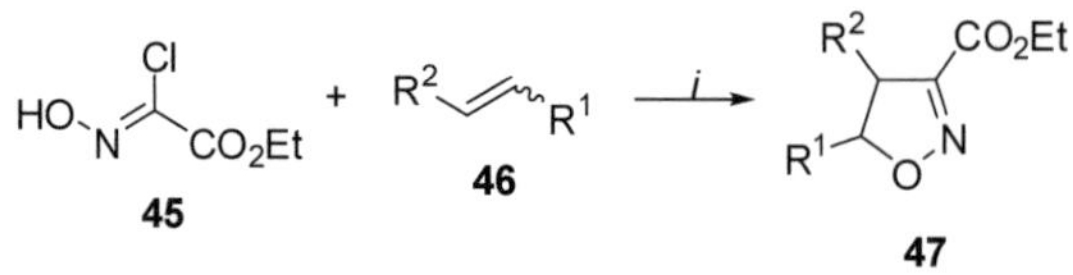

$R^1 = CO_2Et, CO_2Bn, CN, CH_2CN, OC(O)CH_3, SiMe_3, CH_2Cl$;
R^2 = H, *cis*-CO_2Et, *trans*-CO_2Et, *trans*-Me; R^1,R^2 = 2,3-Dihydrofuran, *c*-hexene.
i: $KHCO_3$, [BMIM][BF_4] or [BMIM][PF_6], r.t., 5-12 h (55-95%).

Scheme 10

(84%), when compared with the two other ionic liquids. The conversion in the water-soluble ionic liquid [BMIM][N(CN)$_2$] was very small and thus was disregarded. Steric interactions in the transition state leading to the isomers **49** and **50** are more significant in ionic liquids than they are in molecular solvents. The authors argued that ionic liquid have higher cohesive pressures than molecular solvents [Swiderski et al., 2004]. The amount of reactant and ionic liquid employed was not informed. The authors also performed this reaction in three molecular solvents (acetonitrile, ethyl acetate and THF) and observed similar diastereoisomeric ratios for products **49** and **50** (when R^1 = Me, Et and R^2 = CO_2Et, CH_2OH), however they presented lower molar ratios in regard to the product with R^1 = Ph and R^2 = CO_2Et. The difference between the reaction outcomes in the two ionic liquids is small when compared to the differences between ionic liquids and molecular solvents.

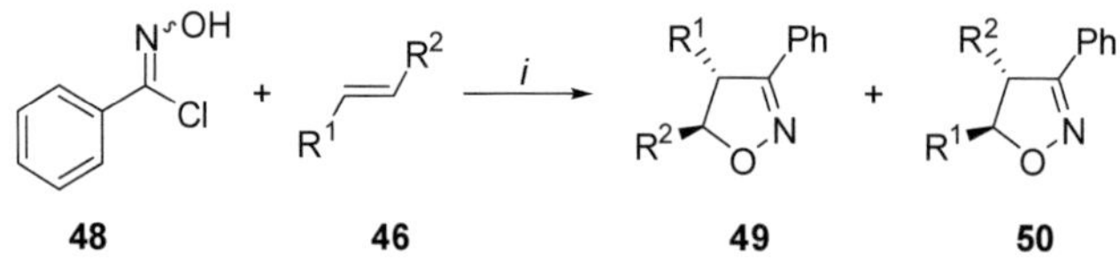

R^1 = Me, Et, Ph; R^2 = CO_2Et, CH_2OH
i: Et_3N, IL, r.t., 12-24 h

Scheme 11

Ionic liquids were found to reduce reaction time and to give better regioselectivity than organic solvents, as presented by Yadav *et al.* [Yadav et al., 2007] in the synthesis of isoxazolidines from 1,3-dipolar cycloaddition reactions of nitrones with electron deficient olefins (Scheme 12). The reaction between aldehyde, *N*-phenylhydroxyl amine and acrylonitrile was carried out in both hydrophilic and hydrophobic ionic liquids, ([BMIM][PF_6] or [BMIM][BF_4]), at room temperature during 4-6 h. The reactants **51**, **52** and **23** and the ionic liquid were used at a molar ratio of 1:1:1.2:10, respectively. Similar results in regard to the reaction rates and yields were obtained in both ionic liquids. The authors believe that the anticipated 1,3-dipoles exhibit enhanced reactivity in ionic liquid thereby reducing the reaction times and improving the yields significantly. Furthermore, the ionic liquids were found to give better regioselectivity than organic solvents since the reaction of C,N-diphenyl nitrone with ethyl acrylate in [BMIM][BF_4] gave the products **53** and **54** in 90% yield at a ratio of 9:1 over 4 h, whereas the same reaction in refluxing benzene gave the desired products **53** and **54** in 68% at a 2:1 ratio after 10 h [Yadav et al., 2007].
In another interesting study that showed the acceleration of cycloaddition reactions in ILs, Dubreuil et al. [Dubreuil & Bazureau, 2000] developed a route to obtain 2-oxazolines **57**, through cycloaddition reaction between imidate **55** and substituted benzaldehyde **56**, which

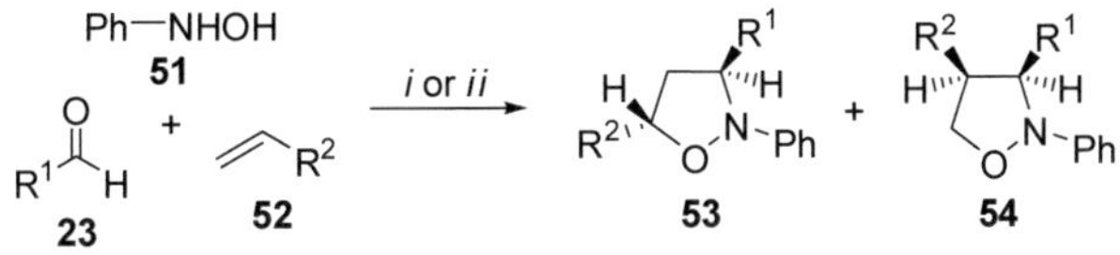

R¹ = Styryl, Ph, 3-Cl-Ph, 3-O_2N-Ph, 4-Cl-Ph, 4-MeO-Ph, 3,4-$(Cl)_2$-Ph, Benzo[3,4]dioxan-2-yl, Fur-2-yl; R² = CN, CO_2Me, COMe
i: [BMIM][BF_4], r.t., 4-5 h (85-93%)
ii: [BMIM][PF_6], r.t., 4.5-6 h (80-92%)

Scheme 12

acts as a dipolarophile (Scheme 13). Initially, the authors investigated the reaction of equimolar amounts of imidate **55** with 2-ethoxybenzaldehyde **56**, in different ionic liquids, maintaining the same temperature (70°C) in all tests. It was observed that the best rate acceleration was in the ionic liquid [EMIM][BF_4], due to the lower time (3 h) required for this reaction when compared with [EMIM][PF_6], which required 10 h for total conversion of the starting material. The addition of 5% of glacial acetic acid as Brønsted catalyst in the ionic liquid increased the reaction rate. However, the product yield was better when the ionic liquid [EMIM][PF_6] was used. This method for the synthesis of oxazolines **57** using molecular solvents has not yet been described in the literature.

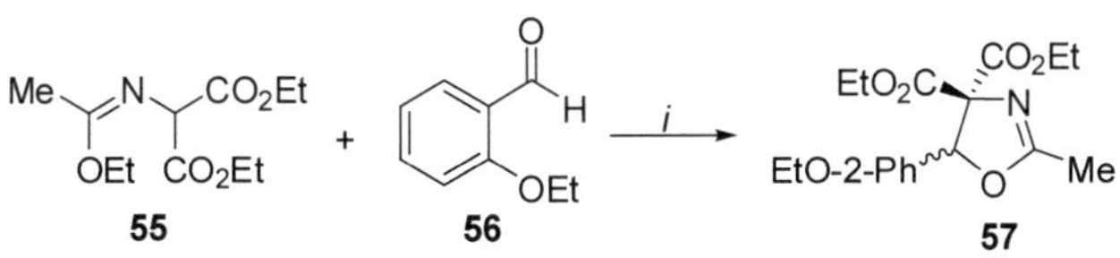

i: [EMIM][BF_4], 70°C, 3 h (70%)

Scheme 13

In a continuation, the authors explored the reactivity of benzaldehyde bound to the ionic liquid, furnishing the ionic liquid dipolarophiles **61** (X = BF_4, NfO), depicted in Scheme 14. The ionic liquid dipolarophiles **61** were reacted with imidate **55**, in equimolar amounts, under the same reaction conditions (70°C), although these reactions required different times. It is noteworthy that the reaction of imidate **55** with ionic liquid dipolarophile **61** was faster than the reaction of **55** with 2-ethoxybenzaldehyde **56** in [EMIM][NfO] ionic liquid. This acceleration observed with **61** is probably due to the intramolecular interaction between the CHO-group of the dipolarophile and the polar 3-methylimidazolium moiety.

Potewar *et al.* [Potewar et al., 2007] claimed the efficacy of ionic liquids to promote cycloaddition reactions to be related to the correlation between the basicity of the anions of the ionic liquids as well as their polarity. They reported a one-pot condensation of sodium azide **65**, substituted amines **64** and triethyl orthoformate **63** in 1-butylimidazolium tetrafluoroborate ([HBIM][BF_4]) at 100°C to afford 1-substituted-1*H*-1,2,3,4-tetrazoles **66**, without any added catalyst (Scheme 15). A variety of amines, such as substituted anilines, heteroaromatic and aliphatic, was employed to investigate the scope of this process. The data obtained revealed that both anilines containing electron-withdrawing and electron-donating groups promoted the cycloaddition reaction in short reaction times (15-35 min) and in good yields (85-93%). The reactants amine, triethyl orthoformate, sodium azide and ionic liquid were used at a molar ratio of 1:1.2:1:3, respectively. The authors assumed that

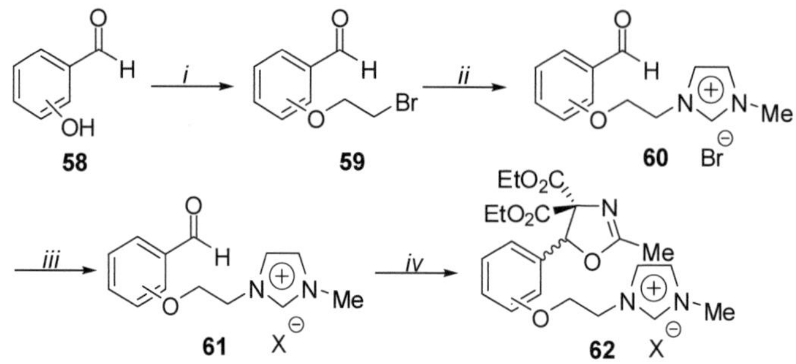

X = BF_4, NfO
i: $Br(CH_2)_2Br$, NaOH, reflux, 48 h (61-67%)
ii: for **61** (position-2) 72 h; for **61** (position-4) 80°C, 15 h (94%)
iii: NH_4BF_4, H_2O, 60°C, 18 h (X = BF_4, Yield: 60-80%)
$C_4F_9SO_3K$, H_2O, 70°C, 18 h (X = NfO, Yield: 53%)
iv: imidate (**55**), 70°C, 6-7 h (conv. 64-80%).

Scheme 14

the nature of the anion would influence the electrophilicity of the imidazolium cation, which in turn has a bearing on the acidity of the ILs. They observed that with the increasing basicity of the anion (increasing pKa of the corresponding acid), there was a progressive increase in the yield. The [HBIM][BF_4] afforded the best results by virtue of its inherent Brønsted acidity. The conventional methods reported for the synthesis of tetrazoles use either acidic conditions involving acids, such as hydrochloric, acetic, trifluoroacetic, and sulfuric, or highly polar solvents, such as 2-methoxyethanol, DMF, or methanol, and require very harsh reaction conditions such as refluxing for 6–24 h [Potewar et al., 2007].

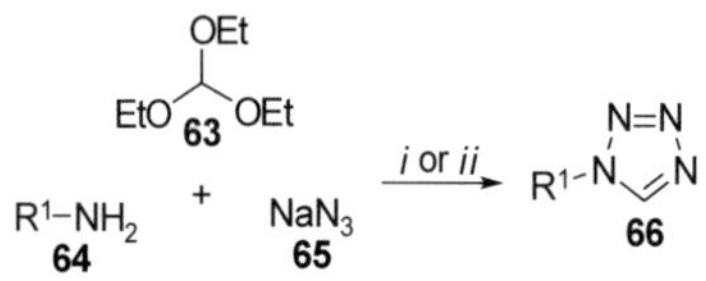

R^1 = Bn, Ph, 4-Me-Ph, 4-MeO-Ph, 4-Cl-Ph, 3-Cl-Ph, 4-F-Ph, 3-Me-Ph, 3-Cl-4-F-Ph, 4-Ac-Ph, Pyrid-2-yl, 4-Me-Pyrid-2-yl, 6-Me-Pyrid-2-yl
i: [HBIM][BF_4], 100°C, 15-35 min (85-93%)
ii: [BBIM][Br]/DMSO, 30°C, 20-90 min (85-90%)

Scheme 15

Our contributions to studies of ionic liquid in heterocyclic synthesis by cycloaddition reactions are quite in beginning, we are preparing a review about heterocyclic synthesis by 1,3-dipolar cycloaddtion and we have performed a cycloaddition reaction between equimolar amounts of oximes **67** and phenyl acetylene **68** in presence of N-chlorosuccinimide and Et_3N in IL to produce isoxazoles **69** and **70** (Scheme 16). The IL was used in the same amount of the reactants. The reaction of oxime **68** (R = H), was also performed in reflux acetonitrile, for 5 h, however the product **69**, **70** was obtained in 70% of yield and in a molar ratio of 3:1 respectively.

In summary, from these examples, we found that the deployment of ILs in 1,3-dipolar cycloaddition reactions has brought about the reduction of reaction times, the increase of yields and better endo/exo selectivity in comparison to the use of molecular solvents. In addition, from these important findings and numerous other results collected in the literature, we believe that the outstanding premises introduced by Chiappe *et al.* [Chiappe et al., 2010] for Diels-Alder reactions can be extended to the 1,3-dipolar cycloaddition reaction as a good approach to explain the solvent effect of ILs. Therefore, rate enhancements of 1,3-dipolar cycloaddition reactions in ILs could be ascribed to the same factors that are found in water: (*i*) increased polarity of the transition state, (*ii*) the hydrophobic effect which aggregates organic reactants raising the energy of the ground state relative to the transition state, thereby lowering the activation energy, and (*iii*) special or enhanced hydrogen bonding effects in the transition state.

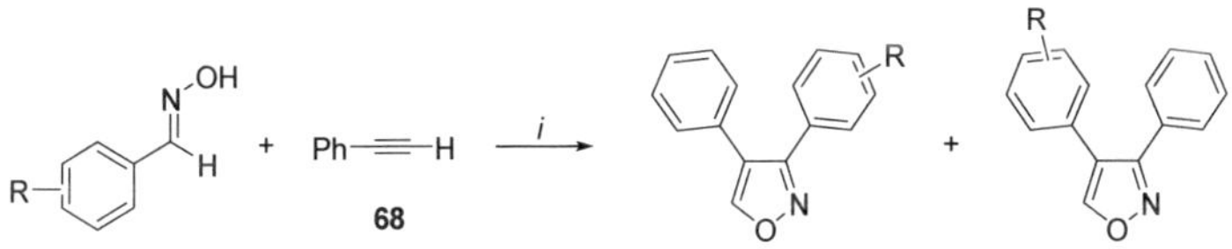

67 68 69 70

i: NCS, Et_3N, [BMIM][BF_4], 90°C, 5h.

R	Molar Ratio 69:70	Yield (%)	R	Molar ratio 69:70	Yield (%)
H	10:1	84	4-Me	3:1	55
2-OH	5:1	42	2-Me	3:1	62
4-Cl	10:1	65	Thien-2-yl	1:1	41
4-OH	3:1	42			

Scheme 16

4. Conclusion

After having examined extensive cyclocondensation and cycloaddition reactions described in the literature, it is necessary to return to the initial question. What is the role of ILs in these reactions? Now, it is clear that to adequately answer this question, the characteristics of each reaction must be considered. In this review, the main effects of the ionic liquids observed in cyclocondensation reactions were to improve the reaction yields and to shorten the reaction time. In cycloaddition reactions, besides the increase of yields and reduction of reaction time a better endo/exo selectivity was observed, in comparison to that found with molecular solvents. One can rationalize that ILs are good solvents for both cyclocondensation and cycloaddition reactions, due to the stabilization of TS of both reactions and, particularly, to the dual effect of catalyst-solvent and liquid-support in cyclocondensations and to the "solvophobic" effect in cycloaddition reactions. Moreover, considering the importance of the cyclocondensation and cycloadditon reactions, the main reactions in heterocyclic synthesis, the information presented here clearly illustrates the substantial advances achieved over the past decade in the use of ionic liquids as solvent in organic reactions. In addition, clear advantages of using ionic liquids, such as increased reaction rates and product yields and the possibility of avoiding complex workup procedures and of reusing these solvents have been demonstrated. Undeniably, ionic liquids enable more efficient reactions to take place when compared with molecular solvents. However, it is necessary to optimize both reactions in order to be able to make a truly accurate comparison. Large increases in reactivity and selectivity have been achieved using this medium for homogeneous reactions, and in some cases, reactions have been shown to only work in the ionic environment and not in molecular solvents. In this chapter, we hope to have given a clear idea of the applicability of ionic liquids in cyclocondensation and cycloaddition reactions. We would like to conclude with an optimistic view for the future expansion of these reactions in ionic liquid media. This positive view comes from the certainty that the results reported here will be the beginning of a great advance in this promising field in the near future.

5. References

Apeloig, Y. & Matzner, E. (1995) Evidence for the Dominant Role of Secondary Orbital Interactions in Determining the Stereochemistry of the Diels- Alder Reaction: The Case of Cyclopropene. *Journal of American Chemical Society* 117, 5375-5376.

Bagley, M. C.; Brace, C.; Dale, J. W.; Ohnesorge, M.; Phillips, N. G.; Xiong, X.; Bower, J. (2002). Synthesis of tetrasubstituted pyridines by the acid-catalysed Bohlmann–Rahtz reaction. *Perkin Transactions 1, 1,* 1663-1671.

Bienaymé, H.; Bouzid, K. (1998). A New Heterocyclic Multicomponent Reaction For the Combinatorial Synthesis of Fused 3-Aminoimidazoles. *Angewandte Chemie International Edition 37,* 16, 2234-2237.

Bruzzone, S.; Malvaldi, M.; Chiappe, C. (2008). Solvation thermodynamics of alkali and halide ions in ionic liquids through integral equations. *Journal of Chemical Physics* 129, 74509.

Chiappe, C. (2007). Nanostructural Organization of Ionic Liquids: Theoretical and Experimental Evidences of the Presence of Well Defined Local Structures in Ionic Liquids. *Monatshefte für Chemie,* 138, 1035-1043.

Chiappe, C.; Malvaldi, M.; Pomelli C. S. (2010). The solvent effect on the Diels–Alder reaction in ionic liquids: multiparameter linear solvation energy relationships and theoretical analysis. *Green Chemistry* 12, 1330-1339.

Chiappe, C.; Pieraccini, D. (2005). Ionic liquids: solvent properties and organic reactivity. *Journal of Physical Organic Chemistry* 18, 4, 275-297.

Comprehensive Heterocyclic Chemistry II; Katritzky, A. R., Ress, C. W., Scriven, E. F. V., Eds.; Pergamon: New York, 1996; Vols. 1-8.

Conti, D.; Rodriquez, M.; Sega, A.; Taddei, M. (2003). 1,3-Cycloaddition of nitrile oxides in ionic liquids. An easier route to 3-carboxy isoxazolines, potential constrained glutamic acid analogues. *Tetrahedron Letters* 44, 5327–5330.

Das, B.; Damodar, K.; Chowdhury, N.; Kumar, R. A. (2007). Application of heterogeneous solid acid catalysts for Friedlander synthesis of quinolines. *Journal of Molecular Catalysis A: Chemical 274,* 148-152.

Desimoni, G.; Faita, G.; Righetti, P.; Toma, T. (1990). Solvent effect as the result of frontier molecular orbital interaction. VI. The Diels-Alder reactions of an N-acyl-o-quinone monoimide behaving as diene or dienophile. *Tetrahedron,* 46, 23, 7951-7970.

Dubreuil, J. F. & Bazureau, J. P. (2000). Rate accelerations of 1,3-dipolar cycloaddition reactions in ionic liquids. *Tetrahedron Letters,* 41, 38, 7351-7355.

Enguehard, C.; Allouchi, H.; Gueiffier, A.; Buchwald, S. L. (2003). Easy Access to Novel Substituted 6-Aminoimidazo[1,2-*a*]pyridines Using Palladium- and Copper-Catalyzed Aminations. *Journal of Organic Chemistry, 68,* 11, 4367-4370.

Evdokimov, N. M.; Magedov, I.; Kireev, A. S.; Kornienko, A.(2006). One-Step, Three-Component Synthesis of Pyridines and 1,4-Dihydropyridines with Manifold Medicinal Utility. *Organic Letters, 8*, 5, 899-902.

Firestone R. A. & Saffar, S. G. (1983). Vibrational Activation. 3. Effect of Solvent Density on the Rate of a Diels-Alder Reaction. Density as an Index of Porosity. *Journal of Organic Chemistry*, 48, 24, 4783-4784.

Fringuelli, F. & Taticchi, A. (2002). *The Diels–Alder reaction: selected practical methods*, John Wiley & Sons, New York.

Frizzo, C. P.; Marzari, M. R. B.; Buriol, L.; Moreira, D. N.; Rosa, F. A.; Vargas, P. S.; Zanatta, N.; Bonacorso, H. G. & Martins, M. A. P. (2009). Ionic liquid effects on the reaction of β-enaminones and tert-butylhydrazine and applications for the synthesis of pyrazoles. *Catalysis Communications*, 10, 15, 1967–1970.

Gajewski, J. J. (1992). A Semitheoretical Multiparameter Approach To Correlate Solvent Effects on Reactions and Equilibria. *Journal of Organic Chemistry*, 57, 5500-5506.

Greave, T. L.; Drummond, C. J. (2008). Protic Ionic Liquids: Properties and Applications. *Chemical Reviews*, 108, 206-237.

Hu, Y.; Wei, P.; Huang, H.; Han, S.-Q.; Ouyang, P.-K. (2006). Synthesis of 2-Aminothiophenes on Ionic Liquid Phase Support using the Gewald Reaction. *Synthetic. Communication 36*, 11, 1543-1548.

Hughes, D. E. & Ingold, C. K. (1935). Mechanism of substitution at a saturated carbon atom. Part IV. A discussion of constitutional and solvent effects on the mechanism, kinetics, velocity, and orientation of substitution. *Journal of Chemical Society*, 244-255.

Huisgen, R. (1984) *1,3-Dipolar Cycloaddition Chemistry*; Padwa, A., Ed.; John Wiley and Sons: New York.

Ibrahim-Ouali, M. (2009). Diels–Alder route to steroids and associated structures. *Steroids* 74, 133-162.

Imade, M.; Hirao, H.; Omoto, K.; Fujimoto, H. (1999). Theoretical Study of Endo Selectivity in the Diels-Alder Reactions between Butadienes and Cyclopropene. *Journal of Organic Chemistry*, 64, 6697-6701.

Ingold, C. K. (1969). *Structure and Mechanism in Organic Chemistry*, 2nd ed; Cornell University Press: Ithaca, NY.

Karthikeyan, G.; Perumal, P. T. (2004). A mild, efficient and improved protocol for the friedländer synthesis of quinolines using lewis acidic ionic liquid. *Journal of Heterocyclic Chemistry*, 41, 1039-1041.

Karthikeyan, G.; Perumal, P. T. (2005). Ionic liquid promoted simple and efficient synthesis of β-enamino esters and β-enaminones from 1,3-dicarbonyl compounds - One-pot, three-component reaction for the synthesis of substituted pyridines. *Canadian Journal of Chemistry* 83, 1746-1751.

Lee, J. W.; Shin, J. Y.; Chun, Y. S.; Jang, H. B.; Song, C. E.; Lee, S. (2010). Toward Understanding the Origin of Positive Effects of Ionic Liquids on Catalysis: Formation of More Reactive Catalysts and Stabilization of Reactive Intermediates and Transition States in Ionic Liquids. *Accounts Chemical Research*, 43, 7, 985–994.

Lütjens, H.; Zickgraf, A.; Figler, H.; Linden, J.; Olsson, R. A.; Scammells, P. J. (2003). 2-Amino-3 benzoylthiophene Allosteric Enhancers of A_1 Adenosine Agonist Binding: New 3, 4-, and 5-Modifications *Journal of Medicinal Chemistry 46*, 10, 1870-1877.

MacFarlane, D. R.; Pringle, J. M.; Johansson, K. M.; Forsyth, S. A. M. Forsyth (2006). Lewis base ionic liquids. *Chemical Communication* 18, 1905–1917.

Martins, M. A. P.; Frizzo, C. P.; Moreira, D. N.; Zanatta, N. & Bonacorso, H. G. (2008) Ionic Liquids in Heterocyclic Synthesis. *Chemical Reviews*, 108, 6, 2015–2050.

Moreira, D. N.; Frizzo, C. P.; Longhi, K.; Zanatta, N.; Bonacorso, H. G. & Martins, M. A. P. (2008). An efficient synthesis of 1-cyanoacetyl-5-halomethyl-4,5-dihydro-1*H*-pyrazoles in ionic liquid. *Monatsheft für Chemie*, 139, 9, 1049-1054.

Moreira, D.N.; Longhi, K.; Frizzo, C.P.; Bonacorso, H.G.; Zanatta, N. & Martins, M.A.P. (2010). Ionic liquid promoted cyclocondensation reactions to the formation of isoxazoles, pyrazoles and pyrimidines. *Catalysis Communications*, 11, 5, 476–479

Muller, P. (1994). Glossary of terms used in physical organic chemistry (IUPAC Recommendations). *Pure Appl. Chem. 66*, 5, 1077-1184.

Olivier-Bourbigou, H.; Magna, L. (2002). *Journal of Molecular Catalysis A: Chemical 182-183*, 419-437.

Otto S. & Engberts, J. B. F. N. Diels–Alder reactions in water. (2000). *Pure Applied Chemistry* 72, 7, 1365–1372.

Palimkar, S.S.; Siddiqui, S. A.; Daniel, T.; Lahoti, R. J.; Srinivasan, K. V. (2003). Ionic Liquid-Promoted Regiospecific Friedlander Annulation: Novel Synthesis of Quinolines and Fused Polycyclic Quinolines *Journal of Organic Chemistry, 68*, 9371-9378.

Pârvulescu, V.I.; Hardacre, C. (2007). Catalysis in Ionic Liquid. *Chemical Reviews, 107*, 6, 2615-2665.

Potewar, T. M.; Siddiqui, S. A.; Lahoti, R. J.; Srinivasan, K. V. (2007). Efficient and rapid synthesis of 1-substituted-1H-1,2,3,4-tetrazoles in the acidic ionic liquid 1-n-butylimidazolium tetrafluoroborate. *Tetrahedron Letters*, 48, 1721–1724.

Ranu, B. C.; Ranjan, J.; Sowmiah, S. (2007) An Improved Procedure for the Three-Component Synthesis of Highly Substituted Pyridines Using Ionic Liquid. *Journal of Organic Chemistry*, 72, 3152-3154.

Rideout, D. C. & Breslow, R. (1980). Hydrophobic Acceleration of Diels-Alder Reactions. *Journal of American Chemical Society, 102*, 7816-7817.

Rosella, C. E.; Harper, J. B. (2009). The effect of ionic liquids on the outcome of nitrile oxide cycloadditions. *Tetrahedron Letters, 50*, 992-994.

Sanin, A. V.; Nenaienko, V. G.; Kuzmin, V. S.; Balenkova, E. S. (1998). Synthesis of trifluoromethyl derivatives of pyrazolidine- and 2-pyrazoline-1-carboxamides and pyrazolidine- and 1-pyrazoline-carbothioamides *Chemistry of Heterocyclic Compounds 34, 5*, 558-567.

Swiderski, K.; McLean, A.; Gordon, C. M.; Vaughn, D. H. (2004). Estimates of internal energies of vaporisation of some room temperature ionic liquids. *Chemical Communications*, 2178–2179.

Wasserscheid, P. & Welton, T. (2008). *Ionic Liquids in Synthesis*, Wiley- VCH, Weinheim.

Welton, T. (1999). Room-Temperature Ionic Liquids. Solvents for Synthesis and Catalysis. *Chemical Reviews, 99*, 8, 2071-2084.

Xia, C.-G.; Sun, W.; Wang, H.-W. (2003). Synthesis of aziridines from imines and ethyl diazoacetate in room temperature ionic liquids. *Tetrahedron Letters,* 44, 11, 2409-2411.

Xie, W. H.; Fang, J. W.; Li, J.; Wang,b P. G. (1999). Aziridine synthesis in protic media by using lanthanide triflates as catalysts. *Tetrahedron, 55*, 45, 12929-12938.

Xu, J.; Stevenson, J. (2000). Drug-like Index: A New Approach To Measure Drug-like Compounds and Their Diversity. *Journal of Chemical Information Computation Science, 40*, 5, 1177-1187.

Yadav, J.S.; Reddy, B.V.S.; Sreedhar, P.; Murthy, Ch.V.S.R.; Mahesh, G.; Kondaji, G.; Nagaiah K. (2007). Three-component coupling reactions in ionic liquids: One-pot synthesis of isoxazolidines. *Journal of Molecular Catalysis A: Chemical* 270, 160–163.

18

Current Knowledge and Potential Applications of Ionic Liquids in the Petroleum Industry

Murillo-Hernández José-Alberto[1] and Aburto Jorge[2]
[1]*Universidad Nacional Autónoma de México,*
[2]*Instituto Mexicano del Petróleo*
Mexico

1. Introduction

The modern petroleum industry is confronting many challenges throughout the chain value, since the production and refining of crude oil to the manufacture of high-value petrochemical products. Conventional light and middle crude oil production reached a maximum and tends to decline. Also, some producing countries are dealing with heaviest and higher content of pollutants (sulfur, nitrogen and metals) crude oils, which are being produced in countries like Canada, China, Mexico, Venezuela and the USA. Such heavy and extra heavy crude oils (HCOs) are more difficult to produce, transport, refine and convert to useful petrochemicals. Sometimes, conventional technologies may be applied but in most cases, new technologies are needed in order to allow economical HCOs production, transportation through pipelines, avoiding pressure drops, pipeline clogging and production stops due to asphaltene and paraffin aggregation and precipitation. Also, the higher content of heteroatoms in HCOs makes them more difficult to refine because of mainly catalyst deactivation that decreases the efficiency of processes like desulfurization, denitrogenation and demetallization. Moreover, the crude oil composition largely varies from one oil field to another among deposits along the countries. Heavy and extra-heavy oil are characterized by an API gravity between $20<API<10$ and $API\leq10$, respectively. The latter oils present a specific gravity near or above 1.0 g/cm^3, i.e. they are as heavier or more than water, making crude oil dehydration, for example, very difficult to reach before refining. Therefore, the modern petroleum industry is facing several problems in petroleum production and refining.

In last years, ionic liquids (ILs) have attracted the attention because of their properties and versatility of potential application in the petroleum industry. ILs may be use to augment the flowability of viscous and bituminous HCOs that could augment oil production and transportation, inhibit the aggregation of asphaltenes and paraffins, and diminish the deactivation of refining catalyst by removing heteroatoms and coke from the fuel and catalyst's surface. Because of the technological, environmental and economical impacts, this work reviews the current knowledge and potential application of ILs in petroleum industry.

2. Petroleum production

According to Speight (1999), crude oil is a mixture of gaseous, liquid and solid hydrocarbon that occur in porous rock deposits called reservoirs and that contains variable quantities of nitrogen-, oxygen-, and sulphur-containing compounds as well as traces of metals. Such complex mixture may be separated in four constituent fractions: saturates (S), aromatics (A), resins (R) and asphaltenes (A); known as SARA fractions (Table 1). Since these major crude oil fractions are obtained by consecutive separation methods (Figure 1), it is difficult to base

	Mexico			**Canada**	**Venezuela**	**China**
Parameter	*Medium*	*Heavy*	*Extra-heavy*	*Bitumen*	*Extra-heavy*	*Extra-heavy*
API gravity	21.3	11.9	9.2	5.9	10	7.2
Molecular weight (g/mol)	314.8	486	507.8	ND	ND	ND
Sulfur content (%)	3.4	5.0	4.8	4.9	3.2	ND
Water content (%)	1.80	0.05	<0.05	0.0	ND	ND
Carbon residue (%)	0.0	0.0	0.0	18.5	ND	ND
SARA analysis (%)						
Saturates	26.5	7.9	15.0	15.1	11.0	36.3
Aromatics	14.7	5.3	19.1	ND	19.0	23.6
Resins	47.6	70.9	46.8	34.1	54.0	32.7
Asphaltenes	11.1	15.9	19.1	16.9	16.0	7.4

Table 1. Properties and composition of some medium, heavy and extra-heavy crude oils from Canada, China, Mexico and Venezuela. Adapted from (Isea, 1987; Speight, 1999; Zhang, 2005; Murillo-Hernández, 2009; Castro, 2009; Riveros, 2006; Strausz, 2010). ND: not determined.

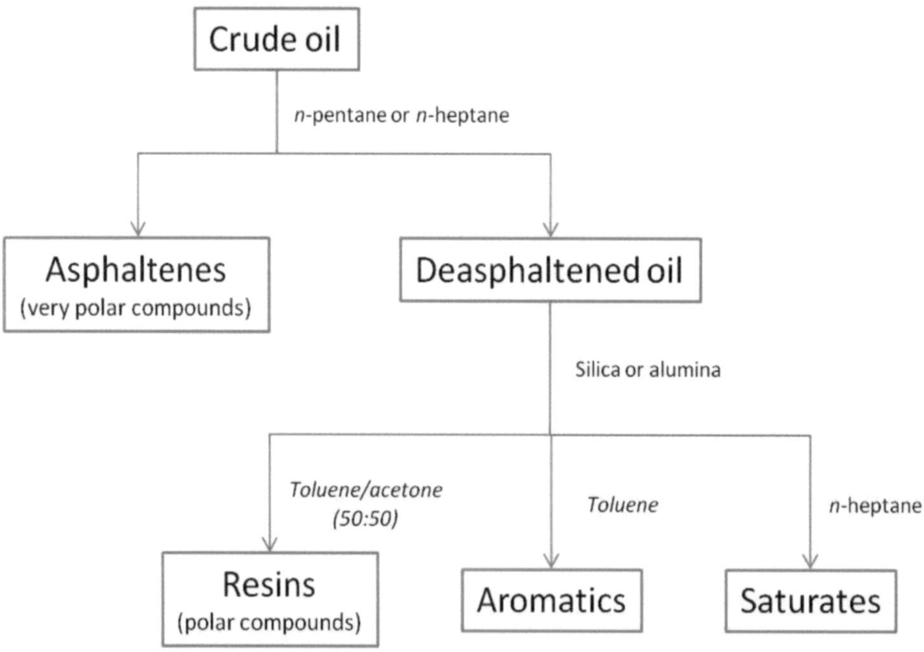

Fig. 1. Simplified fractionation method of crude oil.

such fractions on chemical or structure features. Nevertheless, saturates are found in the n-heptane eluent and are composed by straight or branched acyclic (also called paraffins) and cyclic saturated hydrocarbons ($19 \le C_n \le 40$); aromatics are found in the toluene eluent and we can consider them as hydrocarbons containing one or more aromatic ring systems substituted by other aromatic rings or saturated hydrocarbons. Resins or polar compounds are desorbed with a toluene/acetone eluent; while asphaltenes are the fraction insoluble in *n*-pentane or *n*-heptane and are considered polynuclear aromatic systems and the more polar fraction of crude oil (Speight, 1999; Speight, 2004a; ASTM, 2007; Strausz et al., 2010). Hence, each SARA fraction corresponds to an enormous group of compounds, quantifed in the laboratory by solvent precipitation, adsorption and solvent extraction; that contributes to the physicochemical properties of the crude oil like API gravity and viscosity. Heavy and extra-heavy crude oils are mainly characterized by an API gravity between $20<API<10$ and $API\le 10$, respectively; and high viscosities (>10-10^4 Pa*s). Such crude oils sites are regionally found in countries like Canada, China, Mexico, Venezuela and USA (Figure 2).

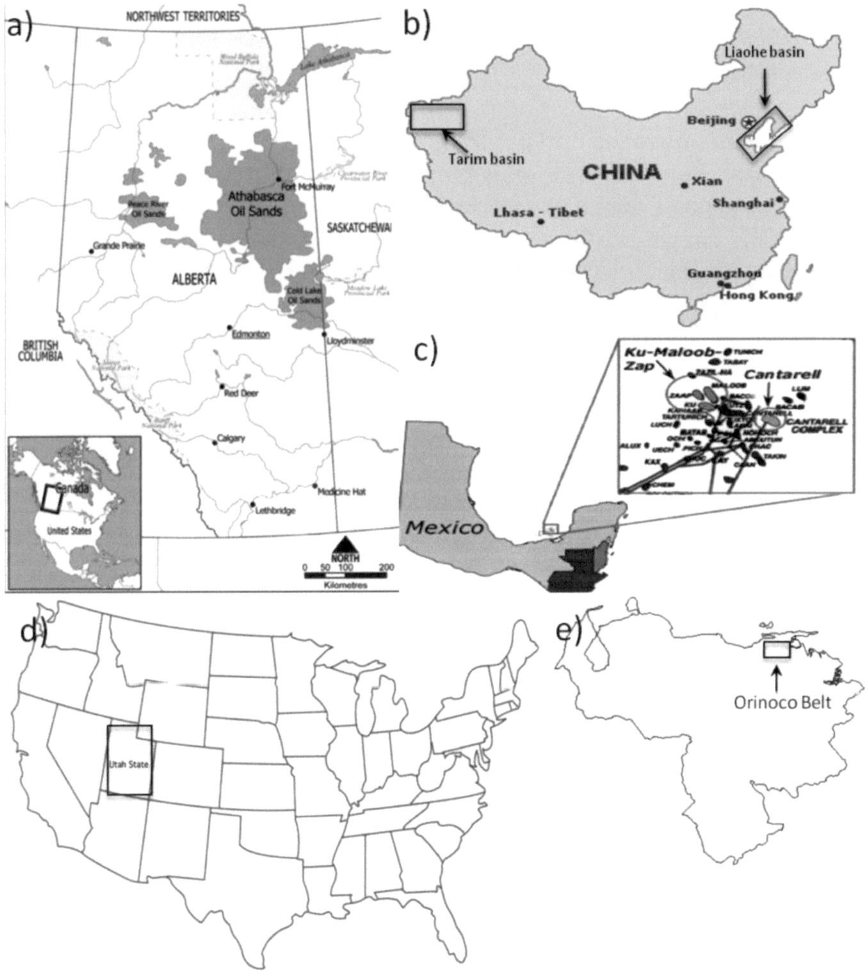

Fig. 2. Production sites of heavy and extra-heavy oil in the world: a) Canada, b) China, c) Mexico, d) USA and e) Venezuela.

Heavy and extra-heavy crude oil production from reservoirs and transport to terminals presents several problems associated with oil production, precipitation of asphaltenes, resins and paraffins (saturated hydrocarbons with straight or branched chains), petroleum's high viscosity and low flowability, formation of water-in-oil (W/O) emulsions, corrosion of production and pipeline infrastructure and decaying of petroleum production. A recent work deals with the evaluation of extra-heavy oil production from Canadian and Utah tar sands using ILs (Painter et al., 2010a, Painter et al., 2010b). They could recover till 90% yield of extra-heavy oil from Canadian tar sands using common ILs like $[Bmim]^+[CF_3SO_3]^-$ and $[Bmmim]^+[BF_4]^-$ and ILs could be recycled till 5 times without noticeable loss of efficiency. Nevertheless, Utah tar sand needed the use of consecutive additions of toluene as cosolvent apparently because of over saturation. These are the first reports concerning the production of crude oil using ILs and some advantages are the elimination or reduction of water use during oil extraction, good separation of sand and clay from crude oil, reduction of corrosion issues, and ease separation of ILs from tar sand. Although, this is only an exploratory study that needs more work in order to address a number of issues like the choice of IL, crude oil/IL ratio, use of cosolvents, the kinetics of the separation process, costs, and environmental concerns.

2.1 Deposition of asphaltenes and paraffins

The stability of asphaltenes in crude oil is proposed to be due to the presence of some polar substances, resins among them, present in crude oil (Chang & Fogler, 1994). Goual and Firoozabadi (2002) suggested that both asphaltene and resins molecules are polar and associated as micelles, thanks to the positively charged asphaltenes that would be dispersed in the crude oil by negatively charged resins through electron donor acceptors and hydrogen bonding interactions (González et al., 2003). In fact, asphaltenes and resins coexist in a petroleum fluid and may be found in the form of monomers or associated as micelles. In the latter form, the micellar core is formed by the self association of asphaltene molecules with adsorbed resins at the surface to form a shell that also contains an oil fraction (Figure 3). Then, resins may be essential in asphaltene aggregation and stabilization because they attach to asphaltene micelles through their polar head and hence stretch their aliphatic groups outward to form a steric-stabilization layer surrounding asphaltene molecules. The formation and properties of such micelles is governed by the relative concentration of asphaltenes and resins. It seems that when resins are desorbed from or exceeds a critical concentration at the micellar core surface, they give rise to the asphaltene phase that aggregates and eventually precipitates (Pacheco-Sánchez et al., 2004; García-Cruz et al., 2005; Pereira et al., 2007). Hence, resins are proposed as the natural dispersants of asphaltenes in crude oil but there some controversy about the aggregation mechanisms (Sedghi & Goual, 2010).

Asphaltene and paraffins aggregation and deposition into the rock reservoir, through the production well clogging pumps, valves and pipes represents a very serious and constant problem in oil production and transport with an enormous economic impact. This is because asphaltene or paraffin aggregation and solid formation may occur in reservoirs and oil wells, provoking the formation of an extremely dense phase that reduces oil extraction and in many cases completely stops oil production. Also, asphaltene and paraffins deposition occurs in destabilized crude oil blends and during oil storage (Thanh et al., 1999; Speight, 1999; Srivastava et al., 2002; Speight 2004a, Speight 2004b; Murgich et al., 2001; Murgichet al., 1998; Mansoori, 1997; Buenrostro-González et al., 2001; Carbognani et al., 1999; Oh &

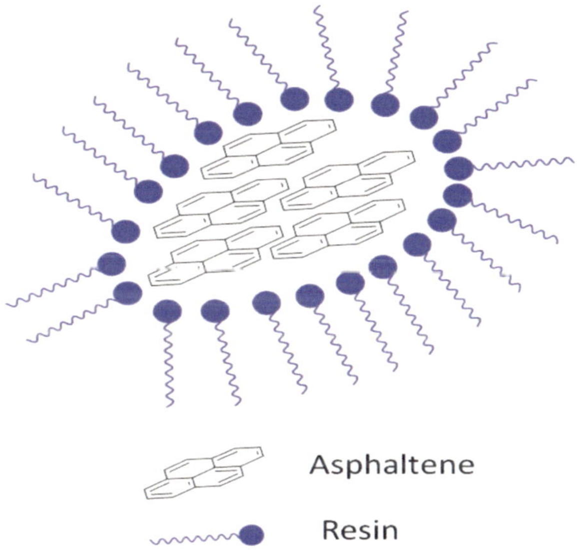

Fig. 3. Schematic representation of an asphaltene-resin micelle presents in crude oil.

Deo, 2002; Leon et al., 2000; Visintin et al., 2008). The factors influencing asphaltene and paraffin aggregation are ruled by the equilibrium disturbance of crude oil's original composition, presence of trapped water in the reservoir (also called formation water), temperature diminution that favors aggregation as well as their chemical modification in the reservoir (Table 2). For example, the stability of asphaltenes in crude oil is influenced by the composition, temperature, and pressure of crude oil; asphaltenes can be destabilized as the pressure approaches toward the bubble point of crude oil (where solubilised gas starts to evaporate from the oil phase). The understanding of these factors and aggregation phenomena are crucial to find practical and effective solutions to the oil industry. It is of commonly use, the addition of aromatic solvents like xylene or cosolvents like amines or sulfonic acid derivatives. Although, the empirical approach may cause damage to the rock reservoir (also called formation damage), increase asphaltene aggregation as well as useless, costly and time-consuming procedures that may compromise oil production. Recently, we have reviewed several conventional and emerging technologies to allow crude oil flowing and transportation by pipeline which somehow reduce asphaltene and paraffin aggregation and precipitation (Aburto et al., 2009).

On the subject of ILs application to the asphaltene problems, Liu et al. (2005) studied the dissolution of asphaltenes from Shengli crude oil in ILs. They mixed the isolated asphaltene sample with ILs and heated them up to different temperatures (50, 80, 135 and 150°C) in order to determine their solubility. It was observed that the ILs based on isoquinolinium cation can dissolve asphaltenes better than pyridinium and imidazolium cations. Also, the ILs capacity to dissolve asphaltenes decreases with the major substitution of the alkyl chain of ILs. The effect of the anion over the capacity to dissolve asphaltenes was found to augment with minor molar volumes, i.e. the anion [Cl]- showed the better dissolution of ILs within a range of 20-27% between 80 to 150°C. Authors proposed that asphaltene dissolution occurs by ILs breaking of asphaltene associations, mainly hydrogen bonds. Now, the dissolution approach to solve asphaltene aggregation and deposition should be carefully studied before practical application since tested ILs are poor soluble in water

Fractions	Factors
Asphaltenes	• Disturbance of the pressure-temperature reservoir equilibrium that alters the natural abundances of crude oil fractions, i.e. gas separation from crude oil in primary recovery, during gas injection into reservoir in secondary recovery or through polymer injection, by instance, into the reservoir during tertiary recovery, also called enhanced oil recovery (EOR). • Injection of light petroleum fractions or light to medium crude oils to the reservoir to facilitate production of heavy and extra-heavy crude oils. • Chemical modification of fractions during recovery processes, especially through thermal recovery processes.
Paraffins	• Ascension of crude oil from the reservoir through the production well and during pipeline transport to the terminal may cause heat loss and paraffin crystallization and sedimentation. • Presence of water in the reservoir that favors the formation of a paraffin shell around water drops. • Blending two or more crude oils

Table 2. Factors influencing asphaltene and paraffins agglomeration and solid formation.

and a solvent may be required to inject them to the reservoir or pipeline. Moreover, dosage, injection and effect on real crude oil sample should be essayed to get more insights about ILs functionality.

Hence, Hu & Guo(2005) studied the effect of IL sand amphiphile molecules in the inhibition of asphaltene precipitation in simulated CO_2-injected reservoir oil. The CO_2 flooding is a promising enhanced oil recovery technique for reservoirs but asphaltene or paraffin (Hu et al., 2004) precipitation and subsequent clogging of the rock formation as well as production and transport equipment may be an important disadvantage (Figure 4). The experiment was conducted in a high-pressure cell where the crude oil and the ILs were mixed and CO_2 was injected as an asphaltene precipitation agent. It was found that inhibition abilities of ILs like $C_n$phol and resins such as $C_n$bsa, and $C_n$bsNa increase with elongating the alkyl chain length between $2 \leq n< 8$ and then remain constant for $n \geq 8$. Nevertheless, the [$C_n$py][Cl] series display an inverse trend and inhibition of asphaltene precipitation auments from $n = 12$ to $n= 4$; and [C4iql][Cl] revealed as the better inhibitor with the minor charge density on the cation. Something important about this work is that the ability of ILs and amphiphiles to inhibit asphaltene deposition varies from one kind of crude oil to another. Indeed, Xinmin, Shengli and MuH are Chinese medium crude oils with an asphaltene content between 1.5 to 3.5% (Hu, 2005). The Xinmin oil has more asphaltenes and a larger specific gravity (0.9206 g/cm^3 at 20°C) than the other two tested oils with similar properties. Non IL could completely inhibit asphaltene precipitation only the benzene sulfonic amphiphiles, but at a high concentration of 4% w/w. Such quantity of an additive to prevent or correct asphaltene deposition into the reservoir or even at a storage tank may cause important operational problems that should be carefully planned: ILs storage, solvent storage and mixing with ILs, ILs injection to the well or tank, ILs effect on water separation, corrosion and crude oil refining.

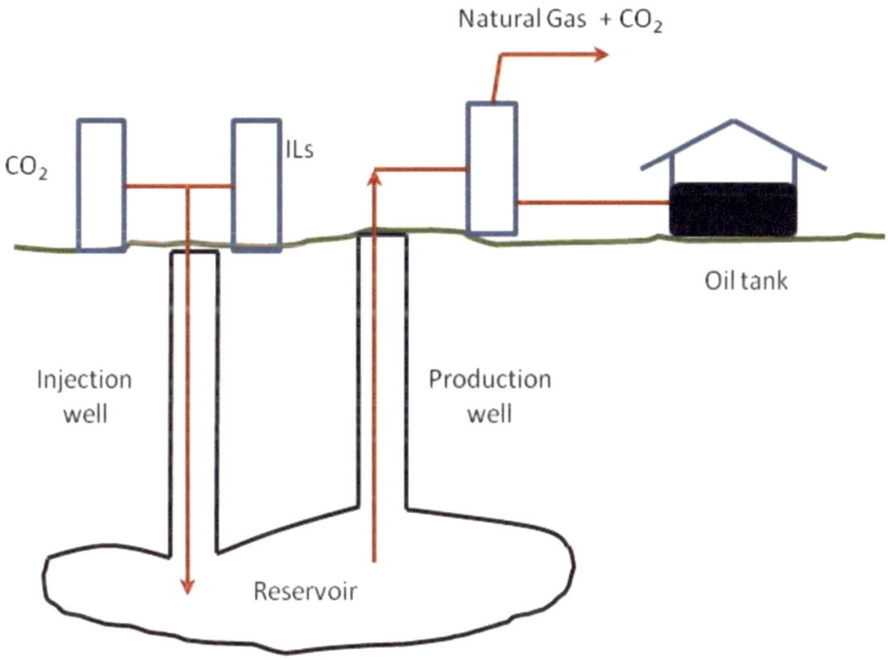

Fig. 4. Scheme of CO_2 flooding and hypothetical ILs utilization for enhanced oil recovery with minor asphaltene precipitation and deposition issues.

Although ILs may be potential asphaltene deposition inhibitors, there still many basic questions that should be answered first prior to industrial essays. An important basic and applicative question is related to IL-IL interactions, ILs dosification and how polar asphaltenes or asphaltene-resin complex interact with ILs. Hence, imidazolium- and pyridinium-based ILs presented a critical aggregation concentration (CAC) in solution, i.e. these ILs aggregate at certain concentration around the molecules of a Mexican heavy crude oil sample (MHCOs) used as fluorescent probe (Figure 5; Murillo-Hernández et al., 2009). The IL molecules surround and clog the fluorophore molecules of MHCOs, which sense a more apolar microenvironment at the CAC value due to the blue shifting of the spectra of the MHCOs at increasing IL concentration. Such observation can be explained if we consider that ILs may interact through their apolar tail with resin-stabilized asphaltenes aggregates in the MHCOs, resulting in an increasing apolar microenvironment around asphaltene molecules (Figure 6a). The CAC value show then the critical concentration of ILs at which a miscelle-like aggregate is formed and varies from 10 to 50 ppm for studied ILs. The resin-stabilized asphaltenes may be further stabilized by this ILs shell and limit asphaltene-asphaltene interaction, aggregation and precipitation. Nevertheless, we must also consider the IL-IL interaction since it forms the outer shell and may favour the formation of loosy or tighty aggregates (Figure 6b). It appears clear here that we need ILs that strongly interact with asphaltenes or resin-asphaltene aggregates in order to stabilize them but slightly or nothing affinity with themselves. The ILs with smaller CAC values should then have a structure that facilitates stronger IL-IL interaction that other ILs with larger CAC values, i.e. $[Emim]^+[Br]^-$ with an ethyl substitution forms at lower concentration a tight micelle-like structure while $[Bmim]^+[AlCl_4]^-$ with a butyl group and a larger anion requires 5 times more

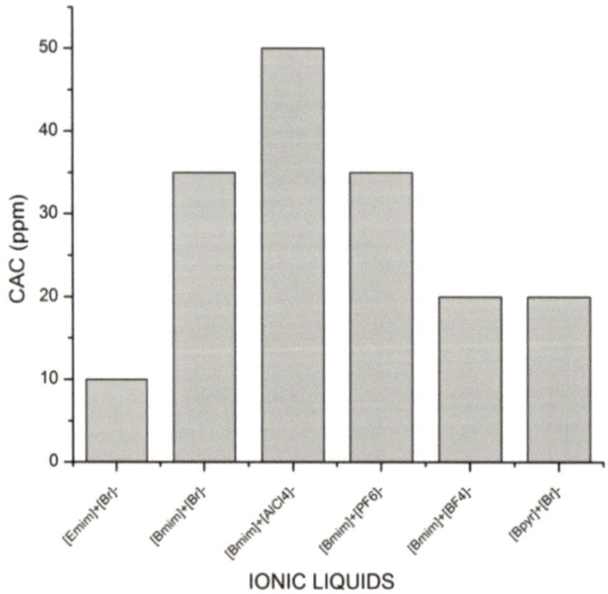

Fig. 5. Values of the critical aggregation concentration of various imidazolium- and pyridinium-based ILs.

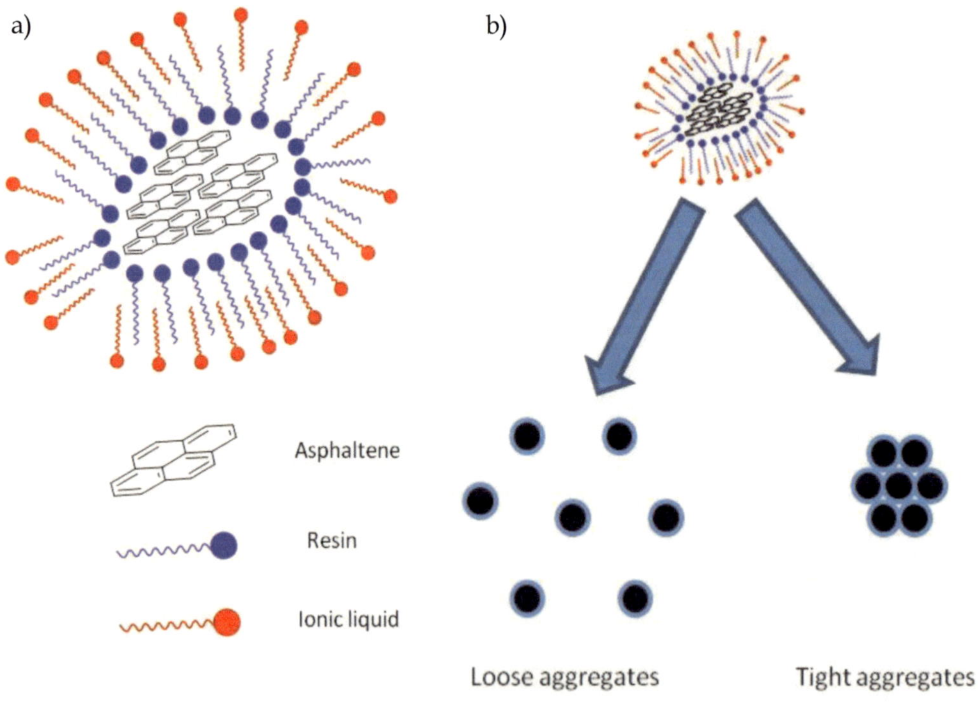

Fig. 6. Proposed scheme of a) the interaction of ILs with resin-asphaltene aggregates and b) further asphaltene aggregates stabilization with ILs.

molecules to aggregate. Additional information about ILs that must be generated is the dissociation constants (K_d, Equation 1), and the surface and interfacial tension values of ILs solutions that may give some insights about their dispersion capacity of crude oils, specifically asphaltenes and/or paraffins.

$$IL + IL \xleftrightarrow{K_d} IL_n \quad (1)$$

We study also the aggregation of a MHCOs by spectrofluorometry and found that it occurs at a predominant polar microenvironment either by incrementing the crude oil concentration or the n-heptane volume (Murillo-Hernández et al., 2009). The addition of an IL modifies such aggregation and displaces it to larger MHCOs concentration or n-heptane volume. Here, ILs set a polar microenvironment around MHCOs aggregates, which stabilized them against further aggregation and precipitation (Figure 6b). The better performance of ILs as inhibitors or stabilizers of asphaltene aggregation was found with those comporting a complex anion, a pyridinium ring, or a shorter alkyl substitution on the cation. Such ILs present also the higher values of the calculated electronic properties. Another important finding is that ILs set a polar microenvironment around MHCOs molecules, which allows MHCOs-ILs-MHCOs interactions, limits MHCOs aggregation, and shifts the aggregation to a larger MHCOs concentration (Figure 7); i.e. ILs like $[Bmim]^+[BF_4]^-$ and $[Bmim]^+[AlCl_4]^-$ shift the MHCOs aggregation from initial 60 to 180 and 330 ppm, respectively. Here, coverage of MHCOs molecules by ILs should be homogenous (mono or multilayer) or a more specific IL-stabilized MHCOs interaction should exist that prevents asphaltene aggregation with poor IL-IL interactions. On the other hand, ILs that not or poorly shift the MHCOs aggregation ($[Bmim]^+[PF_6]^-$ or $[Bmim]^+[PF_6]^-$) may be explained in terms of a heterogeneous or incomplete coverage of MHCOs molecules, to more specific MHCOs-MHCOs interactions that displace MHCOs-IL interaction or to strong IL-IL interactions that favour asphaltene aggregation. Hu & Guo (2005) found that ILs with $[BF_4]^-$ and $[PF_6]^-$ anions have less ability to inhibit asphaltene precipitation because these anions are weakly complexing anions and thus are not expected to break intramolecular associations of asphaltenes because of their low charge densities and symmetrical shapes. But, $[Cl]^-$ anion is a better inihibitor because it possesses the ability to undergo acid-base interactions with asphaltenes, breaking intramolecular interactions and subsequently binding itself with asphaltenes. ILs with a $[Br]^-$ anion have a better MHCOs aggregation shifting when compared to $[PF_6]^-$ but less than $[BF_4]^-$ (Figure 7). It seems then that several factors affect asphaltene stabilization/destabilization, and that general rules do not apply to ILs utilization in crude oils. Further research must take into consideration factors like the chemical properties of asphaltenes and other SARA fractions present in a specific kind of crude oil, spatial variations of properties of the same crude oil, basic and acid sites in asphaltenes and ILs structure (Boukherissa et al., 2009), relative polarity at aggregation point of crude oil and isolated asphaltenes, etc. More insights about asphaltene aggregation and precipitation are needed in order to understand the ruling adsorption mechanisms among asphaltenes, asphaltene-resin aggregates and ILs. Such knowledge may be acquired through experimental adsorption isotherms and the estimation of affinity constants and adsorption capacities.

In regard to paraffin crystallization and deposition, the formation of wax crystals occurs as a consequence of temperature diminution of crude oil. The solids formed are mainly mixtures containing n-paraffins ranging from C20 to C40 and higher. However, the deposition of branched chain paraffins may drag also naphthenic acids, aromatics compounds as well as asphaltenes and resins and the mechanisms are quite complex and poorly understood (Oliveria et al., 2007). The main factors determining organic deposition are petroleum composition, the thermohydraulic profile along the production pipelines and temperature changes among the reservoir, production well, transportation pipelines and storage tanks. The paraffin precipitation also causes a change in the crude oil flow behavior from Newtonian to non-Newtonian, augmenting viscosity and pressure to pump it along the pipelines. Hence, the lowest temperature at which movement of the crude oil specimen is observed under the conditions of the test is called the pour point temperature and corresponds to a phase transition in which the oil goes from a fluid dispersion to a wax-oil gel (ASTM D 5853, 2009). The use of scrapers, hot oil treatments and solvents ease the operation; but the application of remedial measures becomes prohibitive economically in extremely deep production, offshore drilling and ocean floor completions. Certain chemical products are used as flow improvers, crystal modifiers, and pour point reducing agents that diminish apparent viscosity and the pour point of oils having a tendency to wax crystallization through various not still well defined mechanisms, but there is evidence that asphaltenes and naphthenic acids may interfere or favour paraffin crystallization (Oliveira et al., 2007). Despite that mechanisms of paraffin inhibitors are still under discussion, the pour point depressant agent acts most likely by crystal modification and/or dispersion, while asphaltenes probably adsorb to a greater degree to the pipe wall, preventing strong adhesion by the wax (Mendell & Jessen, 1970). Here, ILs could also serve as pour point reducing agents, but the issue is still open and for example, (liquid + liquid) equilibrium data for complex systems are needed in order to approach the technical and economical feasibility of ILs as paraffin inhibitor or extractor (Hansmeier et al., 2010).

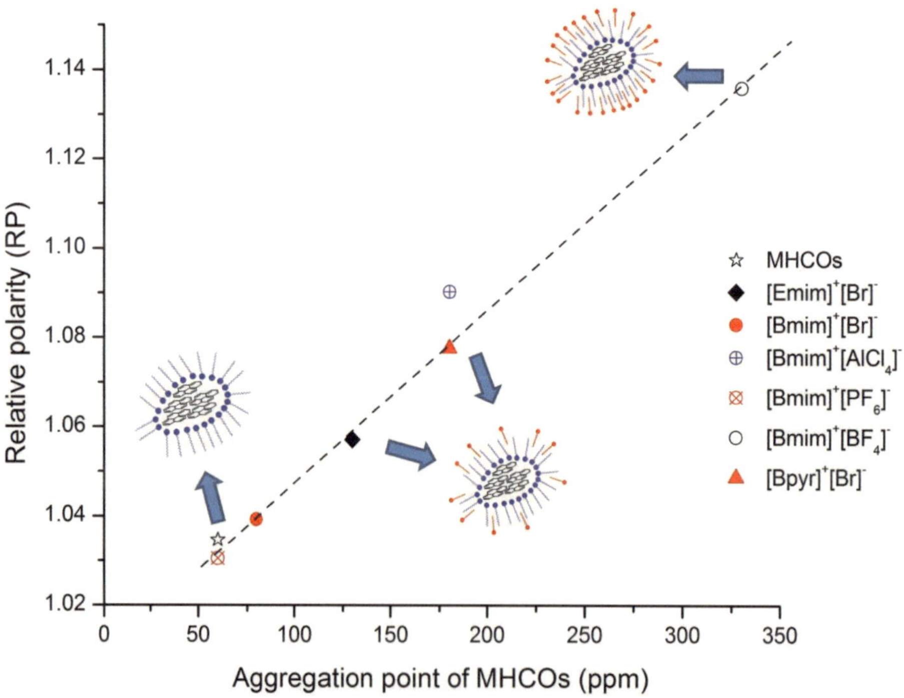

Fig. 7. The relative polarity set by different ILs on the aggregation point of MHCOs and proposed scheme for the interaction of ILs with resin-asphaltene aggregates.

2.2 Crude oil dehydration and desalting

The fluids produced from mature wells are seldom pure crude oil and associated gas, and by far the most abundant material is formation or added water. The first is the water naturally located into the reservoir while the latter comes from production techniques to enhance crude oil recovery. Hence, secondary and tertiary oil recovery processes, some of which involve injecting water or steam into oil bearing formations, have become increasingly common, particularly in the production of heavy oil or bitumen (Sefton & Sinton, 2010). Also, many wells produce large quantities of salt water, and safely disposing of it is a problematic and expensive problem. Additionally, brine may be corrosive and salts tend to precipitate through infrastructure (Speight, 1999). Also, salt-containing water can be present as a continuous or emulsionated phase (Figure 8) and must be separated as soon as possible from crude oil before its storage or refining. Nevertheless, the presence of very stable and tight water-in-oil (W/O) emulsions is a costly problem, resulting in high viscosity, the demand of expensive chemical products, corrosion and equipment with inefficient separations. The crude oil market demands that water in crudes must be removed to a level of less than 0.5% BSW (bottom, solids, water). Therefore, different physicochemical methods have been used to separate water from oil (Fan et al., 2009).

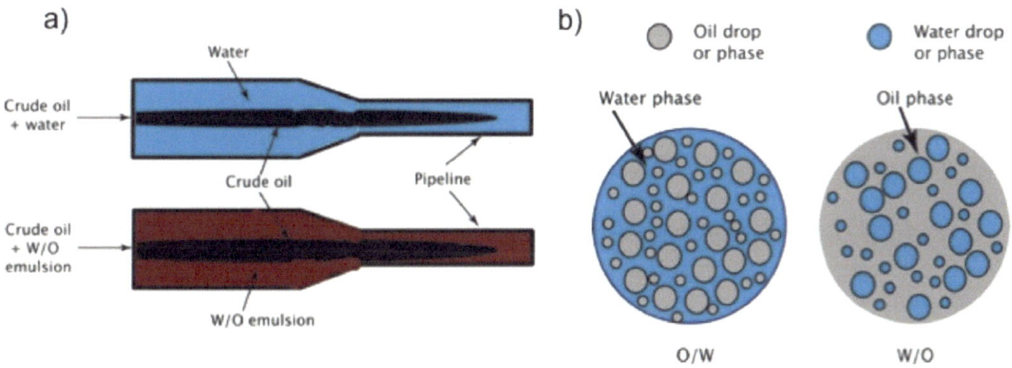

Fig. 8. Scheme for crude oil production with a) continuous and emulsionated water; and b) kinds of emulsions.

The stabilization of the W/O emulsion is mainly attributed to the formation of a rigid interfacial film of asphaltenes, where asphaltenes may accumulate as nanosized aggregates at the W/O interface and have interplay with resins as a solubilizing agent (Sjöblom et al., 2001). W/O emulsions are complex multicomponent, multiphase systems. The effective viscosity of these mixtures is a function of many parameters such as the presence of solids, the presence of emulsifiers, the viscosity of individual phases, the droplet size and distribution, the density of constituents, the age of the emulsion, the shear rate at which the viscosity is sampled, the volume fraction of constituents and the temperature of the emulsion (Farah et al., 2005; Sefton & Sinton, 2010). Hence, crude oil demulsification can be very difficult, non-efficient and it may require high residence times. Therefore, it is still one of the most frequently studied subjects in order to find efficient industrial methods to easily and economically break emulsions and desalts crude oil (Xu et al., 2006).

The self-organization, micelle formation and surface properties of ionic liquids (Luczak et al., 2008) as well as their use to dehydrate and desalt crude oil have been just recently reported (Guzmán-Lucero et al., 2010; Lemos et al., 2010). These research groups work on the demulsification of water-in-medium, heavy and extra-heavy crude oil emulsions assisted by conventional and microwave assisted heating. The latter technology has been known since 1978 and it accelerates the resolution of W/O emulsions as discussed elsewhere (Klaika, 1978; Fortuny et al., 2007). The microwave technology allows for the fast heating of emulsions, favouring the separation of crude oil emulsions into water and oil phases by thermal effects. The latter can be enhanced through the use of specific compounds with high dielectric properties, i.e. that augments the absorption of radiation by the sample and allows for the quicker resolution of the emulsion. Moreover, these additives can not only act in the heating system but also in the mechanisms involved in the process like modification of the interfacial tension between water and oil phases (Lemos et al., 2010). In this direction, ILs have been assayed as additives in the resolution of W/O emulsions together with crude oil desalting. Several ILs with $[Br]^-$, $[Cl]^-$, $[PF_6]^-$ and $[BF_4]^-$ anions were tested on the emulsion rupture using the conventional and microwave-assisted bottle test. It was found that oil microwave-assisted demulsification with ILs depends on ILs dosage and surface activity, W/O interfacial tension, salt content on water phase, the type of crude oil, and irradiation time. Interfacial tension of W/O emulsion diminishes in presence of water soluble ILs, i.e. ILs molecules go to the W/O interphase and weak the interfacial film which allows a more enhanced emulsion rupture and phase separation when microwaves are applied (Lemos et. al., 2010). By the other side, ILs may break W/O emulsions with different types of crude oils, being the faster and higher demulsification efficiency in the order medium > heavy > extra heavy crude oil. It seems to us that high viscosity on extra-heavy crude oils somehow limits the mobility and coalescence of water drops inside the oily matrix, which delays phase separation (Figure 9). Many factors may affect emulsion stability and rupture but relevant parameters that must be deeply study are crude oil viscosity, asphaltenes, paraffins, naphtenic acids, salt content, ILs' surface properties, interfacial phenomena, etc.; in order to assess emulsion breakage factibility.

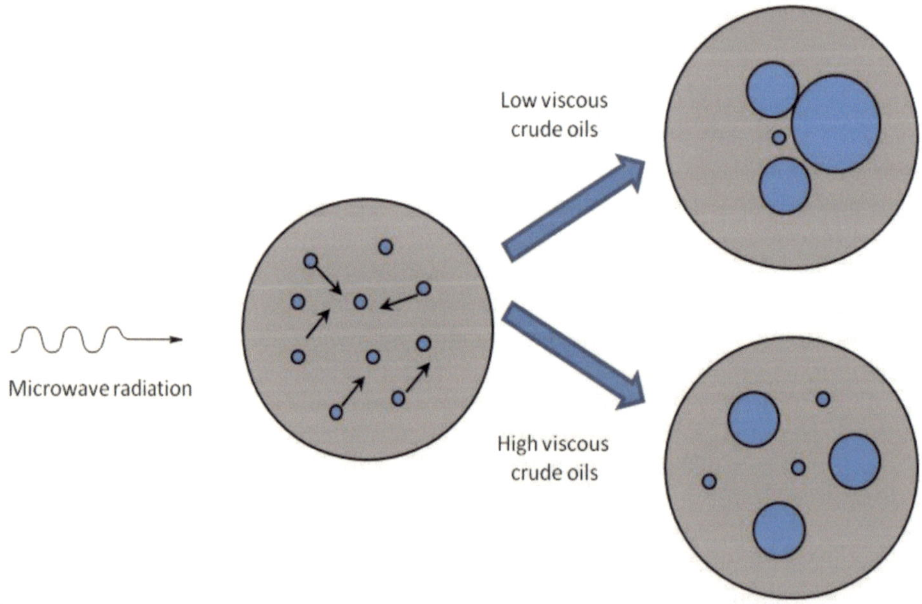

Fig. 9. Scheme for the mobility and coalescence of water drops in W/O emulsions in low and high viscous crude oils under microwave-assisted deselmusification.

3. Petroleum refining

Transportation fuel usage around the world is developing so fast because of their advantages of cost-efficiency and big power. However, SOx emissions are produced during fuel combustion, which brings environmental problems. Therefore, many stringent environment legislations have been issued to limit the sulphur (Esser et al., 2004) and aromatic content of fuels (EPA, 2008). There are two conventional desulfurization methods: hydrodesulfurization (HDS) and non-HDS. In the past few years, most of the sulfur was removed from the fuels through HDS in refineries. Nevertheless, the sulfur- and nitrogen-containing compounds left in fuels like gas oil and diesel after HDS are mainly aromatic compounds such as dibenzothiophene (DBT), carbazole and their alkyl derivatives (Figure 10). Hence, conventional HDS requires higher temperature under higher hydrogen pressures, which induces high operational costs and investments to get the lower sulfur content, reduce HDS catalyst deactivation and meet environmental regulations.

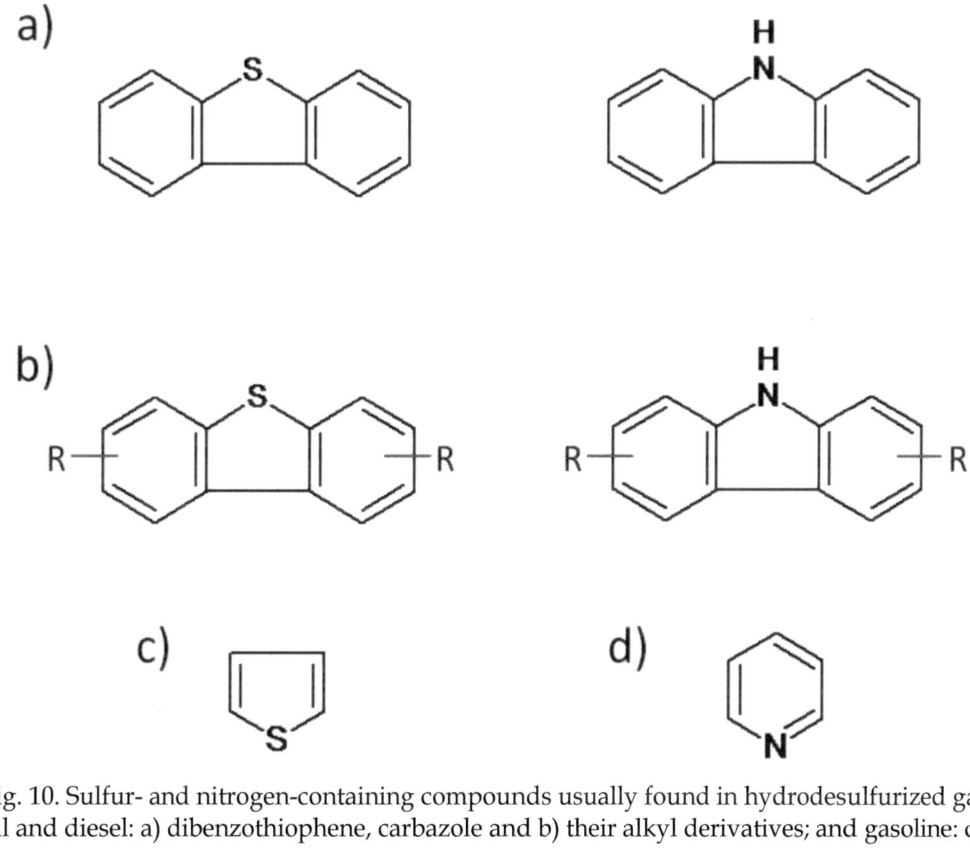

Fig. 10. Sulfur- and nitrogen-containing compounds usually found in hydrodesulfurized gas oil and diesel: a) dibenzothiophene, carbazole and b) their alkyl derivatives; and gasoline: c) thiophene and d) pyridine.

Room temperature ionic liquids (ILs) have been put in a wide range of application as green solvents for their low volatility and excellent thermal stability. In regard to the crude oil refining industry, the usage of ILs has been reported for desulfurization (Huang et al., 2004; Zhang et al., 2004; Mochizuki and Sugarawa, 2008; Xu et al., 2009; Wang et al., 2010), denitrogenation (Zhang et al., 2004; Huh et al., 2009) and dearomatization (Zhang et al.,

2004) of diesel and gasoline fuels as well as reduction of naphthenic acids in petroleum (Shi et al., 2008). In general, desulfurization, denitrogenation and dearomatization have been approached by mere extraction or extraction coupled with catalytic oxidation in ILs (Figure 11). Hence, sulfur removal by mere extraction with ILs displays low efficiencies (Huang et al., 2004). Here, the $[Cu_xCl_y]^-$ containing ILs could remove between 16-37% sulphur in gasoline and removal efficiency augments with the decreasing sulphur content. This was attributed to π-complexation of thiophene by Cu(I) as reported elsewhere (Hernandez and Yang, 2003). Another approach using a $[BF_4]^-$-containing ILs found that sulphur and nitrogen removal slightly augmented with increasing heteroatom content in gasoline (Zhang et al., 2004). Moreover, an important removal of sulphur and aromatic compounds in gasoline occurs first by increasing the IL/fuel ratio and then by multiple extraction cycles with ILs. The aromatic removal, mainly benzene, is an attractive subject of study since the content of benzene cannot exceed 0.62% in gasoline from 2011 in the USA (EPA, 2008).

It is suggested that the main interaction between aromatic and sulphur, nitrogen-containing compounds and ILs and then the extraction mechanism occur with the π-electron density of aromatic compounds as it has been described elsewehere (Su et al., 2004; Gutel et al., 2009). Another approach uses $[SO_4]^-$-based ILs to extract sulphur compounds in order to avoid the possible generation of HF and HCl during hydrolysis (Mochizuki and Sugarawa, 2008). They found a remarkable extraction of dibenzothiophene (40-70%) in *n*-dodecane at an IL/model fuel ratio of 1:1 that increases with extraction time and steps as well as carbon number of ILs. The extraction efficiency of nitrogen-containing compounds increases with larger chain lengths of the alkyl groups in ILs and it was attributed to the increasing free volume left behind by bulky ILs (Huh et al., 2009). Here, the extraction of the basic quinoline may be facilitated by its coordination to the Zn center of ILs but the extraction of neutral indole is ruled by the interaction between the ILs' anion and the H-atom of the secondary amine group as shown by computational calculations.

Recently, Meindersma et al. (2010) have made a review of works dealing with dearomatization as well as some useful contributions. They state that activity coefficients are needed at finite dilutions for practical application and not at infinite dilution as most works present; the distribution coefficients must be calculated on a mass basis, and not mole, since this is the approach used in industry; the work results must be compared with conventional technologies as sulfolane extraction process; and whenever possible, a technical and economical estimation of the process must be done. In this regard, the aromatic/aliphatic selectivities with some ILs are 1.9 times higher when compared to sulfolane, while the investment cost may be reduced by a factor of two. Further research should take in mind the extraction mechanism between heteroatom-containing and aromatics compounds with ILs, removal efficiencies, ILs separation and recycling, use of solvents or cosolvents, applications for separated compounds, economical feasibility, etc. All latter approaches use biphasic extraction where surface area is poor, mass-transfer issues should be important and require a lot of ILs recycling steps. Recently, the use of ILs deposited on a solid, known as supported ionic liquid phase (SILP), incremented significantly the desulfurization efficiency by augmenting surface area, diminishing mass-transfer problems and allowing a continuous process (Kuhlmann et al., 2009). Careful attention must be paid to ILs leaching or saturation during process, but ILs chemical bonding to support as well as ILs-support reactivation are subjects that must be further studied.

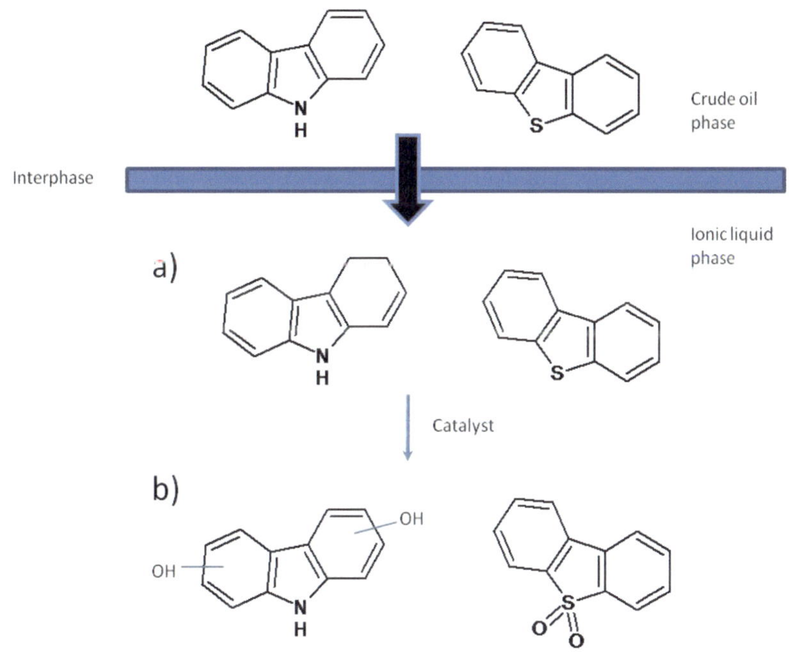

Fig. 11. Scheme for desulfurization and denitrogenation using ILs by a) mere extraction or b) extraction coupled with catalytic oxidation.

Conventional refining processes, as HDS, are typical catalytic ones where it is common the use of supported metals as Al, Co, Mo, V, Ti, etc., to enhance and accelerate fuel desulfurization and denitrogenation. Although, the process results in hydrogen sulfide and ammonia that need to be further treated ahead in Refinery to meet environmental emission regulations. Chemical oxidation in conjunction with ILs extraction can sharply increase, when compared to mere extraction, the removal of sulphur and nitrogen from automotive fuels without producing polluting gases that need to be recovered. For instance, Lo et al. (2003) reported that oxidative desulfurization with H_2O_2-acetic acid combined with water-immiscible IL of 1-butyl-3-methylimidazolium hexafluorophosphate ($[Bmim]^-[PF6]^+$) lead to a high oxidation conversion and extraction of sulphur compounds in model- and light oil but the process took several hours. Moreover, the oxidative desulfurization of fuels using acidic ionic liquids like $[Hmim]$-$[BF_4]^+$ and $[Hnmp]$-$[BF_4]^+$ with H_2O_2 showed that sulfur removal from the model oil could reach over 90% in less than 60 minutes (Lu et al., 2007; Zhao et al., 2007). The reaction time in HDS processes are expressed as LHSV (liquid hourly space velocity) and common values range for nafta are between 6 to 10 h^{-1}. Here, the assays were carried out with a LHSV between 132.14 and 529 h^{-1}, too highs for industrial application but we must consider that it is homogeneous versus conventional heterogeneous catalysis. It is clear here that we need also to standardize reaction criteria for mutual comprehension and set assays conditions near to practical applications. The $[Hnmp]^-[BF_4]^+$ could be recycled up to 7 times without a significant decrease in activity and up to 12 cycles with 99.5% of sulphur removal. Conventional HDS catalyst may be normally used for 10 years, ILs and their recycling should take this into account and economical analysis is crucial to understand the process feasibility using ILs.

Recently, a deep desulfurization system using an extraction and catalytic oxidation desulfurization process and including H_2O_2, V_2O_5, and $[Bmim]^-[BF_4]^+$, under mild conditions, has been proved to be efficient (Xu et al., 2009). Such approach showed high sulfur removal, which was superior to mere extraction with IL and extractive oxidation without catalyst. The advantages of using the extractive and catalytic system containing V_2O_5 and $[Bmim]^-[BF_4]^+$ are that ILs can be recycled seven times without a significant decrease in activity, and is a simple, mild, and environmentally friendly approach for deep desulfurization. Another approach used ozone and hydrogen peroxide as oxidation catalyst and sulphur removal reached 100% between 50 and 80°C (Wang et al., 2010). Nevertheless, these approaches remain still far away from industrial application and more intensive research should be done in order to assay with real diesel or nafta streams, understand the catalysis and thermodynamics of reactions, recycling of sulphur-saturated ILs, finding applications for oxidized heteroatom-containing compounds, and scale up of the extractive and oxidative extraction processes.

Finally, the presence of naphthenic acids in crude oil during refining operations may cause operational issues, such as foaming during crude oil desalting or other operation units as well as carrying cations through the refining process that may cause catalyst deactivation (Shi et al., 2008). They undertook an innovative approach to solve such a problem by forming ILs from the reaction of naphthenic acids and imidazole derivatives directly into *the* crude oil matrix, that allow ILs separation from crude oil. Even if high removal of naphthenic acids, transformed to ILs, was accomplished between 30 to 60°C; it is necessary the use of a cosolvent to facilitate phase separation. This approach requires a further separation process to obtain relatively pure ILs and regenerated cosolvent. Here, the approach is interesting but when applied directly to oil reservoir, forming *in-situ* ILs, may help to all described issues found along the production, extraction and refining of crude oils.

4. Conclusion

ILs have been successfully assayed for the production, transportation and refining of crude oil at lab scale. Nevertheless, their potential application on the field has still to be proven. Many issues have to be solved before ILs may be currently applied in the petroleum industry. The economical, technological and environmental feasibility of large scale production and utilization of ILs must be asserted before any petroleum company accepts their daily use. The effect of the presence of ILs in crude oil during production, transport and refining must be asserted in order to identify operational issues. Therefore, much work is necessary at pilot plant and field scale to identify and fix the possible operational issues, infrastructure dimension and ILs handling. Even if ILs have still to prove their safely utilization in daily and routinely petroleum operations, we believe that they have already open a crucial window to more safe, environmental-friendly and tailor-made solutions to the actual problems of the petroleum industry.

5. References

Aburto, J.; Mar-Juárez, E. & Juárez-Soto, C. Transportation of Heavy and Extra Heavy Crude Oil by Pipeline: A Patent Review for Technological Options. *Recent Patents on Chemical Engineering*, Vol.2, No.2, (2009) 86-97.

ASTM D 2007-03. Standard test method for characteristic groups in rubber extender and processing oils amd other petroleum-derived oils by the clays-gel absorption chromatographic method.

ASTM D 5853. Standard test method for pour point of crude oils. 2009.

Boukherissa M.; Mutelet, F.; Modarressi, A.; Dicko, A.; Dafri, D.; & Rogalski, M. Ionic Liquids as Dispersants of Petroleum Asphaltenes. *Energy & Fuels*, Vol. 23, (2009) 2557–2564.

Buenrostro-Gonzalez, E.; Espinoza-Peña, M.; Andersen, S. I. & Lira-Galeana, C. Characterization of asphaltenes and resins from problematic mexican crude oils. *Pet. Sci. Technol.*, Vol. 19, No.3-4, (2001) 299-316.

Carbognani, L.; Orea, M. & Fonseca, F. Complex nature of separated solid phases from crude oils. *Energy & Fuels*, Vol. 13, (1999) 351-358.

Castro, L.V. & Vazquez, F. Fractionation and characterization of mexican crude oils. *Energy & Fuels*, Vol. 23:, (2009) 1603-1609.

Chang, C.L. & Fogler, H.S. Stabilization of asphaltenes in aliphatic solvents using alkylbenzene-derived amphiphiles. 1. Effect of the chemical structure of amphiphiles on asphaltene stabilization. *Langmuir*, Vol. 10, No.6, (1994) 1749-1757.

Environmental protection Agency (EPA). Control of hazardous air pollutants from mobile sources: early credit technology requirement revision. *Federal Register*, 73(201), 61358-61363, 2008.

Esser, J.; Wasserscheid, P. & Jess, A. Deep desulfurization or oil refinery streams by extraction with ionic liquids. Green Chem., Vol. 6, No. 7, (2004,) 316–322.

Fan, Y.; Simon, S.; Sjöblom, J. Chemical destabilization of crude oil emulsions: effect of nonionic surfactants as emulsion inhibitors. *Energy & Fuels*, Vol. 23, (2009) 4575-4583.

Farah, M.A.; Oliveira, R.C.; Navaes-Caldas, J. & Rajagopal, K. Viscosity of water-in-oil emulsions: variation of temperature and water volume fraction. *Journal of Petroleum Science and Engineering*, Vol.48, (2005) 169-184.

Fortuny, M.; Oliveira, C.B.; Melo, R.L.; Nele, M.; Coutinho, R.C. & Santos, A.F. Effect of salinity, temperature, water content, and pH on the microwave demulsification of crude oils emulsions. *Energy & Fuels*, Vol. 21, No.3, (2007) 1358-1364.

García-Cruz, I.; Martínez-Magadán, J.M.; Salcedo, R. & Illas, F. Electronic structure properties of dibenzofurane and dibenzothiophene derivatives: Implications on asphaltene formation. Energy & Fuels, Vol. 19, No.3, (2005) 998-1002.

González, G.; Neves, G.B.M.; Saraiva, S.M.; Lucas, E.F. & dos Anjous de Sousa, M. Electrokinetic characterization of asphaltenes and the asphaltenes-resins interaction. *Energy & Fuels*,Vol. 17, (2003) 879-886.

Goual, L. & Firoozabadi, A. Measuring asphaltenes and resins, and dipole moments in petroleum fluids. *AICHE Journal*, Vol. 48, No.11, (2002) 2646-2663.

Gutel, T.; Santini, C.C.; Padua, A.A.H.; Fenet, B.; Chauvin, Y.; Conongia-Lopez, J.N.; Bayard, F.; Costa-Gomez, M.F. & Pensado, A.S. Interaction between the p-system of toluene and the imidazolium ring of ionic liquids: a combined NMR and molecular simulation study. *J. Phys. Chem B.*, Vol. 113, (2009) 170-177.

Guzmán-Lucero, D.; Flores, P.; Rojo, T. & Martínez-Palou, R. Ionic liquids as demulsifiers of water-in-crude oil emulsions: study of the microwave effect. *Energy & Fuels*, Vol. 24, (2010), 3610-3615.

Hansmeier, A.R.; Jongmans, M.; Meindersma, G.W. & de Haan, A.B. LLE data for the ionic liquid 3-methyl-*N*-butyl pyridinium dicyanamide with several aromatic and aliphatic hydrocarbons. *J. Chem. Thermodynamics*, Vol.42, (2010) 484-490.

Hernandez-Maldonado, A.J. & Yang, R.T. Desulfurization of liquid fuels by adsorption via π complexation with Cu(I)-Y and Ag-Y zeolites. *Industrial & Engineering Chemical Research*. Vol. 42, No.1, (2003) 123-129.

Hu, Y.F.; Li, S.; Chu, Y.P. & Guo, T.M. Wax precipitation in three Chinese reservoir oils under carbon dioxide (CO_2) injection. *Energy & Fuels*, Vol. 18, (2004) 1183-1186.

Hu, Y. F. & Guo T. M. Effect of the structure of ionic liquids and alkylbenzene-derived amphiphiles on the inhibition of asphaltene precipitation from CO_2-injected reservoir oils. *Langmuir*, Vol. 21, (2005) 8168-8174.

Huang, C.P.; Chen, B.H.; Zhang, J.; Liu, Z.C. & Li, Y.X. Desulfurization of gasoline by extraction with new ionic liquids. *Energy & Fuels*, Vol. 18, (2004) 1862–1864.

Huh, E.S.; Zazybin, A.; Palgunadu, J.; Ahn, S.; Hong, J.; Kim, H.S.; Cheong, M. & Ahn. Zn-containing ionic liquids for the extractive denitrogenation of a model oil: a mechanistic consideration. *Energy & Fuels*, Vol. 23, 2009, 3032-3038.

Isea, A. Geological synthesis of the Orinoco oil belt, Eastern Venezuela.*Journal of Petroleum Geology*. Vol. 10, No.2, (1987) 135-148.

Klaika, W.J. Method and apparatus for controlling fluency of high viscosity hydrocarbon fluids. US Patent 4,067,683, (1978).

Kuhlmann, E.; Haumann, M; Jess, A.; Seeberger, A. & Wasserscheid, P. Ionic liquids in refinery desulfurization: Comparison between biphasic and supported ionic liquid Phase suspension processes. *ChemSusChem*, Vol. 2, (2009) 969-977.

Lemos, R.C.B.; da Silva, E.B.; dos Santos, A.; Guimaraes, R.C.L.; Ferreira, B.M.S.; Guarnieri, R.A.; Dariva, C.; Franceschi, E.; Santos, A.F. & Fortuny, M. Demulsification of water-incrude oil emulsions using ionic liquids and microwave irradiation. *Energy & Fuels*, Vol. 24, (2010), 4439-4444.

Leon, O.; Rogel, E.; Espidel, J. & Torres, G. Asphaltene structural characterization, self-association, and stability behavior. *Energy & Fuels*, Vol. 14, No.1, (2000) 6-10.

Liu, Y.; Hu, Y.; Wang, H.; Xu, Ch.; Ji, D.; Sun, Y. & Guo, T. Ionic liquids : Novel solvents for petroleum asphaltenes. *Chinese J. Chem. Eng.*, Vol. 13, No.4, (2005) 564-567.

Lo, W.H.; Yang, H.Y. & Wei, G.T. One-pot desulfurization of light oils by chemical oxidation and solvent extraction with room temperature ionic liquids. *Green Chem.*, Vol 5, (2003) 639–642.

Lu, L.; Cheng, S.F.; Gao, J.B.; Gao, G.H & He, M. Y. Deep oxidative desulfurization of fuels catalyzed by ionic liquid in the presence of H_2O_2.*Energy & Fuels*, Vol. 21, (2007) 383–384.

Luczak, J.; Hupka, J.; Thörming, J. & Jungnickel, C. Self-organization of imidazolium ionic liquids in aqueous solution. *Colloids and Surfaces A : Physicochemical and Engineering Aspects*, Vol. 329, (2008) 125-133.

Mansoori, G.A. Modeling of asphaltene and other heavy organic depositions. Journal of Petroleum Science and Engineering. Vol 17, No. 1-2, (1997) 101-111.

Meindersma, G.W.; Hansmeier, A.R. & de Haan, A.B. Ionic liquids for aromatic extraction. Present status and future outlook. *Ind. Eng. Chem. Res.*, Vol.49, No.16, (2010) 7530-7540.

Mendell, J.L. & Jessen, F.W. Mechanism of inhibition of paraffin deposition in crude oil systems. *SPE*, (1970) 171-182.

Mochizuki, Y. & Sugarawa, K. Removal of organic sulfur from hydrocarbon resources using ionic liquids. *Energy & Fuels*, Vo. 72, (2008) 3303-3307.

Murgich,J.; Abanero, J. A. & Strausz, O.P. Molecular recognition in aggregates formed by asphaltene and resin molecules from the Athabasca oil sand. *Energy & Fuels*, 1999, Vol.13, No. 2, (1998) 278-286..

Murgich, J.; Rogel, E.; León, O. & Isea, R.A molecular mechanics-density functional study of the adsorption of fragments of asphaltenes and resins on the surface of Fe_2O_3.*Pet. Sci. Technol.*, Vol. 19, (2001) 437-455.

Murillo-Hernández, J.; García-Cruz, I.; López-Ramírez, S.; Durán-Valencia, C.; Domínguez, J.M. & Aburto, J. Aggregation Behavior of Heavy Crude Oil-Ionic Liquid Solutions by Fluorescence Spectroscopy. *Energy & Fuels*, Vol. 23, (2009) 4584-4592.

Oh, K. & Deo, M.D. Effect of organic additives on the onset of asphaltene precipitation. *Energy & Fuels*, Vol. 16, No.3, (2002), 694-699.

Oliveria, G.E.; Mansur, C.R.E.; Lucas, E.F.; González, G. & de Souza, W.F. The effect of asphaltenes, naphtenic acids, and polymeric inhibitors on the pour point of paraffins solutions. *Journal of Dispersion Science and Technology*, Vol.28, (2007) 349-356.

Pacheco-Sánchez, J.H.; Álvarez-Ramírez, F. & Martínez-Magadán, J.M. Morphology of aggregated asphaltene structural models. *Energy & Fuels*, Vol. 18, No.6, (2004) 1676-1686.

Painter, P.; Williams, P. & Lupinsky, A. Recovery of bitumen from Utah tar sands using ionic liquids. *Energy & Fuels*, Vol. 24, (2010a), 2172-2173.

Painter, P.; Williams, P. & Mannebach, E. Recovery of bitumen from oil or tar sands using ionic liquids. *Energy & Fuels*, Vol. 24, (2010b) 1094-1098.

Pereira, J.C.; López, I.; Salas, R.; Silva, F.; Fernández, C.; Urbina, C. & López, J.C. Resins : the molecules responsible for the stability/instability phenomena of asphaltenes. *Energy & Fuels*, Vol. 21, (2007) 1317-1321.

Sedghi, M. & Goual, L. Role of resins on asphaltene stability. *Energy & Fuels*, Vol. 24, (2010) 2275-2280.

Sefton, E. & Sinton, D. Evaluation of selected viscosity prediction models for water in bitumen emulsions. *Journal of Petroleum Science and Engineering*, Vol.72, (2010) 128-133.

Shi, L.J.; Shen, B.X. & Wang, G.Q. Removal of naphthenic acids from Beijiang crude oil by forming ionic liquids. *Energy & Fuels*, Vol. 22, (2008) 4177-4181

Sjöblom J.; Johnsen, E.E.; Westvik, A.; Ese, M.H.; Djuve J.; Auflem, I.H. & Kallevik, H. Demulsifiers in the oil industry. In: *Encyclopedic Handbook of Emulsion Technology*. Sjöblom J. (Ed)., Marcel Dekker, New York, 2001, p. 595-619.

Speight, J. G. (1999). *The Chemistry and Technology of Petroleum*, Marcel Dekker Inc., New York.

Speight, J.G. Petroleum asphaltenes. Part 1 : Asphaltenes, resins and the structure of petroleum. *Oil and Gas Science and Technology- Rev. IFP*, Vol.59, No.5, (2004a) 467-477.

Speight, J.G. Petroleum asphaltenes. Part 2: The effect of asphaltenes and resin constituents on recovery and refining processes. *Oil and Gas Science and Technology- Rev. IFP*, Vol.59, No.5, (2004b) 479-488.

Srivastava, S.P.; Butz, T.; Tiwari, G.B.; Oschmann, H.-J.; Rahimian, I. & Phatak, S.D. Gel formation in vaccum gas oil. I. Role of composition. *Petroleum Science and Technology*, Vol. 20, No.3 & 4, (2002) 269-290.

Strausz, O.P.; Morales-Izquierdo, A.; Kazmi, N.; Montgomery, D.S.; Payzant, J.D.; Safarik, I. & Murgich, J. Chemical composition of Athabasca Bitumen: The saturate fraction. *Energy & Fuels*, DOI: 10.1021/ef100702j

Su, B.M.; Zhang, S. & Zhang, Z.C. Structural elucidation of thiophene interaction with ionic liquids by multinuclear NMR spectroscopy. *J. Phys. Chem. B.*, Vol. 108, (2004) 19510-19517.

Riveros, L.; Jaimes, B.; Ranaudo, M.A.; Castillo, J. & Chirinos J. Determination of asphaltene and resin content in Venezuelan crude oils by using fluorescence spectroscopy and partial least squares. *Energy & Fuels*, Vol. 20, (2006) 227-230.

Thanh, N.X.; Hsieh, M. & Philp, R.P. Waxes and asphaltenes in crude oils. *Organic Geochemistry*. Vol. 30, (1999) 1119-132.

Visintin, R.F.G.; Lockhart, T.P.; Lapasin, R. & D'Antona, P. Structure of waxy crude oil emulsion gels. *Journal of Non-Newtonian Fluid Mechanics*, Vol.149, (2008) 34-39.

Xu, X.; Yang, J.; Jiang, Y. & Gao, J. Effects of process conditions on desalting and demetalization of crude oil. *Petroleum Science and Technology*, Vol. 24, (2006) 1307-1321.

Xu, D.; Zhu, W.; Li, H.; Zhang, J.; Zou, F.; Shi, H. & Yan, Y. Oxidative desulfurization of fuels catalyzed by V_2O_5 in ionic liquids at room temperature. *Energy & Fuels*, Vol. 23, (2009) 5929-5933.

Wang, J.; Zhao, D. & Li, K. Oxidative desulfurization of dibenzothiophene using ozone and hydrogen peroxide in ionic liquid. *Energy & Fuels*, Vol. 24, (2010) 2527-2529.

Zhao, D.S.; Wang, J. L. & Zhou, E.P. Oxidative desulfurization of diesel fuel using a Brønsted acid room temperature ionic liquid in the presence of H_2O_2. *Green Chem.*, Vol. 9, (2007) 1219–1222.

Zhang, S.; Zhang, Q. & Zhang, Z.C. Extractive desulfurization and denitrogenation of fuels using ionic liquids. *Ind. Eng. Chem. Res*, Vol.43, (2004) 614-622.

Zhang, S. & Huang, H. Geochemistry of palaeozoic marine petroleum from the Tarim basin, NW China : parti. Oil family classification. *Organic Geochemistry*, Vol.36, (2005) 1204-1214.

Part 5

Biotechnology

19

Application of Ionic Liquids in Biocatalysis

Maja Habulin, Mateja Primožič and Željko Knez
University of Maribor, Faculty of Chemistry and Chemical Engineering
Slovenia

1. Introduction

Biotransformations have been of enormous economic and social importance throughout the history of mankind (Liese et al., 2000). Biocatalysis may be the most efficient way of producing fine chemicals. Today, several chemicals like pharmaceuticals, amino acids, saccharides and polysaccharides, esters and vitamins are produced by enzymatic biotransformations on industrial scale (Liese et al., 2000).

The production of fine chemicals results in output of considerable volume of waste. Most of wastes are solvents such as water, volatile organic compounds (VOCs) etc. Solvents comprise 2/3 of all industrial emissions and 1/3 of all VOC emissions nationwide. These emissions have been linked to a host of negative effects (global climate change, pollution of air, human illness etc.) (Brennecke & Maginn, 2001). In recent years, green chemistry is become a growing area of research. Therefore the search for new environmental friendly and benign solvents and catalysts which can be easily recycled or reused is of significant interest. The ideal solvent should be chemically and physically stabile, recyclable, and reusable, should have a very low volatility, should allow selective and rapid transformations and should be easy to handle.

For the biocatalysis, there are five main "green" solvent systems: supercritical fluids (SCFs), fluorinated solvents, ionic liquids (ILs), water, and solvent free reactions (Hobbs & Thomas, 2007). Enzymatic reactions could be performed under preferred conditions with minimized yield of the undesired by-products. Meanwhile, low yields, selectivity, and poor solubility of substrates in aqueous medium may require the enzymatic reactions to be carried out in non-aqueous medium (Sureshkumar & Lee, 2009).

SCF is any substance at a temperature and pressure above its critical point. Close to the critical point, small changes in pressure or temperature result in large changes in density, allowing many properties of a SCF to be "fine-tuned" (Jessop & Leitner, 1999). There are several advantages using the SCFs as solvents in chemical synthesis, where all are based on unique thermo-physical properties of SCFs for their mixtures with reactants. The main advantage of biocatalysis in SCFs is the tunability of the properties of the solvent by changes in the pressure and/or the temperature. The application of SCFs enables also design of integrated reaction and separation processes. In mass transfer limited processes the reaction rate can be increased if SCFs are applied due to higher diffusivity and to reduce viscosity of reaction system. SCFs display unique substrate specificity at relative mild reaction conditions.

Environmental benefits of most SCFs in industrial processes are in replacement of environmentally far more damaging conventional organic solvents. An environmental impact is also low energy consumption during operation. Health and safety benefits include the fact that the most important SCFs (supercritical carbon dioxide (SC CO_2) and supercritical water (SC H_2O)) are non-cancerogenic, non-mutagenic, non-flammable, non-toxic, and thermodynamically stable. SC H_2O however cannot be used as medium for biocatalytic reactions because of the high temperatures involved which completely deactivate the enzymes.

Since the first reports on the use of SCFs (Randolph et al., 1985), as reaction media for enzyme-catalyzed reaction several reviews regarding biocatalysis in SCFs have been published (Knez & Habulin, 2002; Habulin et al., 2007a; Knez, 2009; Hobbs & Thomas, 2007; Mesiano et al., 1999; Krishna, 2002; Matsuda et al., 2005). Majority of these biocatalytic reactions have been carried out in SC CO_2.

Enzymes in SCFs could be used in their native form (powder, liquid, whole cells ...) or immobilized on a carrier (resin, sol-gel matrix ...). Enzymes, apart from their form, are not soluble in CO_2. Therefore, biocatalysis in SC CO_2 is always heterogeneous.

In SCFs there are both the direct effects of pressure on enzyme activity which may lead to denaturation, and the indirect effects of pressure on enzymatic activity and selectivity. In the case of SC CO_2 direct effect of pressure on enzyme inactivation is small and the protein structure is expected to be retained on the whole and only local changes may occur. Those local changes may lead to another active state of a protein, which may possess an altered activity, specificity or stability. Pressure is also likely to have an indirect affect on the efficiency of the reaction by changing either the rate constant or the solubility of the reactants. At high pressures solute-solvent interactions increase, resulting in a higher solvent capacity (Habulin & Knez, 2001). However, it has also disadvantages, as sometimes lower catalytic activities in the solvent which have been attributed to the formation of carbonic acid (Habulin et al., 2007a).

Ionic liquids (ILs) are organic salts consisting of ions, which exist in the liquid state at ambient temperatures. In the last 15 years, ILs were recognized as a novel class of solvents for chemical processes. They represent also an exciting new class of reaction solvents for catalysis, which have been used successfully for enzyme-catalyzed reactions (Moniruzzaman et al., 2010; van Rantwijk & Sheldon, 2007; Kragl et al., 2002).

Common ions involved in ILs for biocatalysis are: cations, which are generally bulky, organic with low symmetry, e.g. derivatives of imidazolium, pyridinium, pyrrolidinium, ammonium, sulfonium, phosphonium, …, and anions, which are either organic or inorganic and can be classified in two classes: a.) those which give polynuclear anions, such as $[Al_2Cl_7]^-$, $[Al_3Cl_{10}]^-$, $[Au_2Cl_7]^-$, …, and b.) those that corresponds to mononuclear anions which lead to neutral, stoichiometric ILs, such as tetrafluoroborate, hexafluorophosphate, bis[(trifluoromethyl)sulfonyl]amide, nitrate, trifluoroacetate, methyl-sulfate etc. (Olivier-Bourbigou & Magna 2002).

ILs combine good and tunable solubility properties with no measurable vapour pressure and excellent thermal stabilities. They have rapidly found a place of choice as valuable environmentally benign reaction and separation media. The possible choices of cation and anion that will result in the formation of ILs with different physico-chemical properties are numerous (Brennecke & Maginn, 2001). Furthermore, ILs are compounds which have a potential to be recycled and reused. They provide a medium for performing clean reactions with minimum waste generation.

ILs showed an over-stabilization effect in biocatalysts on the basis of the double role played by these neoteric solvents. First, ILs act as solvents, providing an adequate microenvironment for the catalytic action of the enzyme (mass transfer phenomena and active catalytic conformation); second, ILs may be regarded as liquid immobilization supports, since multipoint enzyme–IL interactions (hydrogen, Van der Waals, ionic, etc.) may occur, resulting in a flexible supramolecular net able to maintain active the protein conformation (De Diego et al., 2005). However, many enzymes were rapidly inactivated in ILs. Some of them remain stable and catalytically active in ILs, even though they are not stable and active in polar organic solvents.

The disadvantages of using ILs as media for enzymatic reactions are their high costs which could present limitation for their application to products of high added value. The question about how "green" are ILs, also appears, since their synthesis involves toxic reagents. The toxicity effect on humans/environment is still not yet clear, so they are rather considered as harmful. For enzyme-catalyzed reactions, ILs can be used as co-solvents in aqueous phase, as two-phase systems together with other solvents and as pure solvents.

One of the problems with the use of ILs for synthesis is the extraction of products. Volatile products could be separated by distillation; on the other hand, the non-volatile products could be separated by solvent extraction with the solvent which is immiscible with the ILs and is environmentally friendly (e.g. SC CO_2).

The volatile and nonpolar SC CO_2 forms different two-phase systems with nonvolatile and polar ILs. The product recovery process with these systems is based on the principle that SC CO_2 is soluble in ILs, but ILs are not soluble in SC CO_2 (Blanchard et al., 2001). Since most of the organic compounds are soluble in SC CO_2, with the high solubility of SC CO_2 in ILs, these products are transferred from the IL to the supercritical phase (Blanchard & Brennecke, 2001).

Recent researches have demonstrated the possibility to carry out integral green biocatalytic processes by combining SC CO_2 and ILs with enzymes (Lozano et al., 2003; Lozano et al., 2004; Lozano et al., 2007a; Miyawaki & Tatsuno, 2008; Knez, 2009; Fan & Qian, 2010), because their different miscibilities produce the two-phase systems that show an exceptional ability to carry out both the biotransformation and the products extraction steps simultaneously.

In our studies, ILs were used as reaction media for lipase-catalyzed kinetic resolution of (*R,S*)-1-phenylethanol with vinyl acetate. Transesterification of chiral substrate, (*R,S*)-1-phenylethanol with vinyl acetate, was performed in different ILs at atmospheric pressure and in SC CO_2/IL biphasic system. Influence of different parameters such as concentration of IL, type of IL on conversion or reaction rate of transesterification were studied. Next, stability of immobilized *Candida antarctica* lipase B (CALB) in selected IL was tested.

2. An overview of biocatalysis in ILs and in SCFs/ILs systems

In the last two decades much research has been done in the field of biocatalysis in ILs. The reason for the rapid increase of performing biotransformations in ILs was in the ability of presenting excellent enzyme activity, stability and selectivity (Yang & Pan, 2005; van Rantwijk & Sheldon, 2007). Additionally, ILs could be "taylor-made" for a specific reaction, simply by selecting appropriate combinations of cations and anions. Probably this is the major attraction of making ILs an alternative to conventional organic solvents. However, the range of ILs suitable for biocatalytic whole-cell applications is still limited and the influence of different anion and cation groups has been investigated by Bräutigam et al. (Bräutigam et al., 2009). The applicability of ionic liquids for their use as second liquid phase in whole-cell biotransformations was evaluated in combination with a recombinant *Escherichia coli* co-expressing a *Lactobacillus brevis* alcohol dehydrogenase gene for the desired asymmetric reduction of prochiral ketones and a *Candida boidinii* formate dehydrogenase for the regeneration of NAD^+ with formate. Ionic liquids, 1-(2-hydroxyethyl)-3-methylimidazolium bis(trifluoromethylsulfonyl)imide ([(E_2OH)mim][NTF]) and *N*-(3-hydroxypropyl)-pyridinium bis(trifluoromethylsulfonyl)imide ([(P_3OH)PYR][NTF]), seem to be toxic to the whole-cell biocatalyst and therefore they are not suited for an application. They resulted in yields below 3 % after a process time of 1 h. Furthermore, it was validated that ionic liquids with hexafluorophosphate ($[PF_6]^-$) and bis(trifluoromethylsulfonyl)imide ($[NTF]^-$) anions are better qualified than corresponding ILs with tris(perfluoroalkyl)trifluorophosphate ($[FAP]^-$) anion. Consequently, they should be preferred for biphasic biotransformations.

Four types of whole-cell biocatalysts: wild-type *Rhizopus oryzae* producing triacylglycerol lipase (w-ROL), recombinant *Aspergillus oryzae* expressing *Fusarium heterosporum* lipase (r-FHL), *C. antarctica* lipase B (r-CALB), and mono- and diacylglycerol lipase from *A. oryzae* (r-mdlB) were used to catalyze methanolysis of soybean oil in the presence of ILs. w-ROL gave very high yield of fatty acid methyl ester (ME) in ionic liquid 1-ethyl-3-methylimidazolium tetrafluoroborate ([emim][BF_4]) or 1-butyl-3-methylimidazolium tetrafluoroborate ([bmim][BF_4]) biphasic systems following a 24 h reaction (Arai et al., 2010).

Numerous types of enzymatic reactions (Habulin et al., 2007b; Contesini & Carvalho, 2006; Lou et al., 2006; Tan et al., 2007; Hernandez-Fernandez et al., 2007) have been carried out using ILs as solvents with similar or enhanced reaction rates and enzyme activities, and with higher operational stabilities and enantioselectivities compared to those observed in organic solvents (Welton, 1999; van Rantwijk et al., 2003; Kragl et al., 2002).

Esterification of glycerol to sinapic acid (SA) in anion $[PF_6]^-$-containing ILs, using a feruloyl esterase (FAE) from *Aspergillus niger* (AnFaeA) as biocatalyst was investigated. Hydrophobic anion ($[PF_6]^-$)-containing ILs were found to be appropriate reaction media for the enzymatic esterification of glycerol to SA, especially when they possess hydrophilic cations (1-[2-(2-methoxyethoxy)-ethyl]-3-methyl-imidazolium cation - $[C_5O_2mim]^+$, 1-(2-hydroxyethyl)-3-methyl-imidazolium cation - $[C_2OHmim]^+$) (Vafiadi et al., 2009).

Lipases, noted for their tolerance of organic solvents, are obvious candidates for the enzymatic synthesis in ILs (Sureshkumar & Lee, 2009). Kurata et al. (Kurata et al., 2010) described a transesterification reaction of methyl caffeate with various alcohols to produce caffeic acid phenethyl ester (CAPE) analogues with a lipase using an IL as the reaction medium. Effect of ILs on immobilized lipase B from *Candida antarctica* – CALB (Novozyme 435) showed that the anion nature is a crucial factor in determining enzyme activity. It was reported that the hydrogen-bond basicities of [bmim][BF_4] and 1-butyl-3-methylimidazolium trifluoromethanesulfonate ([bmim][CF_3SO_3]) are larger than those of 1-butyl-3-methylimidazolium bis(trifluoromethylsulfonyl)imide ([bmim][NTf_2]) and 1-butyl-3-methylimidazolium hexafluorofosfate ([bmim][PF_6]) (Anderson et al., 2002; Kaar et al., 2003). Additionally, it was suggested that the $[BF_4]^-$ and $[CF_3SO_3]^-$ anions are more nucleophilic than the $[NTf_2]^-$ and $[PF_6]^-$ anions, and the $[BF_4]^-$ and $[CF_3SO_3]^-$ anions coordinate more strongly to positively charged sites in the structure of an enzyme.

The immobilized CALB showed excellent storage stability and reusability in $[PF_6]^-$ containing ionic liquids. On account of high enzyme stability and high solubility of product versus substrate, $[PF_6]^-$ containing ionic liquids enabled the synthesis of various lipophilic

derivatives of hydroxycinnamic acid derivatives, such as ferulic acid, with higher conversions and reaction rates than the corresponding $[BF_4]^-$ containing ionic liquids and commonly used organic solvents (Katsoura et al., 2009).
Seventeen types of ILs were screened to test their performance as media to host lipase-catalyzed glycerolysis. It turned out that the reaction rate, triglyceride (TG) conversion, yield of diglyceride (DG), and by-product minimization depended greatly on the structure and property of the ILs applied. The reactions in trioctylmethylammonium bis(trifluoromethylsulphonyl)imide ($[toma][NTf_2]$) and Ammoeng 120 can produce comparable DG yield with those typical conventional solvent systems, but with less by-products (Kahveci et al., 2009).
One reaction system which could greatly benefit from an increased capacity in ILs is the acylation reaction of flavonoids. Flavonoids are naturally occurring bioactive compounds whose application in the food, pharmaceutical and even cosmetics industries could be drastically expanded through improved solubility and miscibility in hydrophobic environments (Lue et al., 2010).
However, there is no "best" ionic liquid for performing biotransformation just as there is no "best" organic solvent in general for carrying out biocatalysis (Vidya & Chadha, 2010).
The main problem associated with the use of ILs as reaction media for biotransformations is the recovery of products from the reaction mixture and recycling of the catalyst (Sheldon, 2005; Blanchard & Brennecke, 2001).
Several techniques of separation and product recovery from ILs exist. An attractive solution is the use SC CO_2 as extraction solvent, whereby the catalyst remains in the IL phase and the product is extracted into the SC CO_2 phase (Sheldon, 2005). Blanchard's group demonstrated the recovery of organic products from ILs by using SC CO_2. They have shown that a wide variety of solutes (alcohols, amides, ketones) can be extracted from $[bmim][PF_6]$ with CO_2 with recovery rates greater than 95 % and without any IL contamination (Blanchard & Brennecke, 2001).
ILs can absorb large quantities of CO_2 at low pressure (0.6 mol fraction at 10 MPa), although the amount of IL dissolved in CO_2 is negligible. This fact not only shows the exceptional ability of SC CO_2 to extract a wide variety of hydrophophic compounds from ILs, but also decreases the viscosity of ILs, thus, improving the mass-transfer phenomena (Lozano et al., 2007b).
Applying enzymes in SC CO_2/ILs biphasic systems has a very short history. The first successful biotransformation in SC CO_2/ILs biphasic systems was reported in the year 2002 (Lozano et al., 2002). The protective effect of ILs towards enzyme deactivation by temperature or CO_2 was demonstrated by the observed increase in synthetic activity of the enzyme when it was assayed in the presence of IL. Studies based on enzymatic kinetic resolution of *rac*-1-phenylethanol with vinyl propionate performed in SC CO_2/ILs biphasic systems have been reported (Lozano et al., 2006; Lozano et al., 2003). Both free and immobilized CALB were able to catalyze specifically the synthesis of (*R*)-1-phenylethyl propionate in SC CO_2/ILs biphasic systems, and excellent activity, stability and enantioselectivity levels were recorded. The suitability of CALB to catalyze the stereoselective transesterification of (*R*)-1-phenylethanol from the racemic mixture has been widely demonstrated (Lozano et al., 2004; Lozano et al., 2003; Eckstein et al., 2002; Suan & Sarmidi, 2004).
When continuous dynamic kinetic resolution (DKR) was performed in SC CO_2/IL biphasic media with simultaneous presence of immobilized CALB the improvement of DKR process was observed. The formation of undesired (*S*)-1-phenylethyl propionate was obtained in

low levels, which could be related to increased mass-transfer limitations for substrates and products in SC CO_2. A commercial solution of free CALB (Novozym 525L) was immobilized onto 12 different silica supports modified with specific side chains, which were assayed for the kinetic resolution of *rac*-1-phenylethanol in both IL/hexane and SC CO_2/IL biphasic media (Lozano, et al., 2007b). Immobilized derivatives coated with ILs [toma][NTf_2] and butyl-trimethylammonium bis(trifluoromethylsulfonyl)imide ([btma][NTf_2]) improved the synthetic activity of the enzyme in SC CO_2 by up to 6 times with respect to the hexane medium. This could be due to the excellent ability of SC CO_2 to transport dissolved solutes through the IL phase, which improves the transfer rate of substrates to the enzyme microenvironment compared with the liquid systems. Next, the suitability of two different ILs based on quarternary ammonium cations associated with the same anion (bis(trifluoromethane)sulfonyl amide [NTf_2]) for CALB-catalyzed ester synthesis in SC CO_2/IL biphasic system was studied (Lozano et al., 2004). The efficiency of the system was depended on both the mass-transfer phenomena between ILs and SC CO_2 immiscible phases and the specificity of the enzyme toward the catalyzed reaction. At lipase-catalyzed kinetic resolution of *rac*-1-phenylethanol with vinyl acetate in 1-butyl-3-methylimidazolium bis(trifluoromethylsulfonyl)amide ([bmim][bta]) using SC CO_2 in a batch-wise process the enzyme fully retained its activity and the enantiomeric discrimination remained uniformly high (Reetz et al., 2002). Garcia et al. (Garcia et al., 2004) explored the catalytic activities of *Fusarium solani pisi* cutinase immobilized on zeolite NaY and CALB immobilized on an acrylic resin (Novozym 435) for the transesterification reaction of *rac*-2-phenyl-1-propanol with vinyl butyrate in SC CO_2/[bmim][PF_6] biphasic system. Although the enantioselectivity of enzymes, cutinase and lipase, towards *rac*-2-phenyl-1-propanol at all the reaction conditions tested was found to be low, the reaction rates, observed in the SC CO_2/[bmim][PF_6] biphasic system, were much higher than for the [bmim][PF_6] on its own. Hernández et al. and de los Rios et al. (Hernández et al., 2006; de los Rios et al., 2007) successfully applied the dynamic membranes with immobilized CALB for butyl propionate synthesis in a recirculating bioreactor in supercritical medium and in SC CO_2/IL biphasic systems at 50 °C and 8 MPa. In SC CO_2/IL systems, the immobilized enzyme coated with different ILs ([bmim][NTf_2], [bmim][PF_6], 1-octyl-3-methylimidazolium hexafluorophosphate - [omim][PF_6] and 1-butyl-2,3-dimethylimidazolium - [bdmim][PF_6]), showed an increase in the selectivity of the process compared with SC CO_2 assayed in the absence of IL. The influence of different co-solvents, such as ethyl methylketone, *n*-heptane, 2-methyl-2-butanol, acetone, [bmim][PF_6] and [bmim][BF_4], on citronellol laurate synthesis in SC CO_2 at 60 °C and 10 MPa in a high-pressure batch stirred-tank reactor was studied (Habulin et al., 2007b). Low ester concentration was obtained in hydrophilic IL [bmim][BF_4], which could be due to desorption of water from the enzyme surface and therefore decrease in the activity of the enzyme occurred. On the contrary, [bmim][PF_6] is a hydrophobic solvent in which higher ester concentration was obtained compared with [bmim][BF_4].

Miyawaki and Tatsuno (Miyawaki & Tatsuno, 2008) studied lipase-catalyzed butanolysis of triolein in an IL methyltrioctylammonium trifluoroacetate ([mtoa][tfa]) and afterwards the product, butyloleate, was selectively extracted from the reaction mixture using SC CO_2 at 35 °C and 8.6 MPa. Although a small amount of IL added seemed inhibitory, a large amount of added IL accelerated the reaction.

Enantiomerically pure alcohols are useful building blocks and chiral auxiliaries for the synthesis of bioactive compounds such as pharmaceuticals, agrochemicals and natural products (Faber, 2000). Recently, a number of microorganisms with lipase activity have been reported for the stereoselective transesterification of racemic-1-phenyl ethanol (Goswami & Goswqami, 2005); e.g. soil isolated bacterial strain *Pseudomonas aeruginosa* catalyses the enantioselective transesterification of 1-phenyl ethanol and its various derivatives with absolute enantioselectivity (Singh et al., 2010).

Paljevac et al. (Paljevac et al., 2009) successfully applied immobilized CALB as chiral biocatalyst for enzyme-catalyzed transesterification of (*R,S*)-1-phenylethanol in SC CO_2 and in SC CO_2/IL two-phase system. An increase in the conversion and in the reaction rate was observed as the temperature and pressure were increased from 40 °C to 80 °C and from 8 MPa to 10 MPa, respectively.

Lipase-catalyzed acylation (kinetic resolution) of chiral substrate, (*R,S*)-1-phenylethanol with vinyl acetate, was performed in ILs as reaction media. Transesterification of chiral substrate, (*R,S*)-1-phenylethanol with vinyl acetate was also performed in SC CO_2/IL biphasic media and was compared with the same reaction, performed at atmospheric pressure. ILs based on the *N,N*-dialkylimidazolium cation were due to the wide spectrum of physico-chemical properties of this class chosen as model reaction media. The reaction was catalyzed by immobilized CALB (Novozym 435).

A comparison of reactions performances obtained in [emim][NTf_2] with those obtained in ILs based on dialkylimidazolium cations associated with mononuclear anions, such as $[BF_4]^-$ and $[PF_6]^-$, was proposed.

3. Materials and methods

3.1 Enzymes and chemicals

Immobilized CALB (Novozym 435) was kindly donated from Novozymes (Bagsvaerd, Denmark). (*R*)-1-phenylethanol and (*S*)-1-phenylethanol were supplied from Sigma-Aldrich (Saint Louis, USA). Vinyl acetate (≥99 %), (*R,S*)-1-phenylethanol (≥98 %), 1-ethyl-3-methyl-imidazolium bromide (≥97 %), and 65 % (gravimetric) hexafluoro-phosphoric acid solution in water were supplied from Fluka (Buchs, Switzerland). Acetone (≥99.8 %), *n*-heptane (≥99 %), anhydrous magnesium sulfate (≥98 %), 1-chlorobutane (≥99 %), dichloromethane (≥99.8 %), ethyl acetate (≥99.5 %), and 1-methylimidazole (≥99 %) were purchased from Merck (Darmstadt, Germany). Decane - Reagent Plus® (≥99 %), *N*-lithiotrifluoromethane-sulfonimide (97 %) and sodium tetra-fluoroborate (98 %) were provided by Aldrich Chemical Co. (Diesenhofen, Germany). Carbon dioxide 2.5 was provided by Messer MG (Ruše, Slovenia). Helium 6.0 was supplied from Linde plin (Celje, Slovenia).

3.2 Synthesis of ionic liquids

3.2.1 1-Butyl-3-methylimidazolium chloride [bmim][Cl]

The ionic liquid [bmim][Cl] was synthesized using 1-methylimidazole and 1-chlorobutane. The reaction mixture was stirred at 60 °C for 70 h in round-bottomed flask fitted with a reflux condenser. Obtained yellow, viscous liquid mixture was cooled to room temperature, washed three times with ethyl acetate portions and the formatted crystals of [bmim][Cl] were dried under vacuum at 60 °C for 24 h (Lewandowski & Galiński, 2004). Acquired ionic liquid served as basic component for synthesis of [bmim][BF_4] and [bmim][PF_6].

3.2.2 1-Butyl-3-methylimidazolium tetrafluoroborate [bmim][BF_4]

The sodium tetrafluoroborate was added to a solution of [bmim][Cl] in acetone. The reaction mixture was stirred at room temperature for 24 h. After stirring, the reaction mixture was filtered through a plug of celite to remove the formed sodium chloride crystals. The solvent was evaporated from hydrophilic ionic liquid [bmim][BF_4] at 40 °C for 24 h (Lozano et al., 2001). Initial water content in the synthesized [bmim][BF_4] was 0.18 % (w/w).

3.2.3 1-Butyl-3-methylimidazolium hexafluorofosfate [bmim][PF_6]

An aqueous solution of hexafluorophosphoric acid was slowly added to a solution of [bmim][Cl] in water and stirred at room temperature for 36 h. The two-phase system was separated, and the lower phase ([bmim][PF_6]) was washed with water portions until the neutral pH value. The light yellow [bmim][PF_6] was dried under vacuum at 80 °C for 24 h (Lewandowski & Galiński, 2004). Initial water content in the synthesized [bmim][PF_6] was 0.04 % (w/w).

3.2.4 1-Ethyl-3-methylimidazolium bis[(trifluoromethyl)sulfonil]amide [emim][NTf_2]

1-Ethyl-3-methylimidazolium bromide and *N*-lithiotrifluoro-methanesulfonimide were mixed in hot water for 1 h. The ionic liquid [emim][NTf_2] was extracted with dichloromethane, and dried under vacuum at 100 °C for 24 h (Lozano et al., 2001). Initial water content in the synthesized [emim][NTf_2] was 0.17 % (w/w).

3.3 Enzyme-catalyzed transesterification of (*R,S*)-1-phenylethanol performed in a batch stirred tank reactor at atmospheric pressure in IL

Enantiomerically pure compound (*R*)-1-phenylethyl acetate production by enzymatic transesterification reaction of (*R,S*)-1-phenylethanol with vinyl acetate as acyl donor was performed. Reaction, catalyzed with immobilized CALB, was performed in a batch stirred-tank reactor (Knez & Habulin, 1992). (*R,S*)-1-Phenylethanol and vinyl acetate were dissolved in ionic liquid, which was used as a reaction medium. The reactor filled up with substrates, was immersed in a water bath, heated to the desired operating temperature and stirred with a magnetic stirrer. The reaction was started by addition of CALB. Aliquots of the sample were periodically withdrawn from the reaction mixture at fixed time intervals, suspended in 0.2 % solution of decane (internal standard, IS) in *n*-heptane. Resulting solution was analyzed by gas chromatography. At least two replicates of experiments were carried out at each operative condition.

3.4 Enzyme-catalyzed transesterification of (*R,S*)-1-phenylethanol performed in a high-pressure variable-volume view cell in two-phase system SC CO_2/IL

CALB-catalyzed transesterification of (*R,S*)-1-phenylethanol with vinyl acetate in two-phase system SC CO_2/IL was performed in a high-pressure variable-volume view cell equipped with stirrer and with two stainless steel cartridge heaters to heat the reaction mixture (Fig. 1). The cell with tunable internal volume between 30 to 60 cm^3 was designed to operate up to 75 MPa and 200 °C. First, the bioreactor was loaded with immobilized CALB and heated to the desired operating temperature. Next, the substrates, (*R,S*)-1-phenylethanol, vinyl acetate, and certain amount of IL were added into the bioreactor. The mass ratio between loaded biocatalyst and substrate was 11.1. Additionally, liquid CO_2 was pumped into the reactor up to the working pressure. The reaction was started when selected pressure was achieved and mixing of reaction mixture was turned on. Transfer of substrates from upper phase through IL to active sites of biocatalyst, where the reaction took place, and transfer of synthesized products from IL phase to SC CO_2 phase was assured by mechanical stirring.

During the reaction, samples of reaction mixture were withdrawn from upper phase to monitor the product evolution. Aliquots of reaction mixture were suspended in 0.2 % solution of decane (internal standard, IS) in *n*-heptane. The enantimers concentrations during the reaction were monitored using gas chromatography. At least two replicates of experiments were carried out at each operative condition.

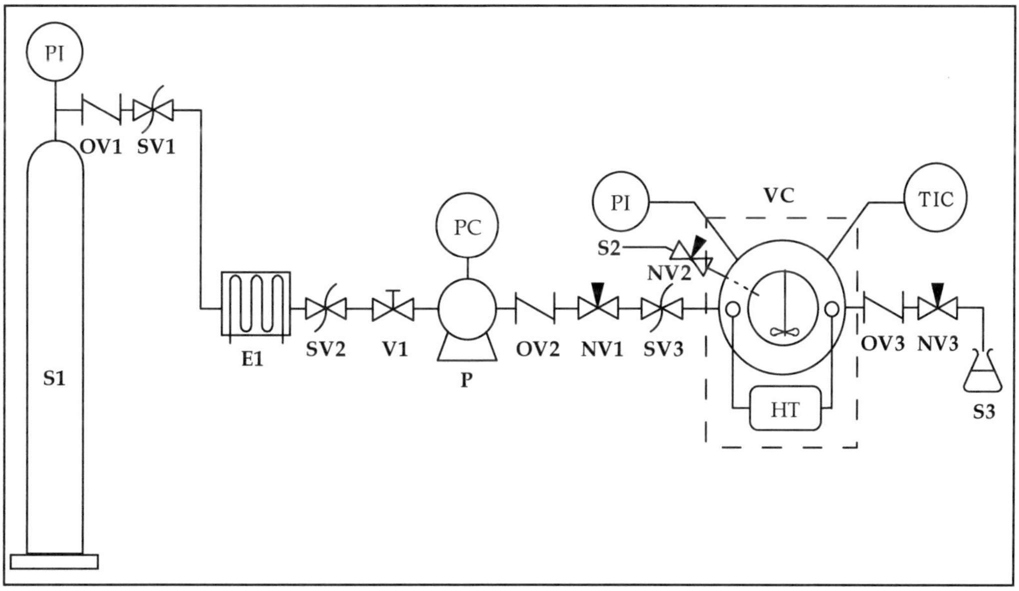

Fig. 1. High-pressure variable-volume view cell: S1 - CO_2 cylinder; E1 - cooler; P - high-pressure membrane pump; VC - variable-volume view cell with mixer; HT - electrical heater; PI - pressure indicator; PC - pressure controller; TIC - temperature indicator and controller; OV1, OV2, OV3 - high-pressure one-way valves; SV1, SV2, SV3 - high-pressure safety valves; NV1, NV2, NV3 - high-pressure needle valves; V1 - high-pressure valves; S2 - liquid charge; and S3 - sample collector.

3.5 Gas chromatography analysis (GC analysis)

Enantiomers content during the reaction was monitored using an HP 5890 series A gas chromatograph equipped with a flame-ionisation detector (FID), using a β-cyclodextrin capillary column (β -DEX 120) with the dimension length × I.D. 30 m × 0.25 mm with 0.25 μm film thickness (Supelco, Schnelldorf, Germany), at following temperature program: 100 °C K hold for 5 min, rise up to 120 °C at rate of 5 °C/min and hold for 11 min. Helium was used as carrier gas. Temperature of injector and detector were maintained at 220 °C and 250 °C, respectively.

The enantiomers of the (*R,S*)-1-phenylethanol and of the product (*R,S*)-1-phenylethyl acetate were baseline separated in the GC analysis. The conversion (*X*) was calculated by applying the below mentioned equation:

$$X = \frac{ee_R}{ee_R + ee_P} \times 100 \tag{1}$$

with

$$ee_R = \frac{[(R)-1-phenylethanol]-[(S)-1-phenylethanol]}{[(R)-1-phenylethanol]+[(S)-1-phenylethanol]} \quad (2)$$

$$ee_P = \frac{[(R)-1-phenylethylacetate]-[(S)-1-phenylethylacetate]}{[(R)-1-phenylethylacetate]+[(S)-1-phenylethylacetate]} \quad (3)$$

where square brackets represent concentration of defined substrate or product.
All samples were analyzed by GC at least twice. The relative deviation was evaluated to be within ± 1 %.

4. Transesterification of (*R,S*)-1-phenylethanol in ILs and in SC CO_2/IL two-phase systems

It is well known that not all ILs are suitable for biocatalysis. ILs containing tetrafluoroborate $[BF_4]^-$, hexafluorophosphate $[PF_6]^-$ and bis[(trifluoromethyl)sulfonyl]amide $[NTf_2]^-$ anions enabled good enzymatic activities, but in ILs containing chloride $[Cl]^-$, nitrate $[NO_3]^-$, trifluoroacetate $[CF_3CO_2]^-$ or acetate $[CH_3CO_2]^-$ anions the activity of the enzymes was observed to decrease (Kaar et al., 2003). Of course, also other parameters, such as polarity and viscosity of ILs affected enzyme activity in ILs, as well, because they can affect both the enzyme activity and mass-transfer limitations, respectively (Lozano et al., 2001).
The influence of three different ILs, based on the *N,N'*-dialkylimidazolium cations associated with mononuclear anions, such as $[BF_4]^-$, $[PF_6]^-$ and $[NTf_2]^-$, on acylation of (*R,S*)-1-phenylethanol with vinyl acetate, was studied. Reactions were performed in a batch stirred-tank reactor at 40 °C and atmospheric pressure. Equimolar ratio of (*R,S*)-1-phenylethanol/vinyl acetate was used. The influence of assayed ILs on conversion of (*R*)-1-phenylethanol to (*R*)-1-phenylethyl acetate after 5 h is presented in Fig. 2.

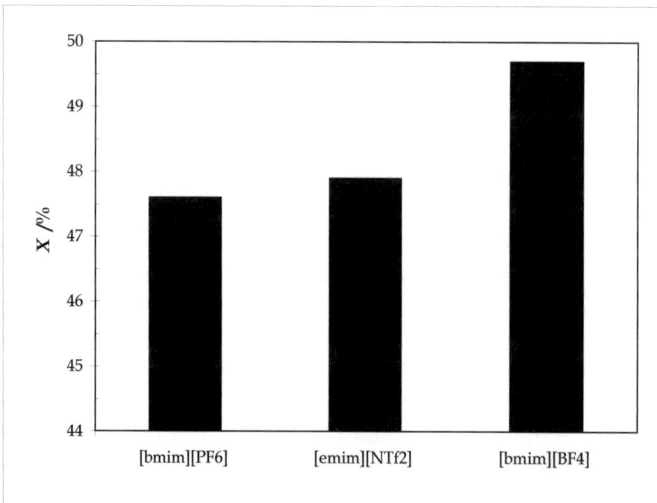

Fig. 2. Influence of three different ILs ($[bmim][PF_6]$, $[bmim][BF_4]$, and $[emim][NTf_2]$) on acylation of (*R,S*)-1-phenylethanol with vinyl acetate after 5 h of reaction performance. Reaction conditions: *c* ((*R,S*)-1-phenylethanol) = 5 mmol, *c* (vinyl acetate) = 5 mmol, *c* (ILs) = 5 mmol, *m* (immobilized CALB) = 100 mg, *T* = 40 °C, *n* = 600 rpm.

The difference in conversion of (*R*)-1-phenylethanol to (*R*)-1-phenylethyl acetate, obtained after 5 h, when the reaction was performed in [emim][NTf_2] and [bmim][PF_6], was very small. Obtained conversions, performed in [bmim][PF_6] and in [emim][NTf_2], were 47.6 % and 47.9 %, respectively. However, the highest conversion for the reaction (49.7 %) after 5 h of reaction performance was obtained in hydrophilic IL [bmim][BF_4]. The assayed ILs, [bmim][BF_4], [bmim][PF_6] and [emim][NTf_2] with log*P* values -2.44, -2.38 and -1.18 (Kaar et al., 2003; Ulbert et al., 2004; Zhao et al., 2008), respectively, proved to be adequate reaction media for lipase-catalyzed acylation of (*R,S*)-1-phenylethanol with vinyl acetate.
Enzyme activity increased with the decrease in log*P* values of ILs and the highest enzyme activity was observed in [bmim][BF_4] with log*P* value -2.44. Immobilized CALB retained its activity in all assayed ILs despite the fact that ILs have low log*P* values (below zero), which seem to suggest that they are highly hydrophilic in nature and would likely inactivate enzymes (Yang & Pan, 2005).
The immobilized CALB retained its activity in assayed ILs also due to their low hydrogen-bond basicity of the enzyme-compatible anions. Namely, the $[BF_4]^-$ spreads its negative charge over four fluorine atoms, the $[PF_6]^-$ over six fluorine atoms and the $[NTf_2]^-$ over five atoms (Park & Kazlauskas, 2003).
ILs are much more viscous than conventional organic solvents. Using them as media for biotransformations, enzyme activity can be controlled by the viscosity of the ILs affecting the mass-transfer limitations. Therefore a lower reaction rate would be expected in an ionic liquid with higher viscosity (Yang & Pan, 2005). Indeed, when the reaction was carried out in [emim][NTf_2], [bmim][BF_4] and [bmim][PF_6] with viscosities of 34 cP, 154 cP and 430 cP (Marsh et al., 2004) respectively, a reduction in the initial reaction rate was corresponding to an increase in the viscosity of the ILs (Fig. 3).
Visual observations confirmed that hydrophobic IL [bmim][PF_6] formed a layer around the enzyme and this could be considered as being included into the media. Because of the limited contact between substrate and the active site of the lipase, lower reaction rate was obtained.

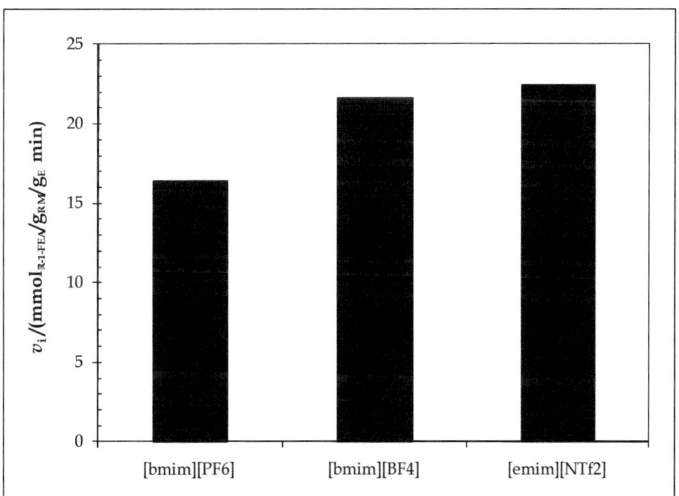

Fig. 3. Influence of different ILs ([bmim][PF_6], [bmim][BF_4], and [emim][NTf_2]), on initial reaction rate. Reaction conditions: *c* ((*R,S*)-1-phenylethanol) = 5 mmol, *c* (vinyl acetate) = 5 mmol, *c* (ILs) = 5 mmol, *m* (immobilized CALB) = 100 mg, *T* = 40 °C, *n* = 600 rpm.

One of the greatest advantages of the immobilized enzymes is the possibility of their reuse for a specific reaction, reducing the process costs. The possibility of enzyme reuse depends on the residual activity of the biocatalyst in the reaction medium. To study the influence of the used IL on residual activity of immobilized CALB, this was reused for transesterification of (*R*)-1-phenylethanol in [bmim][BF_4] for several reaction cycles.
After each reaction cycle, which lasted for 5 h, the biocatalyst was regenerated by filtration and washing with acetone. Changes in the conversion of (*R*)-1-phenylethanol after each reaction cycle in [bmim][BF_4] are presented in Fig. 4. Significant changes in the conversion after each reaction cycle were observed. Conversion, obtained after 5 h, was after 1st reaction cycle 50 %, while after 2nd one it was half of this value and after 5th reaction cycle it was only 3.2 %.
To overcome the problem of enzyme inactivation, another IL should be used as a reaction medium; e.g. Schöfer et al. (Schöfer et al., 2001) reported of only 10 % reduction of enzyme activity of immobilized CALB per reaction cycle in [bmim][NTf_2].
To use the benefits of two unconventional reaction media, transetserification of (*R,S*)-1-phenylethanol, using vinyl acetate as acyl donor, was performed in SC CO_2/[bmim][BF_4] system. CALB was used to catalyze the reaction. At first, the effect of [bmim][BF_4] concentration on lipase activity was studied. IL concentrations were varied between 0 mmol to 50 mmol, with molar ratio of substrates, (*R,S*)-1-phenylethanol and vinyl acetate, of 1:1 at 40 °C and 10 MPa.

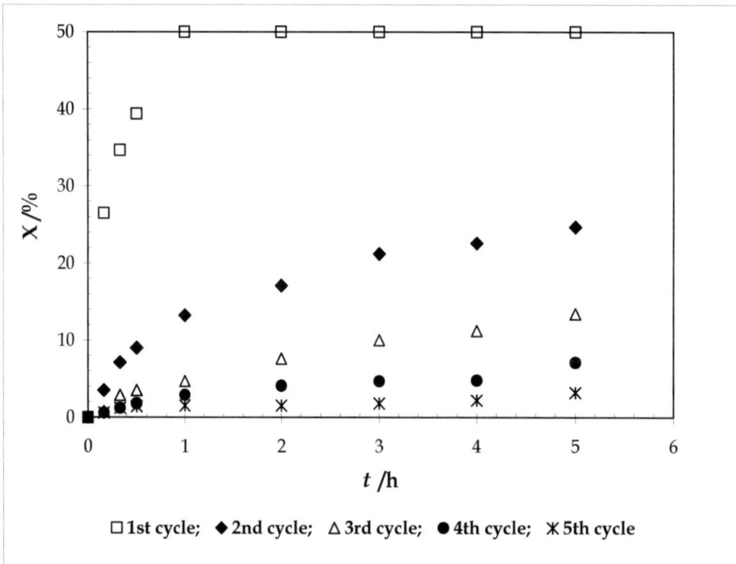

Fig. 4. Stability of immobilized CALB in [bmim][BF_4]. Reaction conditions: *c* ((*R,S*)-1-phenylethanol) = 5 mmol, *c* (vinyl acetate) = 5 mmol, *c* ([bmim][BF_4]) = 5 mmol, *w* (immobilized CALB; w/w of substrates) = 20 %, T = 40 °C, n = 600 rpm.

Using both solvents, SC CO_2 and [bmim][BF_4], the reaction rate increased with the increase of IL concentration. The best results were obtained when the highest amount of [bmim][BF_4] (50 mmol) was used for the transesterification reaction. This is in agreement with the results

obtained at atmospheric pressure which showed that by increasing the IL concentration, the conversion was enhanced (Habulin & Knez, 2009).
Performing the reaction in the system without [bmim][BF_4], which means only in SC CO_2, the highest CALB activity was observed. After 2 h of reaction performance 50 % conversion was obtained. From Table 1 it is obvious that the conversion decreased drastically when [bmim][BF_4] was added into the reaction mixture. This could be related to the presence of external mass transfer limitations, since the reaction was performed in a two-phase system with CALB particles, instead of in one-phase system with CALB particles, as in the case when the reaction was performed only in SC CO_2, without [bmim][BF_4].

$c_{[bmim][BF4]}$ /mmol	X/%	
	2 h	8 h
0.0	50.0	50.0
12.5	27.0	45.0
25.0	24.9	48.9
50.0	35.1	50.0

Table 1. Impact of different [bmim][BF_4] concentrations on conversion of transesterification of (*R,S*)-1-phenylethanol in two-phase system SC CO_2/IL after 2 h and 8 h of reaction.

In this case CALB particles were suspended in [bmim][BF_4] phase, while the substrates were in the supercritical phase. The substrate molecules had to be transferred from SC CO_2 phase to IL phase towards CALB particles first and secondly, products had to be transferred back to SC CO_2 phase.
This thesis could be proved by visual observation of the system. Fig. 5 shows the changes of the phase behaviour for (*R,S*)-1-phenylethanol/vinyl acetate/SC CO_2/[bmim][BF_4] system when the pressure was increased from 0 MPa (without SC CO_2) to 10 MPa at 40 °C.
Since SC CO_2 dissolves quite well in ionic liquids, but ionic liquids do not dissolve in carbon dioxide (Olivier-Bourbigou & Magna, 2002) it was expected that two-phase system with CALB particles would be obtained when the reaction was carried out in SC CO_2/[bmim][BF_4] at 40 °C and 10 MPa. Indeed, liquid reaction bulk (solution of (*R,S*)-1-phenylethanol and vinyl acetate in [bmim][BF_4]) at 0 MPa (one phase system) with CALB particles was with addition of CO_2 transformed to a two-phase system with CALB particles (Fig. 5).
Lower phase (catalytic phase) was rich with [bmim][BF_4], while the upper phase was rich with CO_2 and substrates.
CALB particles were suspended in the lower phase where the reaction took place. When the pressure achieved 10 MPa, the stirring of reaction mixture was started.
Enhanced transfer of substrates from upper phase through [bmim][BF_4] to active sites of the enzyme where the reaction took place and later the transfer of formed products from [bmim][BF_4] phase to SC CO_2 phase was enabled.
Varying the concentration of [bmim][BF_4], initial reaction rates could be enhanced, as well (Fig. 6). They were increasing with increased IL concentration. Since ILs are good solvents for many organic compounds (Sheldon, 2001), higher IL concentration in the reaction mixture could cause higher solubility of substrates (Habulin & Knez, 2009) and consecutively higher conversion and initial reaction rate were obtained.

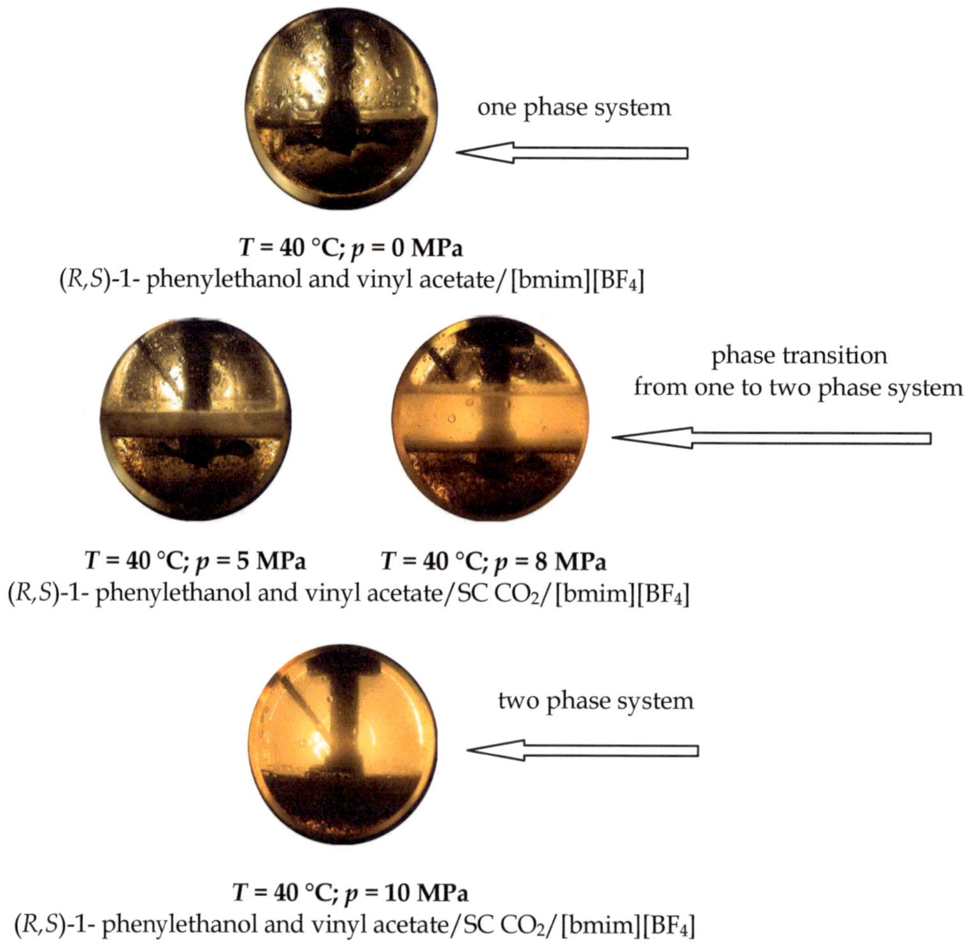

Fig. 5. Phase transitions for the system (*R,S*)-1-phenylethanol and vinyl acetate/SC CO2/[bmim][BF4] with CALB particles at different pressures.

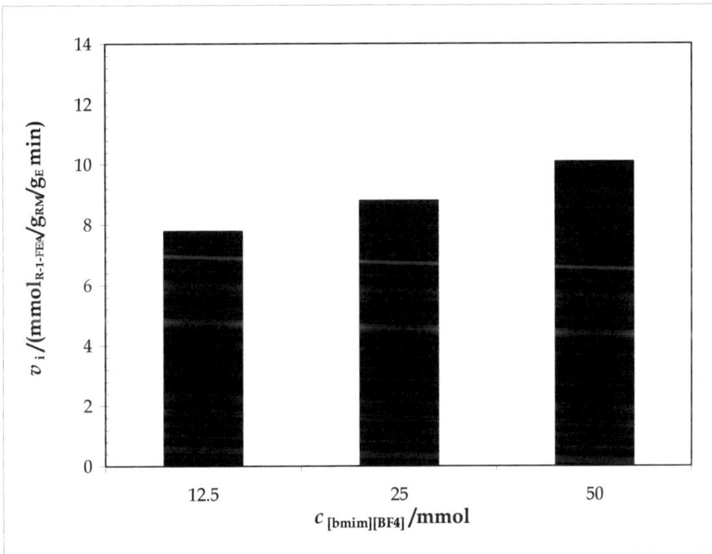

Fig. 6. Influence of different [bmim][BF_4] concentrations on initial reaction rate of (*R,S*)-1-phenylethanol transesterification in SC CO_2. Reaction parameters: *c* ((*R,S*)-1-phenylethanol) = 25 mmol, *c* (vinyl acetate) = 25 mmol, *m* (immobilized CALB) = 0.54 g, $T = 40\,^{\circ}C$, $p = 10$ MPa, $n = 600$ rpm.

5. Outlook

According to a recent review work on biocatalysis in ILs, factors such as polarity and nucleophilicity of the anion, pH, purity of the IL and water content, have a major effect on the activity, the stability and the solubility of enzymes in these non-conventional media (van Rantwijk and Sheldon, 2007). The influence of IL concentration on the CALB activity was found to be a determining factor for the performance of the (*R,S*)-1-phenylethanol kinetic resolution at atmospheric pressure, as well as in SC CO_2/[bmim][BF_4] system. Biocatalysis with CALB, which is an excellent chiral biocatalyst for the stereo-selective acylation of racemic alcohols, gave very high kinetic resolution (*R*)-enantiomer yields and selectivity.

The unique properties of ILs to carry out biotransformations in nonaqueous environments open up new opportunities to develop green industrial processes. However, despite of opportunities afforded by ILs in biocatalysis, before the industrial-scale application of ILs in biotransformation will be feasible, several necessary steps need to be done. First, more attention should be given to synthesis of ILs, especially to new "clean" synthesis methods. The synthesis of safer ILs from natural materials (e.g. carbohydrates, proteins, lipids, and their derivatives) is in the process of development and will provide benefits for the IL industry as well as for the food industry. Therefore, techniques together with corresponding facilities for large-scale production of ILs have to be developed, because the number of ILs presently on the market is still limited (Moniruzzaman et al., 2010). The next problem presents separation of the product from IL. Large scale separation of products from ILs and recovery as well as reuse of ILs represent one of the key technologies that limits the extensive applications of ILs (Kahveci et al., 2009).

One of the possibilities is use of biphasic systems IL/SCF. Lozano et al. proposed two-phase systems IL/SC CO_2 as the first approach to integral green bioprocesses in non-aqueous media, where both the biocatalytic and extraction steps are coupled in an environmentally benign and efficient reaction/separation process (Lozano et al., 2007c).

6. References

Anderson, J.L., Ding, J., Welton, T. & Armstrong, D.W. (2002). Characterizing ionic liquids on the basis of multiple solvation interactions, *J. Am. Chem. Soc.*, 124(47): 14247–14254.

Arai, S., Nakashima, K., Tanino, T., Ogino, C., Kondo, A. & Fukuda, H. (2010). Production of biodiesel fuel from soybean oil catalyzed by fungus whole-cell biocatalysts in ionic liquids, *Enzyme Microb. Tech.*, 46(1): 51-55.

Blanchard, L.A. & Brennecke, J.F. (2001). Recovery of organic products from ionic liquids using supercritical carbon dioxide, *Ind. Eng. Chem. Res.*, 40(1): 287-292.

Blanchard, L.A., Gu, Z. & Brennecke, J.F. (2001). High-pressure phase behavior of ionic liquid/CO_2 systems, *J. Phys. Chem. B*, 105(12): 2437-2444.

Bräutigam, S., Dennewald, D., Schürmann, M., Lutje-Spelberg, J., Pitner, W.R. & Weuster-Botz, D. (2009). Whole-cell biocatalysis: Evaluation of new hydrophobic ionic liquids for efficient asymmetric reduction of prochiral ketones, *Enzyme Microb. Tech.*, 45(4): 310-316.

Brennecke, J.F. & Maginn, E.J. (2001). Ionic Liquids: Innovative Fluids for Chemical Processing, *AIChE Journal*, 47(11): 2384-2389.

Contesini, F.J. & Carvalho, P.D. (2006). Esterification of (*R,S*)-ibuprofen by native and commercial lipases in a two-phase system containing ionic liquids, *Tetrahedron-Asymmetr.*, 17(14): 2069-2073.

De Diego, T., Lozano, P., Gmough, S., Vaultier, M. & Iborra, J.L. (2005). Understanding structure - stability relationships of *Candida antarctica* lipase B in ionic liquids, *Biomacromolecules*, 6(3): 1457−1464.

De los Rios, A.P., Hernandez-Fernandez, F.J., Gomez, D., Rubio, M., Tomas-Alonso F. & Villora, G. (2007). Understanding the chemical reaction and mass-transfer phenomena in a recirculating enzymatic membrane reactor for green ester synthesis in ionic liquid/supercritical carbon dioxide biphasic systems, *J. Supercrit. Fluid.*, 43(2): 303-309.

Eckstein, M., Wasserscheid, P. & Kragl, U. (2002). Enhanced enantioselectivity of lipase from *Pseudomonas sp.* at high temperatures and fixed water activity in the ionic liquid 1-butyl-3-methyl bis((trifluoromethyl)sulfonyl)amide, *Biotechnol. Lett.*, 24(10): 763-767.

Fan, Y. & Qian, J. (2010). Lipase catalysis in ionic liquids/supercritical carbon dioxide and its applications, *J. Mol. Catal. B: Enzym.*, 66(1-2): 1-7.

Faber, K. (2000). *Biotransformations in Organic Chemistry*, Springer-Verlag, Berlin.

Garcia, S., Lourenco, N.M.T., Lousa, D., Sequeira, A.F., Mimoso, P., Cabral, J.M.S., Afonso C.A.M. & Barreiros, S. (2004). A comparative study of biocatalysis in non-conventional solvents: Ionic liquids, supercritical fluids and organic media, *Green Chem.*, 6(9): 466-470.

Goswami, A. & Goswami, J. (2005). DMSO-triggered enhancement of enantioselectivity in Novozyme[435]-catalyzed transesterification of chiral 1-phenylethanols, *Tetrahedron Lett.*, 46(25): 4411–4413.

Habulin, M. & Knez, Ž. (2001). Pressure stability of lipases and their use in different systems, *Acta Chim. Slov.*, 48(4): 521-532.

Habulin, M., Primožič, M. & Knez, Ž. (2007a). Supercritical fluids as solvents for enzymatic reactions, *Acta Chim. Slov.*, 54(4): 667-677.

Habulin, M., Šabeder, S., Paljevac, M., Primožič, M. & Knez, Ž. (2007b). Lipase-catalyzed esterification of citronellol with lauric acid in supercritical carbon dioxide/co-solvent media, *J. Supercrit. Fluid.*, 43(2): 199-203.

Habulin, M. & Knez, Ž. (2009). Optimization of (*R,S*)-1-phenylethanol kinetic resolution over *Candida antarctica* lipase B in ionic liquids, *J. Mol. Catal. B: Enzym.*, 58(1-4): 24-28.

Hernández, F.J., de los Ríos, A.P., Gómez, D., Rubio, M. & Víllora, G. (2006). A new recirculating enzymatic membrane reactor for ester synthesis in ionic liquid/supercritical carbon dioxide biphasic systems, *Appl. Catal. B: Environ.*, 67(1-2): 121-126.

Hernández-Fernandez, F.J., de los Ríos, A.P., Rubio, M., Gómez, D. & Víllora, G. (2007). Enhancement of activity and selectivity in lipase-catalyzed transesterification in ionic liquids by the use of additives, *J. Chem. Technol. Biotechnol.*, 82(19): 882-887.

Hobbs, H.R. & Thomas, N.R. (2007). Biocatalysis in Supercritical Fluids, in Fluorous Solvents, and under Solvent-Free Conditions, *Chem. Rev.*, 107(6): 2786-2820.

Jessop, P.G & Leitner, W. (1999). *Chemical synthesis using supercritical fluids*, Wiley-VCH, Weinheim.

Kaar, J.L., Jesionowski, A.M., Berberich, J.A., Moulton, R. & Russell, A.J. (2003). Impact of ionic liquid physical properties on lipase activity and stability, *J. Am. Chem. Soc.*, 125(14): 4125–4131.

Kahveci, D., Guo, Z., Özçelik, B. & Xu, X. (2009). Lipase-catalyzed glycerolysis in ionic liquids directed towards diglyceride synthesis, *Process Biochem.*, 44(12): 1358-1365.

Katsoura, M.H., Polydera, A.C., Tsironis, L.D., Petraki, M.P., Kostić Rajačić, S., Tselepis, A.D. & Stamatis, H. (2009). Efficient enzymatic preparation of hydroxycinnamates in ionic liquids enhances their antioxidant effect on lipoproteins oxidative modification, *New Biotechnol.*, 26(½): 83-91.

Knez, Ž. & Habulin, M. (1992). Lipase catalysed esterification in supercritical carbon dioxide, In: *Progress in Biotechnology 8, Biocatalysis in Non-Conventional Media,* J. Tramper, M.H. Vermüe, and H.H. Beeftink (Eds.), pp. 401-406, Elsevier, New York.

Knez, Ž. & Habulin, M. (2002). Compressed gases as alternative enzymatic-reaction solvents: a short review, *J. Supercrit. Fluid.*, 23(1): 29-42.

Knez, Ž. (2009). Enzymatic reactions in dense gases, *J. Supercrit. Fluid.*, 47(3): 357-372.

Kragl, U., Eckstein, M. & Kaftzik, N. (2002). Enzyme catalysis in ionic liquids, *Curr. Opin. Biotechnol.*, 13(6): 565-571.

Krishna, S.H. (2002). Developments and trends in enzyme catalysis in nonconventional media, *Biotechnol. Adv.*, 20(3-4): 239-267.

Kurata, A., Kitamura, Y., Irie, S., Takemoto, S., Akai, Y., Hirota, Y., Fujita, T., Iwai, K., Furusawa, M. & Kishimoto, N. (2010). Enzymatic synthesis of caffeic acid phenethyl ester analogues in ionic liquid, *J. Biotechnol.*, 148(2-3): 133-138.

Lewandowski, A. & Galiński, M. (2004). Carbon-ionic liquid double-layer capacitors, *J. Phys. Chem. Solids*, 65(2-3): 281-286.

Liese, A., Seelbach, K. & Wandrey, C. (2000). *Industrial Biotransformations*, Wiley-VCH, Weinheim.

Lou, W.Y., Zong, M.H., Liu, Y.Y. & Wang J.F. (2006). Efficient enantioselective hydrolysis of D,L-phenylglycine methyl ester catalyzed by immobilized *Candida antarctica* lipase B in ionic liquid containing systems, *J. Biotechnol.*, 125(1): 64-74.

Lozano, P., de Diego, T., Guegan, J.-P., Vaultier, M. & Iborra, J.L. (2001). Stabilization of α-chymotrypsin by ionic liquids in transesterification reactions, *Biotechnol. Bioeng.*, 75(5): 563-569.

Lozano, P., De Diego, T., Carrié, D., Vaultier M. & Iborra J.L. (2002). Continuous green biocatalytic processes using ionic liquids and supercritical carbon dioxide, *Chem. Commun.*, 7: 692-693.

Lozano, P., De Diego, T., Carrié, D., Vaultier, M. & Iborra, J.L. (2003). Lipase catalysis in ionic liquids and supercritical carbon dioxide at 150 °C, *Biotechnol. Prog.*, 19(2): 380-382.

Lozano, P., De Diego, T., Gmouh, S., Vaultier, M. & Iborra, J.L. (2004). Criteria to design green enzymatic processes in ionic liquid/supercritical carbon dioxide systems, *Biotechnol. Prog.*, 20(3): 661-669.

Lozano, P., de Diego, T., Larnicol, M., Vaultier, M. & Iborra, J.L. (2006). Chemoenzymatic dynamic kinetic resolution of *rac*-1-phenylethanol in ionic liquids and ionic liquids/supercritical carbon dioxide systems, *Biotechnol. Lett.*, 28(19): 1559-1565.

Lozano, P., De Diego, T., Gmouh, S., Vaultier, M. & Iborra, J.L. (2007a). A continuous reactor for the (chemo)enzymatic dynamic kinetic resolution of *rac*-1-phenylethanol in ionic liquid/supercritical carbon dioxide biphasic systems, *Int. J. Chem. Reactor Eng.*, 5: Article A53.

Lozano, P., De Diego, T., Sauer, T., Vaultier, M., Gmouh, S. & Iborra J.L. (2007b). On the importance of the supporting *Candida antarctica* lipase B in material for activity of immobilized ionic liquid/hexane and ionic liquid/supercritical carbon dioxide biphasic media, *J. Supercrit. Fluid.*, 40(1): 93-100.

Lozano, P., De Diego, T. & Iborra J.L. (2007c). Enzymatic catalysis in ionic liquids and supercritical carbon dioxide biphasic systems, *Chem. Today*, 25(6): 76-79.

Lue, B.M., Guo, Z. & Xu, X. (2010). Effect of room temperature ionic liquid structure on the enzymatic acylation of flavonoids, *Process Biochem.*, 45(8): 1375-1382.

Marsh, K.N., Boxall, J.A. & Lichtenthaler, R. (2004). Room temperature ionic liquids and their mixtures - a review, *Fluid Phase Equilib.*, 219(1): 93-98.

Matsuda, T., Harada, T. & Nakamura, K. (2005). Biocatalysis in supercritical CO_2, *Curr. Org. Chem.*, 9(3): 299-315.

Mesiano, A.J., Beckman, E.J. & Russell, A.J. (1999). Supercritical biocatalysis, *Chem. Rev.*, 99(2): 623-633.

Miyawaki, O. & Tatsuno, M. (2008). Lipase-catalyzed butanolysis of triolein in ionic liquid and selective extraction of product using supercritical carbon dioxide, *J. Biosci. Bioeng.*, 105(1): 61-64.

Moniruzzaman, M., Nakashima, K., Kamiya N. & Goto, M. (2010). Recent advances of enzymatic reactions in ionic liquids, *Biochem. Eng. J.*, 48(3): 295-314.

Olivier-Bourbigou, H. & Magna, L. (2002). Ionic liquids: Perspectives for organic and catalytic reactions, *J. Mol. Catal. A.: Chem.*, 182-183: 419-437.

Paljevac, M., Knez, Ž. & Habulin, M. (2009). Lipase-Catalyzed Transesterification of (*R,S*)-1-Phenylethanol in SC CO_2 and in SC CO_2/Ionic Liquid Systems, *Acta Chim. Slov.*, 56(4): 815-825.

Park, S. & Kazlauskas, R.J. (2003). Biocatalysis in ionic liquids - advantages beyond green technology, *Curr. Opin. Biotechnol.*, 14(4): 432-437.

Randolph, T.W., Blanch, H.W., Prausnitz, J.M. & Wilke, C.R. (1985). Enzymatic catalysis in supercritical fluid, *Biotechnol. Lett.*, 7(5): 325-328.

Reetz, M.T., Wiesenhofer, W., Francio, G. & Leitner, W. (2002). Biocatalysis in Ionic Liquids: Batchwise and Continuous-Flow Processes Using Supercritical Carbon Dioxide as the Mobile Phase, *Chem. Commun.*, (9): 992-993.

Schöfer, S.H., Kaftzik, N., Wasserscheid, P. & Kragl, U. (2001). Enzyme catalysis in ionic liquids: lipase catalysed kinetic resolution of 1-phenylethanol with improved enantioselectivity, *Chem. Commun.*, (5): 425-426.

Sheldon, R.A. (2001). Catalytic Reactions in Ionic Liquids, *Chem. Commun.*, (23): 2399–2407.

Sheldon, R.A. (2005). Green solvents for sustainable organic synthesis: State of the art, *Green Chem.*, 7(5): 267-278.

Singh, M., Singh, R.S., & Banerjee, U.C. (2010). Enantioselective transesterification of racemic phenyl ethanol and its derivatives in organic solvent and ionic liquid using *Pseudomonas aeruginosa* lipase, *Process Biochem.*, 45(1): 25-29.

Suan, C.L. & Sarmidi, M.R. (2004). Immobilised lipase-catalysed resolution of (*R,S*)-1-phenylethanol in recirculated packed bed reactor, *J. Mol. Catal. B: Enzym.*, 28(2-3): 111-119.

Sureshkumar, M. & Lee, C.-K. (2009). Biocatalytic reactions in hydrophobic ionic liquids, *J. Mol. Catal. B: Enzym.*, 60(1-2): 1–12.

Tan, Z.Y., Wu, H. & Zong, M.H. (2007). Novozym 435-catalyzed regioselective benzoylation of 1-beta-D-arabinofuranosylcytosine in a co-solvent mixture of C(4)MIm center dot PF6 and pyridine, *Biocatal. Biotransform.*, 25(5): 408-413.

Ulbert, O., Fráter, T., Bélafi-Bakó, K. & Gubicza, L. (2004). Enhanced enantioselectivity of *Candida rugosa* lipase in ionic liquids as compared to organic solvents, *J. Mol. Catal. B: Enzym.*, 31(1-3): 39-45.

Vafiadi, C., Topakas, E., Nahmias, V.R., Faulds, C. B. & Christakopoulos, P. (2009). Feruloyl esterase-catalysed synthesis of glycerol sinapate using ionic liquids mixtures, *J. Biotechnol.*, 139: 124-129.

van Rantwijk, F., Lau, R.M. & Sheldon, R.A. (2003). Biocatalytic transformations in ionic liquids, *Trends Biotechnol.*, 21(3): 131-138.

van Rantwijk, F. & Sheldon, R.A. (2007). Biocatalysis in ionic liquids, *Chem. Rev.*, 107(6): 2757–2785.

Vidya, P. & Chadha, A. (2010). *Pseudomonas cepacia* lipase catalyzed esterification and transesterification of 3-(furan-2-yl) propanoic acid/ethyl ester: A comparison in ionic liquids vs hexane, *J. Mol. Catal. B: Enzym.*, 65(1-4): 68-72.

Welton, T. (1999). Room-temperature ionic liquids. Solvents for synthesis and catalysis, *Chem. Rev.*, 99(8): 2071-2083.

Yang, Z. & Pan, W. (2005). Ionic liquids: green solvents for nonaqueous biocatalysis, *Enzyme Microb. Technol.*, 37(1): 19–28.

Zhao, H., Baker, G.A., Song, Z., Olubajo, O., Zanders, L. & Campbell, S.M. (2008). Effect of ionic liquid properties on lipase stabilization under microwave irradiation, *J. Mol. Catal. B: Enzym.*, 57(1-4): 149-157.

20

Applications of Ionic Liquids to Increase the Efficiency of Lipase Biocatalysis

Francisc Péter, Cristina Paul and Anca Ursoiu
University „Politehnica" of Timişoara, Faculty of Industrial Chemistry and Environmental Engineering
Romania

1. Introduction

Stabilization of enzymes is a key issue to develop more efficient biocatalysts for industrial, environmental, or biomedical applications (Péter, 2005). In the last decades, important research potential has been focused on enzyme immobilization, explainable by the limited reusability of the native enzymes as industrial catalysts. Enzyme immobilization emerged as a valuable tool to overcome such a drawback, as it allows enzyme recycling and facilitates separation and recovery of the product from the reaction medium. In addition, could often result in increased thermal and operational stability of biocatalysts, thus allowing their employment in wider experimental conditions, compared to the native enzymes (Bucholz et al., 2005; Cao, 2005). Enzyme engineering by immobilization techniques was demonstrated to be compatible with other chemical or biological approaches to improve the enzyme function (Reetz & Jaeger, 1998).

Lipases are enzymes widely used in organic chemistry as biocatalysts for a large number of synthetic reactions. They showed an interesting combination of broad range substrate specificity with high regio- and enantioselectivity, which made them particularly useful in chiral resolution of organic compounds (Whittall & Sutton, 2010). Immobilization of lipases to increase the biocatalytic efficiency and reduce the costs is very well documented (Christensen et al., 2003; Sheldon, 2007; Mateo et al., 2007). The most frequently used lipase immobilization techniques are:

- physical adsorption
- covalent attachment
- entrapment in polymeric matrixes
- cross-linking of enzyme molecules

All these methods target to gather the advantages of immobilization (enhanced stability, repeated or continuous use, easy separation from the reaction mixture, possible modulation of the catalytic properties, prevention of protein contamination in the product), without any significant decline of the enzyme catalytic activity. In the last decades, the utilization of new enzyme carriers was, besides exploitation of the increasing knowledge regarding enzyme structure and mechanism, the most important scientific trend in enzyme engineering. (Bornscheuer, 2003).

In this respect, encapsulation in sol–gel matrices has been proved as one of the most efficient immobilization methods, considering both activity and biocatalyst stability. Due to their nano- or microporous structure, sol–gel materials exhibit valuable properties, such as high surface to volume ratio, large surface area and porosity (Pirozzi et al., 2009). In addition, the sol–gel process allows high compositional and morphological flexibility, by utilization of alkoxyde-type silane precursors holding one or two nonhydrolizable organic functional groups and various additives, to yield organic-inorganic hybrid matrices (Reetz et al., 2003). Sol-gel matrices are formed by hydrolytic polymerization of the silane precursors, mainly alcoxides. The first step is hydrolytic, followed by condensation reactions to yield silica. Silica particles grow progressively as condensation proceeds, leading to the formation of colloidal solutions and gels. Dried at room temperature, the gels form a porous network of hydrated amorphous silica. Conducting the sol-gel process in the presence of a biomolecule will result in its entrapment by the obtained matrix (Pierre, 2004).

Lipase immobilization by the sol-gel method was subject of numerous recent publications (Kiss et al., 2007; Hara et al., 2008; Kawakami et al., 2009; Zarcula et al., 2009; Tomin et al., 2010). The main benefit of this method is the enzyme inclusion throughout the gelation phase, at low synthesis temperature and not adverse reaction medium for the catalytic activity (Ciriminna & Pagliano, 2006). Another important advantage is the possibility to combine the entrapment with adsorption, to obtain supported sol-gel polymers with improved properties. Diverse carriers, as Fe_3O_4 magnetic nanoparticles, non-woven fabric (Chen & Lin, 2003), or Celite diatomaceous earth (Kawakami & Yoshida, 1996; Zarcula et al., 2009), have been tested with promising results.

As protection of the enzyme during the encapsulation process is very important to prevent inactivation caused by gel shrinkage throughout maturation and drying, or by inadequate pore size resulting in slow diffusion rate, several compounds have been tested as immobilization additives. Sol-gel glasses have the ability to entrap large amounts of such additives, resulting in preservation or even enhancement of enzyme activity and selectivity. Most of the tested compounds were of polyhydroxylic nature: sugar, polyethylene glycols with different molecular weights, glycerol, cyclodextrins, polyvinyl alcohol, but also crown ethers, nonionic surfactants, or even proteins have been investigated (Reetz et al., 1996; Reetz et al., 2003; Péter et al., 2005; Cao et al., 2009).

Unexpectedly, ionic liquids (ILs) also proved to be efficient immobilization additives, although they are mainly known as substitute reaction media of volatile organic solvents. Ionic liquid usually contain a bulky cation and a small anion, and are available in large varieties. They gained constantly increasing attention as possible green reaction media of the future, based on their lack of vapour pressure, thermal stability, non-flammability and widely tunable physicochemical properties through the appropriate modification of the cation and/or anion (Van Rantwijk et al., 2003).

ILs with 1,3-dialkylimidazolium cations are generally recognized as the most valuable for biocatalytic applications. Their physical properties cover a broad range of values, but the catalytic properties are mainly influenced by their polarity, hydrophobicity and solvent miscibility. The polarity of ILs based on imidazolium cation is in the range of lower alcohols and formamide, and slightly decreases with increase of the length of alkyl chain linked to the imidazolium ring. However, it is difficult to correlate the IL polarity with the reaction rate of a specific substrate because other parameters, like viscosity, are also influential

(Moniruzzaman et al., 2010). Based on their water miscibility, ILs can be classified as hydrophobic and hydrophilic. This miscibility seems to be influenced by the ability of their anions to form hydrogen bonds, not by the overall polarity of the molecule. From the series of ILs with the same (1-butyl-3-methylimidazolium) cation, those with hexafluorophosphate ($[PF_6]^-$), *bis*(trifluoromethyl)-sulfonylimide ($[Tf_2N]^-$), or perchlorate ($[ClO_4]^-$) anions are water-immiscible, whereas those with tetrafluoroborate ($[BF_4]^-$), bromide ($[Br^-]$), or octylsulphate ($[OcSO_4]^-$) anions are water-miscible (Gorke et al., 2007). The explanation is probably the stronger hydrogen-bond accepting ability of water-miscible IL's anions. For biocatalytic purposes, the hydrophobic ILs are more interesting, as synthetic reactions are usually performed in non-aqueous solvents. Esterification, acylation, or polymerization reactions catalyzed by lipases in hydrophobic ILs have been reviewed (Moon et al., 2006; Sureshkumar & Lee, 2009). Unexpectedly, the hydrophobic ILs are not inherently miscible with hydrophobic organic solvents. The water-immiscible 1-butyl-3-methyl-imidazolium tetrafluoroborate is also immiscible with hexane and toluene, but miscible with acetone (Poole, 2004). Even in case of miscibility, the miscibility gaps of ILs with organic solvents are highly asymmetrical. The IL can dissolve appreciable amount of the solvent, but the solubility of IL in the molecular solvent is low (Weingärtner, 2008). Consequently, it is difficult to presume the solubility of various substrates and reaction products in a given ionic liquid, and experiments could run in a different way as it was expected. Monophasic reaction systems are commonly employed in biocatalytic reactions in IL media, with the main disadvantage of a supplementary product extraction step at the end of reaction. This extraction step could be eliminated using biphasic or multiphasic systems, but the IL and the organic components (product, unreacted substrates) must represent effectively immiscible phases, without negative effect on the reaction yield and rate, an objective not easy to achieve (Fehér et al., 2007).

Beyond their increasing potential as green reaction media in biocatalytic reactions, the ILs also demonstrated to be valuable additives during the sol–gel immobilization process of lipases. Nevertheless, identifying the exact role of ILs in the preparation of xerogel-type materials and the effect of their presence during immobilization on the entrapped enzyme properties is not an easy task. The main issues emerged from the relative scarce number of publications focused on this subject are:

- ILs can protect the enzyme against inactivation by the released alcohol and shrinking of gel during the maturation and drying step of the sol-gel immobilization process (Lee et al., 2006).
- ILs could totally replace the solvents used in the immobilization process, allowing the formation of a structure similar to aerogels in terms of pores size and volume, without supercritical drying, but this technique was not tested in the presence of an enzyme (Dai et al., 2000).
- The presence of ILs as additives increased the gelation time and influenced the gel structure, increasing the average pore radius and resulting in a narrower pore size distribution (Karout & Pierre, 2007).
- ILs can act as a template to form a wormlike mesoporous silica framework, based on hydrogen bond formation between the IL anion ($[BF_4]^-$) and the silanol groups of the silica gel (Zhou et al., 2004), but the formation of such a structure in the presence of an enzyme was not yet demonstrated.

Our work was focused on both possible applications of ionic liquids in lipase biocatalysis: as immobilization additives and as reaction media for the biocatalytic reaction.

Enantioselective acylation reactions of racemic 2-octanol have been used as test reactions for the characterization of the immobilized lipase biocatalysts. Chiral 2-octanol is an important building block for the preparation of liquid crystal materials, as well as an essential intermediate of many optically active pharmaceuticals (Dai & Xia, 2006). The precursor composition was finely tuned to obtain the best catalytic efficiency, by using secondary or tertiary silane mixtures with alkyl- or aryl nonhydrolizable groups, and various ionic liquids as additives. Ionic liquids were tested as well as reaction media, compared with organic solvents. Operational stability of the obtained biocatalysts was studied in several reuse cycles and thermal stability was also tested at temperatures up to 80°C.

2. Experimental

2.1 Materials

Lipases from *Pseudomonas fluorescens* (Amano AK) and *Burkholderia cepacia* (Amano PS) have been purchased from Aldrich. *Candida antarctica* B was produced by C-Lecta (Leipzig, Germany). Methyl-trimethoxysilane (MeTMOS), tetramethoxysilane (TMOS), acetone, n-hexane, 2-propanol, acetonitrile, tetrahydrofuran (THF), toluene, *t*-butanol, sodium fluoride, vinyl acetate, and 2-octanol were products of Merck. Octyl- (OcTMOS), and phenyl-trimethoxysilane (PhTMOS), were obtained from Fluka. All the specified reagents have been purchased at analytical grade and used as purchased. Dodecane (99%, Merck) and decane (>99%, Aldrich) were used as internal standards for quantitative gas-chromatographic analysis. Ionic liquids 1-ethyl-3-methyl-imidazolium tetrafluoroborate [Emim]BF_4, 1-ethyl-3-methyl-imidazolium trifluoroacetate [Emim]$COOCF_3$, 1-hexyl-3-methyl-imidazolium tetrafluoroborate [Hmim]BF_4, 1-butyl-3-methyl-imidazolium hexafluorophosphate [Bmim]PF_6, were purchased from Merck at the highest available purity. 1-Octyl-3-methyl-imidazolium tetrafluoroborate [Omim]BF_4 was product of Fluka.

2.2 Sol-gel immobilization method

A microbial lipase suspension (120 mg/mL) in TRIS/HCl 0.1 M, pH 8.0 buffer, was stirred at 700 rpm for 30 min, centrifuged, and the supernatant used for immobilization. In a 4 mL glass vial, 1 mL of this lipase solution was mixed (magnetic stirrer) with 200 µL ionic liquid or PEG 20,000, followed by addition of 100 µL 1M NaF solution, and 200 µL isopropyl alcohol. This mixture was kept for 30 min under continuous stirring for homogenization, and subsequently a binary or tertiary mixture of silane precursors (total 6 mmoles) was added. The mixture was stirred at room temperature until start of gelation. The obtained gel was kept for 24 h at room temperature to complete polymerization. The bulk gel was washed with isopropyl alcohol (7 mL), distilled water (5 mL), isopropyl alcohol again (5 mL) and finally hexane (5 mL), filtered, dried at room temperature for 24 hrs, and in a vacuum oven at room temperature for another 24 hrs. Finally, it was crushed in a mortar and kept in refrigerator.

2.3 Immobilization by sol-gel entrapment combined with adsorption

The immobilization protocol was identical as described for the simple sol-gel entrapment, until the start of gelation, when 0.5 g Celite 545 was blended with the gelling mixture. Subsequently, the obtained solid preparate was worked-up as described in Subchapter 2.2. A simplified scheme of both immobilization procedures is presented in Fig. 1.

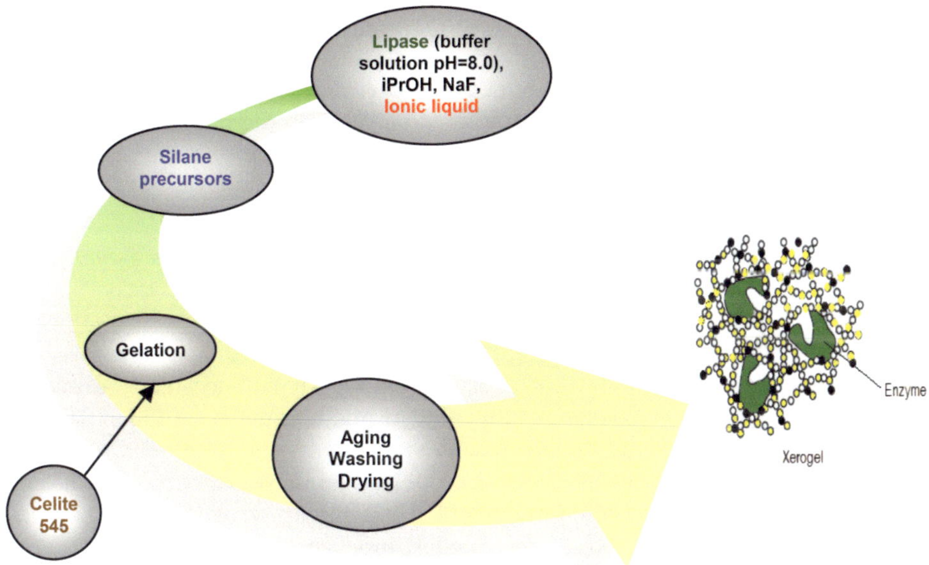

Fig. 1. Immobilization scheme of sol-gel entrapped lipase, simple or combined with adsorption on Celite 545.

2.4 Lipase-catalyzed acylation of 2-octanol

The procedure was the same as previously described (Zarcula et al., 2010). Acylations were performed in 4 mL capacity glass vials, charged with a mixture of 2-octanol (1 mmole), vinyl acetate (3 mmole), reaction medium (organic solvent or ionic liquid, 2 mL) and free (10 mg) or sol-gel immobilized (50 mg) lipase. The solvents, alcohols and ILs used were separately equilibrated to 0.328 water activity at 25°C over saturated $MgCl_2$ solution for 48 hrs, as described in the literature (Bell et al., 2001). The mixture was incubated using an orbital shaker (MIR-S100, Sanyo, Japan) at 300 strokes/min and 40°C (ILW 115 STD incubator, Pol-Eko-Aparatura, Poland). The conversion and enantiomeric excess of the product were assayed by gas-chromatography, on a Varian 450 instrument (Varian Inc., USA) equipped with flame ionization detector, using a 30 m x 0,25 mm Elite-Cyclosil B chiral column with 0.25 mm film thickness (Perkin-Elmer, USA). The analysis conditions were: oven temperature: 50° to 120°C with 10°C/min heating rate, injector temperature 240°C, detector temperature 280°C, carrier gas (hydrogen) flow 1.2 mL/min. The reactions were usually run for 24 hrs. Conversions have been calculated based on the internal standard method. When ILs were used as reaction media, the obtained reaction mixture was extracted 4-times with hexane to isolate the product, the extracts merged and analyzed by gas-chromatography in the same conditions.

Transesterification activities were calculated at 24 hrs reaction time and expressed as the average 2-octyl-acetate amount (in micromole) synthesized per hour by 1 mg of free or immobilized enzyme. The control reaction without enzyme did not give any product in the same conditions. To characterize the overall efficiency of the immobilization process, total activity yield was calculated as % of the total enzymatic activity recovered following immobilization, divided by the total activity of the lipase subjected to immobilization. The enantiomeric excess of the resulted ester product (ee_p) was determined from the two

enantiomers peak area, and the enantiomeric ratio (*E*) values were calculated based on conversion and ee_p values (Chen et al., 1982).
All experiments and samplings were run in duplicate, and the calculated average values are presented in the tables and figures, as the differences between the measured values for the same assay were less than 2% for conversion and 1% in case of enantiomeric excess.

2.5 Enzyme recycle and thermal stability study

The enzyme recycle study was performed at 40°C. The initial reaction system was set up as described for the acylation study (Subchapter 2.3). At the end of every reaction cycle, the product (upper liquid phase of the reaction mixture) was removed with a pipette, the remained solid phase (native or immobilized lipase) was washed two times with 2 mL hexane, centrifuged at 15°C and 5,000 rot/min, and the supernatant decanted. Subsequently, the same amounts of reagents (2-octanol, vinyl acetate, and hexane) as for the initial reaction were added to the reused enzyme, and the reaction run in the same conditions.
For thermal stability, 5 mg native or 25 mg immobilized lipase was incubated in hexane, at specified temperatures for a definite time period. Then, the lipase activity was determined in the acylation reaction of 2-octanol, as described in Subchapter 2.3.

3. Results and discussion

3.1 Influence of the silane precursor ratio in binary and ternary mixtures with ionic liquid additive on the immobilized biocatalyst efficiency

Lipases from three microbial species, *Burkholderia cepacia* (Amano PS), *Pseudomonas fluorescens* (Amano AK) and *Candida antarctica* (CALB-Lecta) were immobilized in sol-gel matrices obtained from binary and ternary silane mixtures in the presence of an IL additive, using our methodology presented in the Experimental part. The test reaction was the enantioselective acylation of 2-octanol by vinyl acetate, in hexane medium. The enantioselectivity in this reaction was based on kinetic discrimination, all studied lipases being (*R*)-selective (Fig. 2), which means that the (*R*)-2-octyl-acetate enantiomer was the fast reacting enantiomer, following the empirical Kazlauskas rule (Kazlauskas et al., 1991; Rottici et al., 1998).

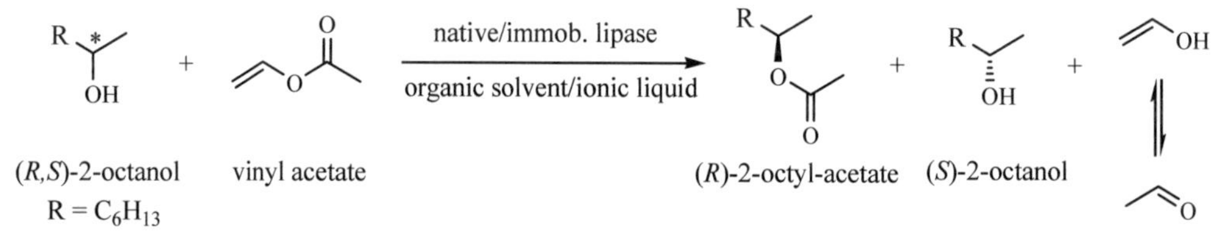

Fig. 2. Reaction scheme of 2-octanol enzymatic acylation

Conversions and product enantiomeric compositions of 2-octanol acylation reactions were measured by chiral GC at 24 hrs reaction time, and used to calculate the immobilized lipase activity, total activity recovery yield, enantiomeric excess (e.e.), and enantiomeric ratio (E) values.
Our previous results demonstrated that the catalytic efficiency of *Pseudomonas fluorescens* lipase in the acylation reaction of secondary alcohols could be enhanced by the presence of

hydrophobic alkyl groups in the hybrid organic-inorganic sol-gel matrix, and utilization of ILs as nonstructural template compounds (Péter et al., 2008; Zarcula et al., 2010). Recently, ternary silane mixtures, containing alkyl-triethoxysilanes, phenyl-triethoxysilane and tetraethoxysilane, have been reported with higher catalytic efficiency than the corresponding binary silane mixtures, for the sol-gel entrapment of Celite-supported *Pseudomonas fluorescens* lipase (Tomin et al., 2010). Based on these results, we used in the immobilization protocol binary and ternary mixtures of methyl-trimethoxysilane (MTMOS), phenyl-trimethoxysilane (PhTMOS), and tetramethoxysilane (TMOS), as well as [Emim]BF_4 or [Omim]BF_4 as additive. It must be pointed out that the total silane precursor amount was the same in all immobilization experiments (6 mmoles).

The immobilization efficiency was excellent for *Burkholderia cepacia* and *Candida antarctica* lipases, as the total activity recovered following immobilization was about 10-fold higher than the total activity of the enzyme subjected to immobilization (Fig. 3). Even in the case of the lowest-performing *Pseudomonas fluorescens* lipase, the recovered total activity value of the entrapped lipase was enough good, up to 90% related to the native lipase.

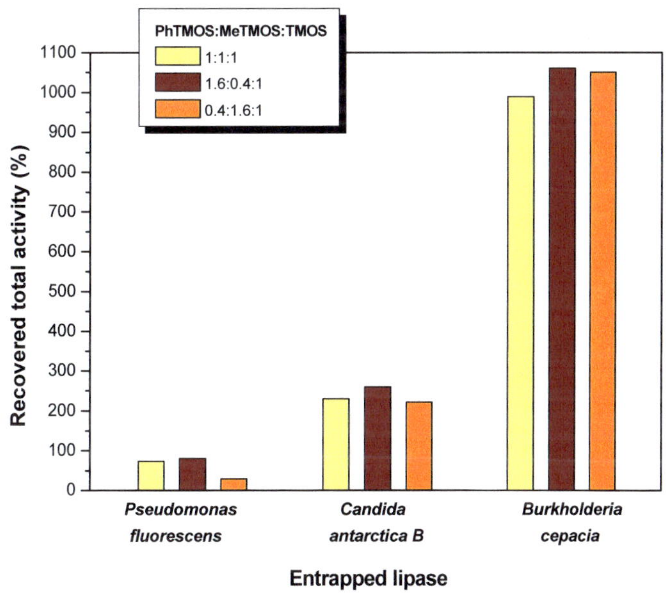

Fig. 3. Influence of silane precursors molar ratio on the relative total activity recovered after sol–gel immobilization of lipases from *Burkholderia cepacia*, *Pseudomonas fluorescens* and *Candida antarctica* B.

Sol-gel entrapped *Burkholderia cepacia* lipase showed higher activity and enantioselectivity, when the immobilization was performed using ternary silane precursor systems, compared to binary mixtures of the same precursors (Table 1).

It is obvious that fine tuning of the sol-gel matrix composition allows maximizing the catalytic efficiency of entrapped enzyme. Among ternary silane precursor mixtures, the highest activity was observed at lower phenyl-group concentration in the precursor silane mixture, but associated with a slightly lower enantioselectivity. The highest enantioselectivity was exhibited by the preparate obtained with equimolar mixture of the

Silane precursors[1] (molar ratio)	Additive	Conversion[2] (%)	Activity[3]	e.e. (%)	E
native enzyme	-	23	0.953	74	8
A:C (1:1)	[Omim]BF_4	38	0.329	79	14
B:C (1:1)	[Omim]BF_4	13	0.114	78	9
A:B:C (1:1:1)	[Omim]BF_4	61	0.493	62	19
	[Emim]BF_4	12	0.105	68	6
A:B:C (1.6:0.4:1)	[Omim]BF_4	53	0.433	73	16
	[Emim]BF_4	20	0.165	73	8
A:B:C (0.4:1.6:1)	[Omim]BF_4	67	0.538	49	15
	[Emim]BF_4	11	0.091	66	5

[1] A = phenyl-trimethoxysilane; B = methyl-trimethoxysilane C = tetramethoxysilane
[2] at 24 hrs reaction time
[3] transesterification activity of 2-octanol substrate, expressed as (μmole h^{-1} mg $catalyst^{-1}$)

Table 1. Catalytic efficiency and enantioselectivity of *Burkholderia cepacia* lipase, immobilized by sol-gel entrapment with binary and ternary silane precursor mixtures in different molar ratios, using an ionic liquid as additive.

three silane precursors, and [Omim]BF_4 as ionic liquid additive. The enantiomeric ratio (E) value of this biocatalyst was more than double, compared to the native enzyme. Therefore, we consider the equimolar mixture of the three silane precursors as the best to obtain high catalytic efficiency of sol-gel entrapped lipase. The presence of a more hydrophobic alkyl chain (octyl vs. ethyl) in the cationic part of the IL yielded an immobilized lipase with higher activity and enantioselectivity, irrespective of silane molar ratio used.

Changing the immobilization technique to sol-gel entrapment combined with adsorption, the same ternary silane precursor mixture yielded the highest activity in different conditions as registrated for the simple sol-gel entrapment (Table 2). The most active preparate was obtained at increased phenyl group concentration in the sol-gel matrix, and using the IL with shorter alkyl chain (ethyl) in the imidazolium moiety. The explanation should be related to the physical properties of the adsorption support, Celite 545 in our case. As hydrophobicity of the matrix is essential for the catalytic efficiency of the immobilized lipase, it is strongly influenced by physical characteristics of the support used for adsorption, overtaking the influence of alkyl chain length in the IL cationic part. The enantioselectivity slightly increased following immobilization, and in this case the same preparate showed the best values for both activity and enantioselectivity. Using the combined method and fine tuning the silane precursor ratio we obtained an immobilized biocatalyst with excellent performances, as the effective transesterification activity for the 2-octanol substrate was 44% compared to the native enzyme (very high, considering the significant enzyme dilution in the matrix), and the total activity next to immobilization showed a more than 20-fold increase.

The sol-gel entrapped lipase from *Pseudomonas fluorescens* (Amano AK) exhibited the lowest activity and enantioselectivity, considering the investigated enzymes. Using a ternary precursor mixture was also in this case beneficial to the catalytic performance of the immobilized enzyme, compared to binary mixtures of the same silanes (Table 3). Similar to *Burkholderia cepacia* lipase, an equimolar phenyl and methyl group concentration in the silica matrix resulted in higher activity of the sol-gel entrapped enzyme, but at this optimal precursor composition we found no significant difference associated with the chain length of alkyl group from the imidazolium moiety of the IL used. The low enantioselectivity of this enzyme for the racemic 2-octanol substrate was not changed by immobilization.

Silane precursors[1] (molar ratio)	Additive	Conversion[2] (%)	Activity[3]	e.e. (%)	E
native enzyme	-	23	0.953	74	8
A:B:C	[Omim]BF_4	29	0.246	76	10
(1:1:1)	[Emim]BF_4	27	0.233	76	10
A:B:C	[Omim]BF_4	40	0.341	75	11
(1.6:0.4:1)	[Emim]BF_4	47	0.421	76	15
A:B:C	[Omim]BF_4	9	0.076	62	5
(0.4:1.6:1)	[Emim]BF_4	12	0.100	69	6

[1]A = phenyl-trimethoxysilane; B = methyl-trimethoxysilane C = tetramethoxysilane
[2] at 24 hrs reaction time
[3]transesterification activity of 2-octanol substrate, expressed as (μmole h^{-1} mg $catalyst^{-1}$)

Table 2. Catalytic efficiency and enantioselectivity of *Burkholderia cepacia* lipase immobilized by sol-gel entrapment combined with adsorption on Celite 435, using ternary silane precursor mixtures in different molar ratios, and an ionic liquid as additive.

Silane precursors[1] (molar ratio)	Additive	Conversion[2] (%)	Activity[3]	e.e. (%)	E
native enzyme	-	47	1.910	53	5
A:C (1:1)	[Omim]BF_4	8	0.065	60	4
B:C (1:1)	[Omim]BF_4	22	0.191	64	5
A:B:C	[Omim]BF_4	45	0.372	53	5
(1:1:1)	[Emim]BF_4	21	0.179	56	4
A:B:C	[Omim]BF_4	41	0.343	56	5
(1.6:0.4:1)	[Emim]BF_4	43	0.349	57	5
A:B:C	[Omim]BF_4	21	0.176	53	4
(0.4:1.6:1)	[Emim]BF_4	23	0.191	53	4

[1] A = phenyl-trimethoxysilane; B = methyl-trimethoxysilane C = tetramethoxysilane
[2] at 24 hrs reaction time
[3] transesterification activity of 2-octanol substrate, expressed as (μmole h^{-1} mg $catalyst^{-1}$)

Table 3. Catalytic efficiency and enantioselectivity of *Pseudomonas fluorescens* lipase, immobilized by sol-gel entrapment with binary and ternary silane precursor mixtures in different molar ratios, and an ionic liquid as additive.

Combining the sol-gel entrapment with adsorption on Celite 545, the influence of relative molar ratio of the three silane precursors showed the same tendency, the highest activity being exhibited by the preparate obtained at equimolar phenyl and methyl group concentration in the matrix. Contrary to the influence observed for *Burkholderia cepacia* lipase, in this case the preparates obtained with [Omim]BF_4 as additive were more active compared to those obtained with [Emim]BF_4, suggesting a more important role of the IL during the combined enzyme entrapment and adsorption (Table 4).

Silane precursors[1] (molar ratio)	**Additive**	**Conversion[2] (%)**	**Activity[3]**	**e.e. (%)**	**E**
native enzyme	-	47	1.910	53	5
A:B:C	$[Omim]BF_4$	37	0.304	58	4
(1:1:1)	$[Emim]BF_4$	2	0.015	27	2
A:B:C	$[Omim]BF_4$	36	0.295	57	5
(1.6:0.4:1)	$[Emim]BF_4$	15	0.131	55	4
A:B:C	$[Omim]BF_4$	10	0.082	42	3
(0.4:1.6:1)	$[Emim]BF_4$	3	0.025	16	1

[1]A = phenyl-trimethoxysilane; B = methyl-trimethoxysilane C = tetramethoxysilane
[2] at 24 hrs reaction time
[3]transesterification activity on 2-octanol substrate, expressed as (μmole h^{-1} mg $catalyst^{-1}$)

Table 4. Catalytic efficiency and enantioselectivity of *Pseudomonas fluorescens* lipase immobilized by sol-gel entrapment combined with adsorption on Celite 435, using ternary silane precursor mixtures in different molar ratios, and an ionic liquid as additive.

The third investigated lipase, from *Candida antarctica* B, showed the highest enantioselectivity in the kinetic resolution of 2-octanol, and an important improvement related to the native enzyme. Regardless to the immobilization method, silane precursor composition, or IL additive used, the reaction reached around 50% conversion in 24 hrs reaction time, at more than 93% enantiomeric excess of the (*R*)-2-octyl acetate product. For this reason, in Table 5 only the results of the combined immobilization method are presented.

Silane precursors[1] (molar ratio)	**Additive**	**Conversion[2] (%)**	**Activity[3]**	**e.e. (%)**	**E**
native enzyme	-	17	0.705	89	21
A:B:C	$[Omim]BF_4$	51	0.427	95	201
(1:1:1)	$[Emim]BF_4$	51	0.426	95	201
A:B:C	$[Omim]BF_4$	51	0.434	95	201
(1.6:0.4:1)	$[Emim]BF_4$	51	0.431	93	114
A:B:C	$[Omim]BF_4$	51	0.418	95	201
(0.4:1.6:1)	$[Emim]BF_4$	47	0.394	95	104

[1]A = phenyl-trimethoxysilane; B = methyl-trimethoxysilane C = tetramethoxysilane
[2] at 24 hrs reaction time
[3]transesterification activity on 2-octanol substrate, expressed as (μmole h^{-1} mg $catalyst^{-1}$)

Table 5. Catalytic efficiency and enantioselectivity of *Candida antarctica* B lipase immobilized by sol-gel entrapment combined with adsorption on Celite 435, using ternary silane precursor mixtures in different molar ratios, and an ionic liquid as additive.

Based on these results, the CALB lipase was selected for the solvent and stability study, as the more serious candidate for a possible industrial application. Although the fine tuning of silane precursor composition and IL additive allows finding the optimal immobilization conditions for every enzyme, the equimolar ratio of the three silanes with $[Omim]BF_4$ as additive can be considered with generic validity for the sol-gel immobilization of microbial lipases.

3.2 Solvent engineering of 2-octanol acylation reaction catalyzed by sol-gel immobilized *Candida antarctica* B lipase

Although many enzymatic synthetic reactions are performing well in organic solvents, they have the disadvantage to be harmful for the environment. Room-temperature ILs are a possible solution to replace volatile organic solvents with an environment-friendly reaction medium, as they are non-volatile, can be used at higher temperatures, and recycled. Besides the environmental advantages, also improvements of reaction rates, conversion, enantioselectivity and regioselectivity have been reported for various biocatalytic reactions (Moniruzzaman et al., 2010). However, it was not possible to correlate the IL structure with the solvent properties, and a careful selection is needed to find the appropriate IL for every enzyme and application. We investigated the enantioselective acylation of 2-octanol in common organic solvents and different ILs, using CALB lipase immobilized by sol-gel entrapment as catalyst. The reactions were carried out at 40°C. As the employed ILs were not miscible with hexane, the product was recovered by repeated extraction with this compound. Such a procedure has a further advantage, the possibility to easily recycle the catalyst and IL which remain in the non-organic phase, but the utilization of an organic solvent invalidates the environment-friendly nature of the process. To overcome such a drawback, extraction with supercritical CO_2 or evaporation at low pressure should be a solution for syntheses at larger scale, which will be investigated in a future work.

Solvent	Conversion[1] (%)	Activity[2] (μmol h^{-1} mg^{-1})	e.e. (%)	E
[Emim]BF_4	54	0.457	83	46
[Hmim]BF_4	51	0.437	89	58
[Omim]BF_4	51	0.440	93	114
[Bmim]PF_6	47	0.402	86	30
[Emim]$COOCF_3$	2	0.015	64	5
tert-Butanol	48	0.402	95	113
Acetone	51	0.431	95	201
Tetrahydrofuran	61	0.529	63	20
Acetonitrile	51	0.433	95	201
Toluene	51	0.424	94	146
iso-Octane	51	0.435	95	201
Hexane	51	0.410	95	201

[1] at 24 hrs reaction time
[2]transesterification activity on 2-octanol substrate, at 24 hrs reaction time

Table 6. Influence of reaction medium on enzymatic acylation of 2-octanol by vinyl acetate, catalyzed by CALB lipase immobilized with PhTMOS:MTMOS:TMOS precursors (1:1:1 molar ratio), and [Omim]BF_4 IL additive

The sol-gel immobilized CALB lipase proved to be once more an excellent catalyst for the investigated reaction, showing high enantioselectivity in all medium- and low-polarity organic solvents tested. This behavior is different as we previously registrated with lipase from *Pseudomonas fluorescens*, which had a significantly lower enantioselectivity in acylation reactions of secondary alcohols and showed a strong influence of the solvent (Zarcula et al, 2010). In ILs, the influence of reaction medium was much more significant. The conversion was very low in [Emim]$COOCF_3$, the single water-miscible of the tested ILs. Higher conversions in water-immiscible than in water-miscible ILs were also reported by other authors for lipase-catalyzed transesterification reactions (Gorke et al, 2007). ILs with BF_4^- anion were more efficient as reaction media than [Bmim]PF_6 , considering both activity and enantioselectivity. From this group of ILs with BF_4^- anion, the best results were obtained in case of a more hydrophobic alkyl chain (octyl) presence in the imidazolium cationic moiety. Previously, we obtained for 2-hexanol acylation with *Pseudomonas fluorescens* lipase in the same reaction media an inverse effect of the IL polarity (Zarcula et al, 2010). The polarities of all ILs tested are in a close range, between 0.67-0.71 on the Reichardt's scale (Reichardt, 2005), and we presume that other properties, like the IL influence on lipase active conformation and/or miscibility with reagents and reaction products, is more important than the intrinsic polarity of the IL. Therefore, are unlikely to predict the solvent behavior of an IL based on its structure, and the selection of the appropriate reaction medium must be done on experimental basis. For the investigated reaction, the enantioselective acylation of 2-butanol, the optimal reaction medium was 1-octyl-3-methyl-imidazolium tetrafluoroborate, yielding the same results as the best organic solvents.

3.3 Operational and thermal stability of sol-gel entrapped lipases with ionic liquid as immobilization additive

The most important enzyme property which can be improved through immobilization is stability. However, such improvement is not an intrinsic result of immobilization, as in some cases the enzyme stability may decrease after immobilization for different reasons, like undesired interactions between the support and the enzyme. It is now generally accepted that tailor-made immobilization protocols are needed to obtain the best catalytic performances. As for industrial application the main requirement is reuse of the enzyme in as many as possible reaction cycles, to make the process economically feasible, the operational stability of the immobilized enzyme is an essential issue. Sol-gel entrapment not only allows an easier separation of the catalyst at the end of every reaction cycle, but also stabilizes the lipase molecules against aggregation and proteolysis due to possible presence of proteases in the crude enzyme. However, depending on pores size, distribution, and location of the enzyme inside the porous matrix, the enzyme stability could also deteriorate, generated by pore collapsing throughout use or colmation caused by different impurities, resulting in increased steric hindrance.

ILs can improve the operational stability of the sol-gel immobilized lipase by two mechanisms:

- increasing the stability of the enzyme itself;
- acting as non-structural templates during the immobilization process.

This study was performed with *Candida antarctica* B lipase (CALB-Lecta), immobilized by sol-gel entrapment in a matrix obtained from a ternary precursor mixture of PhMTOS:MTMOS:TMOS (1.6:0.4:1 molar ratio), using [Omim]BF_4 as immobilization additive. The same enzyme, immobilized by sol-gel entrapment combined with adsorption on Celite, was also studied, using as reference the native lipase (Fig. 3). The reactions were carried out in hexane, at 40°C, for 24 hrs.

As results from Fig. 3, the immobilized biocatalysts showed a remarkable operational stability, as the relative activities (related to the activity measured in the initial reaction with the same catalyst) were above 0.9 for the sol-gel entrapped lipase and 0.8 for the lipase

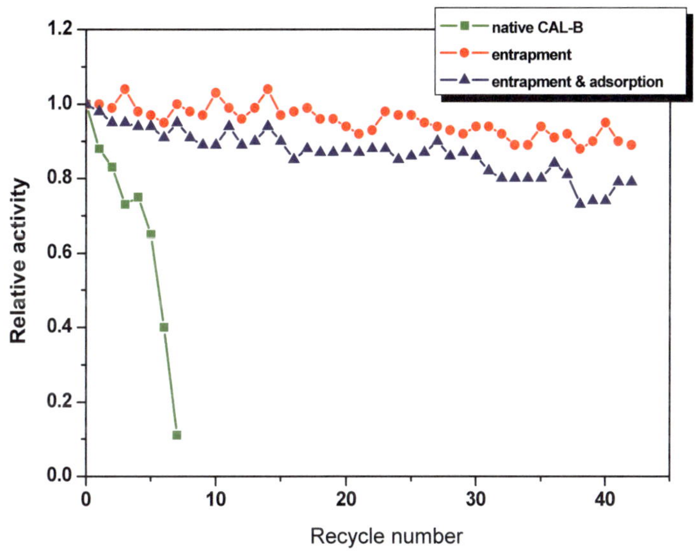

Fig. 3. Operational stability of *Candida antarctica* B lipase, immobilized by sol-gel entrapment and sol-gel entrapment combined with adsorption.

immobilized by the combined method, for more than 40 reuse cycles. In the same time, the native lipase exhibited a very early decline, to less than 20% of the initial activity after only 6 reuse cycles. Such an important difference can be explained by the protective effect of the sol-gel network and IL incorporated in the matrix against any kind of mechanical or chemical inactivation of the enzyme during the chemical reaction and enzyme recovery steps. The enantioselectivity did not change during the whole enzyme recycle process, the enantiomeric excess of the product remaining at the same values, 94-95% (data not shown). Although the sol-gel entrapment does not influence the intrinsic structural stability of the lipase, it results in increased operational stability and manifold reuse possibility.

Thermal stability is another important property of an industrial enzyme, as in many processes a higher temperature than usually employed in enzymatic reactions is needed, to enhance the reaction rate. Thermal inactivation is typically related to unfolding of enzymes, and immobilization could prevent the conformational changes involved in this type of inactivation. The presence of ILs in the immobilization matrix should induce a more prominent stabilizing effect by maintaining the active structure of the enzyme through specific interactions. Additionally, immobilized enzymes are better protected against local temperature increase in the reactor zone close to the heating surface.

Our research was focused on thermal stability of the same immobilized lipases used in the operational stability study. The biocatalysts were maintained at different temperatures for 1 hr, in hexane, and subsequently used for the acylation reaction of 2-octanol by vinyl acetate, in the same solvent, at 40°C.

The results from Table 7 show an excellent thermal stability of immobilized lipases in the organic solvent in the studied temperature range, regardless to immobilization method or lipase nature. Small variation of activity values can be attributed to differences in distribution of the enzyme in the matrix, even if the samples have been run in duplicate.

Temperature (°C)	Activity[1] (µmole h^{-1} mg catalyst^{-1})					
	Pseudomonas fluorescens lipase		*Burkholderia cepacia* lipase		*Candida antarctica* B lipase	
	SG[2]	SGA[3]	SG[2]	SGA[3]	SG[2]	SGA[3]
40	0.343	0.295	0.433	0.341	0.429	0.434
45	0.338	0.299	0.438	0.286	0.418	0.414
50	0.316	0.250	0.395	0.274	0.423	0.406
55	0.315	0.286	0.398	0.289	0.411	0.397
60	0.316	0.282	0.393	0.315	0.424	0.404
65	0.325	0.297	0.413	0.285	0.426	0.413
70	0.335	0.298	0.412	0.291	0.420	0.425
75	0.331	0.310	0.411	0.273	0.424	0.428
80	0.326	0.319	0.415	0.286	0.419	0.430

[1]transesterification activity on 2-octanol substrate, at 24 hrs reaction time
[2]immobilized by sol-gel entrapment
[3]immobilized by sol-gel entrapment with adsorption

Table 7. Influence of temperature on the activity of sol-gel entrapped lipases. The biocatalyst samples were incubated 1 hr in hexane, at temperatures between 40-80°C.

Based on the high thermal stability resulted from this study, the experiments were continued for *Candida antarctica* B lipase in more severe temperature conditions, by incubation in hexane at 80°C, for 5 days. Even in these conditions, the immobilized lipase exhibited remarkable stability, the activity values remaining practically unchanged, while the native enzyme lost about 50% of the initial activity after 5 days (Fig. 4). It was no

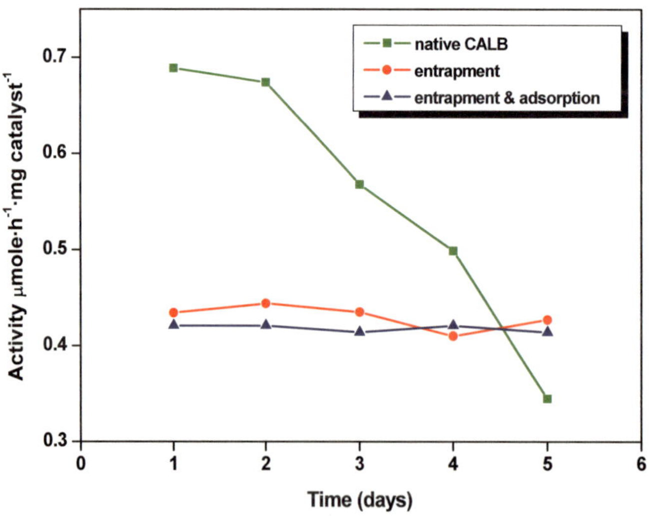

Fig. 4. Influence of temperature on the activity of native and sol-gel entrapped *Candida antarctica* B lipase. The biocatalyst samples were incubated in hexane, at 80°C, and used subsequently for the acylation of 2-octanol.

significant activity difference between the lipase immobilized by simple sol-gel entrapment and entrapment by the combined method. The enantioselectivity of lipase was also not affected, remaining at the same value of 94-95% (*R*)-2-octyl-acetate enantiomeric excess. High thermal stability of lipase from *Candida cylindracea* entrapped in sol-gels from TMOS and propyl-trimethoxysilane deposited on Celite, using a different immobilization protocol, was previously reported (Kawakami & Yoshida, 1996), but the incubation time was not extended beyond 1 hr. Our results demonstrate the possibility to maintain the catalytic properties of sol-gel immobilized enzymes intact for a much longer time period.

4. Conclusions

Sol-gel encapsulation of lipases is an emerging method to obtain biocatalysts with high activity and enantioselectivity for possible industrial applications. Utilization of specific ionic liquids as immobilization additives make possible the design of a more adequate sol-gel matrix to prevent enzyme leakage during operation and preserve the catalytic activity. The fine tuning of silane precursor mixture composition and ionic liquid structure allows developing the best biocatalyst for a specific application. For enantioselective acylation of 2-octanol, the most effective immobilization protocol involved a ternary mixture of PhTMOS:MTMOS:TMOS in equimolar ratio and [Omim]BF_4 as IL additive. Combining the sol-gel entrapment with adsorption on a porous support did not cause any damage of enzyme catalytic properties or stability, moreover we registrated a very important (up to 20-fold) increase of the total immobilized activity related to the total activity of native lipase. Among three microbial lipases investigated, *Burkholderia cepacia* lipase showed the highest activity and *Candida antarctica* B lipase the highest enantioselectivity in the studied reaction. The more hydrophobic [Omim]BF_4 was also the best reaction medium among several ILs tested, with performances similar to organic solvents, but having the advantage of avoiding the operation with toxic volatile compounds in the synthetic process. The obtained biocatalysts demonstrated excellent operational and thermal stability properties, recommending them for scale-up.

Financial support of this research was provided by CNCSIS through PNII-IDEI grant 368/2007. Fellowship of A. Ursoiu was provided by strategic grant POSDRU 2009, project ID 50783, of the Ministry of Labour, Family and Social Protection, Romania, co-financed by the European Social Fund - Investing in People. The authors acknowledge Dr. Luuk van Langen (ViaZym BV, Delft, The Netherlands) for supplying the C-Lecta lipase.

5. References

Bell, G., Halling, P.J., May, L., Moore, B.D., Robb, D.A., Ulijn, R. & Valivety, R.H. (2001). Methods for Measurement and Control of Water in Nonaqueous Biocatalysis in: *Methods in Biotechnology: Enzymes in Nonaqueous Solvents*, Vulfson, E.N., Halling, P.J., Holland H.L. (Eds.), 105-126, Humana Press, Totowa (N.J.)

Bornscheuer, U.T. (2003). Immobilizing enzymes: how to create more suitable biocatalysts, *Angewandte Chemie, International Edition*, 42, 3336 - 3337.

Buchholz, K., Kasche V. & Bornscheuer, U.T. (2005). *Biocatalysts and enzyme technology*, Wiley-VCH Verlag, Weinheim.

Cao, L. (2005). *Carrier-bound immobilized enzymes: principles, applications and design*, Wiley-VCH Verlag, Weinheim.

Cao, X., Yang, J., Shu, L., Yu, B. & Yan, Y. (2009). Improving esterification activity of *Burkholderia cepacia* lipase encapsulated in silica by bioimprinting with substrate analogues, *Process Biochemistry*, 44, 177-182.

Chen, C.-S., Fujimoto, Y., Girdauskas, G. & Sih, C.J. (1982). Quantitative analyses of biochemical kinetic resolutions of enantiomers, *Journal of the American Chemical Society*, 104, 7294-7299.

Chen, J.-P. & Lin, W.-S. (2003). Sol-gel powders and supported sol-gel polymers for immobilization of lipase in ester synthesis, *Enzyme and Microbial Technology*, 32, 801-811.

Christensen, M.W., Andersen, L., Husum, T.L. & Kirk, O. (2003). Industrial lipase immobilization, *European Journal of Lipid Science & Technology*, 105, 318–321.

Ciriminna, R. & Pagliano, M. (2006). Recent Uses of Sol-Gel Doped Catalysts in the Fine Chemicals and Pharmaceutical Industry *Organic Process Research & Development*, 10, 320-326.

Dai, S., Ju, Y.H., Gao, H.J., Lin, J.S., Pennycook, S. J. Barne& C. E. (2000). Preparation of silica aerogel using ionic liquids as solvents, *Chemical Commuications.*, 243–244.

Dai, D.-Z. & Xia L.-M. (2006). Resolution of (*R,S*)-2-octanol by *Penicillium expansum* PED-03 lipase immobilized on modified ultrastable-Y molecular sieve in microaqueous media, *Process Biochemistry*, 41, 1455–1460.

Fehér, E., Major, B., Bélafi-Bakó, K. &Gubicza L. (2007). On the background of enhanced stability and reusability of enzymes in ionic liquids, Biochemical Society Transactions, 35, 1624-1627.

Gorke, J.T., Okrasa, K., Louwagie, A., Kazlauskas, R.J. & Srienc, F. (2007). Enzymatic synthesis of poly(hydroxyalkanoates) in ionic liquids, *Journal of Biotechnology*, 132, 306-313.

Hara, P., Hanefeld, U. & Kanerva, L.T. (2008). Sol–gels and cross-linked aggregates of lipase PS from *Burkholderia cepacia* and their application in dry organic solvents, *Journal of Molecular Catalysis B: Enzymatic*, 50, 80-86.

Karout, A. & Pierre, A.C. (2007). Silica xerogels and aerogels synthesized with ionic liquids, *Journal of Non-Crystallyne Solids*, 353, 2900-2909.

Kazlauskas, R.J. Weissfloch, A.N.E., Rappaport, A.T., & Cuccia L.A. (1991). A rule to predict which enantiomer of a secondary alcohol reacts faster in reactions catalyzed by cholesterol esterase, lipase from *Pseudomonas cepacia,* and lipase from *Candida rugosa*, *Journal of Organic Chemistry*, 56 (8), 2656–2665.

Kawakami, K. & Yoshida, S. (1996). Thermal stabilization of lipase by sol-gel entrapment in organically modified silicates formed on Kieselguhr, *Journal of Fermentation and Bioengineering*, 82 (3), 239-245.

Kawakami, K., Takahashi, R., Shakeri, M. & Sakai S. (2009). Application of a lipase-immobilized silica monolith bioreactor to the production of fatty acid methyl esters, *Journal of Molecular Catalysis B: Enzymatic*, 57, 194-197.

Kiss, C., Zarcula, C., Csunderlik, C. & Péter, F. (2007). Enantioselective acylation of secondary alcohols by biocatalysis with sol-gel immobilized *Pseudomonas fluorescens* lipase, *Revista de Chimie (Bucharest)*, 58 (8), 799-804.

Lee, S.H., Doan, T.T.N., Ha, S.H. & Koo, Y.-M. (2006). Using ionic liquids to stabilize lipase within sol-gel derived silica, *Journal of Molecular Catalysis B: Enzymatic*, 45, 57-61.

Mateo, C., Palomo, J.M., Fernandez-Lorente, G., Guisan, J.M. & Fernandez-Lafuente R. (2007). Improvement of enzyme activity, stability and selectivity via immobilization techniques, *Enzyme and Microbial Technoogy*, 40, 1451-1463.

Moniruzzaman, M., Nakashima, K., Kamiya, N. & Goto, M. (2010). Recent advances of enzymatic reactions in ionic liquids, *Biochemical Engineering Journal*, 48 (3), 295-314.

Moon, Y.H., Lee, S.M., Ha, S.H. & Koo Y.-M. (2006). Enzyme-catalyzed reactions in ionic liquids, *Korean Journal of Chemical Engineering*, 23(2), 247-263.

Péter, F. (2005). Biotransformări enzimatice, Editura Politehnica, Timişoara.

Péter, F., Poppe, L., Kiss, C., Szőcs-Biró, E., Preda, G., Zarcula & C., Olteanu, A. (2005). Influence of precursors and additives on microbial lipases stabilized by sol-gel entrapment, *Biocatalysis and Biotransformation*, 23 (3-4), 251-260.

Péter, F., Zarcula, C., Kakasi-Zsurka, S., Croitoru, R., Davidescu, C., Csunderlik, C. (2008). Solid-phase lipase biocatalysts for kinetic resolutions, *Journal of Biotechnology*, 136S, S374.

Pierre, A.C. (2004). The sol-gel encapsulation of biocatalysts, *Biocatalysis and Biotransformation*, 22, 145-170.

Pirozzi, D., Fanelli, E., Aronne, A., Pernice, P. & Mingione, A. (2009). Lipase entrapment in a zirconia matrix: Sol-gel synthesis and catalytic properties, *Journal of Molecular Catalysis B: Enzymatic*, 59, 116-120.

Poole, C.F. (2004). Chromatographic and spectroscopic methods for the determination of solvent properties of room temperature ionic liquids, *Journal of Chromatography A*, 1037, 49–82.

Reetz, M.T., Zonta, A. & Simpelkamp J. (1996). Efficient immobilization of lipases by entrapment in hydrophobic sol-gel materials, *Biotechnology and Bioengineering*, 49, 527-534.

Reetz M.T. & Jaeger K.-E. (2003). Overexpression, immobilization and biotechnological application of *Pseudomonas* lipases, *Chemistry and Physics of Lipids*, 93, 3–14.

Reetz, M.T., Tielmann, P., Wiesenhöfer, W., Könen, W. & Zonta, A. (2003). Second generation sol-gel encapsulated lipases: robust heterogeneous biocatalysts, *Advanced Synthesis & Cataysis.*, 345, 717-728.

Reichardt, C. (2005). Polarity of ionic liquids determined empirically by means of solvatochromic pyridinium N-phenolate betaine dyes, *Green Chemistry*, 7, 339–351.

Rotticci, D., Hæffner, F., Orrenius, C., Norin, T., & Hult, K. (1998). Molecular recognition of *sec*-alcohol enantiomers by *Candida antarctica* lipase B. *Journal of Molecular Catalysis B: Enzymatic* , 5, 267–272.

Sheldon R.A. (2007). Enzyme immobilization: the quest for optimum performance, *Advanced Synthesis & Catalysis*, 349, 1289-1307.

Sureshkumar, M. & Lee, C.-K. (2009). Biocatalytic reactions in hydrophobic ionic liquids. *Journal of Molecular Catalysis B: Enzymatic*, 60, 1-12.

Tomin, A, Weiser, D., Hellner, G., Bata, Zs., Corîci, L., Péter, F., Koczka, B. & Poppe, L. (2010). Fine-tuning the second generation sol–gel lipase immobilization with ternary alkoxysilane precursor systems, *Process Biochemistry*, doi:10.1016/ j.procbio.2010.07.021.

Van Rantwijk, F., Lau, R.M. & Sheldon, R.A. (2003). Biocatalytic transformations in ionic liquids, *Trends in Biotechnoogy.*, 21, 131-138.

Weingärtner, H. (2008). Understanding ionic liquids at the molecular level: facts, problems, and controversies, *Angewandte Chemie, International Edition*, 47, 654-670.

Whittall, J. & Sutton, P., Editors (2010). *Practical methods for biocatalysis and biotransformations*, John Wiley&Sons, Chichester.

Zarcula, C., Croitoru, R., Corîci, L., Csunderlik, C. & Péter, F. (2009). Improvement of lipase catalytic properties by immobilization in hybrid matrices, *International Journal of Chemical and Biomolecular Engineering*, 2(3), 138-143.

Zarcula, C., Kiss, C., Corîci, L., Croitoru, R., Csunderlik, C. & Péter, F. (2009). Combined sol-gel entrapment and adsorption method to obtain solid-phase lipase biocatalysts, *Revista de Chimie (Bucharest)*, 60, 922-927.

Zarcula C., Corîci L., Croitoru R., Ursoiu A. & Péter F. (2010) Preparation and properties of xerogels obtained by ionic liquid incorporation during the immobilization of lipase by the sol-gel method, *Journal of Molecular Catalysis B: Enzymatic*, 65 (1-4), 79-86.

Zhou, Y., Schattka, J.H., Antonietti, M. (2004). Room-temperature ionic liquids as template to monolithic mesoporous silica with wormlike pores via a sol-gel nanocasting technique, *Nano Letters*, 4 (3), 477-481.

21

Ionic Liquids: Alternative Reactive Media for Oxidative Enzymes

Oscar Rodriguez, Ana P.M. Tavares,
Raquel Cristóvão and Eugénia A. Macedo
Associate Laboratory LSRE/LCM - Faculty of Engineering, University of Porto. Portugal

1. Introduction

During the last decades, biotechnology has attracted a great interest from academic and industrial environments due to the progress made in the discovery of new and efficient biocatalysts for many different applications. Due to the increasing variety of these applications, the aqueous medium became limiting. Biocatalysis in nonaqueous media offers unique capabilities and thus plays a major role in biotransformation technologies. It is well known that when enzymes are introduced in a non-aqueous medium, a number of factors can alter their native structure (critically at the active centre of the protein) and so alter their biological functions, thus causing reversible or irreversible inactivation of the enzymes. In order to overcome these problems, a high number of papers related to biocatalysis in organic solvents are published every year, in several areas of knowledge, investigating the effects of different organic media in enzyme activity, stability, structure and kinetics. It is important to highlight that good results have been reported for several enzymes, involving many different reaction types and a wide variety of organic solvents (Carrea & Riva, 2000). As in aqueous enzymatic reactions, the enzymology in non-aqueous media also presents specific and unique advantages like: reaction enantioselectivity, resistance to contamination by microorganisms, enhanced thermo-stability of the enzyme. Additionally, the solubility of hydrophobic substrates and/or products can be increased with the right selection of the solvent. The latter decreases diffusional barriers for the reactions, thus improving their yields.

The study of enzymatic reactions in non-aqueous media started more than 100 years ago (Halling & Kvittingen, 1999), but did not receive attention until the 1970s with the pioneering works of Berezin and co-workers (Klyosov et al., 1975; Klibanov et al., 1977; Martinek et al., 1981) and today either water and organic solvents are conventional media for enzymes. Klyosov et al. (1975) studied the hydrolysis of p-nitrophenyl esters by means of α-chymotrypsin and related proteins in four different organic solvents (dimethylsulfoxide, dimethylformamide, formamide, and N-methylacetamide). The work of Klibanov et al. (1977) was based on enzymatic reactions in a "water–*water-immiscible* organic solvent" biphasic system with bovine chymotrypsin using chloroform, benzene, ether, acetone, ethanol, dimethylsulfoxide and dioxane as organic solvents. On the other hand, Martinek et al. (1981) investigated the behaviour of α-chymotrypsin, trypsin, pyrophosphatase, peroxidase, lactate dehydrogenase and pyruvate kinase in organic solvents (benzene, chloroform, octane, cyclohexane). After these publications, a growing number of articles in this subject can be found in the open literature. The fundamental findings of this field can be found in deeper detail in a series of (recent) review works (Dordick, 1991; Nikolova & Owen, 1993; Halling, 2000; Castro & Knubovets, 2003; Yu et al., 2010).

2. Oxidative enzymes

Oxidative enzymes are enzymes which catalyse oxidation reactions. The most common types of oxidative enzymes are peroxidases, which use hydrogen peroxide for the oxidation of the enzyme, and oxidases (such as laccases and tyrosinases), which use molecular oxygen. These enzymes have broad substrate specificities and can catalyse the oxidation of a wide range of toxic organic compounds and are normally used in the treatment of organic pollutants. In aqueous medium, oxidative enzymes catalyse the transformation of a large number of phenolic and non-phenolic compounds.

Laccase (EC 1.10.3.2, p-benzenediol:oxygen oxidoreductase) is a multicopper belonging to a small group of enzymes denominated blue oxidases (Mayer & Staples, 2002). It is able to catalyse the oxidation of various aromatic compounds (particularly phenols) with the concomitant reduction of oxygen to water. Laccases are extracellular glycoproteins containing normally 4 atoms of copper, which are distributed into three sites and constitute the active site of the enzyme. A scheme of the reaction mechanism is depicted in Figure 1 (Bourbonnais et al., 1998).

It is possible to find much information about the use of oxidative enzymes in several applied areas in the literature (Xu, 2005; Couto & Herrera, 2006). Particularly, laccase can be used for delignification of wood fibres in the preparation of pulp for paper industry, a large variety of bioremediation processes, treatment of industrial wastes, decolourization of dyes for textile industry, enzyme sensors for drug analysis, etc. Besides these direct reactions, laccases can indirectly react with non-phenolic types of substrates. In that case, they need the presence of a redox mediator: the mediator promotes the enzyme action by increasing its oxidation potential (Husain & Husain, 2008). In contrast, an exhaustive effort to explore the more complex aspects of oxidative enzymes in non-aqueous media is still needed.

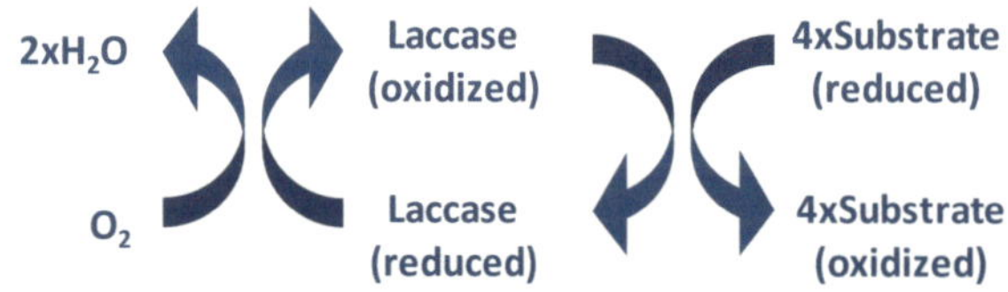

Fig. 1. Scheme of the reaction mechanism (biocatalytic cycle) in laccase-mediated oxidations.

As it comes for peroxidases, also extensive research has been developed towards new possibilities and applications of the enzymes, especially for environmental applications. Peroxidases are heme proteins which use hydrogen peroxide to catalyse one-electron oxidation of a large variety of aromatic compounds (Banci, 1997). A scheme of the reaction mechanism is depicted in Figure 2 (Banci, 1997; Hamid & Rehman, 2009).The most studied peroxidases include: Horseradish peroxidase (HRP, EC 1.11.1.7), Lignin peroxidase (LiP) and Manganese peroxidase (MnP). The most relevant applications for these enzymes are in the removal of recalcitrant phenolic pollutants, the decolourization of Kraft effluents, in biosensor and for some specific organic syntheses (Conesa et al., 2002). The products of the enzymatic reaction can be further polymerized by downstream processes in order to produce insoluble precipitates. These precipitates can be conveniently and easily separated from the reaction medium by sedimentation or filtration (Karan & Nicell, 1997).

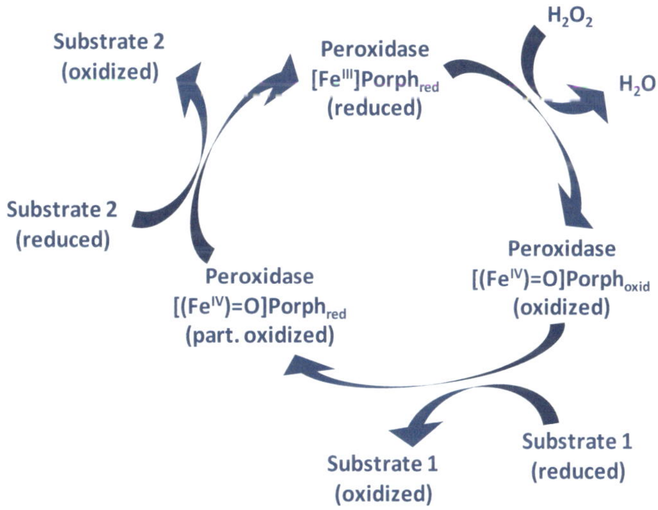

Fig. 2. Scheme of the reaction mechanism (biocatalytic cycle) in peroxidase-mediated oxidations.

Due to the easiness of oxidative enzymes to catalyse the oxidation of phenolic and non-phenolic compounds, and due to the hydrophobicity and low aqueous solubility of many of these chemicals, many researchers have derived their efforts to carry out the reactions in non-aqueous media (organic solvents). Yoshida et al. (1997) studied the oxidation of various phenolics and aromatic amines in organic solvents by means of lignin peroxidase (LiP). It was found that the activity of LiP in organic solvents depends on the nature of solvent and substrate. LiP oxidized the aromatic compounds in the presence of 70% ethylene glycol and aromatic amines were more easily oxidized in 70% aqueous ethylene glycol medium than phenolic compounds. But LiP failed to catalyze the oxidation of veratryl alcohol in the ethylene glycol medium. It was also related to the decrease in the redox potential of LiP in organic solvents. Another study (Azevedo et al., 2001) with horseradish peroxidase (HRP) reported the measurement of enzymatic activity and stability in the presence of organic co-solvents: dimethyl sulfoxide (DMSO), dimethylformamide, dioxan, acetonitrile and tetrahydrofuran. Among the solvents used, DMSO led to the highest activities and stabilities. Additionally, best results were obtained when HRP was immobilized onto silica microparticles. A system based on the use of Manganese Peroxidase (MnP) for the degradation of polycyclic aromatic hydrocarbons (PAHs), where anthracene was selected, was reported by Eibes et al. (2005) with different water miscible organic solvents (acetone, methyl-ethyl-ketone, methanol and ethanol). Comparing the maximum solubilisation of anthracene and the minimum loss of MnP activity, acetone was the best co-solvent for an acetone concentration of 36% (v/v). Other studies show the use of peroxidases for enzyme-catalyzed polymerization reactions due to the advantages over traditional polymerization, namely to improve the control of the polymer structure (Singh & Kaplan, 2002). Further examples of polymerization reactions for syringaldehyde and bisphenol A by peroxidase (An et al., 2010), 4-[(4-phenylazo-phenyimino)-methyl]-phenol by HRP (Turac & Sahmetlioglu, 2010), cardanol by HPR (Won et al., 2004) and phenols by laccase (Mita et al., 2003) are reported in the literature.

3. Enzymes in ionic liquids

The use of ionic liquids as reaction media for enzymatic catalysis has received a boost of attention within the last decade. The reasons for that are easily understood: biocatalysis in non-aqueous media is a subject of interest to expand the possibilities of enzymes to substrates that are not soluble in water, or enhance the yields and process capacity of those with low solubility. Using ionic liquids, it is possible to separate and reuse the catalyst, obtain higher conversions, higher stabilities, and use polar or hydrophilic substrates (Moniruzzaman et al., 2010a). It is a very promissory and environmental friendly alternative to harmful organic solvents. Room-temperature ionic liquids (electrolytes or salts with a low melting point) are compounds that consist only of ions, have a negligible vapour pressure (for industrial applications), present high chemical, electrochemical and thermal stability, tunable polarity and easier product separation and reutilization (Park & Kazlauskas, 2003; van Rantwijk et al., 2003; Wilkes, 2004). Besides this, ionic liquids have arisen during these years as "green" recyclable solvents, awakening the interest of researchers on virtually all fields of chemical engineering, chemistry (organic synthesis, electrochemistry, and so on), physics, polymer science, nanotechnology, biofuel production, or purification of biomolecules, among others. It is straightforward to combine both ideas and thus get into the field of enzymatic catalysis in ionic liquids.

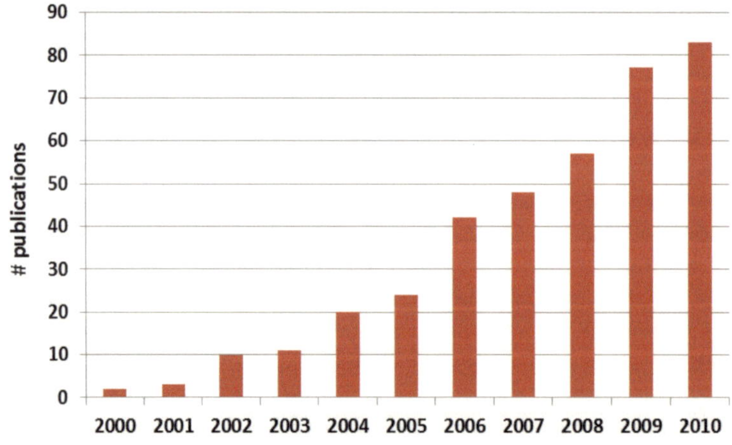

Fig. 3. Number of articles published with keywords "ionic liquid" and either "enzyme" or "biocatalysis". Search done in the Web of Science in January 2011.

The research on enzymatic catalysis in ionic liquids started in 2000 with the work of Erbeldinger et al. (2000). After that, the number of articles found in the open literature regarding "ionic liquid" and "enzymes" or "biocatalysis" increased each year, going up to 377 (search done on January 2011 using the Web of Science©). Just one article could be found before the year 2000, published by Magnuson et al. (1984), where the activity and stability of alkaline phosphatasein were studied in ethylammonium nitrate [$EtNH_3$][NO_3]. The results of this search are presented in Figure 3, where the increasing rate of publication is evident. This trend indicates that ionic liquids are proving to be an attractive media to carry out enzymatic reactions. Just for comparison, when substituting "ionic liquid" for "supercritical" in the previous search, the result is 620 articles published since 1988. Nearly the double of articles produced in about the double of time. Note that supercritical fluids (namely supercritical CO_2) are other attractive non-conventional solvents for biocatalysis. However, these processes are only economically viable when high added-value products are involved, due to the higher costs of elevated pressure. Despite ionic liquids are still rather expensive solvents, scaling-up the productions will certainly reduce their cost.

As detailed in Figure 3, much work has been done throughout this decade with enzymes in ionic liquid media. Excellent review articles have been published during these years collecting the results produced (Kragl et al., 2002; Sheldon et al., 2002, 2003; van Rantwijk et al., 2003; Park & Kazlauskas, 2003; Zhao, 2005; van Rantwijk & Sheldon, 2007; Yang, 2009; Moniruzzaman et al., 2010b; Lozano, 2010). Nevertheless, it is important to highlight that most of these results refer to lipase biocatalysis: lipases provide fast and convenient enantioselective synthetic routes for esterification and transesterification reactions. Besides, they are well suited for reactions in non-aqueous media. Even so, oxidoreductases (such as laccases and peroxidases) also play an important role in biochemical synthesis. These types of enzymes have received less attention in the literature, and this chapter will focus on the research work performed with laccase and peroxidase biocatalysis using ionic liquid as reaction media.

Most biocatalysis in ionic liquids involved no or low water content (Sheldon et al., 2002). Pugin et al. (2004) reported the interest in reaction systems with an ionic liquid and water (wet ionic liquids) due to the enantioselectivity of the biocatalyst and the importance in the use with ionic liquids when compared with organic solvents. These new reaction media consisting of water/ionic liquid present own and unique properties: they may provide a great potential without significant loss of activity due to their excellent catalytic performance, with turnover numbers even above 10^4.

The effect of ions on the enzyme activity, stability and enantioselectivity is another important factor which must be taken into account. Ions can affect the stability of proteins because of the chemical interactions between the proteins and the ions. According to the work of Vrbka et al. (2006) it seems that the ions in the reaction media interact with the charged amino acid groups of the protein as much as they interact with the corresponding ions in bulk solutions. Ions also may cause competitive or non-competitive inhibition of enzymes. The review of Zhao (2005) reports that anions such as PO_4^{3-}, CO_3^{2-}, or SO_4^{2-} (kosmotropic anions) and cations such as Cs^+, Rb^+, or K^+ (chaotropic cations) stabilize enzymes, while chaotropic anions and kosmotropic cations destabilize them. However, the influence of ionic liquids on the enzyme is complex especially when ionic liquids are present as nearly anhydrous solvents. As ionic liquids are composed only by ions, the knowledge about their influence on proteins is fundamental for choosing the most suitable as solvent for a given enzymatic reaction. Specific properties of ionic liquids also affect the enzymatic stability, such as polarity and viscosity (Park & Kazlauskas, 2003). Depending on the structures of cations and anions, ionic liquids can be hydrophobic or hydrophilic (Huddleston et al., 2001). Those with hydrophilic anions such as chloride and iodide are miscible and those with hydrophobic anions (eg. PF_6^- or NTf_2^-) are immiscible in water. The content of water and the hydrophobicity/hydrophilicity are also related to the cation/anion substitution (or functionalization) and is a significant factor considering the applications as reaction medium of a particular ionic liquid. It is important to recall that this substitution/functionlization of the ions allows for modifying (or controlling) their hydrophobicity/hydrophilicity, which can be tuned making ionic liquids the so called "designer solvents".

Cations	**Anions**
1.3-dialkylimidazolium 	Hexafluorophosphate 
1.1-dialkylpyrrolidinium 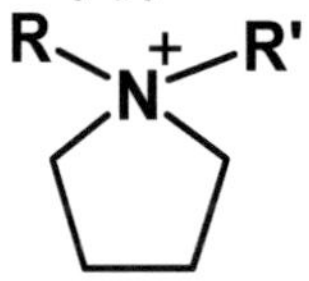	Tetrafluorophosphate 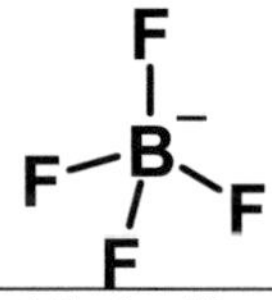
Alkylpiridinium 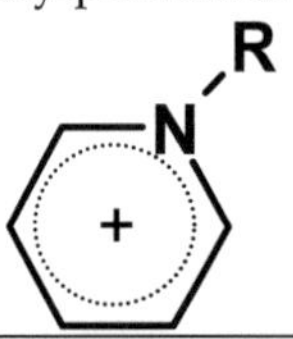	Alkylsulfate 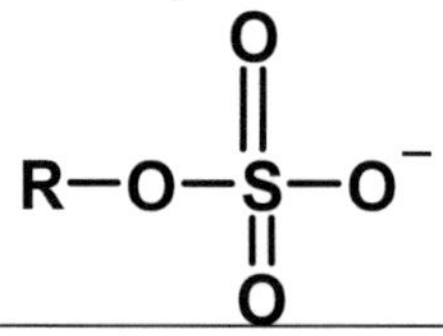
Tetraalkylammonium 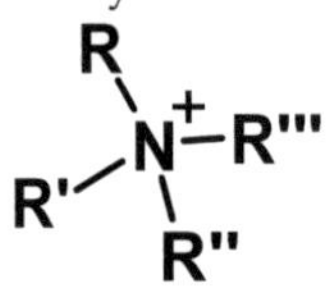	Methanesulfonate 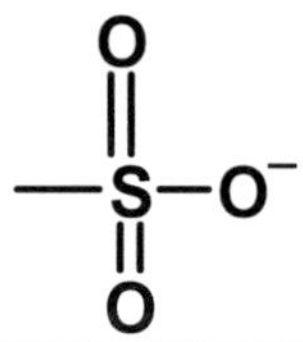
Tetraalkylphosphonium 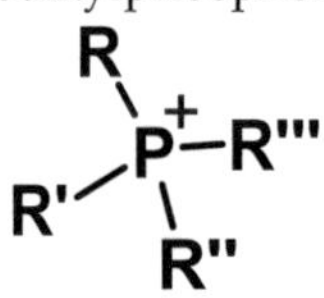	Dicyanamide 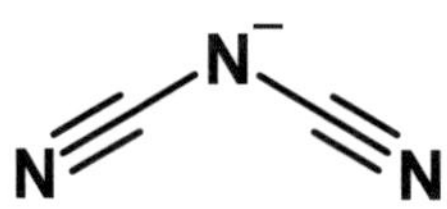
	Halides **Cl^-, Br^-, I^-**

Table 1. Anions and cations of ionic liquids commonly used in biocatalysis.

4. Enzyme activity in ionic liquids: Comparison with aqueous media

Most studies involving enzymatic catalysis are performed by ionic liquids which can be miscible and immiscible with water depending on the suitable choice of the anion and the cation, as described above. This characteristic allows ionic liquids to be designed for specific enzymatic reactions and substrates. In Table 1 it is possible to find the most common structures of ionic liquids used in biocatalysis. In general, ionic liquids consist of a salt where one or both ions are large, and the cation should be (highly) unsymmetrical. These factors tend to reduce the lattice energy of the crystalline form of the salt, and hence lower the melting point (Olivier-Bourbigou, 2010). The review article written by Rantwijk & Sheldon (2007) is focused in the structures and properties of ionic liquids used in biocatalysis. The choice of the cation and the anion is very important in order to avoid the complete denaturation of the enzyme. Either the anion or the cation may deactivate the active centre of the enzyme by interactions with positively/negatively charged residues in the enzyme structure (Park & Kazlauskas, 2003).

Peroxidases and oxidases are very attractive biocatalysts for selective oxidative transformations in ionic liquids. The enzymatic reactions with laccase, horseradish and soybean peroxidases in ionic liquids were started by Hinckley et al. (2002). In their study, 4-methyl-N-butylpyridinium tetrafluoroborate, [4-mbpy][BF_4], and 1-butyl-3-methyl imidazolium hexafluorophosphate, [bmim][PF_6], were used as reaction media at different concentrations for the antracene oxidation by laccase. The previous assay on the catalytic activity of laccase for syringaldazine oxidation showed that the enzyme tolerates moderate concentrations of [4-mbpy][BF_4]. At high concentrations of this ionic liquid (above 50%), laccase activity decreased and then the enzyme precipitated. Similar results of laccase precipitation in ionic liquids were observed with 1-ethyl-3-methylimidazolium 2-(2-methoxyethoxy) ethylsulfate, [emim][$MDEGSO_4$], 1-ethyl-3-methylimidazolium ethylsulfate, [emim][$EtSO_4$], and 1-ethyl-3-methylimidazolium methanesulfonate, [emim][$MeSO_3$] at concentrations of 75% (v/v) or above (Tavares et al., 2008). For the tests with peroxidases Hinckley et al. (2002) observed significant activity when guaiacol was used as substrate, in the presence of 25% (v/v) of the water-miscible ionic liquid [4-mbpy][BF_4]. A similar behavior has been reported (Sgalla et al.,. 2007) for HRP in 1-butyl-3-methylimidazolium tetrafluoroborate ([bmim][BF_4]) and water mixtures. Here, also high activity was obtained for ionic liquid concentrations up to 25%, depending on the pH value. The enzymatic reactions were carried out with non-soluble phenolic compounds with up to 75% of [bmim][BF_4], presenting high yields of dimeric species for 4-phenylphenol (85% dimers). Again, the polymerization of phenol by HRP in [bmim][BF_4] presented a yield of polymerization of 100% with ionic liquid content of 60% (Zaragoza-Gasca et al., 2010). Other study with HRP and [bmim][BF_4] showed the importance of water content in the ionic liquid for the maintenance of the enzyme catalytic activity (Wang et al., 2007): No HRP activity was achieved in the presence of anhydrous ionic liquid, however, this deactivation was reversible with little addition of water. The authors explain that the enzyme structure is dependent on the hydrogen bonding, ionic, hydrophobic, and van der Waals interactions. Consequently, in non-aqueous media these interactions can be broken, which are fundamental for the maintenance of the enzyme activity.

Okrasa et al. (2003) have studied the particular interest in the oxidation products of sulfides using non-aqueous enzymatic reactions with peroxidase and glucose oxidase (GOD), with 1-butyl-3-methyl imidazolium hexafluorophosphate ([bmim][PF_6]). In this study, an interesting transformation of thioanisole to sulfoxide was carried out by the bi-enzymatic system with GOD/peroxidase with 10% of water in ionic liquid. The hydrogen peroxide is produced in situ by GOD (to transform the glucose in gluconic acid) and it is used by peroxidase to oxidize the sulfides.

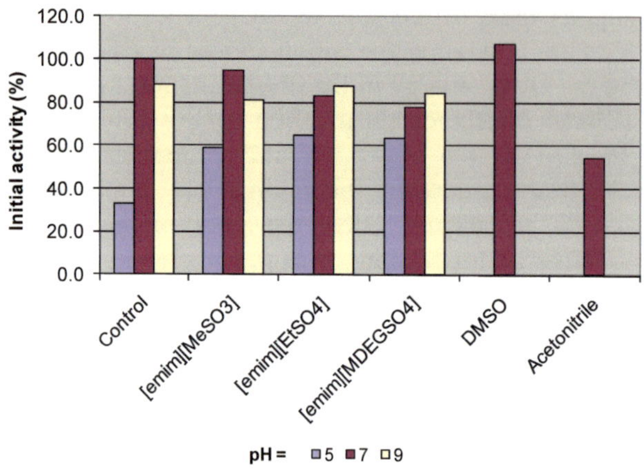

Fig. 4. Comparison of laccase initial activity in aqueous buffer solution (control), ionic liquids, dimethyl sulfoxide and acetonitrile at 50% (v/v) and different pH values.

Up to date, some papers are published in the literature regarding to the enzymatic reactions of laccase in ionic liquids. Most of them are related to the development of biosensors containing laccase and ionic liquids (Liu et al., 2007; Franzoi et al., 2009a,b; Brondani et al., 2009). Tavares et al. (2008) optimized the conditions (pH, ionic liquid concentration) for the activity of commercial laccase (DeniLite base), using ABTS as substrate, in different water-soluble ionic liquids: 1-ethyl-3-methylimidazolium 2-(2-methoxyethoxy) ethylsulfate, [emim][$MDEGSO_4$], 1-ethyl-3-methylimidazolium ethylsulfate, [emim][$EtSO_4$], and 1-ethyl-3-methylimidazolium methanesulfonate, [emim][$MeSO_3$]. As an example, Figure 4 presents a comparison of laccase activity in these ionic liquids (50% v/v) at different conditions. The comparison is extended to an aqueous buffer solution (named control) and organic solvents (dimethyl sulfoxide and acetonitrile, also at 50% v/v). Activities are presented as percentage relative to the initial activity of the enzyme in the buffer with pH 7. Laccase was most active at pH 7.0 - 9.0. The different solvents have little effect in laccase activities at pH 7.0 or 9.0 (both ionic liquids and organic solvents) at the lower concentrations (10 and 25% (v/v)), with a certain reduction in enzyme activity at higher concentrations. Carneiro et al. (2009) also studied the peroxidase activity in two of the previous ionic liquids, [emim][$MDEGSO_4$] and [emim][$EtSO_4$], using once again ABTS as substrate. Promising results of peroxidase activity were obtained at pH 7 for moderate concentrations of ionic liquid (5 and 10 % (v/v)) and good initial stability was obtained when compared to the control sample (an aqueous buffer medium). From both works of the same research group, it is possible to compare the enzymes activities between the buffer solution and ionic liquid media. Figure 5 shows a comparison of initial activities of both enzymes at 50% (v/v) concentration of ionic liquids and different pH values. The initial activity is presented relative to that in the pure aqueous buffer solution (control) for each enzyme. The high activities for laccase and peroxidase show the potential of these enzymes for catalytic reactions in ionic liquids.
Non-traditional ionic liquid media, based on micro emulsions with the nonionic surfactant Triton X-100, have been applied to improve the activity of laccase and lignin peroxidase (LiP) with 1-butyl-3-methylimidazolium hexafluorophosphate ([bmim][PF_6]) and water (Zhou et al., 2008). The results in pure or water-saturated [bmim][PF_6] revealed that both LiP and laccase had insignificant catalytic activity, thus showing the negative effects of [bmim][PF_6]. However, good activities of both enzymes were obtained for [bmim][PF_6]-

based microemulsions and this was attributed to the TX-100 interfacial membrane, as it separates the enzyme from the [bmim][PF_6] bulk phase. In addition, the apparent viscosity of the medium decreases, which reduces diffusional barriers.

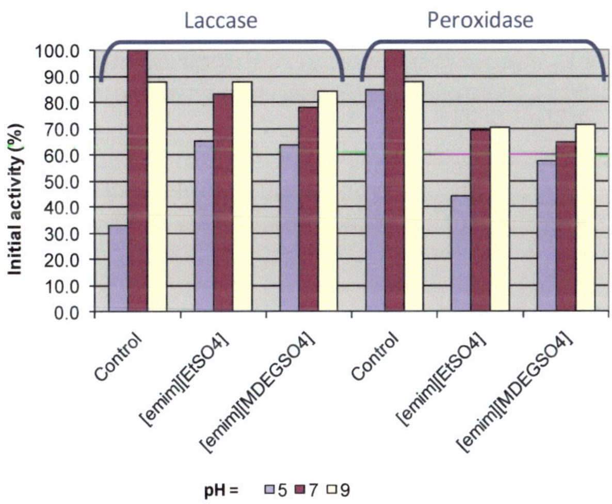

Fig. 5. Comparison of laccase (left side) and peroxidase (right side) initial activities in aqueous buffer solution (control) and ionic liquids, at 50% (v/v) and different pH values.

5. Ezymatic stability in ionic liquids

Enzyme stabilization deserves special attention because of the increasing number of enzyme applications. Enzymatic stability is essential for the enzyme to work at its full potential as catalyst for long periods, and consequently the economic feasibility of applying such enzyme in an industrial process. In addition, stable enzymes permit the use of higher temperatures in biochemical processes, which may have beneficial effects on reaction rates and/or reactant solubilities, while at the same time reduce the risk of microbial contamination. The enzyme stability (storage stability in ionic liquids) is obtained by incubation of the enzyme in an ionic liquid with convenient control of the temperature, and the measurement of the residual activity at certain time intervals. Different methodologies to improve enzyme stability with respect to their use in aqueous (or other non-aqueous) environments have been given in the literature, such as optimization of the reaction medium, chemical modifications of the enzymes, and enzyme immobilization. As discussed above, much of the current focus in enzyme technology involves enhancement of enzyme activity and stability into non-conventional media. The non-aqueous medium such as ionic liquids brings several advantages as enhanced solubility of non-polar substrates and/or products. However, these advantages would often be limited by a lower stability of biocatalysts in these systems. Just as an example, the use of hydrophilic ionic liquids can capture the water molecules which are fundamental to keep the enzyme 3D structure and consequently, its activity. This activity loss is related to the interactions established between the enzyme and the ions, and the rupture of those with water molecules. To overcome these problems, existing strategies for organic solvents have been employed to enhance enzyme stability in ionic liquids with significant results. Below are presented some examples of oxidative enzymes which have demonstrated to provide high stability in different ionic liquids.

Wang et al. (2007) described the stability of free and immobilized HRP in aqueous mixture of 1-butyl-3-methylimidazolium tetrafluroborate ([bmim][BF_4]). The activity was improved by HRP immobilization in agarose hydrogel retaining 80.2% of its initial activity for 10.5 h. The stability of peroxidase in aqueous mixture of [emim][$MDEGSO_4$] and [emim][$EtSO_4$] was well maintained when compared to a phosphate buffer reference medium, for at least seven days of incubation, at optimized conditions: pH 7.0, ionic liquid 10% v/v, room temperature (Carneiro et al., 2009). Chloroperoxidase was highly stable in 1,3-dimethylimidazolium methylsulfate [mmim][$MeSO_4$] or 1-butyl-3-methylimidazolium methylsulfate [bmim][$MeSO_4$] (30%(v/v) co-solvent/citrate buffer), whereas it was deactivated within of 3 hours in the presence of *t*-butyl hydroperoxide or acetone (Sanfilippo et al., 2004).

To improve the stability of HRP in ionic liquids, Das et al. (2007) investigated various anions. Initially, tests with [bmim][Cl], [bmim][CF_3CO_2] and the non-conventional [bmim][$H_2NCH(CH_3)CO_2^-$] with a small amount of water (5% v/v) were evaluated. HRP lost 50% of its initial activity within 24 h in [bmim][Cl] or [bmim][$CF_3CO_2^-$]. On the other hand, [bmim][$H_2NCH(CH_3)CO_2^-$] promoted total deactivation of HRP in only one hour of incubation indicating that the presence of the aminoacid in the ionic liquid structure did not favor the stability of HRP. Further, the water content in ionic liquids was also investigated and HRP activity loss increased with the decrease in water content from 70% to 5%. In more common ionic liquids, a good stability of HRP, for at least 30h of incubation, even at 0% water, was obtained for [bmim][CF_3SO_3] and a similar stability for [bmim][BF_4] was obtained just for high water content (50-70%).

The stability of laccase was also studied in different water soluble ionic liquids [emim][$MDEGSO_4$], [emim][$EtSO_4$], and [emim][$MeSO_3$] at different water contents for up to ten days of incubation. The stabilities were compared with two organic solvents, acetonitrile and dimethyl sulfoxide and with an aqueous buffer solution. The enzyme maintained a high stability at pH 9.0 for all ionic liquids tested, particularly for [emim][$MDEGSO_4$], presenting small activity loss after the incubation for an ionic liquid content of 10%. Increasing this content resulted in a decrease of the enzyme activity for all ionic liquids and organic solvents. Just as an example, laccase activity is presented in Figure 6 as a function of time the

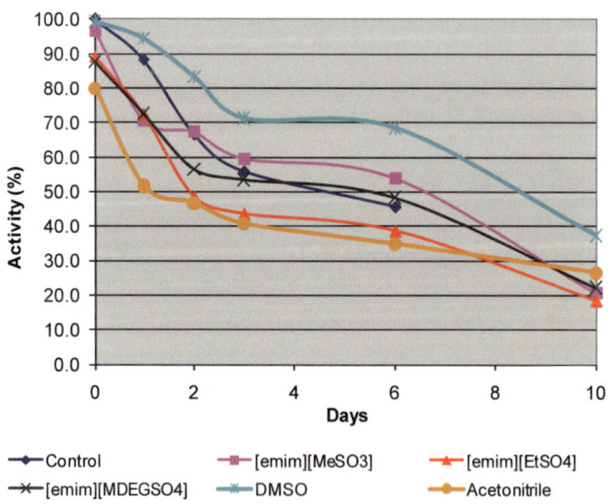

Fig. 6. Evolution of laccase activity with time: initial activity after incubation of the enzyme in aqueous solutions of ionic liquids or organic solvents (25 % v/v; pH = 7). Activities are relative to the initial control value.

above mentioned ionic liquid and organic solvents at 25 % (v/v) and pH 7. All activities are relative to that in pure aqueous buffer solution (named control) whose initial activity is considered 100%. It is clear that all ionic liquids present an initial decrease larger than the control solution (until day 3), but then the decrease is reduced and similar to the control. [emim][$MeSO_3$] provided the best behaviour. Compared to the organic solvents, acentonitrile performed worse but DMSO better than all ionic liquids and control.

6. Kinetics

Most of the previous studies discussed above were concerned with the measurement of activity or stability of enzymes and its reactions, without focusing on how ionic liquids affect the kinetic mechanisms. However, clarification of the role of ionic liquids in the catalytic mechanisms of enzymes is an important factor for the development of efficient biocatalytic processes utilizing ionic liquids as solvents. Many papers regarding enzyme-catalyzed reactions follow the traditional Michaelis–Menten equation, because of its simplicity and easiness to interpret the physical meaning of the equation parameters. The reaction rate can be described on the basis of the concentration of the substrate and of the enzymes. For a single enzyme and single substrate, the Michaelis-Menten parameters can be obtained by non-linear regression, based on the Michaelis-Menten equation:

$$v_0 = \frac{V_{\max} \cdot [S]}{K_m + [S]} \tag{1}$$

where v_0 is the initial reaction rate, $V_{\max}$ is the maximum reaction velocity, K_m is the Michaelis-Menten constant and $[S]$ the substrate concentration. K_m is the value of substrate concentration at which the reaction rate reaches half of its maximum value ($V_{\max}/2$), and can be understood as a measurement of the enzyme affinity for the substrate. Thus, low K_m values indicate that the enzyme attains its maximum catalytic efficiency at a lower substrate concentration. Moreover, the ratio ($V_{\max}/K_m$) can be interpreted as a measurement of the enzyme catalytic efficiency.

The kinetic parameters of laccase were determined in the presence of 4-methyl-N-butylpyridinium tetrafluoroborate, [4-MBP][BF_4], 25% (v/v) and 1-butyl-3-methyl imdizaolium hexafluorophosphate, [bmim][PF_6], both saturated with water, and compared with tert-butyl hydroperoxide (20%v/v) and an aqueous buffer (Hinckley et al., 2002). The authors indicate that the kinetic data revealed a simultaneous decrease of $V_{\max}$ and increase of K_m in the presence of the two ionic liquids. This points to a dramatic reduction in the catalytic efficiency (the $V_{\max}/K_m$ ratio mentioned above). Correspondingly, the catalytic activity of laccase in [bmim][PF_6] was much lower than in aqueous media. These first results of laccase kinetics suggested that the optimization of reaction conditions in ionic liquids may be necessary to obtain higher catalytic activities. Another work can be referred with laccase (Tavares et al., 2008) but different ionic liquids. The ionic liquids used were based on sulfate and sulfonate anions, rather than the tetrafluoroborate: [emim][$MDEGSO_4$], [emim][$EtSO_4$] and [emim][$MeSO_3$], previously referred. Despite the value of $V_{\max}$ decreased for all ionic liquids, that of K_m also decreased (in a lower extent), thus diminishing the reduction in the catalytic effieciency.

Enzymatic inhibition is another point that should be considered in kinetics. An inhibitor is any substance that reduces the reaction rate. If an inhibitor binds reversibly to the same site

than the substrate, the inhibition will be competitive: the inhibitor blocks the active site. Non-competitive inhibition occurs when an inhibitor binds to the enzyme-substrate complex, so the inhibition cannot be overcomed by increasing the concentration of substrate. The inhibitor does not compete with the substrate for the active site of the enzyme, and thus substrate concentration has no influence on the degree of inhibition of the enzyme. The Michaelis-Menten equation can be used to provide evidences of this kind of inhibition: the inhibitor decreases $V_{\max}$ but has no effect on K_m. That is, the inhibitor binding does not affect the enzyme-substrate affinity (K_m), both the substrate (S) and the inhibitor (I) can simultaneously bind to the enzyme, but the resultant E-S-I complex is catalytically inactive. As a result, the inhibitor (I) can hide the adequate positioning of the catalytic center. In that case the reaction of the noncompetitive inhibitor is irreversible and the substrate cannot overcome the inhibitor's impact on the enzyme. Very few studies regarding ionic liquids as possible inhibitors are published in the literature.
An example of such study can be found for peroxidase (Carneiro et al., 2009) with the ionic liquids [emim][$EtSO_4$] and [emim][$MDEGSO_4$]. The Michaelis-Menten parameters for the enzymatic reactions in the presence and absence (buffer solution) of these ionic liquids were determined for a set of different concentrations. The results indicate that K_m values in the absence and in the presence of [emim][$MDEGSO_4$] (for all compositions studied) present close values (in the range 0.009-0.012 mM) while in the presence of [emim][$EtSO_4$] (for all compositions studied) the result was a bit lower (0.0064-0.0079 mM). Nevertheless, the presence of either ionic liquid had a much clearer effect in the values of V_{max}. As the ionic liquid concentration was increased, the V_{max} obtained decreased. The effect of [emim][$EtSO_4$] concentration was higher than that of [emim][$MDEGSO_4$], and so V_{max} was reduced from 3.10 mM/min in the pure buffer to 2.19 mM/min for the first ionic liquid (a reduction of 29 %) and 2.35 mM/min for the latter (a reduction of 24 %). The V_{max} values obtained decrease as the ionic liquid concentration increases. According to these results, K_m value can be considered approximately constant for all concentrations, while V_{max} gradually decreased with the increase in ionic liquid. These facts indicate a non-competitive inhibition mechanism, thus the ionic liquids do not affect the apparent substrate binding to the enzyme but indeed affect the rate of the reaction. The reaction rate was calculated using the following equation:

$$v_0 = \frac{V_{\max}[S]}{([S]+K_m)\left(1+\frac{[I]}{K_i}\right)} \tag{2}$$

where $[I]$ is the concentration of the inhibitor (here, the IL) and K_i is the inhibitory constant. Equation (2) allowed to represent the dependency of the reaction rate on the concentration of both substrate (ABTS) and ionic liquid, with fair agreement with experimental data.

7. Applications of enzymes in ionic liquids

As discussed in the previous sections, performing biocatalytic reactions in ionic liquids can be promising with regards to activity, selectivity and stability towards the target substrates, even presenting a short history. Indeed, the use of enzymes in ionic liquids opens up new possibilities for enzymology in non-aqueous media. Ionic liquids have added benefits for performing different kinds of biotransformations. A large number of biocatalysts have been reported to be active in ionic liquids, and the application of oxidative enzymes has played an important role. The first application of oxidative enzymes in ionic liquids was reported by Hinckley et al. (2002), using laccase, horseradish and soybean peroxidases for veratryl alcohol, anthracene and guaiacol oxidations in the presence of [4-mbpy][BF_4] containing 25% (v/v) of aqueous buffer. Since then, reports have been published using oxidative enzymes. A summary of these applications is depicted in Table 2.

In recent years, the main application found for laccase in ionic liquids is related to the construction of biosensors. As examples, the determination of rutin (Franzoi et al., 2009a) or rosmarinic acid (Franzoi et al., 2009b) using biosensors constructed with laccase and 1-alkyl-3-methylimidazolium bis(trifluoromethylsulfonyl)imide ([$C_x$mim][NTf_2], x = 4, 10, 14), and [bmim][PF_6] or [bmim][BF_4], respectively. The determination of adrenaline in pharmaceutical formulations, using a biosensor constructed with 1-butyl-3-methyl-imidazolium hexafluorophosphate [bmim][PF_6] and laccase, has also been carried out (Brondani et al., 2009).

Enzyme	Ionic Liquid	Application	Reference
Laccase	[4-mbpy][BF_4]	Oxidation of anthracene, veratryl alcohol and guaiacol	Hinckley et al., 2002
Laccase	([$C_x$mim][NTf_2]	Determination of rutin	Franzoi et al., 2009a
Laccase	[bmim][PF_6]; [bmim][BF_4]	Determination of rosmarinic acid	Franzoi et al., 2009b
Laccase	[bmim][PF_6]	Determination of adrenaline	Brondani et al., 2009
Peroxidase	[bmim][NTf_2] [bmim][PF_6]	Oxidation of guaiacol	Laszlo & Compton, 2002
HRP	[bmim][BF_4]	Detection of H_2O_2	Liu et al., 2005; Wang et al., 2007
HRP	[C_2mim]Cl; [C_4mim]Cl; [C_6mim]Cl; [C_8mim]Cl;	Enzyme extraction system	Cao et al., 2008
HRP	[bmim][BF_4]	Polymerization of phenolic compounds	Sgalla et al., 2007; Zaragoza-Gasca et al., 2009
Oxidase-peroxidase	[bmim][PF_6]	Sulfoxidation of thioanisoles	Okrasa et al., 2003
Cloroperoxidase	[mmim][$MeSO_4$]; [bmim][$MeSO_4$]; [bmim][Cl]; [bmim][BF_4]	Oxidation of 1,2-dihydronaphthalene	Sanfilipo et al., 2004
D-amino acid oxidase	[bmim][BF_4]; [mmim][Me_2PO_4]	Deamination of amino acid	Lutz-Wahl et al., 2006
Soybean peroxidase	[bmim][BF_4]; [bmpy][BF_4]	Polymerization of phenols	Eker et al., 2009

Table 2. Examples of application of oxidative enzymes in ionic liquids.

More applications are found for reactions with peroxidases: Immobilized HRP was applied in non-aqueous biosensing for the detection of H_2O_2 with [bmim][BF_4] (Wang et al., 2007). HRP was also used in amperometric biosensing devices, and the extension of the technique to entrapment in a [bmim][BF_4] sol-gel matrix has recently been demonstrated for the detection of H_2O_2 (Liu et al., 2005). A different application for HRP was reported (Cao et al., 2008) showing the enzyme partition behavior in aqueous biphasic systems with the following ionic liquids: [C_2mim]Cl; [C_4mim]Cl; [C_6mim]Cl and [C_8mim]Cl. HRP was also applied for oxidative coupling of water insoluble phenolic compounds in [bmim][BF_4] water mixtures (Sgalla et al., 2007; Zaragoza-Gasca et al., 2009). Besides, soybean peroxidase (SBP) has been used to catalyze the polymerization of phenols in [bmim][BF_4] and 1-butyl-3-methylpyridinium tetrafluoroborate [bmpy][BF_4] (Eker et al., 2009).
The work of Okrasa et al. (2003) demonstrated the oxidase–peroxidase-catalyzed sulfoxidation of thioanisoles to sulfoxide in [bmim][PF_6]. Cloroperoxidase was reported (Sanfilipo et al., 2004) to catalyze the oxidation of 1,2-dihydronaphthalene to (1R,2R)-(+)-dihydroxy-tetrahydro-naphthalene in buffer/ionic liquid mixtures using [mmim][$MeSO_4$], [bmim][$MeSO_4$], [bmim][Cl], and [bmim][BF_4]. The oxidative deamination of the amino acid phenylalanine and cephalosporin-C were catalized by immobilized d-amino acid oxidase in the presence of [bmim][BF_4] and 1,3-dimethylimidazolium dimethylphosphate [mmim][Me_2PO_4] (Lutz-Wahl et al., 2006).

8. Conclusion

Throughout this chapter a revision of the research performed with oxidative enzymes, namely laccase and peroxidase, in ionic liquids has been provided. The focus has been centered in the studies of enzymatic activity and stability as these are critical properties for biocatalytic processes. The results presented allow to conclude that enzyme activity may diminish in the presence of ionic liquids, but such reduction may be balanced with the increase in poorly-soluble substrates and/or products, leading to better performances of teh global bioprocess. In general, the use of ionic liquids as co-solvents to enhance substrate solubility demonstrate the feasibility of this combination, but the type of ionic liquid and its concentration need to be carefully optimized. Works presenting the direct use of an oxidative enzyme in the pure ionic liquid (with very limited water content) are scarce. This fact indicates that direct dissolution of the enzyme in ionic liquids does not seem to be an option. Alternatively, the use of immobilized oxidative enzymes in ionic liquids has received little attention so far. Considering the excellent results that have been obtained with other immobilized enzymes, namely lipases (Lozano, 2010), this approach may be a suitable solution and an excelent path for future research.

9. References

An, E.S.; Cho, D.H.; Choi, J.W.; Kim, Y.H. & Song, B.K. (2010). Peroxidase-catalyzed copolymerization of syringaldehyde and bisphenol A. *Enzyme and Microbial Technology*, Vol. 46, 287-291.

Azevedo, A.M.; Prazeres, D.; Cabral J. & Fonseca, L. (2001). Stability of free and immobilised peroxidase in aqueous–organic solvents mixtures. *Journal of Molecular Catalysis B: Enzymatic*, Vol. 15, 147-153.

Banci, L. (1997). Structural Properties of Peroxidases. *Journal of Biotechnology*, Vol. 53, 253-263.

Bourbonnais, R.; Leech, D. & Paice, M.G. (1998). Electrochemical analysis of the interactions of laccase mediators with lignin model compounds. *Biochimica et Biophysica Acta*, Vol. 1379, 381-390.

Brondani, D.; Scheeren, C.W.; Dupont, J. & Vieira, I.C. (2009). Biosensor based on platinum nanoparticles dispersed in ionic liquid and laccase for determination of adrenaline, *Sensors and Actuators, B: Chemical*, Vol. 140, 252-259.

Cao, Q., Quan, Li; He, C.; Li, Na.; Li, K. & Liu, F. (2008). Partition of horseradish peroxidase with maintained activity in aqueous biphasic system based on ionic liquid. *Talanta*, Vol. 77, 160-165.

Carneiro, A.P.; Rodríguez, O.; Mota, F.L.; Tavares, A.P.M. & Macedo, E.A. (2009). Kinetic and stability study of the peroxidase inhibition in ionic liquids. *Industrial & Engineering Chemistry Research*, Vol.48, 10810-10815.

Carrea, G. & Riva, S. (2000). Properties and synthetic applications of enzymes in organic solvents. *Angewandte Chemie*, Vol. 39, 2226-2254.

Castro, G.R. & Knubovets, T. (2003). Homogeneous biocatalysis in organic solvents and water-organic mixtures. *Critical Reviews in Biotechnology*, Vol. 23, 195-231.

Conesa, A.; Punt, P.J. & van den Hondel, C.A.M.J.J. (2002). Fungal peroxidases: molecular aspects and applications. *Journal of Biotechnology*, Vol. 93, 143–158.

Couto, S.R. & Herrera, J.L.T. (2006). Industrial and biotechnological applications of laccases: A review. *Biotechnology Advances*, Vol. 24, 500–513.

Das, D.; Dasgupta, A. & Das, P. (2007). Improved activity of horseradish peroxidase (HRP) in 'specifically designed' ionic liquid. *Tetrahedron Letters*, Vol. 48, 5635–5639.

Dordick, J.S. (1991). Non-aqueous enzymology. *Current Opinion in Biotechnology*, Vol. 2, 401-407.

Eibes, G.; Lú-Chau, T.; Feijoo, G.; Moreira, M.T. & Lema, J.M. (2005). Complete degradation of anthracene by Manganese Peroxidase in organic solvent mixtures. *Enzyme and Microbial Technology*, Vol. 37, 365-372.

Eker, B.; Zagorevski, D; Zhu, G.; Linhardta, R.J & Dordick, J.S. (2009). Enzymatic polymerization of phenols in room-temperature ionic liquids. *Journal of Molecular Catalysis B: Enzymatic*, Vol. 59, 177–184.

Erbeldinger, M.; Mesiano, A.J. & Russel, A.J. (2000). Enzymatic catalysis of formation of Z-aspartame in ionic liquid - an alternative enzymatic catalysis in organic solvents. *Biotechnology Progress*, Vol. 16, 1129–1131.

Franzoi, A.C.; Migowski, P., Dupont, J. & Vieira, I.C. (2009-a). Development of biosensors containing laccase and imidazolium bis(trifluoromethylsulfonyl)imide ionic liquid for the determination of rutin. *Analytica Chimica Acta*, Vol. 639, 90-95.

Franzoi, A.C.; Dupont, J.; Spinelli, A. & Vieira, I.C. (2009-b). Biosensor based on laccase and an ionic liquid for determination of rosmarinic acid in plant extracts. *Talanta*, Vol. 77,1322-1327.

Halling, P. & Kvittingen, L. (1999). Why did biocatalysis in organic media not take off in the 1930s?. *Trends in Biotechnology*, Vol. 17, 343-344.

Halling, P. (2000). Biocatalysis in low-water media: understanding effects of reaction conditions. *Current Opinion in Chemical Biology*, Vol. 4, 74–80.

Hamid, M. & Rehman, K. (2009). Potential applications of peroxidases. *Food Chemistry,* Vol. 115, 1177–1186.

Hinckley, G.; Mozhaev,V.V.; Budde, C. & Khmelnitsky, Y.L.(2002). Oxidative enzymes possess catalytic activity in systems with ionic liquids. *Biotechnology Letters,* Vol. 24, 2083–2087.

Huddleston, J.G.; Visser, A.E.; Reichert, W.M.; Willauer, H.D.; Broker, G.A. & Rogers, R.D. (2001).Characterization and comparison of hydrophilic and hydrophobic room temperature ionic liquids incorporating the imidazolium cation. *Green Chemistry,* Vol., 3, 156–164.

Husain, M. & Husain, Q. (2008). Applications of Redox Mediators in the Treatment of Organic Pollutants by Using Oxidoreductive Enzymes: A Review. *Critical Reviews in Environmental Science and Technology,* Vol. 38, 1–42.

Karam, J. & Nicell, J.A. (1997). Potential Applications of Enzymes in Waste Treatment. *Journal of Chemical Technology & Biotechnology,* Vol. 69, 141-153.

Klibanov, A.M. (1986). Enzymes that work in organic solvents. *Chemtech,* Vol. 16, 354-359.

Klibanov, A. M.; Samokhin, G.P.; Martinek, K. & Berezin, I. (2000) A new approach to preparative enzymatic synthesis. *Biotechnology and Bioengineering,* Vol.19, 1351-1361 (1977). Special Issue: Special 40th Anniversary Issue: *Biotechnology and Bioengineering,* Vol. 67, 737–747.

Klyosov, A.A.; Van Viet,N. & Berezin, I.V. (1975). The Reactions of a-Chymotrypsin and Related Proteins with Ester Substrates in Non-aqueous Solvents. *European Journal of Biochemistry,* Vol. 59, 3-7.

Kragl, U.; Eckstein, M. & Kaftzik, N. (2002). Enzyme catalysis in ionic liquids. *Current Opinion in Biotechnology,* Vol. 13, 565–571.

Laszlo, J.A. & Compton, D.L. (2002). Comparison of peroxidase activities of hemin, cytochrome c and microperoxidase-11 in molecular solvents and imidazolium-based ionic liquids. *Journal of Molecular Catalysis B: Enzymatic,* Vol. 18, 109–120.

Liu, Y.; Shi, L.; Wang, M.; Li, Z.; Liu, H. & Li, J. (2005). A novel room temperature ionic liquid sol–gel matrix for amperometric biosensor application. *Green Chemistry,* Vol. 7, 655-658.

Liu, Y; Huang, L. & Dong, S. V. (2007). Electrochemical catalysis and thermal stability characterization of laccase-carbon nanotubes-ionic liquid nanocomposite modified graphite electrode. *Biosensors and Bioelectronics,* Vol. 23, 35-41.

Lozano, P. (2010). Enzymes in neoteric solvents: From one-phase to multiphase systems. *Green Chemistry,* Vol. 12, 555–569.

Lutz-Wahl, S.; Trost, E.-M.; Wagner, B.; Manns, A. & Fischer, L. (2006). Performance of d-amino acid oxidase in presence of ionic liquids. *Journal of Biotechnology,* Vol. 124, 163–171.

Magnuson, D.K.; Bodley, J.W. & Evans, D.F. (1984). The activity and stability of alkaline phosphatase in solutions of water and the fused salt ethylammonium nitrate. *Journal of Solution Chemistry,* Vol.13, 583–587.

Martinek, K.; Levashov, A.V.; Klyachko, N.L.; Pantin, V.I. & Berezin, I.V. (1981). The principles of enzyme stabilization. VI. Catalysis by water-soluble enzymes entrapped into reversed micelles of surfactants in organic solvents. *Biochimica et Biophysica Acta,* Vol. 657, 277-294.

Mayer, A.M. & Staples, R.C. (2002). Laccase: new functions for an old enzyme. *Phytochemistry*, Vol. 60, 551-565.

Mita, N.; Tawaki, S.; Uyama, H. and Kobayashi, S. (2003). Laccase-Catalyzed Oxidative Polymerization of Phenols. Macromolecular Bioscience, Vol. 3, 253-257.

Moniruzzaman, M.; Kamiya, N. & Goto, M. (2010a). Activation and stabilization of enzymes in ionic liquids. *Organic and Biomolecular Chemistry* Vol. 8, 2887-2899.

Moniruzzaman, M.; Nakashima, K.; Kamiya, N. & Goto, M. (2010b). Recent advances of enzymatic reactions in ionic liquids. *Biochemical Engineering Journal*, Vol. 48, 295-314.

Nikolova, P. & Ward, O.P. (1993). Whole cell biocatalysis in nonconventional media. Journal of Industrial Microbiology, Vol. 12, 76-86.

Okrasa, K.; Guibé-Jampel, E. & Therisod, M. (2003). Ionic liquids as a new reaction medium for oxidase-peroxidase-catalyzed sulfoxidation. *Tetrahedron: Asymmetry*, Vol. 14, 2487-2490.

Olivier-Bourbigou, H.; Magna, L. & Morvan, D. (2010). Ionic liquids and catalysis: Recent progress from knowledge to applications. *Applied Catalysis A: General*, Vol. 373, 1-56.

Park, S. & Kazlauskas, R.J. (2003). Biocatalysis in ionic liquids -advantages beyond green technology. *Current Opinion in Biotechnology*, Vol., 14, 432-437.

Pugin, B.; Studer, M.; Kuesters, E.; Sedelmeier, G. & Feng, X. (2004). Mixtures of ionic liquids and water as a medium for efficient enantioselective hydrogenation catalyst recycling. *Advanced Synthesis and Catalysis*, Vol. 346, 1481-1486.

Rantwijk, F. & Sheldon, R.A. (2007). Biocatalysis in Ionic Liquids. *Chemical Reviews*, Vol. 107, 2757-2785.

Sanfilippo, C.; D'Antona, N. & Nicolosi, G. (2004). Chloroperoxidase from *Caldariomyces fumago* is active in the presence of an ionic liquid as co-solvent. *Biotechnology Letters*, Vol.26, 1815-1819.

Sgalla, S.; Fabrizi,G.; Cacchi, S.; Macone, A.; Bonamore, A. & Boffi, A. (2007). Horseradish peroxidase in ionic liquids Reactions with water insoluble phenolic substrates. *Journal of Molecular Catalysis B: Enzymatic* Vol. 44, 144-148.

Sheldon, R.A.; Lau, R.M.; Sorgedrager, M.J.; van Rantwijk, F. & Seddon, K.R. (2002). Biocatalysis in ionic liquids. *Green Chemistry*, Vol. 4, 147-151.

Sheldon, R.A.; Lau, R.M. & van Rantwijk, F. (2003). Biocatalytic transformations in ionic liquids. *Trends in Biotechnology*, Vol. 21: 131-138.

Singh, A. & Kaplan, D.L. (2002). Enzyme-Based Vinyl Polymerization. *Journal of Polymers and the Environment*, Vol. 10, 85-91.

Tavares, A.P.M.; Rodriguez, O. & Macedo, E.A. (2008). Ionic liquids as alternative co-solvents for laccase: Study of enzyme activity and stability. *Biotechnology and Bioengineering*, Vol. 101, 201-207.

Turac, E. & Sahmetlioglu, E. (2010). Oxidative polymerization of 4-[(4-phenylazo-phenyimino)-methyl]-phenol catalyzed by horseradish peroxidase. *Synthetic Metals*, Vol. 160, 169-172.

van Rantwijk, F.; Lau, R.M. & Sheldon, R.A. (2003). Biocatalytic transformations in ionic liquids. *Trends in Biotechnology*, Vol. 21, 131-138.

van Rantwijk, F. & Sheldon, R.A. (2007): Biocatalysis in Ionic Liquids. *Chemical Reviews*, Vol. 107, 2757-2785.

Vrbka, L.; Jungwirth,P.; Bauduin, P.; Touraud, D. & Kunz, W. (2006). Specific Ion Effects at Protein Surfaces: A Molecular Dynamics Study of Bovine Pancreatic Trypsin Inhibitor and Horseradish Peroxidase in Selected Salt Solutions. *The Journal of Physical Chemistry B*, Vol. 110, 7036-7043.

Wang, S-F.; Chen, T.; Zhang, Z-L & Pang, D-W. (2007). Activity and stability of horseradish peroxidase in hydrophilic room temperature ionic liquid and its application in non-aqueous biosensing. *Electrochemistry Communications,* Vol. 9, 1337–1342.

Wilkes, J.S. (2004). Properties of ionic liquid solvents for catalysis. *Journal of Molecular Catalysis A: Chemical,* Vol. 214, 11–17.

Won, K.; Kim, Y.H.; An, E.S.; Lee, Y.S. & Song, B.K. (2004). Horseradish Peroxidase-Catalyzed Polymerization of Cardanol in the Presence of Redox Mediators. *Biomacromolecules,* Vol. 5, 1–4.

Xu, F. (2005). Applications of oxidoreductases: Recent progress. *Industrial Biotechnology,* Vol. 1, 38-50.

Yoshida, S.; Watanabe, T.; Honda, Y. & Kuwahara, M. (1997). Effects of water-miscible organic solvents on the reaction lignin peroxidase of *Phanerochaete chrysosporium. Journal of Molecular Catalysis B: Enzymatic,* Vol. 2, 243-251.

Yang, Z. (2009). Hofmeister effects: an explanation for the impact of ionic liquids on biocatalysis. *Journal of Biotechnology,* Vol. 144, 12–22.

Yu, H-L; Ou, L.; Xu, J-H. (2010). New Trends in Non-Aqueous Biocatalysis. *Current Organic Chemistry,* Vol. 14, 1424-1432.

Zaragoza-Gascaa, P.; Villamizar-Galveza,O.J; García-Arrazolaa, R.; Gimenoa,M. & Bárzana, E. (2010). Use of ionic liquid for the enzyme-catalyzed polymerization of phenols. *Polymers Advance Technologies,* Vol. 21, 454–456.

Zhao, H. (2005). Effect of ions and other compatible solutes on enzyme activity, and its implication for biocatalysis using ionic liquids. *Journal of Molecular Catalysis B: Enzymatic,* Vol. 37, 16–25.

Zhou, G-P.; Zhang, Y.; Huang, X-R.; Shi, C-H.; Liu, W-F.; Li, Y-Z.; Qu, Y-B. & Gao, P-J. (2008). Catalytic activities of fungal oxidases in hydrophobic ionic liquid 1-butyl-3-methylimidazolium hexafluorophosphate-based microemulsion. *Colloids and Surfaces B: Biointerfaces,* Vol. 66 146-149.

22

Protease-Catalyzed Synthetic Reactions in Ionic Liquids

Hidetaka Noritomi
Tokyo Metropolitan University
Japan

1. Introduction

Enzymes are very effective and biodegradable catalysts, and act under mild conditions such as room temperature, atmospheric pressure, and around pH 7. Consequently, the applications of enzymes to the organic synthesis have extensively been studied from the standpoint of the development of sustainable synthetic processes (Truesdell, 2005). Enzymes exhibit the high activity and specificity not only in conventional aqueous solutions but also in non-aqueous reaction media (Klibanov, 2001; Noritomi et al., 2007a). Enzymes in non-aqueous media have especially been applied to numerous synthetic processes because of the following benefits, although the enzyme has the low activity in organic solvents compared with that in water: (i) the solubilities of non-polar reactants and products are improved; (ii) the thermostability of enzymes is highly improved; (iii) the stereoselectivity of enzymes is markedly altered; (iv) the thermodynamic equilibria of many processes such the formation of peptide bond by protease are favorable; (v) the immobilization of enzymes is not necessary, and enzymes are easily recycled by recovering them with the filtration or the centrifugation, since enzymes are insoluble in organic solvents; (vi) the product is easily recovered with the evaporation when using the volatile organic solvent as the reaction medium; (vii) the contamination due to the growth of microorganisms is inhibited by organic solvents. Furthermore, it has been found that the activity and selectivity of enzymes can be manipulated by the choice of solvents or enzyme preparation, although they were changed only by protein engineering or enzyme screening prior to the advent of nonaqueous enzymology (Koskinen & Klibanov, 1996; Noritomi et al., 1996; Wescott et al., 1996).

Proteases are stable hydrolytic enzymes with high selectivity, do not need expensive cofactors, and are used as useful synthetic tools for side-directed peptide cleavages, regiospecific ester hydrolyses, or the kinetic resolution of racemates (Bordusa, 2002). In order to improve undesired hydrolysis and substrate solubility in aqueous media, proteases have been used in organic solvents for organic synthesis. Proteases in organic solvents can catalyze water sensitive reactions, such as esterification, transesterification, and peptide synthesis, which are usually impossible in aqueous media.

Ionic solvent that is liquid at room temperature has attracted increasing attention as a green solvent for the chemical processes because of the lack of vapor pressure, the thermal stability, and the high polarity (Welton, 1999; Greaves & Drummond, 2008). Chemical and physical properties of ionic liquids can be changed by the appropriate modification of

Ionic liquid	Structure	m.p. (℃)	Water miscibility
[emim][FSI]	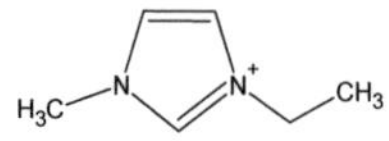 $[(FSO_2)_2N]^-$	-12.9	Partially miscible
[emim][Tf]	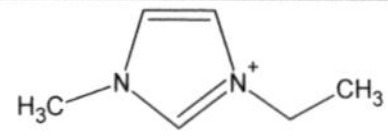 $[CF_3SO_3]^-$	-9	Miscible
[emim][BF_4]	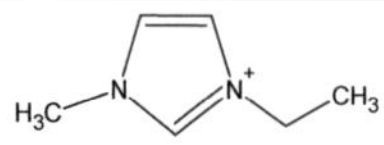 $[BF_4]^-$	14.6	Miscible
[emim][PF_6]	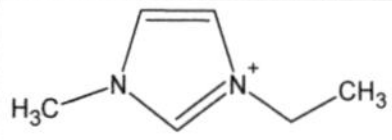 $[PF_6]^-$	62	Partially miscible

Fig. 1. Structures of ionic liquids.

organic cations and anions, which are constituents of ionic liquids. Biotransformation in ionic liquids has increasingly been studied (Moniruzzaman et al., 2010; Yang & Pan, 2005). However, despite such potential capability of proteases, there has been only a limited number of works on protease-catalyzed reactions in ionic liquids.

In this chapter, the effects of ionic liquids on esterification of amino acid with alcohol and peptide synthesis of amino acid ester with amino acid amide catalyzed by proteases are discussed (Noritomi et al., 2007b; Noritomi et al., 2009). The former is the reverse reaction of protease-catalyzed amino acid ester hydrolysis, whereas the latter is the aminolysis of amino acid ester.

2. Protease-catalyzed esterificatin of amino acid in ionic liquids

2.1 Activity of protease in ionic liquids

Figure 1 shows structures and properties of ionic liquids introduced in this chapter. 1-Ethyl-3-methylimidazolium bis(fluorosulfonyl)imide, 1-ethyl-3-methylimidazolium trifluoromethanesulfonate, 1-ethyl-3-methylimidazolium tetrafluoroborate, and 1-ethyl-3-methylimidazolium hexafluorophosphate are abbreviated to [emim][FSI], [emim][Tf], [emim][BF_4], and [emim][PF_6], respectively. Their properties such as melting point and water-miscibility alter by switching from one anion to another.

In order to synthesize amino acid esters by protease in non-aqueous media, there are two reaction processes: the reverse reaction process of protease-catalyzed amino acid ester hydrolysis via thermodynamical control and the transesterification process of amino acid ester with alcohol via kinetical control. The protease activity of the transesterification is much higher than that of the esterification in organic solvents, since esterified amino acids are activated substrates (Kise et al., 1990). On the other hand, acyl enzyme intermediate is formed from neutral carboxylic acid substrate in the esterification process. Consequently, the formation of *N*-acetyl-L-phenylalanine ethyl ester (*N*-Ac-L-Phe-OEt) via protease-catalyzed esterification of *N*-acetyl-L-phenylalanine (*N*-Ac-L-Phe-OH) with ethanol (equation (1)) is enhanced in water-miscible organic solvents, since the apparent pK value of carboxylic acid in *N*-Ac-L-Phe-OH is shifted higher.

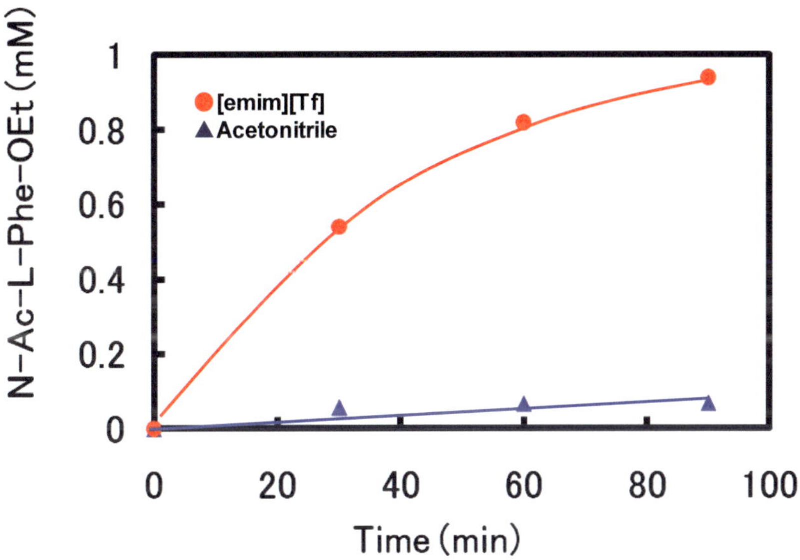

Fig. 2. Time course of subtilisin-catalyzed esterificaton in [emim][Tf] and acetonitrile.

$$N\text{-Ac-L-Phe-OH} + \text{EtOH} \rightleftharpoons N\text{-Ac-L-Phe-OEt} + H_2O \quad (1)$$

Moreover, enzymatic reactions in hydrophilic solvents have the advantage of the solubility of a variety of substrates, including amino acid derivatives, which are poorly soluble in hydrophobic solvents. [emim][Tf] is a water-miscible ionic liquid and suitable for the solubilities of *N*-acetylated amino acids, similar to a hydrophilic organic solvent, such as acetonitrile. Figure 2 shows the time courses of subtilisin-catalyzed esterificaton of *N*-acetyl-L-phenylalanine (*N*-Ac-L-Phe-OH) with ethanol in [emim][Tf] and acetonitrile containing 0.2% (v/v) water at 30 °C. The esterification in [emim][Tf] proceeded with the reaction time, and the concentration of *N*-acetyl-L-phenylalanine ethyl ester (*N*-Ac-L-Phe-OEt) in [emim][Tf] exhibited 14-fold compared to that in acetonitrile when the reaction time reached 90 min. When a hydrophilic solvent is used as a reaction medium, the enzyme molecule directly contacts with the solvent, and thereby its activity is strongly influenced by the nature of the solvent (Klibanov, 2001; Noritomi et al., 2007). From this point of view, subtilisin seems to be stable to [emim][Tf].

Figure 3 shows the initial rates of esterification of *N*-Ac-L-Phe-OH with ethanol catalyzed by free subtilisin and α-chymotrypsin in [emim][Tf] and organic solvents containing 0.2% (v/v) water at 30 °C. In organic solvent systems using subtilisin, the more hydrophobic solvent was used, the greater initial rate was obtained. The hydrophobicity is attributable to the partition of water between enzyme molecules and the bulk organic phase in reaction system (Klibanov, 2001; Zaks & Klibanov, 1988). When the same amount of water is added in different organic solvents, the more hydrophobic the reaction medium is, the more amount of water is located around the enzyme molecule. Consequently, the flexibility of the enzyme molecule, which is essential for catalytic activity, increases in hydrophobic solvents, and thereby high activity is exhibited. The initial rate in [emim][Tf] is superior to the initial rates in all organic solvents, and, for example, is about three times greater than that in octane. On the other hand, the initial rate of α-chymotrypsin in octane is also highest in the organic solvent system. Furthermore, the initial rate of α-chymotrypsin in [emim][Tf] is about fourteen times greater than that in octane. Adding Salts to an aqueous solution containing enzymes is an useful method for improving enzyme stability (Troller & Christian, 1978). Before using the enzyme powder as a suspension in an organic solvent, the inclusion of simple salts during lyophilization or freeze-drying of the enzyme is one of the most effective activation methods. For instance, lyophilized enzyme powder prepared from enzyme solution including excess KCl increases the catalytic efficiency of subtilisin-catalyzed transesterification by 3750-fold (Khmelnisky et al., 1994). Addition of buffer salts or KCl also increases the catalytic activity of lyophilized *Candida Antarctica* lipase 4-fold over that without added salt (Triantafyllou et al., 1997). Moreover, the lyophilizate including KCl enhances thermolysin-catalyzed peptide synthesis in *tert*-amyl alcohol (Bedell et al., 1998). These reports indicate that salts tend to induce activation of enzymes. Furthermore, an amino acid interacts with [emim], and [emim][amino acid] is formed as an ionic liquid (Fukumoto et al., 2005). In the present reaction system, the enhancement of enzymatic activity might be attributable to the interaction between [emim][Tf] and *N*-Ac-L-Phe-OH. As seen in Figs.3, the initial rate of α-chymotrypsin is greater than that of subtilisin in [emim][Tf] system, while the initial rate of α-chymotrypsin is smaller than that of subtilisin in organic solvent system. α-Chymotrypsin tends to prefer the atmosphere of high salt concentration (Martin & Niemann, 1958). As a result, the initial rate of α-chymotrypsin is effectively enhanced compared with that of subtilisin by changing from the organic solvent to [emim][Tf].

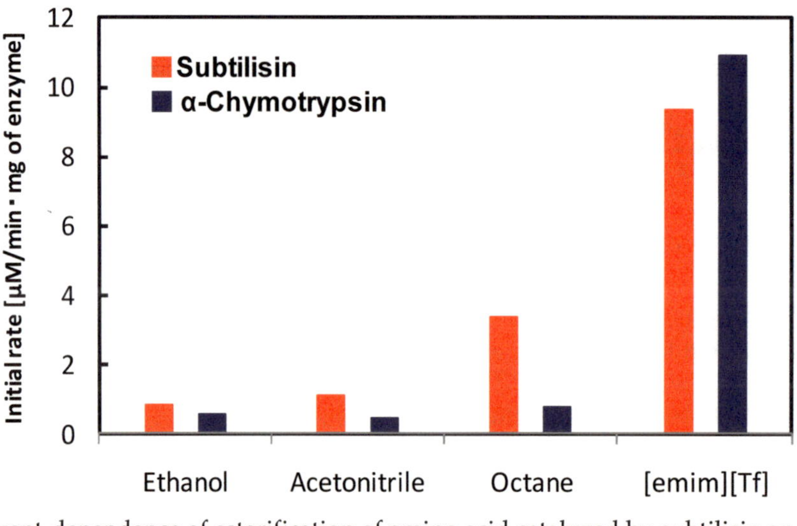

Fig. 3. Solvent-dependence of esterification of amino acid catalyzed by subtilisin and α-chymotrypsin.

2.2 Influence of water content on activity of protease in ionic liquids

A common thread in all studies of enzymes in organic solvents is that the amount of water associated with the enzyme is a key determinant of the properties (e.g. activity, stability, and specificity) that the enzyme exhibits (Klibanov, 2001). Moreover, water can act as a substrate in reactions using hydrolytic enzymes. On the other hand, the esterification of *N*-Ac-L-Phe-OH with ethanol is a reverse reaction of hydrolysis and is therefore thermodynamically controlled. These result in lower product yields in reaction media containing higher water content. However, the initial rate of subtilisin-catalyzed esterification of *N*-Ac-L-Phe-OH with ethanol in [emim][Tf] system at 5 % (v/v) water is enhanced 2.4-fold compared to that at 0.2 %(v/v) water. Similarly, the initial rate in acetonitrile system at 5 %(v/v) water is about seven times greater than that at 0.2 %(v/v) water. When a certain amount of water is added into the non-aqueous enzymatic reaction system, some water is bound to the enzyme, and thereby has a large influence on the enzymatic activity, while the other amount of water is dissolved in the solvent (Klibanov, 2001). Water associated with the enzyme activates the enzyme by increasing the internal flexibility of the enzyme molecule, since water acts as a plasticizer to increase the flexibility (Zaks & Klibanov, 1988). The water content is an influential parameter for the activity in [emim][Tf] system, similar to the case in acetonitirile system.

3. Protease-catalyzed peptide synthesis in ionic liquids

3.1 Protease-catalyzed peptide synthesis

Peptides are molecules of great importance in the pharmaceutical and food fields (Guzman et al., 2007; Gill et al., 1996). Several technologies for peptide synthesis are available: a) chemical synthesis; b) enzymatic synthesis; c) recombinant DNA technology. However, the application of recombinant DNA technology requires a long and expensive research and development phase. The chemical peptide synthesis has some problems such as racemization during peptide bond formation, the requirement of extensive protection of the side chain functionalities of amino acids, and the use of a large excess of coupling reagents and acyl donors. Proteases have been used as a biocatalyst in order to carry out peptide synthesis since 1938 (Bordusa, 2002). The advantages of protease-catalyzed peptide synthesis are freedom from racemization, minimal requirements for carboxyl activation and side-chain protection, mild reaction conditions, and high region- and stereoselectivity.

For example, α-chymotrypsin-catalyzed peptide synthesis of *N*-acetyl-L-tryptophan ethyl ester (*N*-Ac-Trp-OEt) with glycyl glycinamide (Gly-Gly-NH_2) is shown in Fig. 4. When using

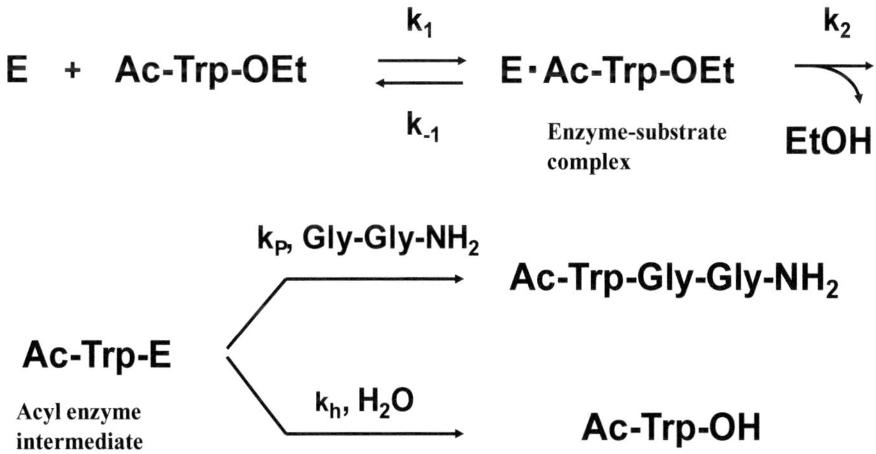

Fig. 4. α-Chymotrypsin-catalyzed peptide synthesis of *N*-acetyl-L-tryptophan ethyl ester (*N*-Ac-Trp-OEt) with glycyl glycinamide (Gly-Gly-NH_2) to *N*-acetyl-L-tryptophan glycyl glycinamide (*N*-Ac-Trp-Gly-Gly-NH_2) and the competing hydrolysis (*N*-Ac-Trp-OH).

amino acid esters as a substrate, the peptide synthesis is a kinetically controlled reaction process. This process involves the competitive distribution of the rapidly formed acyl enzyme intermediate between water (hydrolysis) and another nucleophilic reagent such as an amino acid derivative (peptide synthesis). Since the nucleophilic reaction is rate-determining step, and k_1, k_{-1}, and k_2 are much greater than k_p and k_h (Fersht, 1999), the initial rates are shown as

$$V_p = k_p[\text{Ac-L-Trp-E}][\text{Gly-Gly-NH}_2] \quad (2)$$

$$V_h = k_h[\text{Ac-L-Trp-E}][H_2O] \quad (3)$$

where V_p is the initial rate of peptide synthesis, V_h the initial rate of hydrolysis, k_p the rate constant of peptide synthesis, and k_h the rate constant of hydrolysis. From equations (2) and (3), the selectivity (k_p/k_h) is derivated as the following equation.

$$k_p/k_h = V_p[H_2O]/ V_h[\text{Gly-Gly-NH}_2] \quad (4)$$

3.2 Dependence of α-chymotrypsin-catalyzed peptide synthesis on water content in [emim][FSI]

Figure 5 shows the plot of the initial rates of *N*-acetyl-L-tryptophan glycyl glycinamide (*N*-Ac-Trp-Gly-Gly-NH_2) and *N*-acetyl-L-tryptophan (*N*-Ac-Trp-OH) and the selectivity against the water content in [emim][FSI] at 25 °C. The inherent enzymatic hydrolytic reaction is

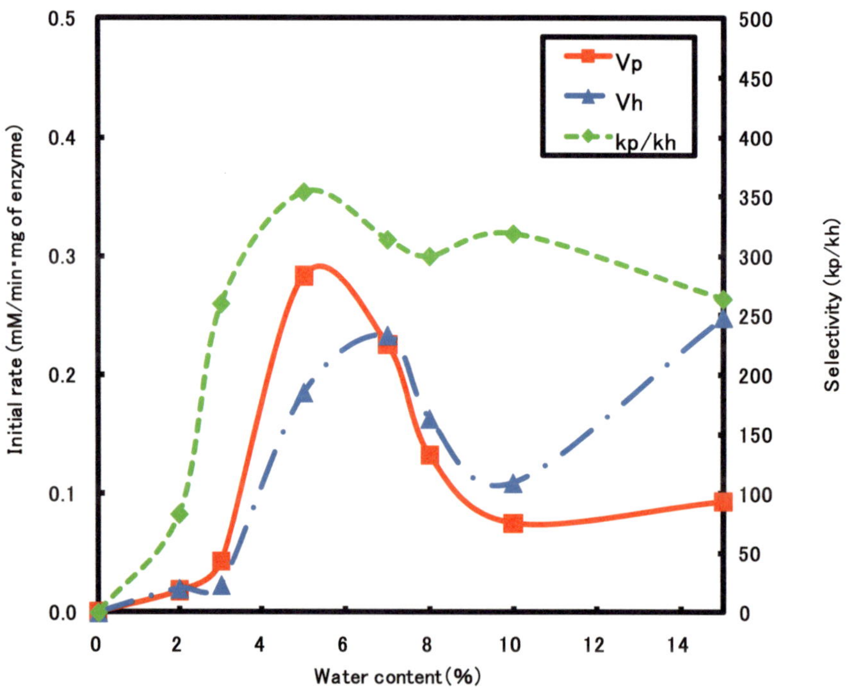

Fig. 5. Effect of water content on α-chymotrypsin-catalyzed peptide synthesis in [emim][FSI].

inhibited by low water content, while the enzymatic peptide synthesis is promoted. The initial rates of *N*-Ac-Trp-Gly-Gly-NH_2 and *N*-Ac-Trp-OH vs. the water content display a bell-shaped curve below 10% (v/v) water. Below 10%(v/v) water, the peptide and hydrolysis profiles are similar to those in acetonitirile and acetonitrile/supercritical carbon dioxide systems (Kise et al., 1988; Noritomi et al., 1995). The relationship between the activity and water content tends to exhibit bell-shaped curve. The optimal water content is due to the balance between kinetic rigidity and thermodynamic stability of enzyme structures, and is called essential water (Klibanov, 1986). The kinetic rigidity is relaxed by increasing water content, while native enzyme structure gradually changes through thermodynamic stability. Thus, the activity increases with an increase in the flexibility of rigid enzyme, and it decreases with an increase in disturbance of enzyme structure. On the other hand, above 10%(v/v) water the hydrolysis increased with an increase in water content. It is suggested that above 10%(v/v) water the amount of water exceeds the solubility of water in the system, the distribution of water is enhanced around the enzyme, and the hydrolysis is enhanced. As the optimum initial rate and selectivity of peptide synthesis were observed at 5% (v/v) water, the enzyme reaction experiments shown below were carried at 5% (v/v) water.

3.3 Peptide synthesis catalyzed by α-chymotrypsin in ionic liquids

Peptide synthesis is advantageous not in hydrophobic organic solvents but in hydrophilic organic solvents due to the solubility of amino acid derivatives (Kise et al., 1990). As seen in Table 1, both the activity and selectivity of peptide synthesis by free α-chymotrypsin in [emim][FSI] were much superior to those in [emim][BF_4] and [emim][Tf], although [emim][BF_4] and [emim][Tf] are more hydrophilic than [emim][FSI] as shown in Fig.1. [emim][FSI] has a good solubility of amino acid derivatives. It has been well known that ionic liquids can dissolve many organic and inorganic compounds, since they exhibit a wide range of intermolecular interactions by the design of ionic liquids (Moniruzzaman et al., 2010). On the other hand, enzymes are generally more stable in hydrophobic solvents than in hydrophilic solvents. Moreover, it has been reported that the activity of *Candida antarctica* lipase B in a transesterification reaction is strongly dependent upon a sort of anion consisting of the same cation when using an ionic liquid as a solvent (Lau et al., 2004).

Solvent	Initial rate (μM/min · mg of enzyme)		Selectivity (-)
	Peptide	Hydrolysate	
[emim][FSI]	280	180	360
[emim][BF_4]	0.031	0.093	77
[emim][Tf]	0	1.8	0

Table 1. Initial rates of *N*-Ac-Trp-Gly-Gly-NH_2 and *N*-Ac-Trp-OH in the peptide synthesis of *N*-Ac-Trp-OEt with Gly-Gly-NH_2 catalyzed by free α-chymotrypsin in several kinds of ionic liquids containing 5% (v/v) water at 25 °C

3.4 The solvent dependence of peptide synthesis catalyzed by α-chymotrypsin

As seen in Fig. 3, enzymatic activity of esterification of amino acid markedly depends upon the nature of solvents. Figure 6 shows the initial rates of peptide synthesis and hydrolysis and the selectivity catalyzed by free α-chymotrypsin in ionic liquid, [emim][FSI] and conventional organic solvents at 5% (v/v) water and 25 °C. The initial rates of peptide synthesis and hydrolysis tended to be dramatically dependent upon a kind of solvents. The reactivity in hydrophilic organic solvents was observed as shown in Fig.6, while that in octane was hardly observed due to low solubility of substrates. Enzymatic reactions in hydrophilic organic solvents have the advantage of the solubility of amino acid derivatives, which are poorly soluble in hydrophobic solvents. The initial rate of peptide synthesis in [emim][FSI] was superior to that in organic solvents, and, for example, was about six times greater than that in tetrahydrofuran. The conformation of enzymes in hydrophilic solvents is remarkably rigid, and disturbs the induced-fit process of enzyme reaction, since essential water is stripped off from enzymes by hydrophilic solvents. As a result, the enzymatic activity in hydrophilic organic solvents is low, compared to that in [emim][FSI]. Moreover, the powder of α-chymotrypsin was finely dispersed in [emim][FSI], and the reaction system was visually transparent, while the enzyme powder was suspended in organic solvent systems. In the heterogeneous catalytic system the apparent rate corresponds upon the rate per unit area of catalysts multiplied by the surface area. The small surface area of enzyme powder resulted in the low initial rate in organic solvents. The sufficient reactivity observed in [emim][FSI] might be attributable to the improvement of the surface area of an enzyme powder due to the fine dispersion.

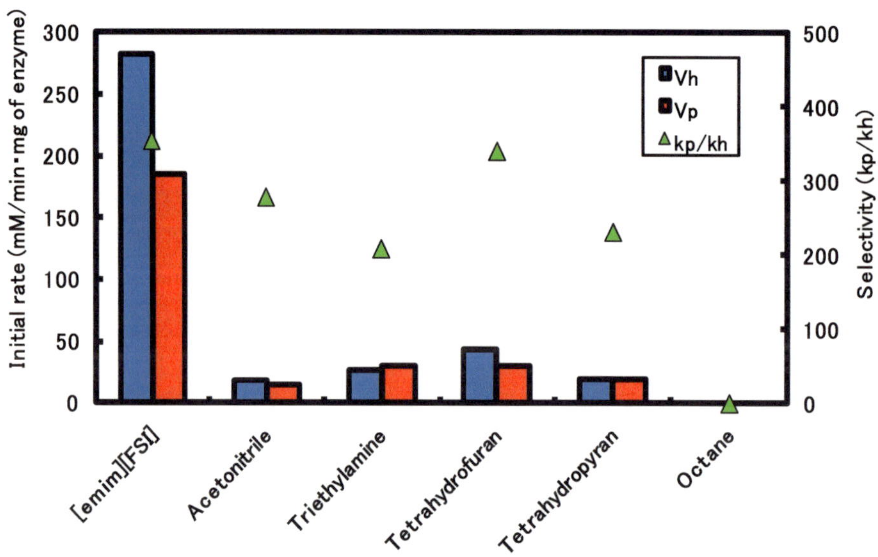

Fig. 6. Solvent-dependence of α-chymotrypsin-catalyzed peptide synthesis.

3.5 The temperature dependence of peptide synthesis catalyzed by α-chymotrypsin in [emim][FSI] and organic solvents

Enzymatic reactions, as well as chemical reactions, obey the Arrhenius correlation between reaction rate constant and temperature, although the temperature range is quite limited. Figure 7 shows the plots of initial rates of peptide synthesis in [emim][FSI] and organic solvents containing 5% (v/v) water against temperature. The initial rate of peptide synthesis in [emim][FSI] exhibited a maximum around 30 °C, and then decreased with an increase in

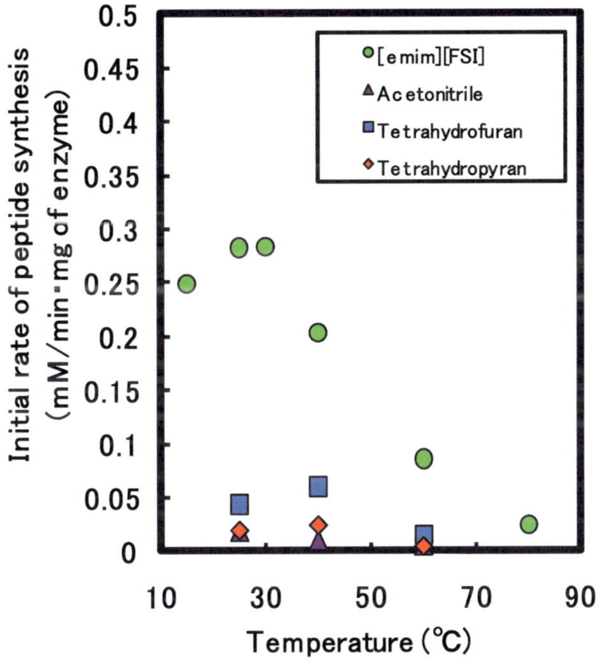

Fig. 7. Temperature-dependence of α-chymotrypsin-catalyzed peptide synthesis.

temperature. Likewise, the initial rates in tetrahydrofuran and tetrahydropyran showed a maximum around 40 °C. According to the Arrhenius equation, the initial rate becomes high at a high temperature. However, the thermal denaturation of enzymes proceeds reversibly and/or irreversibly. Consequently, the profile of activity has an optimal temperature. Comparing [emim][FSI] system with organic solvent systems, the initial rate of peptide synthesis in [emim][FSI] was 6-fold superior to that in tetrahydrofuran at 60 °C, and was seventeenth times greater than that in acetonitrile and tetrahydropyran at 60 °C. On the other hand, thermal denaturation of enzymes in aqueous solutions begins at 45 to 50 °C and is severe. The activity of α-chymotrypsin in aqueous solutions was not observed at 60 °C. As the temperature increases, the atoms in the enzyme molecule have greater energies and a greater tendency to move, acquire sufficient energy to overcome the weak interactions keeping the tertiary structure of the enzyme molecule, and denaturation follows. The thermostability of enzymes in anhydrous media is highly improved, compared with that in aqueous solutions, since the thermal denatuaration of enzymes is due to water (Klibanov, 2001). In other words, the disturbance of the atoms in the enzyme molecule tends to be relaxed, since the enzyme molecule in limited water content is rigid. Furthermore, it has been reported that *Candida antarctica* lipase B is stabilized by ionic liquids in ester synthesis, since ionic liquids maintain the enzyme conformation into the ionic net, and work as both immobilization support and reaction media (Lozano et al., 2001). As shown in Table 2, the activity of α-chymotrypsin in peptide synthesis in [emim][FSI] and [emim][PF_6] was observed, although the initial rates of peptide and hydrolysate at 80 °C were about one-tenth smaller than those at 25 °C in [emim][FSI]. On the other hand, the production of peptides or hydrolysates by enzyme reaction was not shown in [emim][BF_4] and [emim][Tf].

Solvent	Initial rate (μM/min・mg of enzyme)		Selectivity (-)
	Peptide	Hydrolysate	
[emim][FSI]	25	20	290
[emim][PF_6]	22	61	83
[emim][BF_4]	0	0	-
[emim][Tf]	0	0	-

Table 2. Initial rates of *N*-Ac-Trp-Gly-Gly-NH_2 and *N*-Ac-Trp-OH in the peptide synthesis of *N*-Ac-Trp-OEt with Gly-Gly-NH_2 catalyzed by free α-chymotrypsin in several kinds of ionic liquids containing 5% (v/v) water at 80 °C

4. Conclusion

In this chapter the effects of ionic liquids on the catalytic behavior and stability of protease have been described. The activities of protease in both esterification of amino acid and peptide synthesis in ionic liquids were superior to those in conventional organic solvents. The activity in ionic liquid systems was dependent upon the nature of solvent, similar to the case of organic solvent systems, and the tendency in ionic liquid systems seemed to be more sensitive than that in organic solvent systems, since the activity markedly altered by switching from one anion to another of an ionic liquid consisting of the same cation. Concerning the thermostability, the same tendency was exhibited. These indicate that the constituents of ionic liquids directly affect protease, compared to the case of organic solvent systems. For instance, it is apparent that [emim][FSI] functions as a thermal stabilizing agent. When using protease in non-aqueous reaction media, reaction media having the properties of hydrophobicity, good solubility of amino acid derivatives, and the fine dispersion of enzyme are desirable. The conventional organic solvents such as acetonitrile lack these properties, while [emim][FSI] has these properties. Moreover, it is expected that the ionic liquid, which is more suitable for the catalytic behavior and stability of protease, is prepared by tailoring the constituents of ionic liquids.

5. References

Bedell, B.A.; Mozhaev, V.V.; Clark, D.S. & Dordick, J.S. (1998). Testing for diffusion limitations in salt-activated enzyme catalysts operating in organic solvents. *Biotechnol. Bioeng.*, 58, 654-657.

Bordusa, F. (2002). Proteases in organic synthesis. *Chem. Rev.*, 102, 4817-4867.

Fersht, A. (1999). *Structure and Mechanism in Protein Science: A Guide to Enzyme Catalysis and Protein Folding*, W. H. Freeman and Company, NY.

Fukumoto K.; Yoshizawa M. & Ohno H. (2005). Room temperature ionic liquids from 20 natural amino acids. *J. Amer. Chem. Soc.*, 127, 2398-2399.

Gill, I.; Lopez-Fandino, R.; Jorba, X. & Vulfson, E.N. (1996). Biologically active peptides and enzymatic approaches to their production. *Enz. Microbial Technol.*, 18, 162-183.

Greaves, T.L. & Drummond, C.J. (2008). Protic ionic liquids: Properties and applications. *Chem. Rev.*, 108, 206-237.

Guzman, F.; Barberis, S. & Illanes, A. (2007). Peptide synthesis: chemical or enzymatic. *Elect. J. Biotechnol.*, 10, 279-314.

Khmelnitsky, Y.L.; Welsh, S.H.; Clark, D.S. & Dordick, J.S. (1994). Salts dramatically enhance activity of enzymes suspended in organic solvents. *J. Am. Chem. Soc.*, 116, 2647-2648.

Kise, H.; Fujimoto, K. & Noritomi, H. (1988). Enzymatic reactions in aqueous-organic media. VI. Peptide synthesis by α-chymotrypsin in hydrophilic organic solvents. *J. Biotechnol.*, 14, 279-290.

Kise, H.; Hayakawa, A. & Noritomi, H. (1990). Protease-catalyzed synthetic reactions and immobilization-activation of the enzymes in hydrophilic organic solvents. *J. Biotechnol.*, 14, 239-254.

Klibanov, A.M. (1986). Enzymes that work in organic solvents. *CHEMTECH*, 16, 354-359.

Klibanov, A.M. (2001). Improving enzymes by using them in organic solvents. *Nature*, 409, 241-246.

Koskinen, A.M.P. & Klibanov, A.M. (Eds.) (1996). *Enzymatic Reactions in Organic Media*, Blackie A & P, London.

Lau, R.M.; Sorgedrager, M.J.; Carrea G.; van Rantwijk, F.; Secundo, F. & Sheldon, R.A. (2004). Dissolution of *Candida antarctica* lipase B in ionic liquids: effects on substrate and activity. *Green. Chem.*, 6, 483-487.

Lozano, P.; De Diego, T.; Carrie, D.; Vaultier, M. & Iborra, J.L. (2001). Over-stabilization of *Candida antrarctica* lipase B by ionic liquids in ester synthesis. *Biotechnol. Lett.*, 23, 1529-1533.

Martin, R.B. & Niemann, C. (1958). The effect of various salts on the α-chymotrypsin-catalyzed hydrolysis of two acylated α-amino acid esters. *J. Am. Chem. Soc.*, 80, 1481-1486.

Moniruzzaman, M.; Nakashima, K.; Kamiya, N. & Goto, M. (2010). Recent advances of enzymatic reactions in ionic liquids. *Biochem. Eng. J.*, 48, 295-314.

Noritomi, H.; Miyata, M.; Kato, S. & Nagahama, K. (1995). Enzymatic synthesis of peptide in acetonitrile/supercritical carbon dioxide. *Biotechnol Lett.*, 17, 1323-1328.

Noritomi, H.; Almarsson, O.; Barletta, G. L. & Klibanov, A.M. (1996). The influence of the mode of enzyme preparation on enzymatic enantioselectivity in organic solvents and its temperature dependence. *Biotechnol. Bioeng.* 51, 95-99.

Noritomi, H.; Sasanuma, A.; Kato, S. & Nagahama, K. (2007a). Catalytic properties of cross-linked enzyme crystals in organic media. *Biochem. Eng. J.*, 33, 228-231.

Noritomi, H.; Nishida, S. & Kato, S. (2007b). Protease-catalyzed esterification of amino acid in water miscible ionic liquid. *Biotechnol Lett.*, 29, 1509-1512

Noritomi, H.; Suzuki, K.; Kikuta, M. & Kato, S. (2009). Catalytic activity of α-chymotrypsin in enzymatic peptide synthesis in ionic liquids. *Biochem. Eng. J.*, 47, 27-30.

Triantafyllou, A.O.; Wehtje, E.; Adlercreutz, P. & Mattiasson, B. (1997). How do additives affect enzyme activity and stability in nonaqueous media? *Biotechnol. Bioeng.*, 54, 67-76.

Troller, J.A. & Christian, J.H.B. (1978). *Water Activity and Food*, Academic Press Inc., London.

Truesdell, S. (2005). Biotransformations for a green future. *Chem. Eng. Prog.*, 101, 44-47.

Welton, T. (1999). Room-temperature ionic liquids. Solvents for synthesis and calalysis. *Chem. Rev.*, 99, 2071-2083.

Wescott, C.R.; Noritomi, H. & Klibanov, A.M. (1996). Rational control of enzymatic enantioselectivity through solvation thermodynamics. *J. Am. Chem. Soc.*, 118, 10365-10370.

Yang, Z. & Pan, W. (2005). Ionic liquids: Green solvents for nonaqueous biocatalysis. *Enz. Microbial Technol.*, 37, 19-28.

Zaks, A. & Klibanov, A.M. (1988). Enzymatic catalysis in nonaqueous solvents. *J. Biol. Chem.*, 263, 3194-3201.

23

Perdeuterated Pyridinium Ionic Liquids for Direct Biomass Dissolution and Characterization

Nan Jiang and Arthur J Ragauskas
Georgia Institute of Technology
USA

1. Introduction

Fossil fuels, as the primary source of energy, chemicals, and materials for our modern society, are estimated to consist almost 90% of the energy consumption in the world in the resource such as petroleum, coal, and natural gas, particularly for the transportation sector (Simonetti & Dumesic, 2008). However, the use of fossil fuels for the generation of energy is associated with growing concerns related to energy security (the uneven geographical distribution of the fossil fuel reserves) and climate change. Therefore, ever-increasing societal demand for environmental and economic sustainability is placing a renewed focus on the agro-forest industry, due to, among others, the unquestionable environmental advantage associated with the reduction of the net emissions of CO_2 (a greenhouse gas) into the atmosphere (Clark et al., 2006; Ragauskas et al., 2006; Petrus & Noordermeer, 2006; Horvath and Anastas, 2007; Corma et al., 2007; Pu et al., 2008). Unlike fossil fuels, biofuels are widely considered to be "carbon neutral" because any CO_2 produced during fuel combustion can be consumed by subsequent biomass regrowth. In this sense, the aggressive government directives from U.S. and the European Union, are stimulating a gradual shift of our current fossil fuel-based economy towards a more renewable energy-based one to achieve the goals to produce 20% of fuels from renewable sources by 2030 (Serrano-Ruiz et al., 2010).

While 1st-generation biorefineries have successfully utilized readily precessable bioresources such as sucrose, starches and plant oils for bioethanol/biodiesel (Bothast & Schlicher, 2005; Marchetti et al., 2007; Huber & Corma, 2007), important moral and ethical questions have arisen from the use of these edible biomass (sugars, starches, and vegetable oils) for the large-scale production of biofuels. Therefore, more efficient utilization of renewable and low-cost nonedible biomass (lignocellulosic biomass) to maximize sustainable, economic societal development is of extreme significance.

However, intrinsic recalcitrance of biomass, known as resistance of the plant cell walls to break down due to its complicated structure (Cosgrove, 2005; Wyman et al., 2005; Himmel et al., 2007), has limited economical transformation of lignocellulosic biomass into biofuels. To address this challenge, extensive research has been directed towards a better understanding of plant cell wall structures and their constituents (Lewis & Paice, 1989; Jung et al., 1993), consisting of cellulose, hemicelluloses and lignin (Figure 1).

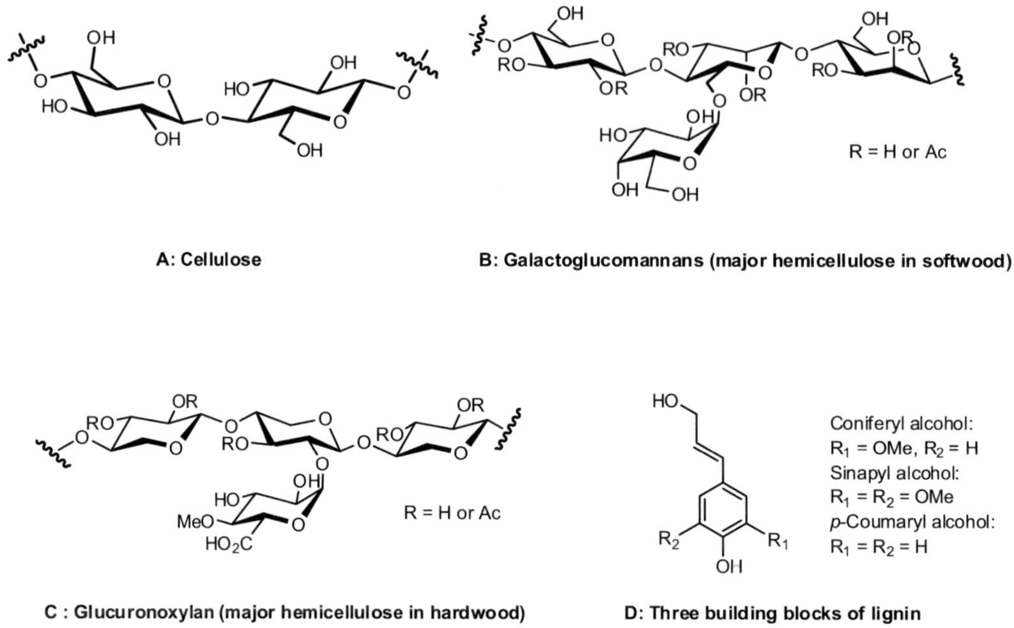

Fig. 1. The structures of three major biopolymers (**A-D**) of the plant cell walls

Among the cell walls' composite materials, lignin (see Fig. 1D), as the second most abundant natural polymer after cellulose and the natural glue for the plant cell walls, is produced by enzyme-mediated radical coupling of the three monolignols (Higuchi, 1985; Boerjan et al., 2003) and widely recognized as the major factor of biomass recalcitrance problem. With recent developments in the genetic engineering of lignin's biosynthetic pathway (Pena & Seguin, 2001; Baucher et al., 2003; Sticklen, 2006), new avenues to rationally designing bio-energy plants with reduced recalcitrance and thus improved processing properties has been opened for the next generation of biorefineries. As a large number of new plants with differing variations in lignin structure are developed the need for the efficient, economical and high-resolution analysis of the plant cell walls, particularly lignin (Davin & Lewis, 2005), becomes a pressing research issue.

Recently, 4:1 mixture of DMSO-d_6 and 1-methylimidazole-d_6 has been reported to dissolve non-derivatized ball-milled plant cell walls for a 2D NMR analysis, which provides a direct and accurate approach for the characterization of lignin and hemicellulose in plant biomass and was applied to samples of loblolly pine, quaking aspen and kenaf (Yelle et al., 2008). This approach is notable due to avoiding the need for lengthy, low-yielding lignin extraction protocols. Key to liquid NMR analysis of plant cell material is the efficient dissolution of ball-milled plant cell walls in a perdeuterated NMR solvent system.

Widely recognized as environmentally-friendly solvents, ionic liquid has been extensively documented for a variety of applications, due to its tunable physicochemical properties, including negligible vapor pressures and thermal/chemical stability over a wide range of temperatures (Welton, 1999). In addition to their role as green alternative reaction media (Wasserscheid & Keim, 2000; Sheldon, 2001), Rogers and co-workers first reported that ionic liquid 1-*n*-butyl-3-methylimidazolium chloride ([Bmim]Cl, see 1a in Fig. 2) can dissolve cellulose without activation (Swatloski et al., 2002). We and others have shown that various imidazolium-based ionic liquids can efficiently dissolve lignin (Pu et al., 2007), wood (Kilpelaeinen et al., 2007; Fort et al., 2007) and other biopolymers (Zhu et al., 2006; Fukaya et al., 2008; Zakrzewska et al., 2010). Despite wide applications of imidazolium-typed ionic liquids (see 1 in Fig. 2) as the solvents for biomass dissolution and functionalization, no perdueterated imidazolium ionic liquids has been reported in biomass NMR characterization due to their tedious multi-step synthesis (Hardacre et al., 2001; Giernoth et al., 2008).

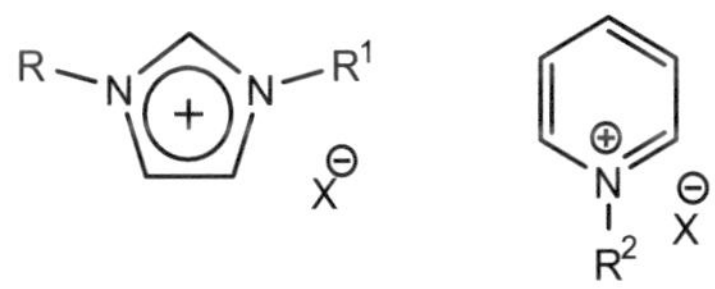

1 **Imidazolium salt** **2** **Pyridinium salt**

R = Me, Allyl; **R**[1] = Me, Et, Bu, Bz, Allyl
X = Cl, OAc, and $(MeO)_2PO_2$
1a : **R** = Me, **R**[1] = Bu, **X** = Cl
2a : **R**[2] = Allyl, **X** = Cl
2b : **R**[2] = Allyl, **X** = Br
2c : **R**[2] = Bu, **X** = Cl
2d : **R**[2] = Cyanomethyl, **X** = Cl
2e : **R**[2] = H, **X** = Cl

Fig. 2. Structures of imidazolium and pyridinium salts for biomass dissolution

Hence, our research in this area is aimed at direct NMR analysis of the plant cell walls via readily accessible perdeuterated ionic liquids, maximizing accuracy and efficiency of characterization by dissolution of Wiley-milled biomass. In this chapter, we shall describe novel perdeuterated ionic liquid for direct dissolution and NMR analysis of the plant cell walls. Our starting point was the use of pyridinium-typed ionic liquids (2 in Fig. 2), which had not been reported for biomass dissolution, despite the easy accessibility of perdeuterated pyridinium salts via commercially available pyridine-d_5.

We started our search for a suitable perdeuterated pyridinium salt for direct biomass dissolution and NMR analysis by screening five pyridinium ionic liquids (see 2a-e in Fig. 2) with ball-milled poplar as the initial sample. Due to the pyridinium salt's high melting point, we also used DMSO-d_6 as the co-solvent to render the sample to be dissolved at a lower temperature and to reduce viscosity of the resulting biomass solution for NMR analysis. For example, 35 mg ball-milled poplar can be dissolved in 1.0 gram 1 : 2 [Apyr]Cl (1-allylpyridinium chloride, 2a) and DMSO-d_6 at 60 °C under nitrogen in 1 h as the suspended poplar particles disappeared and a homogeneous solution formed with increased viscosity and slightly coloration of the solution (entry 1 in Table 1). Further ^1H and ^{13}C NMR analysis of the dissolved poplar solution supported this claim as the spectra data contained siganls readily attributed to lignin and polysaccharides. However, the use of [Apyr]Br (1-allylpyridinium bromide, 2b) to replace [Apyr]Cl proved to be unsuccessful under identical conditions (<5 mg/g, entry 2 in Table 1). In contrast, [Bpyr]Cl (1-*n*-butylpyridinium chloride, 2c) showed similar biomass solubility as [Apyr]Cl (entry 3 in Table 1), and both [Cmpyr]Cl (cyanomethylpyridinium chloride, 2d) and [Hpyr]Cl (pyridinium chloride, 2e) led to excellent results (70 mg/g and 80 mg/g, entries 4 and 5 in Table 1), which is superior to traditional [Bmim]Cl (1-*n*-butyl-3-methylimidazolium chloride, 1a in Fig. 2) under identical conditions (entry 6 in Table 1).

It is therefore clear that pyridinium typed ionic liquids with chloride as the anion show good biomass solubility in the presence of DMSO-d_6 as co-solvent. Furthermore, chloride anion is crucial for our present system, which agrees with Rodgers' speculations that chloride is highly effective in interupting the extensive hydrogen-bonding cellulose network

Entry	Ionic Liquid (IL)	IL : DMSO-d_6	Biomass	Time (h)	Solubility [a] (mg/g)
1	[Apyr]Cl **2a**	1:2	Poplar[b]	1	35
2	[Apyr]Br **2b**	1:2	Poplar[b]	6	<5
3	[Bpyr]Cl **2c**	1:2	Poplar[b]	6	30
4	[Cmpyr]Cl **2d**	1:2	Poplar[b]	1	70
5	[Hpyr]Cl **2e**	1:2	Poplar[b,c]	1	80
6	[Bmim]Cl **1a**	1:2	Poplar[b]	1	40
7	None	-	Poplar[b,c]	6	-
8	[Hpyr]Cl **2e**	1:2	Poplar[d]	6	80
9	[Hpyr]Cl-d_6 **2e'**	1:2	Poplar[b]	1	80
10	[Hpyr]Cl-d_6 **2e'**	1:2	Poplar[d]	6	80
11	[Hpyr]Cl-d_6 **2e'**	1:2	Switchgrass[d]	6	100
12	[Hpyr]Cl-d_6 **2e'**	1:2	Pine[d]	6	60

[a] In all experiments dry biomass dispersed in 1.0 g bi-solvent system consisting of pyridinium salt and DMSO-d_6 was stirred at 60 °C under nitrogen for specific time except where indicated. [b] Sample was the ball-milled biomass. [c] The mixture was heated at 80 °C. [d] Sample was non-ball-milled biomass (20 mesh).

Table 1. Solubility of biomass in ionic liquid/DMSO-d_6 bi-solvent system.

presented in plant cell walls (Swatloski et al., 2002). Control experiments showed that the absence of ionic liquid led to no dissolution of poplar in DMSO-d_6 even under elevated temperature and prolonged heating, as confirmed by direct NMR analysis. Therefore, the presence of pyridinium chloride salt is crucial for the successful dissolution of the ball-milled sample.

A shortcoming of using ball-milled sample is that ball-milling processing has been reported to induce some chemical changes in the structure of lignin (Kondo et al., 1995; Ikeda et al., 2002;), which will affect the accuracy of native lignin characterization. Other shortcoming includes that ball-milled process takes up to 7 days and is less efficient if a large number of samples need to be characterized. In this context, we next investigated the non-ball-milled poplar sample (an average particle size of 20 mesh by Wiley mill, 1041 um) for direct dissolution and characterization in 1 : 2 [Hpyr]Cl/DMSO-d_6 bi-solvent system. Paralleled solubility of Wiley milled sample was obsvered through a prolonged time of 6 hours (entry 8 in Table 1). At this stage, perdeuterated [Hpyr]Cl-d_6 **2e'** was prepared from pyridine-d_5 (see Fig. 3a), and the perdeuterated bi-solvent system was tested for direct dissolution and NMR analysis of ball-milled and Wiley milled poplar sample (entries 9 and 10 in Table 1). Furthermore, besides hardwood poplar, both Wiley-milled grass sample (switchgrass, 20 mesh, average partical size : 1041 um) and Wiley milled softwood sample (pine, 20 mesh, average partical size : 1041 um) showed very good solubility in our current perdeuterated bi-solvent system (entries 11 and 12 in Table 1). Therefore, our current perdeuterated pyridinium chloride-d_6/DMSO-d_6 bi-solvent system proved to be effective for successful dissolution of hardwood (poplar), softwood (pine) and grass (swicthgrass) samples.

^1H NMR Spectra of both ball-milled and Wiley milled poplar solution in [Hpyr]Cl-d_6/DMSO-d_6 system, as shown in Fig. 3b and 3c, serve to illustrate signals from both lignin aromatic/olefinic protons (between 6.0 ppm and 8.0 ppm) and polysaccharides. Similar ^1H NMR spectrum was also found for Wiley-milled switchgrass sample in Figure 3d.

Both ball-milled and Wiley milled poplar samples were then subjected to ^{13}C NMR analysis as presented in Fig. 4a and 4b, which showed paralleled spectra. The signals at δ 61.5, 74.1, 75.8, 76.9, 80.1, and 103.0 ppm were attributed to cellulose and strong signals at δ 57.3 ppm were assigned as the lignin methoxy ^{13}C resonance. Some detailed aliphatic sub-lignin structures could also be identified as C_γ resonance in β-O-4 and β-5 (62.2 and 64.0 ppm), C_β in β-O-4 and C_α resonance in β-5 and β-β (between 80 and 86 ppm). At the same times, signals from sub-lignin aromatic rings, such as resonance of syringyl-like lignin structures at 106.0 ppm for C2/6 and C2, C5 and C6 resonance of guaiacyl-like lignin structures between 110.0 and 120.0 ppm can also be clearly identified in ^{13}C NMR spectra (cellulose, β-O-4, β-β, syringyl-like lignin structures, and guaiacyl-like lignin structures see Fig. 5). In addition, ^{13}C NMR analysis of Wiley milled switchgrass revealed similar spectral data as shown in Fig. 4c.

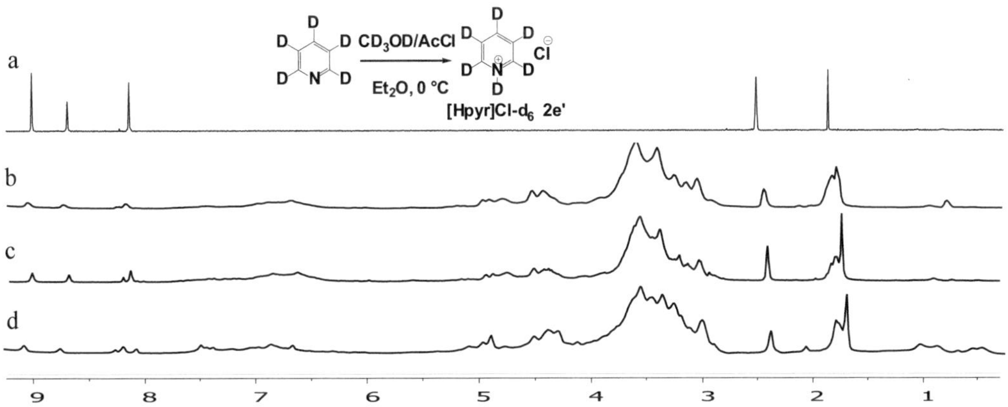

Fig. 3. (a) Preparation and ^1H NMR spectrum of [Hpyr]Cl-d_6 and ^1H NMR spectrum of biomass solution after dissolving biomass sample in 1:2 [Hpyr]Cl-d_6/DMSO-d_6 (1.0 g) at 60 °C: (b) Ball-milled poplar (80 mg) in 1:2 [Hpyr]Cl-d_6/ DMSO-d_6; (c) Wiley milled poplar (80 mg) in 1:2 [Hpyr]Cl-d_6/ DMSO-d_6; (d) Wiley milled switchgrass sample (100 mg).

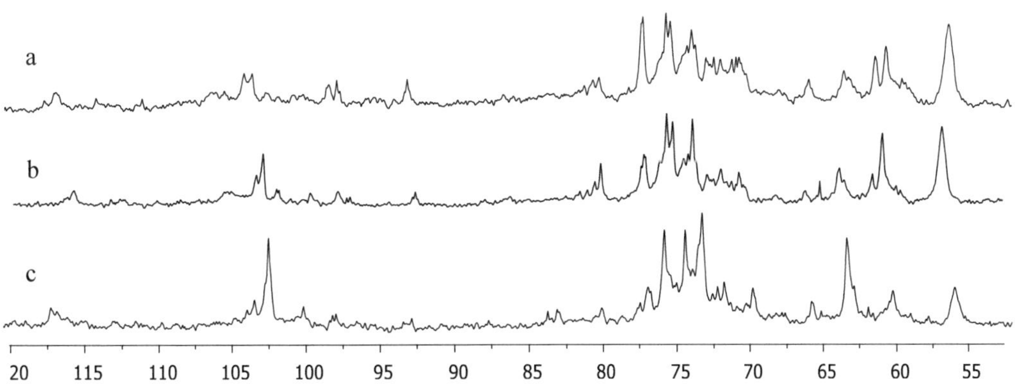

Fig. 4. ^{13}C NMR spectra of sample solution after dissolving biomass sample in 1:2 pyridinium molten salt/DMSO-d_6 (1.0 g) at 60 °C. (a) Ball-milled poplar (80 mg) in 1:2 [Hpyr]Cl-d_6/DMSO-d_6; (b) Wiley milled poplar (80 mg) in 1:2 [Hpyr]Cl-d_6/DMSO-d_6; (c) Wiley milled switchgrass (100 mg) in 1:2 [Hpyr]Cl-d_6/ DMSO-d_6.

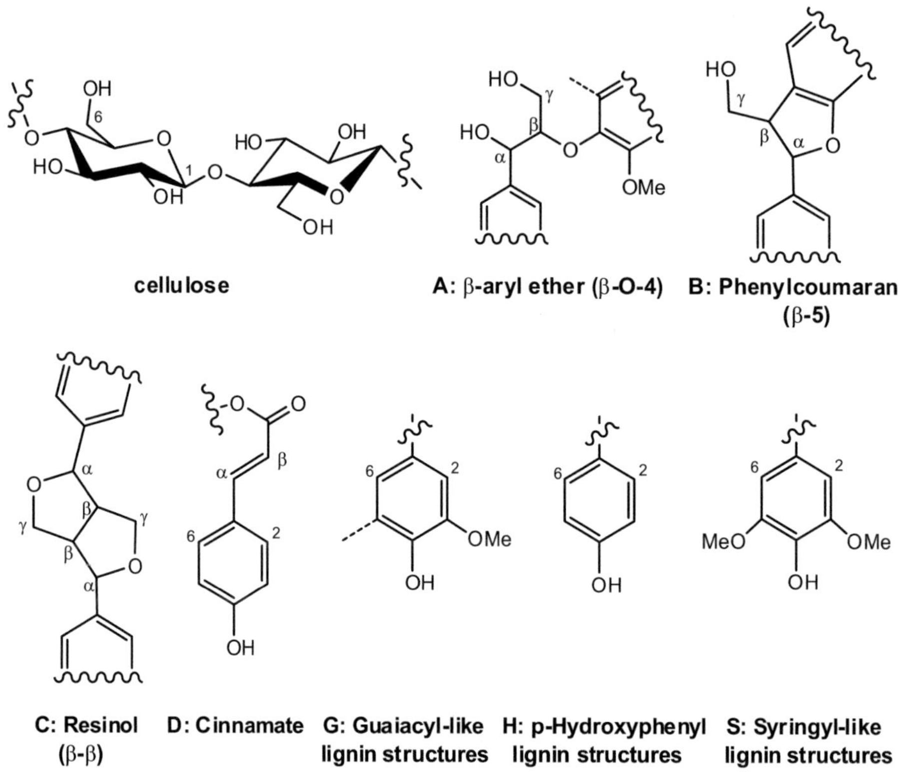

Fig. 5. Structures of cellulose and sub-lignin units

We next used 2D NMR (Heteronuclear Single Quantum Coherence, HSQC) analysis to investigate more detailed sub-lignin structures. As switchgrass lignin contains a typical HGS lignin (H: *p*-hydroxyphenyl, S: syringyl, G: guaiacyl) that can be readily analyzed by HSQC NMR, we summarized HSQC NMR of switchgrass in Fig. 6 to illustrate the aliphatic side chain and the aromatic ring ^{13}C-^{1}H correlations of the lignin component. The main cross signals in the aromatic region correspond to sub-lignin syringyl-like (S) and guaiacyl-like (G) units which appeared separately. The S unit cross peak for the C2,6/H2,6 ($S_{2/6}$) correlation appears at δC/δH 104.7/6.8 ppm. The G units showed different correlations at δC/δH 112.1/6.9, 115.8/6.7, 119.4/6.9 ppm for C2/H2 (G_2), C5/H5 (G_5) and C6/H6 (G_6), respectively, as shown in Figure 6. Furthermore, strong correlation signals at δC/δH 130.5/7.4 and 128.9/7.1 ppm reveal the presence of *p*-hydroxyphenyl (H structure in Fig. 5) unit. In addition, correlation signals at δC/δH 114.0/6.4 ppm (D_β) and 145.9/7.4 ppm (D_α) from *p*-coumarate and ferulate (see D in Fig. 5) can be clearly identified, which agree with the report that the grass lignin contains *p*-coumarate and ferulate units (Ralph et al., 1999). More importantly, the detailed sub-lignin side chain structure (aliphatic region) can be clearly assigned to the cross signal for lignin methoxyl group (δC/δH 57.3/3.7 ppm, the most prominent) and those for β-O-4 substructures at α-, β- and γ-C positions (δc/δH 74.5/4.5, 84.8/3.8, and 60.0/3.6 ppm, respectively, see A in Fig. 5). The presence of phenyl coumaran substructures (B in Fig. 5) can also be confirmed by C-H correlations at δC/δH 63.8/3.9 ppm. Similarly, the Wiley milled poplar sample also affords satisfactory HSQC spectra, as shown in Fig. 7.

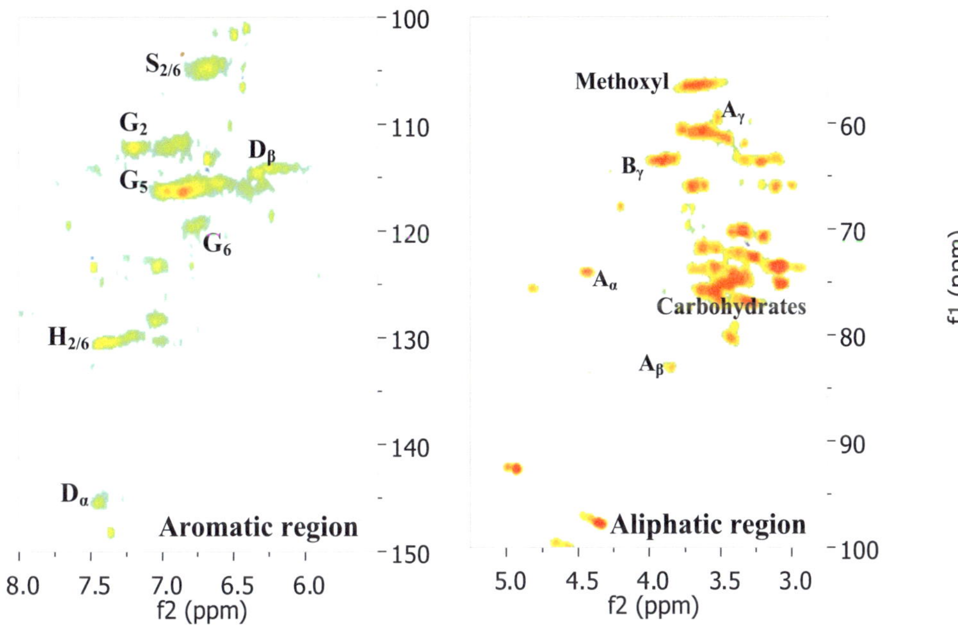

Fig. 6. (see correction 2) 2D HSQC spectra of switchgrass solution after dissolving 100 mg Wiley milled sample in 1:2 [Hpyr]Cl-d_6/DMSO-d_6 (1.0 g) at 60 °C.

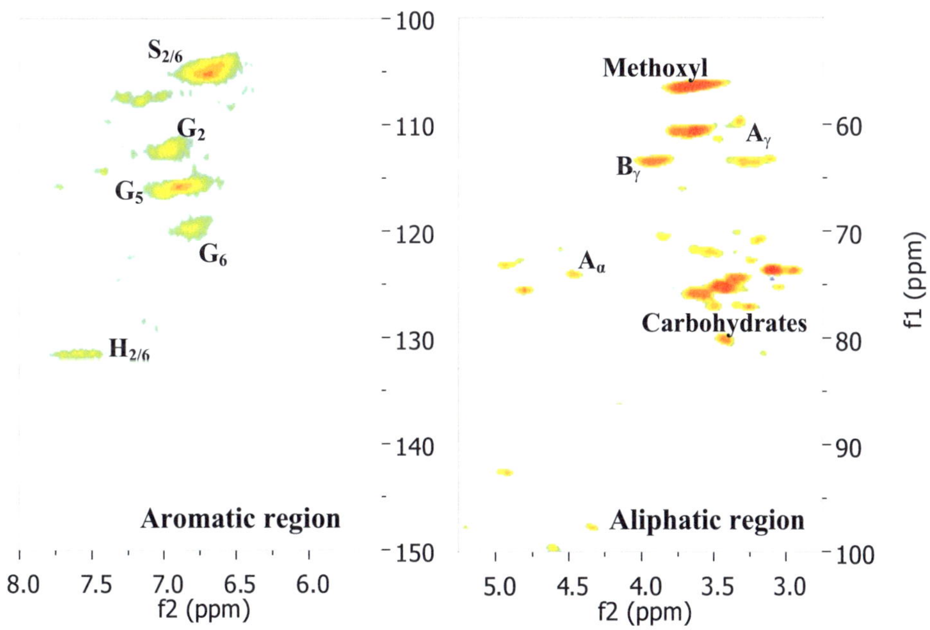

Fig. 7. (see correction 3) 2D HSQC spectra of poplar solution after dissolving 80 mg Wiley milled sample in 1:2 [Hpyr]Cl-d_6/DMSO-d_6 (1.00 g) at 60 °C.

Having successfully developed novel perdeuterated pyridinium chloride-d_6/DMSO-d_6 bi-solvent system for direct biomass dissolution and NMR characterization (Jiang et al., 2009), we further expanded our research to quantitative NMR analysis of the plant cell walls. Our initial efforts were directed toward biomass lignin content determination via ^{1}H NMR analysis.

While many methods based on gravimetric or spectrophotometric analysis (Iiyama & Wallis, 1989; Dence & Lin, 1992; Bose, 1998) have been traditionally used to analyze lignin content in plants with some disadvantages including a relatively large sample size, time-consuming procedure (gravimetric method) and the difficulty in finding an appropriate calibration standard (spectrophotometric methods). Very recently, near-infrared (NIR) spectroscopy method has been further modified to improve the precision of lignin content (Yeh et al., 2004; Hatfield & Fukushima, 2005; Chang et al., 2008; Yao et al., 2010), and high-throughput screening of plant cell wall composition via pyrolysis molecular beam mass spectroscopy can analyze lignin content and determine lignin S/G ratio (Sykes et al., 2009).

While there is a need for assessing a large numbers of "new" plants, precise analytic techniques for efficient lignin content assessment at a microscale are still a pressing research issue. In this sense, NMR methodologies to assess lignin content via dissolving native plant cell walls in perdeuterated ionic liquid system are attractive to serve as a supplemental to detailed structural information.

We started our initial search for a suitable procedure to measure the lignin content via direct biomass dissolution and quantitative ^{1}H NMR analysis with a ball-milled switchgrass sample dissolved in 1:2 [Hpyr]Cl-d_6/DMSO-d_6 system. DMSO-d_6 (99.9 atom%) was used as the co-solvent of [Hpyr]Cl-d_6 to reduce the solvent viscosity. Six dependent experiments were carried out to optimize the conditions for complete biomass dissolution with integration of DMSO as the internal standard, as shown in Table 2. Our experimental results show that the ratio of integration of signals of lignin aromatic/olefinic (signals between 6.0 ppm and 8.0 ppm) and integration of non-deuterated DMSO (2.5 ppm) increased with prolonging stirring at 25 °C, indicating incomplete biomass dissolution (Table 1, entries 1-3). In contrast, biomass dissolution carried out at higher temperature (60 °C) led to comparable ratios, which suggests complete biomass dissolution (Table 1, entries 4-6). Therefore, biomass dissolution carried out at 60 °C for 12 h was chosen as the standard condition for further analysis of lignin content.

Entry	Temperature (°C)	Time (h)	Ratio of integration[a,b] ($Intg_{6.0\text{-}8.0ppm} : Intg_{DMSO}$)
1	25	6	1.35
2	25	12	1.68
3	25	24	1.83
4	60	6	2.18
5	60	12	2.22
6	60	24	2.17

[a] Dried extract-free ball-milled switchgrass cell walls (40.0 mg), 600.0 mg 1 : 2 [Hpyr]Cl-d_6/DMSO-d_6 (DMSO-d_6, 99.9 atom % D) and spinbar (5 mm × 2 mm) in 5 mm NMR tube were stirred at specific temperature for the specific time for the optimization of biomass dissolution. [b] Direct ^{1}H NMR analysis of the resulting biomass solution afford the ratio of integration of signals (6.0 ppm-8.0 ppm, exclusively attributed to lignin aromatic signals) and integration of DMSO (2.5 ppm, non-deuterated DMSO as the internal standard).

Table 2. Optimization of ball-milled switchgrass complete dissolution.

After optimizing the conditions for completely dissolving biomass, we applied the linear extrapolation method for the measurement of lignin content by the addition of a specific amount of isolated switchgrass lignin to the biomass solution to examine integration ratio changes in the quantitative ^1H NMR spectra with non-deuterated DMSO as the internal standard. Four samples of 40.0 mg dried extractive-free ball-milled switchgrass samples and 2.0 mg, 4.0 mg, 6.0 mg, or 8.0 mg of isolated switchgrass lignin were added into a 5 mm NMR tube containing 600.0 mg 1 : 2 [Hpyr]Cl-d_6/DMSO-d_6, respectively. After 12 h stirring at 60 °C for complete biomass dissolution, the four mixtures were subject to quantitative ^1H NMR analysis with non-deuterated DMSO signal as the internal standard. As the signals between 6.0 ppm and 8.0 ppm can be exclusively attributed to lignin aromatic/olefinic region (see Fig. 8), we found a good linear relationship between the ratio of integrations and the added lignin amount (Scheme 1). Further data analysis via regression as modelling approach reveals that the lignin amount in 40 mg dry extractive-free switchgrass sample can be measured as 7.0 mg, and thus the lignin content in the switchgrass sample can be calculated to be 17.4%, which is comparable with traditional Klason lignin content of switchgrass as 17.1%. Therefore, our current method proves to give a comparable result with the traditional Klason lignin content.

Similarly, a ball-milled hardwood sample (poplar) was analyzed by direct dissolution and ^1H NMR analysis for the measurement of lignin content. A good linear relationship between the ratio of integrations and added lignin amount for ball-milled poplar sample, as shown in Scheme 2, was used to calculate the lignin amount in 40 mg poplar sample as 9.3 mg, which affords the poplar lignin content as 23.3%, which is consistent with Klason lignin content (24.1%).

Subsequently, we also investigated the use of Wiley milled samples to analyze biomass lignin content in order to improve efficiency without the complications associated with the ball-milling processes. Due to the incomplete dissolution of a Wiley milled sample in 5 mm

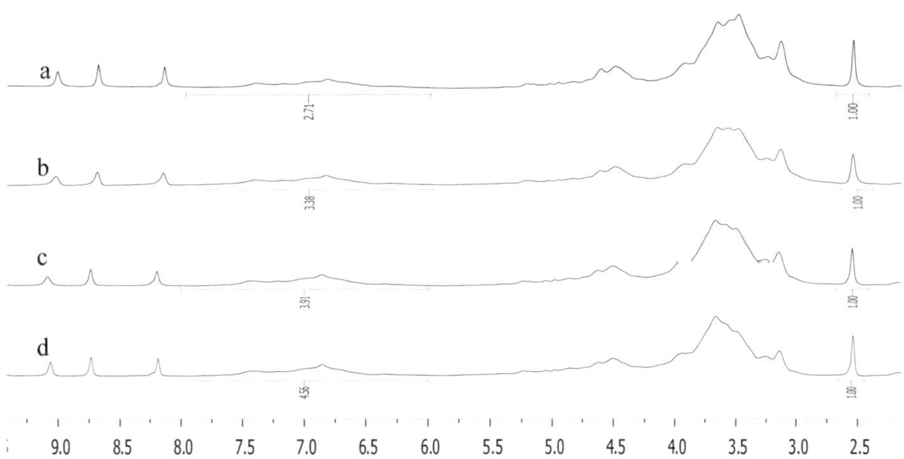

Fig. 8. ^1H NMR analysis of lignin content in ball-milled switchgrass cell walls: a) 40.0 mg dry switchgrass and 2.0 mg isolated switchgrass lignin; b) 40.0 mg dry switchgrass and 4.0 mg isolated switchgrass lignin; c) 40.0 mg dry switchgrass and 6.0 mg isolated switchgrass lignin; d) 40.0 mg dry switchgrass and 8.0 mg isolated switchgrass lignin.

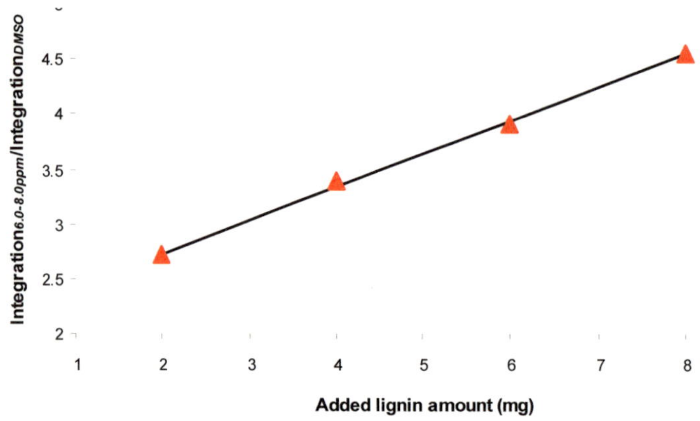

Scheme 1. Linear relationship between ratios of integration of signals and added lignin amount (mg) in ball-milled swichgrass samples ($Y = 2.12 + 0.304 \times X$; lignin content = $0.025 \times 2.12 \div 0.304 = 17.4\%$)

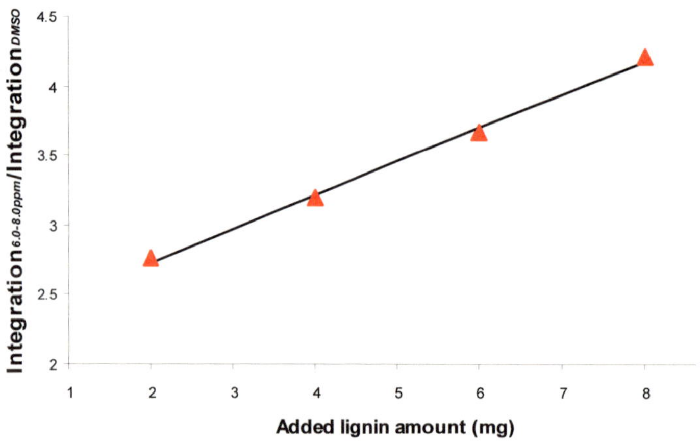

Scheme 2. Linear relationship between ratios of integration of signals and added lignin amount (mg) in 40 mg ball-milled poplar samples ($Y_1 = 2.76$; $Y_2 = 3.19$; $Y_3 = 3.67$; $Y_4 = 4.21$; $Y = 2.25 + 0.2415 \times X$; lignin content = $0.025 \times 2.25 \div 0.2415 = 23.3\%$)

NMR tube, we adopted a modified method involving the use of one signal of non-deuterated pyridinium cation (signals at 9.1 ppm) as the internal standard to analyze lignin content of Wiley milled poplar sample. Thus three samples of 30.0 mg dry extractive-free Wiley-milled poplar sample (20 mesh) and 1.0 mg, 2.0 mg, or 4.0 mg isolated poplar lignin were dissolved in 10 mL vial containing 500.0 mg 1 : 2 [Hpyr]Cl-d_6/DMSO-d_6 via vigorously stirring for 12 h at 60 °C, respectively. After the biomass solution was transferred into 5 mm NMR tube, the vial was rinsed with 2 × 150 mg DMSO-d_6, which is also transferred into the NMR tube for ^{1}H NMR analysis (Figure 9). With data analysis of quantitative ^{1}H NMR spectra confirming a good linear relationship between ratio of integrations and the added lignin amount (see Scheme 3), the corresponding lignin content can be calculated to be 23.3%, which is consistent with Klason lignin content and lignin content analyzed from ball-milled poplar sample.

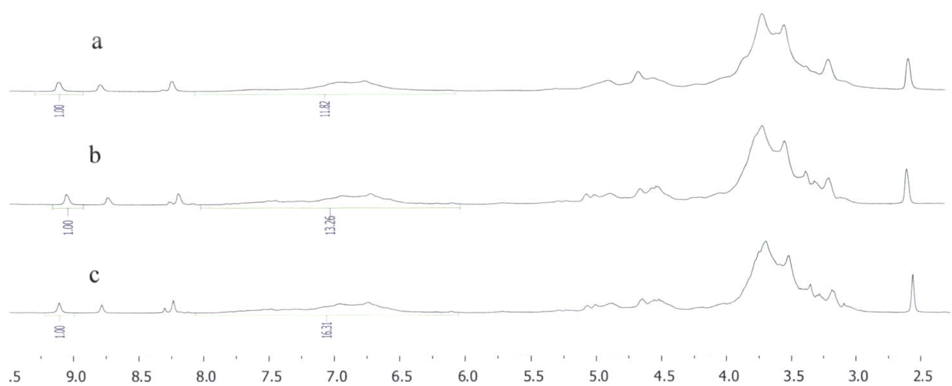

Fig. 9. ^{1}H NMR analysis of lignin content in Wiley milled poplar cell walls: a) 30.0 mg dry poplar and 1.0 mg isolated poplar lignin; b) 30.0 mg dry poplar and 2.0 mg isolated poplar lignin; c) 30.0 mg dry poplar and 4.0 mg isolated poplar lignin

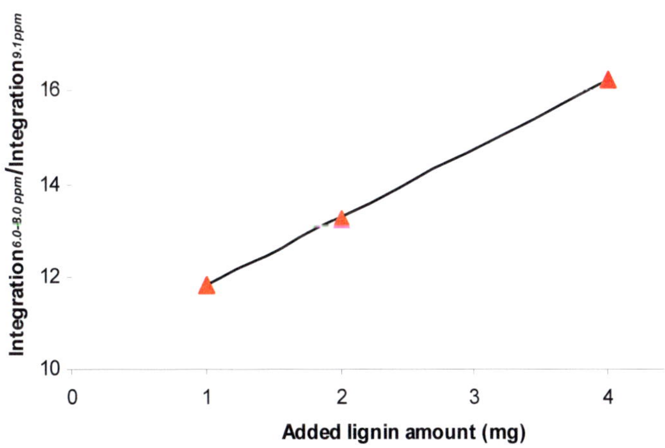

Scheme 3. Linear relationship between ratios of integration of signals and added lignin amount (mg) in Wiley milled poplar samples (Y = 10.325 + 1.4793 × X; lignin content = 0.0333 × 10.325 ÷ 1.4793 = 23.3%)

Having analyzed Wiley milled poplar lignin content, we further analyzed softwood pine lignin content via three samples of 25.0 mg dry extractive-free Wiley-milled pine sample (20 mesh) with addition of 1.0 mg, 2.0 mg, or 4.0 mg isolated pine lignin, which were dissolved in 500.0 mg 1 : 2 [Hpyr]Cl-d_6/DMSO-d_6. After rinsed with 2 × 150 mg DMSO-d_6, the combined biomass solutions was analyzed by ^{1}H NMR. With data analysis of quantitative ^{1}H NMR spectra confirming a good linear relationship between ratio of integrations (signals at 9.1 ppm as the internal standard) and the added lignin amount (see Scheme 4), the corresponding lignin content can be calculated to be 28.9%, which is comparable with Klason lignin content (29.4%).

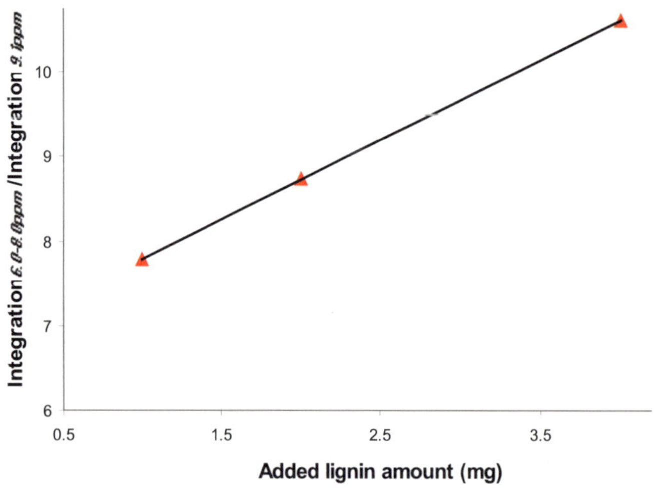

Scheme 4. Linear relationship between ratios of integration of signals and added lignin amount (mg) in 25 mg Wiley milled pine samples ($Y_1 = 7.78$; $Y_2 = 8.74$; $Y_3 = 10.62$; $Y = 6.84 + 0.9457 \times X$; lignin content = $0.04 \times 6.84 \div 0.9457 = 28.9\%$)

Therefore, we have applied perdeuterated pyridinium chloride for a novel methodology of efficient lignin content assessment of biomass samples (ball-milled or Wiley milled) at a microscale (Jiang, 2010), which showed comparable lignin contents as the traditional Klason lignin contents, providing a new venue for rapid assessing the lignin contents in large numbers of 'new' plants during biofuel research. This protocol significantly broadens the application of whole cell NMR analysis as the lignin content acquired by 1D ^{1}H NMR can serve as important supplemental information to the detailed structure of lignin ascertained by 2D HSQC NMR analysis.

In summary, we have successfully developed easy-accessible perdeuterated pyridinium chloride ionic liquid system for direct biomass dissolution and NMR characterization with improved efficiency of characterization by the use of Wiley milled biomass. With regard to future directions, there is still much room for increased applications of this system, such as

HSQC NMR analysis for evaluating biomass pretreatment, lignin HSG ratio determination via quantitative ^{13}C NMR and quantitative ^{31}P NMR analysis of in situ phosphitylated biomass. In addition, novel ammonium typed ionic liquids with good biomass solubility need to be developed due to strong ^{13}C NMR signals from both pyridinium cation and imidazolium cation (between 120.0 ppm and 150.0 ppm), which hinders detailed lignin aromatic ring structural analysis by ^{13}C NMR. Further developing novel ionic liquids and broadening of the scope of current perdeurated pyridinium chloride in whole cell NMR analyses, would similarly enhance better understanding of plant cell walls and recalcitrance.

2. References

Baucher, M.; Petit-Conil, M. & Boerjan, W. (2003). Lignin: genetic engineering and impact on pulping. *Crit. Rev. Biochem. Mol. Biol.*, 38, 305-350.

Boerjan, W.; Ralph, J. & Baucher, M. (2003). Lignin biosynthesis. *Annu. Rev. Plant Biol.*, 54, 519-546.

Bose, S. K.; Wilson, K. L.; Francis, R. C. & Aoyama, M. (1998). Lignin analysis by permanganate oxidation. I. Native spruce lignin. *Holzforschung*, 52(3), 297-303.

Bothast, R. & Schlicher, M. A. (2005). Biotechnological processes for conversion of corn into ethanol. *Appl. Microbiol. Biotechnol.*, 67(1), 19-25.

Chang, X. F.; Chandra, R.; Berleth, T. & Beatson, R. P. (2008). Rapid, Microscale, Acetyl Bromide-Based Method for High-Throughput Determination of Lignin Content in Arabidopsis thaliana. *J. Agric. Food Chem.*, 56, 6825-6834.

Clark, J. H.; Budarin, V.; Deswarte, F. E. I.; Hardy, J. J. E.; Kerton, F. M.; Hunt, A. J., Luque, R.; Macquarrie, D. J.; Milkowski, K.; Rodriguez, A.; Samuel, O.; Tavener, S. J.; White R. J. & Wilson, A. J. (2006). Green chemistry and the biorefinery: a partnership for a sustainable future. *Green Chem.*, 8, 853-860.

Corma, A.; Iborra, S. & Venlty, A. (2007). Chemical Routes for the Transformation of Biomass into Chemicals. *Chem. Rev.*, 107, 2411-2502.

Cosgrove, D. J. (2005). Growth of plant cell wall. *Nat. Rev. Mol. Cell Biol.*, 6, 850-861.

Davin L. B. & Lewis, N. G. (2005). Lignin primary structures and dirigent sites. *Current Opinion in Biotechnology*, 2005, 16(4), 407-415.

Dence, C. W. & Lin, S. Y. (1992) *The determination of lignin In Methods in Lignin Chemistry*, Springer-Verlag: Berlin, Germany, pp. 3361.

Fort, D. A.; Remsing, R. C.; Swatloski, R. P.; Moyna, P.; Moyna G. & Rogers, R. D. (2007). Can ionic liquids dissolve wood? Processing and analysis of lignocellulosic materials with 1-n-butyl-3-methylimidazolium chloride. *Green Chem.*, 9(1), 63-69.

Fukaya, Y.; Hayashi, K.; Wada, M. & Ohno, H. (2008). Cellulose dissolution with polar ionic liquids under mild conditions: required factors for anions. *Green Chem.*, 2008, 10(1), 44-46.

Giernoth, R. & Bankmann, D. (2008). Transitional-metal-free synthesis of perdeuterated imidazolium ionic liquids by alkylation and H/D exchange. *Eur. J. Org. Chem.*, 2881-2886.

Hatfield, R. & Fukushima, R. (2005). Can lignin be accurately measured? *Crop Sci.*, 45, 832–839.

Hardacre, C.; Holbrey, J. D. & McMath, S. E. J. (2001). A highly efficient synthetic procedure for deuterating imidazoles and imidazolium salts. *Chem. Commun.*, 367-368.

Higuchi, T. (1985). *Biosynthesis of lignin In Biosynthesis and Biodegradation of Wood Components*, Academic Press: New York, pp. 141-160.

Himmel, M. E.; Ding, S.-Y.; Johnson, D. K.; W. S. Adney, M. R. Nimlos, Brady, J. W. & Foust, T. D. (2007). Biomass Recalcitrance: Engineering Plants and Enzymes for Biofuels Production. *Science*, 315, 804-807.

Horvath, I. T . & Anastas, P. T. (2007). Innovations and green chemistry. *Chem. Rev.*, 107, 2167–2168.

Huber, W. & Corma, A. (2007). Synergies between bio- and oil refineries for the production of fuels from biomass. *Angew. Chem. Int. Ed.*, 46, 7184–7201.

Iiyama, K. & Wallis, A. F. A. (1989). Effect of acetyl bromide treatment on the ultraviolet spectra of lignin model compounds. *Holzforschung*, 43, 309–316.

Ikeda, T.; Holtman, K.; Kadla, J. F.; Chang, H.-M. & Jamee, H. (2002). Studies on the effect of ball milling on lignin structure using a modified DFRC method. *J. Agr. Food Chem.*, 50(1), 129-135.

Jiang, N.; Pu, Y.; Samual, R. & Ragauskas, A. J. (2009). Perdeuterated pyridinium molten salt (ionic liquid) for direct dissolution and NMR analysis of plant cell walls. *Green Chem.*, 11, 1762-1766.

Jiang, N.; Pu, Y.& Ragauskas, A. J. (2010). Rapid determination of lignin content via direct dissolution and 1H NMR analysis of the plant cell walls. ChemSusChem, (in press).

Jung, H. G.; Buxton, D. R.; Hatfield, R. D. & Ralph, J. (1993). *Forage Cell Wall Structure and Digestibility*, Madison, WI: ASA-CSSA-SSSA, pp. 1-974.

Kilpelaeinen, I.; Xie, H.; King, A.; Granstrom, M.; Heikkinen, S. & Argyropoulos, D. S. (2007). Dissolution of Wood in Ionic Liquids. *J. Agricultural and Food Chem.*, 55(22), 9142-9148.

Kondo, T.; Watanabe, T.; Ohshita, T. & Kyuma, T. (1995). Comparative characterisation of dioxane-soluble lignins released by ball milling and by sheep digestion from forage grasses. *J. Sci. Food Agr.*, 68(3), 383-388.

Lewis, N. G. & Paice, M. G. (1989). *Plant Cell Wall Polymers. Biogenesis and Biodegradation*, Washington, DC, Vol. 399, pp. 1-676.

Marchetti, J. M.; Miguel, V. U. & Errazu, A. F. (2007). Possible methods for biodiesel production. *Renew. Sust. Energ. Rev.*, 11(6), 1300-1311.

Parkas, J.; Brunow, G. & Lundquist, K. (2007). Quantitative lignin analysis based on permanganate oxidation. *BioResources*, 2(2), 169-178.

Pena, L. & Seguin, A. (2001). Recent advances in the genetic transformation of trees. *Trends in Biotechnology*, 19(12), 500-506.

Petrus, L. & Noordermeer, M. A. (2006). Biomass to biofuels, a chemical perspective. *Green Chem.*, 8(10), 861-867.

Pu, Y., Jiang, N. & Ragauskas, A. J. (2007). Ionic liquid as a green solvent for lignin.*J. Wood Chem. Techno.*, 27(1), 23-33.

Pu, Y.; Zhang, D.; Singh, P. M. & Ragauskas, A. J. The new forestry biofuels sector. (2008). *Biofuels, bioproducts & biorefining*, 2, 58-73.

Ragauskas, A. J.; Williams, C. K.; Davison, B. H.; Britovsek, G.; Cairney, J.; Eckert, C. A.; Frederick, W. J.; Hallett, J. P.; Leak, D. J.; Liotta, C. L.; Mielenz, J. R.; Murphy, R.; Templer, R. & Tschaplinski, T. (2006). The path forward for biofuels and biomaterials. *Science*, 311, 484-489.

Ralph, J.; Marita, J. M.; Ralph, S. A.; Hatfield, R. D.; Lu. F.; Ede, R. M.; Peng, J.; Quideau, S.; Helm, R. F.; Grabber, J. H.; Kim, H.; Jimenez-Monteon, G.; Zhang, Y.; Jung, H.-J. G.; Landucci, L. L.; Mackay, J. J.; Sederoff, R. R.; Chapple, C. & Boudet, A. M. (1999). Solution-state NMR of lignins. *Advances in Lignocellulosics Characterization*, D. S. Argyropoulos, Edn., Tappi Press, Atlanta, USA, pp. 55-108.

Serrano-Ruiz, J. C.; West, R. M. & Dumesic, J. A. (2010). Catalytic Conversion of Renewable Biomass Resources to Fuels and Chemicals. *Annu. Rev. Chem. Biomol. Eng.*, 1, 79-100.

Sheldon, R. A. (2001). Catalytic reactions in ionic liquids. *Chem. Commun.*, 2399.

Simonetti D. A. & Dumesic J. A. (2008). Catalytic strategies for changing the energy content and achieving C–C coupling in biomass-derived oxygenated hydrocarbons. *ChemSusChem*, 1, 725–33.

Sticklen, M. (2006). Plant genetic engineering to improve biomass characteristics for biofuels. *Current Opinion in Biotechnology*, 17(3), 315-319.

Swatloski, R. P.; Spear, S. K.; Holbrey, J. D. & Rogers, R. D. (2002). Dissolution of cellulose with ionic liquids. *J. Am. Chem. Soc.*, 124, 4974-4975.

Sykes, R.; Yung, M.; Novaes, E.; Kirst, M.; Peter, G. & Davis, M. (2009). High-Throughput Screening of Plant Cell-Wall Composition Using Pyrolysis Molecular Beam Mass Spectroscopy. *Methods Mol. Biol.*, 581, 169-183.

Wasserscheid P. & Keim, W. (2000). Ionic Liquids—New "Solutions" for Transition Metal Catalysis. *Angew. Chem. Int. Ed.*, 39, 3772-3789.

Welton, T. (1999). Room-Temperature Ionic Liquids. Solvents for Synthesis and Catalysis. *Chem. Rev.*, 99, 2071-2084.

Wyman, C. E.; Dale, B. E.; Elander, R. T.; Holtzapple, M.; Ladisch, M. R. & Lee, Y. Y. (2005). Coordinated development of leading biomass pretreatment technologies. *Bioresour. Technol.*, 96, 1959-1966.

Xie, H.; King, A.; Kilpelainen, I.; Granstrom, M. & Argyropoulos, D. S. (2007). Thorough chemical modification of wood-based lignocellulosic materials in ionic liquids. *Biomacromolecules*, 2007, 8(12), 3740-3748

Yao, S.; Wu, G.; Xing, M.; Zhou, S. & Pu, J. (2010). Determination of lignin content in *Acacia* spp. using near-infrared reflectance spectroscopy. *BioResources*, 5(2), 556-562.

Yeh, T.-F.; Chang, H.-M. & Kadla, J. F. (2004). Rapid Prediction of Solid Wood Lignin Content Using Transmittance Near-Infrared Spectroscopy. *J. Agric. Food Chem.*, 52, 1435-1439.

Yelle, D. J.; Ralph, J. & Frihart, C. R. (2008). Characterization of nonderivatized plant cell walls using high-resolution solution-state NMR spectroscopy. *Magn. Reson. Chem.*, 46, 508-517.

Zhu, S.; Wu, Y.; Chen, Q.; Yu, Z.; Wang, C.; Jin, S.; Ding, Y. & Wu, G. Dissolution of cellulose with ionic liquids and its application: a mini-review. (2006). *Green Chem.*, 8(4), 325-327.

Zakrzewska, M. E.; Bogel-Łukasik, E. & Bogel-Łukasik, R. (2010). Solubility of Carbohydrates in Ionic Liquids. *Energy Fuels*, 24(2), 737-745.

24

Ionic Liquids in the Pretreatment of Lignocellulosic Biomass

Jana Holm and Ulla Lassi
University of Oulu, Department of Chemistry
Finland

1. Introduction

The fossil fuel-based economy is facing several problems and challenges, as stated by the Intergovernmental Panel on Climate Change IPCC (IPCC, 2007). These challenges involve the increasing emissions of CO_2, decreasing reserves and increasing energy prices. A potential solution to the problem could be in the form of lignocellulosic biomass as an alternative and sustainable energy source of the future. It can be used to produce chemicals and biofuels, which do not compete with food production (Huber et al., 2006; Lynd et al., 1999). Extensive research into the conversion of lignocellulosic biomass is currently being undertaken all over the world.

Biomass as a carbon based material is composed of a mixture of organic molecules containing carbon and hydrogen. It usually possesses atoms of oxygen and nitrogen whilst including small quantities of other elements, such as metals. The carbon used to construct biomass is absorbed from the atmosphere as carbon dioxide (CO_2) by plant life, using energy from the sun. Therefore biomass is the most abundant renewable resource available. The major constituents of lignocellulosic biomass are polymeric carbohydrates (cellulose and hemicellulose) and lignin (Zhang & Zhao, 2010). The term "lignocellulosic biomass" is often used to describe the material that composes the plant cell wall, i.e. cellulose, hemicelluloses and lignin. However, the plant cell contains different layers that differ in structure and chemical composition. As a result of the organisation and interaction between these polymeric structures, the plant cell wall is naturally recalcitrant to the biological degradation (da Costa Sousa et al., 2009). Conventional methods to convert lignocellulosic materials to sugars have been in the form of acid hydrolysis or the use of high pressures and temperatures. These methods are either energy-intensive or require the recirculation of acid. There are several solvents which dissolve cellulose, but these usually degrade the cellulose during the process (Holm et al., 2009; Mosier et al., 2005). Although different process steps are necessary in order to convert the biomass to useful products, the intermediate product is mostly glucose.

As indicated above, the pretreatment of lignocellulosic biomass is a significant tool in the practical conversion processes, and it will be considered in this article.

2. Lignocellulosic biomass

Wood, grass, forestry waste, agricultural residues and municipal solid waste are excellent examples of lignocellulosic materials that can be used in the production of lignocellulosic biomass. Typically this contains cellulose (30-50 %), hemicellulose (15-35 %) and lignin (10-30 %) of dry weight (Lynd et al., 2002) and is the most abundant renewable organic material on earth (Bhat & Bhat, 1997), with an estimated global production of around 1.0 x 10^{11} tons/year (Zhang & Zhao, 2010; Sánches & Cardona, 2008). Against the background of this enormous growth of renewable material, plant biomass offers the most potential platform for energy production.

Cellulose is a crystalline polymer that is made up of glucose units. Cellulose fibre are surrounded by intra and intermolecular hydrogen bonds (Zhbankov, 1992) which makes cellulose insoluble in water and the most organic solvents available (Swatloski et al., 2002). Instead hemicellulose, a polymer of five sugars, is completely amorphous. Lignin as an irregular polymer forms a network in which cellulose and hemicellulose fibre are embedded (Huber et al., 2006). Due to the complex structure of lignocellulose, it is resistant to most chemicals and hydrolysis, which form the barrier for its utilisation (Lynd et al., 2002; Zhu et al., 2006).

(A)

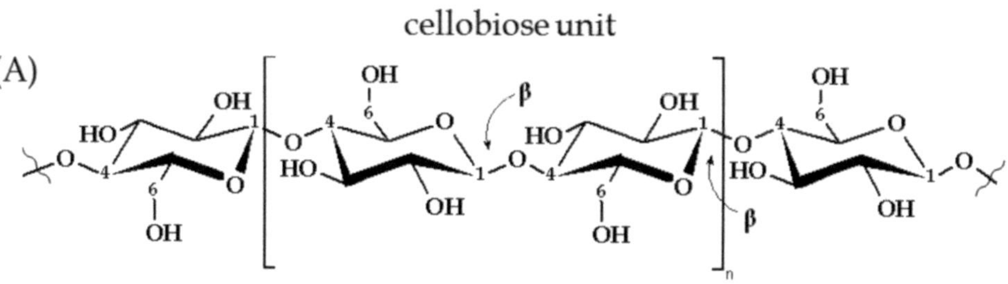

(B)

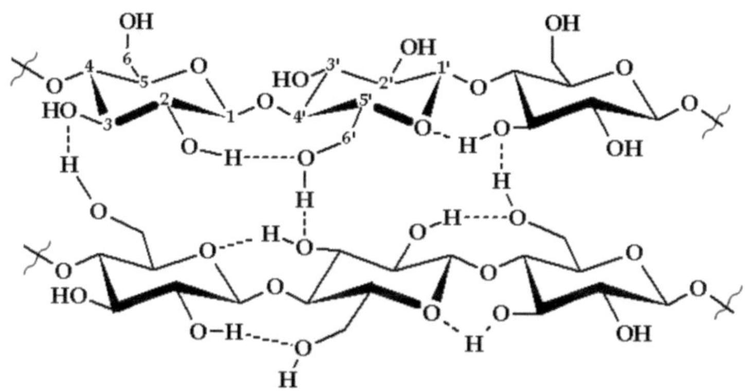

Fig. 1. The cellulose network (A is the cellulose chain and B shows inter and intra H-bonds present in cellulose) adopted from Olivier-Bourdigou et al., 2010.

Cellulose forms the skeleton in which other substances, hemicelluloses (matrix) and lignin (encrust) surround it. Cellulose, hemicelluloses and lignin are closely attached and covalent cross-linkages occur between lignin and the polysaccharides (Rogalinski et al., 2008).

3. Aims of biomass pretreatment

Since different lignocellulosic materials have different physio-chemical characteristics, it is necessary to adopt suitable pretreatment technologies based on the lignocellulosic biomass properties of each raw material (Alvira et al., 2010). Biomass pretreatment is needed for several reasons. The main goal is to alter the structure of cellulosic biomass, and thus to make the cellulose structure more accessible to adsorbates (enzymes and/or chemicals). The lignin seal of the cellulose must be broken and then the crystalline structure of cellulose can be disrupted.

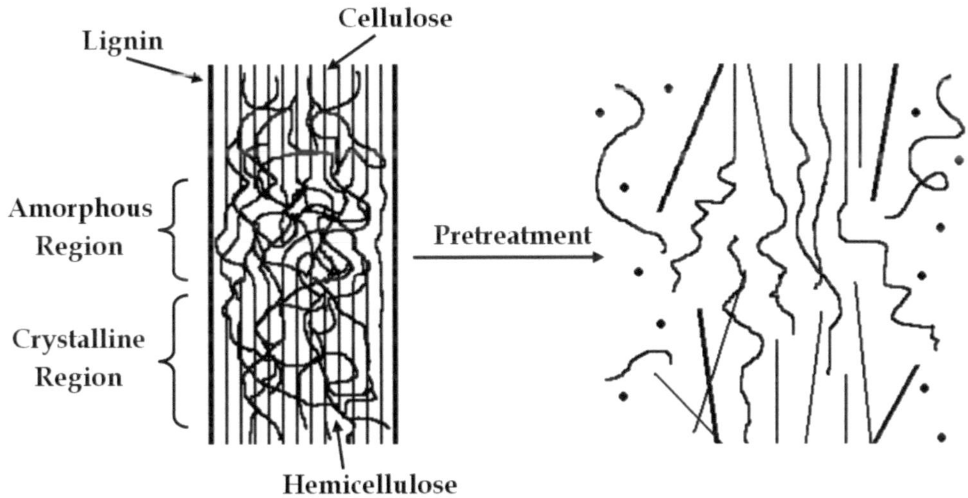

Fig. 2. Goals of the pretreatment adopted from Mosier et al., 2005.

Recently, much research has been undertaken worldwide to convert lignocellulosic biomass (forestry waste, agricultural residues and energy crops) through a sugar platform. Lignocellulose serves as a sustainable feedstock for the production of future valuable products, such as biofuels and chemicals. Pretreatment of biomass is required to decompose the fibre structure of lignocelluloses and to utilise all the components (Kuo & Lee, 2009). Pretreatment alters the structure of cellulosic biomass to make cellulose more accessible to the enzymes that convert carbohydrate polymers. Subsequently, when lignocellulose is separated into its components, it can be hydrolysed to fermentable sugars (monosaccharides) using mineral acids or enzymes. Monosaccharides can then be further converted to valuable bio-based chemicals (Kamm & Kamm, 2004).

The aim of pretreatment is to break the lignin shield and the crystalline structure of cellulose whilst increasing the porosity of cellulose. Pretreatment can be considered as a very important tool for practical cellulose conversion processes. Pretreatment methods are usually categorised into physical, chemical, physiochemical and biological (Zhao et al., 2009) however each of these methods have their own specific drawbacks (Chandra et al., 2007).

The effectiveness of lignocelluloses pretreatment is one of the key factors to a successful conversion of the originally low-cost material into sugars and further into biofuel or biofuel intermediates. The pretreatment of cellulosic materials can affect its physical properties such as its degree of polymerisation, its crystallinity and even the surface area of the substrate accessible in the case of further enzymatic hydrolysis (Olivier-Bourbigou et al., 2010). Pretreatment methods increase the surface area accessible to water and cellulases are expected to generate improvements in hydrolysis kinetics and conversion of cellulose to glucose.

Pretreatment is a significant tool for practical cellulose conversion processes. Several pretreatment procedures for lignocelluloses have been successfully developed (Mosier et al., 2005; Wyman et al., 2005), i.e. after mechanical comminution for chemical and hydrothermal pretreatments have been applied. The most commonly applied methods involved, for example include steam (Cara et al., 2008) or carbon dioxide (Zheng et al., 1998) explosions and hot water treatment (Mosier et al., 2005). Chemical processes involve acid (Lloyd & Wyman, 2005) or alkali treatments (Kaar & Holtzapple, 2000) and organosolv process for example however, all these pretreatment methods have several disadvantages. They are tailored to a specific type of lignocelluloses (Mosier et al., 2005) and some of the solvents cause the degradation of lignocelluloses upon dissolving. These unwanted by-products can inhibit a subsequent hydrolysis or fermentation step (Larsson et al., 1999; Weil et al., 1994).
Cellulose solvents should have most of the features listed below (Olivier-Bourbigou et al., 2010). They have to:

1. be able to dissolve cellulose at low temperatures,
2. be non volatile, non toxic and chemically stable,
3. no decompose cellulose,
4. be easy to regenerate,
5. be recyclable,
6. be cost effective and easily processed
7. non-toxic to enzymatic and microbial fermentation.

As indicated in the list above, solvent use and recovery increases the total cost of pretreatment. Some pretreatment methods are too slow (even weeks), and these methods cannot be considered as "green" since hazardous or toxic compounds are released (Zhu 2008). Therefore, more efficient pretreatment procedures are required. One potential alternative is the use of ionic liquids.

4. Ionic liquids

4.1 General

Ionic liquids (ILs) are a group of new organic salts that are liquids at relative low temperatures (below 100°C). As the name suggest they are completely ionic with most ionic liquids consisting of an organic cation and an inorganic anion. Many ionic liquids are also liquids at room temperature making it an ideal solvent to work with. Compared with traditional molecular solvents, ionic liquids express very interesting properties like broad liquid regions, high thermal stabilities and negligible vapour pressures (Brennecke & Maginn, 2001). Ionic liquids are also called green solvents because no toxic or explosive gases are formed when used (Anderson et al., 2002). Depending on which anion and cation is chosen, their physical and chemical properties such as melting points, viscosity, hydrophobicity and hydrolysis stability are affected (Huddleston et al., 2001). Therefore, optimal ionic liquids for a special application can be designed. Simply by making changes to the structure of either the anion or the cation, or both, properties such as solubility, density, refractive index and viscosity can be adjusted to meet the requirements of the user or experiment. The tuning of properties is possible by varying the length and branching of the alkyl groups that are incorporated to the cation. There is also the potential for task-specific ionic liquids (TSILs) to be produced.
When observing the physical properties of ionic liquids, studies have shown that the melting point is influenced by the charge distribution on the ions, H-bonding ability,

symmetry of the ions and van der Waals interactions. When comparing the melting points between NaCl and 1-propyl- 3-methylimidazolium chloride, 803 °C and 60 °C respectively, it is clear that the reduction in the melting temperature is caused by replacing the small inorganic cations by bulky asymmetric organic cations. Many ionic liquids undergo supercooling which make their melting point rather uncertain. Many RTILs (room temperature ionic liquid) can be supercooled to more than 100 K below the melting point and in many cases they then form a glass (Marsh et al., 2004).

Most of the known ionic liquids are denser than water, with values between 1-1.6 g/cm^{-3}. The density decreases with increasing length of the alkyl chain in a systematic manner (Marsh et al., 2004). In most cases ionic liquids are viscous, similar to that of oils. Their high viscosity is a disadvantage because it will negatively affect the mass transfer and power requirements for mixing heterogeneous liquid-liquid systems. An increase in length of one or both alkyl chains in the cation does not automatically result in a direct increase of the viscosity.

The first published report of dissolution of cellulose in an ionic liquid ([N-ethylpyridinium][Cl] in the presence of a nitrogen base) dates back to 1934 (Graenacher 1934).

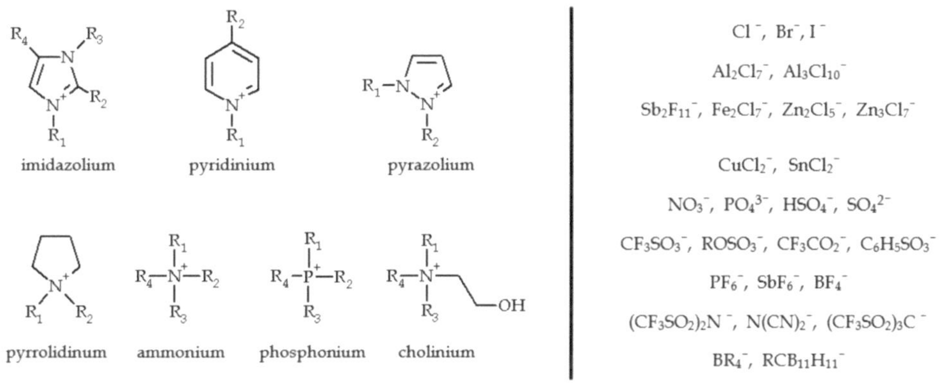

Fig. 3. Main cations and anions described in literature adopted from Olivier-Bourdigou et al., 2010.

Today, there are more than 20 ionic liquids which are known to dissolve cellulose (Barthel & Heinze, 2006). Ionic liquids are also able to dissolve other biomaterials (Fort et al., 2006), such as lignin and lignocelluloses (Kilpeläinen et al., 2007). Dissolved cellulose can be precipitated and separated from lignin and hemicelluloses by the addition of anti-solvents, such as water (Fort et al., 2006). Recent patents also prove that the pretreatment of lignocellulose using ionic liquids (ILs) is of industrial interest, and the commercialisation of such processes is underway (Myllymäki & Aksela, 2008; Zhu, 2008).

The ability of ionic liquids to dissolve cellulose depends on the nature of the native cellulose (its DP and crystallinity) on the operating conditions (temperature, reaction time, initial concentration of cellulose in the IL) and the presence of impurities (water). The use of a non-dried ionic liquid can affect the solubility of cellulose, so much that severely dried ionic liquid is needed to achieve an optimal dissolution (Vitz et al., 2009).

The ionic liquids which are most commonly considered for extraction and synthesis are those with cations based on an imidazolium or pyridinium ring with one or more alkyl groups attached to the nitrogen or carbon atoms. Quaternary ammonium salts have been widely used for electrochemical synthesis (Marsh et al., 2004).

4.2 Impurities in ionic liquids

Ionic liquids may contain several impurities, such as halides, water and volatiles. Volatile impurities, primary originating from starting materials or solvents used in synthesis, can easily be removed from the non-volatile ionic liquid by evaporation.

In ionic liquids possessing halide impurities, these can be removed by halide exchange reactions that precede more or less quantitatively causing residual amounts of halide impurities in the final product. Residual halide can seriously affect the usefulness of the ionic liquid as a solvent for a given chemical reaction (Fukaya et al., 2006). Some physiochemical properties are strongly dependent on the presence of halide impurities, e.g. in the biomass dissolution. The halide can act as a catalyst poison, as a stabilising ligand, nucleophile or reactant, depending on the chemical nature of the reaction (Wagner & Hilgers, 2008).

Protic impurities have to be considered for ionic liquids which have been produced by an exchange reaction using a strong acid and for ionic liquids that are sensitive to hydrolysis. It is easy to control if there are acidic impurities in an ionic liquid that is not fully hydrophilic. The ionic liquid is added to water and the pH is measured in the aqueous phase. If the pH is below seven (acidic solution), the ionic liquid needs to be washed with water until it becomes neutral (Wagner & Hilgers, 2008).

Impurities from incomplete metathesis reactions can also contaminate the ionic liquid. This can occur if the alkali salt that is used in the metathesis reaction shows significant solubility in the formed ionic liquid. This may not be a problem for some catalytic reactions (the alkali cation may not affect the catalytic cycle of a transition metal catalyst), but it is of great relevance for the physicochemical properties of the ionic liquid (Wagner & Hilgers, 2008).

Water is found in all ionic liquids however this may or may not be a problem for some applications. In all cases, the approximate amount of water in the used ionic liquid should be known. The fact is that water in an ionic liquid may not be inert. The presence of water can have significant influence on the physicochemical properties of the ionic liquid, on its stability and on the reactivity of catalysts dissolved in the ionic liquid (Swatloski et al., 2002; Liebert & Heinze, 2008).

Before researching with the aid of a specific ionic liquid, one should be aware of the level of impurities present. A good commercial ionic liquid should be free of organic volatiles, halides (if not a halide of IL) and all ionic impurities. Though, it may be coloured and contain some traces of water.

To remove water, the ionic liquid should be dried at 60 °C over night in a vacuum. The water content should then be checked with IR before use and recorded in any descriptions or documentation that is relevant to the ionic liquid used in experiments.

If a colourless ionic liquid is required, the colour can be removed by chromatography/filtration over silica 60. For this purification method, the ionic liquid is dissolved in a volatile solvent such as CH_2Cl_2 (Wagner & Hilgers, 2008).

Finally, the purity of an ionic liquid is a very important issue when measuring the physical properties.

4.3 Effect of the anion on the dissolution

Whilst attempting to dissolve carbohydrates, many different ionic liquids were explored and it was concluded that good dissolution of cellulose may be obtained using halide based ionic liquids as presented in Table 1. It is well known that higher the anion concentration the better the solubilisation become. Other advantages include the strong electronegativity of the chloride anion and its small size (Dadi et al., 2006). This is especially important since the chloride anion, being a small hydrogen-bond acceptor, is a good selection in cellulose

Substrate	DP (cellulose)	Solvent	Solubility	T (°C)
Cellulose	286	[BMI][Cl]	18 (wt%)	83
Cellulose	593	[BMI][Cl]	13 (wt%)	83
Cellulose	~1000	[BMI][Cl]	3 (wt%)	70
Cellulose	~1000	[BMI][Cl]	10 (wt%)	100
Cellulose	~1000	[BMI][Cl]	25 (wt%)	110
Cellulose	1108	[BMI][Cl]	10 (wt%)	83
Cellulose	6500	[BMI][Cl]	6 (wt%)	80
Cellulose	225	[BMI][Fm]	8 (wt%)	110
Cellulose	250	[BMMI][Cl]	4.5 (wt%)	110
Cellulose	286	[BMI][Cl]	9 (wt%)	80
	593	[BMI][Cl]	6 (wt%)	80
	1198	[BMI][Cl]	4 (wt%)	80
Cellulose	650	[AMI][Cl]	14.5 (wt%)	80
	1600	[AMI][Cl]	8 (wt%)	80
Cellulose	286	[AMI][Br]	12 (wt%)	80
	593	[AMI][Br]	4 (wt%)	80
	1198	[AMI][Br]	4 (wt%)	80
Cellulose	250	[AMI][Fm]	10 (wt%)	60
	250	[AMI][Fm]	22 (wt%)	85
Cellulose	286	[EMI][Cl]	12 (wt%)	80
	593	[EMI][Cl]	6 (wt%)	80
	1198	[EMI][Cl]	4 (wt%)	80
Cellulose	795	[EMI][Ac]	20 (wt%)	80
Cellulose	200-250	[EMI][PO_2(H)(OMe)]	10 (wt%)	45
Cellulose	200-250	[EMI][PO_2(Me)(OMe)]	10 (wt%)	55
Cellulose	200-250	[EMI][$PO_2(OMe)_2$]	10 (wt%)	65
Cellulose	200-250	[EMI][$PO_2(OEt)_2$]	14 (wt%)	100
Cellulose	200-250	[MMI][$PO_2(OMe)_2$]	10 (wt%)	100
Cellulose	286	[BMPy][Cl]	39 (wt%)	105
	593	[BMPy][Cl]	37 (wt%)	105
	1198	[BMPy][Cl]	12 (wt%)	105
Cellulose	200-250	[($HO(CH_2)_2$)MI][Cl]	6.8 (wt%)	70

Table 1. Examples of solubility of cellulose in ILs adopted from Olivier-Bourdigou et al., 2010.

dissolution (Holm, 2010 unpublished data; Swatloski et al., 2002). The halogen anion containing ionic liquids, such as [BMIM][Br] and [BMIM][I], have been found not to be good solvents for cellulose (Zavrel et al., 2009). The fairly high melting point (> 70 °C for [BMIM][Cl]) of ionic liquids containing the chloride anion could be a technical drawback and possibly limit their practical applications in cellulose processing (Wu et al., 2009). Very high dissolution temperatures (mainly above 80 °C) are often required for dissolving cellulose. This may result in cellulose modification by ionic liquids themselves and in thermal decomposition of ionic liquids. Some organohalogenides which have uncertain toxicity and are hazardousness can be produced. In general, high viscosity of [BMIM][Cl] and the high hygroscopicity of halide in ionic liquids make their handling more difficult (Vitz et al., 2009).

Formate, acetate or phosphate based imidazolium ionic liquids have also been studied. They have shown potential to dissolve cellulose under mild conditions such as ionic liquids containing formate anions were found to be good solvents for dissolving cellulose (Olivier-Bourdigou et al., 2010). Formate ionic liquids generally exhibit low thermal stability due to decarboxylation (Fukaya et al., 2008) and they are known to be basic and unsuitable for enzymes (Zhao et al., 2009). Asetate-based ionic liquids were found to be interesting due to their low melting point, lower viscosity and have a less toxic and corrosive character compared with the chloride-based ionic liquids (Feng & Chen, 2008). Futhermore, phosphate-based ionic liquids provide high thermal stability in the 260-290 °C range and low viscosities (Olivier-Bourdigou et al., 2010).

In general, acetate based ionic liquids are less viscous than chloride based ionic liquids and are thermally more stable than formate based ionic liquids (Fukaya et al., 2008).

4.4 Effect of the cation on the dissolution

Studies have shown the existence of weak interaction between cation of ionic liquids and cellulose (El Seoud et al., 2007). The cations interact in the dissolution process while their role in the dissolution mechanism should not be neglected (Dadi et al., 2006). When increasing the length of the alkyl chain on the dialkyl imidazolium cation (with chloride anion) this leads to a decrease in the cellulose solubilisation (Olivier-Bourdigou et al., 2010). The introduction of an allyl group on the imidazolium cation has been shown to provide excellent improvements in dissolving of cellulose. 14.5 % of cellulose can be dissolved using [AMIM][Cl] compared to, 13 % of the same cellulose when [BMIM][Cl] was used (Olivier-Bourdigou et al., 2010). We can also report similar results (Holm et al., 2010 unpublished data). [AMIM][Cl] is a smaller cation because it contains only three carbon atoms in the side chain, while the strong polar of the double bond seems to be an essential part of the cation (Feng & Chen, 2008). Overall, it can be said that relatively small cations are efficient in dissolving cellulose. Zhang et al., 2005 discovered that $[AMIM]^+$ is more powerful in the dissolution of cellulose than $[BMIM]^+$ due to its smaller size. Kosan et al., 2008 published that [EMIM][Cl] was a better choice (i.e. is smaller) than [BMIM][Cl], because of the size of the cation. If the cation is large it decreases the ability to form hydrogen bonds with cellulose (Zhao et al., 2008). Vitz et al., 2009 was able to show [EMIM][Br] to have a poor dissolution effect on the cellulose while Zavrel et al., 2009 indicated that [EMIM][Ac] is the most efficient solvent for cellulose and [AMIM][Cl] is the most suitable ionic liquid for dissolution of hardwood and softwood.

Functional groups in the cation can also be crucial for the dissolution of cellulose. The presence of a hydroxyl end-group in the cation decreases the solubility of cellulose in the corresponding ionic liquid (Mäki-Arvela et al., 2010). The longer-chain substituted ionic liquids ([C_6MIM][Cl] and [C_8MIM][Cl]) appear to be less efficient at dissolving cellulose. This can be due to the reduced effective chloride concentration within these liquids (Swatloski et al., 2002).

4.5 Dissolution process in ionic liquids

The main properties of ILs affecting the dissolution and functional modification of cellulose and carbohydrates are their polar characteristics, the basic form of the ionic liquid anions and their ability to generate H-bonds. The most important contribution is due to the hydrogen´s bonding ability that the IL anion has (chlorine forms an H-bond with the hydroxyl groups of the biomolecule) (Olivier-Bourbigou et al., 2010). In a typical process when dissolving cellulose, the reaction medium has to be heated. Therefore, thermal stability is also an important aspect. Depending on the purity of the IL, the melting points of IL can differ.

In order to dissolve cellulose, the ionic liquid should satisfy the following three conditions: 1) the anion must be a good hydrogen bond acceptor; 2) the cation should be a moderate hydrogen bond donator because the cation has the most moderate activated hydrogen for forming hydrogen bonding with oxygen atoms of the hydroxyls of cellulose and 3) the size of the cation should not be too large (Zhang et al., 2010). The IL solutions become increasingly viscous as the length of the dissolved cellulose oligomer increases (Moulthorp et al., 2005).

Zhang et al., 2005 speculated that the small cation $[AMIM]^+$ favours the attack on the oxygen atom of the hydroxyl group in cellulose and that the less electronic chemical structure

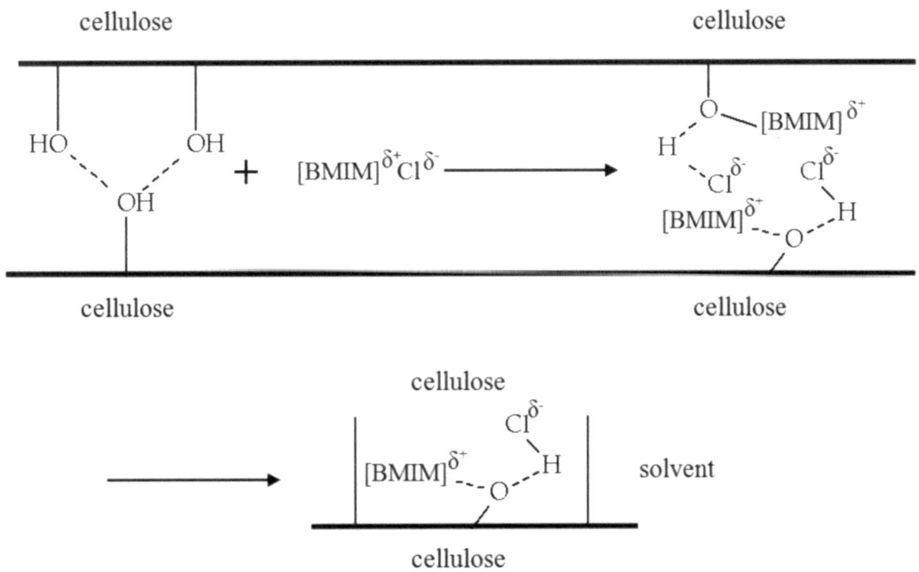

Fig. 4. Dissolution mechanism of cellulose in [BMIM][Cl] adopted from Feng & Chen, 2008.

caused by the alkyl group also enhances the interaction between cations in ionic liquid and oxygen atoms of the hydroxide group in cellulose. Above the critical temperature, the ions in [AMIM][Cl] dissociate to individual $[AMIM]^+$ and Cl^- ions. The free Cl^- ions associate with the cellulose hydroxyl proton and the free cations complex with the cellulose hydroxyl oxygen. This leads to the disruption of hydrogen bonds in cellulose and to the dissolution of cellulose. Fukaya et al., 2006 discovered the same mechanism occurring in their studies.

Demonstrated by Zhang et al., 2010, studies showed that the hydrogen bonding of hydroxyls in cellulose with the $[CH_3COO^-]$ anion and $[AMIM]^+$ cation is the major force behind cellulose dissolution in an ionic liquid. The relatively small acetate anion favours the formation of a hydrogen bond with hydrogen atoms in hydroxyls, while the aromatic protons in the bulky cation imidazolium (especially H2) prefer to associate with the oxygen atoms of hydroxyls with less steric hindrance.

The carbohydrate dissolving ionic liquids usually contain anions such as Cl^-, $HCOO^-$ and OAc^-, which form strong hydrogen bonds with carbohydrates when being dissolved. For this reason, these ionic liquids are more likely to denature the enzymes and thereby preventing a further enzymatic modification of dissolved carbohydrates in ionic liquids. To overcome this problem, development over the past two years has been in the form of synthesizing new glycol-substituted ionic liquids that are capable of dissolving carbohydrates but not denature lipase (Zhao et al., 2008).

When using ionic liquids as cellulose solvents, it is important to investigate if any structural changes occur in the cellulose molecule during the cellulose dissolution, i.e. if any derivatisation or depolymeration of cellulose takes place. Ionic liquids, such as 1-butyl-3-methylimidazolium chloride, 1-methyl-3-methylpyridinium chloride and N-benzyl-N,N-dimethyltetradecylammonium chloride, were found to be non-derivative solvents for cellulose (Heinze et al., 2005). When dissolving dry fibre sludge (fibre sludge is pure cellulose) from a pulp mill in 1-butyl-3-methylimidazolium chloride, the IL was found to be non-derivative (Holm, 2010, unpublished data).

Zavrel et al., 2009 reported that a structural change of cellulose was observed when dissolved cellulose was precipitated with water. This could potentially enhance subsequent hydrolysis. The cellulose regenerated from ionic liquids was essentially found to be amorphous and porous, which was much more prone to degradation by cellulases (Dadi et al., 2006). The celluloses regenerated by rapid precipitation of the dissolved cellulose dopes with an anti-solvent (water) have demonstrated a great improvement on enzymatic hydrolysis kinetics (Kuo & Lee, 2009).

Glucose degradation products observed with acid pretreatment or hydrolysis include hydroxymethyl furfural (HMF) and furfural (Dadi et al., 2006). In the presence of water, HMF produces levulinic and formic acid that inhibit downstream fermentation. For example, if a [BMIM][Cl] pretreated fibre sludge sample is left standing for a long period of time (weeks) before analysis, 5-HMF and levulinic acid can be found in the solution (Holm, 2010 unpublished data).

Following on from this, the dissolved cellulose in the ionic liquid, can be precipitated from its solution by the addition of a non-solvent (anti-solvent), such as water, methanol, ethanol or acetone (Swatloski et al., 2002; Kosan et al., 2008). When the anti-solvent (e.g. water) is added to the solution the ions in the ionic liquid are extracted into an aqueous phase (Mäki-Arvela et al., 2010). The resulting regenerated cellulose can then be separated by filtration or

Fig. 5. Dissolution of dry fibre sludge in [BMIM][Cl].

centrifugation while the ionic liquid can be recovered through distillation of the anti-solvent. The recovery of ionic liquids still requires much more research, although studies on biphasic systems in regards to the recovery have already been performed.

The regenerated cellulose can differ from the native cellulose in both the macro- and micro-structure while the degree of crystallinity can be changed due to the changes in the regeneration conditions. The regenerated cellulose can have the same degree of polymerisation and polydispersity as the native cellulose but this depends on the operating conditions in the treatment (Olivier-Bourdigou et al., 2010). The regenerated cellulose demonstrated improved enzymatic hydrolysis kinetics (Dadi et al., 2006).

During the cellulose regeneration, careful attention should be taken when removing ionic liquid residues as their presence has detrimental effect on the cellulase activity depending on the amount of ionic liquid remaining. Such an inactivation may affect the final concentration levels of the total sugars reduction and glucose after the completion of hydrolysis (Zhao et al., 2009).

5. Hydrolysis of biomass

Much research is being undertaken in the conversion of lignocellulosic biomass into valuable products through a sugarplatform which includes two processes, namely the hydrolysis of lignocellulosic materials to monosaccharides and transforming the gained sugars to bio-based products. The lignocellulosic material, provided by nature, is resistant to hydrolysis due to their structure.

Until now, hydrolysis of lignocelluloses to monosaccharides is usually catalysed either by enzymes or acid catalysts under heterogeneous conditions. Concentrated acids such as H_2SO_4 and HCl have been the preferred choice to treat lignocellulosic materials. They are powerful agents for cellulose hydrolysis but they are also toxic, hazardous and corrosive. The concentrated acid needs to be recovered after hydrolysis.

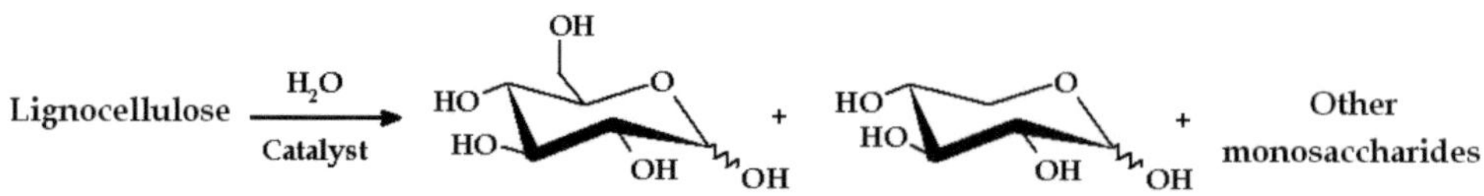

Fig. 6. Hydrolysis of lignocellulosic material adopted from Li et al., 2008.

The main factors that affect the enzymatic hydrolysis of lignocellulosic material can be divided in two groups, namely enzyme-related and substrate-related factors, though many of them are interrelated in the hydrolysis process.

Enzymatic hydrolysis is considered as the most promising method in breaking down cellulosic biomass to fermentable sugars. However, one of the main obstacles to this is the poor enzymatic hydrolysis of cellulose which cannot be explained due to a combination of many factors. There are many reasons behind the poor hydrolysis of cellulose (Zhao et al., 2008). Enzyme factors such as synergism, adsorption, thermal stability and product inhibition influence the cellulose hydrolysis while the degree of polymerisation, crystallinity, accessible surface area, hemicellulose and lignin content, distribution and specific surface area are all substrate related factors that limit the hydrolysis too. Enzyme related factors also include the inhibition of by-products such as cellobiose and D-glucose (Zhao et al., 2009).

Hydrochloric acid/[BMIM][Cl] has been found to be an efficient system for the hydrolysis of lignocellulosic biomass but the reason for this is still unclear. The consensus is that lignocellulosic materials dissolved in [BMIM][Cl] promote the dispersion of the majority of cellulose and hemicelluloses molecules, leading to the substrates being more exposed to H^+ in the homogeneous solution. This in turn should be an advantage when compared to conventional enzymatic or dilute acid hydrolysis under heterogeneous conditions. The result of this is that the physical barrier for hydrolysis has been successfully extinguished through the formation of a solution. The dissociated Cl^- and $[BMIM]^+$ may also weaken the glycosidic linkage contained in the substrates (Swatloski et al., 2002).

Dadi et al., 2006 reported that the cellulose structure and hydrolysis rates were neither affected by the extended cellulose incubation period in [BMIM][Cl], nor by the selection of antisolvents (water, methanol or ethanol). After being regenerated from IL solutions, celluloses exhibited lower crystallinity and higher cellulose adsorption compared with untreated substrates. These regenerated celluloses were hydrolysed by cellulose much faster than untreated substrates (Zhao et al., 2009).

Maleic acid effectively accelerated the hydrolysis of cellulose in [BMIM][Cl]. The catalytic activity on hydrolysis of lignocelluloses in [BMIM][Cl] for studied acids roughly followed the following sequence: hydrochloric acid > nitric acid > sulphuric acid > maleic acid > phosphoric acid. (Li et al., 2008)

6. Conventional methods for pretreatment

Pretreatment is a significant tool for practical cellulose conversion processes. Pretreatment methods are either chemical or physical but in some cases both can be used. Physical pretreatment methods include comminution (mechanical reduction in a particular biomass size), steam explosion and hydrothermolysis. Chemically, acids or bases are used to promote hydrolysis. The most commonly used acid and base is sulphuric acid and sodium hydroxide respectively (Mosier et al., 2005) while cellulose solvents are another type of chemical additive. Solvents such as alkaline peroxide, ozone, organosolv, glycerol, dioxane, and phenol are known to disrupt the cellulose structure and promote hydrolysis. Concentrated mineral acids, ammonia-based solvents, aprotic solvents, metal complexes and wet oxidation are known to reduce cellulose crystallinity, dissolve hemicelluloses and disrupt the linkage between lignin and cellulose. Sulphuric acid has initially been applied to remove hemicelluloses in combination with the breakdown of cellulose to glucose before acid hydrolysis. However, before starting any fermentation process the acid needs to be neutralized. Following from dilute acid pretreatment condensed lignin remains on the surface of crystalline cellulose, which can block the enzyme access to the substrate in sugar production.

Li et al., 2010 investigated the difference between dilute acid and ionic liquid pretreatment of Switch grass, and came to the conclusion that the ionic liquid [C2mim][OAc] and dilute sulphuric acid both remove lignin and hemicelluloses successfully, whilst producing higher glucan content compared with untreated materials. Furthermore, ionic liquid pretreatment removes much more lignin and less hemicellulose than dilute acid pretreatment. Li et al., 2010 results show that the ionic liquid pretreatment resulted in a significant level of delignification. The results also confirm that the dilute acid process removes the majority of hemicelluloses present and small amounts of amorphous cellulose.

7. Conclusion

Lignocellulosic materials which include wood, grass, forestry waste, agricultural residues and municipal solid waste are sustainable energy sources of the future; however, utilizing the unique properties of these materials is a challenge. Due to the different physio-chemical characteristics of lignocellulosic materials a suitable pretreatment technology should be chosen. The implementation of a pretreatment step is necessary to release the components of lignocellulosic materials for usage. The effectiveness of lignocelluloses pretreatment is one of the key factors to a successful conversion of the originally low-cost material into a sugar platform and from there, into biofuels or biofuel intermediates. Cellulose crystallinity, the protective lignin shield and the sheathing of hemicelluloses makes it resistant to enzymatic and chemical hydrolysis.

The dissolution of cellulose using ionic liquids has provided a platform for the utilization of green cellulose in addition to providing a comprehensive application of lignocellulosic materials in a green way. The excellent chemical and physical properties of task-specific ionic liquids make them effective and rapid cellulose solvents. This allows optimal ionic liquids to be designed for a specific application. Compared with traditional solvents ionic liquids express properties such as broad liquid regions, high thermal stabilities and negligible vapour pressure. The viscosity of the ionic liquid is at a disadvantage because it negatively affects the mass transfer. Finally, further research should be preformed on the toxicity of ionic liquids.

8. References

Alvira, P.; Tomás-Pejó, M.; Ballesteros, M. & Negro, M.J. (2010). Pretreatment technologies for an efficient bioethanol production process based on enzymatic hydrolysis: A review. *Bioresource Technology*, Vol. 101, 4851-4861.

Anderson, J.L.; Ding, J.; Welton, T. & Armstrong, D.W. (2002). Characterizing ionic liquids on the basis of multiple salvation interactions. *Journal of the American Chemical Society*, Vol. 124, 14253-14254.

Barthel, S. & Heinze, T. (2006). Acylation and carbanilation of cellulose in ionic liquids. *Green Chemistry*, Vol. 8, 301-306.

Bhat, M.K. & Bhat, S. (1997) Cellulose degrading enzymes and their potential industrial applications. *Biotechnology Advances*, Vol. 15, No. 3-4, 583-620.

Brennecke, J.F. & Maginn, E.J.(2001) Ionic liquids: Innovative fluids for chemical processing. *AIChE Journal*, Vol. 47, No. 11, 2384-2389.

Cara, C.; Ruiz, E.; Ballesteros, M.; Manzanares, P.; Negro, M.J. & Castro, E. (2008) Production of fuel ethanol from steam-explosion pretreated olive tree pruning. *Fuel* Vol. 87, No. 6, 692-700.

Chandra, R.P.; Bura, R.; Mabee, W.E.; Berlin, A.; Pan, X. & Saddler, J.N. (2007) Substrate pretreatment: The key to effective enzymatic hydrolysis of lignocellulosics? *Biofuels* Vol. 108, 67-93.

da Costa Sousa, L.; Chundawat, S.; Balan, V. & Dale, B. (2009). 'Cradle-to-grave' assessment of existing lignocellulose pretreatment technologies. *Current Opinion in Biotechnology*, Vol. 20, 339-347.

Dadi, A.; Varanasi, S & Schall, C. (2006). Enhancement of Cellulose Saccharification Kinetics Using an Ionic Liquid Pretreatment Step. *Biotechnology and Bioengineering*, Vol. 95, No. 5, 904-910.

El Seoud, O.A.; Koschella, A.; Fidale, L.C.; Dorn, S. & Heinze, T. (2007). Applications of ionic liquids in carbohydrate chemistry: a window of opportunies. *Biomacromolecules*, Vol. 8, 2629-2647.

Feng, L. & Chen, Z-I. (2008). Research progress on dissolution and functional modification of cellulose in ionic liquids. *Journal of Molecular Liquids*, Vol. 142, 1-5.

Fort, D.A.; Remsing, R.C; Swatloski, R.P; Moyna, P; Moyna, G. & Rogers, R.D. (2006). Can ionic liquids dissolve wood? Processing and analysis of lignocellulosic materials with 1-*n*-butyl-3-methylimidazolium chloride. *Green Chemistry*, Vol.9, 63-69.

Fukaya, Y.; Hayashi, K.; Wada, M. & Ohno, H. (2008). Cellulose dissolution with polar ionic liquids under mild conditions: required factors for anions. *Green Chemistry*, Vol.10, 44-46.

Fukaya, Y.; Sugimoto, A. & Ohno, H. (2006). Superior Solubility of Polysaccharides in Low Viscosity, Polar and Halogen-Free 1,3-Dialkylimidazolium Formates. *Biomacromolecules*, Vol. 7, No. 12, 3295-3297.

Graenacher, C. (1934). Cellulose Solution. US patent 1,943,176.

Heinze, T.; Schwikal, K. & Barthel, S. (2005). Ionic liquids as reaction medium in cellulose functionalization. *Macromolecular Bioscience*, Vol. 5, 520-525.

Holm, J.; Asikkala, J. & Lassi, U. (2009). Cellulose fractionation from real sludge samples through ionic liquid [BMIM][Cl]. *Chemical Engineering Transactions* Vol. 17, 1645-1650.

Holm, J. (2010) Unpublished data.

Huber, G.W.; Iborra, S. & Corma, A. (2006) Synthesis of transportation fuels from biomass: chemistry, catalysts and engineering. *Chemical Reviews* Vol. 106, No. 9, 4044-4098.

Huddleston, J.G.; Visser, A.E.; Reichert, W.M.; Willauer, H.D.; Broker, G.A. & Rogers, R.D. (2001). Characterization and comparison of hydrophilic and hydrophobic room temperature ionic liquids incorporating the imidazolium cation. *Green Chemistry*, Vol. 3, 156-164.

IPCC (2007). The Fourth Assessment Report of the Intergovernmental Panel on Climate Change. IPCC, Geneva, Switzerland.

Kaar, W.E. & Holtzapple, M.T. (2000). Using lime pretreatment to facilitate the enzymic hydrolysis of corn stover. *Biomass Bioenergy*, Vol. 18, No. 3, 189-199.

Kamm, B. & Kamm, M. (2004). Principles of biorefineries. *Applied Microbiology and Biotechnology*, Vol. 64, 137-145.

Kilpeläinen, I.; Xie, H.; King, A.; Granström, M.; Heikkinen, S. & Argyropoulus, D.S. (2007). Dissolution of wood in ionic liquids. *Journal of Agricultural and Food Chemistry*, Vol. 55, No. 22, 9142-9148.

Kosan, B.; Michels, C. & Meister, F. (2008). Dissolution and forming of cellulose with ionic liquids. *Cellulose*, Vol. 15, 59-66.

Kuo, C-H. & Lee, C-K. (2009). Enhancement of enzymatic saccharification of cellulose by cellulose dissolution pretreatments. *Carbohydrate Polymers*, Vol. 77, 41-46.

Larsson, S.; Palmqvist, E.; Hahn-Hägerdal, B.; Tengborg, C.; Stenberg, K.; Zacchi, G. & Nilvebrant, N-O. (1999). The generation of fermentation inhibitors during dilute acid hydrolysis of softwood. *Enzyme and Microbial Technology*, Vol. 24, 151-159.

Li, C.; Wang, Q. & Zhao Z.K. (2008). Acid in ionic liquid: An efficient system for hydrolysis of lignocellulose. *Green Chemistry*, Vol. 10, 177-182.

Li, C.; Knierim, B.; Manisseri, C.; Arora, R.; Scheller H.V.; Auer, M.; Vogel, K.P.; Simmons, B.A. and Singh, S. (2010). Comparison of dilute acid and ionic liquid pretreatment of switchgrass: Biomass recalcitrance, delignification and enzymatic saccharification. *Bioresourse Technology*, Vol. 101, 4900-4906.

Liebert, T. & Heinze, T. (2008). Interaction of ionic liquids with polysaccharides 5. solvents and reaction media for modification of cellulose. *Bioresources*, Vol. 3, No. 2, 576-601.

Lloyd, T.A. & Wyman, C.E. (2005). Combined sugar yields for dilute sulfuric acid pretreatment of corn stover followed by enzymatic hydrolysis of the remaining solids. *Bioresource Technology*, Vol. 96, No. 18, 1967-1977.

Lynd, L.R.; Weimer, P.J.; van Zyl, W.H. & Pretorius, I.S. (2002) Microbial cellulose utilization: fundamentals and biotechnology. *Microbiology and Molecular Biology Reviews*, Vol. 66., No. 3, 506-577.

Lynd, L.R.; Wyman, C.E. & Gerngross, T.U. (1999) Biocommodity engineering. *Biotechnology progress*, Vol. 15, No. 5, 777-793.

March, K.N.; Boxall, J.A. & Lichtenthaler, R. (2004). Room temperature ionic liquids and their mixtures - a review. *Fluid Phase Equilibria*, Vol. 219, 93-98.

Mosier, N.; Wyman C.; Dale, B.; Elander, R.; Lee, Y.Y.; Holtzapple, M. & Ladisch, M. (2005) Features of promising technologies for pretreatment of lignocellulosic biomass. *Bioresource Technology*, Vol. 96, 673-686.

Moulthorp, J.S.; Swatloski, R.P.; Moyna, G. & Rogers, R.D. (2005). High-resolution ^{13}C NMR studies of cellulose and cellulose oligomers in ionic liquid solutions. *Chemical Communications*, 1557-1559.

Myllymäki, V. & Aksela, R. (2008). Dissolution method. US patent 2008/0023162 A1.

Mäki-Arvela, P.; Anugwom, I; Virtanen, P.; Sjöholm, R. & Mikkola, J.P. (2010). Dissolution of lignocellulosic materials and its constituents using ionic liquids-A review. *Industrial Crops and Products*, Article in press.

Olivier-Bourbigou, H.; Magna, L. & Morvan, D. (2010). Ionic liquids and catalysis: Recent progress from knowledge to applications. *Applied Catalysis A: General,* Vol. 373, 1-56.

Rogalinski, T.; Ingram, T & Brunner, G. (2008). Hydrolysis of lignocellulosic biomass in water under elevated temperatures and pressures. *The Journal of Supercritica Fluids,* Vol. 47, 54-63.

Sánches, Ó.J. & Cardona, C.A. (2008). Trends in biotechnological production of fuel ethanol from different feedstocks. *Bioresource Technology,* Vol.99, 5270-5295.

Swatloski, R.P.; Spear, S.K.; Holbrey, J.D. & Rogers, R.D. (2002). Dissolution of Cellose with Ionic Liquids. *Journal of the American Chemical Society,* Vol. 124, 4974-4975.

Vitz, J.; Erdmenger, T.; Haensch, C. & Schubert, U.S. (2009). Extended dissolutionstudies of cellulose in imidazolium based ionic liquids. *Green Chemistry.* Vol. 11, 417-424.

Wagner, M. & Hilgers, C. (2008). *Ionic liquids in synthesis,* Wiley-VCH, ISBN 978-3-527-31239-9, Weinheim

Weil, J.; Westgate, P.; Kohlmann, K. & Ladisch, M.R. (1994). Cellulose pretreatments of lignocellulosic substrates. *Enzyme and Microbial Technology,* Vol. 16, No. 11, 1002-1004.

Wu, B.; Liu, W.W.; Zhang, Y.M.; & Wang, H.P. (2009). Do We Understand the Recyclability of Ionic Liquids? *Chemistry A European Journal.* Vol. 15, 1804-1810.

Wyman, C.E.; Dale, B.E.; Elander, R.T.; Holtzapple, M.; Landisch, M.R. & Lee, Y.Y. (2005). Coordinated development of leading biomass pretreatment technologies. *Bioresource Technology,* Vol. 96, 1959-1966.

Zavrel, M.; Bross, D; Funke, M.; Büchs, J. & Spiess, A.C. (2009). High-throughput screening for ionic liquids dissolving (ligno-)cellulose. *Bioresource Technology,* Vol. 100, 2580-2587.

Zhang, H.; Wu, J.; Zhang, J. & He, J. (2005). 1-Allyl-3-methylimidazolium Chloride Room Temperature Ionic Liquid: A New and Powerful Nonderivatizing Solvent for Cellulose. *Macromolecules,* Vol. 38, 8272-8277.

Zhang, J.; Zhang, H.; Wu, J.; Zhang, J.; He, J. & Xiang, J. (2010). NMR spectroscopic studies of cellobiose solvation in EmimAc aimed to understand the dissolution mechanism of cellulose in ionic liquids. *Physical Chemistry Chemical Physics,* Vol. 12, 1941-1947.

Zhang, Z. & Zhao, Z.K. (2010). Microwave-assisted conversion of lignocellulosic biomass into furans in ionic liquid. *Bioresource Technology,* Vol. 101, 1111-1114.

Zhao, H.; Baker, G.A.; Song, Z.; Olubajo, O.; Crittle, T. & Peters, D. (2008). Designing enzyme-compatible ionic liquids that can dissolve carbohydrates. *Green Chemistry,* Vol. 10, 696-705.

Zhao, H.; Jones, C.L.; Baker, G.A.; Xia, S.; Olubajo, O. & Person, V.N. (2009). Regenerating cellulose from ionic liquids for an accelerated enzymatic hydrolysis. *Journal of Biotechnology,* Vol. 139, 47-54.

Zhbankov, R.G. (1992). Hydrogen bonds and structure of carbohydrates. *Journal of Molecular Structure,* Vol. 270, 523-539.

Zheng, Y.Z.; Lin, H.M. & Tsao, G.T. (1998). Pretreatment for cellulose hydrolysis by carbon dioxide explosion. *Biotechnology Progress,* Vol. 14, No. 6, 890-896.

Zhu, S.; Wu, Y.; Chen, Q.; Yu, Z.; Wang, C.; Jin, S.; Ding, Y. & Wu, G. (2006). Dissolution of cellulose with ionic liquids and its application: a mini-review. *Green Chemistry,* Vol. 8, 325-327.

Zhu, S.D. (2008). Use of ionic liquids for the efficient utilization of lignocellulosic materials. *Journal of Chemical Technology and Biotechnology,* Vol. 83, No. 6, 777-779.

25

Application of Ionic Liquids in Membrane Separation Processes

Cserjési Petra and Bélafi-Bakó Katalin
Pannon University
Hungary

1. Introduction

Membrane separation processes are one of the most widely researched and fastest growing separation techniques of our century because of their advantages compared to classical (absorption, liquid-liquid extraction, distillation, etc.) processes, such as simple and compact set-up (Stern, 1994), easy operation at ambient temperature and pressure (Porter, 1990), simple up- and downscaling (Atchariyawut et al., 2006), better energy efficiency (Acharya et al., 2008), high purity products (Jönsson & Mathiasson, 1999) and much lower environmental impact (Soni et al., 2009).

They are successfully used in the food and dairy-, chemical-, beauty-, pharmaceutical- and biotechnology industry, in waste water treatment and in medical applications (Nath, 2008). However efficient these methods are nowadays, their continuous improvement is essential to be suitable for the ever changing requirements of the industries and the environmental laws and regulations.

The unique behaviour of ionic liquids, such as low melting point, negligible vapour pressure and tuneable physicochemical properties (Welton, 1999) make them the ideal candidates for membrane development. In liquid phase they can be used as bulk liquid membranes (BLM), emulsion liquid membranes (ELM) or supported liquid membranes (SLM) and after polymerization as non porous solid membrane films.

In the last decade several researchers have studied the application possibilities of ionic liquid based membranes including the transport and separation of solid matters, liquids and gases. In this chapter the most remarkable scientific results of these studies are reviewed and our latest results on gas separation applying supported ionic liquid membranes and polymerized ionic liquid membranes are described.

2. Membrane separation

The most important element of the membrane separation processes is the membrane itself. According to the international nomenclature and definition the **membrane** is a ***permselective barrier between two phases*** (Koops, 1995). It separates the influent (feed) stream into two effluent streams, known as permeate and retentate (Strathmann et al., 2006) as it is shown in Figure 1.

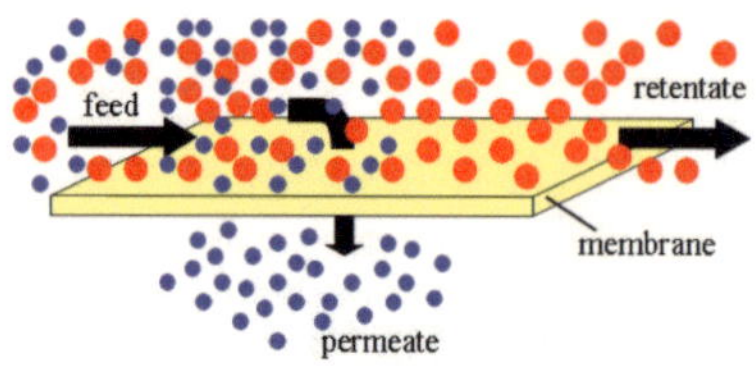

Fig. 1. Theoretical scheme of a membrane process

The separation is based on the difference between the rates of passive transport of the influent molecules. The passive transport occurs as a consequence of a driving force, which is a difference in chemical potential across the membrane (Ulbricht, 2006). It can be the difference in pressure, concentration, temperature or electric potential (Mulder, 1996).

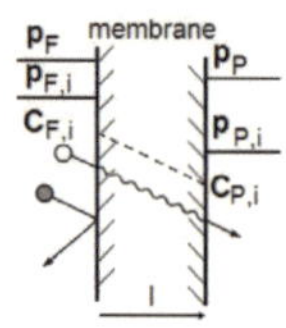

Fig. 2. Scheme of the solution diffusion mechanism

In the case of **non porous solid and liquid membranes** the separation process can be described by the solution diffusion mechanism (Figure 2), which consists of these three independent steps:

1. Molecules dissolve into the membrane material at the high pressure side
2. Molecules diffuse through the membrane according to Fick's law of diffusion
3. Molecules leave the membrane material at the low pressure side (Lashkari et al., 2008).

The rate limiting second step can be described by the following equation:

$$J_i=D_iS_i(p_{i,F}-p_{i,P})/l=P_i(p_{i,F}-p_{i,P})/l \quad (1)$$

where: J_i is the diffusive flux; D_i is the diffusion coefficient; S_i is the solubility coefficient; $p_{i,F}$ is the pressure of the component i on the feed side of the membrane; $p_{i,P}$ is the pressure of the component i on the permeate side of the membrane; l is the membrane thickness; and $P_i=D_iS_i$ is the permeability of the membrane (Baker, 2004).

The capability of the membranes to separate components can be efficiently described by the permeability and the selectivity of the membrane. In a simple component process the ideal selectivity (a^*) can be given by this equation (Stern, 1994):

$$a^*_{i/j}=P_i/P_j=(D_i/D_j)(S_i/S_j) \quad (3)$$

Considering a two component mixture, i and j the selectivity (a) is given by the following equation:

$$a_{i/j}=(C_{P,i}/C_{P,j})/(C_{F,i}/C_{F,j}) \quad (2)$$

where: $C_{P,i}$ and $C_{P,j}$ are the concentrations of the components in permeate; $C_{F,i}$ and $C_{F,j}$ are the concentrations of the components in retentate (Koros & Fleming, 1993).

In general it can be stated that liquid membranes, which are built up of a porous membrane phase and an organic liquid phase held in the pores by capillary forces, have higher selectivity values than the non porous solid membranes because of the higher liquid phase diffusivities

(Chakma, 1995). Due to the incorporation of some carrier (C), which reacts reversibly and selectively with a specific permeate in the liquid membrane, further increase in mass flux and selectivity can be achieved (Ravanchi et al., 2010). This phenomenon is commonly known as facilitated transport (Kocherginsky et al., 2007) and is represented in Figure 3.

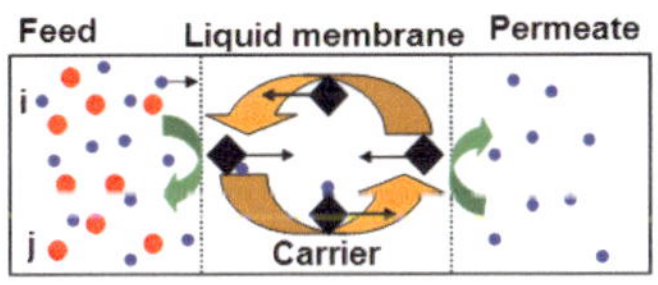

Fig. 3. Facilitated transport of permeants through liquid membranes

3. Ionic liquids

The special properties of ionic liquids, i.e. high thermal and chemical stability, low vapour pressure (Endres et al., 2008), non-flammability (Wasserschied & Welton, 2007) and tuneable physicochemical properties, make them the ideal candidates for the replacement of organic solvents in liquid membranes and hence for the realization of environmentally sound and economical membrane separation processes (Koel, 2008).
Table 1 contains the names, abbreviations and application of the ionic liquids that are most commonly used and investigated in membrane separation processes.

Ionic liquids			
Name	Abbreviation	Compounds to separate	Ref.
MEMBRANE ABSORPTION			
1-butyl-3-methylimidazolium tetrafluoroborate	$[bmim][BF_4]$	propylene/propane	Ortiz et al., 2008
MEMBRANE EXTRACTION			
1-butyl-3-methylimidazolium hexafluorophosphate	$[bmim][PF_6]$	benzene derivates	Visser et al., 2001
		tyramine and 2-methoxy-phenethylamine	Yung et al., 2005
		octane/ethylbenzene	Birdwell et al., 2006
1-alkyl-3-methylimidazolium hexafluorophosphate	$[C_nmim][PF_6]$	short-chain aliphatic carboxylic acids, e.g. acetic acid, butyric acid	Miyako et al., 2003
		penicillin G	Matsumoto et al., 2007
1,3-dialkylimidazolium tetrafluoroborate	$[dC_nim][BF_4]$	tetrahydrofurane from water	Jork et al., 2004
1-alkyl-3-methylimidazolium hexafluorophosphate	$[C_nmim][PF_6]$	phenol, tyrosol and p-hydroxybenzoic acid	Vidal et al., 2004
1-alkyl-3-methylimidazolium tetrafluoroborate	$[C_nmim][BF_4]$		

1-alkyl-3-methyl-imidazolium chloride	$[C_nmim][Cl]$	taurine from sodium sulfate	Gu et al., 2004
1-alkyl-3-methyl-imidazolium bromide	$[C_nmim][Br]$		
1-butyl-3-methyl-imidazolium iodide	[bmim][I]		
1-alkyl-3-methyl-imidazolium octylsulfate	$[C_nmim][OcSO_4]$	organic sulphur compounds from diesel fuel	Eβer et al., 2004
1-alkyl-3-methyl-imidazolium ethylsulfate	$[C_nmim][EtSO_4]$		
1-butyl-3-methyl-imidazolium bis (trifluoro-methylsulfonyl) imide	$[bmim] [NTf_2]$	toluene, nonan-1-ol, acetic acid and hexanoic acid	McFarlane et al., 2005
1-hexyl-3-methyl-imidazolium bis (trifluoro-methylsulfonyl) imide	$[hmim] [NTf_2]$		
1-octyl-3-methyl-imidazolium bis (trifluoro-methylsulfonyl) imide	$[omim] [NTf_2]$		
Trihexyl(tetradecyl)-phosphonium bis (trifluoro-methylsulfonyl) imide	$[thtdp][NTf_2]$		
Trihexyl(tetradecyl)-phosphonium methanesulfonate	$[thtdp][CH_3SO_3]$		
Trihexyl(tetradecyl)-phosphonium dodecylbenzenesulfonate	[thtdp][ddbOTf]		
Tributyl(tetradecyl)-phosphonium dodecylbenzenesulfonate	[tbtdp][ddbOTf]		
1-butyl-1-methyl-pyrrolidinium bis(trifluoro-methylsulfonyl) imide	$[bmpyr][NTf_2]$		
1-butyl-3-methylimidazolium hexafluorophosphate	$[bmim][PF_6]$	phenols	Khachatryan et al., 2005
1-*n*-butylimidazolium hexafluorophosphate	$[bim][PF_6]$	aromatic hydrocarbons	Matsumoto et al., 2005
1-*n*-hexylimidazolium hexafluorophosphate	$[him][PF_6]$		
1-*n*-octyl-3-methylimidazolium hexafluorophosphate	$[omim][PF_6]$		
Diethylmethyl (2-methoxyethyl) ammonium bis(trifluoro-methane sulfon)imid	$[Et_2MeMeON][NTf_2]$		
1-butyl-3-methylimidazolium hexafluorophosphate	$[bmim][PF_6]$	amino acids, e.g. valine, leucine, tyrosine	Wang et al., 2005

1-hexyl-3-methylimidazolium tetrafluoroborate	[hmim][BF_4]		
1-octyl-3-methylimidazolium tetrafluoroborate	[omim][BF_4]	antibiotics, e.g. amoxicillin, ampicillin	Soto et al., 2005
		thiophene from cyclohexane and toluene	Alonso et al., 2007
N-butyl-*N*-methyl-pyrrolidinium bis(trifluoro-methylsulfonyl) imide	[bmpyr][NTf_2]	azo dyes	Vijayaraghavan et al., 2006
1-methyl-3-octylimidazolium chloride	[omim][Cl]	toluene	Chakraborty & Bart, 2007
1-ethyl-3-methyl-imidazolium ethyl sulfate	ECOENG™212	toluene	Chakraborty & Bart, 2007
1-methylimidazole hydrogen sulfate	BASIONIC™ AC39	toluene	Chakraborty & Bart, 2007
1-H-imidazolium ethyl sulfate	TEGO IL IM55	toluene	Chakraborty & Bart, 2007
1-hexyl-3-methylimidazole	hmim	ethanol, butan-1-ol from water	Chapeaux et al., 2008
Tetrahexylammonium dihexylsulfosuccinate	[N_{6666}^+][DHS]	phenols, amines	Egorov et al., 2008
Trioctylmethylammonium salicylate	[$N_{(1)888}^+$][sce]	phenols, amines	Egorov et al., 2008
1-(2-hydroxyethyl)-3-methylimidazolium tetrafluoroborate	[C_2OHmim][BF_4]	ethanol from ethyl acetate	Hu et al., 2008
N-butyl-3-methyl-pyridinium thiocyanate	[bmpy][SCN]	organosulfur compounds from refinery products	Holbrey et al., 2008
N-hexylpyridinium tetrafluoroborate	[hpy][BF_4]	aromatic sulphur compounds from diesel fuel	Gao et al., 2008
N-octylpyridinium tetrafluoroborate	[opy][BF_4]	aromatic sulphur compounds from diesel fuel	Gao et al., 2008
Trihexyl(tetradecyl) phosphonium chloride	Cyphos 101	lactic acid	Matsumoto et al., 2010
Trihexyl(tetradecyl) phosphonium bis(2,2,4-trimethylpentyl-phosphinate	Cyphos 104	lactic acid	Marták et al., 2008
1-alkyl-3-methylimidazolium hexafluorophosphate	[$C_n$mim][PF_6]	phenol, bisphenol A, 4-octylphenol and 4-nonylphenol	Fan et al., 2008
1-alkyl-3-methylimidazolium tetrafluoroborate	[$C_n$mim][BF_4]	phenol, bisphenol A, 4-octylphenol and 4-nonylphenol	Fan et al., 2008
1-butyl-3-methylimidazolium hexafluorophosphate	[bmim][PF_6]	transesterification reaction compounds, e.g. vinyl acetate, ethyl decanoate, 1-butanol, 1-propanol, butyric acid	Hernández-Fernández et al., 2007; de los Ríos et al., 2008 a)
1-octyl-3-methylimidazolium hexafluorophosphate	[omim][PF_6]	transesterification reaction compounds, e.g. vinyl acetate, ethyl decanoate, 1-butanol, 1-propanol, butyric acid	Hernández-Fernández et al., 2007; de los Ríos et al., 2008 a)

1-buty-3-methylimidazolium tetrafluoroborate	[bmim][BF_4]		de los Ríos et al., 2008 a)
1-octyl-3-methylimidazolium tetrafluoroborate	[omim][BF_4]		
1-buty-3-methyl-imidazolium bis(trifluoro-methylsulfonyl) imide	[bmim][NTf_2]		de los Ríos et al., 2010
1-octyl-3-methyl-imidazolium bis(trifluoro-methylsulfonyl) imide	[omim][NTf_2]		
1,3-dialkylimidazolium bis (trifluoromethylsulfonyl) imide	[dC_nim][NTf_2]	benzene from hexane	Poole & Poole, 2009
1-ethylpyridinium bis (trifluoromethylsulfonyl) imide	[epy][NTf_2]		
2-hydroxyethyl-trimethylammonium bis (trifluoromethylsulfonyl) imide	[$C_2OHEtMe_3NO$][NTf_2]		
Trihexyltetradecyl-phosphonium bis (trifluoromethylsulfonyl) imide	[thtdp][NTf_2]	ethyl benzene from octane	
1-ethy-3-methyl-imidazolium bis (trifluoro-methylsulfonyl) imide	[emim][NTf_2]	aromatics from aliphatics	Arce et al., 2009
Tetraalkyl ammonium sulfate	Ammoeng 102		Pereiro & Rodriguez, 2009 a)
Tri-*n*-octylmethyl-ammonium chloride	TOMAC	benzimidazole	Venkantesan & Begum, 2009
3-(trimethoxysilyl)-propylammonium-3-trimethoxysilyl)-propyl carbamate	TMSAC	oil from oil shale and tar sands	Blasucci et al., 2010
3-(triethoxysilyl)-propylammonium 3-(triethoxysilyl)-propyl carbamate	TESAC		
MEMBRANE MICROEXTRACTION			
1-hexadecyl-3-methyl-imidazolium bromide	[hdmim][Br]	polycyclic aromatic hydrocarbons from sediments	Pino et al., 2008
1-butyl-3-methylimidazolium hexafluorophosphate	[bmim][PF_6]	organochlorine pesticides from soil	Khodadoust et al., 2006

1-octyl-3-methylimidazolium hexafluorophosphate	[omim][PF_6]	Sudan dyes, para red from chilli powder	Fan et al., 2009
1-octyl-3-methylimidazolium hexafluorophosphate	[omim][PF_6]	halomethanes, alkyl aromatics	Aguilera-Herrador et al., 2008
1-butyl-3-methyl-imidazolium phosphate	[bmim][PO_4]	aliphatic and aromatic hydrocarbons, e.g. tetradecane, naphthalene	Basheer et al., 2008
1-butyl-3-methyl-imidazolium methylsulfate	[bmim][$MeSO_4$]		
1-butyl-3-methyl-imidazolium octylsulfate	[bmim][$OcSO_4$]		
1-butyl-3-methylimidazolium hexafluorophosphate	[bmim][PF_6]		
1-hexyl-3-methylimidazolium tetrafluoroborate	[hmim][BF_4]	pesticides, aromatic amines	Poole & Poole, 2009
MEMBRANE EXTRACTIVE DISTILLATION			
1-ethy-3-methyl-imidazolium trriiodide	[emim][I_3]	toluene/*n*-heptane	Selvan et al., 2000
1-ethyl-3-methylimidazolium acetate	[emim][ace]	water/ethanol	Seller &Hirsch, 2004
1-ethyl-3-methylimidazolium tetrafluoroborate	[emim][BF_4]		Ge et al., 2008
1-(2-hydroxyethyl)-3-methylimidazolium chloride	[C_2OHmim][Cl]	water/tetrahydrofurane	Hu et al., 2006
1-octylquinolin bis(trifluoromethylsulfonyl) imide	[Ocqnl][NTf_2]	n-hexane/1-hexene	Lei et al., 2006
1-ethyl-3-methyl-imidazolium bis(trifluoro-methylsulfonyl) imide	[emim][NTf_2]	aromatics/alkanes	Arce et al., 2007
1-hexyl-3-methylimidazolium hexafluorophosphate	[hmim][PF_6]	ethanol/*n*-heptane	Pereiro et al., 2006
1,3-dimethylimidazolium methanesulfonate	[dmim][CH_3SO_3]		Zhang et al., 2009
1-ethyl-3-methylimidazolium acetate	[emim][ace]	water/2-propanol	Zhang et al., 2007
1-ethyl-3-methylimidazolium trrifluoromethylsulfonate	[emim][CF_3SO_3]	ethanol/chloroform	Orchilies et al., 2008
1-allyl-3-methyl-imidazolium bromide	[amim][Br]	ethanol/ethyl acetate	Zhang et al., 2008

1-ethyl-3-methylimidazolium tetrafluoroborate	[emim][BF_4]		Li et al., 2009
1-butyl-3-methylimidazolium methanesulfonate	[bmim][CH_3SO_3]	ethanol/*n*-hexane	Pereiro & Rodriguez, 2009 b)
1-hexyl-3-methylimidazolium	[hmim][BTI]	aliphatics/aromatics alkanes/alkenes alcohols/water	Westerholt et al., 2009
1-butyl-3-methyl-pyrrolidinium	[bmpyr][BTI]		
1-ethyl-3-methyl-imidazolium chloride	[emim][Cl]	water/*tert*-butanol	Zhang et al. 2009
MEMBRANE PERVAPORATION			
Tetrapropylammonium tetracyanoborate	[N_{3333}^+][$B(CN)_4$]	1,3-propanediol	Izák et al., 2006
1-ethenyl-3-ethylimidazolium hexafluorophosphate	[eeim][PF_6]	acetone and butan-1-ol from water	Izák et al., 2008
1-ethenyl-3-ethylimidazolium bromide	[eeim][Br]		
MEMBRANE VAPOUR SEPARATION			
1-butyl-3-methyl-imidazolium bis(trifluoro-methylsulfonyl) amide	[bmim][BTA]	olefins/paraffins	Krull et al., 2007
MEMBRANE GAS SEPARATON			
1-ethyl-3-methyl-imidazolium bis (trifluoromethyl - sulfonyl)imide	[emim][NTf_2]	N_2, CO_2, CH_4	Scovazzo et al., 2004
1-ethyl-3-methyl-imidazolium triflate	[emim][CF_3SO_3]		
1-ethyl-3-methylimidazolium dicyandiamide	[emim][dca]		
Trihexyl(tetradecyl)-phosphonium chloride	[thtdp][Cl]		
1,3-dimethylimidazolium bis (trifluoromethyl - sulfonyl)imide	[dmim][NTf_2],	N_2, H_2, O_2, CO	Gan et al., 2006
Methyltrioctylammonium bis (trifluoromethyl - sulfonyl)imide	[$N_{(1)888}^+$][NTf_2]		
N-octylpyridinium bis (trifluoromethyl - sulfonyl)imide	[C_8Py][NTf_2]		
1-(3-aminopropyl)-3-methylimidazolium bis(trifluoromethyl - sulfonyl)imide	[$H_2NC_3H_6mim$][NTf_2]	H_2, CO_2	Myers et al., 20008
N-aminopropyil-3-	[C_3NH_2mim][NTf_2]	CO_2, CH_4	Hanioka et al., 2008

methylimidazolium bis

(trifluoromethyl - sulfonyl)imide			
N-aminopropyil-3-methylimidazolium triflate	$[C_3NH_2mim][CF_3SO_3]$		
1-methyl-3-(3,3,4,4,4-pentylfluorohexyl) imidazolium bis(trifluoromethyl - sulfonyl)imide	$[MpFHim][NTf_2]$,	N_2, O_2, CO_2, CH_4	Bara et al. 2009
1-methyl-3-(3,3,4,4,5,5,6,6,6-nonafluorohexyl) imidazolium bis (trifluoromethyl - sulfonyl)imide	$[MnFHim][NTf_2]$		
1-methyl-3-(3,3,4,4,5,5,6,6, 7,7, 8,8,8-tridecyl-fluorohexyl)imidazolium bis (trifluoromethyl - sulfonyl)imide	$[MtdFHim][NTf_2]$		
1-butyl-3-methylimidazolium tetrafluoroborate	$[bmim][BF_4]$	N_2, CO_2, CH_4, SO_2	Jiang et al., 2009
1-ethyl-3-methylimidazolium tetrafluoroborate	$[emim][BF_4]$		
1-hexyl-3-methylimidazolium tetrafluoroborate	$[hmim][BF_4]$		
1-ethyl-3-methylimidazolium triflate	$[emim][CF_3SO_3]$		
1-butylimidazolim acetate	[bim][ace]	CO_2, SO_2	Luis et al., 2009
1-methylimidazolium acetate	[mim][ace]		
1-butyl-3-methyl-imidazolium bis(perfluoro-ethylsulfonyl)imide	[bmim][BETI]	N_2, CO_2, CH_4	Scovazzo et al., 2009
1-ethyl-3-methylimidazolium dicyandiamide	[emim][dca]		
Trihexyl(tetradecyl) phosphonium bromide	Cyphos 102	N_2, H_2, CH_4, CO_2	Cserjési et al., 2010
Trihexyl(tetradecyl) phosphonium decanoate	Cyphos 103		
Trihexyl(tetradecyl) phosphonium bis (2,4,4-trimethylpentyl) phosphinate	Cyphos 104		
Triisobutylmethyl phosphonium tosilate	Cyphos 106		

Tetrabutyl phosphonium bromide	Cyphos 163		
Triethylsulfonium bis (trifluoromethyl sulfonyl)imide	$[Set_3][NTf_2]$		
1,3-dimethylimidazolium dimethylphosphate	ECOENG™1111P		
1-decyl-3-methylimidazolium tetrafluoroborate	$[dmim][BF_4]$	N_2, CH_4, CO_2	Neves et al., 2010
1-butyl-3-methylimidazolium hexafluorophosphate	$[bmim][PF_6]$		
1-butyl-3-methylimidazolium tetrafluoroborate	$[bmim][BF_4]$		
1-butyl-3-methyl-imidazolium bis (trifluoro-methyl sulfonyl)imide	$[bmim][NTf_2]$		
1-octyl-3-methylimidazolium hexafluorophosphate	$[omim][PF_6]$		

Table 1. Most commonly used ionic liquids in membrane separations

4. Ionic liquid based membranes

4.1 Bulk ionic liquid membranes (BILMs)

In the bulk liquid membrane an organic membrane phase separates two aqueous or gaseous phases from each other. These non miscible phases can be arranged in several ways, five of the most commonly used possibilities are shown in Figure 4 (Clark et al., 2005; Bélafi-Bakó et al., 2000).

Bulk ionic liquid membranes contain ionic liquids as the membrane phase and have higher selectivity due to the tunable nature of the ionic liquids; for a certain separation task the ionic liquids with the best separation properties can be selected. The extraction of the separated compounds can be easily carried out by a solvent that is totally miscible with the ionic liquid and dissolves the compounds (Gubicza et al., 2003).

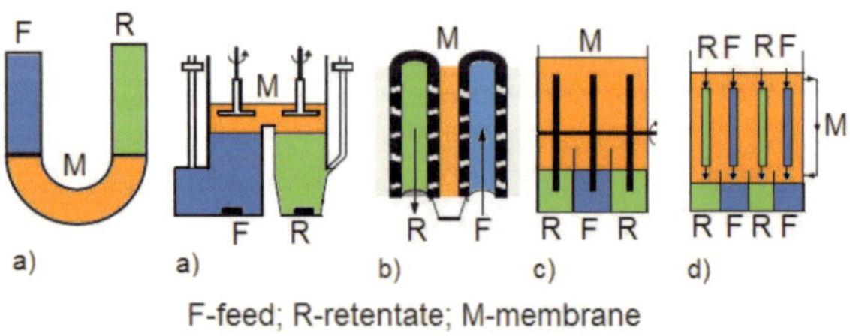

Fig. 4. Most common types of bulk liquid membranes; a) layered BLMs; b) BLM in a rotating disc contactor; c) BLM in a creeping film contactor; d) BLM with immobilised interfaces in a hollow fibre

Several researchers have studied the liquid/liquid extraction properties of bulk ionic liquid membranes, heptane separation from toluene using tetraalkyl ammonium sulfate (Ammoeng 102) ionic liquids was investigated by Pereiro and Rodriguez (Pereiro & Rodriguez, 2009 a)) and the extraction of oil from oil shale and tar sands with the ionic liquids shown in Table 2., by Blasucci and his co-workers (Blasucci, 2010).

Name	Abbreviation
3-(trimethoxysilyl) propylammonium 3-(trimethoxysilyl)-propyl carbamate	TMSAC
3-(triethoxysilyl)-propylammonium 3-(triethoxysilyl-propyl carbamate	TESAC
3-(triethylsilyl)-propylammonium 3-triethylsilyl-propyl carbamate	TEtSAC
3-(tripropylsilyl)-propylammonium 3-(tripropylsilyl)-propyl carbamate	TPSAC

Table 2. Ionic liquids used by Blasucci et al.

Chakraborty and Bart carried out toluene transport experiments in bulk ionic liquid membranes containing 1-methyl-3-octyl imidazolium chloride, 1-ethyl-3-methyl imidazolium ethyl sulfate, 1-methylimidazole hydrogen sulfate, 1H-imidazolium ethyl sulfate, 1-ethyl-4,5-dihydro-3-(2-hydroxyethyl)-2-(8-heptadecenyl) ethyl sulfate. They determined that the stirring speed, the feed phase concentration and the temperature had a strong effect on the permeation and that the permeability of toluene is quite low, while the selectivity is relatively high (Chakraborty & Bart, 2007). Gu and fellow researchers reported the extraction of taurine in 1-propyl-3-
Methylimidazolium-, 1-butyl-3-methylimidazolium-, 1-amyl-3-methylimidazolium, 1-hexyl-3-methylimidazolium chloride, 1-propyl-3-methylimidazolium-, 1-butyl-3-methyl-imidazolium-, 1-amyl-3-methylimidazolium-, 1-hexyl-3-methylimidazolium bromide, 1-butyl-3-methylimidazolium iodide, 1-butyl-3-methylimidazolium hexafluorophosphate and 1-butyl-3-methylimidazolium tetrafluoroborate ionic liquids and achieved 96-98,5% separation yield in a single separation step (Gu et al., 2004).

Propylene separation from propane by absorption was studied by Ortiz et al. in 1-butyl-3-methylimidazolium tetrafluoroborate bulk ionic liquid membranes (Ortiz et al., 2008) and the extractive distillation of benzene/cyclohexane, 1-hexene/n-hexane, 2-propanol/water binary and the ternary systems applying 1-hexyl-3-methyl-imidazolium- and 1-butyl-1-methylpyrrolidinium bis(trifluorpmethylsulfonyl) imide was reported by Westerholt (Westerholt, et al., 2009).

They all concluded that the application of ionic liquids increased the efficiency and repeatability and allowed separation for previously unsolved separation problems.

4.2 Emulsion ionic liquid membranes (EILMs)

Emulsion (or surfactant) liquid membranes are prepared by the intensive mixing of two non miscible phases (phase 1 and phase 2) and then the addition of some strong surfactant as it is shown in Figure 5. In this case phase 2 will be the liquid membrane, which can contain facilitating agents that promote separation through formation of chemical complexes from the agents and the components (Kislik, 2009).

Since the preparation of emulsion liquid membranes can be complicated and the decomposition of the emulsions after the separation is difficult in the case of relatively strong surfactants, the application of this type is limited to laboratory scale.

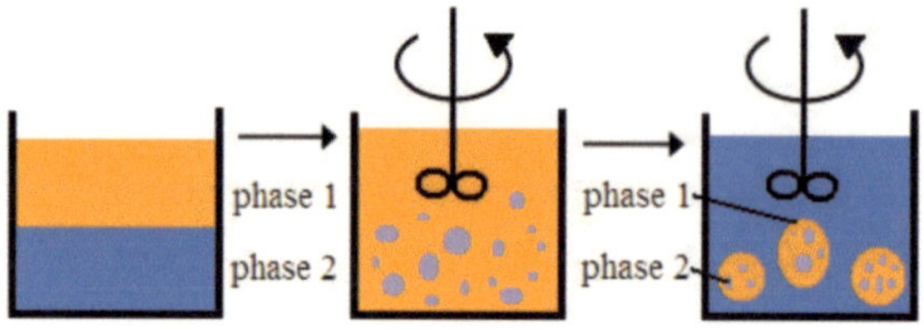

Fig. 5. Formation of emulsion liquid membranes

One of the most important results was achieved by Venkatesan and Begum, who carried out the extraction of benzimidazole by emulsion ionic liquid membrane containing tri-*n*-octyl methyl ammonium chloride as phase 1 and aqueous HCl solution as phase 2 (Venkatesan & Begum, 2009). They studied the effect of surfactant concentration, emulsification time, emulsification speed, carrier concentration, internal phase concentration and the volume ratio of the organic phase (1) to the aqueous phase (2) on emulsion stability and the effect of contact time for extraction, agitation speed, emulsion to external phase ratio, carrier concentration and HCl concentration in the internal aqueous phase on the effect of extraction. It was concluded that this technique is more effective than the conventional liquid-liquid extraction method; more than 97,5% of the benzimidazole present in the aqueous solutions within 12 min.

4.3 Supported ionic liquid membranes (SILMs)

Supported (or immobilized) liquid membranes are also built up of two phases, a supporting porous or non porous membrane and a liquid phase, which is either held in the pores by capillary forces (Jönsson & Mathiasson, 2000) or placed between two membrane sheets (Krull et al., 2008) as it is shown in Figure 6.

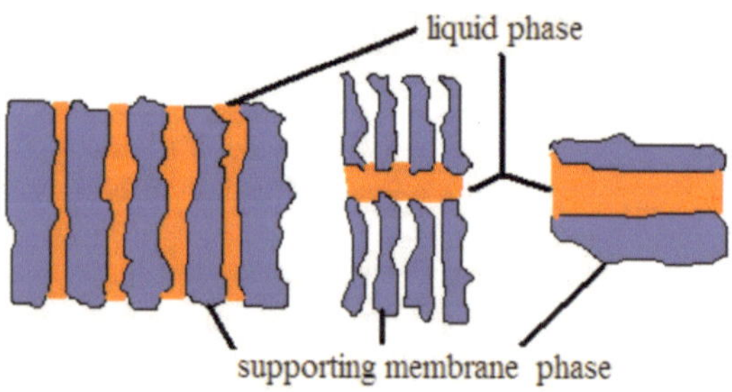

Fig. 6. Types of supported (immobilized) liquid membranes

The separation takes place in the liquid phase, without facilitating agents according to the solution diffusion model or in the presence of such agents through the mechanism of facilitated transport (Kocherginsky et al., 2007).

SLMs offer an elegant solution for separating molecules and have several advantages compared to porous or non porous solid membrane materials (Jönsson & Mathiasson, 1999). With the choice of the right liquid phase high permeability (Ilconich et al., 2007) and selectivity (He, 2008) values can be achieved, expensive, high quality liquids can be used due to the requirement of their small amount (Adebayo & Sarangi, 2008), extraction and stripping can be realized in one step (Poliwoda et al., 2007) and they can be connected to other (analytical) equipments because high pressure is not needed for the separation process (Jönsson & Mathiasson, 1999).

However, SLMs are not stable enough, they suffer from flux decline with time because the liquid phase evaporates or gets easily displaced from the pores due to trans membrane pressure difference (Teramoto et al., 2000). The extension of SLM lifetime was attempted by liquid membrane gelation, surface modification and with the application of ionic liquids (Zhang et al., 2001).

Since ionic liquids are chemically and thermally stable, have non detectable vapour pressure and relatively high viscosity (Keskin et al., 2007) their application as the liquid phase consequently leads to a much more stable membrane, which is often referred to as supported ionic liquid membrane (de los Ríos et al., 2008).

Although the industrial application of supported ionic liquid membranes is still yet to come, laboratory and pilot plant applications have been studied in the last three decades. They were successfully used for liquid-liquid extraction, microextraction, extractive distillation, pervaporation and gas separation.

4.3.1 Liquid-liquid extraction

Miyako et al. successfully realized the transport of organic acids, such as 4-phenoxybutyric acid, 2-phenoxybutyric acid, etc. through 1-alkyl-3-methylimidazolium hexafluorophosphate based SILMs that were stable for 75 hours (Miyako et al., 2003). Marták and co-workers studied lactic acid separation with SILMs prepared with Cyphos 104 and achieved acid purity as high as 95% (Marták et al., 2008). Cassol et al. studied the separation of benzene and alkyl-benzenes from *n*-heptane and *n*-octane (benzene/hexane, toluene/heptane, ethylbenzene/octane and *m*-xylene/*n*-octane) by pyridinium based ionic liquids (Cassol et al., 2007). Matsumoto and co-workers successfully carried out experiments on the separation of lactic acid by Cyphos 101 (Matsumoto et al., 2010), penicillin G by imidazolium based ionic liquids (Matsumoto et al., 2007) and aromatic hydrocarbons i.e. benzene, toluene, p-xylene, hexadecane, methanol, acetonitrile and chloroform by 1-butyl-3-methylimidazolium-, 1-hexyl-3-methylimidazolium and 1-octyl-3-methylimidazolium hexafluorophosphate (Matsumoto et al., 2005). De los Rios and fellow researchers investigated the separation of transesterification products, such as vinyl butyrate, 1-butanol, butyl butyrate, butyric acid (de los Rios et al., 2008 b)), methanol, vinyl acetate, acetic acid, etc. (de los Ríos et al., 2008 a)) applying 1-butyl-3-methylimidazolium-, 1-oxyl-3-methylimidazolium hexafluorophosphate, 1-butyl-3-methylimidazolium-, 1-oxyl-3-methylimidazolium tetrafluoroborate, 1-butyl-3-methyl-imidazolium- and 1-oxyl-3-methylimidazolium bis(trifluoromethylsulfyl)imide based SILMs (de los Ríos et al., 2010). They achieved relatively high average selectivity values for the membranes and high permeabilities for methanol, 1-butanol and butyric acid.

4.3.2 Microextraction

Basheer et al. carried out successful experiments on the microextraction of aliphatic and polycyclic aromatic hydrocarbons, such as naphthalene, tridecane, fluorine, heptadecane, pyrene, etc. from water samples by 1-butyl-3-methyl-imidazolium phosphate, 1-butyl-3-methyl-imidazolium methylsulfate, 1-butyl-3-methyl-imidazolium octylsulfate and 1-butyl-3-methylimidazolium hexafluorophosphate based SILMs, which could be directly connected to GC-MS (Basheer et al., 2008).

4.3.3 Pervaporation

Izák and co-workers studied the transport of 1,3-propanediol from aqueous mixture (Izák et al., 2006) and the removal of acetone and butan-1-ol from water by pervaporation (Izák et al., 2008) using SILMs based on tetrapropylammonium tetracyanoborate, 1-ethenyl-3-ethyl-imidazolium hexafluorophosphate and tetrapropylammonium tetracyano-borate ionic liquids. It was found that the separation factor for propanediol, acetone and butan-1-ol increased from 0.4 to 177, from 2.3 to 7.9 and from 2.2 to 10.9 respectively.

4.3.4 Vapour separation

Krull et al. carried out experiments on the separation of vapourized propylene from propane by 1-butyl-3-methyl-imidazolium bis(trifluoromethylsulfonyl) amide based SILMs and on the stability of the membrane and achieved permeability of 26.5 cm^2hbar and selectivity of 1.67. It was found out that the permeability increased while selectivity decreased with the increase in operation temperature and that the membrane was stable up to 5 bar pressure and 72 h reaction time (Krull et al., 2007).

4.3.5 Gas separation

Scovazzo and co-workers investigated the N_2, CO_2 and CH_4 gas separation properties of SILMs prepared with 1-ethyl-3-methyl-imidazolium bis(trifluoromethyl -sulfonyl)imide 1-ethyl-3-methyl-imidazolium triflate, 1-ethyl-3-methylimidazolium dicyandiamide, trihexyl (tetradecyl) phosphonium chloride (Scovazzo et al., 2004) and 1-butyl-3-methylimidazolium bis(perfluoro-ethylsulfonyl)imide and (Scovazzo et al., 2009). The permeability for CO_2, CH_4 and N_2 were 1500, 920 and 350 Barrers relatively and it was determined that the permeabilities decreased with the increase in trans membrane pressure. Gan et al. used 1-butyl-3-methylimidazoluim-, 1,3-dimethylimidazolium-, methyltrioctylammonium bis (trifluoromethylsulfonyl)imide and *N*-octylpyridinium bis(trifluoromethylsulfonyl)imide based SILMs for the permeation of H_2, N_2, O_2, CO_2, and CO and it was found that, on the contrary to the results of Scovazzo, the permeabilities decreased with the increase in trans membrane pressure (Gan et al., 2006). Ilconich and co-workers carried out experiments on the permeation of He and CO_2 through 1-hexyl-3-methylimidazolium bis(triflluoromethylsulfonyl)imide based SILM between 37°C and 125°C and stated that the permeability for CO_2 and He increased from 744 to 1200 and from 86 to 260 Barrers with the increase in temperature (Ilconich et al., 2007). Hanioka et al. successfully realized the facilitated transport of CO_2 and CH_4 single gases and binary mixture with *N*-aminopropyil-3-methylimidazolium bis(trifluoromethylsulfonyl)imide and *N*-aminopropyil-3-methyl-imidazolium triflate based SILMs. The CO_2/CH_4 selectivity increased with the increase in CO_2 partial pressure and with the increase in temperature from 25°C to 75°C the CO_2 and CH_4 permeability increased from 500 to 1500 and from 100 to 850 Barrers respectively (Hanioka et al., 2008). Jiang and co-workers studied the permeation of N_2, CO_2, CH_4 and SO_2 through SILMs prepared with 1-butyl-3-methylimidazolium-, 1-ethyl-3-methylimidazolium-, 1-hexyl-3-methylimidazolium tetrafluoroborate, 1-ethyl-3-methylimidazolium triflate, 1-butyl-3-methylimidazolium hexafluorophosphate and 1-butyl-3-methylimidazolium bis(trifluoromethylsulfonyl)imide and found out that in the case of SO_2 the permeability increased, while in the other cases no significant change was observed with the increase in trans membrane pressure (Jiang et al., 2009). The permeation of CO_2 and SO_2 through 1-

butylimidazolim acetate and 1-methylimidazolium acetate was studied by Luis et al. (Luis et al., 2009) and the permeation of CO_2, CH_4, H_2S through 1-butyl-3-methylimidazolium tetrafluoroborate based SILM was studied by Park et al. (Park et al., 2009). In 2009 Scovazzo published a gas solubility and diffusivity model and the permeabilities of SILMs prepared from several imidazolium-, ammonium-, phosphonium based and functionalized ionic liquids for N_2, O_2, CO_2 and CH_4 gases (Scovazzo, 2009).
All the scientists concluded that after solving the stability problems of the SILMs, they can perform better or just as well as the polymeric membranes used in the industrial processes.

4.4 Polymerized ionic liquid membranes (PILMs)

The idea of applying polymerized ionic liquids in membrane separation processes is quite young and so far has been mostly used for gas separation (Bara et al., 2007). In this case polymerized ionic liquid solutions can be cast and solid membrane films are achieved (Figure 7), that can be treated as any other commonly used membrane material (Carlisle, 2010).

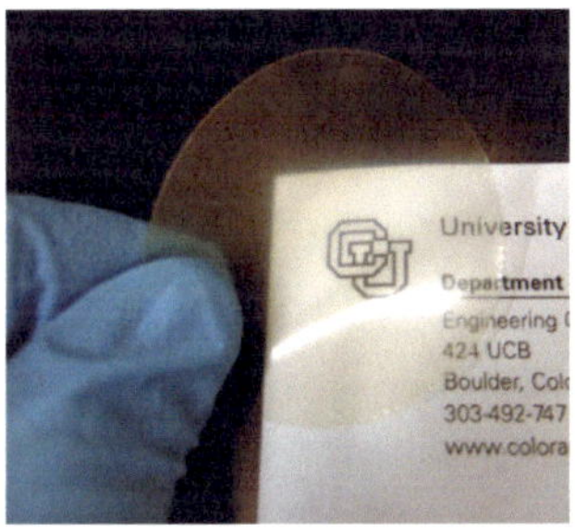

Fig. 7. Picture of a PILM

PILMs combine the advantageous properties of ionic liquids, such as tuneable physicochemical properties, high thermal and chemical stability, high selectivity, etc. and the advantages of dense membranes, i.e. low membrane thickness, high stability and high permeability values.
Noble and co-workers investigated the N_2, CO_2 and CH_4 permeability of PILMs prepared from polymerized ionic liquids with varying length n-alkyl substituents (Bara et al., 2007) and from polymerized imidazolium-based ionic liquids containing polar, oligo(ethyl glycol) and alkyl-terminated nitrile substituents (Bara et al., 2008) and the H_2, N_2, CO_2 and CH_4 permeability of imidazolium ionene based PILMs and ionic liquid composites (Carlisle et al., 2010). The highest N_2, H_2, CO_2 and CH_4 permeability values were as high as 0.83, 5.3, 22 and 1.1 Barrers and the highest CO_2/N_2, CO_2/CH_4 and H_2/CO_2 ideal selectivities were 44, 37 and 6 respectively. Plotting the performance of these membranes on a "Robeson plot" for CO_2/N_2 showed that PILMs perform as well or better than many other polymers for this separation. The CO_2/CH_4 and H_2/CO_2 separations are less impressive when compared to other polymer membranes on a "Robeson plot", but PILMs perform as well or better than ionic liquids do in bulk fluid gas absorptions for that gas pair.
We have prepared four different PILMs using imidazole-based copolymers and anion-changed homopolymers and determined the N_2, H_2, O_2, CO_2 and CH_4 sorption values and the permeabilities and ideal selectivities of the membranes for these gases.

5. Gas sorption and separation properties of polymerized ionic liquid membranes

5.1 Experimental

Materials

High purity nitrogen (N_2), hydrogen (H_2), oxygen (O_2), carbon dioxide (CO_2) and methane (CH_4) gases (99.99%) were provided by Messer Industriegase GmbH (Germany).
To support the ionic liquids Durapore porous hydrophobic polyvinylidene fluoride (PVDF) membrane was used. It has a pore size of 0.22 µm, porosity of 75%, average thickness of 150 mm and was purchased from Millipore Corporation (USA). 1-ethyl-3-methylimidazolium trifluoromethanesulfonate and triethylsulfonium bis(trifluoromethylsulfonyl) imide ionic liquids were purchased from Solvent Innovation GmbH, trihexyltetradecylphosphonium bromide, 2-hydroxyethyl-ammonium formiate, 1-allyl-3-methylimidazolium bis(trifluoromethylsulfonyl) imide and 1-allyl-3-methylimidazolium dicyanamide were procured from IoLiTec GmbH & Co. KG. Vinylimidazole, ethyl-bromide, butyl-bromide, trichloro aceticacid, acrylonitrile and ethanol were provided by Sigma Aldrich Co., sodium dicyanamide and sodium thiocyanate by Fluka and dimethyl sulfoxide by LACH-NER, s.r.o. Asobisisobutironitrile and hydrochinon were synthetidez at Institute of Macromolecular Chemistry AS CR, v.v.i. (IMC) Prague, Czech Republic.

Preparation of polymerized ionic liquid membranes

Monomer synthesis

The synthesis of the two different monomers, 1-vinyl-3-ethyl-imidazolium bromide and 1-vinyl-3-butyl-imidazolium bromide, was carried out in the following way.
In the case of 1-vinyl-3-ethyl-imidazolium bromide first 1.5 moles of vinylimidazole was poured in a 500 ml flask then few grams of hydrochinon was added in order to prevent polymerization and then it was stirred and heated. When its temperature reached 70°C the previously measured 1.4x1.5 moles of ethyl bromide was slowly added to the vinylimidazole drop by drop so that the temperature of the mixture stayed between 70°C and 110°C. Since it is an exothermic reaction the fast decrease in temperature indicates the decrease in reaction rate and in this case the rate of dripping was increased. When all ethyl bromide was added the mixture was heated up to 150°C and was stirred for another 60 minutes so that the reaction can be completed. In order to crystallize the brownish coloured monomer product acetone was added to the solution and the product crystals were dried in a vacuum oven to constant weight at the temperature of 20°C.
The synthesis of 1-vinyl-3-butyl-imidazolium bromide was similar to the synthesis above described with the exception that 1.2*1.5 moles of butyl bromide was slowly added to 1.5 moles of vinylimidazole.

Copolymerisation and membrane preparation

1-vinyl-3-ethyl-imidazolium bromide and acrylonitrile were used to produce copolymer and to prepare membrane from this copolymer (**VEIMBrAN**).
5 g 1-vinyl-3-ethyl-imidazolium bromide and 0.1 g asobisisobutironitrile initiator were dissolved in 90 g dimethyl sulfoxide and 5 g acrylonitrile was added with constant stirring. In order to prevent homopolymerization initiated by oxygen carbon-dioxide was bubbling through the produced yellow solution for a few minutes. Then the solution was kept at the temperature of 70°C for 24 hours so that the polymerization reaction can be finished.

After filtration some of the solution was cast onto a flat glass plate and was kept at 70°C for 24 hours and then at 130°C for 1 hour for complete solvent evaporation. The glass plate was placed into water-acetone (50-50 w/w%) solution and the membrane film was removed carefully and dried further in the vacuum oven at 40°C for 24 hours to remove all detectable traces of the casting solvent.

Homopolymerization and membrane preparation

After the anion of 1-vinyl-3-butyl-imidazolium bromide was exchanged using sodium dicyanamide, trichloro aceticacid and sodium thiocyanate, homopolymerization reactions were carried out and membranes were prepared from the polymers.
85 g, 59 g and 88 g crystal 1-vinyl-3-butyl-imidazolium bromide monomer were dissolved in 100-100 ml distilled water with constant stirring, then 8 g sodium dicyanamide, 11 g trichloro aceticacid and 8 g sodium thiocyanate were added to the three solutions respectively. The brownish coloured products were washed with distilled water several times then were dried at 37°C for 24 hours to evaporate all the water and then the polymer products, 1-vinyl-3-butyl-imidazolium dicyanamide (**VMIMCN**$_2$), 1-vinyl-3-butyl-imidazolium trichloride (**VMIMCl**$_3$) and 1-vinyl-3-butyl-imidazolium thiocyanate (**VBIMTCN**) were formed.
Two grams of the polymers were dissolved in ethanol and cast onto a flat glass plate then kept at 40°C for 24 hours for membrane film formation.

Gas sorption measurements

The sorption measurements are based on the gravimetric method, which is a well established technique for obtaining sorption of pure gases. In this procedure the total amount of adsorbed and absorbed gas is measured at constant temperature with a microbalance while increasing the gas pressure from zero to a final pressure.
The experimental set-up consists of several units; the most important ones are the sorption balance and the computer used for data storage. The two scale beams, the left one holding a known mass and the right one holding the sample, are situated in glass cases, which can be evacuated using a vacuum pump or filled with gas at given pressure using a gas inlet pipe.
Prior to the determination of gas sorption the sample was degassed for 24 hours at room temperature exposing it to vacuum in order to remove all of the absorbed or adsorbed species from the membrane. After desorption the sorption of N_2, H_2, O_2, CO_2 and CH_4 was measured in the following way. The gas was let into the glass cases at given pressure and gas sorption took place for 2 hours then the glass case including the sample was degassed and the gas was let inside. This was repeated in the case of all membrane samples and gases at the pressure of 8 kPa, 16 kPa and 20 kPa respectively.

Gas permeation measurements

Permeability of all PILMs were determined for H_2, N_2, O_2, CO_2 and CH_4 in a self-designed experimental set-up described in our previous work, with an initial driving force of 2 bar and at constant temperature of 30°C, which was maintained with the help of a thermostat (Cserjési et al., 2010). The whole separation system, was vacuumed before the feed gas entered the membrane module then the gas phase pressure was set and with opening one of the cells the gas permeation started. After eight hours of gas permeation from the pressure data the change in the gas volume was determined and permeability and ideal selectivity values were calculated.

5.2 Results and discussion

The N_2, H_2, O_2, CO_2 and CH_4 sorption results for the four PILMs are shown in Table 3.

		Gas sorption (g/g)				
membranes	pressure (kPa)	nitrogen	hydrogen	oxygen	carbon-dioxide	methane
VEIMBrAN	8	0	0.05	0	0.05	0
	16	0	0.15	0	0.1	0
	20	0.05	0.2	0.35	0.35	0.15
$VMIMCN_2$	8	0	0	0	0	0
	16	0	0	0	0	0.08
	20	0	0.03	0.03	0.08	0.23
$VMIMCl_3$	8	0	0	0	0	0
	16	0	0	0.007	0	0.14
	20	0	0.14	0.14	0	0.28
VBIMTCN	8	0	0	0	0	0
	16	0.07	0.14	0,07	0.07	0.07
	20	0.14	0.27	0.14	0.21	0.14

Table 3. Gas sorption of the membranes

All of the PILMs had slightly different but extremely low gas sorption, the highest values are measured in the case of O_2 and CO_2 and the lowest ones in the case of N_2 gas. While the VEIMBrAN membrane exhibits the highest sorption followed by the VBIMTCN membrane, the gas sorption increases with the increase in pressure in all cases.

Table 4 contains the N_2, H_2, O_2, CH_4 and CO_2 permeabilities and the ideal selectivities of the investigated PILMs.

	VEIMBrAN	$VMIMCN_2$	$VMIMCl_3$	VBIMTCN
permeability (Barrer)				
N_2	0.28	0.18	0.23	0.32
H_2	3.1	1.9	2.8	3.8
O_2	4.8	2.3	3.2	4.1
CO_2	11	10.5	13	17
CH_4	2.3	1.8	1.5	2.7
ideal selectivity				
H_2/N_2	11	11	12	12
H_2/CH_4	1.3	1.1	1.9	1.4
O_2/N_2	17	13	14	13
O_2/CH_4	2.1	1.3	2.1	1.5
O_2/H_2	1.5	1.2	1.1	1.1
CO_2/N_2	39	58	57	53
CO_2/CH_4	4.8	5.8	8.7	6.3
CO_2/H_2	3.5	5.5	4.6	4.5
CO_2/O_2	2.3	4.6	4.1	4.1
CH_4/N_2	8.2	10	6.5	8.4

Table 4. Permeability and ideal selectivity values of the PILMs

All PILMs have the highest permeability values for CO_2 and the lowest for N_2 and the membrane permeabilities vary in the range of 0.18-0.32 Barrers for N_2, 1.9-3.8 Barrers for H_2, 2.3-4.8 Barrers for O_2, 1.5-2.7 Barrers for CH_4 and 10.5-17 Barrers for CO_2 respectively. $VMIMCN_2$ membrane has the lowest and VBIMTCN membrane the highest permeability for most of the gases tested. Although, the PILMs show low gas sorption they have relatively high permeability and selectivity values, it can be seen that higher permeabilities belongs to higher gas sorption values. Plotting the performance of the membranes for CO_2/N_2 gas pair on a Robeson plot diagram it can be said that these PILMs perform almost as well as the commercially used polymeric membranes.
In the future PILMs will be prepared based on novel kinds of ionic liquids containing different substituents and their gas permeability and selectivity values will be determined using N_2, H_2, O_2, CO_2 CH_4 and H_2S single gases, binary, tertiary and quaternary gas mixtures.

6. Conclusion

Due to the special features of ionic liquids, such as high thermal and chemical stability, low vapour pressure, non flammability, tunable physicochemical properties, etc. make them perfect candidates for the substitution of organic membrane phase in liquid membranes. Bulk-, emulsion- and supported ionic liquid membranes have higher stability, permeability and selectivity values than conventional liquid membranes and were successfully used in liquid-liquid extraction, microextraction, pervaporation, vapour- and gas separation processes.

7. Acknowledgement

The research work was supported by the Czech-Hungarian Science and Technology Cooperation Programme, grant No. CZ-8/08 the research program "Livable environment and healthier people - Bioinnovation and Green Technology Research at the University of Pannonia, TÁMOP-4.2.2-08/1/2008-0018"

8. References

Acharya, N.K., Kulshrestha, V., Awasthia, K., Jain, A.K., Singh, M., Vijay, Y.K. (2008). Hydrogen separation in doped and blend polymer membranes. *International Journal of Hydrogen Energy*, 33., 1., 327-331., 03603199

Adebayo, A.O., Sarangi, K. (2008). Separation of copper from chalcopyrite leach liquor containing copper, iron, zinc and magnesium by supported liquid membrane. *Separation and Purification Technology*, 63., 2., 392-399., 13835866

Aguilera-Herrador, E., Lucena, R., Cardenas, S., Valcarcel, M. (2008). Direct coupling of ionic liquid based single-drop microextraction and GC/MS. *Analytical Chemistry*, 80., 3., 793-800., 00032700

Alonso, L., Arce, A., Francisco, M., Rodriguez, O., Soto, A. (2007). Gasoline desulfurization using extraction with $[C_8mim][BF_4]$ ionic liquid. *AIChE Journal*, 53., 12., 3108-3115., 15475905

Arce, A, Earle, M.J., Rodriguez, H., Seddon, K.R. (2007). Separation of aromatic hydrocarbons from alkanes using the ionic liquid 1-ethyl-3-methylimidazolium bis{(trifluoromethyl)sulfonyl}amide *Green Chemistry*, 9., 70-74., 14639262

Arce, A, Earle, M.J., Rodriguez, H., Seddon, K.R., Soto, A. (2009). Bis{(trifluoromethyl)sulfonyl}amide ionic liquids as solvents for the extraction of aromatic hydrocarbons from their mixtures with alkanes: effect of the nature of the cation. *Green Chemistry*, 11., 365-372., 14639262

Atchariyawut, S., Feng, C., Wang, R., Jiraratananon, R., Liang, D.T. (2006). Effect of membrane structure on mass-transfer in the membrane gas-liquid contacting process using microporous PVDF hollow fibers. *Journal of Membrane Science*, 285., 1-2., 272-281., 03767388

Baker, R.W. (2004). *Membrane technology and applications*, John Wiley & Sons, ISBN: 0-470-85445-6, Chichester, England

Bara, J.E., Lessmann, S., Gabriel, C.J., Hatakeyama, E.S., Noble, R.D., Gin, D.L. (2007). Synthesis and performance of polymerizable room-temperature ionic liquids as gas separation membranes. *Industrial & Engineering Chemical Research*, 46., 16., 5397-5404., 08885885

Bara, J.E., Gabriel, C.J., Hatakeyama, E.S., Carlisle, T.K., Lessmann, S., Noble, R.D., Gin, D.L. (2008). Improving CO_2 selectivity in polymerized room-temperature ionic liquid gas separation membranes through incorporation of polar substituents. *Journal of Membrane Science*, 321., 1., 3-7., 03767388

Bara, J.E., Gabriel, C.J., Carlisle, T.K., Camper, D.E:, Finotello, A., Gin, Noble, R.D. (2009). Gas separations in fluoroalkyl-functionalized room-temperature ionic liquids using supported liquid membranes. *Chemical Engineering Journal*, 147., 1., 43-50., 13858947

Basheer, C., Alnedhary, A.A., Rao, B.S.M., Balasubramanian, R., Lee, H.K. (2008). Ionic liquid supported three-phase liquid-liquid-liquid microextraction as a sample preparation technique for aliphatic and aromatic hydrocarbons prior to gas chromatography-mass spectrometry. *Journal of Chromatography A*, 1210., 1., 19-24., 00219673

Bélafi-Bakó, K., Gubicza, L., Mulder, M. (2000). *Integration of Membrane Processes into Bioconversions*, Kluwer Academic Publishers, ISBN: 978-0-3064-6437-9, New York

Birdwell, J.F., McFarlane, J., Hunt, R.D., Luo, H., DePaoli, D.W., Schuh, D.L., Dai, S. (2006). Separation of ionic liquid dispersions in centrifugal solvent extraction contactors. *Separation Science and Technology*, 41., 10., 2205-2223., 01496395

Blasucci, V.M., Hart, R., Pollet, P., Liotta, C.L., Eckert, C.A. (2010). Reversible ionic liquids designed for facile separations. *Fluid Phase Equilibria*, 294., 1-2., 1-6., 03783812

Carlisle, T.K., Bara, J.E., Lafrate, A.L., Gin, D.L., Noble., R.D. (2010). Main-chain imidazolium polymer membranes for CO_2 separations: An initial study of a new ionic liquid-inspired platform. *Journal of Membrane Science*, 359., 1-2., 37-43. 03767388

Cassol, C.C., Umpierre, A.P., Ebeling, G., Ferrera, B., Chiaro, S.S.X., Dupont, J. (2007). On the extraction of aromatic compounds from hydrocarbons by imidazolium ionic liquids. *International Journal of Molecular Sciences*, *8.*, 7., 593-605., 14220067

Chakma, A. (1995). Separation of CO_2 and SO_2 from flue gas streams by liquid membranes. *Energy Conversion and Management*, 36., 6-9., 405-410., 01968904

Chakraborty, M., Bart, H.J. (2007). Highly selective and efficient transport of toluene in bulk ionic liquid membranes containing Ag^+ as carrier. *Fuel Processing Technology*, 88., 1., 43-49., 03783820

Chapeaux, A., Sinomi, L.D., Runan, T.S., Stadherr, M.A., Brennecke, J. (2008). Extraction of alcohols from water with 1-hexyl-3-methylimidazolium bis(trifluoromethyl-sulfonyl) imide. *Green Chemistry*, 10., 1301-1306., 14639262

Clark, J.D., Han, B., Bhown, A.S., Wickramasinghe, S.R. (2005). Amino acid resolution using supported liquid membranes. *Separation and Purification Technology*, 42., 3., 201-211., 13835866

Cserjési, P., Nemestóthy, N., Bélafi-Bakó, K. (2010). Gas separation properties of supported liquid membranes prepared with unconventional ionic liquids. *Journal of Membrane Science*, 349., 1-2., 6-11., 03767388

de los Ríos, A.P., Hernández-Fernández, F.J., Rubio, M., Tomás-Alonso, F., Gómez, D., Víllora, G., (2008). Prediction of the selectivity in the recovery of transesterificationb reaction products using supported liquid membranes based on ionic liquids. *Journal of Membrane Science*, 307., 2., 225-232., 03767388 a)

de los Ríos, A.P., Hernández-Fernández, F.J., Tomás-Alonso, F., Rubio, M., Gómez, D., Víllora, G. (2008). On the importance of the nature of the ionic liquids int he selective simultaneous separation of the substrates and products of a transesterification reaction through supported ionic liquid membranes. *Journal of Membrane Science* 307., 2., 233-238., 03767388 b)

de los Ríos, A.P., Hernández-Fernández, F.J., Rubio, M., Gómez, D., Víllora, G. (2010). Highly selective transport of transesterification reaction compounds through supported liquid membranes containing ionic liquids based ont he tetrafluoroborate anion. *Desalination*, 250., 1., 101-104., 00119164

Egorov, V.M., Smirnova, S.V., Pletnev, I.V. (2008). Highly efficient extraction of phenols and aromatic amines into novel ionic liquids incorporating quaternary ammonium cation. *Separation and Purification Technology*, 63., 3., 710-715., 13835866

Endres, F., Abbott, A.P., MacFarlane, D.R. (2008). *Electrodeposition from Ionic Liquids*, Wiley-VCH Verlag GmbH & Co. KGaA, ISBN: 978-3-527-31565-9, Weinheim, Germany

Eβer, J., Wasserscheid, P., Jess, A. (2004). Deep desulfurization of oil refinery streams by extraction with ionic liquids. *Green Chemistry*, 6., 316-322., 14639262

Fan, J., Fan, Y., Pei, Y., Wu, K., Wang, J., Fan, M. (2008). Solvent extraction of selected endocrine-disrupting phenols using ionic liquids. *Separation and Purification Technology*, 61., 3., 324-331., 00032670

Fan, Y., Chen, M., Shentu, C., El-Sepai, F., Wang, K., Zhu, Y., Ye, M. (2009). Ionic liquids extraction of Para Red and Sudan dyes from chilli powder, chilli oil and food additive combined with high performance liquid chromatography. *Analytica Chimica Acta*, 650., 1., 65-69., 18734324

François, Y., Varenne, A., Juillerat, E., Villemin, D., Gareil, P. (2007). Evaluation of chiral ionic liquids as additives to cyclodextrins for enantiomeric separations by capillary electrophoresis. *Journal of Chromatography A*, 1155., 2., 134-141., 00219673

Gan, Q., Rooney, D., Xue, M., Thompson, G., Zou, Y. (2006). An experimental study of gas transport and separation properties of ionic liquids supported on nanofiltration membranes. *Journal of Membrane Science*, 280., 1., 948-956., 03767388

Gao, H., Luo, M., Xing, J., Wu, Y., Li, Y., Li, W., Li, Q., Liu, H. (2008). Desulfurization of fuel by extraction with pyridinium-based ionic liquids. *Industrial & Engineering Chemistry Research*, 47., 21., 8384-8388., 08885885

Ge, Y., Zhang, L., Yuan, X., Geng, W., Ji, J. (2008). Selection of ionic liquids as entrainers for separation of (water+ethanol). *The Journal of Chemical Thermodynamics*, 40., 8., 1248-1252., 00219614

Gu, Y., Shi, F., Yang, H., Deng, Y. (2004). Leaching separation of taurine and sodium sulfate solid mixture using ionic liquids. *Separation and Purification Technology*, 35., 2., 153-159., 13835866

Gubicza, L., Nemestóthy, N., Fráter T., Bélafi-Bakó, K. (2003). Enzymatic esterification in ionic liquids integrated with pervaporation for water removal. *Green Chemistry*, 5 (2003) 236-239., 14639262

Hanioka, S., Maruyama, T., Sotani, T., Teramoto, M., Matsuyama, H., Nakashima, K., Hanaki, M., Kubota, F., Goto, M. (2008). CO_2 separation facilitated by task-specific ionic liquids using a supported liquid membrane. *Journal of Membrane Science*, 314., 1-2., 1-4., 03767388

He, T. (2008). Towards stabilization of supported liquid membranes: preparation and characterization of polysulfone support and sulfonated poly (ether ether ketone) coated composite hollow fiber membranes. *Desalination*, 225., 1., 82-94., 00119164

Hernández-Fernández, F.J., de los Ríos, A.P., Rubio, M., Tomás-Alonso, F., Gómez, D., Víllora, G. (2007). A novel application of supported liquid membranes based on ionic liquids to the selective simultaneous separation of the substrates and products of a transesterification reaction. *Journal of Membrane Science*, 293., 1-2., 73-80., 03767388

Holbrey J.D., Lopez-Martin, I., Rothenberg, G., Seddon, K.R., Silvero, G., Zheng, X. (2008). Desulfurisation of oils using ionic liquids: selection of cationic and anionic components to enhance extraction efficiency. *Green Chemistry*, 10., 87-92., 14639262

Hu, X., Yu, J., Liu, H. (2006). Separation of THF and water by room temperature ionic liquids. *Water Science & Technology*, 53., 11., 245-249., 02731223

Hu, X., Li, Y., Cui, D., Chen, B. (2008). Separation of ethyl acetate and ethanol by room temperature ionic liquids with the tetrafluoroborate anion. *Journal of Chemical & Engineering Data*, 53., 2., 427-433., 00219568

Ilconich, J., Myers, C., Pennline, H., Luebke, D. (2007). Experimental investigation of the permeability and selectivity of supported ionic liquid membranes for CO_2/He separation at temperatures up to 125°C. *Journal of Membrane Science*, 298., 1-2., 41-47., 03767388

Izák, P., Köckerling, M., Kragl, U. (2006). Solute transport from aqueous mixture throught supported ionic liquid membrane by pervaporation. *Desalination*, 199., 1-3., 96-98., 00119164

Izák, P., Ruth, W., Fei, Z., Dysonc, P.J., Kragl, U. (2008). Selective removal of acetone and butan-1-ol from water with supported ionic liquid–polydimethylsiloxane membrane by pervaporation. *Chemical Engineering Journal*, 139., 2., 318-321., 13858947

Jiang, T.F., Gu, Y.L., Liang, B., Li, J.B., Shi, Y.P., Ou, Q.Y. (2003). Dynamically coating the capillary with 1-alkyl-3-methylimidazolium-based ionic liquids for separation of basic proteins by capillary electrophoresis. *Analytica Chimica Acta*, 479., 2., 249-254., 00032670

Jiang, Y., Wu, Y., Wang, W., Li, L., Zhou, Z., Zhang , Z. (2009). Permeability and selectivity of sulfur dioxide and carbon dioxide in supported ionic liquid membranes. *Chinese Journal of Chemical Engineering*, 17., 4., 594-601., 10049541

Jork, C., Seiler, M., Beste, Y.A., Arlt, W. (2004). Influence of ionic liquids on the phase behavior of aqueous azeotropic systems. *Journal of Chemical & Engineering Data*, 49., 4., 852-857., 00219568

Jönsson, J.Å., Mathiasson, L. (1999). Liquid membrane extraction in analytical sample preparation I. Principles. *Trends in Analytical Chemistry*, 18., 5., 318-325., 01659936

Jönsson, J.Å., Mathiasson, L. (2000). Membrane-based techniques for sample enrichment. *Journal of Chromatography A*, 902., 1., 205-225., 00219673

Keskin, S., Kayrak-Talay, D., Akman, U., Hortaçsu, Ö., (2007). A review of ionic liquids towards supercritical fluid applications. *Journal of Supercritical Fluids*, 43., 1., 150-180., 08968446

Khachatryan, K.S., Smirnova, S.V., Torocheshnikova, I.I., Shvedene, N.V., Formanovsky, A.A., Pletnev, I.V. (2005). Solvent extraction and extraction-voltammetric determination of phenols using room temperature ionic liquid. *Analytical and Bioanalytical Chemistry*, 381., 2., 464-470., 16182642

Khodadoust, A.P., Chandrasekaran, S., Dionysiou, D.D. (2006). Preliminary assessment of imidazolium-based room-temperature ionic liquids for extraction of organic contaminants from soils. *Environmental Science & Technology*, 40., 7., 2339-2345., 15205851

Kislik, V.S. (2009). *Liquid membranes: Principles and Application in Chemical Separations and Wastewater Treatment*, Elsevier, ISBN: 978-0-444-53218-3,

Kocherginsky, N.M., Yang, Q., Seelam, L. (2007). Recent advances in supported liquid membrane technology. *Separation and Purification Technology*, 53., 2., 171-177., 13835866

Koel, M. (2008). *Ionic liquids in Chemical Analysis*, CRC Press Taylor & Francis Group, ISBN: 978-1-4200-4646-5, USA

Koops, G.H. (1995). *Nomenclature and symbols in membrane Science and technology*. University of Twente, Membrane Technology Group, ISBN: 90-365-0768-5, The Netherlands

Koros, W.J., Fleming, G.K. (1993). Membrane-based gas separation, Review. *Journal of Membrane Science*, 83., 1., 1-80., 03767388

Krull, F.F. Medved, M., Melin, T. (2007). Novel supported ionic liquid membranes for simultaneous homogeneously catalyzed reaction and vapor separation. *Chemical Engineering Science*, 62., 18-20., 5579-5585., 00092509

Krull, F.F., Fritzmann, C., Melin, T. (2008). Liquid membranes for gas/vapor separations. *Journal of Membrane Science*, 325., 2., 509-519., 03767388

Lashkari, S., Tran, A., Kruczek, B. (2008). Effect of back diffusion and back permeation of air on membrane characterization in constant pressure system. *Journal of Membrane Science*, 324., 1-2., 162-172., 03767388

Lei, Z., Arit, W., Wasserscheid, P. (2006). Separation of 1-hexene and *n*-hexane with ionic liquids. *Fluid Phase Equilibria*, 241., 1-2., 290-299., 03783812

Li, Q., Zhang, J., Lei, Z., Zhu, F., Xing, J. (2009). Isobaric vapor−liquid equilibrium for ethyl acetate+ethanol+1-ethyl-3-methylimidazolium tetrafluoroborate. *Journal of Chemical & Engineering Data*, 54., 2., 193-197., 00219568

Luis, P., Neves, L.A., Afonso, C.A.M., Coelhoso, I.M., Crespo, J.G., Garea, A., Irabien, A. (2009). Facilitated transport of CO_2 and SO_2 through supported ionic liquid membranes (SILMs). *Desalination*, 245., 1-3., 485-493., 00119164

Marszałł, M.P., Markuszewski, M.J., Kaliszan, R. (2006). Separation of nicotinic acid and its structural isomers using 1-ethyl-3-methylimidazolium ionic liquid as a buffer additive by capillary electrophoresis. *Journal of Pharmaceutical and Biomedical Analysis* 41., 1., 329-332., 07317085

Marták, J., Schlosser, Š., Vlčková, S. (2008). Pertraction of lactic acid through supported liquid membranes containing phosphonium ionic liquid. *Journal of Membrane Science*, 318., 1-2., 298-310., 03767388

Matsumoto, M., Inomoto, Y., Kondo, K. (2005). Selective separation of aromatic hydrocarbons through supported liquid membranes based on ionic liquids. *Journal of Membrane Science*, 246., 1., 77-81., 03767388

Matsumoto, M., Ohtani, T., Kondo, K. (2007). Comparison of solvent extraction and supported liquid membrane permeation using an ionic liquid for concentrating penicillin G. *Journal of Membrane Science*, 289., 1-2., 92-96., 03767388

Matsumoto, M., Hasegawa, W., Kondo, K., Shimamura, T., Tsuji, M. (2010). Application of supported ionic liquid membranes to lactic acid recovery. *Desalination and Water Treatment*, 14., 1-3., 37-46., 19443994

McFarlane, J., Ridenour, W.B., Luo, H., Hunt, R.D., DePaoli, D.W., Ren, R.X. (2005). Room temperature ionic liquids for separating organics from produced water. *Separation Science and Technology*, 40., 6., 1245-1265., 01496395

Miyako, E., Maruyama, T., Kamiya, N., Goto, M. (2003). Use of ionic liquids in a lipase-facilitated supported liquid membrane. *Biotechnology Letters*, 25., 10., 805-808., 01415492

Mulder, M. (1996). *Basic principles of membrane technology*, Kluwer Academic Publishers, ISBN: 0-7923-4248-8, Dordrecht, The Netherlands

Myers, C., Pennline, H., Luebke, D., Ilconich, J., Dixon, J.K., Maginn, E.J., Brennecke, J.F. (2008). High temperature separation of carbon dioxide/hydrogen mixtures using facilitated supported ionic liquid membranes. *Journal of Membrane Science*, 322., 1., 28-31., 03767388

Nath, K. (2008). *Membrane Separation Process*, PHI Learning Pvt. Ltd., ISBN: 978-81-203-3532-5, New Delhi

Neves, L.A., Crespo, J.G., Coelhoso, I.M. (2010). Gas permeation studies in supported ionic liquid membranes. *Journal of Membrane Science*, 357., 1-2., 160-170., 03767388

Orchilles, A.V., Miguel, P.J., Vercher, E., Martinez-Andreu, A. (2008). Isobaric vapour-liquid and liquid-liquid equilibria for chloroform+ethanol+1-ethyl-3-methylimidazolium trifluoromethanesulfonate at 100 kPa. *Journal of Chemical & Engineering Data*, 53., 11., 2642-2648., 00219568

Ortiz A., Ruiz, A., Gorri, D., Ortiz, I. (2008). Room temperature ionic liquid with silver salts as efficient reaction media for propylene/propane separation: Absorption equilibrium. *Separation and Purification Technology*, 63., 2., 311-318., 13835866

Pereiro, A.B., Tojo, E., Rodriguez, A., Canosa, J., Tojo, J. (2006). HMImPF$_6$ ionic liquid that separates the azeotropic mixture ethanol+heptane. *Green Chemistry*, 8., 307-310., 14639262

Pereiro, A.B., Rodriguez, A. (2009). Application of the ionic liquid Ammoeng 102 for aromatic/aliphatic hydrocarbon separation. *Journal of Chemical Thermodynamics*, 41., 8., 951-956., 00219614 a)

Pereiro, A.B., Rodriguez, A. (2009). Purification of hexane with effective extraction using ionic liquid as solvent. *Green Chemistry*, 11., 346-350., 14639262 b)

Pino, V., Anderson, J.L., Ayala, J.H., Gonzallez, V., Afonso, A.M. (2008). The ionic liquid 1-hexadecyl-3-methylimidazolium bromide as novel extracting system for polycyclic aromatic hydrocarbons contained in sediments using focused microwave-assisted extraction. *Journal of Chromatography A*, 1182., 2., 145-152., 00219673

Poliwoda, A., Ilczuk, N., Wieczorek, P.P. (2007). Transport mechanism of peptides through supported liquid membranes. *Separation and Purification Technology*, 57., 3., 444-449., 13835866

Poole, C.F., Poole, S.K. (2009). Review. Extraction of organic compounds with room temperature ionic liquids. *Journal of Chromatography A*, 1217., 16., 2268-2286., 00219673

Porter, M.C. (1990). *Handbook of Industrial Membrane Technology*, William Andrew Publishing/Noyes, ISBN: 978-0-8155-1205-9, New Jersey, USA

Ravanchi, M.T., Kaghazchi, T., Kargari, A. (2010). Supported liquid membrane separation of propylene-propane mixtures using a metal ion carrier. *Desalination*, 250., 1., 130-135., 00119164

Scovazzo, P., Kieft J., Finan, D.A., Koval, C., DuBois, D., Noble, R. (2004). Gas separations using non-hexafluorophosphate $[PF_6]^-$ anion supported ionic liquid membranes. *Journal of Membrane Science*, 238., 1-2., 57-63., 03767388

Scovazzo, P., Havard, D., McShea, M., Mixon, S., Morgan, D. (2009). Long-term, continuous mixed-gas dry fed CO_2/CH_4 and CO_2/N_2 separation performance and selectivities for room temperature ionic liquid membranes. *Journal of Membrane Science*, 327., 1-2., 41-48., 03767388

Scovazzo, P. (2009). Determination of the upper limits, benchmarks, and critical properties for gas separations using stabilized room temperature ionic liquid membranes (SILMs) for the purpose of guiding future research. *Journal of Membrane Science*, 343., 1-2., 199-211., 03767388

Seiler, M., Jork, C., Kavarnou, A., Arlt, W., Hirsch, R. (2004). Separation of azeotropic mixtures using hyperbranched polymers or ionic liquids. *AIChE Journal*, 50., 10., 2439-2454., 15475905

Selvan, M.S., Mckinley, M.D., Dubois, R.H., Atwood, J.L. (2000). Liquid-liquid equilibria for toluene+heptane+1-ethyl-3-methylimidazolium triiodide and toluene+heptane+1-butyl-3-methylimidazolium triiodide. *Journal of Chemical & Engineering Data*, 45., 5., 841-845., 00219568

Soni, V., Abildskov, J., Jonsson, G., Gani, R. (2009). A general model for membrane-based separation processes. *Computers and Chemical Engineering*, 33., 3., 644-659., 00981354

Soto, A., Arce, A., Khoshkbarchi, M.K. (2005). Partitioning of antibiotics in a two-liquid phase system formed by water and a room temperature ionic liquid. *Separation and Purification Technology*, 44., 3., 242-246., 13835866

Stern, S.A. (1994). Polymers for gas separation: the next decade, Review. *Journal of Membrane Science*, 94., 1., 1-65., 03767388

Strathmann, H., Giorno, L., Drioli, E. (2006), *An introduction to membrane science and technology*, Ufficio Pubblicazioni e Informazioni Scientifiche, ISBN: 88-8080-063-9, Roma, Italia

Teramoto, M., Sakaida, Y., Fu, S.S., Ohnishi, N., Matsuyama, H., Maki, T., Fukui, T., Arai, K., (2000). An attempt for the stabilization of supported liquid Membrane. *Separation and Purification Technology*, 21., 1-2., 137-144., 13835866

Ulbricht, M. (2006). Advanced functional polymer membranes. *Polymer*, 47., 7., 2217-2262., 00323861

Venkatesan, S., Begum, K.M.M.S. (2009). Emulsion liquid membrane pertraction of benzimidazole using a room temperature ionic liquid (RTIL) carrier. *Chemical Engineering Journal*, 148., 2., 254-262., 13858947

Vidal, S.T.M., Correla, M.J.N., Marques, M.M., M.R., Ismael, Reis, M.T.A. (2004). Studies on the use of ionic liquids as potential extractants of phenolic compounds and metal ions. *Separation Science and Technology*, 39., 9., 2155-2169., 01496395

Vijayaraghavan, R., Vedaraman, N., Surianarayanan, M., MacFarlane, D.R. (2006). Extraction and recovery of azo dyes into an ionic liquid. *Talanta*, 69., 5., 1059-1062., 00399140

Visser, A.E., Holbrey, J.D., Rogers, R.D. (2001). Hydrophobic ionic liquids incorporating *N*-alkylisoquinolinium cations and their utilization in liquid–liquid separations. *Journal of the Chemical Society, Chemical Communications*, 2484-2485., 00224936

Wang, J., Pei, Y., Zhao, Y., Hu, Z. (2005). Recovery of amino acids by imidazolium based ionic liquids from aqueous media. *Green Chemistry*, 7., 196-202., 14639262

Wasserschied, P., Welton, T. (2007). *Ionic Liquids in Synthesis*, VCH–Wiley, ISBN: 978-3-527-31239-9, Weinheim, Germany

Welton, T. (1999). Room-temperature ionic liquids, solvents for synthesis and catalysis. *Chemical Reviews*, 99., 8., 2071-2083., 0009-2665

Westerholt, A., Liebert, V., Gmehling, J. (2009). Influence of ionic liquids on the separation factor of three standard separation problems, *Fluid Phase Equilibria*, 280., 1-2., 56-60., 03783812

Yung, K.K.L., Perera, J.M., Smith, C.D., Stevens, G.W. (2005). The partitioning behavior of tyramine and 2-methoxyphenethylamine in a room temperature ionic liquid-water system compared to traditional organic-water system. *Separation Science and Technology*, 40., 7., 1555-1566., 01496395

Zhang, B., Gozzelino, G., Baldi, G., (2001). Membrane liquid loss of supported liquid membrane based on n-decanol. *Colloids and Surfaces A: Physicochemical and Engineering Aspects*, 193., 1., 61-70., 09277757

Zhang, D.L., Deng, Y.F., Li, C.B. (2008). Separation of ethyl acetate-ethanol azeotropic mixture using hydrophilic ionic liquids. *Industrial & Engineering Chemical Research*, 47., 6., 1995-2001., 08885885

Zhang, L., Qiao, B., Ge, Y., Deng, D., ji, J. (2009). Effect of ionic liquids on (vapor plus liquid) equilibrium behavior of (water+2-methyl-2-propanol). *The Journal of Chemical Thermodynamics*, 41., 1., 138-143., 00219614

Zhang, L.Z., Han, J.Z., Deng, D.S., Ji, J.B. (2007). Selection of ionic liquids as entrainers for separation of water and 2-propanol. *Fluid Phase Equilibria*, 255., 2., 179-185., 03783812

26

Antimicrobial Ionic Liquids

Brendan F. Gilmore
School of Pharmacy, Queen's University Belfast, UK

1. Introduction

First described almost a century ago (Walden, 1914), ionic liquids are a novel class of low temperature (typically <100°C) molten salts, comprised of discrete anions and cations (Seddon, 1997; Scammells et al., 2005; Stark & Seddon, 2007;). Many are liquid at room temperature. The majority of room temperature molten ionic liquids are salts with large nitrogen or phosphorous-bearing cations with alkyl chain substituents and anions such as halides, fluorophosphates, fluoroborates and so on. Over one million simple ionic liquids are theoretically possible, with mixtures of two or more ionic liquids making the possibilities for new reaction media almost limitless. Ionic liquids research has experienced a massive upsurge of interest in the past decade, primarily driven by their application in 'Green' chemistry, for example, as replacements for conventional organic solvents and volatile organic compounds (VOCs) in the chemical industry. Furthermore, ionic liquids have been utilized in multitude of diverse applications from synthetic chemistry (separation/extraction/catalysis) to novel biological applications. The most commonly used and extensively described cations and anions employed in ionic liquids are detailed in **Figure 1** (adapted from Seddon et al, 2000).

The ability to 'tune' the physical, chemical and biological property sets of ionic liquids, by independent modification of the properties of the constituent anions and cations, has been the major driving force behind the huge interest in this rapidly expanding field of chemistry. 'Tuneability' of ionic liquids introduces an unparalleled flexibility in the design of reagents for a particular functional niche, these 'designer solvents' (Earle et al., 2006) are capable of providing a range of new reaction media potentially having greater diversity of character and application than that of the traditional solvents they are designed to replace (Scammells et al., 2005; Earle et al., 2006; Stark & Seddon, 2007). A summary of the physicochemical properties of common ionic liquids is given in **Table 1**. Whilst the majority of industry in this field has, to date, been directed towards 'green' applications, biological issues such as stability, biodegradability, recyclability and toxicity (Scammells et al., 2005) have received relatively little attention. However, these issues have attracted increased scrutiny recently, and the biological properties of ionic liquids, which in themselves are 'tuneable' have become one of the most debated topics in the ionic liquids arena.

Ionic liquids generally have properties such as near-zero vapour pressure (Earle et al., 2006) and thermal stability (Kosmulski et al., 2004). However, by altering the cation and anion, ionic liquids can be specifically created for a purpose or to possess particular properties suited to a given functional niche, and can therefore be described as tunable or designer

Most commonly used cations:

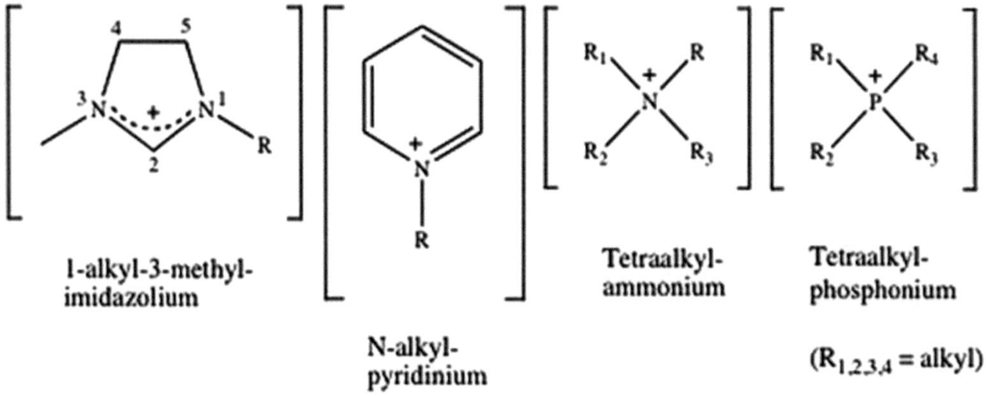

Some possible anions:

Water-insoluble → Water-soluble

$[PF_6]^-$	$[BF_4]^-$	$[CH_3CO_2]^-$, $[CF_3CO_2]^-$
$[(CF_3SO_2)_2N]^-$	$[CF_3SO_2]^-$	Br^-, Cl^-, I^-, $[NO_3]^-$
$[BR_1R_2R_3R_4]^-$		$[Al_2Cl_7]$-

Most commonly used alkyl chains: ethyl, butyl, hexyl, octyl, decyl, dodecyl

Fig. 1. Examples of the most commonly described ionic liquid cations and anions (Adapted from Seddon *et al.*, 2000).

chemicals (Freemantle, 1998; Davis, 2004). Ionic liquids have been validated as ideal replacements for organic solvents in a plethora of chemical processes (Villiagran et al., 2006; Huddleston et al., 2001; Mizuuchi et al., 2008). Perhaps one of the most attractive characteristics of employing ionic liquids in chemical processes is their potential for improving reaction yields, facilitating product recovery and their recyclability without loss of functionality. As a result ionic liquids have found applicability in an impressively diverse range of uses.

Physiochemical property	**Ionic liquid**
Conductivity	Good ionic conductivity compared to organic solvents/electrolyte systems. This is inversely linked to viscosity (Endres & Abedin, 2006)
Viscosity	Generally more viscous than common molecular solvents. Viscosity is determined by van der Waals forces and hydrogen bonding and alkyl chain length in the cation (Endres & Abedin, 2006)
Density	Generally more dense than water (Endres and Abedin, 2006)
Melting point	<100°C
Solubility	Ionic liquids can act as both hydrogen bond acceptors (anion) and donors (cation) and therefore interact with substances with both accepting and donating sites (Dupont & Suarez, 2006). Ionic liquids can be divided into two groups (water-miscible and water-immiscible) according to their solubility in water (Wei & Ivaska, 2008). Examples of water-immiscible ionic liquids include 1-butyl-3-methylimidazolium hexafluorophosphate and 1-decyl-3-methylimidazolium bis(trifluoro-methylsulfonyl)imide. Examples of water-miscible ionic liquids include [1-Butyl-3-methylimidazolium tetrafluoroborate. (Wei and Ivaska, 2008). Miscibility of ionic liquids in water is primarily dependent on the anion present it is also dependent on the structure of the cation (Seddon et al., 2000; Wei & Ivaska, 2008),
Thermal stability	Highly thermally stable (some up to temperatures of 450°C) (Endres & Abedin, 2006)
Chemical stability	Most are stable towards organic and inorganic substances (Dupont & Suarez, 2006)
Electrochemical window (defined as the electrochemical potential range over which the electrolyte is neither reduced or oxidised at an electrode)	Wide electrochemical window (Endres & Abedin, 2006)

Table 1. Physiochemical properties of Ionic liquids

2. Ionic liquids in 'green chemistry'

Green chemistry is defined as the design of chemical products and processes which reduce or eliminate the use and generation of hazardous substances (Anastas & Warner, 1998; Seddon et al., 2005). The design of safe and environmentally benign solvents has become increasingly important in the development of clean manufacturing processes. Conventional organic solvents are often toxic, flammable and volatile which when released into the environment can have potentially devastating effects. Ionic liquids have offered promise as reagents, which have the potential to replace many hazardous volatile organic solvents (including those banned by the Montreal protocol of 1989) (Anastas & Warner, 1998), and have therefore been cited as an important element of green chemistry. Ionic liquids have also been shown to have similar (if not superior and more diverse) properties to the organic/aqueous solvents they could potentially replace (Visser et al., 2000), whilst having negligible vapour pressure thus reducing the likely risk of atmospheric pollution (Fredlake et al., 2004). Attractive physicochemical attributes, improved reaction rates and yields, recyclability and design of ionic liquids lacking inherent biological toxicity all represent approaches for the 'greening' of chemical processes by ionic liquids.

Ionic liquids have already been reported as alternative 'green' solvents for a wide range of reactions (Wasserscheid et al., 2002; Prasad et al., 2005; Tao et al., 2006), however, in addition to possible concerns about the recyclability of ionic liquids there have also been concerns raised over the biodegradability or environmental persistence of ionic liquids (Garcia et al., 2004; Garcia et al., 2005; Stolte et al., 2008). A series of imidazolium compounds were shown to be poorly biodegradable and it was found that bacteria did not use them as a source of carbon under the conditions of the investigation (Romero et al., 2008), making them potentially persistent polluters. In this study, it was also demonstrated that imidazolium based ionic liquids have a wide range of toxicities in this relevant bioassay. Generally, toxicity (EC_{50} value) was found to correlate directly with the length of the *n*-alkyl substituent in the methylimidazolium cation, while the anion has no apparent effect on the EC_{50} value. The authors conclude that the ionic liquids tested are more toxic than conventional organic solvents. In tests against *Vibrio fischeri* and mammalian cell lines, a series of imidazolium ionic liquids of varying alkyl chain length were shown to exhibit significant toxicity (Ranke et al., 2004), once again dependent on the length of the cationic *n*-alkyl substituent.

Many ionic liquids are water-soluble and as such could contribute to pollution of aquatic environments. For example, it has been demonstrated that imidazolium, pyridinium and pyrrolidinium ionic liquids had LC_{50} > 100 mg/L against *Danio rero* (zebra fish), and as such can be regarded as non-lethal (Pretti et al., 2006). On the other hand, however, the ammonium based ionic liquids had LC_{50} values remarkably lower than that reported for organic solvents and yet proved fatal when zebrafish were exposed to them. Ecotoxicological tests on several ionic liquids have revealed that imidazolium and pyridinium ionic liquids exhibit significant toxicity towards the freshwater algae *Pseudokirchneriella subcapitata* (Pham et al., 2008), while imidazolium ionic liquids are toxic to the freshwater crustacean *Daphnia magna* (Wells & Coombe et al., 2004) and *Caenorhabditis elegans* (Swatloski et al., 2004). A number of recent studies have also demonstrated the potential of certain ionic liquids to exhibit excellent antimicrobial activity, discussed below, thus presenting the exciting possibility that ionic liquids could have application as biocidal agents in the control of microorganisms in the environment for contamination and infection control.

3. Methods for evaluating the antimicrobial activity of ionic liquids

A number of methods exist for the accurate determination of microbial susceptibility to antimicrobial/antibiotic compounds. Such methods yield vital data regarding fundamental sensitivity or tolerance to a given antimicrobial biocide or antibiotic and are therefore vital to the successful treatment and management of microbial infections. Furthermore, such tests are useful for determining relative potency of an antimicrobial agent across a range of species and for identifying antimicrobial synergies. The basic testing procedures, which have been used in the assessment of the antimicrobial activity of ionic liquids are considered briefly below. Whilst the majority of these tests have relied on basic planktonic susceptibility assays (minimum inhibitory concentrations (MIC) or minimum bactericidal/fungicidal concentration (MBC/MFC)) or agar diffusion techniques, the importance of evaluation of antimicrobial activity against microbial biofilms is also discussed. In our group, we have pioneered the use of high throughput screening of ionic liquids against clinically relevant microorganisms grown as biofilms, by determination of minimum biofilm eradication concentration (MBEC) (Carson et al., 2009; Busetti et al., 2010).

3.1 Agar diffusion tests

The agar diffusion technique (also known as the Kirby-Bauer test (Bauer et al., 1966) but described somewhat earlier by Abraham am co-workers in 1941 (Abraham et al., 1941)) is a simple and commonly employed technique for determination of MIC on solid media. The basic method requires antibiotic/biocide impregnated discs to be placed on the surface of agar plates seeded or spread with the appropriate test strain of bacteria or fungi. Antimicrobial agent(s) may also be added (as a solution) to wells punched in the agar. The diffusion of antimicrobial agent into the surrounding agar results in inhibition of growth around the reservoir/source and gives rise to zones or clearance where (for sensitive organisms) microbial growth is inhibited. Generally, the diameter of these zones of inhibition or clearance increases with increasing concentration of antimicrobial agent, and this may be measured to determine qualitatively the relative degree of toxicity. The MIC may also be determined from the zero intercept of a linear regression of the squared size of these zones of inhibition, x, versus the natural logarithm of the antibiotic concentration (Bonov et al, 2008). This is described in the equation below, where D is the diffusion coefficient (assumed to be independent of concentration) and t the time over which antibiotic diffusion occurs (incubation time):

$$\ln(MIC) = \ln(c) - \frac{x^2}{4Dt}$$

The technique is also useful for empirical determination of antimicrobial activity of a given compound or assessing relative antimicrobial potency by measuring zones of inhibition of bacterial or fungal growth around the antimicrobial site of application. Recently, this method has been championed by Stephens and co-workers (Rebros et al., 2009; Wood & Stephens, 2010) as a simple method for rapid determination of relative toxicity of ionic liquids. This simple, inexpensive method has been suggested as a basic requirement in the toxicological assessment of ionic liquids and, since it requires neither specialist equipment nor advanced microbiological techniques, may be performed routinely in laboratories conducting research into ionic liquids with minimum microbiological expertise. However, the method is not without inherent limitations and consequently, care must be taken in interpretation of the data obtained. For example, it is well established that some antibiotics deviate from the behaviour described above by interacting with components of the growth media; similar effects might be expected with some ionic liquids especially those having hydrophobic or amphipathic character. Interaction of ionic liquids and other ionic components of the growth media (dissolved salts, nutrients etc.), chemical reactivity of the reagent and interaction with the agar itself may all result in erroneous data. Furthermore, the method is unlikely to be of any practical use for ionic liquids which are immiscible with water, since agar is >98% water, and thus water miscibility will have a significant effect on the extent of diffusion through the medium. Despite this, the use of agar diffusion assays will provide basic toxicity information for a large number of ionic liquids and provides a rapid, high-throughput 'first look' in the hierarchical screening of antimicrobial activity of ionic liquids.

3.2 Dilution tests

Dilution tests are routinely used for the determination of the two most fundamental parameters in antimicrobial susceptibility testing; the minimum biofilm eradication concentration (MIC) and the minimum bactericidal/fungicidal concentration (MBC/MFC), sometimes referred to as the minimum lethal concentration (MLC). Dilution tests usually involve the use of liquid media but agar may also be used (as discussed above). Doubling dilutions of the antimicrobial agent are prepared and added to a defined inoculum of test microorganism taken from the logarithmic phase of growth, such that a final inoculum of 5 x 10^5 colony forming units (CFU or viable cells)/ml is achieved. Following incubation at 35°C ± 2.5°C overnight (18 hours), the MIC is determined as the concentration of antimicrobial contained in the first clear tube/well. Therefore the MIC is defined as the minimum concentration of antimicrobial agent that inhibits the growth of an overnight culture of microorganism. The conditions used for the test and appropriate control tests (which must be included) are most commonly obtained either from the Clinical and Laboratory Standards Institue (CLSI) formerly the National Committee for Clinical Laboratory Standard (NCCLS) (NCCLS document M27-A, 1997; NCCLS document M7-A5) or the British Society for Antimicrobial Chemotherapy (Andrews, 2001).

The MBC is the lowest concentration (in mg/L) of antimicrobial that results in ≥99.9% killing of the bacterium under test. The 99.9% cut-off is an arbitrary *in vitro* value with 95% confidence limits that has uncertain clinical relevance. MBCs are determined by spreading 0.1-ml (100-ml) volumes of all clear (no growth) tubes from a dilution MIC test onto separate agar plates (residual antimicrobial in the 0.1-ml sample is 'diluted' out over the plate). After incubation at 35°C overnight (or longer for slow-growing bacteria), the numbers of colonies growing on each plate are recorded. The first concentration of drug that produces <50 colonies after subculture is considered the MBC. Minimum fungicidal concentrations are determined in the same manner, however, different growth media is necessary (e.g. use of RPMI 1640 plus 2% dextrose) and the inoculum density (yeast cells or spores) is reduced (c. 10^4 CFU/ml).

3.3 Evaluating biofilm susceptibility to antimicrobial agents

Both the MIC and MBC/MFC evaluations are suspension tests, which test the susceptibility of planktonic (free floating) microorganisms grown under optimum conditions to a given antimicrobial challenge. However, evaluation of the antimicrobial susceptibility of microbial biofilms is now recognized as a more physiologically relevant assay. A biofilm may be defined as *'a microbially derived sessile community characterized by cells that are irreversibly attached to a substratum or interface or to each other, are embedded in a matrix of extracellular polymeric substances that they have produced and exhibit altered phenotype with respect to growth rate and gene transcription'* (Donlan & Costerton, 2002). Biofilms represent the predominant mode of growth of microorganisms and also the most persistent, phentypically resistant mode of growth, with increased tolerance to antimicrobial challenge. Irrespective of site, biofilm formation follows a series of discreet events, summarized below in Figure 2.

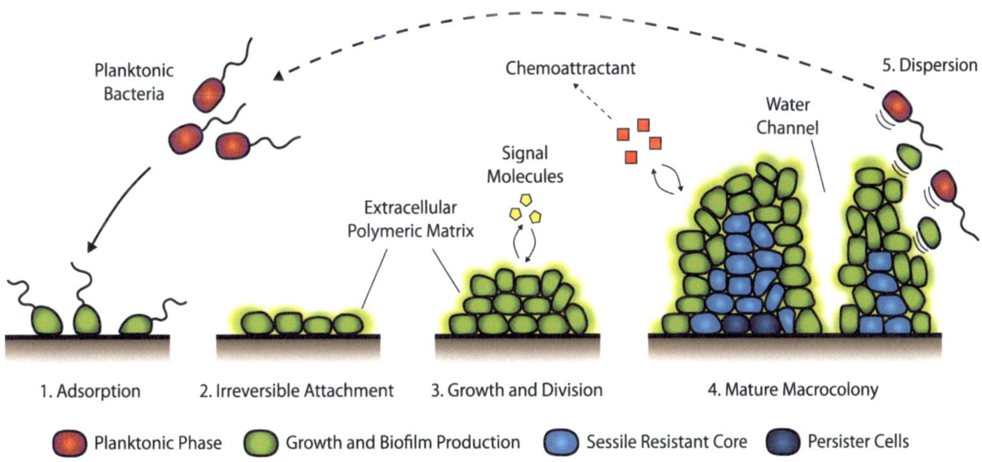

Fig. 2. Microbial Surface Colonisation; Main Stages in Surface attachment and biofilm formation (Adapted from Harrison et al., 2005)

As the importance of microbial biofilms in medicine, industry and agriculture has become clear, a huge amount of industry has been invested into studying their growth and control. As a result of this a number of *in vitro* models have been developed to facilitate elucidation of the mechanisms central to this important microbiological process. Each model has relative advantages and disadvantages, depending on the aspect of biofilm physiology the models were designed study. These are expertly reviewed in (McBain, 2009; Coeyne & Nelis, 2010). However, to date the only model used for the study of biofilm susceptibility of ionic liquids is that employed in our group, namely the Calgary Biofilm Device (commercially available from Innovotech Inc., Edmonton, AB, Canada as the MBEC Assay). The MBEC assay, developed in 1999 by Ceri and co-workers (Ceri et al., 1999) was developed specifically to evaluate biofilm susceptibility to antimicrobials. Essentially, the device consists of a 96-well plate and a lid bearing 96 polycarbonate pegs, each peg protrudes into the 96 wells and provides a surface onto which the bacteria/fungi may attach and form a biofilm, as shown below in Figure 3.

Shear forces (provided by gyration of the plate) stimulates microbial attachment and biofilm formation, the density of which may be determined by sonication of the biofilm back into fresh growth media followed by enumeration *via* standard plate counting. Biofilms grown on pegs may then be transferred to a 96-well challenge plate, set up in the same manner as an MIC assay with varying concentrations of antimicrobial agent(s) alone or in combination.

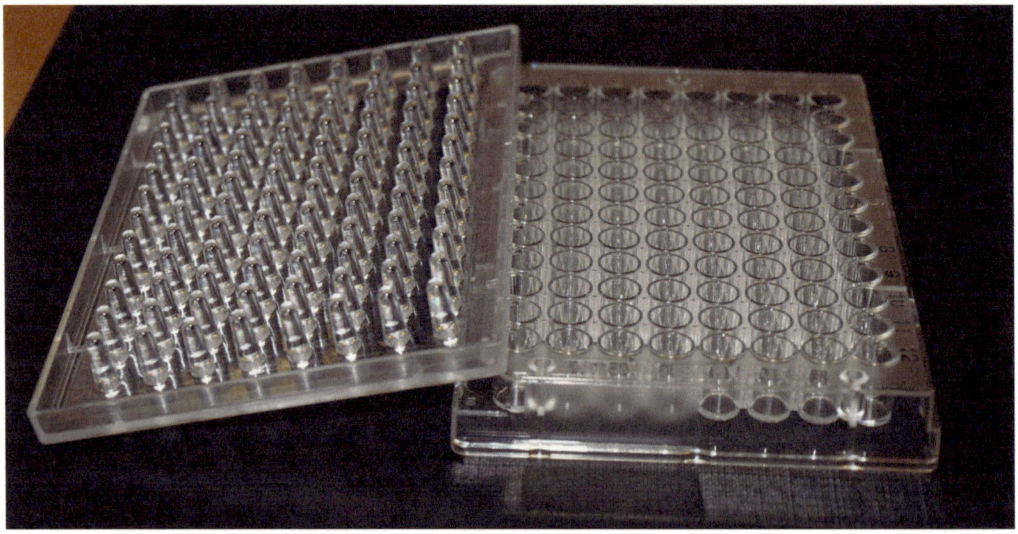

Fig. 3. The commercially available Calgary Biofilm Device/MBEC Assay Plate

Following antimicrobial exposure bacteria would again be sonicated from the pegs and counted to determine the biofilm MIC (BMIC), biofilm bactericidal concentration (BMBC) and biofilm eradication concentration (MBEC) in a highly standardized and reproducible assay based on existing MIC technology.

4. Antimicrobial and antibiofilm activities of ionic liquids

The toxicity shown in the studies highlighted previously raises issues over the validity of the classification of ionic liquids as 'green' compounds. However, toxicity itself a tuneable property which may be of utility in a number of other applications, for example, in the development of antiseptics, disinfectants and anti-fouling reagents (Pernak et al., 2004a; Pernak et al., 2004b; Pernak et al., 2007a; Fischmeister et al., 2007). The antimicrobial activities of five new groups of choline-like quaternary ammonium chloride ionic liquids were evaluated against a range of Gram positive and Gram negative bacteria (Pernak & Chwala, 2003). The ionic liquids tested all showed good antimicrobial activity, and confirmed that lipophlicity was the main factor in determining antimicrobial activity. Compounds with an alkyl chain substituent of 12 carbon atoms on the cation all exhibited the highest antimicrobial activity across all groups of ionic liquids tested, for a range of test microorganisms.

In a similar study, a series of 3-alkoxymethyl-1-methylimidazolium ionic liquids bearing [Cl], [BF_4] and [PF_6] anions were tested against a range of bacterial species, as well as fungi (Pernak et al., 2003). This study demonstrated that shorter cationic alkyl chain substituents resulted in reduced antimicrobial activity compared to the imidazolium compounds containing 10, 11, 12 and 14 carbon atoms in their alkoxy group, confirming earlier findings (Pernak et al., 2004a). Again, the imidazolium ionic liquids with alkoxy substituents of twelve carbon atoms were the most active against the bacteria and fungi tested. Another study showed that 1, 3 - (dialkloxymethyl)-substituted imidazolium ionic liquids (Pernak et al., 2004b) also exhibited broad-spectrum antimicrobial activity against various bacterial rods, cocci and fungi.

Pyrrolidinium ionic liquids with varying alkyl chain substituents were shown to possess good antimicrobial activity against rods, cocci and fungi (Demberelnyamba et al., 2004). Compounds exhibiting the greatest antimicrobial activity were those having 14 carbon atoms in the alkyl chain. In a recent study, Pernak and co-workers tested a range of trigeminal tricationic ionic liquids for antimicrobial activity (Pernak et al., 2007b), it was found that, in addition to their broad spectrum antimicrobial activity, their potency was much better than the commercially available benzalkonium chloride. A further study on chiral ammonium-based ionic liquids (Pernak & Feder-Kubis, 2005) revealed that compounds with 11 carbons in the alkyl substituent showed the most activity against a range of bacteria and fungi. In a study carried out on a number of ionic liquids with varying anions (Docherty & Kulpa, 2005), it was found that improved antimicrobial activity resulted from increasing alkyl group chain length as well as increasing the number of alkyl groups substituted on the cation ring. Varying the anion present in the compound did not significantly alter toxicity. Recently, the antimicrobial activity of multifunctional long-alkyl-chain quaternary ammonium azolate based ionic liquids has been described (Walkiewicz et al., 2010). These ionic liquids, based on didecylmethylammonium, benzalkonium, domiphen and hexadecyltrimethlammonium cations combined with benzotriazole, 1,2,4-triazolate, 4-nitroimidazolate or 2-methyl-4-nitroimidazolate anions all exhibited excellent, broad spectrum anti-bacterial and antifungal activity, which was comparable or superior to that of the original benzalkonium chloride (Walkiewicz et al., 2010).

According to the studies discussed above, a general feature common to the ionic liquids is a dependency on substituent alkyl chain length for antimicrobial potency, indicating a general mechanism for antimicrobial activity. Other studies have indicated that the mechanism of antimicrobial activity of ionic liquids is *via* membrane disruption. This seems likely given the structural similarity between ionic liquids and antimicrobial agents whose mechanism is more fully elucidated (Li et al., 1998; Pernak et al., 2001). Many ionic liquids have a similar structure to cationic surfactants whose primary mode of action membrane-bound protein disruption (Bernot et al., 2005). Another suggested mechanism of toxicity and antimicrobial activity is the inhibition of the enzyme acetylcholinesterase, as illustrated in studies of the inhibitory effects of imidazolium and pyridinium ionic liquids which were shown to inhibit purified enzyme with EC_{50} levels as low as 13 μM (Stock et al., 2004).

4.1 Antibiofilm activity of 1-alkyl-3-methylimidazolium chloride and 1-alkylquinolinium bromide ionic liquids

All microbiological toxicity studies conducted to date have described antimicrobial activity against planktonic, or free swimming, microbial phenotypes. However, the predominant mode of growth of both pathogenic and environmental microorganisms, is as highly-ordered surface-adhered communities encased within a self-produced protective extracellular polymeric matrix (glycocalyx), collectively known as a biofilm (Donlan & Costerton, 2002; Hall-Stoodley et al., 2004). A general characteristic of biofilm communities is that they tend to exhibit significant tolerance/resistance to antibiotics and antimicrobial/biocidal challenge compared with planktonic bacteria of the same species (Ceri et al., 1999; Stewart & Costerton, 2001; Gilbert et al., 2002; Stewart, 2002). Therefore, significant limitations exist when attempting to extrapolate planktonic culture susceptibility data to environmental or clinical scenarios where the majority of microbial growth is as biofilms. This is illustrated by the NIH estimation that up to 80% of all chronic human infections are biofilm-mediated and that 99.9% of bacteria in aquatic ecosystems live as biofilm communities (Lewis, 2001; Costerton & Wilson, 2004). Indeed, it has been demonstrated that there is often no correlation between planktonic susceptibility to antimicrobials (MIC values) and biofilm susceptibility of the same species and strain to those same antimicrobial agents (Smith et al., 2003).

Biofilms are a major survival strategy for microbial populations in the face of environmental stresses and have been linked to a host of industrially and clinically relevant complications; from chronic plant, animal and human infections, to failure of implanted medical devices and microbially-influenced biocorrosion. Therefore, knowledge of the antibiofilm activity of ionic liquids is both environmentally and clinically relevant.

In a recent study, Carson and colleagues reported for the first time the *in vitro* antibiofilm activity of a library of 1-alkyl-3-methylimidazolium chloride ionic liquids (the general structure is given below in Figure 4) against a panel of clinically relevant pathogenic bacteria (including MRSA) and fungi using the Calgary Biofilm Device (CBD), a high-throughput micro-titre plate-based technology for screening antimicrobial susceptibility of microbial biofilms, which permits the determination of minimum biofilm eradication concentration (MBEC), or the concentration of an antimicrobial agent required to kill a microbial biofilm. This study illustrated that antibiofilm activity of these ionic liquids was also dependent on alkyl chain length, with the MBEC value decreasing (increased antibiofilm potency) with increasing alkyl chain length. Ionic liquids [$C_n$mim]Cl where $n \geq 10$ exhibited potent, broad spectrum antimicrobial activity. In general, of the compounds tested in this series, [$C_n$mim]Cl where $n = 14$, exhibited greatest antibiofilm activity against all microbial biofilms. The data from this study indicate that Gram positive microbial biofilms (in keeping with planktonic cultures) are generally more susceptible to 1-alkylmethylimidazolium ionic liquids than Gram negative bacterial biofilms, whilst *Candida tropicalis* biofilms exhibited a similar susceptibility profile to these reagents as the representative Gram positive organisms tested in this study (Carson et al., 2009).

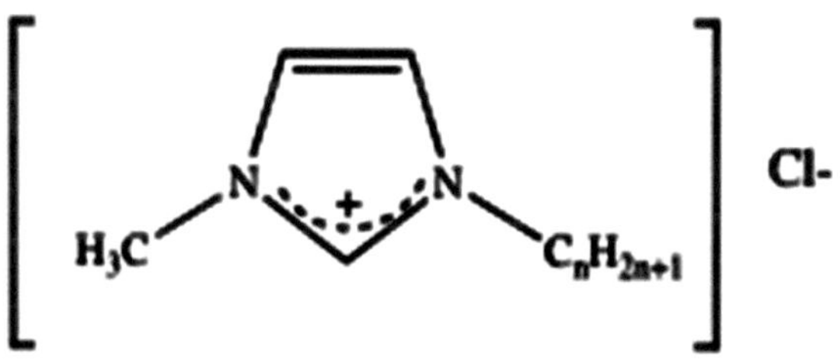

Fig. 4. General structure of 1-alkyl-3-methylimidazolium chlorides [$C_n$mim]Cl which exhibited antibiofilm activity across a range of clinically relevant pathogens.

In a further study from the same group, Busetti and co-workers, described the antimicrobial and antibiofilm activities of a range of 1-alkylquinolinium bromide ionic liquids (Busetti et al., 2010). In general, these ionic liquids are the most potent antibiofilm ionic liquids tested so far, having a superior microbiological toxicity to the 1-alkyl-3-methylimidazolium ionic liquids against both planktonic and biofilm cultures of a range of bacteria and fungi commonly implicated in nosocomial and device associated infections, including *Staphylococcus epidermidis, Pseudomonas aeruginosa, Klebsiella aerogenes* and *Bacillus cereus*. In keeping with the observations from our previous studies, the antimicrobial activity is dependent on the length of alkyl chain substituent, with compounds having alkyl chain lengths of 12-14 carbon atoms exhibiting greatest antimicrobial potency. The general structure of the 1-alkylquinolinium bromides is given below in Figure 5.

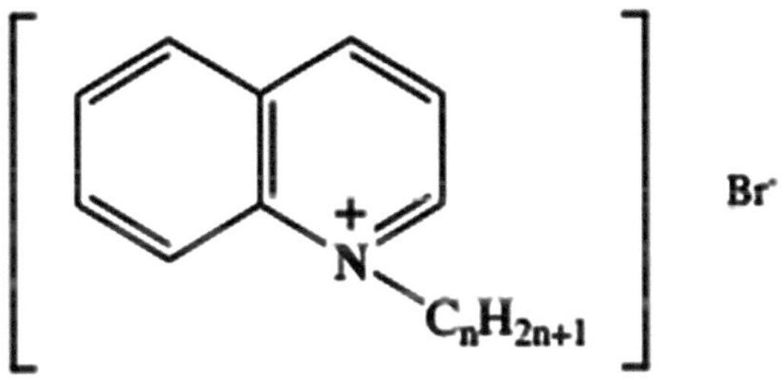

Fig. 5. General structure of 1-alkylquinolinium bromides with demonstrated antimicrobial and antibiofilm activities

These important studies not only highlight the potential environmental effects of ionic liquids to the microbial ecosystem (which are susceptible to their antimicrobial activities even in their predominant environmental mode of growth as biofilms) but also opens up the real possibility of employing ionic liquids for a plethora of beneficial uses as antimicrobials in disinfectants, preservatives, antiseptics and development of anti-infective medical device surfaces for use in healthcare and as antibiofouling reagents for a host of industrial applications.

The challenges that remain in bringing the first ionic liquid based biocides to the clinical setting as either disinfectants, antiseptics, sterilants for medical devices/instruments or preservatives include demonstrating suitably rapid rates of kill, for example high level disinfectants would typically be required to be sporocidal in <7 minutes, and (as an industry 'rule of thumb') reduce the original (vegetative) bioburden of 5 microbial species by 5 log reductions (99.999% kill) within 5 minutes. The factors that attenuate the activity of ionic liquids as disinfectants have not, as yet, received sufficient attention. Furthermore, the toxicological profile of these compounds is, for the present, not elucidated (although ongoing work in our laboratory is aimed at addressing this lacuna in our knowledge). Despite this, ionic liquids appear to hold great promise in the future development of biocides for use in clinical and industrial infection and contamination control.

5.0 Potential regulatory challenges to antimicrobial applications of ionic liquids

As with all novel compounds with potential application as biocidal products, legislative and commercial barriers to their entry to the market exist. In the following sections, the current European legislation regarding biocidal products and new chemical entities coming onto the EU market are discussed. These legislative instruments are necessary in safeguarding public safety and will no doubt prove burdensome in bringing ionic liquids from the bench to various in-use settings where we predict they will be usefully employed as biocides.

5.1 The Biocidal Products Directive (98/8/EC)

Directive 98/8/EC of the European Parliament and of the Council on the placing on the market of biocidal products was adopted in 1998. The Biocidal Products Directive aims to harmonise both the issues of biocide manufacture and use, and the European market for biocidal products. Furthermore, the directive aims to provide a high level of protection for humans, animals and the environment. The scope of the directive is wide, covering some 23 different product types including disinfectants (classified by use in given areas), chemicals for preservation of products and material (such as timber), non-agricultural pesticides and anti-fouling agents used to prevent hull-fouling on vessels. However, medicines and cosmetics fall outside the remit of the directive, potentially allowing ionic liquids to be used as preservatives in the first instance.

The basic objective of the directive is to produce a list of active biocidal products that are licensed for use across all member states, known as 'Annex I'. Active substances must be assessed and any decision on their inclusion in Annex I will be taken at Community level. Only products containing active substances listed in Annex I will be authorised for use in the EC. The two-tier system mandated by the directive requires that firstly, active substances must be assessed and a decision reached as to their suitability for inclusion in Annex I and secondly, the producers and formulators responsible for the placing of the market of the biocidal products and their active substances must apply for authorisation of the biocidal product. In each member state, it is the responsibility of the national competent authority (in the GB this is the Health and Safety Executive) to authorise products containing active substances included in Annex I. The principle of mutual recognition outlined in Article 4 of the directive means that once a product containing an Annex I active substance has been authorised by one member state it can be recognised in as an authorised product in other member states.

The legislation related to the directive came into force in September 2000, with guidance on deadlines for identification and notification of active substances. In June 2009, based on experience of working under Directive 98/8/EC, the European Commission adopted a proposal for a Regulation concerning the placing on the market and use of biocidal products (COM(2009)267), which is intended as a full revision of the existing directive which it will repeal and replace. This revision is a response to a 2008 report on the implementation of the directive, which highlighted the inherent weaknesses of the original directive, primarily the complexity of the legal framework and the high costs associated with compliance (especially the cost of compiling a dossier in support of inclusion of an active substance in Annex I). The proposed new regulation is scheduled to enter into force on January 1st 2013.

Although there is unlikely to be any direct impact on the pharmaceutical industry *per se*, the biocidal products directive and proposed revision *is likely to present a significant restriction to bringing new disinfectants to market*. This is likely to have some impact on the use of new ionic liquids for biocidal applications. The experience under the original directive indicated that expense was a major issue in supporting the addition of an active substance to Annex I. It remains to be seen if the revised regulation will streamline the process and reduce costs associated.

5.2 Registration, Evaluation, Authorisation and Restriction of Chemicals (REACH)

EC Regulation (EC1907/2006) Registration, Evaluation, Authorisation and Restriction of Chemicals (REACH) entered into force on June 1st 2007, replacing a number of European Directives and Regulations within a single legislative framework. The former EC legislative framework for chemical substances was a collection of numerous different directives and regulations, which had developed historically. Part of the problem, which REACH sought to address, was the different rules governing 'existing' and 'new' chemicals. This distinction between existing and new chemical substances was introduced under regulation (EC) 793/93 based on a cut-off date of 1981. Chemical substances which were reported as being on the EC market between 1st January 1971 and 18th September 1981 were called 'existing' chemicals (~100,000, listed in the European Inventory of Existing Commercial Chemical Substances (EINECS)); those introduced to the market after the cut-off date were classified as 'new' chemicals (~3800). Whilst new chemicals must undergo rigorous testing before being placed on the market, there were no such regulations for existing chemicals, as a result there is generally insufficient publically accessible information available to accurately assess and control these substances effectively.

Further issues which led to the drafting of this Regulation include the pre-REACH allocation of responsibilities whereby public authorities were responsible for undertaking risk assessments of substances, with no such responsibilities on downstream users (manufacturers, importers, end users). Manufacturers and importers were required to provide information on use of the substances, but downstream users (industrial users, formulators) were not. In addition, the threshold for notification and testing of new chemical substances could be as low as 10 kg per year. This has been regarded as a significant barrier to innovation within the EU chemical industry, since the resultant trend has been away from developing new chemicals and towards using exiting agents.

The aims of REACH are:

i. To provide a high level of protection of human health and the environment from the use of chemicals
ii. To give those who place chemical substances on the market (manufacturers and importers) responsibility for understancing and managing the risks associated with their use
iii. To allow free movement of substances on the EU market
iv. To enhance the innovation in and competitiveness of the EU chemicals industry
v. To promote the use of alternative methods (other than animal studies) for assessment of hazardous properties of substances, e.g. QSAR studies.

REACH is based on the concept that the chemical industry itself is best placed to ensure that chemicals placed on the market in the EU do not adversely affect human health or the environment. REACH creates a single system for both new and existing chemical substances. Substances are now described as 'non-phase-in' substances (i.e. those not produced or marketed prior to the entry into force of REACH) and 'phase-in' substances (i.e. those substances listed in EINECS, or those that have been manufactured in the Community, but not placed on the Community market, in the last 15 years or the so called "no longer polymers" of Directive 67/548). A major part of REACH is the requirement for manufacturers or importers of substances to register them with a central European Chemicals Agency (ECHA), with a standard set of data to be submitted for each substance. If the substance is not registered, data will not be available and the substance will no longer be able to be legally manufactured or supplied.

REACH proposes a number of benefits over the existing patchwork of legislation and regulations. Primarily, by creating parity for 'existing' and 'new' chemical substances with respect to risk management and the making available of data for all substances in relation to this. Furthermore, it simplifies the existing EU level regulation by replacing 40 existing pieces of legislation and creates a single system for all chemicals, removing the distinction of 'existing' and 'new' substances. REACH will result in better risk characterisation of chemicals and mandates improved information flows in the supply chain. REACH will close the knowledge gap for over 30,000 existing substances, providing information on both acute and long-term toxicity. REACH provisions are intended to be phased in over a period of 11 years. Manufacturers of biocides and biocidal products are examining closely the likely impact of REACH and it is generally regarded that REACH will impose a significant burden, but one which must be borne for commercial reasons. In this respect, the bringing onto the market of ionic liquid based biocidal products will require (i) demonstration that these reagents are more effective or have superior properties and in-use characteristics compared with existing biocides and (ii) significant financial investment to satisfy the compliance programmes required by REACH. That said, there exists a pressing need for new and effective biocides, in the face of increasing emergence of resistance to most conventional biocides. Ionic liquids have numerous attractive properties, which render them excellent candidates for biocidal applications, and, to date, no reports of resistance have been published.

6. Conclusion

Since ionic liquids are tunable and designer chemicals, they have been used in a wide variety of applications. Many have been developed for use as solvents in industrial chemistry, and generally their use confers a number of advantages over using other solvents; superior reaction rates, recyclability of reactants and catalysts and improved product recovery. Having negligible vapour pressure, it has been suggested that ionic liquids will not contribute to air pollution and are thus green alternative to conventional organic solvents which are generally volatile, flammable and toxic. As a result of this, many studies have been conducted on the use of ionic liquids as novel green solvents to replace established solvents for particular reactions.

However, toxicity of ionic liquids has been demonstrated by a number of groups, including our own, in a variety of environmental niches, and raises questions over their 'green' credentials. Nonetheless the toxicity of ionic liquids is in itself a property which can be tuned and exploited for other beneficial uses, for example in developing novel antimicrobials. We have demonstrated the utility of ionic liquids as antibiofilm agents; biofilms are complex, organised communities of bacteria which have been shown to have greater tolerance and resistance to antimicrobials, accounting for the majority of chronic and acute infections as well as the majority of bacterial communities in aquatic environments. In summary, the biological properties of ionic liquids may yet prove their most exciting and the benefits of rationally designed, bespoke ionic liquid-based antimicrobials to human health has yet to be harnessed.

7. References

Abraham, E.P., Gardner, A.D., Chain, E.B., Heatley, N.G., Fletcher, C.M., Jennings, M.A., Florey, H.W. (1941) Further observations on penicillin. Lancet, ii: 177-189.

Andrews, J. M. (2001) Determination of minimum inhibitory concentrations. *J Antimicrob Chemother*, 48 *Suppl. S1*. 5-16.

Anastas, P.T. and Warner, J.C. (1998) *Green Chemistry: Theory and Practice*; Oxford University Press: New York, 1998, p.30.

Bauer, A.W., Kirby, W.M.M., Sherris, J.C., Turck, M. (1966) Antibiotic susceptibility testing by a standardized single disk method. *Am. J. Clin. Pathol.* 45, 493-496.
Bonev, B., Hooper, J., Parisot, J. (2008) Principles of assessing bacterial susceptibility to antibiotics using the agar diffusion method, *J. Antimicrob. Chemother.*, 61, 1295-1301.
Bernot R. J., Kennedy, E. E., Lamberti, G. A., (2005) Effects of ionic liquids on the survival, movement and feeding behaviour of the freshwater snail, *Physa acuta. Environ. Toxicol. Chem.*, 24, 1759-1765.
Busetti, A., Crawford, D.E., Earle, M.J., Gilea, M., Gilmore, B.F., Gorman, S.P., Laverty, G., Lowry, A.F., McLaughlin, M., Seddon, K.R. (2010) Antimicrobial and antibiofilm activities of 1-alkylquinolinium bromide ionic liquids. *Green Chem.*, 12, 420-425.
Carson, L., Chau, P. K. W., Earle, M. J., Gilea, M. A., Gilmore, B. F., Gorman, S. P., McCann, M. T., Seddon K. R. (2009) Antibiofilm activities of 1-alkyl-3-methylimidazolium chloride ionic liquids. *Green Chem.*, 11(4) 492-497.
Ceri, H., Olson, M. E., Stremick, C., Read, R. R., Morck, D., Buret, A. (1999) The Calgary Biofilm Device: new technology for rapid determination of antibiotic susceptibilities of bacterial biofilms. *J Clin Microbiol.*, 37, 1771-1776.
Coeyne, T. & Nelis, H.J. (2010) *In vitro* and *in vivo* model systems to study microbial biofilm formation. *J Microbiol Methods,* Sept 16 - ahead of print.
Costerton, J. W. & Wilson, M. (2004) Introducing *Biofilms. Biofilms*, 1, 1-4.
Davis, J.H. (2004) Task-specific ionic liquids. *Chem. Lett.*, 33, 1072-1077.
Demberelnyamba, D., Kim, K., Choi, S., Park, S., Lee, H., Kim, C., Yoo, I. (2004) Synthesis and antimicrobial properties of imidazolium and pyrrolidinium salts. *Bioorg. Med. Chem.*, 12, 853–857.
Docherty, K. & C.F. Kulpa Jr. (2005), Toxicity and antimicrobial activity of imidazolium and pyridinium ionic liquids *Green Chem.*, 7, 185–189.
R. M. Donlan and Costerton, J. W. (2002) Biofilms: Survival mechanisms of clinically relevant microorganisms. *Clin. Microbiol. Rev.*, 15, 167-193
Dupont J. & Suarez, P.A.Z. (2006) Physico-chemical processes in imidazolium ionic liquids *Phys. Chem. Chem. Phys.*, 8, 2441-2452.
Earle, M.J., Esperancüa, J.M.S.S., Gilea, M.A., Canongia Lopes, J.N., Rebelo, L.P.N., Magee, J.W., Seddon, K.R., Widegren, J.A. (2006) The distillation and volatility of ionic liquids *Nature,* 439, 831.
Endres, F. & El Abedin, S.Z. (2006) Air and water stable ionic liquids in physical chemistry *Phys. Chem. Chem. Phys.*, 8, 2101-2116.
Fischmeister, C., Griffin, S.T., Rogers, R.D. (2007) Synthesis and properties of chiral imidazolium ionic liquids with a (1*R*,2*S*,5*R*) (-)-menthoxymethyl substituent *New J. Chem.*, 31, 879-892.
Fredlake, C. P., Crosthwaite, J. M., Hert, D. G., Aki, S. N. V. K., Brennecke, J. F. (2004) Thermophysical Properties of Imidazolium-Based Ionic Liquids, *J. Chem. Eng. Data.*, 49, 954-964.
Freemantle, M. (1998) Designer solvents - Ionic Liquids may boost clean technology development. *Chem. Eng. News,* 76, 32.
Gathergood, N., Garcia, M.T., Scammells P.J. (2004) Biodegradable ionic liquids: Part I. Concept, preliminary targets and evaluation. *Green. Chem.*, 6, 166–175.
Garcia, M.T., Gathergood, N., Scammells, P.J. (2005) Biodegradable ionic liquids Part II. Effect of the anion and toxicology *Green. Chem.*, 7, 9–14.

Gilbert, P. T., Maira-Litran, T., McBain, A. J., Rickard, A. H., Whyte, F. (2002) The physiology and collective recalcitrance of microbial biofilm communities *Adv. Microbial. Phys.*, 46, 203-256.

Hall-Stoodley, L., Costerton, J. W., Stoodley, P. (2004) Bacterial biofilms: from the natural environment to infectious diseases. *Nat. Rev. Microbiol.*, 2, 95108.

Harrison, J.J., Turner, R.J., Marques, L.R. & Ceri, H. (2005) Biofilms: A new understanding of these microbial communities is driving a revolution that may transform the science of microbiology. *American Scientist*, 93, 508-515.

Huddleston J.G., Visser, A.E., Reichart, W.M., Willauer, H.D., Broker, G.A., Rogers, R.D. (2001) Characterization and comparison of hydrophilic and hydrophobic room temperature ionic liquids incorporating the imidazolium cation. *Green Chem.*, 3, 156-164.

Kosmulski, M., Gustafsson, J., Rosenholm, J.B. (2004) Thermal stability of low temperature ionic liquids revisited *Thermochim. Acta.*, 412, 47-53.

Li, G., Shen, J., Zhu, Y. (1998) Study of pyridinium-type functional polymers. II. Antibacterial activity of soluble pyridinium-type polymers. *J. Appl. Polym. Sci.*, 67, 1761-1768.

Lewis, K. (2001) Riddle of biofilm resistance. *Antimicrob. Agents Chemother.*, 45, 999-1007.

McBain A. (2009). In vitro biofilm models: an overview. *Advances in Applied Microbiology*, 69, 99-132.

Mizuuchi, H., Jaitely, V., Murdan, S., Florence, A.T. (2008) Room temperature ionic liquids and their aqueous mixtures as potential drug solvents. *Eur. Journ. Pharm. Sci.*, 33, 326-331

National Committee for Clinical Laboratory Standards (1997) Reference method for broth dilution antifungal susceptibility testing of yeasts; approved standard. NCCLS document M27-A, Wayne, PA.

National Committee for Clinical Laboratory Standards (2000) Methods for dilution antimicrobial susceptibility tests for bacteria that grow aerobically; approved standard, 5th edn. NCCLS document M7-A5, Wayne, PA.

Pernak, J., Rogoza, J., Mirska, I. (2001) Synthesis and antimicrobial activities of new pyridinium and benzimidazolium chlorides. *Eur. J. Med. Chem.*, 36, 313-320.

Pernak, J. & Chwala, P. (2003) Synthesis and anti-microbial activities of choline-like quaternary ammonium chorides. *Chem. Eur. J.*, 38, 1035-1042.

Pernak, J., Sobaszkiewicz, K., Mirska, I. (2003) Anti-microbial activities of ionic liquids *Green Chem.*, 5, 52–56.

Pernak, J., Goc, I., Mirska, I. (2004a) Anti-microbial activities of protic ionic liquids with lactate anion. *Green Chem.*, 6, 323-329.

Pernak, J., Sobaszkiewicz, K. Foksowicz-Flaczyk, J. (2004b) Ionic Liquids with Symmetrical Dialkoxymethyl-Substituted Imidazolium Cations. *Chem. Eur.*, 10, 3479-3485.

Pernak J. & Feder-Kubis, J. (2005) Synthesis and properties of chiral ammonium-based ionic liquids. *Chem. Eur. J.*, 11, 4441- 4449.

Pernak, J., Syguda, A., Mirska, I., Pernak, A., Nawrot, J., Pradzynska, A., Griffin, S.T., Rogers, R.D. (2007a) Choline-Derivative-Based Ionic Liquids *Chem. Eur. J.*, 2007a, 13, 6817-6827.

Pernak, J., Skrzypczak, A., Lota, G., Frackowiak, E. (2007b) Synthesis and Properties of Trigeminal Tricationic Ionic Liquids. *Chem. Eur. J.*, 13, 3106-3112.

Pham, T. P. T., Cho, C.W., Min, J., Yun, Y. S. (2008) Alkyl-chain length effects of imidazolium and pyridinium ionic liquids on photosynthetic response of *Pseudokirchneriella subcapitata. J. Biosci. Bioeng.*, 105, 425-428.
Prasad, A.K., Kumar, V., Malhotra, S., Ravikumar, V.T., Sanghvi Y.S., Parmar, V.S. (2005) 'Green' methodology for efficient and selective benzoylation of nucleosides using benzoyl cyanide in an ionic liquid *Bioorg. Med. Chem.*, 13, 4467-4472.
Pretti, C., Chiappe, C., Pieraccini, D., Gregori, M., Abramo, F., Monni, G., Intorre, L. (2006) Acute toxicity of ionic liquids to the zebrafish (*Danio rerio*) *Green Chem.*, 8, 238–240.
Ranke, J., Mölter, K., Stock, F., Bottin-Weber, U., Poczobutt, J., Hoffman, J., Ondrussckka, B., Jastorff, B. (2004) Biological effects of imidazolium ionic liquids with varying chain lengths in acute *Vibrio fischeri* and WST-1 cell viability assays. *Ecotoxicology and Environmental Safety*,58, 396–404.
Rebros, M., Gunaratne, H.Q.N., Ferguson, J., Seddon, K.R., Stephens, G. (2009) A high throughput screen to test the biocompatibility of water-miscible ionic liquids, *Green Chem.*, 11, 402-408
Romero, A., Santos, A., Tojo, J., Rodriguez, A. (2008) Toxicity and biodegradability of imidazolium ionic liquids. *J. Hazard. Mater.*, 151, 268–273.
Scammells, P.J., Scott, J.L., Singer, R.D. (2005) Ionic Liquids: The Neglected Issues. *Aust. J. Chem.* 58, 155-169.
Seddon K. R. (1997) Ionic liquids for clean technology. *J. Chem. Technol. Biotechnol.*, 68, 351-356.
Seddon, K.R., Stark, A., Torres, M.J. (2000) Influence of chloride, water, and organic solvents on the physical properties of ionic liquids. *Pure Appl. Chem.*, 72, 2275-2287.
Smith, A. L., Fiel, S. B., Mayer-Hamblett, N., Ramsey, B., Burns, J. L. (2003) Susceptibility Testing of Pseudomonas aeruginosa Isolates and Clinical Response to Parenteral Antibiotic Administration : Lack of Association in Cystic Fibrosis. *CHEST*, 123, 1495-1502.
Stark, A. & Seddon, K. R. (2007) 'Ionic Liquids' in *'Kirk-Othmer Encyclopaedia of Chemical Technology'*, ed. A. Seidel, John Wiley & Sons, Inc., Hoboken, New Jersey, 2007, vol. 26, pp. 836–920.
Stewart, P. S. (2002) Mechanisms of antibiotic resistance in bacterial biofilms. *Int. J. Med. Microbiol.*, 292, 107-113.
Stewart, P. S. & Costerton, J. W. (2001) Antibiotic resistance of bacteria in biofilms. *Lancet.*, 358, 135-138.
Stock, F., Hoffmann, J., Ranke, J., Stormann, R., Ondruachka B., Jastorff, B. (2004) Effects of ionic liquids on the acetylcholinesterase – a structure–activity relationship consideration. *Green Chem.*, 6, 286-290.
Stolte, S., Abdulkarim, S., Arnig, J., Blomeze-Nienstedt, A., Bottin-Weber, U., Matzke, M., Ranke, J., Jastorff, B., Thorning, J. (2008) Primary biodegradation of ionic liquid cations, identification of degradation products of 1-methyl-3-octylimidazolium chloride and electrochemical wastewater treatment of poorly biodegradable compounds. *Green Chem.*, 10, 214-224.
Swatloski, R. P., Holbrey, J. D., Memon, S. B., Caldwell, G. A., Caldwell K. A., Rogers, R. D. (2004) Using *Caenorhabditis elegans* to probe toxicity of 1-alkyl-3-methylimidazolium chloride based ionic liquids, *Chem. Commun.*, 668-669.

Tao, G., He, L., Lui, W., Xu, L., Xiong, W., Wang, T., Kou, Y. (2006) Preparation, characterization and application of amino acid-based green ionic liquids. *Green Chem.*, 8, 639-646.

Villiagran, C., Aldous, L., Lagunas, M.C., Compton, R.G., Hardacre, C. (2006) Electrochemistry of phenol in bis{(trifluoromethyl)sulfonyl}amide ($[NTf_2]^-$) based ionic liquids. *J. Electoanalytical Chem.*, 588, 27-31.

Visser, A. E., Swatloski, R. P., Rogers, R. D. (2000) pH-Dependent Partitioning in Room Temperature Ionic Liquids Provides a Link to Traditional Solvent Extraction Behavior. *Green Chem.*, 2000, 2, 1-4.

Walden, P. (1914) Molecular weights and electrical conductivity of several fused salts. *Bull Acad Sci St. Petersbourg.* 405-422.

Wasserscheid, P., van Hal R., Bösmann, A. (2002) 1-*n*-Butyl-3-methylimidazolium ([bmim]) octylsulfate—an even 'greener' ionic liquid, *Green Chem.*, 2002, 4, 400-404.

Wei, D. & Ivaska, A. (2008) Applications of ionic liquids in electrochemical sensors *Anal. Chim. Acta*, 607, 126-135.

Wells A. & Coombe, V.T. (2006) On the Freshwater Ecotoxicity and Biodegradation Properties of Some Common Ionic Liquids, *Org. Process Res. Dev.*, 2006, 10, 794-798.

Wood, N. & Stephens, G. (2010) Accelerating the discovery of biocompatible ionic liquids. *Phys. Chem. Chem. Phys.*, 12, 1670-1674.

Part 6

Electrochemistry

27

Application of Electrochemical Impedance Spectroscopy (EIS) and X-ray Photoelectron Spectroscopy (XPS) to the Characterization of RTILs for Electrochemical Applications

J. Benavente[1] and E. Rodríguez-Castellón[2]
[1]*Grupo de Caracterización Electrocinética en Membranas e Interfases. Departamento de Física Aplicada I. Facultad de Ciencias. Universidad de Málaga*
[2]*Departamento de Química Inorgánica. Facultad de Ciencias. Universidad de Málaga. Spain*

1. Introduction

Ionic liquids (ILs) are low temperature molten salts, that is, a salt in the liquid state. ILs used to present a very low vapour pressure and this property makes of the ILs key materials for the development of a wide variety of emerging technologies. The stability of ILs at high temperatures (several hundred degrees), low combustibility, and even the relatively high viscosity of some of them compared to conventional solvents, are characteristic of interest for some applications. Due to the large diversity of ILs components, they may present wide structural variations which can be used to design the IL with more adequate properties for a particular application. These applications might include new types of lubricants and fluids for thermal engines, electrodeposition, energy and CO_2 capture devices, biomimetics, double layer capacitors,.....in fact, the scientific and technological importance of the ILs spans nowadays to a wide range of applications [1-5].

Among the energy devices, polymer-electrolyte membranes for fuel cell application are under development as a way to reduce global warming and energy cost and ILs incorporation in the structure of Nafion, a typical membrane for fuel cell use, is under study [6-8]. Since transport properties of porous and dense membranes can be modified with the addition of substances which could favour/reject the pass of some of the particles or ions in a mixture, the incorporation of a particular IL in the structure of a membrane may increase its selectivity and/or specificity.

Chemical characterization and determination of electrical parameters for different ILs as well as the changes associated to water incorporation, a subject of interest for different electrochemical applications, is considered in this work. Moreover, due to the importance that membrane separation technology has nowadays, modification of membranes with different structures by incorporation of RTILs or IL-cations and their effect on mass and charge transport is also presented.

Ionic liquids, membranes and membranes/IL-modified samples were chemically characterized by X-ray photoelectron spectroscopy (XPS). This technique allows the determination of the surface chemical composition of a given sample and other properties related to the structure and chemical environment in which the atom lies within the solid and, in the case of membranes, it is commonly used to study chemical changes in polymer matrix [9-11]. Impedance spectroscopy (IS) measurements were performed for electrical characterization of both ILs and IL-modified membranes by analyzing the impedance plots and using equivalent circuits as models [12-16]. Time evolution of the IS plots was used as a way for monitoring both water diffusion in the ILs and IL inclusion in the membranes (or its loose from them), but also to show interfacial effects depending on the external conditions of the studied systems. Moreover, a comparison of the electrochemical parameters (ion transport numbers, water diffusion coefficient, electrical resistance and capacitance) obtained for fresh and aged samples of a IL-supported membrane or for original and IL-modified cellulosic membranes is also presented as examples of the interest of the studied systems and potentiality of the techniques used.

2. Experimental

2.1 Ionic liquids

The following room temperature ionic liquids were studied: 1-*n*-butyl-3-methylimidazolium hexafluoro-phosphate or $[C_4MIM^+][PF_6^-]$; 1-*n*-octyl-3-methylimidazolium hexafluoro-phosphate or $[C_8MIM^+][PF_6^-]$, 1-*n*-decyl-3-methylimidazolium tetrafluoroborate or $[C_{10}MIM^+][BF_4^-]$, 1-*n*-butyl-3-methylimidazolium tetrafluoroborate or $[BMIM^+][BF_4^-]$ and n-dodecyltriethylammonium chloride or $[DTA^+][Cl^-]$. Both commercial (Solchema, Portugal) and prepared ILs following reported procedures [17-19] were used; in the case of the $[DTA^+][Cl^-]$ (solid at room temperature) a 40% (w/w) solution with deionised water was prepared.

2.2 Membranes

The electrochemical characterization of the RTILs-modified membranes was performed with different kinds of flat membranes with porous and dense structure: i) a porous polyvinylidene fluoride (PVDF) membrane (FP-Vericel, Pall, USA) with nominal pore size of 0.2 μm and a thickness of 125 μm; ii) a dense perfluorinated proton-exchange Nafion-112 membrane in protonated form from Dupont, USA; iii) a dense but highly hydrophilic cellulosic membrane with 0.06 kg/m^2 of regenerated cellulose (sample RC-6) from Cellophane Española, S.A. (Burgos, Spain).

The supported liquid membrane (SLM) was obtained placing the PVDF porous support in a desiccator under vacuum for 1 hour, then (still under vacuum) the ionic liquid was released onto the membrane surface from a syringe (100 μl of ionic liquid per cm^2 of membrane area). The SLM is named by the RTIL used for its preparation and a detailed explanation of SLM preparation is presented in ref [11].

ILs incorporation into the structures of two dense membranes was performed by contacting the membranes with aqueous solutions of the ILs for different periods of time. A Nafion-112 membrane (Dupont, USA) was placed in the test-cell B (see Fig. 1) and both half-cell were filled with a 40% (w/w) aqueous solution of $[DTA^+][Cl^-]$ and cation incorporation into the Nafion structure was obtained by proton-exchange (H^+/DTA^+) according to previous studies [20], while a piece of the hydrophilic cellulosic membrane was immersed for 1 week in a 50 % water solution of $[BMIM^+BF_4^-]$ and dried at room temperature conditions for another week; these ILs-modified membranes will be hereafter named as Nafion/DTA and CR-6/$BMIMPB_4$, respectively.

2.2 X-ray photoelectron spectroscopy (XPS) measurements

$[C_8MIM^+][PF_6^-]$, $[C_8MIM^+][PF_6^-]$ and $[C_{10}MIM^+][BF_4^-]$ RTILs and the surfaces of the original and IL-modified membranes were chemical characterized by XPS analysis. The high-resolution spectra were recorded with a Physical Electronics PHI 5700 spectrometer by using a concentric hemispherical analyzer operating in the constant pass energy mode at 29.35 eV, with 720 μm diameter analysis area, and MgK_α X-ray as an excitation source ($h\nu$ = 1253.6 eV). Accurate ±0.1 eV binding energies were determined with respect to the position of the adventitious C 1s peak at 284.8 eV, and the residual pressure in the analysis chamber was maintained below 5×10^{-7} Pa during data acquisition. A PHI ACCESS ESCA-V6.0F software package was used for acquisition and data analysis [21]. Atomic concentration (A.C.) percentages of the characteristic sample elements were determined after subtraction of a Shirley-type background taking into account the corresponding area sensitivity factor [22] for the different measured spectral regions. In the study performed with original and ILs- modified membranes, an irradiation time less than 20 minutes was used to minimize possible X-ray induced damage in the polymer structure [23].

2.3 Impedance spectroscopy measurements

Two slightly different test cells were used for impedance spectroscopy (IS) measurements carried out with different ILs and membrane/electrolytes (ILs or NaCl aqueous solutions) systems. Fig. 1 shows a scheme of the two open electrochemical cells used for electrochemical characterization, which consist of two glass semi-chambers with one

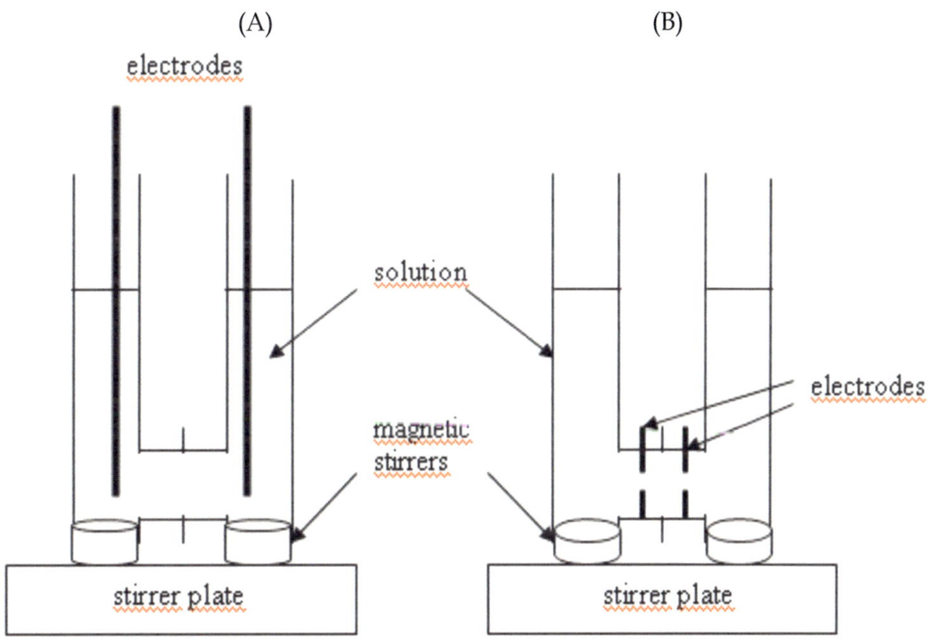

Fig. 1. Test cells for impedance spectroscopy and membrane potential measurements. In the case of RTILs and NaCl solutions both semi-cameras (in cells A or B) are filled with the electrolytes, while membranes are placed in the middle of the semi-cameras for membrane characterization.

electrode in each one, which were located in position (A) or position (B). The main difference between both cells is the direct contact between the electrode and the liquid open surface for test-cell A, which does not exist in the case of test-cell B. IS measurements were carried out with both semi-cell were filled with the studied RTILs.
Membrane electrical characterization was performed in test-cell A, with the membranes placed between both chambers supported by rubber rings and ILs or NaCl solutions filling both half-cells (electrode/electrolyte/membrane/electrolyte/electrode system) [16]. In the case of dry samples, the test-cell consists of a Teflon support on which two Pt electrodes were placed and screwed down (system electrode/membrane/electrode). In all cases, the electrodes were connected to a Frequency Response Analyzer (FRA, Solartron 1260) and measurements were recorded for 100 data points with a frequency ranging between 1 Hz and 10^7 Hz, at a maximum voltage of 0.01 V. Impedance data were corrected by parasite capacitances.

2.4 Electrochemical characterization of membranes.

Membrane potential (MP) measurements were performed in test-cell A using aqueous NaCl solutions and reversible Ag/AgCl electrodes connected to a high impedance voltmeter (Yokohama 7552, 1GΩ input resistance). The concentration of the NaCl solution was kept constant at one side of the membranes (c_c = 0.01 M), while the concentration of the NaCl solution at the other membrane side (c_v) was gradually changed from 0.002 M to 0.1 M [24].
Water diffusion coefficient through RC-6 and RC-6/BMIM membranes was determined by diffusion experiments with tritiated water (TOH) carried out in-cell A. One of the half-cells (donor chamber) contained 30 μl tritium/15 mL of distilled water while the other half-cell (receiving chamber) was filled with distilled water at t = 0; time variation of the tritium activity in the donor and receiving chambers was determined at different time instances by taken samples of 50 μl which were analyzed in a Beckman LS6500 scintillation counter [25].

3. Results and discussion

3.1 Electrical and chemical characterizations of ionic liquids.

Electrical characterization of pure and water-containing ILs was carried out by analyzing the impedance spectroscopy (IS) plots obtained for medium range frequencies (between 1 Hz and 10 MHz). IS is a non-destructive a.c. technique for electrical characterization of solid and liquid systems, but it can also be used for the determination of interfacial effects [16].
When a linear system is perturbed by a small alternating voltage $v(t) = V_o \sin\omega t$, its response, the electric current is also a sine wave, $i(t) = I_o \sin(\omega t+\phi)$, where V_o and I_o represent the maximum voltage and intensity, respectively, while $\omega=2\pi f$ is the angular frequency and ϕ the phase angle. A transfer function, the admittance function, can be defined as: $Y^*(\omega) = |Y(\omega)| e^{j\phi}$, where $|Y(\omega)|$ represents the amplitude, and the impedance function, $Z(\omega)$, is the inverse of the admittance: $Z(\omega)=[Y^*(\omega)]^{-1}$; the impedance is expressed as a complex number: $Z = Z_{real} + j\, Z_{img}$, where Z_{real} and Z_{img} are the real and imaginary parts of the impedance, respectively. The admittance for a parallel resistance-capacitor (RC) circuit is given by the sum of conductance (1/R) and capacitance (C) contributions, where the resistance (R) represents the dissipative component of the dielectric response while the capacitance describes the storage component. The impedance function for that circuit is: $1/Z^* = (1/R) + (j\omega C)$, and it can be separated into real and imaginary parts by algebra rules, which are related with the electrical parameters of the system by the following expressions:

$$Z_{real} = (R/[1 + (\omega RC)^2]) \ ; \qquad Z_{img} = -(\omega R^2C/[1 + (\omega RC)^2]) \qquad (1)$$

These expressions correlate impedance components, which are determined from experimental values using impedance plots, with the electrical parameters of the system.
The analysis of the impedance data can be carried out by the complex plane method by using the Nyquist plot ($-Z_{img}$ vs Z_{real}). The equation for a parallel (RC) circuit gives rise to a semi-circle in the $Z^*(\omega)$ plane as those shown in Fig. 2.a (obtained for C_8MIMPF_6 and water-saturated/C_8MIMPF_6 systems); this semicircle has intercepts on the Z_{real} axis at R_∞ ($\omega \rightarrow \infty$) and R_o ($\omega \rightarrow 0$), being (R_o-R_∞) the resistance of the system; the maximum of the semi-circle equals $0.5(R_o-R_\infty)$ and occurs at such a frequency that $\omega RC=1$, being $\tau=RC$ the relaxation time [26]. A comparison of the semicircles presented in Fig. 2.a shows the strong effect of water on the electrical parameters associated the C_8MIMPF_6 ionic liquid, being the electrical resistance of the water-saturated/C_8MIMPF_6 approximately 30 % of that for the dry IL. Moreover, the data drawn in Fig. 2.b (impedance real part as a function of the frequency) also show differences in the liquid-electrode interface (f < 1000 Hz) between both systems.

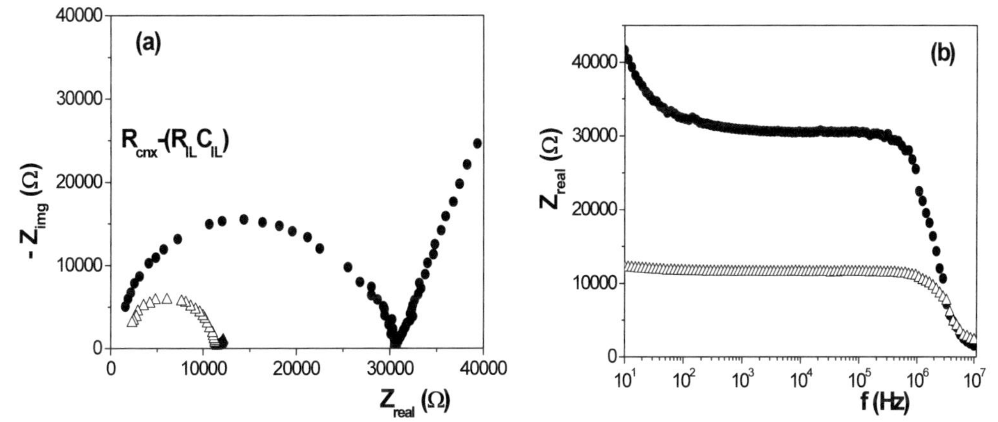

Fig. 2. Impedance plots for the ionic liquid C_8MIMPF_6 (•) and a water saturated mixture of C_8MIMPF_6 (Δ). (a) Nyquist and (b) Bode plots.

The reduction of the IL electrical resistance (or the conductivity increase) with water addition might be a point of interest for measurements carried out with ILs in electrochemical applications due to room humidity (open cells) or contact with aqueous solutions. According to Rivera-Rubero and Baldelli water is often present as a contaminant (up to ~ 0.2 mol fraction) in hydrophilic and hydrophobic ILs, highly affecting the bulk physical properties as well as the surface of hydrophobic ones [27]. In this context, it is interesting to remark the interfacial effects found when IS measurements at different time instances were performed with C_8MIMPF_6, in the open test-cell A and laboratory humidity of 50 %, shown in Fig. 3.a, where a new relaxation can be observed in the interfacial region ($f_{max} \approx 300$ Hz), but this effect hardly appears in measurements performed with test-cell B; in both cases bulk liquid contribution is practically unaffected. Results in Fig. 3.a could indicate that the electrodes also act as a pathway for water transport (no diffusion through the IL), which may cover the electrode measuring surface and, consequently, to modify the electrode/IL interface.

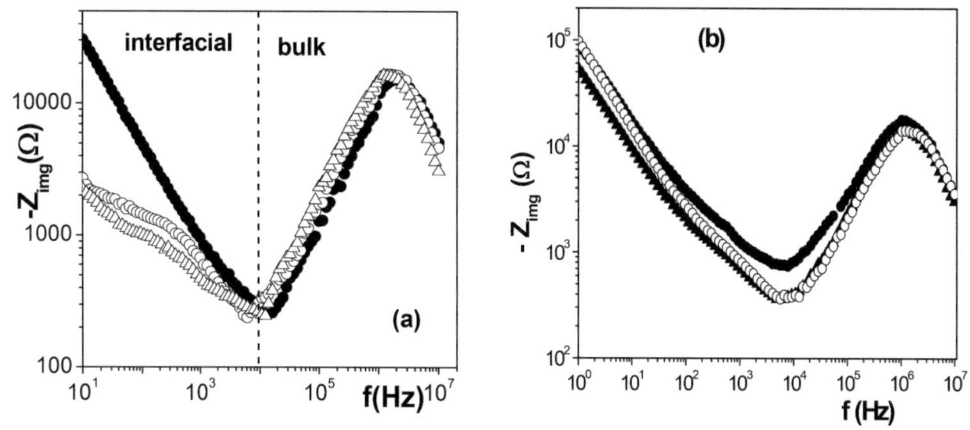

Fig. 3. Effect of water adsorption in the impedance plots measured for the RTIL C_8MIMPF_6 at different times: t = 0 h (•), t = 6 h (o) and t = 8 h (Δ). (a) in test-cell A; (b) in test-cell B.

To avoid that effect and clarify the electrical modifications caused by water diffusion into the IL, a more detailed study was performed by covering both free surfaces of the IL C_8MIMPF_6 with distilled water (measures carried out in test-cell B), and Fig. 4 shows time evolution of the impedance plots due to both water diffusion into the IL and water mixture/content. As can be observed, water diffusion (time evolution) seems to affect more to the real part of the impedance (Fig. 4.a) than the imaginary part (Fig. 4.b); however, water content modifies both real and imaginary impedance parts, shifting to higher frequencies the maximum frequency.

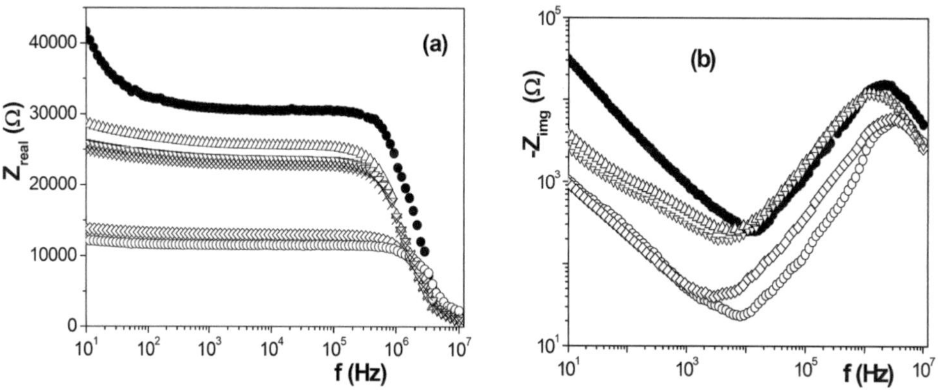

Fig. 4. Modification of impedance plots with water content in the ionic liquid C_8MIMPF_6. (•) no-water content, (Δ) 25 % of water on the IL surfaces (no-mixing) at t = 0 h and (∇) at t = 24 h; (x) 35 % of water on the IL surfaces (no-mixing) at t = 0, (◊) 35 % water mixture, (o) water saturated IL.

The fit of the impedance data using a non-linear program allows the determination of the electrical resistance of the ionic liquid C_8MIMPF_6 at different water content and the values are shown in Fig. 5.a; for comparison, results obtained with C_4MIMPF_6, another RTIL, are also indicated in Fig. 5.a [28]. These results show a direct relationship between the IL electrical resistance and the size of the alkali chain and agree with lower conductivity of C_4MIMPF_6 when compared with C_8MIMPF_6 already reported in the literature [29]. Among other different physicochemical parameters of liquid systems, the viscosity and its dependence with water content is of higher interest. Measurements of $[C_8MIM][PF_6]$ at different water contents also showed a decrease of viscosity values with increasing water content [30], which is due to the reduction of the electrostatic attraction between the ions associated to the presence of the water molecules, lowering the energy of the system and, as a result of this, its viscosity. Combining the Stokes-Einstein equation for the diffusion coefficient of a particle with the Nernst-Einstein law for the total conductivity is possible to obtain a linear relationship between the electrical resistance and the apparent viscosity (η) for a particular IL [31], which allows the determination of η value for certain water content. Fig. 5.b shows the variation of apparent viscosity with water percentage for the ILs C_8MIMPF_6 and C_4MIMPF_6, where the lower viscosity of this latter can be observed [28].

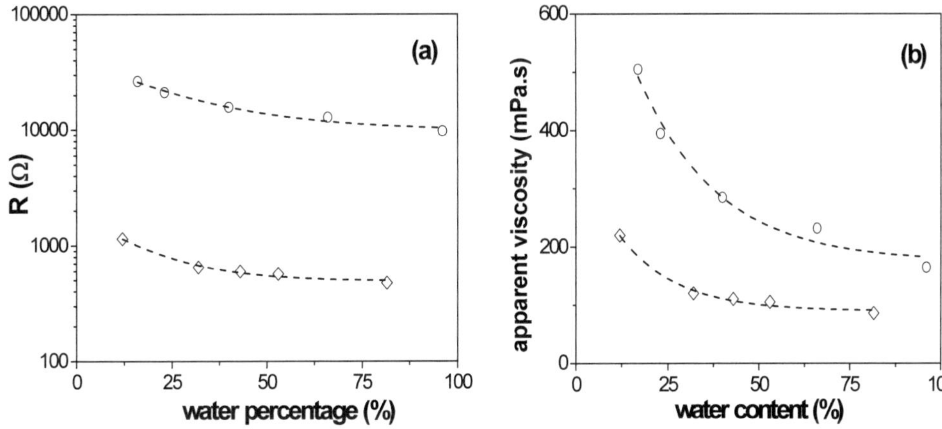

Fig. 5. Variation of C_8MIMPF_6 ionic liquid electrical resistance (a) and apparent viscosity (b) with water percentage (from ref. [28]).

Chemical characterization of $[C_4MIM^+][PF_6^-]$, $[C_8MIM^+][PF_6^-]$ and $[C_{10}MIM^+][BF_4^-]$ room temperature ionic liquids was carried out by X-ray photoelectron spectroscopy (XPS) analysis. XPS technique consists in the irradiation of a sample with X-rays under vacuum and the measure of the kinetic energy ($E_{kinetic}$) of the photoelectrons ejected from the sample's surface. The emitted electrons binding energy (B.E.) can be calculated as: $E_{binding} = E_{photon} - E_{kinetic}$, where E_{photon} is the energy of the X-ray incident radiation. Since the electrons of each chemical element have a characteristic B.E., it is possible to identify which elements are present in the surface sample (a thin layer of 30-50 Å) and their relative atomic concentration percentages, A. C. (%) (except hydrogen and helium). Additionally, it may also be possible to know the chemical state of the elements based on small shifts in the binding energies [10].

Table 1 shows the A.C. percentages of the characteristic RTILs elements found on the surface of the studied samples, but small percentages of other elements (silicon, oxygen,...) were also detected and attributed to impurities/environmental contamination [10,32]; particularly, a small percentage of oxygen (no included in Table 1) was also found in the three ILs which is associated to the presence of water due to their hygroscopic character.

Since the values shown in Table 1 correspond to relative percentages and, therefore, can not be easily compared between different samples, the ratio between the different elements detected is also presented, as well as the expected theoretical ratios (in brackets).

Ionic Liquid	%C	%F	%N	%P	C/F	N/P	F/P	F/N	C/P	C/N
[C_4MIM][PF_6]	48.2	30.0	9.8	5.4	1.6(1.3)	1.8 (2)	5.5 (6)	3.1 (3)	9.0 (8)	5.0 (4)
[C_8MIM][PF_6]	63.1	21.6	9.3	3.7	2.9 (2)	2.5 (2)	5.8 (6)	2.3 (3)	16.9 (12)	6.8 (6)
[C_{10}MIM][BF_4]	60.3	12.8	6.1	3.8	4.7(3.5)	1.6 (2)	3.3 (4)	2.1 (2)	15.7 (14)	9.9 (7)

Table 1. Surface composition by XPS analysis of the ionic liquids [C_4MIM][PF_6], [C_8MIM][PF_6] and [C_{10}MIM][BF_4] (theoretical ratios in brackets).

Because the presence of external contaminants does not affect the ratios between the elements of interest, they will be used for comparison between the different ILs as well as with the theoretical ones. The N/P, F/P and F/N ratios found for the [C_nMIM][PF_6] (n = 4 or 8) ionic liquids show good agreement with the theoretical ones, as well as the F/B and F/N ratios in the case of [C_{10}MIM][BF_4], while the slightly higher differences obtained for C/F, C/P and C/N may be associated to carbon contamination.
A comparison of the carbon core level (C 1s) spectra obtained for the three RTILs is shown in Fig. 6, where two different peaks, at binding energies of 286.5 eV (assigned to carbons of the imidazolonium head) and 285.0 eV (assigned to the alkyl chain and adventitious carbon) can be observed; differences in the relative contributions of the two peaks seem to be dependent on the length of the RTILs' alkyls chain [22].

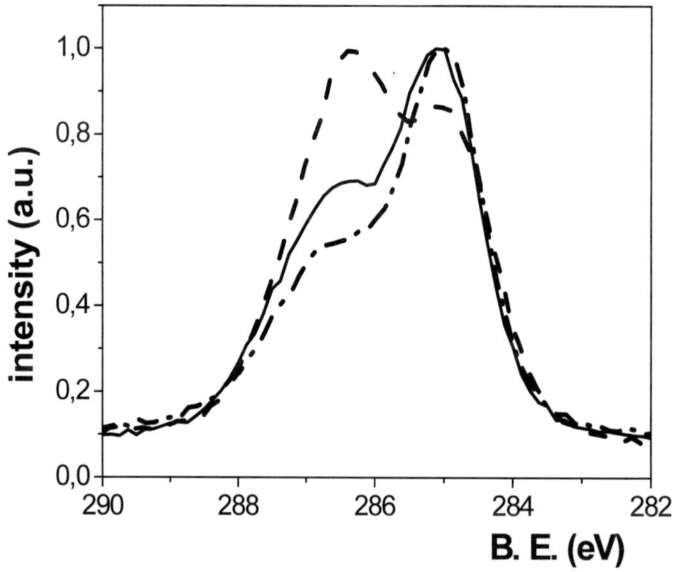

Fig. 6. C is core level spectra for the ionic liquids [C_4MIM][PF_6] (solid line) [C_8MIM][PF_6] (dashed line) and [C_{10}MIM][BF_4] (dashed-dot line).

3.2 Electrical and chemical characteristics of membranes containing ionic liquids

Nowadays, membranes are used in different electrochemical applications and as was already indicated, the incorporation of room temperature ionic liquids into the membrane matrix may change its transport properties (mass/charge transport and selectivity) depending on the membrane structure and characteristics [1,33]. For that reason, changes in ion transport associated to the modification by ILs of porous and dense membranes are presented in this section.

3.2.1 Supported liquid membranes

Supported liquids membranes (SLMs) basically consist of an organic solvent immobilised in the pores of a support membrane [34]. SLMs has important transport advantages due to the high liquids diffusion rates and selectivity, but stability problems caused by the possible loss of the organic liquid from the porous support (or the formation of emulsions in the pores) have reduce their application in industrial separation processes [35-36]. Among the approaches suggested for improving SLMs stability, the use of RTILs as organic phase was also considered [28]. Among the different RTILs, those based in imidazolonium cation (1-n-alkyl-3-methylimidazolonium) seems to be particularly adequate due to their relatively high viscosity and reduced solubility for various solvents (including water), which are fundamental requirements for the durability of the organic phase of SLMs.

Chemical and electrical characterizations of a fresh SLM (sample *f*) and a 4 year old membrane stored without any special preservation procedure (sample *a*) were performed to see age effect on the structure and transport behaviour of the SLM (a porous PVDF matrix with pores filled with C_8MIMPF_6) [37].

A.C. (%) of the characteristic elements found on the surfaces of the porous PVDF support, and fresh (f) and aged (a) samples of the SLM containing $[C_8MIM^+][PF_6^-]$ into its pores are indicated in Table 2 (oxygen percentage and other impurities are not included, then total percentages differ from 100%). The coverage of the PVDF support by the IL is detected by a reduction in the percentage of fluorine and, more significant, the presence of phosphorous, an IL characteristic element, on the surface of the SLM fresh sample; however, the results obtained for the SLM aged sample show a slight increase of carbon (7 %) attributed to surface contamination, a reduction of 50 % in the A.C. of fluorine but an increase in the percentage of nitrogen and phosphorous, which might be taken as an indication of the re-organization of the IL rather than its lost.

Fig. 7.a shows the C 1s spectra for the PVDF-support, $[C_8MIM][PF_6]$ (f) and $[C_8MIM][PF_6]$ (a) samples. The two clear peaks showed by the PVDF support are ascribed to the $-CF_2-$ (at 291.2 eV) and the C* carbon in CF_2-C^*H_2- bonds (at 286.7 eV due to the neighbourhood of the $-CF_2$ group) [38]; moreover, a shoulder at 285.0 eV ascribed to adventitious carbon can also be observed, which support the slightly higher experimental C/F ratio obtained and indicated in Table 2. However, when fresh and aged SLM samples are compared, only slight differences associated to C=O bond (287.5 eV) attributed to membrane oxidation can be observed.

To ensure the presence of the IL not only on the membrane surface but also filling the pores of the aged sample, attenuated total reflectance Fourier transform infrared (ATR/FTIR) technique was used due that depth penetration of the infrared waves in this type of membranes is around 1.5 μm [39]. Fig. 7.b shows a comparison of the ATR-FTIR spectra in the range 3300-2800 cm^{-1} for the $[C_8MIM][PF_6]$ aged membrane and the PVDF support where the presence of the RTIL is clearly demonstrated by the appearance of the bands assigned at imidazolium derivates when a comparison with the polymeric support is carried out; particularly, the absorption bands at 2928 and 2856 cm^{-1} correspond to asymmetric and symmetric stretching vibrations of methylene groups, and other bands assigned to asymmetric and symmetric stretching vibrations of methyl groups can also be appreciated at 2956 and 2873 cm^{-1} [37].

Membrane	C (%)	F (%)	N (%)	P (%)	C/F
PVDF support	54.1	41.9	0.9	---	1.29
$[C_8MIM^+][PF_6^-]$ (f)	57.4	24.1	3.7	2.5	2.38
$[C_8MIM^+][PF_6^-]$ (a)	61.5	10.5	6.9	4.0	5.86

Table 2. Atomic concentration percentage of the characteristic elements present on the surface of the PVDF support, $[C_8MIM^+][PF_6^-]$ fresh (f) and aged (a) RTIL supported membranes

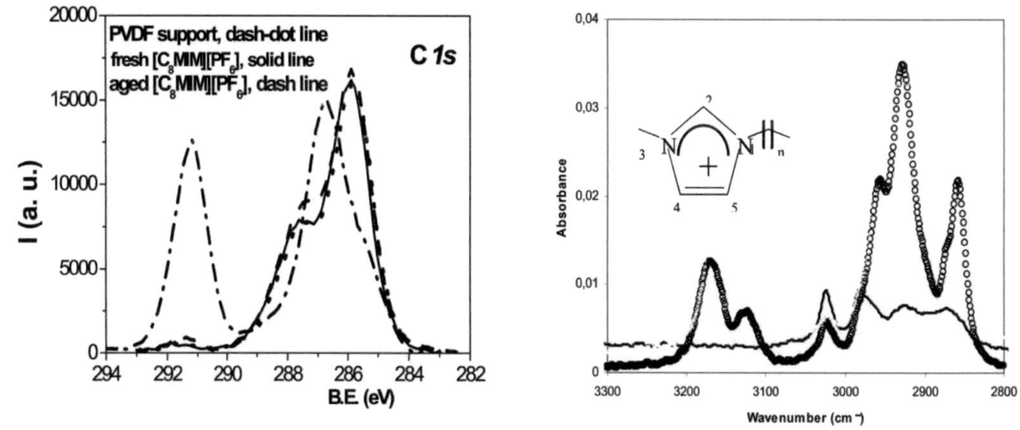

Fig. 7. (a) C1s core level spectra for $[C_8MIM][PF_6]$ fresh (solid line) and aged (dash line) membranes, and the PVDF porous support (dash-dot line). (b) ATR-FTIR spectra for aged $[C_8MIM][PF_6]$ SLM (o) and PVDF support (solid line).

Age effect on the transport across the SLM was determined by considering modification in the ion transport number (t_i), an electrochemical characteristic parameter which represents the ratio between the electric current transported by the ion *i* with respect to the total current crossing the membrane ($t_i = I_i/I_T$), then for single salts: $t_+ + t_- = 1$. Fig. 8.a shows a comparison of the electromotive force, ΔE_{med}, measured at both sides of fresh and aged SLM samples versus concentration ratio; rather similar ΔE_{med} values were obtained for both samples at low concentrations (c_v <0.02 M), but differences at high concentrations can be observed. Thermodynamic arguments lead to the following relationship between the electromotive force measured by the electrodes placed in the two half-cells, ΔE_{med}, the cation transport number, t_+, and the NaCl concentration of the solution filling each half-cell (c_1 and c_2) [40]:

$$\Delta E_{med} = -(2RT/F) \int_{c1}^{c2} t_+ \, dc \qquad (2)$$

where R and F are gas and Faraday constants, T is the thermodynamic temperature of the system. For an ideal cation-exchange membrane ($t_+ = 1$), the ΔE_{med} reaches the maximum value [40]: $\Delta E_{max} = -(2RT/F)\ln(c_2/c_1)$, and the cation transport number in the membrane for a given pair of solutions can be obtained as:

$$t_+ = \Delta E_{med}/\Delta E_{max} \qquad (3)$$

Cation transport number across fresh and aged membranes were determined by using Eq. (3) and Fig. 8.b shows a comparison of t_+ values as a function of the NaCl average concentration ($\langle c_{NaCl}\rangle = (c_1 + c_2)/2$) at the highest concentrations (c_2 > 0.02 M, most significant differences). As can be observed, a practically constant cation transport value ($\langle t_+ \rangle$ = (0.416 ± 0.014)) for the whole interval of concentration was obtained with the fresh sample, but t_+ values decrease with the increase of concentration for the aged sample. This result could be due to modification of the PVDF-RTILs interactions as a result of age, with could facilitate water (or NaCl aqueous solution) transport through the RTIL [37], although the partial lost of the RTIL from the pores of the aged membrane could also be considered. It should be remark that modification of Na^+ transport for the aged sample is only significant at high concentrations and, consequently, when a relatively high osmotic pressure is acting on the RTILs placed into the pores (between 1 and 4.5 bar), but the system is stable at lower pressures.

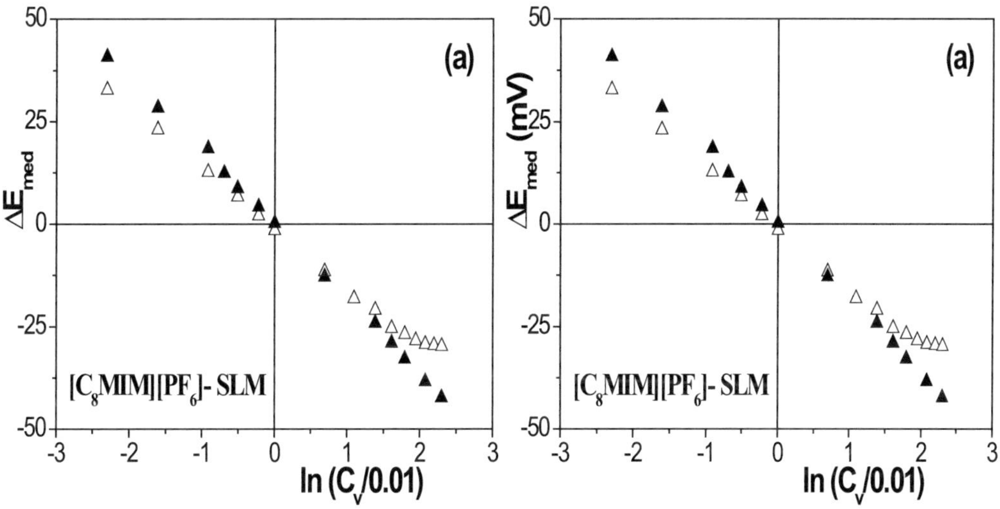

Fig. 8. (a) Variation of the measured potential at both membrane sides with NaCl concentration ratio; (b) Variation of cation transport number with NaCl average concentration. (▲) fresh $[C_8MIM][PF_6]$ membrane, (Δ) aged $[C_8MIM][PF_6]$ membrane.

3.2.2 Modification of dense membranes by inclusion of a IL or a IL-cation.

Two dense commercial polymeric membranes, a cation-exchange polytetrafluoroethylene backbone with sulfonic groups (Nafion) and a hydrophilic regenerated cellulose (RC), were modified with by inclusion of a IL-cation or a IL, respectively, into the polymer structure. Nafion membranes are widely used in electrochemical applications, mainly for fuel cell studies (PEMFCs) due to their good mechanical, chemical and thermal stability up to temperatures of 80 °C, while cellulose and is derivatives are common membrane materials for different separation processes (hemodialysis, reverse osmosis, microfiltration,…) [6,41].

Although IL-modification of both dense membranes was performed by immersion of the samples in the IL or in an IL-water mixture, modification mechanism differs depending on the type of membrane. In the case of Nafion membranes the mechanism consists in the H^+/IL-cation exchange [20], while the IL-water mixture is embeded into the structure of the highly hydrophilic RC membrane.

Modification of the Nafion-protonated membrane was performed by immersion in a solution of the IL for a specific period of time and both conductivity and pH of the IL solution were measured at different time intervals [42]. The system presented in this paper correspond to n-dodecyltriethylammonium (DTA^+) which is solid at room temperature, and a 40% (w/w) IL aqueous solution with deionised water was prepared. These results showed an increase of DTA^+ incorporation along the time associated to the H^+/DTA^+ exchange, with a maximum value around 88 % at 20 h and a stable degree of incorporation percentage of 66 % between 24 h and 48 h [42].

Time evolution of IS measurements for the system *electrode/ DTA^+-water solution/ Nafion membrane/DAT^+-water solution /electrode* were carried out to monitory electrical changes in the Nafion membrane related with DTA^+ incorporation and Fig. 9.a shows the increase of the real part of the impedance (directly related to the electrical resistance) with time as a result of DTA^+ incorporation and, consequently, proton-content reduction; moreover, Z_{real} also presents higher values around 20 h (higher DTA^+ content) and a slight reduction and stabilization at higher contact times was also found in agreement with solution modification measurements previously indicated. XPS analysis of Nafion-protonated sample for different contact time with the DTA^+-water solution was also carried out and time evolution of the A.C. (%) for two characteristic elements, carbon and fluorine, is shown in Fig. 9.b. A decrease of fluorine A.C. (%) and an increase of carbon A.C. (%) with the increase of Nafion/IL-cation contact time can be observed. This is also a confirmation of the increase of DTA^+ incorporation into the Nafion structure.

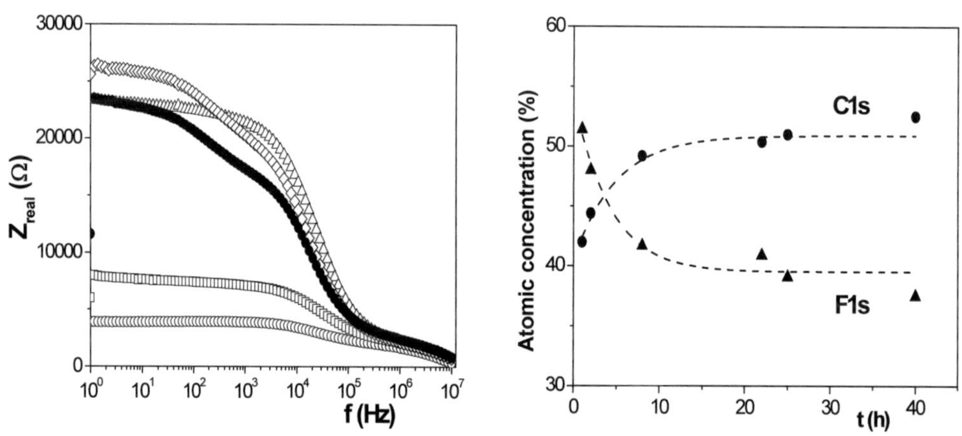

Fig. 9. (a) Time evolution of the Bode plot (Zreal vs frequency) for the Nafion membrane associated to the DTA^+ IL-cation incorporation by cation-exchange mechanism. (o) t = 0, (□) t = 1 h, (Δ) t = 6 h, (∇) t = 19 h, (◊) t = 26 h, (•) t = 40 h. (b) Time evolution of the carbon and fluorine atomic concentration percentages in the Nafion/DTA^+ system determined by XPS analysis.

Interfacial effect for the Nafion/DTA⁺-water solution system after a certain time of contact can also be observed if the - Z_{img} vs frequency plot is considered as is shown in Fig. 10.a. The impedance plot shows a new relaxation process in the interfacial region after 10 h of contact between the Nafion membrane and the DTA⁺-water solution [43], which might be related with the hydrophobic character of the polytetrafluoroethylene backbone of Nafion. To check this point, impedance measurements with the Nafion membrane immerse in the room temperature ionic liquid $BMIMBF_4$ (no water solution was necessary) at different times were also performed. No interfacial relaxation process was detected in this system as can be observed in Fig. 10.b. However, physicochemical differences between both cations could also be considered to explain the different electrical behaviour of both membrane systems.

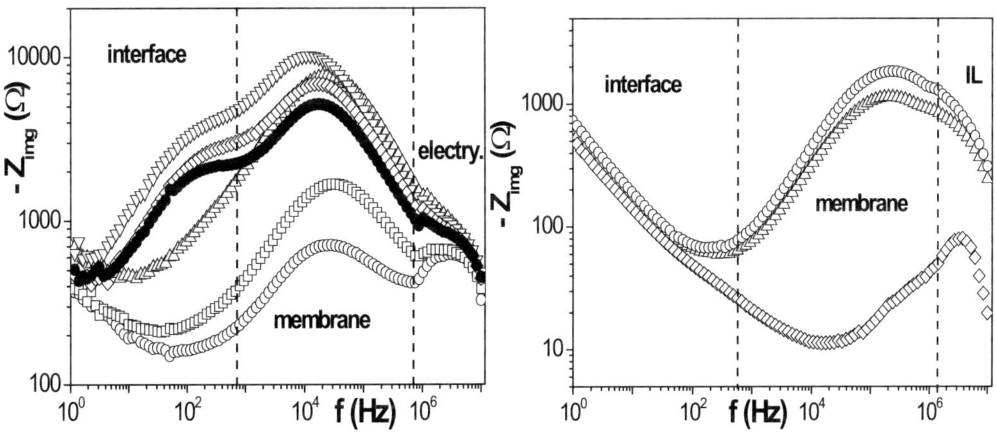

Fig. 10. (a) Time evolution of the Bode plot (- Zimg vs frequency) for the Nafion membrane associated to IL-cation incorporation. (a) Nafion/DTA⁺ system: (o) t = 0, (□) t = 1 h, (Δ) t = 6 h, (∇) t = 19 h, (◊) t = 26 h, (•) t = 40 h. (b); (b) Nafion/BMIM⁺ system: (□) t = 1 h, (Δ) t = 5 h, (◊) t = 20 h.

The small changes caused by the incorporation of BF_4^+ in the Nafion membrane and the reduction in water lost at temperatures higher than 100° C reported by Neves et al [42] makes of this RTIL a good candidate for inclusion in membranes with applications in low temperature fuel cells. In fact, these results have shown the possibility of electrical and chemical modifications of a typical commercial membrane for electrochemical applications with incorporation of ILs by ion-exchange mechanism, which strongly depend on the IL selected, opening a wide range of possibilities for particular membrane applications.

The inclusion of $BMIMBF_4$ in the structure of the regenerated cellulose RC-6 membrane and its effect on electrical and transport parameters was determined by comparing changes in electrical resistance and cation transport number determined with original and IL-modified membranes in contact with NaCl solutions at different concentrations (electrode/NaCl solution/membrane/NaCl solution/electrode system), moreover XPS analysis and IS measurements with dry samples were also performed for a more complete characterization. Fig. 11.a shows impedance plots (Z_{real} vs frequency and –Z_{img} vs frequency plots) for RC-6 y

RC-6/BMIM membranes in contact with a 0.001 M NaCl solution, where clear differences between both samples can be observed. Since a unique relaxation process for the whole membrane system (membrane plus electrolyte placed between the membranes and the electrodes) was obtained, separate membrane characterization from IS measurement is not possible, but the results show higher values for the RC-6/BMIM membrane electrical resistance (Z_{real} vs f plot in the left axis of Fig. 11.a) and a shift to lower frequency in the - Z_{img} vs f plot (right axis in Fig. 11.a) associated to a denser structure [16], which seems to be caused by a reduction in the swelling degree of the cellulosic membrane as a result of the IL inclusion. Fig. 11.b shows the variation with the NaCl concentration of the membrane system electrical resistance (R_{ms}) determined from the analysis of the impedance data; the strong reduction for R_{ms} values with solution increase obtained is associated to the electrolyte-concentration dependence, but differences between RC-6 and RC-6/BMIMBF$_4$ membranes are only significant at low concentrations, since at high concentrations the charges in solution can screen the membrane electrical properties; however, higher capacitance values (around 18 %) were obtained for the RC-6/BMIMBF$_4$ membrane and whole interval of NaCl concentration.

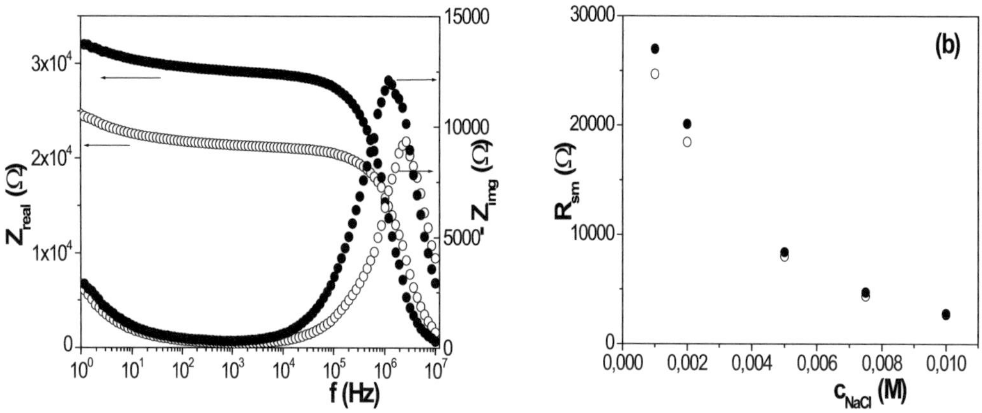

Fig. 11. (a) Impedance plots (Z_{real} vs f, left axis; - Z_{img} vs f, right axis). (b) Variation of the membrane system electrical resistance with NaCl concentration. (o) membrane RC-6/w, (•) membrane RC-6/BMIMBF$_4$.

However, dry samples show opposite behaviour as can be observed in Fig. 12 where Z_{real} vs frequency plots are compared. Since these measurements correspond to the system electrode/membrane/electrode, R_m values for the (dry) membrane matrix are obtained from these measurements (no electrolyte exists), which allows the determination of the membrane conductivity, a material characteristic parameter: $\sigma_m = \Delta x_m/S_m.R_m$, where S_m and Δx_m represent the membrane area and thickness, respectively. The following values were obtained:

RC-6 membrane: $\sigma_m = 1.0x10^{-5}$ $(\Omega.m)^{-1}$.

RC-6/BMIM$^+$ membrane: $\sigma_m = 1.8x10^{-3}$ $(\Omega.m)^{-1}$.

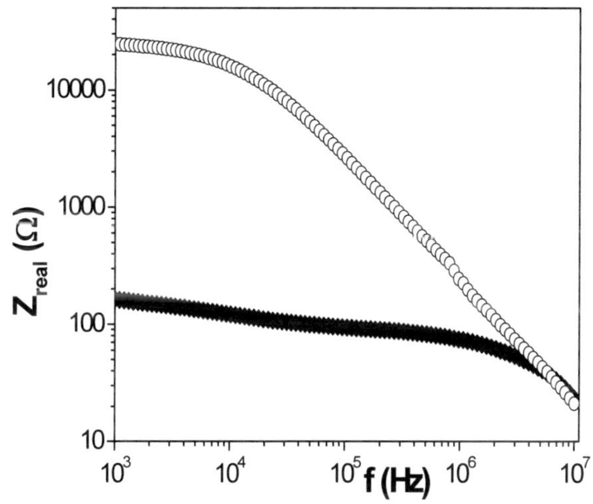

Fig. 12. A comparison of Z_{real} versus frequency plot for dry RC-6 (◊) and RC-6/$BMIMBF_4$ (♦) membranes.

According to these results, the presence of the ionic liquid $BMIMBF_4$ impregnating the free space among the cellulose chains in the solid matrix significantly increases the membrane conductivity.
The membrane impregnation assumption is supported by the values obtained from the XPS analysis carried out at different take off angles according to the A.C. (%) determined, which are indicated in Table 3. These results show and increase in fluorine and nitrogen, both IL elements, in the RC-6/$BMIMBF_4$ membrane with the increase of the take off angle, that is, when a deeper analysis is carried out.

	RC-6/w					RC-6/$BMIMBF_4$				
Φ (°)	C 1s (%)	O 1s (%)	Si 2p (%)	N 1s (%)		C 1s (%)	O 1s (%)	Si 2p (%)	N 1s (%)	F 1s (%)
20	62.9	19.9	16.9	0.3		65.1	28.1	4.1	1.2	1.5
45	64.0	20.2	15.3	0.5		61.9	32.8	2.4	1.4	1.5
75	65.4	20.9	13.1	0.6		62.4	32.3	1.8	1.7	1.9

Table 3. Atomic concentration percentages of the elements found on the surfaces of the RC-6/w and RC-6/LL membranes at different take off angles (Φ)

The effect of IL on the ion transport was also determined from membrane potential measurements and Fig. 13.a shows the membrane potential for RC-6/w and RC-6/$BMIMBF_4$ samples as a function of the ratio of the NaCl solution concentrations at both membrane sides. For NaCl concentrations lower than 0.01 M membrane potentials for both membranes are very similar and they slightly differ from the values associated to an ideal cation-exchange membrane, but clear differences are obtained at higher concentrations (0.02

M ≤ C_{NaCl} ≤ 0.2 M). This behaviour is common for weak charged membranes [25,41] and is associated to the Donnan co-ion exclusion at low concentrations (when the solution concentration is lower than the effective membrane fixed concentration), but this effect is practically neglected at high solution concentrations, when the number of solution charges are able to screen membrane fixed charge, and the membrane potential mainly corresponds to a diffusion potential due to the different mobility of Na^+ and Cl^- in the membrane.

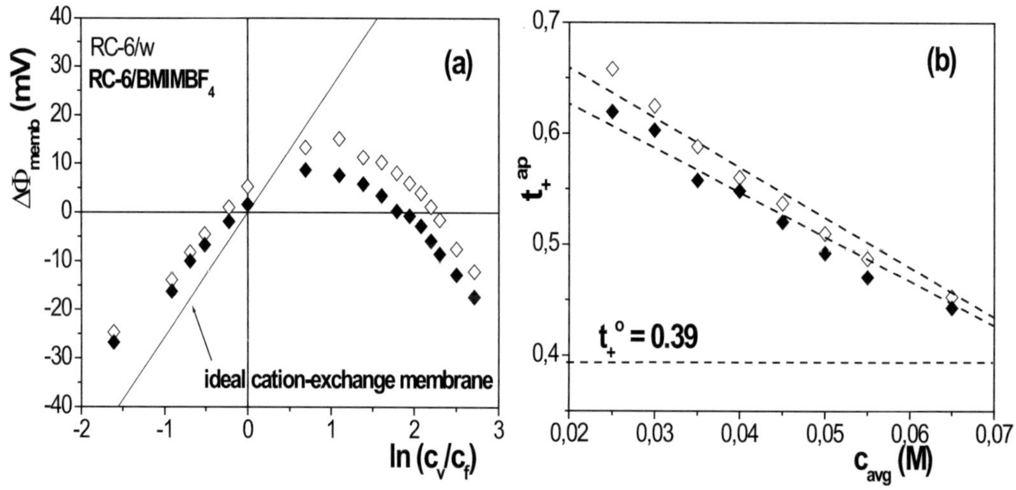

Fig. 13. (a) membrane potential vs NaCl concentration ratio. (b) Apparent cation transport number vs average NaCl concentration. (◊) membrane RC-6/w, (♦) membrane RC-6/$BMIMBF_4$.

Cation transport number across RC-6/w and RC-6/$BMIMBF_4$ membranes was determined by Eq. (3) and Fig. 13.b shows its variation with the average concentration ($c_{avg} = (c_v + c_c)/2$) for both membranes. Cation transport numbers determined by Eq. (2) are usually named "apparent transport number", t_+^{ap}, since water transport is not considered, which is a rather good approximation for most dense membranes, but it can slightly differ from true membrane transport number, t_+^m, in the case of highly hydrophilic membranes as the RC samples. Scatchard obtained the following relationship between both apparent and true cation transport numbers in a membrane [44]:

$$t_+^{ap} = t_+^m - 0.0018\, t_w\, (C_{avg}) \qquad (4)$$

where t_w represent the water transport number. From the slopes of the straight lines shown in Fig. 13.b, taking into account Eq. (4), the following values for t_+^m and t_w in RC-6/w and RC-6/LL membranes were obtained:

membrane RC-6/w: $t_+^m = 0.66 \pm 0.02$, $t_w = 255 \pm 15$

membrane RC-6/$BMIMBF_4$: $t_+^m = 0.62 \pm 0.02$, $t_w = 220 \pm 12$

These results show a reduction of 6 % in the transport of Na^+ ions and 14 % in the water transport associated as a result of the presence of the $BMIMBF_4$ ionic liquid.
To check that result, water diffusion measurements were performed using tritied water with a given activity (A_f) at one side of the membrane (donor chamber) and distilled water at the other membrane side (receiving chamber) [25, 45]. Taking into account the mass continuity: $A_f^o = A_f^t + A_r^t = cte$, where A_f^o is the initial activity of tritiaed water in the donor chamber (t = 0) while A_f^t and A_r^t the water activities in donor and receiving chambers at time t, the following expression is obtained [46]:

$$\ln([1 - (2A_r/A_f)] = - 2[S/(V_o.\Delta x)].P_w.t \qquad (5)$$

where V_o and P_w are the chamber volume and water permeability, respectively. Fig. 14 shows variation of solutions activities with time, and P_w value for each membrane was obtained from the slopes of those straight-lines by using Eq. (4). Water diffusion coefficient was obtained by [41]: $D_w = P_w.\Delta x_m$, and the following values were obtained: $D_w^{RC\text{-}6/w} = 2.1x10^{-10}$ m^2/s and $D_w^{mRC\text{-}6/LL} = 1.7x10^{-10}$ m^2/s, which indicates a reduction of 15 % in water diffusion from direct measurements. This result agrees with those previously obtained and confirms a diminution of the free space among the cellulose chains associated to the presence of the IL, reducing the transport of both water and ions through the IL-modified cellulosic membrane.

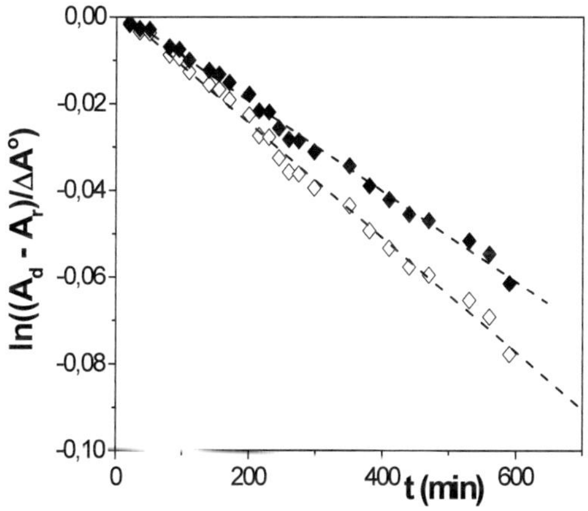

Fig. 14. Variation of tritiaed water activity in donor and receiving chambers as a function of time for RC-6/w (◊) and RC-6/$BMIMBF_4$ (♦) membranes.

Moreover, these results show the possibility of easy membrane modification by its immersion in an IL-water solution, which hardly modifies the transport of ions across the original membrane but reduces the mass transport, which is a requirement of interest for energy applications of membranes (reduction of cross flow in fuel cells) and electrochemical devices.

4. Conclusions

Electrical and chemical surface characterizations of different imidazolonium-bases RTILs carried out by impedance spectroscopy and XPS measurements was presented. The reduction in the electrical resistance of the ILs the increase of water content were correlated with the reduction in viscosity of the ILs.
The modification of various kinds of polymeric membranes, with different structures (porous and dense), materials (PVDF, RC and Nafion) and process applications (filtration, dialysis and electrodialysis), by inclusion of ILs in the pores as organic phase of supported liquid membranes or into the structure of dense samples (by ion-exchange or embeded into the membrane matrix) and its effect on transport parameters was also presented. The variety of ILs made possible to choose that more adequate for a specific membrane process, which opens its use in a wide variety of applications, particularly related to electrochemical systems.

5. Acknowledgements

We thank to Comisión Interministerial de Ciencia y Tecnología (CICYT, Project MAT/2007-65065, Spain) for financial support.

6. References

[1] M. Armandi, F. Endres, D.R. MacFarlane, H. Ohno, B. Scrosati; *Nature Materials*, 8 (2009) 621.
[2] J.F. Brennecke, E.J. Magin; *AICHEJ*, 47 (2001) 2384.
[3] J.D. Figueroa, T. Fout, S. Plasynski, H. McIlvried, R.D. Srivastava; *Int. J. Greenhouse Control*, 2 (2008) 9.
[4] J. Ranke, S. Stolte, R. Strmann, J. Arning, B. Jastorff; *Chemical Reviews*, 107 (2007) 2183.
[5] K. Yuyama, G. Masuda, H. Yoshida, T. Sato; *J. Power Sources*, 162 (2006) 1401.
[6] V. S. Silva, B. Ruffmann, S. Vetter, M. Boaventura, A. M. Mendes, L. M. Madeira, S. P. Nunes; *Electrochimica Acta*, 51 (2006) 79.
[7] M. Doyle, S. Choi, G. Proulx; *J. Electrochem. Soc.*, 147 (2000) 34.
[8] C. Schmidt, T. Glück, G. Schmidt-Naake; *Chem. Eng. Technol.*, 31 (2008) 13.
[9] D. Susac, M. Kono, K.C. Wong, K.A.R. Mitchell; *Appl. Surf. Sci.* 174 (2001) 43.
[10] M. J. Ariza, J. Benavente, E. Rodríguez-Castellón; in "Handbook of Membranes: Properties, Performance and Applications" (Ed. S. V. Gorley). Nova Science Publishers, Inc., New York, 2009.
[11] R. Fortunato, C.A.M. Afonso, J. Benavente, E. Rodriguez-Castellón, J.G. Crespo; *J. Membr. Sci.*, 256 (2005) 216.
[12] K. Sakai; *J. Membr. Sci.*, 96 (1994) 91.
[13] J. Mijovíc, F. Bellucci; *Trends in Polym. Sci.*, 4 (1996) 74.
[14] J. Benavente, J.M. García, R. Riley, A.E. Lozano, J. de Abajo; *J. Membr. Sci.*, 175 (2000) 43.
[15] V. Compañ, E. Riande, F.J. Fernandez-Carretero, N.P. Berezina, A.A.-R. Sytcheva; *J. Membr. Sci.*, 318 (2008) 255.

[16] J. Benavente; Electrical Characterization of Membranes, in Monitoring and Visualizing Membrane-Based Process, Eds. C. Güell, M. Ferrando, F. López. Wiley-VCH, 2009.
[17] P.A.Z. Suarez, J.E.L. Dullius, S. Einloft, R.F. de Sousa, J. Dupont; *Polyhedron*, 15(7) (1996) 1217.
[18] J.D. Holbrey, K.R. Seddon; J. Chem. Soc., *Dalton Trans.*, 13 (1999), 2133.
[19] A.E. Visser, R.P. Swatloski, R.D. Rogers; *Green Chemistry*, 2 (2000) 1.
[20] T. Schäfer, R. E. Di Paolo, R. Franco, J. G. Crespo; *Chemical Communications* (2005) 2594.
[21] D. Brigg, M. P. Seah, Practical Surface Analysis: Auger and X-Ray Photoelectron Spectroscopy, vol.1, 2º ed., John Wiley & Sons, Chichester, 1995.
[22] J. F. Moulder, W. F. Stickle, P. E. Sobol, K. D. Bomben; Handbook of X-Ray Photoelectron Spectroscopy, ed. J. Chastain. Perkin-Elmer Corporation, Minneapolis, 1992.
[23] M.J. Ariza, P. Prádanos, R. Rico, E. Rodriguez-Castellón, J. Benavente; *Surf. Interface Anal.*, 35 (2003) 360.
[24] R. de Lara, M.I. Vázquez, P. Galán, J. Benavente; *J. Membr. Sci.*, 273 (2006) 25.
[25] J.D. Ramos, C. Milano, V. Romero, S. Escalera, M.C. Alba, M.I. Vázquez, J. Benavente; *J. Membr. Sci.*, 352 (2010) 153.
[26] J.R. Macdonals; *Impedance Spectroscopy*, Wiley, New York, 1987.
[27] S. Rivera-Rubero, S. Baldelli; *J. Am. Chem. Soc.*, 126 (2004) 11788.
[28] R. Fortunato, L.C. Branco, C.A.M. Afonso, J. Benavente, J.G. Crespo; *J. Membr. Sci.*, 270 (2006) 42.
[29] D.L. Compton, J.A. Laszlo; *J. Electroanal. Chem.*, 520 (2002) 71.
[30] R. Fortunato, C.A.M. Afonso, M.A. Reis, J.G. Crespo; *J. Membr. Sci.*, 242 (2004) 197.
[31] K.B. Oldham, J.C. Myland; Fundamental of Electrochemical Science, Academic Press Incorporated, 2004.
[32] J.T.F. Keurentjes, J.G. Harbrecht, D. Brikman, H.H. Hanemaaijer, M.A. Cohen, H. van´t Riet; *J. Membr. Sci.*, 47 (1989) 333.
[33] L.A. Neves, I.M. Coelhoso, J.G. Crespo; *J. Membr. Sci.*, 360 (2010) 363.
[34] M.Oleinikova, M.Muñoz, J.Benavente, M.Valiente; *Langmuir*, 16(2) (2000) 716.
[35] P. Scovazzo, J. Kieft, D.A. Finah, C. Koval, D. DuBois, R. Noble; *J. Membr Sci.*, 238 (2004) 57.
[36] E. Miyako, T. Maruyama, N. Kamiya, M. Goto; *Chem. Commun.*, 23 (2003) 2926.
[37] S. Bijani, R. Fortunato, M.V. Martínez de Yuso, F. A. Heredia-Guerrero, E. Rodríguez-Castellón, I. Coheloso, J. Crespo, J. Benavente; *Vacuum*, 83 (2009) 1283.
[38] M.D. Duca, C.L. Plosceanu, T.J. Pop; *Appl Polym Sci.*, 67 (1998) 2125.
[39] Y. Jeon, J. Sung, C. Seo, H. Lim, H. Cheong, M. Kang, B. Moon, Y. Ouchi, D.J. Kim; *J. Phys. Chem. B*, 112 (2008) 4735.
[40] F. G. Helfferich; Ion Exchange, McGraw-Hill Book Company, New York, 1962.
[41] M. Mulder, Basic Principles of Membrane Technology, Kluwer Acad. Publishers, Dordrecht, The Netherland, 1992.
[42] L. Neves, J. Benavente, I.M. Coelhoso, J.G. Crespo; *J. Membr. Sci.*, 347 (2010) 42.
[43] J. Benavente, L. Neves, I. Coelhoso, J.G. Crespo; 2º Interface and Colloid Iberic Conference. Coimbra (Portugal), 2007.
[44] G.J. Scartchard; *J. Amer. Chem. Soc.*, 75 (1953) 2883.

[45] S. Escalera, S. Bijani, P. Galán, J. Benavente; XI Congreso Nacional de Materiales, Zaragoza (Spain), 2010.
[46] D.L. Gilbert, T. Okano, T. Miyata, S.W. Kim; *Int. J. Pharm.*, 47 (1988) 79.

28

Ionic Liquids for the Future Electrochemical Applications

Yu-Sheng Liu and Ge-Bo Pan
Suzhou Institute of Nano-tech and Nano-bionics, CAS
P. R. China

1. Introduction

Room temperature ionic liquids (RTILs) are the salts with weak interactions between the ions, which allows them to be liquid at room temperature (typically below 100°C). The scientific and technological importance of RTILs nowadays have spanned a wide range of applications, owing to their unique physicochemical properties, such as thermal and chemical stability, low melting point, negligible volatility, flame retardancy, high ionic conductivity, moderate viscosity, high polarity, and solubility (affinity) with many compounds (Fischer et al.; 1999; Welton, 1999; Earle & Seddon, 2000; Wasserscheid & Keim, 2000; Wasserscheid & Welton, 2003). For instance, RTILs are good solvents for a wide range of organic and inorganic compounds, which makes them an attractive solvent alternative for environmentally friendly processes. To date, there have been many reviews dealing with these topics. In particular, the easy modification of the cation and anion in RTILs is an advantage for the development of task-specific RTILs for organic synthesis, extraction, dissolution, etc. (Blanchard et al.,1999; Earle et al.,1999; Kubo et al., 2002). Electrochemical processes have been another important application area for RTILs since their early development. Key advantages of RTILs over common aqueous or organic media in this field are their wide electrochemical window (up to 6 V in some cases), high conductivity, and vanishingly low vapor pressure. These features not only facilitate investigations into metal electrodeposition (Hamelin et al., 1987; Schmickler, 1996; Freyland et al., 2003; Mann et al., 2009), electrocapacitor, and electrocatalysis in a less demanding manner, but also open up new possibilities for increased reactivity of processes and/or stability of reactants/products in ionic liquids (Hamelin et al., 1987; Dom & Mar, 2008).
The subject of this chapter covers the newest aspects of ionic liquids in applications where their ion conductivity is exploited. Our goal in this chapter is to survey the recent key developments and issues within ionic liquid research in these areas. First, the current state of the knowledge of the ionic liquid/electrode interfaces, which is vital for applications and whose studies are still in its infancy, has been reviewed. As the range of available surface techniques and systematic, investigations increase, our understanding will improve which lead to advances on this field. Then, updated researches on the electrochemical applications of ionic liquids are exploited. In particular, their potentials as electrochemical solvents for metal/semiconductor electrodeposition, batteries and fuel cells are reviewed where conventional medias fail. Finally, the applications of ionic liquids in electrosynthesis, electrocatalysis, and electrochemical biosensing are briefly discussed. We conclude this

chapter with personal perspectives on the directions to which future research on this field might be directed.

2. Ionic liquid / electrode interfaces

During the last years, the electrochemical applications of RTILs have been highly explored (Buzzeo et al., 2004; Zhang & Bond, 2005; Endres & Abedin, 2006) from both experimentally and theoretically. It has been recognized that the ionic liquid/electrode interfaces greatly influence the electrochemical processes. Thus, it is necessary to study these interfacial systems, especially to obtain a molecular level perspective. Moreover, RTILs provide an interesting system in physical chemistry because they are composed purely of ions yet are liquid at room temperature. While the surface chemistry of aqueous electrolytic systems has been under extensive study for nearly a century, the study of neat ions in the liquid state is very new and highly investigated. In addition, there are various experimental techniques, e.g. electrochemical, spectroscopic and microscopic instruments available to study surfaces.

2.1 Electrochemical studies

To obtain a better picture of the structure of the ionic liquid/electrode interface, different models including Helmholtz, Gouy Chapman-Stern, and multilayer (Fig. 1) have been developed to describe and predict the experimental behavior of electrochemical systems observed in electrochemical studies, such as interfacial capacitance and electrocapillary measurements (Aliaga et al., 2007). Among the three models, the Helmholtz and the multilayer conceptions seemed to best describe the structure at the interfaces, although they both failed at some point (Inman & Lovering, 1983). In addition, as far as the studies of ionic liquid/electrode interface are concerned, techniques such as EIS, differential capacitance measurements, and voltammetry at DME (dropping mercury electrode) were found to be useful.

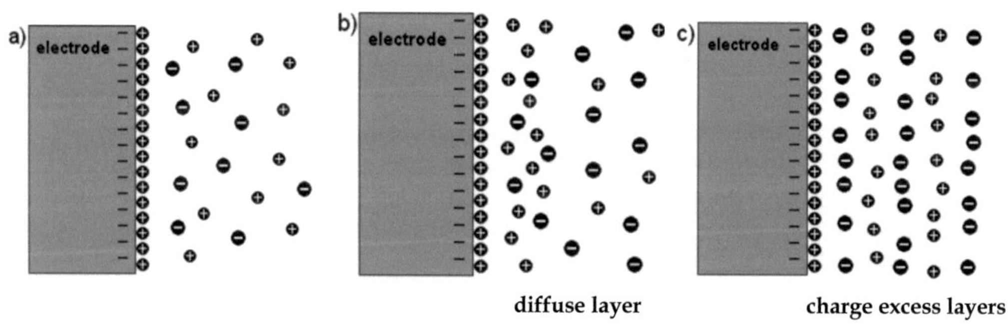

Fig. 1. Models of the electrical double layer. (a) Helmholtz, (b) Gouy Chapman–Stern, (c) multilayer.

Note that the number of papers concerning research of the ionic liquid/electrode interface is limited. Most of the efforts have been aimed at the study of the ionic liquid/platinum interface as a method of proving the structure of the adjacent ionic environment and the study of the ionic liquid/carbon interface with an emphasis on the design of electrochemical double layer capacitors. The results have revealed that different structures may form at the surface of the electrode depending on the nature of the ions that constitute an ionic liquid. Subsequently, anions such as dicyanamide, tend to form multilayers of adsorbed ions due to their coordinating properties and others will simply assemble as monolayers. For instance, Baldelli (2005) studied the structure of the [BMIM][BF_4]/platinum interface by EIS. The works showed that the ionic liquid corresponds to a Helmholtz layer with a double layer thickness of ~0.5 nm.

2.2 Spectroscopic studies.

Surface sensitive vibrational spectroscopic techniques such as surface enhanced Raman scattering (SERS), sum-frequency generation (SFG), surface-enhanced infrared absorption spectroscopy (SEIRAS), and in situ Fourier transform infrared reflection absorption spectroscopy (FT-IRAS) have been used to investigate the structure of the ionic liquid-electrode interface (Baldelli, 2005). The advantages of these techniques are their capabilities to perform experiments at controlled electrochemical conditions. On the other hand, by relating spectroscopic information with previous electrochemical measurements, a more complete description of the electrified interface, such as double layer thickness, orientation of the ions at the surface, and identification of the ionic species adsorbed, can be obtained. Note that this is very important since the structure of the interfacial layer dramatically influences the electron transfer reactions.

Recently, Vianney et al. (2006) carried out the SERS at the $BMIMPF_6$/silver electrode interface. It has been shown that the $BMIM^+$ cations adsorb on the silver electrode at potentials more negative than -0.4 V *vs* a Pt quasi-reference electrode (Fig.2). The potential dependence of the SERS intensities of Py adsorbed on a silver electrode in $BMIMPF_6$ has also been investigated by Rivera-Rubero and Baldelli (2004). Their results have shown that at potentials less negative than -0.8 V Py adsorbs at an end-on configuration forming an Ag-N bond. The simulation results further confirmed this assumption. It was concluded that Py molecules lie flat on the electrode surface from the -0.9V to 1.4V potential range and Py is replaced by the $BMIM^+$ at potentials < -1.4 V. In the previous studies (Reed et al., 2007), ionic molecular dynamics simulation model for an electroactive interface in which a metallic electrode is maintained at a preset electrical potential was described.

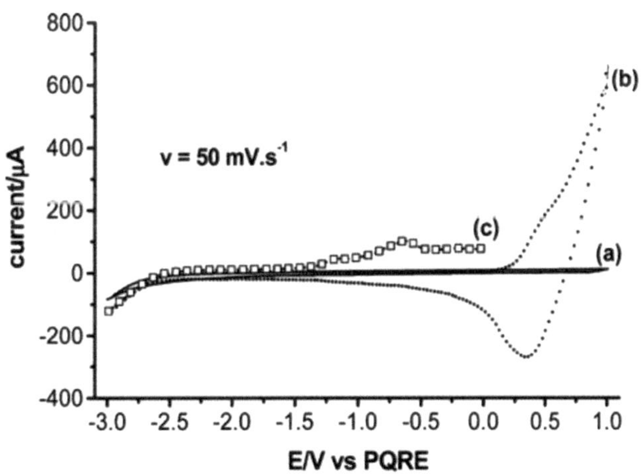

Fig. 2. Cycle voltammograms of a Au (a) and Ag (b) electrodes in $BMIPF_6$; (c) the voltammogram of the Ag electrode after staying at -3.0 V for 60 s and then scanning the potentials in the positive.

In brief, the spectroscopic surface studies of the ionic liquid/electrode interface of compounds based on the imidazolium cation using techniques such as SFG, SERS, FT-IRAS and SEIRAS agree with the fact that the structure of electrified interface can be specifically probed and is strongly influenced by the applied potentials. In all the cases, the ring planes of the cation tends to align themselves parallel to the surface as the excess charge becomes negative, and is repelled when it turns positive, to allow for the adsorption of the anion. The anionic species may adsorb on the surface of the electrode in single or multilayers depending on the nature of the anion, as probed by the relative increase of their surface concentration as a function of the applied potential.

2.3 Microscopy studies.

Scanning tunneling microscopy (STM) and atomic force microscopy (AFM), which can provide atomic and sub-molecular resolution images of adsorbed targets, were used in the investigation of the electrified ionic liquid-metal interface as in situ techniques. Recently, we have studied the structuring of adsorbed ions at the electrified [BMIM][PF_6]/Au(111) interface by means of STM under potential control at 0 V (Pan & Freyland, 2006). The formation of ordered PF_6^- adlayers was observed in a potential range from -0.5 to 0.5 V *vs* a Pt reference electrode. The results indicate that the anion molecules formed Moiré-like patterns at potentials above -0.2 V and adopted a ($\sqrt{3} \times \sqrt{3}$) structure at approximately -0.45 V although the Moiré patterns often co-existed with uncompressed structures (Fig.3). Mao et al. (2003) have observed the drastic long-range surface restructuring with the appearance of small pits. The enlargement of the pits has resulted in forming wormlike structures.

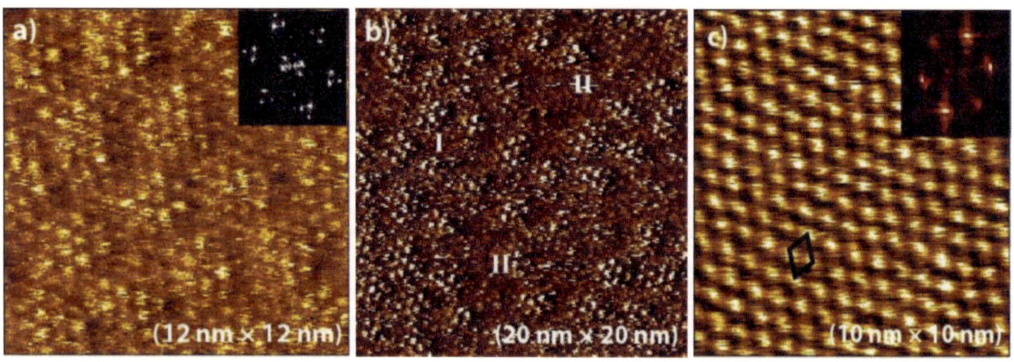

Fig. 3. Typical STM images of a Au(111) electrode recorded under different potential controls in pure ionic liquid of [BMIM][PF_6]. (a) E = 0.5 V; (b) E = 0.0 V; (c) E = -0.25 V.

The use of high-resolution techniques such as STM and AFM for the study of the ionic liquid/electrode interface is then evident. The spatial arrangement of the adsorbed ions in ordered layers on the surface of metallic electrodes was probed and ordered arrays of ions were observed, which is also supported by the SFG results. In addition, the surface reconstruction effect that the ionic liquids have on metals when the electric field is applied at the surface of the electrode, which seems due to the charge transfer from the metal to the adsorbed ions.

3. Electrochemical applications

RTILs can provide exciting opportunities for overcoming the limitations encountered in traditional chemical processes, in particular in electrochemistry. The nonvolatile, noncombustible, and heat resistance nature of RTILs is observed for applications in electrodeposition, batteries, fuel cells, electrosynthesis, electrocatalysis, and electrochemical biosensing.

3.1 Electrodeposition

Electrodeposition is essential for a variety of industries including electronics, optics, sensors, automotive and aerospace. Most of conventional solutions, e.g., water, suffer from the drawback that it has a relatively narrow potential window, and hence the deposition of metals with large negative reduction potentials is hindered by poor current efficiencies. RTILs are superior media for the electrodeposition of metals and semiconductors, and have an unprecedented potential to revolutionize electroplating. The use of ionic liquids heralds not only the ability to electrodeposit metals that have hitherto been impossible to reduce in aqueous solutions but also the capability to engineer the redox chemistry and control metal nucleation characteristics.

Quite a large number of ionic liquids, especially those based on NTf_2, BF_4 and PF_6 anions, are stable in some cases to below Li/Li^+ reductive region. Therefore, semiconductors that previously could not be deposited from conventional water baths by turning to ionic liquids or be directly electroplated. Recently, the photoluminescent Si_xGe_{1-x} with a band gap of at least 1.5–3.2 eV has been successfully electrodeposited from pure ionic liquid containing silicon and germanium halides (Fig.4) (Al-Salman et al., 2008). During the deposition, different colors from orange to green are observed in the visible spectrum. These are due to a quantum size effect of the semiconductor particles with sizes between 2 and 20 nm. The results show that the ionic liquid allows deposits very pure and photoluminescence effects can be seen. In addition, the material absorbs visible light and may open the way to a simple electrochemical fabrication of inexpensive solar cells.

The electrodeposition of metals and alloys using a variety of ionic liquids has also been demonstrated by previous studies (Abbott et al., 2008). Electroplating electronegative metals, e.g. Al, Ta, Nb, Mo, W; direct electroplating of metals on water sensitive substrate materials such as Al and Mg can be achieved, removal of hydrogen embrittlement from the substrate; alloy deposition is easier to achieve; the possibility exists to develop novel immersion plating baths; potential energy savings compared with aqueous solutions, replacement of many hazardous and toxic materials currently using and access to novel deposit morphologies. The electrolytic deposition of nickel is also demonstrated using a solution of the metal chloride salt separately in either a urea or ethylene glycol/choline chloride based ionic liquid (Haerens, 2009).

Moreover, it has been shown that the deposition kinetics and thermodynamics of ionic liquids differ from them in the aqueous processes, which results in different deposit morphologies. Dobbs et al. (2006) have produced silver and gold nanoparticles by electrodeposition from ionic liquid-crystal precursors, illustrated by the electrolysis of imidazolium liquid crystals containing dicyanoargentate and dicyanoaurate anions. We have recently studied the Pb and Sb deposition on Au (111) from a Lewis acidic ionic liquid of MBIC-$AlCl_3$. A quasi-equilibrium process has been observed and the results indicate that Pb UPD is kinetically

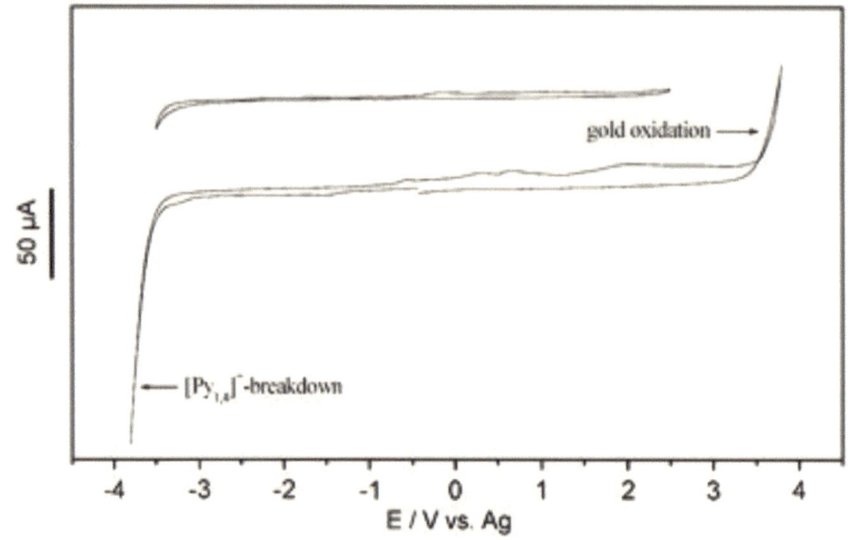

Fig. 4. CVs of pure $[Py_{1,4}]Tf_2N$ on Au(111) showing the electrochemical window of this ionic liquid. Scan rate: 10 mV s^{-1}, 25 °C.

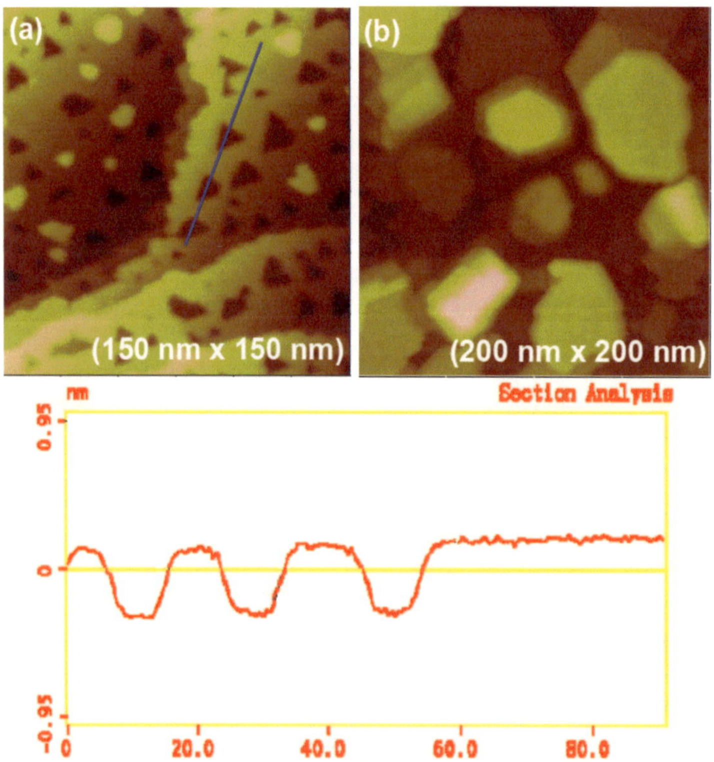

Fig. 5. Typical large-scale STM images of Pb electrodeposition on Au(111) in MBIC-$AlCl_3$ + 5mM $PbCl_2$ solution. (a) E = 0.5 V; (b) E = 0.27 V.

controlled (Fig. 5). In the OPD range, the nucleation of 3D Pb crystal occurs (Pan et al., 2010, a, b). As for the Sb deposition, 2D-nanostripe architectures have been observed. In addition, the ionic liquids have the ability to obtain high concentrations of aluminium in a highly conducting aprotic medium for aluminium deposition (Buzzeo et al., 2004).

For ionic liquids, anions such as BF_4^- and PF_6^- were initially used quite extensively because of their wide potential window, however, slow hydrolysis by water, yielding HF, has led to an increase in the use of water stable anions such as $(CF_3SO_2)_2N^-$ (Ramesh & Ang, 2010). As a result, other anions such as $(CN)_2N^-$ p-toluenesulfonate and methylsulfonate have recently been extensively used for metal deposition (Borra et al., 2007). Besides anions, ionic liquids with alternative cations such as those derived from biodegradable imidazoles, lactams, amino acids and choline have been prepared and used for metal deposition (Haerens et al.,1987). This area is still needed to be further investigated although Endres et al. (2006) have shown that the cation appears to control the morphology of aluminium deposited from various triflamide based ionic liquids.

3.3 Batteries and fuel cells

RTILs are important in the field of energy, especially, lithium batteries and fuel cells, which greatly benefit from the switch of non-volatile, non-flammable, ionic liquid-based electrolytes. They have the excellent energy efficiency of all known electrochemical storage systems. It has been found that ionic liquids based on the NTf_2, BF_4 and PF_6 anions, exhibiting wide and stable electrochemical windows, in some cases, can reach in some cases to below Li/Li^+ reductive potential region.

Lithium ion (Li-ion) batteries have been utilized as power supplies of electronic devices such as cell phones and laptop computers. However, Li-ion batteries consist of conventional electrolytes, such as ethylene carbonate, dimethyl carbonate, and diethyl carbonate, may suffer thermal runaway and cell rupture if over heated or overcharged. A significant research effort into RTILs-based electrolytes have been thus made to pursue a safe battery in the last years (Matsumoto et al., 2005; Sakaebe et al., 2007; Noto et al., 2010). An electrolyte of lithium batteries based on the ionic liquid PMIMTFSI complexed with lithium bis LiTFSI at a molar ratio of 1:1 has been reported by Kim et al. (2010). The electrolyte shows a high ionic conductivity at room temperature (Fig.6). Ionic liquids like $BMIMBF_4$ or $BMIMPF_6$ and $PyBF_4$ were also mixed with organic solvents for the use of batteries such as butyrolactone and acetonitrile (Diaw et al., 2005). In recently studies (Lee et al., 2010), Lithium salt ($LiBF_4$ or $LiPF_6$) was added to the ionic liquids mixtures for possible application in the field of energy storage (batteries or supercapacitors), electrolytes are stable toward oxidation and exhibit a vitreous phase transition.

RTILs have also been widely used in fuel cells. Although polymer-electrolyte-membrane fuel cells have been used very long time, they have not reached large-scale development as some issues are still unresolved. These issues may lead to an extent by switching to ionic-liquid-based polymer membranes. The diethylmethylammonium trifluoromethanesulfonate ([dema][TfO]) ionic liquid, functions as a proton conductor and is suitable for use as an electrolyte in H_2/O_2 fuel cells, which can be operated at temperatures higher than 100 °C under non-humidified conditions. In earlier reports (Su et al., 2009), in order to fabricate a polymer electrolyte fuel cell, matrix polymers for [dema][TfO] are explored and sulfonated polyimides, in which the sulfonic acid groups are found to be highly compatible with [dema][TfO]. Polymer electrolyte membranes for non-humidified fuel cells are prepared by the solvent casting method using [dema][TfO]. The composite membranes have good thermal stability (>300 °C) and ionic conductivity (>10^{-2} Scm^{-1} at 120 °C when the [dema][TfO] content is higher than 67 wt%) under anhydrous conditions(Fig. 7).

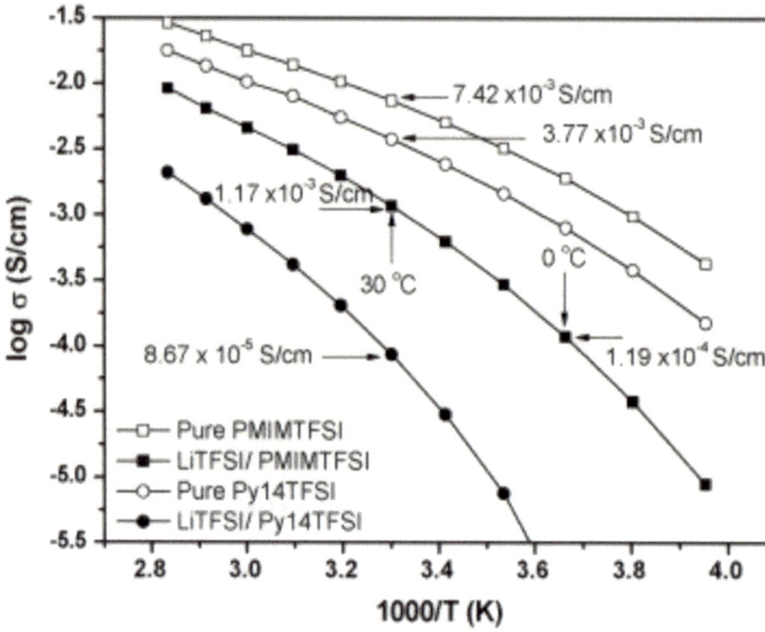

Fig. 6. Ionic conductivity as a function of temperature for pure PMIMTFSI, $Py_{14}TFSI$ and the two corresponding liquid electrolytes obtained after complexation with LiTFSI at molar ratio of 1:1

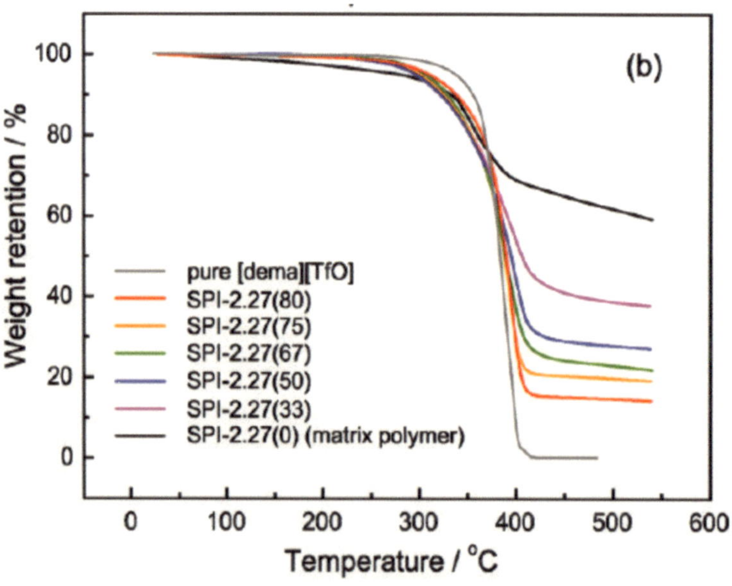

Fig. 7. Thermogravimetric curves for [dema][TfO] and composite membranes

3.4 Electrosynthesis and electrocatalysis

RTILs have also been utilized in the fields of electrosynthesis and electrocatalysis. These features open up new possibilities for increased reactivity of processes and stability of reactants/products in ionic liquids (Zhang et al., 2008). A potential advantage in the use of ionic liquids during electrosynthesis is their unique solvent capabilities. Many advances made in the field to implement electrochemical methodology at preparative-scales. The electrosynthetic processes would inevitably have the limitations associated with organic electrosynthesis. Since several years, RTILs are increasingly being applied for electrochemical and electrosynthetic purposes.

A vast array of chemical processes using ionic liquids at electrodes, the obvious advantages of adopting ILs as 'clean' reagents for synthetic processes compared with the organic electrochemical reactions. Due to its low cost, ease of fabrication, high sensitivity, RTILs can be used as the supporting electrolyte, or the binder in the carbon paste electrode, or the modifier on the chemically modified electrodes (Sun et al., 2007). Indeed, it has already been demonstrated that a wide range of organic syntheses, including catalysed reactions, can be carried out in these alternative non- volatile, non-flammable solvents. Ionic liquids are attractive according to their easy access, thermal stability, low vapor pressure and simple product recovery (Liu et al., 2010). Makeli et al. (2006) applied n-octylpyridinum hexafluorophosphate (OPFP) as binder to fabricate a carbon composite electrode, which provided a remarkable increase in the electron transfer rate and decreased the overpotentials of some organic substances (Fig.8). Besides, the electrochemical reduction of benzaldehyde at a Pt microelectrodes in 1-butyl-1-methyl pyrrolidinium triflimide ([Bmpyr][NTF_2]) ionic liquid is reported by Doherty and Brooks.(2003), the reaction process is shown in scheme 1.

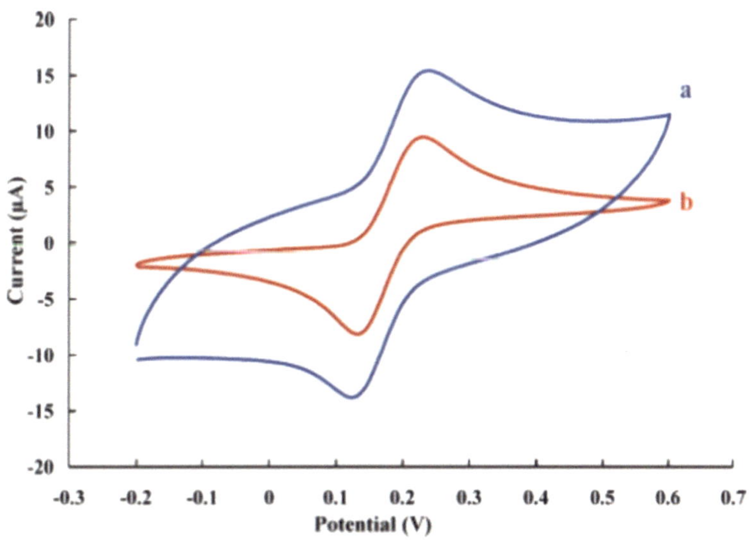

Fig. 8. Cyclic voltammograms of 1 mM catechol in phosphate buffer pH 7, on CILE (a) before heating and (b) after heating. The scan rate was 100 mV.s^{-1}.

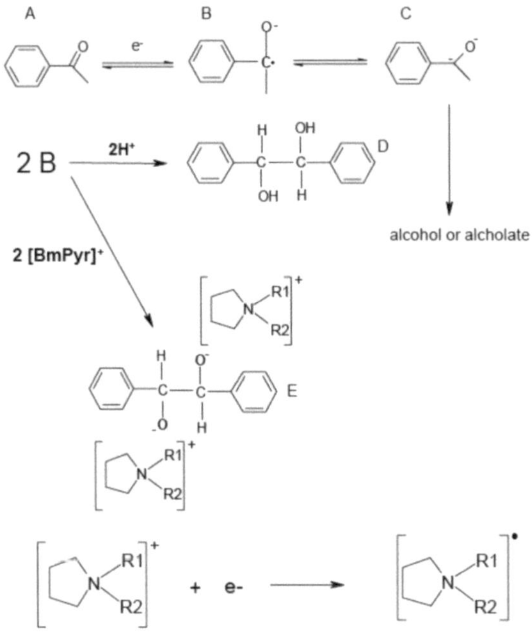

Scheme 1. Proposed reaction sequence to account for the three reduction process.

Due to the specific characteristics such as good electronic and mechanical properties, RTILs can also be used in the process of electrocatalysis. IL/CNTs modified electrodes had been applied in the field of electroanalysis. By incorporating IL/CNTs on the surface of modified electrode, the electron transfer rate of electroactive substances are greatly improved. Sun et al. (2007) reported an electrocatalysis of hemoglobin on multi-walled carbon nanotubes (MCNTs) modified carbon ionic liquid electrode (Fig.9) with hydrophilic $EMIMBF_4$ as modifier. The as-abstained carbon ionic liquid electrode showed excellent electrocatalytic activity to the reduction of trichloroacetic acid (TCA) and hydrogen peroxide.

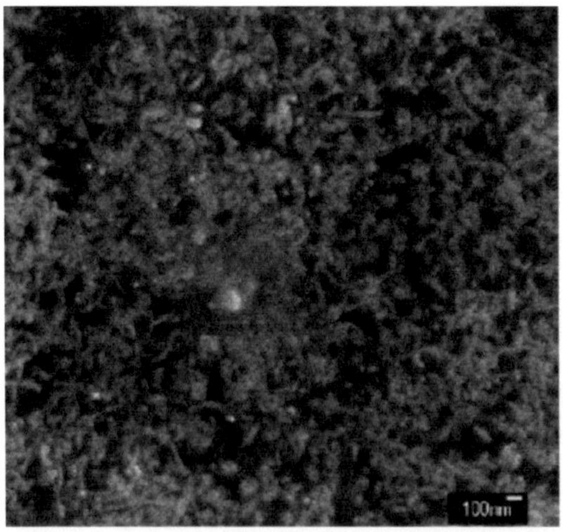

Fig. 9. SEM image of MWCNTs/IL carbon electrode.

3.5 Electrochemical biosensing

Bioscience is an interesting and emerging area where ionic liquids are beginning to play an important role. In general, water has been believed to be the unique solvent for biomolecules and biosystems. However, for long-life biodevices, there are a number of drawbacks to this situation, such as the volatility of water, the limited temperature range and the very narrow pH range for protein stability. On the other hand, immobilization of electron transfer mediators to electrode surfaces is a key step for design, fabrication and performance of the sensors and biosensors. In conventional methods, however, most of these electrodes presented quasi-reversible electrochemical behavior with poorly defined cyclic voltammograms with large background currents and low electrocatalytic activity. The use of RTILs might overcome these drawbacks and proved to be efficient pasting binder in place of non conductive organic binders of the preparation of electrodes.

Recently, a colloidal gold-modified carbon ionic liquid electrode was constructed by mixing colloidal gold-modified graphite powder with a solid RTIL n-octyl-pyridinium hexafluorophosphate ($OPPF_6$) (Ren et al., 2010). The electrode has shown the good bioactivity and excellent stability. This sensor was capable of distinguishing the complementary target DNA at low concentration from the three-base mismatched DNA at higher concentration (Fig.10). The response time of the biosensor is fast (within 10 s), and the life time is over two months. By using a hydrophilic ionic liquid $EMIMBF_4$ as the modifier, a new carbon ionic liquid electrode (CILE) was fabricated and further modified with MWCNTs to get the MWCNTs/CILE. The proposed electrode showed the potential application in the third generation reagentless biosensor (Choi et al., 2009; Ding, et al., 2009).

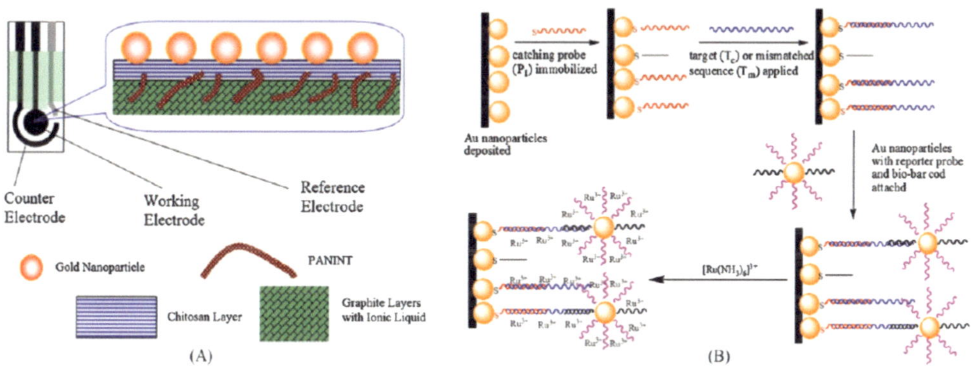

Fig. 10. The pattern of the PANINT-IL/chitosan/AuNP SPE and the structure of the working electrode. (B) Procedure for the detection of DNA with RuHex as electrochemical probe.

On the other hand, nanomaterials have been widely used in the electrochemical biosensors. Among the nanomaterials used for protein film electrochemistry, ionic liquids had been extensively studied. Due to the specific characteristics such as good electronic and mechanical properties, electric conductivity and biocompatibility, RTILs electrodes had been applied in the field of electroanalysis. For example, a DNA-modified carbon paste electrode (DNACPIE) was designed by using a mixture of the ionic liquid 1-butyl-3-methylimidazolium hexafluorophosphate and paraffin oil as the binder (Xi et al., 2010). The electrode exhibits higher sensitivity compared to DNA modified carbon paste electrode without ionic liquid and better selectivity compared with electrodes without DNA. Adenine showed an irreversible adsorption-controlled oxidation reaction with enhanced electrochemical response, which was due to the presence of high conductive MWCNTs on the CILE surface. The electrode showed good stability and selectivity, and was further applied to milk powder samples with satisfactory results. Salimi et al. (2010) recently have reported a new carbon nanotubes-ionic liquid and chloropromazine modified electrode (Fig.11) biosensor for determination of NADH and fabrication of Ethanol. The nanocomposite modified electrode displays excellent electrocatalytic activity toward oxidation of NADH. The proposed electrode showed the potential application in the third generation reagentless biosensor.

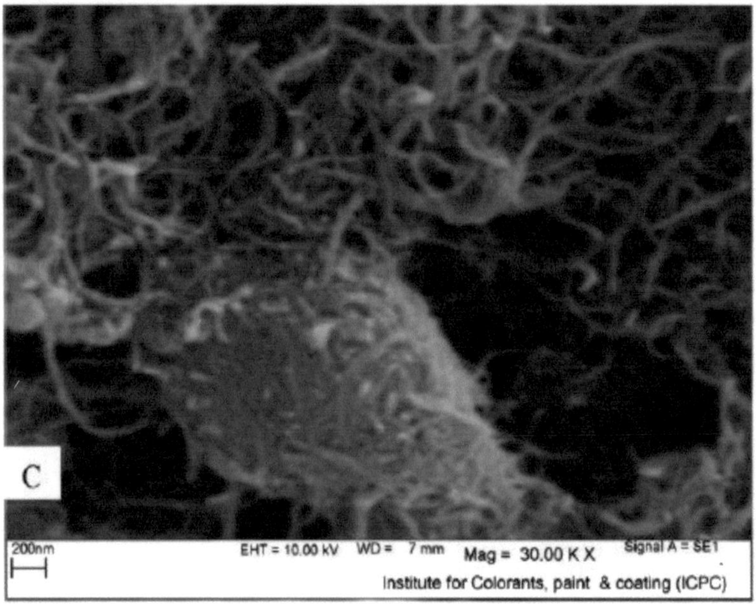

Fig. 11. SEM images of MWCNTs-IL modified electrode

4. Conclusions

RTILs have wide electrochemical windows and high conductivities, they are of current interest the research field of electrochemically. RTILs hold a great promise for various electrochemical applications, even for broad electrochemical applications. Previous attempts have revealed that the RTILs have advantageous for electrochemical applications, in particular, in electrosynthesis, electocatalysis, electrodeposition and batteries and fuel cells. Nonetheless, there remain many aspects of the behavior of ionic liquids that need urgent investigation if the applications discussed here are to be better understood and developed. Examples are fundamental electrochemical issues such as the structure of the double layer at an electrode in an ionic liquids; the speciation of solute ions such as metal ions; transference numbers and how these are influenced by speciation; ion association and its effect on thermodynamic and transport properties; and interactions of ionic liquids with solutes and interfaces. For ionic liquids, there are essentially advanced properties to the use of electrochemistry and energy devices. It is taken into account in the trends in this novel materials developed for practical applications. In the future, more developments in ionic liquids are expected to be seen. The simple dissolution of a lithium salt always results in electrolytes in which only a fraction of the current is actually carried by Li^+ ions. It would be interesting to see whether the incorporation of poly(Li salts), that is, lithium salts with a polyanionic chain, through specific interaction with the fixed negative charge of polarizing Li^+ ions, could change the flux balance. All of these fundamental aspects of the physical chemistry and electrochemistry of ionic liquids remain to be thoroughly explained, and they promise to further improve the potential of their various electrochemical applications. We can easily expect wider applications in electrochemical fields of using these remarkable materials.

5. References

Aliaga, C.; Santos, C.S. and Baldelli, S. (2007). Surface chemistry of room-temperature ionic liquids, *Phys. Chem. Chem. Phys.*, 9, 3683–3700.

Abbott, A. P.; Ttaib, K.; Ryder S. and Smith, E. L. 2008. Electrodeposition of nickel using eutectic based ionic liquids, *Transactions of the Institute of Metal Finishing*, 86, 234-238.

Al-Salman, R.; Abedin, S. Z. and Endres, F. (2008). Electrodeposition of Ge, Si and Si_xGe1_{-x} from an air and water-stable ionic liquid, *Phys. Chem. Chem. Phys.*, 10, 4650-4657.

Blanchard, L.A,. Hancu, D.; Beckman, E. J. and .Brennecke, J. F. (1999.) Green processing using ionic liquids and CO_2, *Nature*, 399, 28–29.

Buzzeo, M. C.; Evans, R. G.; Compton, R.G. 2004. Non-haloaluminate room-temperature ionic liquids in electrochemistry: A review, *ChemPhysChem*, 5, 1106-1120.

Baldelli, S. (2005). Probing Electric Fields at the Ionic Liquid-Electrode Interface Using Sum Frequency Generation Spectroscopy and Electrochemistry, *J. Phys. Chem. B*, 109, 13049-13051.

Borra, E.F., (2007). Seddiki, O.; Angel, R.; Eisenstein, D., Hickson, P.; Seddon, K.R. , Worden, S.P. Deposition of metal films on an ionic liquid as a basis for a lunar telescope, *Nature*, 447, 979-981.

Choi, B.G.; Park, H.; Park, T.J.; Kim, D.H. Lee, S.Y.; Hong, W.H. 2009. Development of the electrochemical biosensor for organophosphate chemicals using CNT/ionic liquid bucky gel electrode, *Electrochemistry Communications* 11, 672–675.

Doherty, A.P. and Brooks, C.A. (2004). Electrosynthesis in room-temperature ionic liquids: benzaldehyde reduction, *Electrochim. Acta*, 49, 3821–3826.

Diaw, M.; Chagnes, A.; Carré, Willmann, B.P. and Lemordant, D. (2005). Mixed ionic liquid as electrolyte for lithium batteries, *J. Power Sources*, 146, 682–684.

Dobbs, W.; Suisse, J.M.; Douce, L. and Welter, R. (2006). Electrodeposition of Silver Particles and Gold Nanoparticles from Ionic Liquid-Crystal Precursors, *Angew. Chem. Int. Ed.* 45, 4179–4182.

Dom, P.; Mar, N.D. (2008). "Nonsolvent" Applications of Ionic Liquids in Biotransformations and Organocatalysis, *Angew. Chem. Int. Ed.* 47, 6960 - 6968.

Ding, C.; Zhao,F.; Ren, R. Lin, J. (2009). An electrochemical biosensor for α-fetoprotein based on carbon paste electrode constructed of room temperature ionic liquid and gold nanoparticles, *Talanta*, 78 1148–1154.

Earle, M. J.; McCormac, P. B. and Seddon, K. R. (1999). Diels–Alder reactions in ionic liquids. A safe recyclable alternative to lithium perchlorate–diethyl ether mixtures, *Green Chem.*, 1, 23–25.

Earle, M.J. and Seddon, K.R. (2000). Ionic liquids. Green solvents for the future, *Pure Appl. Chem.*, 72, 1391–1398.

Endres, F. and Abedin, S.Z. (2006). Air and water stable ionic liquids in physical chemistry, *Phys. Chem. Chem. Phys.* 8, 2101-2116.

Fischer, T.; Sethi, A.; Welton T. and Woolf, J. (1999). Diels-Alder reactions in room-temperature ionic Liquids,*Tetrahedron Lett.*, 40, 793–796.

Freyland , W.; Zell, C.A.; Abedin, Z.E. and Endres, S.F. (2003). Nanoscale electrodeposition of metals and semiconductors from ionic Liquids, *Electrochim. Acta,* 48, 3053-3061.

Hamelin, A.; Stoicoviciu, L.; Silva, F.J. (1987). The temperature dependence of the double-layer properties of gold faces in perchloric acid solutions, open: Part I. the (210) gold face, *Electroanal. Chem.,* 229,107–124.

Haerens, K.; Matthijs, E. Binnemansc, K. and Bruggenb, B.V. 2009. Electrochemical decomposition of choline chloride based ionic liquid analogues, *Green Chem.*, 11, 1357–1365.

Inman, D. and Lovering, D.G. (1983). *Comprehensive Treatise of Electrochemistry,* Plenum Press, New York , vol. 7, pp. 543-593.

Kisza, A. (2002). The capacitance of the electric double layer of electrodes in molten salts, *J. Electroanal. Chem. 534,* 99-106.

Kubo, K.; Shirai, M.; and Yokoyama, C. (2002). Heck reactions in a non-aqueous ionic liquid using silica supported palladium complex catalysts, *Tetrahedron Lett.*, 43, 7115–7118.

Kim, J.K.; Matic, A.; Ahn, J.H.and Jacobsson, P. (2010). An imidazolium based ionic liquid electrolyte for lithium batteries, *J. Power Sources*, 195, 7639–7643.

Lin, L.G.; Wang, Y.; Yan, J.W.; Yuan, Y.Z.; Xiang, J. and B. Mao, W. 2003. An in situ STM study on the long-range surface restructuring of Au(111) in a non-chloroaluminumated ionic liquid, *Electrochem. Commun.*, 5, 995-999

Lee, S.Y.; Yasuda, T. Watanabe, M. (2010). Fabrication of protic ionic liquid/sulfonated polyimide composite membranes for non-humidified fuel cells, *J. Power Source,* 195, 5909–5914.

Liu, X.; Zeng, X.; Mai, N.; Liu, Y.; Kong, B.; Li Y.; Wei, W. and Luo, S. 2010. Amperometric glucose biosensor with remarkable acid stability based on glucose, *Biosens. Bioelectron.*, 25, 2675-2679.

Matsumoto, H.; Sakaebe, H. and Tatsumi, K. (2005). Preparation of room temperature ionic liquids based on aliphaticonium cations and asymmetric amide anions and their electrochemical properties as a lithium battery electrolyte, *J. Power Sources* ,146, 45–50.

Maleki, N.; Safavi, A. and Tajabadi, F. High-performance carbon composite electrode based on an ionic liquid as a binder (2006). *Anal.Chem.* 78, 3820-3826.

Mann, O.; Pan, G.B. and Freyland, W. (2009). Nanoscale electrodeposition of metals and compound semiconductors from ionic liquids, *Electrochimica Acta* ,54, 2487–2490.

Noto,V.D.; Negro, E.; Sanchez, J.Y. and Iojoiu, C. (2010). Structure-Relaxation interplay of a new nanostructured membrane based on tetraethylammonium trifluoromethanesulfonate ionic liquid and neutralized nafion 117 for high-temperature fuel cells, *J. Am. CHem. SOC.*, 132, 2183–2195.

Pan, G.B. and Freyland, W. (2006). 2D phase transition of PF_6 adlayers at the electrified ionic liquid/Au(1 11) interface, *Chemical Physics Letters*, 427, 96–100.

Rivera-Rubero, S. and Baldelli, S. (2004). Surface spectroscopy of room-temperature ionic Liquids on a platinum electrode: a sum frequency generation study, *J. Phys. Chem. B, 108*, 15133-15140.

Reed, S.K. Lanning, O.J., and Madden, P.A. (2007). Electrochemical interface between an ionic liquid and a model metallic electrode, *J. Chem. Phys.*, 126, 084704.

Ramesh, S. and Ang, G. P. (2010). Impedance and FTIR studies on plasticized PMMA-$LiN(CF3SO_2)_2$ nanocomposite polymer electrolytes, *Ionics* 16, 465–473.

Ren, R.; Leng, C., Zhang, S. 2010. A chronocoulometric DNA sensor based on screen-printed electrode doped with ionic liquid and polyaniline nanotubes, *Biosens. Bioelectron.*, 25, 2089–2094.

Schmickler, W. (1996). Electronic Effects in the Electric Double Layer, *Chem. Rev.* 96, 3177–3200.

Sun, W; W. D. Gao R.; J. K. (2007). Direct electrochemistry and electrocatalysis of hemoglobin in sodium alginate film on a $BMIMPF_6$ modified carbon paste electrode, *Electrochemistry Communications*, 20,1159–1164.

Su,Y.Z.; Fu,Y.C.; Yan, J.W.; Chen, Z.B. and Mao, B.W. (2009). Double Layer of Au(100)/Ionic Liquid Interface and Its Stability in Imidazolium-Based Ionic Liquids, *Angew. Chem. Int. Ed.* 48, 5148 -5151.

Salimi, A.; Lasghari, S. and Noorbakhash, A.(2010).Carbon nanotubes-Ionic Liquid and chloropromazine modified Electrode for determination of NADH and fabrication of ethanol biosensor, *Electroanalysis*, 22, 1707 - 1716

Sakaebe, H.; Matsumoto, H. and Tatsumi, K. 2007. Application of room temperature ionic liquids to Li batteries, *Electrochimica Acta*, 53, 1048–1054.

Vianney O.; Santos, Jr.; Melquizedeque B.; Alves, M.S.; Carvalho, Paulo, A.Z.S., and Joel C.R. (2006). *J. Phys. Chem. B*, 110, 20379-20385.

Welton, T. (1999). Room-Temperature Ionic Liquids. Solvents for Synthesis and Catalysis , *Chem. Rev.*, 99, 2071–2083.

Wasserscheid, P. and Keim, W. (2000). catalysis with polymer-stabilized colloids in biphasic systems, *Angew. Chem., Int. Ed.* , 39, 3772–3789.

Wasserscheid, P. and Welton,T. (2003). *Ionic Liquids in Synthesis*,Wiley-VCH, Weinheim, pp. 103–126.

Wang, F.X.; Pan, G.B,; Liu,Y.D.; Xiao,Y. (2010). Pb deposition onto Au(111) from acidic chloroaluminate ionic liquid, *Chemical Physics Letters*, 488,112–115, a.

Wang, F.X.; Pan, G.B,; Liu, Y.D.; Xiao,Y. (2010). Electrodeposition of Sb on Au(111) from an acidic chloroaluminate ionic Liquid: An in Situ STM Study, *J. Phys. Chem.* C 114, 4576–4579, b.

Xi, M. ; Duan, Y.; Li ,X.; Qu, L.; Sun, W.and Jiao, K. (20100. Carbon electrode modified with ionic liquid and multi-walled carbon nanotubes for voltammetric sensing of adenine, *Microchim Acta*, 170, 53–58.

Zhang, J.; Bond, A.M. (2005). Practical considerations associated with voltammetric studies in room temperature ionic liquids, *The Analyst*, 130,1132-1147.

Zhang, X.Z.; Jiao, K.; Wang, X.L. (2008). Paste Electrode Based on Short Single-Walled Carbon Nanotubes, and Room Temperature Ionic Liquid:Preparation, Characterization and Application in DNA Detection, *Electroanalysis*, 12, 1361 - 1366.

29

Application of Room Temperature Ionic Liquids in Electrochemical Sensors and Biosensors

Farnoush Faridbod,[1] Mohammad Reza Ganjali,[2,*]
Parviz Norouzi,[2] Siavash Riahi,[3,2] and Hamid Rashedi[4]
[1]*Endocrinology and Metabolism Research Center, Tehran University of Medical Sciences,*
[2]*Center of Excellence in electrochemistry, Faculty of Chemistry, University of Tehran,*
[3]*Institute of Petroleum Engineering, Faculty of Engineering, University of Tehran,*
[4]*Department of Chemical Engineering, Faculty of Engineering, University of Tehran,*
Tehran,
Iran

1. Introduction

Ionic Liquids (ILs) are a new class of purely ionic, salt-like materials that are liquid at unusually low temperatures. The official definition of ILs uses the boiling point of water as a point of reference: "Ionic Liquids are ionic compounds which are liquid below 100 °C". In particular, salts that are liquids at room temperature are called room-temperature ionic liquids (RTILs). RTILs, also known as organic liquid, molten, or fused salts, are a class of non-molecular ionic solvents with low melting points. The accepted definition of an RTIL is any salt that has a melting point lower than ambient temperature.

Ionic liquids (ILs) can be composed of a large number of cations and anions. Most common RTILs are composed of unsymmetrically substituted nitrogen-containing cations (e.g., imidazolium, pyridinium, pyrrolidinium) with organic or inorganic anions (e.g., Cl^-, PF_6^-, BF_4^-), Fig. 1.

First RTIL was reported by Wilkes et al. in 1982 (Wilkes et al., 1982). It was based on the 1-alkyl-3-methylimidazolium cation. Thereafter, many ILs containing a variety of cations and anions of different sizes have been synthesized for specific applications.

At the same time as ionic liquids become commercially available more and more (more than 350 ILs are now commercially available (Koel, 2009)), they show the interesting perspectives in different fields of researches such as catalysis, materials science, sensors, biosensors and separation technology. Since these techniques are in developing, there is always a need to design and synthesis many new ILs. An estimation predicts the number of possible ILs on order of 10^{18} (Koel, 2009).

A variety of cations and anions can be form ILs. The most common classes of ILs are imidazolium, pyridinium, pyrrolidinium, quaternary ammonium, and tetraalkylphosphonium ILs (Fig. 1). Of these, the most popular in experimental laboratory worldwide are undoubtedly 1,3-dialkylimidazolium salts, primarily due to the attractive and suitable physical properties.

The structure of IL and its interaction with the environment is extremely important for evaluating and selecting ILs for special application.

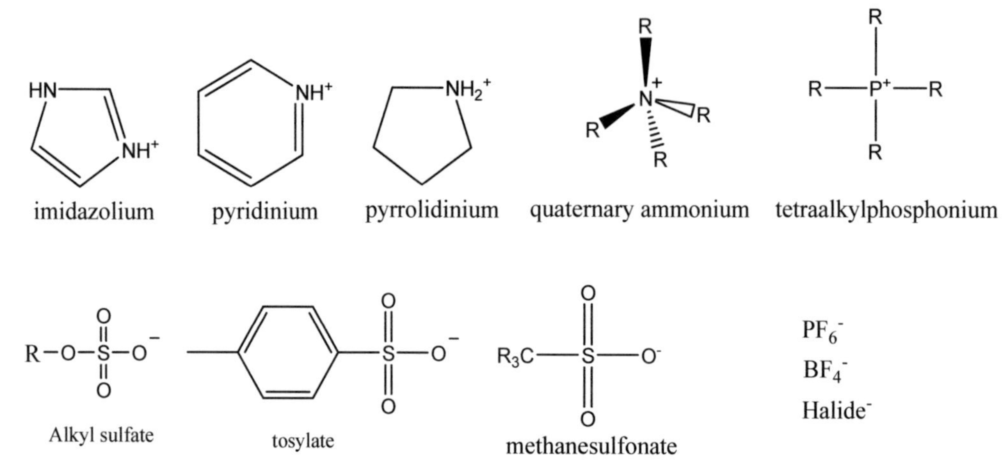

Fig. 1. The most common cations and anions used in construction of ILs

Electrochemical sensors and biosensors are the most rapidly growing class of chemical sensors, in analytical chemistry. Since the 1930s, these devices have found the most practical applicability in different fields due to the low cost, simplicity, portability, and high selectivity. A chemical sensor is a device that provides continuous information about some specific chemical properties of its environment. An ideal chemical sensor provides a certain type of response that is directly related to the quantity of a specific chemical species. Each chemical sensor composed of a transducer which is a device that provides continuous information about its environment, and converts the chemical response into a signal that can be detected by modern instrumentations. The second part which is the most important part of a chemical sensor is a chemically selective material, or the recognition element, which recognizes and differentiates between the response of the analyte and that of its environment. According to the type of signal transduction, chemical sensors can be classified as electrical, optical, mass or thermal sensors. Due to the considerable delectability, simplicity and low cost, electrochemical sensors receive more interest in comparison with optical, mass and thermal sensors which can be applied in a wide range of clinical, industrial, environmental and agricultural analyses.

An electrochemical biosensor is an analytical device which converts a biological response into an electrical signal. The term 'biosensor' is often used to cover sensor devices used to determine the concentration of substances and other parameters of biological process. Transduction of the biological or chemical signal in to the electrical signal can be done by amperometry, potentiometry and conductometry.

ILs due to their interesting properties have recently found various applications in construction of electrochemical sensors and biosensors in order to modify the responses. This chapter summarizes the properties of ILs and discusses their importance in electrochemistry. Then, it review ILs advantageous in construction of electrochemical sensors and biosensors.

2. General properties of ionic liquids

In general, ILs have some unique properties, such as a low vapor pressure, good thermal stability, high polarity, tunable viscosity and an ability to dissolve many compounds, a wide electrochemical window, the ability to dissolve many compounds high conductivity, high heat capacity and solvents available to control reactions. Although they have wide range of polarity and hydrogen-bonding ability, they are liquid from 180 K to 600 K.

The physical and chemical properties of ILs depend mostly on the nature and size of both their cation and anion constituents. The main structural factors of the cation are symmetry, charge density, number of carbon atoms in the alkane substituent and its flexibility, the rotational symmetry of the head ring, the cyclic and branched structures, and the functional tail group. Similar structural factors have an influence on the properties of anions, including charge delocalization either by a large volume of the central atom, or by the presence of the perfluoroalkyl chain.

ILs are mostly denser than water with values ranging from 1 for typical ILs to 2.3 g cm^{-3} for fluorinated ILs. Density strongly depends on the size of the ring in the cation, on the length of the alkyl chain in the cation, on the symmetry of ions and on the interaction forces between the cation and the anion. The ILs with aromatic head ring, in general, present greater densities than pyridinium head ring ILs and than do imidazolium ring ILs. By increasing the symmetry of the cations, density will increase. ILs with functional group reveals higher densities than those of alkyl chains.

Transport properties play an important role in chemical reactions, electrochemistry, and liquid–liquid extraction. This concerns mainly the viscosity of ILs and their solutions with molecular solvents. Compared with typical organic solvents, ILs are much more viscous. Viscosity of ILs, typically at the level of 10–500 cP at room temperature, is much higher than water ($\eta(H_2O)$ = 0.89 cP at 298.15 K) and aqueous solutions. The viscosity of ILs is determined by van der Waals forces and hydrogen bonded structures. Also, electrostatic forces can affect viscosity. The viscosities of the same class of ILs (with the anion held constant) increase as the alkyl group is lengthened. Increasing the alkyl chain length from butyl to octyl of a series of 1-alkyl-3-methylimidazolium cations, increases the hydrophobicity and the viscosity of the IL, while densities decrease. This is due to the stronger van der Waals forces between cations, leading to an increase in the energy required for molecular motion (Endress & Abedin, 2006). It is expected that the replacement of the alkyl chain by a hydroxyl functional group would increase the viscosity by increasing the H-bonding. As previously reported by Okoturo and Van der Noot viscosity temperature dependence in ILs is more complicated than in most molecular solvents. Most of ILs do not follow the typical Arrhenius behaviour. Most temperature studies fit the viscosity values into the Vogel–Tammann–Fulcher (VTF) equation, which adds an additional adjustable parameter (glass transition temperature) to the exponential term. In general, all ILs show a significant decrease in viscosity as the temperature increases (Okoturo and Van der Noot, 2004).

As a type of substances, ILs have been defined to have melting points (m.p.) below 373 K, however, most of them are liquid at room temperature. A typical ILs, e.g., 1-ethyl-3-methylimidazolium ethylsulfate has m.p. of <-20 °C) while a typical inorganic salt e.g. NaCl has m.p. of about 801°C. In this IL, the charge of the cation as well as the charge of the anion is distributed over a larger volume of the molecule by resonance. As a consequence, the solidification of the IL will occurred at lower temperature. However, in some long aliphatic side chains, a glass transition is observed instead of a melting point. In general, salts with a halogen anion, reveal to a higher melting temperature which are known as the precursors of ILs. Both, cations and anions have influence on the lowering melting points of ILs. Typically, the increase in anion size and its asymmetric substitution leads to a decrease in the melting

point. Also, the size and symmetry of the cation have an important impact on the melting points. For the longer alkane chain as C_{10}, the melting point increases. The short chain alkyl substituents in 1,3-dialkylimidazolium salts decreases the melting temperature.
Most of ILs have a glass transition temperature about 200 K (Domanska 2009). It is very interesting that also for the glass transition temperatures the structure of cations is important. However, the changes of glass transition temperatures with changing the length of the alkyl chain are much smaller than the melting temperatures. Common ILs are thermally stable up to 700 K. Thermal stability is limited by the same factors that contribute to the melting temperature. Branching the alkyl chain decreases the thermal stability of imidazolium ILs. Thermal stability increases with increasing anion charge density if the cationic charge density is also high.
Many synthetic reactions are faster in RTILs, and this saves time and or energy (Anderson et al., 2006). Compare to traditional organic solvents, a few volatile organic compounds are produced due to the little vapour pressure of ILs. The potential for explosions is minimized in RTIL because many RTILs have little or no flammability and no flash point. In many cases, products can be extracted from the RTIL after reaction, and the RTIL can be recovered and recycled.

3. The importance of ionic liquids in electrochemistry

Among the most important characteristics of ionic liquids, ionic conductivity, the width of the electrochemical potential window, viscosity, hydrophobicity and non-volatility cause the use of these solvents in electrochemical devices. In general, an ideal electrolyte should has high ionic conductivity ($>10^{-4}$ S/cm), fast ion mobility during redox reactions ($>10^{-14}$ $m^2/V.s$), large electrochemical potential windows (>1 V), and low volatility. RTILs exhibit many of these properties and characteristics.
Non-flammability, high ionic conductivity, electrochemical and thermal stability of ILs are a unique properties for used as an electrolyte in electrochemical devices like in batteries, capacitors, fuel cells, photovoltaics, actuators, and electrochemical sensors (Wei & Ivaska, 2008).
Most properties of ILs relevant to electrochemistry are centrally based on the following three properties: conductivity, viscosity, and electrochemical potential windows.

3.1 Large electrochemical window

One of the very important properties of ILs is their wide electrochemical potential window, which is a measure for their electrochemical stability against oxidation and reduction processes. In fact, the electrochemical potential window is a voltage range between which the electrolyte is not oxidized or reduced. This value, on the one hand, characterizes the electrochemical stability of ionic liquids, i.e., the limits of the window correspond to the start and the end of the electrochemical decomposition of the involved ions. On the other hand, the width of the electrochemical window governs the range of potentials available for the electrochemical processes not affecting the solvent.
The electrochemical potential window is sensitive to impurities. Halides are oxidized much easier than organic anions. In organic anions the negative charge is delocalized over larger volume. Thus, contamination with halides cause to lower electrochemical stabilities.

3.2 Ionic conductivity

The conductivity of an ionic liquid mainly depends on the mobility of its cation because in general the diffusion coefficients of ILs cations are higher than anions. Ionic liquids based on imidazolium and pyridinium cations have the highest ionic conductivity (~1 and 10^{-1} S/m, respectively) (Every et al., 2000). Typical RTILs have conductivities of >10^{-2} S/cm which are often useless as electrolyte. This can be problematic because IL electrolyte ions also migrate along the potential gradient. However, ILs possessing a zwitterionic structure in which the cation and anion are not expected to migrate with the potential gradient are useful in construction of an electrochemical cells. They showed much lower ionic conductivities in the range of 10^{-5}–10^{-7} S/cm. (Wilkes et al., 1982; Anderson et al., 2006). The electrochemically most stable materials having comparable small conductivities like N-butyl-N-methylpyrrolidinium bis(trifluoromethylsulfonyl)imide, triethylsulphonium bis(trifluoromethylsulfonyl)imide, and N-methyl-N-trioctylammonium bis(trifluoromethylsulfonyl)imide. These materials are good electrolytes for use in batteries, fuel cells, metal deposition, and electrochemical synthesis of nano-particles.

The ILs showing the highest conductivities, e.g. 1-ethyl-3-methylimidazolium thiocyanate and dicyanamide exhibit the lowest electrochemical stabilities. Nevertheless, these materials are good candidates for use in any application where a high conductivity combined with thermal stability and non-volatility is necessary.

When conductivity and electrochemical stability are both required in an application, e.g., in supercapacitors, sensors and biosensors, imidazolium-based ILs with stable anions e.g., tetrafluoroborate or trifluoromethylsulfonate are applied. Table 1 summarized the electrochemical conductivity of some common ionic liquids which can be suitable for electrochemical sensors and biosensors.

Entry	Name	Electrical conductivity (S/m) In 25 °C *
1	1-hexyl-3-methylimidazolium bis[(trifluoromethyl)sulfonyl]imide	0.26
2	1-butyl-3-methylimidazolium bis[(trifluoromethyl)sulfonyl]imide [C_4mim] [NTf_2]	0.40
3	1,3-dimethylimidazolium bis[(trifluoromethyl)sulfonyl]imide	0.74
4	1-butyl-1-methylpyrrolidinium bis[(trifluoromethyl)sulfonyl]imide	0.22
5	1-butyl-3-methylimidazolium hexafluorophosphate; [BMIM][PF_6]	0.14
6	1-butyl-3-methylimidazolium tetrafluoroborate ([BMIM][BF_4])	0.35
7	1-butyl-3-methylimidazolium methylsulfate	0.21
8	1-butyl-3-methylimidazolium trifluoroacetate	0.31

*Data are according to "Ionic Liquids Database - (ILThermo); http://ilthermo.boulder.nist.gov/ILThermo/pureprp.uix.do".

Table 1. Electrical conductivity of some ILs which can be used in construction of electrochemical sensors and biosensors

As it can be seen from Table 1, effect of ILs cation on electrical conductivity of the ILs is more than its anion. Nos. 3 to 5 have the same cations but the difference in the electrical conductivity are not too much. While, Nos. 1 to 4 the difference is more.

3.3 Hydrophobicity

Miscibility with water is often understood by hydrophobicity. The hydrophobicity mainly depends on the composition of ILs. From the point of view of hydrophobicity (solubility in water), ILs can be divided into two groups. Water-immiscible or hydrophobic IL like 1-butyl-3-methylimidazolium hexafluorophosphate ([BMIM][PF_6]) and 1-decyl-3-methylimidazolium bis(trifluoro-methylsulfonyl)imide ([DMIM][Tf_2N]) and water-miscible or hydrophilic ILs such as [BMIM][BF_4] (Wei & Ivaska, 2008). The first group due to the immiscibility with water are a good candidate for using in construction of electrochemical sensors and biosensors because these electrochemical devices contact water for a long period of operation. The second group are unstable in aqueous solutions and are not suitable for using in devices which are in contact with water.

The miscibility of ILs in water is strongly dependent on their anions (Wei & Ivaska, 2008). Cl^-, Br^-, I^-, NO_3^-, CH_3COO^- and CF_3COO^- are anions that make the ILs miscible with water. ILs composed of anions such as PF_6^- and Tf_2N^- are in immiscible with water. Miscibility of water of ILs based on anions such as BF_4^- and $CF_3SO_3^-$ is dependent on the structure of the cations, even though they in general are miscible with water. The miscibility will decrease with the increase in the cation chain length which is due to the increased surface activity of the longer chain cations (Fitchett et al. 2005a,b; Wei & Ivaska, 2008).

3.4 Viscosity

Since ILs have much higher viscosities than normal electrochemical supporting electrolytes, they have an effect on the diffusion coefficients of species. A previous reported study compares the diffusion coefficients of a neutral molecule and the radical cation produced after an electrochemical reaction in an IL and acetonitrile. The diffusion coefficient of the radical cation was consistently about half that of the neutral molecule. In contrast, in acetonitrile, the ratio of the diffusion coefficients was nearly 60% higher, an indication that viscosity and charge have considerable effects on the transport of diffusing species in IL solutions. Table 2 listed viscosity of some common ILs suitable for the electrochemistry.

Entry	Name	Viscosity (p) In 25 °C *
1	1,2-dimethyl-3-propylimidazolium tetrafluoroborate	0.377
2	1-butyl-3-methylimidazolium bis[(trifluoromethyl)sulfonyl]imide [C_4mim] [NTf_2]	0.069
3	1,3-dimethylimidazolium bis[(trifluoromethyl)sulfonyl]imide	0.047
4	1-butyl-1-methylpyrrolidinium bis[(trifluoromethyl)sulfonyl]imide	0.074
6	1-octyl-3-methylimidazolium bis[(trifluoromethyl)sulfonyl]imide	0.096
5	1-octyl-3-methylimidazolium tetrafluoroborate; [C_8mim] [BF_4]	0.439
7	1-butyl-3-methylimidazolium methylsulfate	0.21
8	1-butylpyridinium tetrafluoroborate	0.163

*Data are according to "Ionic Liquids Database - (ILThermo); http://ilthermo.boulder.nist.gov/ILThermo/pureprp.uix.do".

Table 2. The viscosity of some ILs suitable for electrochemistry

4. Ionic liquids used in electrochemical sensors

Some studies revealed that ILs can be useful as electrolytes in batteries, electrochemical cells, and electroplating. However, they have been recently used also in construction of electrochemical sensors and biosensors. An electrochemical sensor is an analytical device which converts a chemical response into an electrical signal. Most of the sensors which are used in electrochemical measurements and have ability to modify with ionic liquids are ion selective electrodes based on polymeric membrane, carbon paste electrodes, and all solid state electrodes.

4.1 Ion selective liquid membrane sensors

Ion selective electrodes (ISEs) have been developed over four decades as sensitive, inexpensive and handy electrochemical sensors to selectively determine the concentration (activity) of ions in aqueous media. Typical ion selective membrane sensors are composed of a hydrophobic plasticized polymeric membranes or films that are doped with one ionophore in addition to a lipophilic ion-exchanger that plays an important role to the sensor response. The membrane matrix should acts as a solvent of low viscosity for all active sensing materials in the membrane. Therefore, it is required to use a plasticizer that can reduce the glass transition temperature of the polymer to below room temperature and increase the elasticity of the polymeric membrane and helps providing mechanical stability. Plasticizer also can improve the solubility of the sensing materials in the membrane (Peng et al., 2008).

According to the mentioned above, ILs can be excellent materials to prepare ISEs membranes because they have polymer plasticizing ability and ionic nature. Hence, they can be used as an ionic additive and plasticizer at the same time.

One of the key factors which helps the ions extract to the liquid membrane is a plasticizer. It is well known that the selectivity and working concentration range of the membrane sensors are affected by the nature and amount of the plasticizer used. This is due to the influence of the plasticizer on the dielectric constant of the membrane phase, the mobility of the ionophore molecules. Dielectric constant of the plasticizer can affect the selectivity manner of the ion selective membrane electrode. For example, for the extraction of polar ions, a high polar plasticizer or solvent mediator is required, and in contrast, for extraction of a lipophilic cation a low polarity plasticizer is better. Sometimes, very polar plasticizers lead to the extraction of the polar interfering ions which may have negative effects on the selectivity behaviour of the sensor, thus, a low polarity solvent mediator is more suitable. A variety of dielectric constant and polarity of the ILs offers a various number of plasticizer choices. However, it should be noted that hydrophobic (immiscible) ionic liquids are suitable for use in liquid membrane.

Recently, tetraphenylborate derivatives are used as cationic additives and lipophilic tetraalkylammonium salts are applied as anion exchangers. The hydrophilic counter ions of these lipophilic additives are exchanged with the primary ion when the ISE is conditioned in specific aqueous solution.

The incorporation of lipophilic ILs in the ion-selective membrane electrode as an ion-exchanger diminishes the ohmic resistance and enhances the response behaviour and selectivity manner and also, in the case of the poor extraction capability, increases the sensitivity of the membrane electrodes.

In 2005, Coll et al. (Coll et al., 2005) reported using the hydrophobic IL ([BMIM][PF_6]) to prepare the poly(vinyl chloride) (PVC) membrane. A high selective response to sulfate anion was observed. They showed that ILs can be used as ionic additives in conventional ISE membranes. The ionophore used in this electrode polyazacycloalkane.
In 2006, Shvedene et al. (Shvedene et al., 2006) have used two ILs in different polymer membranes both as the plasticizer and the ion-exchange additive. The compounds 1-butyl-2,3-dimethylimidazolium bis(trifluoro-methylsulfonyl)imide ([BDMIM][Tf_2N]) and dodecylethyldiphenylphosphonium bis(trifluoro-methylsulfonyl)imide ([DEDPP] [Tf_2N]) were used to plasticize the PVC and poly(methyl methacrylate) (PMMA) membranes, respectively. Good and stable response to relatively hydrophobic cations and anions were obtained with the proposed membrane compositions.
In 2008, Nishi et al. showed that hydrophobic RTILs can be used as a non-volatile ionic medium for ion-selective liquid membrane sensors. They used a hydrophobic RTIL, trioctylmethylammonium bis(nonafluorobutylsulfonyl)imide ([$TOMA^+$][$C_4C_4N^-$]) in a K^+ ion selective liquid membrane sensor (Nishi et al., 2008). The phase-boundary potential at the interface between an aqueous KCl solution and ILs, shows the Nernstian response to K^+. Dicyclohexano-18-crown-6 ($DCH_{18}C_6$) was used as an ionophore in the membrane. The complex formation constant of K^+ with $DCH_{18}C_6$ in ([$TOMA^+$][$C_4C_4N^-$] is estimated to be on the order of 10^9 from the upper detection limit using a partition equilibrium model in the presence of a neutral ionophore.
In 2008, Peng et al. (Peng et al., 2008) also used RTILs as both ion-exchanger and plasticizer for PVC-based ion-selective membranes. 1-Methyl-3-octylimidazolium chloride (MOImCl) and trihexyltetradecylphosphonium chloride (THTDPCl) can plasticize PVC to form flexible ion-sensing membranes. PVC–MOImCl membrane without additional ionophore and ion-exchanger demonstrated Nernstian response to sulfate ion with slope of 29.1 mV/decade in the concentration range of 10^{-5} to 10^{-1} M. PVC-MOImCl-based electrode have fast response time within 10 s and wide pH independent range of 3–10. PVC–THTDPCl membrane exhibited stable and Nernstian response to different anions and the selectivity followed the Hofmeister series.

4.2 Carbon Paste Electrodes (CPEs)

Most of potentiometric carbon paste electrodes reported are based on incorporation of a sensing material into the carbon paste. The carbon paste usually consists of graphite powder dispersed in a non-conductive mineral oil such as paraffin.
Incorporation of mineral oil gives CPEs some disadvantages. Mineral oil is not component-fixed since it is derived from refining of petroleum and processing of crude oil. As a result, contaminants or matrix components may unpredictably effect on detection and analysis. In addition, the mechanical stability of CPEs rests somewhere between that of liquid membrane electrodes and solid state electrodes.
Carbon paste electrodes (CPEs) have attracted attention as ion selective electrodes mainly due to their improved renewability, stable response, and low ohmic resistance when compared to membrane electrodes (Javanbakht et al. 2007; Norouzi et al., 2010; Ganjali et al., 2009a, 2010).
Recently, RTILs have been widely used in construction of carbon paste electrodes (CPEs). These sensors are, in turn, called carbon ionic liquid electrodes (CILEs) (Safavi et al., 2007; Ganjali et al. 2009b,c; Faridbod et al. 2010). RTILs are a good choice as binders in carbon

paste electrodes due to their chemical stability, low vapor pressure, low toxicity, low melting temperature, high ionic conductivity and good electrochemical and thermal stability (Maleki et al., 2006). Using room temperature ionic liquids instead of paraffin oil in the carbon paste yields more efficient extraction of ions with high charge density into the carbon paste surface. This is due to the much higher dielectric constant of the ionic liquids as binder compared to paraffin oil (Ganjali et al. 2009b,c; Faridbod et al. 2010).

Due to its mechanical strength, the IL CPE could be applied as an effective flow-through detector in flowing streams, and since a mixture of IL and graphite is easily moldable, the fabrication of different electrode geometries is completely feasible. The favorable electrochemical response, high reversibility, sensitivity, and selectivity observed for these electrodes toward molecules together with its resistance to electrode fouling make it an excellent candidate for the construction of a new generation of sensors.

Recently, carbon nanotubes (CNTs) have also been added to the carbon paste (Ganjali et al. 2009b,c; Faridbod et al. 2010). CNTs have very interesting physicochemical properties, such as an ordered structure with a high aspect ratio, ultra-light weight, high mechanical strength, high electrical conductivity, high thermal conductivity, metallic or semi-metallic behavior and high surface area (Ajayan 1999). The combination of these characteristics makes CNTs unique materials with the potential for diverse applications.

Our research group, recently used [bmim]BF_4 in order to modify the response of an erbium carbon paste potentiometric electrode (Faridbod et al., 2010). The general procedure to prepare the carbon paste electrode was as follows: Different amounts of the ionophore [5-(dimethylamino) naphthalene-1-sulfonyl-4-phenylsemicarbazide] (NSP), along with an appropriate amount of graphite powder, ionic liquid and MWCNTs were thoroughly mixed. The resulting mixture was transferred into a glass tube with 5mm i.d. and a height of 3 cm. After homogenization of the mixture, the paste was carefully packed into the tube tip to avoid possible air gaps, which often enhance the electrode resistance. A copper wire was inserted into the opposite end of the CPE to establish electrical contact. The external surface of the carbon paste was smoothed with soft paper. A new surface was produced by scraping out the old surface and replacing the new carbon paste. The electrode was finally conditioned for 48 h by soaking it in a 1.0×10^{-3} mol L^{-1} $Er(NO_3)_3$ solution. Using RTILs instead of paraffin oil in the carbon paste yield smore efficient extraction of Er(III) (which is a cation with high charge density) into the CPE. This is due to the much higher dielectric constant of the RTIL binder when compared to paraffin oil. Using MWCNTs in the carbon paste improves the conductivity and, therefore, conversion of the chemical signal to an electrical signal. Carbon nanotubes have many properties that make them ideal as components in electrical circuits, including their unique dimensions and their unusual current conduction mechanism. The performance of the Er(III) carbon paste sensor can be greatly improved by using RTIL instead of mineral oil (paraffin), and also by using MWCNTs as enhanced signal transducers. The modified CPEs show better potentiometric response than unmodified CPEs in terms of sensitivity, Nernstian slope, linear range, and response stability. These are important characteristics of ion selective electrodes. The proposed sensor exhibits a long lifetime (about three months). The best performance for the modified sensor was obtained with an electrode composition of 20% [bmim]BF4, 20% NSP, 45% graphite powder and 15% MWCNT. This particular sensor formulation exhibits a Nernstian response (19.8±0.3mVdecade−1) toward Er(III) ions in the range of 1.0×10−7 to 1.0×10−1 mol L−1 with a detection limit of 5.0×10−8 mol L−1. The proposed modified Er(III) sensor can be used over the pH range from 3.5 to 9.0.

In addition to potentiometric measurements, carbon paste modified with ILs have more been used in other electrochemical sensors, particularly voltammetric sensor. For the first time, in 2005, Liu et al. introduced using imidazolium salt functionalized by polyelectrolyte in carbon paste electrodes (Liu et al., 2005). In voltammetric carbon paste sensors, as well as advantages mentioned in potentiometric carbon paste sensors, addition of traces of the ILs increased the electrocatalytic activity of the electrode. Maleki et al. used N-octylpyridinium hexafluorophosphate (OPFP) as a binder in a carbon paste electrode (Maleki et al., 2006). This type of electrode has lower ohmic resistance than CPE and gives very reproducible and sensitive voltammetric results. The electrocatalytic activity of the ILs 1-octyl-3-methylimidazolium hexafluorophosphate [OMIM][PF_6] in carbon paste electrode was investigated by using the redox probe $Fe(CN)_6^{3-/4-}$ (Maleki et al., 2007). Trace amount of chloride has been measured by linear sweep, square wave and cathodic stripping voltammetry using [BMIM][BF_4], [BMIM][Tf_2N] and [BMIM][PF_6] (Villagran et al., 2004). In another report a carbon paste base ILs was used in a flow-injection system with a voltammetric detector (Shen et al. 2007). The electroactive compounds to be determined can diffuse better from the eluent into the thin layer of the ionic liquid on the surface of a carbon paste electrode as a working electrode.

4.3 All solid state sensors

All-solid-state sensors are a kind of potentiometric sensors which are more durable and can be miniaturized. The potential of the sensor to be miniaturized is an additional requirement that can give a lot of new applications to this device. Solid state ISEs are based on conducting polymers as the transduction layer in the electrode construction. Although the fabrication of all-solid-state sensor, which do not require any internal filling solutions, is one way to achieve durable sensors, the design of a proper solid contact between the ion-selective membrane and the electronic conductor is a difficult challenge in the way of obtaining reliable all solid-state electrodes.

Application of ILs in construction of all-solid-state sensor has recently been reported by Maminska et al. (Maminska et al., 2006). They used 1-dodecyl-3-methylimidazolium chloride [DMIM][Cl] in PVC membranes. In this way, the electrode showed good potential stability and reproducibility. In 2007, Kakiuchi et al. have also introduced a new solid-state reference electrode (Kakiuchi et al., 2007). A new type of Ag/AgCl reference electrode is consists of a Ag/AgCl electrode coated with a AgCl-saturated with a hydrophobic ionic liquid ionic liquid, 1-methyl-3-octylimidazolium bis(trifluoromethylsulfonyl)imide ([C_8mim^+][$C_1C_1N^-$]), instead of the internal aqueous solution. The [C_8mim^+][$C_1C_1N^-$] phase plays dual roles, that is, as a medium dissolving AgCl and an ionic-liquid-type salt bridge upon contact with an aqueous solution. This new class of reference electrodes opens the way for a variety of miniaturized and solid-state reference electrodes.

There is another report in 2009 (Chernyshov et al. 2009) in which ILs that melt slightly above room temperature (and may be called low-melting ionic solids, LMISs) were used as sensing materials for the detection of ions in aqueous solutions. In this work, Chernyshov et al. used a simple method to prepare solid-contact ion-sensitive electrodes based on the consecutive melting and further solidification of the LMISs. Indeed, instead of using plasticizer, ion-exchanger polymeric matrix, they used LMIS in the potentiometric sensor. A potentiometric response toward a number of anions and the possibility of altering the selectivity by incorporating additional ionophores into the LMIS matrix was studied in this work.

5. Ionic liquids used in electrochemical biosensors

An electrochemical biosensor is an analytical device which converts a biological response into an electrical signal. In general, it is difficult to exchange the electron between an enzyme and solid surface of the electrodes directly. This is because of the inaccessibility of its redox center and loss of bioactivity of the enzyme due to the conformational changes by adsorption on the electrode surface. ILs have shown good compatibility with biomolecules and enzymes, and even whole cells. Thus, ILs can be used in electrochemical biosensors typically as both binder and conductor. Common ILs advantages observed when they incorporate into the biosensors include higher conductivity, good biocatalytic ability (Park, S. & Kazlauskas, 2003), long-term stability (including stability at high temperature), superior sensitivity, improved linearity, better selectivity, and the ability to fabricate third-generation biosensors with direct (without using mediator) electron transfer between protein and electrode.

ILs have a wide electrochemical potential window. The difference between the potentials of their anodic (E_a) and cathodic (E_c) decomposition is usually greater than 3 V (Ohno, 2005), while for aqueous electrolytes is about 1.2 V. Because of this remarkable property of ionic liquids, they find wide use in electrochemical biosensors.

Recently, some authors have reported increased stability of enzymes in ILs compared with stability in some organic solvents (Lozano et al., 2001; Laszio, et al. 2002; Park, S. & Kazlauskas, 2003; Persson & Bornscheuer, 2003). ILs were also found to act as agents to stabilize proteins effectively at high temperatures (Baker et al., 2004). [BMIM][BF_4], [BMIM][PF_6], [OMIM][PF_6], 1-(2-hydroxyethyl)-3-methylimidazolium tetrafluoroborate [HEMIM][BF_4], and Butylpyridinium hexafluorophosphate ([BP][PF_6]) was widely used in construction of electrochemical biosensors due to their very good biocompatibility.

Most investigations to date have focused on the amperometrically determined biocatalytic activity of common enzymes such as glucose oxidase (GOx), horseradish peroxidase (HRP), or various other heme proteins (e.g., Hb, Mb) incorporated into electrodes. Laszlo and Compton have also reported the catalysis of hemin activated by an electron acceptor in IL solutions and it was found that the activity of hemin increased with the enhanced amount of IL in the methanol-IL system (Laszlo & Compton, 2002). Dramatic enhanced activity and thermal stability of horseradish peroxide (HRP) were obtained when it was immobilized it in the [BMIM][BF_4] based sol–gel matrix (Liu et al. 2005a). Direct electrochemical response of HRP (Liu et al. 2005b), myoglobin (Ding et al., 2007) and Hb (Sun et al., 2007) have been observed on IL modified electrodes. Direct electrochemical reduction of hemin has been studied by cyclic voltammetry and chronocoulometry in the ILs, [BMIM][PF_6] and [OMIM][PF_6] (Compton & Laszlo, 2002).

Nafion films have been used widely in construction of electrochemical biosensors. Nafion, due to its easy fabrication, good electrical conductivity, high chemical stability and good biocompatibility, has been used as a protective coating material for enzyme immobilization. Mixture of nafion with ILs can improved the coating ability and stability. Nafion–[BMIM][PF_6] composite film has been reported to help immobilization of enzyme HRP on the glassy carbon electrode (Chen et al. 2007). Paraffin can also be replaced by [BMIM][PF_6] as binder in H_2O_2 and nitrite carbon paste biosensor (Sun et al., 2007).

Yu et al also reported that the water-miscible imidazolium-based ILs can interact with glassy carbon electrode and form molecular films on the electrode surface (Yu et al. 2005). Various approaches have been investigated to use ionic liquids with carbon nanotube in biosensors. One interesting approach involves the use of multiwall carbon nanotube–IL (MWCN-IL) modified glassy carbon electrodes (GCEs). In this approach, MWCNs was thoroughly mixed with the ILs, by grinding them together in a mortar to create a gel-like paste which was then applied to the surface of a cleaned GC electrode. Using a platinum wire and a saturated calomel electrode as auxiliary and reference electrodes, respectively, cyclic voltammograms (CVs) can be measured. An immediate advantage of the MWCN-IL is a larger peak current with smaller peak separations, an indication of faster electron transport to the electrode surface.

In 2010, a nano-composite material consisting of amine functionalized multi-walled carbon nanotubes and a room temperature ionic-liquid, 1-butyl-3-methylimidazolium tetrafluoroborate, was reported to use in construction of a novel catalase based biosensor for the measurement of hydrogen peroxide. The modified electrode exhibited a quasi-reversible cyclic voltammogram corresponding to the Fe(II)/Fe(III) redox couple in the heme prosthetic group of catalase with a formal potential of −460 mV in 0.1 M phosphate buffer solution at pH=7.0. The nano-composite film showed an obvious promotion of the direct electron transfer between catalase and the underlying electrode. The apparent charge transfer rate constant and transfer coefficient for electron transfer between the electrode surface and enzyme were reported as 2.23 s^{-1} and 0.45, respectively. The immobilized catalase exhibited a relatively high sensitivity (4.9 nA/nM) toward hydrogen peroxide (Rahimi et al., 2010).

Table 3, summarized some important ILs which are immiscible in water and they have been widely used in construction of electrochemical sensors and biosensors.

6. Conclusion

In this chapter, besides a brief discussion of the properties of ILs, the application of ILs in electrochemical sensors and biosensors are reviewed. ILs are liquids with non-volatility, high ion conductivity, the ability to dissolve many compounds, thermal stability, high viscosity, high polarity, and low vapor pressure. Ionic liquids have many applications, such as powerful solvents and electrically conducting electrolytes. Nowadays, electrochemical sensors and biosensors can be powerful tools for analysis of different species. Due to the some special characterization of ionic liquids such as wide potential windows (a voltage range between which the electrolyte is not oxidize or reduced.) and high electrical conductivity, hydrophobicity and the insolubility in water, the extraction and plasticizing ability, they are used in construction of electrochemical sensors and biosensors. They can be applied for improvement and modification of the composite materials in an electrochemical sensor or biosensor. Using room temperature ionic liquids instead of paraffin oil in the carbon paste electrodes yields more efficient extraction of ions with high charge density into the carbon paste surface. This is due to the much higher dielectric constant of the ionic liquids as binder compared to paraffin oil. Also, in the case of voltammetric measurements, the electroactive compounds to be determined can better diffuse from the eluent into the thin layer of the ionic liquid on the surface of a carbon paste electrode. In addition, ILs can be used in electrochemical biosensors typically as both binder and conductor. Common ILs advantages observed when they incorporate into the biosensors include higher conductivity, good biocatalytic ability, long-term stability, superior sensitivity, improved linearity, better selectivity, and the ability to direct (without using mediator) electron transfer between protein and electrode.

Name	Structure	M.P. (°C)	Viscosity (cP) in 20°C	Application	Reference
1-butyl-3-methylimidazolium hexafluorophosphate ([BMIM][PF_6])	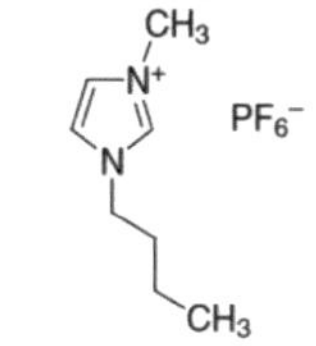	11	272.7	ISE	Coll et al., 2005
1 butyl 2,3-dimethylimidazolium bis(trifluoro-methylsulfonyl)imide ([BDMIM][Tf_2N])	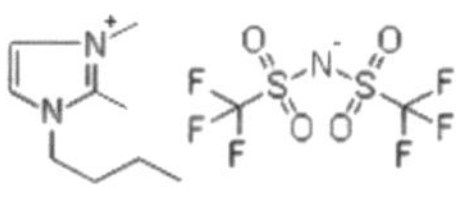	-15	-	ISE	Shvedene et al., 2006
1-Methyl-3-octylimidazolium chloride (MOImCl)	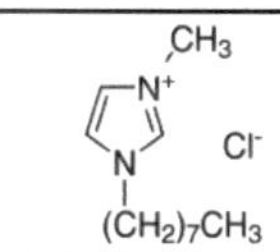	0	-	ISE	Peng et al., 2008
1-dodecyl-3-methylimidazolium chloride [DMIM][Cl]	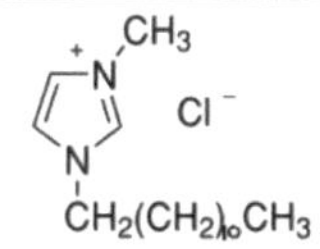	102	-	All-solid-state	Maminska et al., 2006
1-octyl-3-methylimidazolium hexafluorophosphate [OMIM][PF_6]	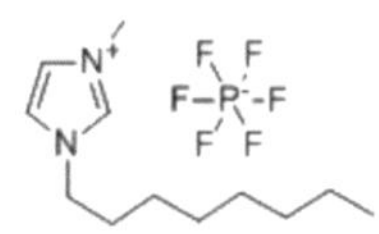	-	807.1	Voltammetric carbon paste sensor	Maleki et al., 2007
1-butyl-3-methylimidazolium tetrafluoroborate ([BMIM][BF_4])	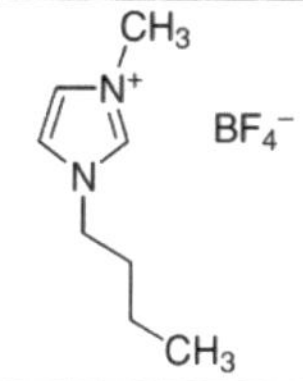	-71	99.9	1-CPE 2- Biosensor	Ganjali et. al, 2009b,c; Faridbod et al., 2010; Rahimi et al., 2010

Table 3. Some common ILs which are immiscible in water and can be used in electrochemical sensors and biosensors

7. References

Ajayan, P.M. (1999). Nanotubes from Carbon. *Chem. Rev.*, 99, 7, 1787-1800.

Anderson, J. L.; Armstrong, D. W. & Wei, G. (2006). Ionic Liquids in Analytical Chemistry, *Anal. Chem.*, 78, 9, 2893-2902.

Baker, S.N.; McCleskey, T.M.; Pandey, S. & Baker, G.A.; (2004). Fluorescence study of protein thermostability in ionic liquids. *Chem. Commun.* 940-941.

Chen, H.; Wang, Y.; Liu, Y.; Wang, Y.; Qi, L. & Dong, S. (2007). Direct electrochemistry and electrocatalysis of horseradish peroxidase immobilized in Nafion-RTIL composite film. *Electrochem. Commun.* 9, 3, 469-474.

Chernyshov, D.V.; Egorov, V. M.; Shvedene, N. V. & Pletnev, I. V. (2009). Low-Melting Ionic Solids: Versatile Materials for Ion-Sensing Devices. *Appl. Mater. Interfaces*, 1, 9, 2055–2059.

Coll, C.; Labrador, R.H.; Manez, R.M.; Soto, J.; Sancenon, F.; Segui, M.J. & Sanchez, E. (2005). Ionic liquids promote selective responses towards the highly hydrophilic anion sulfate in PVC membrane ion-selective electrodes*Chem. Commun.*, 3033-3035.

Compton, D.L. & Laszlo, J.A. (2002). Direct electrochemical reduction of hemin in imidazolium-based ionic liquids. *J. Electroanal. Chem.*, 520, 71-78.

Ding, S.; Xu, M.; Zhao, G. & Wei, X. (2007). Direct electrochemical response of Myoglobin using a room temperature ionic liquid, 1-(2-hydroxyethyl)-3-methyl imidazolium tetrafluoroborate, as supporting electrolyte. *Electrochem. Commun.* 9, 2, 216-220.

Domanska, U. (2009). Ionic Liquids in Chemical Analysis, Edited by Mihkel Koel , CRC Press, Chapter 1, Pages 1-60, ISBN: 978-1-4200-4646-5, eBook ISBN: 978-1-4200-4647-2, DOI: 10.1201/9781420046472.ch1

Endress, F. & El Abedin, S.Z. (2006). Air and water stable ionic liquids in physical chemistry. *Phys. Chem. Chem. Phys.*, 8, 2101-2116.

Evans, R. G.; Klymenko, O.V.; Price, P.D.; Davies, S.G.; Hardacre, C. & Compton, R.G. (2005). An electrochemical and ESR spectroscopic study on the molecular dynamics of TEMPO in room temperature ionic liquid solvents. *Chem. Phys. Chem.* 6, 526–533.

Every, H.; Bishop, A.G.; Forsyth, M. & MacFarlane D.R. (2000). Ion diffusion in molten salt mixtures. *Electrochim. Acta*, 45, 1279-1284.

Faridbod, F.; Ganjali, M. R.; Larijani, B. & Norouzi, P. (2009). Multi-walled carbon nanotubes (MWCNTs) and room temperature ionic liquids (RTILs) carbon paste Er(III) sensor based on a new derivative of dansyl chloride. *Electrochim. Acta*, 55, 1, 234-239.

Fitchett, B.D.; Rollins, J.B. & Conboy, J.C. (2005a). Interfacial Tension and Electrocapillary Measurements of the Room Temperature Ionic Liquid/Aqueous Interface. *Langmuir* 21, 12179-12186.

Fitchett, B. D.; Rollins, J. B. & Conboy, J. C. (2005b). 1-Alkyl-3-Methylimidazolium Bis(perfluoroalkylsulfonyl)imide Water-Immiscible Ionic Liquids: Electrochemical Properties at Liquid-Liquid Interfaces. *J. Electrochem. Soc.*, 152, E251-E258.

Ganjali, M. R.; Motakef-Kazemi, N.; Norouzi, P.; & Khoee, S. (2009a). A Modified Ho^{3+} Carbon Paste Electrode Based on Multi-walled Carbon Nanotubes (MWCNTs) and Nanosilica. *Int. J. Electrochem. Sci.*, 4, 7, 906-913.

Ganjali, M. R.; Khoshsafar, H.; Shirzadmehr, A.; Javanbakht, M. & Faridbod, F. (2009b). Improvement of Carbon Paste Ion Selective Electrode Response by Using Room Temperature Ionic Liquids (RTILs) and Multi-Walled Carbon Nanotubes (MWCNTs). *Int. J. Electrochem. Sci.*, 4, 3, 435-443.

Ganjali, M. R.; Khoshsafar, H.; Faridbod, F.; Shirzadmehr, A.; Javanbakht, M. & Norouzi, P. (2009c). Room Temperature Ionic Liquids (RTILs) and Multiwalled Carbon Nanotubes (MWCNTs) as Modifiers for Improvement of Carbon Paste Ion Selective Electrode Response; A Comparison Study with PVC Membrane. *Electroanalysis*, 21, 19, 2175-2178.

Ganjali, M. R.; Motakef-Kazami, N.; Faridbod, F.; Khoee, S. & Norouzi, P. (2010). Determination of Pb^{2+} ions by a modified carbon paste electrode based on multi-walled carbon nanotubes (MWCNTs) and nanosilica. *J. Hazard. Mater.* 173, 1-3, 415-419.

Kakiuchi, T. ; Yoshimatsu, T. & Naoya, N. (2007). New class of Ag/AgCl electrodes based on hydrophobic ionic liquid saturated with AgCl. *Anal. Chem.* 79, 18, 7187-7191.

Koel, M. (2009). Ionic liquids in chemical analysis, In: *Analytical Chemistry Series*, first edition, CRC Press Taylor & Francis Group, ISBN-13: 978-1-4200-4646-5, USA.

Javanbakht, M.; Ganjali, M. R.; Norouzi, P.; Badiei, A.; Hasheminasab, A. & Abdouss, M. (2007). Carbon paste electrode modified with functionalized nanoporous silica gel as a new sensor for determination of silver ion. *Electroanalysis*, 19, 12, 1307-1314.

Laszlo, J.A. & Compton, D.L. (2002). Comparison of peroxidase activities of hemin, cytochrome c and microperoxidase-11 in molecular solvents and imidazolium-based ionic liquids. *J. Mol. Catal. B: Enzym.*, 18, 109-120.

Liu, H.; He, P.; Sun, C.; Shi, L.; Liu, Y.; Zhu, G.; Li, J. (2005). An ionic liquid-type carbon paste electrode and its polyoxometalate-modified properties. *Electrochem. Commun.*, 7, 12, 1357-1363.

Liu, Y.; Wang, M.; Li, J.; Li, Z.; He, P.; Liu, H. & Li, J. (2005a). Highly active horseradish peroxidase immobilized in 1-butyl-3-methylimidazolium tetrafluoroborate room-temperature ionic liquid based sol–gel host materials. *Chem. Commun.*, 13, 1778-1780.

Liu, Y.; Shi, L.; Wang, M.; Li, Z.; Liu, H. & Li, J. (2005b). A novel room temperature ionic liquid sol-gel matrix for amperometric biosensor application. *Green Chem.*, 7, 655-658.

Lozano, P.L.; Diego, T.D.; Carrie, D.; Vaultier, M. & Iborra, J.L. (2001). Stabilization of a-chymotrypsin by ionic liquids in Transesterification. *Biotechnol. Bioeng.* 75, 563-569.

Maleki, N.; Safavi, A. & Tajabadi, F. (2006). High-performance carbon compo- site electrode based on an ionic liquid as a binder. *Anal. Chem.*, 78, 3820-3826.

Maleki, N. ; Safavi, A. & Tajabadi, F. (2007). Investigation of the role of ionic liquids in imparting electrocatalytic behavior to carbon paste electrode.*Electroanalysis*, 19, 21, 2247-2250.

Maminska, R.; Dybko, A. & Wroblewski, W. (2006). All-solid-state miniaturised planar reference electrodes based on ionic liquids. *Sens. Actuators B*, 115, 552-557.

Nishi, N.; Murakami, H.; Yasui, Y. & Kakiuchi, T. (2008). Use of Highly Hydrophobic Ionic Liquids for Ion-selective Electrodes of the Liquid Membrane Type. *Anal. Sci.*, 24, 10, 1315-1320.

Norouzi, P.; Rafiei-Sarmazdeh, Z.; Faridbod, F.; Adibi, M. & Ganjali, M. R. (2010). Er^{3+} Carbon Paste Electrode Based on New Nano-composite. *Int. J. Electrochem.* Soc., 5, 3, 367-376.

Ohno H., Electrochemical Aspects of Ionic Liquids, Ed., Hoboken: Wiley, 2005.

Okoturo, O.O. & Van der Noot, T.J. (2004). Temperature dependence of viscosity for room temperature ionic liquids, *J. Electroanal. Chem.*, 568, 167-181.

Park, S. & Kazlauskas, R. J. (2003). Biocatalysis in ionic liquids–advantages beyond green technology. *Current Opinion in Biotechnology*, 14, 432–437.

Penga, B.; Zhub, J.; Liua, X. & Qina, Y. (2008). Potentiometric response of ion-selective membranes with ionic liquids as ion-exchanger and plasticizer. *Sens. Actuators B*, 133, 308–314.

Persson, M. & Bornscheuer, U.T. (2003). Increased stability of an esterase from Bacillus stearothermophilus in ionic liquids as compared to organic solvents. *J. Mol. Catal. B: Enzym.*, 22, 21-27.

Rahimi, P.; Rafiee-Pour, H. A.; Ghourchian, H.; Norouzi, P. & Ganjali, M. R. (2010). Ionic-liquid/NH2-MWCNTs as a highly sensitive nano-composite for catalase direct electrochemistry. *Biosens. Bioelectron.*, 25, 1301-1306.

Safavi, A.; Maleki, N.; Honarasa, F.; Tajabadi, F. & Sedaghatpour, F. (2007). Ionic Liquids Modify the Performance of Carbon Based Potentiometric Sensors. *Electroanalysis* 19, 5, 582-586.

Shen, Y. ; Zhang, Y.; Qui, X.; Guo, H.; Niu, L. & Ivaska, A. (2007). Polyelectrolyte functionalized ionic liquid for electrochemistry in supporting electrolyte free aqueous solutions and application in amperometric flow injection analysis. *Green Chem.*, 9, 746-753.

Shvedene, N.V.; Khrenova, M.G.; Chernyshov, D.V.; Formanovsky, A.A.; Baulin, V.E.; Pletnev, I.V. (2006). Ionic liquids plasticize and bring ion-sensing ability to polymer membranes of selective electrodes. *Electroanalysis* 18, 13-14, 1416–1421.

Sun, W.; Wang, D.; Gao, R.; Jiao, K. (2007). Direct electrochemistry and electrocatalysis of hemoglobin in sodium alginate film on a BMIMPF6 modified carbon paste electrode. *Electrochem. Commun.*, 9, 5, 1159-1164.

Villagran, C.; Banks, C.E.; Hardacre, C. & Compton, R.G. (2004) Electroanalytical determination of trace chloride in room-temperature ionic liquids. *Anal. Chem.* 76, 1998-2003.

Wei, D. & Ivaska, A. (2008). Applications of ionic liquids in electrochemical sensors. *Anal. Chim. Acta*, 607, 126-135.

Wilkes, J. S.; Levisky, J.A.; Wilson, R.A. & Charles L.H. (1982). Diaklyimidazolium chloroaluminate melts: A new class of room-temperature ionic liquids for electrochemistry, spectroscopy, and synthesis. *Inorg. Chem.* 21, 1263–1264.

Yu, P.; Lin, Y.Q.; Xiang, L.; Su, L.; Zhang, J. & Mao, L.Q. (2005). Molecular Films of Water-Miscible Ionic Liquids Formed on Glassy Carbon Electrodes: Characterization and Electrochemical Applications. *Langmuir*, 21, 20, 9000-9006.

Electrochemical Studies on Uranyl(VI) Species in 1-Butyl-3-methylimidazolium Based Ionic Liquids and Their Application to Pyro-Reprocessing and Treatment of Wastes Contaminated with Uranium

Yasuhisa Ikeda[1], Noriko Asanuma[2] and Yusuke Ohashi[3]
[1]*Research Laboratory for Nuclear Reactors, Tokyo Institute of Technology,*
[2]*Department of Energy Science and Engineering, Tokai University,*
[3]*Ningyo-toge, Environment Engineering Center, Japan Atomic Energy Agency,*
Japan

1. Introduction

Room temperature ionic liquids (ILs) have been paid attention as environmentally benign media, because they have attractive properties such as thermal stability, nonflammability, high ionic conductivity, and wide electrochemical potential windows (Earle & Seddon, 2000; Rogers et al., 2000; Wasserscheid & Welton, 2003). In the nuclear industry field, ILs are expected to be applied as media for reprocessing of spent nuclear fuels and treatment of radioactive wastes contaminated with radioactive nuclides (Bladley et al., 2002; Rogers et al., 2002; Giridhar et al., 2006; Cocalia et al., 2006; Giridhar et al., 2007; Binnemans, 2007).
In this chapter, our feasibility studies on applications of ILs as the media of pyro-reprocessing processes and the treatment method of radioactive wastes contaminated with uranium will be introduced.

2. Investigation on application of ILs as media of the pyro-reprocessing processes

We studied electrochemical properties of uranyl species in 1-butyl-3-methylimidazolium (BMI) based ILs (BMICl, $BMIBF_4$, and BMINfO (NfO = nonafluorobutanesulfonate)) to examine their feasibility as alternatives to conventional molten salts as media for pyro-reprocessing processes for spent nuclear fuels.
BMICl (Kanto Chemical Co., Inc.) was used without further purification and $BMIBF_4$ (Kanto) was purified by using activated carbon. BMINfO was synthesized as follows: 1-Butyl-3-methylimidazole was dissolved into tetrahydrofuran (THF) and stirred vigorously. After that, 1-buromo butane was dropped slowly with a dropping funnel, and the resulting solution was refluxed. After refluxing, THF phase was separated and stirred with ethylacetate (EA). Crude 1-butyl-3-methylimidazolium bromide (BMIBr) was obtained by removing THF and EA *in vacuo*. The crude BMIBr was dissolved into distilled water and

stirred vigorously. To this solution, KNfO was added. The resulting solution was refluxed with stirring at 70 °C. The BMINfO phase was separated from aqueous one and mixed with activated carbon for removing organic impurities. After filtration, the filtrate was mixed with the distilled water for striping inorganic impurities. In order to remove water and volatile impurities, all ILs used were kept for more than 3 h under reduced pressure at 120 °C.

Sample solutions for electrochemical experiments were prepared by dissolving $Cs_2UO_2Cl_4$ or $UO_2Cl_2{\cdot}nH_2O$ (n = 1~3) into ILs. $Cs_2UO_2Cl_4$ and $UO_2Cl_2{\cdot}nH_2O$ (n = 1 ~ 3) were synthesized according to the reported procedures (Cordfunke, 1969; Denning et al., 1979). Cyclic voltammograms (CV) of sample solutions controlled at appropriate temperatures were measured by using an electrochemical analyzer (BAS, ALS model 660B) in glove box under an Ar atmosphere. A glassy carbon and a Pt wire were used as a working electrode and a counter electrode, respectively. As a reference electrode, an Ag/AgCl electrode was used and connected with a cyclic voltammetry cell by a liquid junction filled with $BMIBF_4$ or BMINfO. All potentials reported here are *vs.* Ag/AgCl. In the CV measurements, potential was swept to cathodic direction initially.

2.1 Electrochemical study on uranyl chloride in BMICl

Figure 1(a) shows the UV-visible absorption spectrum of the solution prepared by dissolving $Cs_2UO_2Cl_4$ into BMICl at 80°C. This absorption spectrum is found to exhibit remarkable vibrational fine structure, which is similar to that of $[UO_2Cl_4]^{2-}$ in $AlCl_3$/EMIC (EMIC = 1-ethyl-3-methylimidazolium chloride) (Dai et al., 1997), $BMITf_2N$, $MeBu_3NTf_2N$, and C_4minTf_2N (Tf_2N = bis(trifluoromethanesulfonyl)imide, $MeBu_3N$ = tri-*n*-butylmethyl-ammonium, C_4min = 1-hexyl-3-methylimidazolium) (Sornein et al., 2006; Nockemann *et al.*, 2007). Nockemann et al. have reported that the fine structure is typical for the $[UO_2Cl_4]^{2-}$ with D_{4h} coordination symmetry (Nockemann et al., 2007). The molar absorption coefficient (ε) at 429.8 nm (maximum peak) is 14.5 M^{-1} cm^{-1} (M = mol dm^{-3}), and is almost same as those reported previously (Sornein et al., 2006; Nockemann et al., 2007). We also measured the UV-visible absorption spectrum of the solution obtained by dissolving $UO_2Cl_2{\cdot}nH_2O$ in BMICl at 80 °C (see Fig. 1(b)), where the concentration of uranium was determined by ICP-AES. A similar absorption spectrum to that in Fig. 1(a) was observed. The ε value of maximum peak at 428.6 nm was 16.3 M^{-1} cm^{-1}. These results indicate that the uranyl species in solutions prepared by dissolving $Cs_2UO_2Cl_4$ or $UO_2Cl_2{\cdot}nH_2O$ into BMICl is $[UO_2Cl_4]^{2-}$. Slight differences in the ε values and the wavelength of peak maxima in Fig. 1(a) and (b) might be due to the effects of Cs^+ ions in BMICl system obtained by dissolving $Cs_2UO_2Cl_4$ and residual water in the sample solutions.

Based on spectrophotometric data, to examine the electrochemical behavior of $[UO_2Cl_4]^{2-}$ in BMICl, CVs of the sample solutions prepared by dissolving $Cs_2UO_2Cl_4$ or $UO_2Cl_2{\cdot}nH_2O$ into BMICl (abbreviated as $Cs_2UO_2Cl_4$/BMICl system and $UO_2Cl_2{\cdot}nH_2O$/BMICl system) were measured at 80 °C in the potential range of –1.0 ~ –0.4 V at various scan rates (v = 10 ~ 50 mV s^{-1}). A typical result for the $Cs_2UO_2Cl_4$/BMICl system is shown in Fig. 2 (a). As seen from this figure, two peaks corresponding to one redox couple were observed around –0.72 (E_{pc}) and –0.65 V (E_{pa}). The potential differences between two peaks (ΔE_p) are 75 and 81 mV at 10 and 50 mV s^{-1}, respectively, and close to the theoretical value (67 mV) for the reversible one electron transfer reaction at 80 °C. Furthermore, the values of (E_{pc} + E_{pa})/2 is constant, –0.687 ± 0.005 V, regardless of v (see Table 1). Similar results were also obtained from the CVs

for the $UO_2Cl_2 \cdot nH_2O$/BMICl system at 80 °C as shown in Fig. 2 (b), that is, one redox couple was observed around -0.73 (E_{pc}) and -0.66 V (E_{pa}), the ΔE_p values are 69 mV at 10 mV/s and 77 mV at 50 mV/s, the values of $(E_{pc} + E_{pa})/2$ is constant, -0.693 ± 0.001 V (see Table 1). From these results, it is suggested that $[UO_2Cl_4]^{2-}$ in BMICl is reduced to $[UO_2Cl_4]^{3-}$ quasi-reversibly and that the formal redox potential (E^o) is -0.690 V in the present system.
Hence, it should be concluded that BMICl is not applicable as the medium of the pyro-reprocessing process, because the uranyl species in BMICl are not reduced to UO_2.

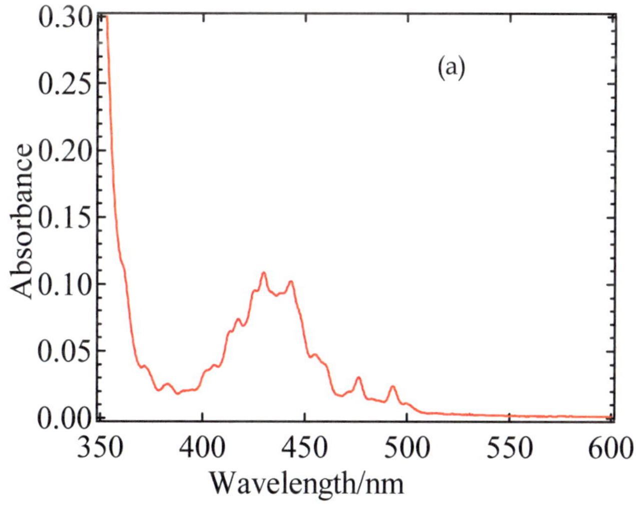

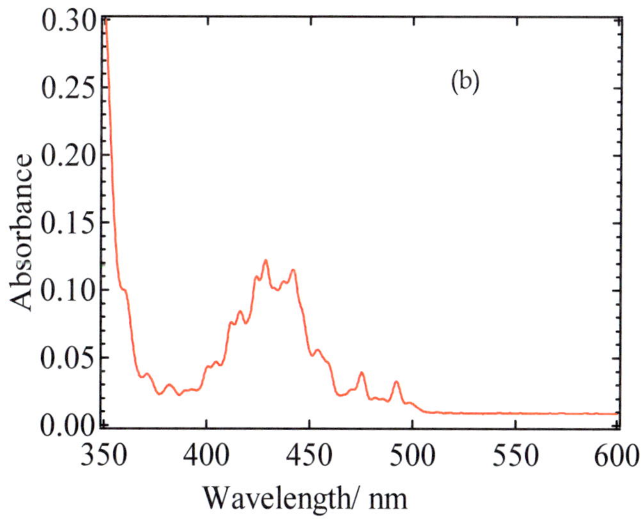

Fig. 1. UV-visible absorption spectra of the solutions prepared by dissolving uranyl chloride complexes into BMICl at 80 °C. (a): Complex = $Cs_2UO_2Cl_4$; $[UO_2^{2+}] = 1.47 \times 10^{-2}$ M. (b): Complex = $UO_2Cl_2 \cdot nH_2O$; $[UO_2^{2+}] = 1.47 \times 10^{-2}$ M.

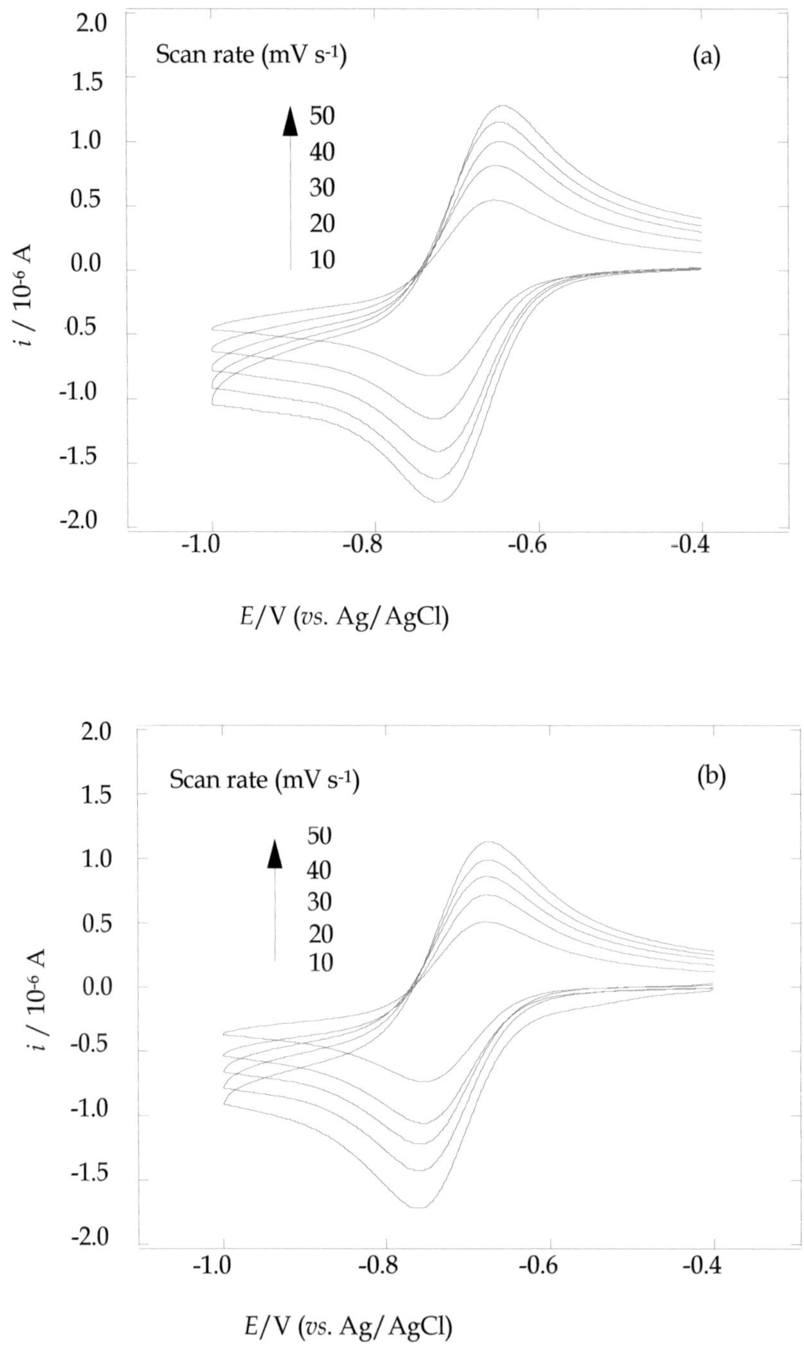

Fig. 2. Cyclic voltammograms of the solutions prepared by dissolving uranyl chloride complexes into BMICl measured in the potential range from –0.1 to –0.4 V at different scan rates (v = 10 ~ 50 mV s^{-1}). (a): Complex = $Cs_2UO_2Cl_4$; [UO_2^{2+}] = 1.47 x 10^{-2} M. (b): Complex = $UO_2Cl_2{\cdot}nH_2O$; [UO_2^{2+}] = 1.47 x 10^{-2} M. Temp. = 80 °C. Initial scan direction : cathodic.

System	ν / (mV•s⁻¹)	E_{pc} / V	E_{pa} / V	ΔE/V	i_{pc} / A	i_{pa}/ A
(a)	10	−0.729	−0.654	0.075	-7.21×10^{-7}	6.59×10^{-7}
	20	−0.727	−0.653	0.074	-1.08×10^{-6}	9.38×10^{-7}
	30	−0.725	−0.646	0.079	-1.32×10^{-6}	1.13×10^{-6}
	40	−0.726	−0.647	0.079	-1.52×10^{-6}	1.29×10^{-6}
	50	−0.723	−0.642	0.081	-1.68×10^{-6}	1.42×10^{-6}
(b)	10	−0.752	−0.682	0.070	-7.36×10^{-7}	4.37×10^{-7}
	20	−0.752	−0.678	0.074	-1.06×10^{-6}	5.95×10^{-7}
	30	−0.754	−0.679	0.075	-1.22×10^{-6}	7.09×10^{-7}
	40	−0.759	−0.679	0.080	-1.43×10^{-6}	8.15×10^{-7}
	50	−0.762	−0.676	0.086	-1.72×10^{-6}	9.18×10^{-7}

(a): $[UO_2^{2+}] = 1.47 \times 10^{-2}$ M. (b): $[UO_2^{2+}] = 1.47 \times 10^{-2}$ M

Table 1. Cyclic valtammetric data for solutions prepared by dissolving $Cs_2UO_2Cl_4$ (a) and $UO_2Cl_2{\cdot}nH_2O$ (b) into BMICl

2.2 Electrochemical study on uranyl chloride in $BMIBF_4$

Adding $UO_2Cl_2{\cdot}nH_2O$ to $BMIBF_4$, precipitates were formed. Hence, supernatant solutions containing UO_2^{2+} were used for electrochemical experiments. The UV-visible absorption spectrum of supernatant is shown in Fig. 3, and is found to be similar to those of $[UO_2Cl_4]^{2-}$ in Fig. 1. This result suggests that the uranyl species in the supernatant exists as $[UO_2Cl_4]^{2-}$. The CV measurements were continuously repeated five times in the range of −1.0 ~ 1.0 V. The results are shown in Fig. 4. An irreversible reduction peak was observed around −0.7 V and gradually decreased with the repetition of the potential sweep. By wiping off the surface of working electrode, the reduction peak was appeared again. This suggests that the surface of the electrode is covered by insoluble film produced by redox reaction. Similar phenomenon was reported by Chagnes et al., that is, they observed the formation of a blocking film on the graphite electrode in CV measurements in $BMIBF_4$ (Chagnes et al., 2005).

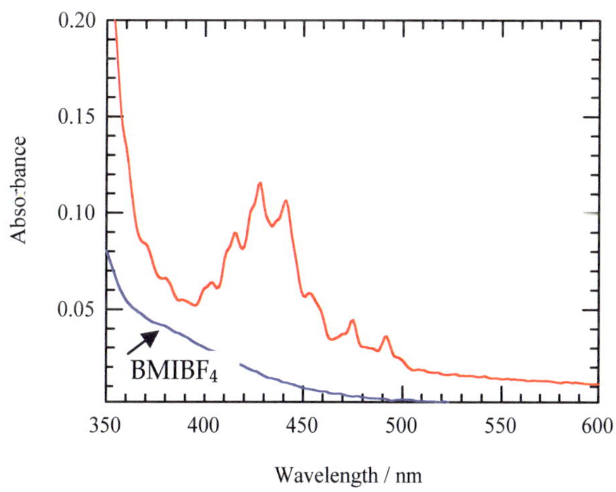

Fig. 3. UV-visible absorption spectrum of the supernatant solution prepared by adding $UO_2Cl_2{\cdot}nH_2O$ into $BMIBF_4$ at 80 °C. $[UO_2^{2+}] = 5.1\times10^{-3}$ mol/kg.

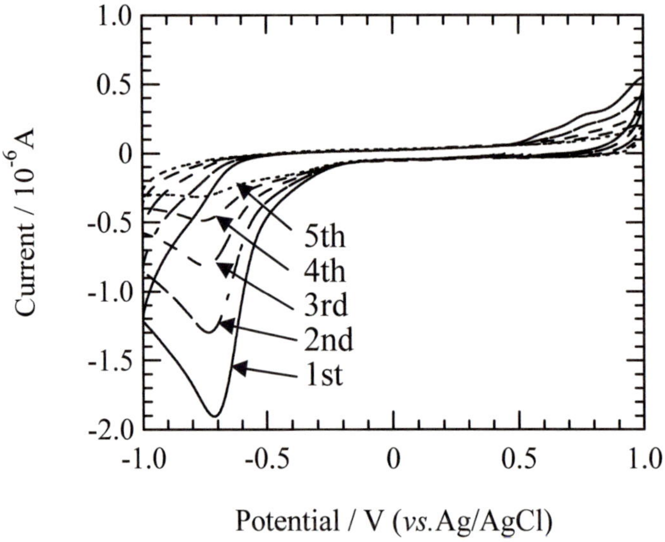

Fig. 4. Cyclic voltammograms of uranyl species in $BMIBF_4$ (80°C, $[UO_2^{2+}] = 5.1x10^{-3}$ mol/kg, sweep rate = 50 mV/s).

Judging from the above results, $BMIBF_4$ is concluded to be not applicable as the medium of the pyro-reprocessing process.

2.3 Electrochemical study on uranyl chloride in BMINfO

Figure 5 shows the UV-visible absorption spectrum of the solution prepared by dissolving $UO_2Cl_2 \cdot nH_2O$ into BMINfO at 80°C. As seen from this figure, the absorption spectrum is different from those of $[UO_2Cl_4]^{2-}$ shown in Fig. 1. Although we do no have exact data on the structure yet, it seems likely that the uranyl species in BMINfO is present as $[UO_2Cl_2(NfO)_n]^{n-}$.

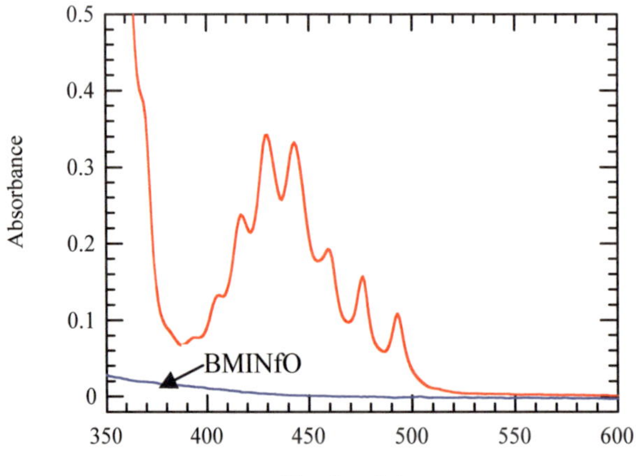

Fig. 5. UV-visible absorption spectrum of the solution prepared by dissolving $UO_2Cl_2 \cdot nH_2O$ into BMINfO at 80°C. $[UO_2^{2+}] = 8.7 \times 10^{-3}$ mol/kg.

Figure 6 shows CVs of neat BMINfO and uranyl species in BMINfO. As seen from this figure, three irreversible reduction peaks (i, ii, iii) and a sharp oxidation one (iv) appear in the range of −0.6 ~ −0.2 V and around 0.85 V, respectively. It is known that a sharp oxidation peak (iv) is due to the oxidative dissolution of reduction products deposited on the electrode (Shirai et al., 1998). Therefore, the sharp oxidation peak at 0.85 V is considered to corresponds to oxidative dissolution of U(IV) compounds deposited on the electrode. Thus, the reduction peaks should be assigned to multi step reduction of U(VI) to U(IV) as follows.

i. U(VI) + e → U(V)
ii. U(V) + e → U(IV)
iii. U(VI) + 2e → U(IV)

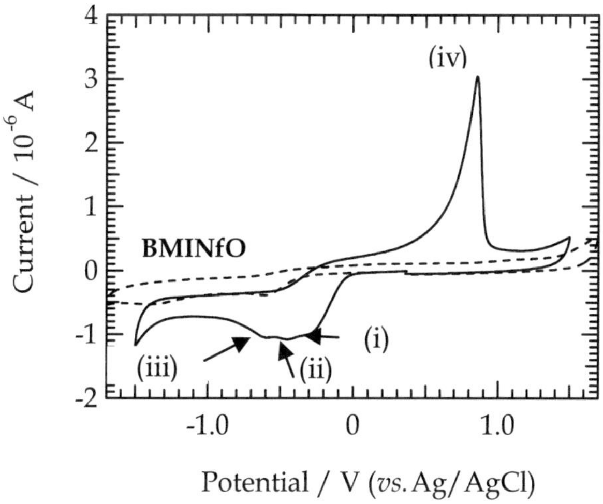

Fig. 6. Cyclic voltammograms of neat BMINfO and uranyl species in BMINfO (80°C, [UO_2^{2+}] = 8.7 x 10^{-3} mol/kg, sweep rate = 50 mV/s).

2.4 Bulk electrolysis of uranyl species in BMINfO

Judging from the results of the CV measurements described above, it might be possible to recover UO_2 by electrochemical reduction of UO_2^{2+} in BMINfO. Thus, bulk electrolysis of UO_2^{2+} (0.3 mol·kg^{-1}) in BMINfO was carried out at −1.0 V by using cell for bulk electrolysis shown in Fig. 7. As a result, the deposits were produced on a carbon electrode as cathode and a part of such deposits was fallen to the bottom of the electrolysis cell. The photograph of recovered deposits is shown in Fig. 8. After the electrolysis, the IL on the carbon electrode was washed away with acetone and dichloromethane, and then the surface of the carbon electrode was analyzed by the scanning electron microscope (SEM) and the energy dispersive X-ray spectrometer (EDX). The micrograph and distribution of elements on the electrode surface are shown in Fig. 9. The distribution of carbon is attributable to the carbon electrode. Consequently, it was found that the deposits are uranium compounds including chlorine components such as uranium oxides and uranium oxychlorides.

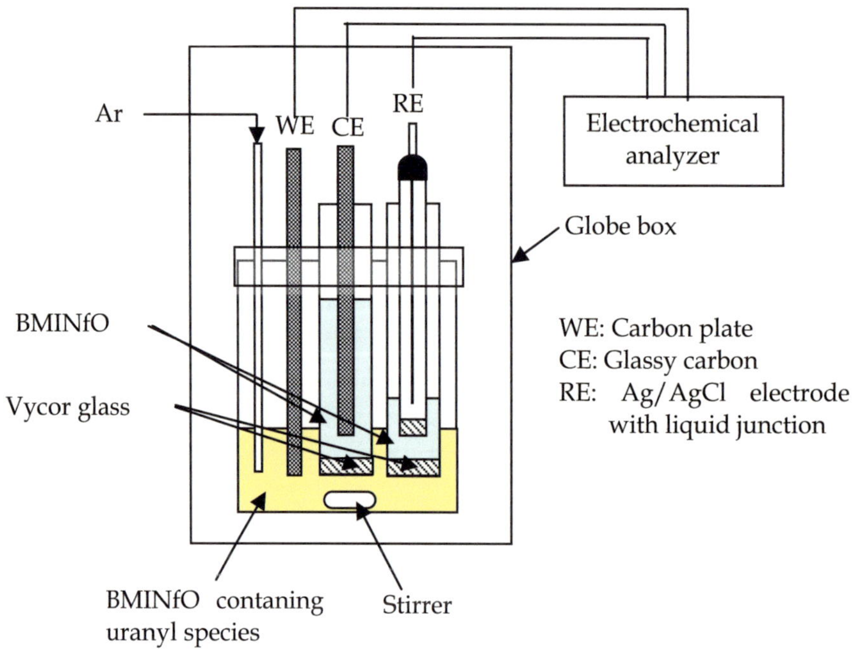

Fig. 7. Cell for bulk electrolysis.

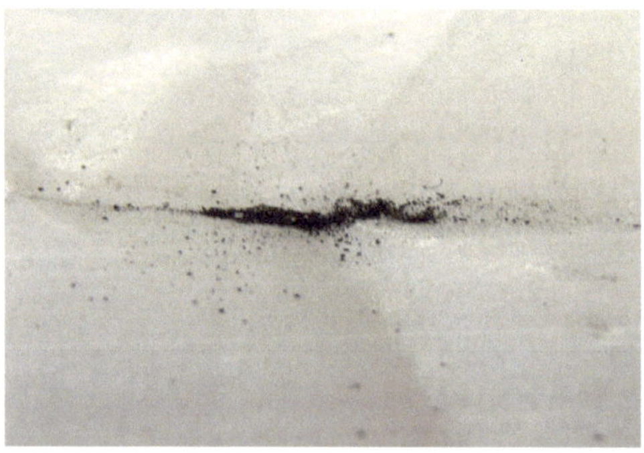

Fig. 8. A photograph of the deposits recovered by electrochemical reduction of uranyl species in BMINfO.

2.5 Summary of electrochemical properties of uranyl species in BMI based ionic liquids

Electrochemical properties of uranyl species in BMICl, $BMIBF_4$, and BMINfO were examined by using cyclic voltammetry. And based on such investigations, the applicability of ILs as the media of pyro-reprocessing processes was also examined. The results are summarized as follows.

- In BMICl, the reversible redox couple was observed in CV. This suggests that the redox couple corresponds to the redox couple of UO_2^{2+}/UO_2^{+} with one electron transfer.
- In $BMIBF_4$, an irreversible reduction peak was observed around −0.7 V and gradually decreased by the repetition of the potential sweep. This phenomenon is caused by the formation of insoluble reduction products on the surface of the electrode.

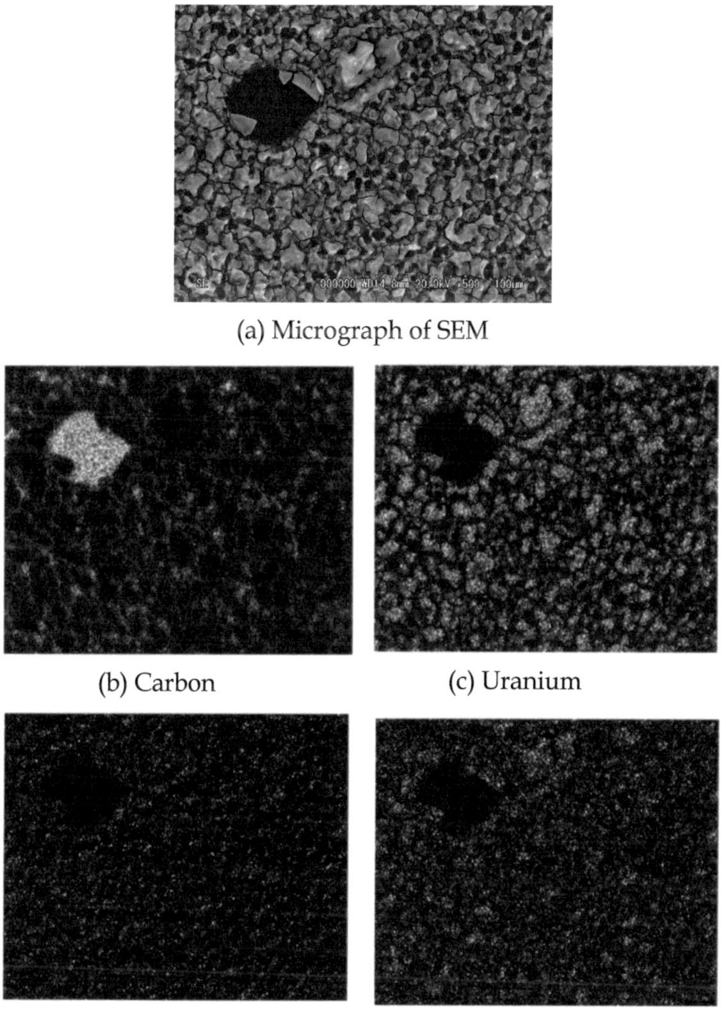

(a) Micrograph of SEM

(b) Carbon (c) Uranium

(d) Oxygen (e) Chlorine

Fig. 9. Micrograph of SEM and distribution of elements analyzed by EDX for the electrode surface after the bulk electrolysis of uranyl species in BMINfO.

- In BMINfO, three irreversible reduction peaks and a sharp oxidation one were observed in the range of −0.6 ~ −0.2 V and around 0.85 V, respectively. This suggests that the redox reactions consist of the multi step reduction of U(VI) to U(IV) and the oxidative dissolution of U(IV) as reduction products.
- Electrochemical reduction of uranyl species in BMINfO was performed by bulk electrolysis. As a result, deposits were observed on the cathodic electrode. From the SEM-EDX analyses, it was confirmed that the deposits are uranium compounds including chlorine components such as uranium oxides and uranium oxychlorides.
- These results indicate that the uranyl species in IL can be recovered electrolytically as uranium compounds. Hence, from the electrochemical viewpoint it is expected that ILs can be used as media of pyro-reprocessing processes.

- A new pyro-reprocessing method shown in Fig. 10 should be proposed. This method consists of three processes, i.e., first one is the dissolution of spent nuclear fuels using oxidant such as Cl_2, second one is the recovery of UO_2 by electrochemical reduction, and third one is the electrochemical deposition of UO_2/PuO_2 mixed oxide.

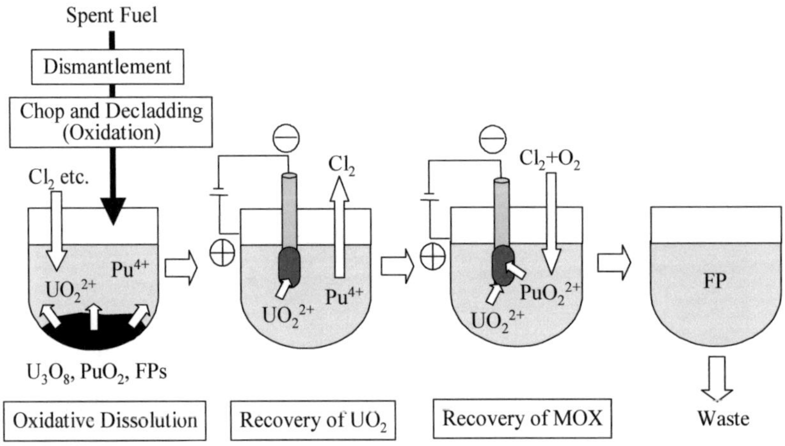

Fig. 10. Schematic diagram of proposed processes of the pyro-reprocessing by using IL as media

3. Investigation on application of ILs as electrolytic media for treating wastes contaminated with uranium

Most of metal and bed materials generated from uranium enrichment facilities or uranium refining and conversion plants are contaminated by uranium fluorides such as UF_4. These wastes are mainly classified as the medium-level wastes. Hence, it is desired to recover uranium as much as possible from such wastes. Moreover, if these wastes are decontaminated up to the clearance level, the resulting decontaminated materials should be reused. As one of effective decontamination methods of metal wastes, wet chemical decontamination processes using inorganic or organic acids have been developed (Ikeda et al., 2002; Enda et al., 2006). However, from such wet processes, a relatively large amount of secondary wastes should be generated with treating spent acid solutions, because base metal part of wastes is dissolved by acid with the dissolution of contaminated part. And also it is reported that uranium of spent adsorbents can be recovered by electrolysis in sodium chloride molten salt (Amamoto et al., 2005). However, this method must be performed under high temperature (672°C). Decontamination methods carried out under milder conditions must be preferable. Ionic liquids are expected to meet such demands. Hence, we investigated the solubility of UF_4 in ILs and the electrochemical properties of uranium species dissolved into ILs

Uranium tetrafluoride (UF_4) was synthesized from yellow cake according to the reported method (Higgins et al., 1958). Synthesis of UF_4 was confirmed using a X-Ray diffractometer (Rigaku, RAD-rPC). Impurities in UF_4 were detected using an ICP-MS (Thermo electron Co., ELEMENT). Purity of UF_4 was 97 %. BMICl was used as an IL. Water containing BMICl was removed by heating *in vacuo*. Water content in BMICl after drying was determined to be 0.1 wt % using a Karl Fischer moisture content meter (Metrohm, 737 KF Coulometer).

3.1 Dissolution behavior of UF_4 powders

Dissolution experiments were carried out at 100°C under the atmosphere with stirring at 100 rpm. The UF_4 powders (0.1 g, 3.2×10^{-4} mol) were weighed accurately and dissolved in BMICl solution (1.0 ml) in a beaker. After dissolution, the residual solid phases were filtered off and the uranium concentrations in the filtrates were measured ICP-MS. Figure 11 shows a plot of dissolution ratios *vs.* time. As seen from this figure, the UF_4 powders do not dissolve in BMICl easily.

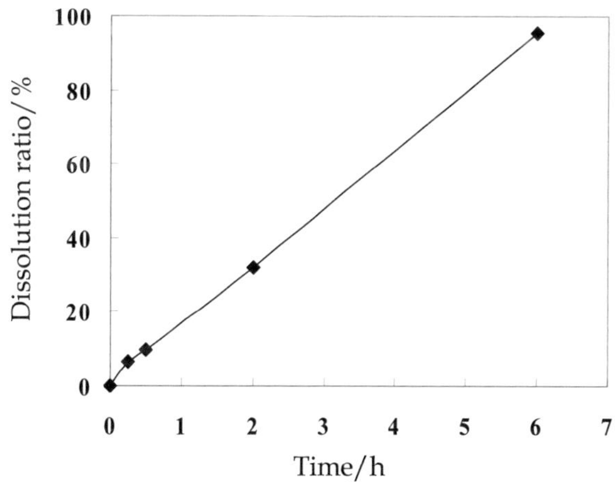

Fig. 11. A plot of dissolution ratios *vs.* time for the dissolution of UF_4 powders (0.1 g) in BMICl (1.0 ml) at 100 °C.

The color of the dissolution solution was green after about 1 h and the powders were completely dissolved after around 6 h. The color of BMICl solution after complete dissolution of UF_4 was yellowish green. However, its absorption spectrum did not show characteristic bands assigned to the U^{4+} species (Rodden, 1964). The yellowish green solution was further heated for 10 h under the atmosphere. As a result, the color of the solution changed from yellowish green to yellow. Figure 12 shows the absorption spectrum of the resulting solution and is similar to that of $UO_2Cl_4^{2-}$ shown in Fig. 1. The ε value at maximum peak of 422 nm is 13.1 M^{-1} cm^{-1} and almost the same as that (about 14 M^{-1} cm^{-1}) at maximum peak of 429 nm reported (Sornein et al., 2006). This result indicates that the species generated with the dissolution of UF_4 powders are oxidized to uranyl(VI) by O_2 under the present conditions. The relative slow dissolution of UF_4 in BMICl should be due to that the oxidation process of U(IV) with O_2 is slow.

Wipff *et al.* have reported that in BMICl dissolving uranyl triflate (TfO^-) or uranyl perchlorate, Cl^- ions interact with uranyl(VI) more strongly than ClO_4^- and TfO^-, and that the uranyl(VI) species mainly exist as $UO_2Cl_4^{2-}$ (Gaillard et al., 2007). And also they have proposed from the results of molecular dynamics (MD) and quantum mechanical (QM) calculations that F^- ions coordinate to uranyl(VI) more strongly than Cl^-(Gaillard et al., 2007; Chaumont & Wipff, 2005). Based on these reports, it is suggested that the uranyl(VI) species with the mixed ligands of F^- and Cl^- should be formed in the BMICl solution dissolving UF_4.

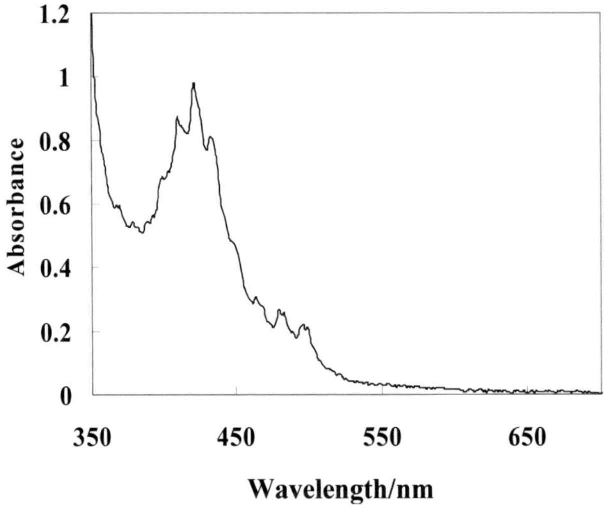

Fig. 12. UV-visible absorption spectrum of the solution prepared by dissolving UF_4 into BMICl ($[UO_2^{2+}] = 8.0 \times 10^{-2}$ M)

3.2 Elecrochemistry of sample solutions prepared by dissolving UF_4 into BMICl

The cyclic voltammograms of the sample solutions prepared by dissolving UF_4 (0.52 g) into BMICl (30 ml) were measured at 80 °C in the potential range -2.0 – 0.95 V at 50 mV/s. In these experiments, a glassy carbon wire, a Pt wire, a Ag/AgCl electrode (BAS, RE-1B) with a liquid junction filled with BMICl were used as working, counter, and reference electrode, respectively. The result is shown in Fig. 13. As seen from this figure, one uncoupled reduction peak and one uncoupled oxidation peak are observed around -0.93 and 0.18 V, respectively, and the current value of the oxidation peak is smaller than that of the reduction peak. This result is different from that of $Cs_2UO_2Cl_4$ system mentioned in 2.1. In the BMICl system dissolving $Cs_2UO_2Cl_4$, the uranyl(VI) species were confirmed to be present as $UO_2Cl_4^{2-}$, and one quasi-reversible redox couple assigned as $UO_2Cl_4^{2-} + e^- = UO_2Cl_4^{3-}$ was observed around -0.72 and -0.65 V. These support the above suggestion that the uranyl(VI) species with the mixed ligands of F^- and Cl^- are formed in the BMICl solution dissolving UF_4, and suggest that the reduction product of the uranyl(VI) complexes with the mixed ligands of F^- and Cl^- are less stable than that of $UO_2Cl_4^{2-}$, *i.e.*, $UO_2Cl_4^{3-}$. Sornein et al. have reported that the uncoupled reduction peak should correspond to the following reduction processes (Sornein et al., 2006).

$$\text{Uranyl(VI)} + e^- \rightarrow \text{Uranyl(V)}$$

$$\text{Uranyl(V)} + e^- \rightarrow UO_2$$

From these results, it is expected that the uranium component can be recovered electrolytically from the solutions generated in the decontamination treatments of the wastes contaminated with UF_4 in BMICl.

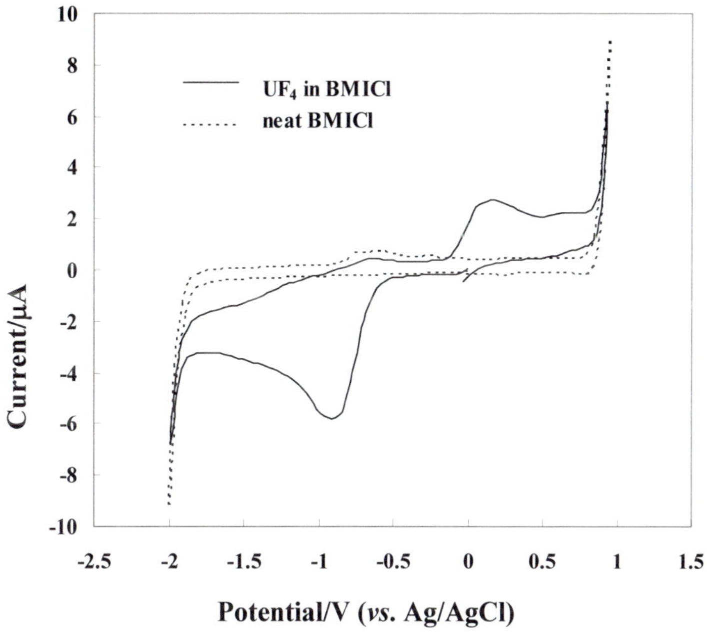

Fig. 13. Cyclic voltammograms of the solution prepared by dissolving UF_4 in BMICl ($[UO_2^{2+}]$ = 8.0×10^{-2} M) and neat BMICl at 80 °C. Initial scan direction: cathodic.

3.3 Application to decontamination of the steel wastes contaminated with uranium

Samples of steel waste were prepared from the dismantled carbon steel cylinders which had been used for storing UF_6 (see Fig. 14). The chemical forms of uranium species adhered on the steel wastes were confirmed to be UF_4 by measuring XRD. Chemical forms of iron species on surfaces of contaminated steel wastes were confirmed to be FeF_3 and Fe_2O_3 by X-ray Photoelectron Spectroscopy (JEOL, JPS-9000MC) using Mg Kα radiation of 1253.6 eV.

Fig. 14. A photograph of sample of steel waste

The contaminated steel wastes were cut into the quarter sector (28mm Φ × 6 mm thick, central angle 90°), and soaked in BMICl (2.0 ml) at 100 °C under the atmosphere. After decontamination, the ILs remained on surfaces of steel wastes were washed off with ethanol. Uranium concentrations (Bq/g) of decontaminated steels were evaluated as the ratio of radioactivity due to U of samples to total weight of samples.

Figure 15 shows a plot of U concentrations (Bq/g) against time in the dissolution of adhered uranium by soaking the contaminated steel waste into BMICl. The U concentrations are found to drop below the temporary proposed clearance level (1.0 Bq/g) within 3 h under the present conditions (IAEA, 2004).

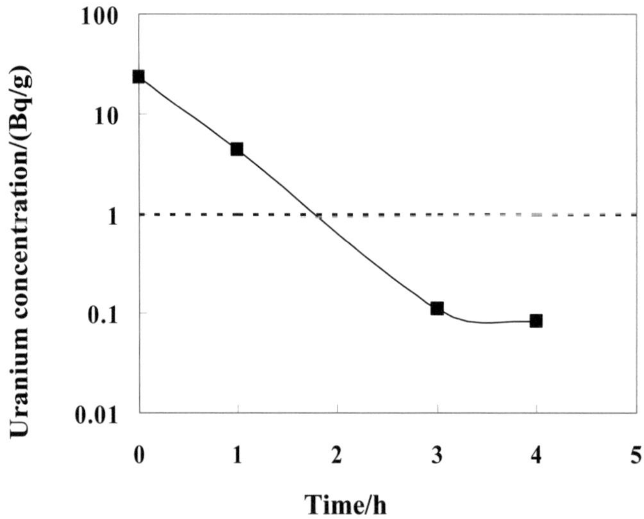

Fig. 15. A plot of uranium concentrations of steel waste *vs.* soaking time for the dissolution in BMICl at 100 °C under the atmosphere

Furthermore, the XPS spectra for the top surface of the steel waste were measured after decontamination treatment. As a result, the peaks due to UF_4 and FeF_3 were found to disappear. This indicates that the FeF_3 component is also dissolved with the dissolution of UF_4 in BMICl.

As mentioned in 3.2, it is suggested that the uranium component can be recovered electrolytically from the BMICl solution dissolving UF_4. Hence, it should be possible to recover only uranium component from the solutions after decontamination of the steel wastes in BMICl by controlling electrolytic potential.

3.4 Summary for application of ILs to the treatment of wastes contaminated with uranium

Dissolution behaviour of UF_4 in BMICl and the electrochemical properties of dissolved uranium species were investigated. Based on such basic studies, the feasibility of decontamination of steel wastes contaminated uranium using BMICl as medium was also examined. The results are summarized as follows.

- UF_4 can be dissolved completely in BMICl by heating under the atmosphere.

- From the UV-visible absorption spectra of dissolution solutions, it was found that the dissolved uranium species are oxidized to uranyl(VI) by O_2 and that the resulting uranyl(VI) species are the complex with the mixed ligands of F^- and Cl^-.
- The CV measurements suggest that the resulting uranyl(VI) species with the mixed ligands of F^- and Cl^- are reduced to UO_2 electrochemically.
- The steel wastes contaminated with UF_4 can be decontaminated below the temporarily proposed clearance level (1.0 Bq/g) within 3 h by soaking in BMICl at 100 °C.
- It should be possible to recover only uranium component from the solutions after decontamination of the steel wastes in BMICl by controlling electrolytic potential.

4. References

Earle, K.J.; Seddon, K.R. (2000). Ionic liquids. Green solvents for the nature. *Pure Appl. Chem.*, 72, 1391-1398

Rogers, R.D.; Seddon, K.R.; Volkov, S. (2000). *Green Industrial Applications of Ionic Liquids*, Kluwer Academkic Publishers, Dordrecht

Wasserscheid, P. & Welton, T. (2003). *Ionic Liquids in Synthesis*, Wiley-VCH, Weinheim

Bradley, A.E.; Hatter J.E.; Nieuwenhuyzen, M.; Pinter, W.R.; Seddon, K.R. & Thied, R.C. (2002). Precipitation of a dioxouranium(VI) species from a room temperature ionic liquid medium. *Inorg. Chem.*, 41, 1692-1694

Rogers, R.D. & Seddon, K.R. (2002). *Ionic Liquids: Industrial Applications to Green Chemistry*, ACS Symposium Series 818, American Chemical Society, Washington, DC

Giridhar, P.; Venkatensan, K.A.; Subramaniam, S.; Srinivasan, T.G. & Rao, P.R.V. (2006). Electrochemical behaviour of uranium(VI) in 1-butyl-3-methylimidazolium chloride and in 0.05 M aliquat-336/chloroform. *Radiochim. Acta*, 94, 415-420

Cocalia, V.A.; Gutowski, K.E. & Rogers, R.D. (2006). The coordination chemistry of actinides in ionic liquids: a review of experiment and simulation. *Coord. Chem. Rev.*, 250, 755-764

Giridhar, P.; Venkatesan, K.A.; Srinivasan, T.G. & Rao, P.R.V. (2007). Electrochemical behavior of uranium(VI) in 1-butyk-3-methylimidazolium chloride and thermal characterization of uranium oxide deposit. *Electrochim. Acta*, 52, 3006-3012

Binnemans, K. (2007). Lanthanides and actinides in ionic liquids. *Chem. Rew.*, 107, 2592-2614

Cordfunke, R.G. (1969). *The Chemistry of Uranium*, Elsevier Publishing Co., Amsterdam

Denning, R.G.; Snellgrove, T.R. & Woodwork, D.R. (1979). The electronic structure of uranyl ion Part I. The electronic spectrum of $Cs_2UO_2Cl_4$. *Mol. Phys.*, 32, 419-442

Dai, S.; Shin, Y.S.; Toth, L.M. & Barnes, C.E. (1997). Comparative UV-Vis studies of uranyl chloride complex in two basic ambient-temperature melt systems: the observation of spectral and thermodynamic variations induced via hydrogen bonding. *Inorg. Chem.*, 36, 4900-4902

Sornein, M.-O.; Cannes, C.; Naour, C.L.; Langarde, G. ; Simoni, E. & Berthet, J.-C. (2006). Uranyl complexation by chloride ions. Formation of a tetrachlorouranium(VI) complex in room temerature ionic liquids [Bmim][Tf_2N] and [$MeBu_3N$][Tf_2N]. *Inorg. Chem.*, 45, 10419-10421

Nockemann, P.; Servaes, K.; Deun, R.V.; Hecke, K.V.; Meervelt, L.V.; Binnemans, K. & Walrand, C.G. (2007). Speciation of uranyl complexes in ionic liquids by optical spectroscopy. *Inorg. Chem.*, 46, 11335-

Chagnes, A.; Diaw, M.; Carré, B.; Willmann, P.; Lemordant, D. (2005). Immidazolium-organic solvent mixtures as electolytes for lithium batteries. J. Power Sources, 145, 82- 88

Shirai, O.; Iwai, T.; Suzuki, Y.; Sakamura, Y.; Tznaka, H. (1998). Electrochemical behavior of actinide ions in LiCl-KCl eutectic melts. *J. Alloys Com.*, 271-273, 685-688

Ikeda, A.; Aida, M.; Fujii, Y.; Kataoka, S.; Annen, S. & Sato, J. (2002). Ion exchange separation for decontamination of centrifuge enrichment plant. *J. Nucl. Sci. Technol.*, 39, 1099-1105

Enda, M.; Sato, T.; Sakai, H.; Natsui, K. & Ishikura, T. (2006). Development of chemical decontamination technique using formic acid. *Trans. At. Energy Soc. Japan*, 15, 364-373

Amamoto, I.; Terai, T.; Oobayashi, H. & Fujita, R. (2005). Separation and recovery study of uranium from spent NaF (fillers). *J. Phys. Chem. Solids*, 66, 602-607

Higgins, I.R.; Robert, J.T. & Hancher, C.W. (1958). *Ind. Eng. Chem.*, 50, 285-292

Rodden, C.J. (1964). *Analysis of Essential Nuclear Reactor Materials*, U.S.A.E.C., GPO, Washington, D.C.

Gaillard, C.; Chaumont, A.; Billard, I.; Hennig, C, Ouadi, A. & Wipff, G. (2007). The competition between ionic liquid anions, uranyl counterions, and Cl- anions investigated by extended X-ray absorption fine structure and UV-visible spectroscopies and molecular dynamics simulations. *Inorg. Chem.*, 46, 4815-4826

Chaumoun, A. & Wipff, G. (2005). Solvation of fluoro and mixed fluoro/chloro complexes of Eu^{III} in the [BMI][PF_6] room temperature ionic liquid. A theoretical study. *Phys. Chem. Chem. Phys.*, 7, 1926-1932

International Atomic Energy Agency. (2004). *Application of the concepts of exclusion, exemption and clearance*, IAEA Safety Standards Series, No. RS-G-1.7